Encyclopedia of

Plant Physiology

New Series Volume 14B

Editors

A. Pirson, Göttingen
M. H. Zimmermann, Harvard

Nucleic Acids and Proteins in Plants II

Structure, Biochemistry and Physiology of Nucleic Acids

Edited by

B. Parthier and D. Boulter

Contributors

H. J. Bohnert W. Bottomley J. A. Bryant E. J. Crouse
T. A. Dyer G. L. Farkas R. B. Flavell G. Galling
D. Grierson K. W. Henningsen L. Hirth W. Nagl
L. Neeleman H. L. Sänger J. Schell J. M. Schmitt
H. G. Schweiger B. M. Stummann L. van Vloten-Doting
C. Wasternack R. Wollgiehn

With 173 Figures

Springer-Verlag Berlin Heidelberg New York 1982

Professor Dr. BENNO PARTHIER
Akademie der Wissenschaften der DDR
Institut für Biochemie der Pflanzen Halle
Weinberg 3, Postfach 250
401 Halle (Saale)/GDR

Professor Dr. DONALD BOULTER
University of Durham
Department of Botany
Science Laboratories, South Road
Durham, DH1 3LE/UK

ISBN-13: 978-3-642-68349-7 e-ISBN-13: 978-3-642-68347-3
DOI: 10.1007/978-3-642-68347-3

Library of Congress Cataloging in Publication Data. Main entry under title: Nucleic acids and proteins in plants. (Encyclopedia of plant physiology; new ser.; v. 14, pt. A–B). Bibliography: p. Includes index. Contents: pt. A. Structure, biochemistry, and physiology of proteins / edited by D. Boulter and B. Parthier pt. B. Structure, biochemistry, and physiology of nucleic acids / edited by B. Parthier and D. Boulter. 1. Nucleic acids. 2. Plant proteins. 3. Botanical chemistry. I. Boulter, D. II. Parthier, Benno. III. Series. QK711.2.E5 new ser., vol. 14, pt. A, etc. 81-18256 [QK898.N8] 581.1s [581.19′24].

Typesetting, printing and bookbinding: Universitätsdruckerei H. Stürtz AG, Würzburg.
2131/3130-543210

Contents

1 Nuclear Chromatin
W. NAGL (With 16 Figures)

2 Chromosomal DNA Sequences and Their Organization
R.B. FLAVELL (With 11 Figures)

3 DNA Replication and the Cell Cycle
J.A. BRYANT (With 4 Figures)

6 RNA Sequences

T.A. Dyer (With 4 Figures)

7 RNA Processing and Other Post-Transcriptional Modifications

D. Grierson (With 14 Figures)

8 Ribonucleases and Ribonucleic Acid Breakdown

G.L. FARKAS

9 Metabolism of Pyrimidines and Purines

C. WASTERNACK (With 10 Figures)

12 Biology, Structure, Functions and Possible Origin of Viroids

H.L. Sänger (With 36 Figures)

13 The Ti-Plasmids of Agrobacterium tumefaciens

J. SCHELL (With 3 Figures)

14 Organization and Expression of Plastid Genomes

H.J. BOHNERT, E.J. CROUSE, and J.M. SCHMITT (With 5 Figures)

15 The Biosynthesis of Chloroplast Proteins

W. BOTTOMLEY and H.J. BOHNERT (With 8 Figures)

List of Contributors

H.J. BOHNERT
EMBL
Postfach 102209
D-6900 Heidelberg/FRG

Present address:
MPI für Züchtungsforschung
Postfach
D-5000 Köln-Vogelsang/FRG

W. BOTTOMLEY
C.S.I.R.O.
Division of Plant Industry
P.O. Box 1600
Canberra City, A.C.T. 2601/Australia

J.A. BRYANT
Dept. of Plant Science
University College
P.O. Box 78
Cardiff CFI IXL/UK

E.J. CROUSE
Institut de Biologie
Moléculaire et Cellulaire
15, rue Descartes
F-67084 Strasbourgh/France

T.A. DYER
Plant Breeding Institute
Maris Lane
Trumpington, Cambridge CB2 2LQ/
UK

G.L. FARKAS
Institute of Plant Physiology
Hungarian Academy of Sciences
P.O. Box 521
H-6701 Szeged/Hungary

R.B. FLAVELL
Plant Breeding Institute
Maris Lane
Trumpington, Cambridge CB2 2LQ/
UK

G. GALLING
Botanisches Institut der Technischen
Universität
D-3300 Braunschweig/FRG

D. GRIERSON
Dept. of Physiology and
Environmental Science
University of Nottingham School of
Agriculture
Sutton Bonington
Loughborough, Leicestershire, LEI2
5RD/UK

K.W. HENNINGSEN
The Royal Veterinary and
Agricultural University
Dept. of Genetics
Bülowsvej 13
DK-1870 Copenhagen V./ Denmark

L. HIRTH
Institut de Biologie
Moléculaire et Cellulaire
15, rue Descartes
F-67084 Strasbourgh/France

W. NAGL
Institut für Zellbiologie
der Universität Kaiserslautern
Postfach 3029
Erwin-Schrödinger-Straße
D-6750 Kaiserslautern/FRG

L. NEELEMAN
Dept. of Biochemistry
State University Leiden
P.O. Box 9505
2300 RA Leiden/The Netherlands

H.L. SÄNGER
Max-Planck-Institut für Biochemie
Abteilung Viroidforschung
D-8033 Martinsried b. München/
FRG

J. Schell
Max-Planck-Institut für Züchtungsforschung
Egelspfad
D-5000 Köln 30 Vogelsang/FRG

J.M. Schmitt
Botanisches Institut der Universität Würzburg
Mittlerer Dallenbergweg 64
D-8700 Würzburg/FRG

H.G. Schweiger
Max-Planck-Institut für Zellbiologie
D-6802 Ladenburg b. Heidelberg/FRG

B.M. Stummann
The Royal Veterinary and Agricultural University
Dept. of Genetics
Bülowsvej 13
DK-1870 Copenhagen V./Denmark

L. van Vloten-Doting
Dept. of Biochemistry
State University Leiden
P.O. Box 9505
2300 RA Leiden/The Netherlands

C. Wasternack
Martin-Luther-Universität Halle-Wittenberg
Sektion Biowissenschaften
Neuwerk 1
4020 Halle (Saale)/GDR

R. Wollgiehn
Akademie der Wissenschaften der DDR
Institut für Biochemie der Pflanzen
Weinberg 3, Postfach 250
401 Halle (Saale)/GDR

List of Abbreviations

A	adenosine (likewise: C, cytidine; G, guanosine; I, inosine; U, uridine; T, thymidine; ψ, pseudouridine)
i^6A	isopentenyl-adenosine
ms^2i^6A	2-methyl-thio-iso pentenyl-adenosine
Aa-RS	amino acyl-tRNA synthetases; (Thr-RS, threonine-tRNA synthetase, other amino acids correspondingly)
ABA	abscisic acid
ADP	adenosine 5′-diphosphate (likewise CDP, GDP, UDP)
AMP	adenosine 5′-monophosphate, adenylic acid (likewise CMP, GMP, UMP, TMP)
cAMP	cyclic 3′,5′AMP
Ap	adenosine 3′-monophosphate (likewise Cp, Gp, Up, Tp) (in polynucleotide chains)
ATP	adenosine 5′-triphosphate (likewise CTP, GTP, UTP, TTP)
BA	benzyladenine
bp	basepairs
BSA	bovine serum albumin
2C	diploid (likewise 4C, tetraploid, etc.)
2,4-D	dichlorophenoxyacetic acid
dA	deoxyadenosine (likewise: dC, dG etc.)
DCMU	3-(3′,4′-dichlorophenyl)-1,1-dimethylurea
DEAE	diethylaminoethyl
cDNA	complementary DNA
ctDNA	chloroplast DNA
EC	energy charge
EDTA	ethylene-diamintetra-acetic acid
EF	elongation factor in ribosomal translation (e.g., EF-1; EFT_u, EFT_s, EFG)
EGTA	ethyleneglycolbis-tetra-acetic acid
eIF	eukaryotic initiation factor
ER	endoplasmic reticulum
FMN	flavin mononucleotide
G0, G1, G2	phases of the cell cycle
GA	gibberellin
GA_3	gibberellic acid
Kbp	kilobase pairs
Kd	kilodalton
M_r	molecular mass
mRNA	messenger RNA
mRNP's	messenger ribonucleic acid particles
MW	molecular weight (mol. wt.)
N	nucleoside (usually in connection with p, pp or ppp) and chemical symbols for nitrogen
NAA	naphthaleneacetic acid
NAD	nicotinamide adenine dinucleotide (oxidized form)
NADH	nicotinamide adenine dinucleotide (reduced form)
NTP	nucleoside triphosphate
pA	adenosine 5′-monophosphate (in polynucleotide chains); likewise pC, pG, pU, etc.
PAL	phenylalanine ammonia lyase
pBR322	plasmid of E. coli frequently used for gene transformation

P_i	inorganic phosphate
poly(A)	polyadenylic acid, polyadenylated 3′ terminus of (virus or messenger) RNA; likewise: poly(U), poly(dA), poly(dC), poly(dA-T), etc.
poly(A)$^+$RNA	RNA containing terminal poly(A)
poly(A)$^-$RNA	RNA lacking terminal poly(A)
rRNA	ribosomal RNA
$tRNA^{Val}_{Euglena}$	valine-specific transfer RNA of Euglena (correspondingly for other amino acids and other species)
RuBPCase (RuBPC)	ribulose-1,5-bisphosphate carboxylase
S	sedimentation coefficient in Svedberg units
SDS-PAGE	sodium dodecylsulphate polyacrylamide gel electrophoresis
Thr-RS	threonine-tRNA synthetase (other amino acids correspondigly)
T_m	melting point

Virus Species Mentioned in Chapters 10 and 11

AMV	alfalfa mosaic virus
BMV	brome mosaic virus
BNYVV	beet necrotic yellow vein virus
BPMV	bean pod mottle virus
BSMV	barley stripe mosaic virus
CaMV	cauliflower mosaic virus
CarMV	carnation mottle virus
CCMV	cowpea chlorotic mottle virus
CcTMV	cowpea strain of TMV
CMV	cucumber mosaic virus
CPMV	cowpea mosaic virus
EAMV	Echtes Ackerbohnen-Mosaik-Virus
EMC	encephalomyocarditis virus
EMV	eggplant mosaic virus
OMV	okra mosaic virus
PeMV	pepper mottle virus
PMV	papaya mosaic virus
PVX	potato virus X
RCMV	red clover mosaic virus
RRV	raspberry ringspot virus
SBMV	southern bean mosaic virus
SqMV	squash mosaic virus
STNV	satellite tobacco necrosis virus
TBRV	tomato black ring virus
TEV	tobacco etch virus
TMV	tobacco mosaic virus
TNV	tobacco necrosis virus
TRosV	turnip rosette virus
TRSV	tobacco ringspot virus
TRV	tobacco rattle virus
TSV	tobacco streak virus
TVMV	tobacco vein mottling virus
TYMV	turnip yellow mosaic virus
VSV	vesicular stomatitis virus

1 Nuclear Chromatin

W. NAGL

1 Introduction

Chromatin is the virtual genetic material of eukaryotes which essentially exhibits two molecular components and functional aspects: the DNA encoding the genetic information, and the chromosomal proteins controlling DNA packaging and thus gene activity. Although massive progress has been achieved in the elucidation of chromatin composition and structure during the last years, the ultimate solution to the twin problems of gene control and DNA packaging is likely to be complex. The complications we already find in prokaryotes should warn us against over-simple answers (CRICK 1979). This review will deal with evidence obtained by biochemistry, biophysics, and electron microscopy on the structure of chromatin at several levels of order, the functional significance of structural changes, as well as some speculations about the regulation of such changes. Some phylogenetic aspects will be briefly discussed and related to ontogenetic aspects of differentiation.

Compared to the increasing information on the basic and higher-order structure of animal chromatin, only little is known about plant chromatin. Therefore, some data will be taken from studies in animals. This may be permissible, because it seems that the subunit structure of chromatin is the same in plants and animals (McGHEE and ENGEL 1975). Higher-order structures, however, such as condensed chromatin in interphase nuclei and mitotic chromosomes, might be quite different in animals and plants, and in organisms with a large and with a small genome. Generalization should, therefore, be avoided until more is known about those aspects.

2 Chemistry of Chromatin

Methods of chromatin isolation from animal tissues have been elaborated since 1959 (reviewed by BONNER 1979a, b, FELLENBERG 1974). Gain of chromatin from isolated plant nuclei was described by TOWILL and NOODÉN (1973). In principle, the nuclei are lysed with detergent, the chromatin pelleted by centrifugation, resuspended in low ionic strength buffer, and purified by centrifugation through 1.7 M sucrose. Solubilization is obtained by shearing. Important prob-

---

Abbreviations. H1, H2A, H2B, H3, H4, histones (animal cells); PH1, PH2, plant histones corresponding to H2A, H2B

lems are to avoid cytoplasmic contaminations, to inhibit the powerful serine protease, e.g., by phenylmethanesulfonyl fluoride (PMSF) or di-isopropylfluorphosphate (DFP), and to remove ribonuclease by the precipitation of purified chromatin from low ionic strength buffer (BONNER 1979a).

Composition of chromatin depends on both its definitions and its purity. Some researchers identify chromatin with the DNA-histone (nucleohisteone) complex, others acknowledge a concept that chromatin includes everything within the nuclear envelope. In this essay I shall use the term chromatin in a sense that covers the structural components of chromatin such as DNA, histones, and the nonhistone proteins of the "scaffold" (PAULSON and LAEMMLI 1977), i.e., the residual proteins of the older literature. I shall not include acidic chromosomal proteins which may be of regulatory nature, nor the nuclear enzymes and the various species of RNA. All these parts will be briefly discussed only in relation to chromatin function. According to the given definition, chromatin is composed of DNA and histones in a ratio of about 1:1 and small amounts of nonhistone proteins. The total content of nuclei possesses a nonhistone chromosomal protein ratio to DNA of 0.6:1 (but some variation occurs due to different activity of the nuclei), and an RNA:DNA ratio of 0.1:1. Chromatin which shows a composition substantially different from the proportions outlined above may be suspected of contamination (BONNER 1979a). Criteria for purity of isolated chromatin have been outlined by BONNER et al. (1978).

The organization of eukaryotic DNA is reviewed by FLAVELL in Chapter 2, this volume. Therefore, only some aspects of the histones and scaffold proteins have to be added. Histones are basic proteins, possessing approximately 24 M% basic amino acids, and can be separated by polyacrylamide gel electrophoresis into a very lysine-rich fraction, slightly lysine-rich fractions, and arginine-rich fractions. Five histone classes are found in all of the animal species: H1 (very lysine-rich), H2A and H2B (slightly lysine-rich), and H3 and H4 (arginine-rich). These classes also occur with some slight alterations in lower eukaryotes, but not in prokaryotes.

The primary structure of histone H4 is almost totally conserved and exhibits nearly the same amino acid sequence in peas and cows (an evolutionary history of 600 million years), that of H3 is also quite conserved (DE LANGE et al. 1969, 1973). Histones H2A and H2B are somewhat less conserved, and H1, which is the most different in its properties from the other histones, is the least conserved and exhibits many variants (COLE 1977). Cysteine is restricted to H3 in nearly all organisms. With the exception of yeast, a cysteine in position 110 has been preserved evolutionarily. One of the few mutations that have occurred in H3 concerns residue 96 which is alanine or serine in *Pisum*, or an additional cysteine in mammals more advanced than rodents. These cysteine residues are completely protected in native chromatin and core particles, but come unmasked simultaneously during a salt-induced dissociation (BODE and STANDT 1978). If cysteines 96 or 110 become accessible during any phase of the cell cycle (e.g., in connection with chromosome condensation), their exposure must be triggered by histone modification or another structural alteration of the chromatin subunit. A specific H3 phosphorylation actually occurs just before metaphase, and this might bring about the rearrangements required (GURLEY et al. 1974, COLE 1977; for details see Sect. 5).

In plants, only the two arginine-rich histones H3 and H4 are nearly identical to those of animals. The H1 fractions of plants show a slower migration on polyacrylamide-SDS gels and a more complex electrophoretic patterns than

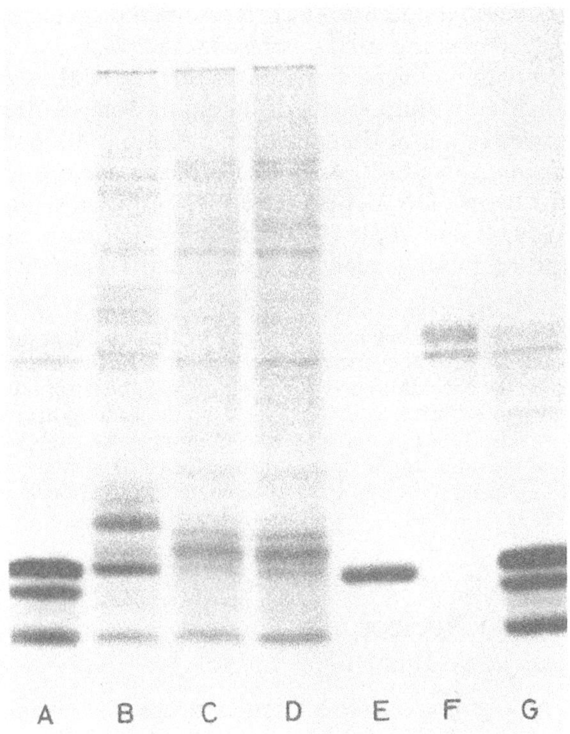

Fig. 1. Electrophoretic mobility of plant histones compared to calf thymus histones. Plant histones were isolated from nuclei of *Brassica* and *Glycine,* and run on a 15% polyacrylamide SDS gel for 6 h at 140 V: *A* total histones from calf thymus; *B* total histones from *Brassica: C* and *D* total histones from *Glycine; E* histone H2B (calf); *F* histone H1 (calf); *G* total histones from calf thymus. (LEBER and HEMLEBEN 1979a, modified)

the corresponding animal histone. No correspondence can be found between the faster-migrating plant histones now called PH1 and PH2 (NADEAU et al. 1974) and the animal histones H2A and H2B. They appear to be species-specific (LEBER and HEMLEBEN 1979a; for an example see Fig. 1). H1 is species- and tissue-specific. Unique histone fractions have been described for meiotic cells, erythrocytes and sperm cells, where they become replaced by protamines (for reviews see STEIN et al. 1978, ISENBERG 1978, 1979; for further aspects of plant histones refer to SPIKER and KRISHNASWAMY 1973, BRANDT and VAN HOLT 1975, SPIKER 1975, 1976, GIGOT et al. 1976).

The nonhistone chromosomal proteins can be studied, for instance, after removal of histones by 0.2 or 0.4 M H_2SO_4 followed by treatment of the DNA-containing nonhistone proteins with SDS and SDS chromatography. The SDS complexes of the nonhistone chromosomal proteins, electrophoresed on polyacrylamide gels, exhibit a wide variety of molecular weights, from about 225,000 down to the lower limit of resolution of such gels, namely of about 10,000–15,000. The major nonhistone chromosomal protein components are similar in a wide variety of chromatins (ELGIN and BONNER 1970), while others exhibit a high species- and tissue-specificity (OLSON and BUSCH 1974, BONNER 1979a, b). Two major proteins of the common nonhistone category are hnRNA packaging proteins (MARTIN et al. 1973), others represent tubulin, actin, and other components of the actomyosin system (DOUVAS et al. 1975). Fifty percent of the total nonhistone chromosomal protein consists of structural components. The remaining 50% represents enzymes and acidic proteins involved in the regulation of gene activity (see SEVALL et al. 1978, WANG and KOSTRABA 1978).

Minor components of the nonhistone proteins are evidently involved in the organization of domains in chromatin and in the formation of a chromosome skeleton (see Sect. 5). These proteins are protected from protease attacks, and the bound DNA is protected from nuclease digestion (JEPPERSEN and BANKIER 1979, RAZIN et al. 1979). It seems that such proteins are covalently bound to certain DNA sequences (KRAUT and WERNER 1979).

There are a number of phenomena which are of great interest in chromatin research, but which are beyond the scope of this chapter. Some of them are the attachment of chromatin to the nuclear envelope and to the nuclear matrix, which evidently help to keep the chromatin arranged. Moreover, the chromatin is apparently more than a nucleoprotein complex, as it contains also minor components such as sphingomyelin and carbohydrate. Glycoproteins, glycosaminoglycans, and saccharides indicate that chromatin plays a role in the metabolism of carbohydrates. This is consistent with the finding of cytidine monophosphate-sialic-acid synthetase in nuclei (for a review see STODDART 1979).

3 The Nucleosome

There is now overwhelming evidence that chromatin of eukaryotes is organized as a regular chain of repeated subunits, termed ν-bodies (OLINS and OLINS 1974) or nucleosomes (OUDET et al. 1975). Dinophyceae do not show nucleosomes in their mesokaryotic nucleus, but in the eukaryotic one if present (BODANSKY et al. 1979, RIZZO and BURGHARDT 1980). Nucleosomes from lower fungi were described by HOZIER and KAUS (1976), JOHNSON et al. (1976), MORRIS (1976), NOLL (1976), VOGT and BRAUN (1976), BAKKE et al. (1978), GRAINGER and OGLE (1978), and from angiosperms by GIGOT et al. (1976), NAGL (1976b), PHILIPPS and GIGOT (1977), MORENO et al. (1978), YAKURA et al. (1978), LEBER and HEMLEBEN (1979a, b), and LUTZ and NAGL (1980). The nucleosome core particle is formed by an octamer of each of the histones H2A, H2B, H3 and H4 (KORNBERG 1974) and 140 bp of DNA, while the core particles are connected by a linker DNA segment, which is variable in length and which is associated with histone H1 (for reviews see Cold Spring Harbor Symposium on Quantitative Biology 1977, LI and ECKHARDT 1977, BUSCH 1978, CALLAN and KLUG 1978, FELSENFELD 1978, STEIN et al. 1978, NICOLINI 1979a, b, SONNENBICHLER 1979, TAYLOR 1979). Extraction of H1 leads to unfolding of the linker DNA so that the characteristic "beads-on-a-string" structure becomes visible in the electron microscope (Fig. 2). Addition of one molecule H1 per octamer core leads to the binding of 160 bp of DNA. Binding of a second H1 molecule per core particle causes a dramatic structural compaction into polynucleosome chains (e.g., NELSON et al. 1979; see Sect. 4). Nucleosomal histones represent a digestion barrier to nucleases, so that nucleosomes can be isolated upon mild digestion with micrococcal nuclease (e.g., MARALDI et al. 1979, and many others).

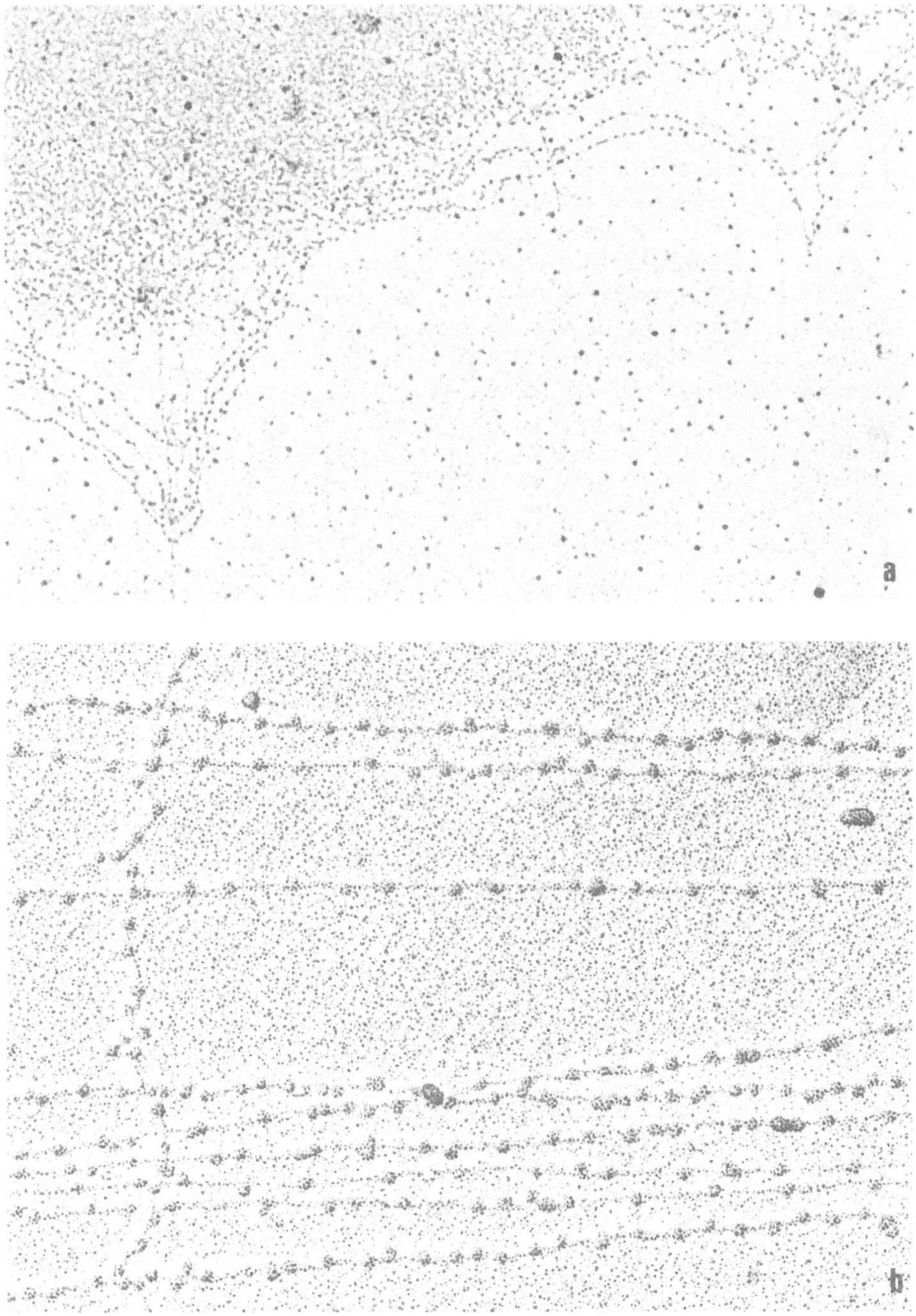

Fig. 2 a, b. Electron micrographs of nucleosomes from onion cell nuclei of *Allium cepa*. **a** Surface of spread nucleus, some nucleosome fibers are extruding (× 46,000). **b** Nucleosome fiber (typical "beads-on-a-string" structure), rotary shadowed with platinum/palladium (× 100,000)

3.1 The Nucleosome Core Particle

Nucleosomes obtained by micrococcal nuclease digestion can be stored, probably indefinitely, frozen in 0.2 mM EDTA (pH 7.0) at $\leqq -25$ °C (OLINS et al. 1976), but their preparation from plant material for electron microscopic visualization is a rather critical procedure (LUTZ and NAGL 1980). This difficulty may be the reason for some yet unexplained differences in the results on chromatin obtained by either nuclease digestion and gel electrophoresis of the DNA, or spreading and electron microscopy of chromatin.

The nucleosome core can crystallize and both ultrastructural and X-ray diffraction analyses indicate that it is a flat particle ("platysome"; CRICK 1979) of $110 \times 110 \times 57$ Å (FINCH et al. 1977), divided into two half-nucleosomes. This fact possibly allows strand separation during DNA replication and transcription without requiring histone displacement (WEINTRAUB et al. 1976). The histone core of the nucleosome consists of the tetramer (2H3–2H4) and the two dimers $(H2A\text{-}H2B)_2$ (Fig. 3b). These histone complexes are arranged in the core particle as spatially separated groups (two heterotypic tetramers) and held together by interactions between structured and apolar central carboxyl regions (for reviews and other models see WORCEL 1977, BRADBURY 1978, CARTER 1978, MIRZABEKOV et al. 1978). Cross-linking sites of the core histones have been directly identified (e.g., SUDA and IWAI 1979). The H3-H4 tetramer seems to be necessary for the appearance of the nucleosome structure (CAMERINI-OTERO et al. 1976) and for the process of self-assembly in vitro (RUIZ-CARILLO and JORCANO 1979, JORCANO and RUIZ-CARILLO 1979). H2A and H2B are considered as stabilizing factors both of the core particle and of the linker DNA. According to CAMERINI-OTERO and FELSENFELD (1977) there exists a close contact between H3 histones on the dyad axis of the core particle. The ends of the core DNA segment are associated with the (2H3–2H4) tetramer. Furthermore the histones H2A and H2B are neighbors along the DNA. The histone core essentially contains

▷

Fig. 3A–C. The nucleosome core particle. **A** Space-filling model of the path of the DNA in the nucleosome core. The rear half nucleosome, from base pair 70 to 140, is *shadowed*. (Modified and redrawn for MARALDI et al. 1979). **B** Drawings of the histone localization in the nucleosome core. The H4-H3 tetramer is localized in the bottom, while the two H2A-H2B dimers are in the top of the nucleosome. The *continuous thick line* indicates the path of the DNA from the top. H4 histones are localized in the outer coils of the DNA from about 0 to 40 bp and from about 100 to 140 bp. H3 histones bind the two H4 histones and are superimposed to the inner coil of DNA in correspondence with the bp 60–80. H2A histones are localized in the inner coil of DNA from about 40 to 60 bp and from 80 to 100 bp. H2B histone. linked with H2A, binds to opposite DNA coils, in correspondence with 30 and 110 bp positions. (Modified from MARALDI et al. 1979). **C** Core particle model showing the arrangement of the histone heads, histone tails, and DNA double helix. According to this model, the tails of the (2H3-2H4) tetramer encompass the $(H2A\text{-}H2B)_2$ complex, at the same time fixing the start and end of the 140 bp DNA superhelix. On the other hand, the tails of the $(H2A\text{-}H2B)_2$ complex encompass the (2H3-2H4) tetramer. Unlike the H3, H4, H2B and H2A, the histones H3′, H4′, H2B′, and H2A′ lie below the drawing plane; the *straight lines* mark the dyad axis lying in the drawing plane and passing through bp 70. (Redrawn from ZINKE 1979)

the apolar regions of the histone molecule, while the very basic N-terminal regions are not involved in the core complex formation and are major sites of interaction with the DNA phosphate groups (Fig. 3c). Besides the apolar, stereochemical contacts between the hydrophobic C-terminals of the histones, series of alternately charged residues in the central regions of all four core histones, perfectly aligned and spaced to form intermolecular salt bridges, may hold the core molecules together and determine the diad axis of symmetry (OHLENBUSCH 1979). Models and references were given by TRIFONOV (1978), MIRZABEKOV et al. (1978), and ZINKE (1979).

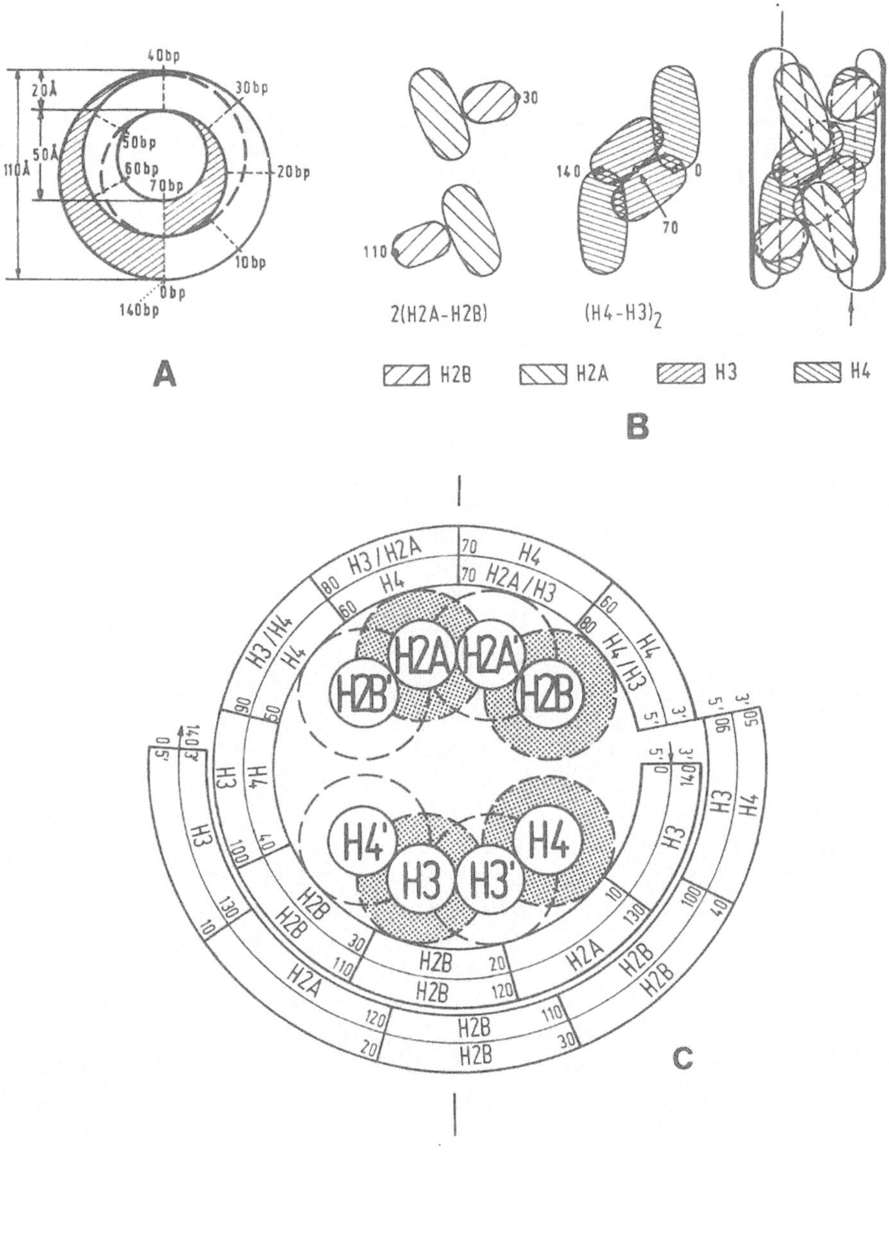

The influence of the ionic strength on the conformational state of the core particle was discussed in detail by ZINKE (1979); see also ZAMA et al. 1978). Conformational changes of the nucleosome may be an important step in the control and process of DNA replication and transcription (see below). Under in vivo ionic strength conditions (corresponding to about 150 mM NaCl), nucleosomes do not assemble in vitro. Recently, a so-called nucleosome-assembly factor was found (LASKEY et al. 1978). STEIN et al. (1979) demonstrated that acidic polypeptides can assemble nucleosomes at physiological ionic strength, and GERMOND et al. (1979) reported the same assembling activity for a nicking-closing enzyme. Hence it is very likely that certain proteins may also in vivo be involved in the maintenance of chromatin organization. Such proteins may inhibit an unorganized binding of differently charged molecules. The use of high salt concentrations during in vitro studies may, therefore, cause some changes in chromatin structure, such as nucleosome sliding, etc. (e.g., LEVY-WILSON 1979, WEISCHET 1979). Moreover, in vitro-reconstituted chromatin exhibits a considerably shorter DNA length per nucleosome (STEINMETZ et al. 1978, NOLL et al. 1980).

Histone H1 is involved in the interactions with the linker DNA (NOLL 1976). This internucleosome spacer exhibits a variable length between 10 and 70 bp (NOLL 1976, LOHR et al. 1977; reviewed by KORNBERG 1977, NAGL 1977a, THOMAS 1978), probably depending on the amount and conformational state of H1. The developmental and tissue-specific spacer lengths (e.g., SPADAFORA et al. 1976a, b, COMPTON et al. 1976, MORRIS 1976) may also be related to the transcriptional activity of a nucleus.

3.2 The Nucleosomal DNA

The histone core induces a tertiary structure in the DNA, the left-handed superhelix. The DNA is folded to about one-seventh its length, a value deduced from electron microscopic measurements of minichromosomes of SV40 and adenovirus 2 (GRIFFITH 1975, GERMOND et al. 1975) and theoretical calculations (CARLSON and OLINS 1976). The exact manner in which DNA is folded over the surface of the core is not yet known, but several models have been put forward (see Fig. 3). Some of the models display about $1^3/_4$ coils of DNA around a nucleosome core (NOLL 1977, FINCH et al. 1977, NOLL et al. 1980), other authors suggest that two concentric coils of DNA surround the core, an inner with 80 Å and an outer of 150 Å external diameters (BAUDY et al. 1976). MARALDI et al. (1979) showed the path of DNA in such a space-filling model, and summarized the evidence for it as obtained by various biochemical and biophysical methods (Fig. 3a). Many studies were performed with nucleases of different specificity, leading to cuts at defined sites of the linker DNA and/or nucleosomal DNA, but this topic is beyond the scope of this chapter (refer to FINCH et al. 1977, DOENECKE 1979, PRUNELL et al. 1979; examples of digested plant chromatin are given in Figs. 4 and 5; Table 1 summarizes the main results of nuclease experiments). The fundamental findings are that mild digestion of chromatin leads to DNA fragments of about 200 bp (and multiples thereof) due to cuts between nucleosomes. Longer digestion leads to degradation of the linker DNA. DNA fragments isolated from "trimmed" nucleosomes (i.e., core particles) have a length of 143 bp (e.g., THOMA et al. 1979), while fragments isolated from "sealed off" nucleosomes (i.e., core particles plus H1) are 166 bp

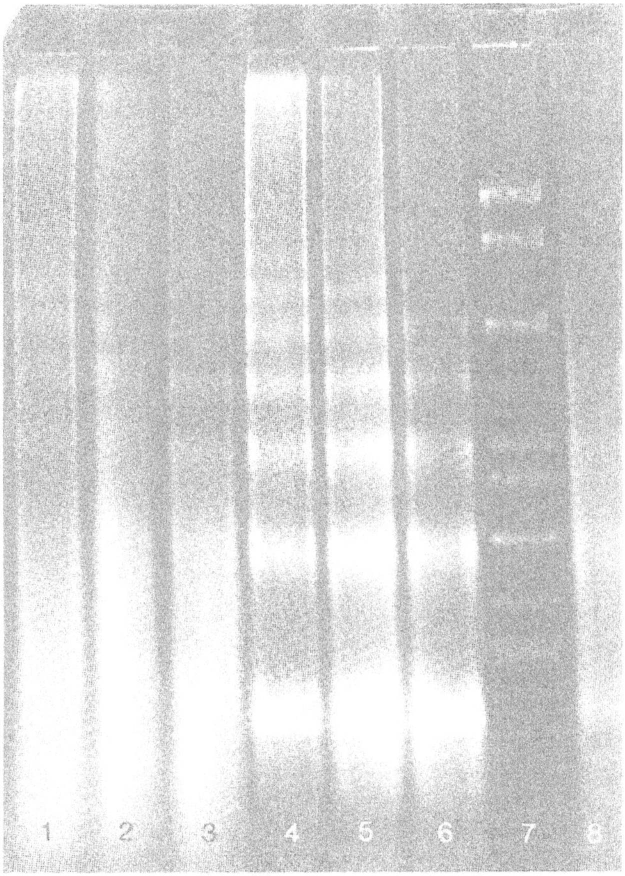

Fig. 4. Nucleosomal DNA pattern of chromatin from *Matthiola incana* seedlings (slots *1–3*), *Matthiola* flower petals (slots *4–6*) and *Brassica pekinensis* seedlings (slot *8*) as obtained by electrophoresis on 2.5% agarose gels. Staphylococcal nuclease digest was for 3 min (slots *1* and *4*), 6 min (slots *2* and *5*), 10 min (slots *3* and *6*) and 5 min (slot *8*). Molecular weight markers were DNA fragments of 1686 (*top*), 1320, 881, 535, 462, 357, 271, 230, 215, 180, 144, 133, 84, 38 and 34 base pairs of Hae III digested λdvl DNA. Slot *7* molecular weight markers. (Modified from LEBER and HEMLEBEN 1979b)

in length. An important structural feature of the core particle is the limited sensitivity of its DNA to nuclease digestion which is periodically varying along the DNA with a period close to 10 bp. One existing explanation of the phenomenon is that periodical distribution of sensitive sites along the chromatin in DNA results from periodical variation of exposure of sugar phosphate bonds to the surroundings due to the helical structure of DNA, folded around the histone core (NOLL 1977). Recent measurements of lengths of the digestion fragments of nucleosomal DNA using sequenced standards lead to the conclusion that the average distance between adjacent sensitive sites in chromatin DNA is noninteger: 10.3–10.4 bp, but that some sites are less sensitive (e.g., TRIFONOV and

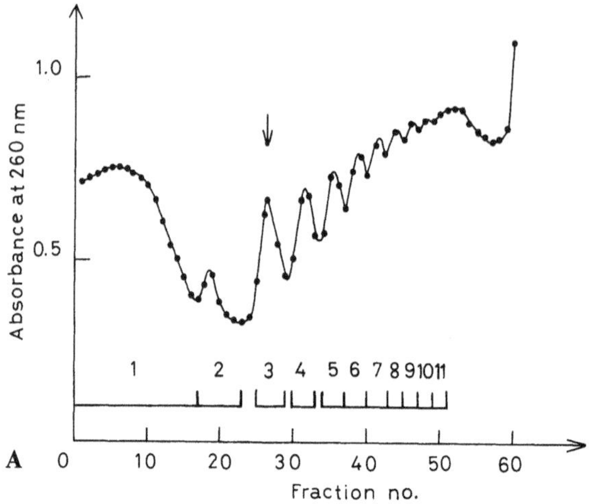

Fig. 5A, B. Size distribution of DNA fragments derived from fractionated chromatin subunits. Nuclease digested *Matthiola incana* chromatin (4 min digest) was fractionated on an isokinetic sucrose gradient (**A**). DNA was isolated from the pooled fractions (see *inserts*) and separated on 2.5% agarose gel (**B**) together with Bsp-digested λdvl DNA. *Arrow* points to positions of the nucleosome monomer. (Modified from LEBER and HEMLEBEN 1979a)

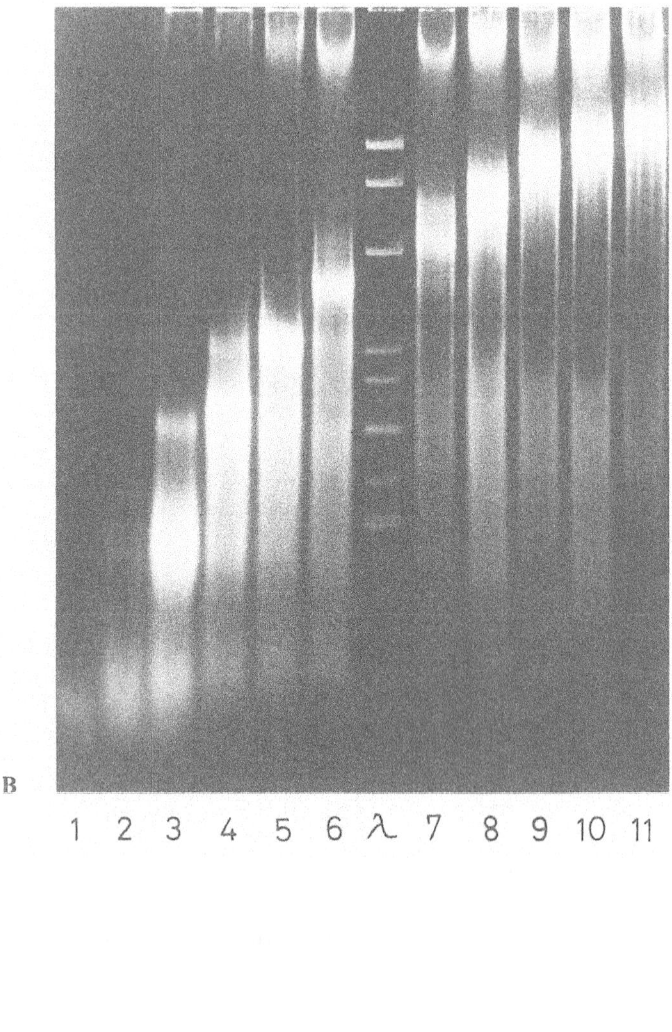

Table 1. Results of chromatin digestion with various nucleases

Type of nuclease	Digestion of native chromatin	Digestion of core particles
Streptodornase, endogeneous nuclear nuclease, *Micrococcus* nuclease	Nucleosomes (oligomeres, monomeres, core particles, depending on duration of treatment)	140 bp DNA fragment, later staggered cuts every 10.4 bp
Ca^{2+}/Mg^{2+}-dependent endogeneous nuclease	Nucleosomes (oligomeres, monomeres)	No cuts
Pancreatic DNase (DNase I)	Favored digestion of active genes	Staggered cuts on complementary strands every 10.4 bp (and multiples thereof)
DNase II	200, 400, 600 bp pieces	No cuts
DNase II $+Mg^{2+}$	100, 200, 300 bp pieces	Cuts at sites on and between nucleosomes, staggered cuts every 10.4 bp

BETTECKEN 1979, PRUNELL et al. 1979). This can be explained on the basis of smooth bending of DNA in chromatin (TRIFONOV and SUSSMAN 1979). But it is still a question whether the double helix can be bent around the nucleosome core without severe alterations in its conformation. CRICK and KLUG (1975) postulated that DNA is wound around the outer surface of the histone core not by continuously deforming the DNA, but by kinking DNA every 20 bp ("kinky helix"; see also SOBELL et al. 1976, VARSHAVSKY 1976, YATHINDRA 1979). On the contrary, LI 1976) postulated a distorted helix, which is compatible with the experimental data and the theoretical considerations of DNA chemophysics (see also SUSSMAN and TRIFONOV 1978, MIRZABEKOV and RICH 1979, SHINDO et al. 1980; for a general discussion of the DNA backbone refer to COHEN 1980).

Since the interactions between the DNA and the histones are electrostatic ones and therefore not sequence-specific, the molecular assembly of the chromatin fiber must be a function solely of the general, stereochemical and ionic, histone-histone and histone-DNA associations (WORCEL 1977, STEINMETZ et al. 1978, NOLL et al. 1980), although some proteins may control the assembly process. This is also consistent with the idea that histones represent general and unspecific repressors of gene activity by packaging the DNA into higher order structures. But while some authors find that nucleosomes are distributed in a random fashion over the genome (LACY and AXEL 1975, BOKHON'KO and REEDER 1976, DUERKSEN and PAUL 1976, FITTLER and ZACHAU 1979, PRUNELL 1979), some others report that nucleosomes in satellite DNA-containing chromatin are in tact with the DNA repeat unit (MAIO et al. 1977, BROWN et al. 1979).

3.3 Transcription, Replication, and Nucleosomes

A further controversial issue is the organization of "active chromatin". The histone-DNA ratio of the expressed portion of the genome is approximately the same, perhaps slightly lower than that of whole chromatin, but the non-histone-DNA ratio is drastically increased (reviewed by CAMERON and JETER 1975). Chromatin is converted into about 200 bp long pieces of DNA, when attacked with DNase II. These fragments contain eight molecules of the core histones and one molecule of H1. The template active portion of the genome is also converted into 200 bp long units of DNA, containing histones in the same proportion. Thus, histones and nucleosomes are apparently not disassembled during transcription in both animals (MATHIS and GOROVSKY 1976, REEVES 1976, 1978, GOTTESFELD and MELTON 1978, LILLEY 1978, PEDERSON 1978, BONNER 1979b) and plants (LEBER and HEMLEBEN 1979a, b). On the other hand, the template active chromatin shows a melting point as pure DNA also does. This fact indicates that histones in this fraction do not interact with DNA as they do in inactive, or bulk of, chromatin. Other types of biophysical studies (e.g., circular dichroism spectrum), attack by nuclease (see the following paragraph), and electron microscopic investigations have also shown that the DNA of template active chromatin is more like pure DNA (FRANKE et al. 1976, LAIRD et al. 1976, FRANKE et al. 1979, SCHEER et al. 1979, DOENECKE 1979). Possible mechanisms of chromatin activation by histone modification, acidic protein binding, and changes in nucleosome conformation are discussed and reviewed by WEINTRAUB and GROUDINE (1976).

Selective digestion of transcriptionally active chromatin is optimally obtained by use of DNase I (pancreatic) (WEINTRAUB and GROUDINE 1976, GAREL and AXEL 1976, STALDER and BRAUN 1978).

Also DNase II cleaves preferentially in such regions of chromatin as are activated for RNA synthesis (GOTTESFELD 1978), and under certain conditions. DNase II cleaves the nucleosomal DNA internally in a way that yields DNA fragments of half-nucleosomal size (ALTENBURGER et al. 1976). Thus, nucleosomes in transcriptionally active portions of the genome are apparently different from those in inactive chromatin, probably in an altered conformation due to histone modification and nonhistone protein binding (GOTTESFELD et al. 1975, ref. in NAGL 1977a, GIRI and GOROVSKY 1980, JOHNSON et al. 1979). A structural difference between nucleosomes of active and inactive chromatin was found in an electron microscopic study by FRANKE and SCHEER (1978) and JOHNSON et al. (1979), for higher order structures see GREIMERS and DELTOUR (1981). FRANKE et al. (1976) found no nucleosomes in active ribosomal genes, and only a few nucleosomes in active mRNA genes. There is a technical problem, as RNA polymerase and nucleosomes are of the same size, and therefore it is not always easy to make the distinction, but the histones can be removed with SDS, while the RNA polymerases remain bound to DNA (e.g., FRANKE et al. 1977; for the preparation and appearance of transcriptionally active chromatin see also MILLER and BAKKEN 1972; SPRING et al. 1974, FRANKE et al. 1979, MCKNIGHT et al. 1979, SCHEER et al. 1979). Thus, in general, the electron microscopic investigations indicate that there is a negative correlation between the number of nucleosomes and the rate of transcription, while the digestion experiments do not indicate such a relationship (e.g., LEBER and HEMLEBEN 1979a, b).

Chromatin is known to be soluble in low ionic strength buffer. In 0.15 to 0.3 M NaCl or KCl solution, however, chromatin becomes almost completely insoluble. This is the range of ionic strength found in the nucleus of higher organisms. Therefore, in the nucleus the chromatin must be insoluble. The template expressed portion of chromatin is soluble in solutions of the above ionic strength and we must therefore envisage the chromatin of interphase nuclei as consisting of an aggregated, precipitated mass of nonexpressed nucleosomal chromatin from which protrude a few loops of soluble chromatin which is in the expressible form, that is, in a form transcribable by RNA polymerase (BONNER 1979b).

The physical properties of replication-active chromatin are also only poorly understood. Chromatin assembling apparently follows DNA replication immediately, as visible in electron micrographs of replicating chromatin (McKNIGHT and MILLER 1977, McKNIGHT et al. 1979), and by nuclease digestion (LEVY and JACOB 1978). If protein synthesis is inhibited during the DNA replication period, the results obtained indicate a conservative distribution of the old histones (WEINTRAUB 1974), but this result was contradicted by other authors (reviewed by NAGL 1977a). For instance, LEFFAK et al. (1977) found that nucleosomes remain bound to the conservative DNA strand, while the most recently formed nucleosomes are associated only with the newly polymerized DNA strands. WEINTRAUB (1979) reported that actively transcribed genes are assembled into nucleosome-like structures close to the replication fork and that they also acquire a conformation characteristic of actively transcribed nucleosomes (i.e., DNase sensitivity). Interesting aspects to the problem have been contributed by ZINKE (1979). According to his model the dynamics of chromatin structure can be explained by means of the balance between hydrophobic and electrostatic interactions of the histones and the DNA.

Apart from the question of nucleosomal organization of replicating chromatin, no changes in nucleosomes occur during the rest of the cell cycle. The basic subunit structure is retained even by mitotic chromosomes, although significant morphologic changes occur within the chromatin fiber (RATTNER et al. 1975, HOZIER and KAUS 1976, WIGLER and AXEL 1976). This leads to aspects of higher-order structures of chromatin.

4 Higher-Order Coiling: Chromatin Fibers

The DNA helix is coiled in the nucleohistone fiber in a nonrandom, precise manner. At one end of the coiling spectrum we have the nucleosome and at the other end the metaphase chromosome, with higher-order DNA coils bridging the gap between these two levels of DNA organization. We do not know exactly how DNA is packed into higher-order chromatin structures, nor do we understand the various levels of chromatin condensation between the nucleosome level and the level of the mitotic chromosome. Nevertheless, several factors are known to be involved in these processes, and some of the steps have been elucidated.

Since chromatin usually is, by its very nature, only imperfectly ordered, X-ray and neutron diffraction observations permit only limited and indirect interpretation, and the obvious method is that of electron microscopy. Although methodical aspects will not be reviewed in this chapter, it must be pointed out that structural analysis of chromatin at various levels of condensation normally needs spreading or sectioning of chromatin after pretreatments under unphysiological conditions, i.e., after induction of artificial changes. The salt concentration during isolation strongly influences the appearance of chromatin in the electron microscope due to changes in (1) charge repulsion at low ionic strength, (2) the decrease of the electrostatic interactions at high ionic strength, and (3) the stabilization of the particle due to hydrophobic associations at moderate ionic strength (ZINKE 1979). The appearance of chromatin is, therefore, significantly different after different extraction methods and the use of different buffers and chelating agents, etc. (e.g., POPENKO and VENGEROV 1978; one-step method for root tip nucleosomes was reported by MORENO et al. 1978, and a reliable method for higher plant nucleosomes was published by LUTZ and NAGL 1980). Chromatin prepared for ultrathin sections only very rarely disperses to the nucleosome level in vivo (for *Helianthus,* see NAGL 1976b) but can also be decondensed before fixation (e.g., ERENPREIS et al. 1978, CAMERON et al. 1979, POLYAKOV et al. 1979). FUGE (1980) described nucleosome-like particles in chromosomes of the crane fly, *Pales ferruginea,* during the first meiotic anaphase. This interesting finding is consistent with the biochemical data on histone phosphorylation during chromosome condensation (see Sect. 5). Evidently, the formation of the highest-order structures requires some uncoiling at the lowest-order level, probably because the condensing proteins need certain binding sites.

In the presence of histone H1 the linker DNA is coiled up, as this histone may form cross-links between core particles. Then the "beads-on-a-string" structure is transformed into a chromatin filament of 10 nm in diameter. Probably there are specific binding sequences for H1, as RENZ (1975) reported that lymphocyte DNA but not *E. coli* DNA shows preferential binding of H1 (this idea is, however, a contradiction to the nonspecific binding of the core histones to DNA). Analyzing DNA fragments of different size, he found that most of the 2×10^6 fragments had at least one binding site, whereas most of the 2×10^5 fragments had none. In addition, BLUMENFELD et al. (1977) could show that specifically modified H1 bound preferentially to repetitive DNA sequences in heterochromatin.

The 10 nm filament has long been known from electron microscopy and has sometimes been named the "unit fiber" or the "elementary fiber", in which the DNA is in its tertiary structure or super-coil (Fig. 6; unfortunately both terms have been also used for different structures).

The compaction of the beaded nucleosome fibril into the 10 nm fiber and vice versa can be studied by dialysis against different NaCl concentrations. Such dissociation and reassociation studies are manifold, and only the principle will be reported here. The dissociation of histone types from chromatin is a function of NaCl concentration and takes place in three distinct changes (BURTON et al. 1979). Lowering the ionic strength first leads to two transitions in $s_{20,w}$ and $D_{20,w}$ which indicate the re-assembling of the nucleosomes (GORDON

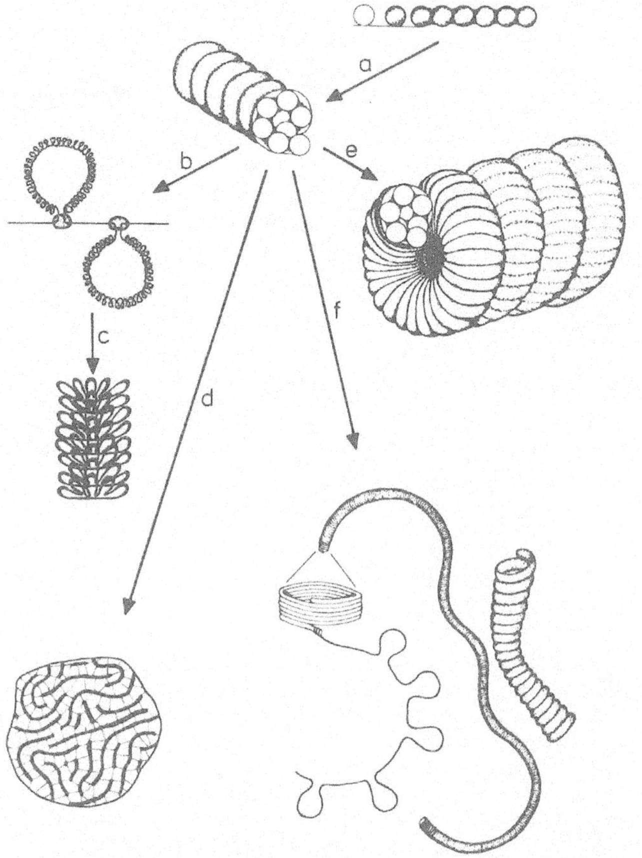

Fig. 6. From the level of the nucleosome to the levels of mitotic chromosomes and condensed interphase chromatin. *a* Supercoiling of the 10 nm nucleosome fibril to the 25–35 nm solenoidal fiber; *b* folding of the soilenoidal fiber into domains by binding of nonhistone chromosomal proteins. (According to NAGL 1979a); *c* polymerization of the nonhistone proteins to form a chromosome scaffold (radial loop model of MARSDEN and LAEMMLI 1979); *d* interphase chromatin condensation by cross-linking of the 10 nm or 30 nm fibers; *e* and *f* helical coil models of the mitotic chromosome; *e* the solenoidal fiber is once more coiled to form the interphase chromosome tubes (∅ 200 nm); a further wind would lead to a metaphase chromosome 600 nm in width. (According to SEDAT and MANUELIDIS 1977); *f* higher levels of organization in chromosomes according to BAK et al. (1979a): The solenoid fiber is coiled to the 400 nm "unit fiber" (i.e., the earthworm-like structure); the *lower part* of the drawing indicates decondensation into the 10 nm fibril as occurring in lampbrush chromosomes. *At right* the final coiling of the "unit fiber" generating the maximally contracted metaphase chromatid is shown

et al. 1978), and then allow the addition of H1 and the formation of the 10 nm fiber (BURTON et al. 1979). The multifunctional role of H1 was discussed by GRIFFITH and CHRISTIANSEN (1977).

A quaternary structure of DNA, a coiled super-coil, was expected by BRAM et al. (1975) on the basis of diffraction studies, and by electron microscopists who regularly found a chromatin fiber, some 25 to 35 nm in diameter (e.g., DUPRAW 1970, RATTNER and HAMKALO 1978a, b, 1979).

H1-depleted chromatin and pure DNA spontaneously take up a superstructure at sodium acetate concentrations of 100 mM and above. Presumably the cations condensate H1-depleted chromatin by inducing DNA superstructures, probably by binding to phosphate groups that are not neutralized by histones (THOMA et al. 1979). However, the condensation is not so pronounced or, more important, so definite as when H1 is present. In the formation of higher-order structures, unique regions of DNA on neighboring nucleosomes come close together so that an H1 polymer may be formed in the center of the superhelical nucleosome filament (THOMA et al. 1979).

There is no general agreement about how the 25 nm fiber is organized, but several findings point in the following direction. Under certain salt concentrations and the presence of divalent cations, nucleosomes are able to form super-beads (or super-nucleosomes). Such super-structures may finally form a compact or solenoidal, knobby 25–35 nm fiber. According to other authors, the 10 nm fibril is directly converted into a solenoidal 25 nm fiber by continuous super-coiling (Fig. 6; FINCH and KLUG 1976, BASU 1979; see also HOZIER et al. 1977, CAMPBELL et al. 1978, STRÄTLING et al. 1978, SCHEER and ZENTGRAF 1978, NICOLINI 1979a, b).

Recent models consider that super-beads and the 25 nm fiber are two configurations of the same level of higher-order organization, a discontinuous (clustered) one and a continuous one (OLINS 1977, BURTON et al. 1979, RENZ 1979, VENGEROV et al. 1978, KIRYANOV et al. 1976). The actual diameter of the fiber, and the presence of a solenoidal or a compact fiber, largely depend on the length of the chromatin repeat length, which is variable in the linker region (WORCEL 1977), or on how tightly the nucleosomes become packed, which is dependent on the Mg^{2+} concentration (BASU 1979), or on both so that different ratios of DNA packaging are achieved (WORCEL and BENYAJATI 1977, WORCEL 1977). In a systematic electron microscopic study, THOMA et al. (1979) found that with increasing ionic strength, chromatin folds up progressively from a filament of nucleosomes (10 nm fiber) at 1 mM monovalent cations through some intermediate higher-order helical structures with a fairly constant pitch but increasing numbers of nucleosomes per turn, until finally at 60 mM (or else in 0.3 mM Mg^{2+}) a thick fiber of 25 nm diameter is formed, corresponding to a structurally well-organized but not perfectly regular super-super-helix or solenoid of pitch 11 nm as described by FINCH and KLUG 1976 (SEE ALSO CAMPBELL et al. 1978, NICOLINI 1979b). It is concluded that H1 is bound to the unique region on a complete two-turn, 166 bp nucleosome. In the formation of higher-order structures, either the H1 molecules may polymerize (THOMA et al. 1979), or Mg^{2+} ions act as neutralizers of repulsion forces between negatively charged DNA phosphates (JERZMANOWSKI et al. (1979). In any case, formation of superstructures is controlled by the increase of ionic strength in the nucleus (see also JERZMANOWSKI and STARON 1980).

Table 2 shows the main requirements for chromatin assembly and condensation into the 25–35 nm fiber. The higher orders of packaging are not as clear as those discussed, and will be presented separately for mitotic condensation of chromosomes and interphase chromatin condensation.

Table 2. Coiling hierarchy in compaction of DNA in metaphase chromosomes according to SEDAT and MANUELIDIS (1977), BAK et al. (1979a)

	Level of condensation	Diameter (nm)	Molecular requirement	Compaction of DNA	
				Partial	Total
I:	DNA Nucleosome fibril	3 10	H1	7	7
II:	Nucleosome fibril Solenoid fiber	10 25–50	Mg^{2+} (Ca^{2+}, in vitro)	6/12	42/84
III:	Solenoid fiber Tube/unit fiber	30 200/400	Mg^{2+}, pH1	17.8/40	748/3,360
IV:	Tube/unit fiber Chromosome	200/400 600	NHP	6.3/5	4710/16,800

pH1 = phosphorylation of histone H1, NHP = nochhistone chromosomal proteins. For some other comparative data see Table 12 in NAGL (1976a), Table 1 in SEDAT and MANUELIDIS (1977, MARSDEN and LAEMMLI (1979).

5 Domains and Mitotic Chromosomes

Looking at solenoids and super-nucleosomes we are still far away from the packaging ratio of DNA in a metaphase chromosome being in the order of 5,000 or 10,000:1. (DUPRAW 1970, NAGL 1976a, SEDAT and MANUELIDIS 1977). Three different approaches exist to explain the next steps of chromatin condensation, and thus DNA packaging: folding, coiling, and looping of the chromatin fiber.

The folded-fiber model as originally suggested by DUPRAW (1970) is based on the well-known "spaghetti-chromosomes" as seen in the electron microscope after whole-mount preparation (see also COMINGS 1974). As, however, this appearance is very probably an artifact, the representative of the folded-fiber model reduced the folding concept to the level between the solenoidal fibril and the coiling of mitotic chromosomes as visible in the light microscope (LEWIN 1974).

More elegantly the coiling models explain the condensation from the level of nucleosomes to the level of chromosomes (PERSON and SUZUKI 1968), but there still exist, however, considerable differences between the results of different authors. While SEDAT and MANUELIDIS (1977) suggest that the solenoid fiber is further coiled to a super-solenoidal tube of 200 nm diameter, and this tube again to a metaphase chromosome of about 0.6 μm, BAK and ZEUTHEN (1977) found that the solenoid fiber is supercoiled to a supersolenoid, a tube some 400 nm in diameter with a wall made up of the solenoidal structure of 30 nm diameter. This tube was termed "unit fiber". In this "unit fiber" the DNA is contracted in the order of 1,300–1,500. The "unit fiber" is then further coiled to the chromosome by a factor of about 5 (Fig. 6; see also BAK et al. 1979a, b, ZEUTHEN et al. 1979). LERMAN and DEGTYAREV (1978) assumed that the formation of such regular structures requires a molecular basis, i.e., distinct protein-binding DNA sequences at identical distances. The authors argued that the necessity of this regular distribution of certain sequences involved in chromatin coiling could have made the splitting of genes in eukaryotes necessary.

Another attractive hypothesis is that of radial loops in chromosomes (Fig. 6). Already in 1969, SONNENBICHLER observed subunits of chromosomes from which loops were projecting. More recently, ADOLPH et al. (1977), PAULSON and LAEMMLI (1977) spread chromosomes after extraction of most of the proteins and could show that there remained a chromosome scaffold built up of residual proteins, from which many long loops of DNA protruded. The residual proteins are apparently firmly bound to DNA (perhaps covalently) and must polymerize to the chromosome scaffold during mitotic chromosome condensation (see JEPPERSEN and BANKIER 1979, KRAUT and WERNER 1979, RAZIN et al. 1979). MARSDEN and LAEMMLI (1979) showed that metaphase chromosomes released radial loops of chromatin fibers or DNA, the diameter and length of which depended on how many of the histones and nonhistone proteins are removed.

There is one aspect of chromatin condensation which is common to all three models: the domain. Domains have been detected in several ways, as regions of supercoiling, as regions defined by enzymatic excision of intact chromatin, as lengths of DNA seen attached to the scaffolding of the chromosome, as super-coiled discs of DNA after gentle lysis of cells and spreading of nucleids, and as chromatin domains directly shown in the electron microscope (COOK and BRAZELL 1975, BENYAJATI and WORCEL 1976, ADOLPH et al. 1977, IGO-KEMENES and ZACHAU 1978, MCCREADY et al. 1979, MULLINGER and JOHNSON 1979, OKADA and COMINGS 1979). One might speculate that the domains are the basis of longitudinal chromosome differentiation as known to exist from the chromomere pattern at meiosis, the banding pattern of giant chromosomes, and the fluorescence and Giemsa bands which can be induced in metaphase chromosomes (e.g., COMINGS 1978).

Evidence is now increasing that the quinacrine and Giemsa-G bands and the complementary pattern of R bands along chromosomes are a reflection of the predominance of GC- and AT-rich regions (e.g., COMINGS and DRETS 1976, SCHWEIZER 1976, VAN DE SANDE et al. 1977). Besides these nonpolymorphic, species-specific and highly conservative bands, Giemsa-C bands can be induced in constitutive heterochromatin, which may occur in a polymorphic condition both within and between individuals, and which are highly variable between species (SCHWEIZER and EHRENDORFER 1976, BENNETT et al. 1977, GREILHUBER and SPETA 1978). But let us bring our attention back to the quinacrine (Q) and G bands. Although many of the possible bands cannot be discerned now, because their resolution is reduced by chromosome condensation, and the fusion of several possible bands is reduced to the fewer visible ones, it has been suggested that the Q and G bands actually may correspond to (fused) chromomeres (SCHNEDL 1974). The latter represent nothing else than regions of the extended chromosomes which are more condensed than the remainder. This could indicate that AT-rich DNA sequences are involved in chromosome condensation in general.

Unfortunately, G bands cannot be induced in plant chromosomes, and Q bands can so far be induced only in *Lilium* (HOLM 1976). The reason for this probably is the higher condensation of plant chromosomes (about 3.5 to 6.5 times more than human chromosomes: GREILHUBER 1977), and this again might be caused by the higher proportion of repetitive DNA (and thus also of condensation sequences) in plants than in animals (NAGL 1976a). A corollary of the above considerations is that one difference of plant chromosomes versus mammalian chromosomes must be sought in the higher-order arrangement of the chromatin fiber (SCHWEIZER 1979). Unfortunately, the organization of plant chromosomes is currently a rather neglected field. Another neglected difference between plant and animal chromatin is shown in the following section (Sect. 6) on interphase nuclei.

The control of condensation of the solenoid fiber to higher-order structures seems to depend on histone H1 phosphorylation. It has been known for some years that certain residues of H1 become phosphorylated during mitotic condensation (LAKE and SALZMAN 1972, BRADBURY et al. 1973a, b, 1974a, b, GURLEY et al. 1973, 1974; BALHORN et al. 1975, ELGIN and WEINTRAUB 1975). H1 phosphorylation was found to strongly support its cross-linking capacity (MATTHEWS and BRADBURY 1978). WORCEL (1977) suggested that variations in the basic subunit, the nucleosome, are responsible for the changes in the higher-order structures. Variant inner core histones and a modified H1, and specific nonhistone proteins becoming attached to the nucleosome may alter its configuration. But also a polymerization of nonhistone chromosomal proteins to the scaffold seems reasonable [as already suggested by SUTTON (1972) in a somewhat different manner].

6 Interphase Chromatin: Heterochromatin

The chromatin of interphase nuclei and noncycling nuclei is normally divided into euchromatin and heterochromatin. These terms, however, are used in a different sense by different authors. While some understand them more in a functional sense (as template active and template inactive chromatin), heterochromatin is regarded by others in the original structural sense as those portions of the chromosomes which remain condensed during interphase, which are rich in repetitive DNA, which do not carry genes, which replicate asynchronously when compared with euchromatin, etc. The latter definition applies to constitutive or karyotypical heterochromatin, which can be made visible in the karyotype by the C-banding technique (for further description and for references see YUNIS and YASMINEH 1971, LIMA-DE-FARIA 1975, 1979, NAGL 1976a, MELLO 1978). In female mammals, the nuclei exhibit a sexchromatin body, which was defined as facultative heterochromatin by SMITH (1945). This body represent one of the two X chromosomes of females, which evidently becomes inactivated due

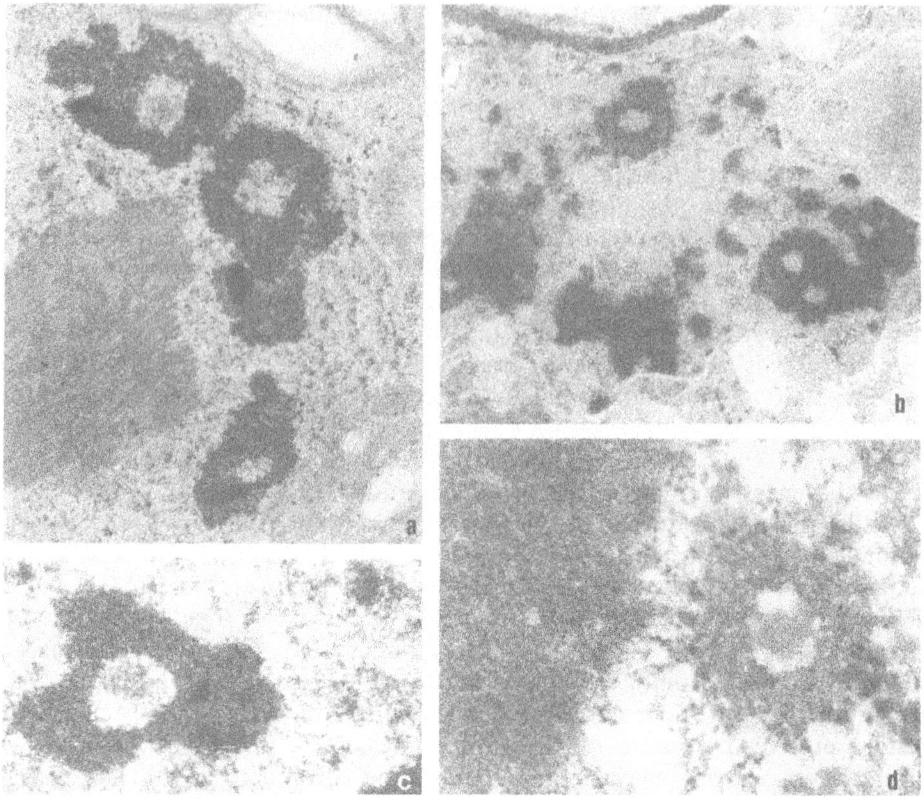

Fig. 7a–d. Organization of heterochromatin in plants during interphase. Note the central hole in many chromocenters; this can be taken as indication of a super-solenoidal structure of interphase heterochromatin. **a** *Rhinanthus minor* (×15,000); **b** *Hedera helix* (×20,000); **c** *Allium flavum* (×25,000); **d** *Anthemis altissima*. (×80,000)

to a dose compensation effect (LYON 1968). Although the condensed X chromosome is late replicating, it possesses the same DNA nucleotide sequence as the active one, so that it cannot be regarded as constitutive heterochromatin. In certain vertebrate nuclei also euchromatin as well as constitutive and facultative heterochromatin can be in the condensed state. This condensed euchromatin does not, of course, exhibit any of the characteristics of real heterochromatin except that it appears condensed and inactive (the condensation mechanisms in euchromatin and heterochromatin may be quite different, see the following paragraphs). The increasing use of these terms synonymously and indiscriminately in the literature has led to a serious degree of conceptual and operational confusion (NAGL 1979a, b). In this essay only the constitutive heterochromatin is considered and further called heterochromatin. This does not exlude the fact that the term heterochromatin still represents a super-conception covering various types of chromatin (COMINGS, D.E., IN MELLO 1978; see also the differential banding behavior of heterochromatic segments: SCHNEDL 1974, NAGL 1976a).

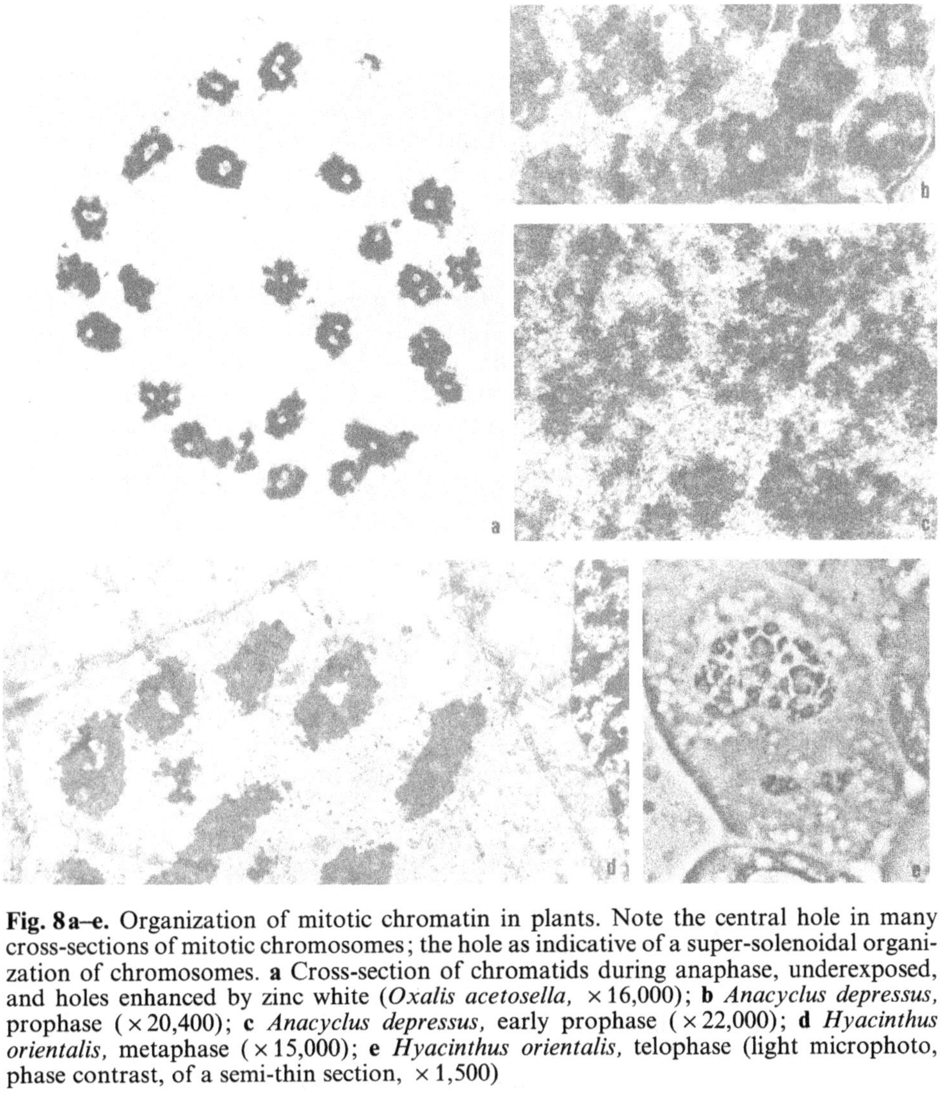

Fig. 8a–e. Organization of mitotic chromatin in plants. Note the central hole in many cross-sections of mitotic chromosomes; the hole as indicative of a super-solenoidal organization of chromosomes. **a** Cross-section of chromatids during anaphase, underexposed, and holes enhanced by zinc white (*Oxalis acetosella*, ×16,000); **b** *Anacyclus depressus*, prophase (×20,400); **c** *Anacyclus depressus*, early prophase (×22,000); **d** *Hyacinthus orientalis*, metaphase (×15,000); **e** *Hyacinthus orientalis*, telophase (light microphoto, phase contrast, of a semi-thin section, ×1,500)

If heterochromatin actually remains in a mitotically coiled state at interphase, there should be no difference between the organization of the metaphase chromosome and the chromocenter (i.e., a piece of interphase heterochromatin). Actually, there is often a close structural similarity between mitotic chromosomes and chromocenters at both the light and ultrastructural level, particularly with respect to the appearance of a super-solenoidal organization (Figs. 7, 8; NAGL unpublished). CARNIEL (1979) described band-like structures in chromocenters of *Lupinus polyphyllus*, which somewhat resemble meiotic poly-complexes. Their origin and functional significance are not known, but they show that interphase heterochromatin may represent various classes or types of constitutively condensed chromatin.

As already discussed, the permanently condensed state of heterochromatin ultimately depends on the primary sequence of the heterochromatic DNA, but other chromosomal constituents must mediate this condensation. Heterochromatin is known to be rich in highly repetitive DNA, although this DNA is not necessarily discernible as satellite DNA. HSIEH and BRUTLAG (1979) isolated a protein from *Drosophila* embryos that preferentially binds to a limited region of supertwisted satellite DNA.

Understanding the function of heterochromatin is as difficult as understanding its structure. BRUTLAG et al. (1977) pointed out that the misconception that satellite DNA has no essential function is based on the lack of genes in heterochromatin and on the viability of individuals with large heterochromatic deletions. The authors reviewed the literature, which indicates that heterochromatin does have a function in the germ line, and that variations in satellite DNA could lead to speciation (see also PEACOCK et al. 1977). NAGL (1978, 1979b) reviewed indications for a role of quantitative and spatial variation of heterochromatin (and chromatin in general) in differentiation and speciation, i.e., in ontogenesis and phylogenesis. LIMA-DE-FARIA (1975) attributed the highest level in the control hierarchy of the karyotype and the chromosomes to heterochromatin and satellite DNA. JOHN and MIKLOS (1979) emphasized that the predominant feature of satellite DNA (and heterochromatin) is its variation in several respects. At least some types of satellite DNA appear to have no somatic functions but do play a role in the germ line, particularly in the regulation of recombination between homologous chromosomes (see also MIKLOS and JOHN 1979).

7 Interphase Chromatin: Euchromatin

Euchromatin organization evidently underwent striking evolution and cannot be compared, therefore, between distinct taxa. Particular structural dynamics are found in the euchromatin of higher animals, whereas nuclear structure in plants is evidently controlled by a different mechanism (NAGL 1979c). Overgeneralization of results obtained in mammalian nuclei has for a long time obscured specific features of plant chromatin. One of the main problems in studying chromatin by means of electron microscopy can be seen in the fact that condensed chromatin of the same electron density covers at least four different classes of chromatin: constitutive heterochromatin, facultative heterochromatin (in female mammals), tissue-specifically condensed euchromatin (in vertebrates only), and species-specifically condensed euchromatin (in plants only (NAGL 1979c).

The main difference between euchromatin in plants and higher animals is, therefore, the species-specific degree of interphase condensation in the former, and the tissue-specific degree and pattern of interphase condensation in the latter. Actually, it has long been known that the nuclear structure is specific for many plant species, genera, and sometimes families (reviewed by TSCHERMAK-WOESS 1963). At the ultrastructural level, three types of chromatin organization have been found: the diffuse, the chromomeric, and the chromonematic (or reticulate) (Fig. 9; NAGL 1979, NAGL and FUSENIG 1979, LAFONTAINE 1974). Nuclei with diffuse euchromatin always possess chromocenters, so that this type is also known as "chromocenter type". Nuclei with chromomeric and chromonematic euchromatin organization may also possess chromocenters. In general, the structural complexity is positively correlated with increasing 2C DNA content of the species (NAGL and FUSENIG 1979). In detail, the proportion of euchromatin which is species-specifically condensed (in the form of chromomeres or chromonemata) is positive correlated with the 2C DNA value, indicating that any excess DNA which may originate during evolution becomes condensed and thus "silenced" (NAGL 1979d). This could recently be proven by [^{3}H]-uridine autoradiography: The absolute amount of RNA synthesized is similar in different plants, because the transcription rate decreases with increasing 2C DNA content and proportion of condensed chromatin (Fig. 10). [^{3}H]-uridine is mainly incorporated into decondensed chromatin (Fig. 11a), and the probability of finding transcription complexes in spread chromatin (Fig. 11b) is very rare in species with chromonematic, DNA-rich nuclei (W. NAGL and U. SOKOL, unpublished).

If the proportion of condensed chromatin is estimated quantitatively, the high species-specificity of chromatin organization in plants allows identification of a species by this parameter. NAGL (1980) used electron microscopic morphometry of thin sections of noncycling nuclei and statistics for the diagnosis of species and hybrids [for several *Microseris* species and allopolyploid hybrid species see NAGL and BACHMANN (1980), for the somatic hybrid *Arabidobrassica* see NAGL and HOFFMANN (1980)]. Moreover, the close, although genus- and family-specific correlation between 2C DNA content and percentage of condensed chromatin (Table 3) is significant and allows prediction of the 2C value from the nuclear texture and vice versa. Also the highest 2C value in a given genus or family can be predicted, i.e., the value which corresponds to 100% condensed chromatin (NAGL 1980).

At the light microscope level, species-, genus-, and family-specific nuclear structures have been well known for a long time (reviewed by TSCHERMAK-WOESS 1963). The morphological features can be digitalized by either scanning cytophotometry (Fig. 12) or by texture analyzing systems.

In spite of species-specificity, the nuclear structure and particularly the nuclear ultrastructure show some variation in an individual plant. Most of this variation can be attributed to cell cycle-specific changes, particularly in chromonematic nuclei (Figs. 13, 14; NAGL 1970b, 1977b; KWIATKOWSKA and MASZEWSKI 1979, SANS and DE LA TORRE 1979, for animals see SETTERFIELD et al. 1978, MOSER and MÜLLER 1979, NICOLINI 1979a, b, SAWICKI 1979; for a review see NAGL 1977c). Some structural variation is caused by extreme dehydration of nuclei (e.g., in resting seed, dormant buds and roots; LAI and SRIVASTAVA 1976, SAUTER and ULRICH 1977, DELTOUR et al. 1979, RISUENO and DE LA ESPINA 1979), or by hydration and accumulation of acidic proteins in very active nuclei

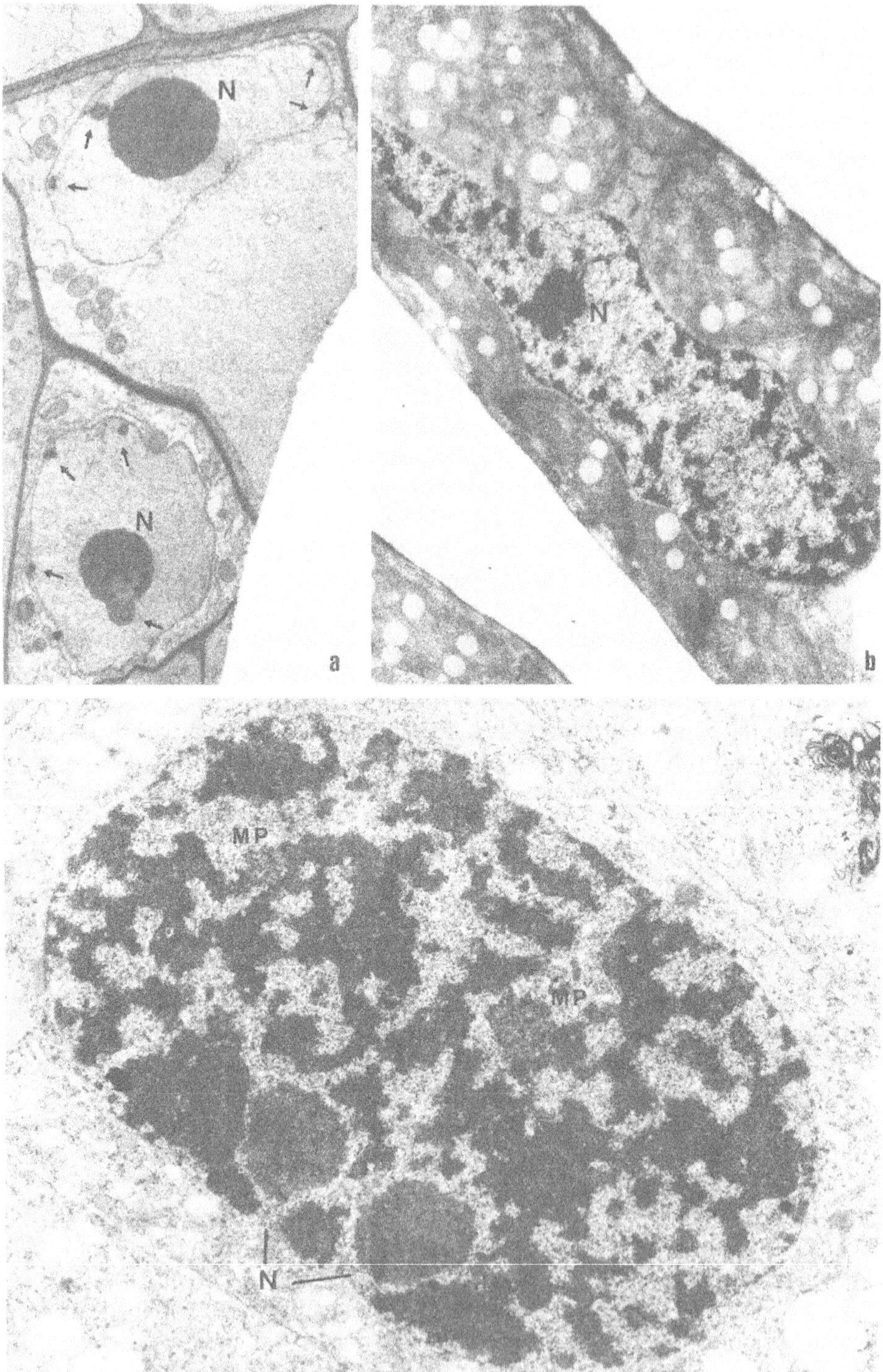
N
N
a
N
b
MP
MP
N
c

of the gametophyte and embryo suspensor. In such cases either the proportion of condensed chromatin (but probably not its total amount) may be altered, or the gross morphology may be changed, particularly if a nucleus also becomes highly endopolyploid (e.g., origin of polytene chromosomes; NAGL 1969a, b, 1970a). But all those changes never reach the high structural variability that is visible in nuclei of mammalian cells.

In mammals and other vertebrates, mitotically cycling nuclei often show a diffuse structure during interphase with some small differences between G1 and G2, while the chromatin undergoes tissue-specific condensation during differentiation (DERENZINI et al. 1977, 1979, BRASCH and SINCLAIR 1978, JEANNY and GONTCHAROFF 1978, AL-HAMDANI et al. 1979, BARSOTTI and MARINOZZI 1979, SCHMALENBERGER 1980). It seems that large portions of the genome become irreversibly silenced; i.e., cell determination occurs by inactivation of certain chromatin domains. Even if the cell activity is enhanced by hormonal stimulation etc., the decondensed chromatin is the only decondensing part of the nucleus, thus increasing the total nuclear volume, while the condensed chromatin is not affected (BRASCH 1978, KIEFER and SANDRITTER 1978). The amount and pattern of condensed chromatin may even vary extremely between different cell types of identical tissue as in the Guinea pig retina (NAGL 1979a, SCHMALENBERGER and NAGL 1979, SCHMALENBERGER 1980).

It was shown in several papers that condensed chromatin is inactive with respect to RNA synthesis (HSU 1962, ALLFREY et al. 1963, SCHMALENBERGER and NAGL 1979). The tissue-specific condensation pattern and thus inactivation of parts of the genome are understandable in the light of suggestions that the chromosomes occupy distinct regions of the nuclear cavity during interphase (BROWN et al. 1979; CREMER et al. 1979). This position may be fixed by attachment of telomeres, or centromeres, and/or other chromosomal regions to the nuclear membranes (BRASCH and SETTERFIELD 1974, and others) or by interaction with the nuclear matrix (COMINGS and WALLACK 1978, WUNDERLICH 1978, BEREZNEY 1979).

The control of chromatin condensation is exerted by modification of histones and nonhistone proteins. Particularly the phosphorylation of H1 at a residue different from that which is phosphorylated during mitotic chromosome condensation, causes chromatin decondensation in working nuclei (LOUIE and DIXON 1973, ORD and STOCKEN 1979, and many others, for reviews see BUSCH 1978, NAGL 1976a).

On the other hand, condensation of chromatin during the maturation of avian red blood cells and during spermiogenesis is brought about by a specialized histone H5 and protamines, respectively, which are phosphorylated after their synthesis and subsequently dephosphorylated. Cross-linking proteins are known

Fig. 9a–c. Electron micrographs of thin sections showing the species-specific chromatin organization in plant interphase (working) nuclei. **a** *Arabidopsis thaliana,* chromocenter-type nuclei (euchromatin is diffuse, completely decondensed, heterochromatin forms chromocenters: *arrows*). **b** *Dictamnus albus,* chromomeric organization (euchromatin is partly decondensed, partly condensed into chromomeres; heterochromatin is not visible in this nucleus). **c** *Allium carinatum,* chromonematic nucleus with chromocenters (most of the euchromatin is condensed into chromonemata, the heterochromatin forms chromocenters of the same electron density as the chromonemata. *MP* "micropuff"; *N* nucleolus. (**a**, **b** ×6,000; **c** ×13,000). (**a** Modified from NAGL 1979c; **b** original; **c** modified from NAGL and FUSENIG 1979)

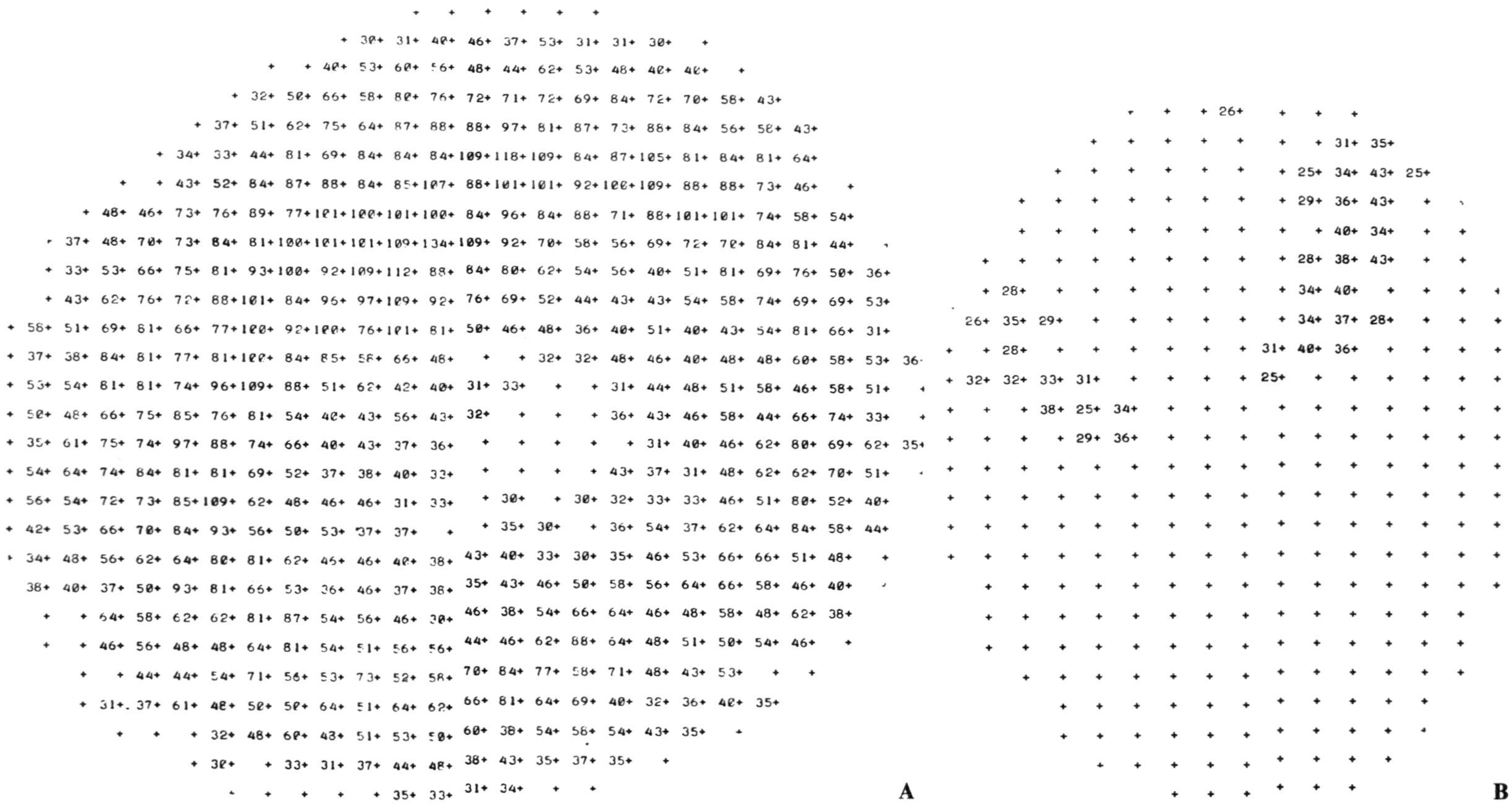

Fig. 10 A, B. Computer drawings of DNA extinction values as obtained by Feulgen scanning cytophotometry. **A** *Allium cepa* nucleus with chromonematic structure; the individual chromonema strands are not separated, but can be visualized by connecting the highest extinction values which each other. (Nagl 1979d). **B** Nucleus of *Rhinanthus minor* with chromocentric structure; only the heterochromatin (chromocenters) exhibit measurable extinction. (Original)

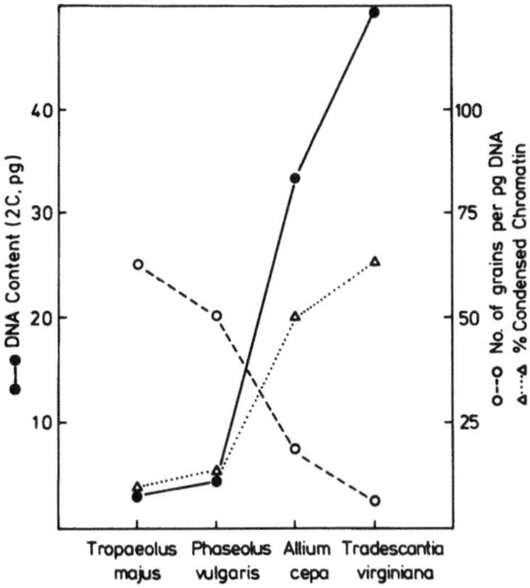

Fig. 11. Relationship between 2C nuclear DNA content, the proportion of chromatin in the condensed state, and [^{3}H]-uridine incorporation (given as number of silver grains per pg DNA per h). The transcription rate is lower in species with high 2C values and a high percentage of condensed chromatin, so that the average transcription rate is similar in all plant species. (Data were obtained in root tips; unpublished results of W. NAGL and U. SOKOL)

Table 3. Relationship between basic nuclear DNA content (2C value) and proportion of condensed chromatin per working nucleus (as measured in electron micrographs). (According to NAGL 1980)

Taxon	Coefficient of correlation	Slope of regression line (% c.c./pg DNA)	Cross with y axis
Anthemis	0.87	3.50	44.72
Luzula	0.98	9.36	0.68
Microseris	0.89	12.66	−4.22
Cruciferae	1.00	4.36	2.61
Rutaceae[a]	1.00	3.89	2.30
Angiosperm sample	0.76	2.63	14.82

[a] Unpublished data

to give rise to the formation of macrocomplexes from DNA or core particles. From the viewpoint of electrostatic attraction, aggregation should be best promoted by those proteins which have a high specific positive charge (i.e., the net positive charge per amino acid residue, which indicates the degree of overall basicity of the polypeptide and hence its chromatin condensation ability; see Table 2 in WAGNER et al. 1978). Thus protein phosphorylation which reduced this value should also reduce the tendency of aggregate formation. However, in all instances investigated, including those of the recent literature, a protein phosphorylation promotes the formation of large complexes with a high internal order. BODE et al. (1979) suggested that the reduction of electrostatic affinities, which is achieved in vivo by protein phosphorylation, serves to enhance complex size, probably by permitting a thermodynamically controlled and ordered association of DNA segments.

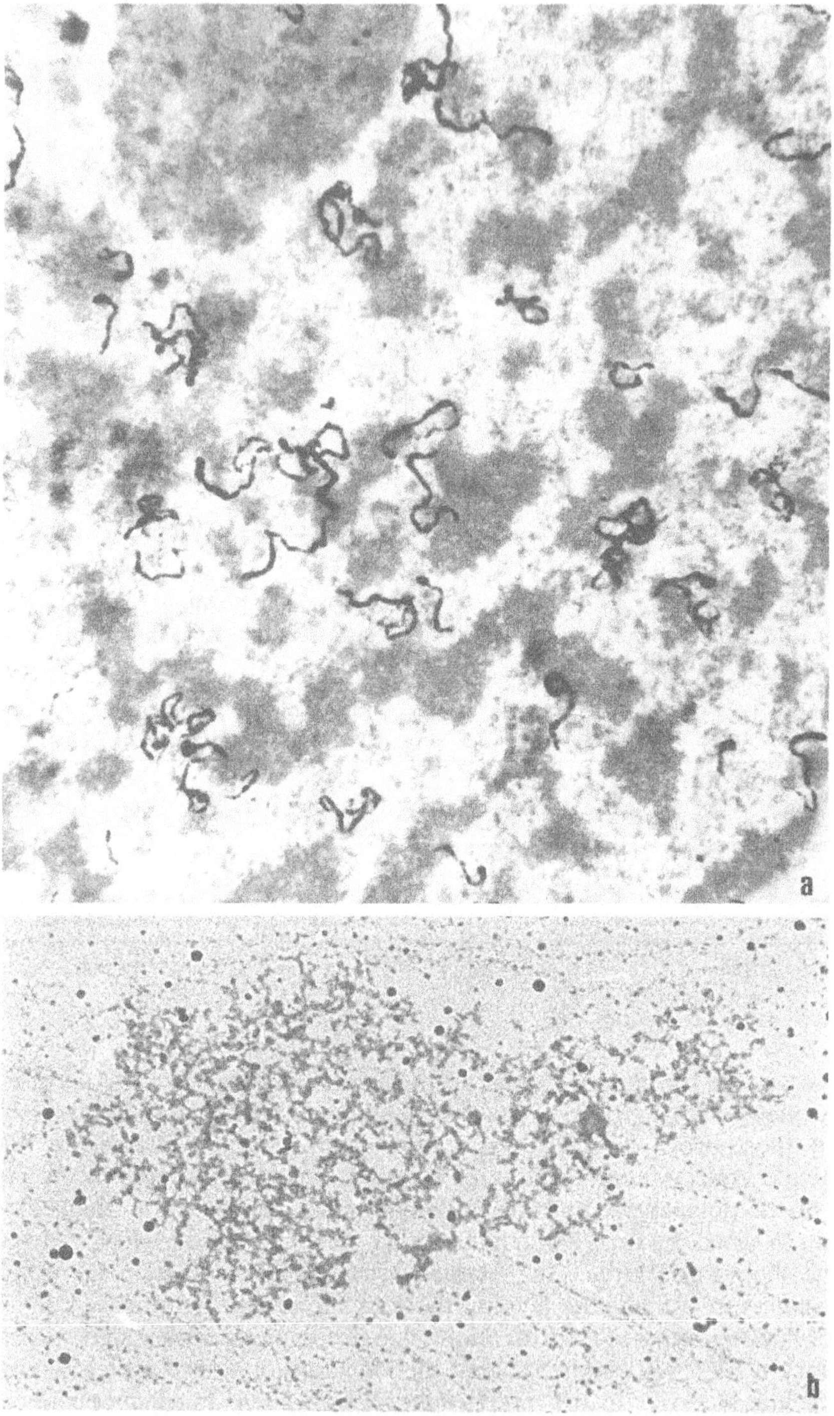
a
b

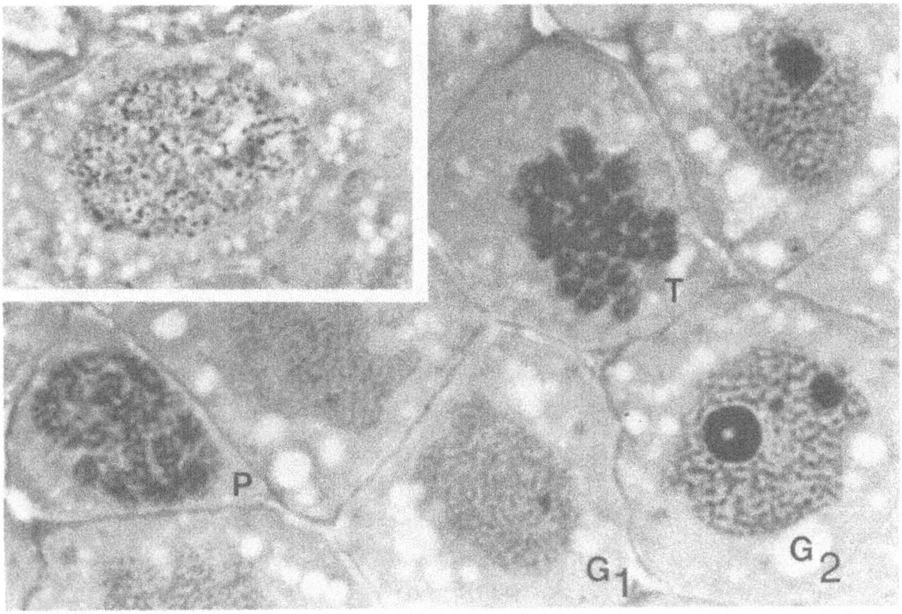

Fig. 13. Cell cycle-dependent changes in nuclear structure (semi-thin section of the root tip meristem of *Hyacinthus*). Note fine chromatin in G_1 nucleus, coarser chromatin in G_2 nucleus, evidence of coils in prophasic nucleus (*P*) and holes in telophasic chromosomes (*T*). The insert shows a nucleus in S phase; note label after [^{3}H]-thymidine incorporation. The structure of the chromatin threads is somewhat intermediate between the G_1 and G_2 structure. (Phase contrast, ×1,500; Original)

Condensation is often the result of poly-(ADP-ribose) polymerase activity. This chromatin-bound enzyme transfers the adenosine diphosphate ribose moiety of NAD (nicotineamide adenine dinucleotide) to the nuclear proteins (particularly H1), adds successively ADP-ribose to form a homopolymer, and liberates nicotinamide. Both ends of the homopolymer become linked to chromosomal proteins, so that cross-linking takes place. Structurally, therefore, this kind of chromatin condensation is completely different from the chromosome condensation at mitosis, although it cannot be distinguished from the latter in electron microscopic sections. For effects of poly-(ADP-ribose) polymerase activity for chromatin condensation and differentiation see e.g. BERGER et al. 1979, CAPLAN et al. 1979, FARINA et al. 1979, KITAMURA et al. 1979; for some controversal aspects see JUMP et al. 1979; review by NAGL 1979e).

Fig. 12a, b. Transcription of chromatin in the DNA-rich species, *Hyacinthus orientalis*. **a** Electron microscopic autoradiography of an ultrathin section through a root-tip nucleus. The label is exclusively located over decondensed chromatin. (×20,000). **b** Complex transcription product as rarely seen in spread chromatin of otherwise nucleosomal organization (×20,000; Originals)

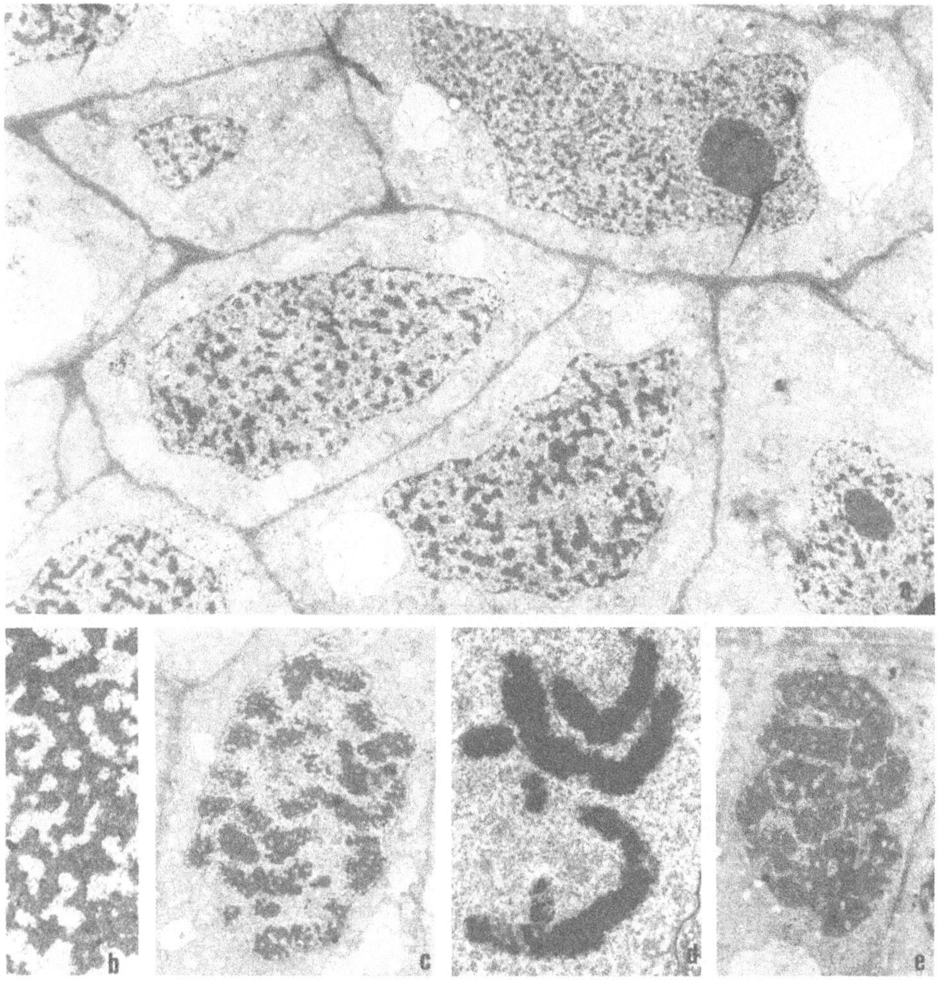

Fig. 14a–e. Some cell-cycle-dependent changes in nuclear ultrastructure (root tip meristem of *Hyacinthus orientalis*). **a** Interphase nuclei in mid-G_1 and mid-G_2; **b** part of an interphase nucleus in late G_2; **c** prophase; **d** metaphase; **e** late telophase. (× 3,200; Original)

DNA superpacking as visible in the pattern of chromatin condensation has also been referred to as quinternary structure. Models for mammalian nuclei have been published by NICOLINI (1979b) and COMINGS (1980).

8 Chromatin Organization and Genome Organization

The correlation between 2C nuclear DNA content and the proportion of euchromatin in the condensed state suggests that the gross organization of chromatin in plant nuclei is determined by the DNA itself (see also Sect. 7). The molecular

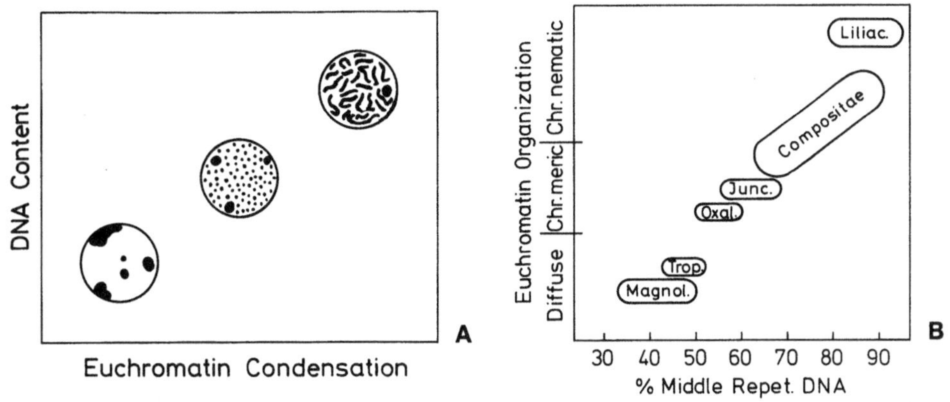

Fig. 15A, B. Diagrams to illustrate some of the basic relationships between genome organization and chromatin organization in plants. **A** There is a general trend that the more euchromatin is species-specifically condensed, the higher the 2C DNA value of the species: species with little DNA exhibit diffuse, decondensed euchromatin, those with intermediate amounts of DNA display a chromomeric euchromatin organization, and in species with high 2C values, most of the euchromatin is condensed into chromonemata. On the contrary, the probability of finding chromcentric heterochromatin is highest in species with low DNA content. (Modified from NAGL 1979c). **B** Family-specific organization of euchromatin as related to the proportion of middle repetitive DNA. (Original)

determinant could be, for instance, the nucleotype (BENNETT 1973). In this case the DNA amount per se, independent of the nucleotide sequence, would exert a condensation effect. It is more likely that a certain repetitive DNA sequence, which is regularly interspersed with other coding and noncoding sequences, and which binds H1 or another "condensing protein" is involved in determining the degree of interphase chromatin condensation (NAGL 1976a, 1979d, BARLOW 1977, NAGL and BACHMANN 1980). In any case the relationship between genome organization and chromatin organization is rather a complex one and is not yet well understood (see also BLUMENFELD et al. 1977, LERMAN and DEGTYAREV 1978, FUHRMANN and NAGL 1979).

Figure 15 shows how the available data can be used to find some correlations between DNA organization and chromatin structure in plants. It must be emphasized that this is rather a rough model based on limited data, but it can show the probability with which total nuclear DNA content and euchromatin condensation are correlated. Highly repetitive DNA and heterochromatin, on the other hand, are found with less probability in species with high 2C values, but due to the high variability and polymorphism of heterochromatin, no clear correlation can be detected (for further discussion see INGLE et al. 1975, TIMMIS et al. 1975, NAGL and FUSENIG 1979).

Looking at more closely related species and general, DNA and chromatin organization may be understood in relation to the process of evolution. Variation in satellite DNA and other highly repetitive DNA, as well as in the heterochromatin content, is often associated with, or even the basis of, speciation (reviewed by NAGL 1978; for annual plants see NAGL 1974). Particularly clear

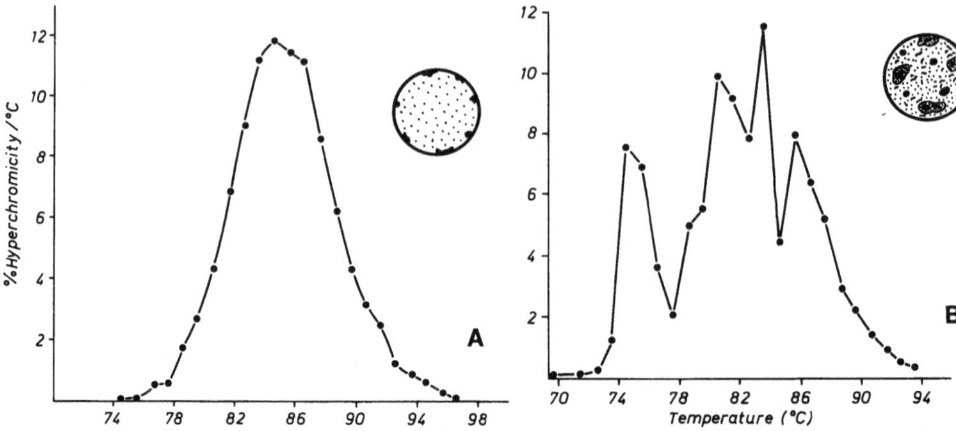

Fig. 16A, B. Comparison of DNA heterogeneity and chromatin structural complexity in two orchids. The DNA of the rather conservative genus *Brassica* (**A**) shows only one peak in derivate melting profiles, the euchromatin is diffuse. The DNA of the rather progressive genus *Cymbidium* (**B**) exhibits several thermic satellite peaks, and chromatin organization is of a rather complex type. (In part according to CAPESIUS and NAGL 1978)

parallel changes of the DNA composition and the chromatin texture were observed in orchids (Fig. 16; CAPESIUS and NAGL 1978). BRITTEN and DAVIDSON (1971) suggested that the origins of evolutionary novelty are closely related to the evolution of repetitive, i.e., control DNA, while the gene number is fairly constant. In orchids, progressive genera actually show inhomogeneous DNA composition and complex nuclear structures with much heterochromatin, while taxa with small morphological variation and slow speciation display homogeneous DNA composition and homogeneous nuclear texture (TANAKA 1971, CAPESIUS and NAGL 1978).

It seems that ontogenetic changes in DNA content and chromatin structure depend somewhat on the phylogenetic state of the genome (NAGL 1977d, 1978, see also GALAU et al. 1976). Evidence for such a relationship was also obtained in some orchids (NAGL and CAPESIUS 1977), but this aspect will be discussed in connection with DNA endoreplication and differential DNA replication in Chapter 4, this volume.

9 Conclusions

a) The universal subunit of chromatin in eukaryotes is the nucleosome core particle which represents an octamer of the histones H2A (PH1), H2B (PH2), H3 and H4, and 143 base pairs of DNA, held together by electrostatic and hydrophobic forces.

b) The multifunctional and variable histone H1 connects core particles so that a 10 nm chromatin fiber is formed.
c) Higher-order structures such as occur in thick chromatin fibrils are induced by divalent cations resulting in super-nucleosome and/or solenoidal fiber formation.
d) The solenoid is either super-coiled again to form the chromatid, or locally bound by nonhistone proteins to form domains and loops, and upon protein polymerization, the chromatid. Giemsa bands may be the expression of domains (or groups of them). Mitotic chromosome condensation requires H1 phosphorylation so that binding sites are liberated for the formation of higher-order structures.
e) Interphase chromatin condensation is species-specific in plants, governed by genome size and organization, and tissue-specific (functional) in mammals, brought about by phosphorylation of a different H1 residue and by ADP-ribosylation (cross-linking of nucleosomes or solenoidal fibers).
f) The correlation between 2C nuclear DNA content and proportion of chromatin in the condensed state in plants allows species and hybrid diagnosis, and the prediction of nuclear structure from the DNA content (and vice versa).
g) Nuclear ultrastructure and chromatin organization in plants is an expression of the evolutionary state of a species rather than of the functional state of a cell.

Acknowledgment. I thank Mrs. Ulrike Sokol and Mrs. Ursula Gille for preparing the line drawings, and Dr. Harald Fuge for reading the manuscript.

Note Added in Proof: Additional information about species-specific, tissue-specific, and cell cycle-specific chromatin condensation is available in a review by NAGL (1982); plant polytene chromosomes were reviewed by NAGL (1981). Loss of DNA and euchromatin in senescing leaf cells was reported by DRUMM and NAGL (1982). The possible role of photon emission from DNA in dependence of the degree of chromatin condensation and DNA packaging was discussed by RATTEMEYER et al. (1981).

References

Adolph KW, Cheng SM, Laemmli UK (1977) Role of nonhistone proteins in metaphase chromosome structure. Cell 12:805–816
Al-Hamdani MM, Atkinson ME, Mayhew TM (1979) Ultrastructural morphometry of blastogenesis I: Transformation of small lymphocytes stimulated in vivo with dinitrochlorobenzene. Cell Tissue Res 200:495–509
Allfrey VG, Littau VC, Mirsky AE (1963) On the role of histones in regulating RNA synthesis in the cell nucleus. Proc Natl Acad Sci USA 49:414–421
Altenburger W, Hörz W, Zachau HG (1976) Nuclease cleavage of chromatin at 100 nucleotide intervals. Nature (London) 264:517–522
Bak AL, Zeuthen J (1977) Higher-order structure of mitotic chromosomes. Cold Spring Harbor Symp Quant Biol 42:367–377
Bak AL, Bak, P, Zeuthen J (1979a) Higher levels of organization in chromosomes. J Theor Biol 76:205–217
Bak P, Bak AL, Zeuthen J (1979b) Characterization of human chromosomal unit fibers. Chromosoma 73:301–315

Bakke AC, Wu J-R, Bonner J (1978) Chromatin structure in the cellular slime mold *Dictyostelium discoideum*. Proc Natl Acad Sci USA 75:705–709

Balhorn R, Jackson V, Grauner D, Chalkley R (1975) Phosphorylation of the lysine-rich histones throughout the cell cycle. Biochemistry 14:2504–2511

Barlow PW (1977) Determinants of nuclear chromatin structure in angiosperms. Ann Sci Nat Bot Biol Veg 18:193–206

Barsotti P, Marinozzi V (1979) A morphological study on the early changes in hepatocyte nuclei induced by low doses of α-amanitin. J Submicrosc Cytol 11:13–23

Basu S (1979) Evidence for superstructures of wet chromatin. In: Nicolini CA (ed) Chromatin structure and function, Part B. Plenum, New York, pp 515–540

Baudy P, Bram S, Vastel D, Lepault J, Kitzis A (1976) Chromatin subunit small angle neutron scattering: a DNA-rich coil surrounds a protein-DNA core. Biochem Biophys Res Common 72:176–183

Bennett MD (1973) Nuclear characters in plants. Brookhaven Symp Biol 25:344–366

Bennett MD, Gustafson JP, Smith JB (1977) Variation in nuclear DNA in the genus *Secale*. Chromosoma 61:149–176

Benyajati C, Worcel A (1976) Isolation, characterization, and structure of the folded interphase genome of *Drosophila melanogaster*. Cell 9:343–407

Berezney R (1979) Dynamic properties of the nuclear matrix. In: Busch H (ed) The cell nucleus Vol VII. Academic Press, London New York, pp 413–456

Berger NA, Petzold SJ, Berger SJ (1979) Association of (ADP-ribose) synthesis with cessation of DNA synthesis and DNA fragmentation. Biochim Biophys Acta 564:90–104

Blumenfeld M, Orf JW, Sina BJ, Kreber RA, Callahan MA, Snyders LA (1977) Satellite DNA, Hl histone, and heterochromatin in *Drosophila virilis*. Cold Spring Harbor Symp Quant Biol 42:273–276

Bodansky S, Mintz LB, Holmes DS (1979) The mesokaryote *Gyrodinium cohnii* lacks nucleosomes. Biochem Biophys Res Commun 88:1329–1336

Bode J, Standt U-D (1978) Structural aspects of histone complexes and nucleosomes revealed by the accessibility of cysteine side chains. Z Naturforsch 33c:884–890

Bode J, Willmitzer L, Arfmann H-A, Wehling K (1979) Condensation of DNA and chromatin by histones H1, H5 and protamines. The influence of a protein phosphorylation. Int J Biol Macromolec 1:73–78

Bokhon'ko A, Reeder RH (1976) The subunit structure of mouse satellite chromatin. Biochem Biophys Res Commun 70:146–152

Bonner J (1979a) Properties and composition of isolated chromatin. In: Nicolini CA (ed) Chromatin structure and function, Part A. Plenum, New York, pp 3–13

Bonner J (1979b) Expressed and nonexpressed portions of the genome: their separation and their characterization. In: Nicolini CA (ed) Chromatin structure and function, Part A. Plenum, New York, pp 15–23

Bonner J, Chalkley GR, Dahmus M, Fambrough D, Fujimara F, Huang RCC, Hubermann J, Jensen R, Marushige K, Ohlenbusch H, Oliviera B, Widholm J (1968) Isolation and characterization of chromosomal proteins. Methods Enzymol 12B:3–65

Bradbury EM (1978) Histone interactions, histone modifications and chromatin structure. Philos Trans R Soc London Ser B 283:291–293

Bradbury EM, Carpenter BC, Rattle HWE (1973a) Magnetic resonance studies of deoxyribonucleoprotein. Nature (London) 241:123–126

Bradbury EM, Cary PD, Crane-Robinson C, Rattle HWE (1973b) Conformations and interactions of histones and their role in chromosome structure. Ann N Y Acad Sci 222:266–289

Bradbury EM, Inglis RJ, Matthews HR (1974a) Control of cell division by very lysine-rich histone (F1) phosphorylation. Nature (London) 247:257–261

Bradbury EM, Inglis RJ, Matthews HR, Langan TA (1974b) Molecular basis of control of mitotic cell division in eukaryotes. Nature (London) 249:553–556

Bram S, Butler-Browne G, Baudy P, Ibel K (1975) Quaterny structure of chromatin. Proc Natl Acad Sci USA 72:1043–1045

Brandt WF, Van Holt C (1975) Isolation and characterization of the histones from cycad pollen. FEBS Lett 51:84–87
Brasch K (1978) Morphometric studies of chromatin structure and nuclear organization: hepatic polyploidy and mammalian erythropoiesis. 9th Int Congr Electr Microsc, Toronto, Vol II, p 208
Brasch K, Setterfield G (1974) Structural organization of chromosomes in interphase nuclei. Exp Cell Res 83:175–185
Brasch K, Sinclair GD (1978) The organization, composition and matrix of hepatocyte nuclei exposed to α-amanitin. Virchows Arch B 27:193–204
Britten RJ, Davidson EH (1971) Repeated and non-repeated DNA sequences and a speculation on the origins of evolutionary novelty. Q Rev Biol 46:111–138
Brown DB, Stack SM, Mitchell JB, Bedford JS (1979) Visualization of the interphase chromosomes of *Ornithogalum virens* and *Muntiacus muntjak*. Cytobiologie 18:398–412
Brown FL, Musich PR, Maio JJ (1979) The repetitive sequence structure of component α DNA and its relationship to the nucleosomes of the African green monkey. J Mol Biol 131:777–799
Brutlag D, Carlson M, Fry K, Hsieh TS (1977) DNA sequence organization in *Drosophila* heterochromatin. Cold Spring Harbor Symp Quant Biol 42:1137–1146
Burton DR, Butler MJ, Hyde JE, Phillips D, Skidmore CJ, Walker IO (1979) The interaction of histones with DNA: equilibrium binding studies. In: Nicolini CA (ed) Chromatin structure and function. Plenum, New York, pp 137–165
Busch H (ed) (1978) The cell nucleus, Vol IV, V: Chromatin. Academic Press, London New York
Callan HG, Klug A (eds) (1978) Structure of eukaryotic chromosomes and chromatin. Philos Trans R Soc London Ser B 283:231–416
Camerini-Otero RD, Felsenfeld G (1977) Supercoiling energy and nucleosome formation: the role of the arginine-rich kernel. Nucl Acids Res 4:1159–1181
Camerini-Otero RD, Sollner-Webb B, Felsenfeld G (1976) The organization of histones and DNA in chromatin: evidence for an arginine-rich histone kernel. Cell 8:333–347
Cameron LJ, Jeter JR (eds) (1975) Acidic proteins of the nucleus. Academic Press, London New York
Cameron IL, Pavlat WA, Jeter JR Jr (1979) Chromatin substructure: an electron microscopic study of thin-sectioned chromatin subjected to sequential protein extraction and water swelling procedures. Anat Rec 194:547–562
Campbell AM, Cotter RI, Pardon JF (1978) Light scattering measurements supporting helical structures for chromatin in solution. Nucl Acids Res 5:1571–1580
Capesius I, Nagl W (1978) Molecular and cytological characteristics of nuclear DNA and chromatin for angiosperm systematics: DNA diversification in the evolution of four orchids. Plant Syst Evol 129:143–166
Caplan AI, Niedergang C, Okazaki H, Mandel P (1979) Poly(ADP-ribose) levels as a function of chick limb mesenchymal cell development as studied *in vitro* and *in vivo*. Dev Biol 72:102–109
Carlson RD, Olins DE (1976) Chromatin model calculations – arrays of spherical nu-bodies. Nucl Acids Res 3:89–100
Carniel K (1979) Über bandartige Strukturen in den Chromozentren von *Lupinus polyphyllus*. Plant Syst Evol 131:235–242
Carter CW Jr (1978) Histone packing in the nucleosome core particle of chromatin. Proc Natl Acad Sci USA 75:3649–3653
Cohen JS (1980) DNA: is the backbone boring? Trends Bioch Res 5:58–60
Cold Spring Harbor Symp Quant Biol, Vol. 42: Chromatin (1977)
Cole RD (1977) Special features of the structures of H1 histones. In: Ts'o POP (ed) The molecular biology of the mammalian genetic apparatus. North-Holland, Amsterdam, pp 91–104
Comings DE (1974) The structure of human chromosomes. In: Busch H (ed) The cell nucleus, Vol. I. Academic Press, London New York, pp 537–563

Comings DE (1978) Mechanisms of chromosome banding and implications for chromosome structure. Annu Rev Genet 12:25–46
Comings DE (1980) Arrangement of chromatin in the nucleus. Human Genetics 53:131–144
Comings DE, Drets ME (1976) Mechanisms of chromosome banding IX. Are variations in DNA base composition adequate to account for quinacrine, Hoechst 33258 and daunomycin banding? Chromosoma 56:199–211
Comings DE, Wallak AS (1978) DNA-binding properties of nuclear matrix proteins. J Cell Sci 34:233–246
Compton JL, Bellard M, Chambon P (1976) Biochemical evidence of variability in the DNA repeat length in the chromatin of higher eukaryotes. Proc Natl Acad Sci USA 73:4382–4386
Cook PR, Brazell IA (1975) Supercoils in human DNA. J Cell Sci 19:261–279
Cremer C, Cremer T, Zorn C, Cioreanu V (1979) Partial UV-irradiation of Chinese hamster cell nuclei and detection of unscheduled DNA synthesis in interphase and metaphase. A tool to investigate the arrangement of interphase chromosomes in mammalian cells. Hoppe-Seyler's Z Physiol Chem 360:244–245
Crick FHC (1979) Aspects of chromatin structure. In: Engberg J, Klenow H, Leick V (eds) Specific eukaryotic genes. Munksgaard, Copenhagen, pp 17–21
Crick FCH, Klug A (1975) Kinky helix. Nature 255:530–533
DeLange RJ, Fambrough DM, Smith EL, Bonner J (1969) Calf and pea histone 4. II. Complete amino acid sequence of pea seedling histone 4; comparison with the homologous calf thymus histobe. J Biol Chem 244:5669–5679
DeLange RJ, Hooper J, Smith EL (1973) Histone 3. III. Sequence studies on the cyanogen bromide peptides; complete amino acid sequence of calf thymus histone 3. J Biol Chem 248:3261–3274
Deltour R, Gautier A, Fakan J (1979) Ultrastructural cytochemistry of the nucleus in *Zea mays* embryos during germination. J Cell Sci 40:43–62
Derenzini M, Lorenzoni E, Marinozzi V, Barsotti P (1977) Ultrastructural cytochemistry of active chromatin in regenerating rat hepatocytes. J Ultrastruct Res 59:250–262
Derenzini M, Pession-Brizzi A, Bonetti E, Novello F (1979) Relationship between ultrastructure and function of hepatocyte chromatin: a study with adrenalectomized rats after cortisol administration. J Ultrastruct Res 67:161–179
Doenecke D (1979) Analysis of accessible site in modified nucleosomes. Plant Syst Evol Suppl 2:201–219
Douvas AS, Harrington CA, Bonner J (1975) Major nonhistone proteins of rat liver chromatin. Preliminary identification of myosin, actin, tubulin, and tropomyosin. Proc Natl Acad Sci USA 72:3902–3906
Drumm N, Nagl W (1982) Loss of DNA and euchromatin in senescing leaf cells of *Allium*. Mech Ageing Devel 18:103–110
Duerksen JD, Paul IJ (1976) Satellite DNA sequence content of polylysine-titritable and nuclease-resistant fraction of mouse liver hepatoma chromatin. Nucl Acids Res 3:2277–2291
DuPraw EJ (1970) DNA and chromosome. Holt, Rinehart, Winston, New York
Elgin SCR, Bonner J (1970) Limited heterogeneity of the major nonhistone chromosomal proteins. Biochemistry 9:4440–4447
Elgin SCR, Weintraub H (1975) Chromosomal proteins and chromatin structure. Annu Rev Biochem 44:725–764
Erenpreis JG, Zirne RA, Demidenko OJ (1978) Electron microscopic observation of nucleosomes in ultrathin sections of heparin-treated nuclei. Experientia 34:882–883
Farina B, Menella MRF, Leone E (1979) Nucleic acids, histones and spermiogenesis: the poly(ADP ribose) polymerase system. In: Salvatore F, Marino G, Volpe P (eds) Macromolecules in the Functioning Cell. Plenum, New York, pp 283–300
Fellenberg G (1974) Chromosomale Proteine. Ulmer, Stuttgart
Felsenfeld G (1978) Chromatin. Nature (London) 271:115–122
Finch JT, Klug A (1976) Solenoidal model for superstructure in chromatin. Proc Natl Acad Sci USA 73:1897–1901

Finch JT, Lutter LC, Rhodes D, Brown RS, Rushton B, Levitt M, Klug A (1977) Structure of nucleosome core particles of chromatin. Nature (London) 269:29–36
Fittler F, Zachau HG (1979) Subunit structure of α-satellite DNA containing chromatin from African green monkey cells. Nucl Acids Res 7:1–13
Franke WW, Scheer U (1978) Morphology of transcriptional units at different states of activity. Philos Trans R London Ser B 283:333–342
Franke WW, Scheer U, Trendelenburg MF, Spring H, Zentgraf H (1976) Absence of nucleosomes in transcriptionally active chromatin. Cytobiologie 13:401–434
Franke WW, Scheer U, Trendelenburg M, Zentgraf H, Spring H (1977) Morphology of transcriptionally active chromatin. Cold Spring Harbor Symp Quant Biol 42:755–772
Franke WW, Scheer U, Spring H, Trendelenburg MF, Zentgraf HW (1979) Organization of nucleolar chromatin. In: Busch H (ed) The cell nucleus, Vol. VII. Academic Press, London New York, pp 49–95
Fuge H (1980) Local decondensation of chromatin at the surface of meiotic chromosomes during anaphase movement. Eur J Cell Biol
Fuhrmann B, Nagl W (1979) Chromatin organization and repetitive DNA in *Anacyclus* and *Anthemis* (Asteraceae). Plant Syst Evol Suppl 2:235–245
Galau GA, Chamberlin ME, Hough BR, Britten RJ, Davidson EH (1976) Evolution of repetitive and nonrepetitive DNA. In: Ayala FJ (ed) Molecular evolution. Sinauer, Sunderland, pp 220–224
Garel A, Axel R (1976) Selective digestion of transcriptionally active ovalbumin genes from oviduct nuclei. Proc Natl Acad Sci USA 73:3966–3970
Germond JE, Hirt B, Oudet P, Gross-Bellard M, Chambon P (1975) Folding of the DNA double helix in chromatin-like structures from Simian virus 40. Proc Natl Acad Sci USA 72:1843–1847
Germond, JE, Rouvière-Yaniv J, Yaniv M, Brutlag D (1979) Nicking-closing enzyme assembles nucleosome-like structures *in vitro*. Proc Natl Acad Sci USA 76:3779–3783
Gigot C, Philipps G, Nicolaieff A, Hirth L (1976) Some properties of tobacco protoplast chromatin. Nucl Acids Res 3:2315–2329
Giri CP, Gorovsky MA (1980) DNase I sensitivity of ribosomal genes in isolated nucleosome core particles. Nucl Acids Res 8:197–214
Gordon VC, Knobler CM, Olins DE, Schumaker VN (1978) Conformational changes of the chromatin subunit. Proc Natl Acad Sci USA 75:660–663
Gottesfeld JM (1978) Organization of transcribed regions of chromatin. Philos Trans R Soc London Ser B 283:343–357
Gottesfeld JM, Melton DA (1978) The length of nucleosome-associated DNA is the same in both transcribed and nontranscribed regions of chromatin. Nature (London) 273:317–318
Gottesfeld JM, Murphy RF, Bonner J (1975) Structure of transcriptionally active chromatin. Proc Natl Acad Sci USA 72:4404–4408
Grainger RM, Ogle RC (1978) Chromatin structure of the ribosomal RNA genes in *Physarum polycephalum*. Chromosoma 65:115–126
Greilhuber J (1977) Why plant chromosomes do not show G-bands. Theor Appl Genet 50:121–124
Greilhuber J, Speta F (1978) Quantitative analysis of C-banded karyotypes, and systematics in the cultivated species of the *Scilla sibirica* group (Liliaceae). Plant Syst Evol 129:63–109
Greimers R, Deltour R (1981) Organization of transcribed and nontranscribed chromatin in isolated nuclei of Zea mays root cells. Eur J Cell Biol 23:303–311
Griffith JD (1975) Chromatin structure: deduced from a minichromosome. Science 187:1202–1203
Griffith JD, Christiansen G (1977) The multifunctional role of histone H1, probed with the SV 40 minichromosome. Cold Spring Harbor Symp Quant Biol 42:215–226
Gurley LR, Walters RA, Tobey RA (1973) Histone phosphorylation in late interphase and mitosis. Biochem Biophys Res Commun 50:744–750

Gurley LR, Walters RA, Tobey RA (1974) Cell cycle-specific changes in histone phosphorylation associated with cell proliferation and chromosome condensation. J Cell Biol 60:356–364

Holm PB (1976) The C and Q banding patterns of the chromosomes of *Lilium longiflorum* (Thunb). Carlsb Res Comm 41:217–224

Hozier JC, Kaus R (1976) Subunit structure of chromosomes in mitotic nuclei of *Physarum polycephalum*. Chromosoma 57:95–102

Hozier J, Renz M, Nehls P (1977) The chromosome fiber: evidence for an ordered superstructure of nucleosomes. Chromosoma 62:301–317

Hsieh TS, Brutlag DL (1979) A protein that preferentially binds *Drosophila* satellite DNA. Proc Natl Acad Sci USA 76:726–730

Hsu TC (1962) Differential rate in RNA synthesis between euchromatin and heterochromatin. Exp Cell Res 27:332–334

Igo-Kemenes T, Zachau HG (1977) Domains in chromatin structure. Cold Spring Harbor Symp Quant Biol 42:109–118

Ingle J, Timmis JN, Sinclair J (1975) The relationship between satellite DNA, ribosomal RNA gene redundnancy, and genome size. Plant Physiol 55:496–501

Isenberg I (1978) Protein-protein interactions of histones. In: Busch H (ed) The cell nucleus, Vol. IV. Academic Press, London New York, pp 135–154

Isenberg I (1979) Histones. Annu Rev Biochem 48:159–191

Jeanny JC, Gontcharoff M (1978) Etude en microscopie électronique et par cytophotométrie à balayage de la structure et de la distribution de la chromatine dans les noyaux des cellules cartilagineuses de *Triturus cristatus* agés au cours de la régénération. Wilh Roux' Arch Dev Biol 184:195–211

Jeppersen PGN, Bankier AT (1979) A partial characterization of DNA fragments protected from nuclease degradation in histone-depleted metaphase chromosomes of the Chinese hamster. Nucl Acids Res 7:49–67

Jerzmanowski A, Staron K (1980) Mg^{2+} as a trigger of condensation-decondensation transition of chromatin during mitosis. J Theor Biol 82:41–46

Jerzmanowski A, Staron K, Tyniec-Kroenke B, Toczko K (1979) Chromatin condensation. Possible dehydrating and stabilizing factors. Biochim Biophys Acta 565:356–364

John B, Miklos GLG (1979) Functional aspects of satellite DNA and heterochromatin. Int Rev Cytol 58:1–114

Johnson EM, Littau VC, Allfrey VG, Bradbury EM, Matthews HR (1976) The subunit structure of chromatin from *Physarum polycephalum*. Nucl Acids Res 3:3313–3329

Johnson EM, Campbell GR, Allfrey VG (1979) Different nucleosome structures on transcribing and nontranscribing ribosomal gene sequences. Science 206:1192–1194

Jorcano JL, Ruiz-Carillo A (1979) H3-H4 Tetramer directs DNA and core histone octamer assembly in the nucleosome core particle. Biochemistry 18:768–774

Jump DB, Butt TR, Smulson M (1979) Nuclear protein modification and chromatin substructure. 3. Relationship between poly(ADP)ribosylation and different functional forms of chromatin. Biochemistry 18:983–990

Kiefer G, Sandritter W (1978) Anwendungsmöglichkeiten der automatischen Kernbildanalyse. Acta Histochem Suppl 20:193–201

Kiryanov GI, Manamshjan TA, Polyakov V Yu, Fais D, Chentsov Jus (1976) Levels of granular organization of chromatin fibers. FEBS Lett 67:323–327

Kitamura A, Tanigawa Y, Yamamoto T, Kawamura M, Doi S, Shimoyama M (1979) Glucocorticoid-induced reduction of poly(ADP-ribose) synthetase in nuclei from chick embryo liver. Biochem Biophys Res Commun 87:725–733

Kornberg RD (1974) Chromatin structure: a repeating unit of histones and DNA. Science 184:868–871

Kornberg RD (1977) Structure of chromatin. Annu Rev Biochem 46:931–954

Kraut W, Werner D (1979) Analysis of the most tightly bound proteins in eukaryotic DNA. Biochim Biophys Acta 564:390–401

Kwiatkowska M, Maszewski J (1979) Changes in the content of condensed chromatin during the cell cycle in antheridial filaments of *Chara vulgaris* L as related to DNA

and RNA synthesis, ^{3}H-actinomycin D binding and RNA polymerase activity. Protoplasma 98:363–367
Lacy E, Axel R (1975) Analysis of DNA of isolated chromatin subunits. Proc Natl Acad Sci USA 72:3978–3982
Lafontaine JG (1974) Ultrastructural organization of plant cell nuclei. In: Busch H (ed) The cell nucleus Vol I. Academic Press, London New York, pp 149–185
Lai V, Srivastava LM (1976) Nuclear changes during differentiation of xylem vessel elements. Cytobiologie 12:220–243
Laird CD, Wilkinson LE, Foe VE, Chooi WY (1976) Analysis of chromatin- associated fiber arrays. Chromosoma 58:169–192
Lake RS, Salzman NP (1972) Occurrence and properties of a chromatin associated f1-histone phosphokinase in mitotic Chinese hamster cells. Biochemistry 11:4817–4826
Laskey RA, Honda BM, Mills AD, Finch J (1978) Nucleosomes are assembled by an acidic protein which binds histones and transfers them to DNA. Nature (London) 275:416–420
Leber B, Hemleben V (1979a) Nucleosomal organization in active and inactive plant chromatin. Plant Syst Evol Suppl 2:178–199
Leber B, Hemleben V (1979b) Structure of plant nuclear and ribosomal DNA containing chromatin. Nucl Acids Res 7:1263–1281
Leffak IM, Grainger R, Weintraub H (1977) Conservative assembly and segregation of nucleosomal histones. Cell 12:837–845
Lerman MI, Degtyarev SV (1978) Periodically interspersed repetitive sequences may govern higher order DNA coiling in chromatin and chromosomes. Mol Biol Rep 4:117–120
Levy A, Jakob KM (1978) Nascent DNA in nucleosome-like structures from chromatin. Cell 14:259–267
Levy-Wilson B (1979) Effects of nuclear disruption on the macromolecular composition of nucleosome subfractions. Nucl Acids Res 7:2239–2254
Lewin B (1974) Gene expression – 2-eucaryotic chromosomes. Wiley and Sons, London New York
Li HJ (1976) Chromatin structure. Int J Biochem 7:181–185
Li HJ, Eckhardt RA (1977) Chromatin and chromosome structure. Academic Press, London New York
Lilley DMJ (1978) Active chromatin structure. Cell Biol Int Rep 2:1–10
Lima-de-Faria A (1975) The relation between chromomeres, replicons, operons, transcription units, genes, viruses and palindromes. Hereditas 81:249–284
Lima-de-Faria A (1979) Prediction of gene location and classification of genes according to the chromosome field. In: Engberg J, Klenow H, Leick V (eds) Specific eukaryotic genes. Munksgaard, Copenhagen, pp 25–38
Lohr D, Tatchell K, van Holde KE (1977) on the occurrence of nucleosome phasing in chromatin. Cell 12:829–836
Louie AJ, Dixon GM (1973) Kinetics of phosphorylation and dephosphorylation of testis histones and their possible role in determining chromosomal structure. Nature (New Biol) 243:164–168
Lutz C, Nagl W (1980) A reliable method for the preparation and electron microscopic visualization of nucleosomes in higher plants. Planta 149:408–410
Lyon MF (1968) Chromosomal and subchromosomal inactivation. Annu Rev Genet 2:31–52
Maio JJ, Brown FL, Musich PR (1977) Subunit structure of chromatin and the organization of eukaryotic highly repetitive DNA: recurrent periodicites and models for the evolutionary origin of repetitive DNA. J Mol Biol 117:637–655
Maraldi NM, Capitarsi S, Cocco L, Manzoli FA (1979) Histones assembly and their structural role for nucleosome core. In: Nicolini CA (ed) Chromatin structure and function. Plenum, New York, pp 371–388
Marsden MPF, Laemmli UK (1979) Metaphase chromosome structure: Evidence for a radial loop model. Cell 17:849–858

Martin T, Billings P, Levy A, Ozarslan S, Quinlan T, Swift H, Urber L (1973) Some properties of RNA-protein complexes from the nucleus of eukaryotic cells. Cold Spring Harbor Symp Quant Biol 38:921–932
Mathis DJ, Gorovsky MA (1976) Subunit structure of rDNA-containing chromatin. Biochemistry 15:750–755
Matthews HW, Bradbury EM (1978) The role of H1 histone phosphorylation in the cell cycle. Turbidity studies of H1-DNA interaction. Exp Cell Res 111:343–351
McCready SJ, Akgrigg A, Cook PR (1979) Electron microscopy of intact nuclear DNA from human cells. J Cell Sci 39:53–62
McGhee JD, Engel JD (1975) Subunit structure of chromatin is the same in plants and animals. Nature (London) 254:449–450
McKnight SL, Miller OL Jr (1977) Electron microscopic analysis of chromatin replication in the cellular blastoderm *Drosophila melanogaster* embryo. Cell 12:795–804
McKnight SL, Martin KA, Beyer AL, Miller OL Jr (1979) Visualization of fractionally active chromatin. In: Busch H (ed) The cell nucleus Vol VII. Academic Press, London New York, pp 97–122
Mello MLS (1978) Heterocromatina. Cien Cult 30:290–303
Miklos GLG, John B (1979) Heterochromatin and satellite DNA in man: properties and prospects. Am J Hum Genet 31:264–280
Miller OL, Bakken AH (1972) Morphological studies of transcription. Karol Symp Res Meth Reprod Endocrinol 5:155–177
Miller OL Jr, Beatty BR (1969) Protrait of a gene. J Cell Physiol 74 Suppl 1:225–232
Mirzabekov AD, Rich A (1979) Asymmetric lateral distribution of unshielded phosphate groups in nucleosomal DNA and its role in DNA bending. Proc Natl Acad Sci USA 76:1118–1121
Mirzabekov AD, Shick VV, Belyavsky AN, Bavykin SG (1978) Primary organization of nucleosome core particle of chromatin: Sequence of histone arrangement aling DNA. Proc Natl Acad Sci USA 75:4184–4188
Moreno ML, Sogo JM, de la Torre C (1978) Higher plant nucleosomes. A micromethod for isolating and dispersing chromatin fibers. New Phytol 81:681–683
Morris NR (1976) Nucleosome structure in *Aspergillus nidulans*. Cell 8:357–363
Moser GC, Müller HJ (1979) Cell cycle dependent changes of chromosomes in mouse fibroblasts. Eur J Cell Biol 19:116–119
Mullinger AM, Johnson RT (1979) The organization of supercoiled DNA from human chromosomes. J Cell Sci 38:369–389
Nadeau P, Palotta D, Lafontaine JG (1974) Electrophoretic study of plant histones: Comparison with vertebrate histones. Arch Biochem Biophys 161:171–177
Nagl W (1969a) Puffing of polytene chromosomes in a plant (*Phaseolus vulgaris*), Naturwissenschaften 56:221–222
Nagl W (1969b) Correlation of structure and RNA synthesis in the nucleolus-organizing polytene chromosomes of *Phaseolus vulgaris*. Chromosoma 28:85–91
Nagl W (1970a) Temperature-dependent functional structures in the polytene chromosomes of *Phaseolus*, with special reference to the nucleolus organizers. J Cell Sci 6:87–107
Nagl W (1970b) The mitotic and endomitotic nuclear cycle in *Allium carinatum*. II. Relations between DNA replication and chromatin structure. Caryologia 23:71–78
Nagl W (1974) Role of heterochromatin in the control of cell cycle duration. Nature (London) 249:53–54
Nagl W (1976a) Zellkern und Zellzyklen. Ulmer, Stuttgart
Nagl W (1976b) Chromatin subunits visualized in common ultrathin section of *Helianthus* nuclei. Experientia 32:703–704
Nagl W (1977a) Organization and replication of the eukaryotic chromosome. Progr Bot 39:132–152
Nagl W (1977b) Early and late DNA replication in respectively condensed and decondensed heterochromatin of *Allium cepa*. Protoplasma 91:389–407
Nagl W (1977c) Nuclear structures during cell cycles. In: Rost TL, Gifford EM Jr

(eds) Mechanisms and control of cell division. Dowden, Hutchison and Ross, Stroudsburg, pp 147–193
Nagl W (1977d) The evolution of chromosomal DNA redundancy: Ontogenetic lateral, versus phylogenetic tandem changes. Nucleus 20:10–27
Nagl W (1978) Endopolyploidy and polyteny in differentiation and evolution. Elsevier North-Holland, Amsterdam
Nagl W (1979a) Chromosomen. Parey, Berlin
Nagl W (1979b) Search for the molecular basis of diversification in phylogenesis and ontogenesis. Plant Syst Evol Suppl 2:3–25
Nagl W (1979c) Condensed interphase chromatin in plant and animal cell nuclei: fundamental differences. Plant Syst Evol Suppl 2:247–260
Nagl W (1979d) Nuclear ultrastructure: Condensed chromatin in plants is species-specific (karyotypical), but not tissue-specific (functional). Protoplasma 100:53–71
Nagl W (1979e) Replication of the eukaryotic chromosome. Progr Bot 41:161–172
Nagl W (1980) Species and hybrid diagnosis in plants by means of quantitative light and electron microscopic morphometry of chromatin texture. Microsc Acta Suppl. 4:19–25
Nagl W (1981) Polytene chromosomes of plants. Int Rev Cytol 73:21–53
Nagl W (1982) Condensed chromatin: species-specificity, tissue-specificity, and cell cycle-specificity, as monitored by scanning cytometry. In: Nicolini C (ed) Cell Growth, Plenum, New York, pp 171–218
Nagl W, Bachmann K (1980) Condensed chromatin in diploid and allopolyploid *Microseris* species with different genome size: a quantitative electron microscopic study. Theor Appl Genet 57:107–111
Nagl W, Capesius I (1977) Repetitive DNA and heterochromatin as factors of karyotype evolution in phylogeny and ontogeny of orchids. Chromosomes Today 6:141–152
Nagl W, Fusenig HP (1979) Types of chromatin organization in plant nuclei. Plant Syst Evol Suppl 2:221–233
Nagl W, Hoffmann F (1980) "Arabidobrassica". Evidence for intergeneric somatic hybrid nature from electron microscopic morphometry of chromatin. Eur J Cell Biol 21:227–228
Nelson PP, Albright SC, Wisemann JM, Garrard WT (1979) Reassociation of histone H1 with nucleosomes. J Biol Chem 254:11751–11760
Nicolini CA (ed) (1979a) Chromatin structure and function. Plenum, New York
Nicolini CA (1979b) Chromatin structure, from Ångström to micron levels, and its relationship to mammalian cell proliferation. In: Nicolini CA (ed) Chromatin structure and function. Part B. Plenum, New York, pp 613–666
Noll M (1976) Differences and similarities in chromatin structure of *Neurospora crassa* and higher eukaryotes. Cell 8:349–355
Noll M (1977) DNA folding in the nucleosome. J Mol Biol 116:49–71
Noll M, Zimmer S, Engel A, Dubochet J (1980) Self-assembly of single and closely spaced nucleosome core particles. Nucl Acids Res 8:21–42
Ohlenbusch HH (1979) Structural complementarity in chromatin subunits. Naturwissenschaften 66:206–207
Okada TA, Comings DE (1979) Higher order structure of chromosomes. Chromosoma 72:1–14
Olins AL (1977) V-Bodies are closely-packed in chromatin fibers. Cold Spring Harbor Symp Quant Biol 42:325–329
Olins AL, Olins DE (1974) Spherical chromatin units (v bodies). Science 183:330–332
Olins AL, Carlson RD, Wright EB, Olins DE (1976) Chromatin V-bodies: isolation, subfractionation and physical characterization. Nucl Acids Res 3:3271–3291
Olson MOJ, Busch H (1974) Nuclear proteins. In: Busch H (ed) The cell nucleus Vol III. Academic Press, London New York, pp 211–268
Ord MG, Stocken LA (1979) Nucleosomes from normal and regenerating rat liver. Biochem J 178:173–185
Oudet P, Gross-Bellard M, Chambon P (1975) Electron microscopic and biochemical evidence that chromatin structure is a repeating unit. Cell 4:281–300

Paulson JR, Laemmli UK (1977) The structure of histone-depleted metaphase chromosomes. Cell 12:817–828

Peacock WJ, Lohe AR, Gerlach WL, Dunsmuir P, Dennis ES, Appels R (1977) Fine structure and evolution of DNA in heterochromatin. Cold Spring Harbor Symp Quant Biol 42:1121–1135

Pederson T (1978) Chromatin-structure and gene transcription – nucleosomes permit a new synthesis. Int Rev Cytol 55:1–22

Person C, Suzuki DT (1968) Chromosome structure – a model based on DNA replication. Can J Genet Cytol 10:627–647

Philipps G, Gigot C (1977) DNA associated with nucleosomes in plants. Nucl Acids Res 4:3617–3626

Polyakov V Yu, Kirýanov GI, Manamshyan TA, Fais D, Chentsov YuS (1979) Ultrastructure of DNP fibrils and interchromatin granules in isolated nuclei from rat liver cells. Tsitologiya 21:514–519

Popenko VI, Vengerov YuYu (1978) The effects of various conditions of chromatin isolation on the nucleosomal structure of the isolated chromatin. Mol Biol Rep 4:45–50

Prunell A (1979) Random arrangement of nucleosomes on DNA in chromatin. FEBS Lett 107:285–287

Prunell A, Kornberg RD, Lutter L, Klug A, Levitt M, Crick FHC (1979) Periodicity of DNase I digestion of chromatin. Science 204:855–858

Rattemeyer M, Popp, FA, Nagl W (1981) Evidence of photon emission from DNA in living systems. Naturwiss 68:572–573

Rattner JB, Hamkalo BA (1978a) Higher order structure in metaphase chromosomes. I. The 250 fiber. Chromosoma 69:363–372

Rattner JB, Hamkalo BA (1978b) Higher order structure in metaphase chromosomes. II. The relationship between the 250 fiber, superbeads and beads-on-a-string. Chromosoma 69:373–379

Rattner JB, Hamkalo BA (1979) Nucleosome packing in interphase chromatin. J Cell Biol 81:453–457

Rattner JB, Branch A, Hamkalo BA (1975) Electron microscopy of whole mount metaphase chromosomes. Chromosoma 52:329–338

Razin SV, Mantieva VL, Georgiev GP (1979) The similarity of DNA sequences remaining bound to scaffold upon nuclease treatment of interphase nuclei and metaphase chromosomes. Nucl Acids Res 7:1713–1735

Reeves R (1976) Ribosomal genes of *Xenopus laevis*: Evidence of nucleosomes in transcriptionally active chromatin. Science 194:529–532

Reeves R (1978) Nucleosome structure of *Xenopus* oocyte amplified ribosomal genes. Biochemistry 17:4908–4915

Renz M (1975) Preferential and cooperative binding of histone to chromosomal mammalian DNA. Proc Natl Acad Sci USA 72:733–736

Renz M (1979) Heterogeneity of the chromosome fiber. Nucl Acids Res 6:2761–2767

Risueno MC, de la Espina SMD (1979) Ultrastructural and cytochemical study of the quiescent root meristematic cell nucleus. J Submicrosc Cytol 11:85–98

Rizzo PJ, Burghardt RC (1980) Chromatin structure in the unicellular algae *Olisthodiscus luteus, Crypthecodinium cohnii*, and *Peridium balticum*. Chromosoma 76:91–99

Ruiz-Carillo A, Jorcano JL (1979) An octamer of core histones in solution: central role of the H3–H4 tetramer in the self-assembly. Biochemistry 18:760–768

Sans J, de la Torre C (1979) Proliferative index estimated by chromatin pattern analysis. Cell Biol Int Rep 3:227–235

Sauter J, Ulrich J (1977) Cytophotometric investigation of DNA and RNA content in nuclei of active Strasburger cells in *Pinus nigra* var. *austriaca* (Hoess) Badoux. Planta 137:5–11

Sawicki W (1979) Chromatin pattern *in situ*: Dependence upon cell cycle, preimplantation development and cellular aging *in vitro*. In: Nicolini CA (ed) Chromatin structure and function, Part B. Plenum, New York, pp 667–681

Scheer U, Zentgraf H (1978) Nucleosomal and supranucleosomal organization of transcriptionally inactive rDNA circles in *Dytiscus* oocytes. Chromosoma 69:243–254
Scheer U, Spring H, Trendelenburg MF (1979) Organization of transcriptionally active chromatin in lampbrush chromosomes. In: Busch H (ed) The cell nucleus Vol VII. Academic Press, London New York, pp 4–47
Schmalenberger B (1980) Differentiation of the Guinea pig eye: Nuclear ultrastructure, template activity and DNA content. Protoplasma 103:377–391
Schmalenberger B, Nagl W (1979) Different DNA content, chromatin condensation, and transcription activity in retina cell nuclei of the Guinea pig. Plant Syst Evol Suppl 2:119–125
Schnedl W (1974) Banding patterns in chromosomes. Int Rev Cytol 4:237–272
Schweizer D (1976) Reverse fluorescent chromosome banding with chromomycin and DAPI. Chromosoma 58:307–324
Schweizer D (1979) Fluorescent chromosome banding in plants: Applications, mechanisms, and implications for chromosome structure. Proc 4th John Innes Symp. John Innes Inst, Norwich, VK, pp 61–72
Schweizer D, Ehrendorfer F (1976) Giemsa banded karyotypes, systematics and evolution in *Anacyclus* (Asteraceae, Anthemideae). Plant Syst Evol 126:107–148
Sedat J, Manuelidis L (1977) A direct approach to the structure of eukaryotic chromosomes. Cold Spring Harbor Symp Quant Biol 42:331–350
Setterfield G, Sheinin R, Dardick I, Kiss G, Dubsky M (1978) Structure of interphase nuclei in relation to the cell cycle. J Cell Biol 77:246–263
Sevall JS, Jagodzinski LL, Tsai S, Castro CE, Lee D (1978) Nonhistone proteins and gene organization. In: Busch H (ed) The cell nucleus Vol IV. Academic Press, London New York, pp 319–343
Shindo H, McGhee JD, Cohen JS (1980) ^{31}P-NMR studies of DNA in nucleosome core particles. Biopolymers 19:523–537
Smith SG (1945) Heteropycnosis as a means of diagnosing sex. J Hered 36:194–196
Sobell HM, Tsai CC, Gilbert SG, Jain SC, Sakore TD (1976) Organization of DNA in chromatin. Proc Natl Acad Sci USA 73:3069–3072
Sonnenbichler J (1969) Substructures of chromatin. Nature (London) 223:205–206
Sonnenbichler J (1979) Advances in chromatin reasearch. Naturwissenschaften 66:244–250
Spadafora C, Bellard M, Compton JL, Chambon P (1976a) The DNA repeat lengths in chromatins from sea urchin sperm and gastrula cells are markedly different. FEBS Lett 69:281–285
Spadafora C, Noviello L, Geraci G (1976b) Chromatin organization in nuclei of sea urchin embryos. Comparison with the chromatin organization of the sperm. Cell Differ 5:225–231
Spiker S (1975) An evolutionary comparison of plants histones. Biochim Biophys Acta 400:461–467
Spiker S (1976) Expression of parental histone genes in the intergeneric hybrid *Triticale* hexaploide. Nature (London) 259:418–420
Spiker S, Krishnaswamy L (1973) Constancy of wheat histones during development. Planta 110:71–76
Spring H, Trendelenburg MF, Scheer U, Franke WW, Herth W (1974) Structural and biochemical studies of the primary nucleus of two green algal species, *Acetabularia mediterranea* and *A. major*. Cytobiologie 10:1–65
Stalder J, Braun R (1978) Chromatin structure of *Physarum polycephalum* plasmodia and amoebae. FEBS Lett 90:223–227
Stein G, Stein J, Kleinsmith LJ (eds) (1978) Chromatin and chromosomal protein research. Meth Cell Biol Vol 16. Academic Press, London New York
Stein A, Whitlock JP Jr, Bina M (1979) Acidic polypeptides can assemble both histones and chromatin *in vitro* at physiological ionic strength. Proc Natl Acad Sci USA 76:5000–5004
Steinmetz M, Streeck RE, Zachau HG (1978) Closely spaced nucleosome cores in recon-

stituted histone-DNA complexes and histone-H1-depleted chromatin. Eur J Biochem 83:615–628
Stoddard RW (1979) Nuclear glycoconjugates and their relation to malignancy. Biol Rev 54:199–235
Strätling WH, Müller U, Zentgraf H (1978) Supranucleosomal structure of chromatin. Cell Biol Int Rep 2:495–499
Suda M, Iwai K (1979) Identification of suberimate cross-linking sites of four histone sequences in H1-depleted chromatin. Histone arrangement in nucleosome core. J Biochem 86:1659–1670
Sussman JL, Trifonov EN (1978) Possibility of nonlinked packing of DNA in chromatin. Proc Natl Acad Sci USA 75:103–107
Sutton WD (1972) Chromatin packing, repeated sequences and gene control. Nature New Biol 237:70–71
Tanaka R (1971) Types of resting nuclei in Orchidaceae. Bot Mag 84:118–122
Taylor JH (1979) Chromosome structure. Academic Press, London New York
Thoma F, Koller Th, Klug A (1979) Involvement of histone H1 in the organization of the nucleosome and of salt-dependent superstructures of chromatin. J Cell Biol 83:403–427
Thomas JO (1978) Chromatin structure. In: Clark BFC (ed) International review of biochemistry of nucleic acids, Vol 17/II. Univ Park Press, Baltimore, pp 181–232
Timmis JN, Deumling B, Ingle J (1975) Localization of satellite DNA sequences in nuclei and chromosomes of two plants. Nature (London) 257:152–155
Towill LE, Noodén LD (1973) An electrophoretic analysis of the acid-soluble chromosomal proteins from different organs of maize seedlings. Plant Cell Physiol 14:851–863
Trifonov E (1978) The helical model of the nucleosome core. Nucl Acids Res 5:1371–1380
Trifonov E, Bettecken T (1979) On the orientational nature of sensitivity of nucleosomal DNA to nuclease digestion. In: Balaban M (ed) Molecular mechanisms of biological recognition. Elsevier North-Holland, Amsterdam, pp 233–234
Trifonov EN, Sussman JL (1979) Smooth bending of DNA in chromatin. In: Balaban M (ed) Molecular mechanisms of biological recognition. Elsevier North-Holland, Amsterdam, pp 227–232
Tschermak-Woess E (1963) Strukturtypen der Ruhekerne von Pflanzen und Tieren. Protoplasmatologia VI/1. Springer, Vienna, New York
Van de Sande JH, Liu CC, Jorgenson KF (1977) Reverse banding on chromosomes produced by a guanosine-cytosine specific DNA binding antibiotic: olivomycin. Science 195:400–402
Varshavsky AJ (1976) Structural and functional organization of eukaryotic chromosomes. Biol Zentralbl 95:301–316
Vengerov YuYu, Popenko VI, Lang H, Tikonenko AS (1978) Organization of nucleosomes in chromatin fibres. Biol Zentralbl 97:29–38
Vogt VM, Braun R (1976) Repeated structure of chromatin in metaphase nuclei of *Physarum*. FEBS Lett 64:190–192
Wagner KG, Bode J, Hofmann KW, Willmitzer L (1978) Studies on the mechanism of DNA condensation by basic chromosomal proteins. Stud Biophys 67:85–86 microfiche 2/7–16
Wang TY, Kostraba NC (1978) Proteins involved in positive and negative control of chromatin function. In: Busch H (ed) The cell nucleus Vol IV. Academic Press, London New York, pp 289–317
Weintraub H (1974) The assembly of newly replicated DNA into chromatin. Cold Spring Harbor Symp Quant Biol 38:247–256
Weintraub H (1979) Assembly of an active chromatin structure during replication. Nucl Acids Res 7:781–792
Weintraub H, Groudine M (1976) Transcriptionally active and inactive conformations of chromosomal subunits. Science 193:848–856
Weintraub H, Worcel A, Alberts B (1976) A model for chromatin based upon two symetrically paired half-nucleosomes. Cell 9:409–417

Weischet WO (1979) On the *de novo* formation of compact oligonucleosomes at high ionic strength. Evidence for nucleosomal sliding in high salt. Nucl Acids Res 7:291–304

Wigler MH, Axel R (1976) Nucleosomes in metaphase chromosomes. Nucl Acids Res 3:1463–1471

Worcel A (1977) Molecular architecture of the chromatin fiber. Cold Spring Harbor Symp Quant Biol 42:313–324

Worcel A, Benyajati C (1977) Higher order coiling of DNA in chromatin. Cell 12:83–100

Wunderlich F (1978) Die Kernmatrix: Dynamisches Proteingerüst im Zellkern. Naturwiss Rundsch 31:282–288

Yakura K, Fukuei K, Tanifuji S (1978) Chromatin Subunit structure in different tissues of higher plants. Plant Cell Physiol 19:1381–1390

Yathindra N (1979) DNA flexibility and tertiary folding. A model for superfolding of DNA in chromatin. Curr Sci 48:753–756

Yunis JJ, Yasmineh WG (1971) Heterochromatin, satellite DNA, and cell function. Science 174:1200–1209

Zama M, Olins DE, Prescott B, Thomas GJ Jr (1978) Nucleosome conformation: pH and organic solvent effects. Nucl Acids Res 5:3881–3897

Zeuthen J, Bak P, Bak AL (1979) Chromosomal unit fibers in *Drosophila*. Chromosoma 73:317–326

Zinke M (1979) New model of DNA and higher histone organization in chromatin – explanation of conformational transitions in chromatin structure and their dissociation. Stud Biophys 75:107–130

2 Chromosomal DNA Sequences and Their Organization

R.B. FLAVELL

1 Introduction

One of the most bewildering features about the nuclear DNA of higher plants is the variability in amount and type between different species. The haploid DNA contents of angiosperms range from about 0.5 to over 200 pg (BENNETT and SMITH 1976). Sometimes the variation is extensive within a genus. For example, there is a threefold variation in haploid DNA content within the genus *Lathyrus* without any variation in chromosome number (REES and HAZARIKE 1969) and a tenfold variation in the genus *Crepis* (JONES and BROWN 1976). The amount of DNA in Angiosperms at the lowest end of the range, i.e., about 0.5 pg is approximately five times the amount in *Drosophila melanogaster*, a fact which immediately provokes the conclusion that higher plants contain a considerable excess of DNA over what is needed to specify their biological complexity (HINEGARDNER 1976). The large amounts of DNA in plants and the extensive variation in DNA content between species pose many problems of a technical as well as an intellectual nature for research on chromosomal DNA sequences.

In the mid-1960's the presence of substantial amounts of repeated sequence DNA in plant chromosomes was discovered (BRITTEN and KOHNE 1968) and it was soon recognised that this kind of DNA was predominantly, but not exclusively, responsible for the large amount of DNA in plants and the substantial variations in DNA content between species (FLAVELL et al. 1974). The structure, chromosomal organisation and evolution of families of repeated sequences has been the subject of considerable study in plant genomes ever since, not only because these kinds of sequences are present in such large amounts and biologically extremely interesting, but also because they are technically simpler to isolate. The non-repeated DNA includes most of the genes, but the coding sequences occupy only a small fraction of the non-repeated DNA (WALBOT and GOLDBERG 1980, THOMPSON and RUBENSTEIN 1980).

Genes are usually recognised and in part defined by their function or RNA product. With the exception of reiterated genes, such as those specifying the ribosomal RNA's, the function of repeated sequences is a matter of considerable

---

Abbreviations. C_0t, unit of DNA reassociation after melting (definition see p. 47); HAP, hydroxyapatite. Restriction enzymes: Bam HI, *Bacillus amyloliquefaciens*; Bgl II, *Bacillus globigii*; Eco RI, *Escherichia coli* RY 13; Hae III, *Haemophilus aegypticus;* Hind III, *Haemophilus influenzae* R_d; Hpa II, *Haemophilus parainfluenzae;* Pst I, *Providencia stuartii*; Sal I, *Streptomyces albus*; Sma I, *Serratia marcescens*; Xba I, *Xanthomonas badrii*

speculation (Orgel and Crick 1980, Doolittle and Sapienza 1980, Cavalier-Smith 1978, 1980, Dover 1980, Flavell 1980, 1981, Hinegardner 1976). Thus most repeated sequences cannot be defined by their function. Neither can most of the non-repeated DNA. Therefore, nearly all the nucleotide sequences or groups of sequences that have been recognised in plant genomes have been defined by physical properties, e.g., length, copy number, base composition, organisation and renaturation rate, rather than by function. In this review I wish to summarize (a) the methods that have been commonly used to elucidate these properties and (b) our current knowledge of plant genome structure. Other reviews on the analysis of plant chromosomal DNA sequences, focussing especially on DNA renaturation, have been published recently and should also be consulted (Walbot and Goldberg 1980, Flavell 1980, Thompson and Rubenstein 1980).

2 Genome Analysis by Renaturation Kinetics

The assay used most extensively to characterise the general features of the genomes of higher organisms is that of renaturation kinetics. By following the reassociation of denatured DNA sequences over a range of C_0t values (C_0t is the product of initial concentration of single-stranded DNA in moles of nucleotide per litre and the time of reassociation in seconds), the total DNA content, the proportion of repetitive DNA and the average reiteration frequencies of repeated DNA classes can be determined. This kind of analysis was introduced by R.J. Britten and his colleagues. The mathematics and theory of DNA reassociation kinetics are described in several papers and reviews (Britten and Kohne 1968, Wetmur and Davidson 1968, Smith et al. 1975, Britten and Davidson 1976) and so will not be repeated in detail here. The technical details have also been well reviewed by Britten et al. (1974) to which the reader should refer. In this chapter I review only the key features of genome analysis by renaturation kinetics, drawing attention to special aspects in the design or interpretation of experiments that have often been overlooked.

DNA renaturation studies are first carried out using (1) short fragments (200 to 400 nucleotides) to ensure that neighbouring sequences are separated from one another as much as possible and that the efficiency of collision of different sequences is maximised and (2) a temperature and salt concentration which gives a maximum rate of renaturation while allowing duplexes with only a moderate number of mismatched bases to be stable. These conditions, which are close to 20 to 25 °C below the average melting temperatures of sheared native DNA fragments, are often approximately 60 °C in 0.18 M Na^+ for plant DNA's. They allow the formation of duplexes with up to approximately 25% mismatched bases because for every degree below the mean melting temperature of native DNA, duplexes with an additional 1% mismatched bases are stable (Britten et al. 1974). This point is important because altering the renaturation conditions not only alters the rate of renaturation, but also the proportion of related sequences able to form stable duplexes (BENDICH and MCCARTHY 1970a, BENDICH and ANDERSON 1977, FLAVELL and SMITH 1976). The extent of fragment reassociation can be determined by chromatography on hydroxyapatite (HAP), by measurement of hypochromicity at a wavelength of 260 mμ in a spectrophotometer, or by estimating the proportion of DNA resistant to the nuclease S_1 which degrades only single-stranded DNA. Measurements of hypochromicity and S_1 nuclease-resistant DNA determine the proportion of bases which are paired. In contrast, HAP chromatography measures the proportion of fragments which contain regions of renatured DNA of approximately 30 base pairs or more. When randomly sheared fragments reanneal, only 55%, on average, of each fragment overlaps in sequence with a complementary fragment (SMITH et al. 1975). Consequently, the DNA which binds to HAP in the renatured fraction is an overestimate of the amount of renatured DNA. An accurate estimate of the proportion of renatured DNA is gained by hypochromicity measurements or by digestion with S_1 nuclease under conditions where the nuclease degrades single-stranded DNA but not mismatched regions within duplex DNA (DAVIDSON et al. 1973, FLAVELL and SMITH 1976, THOMPSON and RUBENSTEIN 1980, WALBOT and GOLDBERG 1980).

A homogenous family of sequences renatures with second-order kinetics which approximate to the equation $\frac{C_t}{C_0}=\frac{1}{1+K_2C_0t}$, where C_t is the concentration of single-stranded DNA remaining in the reaction, C_0 is the initial concentration of DNA and K_2 is the second-order rate constant (BRITTEN and KOHNE 1968, BRITTEN and DAVIDSON 1976). When half the family has renatured $\frac{C_t}{C_0}=\frac{1}{2}$ or $\frac{1}{2}=\frac{1}{1+K_2(C_0t)}$. This simplifies to $K_2(C_0t)=1$ showing that C_0t at half renaturation ($C_0t\frac{1}{2}$) is inversely proportional to the rate of the reaction. The C_0t value at which a particular family of sequences in a complex genome renatures is clearly dependent upon the concentration of the family in the genome. The reassociation of sequences present in one copy per haploid genome is inversely proportional to the total genome size of the species. Sequences which renature more quickly are in a proportionally higher concentration (BRITTEN and KOHNE 1968).

The rate at which DNA fragments become resistant to S_1 nuclease in a renaturation reaction are defined better by the equation $\frac{C_t}{C_0}=\left[\frac{1}{1+K_2(C_0t)}\right]^{0.44}$ (SMITH et al. 1975) because the rate of renaturation of single-stranded "tails" on fragments that contain duplex regions is slower than the rate for fragments lacking renatured regions.

In order to determine what fraction of the total DNA renatures with single copy kinetics it is necessary to know the haploid DNA content of the species and the renaturation kinetics of a genome of known size under the same experimental conditions. This is best achieved by including a known amount of labelled bacterial or other defined DNA in the reannealing plant DNA as an internal standard and determining the renaturation kinetics of the bacterial DNA and plant DNA simultaneously (MURRAY et al. 1978, 1979). The bacterial

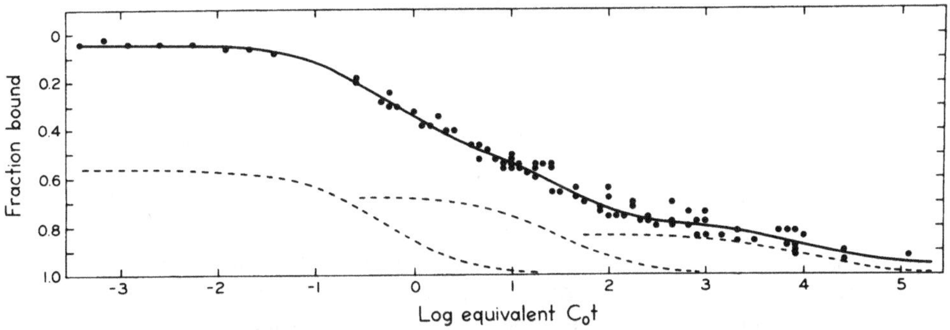

Fig. 1. Reassociation kinetics of pea DNA fragments sheared to a modal single strand length of 300 nucleotides. The fraction of fragments able to bind to hydroxyapatite (i.e., containing duplex regions) is plotted as a function of C_0t. The *dashed lines* represent elements of the best three-component model describing the data. (MURRAY et al. 1978)

DNA should be sheared to the same size as the plant DNA, as renaturation rates are dependent on fragment length (BRITTEN et al. 1974).

The renaturation kinetics of short DNA fragments of pea DNA (MURRAY et al. 1978) are shown as an example in Fig. 1. The proportion of renatured DNA is plotted against C_0t value on a log scale and was determined by HAP chromatography. The simplest mathematical description of the data is given by three second-order curves, fitted to the data by least-squares procedures (PEARSON et al. 1977). The 15% of the genome which renatures with slowest kinetics comprises single- or few-copy DNA sequences. About 85% of the fragments (by weight) renature more rapidly, the renaturation curves covering over a 10,000-fold range of C_0t values. These fragments consist predominantly of repetitive sequences which belong to families with more than a 1,000-fold variation in repetition frequency. Although the renaturation of complex genomes can frequently be fitted to a small number of second-order curves by least-squares analyses, this is probably an oversimplification. Analysis of individual component families usually uncover more heterogeneity in repetition frequency within a genome than implied by best-fit second-order curves to total genome renaturation curves.

When analyses on DNA purified from leaves are carried out, contamination with chloroplast DNA cannot be ignored (MURRAY et al. 1979). Chloroplast DNA can occupy up to 21% of total DNA extracted from plants with small nuclear genomes (SCOTT and POSSINGHAM 1980). Consequently, it is necessary either to use DNA purified from roots or to determine the proportion of chloroplast DNA and its renaturation kinetics and correct the renaturation curves for its presence (PREISLER and THOMPSON 1978, PALMER et al. 1979, LAMPPA and BENDICH 1979).

3 Proportions of Repeated and Non-Repeated DNA

From renaturation kinetics the proportions of single-(or few-)copy DNA and repeated DNA have been determined in a diverse range of species, monocots and dicots, of different genome sizes. These include a range of *Vicia* species (CHOOI 1971, STRAUS 1972), tobacco (ZIMMERMAN and GOLDBERG 1977), pea (MURRAY et al. 1978), mung bean (MURRAY et al. 1979), parsley (KIPER and HERZFELD 1978), pearl millet (WIMPEE and RAWSON 1979), soybean (GOLDBERG 1978, GURLEY et al. 1979, PELLEGRINI and GOLDBERG 1979), Jerusalem artichoke (NZE-EKEKANG et al. 1974), anemone (CULLIS and SCHWEIZER 1974), a range of *Lathyrus* species (NARAYAN and REES 1976), lily (SMYTH and STERN 1973), cotton (WALBOT and DURE 1976, WILSON et al. 1976), a range of Cichorieae species (BACHMANN and PRICE 1977), daffodil, deer fern and parsley fern (BENDICH and ANDERSON 1977), wheat (BENDICH and MCCARTHY 1970a, b, MITRA and BHATIA 1973, SMITH and FLAVELL 1975, FLAVELL and SMITH 1976, RANJEKAR et al. 1976), rye (BENDICH and MCCARTHY 1970a, b, SMITH and FLAVELL 1977, RANJEKAR et al. 1974), a range of conifer species (MIKSCHE and HOTTA 1973), a range of pteridophytes (BOUCHARD and SWIFT 1977), oats (RIMPAU et al. 1980, BENDICH and MCCARTHY 1970a), barley (BENDICH and MCCARTHY 1970a, RANJEKAR et al. 1976, BENDICH and ANDERSON 1977, RIMPAU et al. 1980), maize (CHILTON and MCCARTHY 1973, HAKE and WALBOT 1980), some *Osmunda* species (STEIN and THOMPSON 1975), *Atriplex* species (BELFORD and THOMPSON 1979) two *Poa* species, sunflower, nasturtium, tulip, hyacinth, carrot, sugar beet, shepherd's purse, *Lamium, Senecio, Stellaria media* (FLAVELL et al. 1974), onion (FLAVELL et al. 1974, STACK and COMINGS 1979), peanut (DHILLON et al. 1980), flax (CULLIS 1973, 1977, FLAVELL et al. 1974). Not all these published results are comparable because the renaturation analyses were carried out under different conditions. A few genomes have been examined in considerable detail. Figure 2 shows the results for those genomes which were analysed in similar renaturation conditions (60 °C, 0.18 Na^+) using HAP to determine renatured DNA, and whose lengths of single-copy DNA sequences have also been estimated. Between 20 and 35% of the chromosomal DNA of species with haploid genomes containing more than 2 pg behaves as single- or few-copy DNA. In species with haploid genomes containing less than around 2 pg the *proportion* of single-copy DNA appears to be inversely related to genome size (see also FLAVELL et al. 1974). The absolute amount of single-copy DNA usually (but not always) decreases with decreasing genome size but the rate of decrease is less in genomes smaller than 2 pg. The values are, of course, a reflection of the renaturation conditions. Under more stringent incubation conditions the estimated proportion of single-copy DNA increases, while under more relaxed conditions it decreases (BENDICH and MCCARTHY 1970A, BENDICH and ANDERSON 1977, FLAVELL et al. 1979; THOMPSON and MURRAY 1980). This strongly suggests that much of the single-copy DNA in larger plant genomes belongs to families of repeated DNA but has diverged so much that it does not recognise evolutionarily related sequences at $T_m - 25$ °C. In genomes with only small *amounts* of single-copy DNA the proportion of diverged repeated DNA would

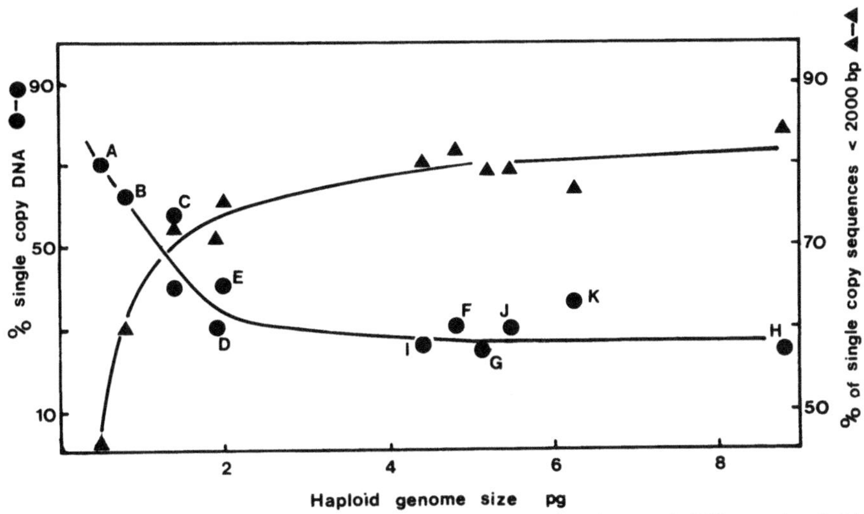

Fig. 2. Proportion and sizes of single copy DNA sequences in plants of different haploid DNA content. The species indicated are ones in which the proportion of repeated sequence DNA has been assayed by renaturation kinetics at 60 °C in 0.12 M phosphate buffer, in which the lengths of single-copy DNA sequences have been estimated and whose haploid genome size has been reliably measured. The species are as follows: *A* mung bean (Murray et al. 1979); *B* cotton (Walbot and Dure 1976); *C* soybean (Goldberg 1978, Gurley et al. 1979); *D* parsley (Kiper and Herzfeld 1978); *E* tobacco (Zimmerman and Goldberg 1977); *F* pea (Murray et al. 1978); *G* wheat (Flavell and Smith 1976; Rimpau et al. 1978). *H* rye (Smith and Flavell 1977, Rimpau et al. 1978); *I* oats (Rimpau et al. 1980); *J* barley (Rimpau et al. 1980); *K* maize (Hake and Walbot 1980). Because estimates of single-copy DNA and their length are dependent on many experimental parameters (see text), the values should be interpreted carefully

be expected to be less than in large genomes. This has been confirmed for the mung bean genome (Thompson and Murray 1980).

The amounts of single-copy DNA determined at $T_m - 25$ °C are very large compared with the amount of coding DNA transcribed and translated. For example, in a species with 30,000 genes (Kiper et al. 1979, 1980, Goldberg et al. 1978) each containing 1,000 base pairs, approximately 3×10^7 base pairs would be in the genes. A plant genome of about 2 pg contains approximately 2×10^9 base pairs. Thus the sequences transcribed and translated probably constitute only about 1% of the total, DNA whereas the single-copy DNA, assayed at $T_m - 25$ °C, constitutes 20 to 40% (Fig. 2).

The proportions of single-copy DNA in plant genomes are generally lower than in animal genomes of a comparable size (Thompson et al. 1979).

4 Single-Copy and Repeated DNA Interspersion Patterns

The patterns of interspersion of different kinds of sequences have now been studied in a small number of plant genomes by the methods first introduced

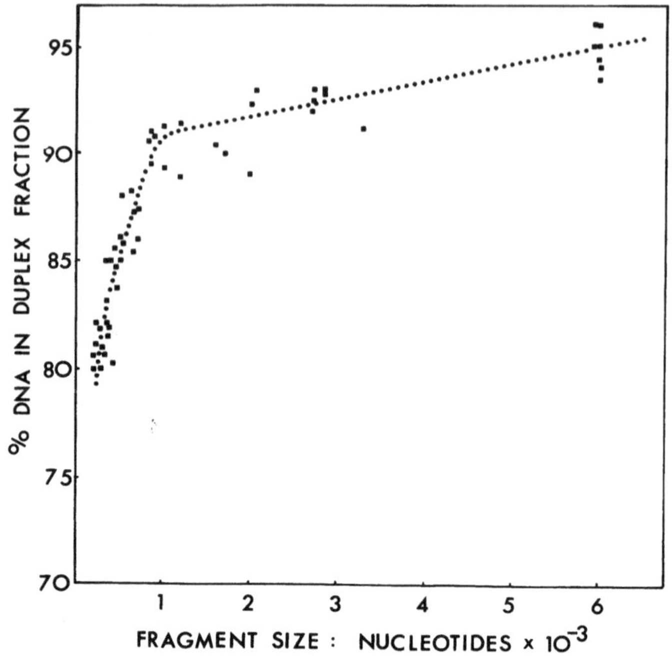

Fig. 3. Interspersion of single- or few-copy DNA sequences with repeated DNA. This graph shows the proportion of wheat DNA in the duplex fraction of the hydroxyapatite chromatography, when DNA fragments of different mean length were renatured to a C_0t value which allows only repeated sequences to renature. The additional DNA in the duplex fraction with increasing fragment length is denatured single-copy DNA covalently attached to renatured repeated DNA (see text). (Results taken from FLAVELL and SMITH 1976)

by DAVIDSON et al. (1973) and GRAHAM et al. (1974). These patterns define the lengths and distribution of sequences and so have contributed important knowledge about eukaryote genome structure. The experimental approach is based upon the ability of HAP to bind DNA molecules which are only partially double-stranded. For example, when single-copy DNA is adjacent to repeated DNA in the genome, and long fragments are renatured to C_0t values which allow only repeated DNA to reassociate, then the duplex DNA retained on HAP will also contain the single copy DNA in the form of single-stranded "tails". The additional DNA included in the HAP duplex fraction with very long fragments gives an estimate of the proportion of single-/low-copy DNA interspersed with repeated DNA and also an estimate of the mean length of the interspersed low copy DNA sequences. In practice, labelled "tracer" DNA's of different lengths are usually incubated with large excesses of short, unlabelled "driver" DNA to prevent self-reannealing of labelled DNA. This method also keeps to a minimum the concentration of long DNA fragments which form complex networks or hyperpolymers when each fragment reassociates with more than one other fragment (BRITTEN et al. 1974, THOMPSON 1976, FLAVELL and SMITH 1977). The results of an experiment using increasing lengths of tracer wheat DNA are shown in Fig. 3 (FLAVELL and SMITH 1976). The proportion binding to HAP after renaturation at 60 °C in 0.18 M Na^+ to a C_0t value of 50 increases up to a fragment length of about 1,000 bases. Such a curve shows that the proportion of DNA behaving as repeated under these conditions is about 75% (extrapolation of curve back to y axis) and that about 15% of the genome consists of low-copy DNA sequences shorter than 1,000 base pairs interspersed between repeated sequences. The failure of 4 or 5% of the total tracer DNA to be included in the HAP duplex fraction even with very long fragments is probably due to a small proportion of the labelled DNA being unable to take part in a hybridisation reaction. The small increase in the proportion of DNA in the duplex fraction using fragments between 1,000 and 6,000 bases probably implies that a small proportion of the few-copy sequences are much longer than 1,000 base pairs (FLAVELL and SMITH 1976).

Several important factors complicate this experimental design and the interpretation of results. One is that the renaturation rate is related to fragment length (WETMUR and DAVIDSON 1968, HINNEBUSCH et al. 1978). Thus a C_0t value which does not allow renaturation of single-copy DNA using short fragments may allow renaturation of single-copy DNA using much longer fragments. This problem is particularly severe for genomes where there is not a large kinetic distinction between the reassociation of repetitive and single-copy DNA as is often found with small genomes, e.g., mung bean (MURRAY et al. 1979). To overcome this problem it is best to use only repeated DNA as the "driver", or to study the renaturation kinetics of the whole genome using tracer fragments of different lengths and to determine the proportion of the genome which renatures with single-copy kinetics for each fragment length (MURRAY et al. 1979, WALBOT and GOLDBERG 1980). The single-copy kinetics can be corrected for the effect of fragment length and then the proportion of single-copy DNA in fragments renaturing with repeated DNA kinetics calculated for each size of fragment. This gives a more accurate estimate of the interspersion of short low-copy sequences with repeats.

Another important factor is that the interspersion patterns obtained are affected by the renaturation conditions because the proportion of DNA which renatures as repeated DNA is dependent upon the renaturation conditions (FLAVELL and SMITH 1976, FLAVELL et al. 1979). This fact has been inadequately considered in discussions of published results.

From renaturation assays using tracer fragments of increasing length the proportions and lengths of the single-copy sequences have been estimated in a small number of plant genomes (Fig. 2). The results for the wheat genome described above are similar to those for other large genomes with most (>75%) of the single-copy DNA being in segments shorter than approximately 2,000 base pairs interspersed between repeated sequences (FLAVELL 1980, THOMPSON and RUBENSTEIN 1980, WALBOT and GOLDBERG 1980). The average lengths of the interspersed repeated sequences have been measured separately in very few studies but it can be deduced that they are generally short, i.e., less than 1,000 base pairs (WALBOT and GOLDBERG 1980). This interspersion pattern for the single-copy DNA, which predominates in plant and animal genomes with genome sizes greater than 2 pg analysed under renaturation conditions equivalent to about $T_m - 25\,°C$, has been called the "short period" interspersion pattern.

The smaller the genome size below 2 pg, the larger the proportion of single-copy sequences which appear longer than 4,000 base pairs. This relationship is also shown in Fig. 2. The sequence organisation in which longer single-copy sequences are interspersed with longer repeats (e.g., 1,000 base pairs in cotton (WALBOT and DURE 1976) and mung bean (MURRAY et al. 1979) has been termed the "long period" interspersion pattern. It is also found in small animal genomes such as *Drosophila* (CRAIN et al. 1976b), honeybee (CRAIN et al. 1976a), *Chironomus* (WELLS et al. 1976) and some birds (ARTHUR and STRAUS 1978, EPPLEN et al. 1978).

The distinction between short and long interspersion patterns and the biological significance of the patterns has probably been overemphasized. The dependence of the patterns on genome size and the renaturation conditions, together with the considerable variation in the amounts of DNA with such interspersion patterns in closely related species reduces the likelihood that much of the interspersed single-copy/repeated DNA is vital for gene expression (DAVIDSON and BRITTEN 1979).

Estimation of the proportion of "tracer" DNA in the HAP duplex fraction after renaturation of tracers of different lengths can also indicate if repeated sequences, present in very different numbers of copies in a genome, are interspersed. For example, if repeats in a moderate frequency are interspersed with high frequency repeats, then they will appear in the HAP duplex fraction with the kinetics of the high frequency repeats when longer "tracer" fragments are used. Results from studies on a number of genomes suggest that interspersed repeats may be common in plants (KIPER and HERZFELD 1978, GURLEY et al. 1979, but see GOLDBERG 1978, PELLEGRINI and GOLDBERG 1979, HAKE and WALBOT 1980). This experimental approach has also been used to investigate the organisation in wheat, rye, oat and barley of sequences which are unique to each genome or are common to two or more of the genomes (RIMPAU et al. 1978, 1980). Sequence interspersion maps were produced for each of the genomes and they showed that short species-specific repeats are often found interspersed with repeats common to two or more of the species. The similarity of these interspersion patterns to the "short period" single copy-repeat pattern suggest that repeated sequences from different families are rearranged into new permutations during evolution by mechanisms similar to those responsible for the short period single-copy/repeat pattern, i.e., the rearrangement of short pieces of DNA around the genome involves all kinds of sequences (FLAVELL et al. 1979, 1981, FLAVELL 1980, THOMPSON and MURRAY 1980). The interspersion of different repeats is considered further in Section 8.

5 Reverse Repeats

All higher plant chromosomes contain duplicate sequences in close proximity but different orientation (THOMPSON and RUBENSTEIN 1980, WALBOT and GOLD-

BERG 1980, FLAVELL 1980). These reverse repeats renature with first-order kinetics when on the same fragment, to form a hairpin structure, visible in the electron microscope (WILSON and THOMAS 1974). Such renaturation is independent of DNA concentration and so occurs in very low DNA concentrations, i.e., at very low C_0t values. The DNA in the HAP duplex fraction at very low C_0t values in Fig. 1 contains the reverse repeats. It comprises less than 5% of the DNA in the genomes studied (FLAVELL 1980). The lengths of the duplex regions in the hairpin structures, which provides an estimate of the length of the reverse repeats, range from 50 bp (the shortest length easily visible in the electron microscope) to over 900 base pairs. Twenty percent of the molecules scored by BAZETOUX et al. (1978) in wheat DNA had a single-stranded terminal loop containing 300 to 1,000 base pairs. This loop length is the distance apart of the reverse repeats in the chromosomes. The other 80% showed no loop, indicating that the reverse repeats lie extremely close together.

The number of reverse repeats in a genome appears to be related to the size of the genome. In wheat over 10^6 occur per haploid genome (BAZETOUX et al. 1978), while in cotton there are about 40,000 (WALBOT and DURE 1976). Some clustering of pairs of reverse repeats occurs in the chromosomes but they appear to be found at least every 5,000 to 10,000 base pairs through 20% of each genome studied. It is quite possible that repeats are in opposite orientations much more frequently than implied by these figures. If they are a long distance apart (or on different chromosomes) they would not be recognised. This possibility is supported by the observation that more reverse repeats with longer hairpin loops are found when longer DNA molecules are renatured. Reverse repeats may have arisen by related sequences becoming closely linked by the mechanism which translocates short pieces of DNA around the genome or alternatively by a precise duplication-inversion process. This process might leave the reverse repeats adjacent initially, but later members of some pairs could become separated by the insertion of unrelated DNA between them. The function, if any, of reverse repeats is unknown. A pair is probably necessary to complete replication of the end of a chromosome (CAVALIER-SMITH 1974).

6 Genome Analysis by Equilibrium Centrifugation in Heavy Salt Gradients

DNA sequences band at specific positions in gradients of heavy salts, such as CsCl. The positions are dependent on the buoyant density and other structural features of the DNA sequences. Buoyant density for most sequences is related to base composition (SCHILDKRAUT et al. 1962). Where a sufficient quantity of the genome consists of clustered arrays of a sequence with a non-representative base composition then DNA can often be recognised in the analytical ultracentrifuge as a separate band or "satellite". Satellites are common in DNA's from dicots with small genomes but rare in monocots with larger genomes. They can constitute up to 45% of the total DNA (INGLE et al. 1973). Clustered arrays with non-representative base compositions often contain short repeating units. Arrays of repeats which do not separate from the bulk of the DNA in such gradients can sometimes be induced to do so by incorporation of certain drugs (e.g., actinomycin D) or heavy metal ions, e.g., Ag^+, Hg^{2+} into the gradients. These molecules can interact preferentially with certain kinds of sequences (Ag^+ to A-T rich DNA and Hg^{2+} to G-C rich DNA) to alter differentially their buoyant density (NANDI et al. 1965). Equilibrium centrifugation in heavy salt gradients has been the means of purifying arrays of repeated sequences from many plant genomes, e.g., muskmelon (BENDICH and TAYLOR 1977), tomato (CHILTON 1975), barley (DENNIS et al. 1980, RANJEKAR et al. 1978), soybean (GURLEY et al. 1979), rye (APPELS et al. 1978), orchid (CAPESIUS et al. 1975) and wheat (RANJEKAR et al. 1978, HUGUET and JOUANIN 1972; DENNIS et al. 1980), melon, cucumber, runner bean, turnip, orange, flax, *Oenothera, Aquilegia alpina, Lagenaria vulgaris, Luffa cylindrica,* pumpkin, marrow, witch hazel, watermelon, Chinese cabbage, French bean, mung bean, lemon grapefruit, potato and marigold (INGLE et al. 1973, 1975). Satellite DNA's which appear as homogeneous peaks in the analytical ultracentrifuge frequently contain more than one family of sequences, so further analyses using restriction endonucleases are necessary. It is also necessary to ascertain the origin of the satellites because chloroplast and mitochondrial DNA's frequently band at densities different from nuclear DNA (KADOURI et al. 1975; cf. Chapter 14, this Vol.).

7 Genome Analysis Using Restriction Endonucleases

7.1 Analysis of Whole Genome Digests After Electrophoresis and Ethidium Bromide Staining

Restriction endonucleases (NATHANS and SMITH 1975) offer great opportunities for analysing complex genomes and they are essential tools in cloning protocols. The special value of restriction endonucleases is that they produce a defined set of fragments which when separated by electrophoresis provide a characteristic pattern or "fingerprint" of the genome. This is because the type II restriction endonucleases which produce these fingerprints cleave DNA at or near to specific nucleotide sequences. Over 100 such enzymes are now known and the reader should consult an up to date review because the number known and commercially available is expanding rapidly. Most, but certainly not all recognition sites of type II restriction endonucleases contain four or six base pairs.

A photograph of rye DNA taken under UV light after cleavage with different restriction endonucleases, fractionation by agarose electrophoresis and staining with ethidium bromide is shown in Fig. 4. With some of the enzymes, a characteristic pattern of bands is seen superimposed upon a smear of fluorescence. The bands contain repeated sequences. Other repeated sequences and single-copy sequences are in the background smear. Such complex products of restriction endonuclease digestion are typical for higher plant genomes.

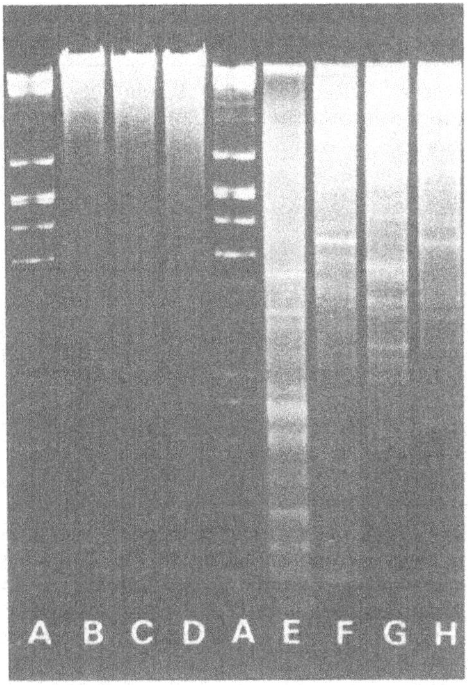

Fig. 4. Agarose gel electrophoresis of rye DNA digested with renaturation endonucleases. Rye nuclear DNA was digested to completion with restriction endonucleases: *B* Sma I; *C* Hpa II; *D* Pst I; *E* Hind III; *F* Bgl II; *G* Bam HI; *H* Eco RI. Tracks labelled *A* contain size markers. Electrophoresis was from top to bottom. The DNA was stained with ethidium bromide and photographed under short wave ultraviolet light

HAMER and THOMAS (1974) showed for *Drosophila* DNA that the size distribution of DNA fragments produced by restriction endonuclease digestion fits that predicted from random fragmentation theory when the base composition of the genome is taken into account. Similar analyses have been carried out on rye genome digests by BEDBROOK et al. (1978) and BEDBROOK and GERLACH (1980). The results are summarised in Table 1. The average fragment sizes after restriction with Sma I, Hpa II and Pst I are considerably higher than predicted (see also Fig. 4, tracks B, C and D). The failure of Sma I and Hpa II to cleave plant DNA frequently is due to the extensive methylation of cytosine residues in the configuration CpG. These enzymes with the recognition sequence CCCGGG and CCGG respectively do not cleave if the C residue to the 3′ side of the cutting site is methylated (MANN and SMITH 1977). Twenty to 25% of the cytosine residues in most higher plant genomes are methylated (BEDBROOK et al. 1978).

Restriction endonucleases Eco RI, Bam HI, Bgl II and Hind III produce fragments of mean length similar to that predicted (Table 1), but each enzyme produces considerably more fragments of much higher molecular weight and much lower molecular weight than predicted by random fragmentation theory (BEDBROOK and GERLACH 1980). These departures are almost certainly because plant genomes contain a high proportion of repeated sequences of relatively low complexity. Digestion of rye DNA with the enzyme Hae III which recognises only four bases (GGCC) produced fragments with sizes close to those expected for a random distribution of GGCC sequences in the genome.

Table 1. Comparison of theoretical and empirical frequencies for the occurrence of restriction endonuclease targets in rye DNA (BEDBROOK and GERLACH 1980)

Restriction enzyme	Recognition sequence	Probability of occurrence of sequence in genome p	Mean segment length determined from p	Calculated mean segment length determined from restriction enzyme digests	Calculated number of "different" fragments produced by digestion of *S. cereale* DNA
Eco RI	G'AA*TTC	2.887×10^{-4}	3,464	3,400	2.30×10^{6}
Bam HI	G'GATC*C	1.948×10^{-4}	5,132	3,311	2.35×10^{6}
Bgl II	A'GATCT	2.887×10^{-4}	3,464	2,512	3.10×10^{6}
Hae III	GG'C*C	25.860×10^{-4}	387	400	1.95×10^{7}
Hpa II	CC*'GG	25.860×10^{-4}	387	>10,000	$<7.80 \times 10^{5}$
Hind III	A*'AGCTT	2.887×10^{-4}	3,464	2,400	3.25×10^{6}
Pst I	CTGCA'G	1.948×10^{-4}	5,132	8,128	9.50×10^{5}
Sma I	CCC*'GGG	1.315×10^{-4}	7,605	>10,000	$<7.80 \times 10^{5}$

Rye DNA was digested with restriction enzymes and fractionated by gel electrophoresis as described in the legend to Fig. 4. Traces of photographic negatives of the gel were made using a scanning densitometer. Cumulative size distributions were obtained by measuring the fluorescence within specific fragment size classes determined from molecular weight markers. The integral of the weight over the range of fragment sizes was plotted against fragment size. The plots were tested for conformance to the Kuhn approximation ($wt \cong tp^2 e^{-pt}$) where wt is the weight fraction of DNA found in segments t base pairs long. The value for p (probability of a double-strand cleavage) was determined from the base composition of rye DNA and the base composition of the enzyme recognition sequence.

Consideration of digestion patterns of plant DNA's as described for rye is of practical as well as of theoretical importance for the following reasons. (1) After digestion with a restriction endonuclease and electrophoresis a proportion of the genome is commonly found to be at the top of the gel uncleaved. This fraction usually contains repeated sequence families of low sequence complexity. Its maintenance in a high molecular form after restriction provides a way of purifying it from other sequences (see BEDBROOK et al. 1980a for purification of repeats from the rye genome using this property) but also prevents it being inserted into a cloning vehicle. The high proportion of large fragments remaining after cleavage with many of the restriction endonucleases also makes unlikely their recovery from certain kinds of cloning experiments. (2) Enzymes which cut the DNA in question only rarely can be very useful in purification protocols if the desired sequences are released in low molecular weight fragments. Alternatively, these enzymes are useless if the sequences remain uncleaved with the bulk of the DNA. (3) To have the maximum chance of inserting almost every DNA sequence into a cloning vehicle it is desirable to use a restriction endonuclease that is capable of cleaving all the DNA to low molecular weight fragments as predicted by the random fragmentation theory. Such an enzyme is likely to be one which recognises only four bases such as Hae III. (4) It is possible to calculate the total number of different fragments produced by different restriction enzymes of a size suitable for insertion into cloning vehicles and so assess the number of clones required to provide say a 90% chance of cloning every fragment capable of being cloned. Estimates of the total number of different rye fragments produced by eight restriction endonucleases are shown in Table 1.

7.2 Interspecies Comparisons of Major Families of Repeated Sequences

It was noted above that the banding pattern produced after gel electrophoresis of DNA digested by a restriction endonuclease is useful because it is a "fingerprint" of the genome. The drawback with complex genomes is that only a few percent of the total DNA are represented in clearly identified bands, visible by ethidium bromide staining, although digests with enzymes that cleave DNA frequently, e.g., Hae III, contain many more bands than digests with enzymes which produce fewer fragments (see Table 1). In spite of this drawback the banding patterns produced in total DNA digests are useful for comparative studies between related species. For example, a range of *Secale* species have been shown to have very similar restriction banding patterns, clearly identifying them as *Secale* species, distinct from species belonging to other genera of the Gramineae (Jones and Flavell 1982b). However, some major band differences can be seen between digests of different *Secale* DNA's. Recognition of these differences opened up a research programme which led to the description of more than 50% of the DNA difference between *Secale silvestre* and *Secale cereale* (Bedbrook et al. 1980a).

7.3 Sequence Analysis After Transfer to Nitrocellulose, Using Radioactive Probes

The drawbacks of being able to assay only the repeated fragments which appear as bands on gels after ethidium staining are overcome when the DNA is transferred to nitrocellulose by the method of Southern (1975) and "probed" by hybridisation with a radioactive RNA or DNA molecule. The fingerprint of the fragments on the gel homologous to the probe nucleic acid is then revealed by autoradiography. The sensitivity of this method is such that fragments carrying sequences present in only one copy in the genome can be detected if sufficient DNA is applied to the gel and sufficient probe nucleic acid, radioactively labelled to high specific activity, is used (Wahl et al. 1979). This technique not only provides estimates of the sizes of fragments carrying a particular sequence, but also enables different fragments carrying the same sequence to be identified. Using DNA digested with different restriction endonucleases, it is also possible to map the cleavage sites for restriction endonucleases within a segment of DNA or a gene. It is also often possible to deduce the arrangement of different sequences or segments of DNA with respect to one another. As examples of the value of such techniques, I describe (1) how the organisation of a repeated sequence family in the rye genome was uncovered and (2) how a physical map of the repeating units containing the genes specifying the 18S and 25S ribosomal

RNA genes in soybean was obtained. It is important to note that this powerful method of genome analysis can be used only when a pure radioactive probe is available.

7.4 Analysis of Repeated DNA Families Using Restriction Endonucleases

Figure 5 shows rye DNA digested with Hae III (track A) Bam HI (track B) and Eco RI (track C), fractionated by electrophoresis on agarose, transferred to nitrocellulose (SOUTHERN 1975) and hybridised with [^{32}P]-labelled RNA complementary to a piece of repeated sequence DNA purified from the rye genome by molecular cloning (BEDBROOK et al. 1980b). The autoradiograph shows that after digestion with Hae III the DNA fragments homologous to the probe appear as a "ladder" where the DNA bands have lengths which are multiples

Fig. 5. Hybridisation of a cloned repeated sequence from rye to rye DNA digested with Hae III (track *A*), Bam HI (*B*) and Eco RI (*C*). After digestion with the restriction endonuclease, the DNA was fractionated by electrophoresis on agarose and transferred from the gel to a sheet of nitrocellulose by the method of SOUTHERN (1975). The filter was hybridised to a [^{32}P]-labelled cRNA probe prepared in vitro by *E. coli* RNA polymerase using as template a cloned example of a 120 bp repeated sequence from rye. Stable hybridisation was revealed by autoradiography after extensive washing. The 2.2 kilobase repeats in the Eco RI digest are *arrowed*. (Results taken from BEDBROOK et al. 1980b)

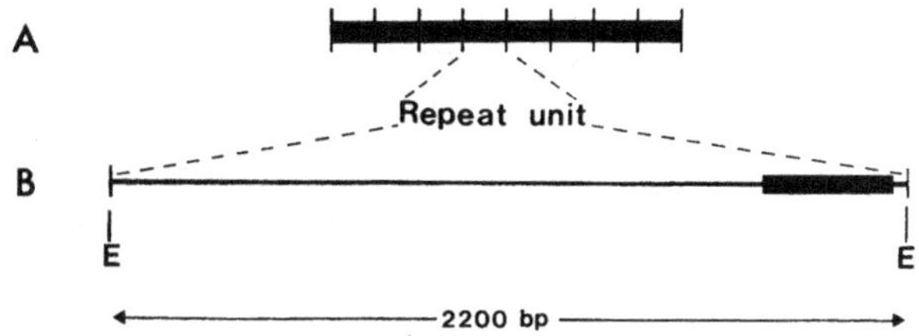

Fig. 6 A, B. The organisation of related rye sequences in two different repeating units. **A** Where the repeating unit is 120 bp as defined by the restriction endonuclease Hae III. **B** Where the repeating unit is 2.2 Kbp and contains a small number of 120 bp repeats (*thickened line*) and other unrelated sequences (*thin line*) which have had a different evolutionary history (see text). (Results taken from BEDBROOK et al. 1980b)

of 120 base pairs. This ladder pattern, in terminal or incomplete digests, is typical for a family of repeated sequences where the unit repeats, in this case 120 base pairs long, are tandemly arranged in the chromosomes. The ladder is created by dimers, trimers, tetramers etc. of the unit repeat. In complete digests the ladder is created by the loss of restriction sites by mutation. Loss of restriction sites is implied by the monomer being in highest concentration and the concentration of the multimers decreasing with increasing size.

A "reverse ladder" with a periodicity of 120 base pairs is seen in the tracks B and C in which rye DNA is digested with Bam HI and Eco RI. Here it can be inferred that most of the family is not cleaved, or only rarely, by Bam HI and Eco RI because most of the probe hybridised to high molecular weight fragments. The ladder is termed "reverse" because the monomer is in lowest concentration. Again the ladder implies that many monomer units are tandemly arranged in the chromosomes. The reverse ladder is probably created by the generation of restriction sites for Eco RI by mutation.

Also of note in tracks B and C are major bands of fragments in very high stoichiometry relative to the ladder. Representative fragments of the major band in the Eco RI digest have been cloned and their structure studied. They are unit repeats 2,200 base pairs long of a family which contains not only sequences homologous to the 120 base pair unit repeats, but also other sequences unrelated to the 120 base pair repeat (BEDBROOK et al. 1980b). The two kinds of repeating unit are drawn in Fig. 6.

These sorts of hybridisation patterns, viz. ladders and bands of varying intensities over a wide range of fragment sizes, appear typical for the repeated sequence families of higher plant genomes digested with restriction endonucleases. The patterns indicate (1) that repeats are commonly organised in tandem arrays and (2) that families of different complexity and concentration are sometimes related to one another in parts of their sequence.

7.5 Derivation of a Physical Map of the rDNA Repeat Unit in Soybean

The genes coding for cytosolic 18S+25S rRNA are highly repeated in plant genomes. This makes rRNA/DNA hybridisation reactions easy to perform. The

Table 2. Hybridisation of 18S+25S rRNA's to restriction fragments of ribosomal DNA (VARSANYI-BREINER et al. 1979)

Digestion	Fragment Size (Kbp)	Relative Intensity of Hybridization			
		25S rRNA-probe	25S rRNA-5′ probe	18S rRNA-probe	18S rRNA-5′ probe
Eco RI	4.0	+	−	+++	+++
	3.8	+++	+++	+++	−
Eco RI+Xba I	3.8	+++	++	++	+
	2.6	+	−	+	+++
	1.4	−	−	+++	+
Eco RI+Sac I	3.7	+	−	++	+++
	2.0	+++	++	++	+
	1.6	++	+	+	−
Eco RI+Bgl II	4.0	+	−	+++	+++
	2.3	+++	++	++	+
	1.5	+	+	−	−

+ → +++ implies increasing intensity of hybridisation.

experiments described here, as an example of how to obtain a physical map of a complex gene, were carried out by VARSANYI-BREINER et al. (1979). When soybean DNA was cleaved with the restriction endonuclease Xba I, fractionated by electrophoresis through agarose, transferred to nitrocellulose and hybridised with 18S+25S ribosomal RNA, [^{32}P]-labelled with ATP+polynucleotide kinase, only one band of hybridisation was observed. This corresponds to a fragment length of 7.8 Kbp and implies that repeating units of the genes contain 7.8 Kbp with only one site for Xba I. Digestion with Eco RI produced two fragments, 3.8 Kbp and 4.0 Kbp which implies there are two Eco RI sites in each 7.8 Kbp repeat unit. Partial digests with Eco RI produced only 3.8, 4.0, 7.8, 11.8 and 15.6 Kbp etc. hybridisation bands, as predicted for essentially homogeneous repeating units arranged in tandem.

In double digests with Eco RI and Xba I, the 3.8 Kbp fragment stayed intact, but the 4.0 Kbp fragment was cleaved into two fragments 1.4 Kbp and 2.6 Kbp long. The order of the three Xba I and Eco RI fragments relative to one another was easily determined by digestion with other pairs of restriction endonucleases. A map derived in this way showing restriction sites for five different endonucleases is shown in Fig. 7 and the lengths of the various restriction fragments are given in Table 2. To position the 18S and 25S coding sequences within the 7.8 Kbp unit, 18S and 25S rRNA were purified from polysomes by sedimentation through 10–30% (w/v) sucrose gradients. Each purified RNA was hybridised separately to the rDNA unit cleaved with the different combinations of restriction endonucleases shown in Table 2. For simplicity, the hybridisations were carried out to a cloned unit repeat instead of to a whole genome digest (VARSANYI-BREINER 1979). The hybridisation results in Table 2 enabled the genes

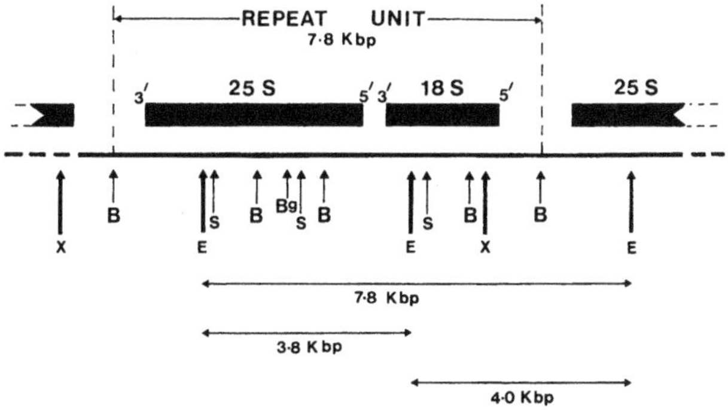

Fig. 7. Organisation of the 18S+25S ribosomal RNA gene repeating units. The figure, drawn from VARSANYI-BREINER et al. (1979), shows a physical map of one repeat unit taken from a tandem array of similar units. The approximate positions of the 18S and 25S coding sequences are shown, together with the positions of the cutting sites of the restriction endonucleases used to determine the positions of the coding sequences. *E* cutting site for Eco RI; *B* for Bam HI; *Bg* for Bgl II; *S* for Sal I; *X* for Xba I. Also shown are the Eco RI fragments described in the text and Table 2 to illustrate how the maps were constructed

to be positioned on the restriction map as shown in Fig. 7. To discover the 5′ end of the coding sequence, 18S and 25S rRNA's were labelled at the 5′ end with [^{32}P]-labelled ATP+polynucleotide kinase and after partial alkaline hydrolysis were hybridised to restriction digests of rDNA. Only one fragment in each digest hybridised very strongly to each probe. This enabled the 5′ end of each gene to be determined unambiguously. Thus the direction of transcription is 18S→25S, as shown in Fig. 7. Further features of rDNA in plants are given in Sect. 9.

8 The Properties and Arrangements of Repeated Sequences

Families of repeated sequences are detected by their more rapid renaturation in vitro than predicted for single-copy DNA by the genome size. Sequences belonging to the same family are not necessarily identical. They commonly differ owing to the accumulation of mutations. This is evident from the heterogeneity in restriction endonuclease targets (Fig. 5; BEDBROOK et al. 1980a, b) and from the reduced thermal stability of renatured repeats. This latter parameter is conveniently assayed by hybridising a radioactively labelled member of the family purified by molecular cloning, to a random sample of all the other members in the genome and then measuring the thermal stability of the duplexes by melting them off hydroxyapatite (HAP) in low molarity phosphate buffer (BEDBROOK et al. 1980a, b). Where essentially all members of the family can be purified from the bulk of the DNA, for example by centrifugation in heavy salt gradients, then the thermal stability of renatured duplexes can be determined by following the hyperchromic increase in optical density at 260 nanometers

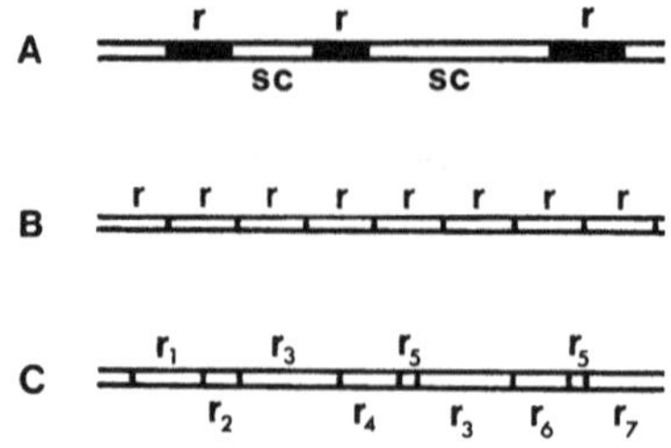

Fig. 8 A–C. General arrangements of repeats. **A** Where short repeats 50 → 1,000 bp are interspersed with non-repeats 2,000 → 4,000 bp. **B** Where essentially identical sequences are tandemly arrayed. **C** Where unrelated short repeats are interspersed in different permutations

as the DNA is melted (BENDICH and TAYLOR 1977). This is best carried out in 2.4 M tetraethylammonium chloride, which eliminates the effect of base composition on melting temperature. On average, a 1% decrease in thermal stability implies about a 1% mismatching of bases in the renatured DNA (ULLMAN and MCCARTHY 1973).

Different families of DNA vary in the extent of sequence divergence within their members (BEDBROOK et al. 1980a). This may reflect the time interval since the family was first formed, i.e., the length of time available for it to have accumulated mutations, or may reflect conservation of certain sequence variants. Interpretations of sequence heterogeneity are complicated by the fact that members of families are often reamplified during evolution (FLAVELL et al. 1977, PREISLER and THOMPSON 1981a). Such quantitative changes prohibit simple analyses of the rates of accumulation of mutations in families of repeats. The shape of DNA melting curves in relation to family homogeneity versus heterogeneity due to reamplifications has been discussed by BENDICH and ANDERSON (1977) and more recently by PREISLER and THOMPSON (1981a, b). Melting profiles of total renatured repeated sequence DNA from plant genomes suggest that most families of repeated sequences are very heterogeneous.

The arrangement of repeated sequences in the chromosomes can be simplified into at least three kinds of pattern. In the first, shown in Fig. 8, the repeats are interspersed with single or few copy sequences as discussed in Sect. 4. These repeats are predominantly short (<1,000 base pairs) in genomes larger than about 2 pg (Fig. 2), but longer in smaller genomes. The size estimates have been inferred from electron microscopy, hyperchromicity measurements and/or molecular weight distributions after S_1 nuclease treatment of renatured repeats (e.g., WALBOT and GOLDBERG 1980, BENDICH and WARD 1980, FLAVELL and SMITH 1976, MURRAY et al. 1978). Analyses on the soybean genome have suggested that the repeats interspersed with single-copy DNA belong predominantly to families which have fewer repeats than average (WALBOT and GOLDBERG 1980).

The repeated DNA not interspersed in small segments with single-copy DNA lies in long arrays. Two kinds of clustered repeats have been described (Fig. 8) (a) those in which closely related repeats are tandemly arrayed and (b) those which are permutations of different repeats. Evidence for tandem arrays of related repeats comes from (1) "satellite" DNA where the method of isolation, as well as studies on the purified satellites, imply that the constituent repeats are few in type and in tandem arrays. (2) the demonstration of "ladders"

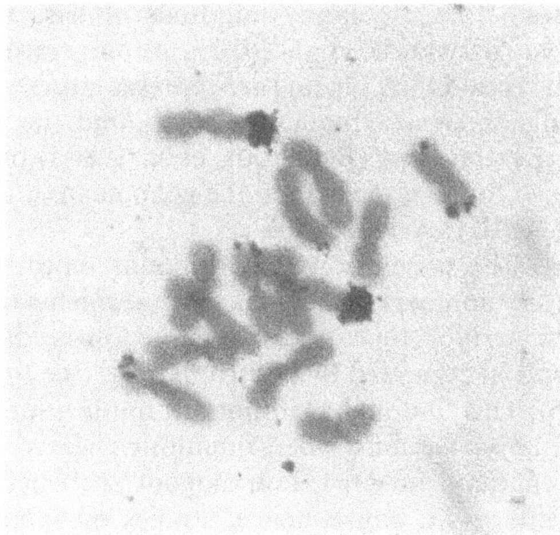

Fig. 9. Localisation of a family of repeated sequences in the karyotype by in situ hybridisation. [^{3}H]-labelled RNA complementary to a DNA sequence purified from a rye family with a repeating unit of 480 base pairs (BEDBROOK et al. 1980a) was hybridised to denatured metaphase chromosomes of *Secale anatolicum* (JONES and FLAVELL 1982b). The sites of hybridisation were revealed by autoradiography. This photograph illustrates how individual pairs of chromosomes can be distinguished by tandem arrays of repeats they possess

by hybridisation of labelled probes to DNA digested with a restriction endonuclease and fractionated by electrophoresis (see Fig. 5), (3) the resistance of long molecules of repeated sequence DNA renatured in vitro to digestion by S_1 nuclease (S_i nuclease degrades single-stranded discontinuities in renatured DNA) and (4) hybridisation of radioactive repeated sequences to metaphase chromosomes at specific sites in large amounts, as expected only for sequences highly clustered (reviewed in HUTCHINSON et al. 1981).

A photograph of metaphase chromosomes of *Secale anatolicum* after hybridisation with a [^{3}H]-labelled repeat from *Secale cereale* (rye) is shown in Fig. 9. The location of the sites of hybridisation, i.e., the positions of the tandem arrays of the repeats in the chromosomes, was revealed by autoradiography. Two pairs of chromosomes have minor sites of hybridisation at their telomeres, while a third pair possesses a very large number of copies of the repeat at its telomeres. The technique of in situ hybridisation is especially useful because it not only provides information on where in the karyotype specific sequences are localised, but also often enables the chromosomes to be distinguished from each other (HUTCHINSON et al. 1981, GERLACH and PEACOCK 1980, JONES and FLAVELL 1982a). In species where the chromosomal distribution of several different tandem arrays of sequences present in very high copy numbers has been investigated, e.g., rye (APPELS et al. 1978, BEDBROOK et al. 1980a, JONES and FLAVELL 1982a) and wheat (GERLACH and PEACOCK 1980, DENNIS et al. 1980), the different tandem arrays are often clustered together in regions of constitutive heterochromatin. These regions can be telomeric, centromeric or interstitial.

Although tandem arrangement of repeating units has been shown only for families with a very large number of repeats, the arrangement is probably also common for some families with relatively few repeats.

The number of repeats per family can vary between two and several million. The repeating units in tandem arrays can vary in length from six base pairs as in the satellite DNA purified from wheat in Ag^+—Cs_2SO_4 gradients (DENNIS et al. 1980) to many hundreds of base pairs as shown for tandem arrays in rye (BEDBROOK et al. 1980a, b) and muskmelon (BENDICH and TAYLOR 1977). If cereal DNA repeats are representative of plant repeats, then the longer repeating units are frequently compound, i.e., part of the repeat contains a sub-repeating unit (BEDBROOK et al. 1980a) or the unit contains segments of DNA also found elsewhere in the genome in different conformations (BEDBROOK et al. 1980b, FLAVELL 1980).

The conclusion that different repeating units present in a wide range of copy numbers sometimes share segments of homology comes from hybridisation of purified repeats to whole genomes, digested with restriction endonucleases and fractionated by electrophoresis (see Fig. 5 and 6 and BEDBROOK et al. 1980a, b). This finding has important implications for many of the methods of genome analysis because when randomly sheared fragments of such repeating units are renatured in vitro, homologous sequences in different repeating units reassociate. As a consequence, studies on renatured repeated sequence duplexes to determine their thermal stability or resistance to S_1 nuclease may not necessarily give results relevant to the sequence diversity within a single family of repeating units. Also, estimates of the number of copies of sequences from renaturation kinetics may be overestimates of the number of repeating units of a particular kind. Partial homology between different repeating units may also give rise to an erroneous conclusion, on the basis of renaturation experiments comparing short and long fragments, that repeated sequences belonging to families with very different numbers of repeats are interspersed in the chromosomes (GURLEY et al. 1979, HAKE and WALBOT 1980, KIPER and HERZFELD 1978). The apparent interspersion may in many cases have been created by renaturation of members of different repeating units. The proportion of repeating units in plant genomes which are interrelated in this way is not known. It is likely that larger genomes have a higher proportion of families with partially related, compound repeating units. PREISLER and THOMPSON (1981a, b) have concluded from melting curves of renatured fragments that the large pea genome contains a much greater proportion of repeated sequence units which are related to one another than the small mung bean genome. It seems probable that the longer the repeating unit, the more likely that it shares some homology with another repeating unit.

The finding that some different repeating units share homology implies that segments of DNA evolve by the recombination of short pieces of repeated DNA into new permutations (Fig. 8c). Comparison of the organisation of repeated sequences in wheat, barley, oat and rye also showed that short repeats are rearranged into many different permutations during evolution (RIMPAU et al. 1978, 1980, FLAVELL et al. 1981). The interspersion of short repeats with single-copy DNA also implies the rearrangement of short segments of DNA into new configurations. Indeed, it appears that a very high proportion of each higher plant genome has evolved by amplification and sequence rearrangement as well as by mutation, duplication and deletion. Because related sequences are found on all chromosomes, the rearrangements must also include inter- as well as intra-chromosomal translocation (FLAVELL et al. 1981). This has been discussed extensively by FLAVELL (1980), FLAVELL et al. (1979) and THOMPSON and MURRAY (1980).

9 The Structure of Nuclear Genes

The approximate number of different genes in higher plants still needs further study. Estimates for parsley (KIPER et al. 1979), barley (KIPER et al. 1980) and tobacco (GOLDBERG et al. 1978) have been made from the kinetics of renaturation of DNA to an excess of polysomal RNA and of single-copy DNA to an excess of polysomal RNA. The former method gave estimates of 10,000 to 15,000 different types of RNA sequence, while the latter gave estimates around 25,000 to 35,000. The genes are dispersed throughout the DNA of the genome if the recombination mapping available for a few species is representative. However, a few genes are highly reiterated and organised in tandem arrays, e.g., the 18S, 5.8S and 25S ribosomal RNA genes (Fig. 7) and the 5S ribosomal RNA genes (Fig. 10). Other genes are also represented in many copies, e.g., storage protein and leghaemoglobin genes (BAULCOMBE and VERMA 1978). Members of closely related families of genes are probably very close to one another in the chromosomes.

Where radioactive probes of a purified mRNA or cDNA are available, then restriction maps of nuclear genes can be determined as described in Section 7.5. This allows certain structural features to be determined, whether the coding sequence possesses intervening sequences, whether multiple copies occur and whether all the copies are identical or different. For more detailed analyses, particularly concerning the sequences surrounding the genes, it is essential to purify the gene sequences from the chromosomes. The techniques of molecular cloning make this possible. The ribosomal RNA genes (GERLACH and BEDBROOK 1979, GERLACH and DYER 1980, VARSANYI-BREINER et al. 1979), a gene for the storage protein phaseolin from *Phaseolus vulgaris* (French bean) (SUN et al. 1981) and for leghaemoglobin from soybean (SULLIVAN et al. 1981) were among the first to be purified using these techniques. The cloned DNA's enabled the structures and sequences of the isolated genes to be determined and compared with other members of the multiple gene families.

The DNA's used for cloning the 5S and 18S+5.8S+25S cytosolic ribosomal RNA genes from wheat and barley were first enriched for these genes by centrifugation in CsCl-actinomycin D gradients. The genes band at a higher density than the bulk of the nuclear DNA. This enrichment (about 30- to 100-fold) meant that correspondingly fewer recombinant clones had to be screened. The 5S rRNA repeating units were shown to contain a single site for Bam HI by hybridisation of [^{32}P]-labelled 5S RNA to restricted nuclear DNA (APPELS et al. 1980). Consequently Bam HI fragments of the enriched DNA were cloned into the vector pBR 322. The 18S, 5.8S and 25S RNA gene units contained a single Eco RI site (APPELS et al. 1980) and were cloned into the Eco RI site of the gene specifying chloramphenicol resistance of the vector pACYC 184. The nuclear DNA fragments containing the rRNA genes of soybean and the phaseolin genes of French bean were not enriched first, but selected from a large genomic bank of fragments cloned in bacteriophage lambda vectors.

The genes specifying the 5S ribosomal RNA genes in the wheat variety Chinese Spring are found in repeating units of two size classes viz. approximately

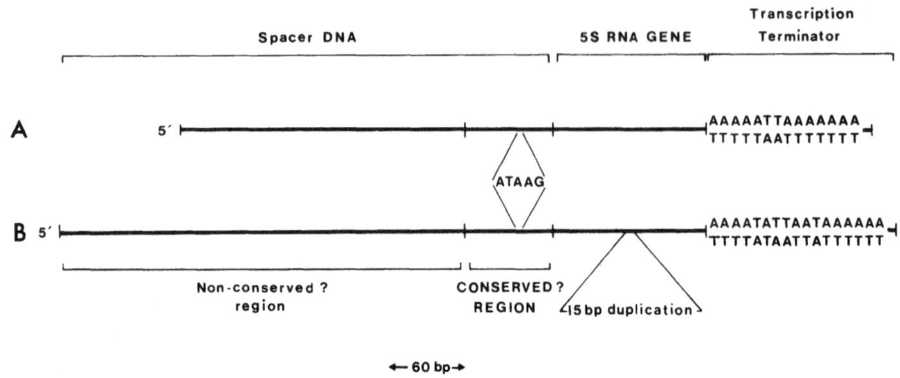

Fig. 10 A, B. The structure of the repeating units containing the 5S ribosomal RNA genes of wheat. The details have been redrawn from the complete nucleotide sequences of the units (GERLACH and DYER 1980). The terms *conserved* and *non-conserved* refer to comparisons between the sequences of the 410 base pair unit (**A**) and the 500 base pair unit (**B**)

500 and 410 base pairs long (APPELS et al. 1980). Three units of the shorter class and one of the longer class were isolated and sequenced (GERLACH and DYER 1980). The structure of the sequenced units is shown in Fig. 10. The sequenced member of the 500 base pair class contained a 15 base pair duplication within the coding sequence. There may be many 5S RNA genes with this duplication in this wheat variety. In *Xenopus* 5S RNA genes, the region homologous to that containing the duplication in some wheat genes, is very important in regulating transcription (SAKONJU et al. 1980, BOGENHAGEN et al. 1980). These genes may therefore be incapable of being transcribed in wheat.

In each 5S RNA gene repeat unit, there is an array of A-T base pairs preceded by a region of weak dyad symmetry immediately following the sequence specifying 5S RNA. This array, with the T residues predominating in the non-coding strand, dictates in part the termination of transcription. It appears more similar to the terminator sequences in yeast than those in *Xenopus* and *Drosophila*. The 70 base pairs before the beginning of the coding sequence are very similar in the cloned representatives of both the 410 and 500 base pair size classes, in contrast to most of the spacer DNA which has diverged considerably between repeating unit classes. These 70 base pairs are probably conserved to determine the correct initiation of transcription of 5S RNA. The sequence ATAAG occurs 29 to 25 base pairs before the transcription start point and is closely related to the AGAAG found in *Xenopus* and identical with the ATAAG sequence found in *Drosophila* 5S DNA at a similar location. This sequence is probably important for correct transcription and appears to be a common feature of nuclear genes in higher eukaryotes.

The heterogeneity in length between the different repeating units is due to variation in the non-transcribed spacer whose structure appears to have had a complex evolution.

The 18S, 5.8S and 25S rRNA genes are organised in most higher eukaryotes as shown in Fig. 7 and transcribed into a single RNA molecule which is subse-

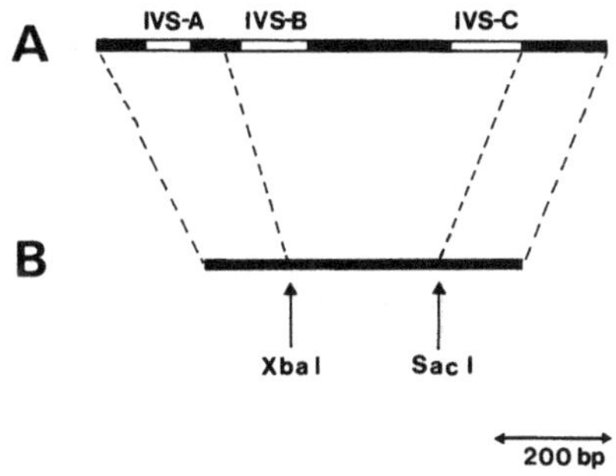

Fig. 11 A, B. The presence of intervening sequences in a phaseolin gene from *Phaseolus vulgaris*. **A** Segment of the chromosomal gene sequence showing three intervening sequences 1VS A, 1VS B and 1VS C. **B** Segment of the cDNA sequence lacking the intervening sequences, prepared from mRNA by reverse transcriptase. The molecules A and B were compared by complete nucleotide sequencing. (Redrawn from SUN et al. 1981)

quently processed into the mature RNA species (LEAVER and KEY 1970, COX and TURNOCK 1973). The 18S and 25S RNA's have almost identical lengths (2 Kbp and 4 Kbp respectively) in unrelated plant species and very similar sequences. The repeating unit lengths, however, differ markedly from 7.8 Kbp in soybean to over 10 Kbp in some Gramineae species. Repeating units with different lengths have been reported in wheat (GERLACH and BEDBROOK 1980) but not soybean (VARSANYI-BREINER et al. 1979). As in the 5S RNA repeating units, the heterogeneity is probably confined to the spacer regions. The number of ribosomal RNA genes varies at least up to threefold within many plant species (FLAVELL and SMITH 1974, RAMIREZ and SINCLAIR 1975, CULLIS and DAVIES 1975).

One particular feature of the rRNA genes is that within a plant there appears to be heterogeneity in the sites of cytosine methylation. This is concluded from the observation that restriction endonucleases which fail to cut DNA when the cytosine in its recognition sequence is methylated, cut some but not all rRNA genes (GERLACH and BEDBROOK 1980, VARSANYI-BREINER et al. 1979). This may be related to whether the rRNA genes function.

No intervening sequences or introns have been detected in plant rRNA genes but they are present in at least one of the phaseolin genes of French bean (SUN et al. 1981) and leghaemoglobin genes of soybean (SULLIVAN et al. 1981). A comparison of the chromosomal sequence for a phaseolin protein and the cDNA sequence synthesised from mRNA is shown in Fig. 11. Three regions are found in the chromosomal sequence that are not in the cDNA, suggesting the presence of three intervening sequences which are spliced out of the primary transcript in production of the mRNA, assuming that this particular gene is transcribed and translated in vivo.

It remains to be seen how common in plant genes are intervening sequences, duplications, A-T rich termination signals and conserved sequences 25 to 30 base pairs before the initiation site for regulating transcription. Only when more genes and mutant variants have been studied in vitro and in vivo will the significance of the structural features of genes be fully understood.

References

Appels R, Driscoll C, Peacock WJ (1978) Heterochromatin and highly repeated DNA sequences in rye (*Secale cereale*). Chromosoma 70:67–89

Appels R, Gerlach WL, Dennis ES, Swift H, Peacock WJ (1980) Molecular and chromosomal organisation of DNA sequences coding for ribosomal RNAs in cereals. Chromosoma 78:293–311

Arthur RR, Straus NA (1978) DNA sequence organisation in the genome of the domestic chicken (*Gallus domesticus*) Can J Biochem 56:257–263

Bachmann K, Price HJ (1977) Repetitive DNA in Cichorieae (compositae). Chromosoma 61:267–275

Baulcombe D, Verma DPS (1978) Preparation of a complementary DNA for leghaemoglobin and direct demonstration that leghaemoglobin is encoded by the soybean genome. Nucl Acid Res 5:4141–4153

Bazetoux S, Jouanin L, Huguet T (1978) Characterisation of inverted repeated sequences in wheat nuclear DNA. Nucl Acid Res 5:751–769

Bedbrook J, Gerlach WL (1980) Cloning of repeated sequence DNA from cereal plants. In: Setlow J, Hollaender A (eds) Genetic engineering, principles and methods Vol II, Plenum, New York

Bedbrook J, Gerlach WL, Thompson RD, Jones J, Flavell RB (1978) Molecular cloning of higher plant DNA. In: Rubenstein I, Gengenbach B, Phillip RL (eds) Genetic improvement of crops. Univ Minnesota Press, pp 93–114

Bedbrook JR, Jones J, O'Dell M, Thompson R, Flavell RB (1980a) A molecular description of telomeric heterochromatin in *Secale* species. Cell 19:545–560

Bedbrook JR, O'Dell M, Flavell RB (1980b) Amplification of rearranged sequences in cereal plants. Nature (London) 288:133–137

Belford HS, Thompson WF (1979) Single copy DNA homologies and the phylogeny of *Atriplex*. Carnegie Inst Washington Year Book 78:217–223

Bendich AF, Anderson RS (1977) Characterisation of families of repeated DNA sequences from four vascular plants. Biochemistry 16:4655–4663

Bendich AJ, McCarthy BJ (1970a) DNA comparisons among barley, oats, rye and wheat. Genetics 65:545–566

Bendich AJ, McCarthy BJ (1970b) DNA comparisons among some biotypes of wheat. Genetics 65:567–573

Bendich AJ, Taylor WC (1977) Sequence arrangement in satellite DNA from the muskmelon. Plant Physiol 59:604–609

Bendich AJ, Ward BL (1980) On the evolution and functional significance of DNA sequence organisation in vascular plants. In: Leaver CJ (ed) Genome organisation and expression in plants. Plenum, New York, pp 17–30

Bennett MD, Smith JB (1976) Nuclear DNA amounts in Angiosperms. Proc R Soc London Ser B 274:227–274

Bogenhagen DF, Sakonju S, Brown DD (1980) A control region in the center of the 5S RNA gene directs specific initiation of transcription. II The 3′ border of the region. Cell 19:27–35

Bouchard RA, Swift H (1977) Nature of the heterogeneity in mispairing of reannealed middle repetitive fern DNA. Chromosoma 61:317–333

Britten RJ, Davidson EH (1976) Studies on nucleic acid reassociation kinetics: Empirical equations describing DNA reassociation. Proc Natl Acad Sci USA 73:415–419

Britten RJ, Kohne DE (1968) Repeated sequences in DNA. Science 161:529–540

Britten RJ, Graham DE, Neufeld BR (1974) An analysis of repeating DNA sequences by reassociation. In: Grossman L, Moldave K (eds) Methods in enzymology 29E. Academic Press, London New York, pp 363–418

Capesius I, Bierweiler B, Bachmann K, Rucker W, Nagl W (1975) An A+T rich satellite DNA in a monocotyledonous plant, *Cymbidium*. Biochim Biophys Acta 395:67–73

Cavalier-Smith T (1974) Palindromic base sequences and replication of eucaryotic chromosome ends. Nature (London) 250:467–470

Cavalier-Smith T (1978) Nuclear volume control by nucleoskeletal DNA, selection for cell volume and cell growth rate and the solution of the DNA C-value paradox. J Cell Sci 34:247–278
Cavalier-Smith T (1980) How selfish is DNA? Nature (London) 285:617–618
Chilton MD (1975) Ribosomal DNA in a nuclear satellite of tomato. Genetics 81:469–483
Chilton MD, McCarthy BJ (1973) DNA from maize with and without B chromosomes: a comparative study. Genetics 74:605–614
Chooi WY (1971) Comparison of the DNA of six Vicia species by the method of DNA-DNA hybridisation. Genetics 68:213–230
Cox BJ, Turnock B (1973) Synthesis and processing of ribosomal RNA in cultured plant cells. Eur J Biochem 37:367–376
Crain WR, Davidson EH, Britten RJ (1976a) Contrasting patterns of DNA sequence arrangement in *Apis mellifera* (honeybee) and *Musca domestica* (housefly). Chromosoma 59:1–12
Crain WR, Eden FC, Pearson WR, Davidson EH, Britten RJ (1976b) Absence of short period interspersion of repetitive and non-repetitive sequences in *Drosophila* DNA. Chromosoma 56:309–326
Cullis CA (1973) DNA differences between flax genotypes. Nature (London) 243:515–516
Cullis CA (1977) Molecular aspects of the environmental induction of heritable changes in flax. Heredity 38:129–154
Cullis CA, Davies DR (1975) Ribosomal DNA amounts in *Pisum sativum*. Genetics 81:485–492
Cullis CA, Schweizer D (1974) Repetitious DNA in some *Anemone* species. Chromosoma 44:417–421
Davidson EH, Britten RJ (1979) Regulation of gene expression: possible role of repetitive sequences. Science 204:1052–1059
Davidson EH, Hough BR, Amenson CS, Britten RJ (1973) General interspersion of repetitive with non repetitive sequence elements in the DNA of *Xenopus*. J Molec Biol 77:1–23
Dennis ES, Gerlach WL, Peacock WJ (1980) Identical polypyrimidine-polypurine satellite DNAs in wheat and barley. Heredity 44:349–366
Dhillon SS, Rake AV, Miksche JP (1980) Reassociation kinetics and cytophotometric characterisation of peanut (*Arachis hypogaea* L) DNA. Plant Physiol 65:1121–1127
Doolittle WF, Sapienza C (1980) Selfish genes, the phenotype paradigm and genome evolution. Nature (London) 284:601–603
Dover GA (1980) Ignorant DNA? Nature (London) 285:618–620
Epplen JT, Leipoldt M, Engel W, Schmidtke J (1978) DNA sequence organisation in avian genomes. Chromosoma 69:307–321
Flavell RB (1980) The molecular characterisation and organisation of plant chromosomal DNA sequences. Annu Rev Plant Physiol 31:569–596
Flavell RB (1981) Molecular changes in chromosomal DNA organisation and origins of phenotypic variation. Chromosomes Today. Bennett, MD, Bobrow, M and Hewitt, eds. George Allen & Unwin, London. pp 42–54
Flavell RB, Smith DB (1974) Variation in nucleolar organiser rRNA gene multiplicity in wheat and rye. Chromosoma 47:327–334
Flavell RB, Smith DB (1976) Nucleotide sequence organisation in the wheat genome. Heredity 37:231–252
Flavell RB, Smith DB (1977) Hyperpolymer formation during renaturation of DNA from genomes with different sequence organisation. Nucl Acid Res 4:2429–2444
Flavell RB, Bennett MD, Smith JB, Smith DB (1974) Genome size and the proportion of repeated sequence DNA in plants. Biochem Genet 12:257–269
Flavell RB, Rimpau J, Smith DB, O'Dell M, Bedbrook JR (1979) The evolution of plant genome structure. In: Leaver CJ (ed) Plant genome organisation and expression. Plenum, New York, pp 35–47
Flavell RB, O'Dell M, Hutchinson J (1981) Nucleotide sequence organisation in plant chromosomes and evidence for sequence translocation during evolution. Cold Spring Harbor Symp Quant Biol vol XLV pp 501–508

Gerlach WL, Dyer TA (1980) Sequence organisation of the repeating units in the nucleus of wheat which contain 5S rRNA genes. Nucl Acid Res 8:4851–4865
Gerlach WL, Peacock WJ (1980) Chromosomal locations of highly repeated DNA sequences in wheat. Heredity 44:269–276
Goldberg RB (1978) DNA sequence organisation in the soybean plant. Biochem Genet 16:45–68
Goldberg RB, Hoschek G, Kamalay JC (1978) Sequence complexity of nuclear and polysomal RNA in leaves of the tobacco plant. Cell 14:123–131
Graham DE, Neufeld BR, Davidson EH, Britten RJ (1974) Interspersion of repetitive and non-repetitive DNA sequence in the sea urchin genome. Cell 1:127–137
Gurley WB, Hepburn AG, Key JL (1979) Sequence organisation of the soybean genome. Biochim Biophys Acta 561:167–183
Hake S, Walbot V (1980) The genome of *Zea mays,* its organisation and homology to related grasses. Chromosoma 79:251–270
Hamer DH, Thomas CA (1974) The cleavage of *Drosophila melanogaster* DNA by restriction endonucleases. Chromosoma 49:243–267
Hinegardner R (1976) Evolution of genome size. In: Ayala FJ (ed) Molecular evolution. Sinauer, Sunderland, pp 179–199
Hinnebusch AG, Clark VE, Koltz LC (1978) Length dependence in reassociation kinetics of radioactive tracer DNA. Biochemistry 17:1521–1529
Huguet T, Jouanin L (1972) Wheat DNA: Study of the heavy satellite in Ag^+-Cs_2SO_4 density gradient. Biochem Biophys Res Commun 46:1169–1174
Hutchinson J, Flavell RB, Jones J (1981) Physical mapping of plant chromosomes by *in situ* hybridisation. In: Setlow J, Hollaender A (eds) Genetic engineering Vol III, Plenum, New York
Ingle J, Pearson GC, Sinclair J (1973) Species distribution and properties of nuclear satellite DNA in higher plants. Nature New Biol 242:193–197
Ingle J, Timmis JN, Sinclair J (1975) The relationships between satellite deoxyribonucleic acid, ribosomal ribonucleic and gene redundancy and genome size in plants. Plant Physiol 55:496–501
Jones JDG, Flavell RB (1982a) The mapping of highly repeated DNA families and their relationship to C bands in chromosomes of *Secale cereale.* Chromosoma, in press
Jones JDG, Flavell RB (1982b) The structure, amount and localisation of defined repeated sequences in species of the genus *Secale.* Chromosoma, in press
Jones RN, Brown LM (1976) Chromosome evolution and DNA variation in *Crepis.* Heredity 36:91–104
Kadouri A, Atsmon D, Edelman M (1975) Satellite-rich DNA in cucumber: Hormonal enhancement of synthesis and subcellular identification. Proc Natl Acad Sci USA 72:2260–2264
Kiper M, Herzfeld F (1978) DNA sequence organisation in the genome of *Petroselinum sativum.* Chromosoma 65:335–351
Kiper M, Bartels D, Herzfeld F, Richter G (1979) The expression of a plant genome in hnRNA and mRNA. Nucl Acid Res 6:1961–1978
Kiper M, Bolte M, Herzfeld F (1980) Reiteration frequency of genes coding for abundant and rare messenger RNA in greened barley seedlings. Heredity 45:233–243
Lamppa GK, Bendich AJ (1979) Chloroplast DNA sequence homologues among vascular plants. Plant Physiol 63:660–668
Leaver CJ, Key JL (1970) Ribosomal RNA synthesis in plants. J Molec Biol 49:671–680
Mann MB, Smith HO (1977) Specificity of Hpa II and Hae III DNA methylases. Nucl Acid Res 4:4211–4221
Miksche JP, Hotta Y (1973) DNA base composition and repetitious DNA in several conifers. Chromosoma 41:29–36
Mitra R, Bhatia CR (1973) Repeated and non-repeated nucleotide sequences in diploid and polyploid wheat species. Heredity 31:251–262
Murray MG, Cuellar RE, Thompson WF (1978) DNA sequence organisation in the pea genome. Biochemistry 17:5781–5790

Murray MG, Palmer JD, Cuellar RE, Thompson WF (1979) Deoxyribonucleic acid sequence organisation in the mung bean genome. Biochemistry 18:529–5266
Nandi VS, Wang JC, Davidson N (1965) Separation of deoxyribonucleic acids by Hg(II) binding and Cs_2SO_4 density gradient centrifugation. Biochemistry 4:1687–1696
Narayan RKJ, Rees H (1976) Nuclear DNA variation in *Lathyrus*. Chromosoma 54:141–154
Nathans D, Smith HO (1975) Restriction endonucleases in the analysis and restructuring of DNA molecules. Annu Rev Biochem 44:273–293
Nze-ekekang L, Patellon M, Schafer A, Kovoor A (1974) Repetitive DNA of higher plants. Separation of fractions of different complexity from Jerusalem artichoke tissue by reassociation kinetics. J Exp Bot 25:320–329
Orgel LE, Crick FHC (1980) Selfish DNA: the ultimate parasite. Nature (London) 284:604–607
Palmer JD, Murray MG, Thompson WR (1979) Studies on chloroplast DNA of mung bean and pea. Carnegie Inst Washington Yearb 78:226–231
Pearson WR, Davidson EH, Britten RJ (1977) A program for least squares analysis of reassociation and hybridisation data. Nucl Acid Res 4:1727–1735
Pellegrini M, Goldberg RB (1979) DNA sequence organisation in soybean investigated by electron microscopy. Chromosoma 75:309–326
Preisler RS, Thompson WF (1978) Distribution of nucleotide sequence divergence among families of repetitive sequences in mung bean DNA. Carnegie Inst Washington Yearb 77:323–330
Preisler RS, Thompson WF (1981a) Evolutionary sequence divergence within repeated DNA families of higher plant genomes. I. Analysis of reassociation kinetics. J Molec Evol 17:78–84
Preisler RS, Thompson WF (1981b) Evolutionary sequence divergence within repeated DNA families of higher plant genomes. II. Analysis of thermal denaturation. J Molec Evol 17:85–93
Ramirez SA, Sinclair JH (1975) Intraspecific variation of ribosomal gene redundancy in *Zea mays*. Chromosoma 80:495–504
Ranjekar PK, Lafontaine JG, Pallotta D (1974) Characterisation of repetitive DNA in rye (*Secale cereale*). Chromosoma 48:427–440
Ranjekar PK, Pallotta D, Lafontaine JG (1976) Analysis of the genome of plants. II. Characterisation of repetitive DNA in barley (*Hordeum vulgare*) and wheat (*Triticum aestivum*). Biochem Biophys Acta 425:30–40
Ranjekar PK, Pallotta D, Lafontaine JG (1978) Analysis of plant genomes. III. Denaturation and reassociation properties of cryptic satellite DNAs in barley (*Hordeum vulgare*) and wheat (*Triticum aestivum*). Biochim Biophys Acta 520:103–110
Rees H, Hazarika MH (1969) Chromosome evolution in *Lathyrus*. Chromosomes Today 2:157–165
Rimpau J, Smith DB, Flavell RB (1978) Sequence organisation analysis of the wheat and rye genomes by interspecies DNA/DNA hybridisation. J Molec Biol 123:327–359
Rimpau J, Smith DB, Flavell RB (1980) Sequence organisation in barley and oats chromosomes revealed by interspecies DNA/DNA hybridisation. Heredity 44:131–149
Sakonju S, Bogenhagen DF, Brown DD (1980) A control region in the centre of the 5S RNA gene directs specific initiation of transcription. I. The 5′ border of the region. Cell 19:13–25
Schildkraut CL, Marmur J, Doty P (1962) Determination of the base composition of deoxyribonucleic acid from its buoyant density in CsCl. J Molec Biol 4:430–443
Scott NS, Possingham JV (1980) Chloroplast DNA in expanding spinach leaves. J Exp Bot 31:1081–1092
Smith MJ, Britten RJ, Davidson EH (1975) Studies on nucleic acid reassociation kinetics. Reactivity of single stranded tails in DNA-DNA renaturation. Proc Natl Acad Sci USA 72:4805–4809
Smith DB, Flavell RB (1975) Characterisation of the wheat genome by renaturation kinetics. Chromosoma 50:223–242

Smith DB, Flavell RB (1977) Nucleotide sequence organisation in the rye genome. Biochim Biophys Acta 474:82–97

Smyth DR, Stern H (1973) Repeated DNA synthesized during pachytene in *Lilium henryi*. Nature (London) 245:94–96

Southern EM (1975) Detection of specific sequences among DNA fragments separated by gel electrophoresis. J Molec Biol 98:503–517

Stack SM, Comings DE (1979) The chromosomes and DNA of *Allium cepa*. Chromosoma 70:161–181

Stein DB, Thompson WF (1975) DNA hybridisation and evolutionary relationships in three *Osmunda* species. Science 189:888–890

Straus N (1972) Reassociation of bean DNA. Carnegie Inst Washington Yearb 71:257–259

Sullivan D, Brisson N, Goodchild B, Verma DPS, Thomas DY (1981) Molecular cloning and organisation of two leghaemoglobin genomic sequences of soybean. Nature (London) 289:516–518

Sun SM, Slightom JL, Hall TC (1981) Intervening sequences in a plant gene – comparison of the partial sequence of cDNA and genomic DNA of French bean phaseolin. Nature (London) 289:37–41

Thompson WF (1976) Aggregate formation from short fragments of plant DNA. Plant Physiol 57:617–622

Thompson WF, Murray MG (1980) Sequence organisation in pea and mung bean DNA and a model for genome evolution. In: Davies DR, Hopwood DA (eds) Fourth John Innes Symp. John Innes Inst, Norwich, UK, pp 31–45

Thompson WF, Murray MG, Cuellar RE (1979) Contrasting patterns of DNA sequence organisation in plants. In: Leaver CJ (ed) Genome organisation and expression in plants. Plenum, New York

Thompson WF, Rubenstein I (1980) The nuclear genome. Organisation and function. In: Stumpf PK, Conn CE (eds) The biochemistry of plants Vol. 6. Academic Press, London New York

Ullman JS, McCarthy BJ (1973) The relationship between mismatched base pairs and the thermal stability of DNA duplexes. Biochim Biophys Acta 294:405–415

Varsanyi-Breiner A, Gusella JF, Keys C, Housman DE, Sullivan D, Brisson N, Verma DPS (1979) The organisation of a nuclear DNA sequence from a higher plant: Molecular cloning and characterisation of soybean ribosomal DNA. Gene 7:317–334

Walbot V, Dure L (1976) Development biochemistry of cottonseed embryogenesis and germination. J Molec Biol 101:503–536

Wahl MW, Stern M, Stark GR (1979) Efficient transfer of large DNA fragments from agarose gels to diazobenzyloxymethyl-paper and rapid hybridisation by using dextran sulfate. Proc Natl Acad Sci USA 76:3683–3687

Walbot V, Goldberg RB (1980) Plant genome organisation and its relationship to classical plant genetics. In: Hall TC, Davies JW (eds) Nucleic acids in plant Vol I, CRC Press, Boca Raton, pp 3–40

Wells R, Royer HD, Hollenberg CP (1976) Non *Xenopus*-like DNA sequence organisation in the *Chironomus tentans* genome. Mol Gen Genet 147:45–51

Wetmur JG, Davidson N (1968) Kinetics of reassociation of DNA. J Molec Biol 31:349–370

Wilson JT, Kattermann FRH, Endrizzi JE (1976) Analysis of repetitive DNA in three species of *Gossypium*. Biochem Genet 14:1071–1075

Wilson DA, Thomas CA (1974) Palindromes in chromosomes. J Molec Biol 84:115–144

Wimpee CF, Rawson JRY (1979) Characterisation of the nuclear genome of pearl millet. Biochim Biophys Acta 562:192–206

Zimmerman JL, Goldberg RB (1977) DNA sequence organisation in the genome of *Nicotiana tabacum*. Chromosoma 59:227–252

3 DNA Replication and the Cell Cycle

J.A. Bryant

1 Introduction

The replication of the genetic material, DNA, and its subsequent distribution to two daughter cells are perhaps the most fundamental processes carried out by living organisms, ensuring that each cell generation contains the same set of information as the previous cell generation. At the most basic level, as seen in asexually-reproducing unicellular organisms, DNA synthesis and cell division are the processes leading directly to the next generation of organisms. In multicellular organisms these processes take on further significance, since cell division provides the basic material, the cells, for construction of the organism and for the differentiation and organisation of cells in tissues and organs.

Within the higher plant, the localisation and organisation of dividing cells in particular regions, the primary and secondary meristems, is very important

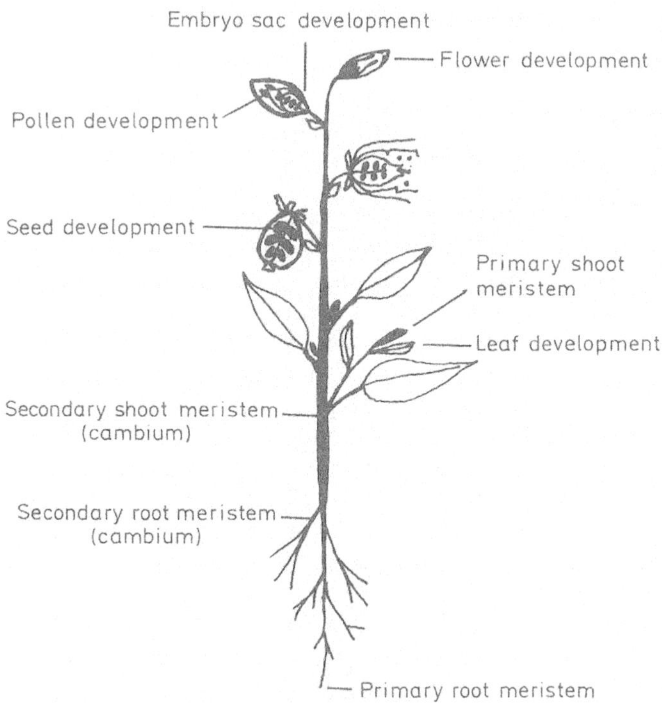

Fig. 1. Regions of higher plants containing dividing cells. (Based on Fig. 1 in Dyer 1976)

in the generation of form (i.e., in provision of cells in the right places) and in maintaining the "unlimited" growth potential exhibited by higher plants. Further, specific phases in the life of a higher plant, such as floral morphogenesis, may be associated with changes in the pattern and rate of cell division (BERNIER 1971, FRANCIS and LYNDON 1979). Figure 1, showing some of the regions of mitotic activity in higher plants, illustrates the fundamental role of cell division in the generation and maintenance of form and function.

Obviously, in dealing with mitotic cell division, we are ultimately dealing with the whole range of processes which lead to the formation of two daughter cells from one parent cell. However, since the duplication of DNA is normally a pre-requisite for mitosis and cytokinesis, and since the DNA contains the information necessary for subsequent development of the cells (given the right conditions for expression of the information) it is natural to regard duplication of DNA as the basic process in mitotic cell division. This is particularly true in the context of a volume such as this, with its emphasis on nucleic acids. The aim of this chapter, therefore, is to describe, in somewhat general terms, the mitotic cell division cycle, whilst dealing in some detail with DNA replication. Wherever possible I have used material obtained from work with plants (including algae, fungi and slime-moulds), but from time to time it has been necessary to refer to other types of eukaryotic organisms and to prokaryotic organisms in order to present as complete a picture as possible.

2 Phases of the Cell Cycle

Mitosis was first described in detail at the cytological level by STRASBURGER in 1880, and for many years after Strasburger's work, the cell division cycle was defined in terms of what can be seen with the microscope. Mitosis was thus divided, somewhat arbitrarily, into four phases: prophase (in which the chromosomes become visible), metaphase (in which the chromosomes become aligned at the equator of the dividing cell), anaphase (in which the two chromosome sets move to opposite poles of the dividing cell), and telophase (in which the new nuclear membranes are formed, and the chromosomes disappear). Telophase is normally accompanied by, or immediately followed by formation of a cross-wall between the two daughter nuclei, thus completing the process of cytokinesis. Concentration on these visible events led to the remainder of the cell division cycle being known as interphase, or even more misleadingly, resting phase. However, despite the misleading implications of the term "interphase", the importance of this phase, as the phase in which the hereditary material is duplicated, was recognised by the early 1920's, although direct experimental verification of this was not achieved until 1950 (SWIFT 1950). These observations were extended by WALKER and YATES (1952) working with animal cells and by HOWARD and PELC (1953), working with plant cells, to show that not only is DNA replicated in interphase, but that DNA replication is confined to a specific period within interphase. HOWARD and PELC redescribed the process

Table 1. Duration of the cell cycle in higher plants

Species	Duration of the phases of the cycle, h					References
	G1	S	G2	M	Total	
Two populations of *Acer* cells:						KING and STREET (1973)
population 1	4	7	8	2	21	
population 2	26	8	8	3	45	
Zea (root tips):						CLOWES and JUNIPER (1968)
Quiescent centre	151	9	11	3	174	
Meristem, just behind Q.C.	2	11	7	2	22	
200 μm behind Q.C.	4	9	6	4	23	
Tradescantia (root tips):						WIMBER (1966)
Grown at 30 °C	2	10	2	2	16	
Grown at 21 °C	6	11	2	2	21	

of cell division in terms of what happens to the DNA and it is to these workers that we owe our concept of the cell cycle. HOWARD and PELC defined four phases of the cell cycle, G1 (gap 1, between one mitosis and the replication of DNA for the next mitosis), S phase (the period of DNA replication), G2 (gap 2, between DNA replication and mitosis) and M (mitosis, the distribution of the duplicated DNA to two daughter nuclei). The duration of the phases of the cell cycle in a number of plants is shown in Table 1.

Since the introduction of the concept of the cell cycle, our understanding of the events in each of the phases has increased enormously. Within G1, many investigators find it useful to distinguish a sub-phase, G0, in which the cell is not yet "committed" to enter S phase (see Sect. 6). G1 proper is then defined as the period in which the cell is specifically preparing for S phase, for example by synthesising the enzymes necessary for DNA replication. Within S phase we know that, in addition to DNA, histones are synthesised (WOODARD et al. 1961, BLOCK et al. 1967). Within G2, the plane of division is established (see DYER 1976) and the cell synthesises proteins necessary for mitosis (see BRYANT 1976). Many of these processes are not open to cytological investigation, and our knowledge of them has come from application of the techniques of biochemistry and molecular biology to the cell cycle.

3 Methodology

The data discussed in Sect. 4 to 7 have been obtained from a variety of experimental techniques and from a large number of different experimental systems.

In order to avoid giving an excessive amount of experimental detail in Sect. 4 to 7, the basic methodologies used in investigating the cell cycle are described briefly here.

As is emphasised in Sect. 1, mitotic cell division occurs during many phases of plant development, and in many regions within the plant. Although each of these regions of the plant is amenable to some study, for many of them it is very difficult to combine detailed cytological investigations with detailed biochemical studies. We find therefore that within the higher plant, a disproportionate amount of attention has been paid to the root apical meristem (particularly in respect of the reinitiation of DNA synthesis and cell division during germination). Further, many of the experimental systems in current use do not actually relate directly to a growing multicellular plant. Such systems include cultures of unicellular lower plants such as yeasts and algae, and cultured cells from higher plants, grown either as callus or in suspension culture. These systems have been, and still are very useful, particularly for biochemical studies (and hence receive attention in this chapter), but of necessity cannot provide direct information on the coordination and regulation of cell cycle activity in the growing higher plant.

3.1 Experimental Systems

3.1.1 Synchronous Populations of Cells

Synchronous populations of cells, in which all the cells progress through the cell cycle together, are very rare in nature (although the plasmodial stage of the true slime-mould, *Physarum polycephalum,* provides a particularly spectacular example of natural synchrony). For most purposes therefore, synchronous cell populations must be derived from non-synchronous populations. The techniques for obtaining synchronous populations have progressed very little during the past five years, and have been reviewed elsewhere (YEOMAN and AITCHISON 1976, BRYANT 1976). Only a brief account will therefore be given here.

There are two main methods for obtaining synchronous populations: selection and induction. In selection synchrony, cells at a particular stage of the cell cycle are selected by size, density or mass. The selected cells are then used to initiate a new culture which retains synchrony for several generations. The technique has been successfully used for unicellular lower plants, including yeasts (e.g., HALVORSON et al. 1971) and algae (HOPKINS et al. 1970). The method clearly cannot be used for aggregates of cells, nor unfortunately can it be used for cells of higher plants in suspension culture, since there is too much variation between cells at a particular stage of the cell cycle.

In induction synchrony, cells are held at a particular stage of the cell cycle by (1) addition of inhibitors: e.g., hydroxyurea prevents entry of cells into S phase, thus accumulating the cells in G1, or (2) omission from the growth medium of essential nutrients or growth factors, such as sucrose or kinetin, or (3) manipulating the cells' growth environment, e.g., by lowering or raising the temperature or by incubation in darkness. Removal of the inhibitor, or addition of the essential nutrient or growth factor, or returning the cells to a permissive growth environment results in the cells in the culture dividing synchronously for a limited number of generations. For synchrony induced by manipulation of environmental factors, the synchrony can be maintained for a long time by alternating the permissive and non-permissive environmental conditions (see, e.g., TERRY and EDMUNDS 1970). These methods have been used with success for cultures of green algae (e.g. PADILLA and COOK 1964, TAMIYA 1966, TERRY and EDMUNDS 1970), but for cells of higher plants, the degree of synchrony achieved is not great (ROBERTS and NORTHCOTE 1970, JOUANNEAU 1971). However, a variant of the technique, known as the starvation and growth technique, has been used successfully for suspension cultures of sycamore (KING and STREET 1973) and soybean (BREWIN and NORTHCOTE 1973). In this method, cell cultures are allowed to deplete their growth medium and enter stationary phase. The cells are then inoculated at low density into fresh medium. After a lag of several hours, the cells enter S phase synchronously, and the synchrony is maintained for several cell division cycles. However, the technique suffers from the disadvantage that the cell density is low enough to make certain biochemical investigations difficult. For example, there has to date been no successful investigation of DNA polymerase activity in higher plant cells synchronised in this way.

The techniques described above are used with single-celled organisms, or with suspension cultures of cells from higher plants. However, some degree of synchrony can also be induced in cell aggregates. The cells in the meristems of cultured roots are partially synchronised by a combination of carbohydrate depletion and the use of inhibitors (KOVAKS and VAN'T HOF 1970). Some tissue explants, such as those from artichoke tubers, exhibit an acceptable level of synchrony for two or three cell division cycles when cultured under suitable conditions (YEOMAN and AITCHISON 1976). Although the degree of synchrony may not be as high as in synchronous suspension cultures, these systems have the advantage that they can provide enough material for biochemical investigations (YEOMAN and AITCHISON 1976).

3.1.2 Non-Synchronous Populations of Cells

As has been emphasised, synchronous populations are very rare in nature, and non-synchronous populations are therefore more typical. Non-synchronous populations of cells can be studied, and indeed are. In batch cultures of unicellular organisms, or in batch suspension cultures of higher plant cells, comparisons may be made between exponential phase and stationary phase cultures, whilst in higher plants comparisons may be made between meristematic and non-meristematic regions. As mentioned earlier, one very popular system is the angiosperm root, with its apical meristem giving rise to non-dividing cells. In particular, much attention has been paid to the re-initiation of DNA replication and cell division in the root meristem during germination. However, study of germination suffers from the disadvantage that biochemical changes specifically related to the cell division cycle may be masked by biochemical changes relating to the process of rehydration.

3.2 Techniques

Useful techniques for study of DNA replication and cell division may be broadly divided into cytological, genetic and biochemical. Ideally, study of the cell cycle in any system should involve all three types of technique. However, as has already been mentioned, in practice particular experimental systems lend themselves to study by particular techniques and so the ideal situation is rarely, if ever, achieved.

3.2.1 Cytological Techniques

Cytological techniques for study of the cell cycle include the use of autoradiography to locate and estimate the frequency of cells which incorporate precursors into DNA (i.e., which are in S phase) and counting mitotic figures to locate and estimate the frequency of cells in division. A combination of these two techniques may be used to follow the progress of cells through the cell cycle, and to estimate the lengths of the phases of the cell cycle (HOWARD and PELC 1953, VAN'T HOF 1968, MITCHISON 1971).

DNA fibre autoradiography, the autoradiography of individual radioactively labelled DNA molecules, was first developed by CAIRNS (1964), working with bacterial DNA. For plant chromosomes, the technique has been modified and used to great effect by VAN'T HOF and his associates (see VAN'T HOF and BJERKNES 1979). Fibre autoradiography is extremely useful in locating the replication sites along the length of DNA molecules (Fig. 2). If used with DNA from cells which have been labelled for different lengths of time the technique gives information on the rate of chain growth (often referred to as "replication fork rate").

The difficulty of obtaining enough material from certain experimental systems for study of the enzymes involved in DNA replication has led to the development of autoradiographic techniques for detection and assay of DNA polymerase (CHEVALLIER and PHILIPPE 1977, OLSZEWSKA and KONONOWICZ 1979). These techniques are still in an early phase of development, and may not distinguish between replicative and repair polymerase activities. However, there is hope that further sophistication may lead to the ability to assay enzymes in, for example, the floral apex, or in the quiescent centre of the root meristem which cannot be investigated by conventional biochemical techniques.

Finally, under the heading of cytological techniques, mention must be made of the technique which first indicated the occurrence of the cell division cycle, namely investigation of chromatin with the light microscope (and of course, more recently with the electron microscope). These basic techniques can still yield much information about the cell cycle, particularly when used in combination with biochemical techniques, as, for example, in the study of chromatin condensation in prophase.

3.2.2 Genetic Techniques

Genetic techniques, involving mutants showing deficiencies in cell cycle activity, have been extremely useful in investigation of the cell cycle in bacteria. Condi-

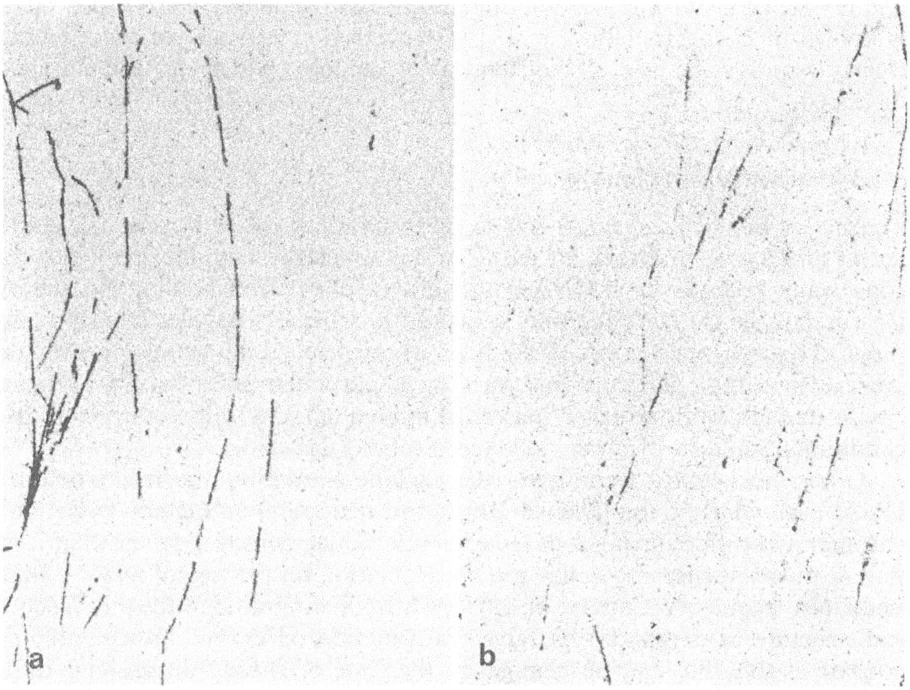

Fig. 2a, b. Fibre autoradiography of replicating DNA in meristems of cultured roots of sunflower (*Helianthus annuus*, var. Russian Mammoth). Cultured roots were supplied with [^{3}H]-thymidine for the times indicated. DNA fibres were prepared from the meristems and spread out across microscope slides. Slides were then dipped in photographic emulsion in order to make autoradiographs. **a** Labelled with high specific activity [^{3}H]-thymidine for 90 min. **b** Labelled with high specific activity [^{3}H]-thymidine for 30 min, followed by low-specific activity [^{3}H]-thymidine for 60 min. In this picture, the bidirectional nature of DNA replication is clearly seen, with the less intense graining (from the low specific activity labelling) seen on either side of the more intensely grained areas, labelled during the initial pulse with high specific activity [^{3}H]-thymidine. (Autoradiographs kindly supplied by Dr. J. VAN'T HOF, Brookhaven Institute, New York)

tional mutants, in which, for example, a deficiency in an enzyme activity is exhibited at one temperature (e.g., 37 °C), but not at another temperature (e.g., 25 °C) have been particularly useful, and have provided direct evidence on the involvement of a number of proteins in DNA replication (KORNBERG 1974, ALBERTS and STERNGLANZ 1977). In higher plants, the problem of detecting mutants is compounded by the diploid nature of the genome, and the genetic approach to the cell cycle has not yet proved very useful. However, in lower plants such as algae and fungi, which may of course be grown as haploid organisms, a number of useful mutants have been detected. Several cell cycle mutants of the fission yeast *Schizosaccharomyces pombe* are known (NURSE et al. 1976) and one of these is a conditional mutant exhibiting a deficiency in DNA ligase activity (NASMYTH 1977). A conditional mutant with a deficiency in DNA ligase activity is known in yeast, *Saccharomyces cerevisiae* (JOHNSTON and NASMYTH 1978) and a conditional mutant with a defective DNA polymerase is known in *Ustilago* (JEGGO et al. 1973). Detection of many of these mutants clearly requires the use of biochemical techniques, which are therefore now discussed.

3.2.3 Biochemical Techniques

The major biochemical approach to understanding the cell cycle has been to detect and assay, and then to purify and characterise any enzyme which may conceivably be involved in DNA synthesis and cell division. With plants, detailed information is available for only a limited number of enzymes from a limited range of species (see BRYANT 1980). Even where detailed information is available, the evidence that particular enzymes have particular roles in vivo is mainly circumstantial, with the exception of the limited amount of direct evidence from conditional mutants of lower fungi (see Sect. 3.2.2).

Other biochemical techniques used include extraction and fractionation of DNA, particularly pulse-labelled DNA, and extraction and characterisation of chromatin at different stages of the cell cycle. Thus, density gradient centrifugation of pulse-labelled DNA has given information on the size of newly synthesised DNA fragments (CRESS et al. 1978), whilst investigation of the physicochemical structure of chromatin has given information on chromatin assembly (JALOUZOT et al. 1980) and on changes in the state of chromatin proteins during the cell cycle (e.g., BRADBURY et al. 1974). Finally, the technique of molecular cloning, using recombinant DNA, has recently been applied to the study of the cell cycle (BEACH et al. 1980a). Fragments of yeast DNA obtained by digestion of the DNA with restriction nucleases have been inserted into a recombinant plasmid lacking DNA replication origins. A limited number of transformed plasmids are obtained which will replicate within a yeast cell, and which are therefore presumed to contain a replication origin within the fragment of yeast DNA inserted. This technique thus gives the potential for detailed investigation of the replication origins of eukaryotic chromosomes.

4 Biochemistry of DNA Replication

4.1 General Features

Much of what we know about the biochemistry of DNA synthesis in eukaryotes is based on work originally done with prokaryotes, and in particular, with *E. coli*. The autoradiographic work of CAIRNS (1964) showed that DNA synthesis starts at a particular point, the replication origin, and proceeds bidirectionally from that point, by the progression of two replication forks working in opposite directions round the circular DNA molecule (Fig. 3). A detailed understanding of the enzymology of DNA replication in *E. coli* has been reached through the use of conditional mutants. For a detailed account of this work, the reader

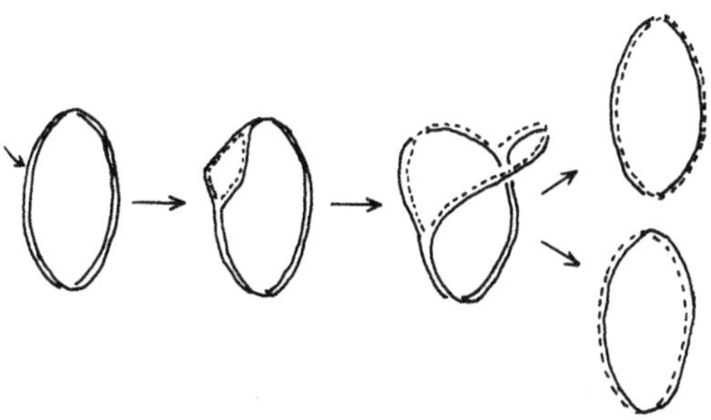

Fig. 3. Diagrammatic representation of DNA replication in *E. coli*. Note the two replication forks moving in opposite directions away from the origin. The newly synthesised DNA strands are indicated by the *dashed lines*

is referred to the work of KORNBERG (1974, 1980) and of ALBERTS and STERNGLANZ (1977). Briefly, the enzymes thought to be directly involved in DNA synthesis in *E. coli* are: (1) Gyrase, which negatively supercoils the DNA, thereby "energising" it. (2) Unwinding enzyme, which is involved in strand separation. (3) DNA-binding protein. (4) Primase or RNA polymerase. (5) DNA polymerase. (6) (Ribo)nuclease (perhaps the exonuclease activity which is part of the DNA polymerase molecule), for removing primers. (7) DNA ligase, for joining nascent DNA fragments. It must be emphasised again here that useful conditional mutants of eukaryotes are somewhat rare, and our current understanding of the enzymology of DNA replication in eukaryotes is at least partly based on our knowledge of these prokaryotic enzymes.

The chromosomes of plants are different in several respects from those of prokaryotes. Plant chromosomes contain large amounts of DNA (BENNETT and SMITH 1976), with the DNA in each chromosome being envisaged as a long linear molecule. The DNA is associated with histones to form a nucleohistone complex with the DNA being wound round the outside of beads of histones known as nucleosomes (see NAGL, Chap. 1, this Vol. for details).

The technique of fibre autoradiography has shown that in eukaryotes, DNA replication is initiated at many hundreds, or even thousands of points along the length of a DNA molecule (CALLAN 1973, VAN'T HOF and BJERKNES 1979). Synthesis proceeds bidirectionally from each origin, in a manner analogous to that described for *E. coli*. The length of DNA synthesised from each origin is known as a replicon. The studies carried out by VAN'T HOF and his associates (VAN'T HOF 1975, 1976a, b, VAN'T HOF and BJERKNES 1977, 1979, VAN'T HOF et al. 1978a, b), and by LARK and his associates (CRESS et al. 1978), have established that in a range of higher plants, the replicons are mainly between 20 and 30 μm (5.9 to 8.8×10^4 nucleotide pairs) in length. In a given species, there appears to be only a limited variation in replicon size (which suggests that replication origins are more or less evenly spaced). In lower eukaryotes, such as the slime-mould *Physarum polycephalum*, there appears to be much more variation within a species in terms of replicon size (FUNDERUD et al. 1978), although the average replicon size for a range of lower eukaryotes is again between 20 and 30 μm (see VAN'T HOF and BJERKNES 1979). In higher plants, the replication forks more outward from the origins at ca. 10 μm/h (VAN'T HOF and BJERKNES 1979). Knowledge of the average replicon size for a given species, and of the total genome size for that species enables calculation of the number of replicons per genome. For example, the diploid genome of *Arabidopsis thaliana* contains ca. 5,200 replicons (VAN'T HOF et al. 1978b) whilst that of *Crepis capillaris* contains ca. 60,000 (VAN'T HOF and BJERKNES 1979).

The rates of replication fork movement observed in higher plants, although up to several hundred times slower than observed in prokaryotes (KORNBERG 1974, ALBERTS and STERNGLANZ 1977) mean that DNA synthesis within a replicon is completed in only a fraction of the time taken to complete the S phase. This apparent paradox is resolved by the observation that different replicons are active at different times during the S phase. Evidence obtained with *Arabidopsis thaliana* indicates that the replicons are organised into groups or "families". In each family, each replicon is active in DNA synthesis at the same time, but the different families are active at different times within the S phase. In the very small genome of *Arabidopsis* there are two distinct families of replicons (VAN'T HOF et al. 1978b), but it is likely that the genomes of most higher plants contain around 25 replicon families.

Our understanding of the biochemistry and enzymology of DNA replication in eukaryotes has been hindered by the scarcity of suitable conditional mutants, and for plants in particular, by lack of data on the occurrence and properties of enzymes. There are therefore still enormous gaps in our knowledge. Two major examples of this are firstly, we have no ideas as to what actually happens, biochemically, at the origins (initiation sites) of each replicon, and secondly that the process of unwinding and strand separation is very poorly understood. It has been argued by many investigators that the replication origins represent sites at which endodeoxyribonuclease nicks one of the DNA strands to assist unwinding (see BRYANT 1980). However, it is both topologically and energetically unlikely that several μm of DNA could be unwound from one nick (see ALBERTS and STERNGLANZ 1977), and it is therefore probable that the progress of the replication forks away from the origins involves further nicking (and possibly re-sealing) of the DNA strands. Nicking, if it does in fact occur, could be carried out by the same enzyme as is responsible for the actual unwinding process (as in prokaryotes), or by a separate endonuclease.

Following unwinding, the DNA is probably held in its single-stranded configuration by a DNA-binding protein (but see Sect. 4.2.3), allowing access to the template of the enzymes involved in synthesis of the new strands. DNA polymerase cannot initiate DNA synthesis de novo, and needs a 3′-OH terminus on which to join the incoming deoxyribonucleotide (KORNBERG 1974). It is very likely that the 3′-OH terminus is provided by the synthesis of a short RNA primer, hydrogen-bonded to the template strand. To date, RNA primers have not been detected in green plants, but the presence of very short pieces of RNA, covalently linked to nascent DNA fragments, has been convincingly demonstrated on the slime-mould, *Physarum polycephalum* (WAQAR and HUBERMAN 1975) and in mammalian cells in culture (BRUN and WEISSBACH 1978, TSENG et al. 1979). The nascent DNA fragments synthesised on the RNA primers are termed Okazaki fragments. Analysis of pulse-labelled DNA from

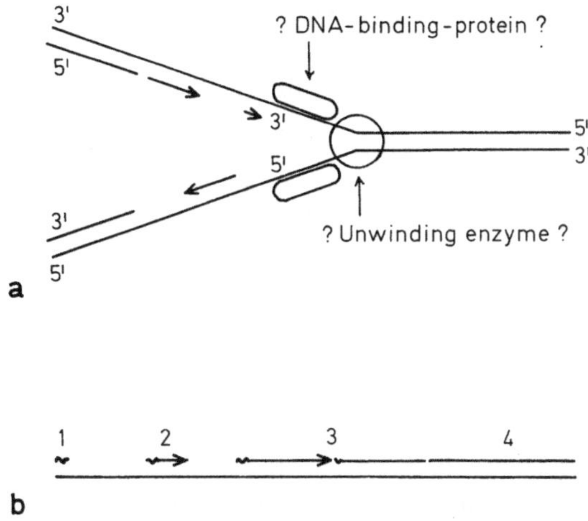

Fig. 4 a, b. Diagrammatic representation of DNA replication in eukaryotes. **a** A replication fork. Note the discontinuous synthesis of DNA in both daughter strands leading to the formation of Okazaki fragments. The question marks surrounding the terms "unwinding enzyme" and "DNA-binding protein" indicate the uncertainties about the roles of these enzymes in eukaryotic organisms. If these enzymes are involved, their most probable sites of action are as indicated. **b** Synthesis of Okazaki fragments. *1* Primer synthesis. *2* Chain extension. *3* Primer excision and gap filling. *4* Ligation of Okazaki fragments

soybean (CRESS et al. 1978) and *Physarum* (FUNDERUD et al. 1978) by sedimentation in density gradients indicates that the Okazaki fragments are ca. 200 nucleotide pairs long (an order of magnitude shorter than in prokaryotes). Sedimentation analysis coupled with fibre autoradiography of DNA labelled for increasing lengths of time shows that Okazaki fragments are joined to give replicon-sized pieces, and at a later stage still, the replicon-sized pieces are joined to give complete daughter strands (CRESS et al. 1978, FUNDERUD et al. 1978).

In order for these joining processes to occur, three things must happen. Firstly, the RNA primers must be removed. In prokaryotes, this is thought to be mediated by the $5' \rightarrow 3'$ exonuclease activity which is exhibited by DNA polymerase (ALBERTS and STERNGLANZ 1977). Some eukaryotic DNA polymerases exhibit exonuclease activity (see Sect. 4.2.5), but many do not, and in eukaryotes primer excision is thought to be mediated by a specific ribonuclease, ribonuclease H (Sect. 4.2.6). Secondly, the gaps left after removal of the primers must be filled, presumably by DNA polymerase. Thirdly, the adjacent DNA fragments are joined by DNA ligase (Sect. 4.2.7).

After synthesis, up to 25% of the cytosine residues in plant DNA are modified by methylation to form 5-methyl cytosine. In animals, there is good evidence that repetitive DNA sequences are much more heavily methylated than unique sequences, and that the methylation enzymes recognise particular sequences within the repetitive DNA (HARBERS and SPENCER 1975). Whether this is true for plants remains to be discovered, although what little evidence there is suggests that it may be so (see BRYANT 1980).

The overall process of DNA replication is illustrated diagramatically in Fig. 4. It is clear that the process requires the sequential and coordinated action of several different enzymes. These enzymes are now described in detail.

4.2 Enzymology of DNA Replication

4.2.1 Endodeoxyribonuclease

The involvement of deoxyribonuclease in DNA replication is suggested by the observation that in a number of plants, including *Tradescantia* (KLIGMAN and TAKATS 1975), *Pisum sativum* (JENNS and BRYANT 1978) and the green algae, *Chlorella* (SCHÖNHERR et al. 1970) and *Euglena* (WALTHER and EDMUNDS 1970) increases in deoxyribonuclease activity are associated with periods of DNA synthesis. In *Tradescantia* and pea, the deoxyribonuclease is known to be located in chromatin and to have an endo-nucleolytic mode of action. Further, in pea, the activity of the enzyme is increased threefold by 8 mM calcium. These observations thus parallel those made on a variety of animal cells, in which nuclear or chromatin-associated calcium-dependent endonucleases are associated with DNA synthesis (see BRYANT 1980).

Although these data provide evidence for a link between endodeoxyribonuclease and DNA replication, a specific role for the endonuclease has yet to be unequivocally demonstrated. In *Tradescantia,* the enzyme is inhibited by Actinomycin D (at concentrations which do not directly affect DNA). This inhibition prevents the nuclease from "priming" the DNA for use by DNA polymerase. These observations, coupled with the observation that the endonuclease prefers poly(dA-T) as a substrate, have led KLIGMAN and TAKATS (1975) to suggest that in vivo, DNA replication may require the nicking of the DNA, perhaps in A-T rich regions. This is again paralleled by observations on mammalian nuclei, where withdrawal of calcium prevents endonuclease activity, which in turn prevents the nuclei bringing about DNA synthesis in vitro (BURGOYNE et al. 1970a, b).

The evidence that endodeoxyribonuclease has a role in DNA replication is clearly not extensive. However, even if we accept for the present that endonuclease does have a role in DNA replication, we do not clearly understand what that role might be. Currently, two possible roles are envisaged. Firstly, it is possible that endonuclease is involved in initiation (as described in Sect. 4.1). This view is consistent with the suggestion made by KLIGMAN and TAKATS (1975) that nicking of DNA at specific sites (e.g., A-T rich regions) is a prerequisite for DNA replication. Recent developments have made the testing of this hypothesis a real possibility. The identification of yeast replication origins, by their ability to restore replicative activity to recombinant plasmids from which replication origins have been excised (BEACH et al. 1980a), and the isolation by density-gradient centrifugation of origin-enriched DNA, after density labelling (with BUdR) of *Physarum* during the first few minutes of S phase (BEACH et al. 1980b), raise the hope that enough "origin-DNA" can be obtained (e.g., by molecular cloning in *E. coli* plasmids) to test the sequence specificity of the putative "initiation endonucleases".

The second role envisaged for the endonuclease is in unwinding. As has already been pointed out, the process of unwinding almost certainly involves nicking the DNA. This could be mediated by endonuclease activity working either immediately before an unwinding enzyme or as part of an unwinding enzyme. Unfortunately, there are no data concerning this possible role, and our consideration of the unwinding enzyme in the next section does not help either.

4.2.2 DNA-Unwinding Enzyme

DNA-unwinding enzymes have been isolated from a number of eukaryotic organisms. One of these is yeast, *Saccharomyces cerevisiae* (DURNFORD and CHAMPOUX 1978), but none, unfortunately, is a green plant. The enzyme brings about a relaxation of positively or negatively supercoiled DNA, via a transient nicking and re-sealing of one of the DNA strands. However, although the DNA in chromatin is negatively supercoiled by virtue of being wound round the nucleosomes (FELSENFELD 1978), the activity which the enzyme exhibits in vitro (i.e., relaxation of supercoils) will not bring about strand separation. It is of course possible that in vivo the enzyme promotes strand separation (as well as relaxation of supercoils), as suggested by DURNFORD and CHAMPOUX (1978), but there is no direct evidence for this view.

4.2.3 DNA-Binding Proteins

The term "DNA-binding proteins" refers to proteins which bind preferentially to single-stranded DNA (ALBERTS and STERNGLANZ 1977). Unfortunately, the nomenclature concerning these proteins is somewhat confused. For example, the DNA-binding protein isolated from *Ustilago maydis* has been referred to as a DNA-unwinding protein (BANKS and SPANOS 1975), although it certainly does not carry out the type of unwinding activity described in Section 4.2.2. Whether or not DNA-binding proteins promote strand separation is not clear. What is clear is that, given a short stretch of single-stranded DNA, the DNA-binding protein will maintain it in that state. For example, the protein from *Ustilago* reduces the T_m of DNA from 60 to 12.5 °C. However, the presence of a short stretch of single-stranded DNA pre-supposes the existence of an enzyme which promotes strand separation, and that, as yet, has not been convincingly demonstrated.

A further problem is that in the one higher plant in which a DNA-binding protein has been detected, namely *Lilium,* it has only been found in meiotic cells, particularly developing pollen grains (HOTTA and STERN 1971, 1979). In pollen grains, the enzyme shows a peak of activity not in the meiotic S phase, but at the time of crossing over. A similar situation exists in mammalian spermatocytes (HOTTA and STERN 1979). These data suggest that DNA-binding protein is involved in recombination, rather than in replication.

4.2.4 RNA Polymerase

The need to synthesise oligonucleotide primers introduces the need for an RNA polymerase or primase. To date, no specific priming RNA polymerase has been detected in eukaryotes, although there is a very limited amount of circumstantial evidence to implicate RNA polymerase I in priming. In mammalian cells, primer synthesis is not inhibited by α-amanitin, but is inhibited by antibodies raised against RNA polymerase I (BRUN and WEISSBACH 1978). In yeast, RNA polymerase I is claimed to possess ribonuclease-H activity (see Sect. 4.2.6) and could thus mediate both primer synthesis and primer removal (HUET et al. 1977). However, before assigning the role of primer synthesis to RNA polymerase I, it must be pointed out that the "normal" role of polymerase I is the synthesis of ribosomal RNA within the nucleolus (see Chap. 5, this Vol.), and it seems unlikely, although not impossible, that it should also be involved in DNA replication.

4.2.5 DNA Polymerase

Of all the enzymes involved in DNA synthesis, DNA polymerase has been the most extensively studied. All organisms which have been investigated possess two or three types of DNA polymerase (in addition to those occurring in chloroplasts and mitochondria). In prokaryotes, the use of conditional mutants has led to some understanding of the roles of the different enzymes. In *E. coli*, for example, there are three DNA polymerases, two of which are involved in DNA replication (one in the synthesis of Okazaki fragments and the other possibly in gap filling), and at least one of which is involved in DNA repair. Prokaryotic DNA polymerases are high molecular weight enzymes, and many of them possess $5' \rightarrow 3'$ and $3' \rightarrow 5'$ exonuclease activity, enabling them to function in primer removal and in proof-reading or removal of mismatched bases (KORNBERG 1974, 1980, ALBERTS and STERNGLANZ 1977).

As a background to our discussion of plant DNA polymerases, mention must be made of the DNA polymerases of vertebrates, since the properties of the vertebrate enzymes are an aid in interpreting data on the plant enzymes. Cells of vertebrates (and possibly of higher invertebrates) contain three distinct, very well-defined types of DNA polymerase, termed α, β and γ. The properties of these polymerases have been reviewed a number of times in recent years (BRUN and CHAPEVILLE 1977, KEIR et al. 1977, WEISSBACH 1977, BRYANT 1980), and the reader is referred to these reviews for further details. Here, as a help to our understanding of the DNA polymerases of plants, the properties of the polymerases of vertebrates are summarised in Table 2. It is clear from the data in the table that the DNA polymerases of vertebrates are somewhat different from their prokaryotic counterparts. In particular, in relation to the enzymology of DNA replication, the polymerases of vertebrates do not possess exonuclease activity and therefore other enzymes are needed for proof-reading and primer removal. One further important point is that assignment of roles to the DNA polymerases of vertebrates is based on circumstantial, but none the less convincing evidence, rather than on the use of mutants.

The pattern of multiple DNA polymerases in algae and fungi is somewhat different from that of vertebrates. The DNA polymerases of *Neurospora* (JOESTER et al. 1978); *Saccharomyces* (WINTERSBERGER and WINTERSBERGER 1970, HELFMAN 1973, E. WINTERSBERGER 1974, U. WINTERSBERGER 1974), *Ustilago* (BANKS et al. 1976, BANKS and YARRANTON 1976) *Euglena* (MCLENNAN and KEIR 1975a–d, 1977) and *Chlamydomonas* (ROSS and HARRIS 1978a, b) have

Table 2. Properties of DNA polymerases from vertebrates. (Data mainly taken from Brun and Chapeville 1977)

Property	DNA polymerase α	DNA polymerase β	DNA polymerase γ
Cellular location:			
(1) Aqueous extraction	Cytoplasm	Mainly Nucleus	Nucleus, cytoplasm
(2) Non-aqueous extraction	Nucleus	Nucleus	Nucleus
Molecular weight	100–150,000	45,000	110,000
Inhibition by:			
N-ethyl-maleimide	Yes	No	Yes
pH optimum	7.5	8.5–9.0	7.5
Di-valent cation preferred (with gapped DNA as template-primer)	Mg^{2+}	Mg^{2+}	Mg^{2+}
Requirement for K^+/Na^+	Low	High	High
Activity with poly(A)-oligo(dT), as percentage of that with gapped DNA	0	50–100	2,000
Associated nuclease	None	None	None
Possible function	DNA replication	DNA repair	DNA replication

been studied in some detail. The major difference between the DNA polymerases of fungi and algae and the DNA polymerases in vertebrates is that lower eukaryotes do not possess an enzyme similar to DNA polymerase β, i.e., they do not have a low molecular weight, chromatin-bound enzyme which is resistant to the SH-group reagent, N-ethyl-maleimide, and which is stimulated by relatively high concentrations of monovalent cations. In fact, these organisms possess *two* major *soluble* DNA polymerases, termed A and B or I and II, the names simply reflecting their order of elution from ion-exchange columns. In *Euglena* and *Chlamydomonas,* a third, very minor form has been detected, but it is not clear whether this is a genuinely distinct enzyme, or merely a sub-population of one of the major forms (McLennan and Keir 1977, Ross and Harris 1978a, b). The properties of the major DNA polymerases from *Neurospora, Saccharomyces, Ustilago, Euglena* and *Chlamydomonas* are shown in Table 3. It is clear from the table that these DNA polymerases from fungi and algae have a general resemblance to DNA polymerase α of vertebrates in being high molecular weight enzymes which are inhibited by N-ethyl-maleimide. However, some of these enzymes also have similarities to DNA polymerase γ of vertebrates, including the relatively high KCl optimum of the *Ustilago* DNA polymerases and of polymerase B of *Chlamydomonas,* and the ability of polymerase I of *Ustilago* (and to a lesser extent, *Euglena* polymerase B and *Chlamydomonas* polymerase A) to use poly(A)·oligo(dT) as a template-primer (although this is not the *pre-*

Table 3. Properties of fungal and algal DNA polymerases

Organism	Poly-merase	Mol. wt.	Optima, mM KCl	Optima, mM Mn	Optima, mM Mg	Mn or Mg pre-ferred	Inhibi-tion by by NEM[a]	Use of poly(A)· ·oligo(dT)	Associated nuclease activity
1. *Neurospora crassa*	A	147,000	5–60	0.2	4	Mg		No	None
	B	110,000	45	0.5	6	Mg		No	None
2. *Saccharomyces cerevisiae*	A	90/180,000	50+	1.5	10	Mg	Yes	No	None
	B	90/180,000	50+	1.5	25	Mg	Yes	No	Exo, 5′→3′ ?
3. *Ustilago maydis*	I	80/100,000	120	0.1	6–12	Mn	Yes	Yes	Exo, 3′→5′
	II	80/100,000	120		6–12		Yes		Exo, 3′→5′
4. *Chlamydomonas reinhardii*	A	90/100,000	20–75	0.2	2	Mg	Yes	Slight	None
	B	200,000	150–200	0.2	2	Mg	Yes	No	Yes
5. *Euglena gracilis*	A	190,000	25	0.2	2	Mn	Yes	No	No
	B	240,000	25	0.2	2	Mn	Yes	Slight	RNase H, exo (5′→3′ ?)

[a] NEM = N-ethyl maleimide

References: 1. Joester et al. 1978
2. Wintersberger and Wintersberger 1970, Helfman 1973, Wintersberger E 1974, Wintersberger U 1974
3. Jeggo et al. 1973, Banks et al. 1976, Banks and Yarranton 1976
4. Ross and Harris 1978a, b
5. McLennan and Keir 1975a–d 1977

ferred template-primer). Further, a number of these DNA polymerases possess nuclease activity (a feature not shown by any of the DNA polymerases of vertebrates). In those organisms in which the nuclease activity has been characterised, it is exonucleolytic, but the direction of exonucleolytic hydrolysis varies between organisms. In *Ustilago,* both polymerases show nuclease activity which appears to be 3′→5′ in direction (and could therefore have a proof-reading function: Yarranton and Banks 1977), whilst the exonuclease activities of the polymerase B of *Saccharomyces* and of polymerase B of *Euglena* seem to be 5′→3′ nucleases (and could therefore function in primer excision). The exonuclease activity of the polymerase B of *Euglena* exhibits a substrate preference for RNA hydrogen-bonded to DNA (i.e., may be classified as ribonuclease H), which lends further support to the view that the exonuclease could function in primer removal (but see Sect. 4.2.6). However, nuclease activity does not seem to be a universal feature of these DNA polymerases from lower plants. Both the DNA polymerases of *Neurospora* have been assayed for endo- and exonuclease activity, and neither has been detected.

Studies of changes in DNA polymerase activity in relation to the growth stage of the cell have shown that in *Chlamydomonas* and *Euglena,* the activity of polymerase B is much better correlated with DNA synthesis than is the activity of polymerase A. In *Saccharomyces,* the activity of polymerase A shows the better correlation with DNA synthesis. Direct evidence for the involvement of DNA polymerase I of *Ustilago* in DNA replication has been obtained with the conditional mutant strain, *Pol 1-1* (JEGGO et al. 1973). Thus, although the two DNA polymerases of a particular alga or fungus may well be somewhat similar in properties, they apparently differ in function, with one of the pair being functionally similar to DNA polymerase α of vertebrates. Interestingly, there seems to be no correlation between a putative role in DNA replication and the possession of nuclease activity. Thus, in *Chlamydomonas* and *Euglena,* the DNA polymerase associated with DNA replication possesses nuclease activity and it has already been noted that the properties of the *Euglena* nuclease are such that it could act as a primer remover. In *Saccharomyces* the putative replicative DNA polymerase does not possess nuclease activity, whilst the other DNA polymerase possesses exonuclease activity. In *Ustilago,* both polymerases possess exonuclease activity, but since the exonucleases are $3' \rightarrow 5'$ in direction, they are more likely to be involved in proof-reading than in primer-removing.

Whilst the patterns of multiple DNA polymerases in vertebrates, and in fungi and algae, are relatively well characterised, the nature of the DNA polymerases of higher plants is still a matter of some confusion. Some investigators (CHANG 1976, GARDNER and KADO 1976, MCLENNAN and KEIR 1977) have suggested that higher plants resemble algae and fungi: i.e., that they do not possess a DNA polymerase β-like enzyme, but that they do possess one or more enzymes which have some similarities to polymerase α and polymerase γ. Other investigators (SRIVASTAVA 1974, TYMONKO and DUNHAM 1977, STEVENS et al. 1978, BRYANT 1980) have suggested that higher plants contain DNA polymerases of direct equivalence to the DNA polymerase-α, β and γ of vertebrates.

Table 4 shows that a number of higher plants covering a relatively wide taxonomic range have been shown to possess a DNA polymerase which has some properties in common with the DNA polymerase α. The properties of these enzymes are also shown in Table 4. The similarities to DNA polymerase α include the high molecular weight, the low KCl optimum (and, for the enzymes from pea and rice at least, marked inhibition by higher concentrations of KCl), the inhibition by N-ethyl-maleimide, and the inability to use poly(A)·oligo(dT) as a template-primer.

Although it is possible to use the data from Table 4 to build up an overall picture of an enzyme which is similar to the DNA polymerase α of vertebrates, there are some anomalies. Firstly, the enzyme detected by CASTROVIEJO et al. (1975, 1979) in wheat embryos differs in molecular weight and pH optimum from the enzyme detected by MORY et al. (1972, 1974, 1975) in germinated wheat seedlings. It is possible, of course, that the two apparently different enzymes are in fact different aggregation states of the same enzyme, particularly in view of their molecular weights, but there is no direct evidence for this view. Secondly, the soluble enzyme from periwinkle is somewhat different from the others. Although it exhibits a relatively low KCl optimum, it is not actually inhibited by high concentrations of KCl. Its preferred template in vitro is denatured DNA, whereas "activated" or "gapped" double-stranded DNA is the

Table 4. Higher plant DNA polymerases which resemble DNA polymerase α of vertebrates

Species	Mol. wt.	Sedimentation coefficient	Optima KCl mM	Optima Mg, mM	Optima pH	Inhibition by NEM	Use of poly(A)·oligo(dT)	Associated nuclease activity
1. *Beta vulgaris* (Beet)	113,000	7S		15	7.5	Yes		
2. *Vicia faba* (Broad bean)				4	7.5			
3. *Phaseolus aureus* (Mung bean)				6.6	7.2			
4. *Pisum sativum* (Pea)			0–25	15	8.1	Yes		None
5. *Glycine max* (Soya bean)				6	7.5			
6. *Daucus carota* (Carrot)				6	7.5			
7. *Vinca rosea* (Periwinkle)	105,000		50	6–15	7.5	Yes	No	Exo
8. *Nicotiana tabacum* (Tobacco)		7–10S	50[a]	7	8.3	Yes	No	
9. *Triticum* (Wheat) a	210–240,000		50	5	7.6			
b[b]	110,000		0	5	7	Yes	Only with Mn	Exo, 5′→3′ and 3′→5′
10. *Zea mays* (Maize)				5	8.4			
11. *Oryza sativa* (Rice)	180,000	75	0	5–10	7.2	Yes	No	? Exo, 3′→5′

[a] With NaCl

[b] The enzyme referred to here is the one designated pol B by CASTROVIEJO et al. 1975, 1979

References:
1. TYMONKO and DUNHAM 1977
2. HOVEMANN and FOLLMAN 1979
3. SCHWIMMER 1966
4. STEVENS and BRYANT 1978
5. HOVEMANN and FOLLMAN 1979
6. SAWAI et al. 1978
7. GARDNER and KADO 1976
8. SRIVASTAVA 1974
9a. MORY et al. 1972, 1974, 1975
9b. CASTROVIEJO et al. 1975, 1979
10. STOUT and ARENS 1970
11. AMILENI et al. 1979

normally preferred template in vitro. Finally, it, and the putative α-like enzyme detected by CASTROVIEJO et al. (1979) in wheat, possess exonuclease activity. This feature is not shown by the other soluble enzymes, although it should be pointed out that the nuclease assays used, for example by STOUT and ARENS (1970) or by STEVENS and BRYANT (1978), may not have been sensitive enough to detect low levels of exonuclease activity.

If the evidence concerning DNA polymerase α in higher plants is somewhat unclear, then that concerning polymerase β is even more so. Firstly, CHANG (1976) and CASTROVIEJO et al. (1975, 1979) failed to detect enzyme activity in chromatin or nuclei from wheat embryos, and GARDNER and KADO (1976) failed to detect any DNA polymerase activity in periwinkle chromatin. Further, CHANG (1976) was unable to detect a low-molecular weight enzyme in whole cell extracts from wheat embryos. On the other hand, a number of investigators have detected DNA polymerase activity tightly bound to chromatin in various higher plants. These enzymes, and their properties, are listed in Table 5. It is obvious that these chromatin-bound DNA polymerases are not nearly as well characterised as the soluble DNA polymerases. Indeed, the only feature known to be shown by *all* these chromatin-bound DNA polymerases is that they are chromatin-bound. Of the other characteristics of the enzymes, some are certainly similar to those of polymerase β of vertebrates. These include firstly, the high KCl optimum of the pea and soybean enzyme. Secondly, in the three plants for which data are available (pea, soybean and beet) the putative polymerase β is certainly less sensitive to N-ethyl-maleimide than the polymerase α from the some plants. Finally, in the two instances where molecular weights and/or sedimentation coefficients have been determined, the values are very similar to those obtained for DNA polymerase β of vertebrates.

Thus a number of higher plants have been shown to possess an enzyme which resembles DNA polymerase β in being bound to chromatin. In soybean, pea, beet and tobacco it has been shown to be a distinct, separate enzyme, with different properties from the soluble α-like polymerase. Detailed characterisations have not been carried out for any of these chromatin-bound enzymes, but the properties so far reported suggest a similarity to DNA polymerase β of vertebrates. The failure to detect such an enzyme in wheat embryos (CHANG 1976, CASTROVIEJO et al. 1975, 1979) and in cultures of periwinkle (GARNDER and KADO 1976) is puzzling, but in view of the occurrence of chromatin-bound polymerases in a relatively wide range of plants, it seems likely that its apparent absence in wheat embryos and periwinkle may be attributed to the techniques employed by the investigators concerned (see STEVENS et al. 1978). This view is supported by the recent detection of DNA polymerase in nuclei from wheat coleoptiles (BRYANT 1980).

If the pattern of multiple DNA polymerase activity in higher plants does in fact resemble that of vertebrates, then higher plants should possess, in addition to the two enzymes described above, a third DNA polymerase, polymerase γ, which shows a marked preference for poly(A)·oligo(dT) as a template-primer. Such an enzyme has been reported from five plants, viz, tobacco (SRIVASTAVA 1974), wheat (CASTROVIEJO et al. 1975, 1979), rice (AMILENI et al. 1979), spinach (SALA et al. 1980) and turnip (DUNHAM and BRYANT 1981). No detailed charac-

Table 5. Chromatin-bound DNA polymerases in higher plants

Species	Mol. wt.	Sedimentation coefficent	Optima KCl, mM	Mg, mM	pH	Inhibition by NEM
1. *Beta vulgaris* (Beet)	50,200	3.4S		10	8.0	No
2. *Pisum sativum* (Pea)			150–200	5	7.25	No[a]
3. *Glycine max* (Soybean)			100	10	8.5	>1 mM
4. *Nicotiana tabacum* (Tobacco)		3.4S				
5. *Tradescantia palidosa*			40 ?	5	7.6	

[a] Crude preparations are partially inhibited

References: 1. TYMONKO and DUNHAM 1977
2. STEVENS et al. 1978
3. D'ALESANDRO et al. 1980
4. SRIVASTAVA 1974
5. WEVER and TAKATS 1970

terisation has been carried out for the γ-like polymerases of tobacco, rice and turnip except that they are known to make up only a small percentage of total polymerase activity. The γ-like polymerases of wheat and spinach have, however, been partially characterised. In wheat, its preference for poly(A)·oligo(dT) as a template-primer is very marked. It is not affected by changes in KCl concentration (at least up to 200 mM) but is inhibited by N-ethyl-maleimide. It has a molecular weight of 150,000. These properties are certainly somewhat similar to those of DNA polymerase γ of vertebrates. However, this particular polymerase makes up ca. 30% of the total polymerase activity as compared with 1%–5% in vertebrates (and in rice, spinach, turnip and tobacco). In view of this, and of their failure to detect a chromatin-bound DNA polymerase in wheat, CASTROVIEJO et al. (1975, 1979) suggest that this polymerase γ-like enzyme is actually one of a group of polymerases of the lower plant type, the other main member of the group being the polymerase α-like enzyme referred to in Table 4. In spinach, on the other hand, the γ-polymerase makes up only a small percentage of the total and has a molecular weight of 105,000 (SALA et al. 1980). It is very much enriched in DNA polymerase preparations made from chloroplasts. This has led SALA et al. (1980) to propose that polymerase γ is chloroplastic. Clearly, before definite conclusions can be reached on the existence of DNA polymerase γ in higher plants, it must be looked for in far greater numbers of species, and its subcellular location studied.

In addition to their location and their physicochemical properties, the DNA polymerases of vertebrates are characterised by their relationship to periods of DNA replication. DNA polymerase α, and probably also DNA polymerase γ, exhibit their highest activities in cells undergoing DNA replication, whilst the activity of DNA polymerase β shows little or no correlation with DNA replication. Unfortunately, with higher plants, a number of investigators have confined their attention to only one of the DNA polymerases in respect of relationship to periods of DNA replication, and so only an incomplete picture is available. However, even from the relatively little evidence available, it is possible to state that the activity of the soluble α-like polymerase shows a marked correlation with DNA replication, whilst that of the chromatin-bound DNA polymerase shows no such correlation (SRIVASTAVA 1974, BRYANT 1980, BRYANT et al. 1981). Further, in rice, the inhibitor, aphidicolin, which is specific for the α-type polymerase, very markedly inhibits replicative DNA synthesis in the nucleus (SALA et al. 1981). These data provide further support for the view that higher plant DNA polymerases resemble those of vertebrates, rather than those of lower plants.

4.2.6 Ribonuclease H

The existence of oligoribonucleotide primers at the 5′ end of the Okazaki fragments implies the existence of an enzyme to remove the primers. That enzyme is believed to be ribonuclease H, an enzyme which degrades RNA in DNA-RNA hybrids. However, although the enzyme's existence is well established, our views on its possible role in DNA replication are largely based on correlations. For example, in batch suspension cultures of carrot cells, ribonuclease H activity closely parallels that of the soluble DNA polymerase (polymerase α), which in turn is closely correlated with DNA synthesis (SAWAI et al. 1978). Similarly, in mammalian cells, ribonuclease H activity is correlated with DNA synthesis (SAWAI and TSUKADA 1977). Carrot is the only higher plant in which the enzyme has been detected, whilst the enzymes from yeast (WYERS et al. 1973) and from *Ustilago* (BANKS 1974) are relatively well characterised. Interestingly, and perhaps unexpectedly, ribonuclease H is endonucleolytic in action, which clearly distinguishes it from the exo-nucleolytic ribonuclease H activity which is part of the DNA polymerase B in *Euglena* (Table 3).

4.2.7 DNA Ligase

DNA ligase is an enzyme which joins the fragments of nascent DNA by forming a phosphodiester linkage between the 3′-OH terminus of one fragment and 5′-phosphate terminus of the next. This has been directly demonstrated in yeast (JOHNSTON and NASMYTH 1978, FABRE and ROMAN 1979) and in fission yeast (NASMYTH 1977) in both of which conditional mutants with deficiencies in DNA ligase fail to join the Okazaki fragments under non-permissive conditions. The enzyme has not been extensively investigated in green plants, but it is known to occur in pea, soybean, spinach, cucumber (KESSLER 1971), carrot (TSUKADA and NISHI 1971) and *Lilium* (HOWELL and STERN 1971). The enzyme has not been well characterised in any of these plants, and even its subcellular distribution is uncertain. KESSLER (1971) found the bulk of the activity to be associated with the nuclei, whilst the ligases studied by TSUKADA and NISHI (1971) and HOWELL and STERN (1971) were "soluble" (although neither of these groups of workers assayed the nuclear fraction). It is possible that higher plants resemble vertebrates, in which there are two major DNA ligases, one nuclear and one soluble, and in which the nuclear enzyme predominates in cells undergoing DNA replication (SÖDERHALL and LINDAHL 1976). In yeast, however, there appears to be only one DNA ligase, and data obtained from conditional mutants suggest that it functions in DNA replication, DNA repair and in recombination (FABRE and ROMAN 1979).

4.2.8 DNA Methylase

The post-synthetic methylation of cytosine residues is carried out by an enzyme known as DNA methylase, which uses S-adenosyl methionine as a methyl donor (KALOUSEK and MORRIS 1969). Very little is known about the enzyme, and it has only been detected in one plant, the garden pea (KALOUSEK and MORRIS 1969). The most interesting feature of the enzyme is that it is able in vivo to methylate cytosine residues at specific sites within the DNA (see Sect. 4.1). What controls this site specificity is at present completely unknown.

5 DNA Replication and Chromatin Structure

5.1 General Features

The DNA in nuclei of eukaryotic organisms is organised in a DNA-protein complex, chromatin, the structure of which is similar in all types of eukaryotes, with the possible exception of the dinoflagellates. The main structural features of chromatin are dealt with in detail by NAGL, Chapter 1, this Volume, but are summarised briefly here in order to provide the basic information for consideration of the relationship between DNA replication and chromatin structure. The DNA in chromatin is wound round discoid beads of histone, each bead consisting of two molecules each of histones H2a, H2b, H3 and H4. The amount of DNA wound round each bead is about 145 base pairs in length, and the length of DNA between the beads is 55 to 60 base pairs, giving a structural repeat length of around 200 base pairs. Associated with this structure are firstly histone H1 (probably located in the interbead regions, and present at a concentration of one molecule per nucleosome), and secondly an array of non-histone proteins which do not determine chromatin structure, but which may have dramatic effects on chromatin function. In consideration of the relationship between the structural features of chromatin and DNA replication, there are three unresolved problems: (1) the organisation of chromatin at the initiation sites (origins); (2) the movement of the replication fork; (3) chromatin reassembly.

5.2 Chromatin Organisation at the Replication Origins

The more-or-less regular spacing of replication origins (initiation sites) along the length of a DNA molecule implies that there is some more-or-less regularly spaced feature of chromatin which allows the action of enzymes (endonucleases?) involved in initiation. Since we do not know what is actually involved in initiation, this is a difficult problem to resolve. It is possible that the spacing of replication origins depends on some higher order feature of chromatin structure. On the other hand, it is equally possible that a replication origin is a specific DNA sequence. If this is so, then it will be revealed by sequence analysis of replication origins which have been bulked up by molecular cloning (see Sect. 4.2.1). Finally, it is obviously possible that both DNA sequence and a higher order structural feature are involved in determining the spacing of replication origins, perhaps with the higher order structural feature being dependent on the DNA sequence.

5.3 Movement of the Replication Fork

The general observation that replication fork movement (i.e., chain extension) is much slower in eukaryotes than in prokaryotes has led to the suggestion that in eukaryotes, DNA synthesis has to progress over bound nucleosomes, i.e., that the DNA is still wound round the histone beads as replication is occurring (ALBERTS and STERNGLANZ 1977). In the opinion of the present author, this seems unlikely. It does seem possible, however, that DNA replication progresses in nucleosome-sized jumps. The following evidence favours such a view. Firstly, the Okazaki fragments in plants (and other eukaryotes) are about 200 base pairs long (see Sect. 4.1). This is similar to the length of DNA associated with the nucleosome. Secondly, DNA *in chromatin* is broken into lengths of 200 base pairs by many endonucleases, suggesting that there is one particularly vulnerable cleavage site in the stretch of DNA between one histone bead and the next (KORNBERG 1977, FELSENFELD 1978). Of course, in order to release a piece of DNA 200 base pairs long, both strands must be broken, but this does not exclude the possibility that under appropriate conditions, with specific enzymes, a single-strand nick could be made at this vulnerable cleavage site. Thirdly, there is a limited amount of evidence that DNA is free from histones during synthesis of the Okazaki fragments, and for a short while afterwards (see Sect. 5.4). These data are consistent with the following, highly speculative model. The size of the Okazaki fragment is determined by the length of DNA associated with the nucleosome. As DNA synthesis progresses, an endonuclease (or an unwinding enzyme) nicks one strand of the DNA at the next internucleosome cleavage site allowing unwinding to start (by an as yet unknown mechanism: see Sects. 4.1 and 4.2.1). Unwinding causes the DNA to dissociate from the histone bead, allowing access of the priming enzyme to the separated strands; synthesis of the Okazaki fragment follows. The whole process is then repeated at the next nucleosome. This model clearly requires testing, but it does explain some otherwise slightly puzzling features.

5.4 Re-Assembly of Chromatin

The synthesis of histone is limited to the S phase of the cell cycle (WOODARD et al. 1961, BLOCH et al. 1967) and it appears that histone synthesis keeps in step with DNA synthesis (WEINTRAUB 1972). At any one moment within the S phase, therefore, it may be envisaged that there is a small pool of histones, some newly synthesised, and some recently released from the DNA actually involved in synthesis. From this pool, the new nucleosomes are re-assembled. There is no evidence on chromatin re-assembly from green plants, and evidence from other systems is very limited. Analysis of replicating chromatin in *Drosophila* suggests that nascent DNA associates first with the arginine-rich histones H3 and H4, and then with histones H2a and H2b (WORCEL et al. 1978). Probing the structure of *Physarum* chromatin with endonuclease during S phase also gives evidence for a two-stage assembly process. Nascent DNA is very vulnerable to nuclease for a few minutes after synthesis. It then becomes less vulnerable, before finally attaining the same degree of protection as in mature chromatin (JALOUZOT et al. 1980). Whether this association of DNA with histones takes place before or after ligation of the nascent DNA fragments is not clear. What is clear, however, is that the daughter chromosomes (chromatids) which appear in prophase are essentially complete by the end of S phase.

6 Relationship Between DNA Replication and Cell Division

Sections 4 and 5 of this chapter have concentrated extensively on DNA replication, with a tendency to deal with it as an isolated process. However, as is implied in the title of this chapter, DNA replication is normally part of the cell division cycle. What then is the relationship between DNA replication and the remainder of the cell cycle? Three groups of observations are pertinent to this question.

Firstly, it should be noted that the onset of S phase cannot be regarded as a transition from the non-dividing to the dividing state. Preparation for S-phase starts some time during the G1 phase of the cell cycle. In the terminology used in Section 2 of this chapter, the start of this preparation may be thought of as the transition from G0 to G1, although in reality there is likely to be no *readily identifiable* event which marks this transition and in cells going through rapid, successive cycles, G0 may be non existent. From work with germinating seeds (MORY et al. 1972, 1975, ROBINSON and BRYANT 1975, SUTCLIFFE and BRYANT 1977, JENNS and BRYANT 1978, SCHIMPFF et al. 1978, HOVEMANN and FOLLMANN 1979, D'ALESANDRO et al. 1980) and with storage tissue explants (YEOMAN and AITCHISON 1976), there is evidence that the preparative period involves the synthesis (or activation) of the enzymes which make the precursors for DNA and of the enzymes involved in DNA replication. However, as has been pointed out (Sect. 3.1. and BRYANT 1976) the first cell division cycle during germination, or following tissue excision, may well not be typical. A limited amount of evidence from systems which may be more typical, such as synchronous cultures of algae (SHEN and SCHMIDT 1966, SCHMIDT 1969) and yeast (ECKSTEIN et al. 1967) and roots of higher plants (KOVACS and VAN'T HOF 1970, VAN'T HOF and KOVACS 1972) again suggests that synthesis or activation of proteins, including enzymes, which are required for the S phase, occurs during G1. There is also evidence from fission yeast that protein synthesis in general (as well as synthesis of specific enzymes) is a requirement for onset of DNA replication, since in this organism, progress through the cell cycle is dependent on cell size, which in turn is at least partly dependent on accumulation of protein (NURSE and THURIAUX 1977, NASMYTH et al. 1979). In those cells which do not exhibit a G1 phase during successive divisions, such as *Physarum,* it is assumed that these preparative evens take place during G2 (and perhaps M) of the previous cell cycle. This approach to the cell cycle thus sees the S phase not as the onset of cell division, but as one event (albeit a dramatic one) in a process initiated much earlier.

The second observation to be noted is that, in plant cells at least, for a given species of plant, the major source of variation in the time taken to complete the cell cycle is the length of the G1 phase (or, more strictly, G0 plus G1) (Table 1 and BRYANT 1976). Under most conditions the time occupied by S+G2+M is more or less constant. This clearly implies that normally the phase of DNA replication (i.e., S) is part of a train of events which, once set in motion, proceeds at the same rate under a variety of conditions. Again, however, it must be emphasised that we must not automatically regard the initiation of S phase as the start of that train of events.

This leads to consideration of the third type of observation. Although we may regard a linkage between DNA replication and mitosis as the norm, it is clearly a norm from which there are frequent exceptions amongst higher plants. The linkage is broken, for example, in the many types of truncated cell cycle which occur in differentiated plant cells (see NAGL, Chap. 4, this Vol. and BRYANT 1976). Indeed, the existence in higher plants of significant numbers of mature cells with 4C (rather than 2C) amounts of DNA (see BRYANT 1976), and the observation that withdrawal of carbohydrate from cultured root meristems of pea and broad bean leads to the establishment of two populations of cells, one held in G1 and one held in G2, has led VAN'T HOF and his associates to propose that there are two major transitions within the cell cycle (VAN'T HOF and KOVACS 1972). One of these transitions is envisaged as being in late G1 (and therefore does not correspond to the postulated G0-G1 transition discussed earlier, and the other in late G2. VAN'T HOF refers to these transitions as principal control points, although he would be first to point out that no specific controlling factors acting at these transition points have been identified. The existence of separate controls operating on S and M has also been suggested by NURSE and his associates, from their work on the control of the cell cycle by cell size in fission yeast (NURSE and THURIAUX 1977, NASMYTH et al. 1979). Detailed examination of the various types of truncated cycle which occur in higher plants in fact suggests that there may be many other transition points (BRYANT 1976) in addition to those discussed already, and doubtless as our knowledge of the molecular events of the cell cycle increases, many more transition points will become apparent.

So we may regard the cell division cycle as a series of steps, each of which is dependent on the previous step, and which preconditions the cell for the next step. Under most circumstances, the whole process goes to completion, but may be halted at particular stages under particular circumstances. In higher plants, the circumstances leading to the occurrence of particular types of truncated cycle may be inherent in the differentiation pattern of particular types of cell (see BRYANT 1976). In unicellular organisms (and in higher plant cells grown in suspension culture) truncated cell cycles are very rare, but blocks can be imposed at certain transition points by the manipulation of conditions or by the use of inhibitors (see Sect. 3.1).

7 Regulation of the Cell Cycle

Consideration in the previous section of the relationship between DNA replication and the remainder of the cell cycle leads naturally to consideration of the regulation of the cell cycle. Unfortunately, in a chapter of limited length, space permits only a limited discussion of this large and complex topic. Two facets of regulation are therefore briefly considered, the first being of general application, and the second relating more specifically to multicellular organisms.

Firstly, then, we can investigate the internal regulation of the cell cycle, i.e., the control of each of many biochemical steps which go to make up the cycle, particularly in cells which are going through successive cell division cycles. Although, in the main, investigations of this type have concentrated on the biochemical mechanisms themselves, the very detailed analysis of the cell cycle in fission yeast carried out in Mitchison's laboratory indicates that there may be general controls, such as cell size, working to give an overall control of the biochemical mechanisms (FANTES 1977, NURSE and THURIAUX 1977, NASMYTH et al. 1979). Secondly, in the context of multicellular organisms, we can investigate what controls the distribution and organisation of dividing cells, restricting division activity to specific regions or to specific growth phases. In the simplest terms this problem becomes one of searching for the factors which prevent cells replicating DNA and dividing as they grow away from the meristem, or for factors which lead to the re-initiation of DNA replication and mitosis, for example, after wounding. We may define such factors as physiological or even morphogenic, since they clearly relate to form and function. Ultimately, however, such factors must act on the biochemical events of the cell cycle, but at present, as will become apparent, there is little information as to how the physiological or morphogenic factors interact with the biochemical control mechanisms.

Before going on to discuss control mechanisms in slightly more detail, one further point must be made. In the preceding paragraph, distinction was made between the internal regulation of DNA replication and division, particularly as exemplified by single-celled organisms, and the regulation of the cessation or re-initiation of these processes, particularly as exemplified by multicellular organisms, such as higher plants. It hardly needs saying that a population of fission yeast cells going through successive cycles is very different from a tissue explant reinitiating division in response to wounding. However, there is a tendency in the literature (including this chapter!) to assume that information gained in one type of system will be relevant to all other systems: for example, that the knowledge of the events leading to the reinitiation of DNA replication in tissue explants gives us knowledge about the G1 phase in general. This assumption may well turn out to be true, but as yet we do not know (but see Sect. 7.1.5).

7.1 Biochemical Aspects of Regulation

In dealing with biochemical aspects of regulation, I wish to confine my attention to changes in the activity of a limited number of enzymes which are believed to have fundamental roles in the cell cycle. Two general levels of regulation of enzyme activity may be envisaged; these are known as "coarse" and "fine" control (see AP REES et al. 1976). In coarse control, regulation is achieved by regulating the actual amount of enzyme present. This is obviously rather a general type of control mechanism, and simply defines the maximum rate at which a particular reaction can proceed. In fine control the activity of preexisting enzymes is modulated in some way, so that the rate of a reaction can be regulated by factors other than the amount of enzyme present. The four examples chosen will illustrate these two types of control.

7.1.1 Ribonucleotide Reductase

It has been argued by several investigators that the most economical point at which to regulate DNA synthesis is at the level of synthesis of the precursors (see HOVEMANN and FOLLMANN 1979). The effectiveness of hydroxyurea (an inhibitor of ribonucleotide reductase) and folate antagonists (which inhibit dTMP synthetase) as inhibitors of DNA synthesis (see BRYANT 1980), certainly indicates that DNA synthesis could in theory be regulated by regulating the amounts of available precursors. HOVEMANN and FOLLMANN (1979) have shown that in germinating broad bean (*Vicia faba*) seeds, the first detectable indication of the initiation of DNA replication is a dramatic increase in the activity of ribonucleotide reductase; in cultured soybean cells, the activity of the reductase shows an equally dramatic decrease as the cultures pass from the exponential to the stationary phase. HOVEMANN and FOLLMAN (1979) suggest that the acquisition of the ability to synthesise deoxyribonucleotides is a major control point (perhaps the G0/G1 transition), but they do not make any suggestions as to how the presence of deoxyribonucleotides might initiate DNA replication. It is clear, however, that this step is under very tight regulation. Not only is the amount of enzyme regulated, but the activity of the enzyme is inhibited by the deoxyribonucleoside triphosphates, thus preventing excessive accumulation of the deoxyribonucleotides.

7.1.2 Endodeoxyribonuclease

The possibility that endonuclease activity is involved in initiation of DNA replication replication and/or in unwinding the double helix, was discussed extensively in Sections 4.1 and 4.2.1. Certainly, in synchronised cultures of *Chlorella* (SCHÖNHERR et al. 1970) and *Euglena* (WALTHER and EDMUNDS 1970) and in germinating peas (JENNS and BRYANT 1978), elevated levels of the enzyme are associated with DNA synthesis. Further, the enzyme is under a relatively complex control system, being associated, in both *Chlorella* and pea, with an inhibitor during periods when DNA is not being synthesised. At present, the identity of the inhibitor is unknown, although earlier ideas that it might be poly(ADP-ribose) have not been confirmed (see BRYANT 1980 for detailed discussion of this point). However, even without knowledge of the identity of the inhibitor (and indeed without knowledge of the actual role of the endonuclease!) we can suggest that regulation of the enzyme involves release from the inhibitor, allowing access of the enzyme to the DNA.

7.1.3 DNA Polymerase

A relatively large amount of work has been carried out on DNA polymerase. Like many of the enzymes involved in DNA synthesis, the activity of the putative replicative DNA polymerase fluctuates in parallel with the cell's ability to replicate DNA. Thus, DNA polymerase activity increases prior to the re-initiation of DNA synthesis during germination (MORY et al. 1972, 1975, ROBINSON and BRYANT 1975, HOVEMANN and FOLLMANN 1979, D'ALESANDRO et al. 1980) and in explants of storage tissue (DUNHAM and CHERRY 1973, YEOMAN and AITCHISON 1976). In suspension cell cultures of carrot (SAWAI et al. 1978), rice (AMILENI et al. 1979) and soybean (HOVEMANN and FOLLMAN 1979), the soluble, α-like polymerase is the major DNA polymerase in exponential-phase cultures, and its activity falls markedly as the cells enter stationary phase. In pea roots, the loss of DNA replicative activity during cell maturation is associated with a loss of soluble DNA polymerase activity (BRYANT et al. 1981). Such data should not be taken to indicate that DNA polymerase activity necessarily regulates DNA replication. What the data do show is that the amount of enzyme available for making DNA can be regulated, and that this regulation is linked in some way to the growth state of cell.

7.1.4 Histone H1 Phosphokinase

Little attention has been given in this chapter to changes in chromatin structure during the cell cycle. This is because, despite a large catalogue of known changes (see MATSUMOTO et al. 1980), it is at present difficult to show the functional significance of most of these changes. One exception to this, however, concerns the phosphorylation of histone H1. Histone H1 does not form part of the nucleosomes, but instead is more loosely associated with the DNA, possibly in the short regions of DNA which are not wrapped round the beads (KORNBERG 1977, FELSENFELD 1978). BRADBURY has proposed that phosphorylation of histone H1 acts as a trigger for chromatin condensation at the beginning of prophase: the phosphorylation of histone is thought to alter the charge distribution in chromatin, leading to increased inter- and intra-molecular attractions (see BRADBURY et al. 1974). In *Physarum,* phosphorylation of histone reaches a peak right at the end of the G2 phase, and the enzyme responsible for the phosphorylation, histone H1 phosphokinase, can only be detected in late G2. Support for the view that phosphorylation controls chromatin condensation comes from findings made by MATSUMOTA et al. (1980), working with mouse cancer cells in culture: in conditional mutant cell lines in which there is a deficiency in histone phosphorylation, there is also a deficiency in chromatin condensation during S phase.

7.1.5 General Aspects of Biochemical Regulation

The four biochemical examples cited serve to illustrate that changes in the state of a higher plant cell (e.g., from non-proliferative to proliferative) are certainly associated with changes in the amount of biochemical machinery involved at various stages of the cell cycle. Further, in a number of lower plants, in which biochemical events have been investigated through several successive cell cycles, it appears that the amount of biochemical machinery can fluctuate through the cell cycle, in a manner which is consistent with the data obtained from higher plants. It thus appears that study of higher plants undergoing a change of state with respect to DNA replication may be relevant to populations of single cells going through successive cell division cycles, and vice versa.

Finally, it must be emphasised again that most of the studies of enzymes are studies of correlations between enzyme activities and particular aspects of the cell cycle. The fact that a given enzyme fluctuates during the cell cycle does not necessarily imply that the enzyme regulates cell cycle activity, except in the rather general sense that in the absence of an enzyme mediating a particular process, the process will not take place.

7.2 Physiological Aspects of Regulation

It has already been suggested that the physiological regulation of DNA synthesis and cell division must be ultimately mediated via biochemical regulation. In other words, the physiological regulatory factors must affect the amount of, or the activity of, the components of the biochemical machinery, including enzymes, which are necessary for DNA replication and cell division. What then are these physiological factors? Unfortunately this is a question to which it is as yet impossible to give more than a very general answer.

Firstly, several "cell division factors" have been isolated from dividing cells of a variety of organisms. These factors have the effect, either of stimulating non-proliferative cells to enter S phase and then complete the cell division cycle, or, if applied to cells which are already progressing through the cell cycle, of accelerating their progress. We have advanced very little in characterising these factors since this author last reviewed this topic (BRYANT 1976), and for this reason, little space is devoted to them here. One feature worth mentioning, however, is that many of them are very non-specific in terms of the cell type which they stimulate, which implies that some aspects of regulation must be of more or less universal occurrence.

Secondly, for higher plants, plant growth substances (plant hormones) have been implicated in the regulation of DNA replication and cell division. In storage tissue explants from a variety of species, plant growth substances (generally auxin and/or cytokinin) are necessary for the re-initiation of cell cycle activity (SETTERFIELD 1963, BRYANT 1976, YEOMAN and AITCHISON 1976, MINOCHA 1979), and in callus and suspension cultures, plant growth substances are certainly required for the maintenance of cell cycle activity (KING and STREET 1973, YEOMAN and AITCHISON 1973). In explants of artichoke, there is evidence that the stimulation of cell cycle activity by auxin involves the synthesis (or activation) of the enzymes involved in DNA replication: i.e., that the physiological regulation has a direct effect on the amount of enzyme available to mediate cell cycle activity (YEOMAN and AITCHISON 1976). There is also some evidence that plant growth substances can have an inhibitory effect on cell cycle activity: in two species of *Lemna* (VAN OVERBEEK et al. 1967, STEWART and SMITH 1972) and in wheat embryos (CHEN and OSBORNE 1970) abscisic acid is a very effective inhibitor of DNA synthesis. However, there is at least one report of abscisic acid as a stimulator of cell cycle activity. In artichoke explants stimulated to reinitiate DNA replication and cell division by auxin, abscisic acid causes a large increase in the precentage of cells which participate in DNA replication and cell division (MINOCHA 1979). So there is a good deal of evidence, some of it a little puzzling, from experiments involving exogenous application of plant growth substances, that plant growth substances may regulate DNA synthesis and cell division. However, no-one has yet succeeded in demonstrating that *endogenous* plant growth substances normally regulate cell cycle activity in vivo. There is a very limited amount of evidence to implicate auxin and cytokinin in the stimulation of cell division during nodule formation in pea roots infected with *Rhizobium* (LIBBENGA et al. 1973), but clearly we are a long way from being able to state definitely that plant growth substances regulate DNA replication and cell division during normal plant growth and development.

Nutrients have also been implicated in the physiological control of DNA replication and cell division. SETTERFIELD (1963) has suggested that calcium ions may be specifically required for cell division, and the work of VAN'T HOF and his associates (KOVACS and VAN'T HOF 1970, VAN'T HOF and KOVACS 1972) suggests that high carbohydrate status is necessary for the onset both of S phase and mitosis. However, as with plant growth substances, we are still far from being able to state definitely that the nutrient status of individual cells or groups of cells is a factor in regulating cell cycle activity in a growing higher plant. Further, whereas in experiments with plant growth substances, some progress has been made in analysing the biochemical (or enzymological) effects, albeit in "artificial" systems, there has been no work on the response of the cell cycle enzymes to manipulation of, for example, carbohydrate supply in cultured pea roots.

8 Concluding Remarks

It will be apparent by now, that even with a range of sophisticated biochemical techniques at our disposal, our analysis of the cell cycle is still in the descriptive phase, particularly in higher plants. Much of the biochemistry of the cell cycle is still unknown: what actually happens to bring a cell from the G0 to the G1 state? What happens at the initiation sites (replication origins)? How is helix unwinding brought about? The descriptive picture needs to be more complete before we can achieve an understanding of the regulation of the cell cycle at the biochemical level. The more widespread use of conditional mutants, and the cloning of specific DNA sequences will doubtless be aids in extending our knowledge of the cell cycle and its regulation.

When we consider regulation at levels higher than the biochemical level, as is seen for example, in higher plants, we are even further from an adequate understanding. Detailed analyses of the distribution of plant growth substances and nutrients within a plant may well help us to decide whether they may regulate the cell cycle in vivo. Even then, however, we would be a long way from comprehending the means whereby these physiological factors have their effects on the biochemical machinery of the cell cyle.

Acknowledgements. I am grateful to Drs. Valgene Dunham, Chris Ford and Jack Van't Hof and to Professor Sydney Shall for helpful and stimulating discussion, to Professor Sydney Shall and to Dr. Francesco Sala for making some of their data available prior to publication, and to Dr. Jack Van't Hof for the autoradiographs used in Fig. 2. Work on DNA synthesis in my laboratory has been generously supported by the Agricultural Research Council, the Royal Society, the Science Research Council and Shell Research Ltd.

Notes Added in Proof:

1. Methylation of cytosine residues in plant DNA has recently been shown to be largely confined to the sequences C.p.G. (and its complement G.p.C.) and C.p.A.p.G. (and its complement G.p.T.p.C.) (Gruenbaum Y, Naveh-Mony T, Cedar H and Razin Y, 1981). Nature 292:860–862
2. A DNA topoisomerase (DNA unwinding enzyme) has been extracted from wheat and extensively characterised. Its properties are similar to those of the yeast enzyme (Dynan WS, Jendrisak JJ, Hager DA and Burgess RR, 1981). J. Biol. Chem. 256:5860–5865.

References

Alberts B, Sternglanz R (1977) Recent excitement in the DNA replication problem. Nature (London) 269:655–661

Amileni A, Sala F, Cella R, Spadari S (1979) The major DNA polymerase in cultured plant cells: partial purification and correlation with cell multiplication. Planta 146:521–528

Ap Rees T, Fuller WA, Wright BW (1976) Pathways of carbohydrate oxidation during thermogenesis by the spadix of *Arum maculatum*. Biochim Biophys Acta 437:22–35

Banks GR (1974) A ribonuclease H from *Ustilago maydis*. Properties, mode of action and substrate specificity of the enzyme. Eur J Biochem 47:499–507

Banks GR, Spanos A (1975) The isolation and properties of a DNA-unwinding protein from *Ustilago maydis*. J Mol Biol 93:63–77

Banks GR, Yarranton GT (1976) A DNA polymerase from *Ustilago maydis*. 2. Properties of the associated deoxyribonuclease activity. Eur J Biochem 62:143–150

Banks GR, Holloman WK, Kairis WT, Spanos A, Yarranton GT (1976) A DNA polymerase from *Ustilago maydis*. 1. Purification and properties of the polymerase activity. Eur J Biochem 62:131–142

Beach D, Piper M, Shall S (1980a) Isolation of chromosomal origins of replication in yeast. Nature (London) 284:185–187
Beach D, Piper M, Shall S (1980b) Isolation of newly-initiated DNA from the early S phase of the synchronous eukaryote, *Physarum polycephalum*. Exp Cell Res 129:211–221
Bennett MD, Smith JB (1976) Nuclear DNA amounts in angiosperms. Phil Trans R Soc London Ser B 274:227–274
Bernier G (1971) Structural and metabolic changes in the shoot apex in transition to flowering. Can J Bot 49:803–819
Block DP, MacQuigg RA, Brack SD, Wu J-R (1967) The synthesis of deoxyribonucleic acid and histone in the onion root meristem. J Cell Biol 33:451–468
Bradbury EM, Inglis RJ, Matthews HR (1974) Control of cell division by very lysine rich histone (fl) phosphorylation. Nature (London) 247:257–261
Brewin NJ, Northcote DH (1973) Variations in the amounts of 3′, 5′-cyclic AMP in plant tissues. J Exp Bot 24:881–888
Brun G, Chapeville F (1977) Multiplicity of animal cell deoxyribonucleic acid polymerase. Biochem Soc Symp 42:1–16
Brun G, Weissbach A (1978) Initiation of HeLa cell DNA synthesis in a subnuclear system. Proc Natl Acad Sci USA 75:5931–5935
Bryant JA (1976) The cell cycle. In: Bryant JA (ed) Molecular aspects of gene expression in plants. Academic Press, London New York, pp 177–216
Bryant JA (1980) Biochemical aspects of DNA replication, with particular reference to plants. Biol Rev 55:237–284
Bryant JA, Jenns SM, Francis D (1981) DNA polymerase activity and DNA synthesis in roots of pea (*Pisum sativum L.*) seedlings. Phytochemistry 20:13–15
Burgoyne LA, Waqar MA, Atkinson MR (1970a) Calcium-dependent priming of DNA synthesis in isolated rat liver nuclei. Biochem Biophys Res Commun 39:254–259
Burgoyne LA, Waqar MA, Atkinson MR (1970b) Initiation of DNA synthesis in rat thymus: correlation of calcium-dependent initiation in thymocytes and in isolated thymus nuclei. Biochem Biophys Res Commun 39:918–922
Cairns J (1964) The chromosome of *Escherichia coli*. Cold Spring Harbor Symp Quant Biol 28:43–46
Callan HG (1973) Replication of DNA in eukaryotic chromosomes. Br Med Bull 29:192–195
Castroviejo M, Tarrago-Litvak L, Litvak S (1975) Partial purification and characterization of two cytoplasmic DNA polymerases from ungerminated wheat. Nucl Acids Res 2:2077–2090
Castroviejo M, Tharaud D, Tarrago-Litvak L, Litvak S (1979) Multiple deoxyribonucleic acid polymerases from quiescent wheat embryos. Purification and characterization of three enzymes from the soluble cytoplasm and one from purified mitochondria. Biochem J 181:183–191
Chang L (1976) Phylogeny of DNA polymerase-β. Science 191:1183–1185
Chen D, Osborne DJ (1970) Hormones in the translational control of early germination in wheat embryos. Nature (London) 266:1157–1160
Chevaillier Ph, Philippe M (1977) *In situ* detection and characterization of DNA polymerase activities in the nucleus of eukaryotic cells. Chromosoma 63:385–399
Clowes FAL, Juniper BE (1968) Plant cells. Blackwell, Oxford
Cress DE, Jackson PJ, Kadouri A, Chu YE, Lark KG (1978) DNA replication in soybean protoplasts and suspension-cultured cells: comparison of exponential and fluoro deoxyridine synchronized cultures. Planta 143:241–253
D'Alesandro MM, Jaskot RH, Dunham VL (1980) Soluble and chromatin-bound DNA polymerases in developing soybean. Biochem Biophys Res Commun 94:233–239
Dunham VL, Bryant JA (1981) DNA polymerases of turnip (*Brassica rapa*). Proc 14th FEBS Meet Abstr., Biochem Soc Trans 9:230
Dunham VL, Cherry JH (1973) Multiple DNA polymerase activity solubilized from higher plant chromatin. Biochem Biophys Res Commun 54:403–410
Durnford JM, Champoux JJ (1978) The DNA untwisting enzyme from *Saccharomyces cerevisiae*. J Biol Chem 253:1086–1089

Dyer AF (1976) The visible events of mitotic cell division. In: Yeoman MM (ed) Cell division in higher plants. Academic Press, London New York, pp 49–110
Eckstein H, Paduch V, Aitz H (1967) Synchronized yeast cells. 3. DNA synthesis and DNA polymerase after inhibition of cell division by X-rays. Eur J Biochem 3: 224–231
Fabre F, Roman H (1979) Evidence that a single DNA ligase is involved in replication and recombination in yeast. Proc Natl Acad Sci USA 76:4586–4588
Fantes P (1977) Control of cell size and cycle time in *Schizosaccharomyces pombe*. J Cell Sci 24:51–67
Felsenfeld G (1978) Chromatin. Nature (London) 271:115–122
Francis D, Lyndon RF (1979) Synchronisation of cell division in the shoot apex of *Silene* in relation to flowering. Planta 145:151–157
Funderud S, Andreassen R, Haugli F (1978) DNA replication in *Physarum polycephalum:* bidirectional replication of DNA within replicons. Nucl Acids Res 5:713–722
Gardner JM, Kado CI (1976) High molecular weight deoxyribonucleic acid polymerase from crown gall tumour cells of periwinkle (*Vinca rosea*). Biochemistry 15:688–696
Halvorson HO, Carter BLA, Tauro P (1971) Synthesis of enzymes during the cell cycle. Adv Microb Physiol 6:47–106
Harbers H, Spencer JH (1975) Distribution of 5-me-cytosine in pyrimidine oligonucleotides of mouse-L-cell satellite DNA and mainband DNA. Biochem Biophys Res Commun 66:738–746
Helfman WB (1973) The presence of an exonuclease in highly purified DNA polymerase from bakers yeast. Eur J Biochem 32:42–50
Hopkins HA, Sitz TO, Schmidt RR (1970) Selection of synchronous *Chlorella* cells by centrifugation to equilibrium in Ficoll. J Cell Physiol 70:231–234
Hotta Y, Stern H (1971) A DNA-binding protein in meiotic cells of *Lilium*. Dev Biol 26:87–99
Hotta Y, Stern H (1979) The effect of dephosphorylation on the properties of a helix-destabilising protein from meiotic cells, and its partial reversal by a protein kinase. Eur J Biochem 95:31–38
Hovemann B, Follmann H (1979) Deoxyribonucleotide synthesis and DNA polymerase activity in plant cells (*Vica faba* and *Glycine max.*). Biochim Biophys Acta 561:42–52
Howard A, Pelc SR (1953) Synthesis of desoxyribonucleic acid in normal and irradiated cells and its relation to chromosome breakage. Heredity 6:261–273
Howell SH, Stern H (1971) The appearance of DNA breakage and repair activities in the synchronous meiotic cycle of *Lilium*. J Mol Biol 55:357–378
Huet J, Buhler JM, Sentenac A, Fromageot P (1977) Characterization of ribonuclease H activity associated with yeast RNA polymerase A. J Biol Chem 252:8848–8855
Jalouzot R, Briane D, Ohlenbusch H, Wilhen ML, Wilhen FX (1980) Kinetics of nuclease digestion of *Physarum polycephalum* nuclei at different stages of the cell cycle. Eur J Biochem 104:423–431
Jeggo PA, Unrau P, Banks GR, Holliday R (1973) DNA polymerase mutants of *Ustilago maydis*. Nature New Biol 242:14–16
Jenns SM, Bryant JA (1978) Correlation between deoxyribonuclease activity and DNA replication in the embryonic axes of germinating peas (*Pisum sativum* L.). Planta 138:99–103
Joester W, Joester KE, Van Dorp B, Hofschneider PH (1978) Purification and properties of DNA-dependent DNA polymerases from *Neurospora crassa*. Nucl Acids Res 5:3043–3055
Johnston LH, Nasmyth KA (1978) *Saccharomyces cerevisiae* cell cycle mutant *cdc 9* is defective in DNA ligase. Nature (London) 274:891–893
Jouanneau JP (1971) Controle par les cytokinines de la synchronisation des mitoses dans les cellules de tabac. Exp Cell Res 67:329–337
Kalousek F, Morris NR (1969) Deoxyribonucleic acid methylase activity in pea seedlings. Science 164:721–722
Keir HM, Craig RK, McLennan AG (1977) Variation of deoxyribonucleic acid polymerases in the cell cycle. Biochem Soc Symp 42:37–54

Kessler B (1971) Isolation, purification and distribution of a DNA ligase from higher plants. Biochem Biophys Acta 240:496–505
King PJ, Street HE (1973) Growth patterns in cell cultures. In: Street HE (ed) Plant tissue and cell culture. Blackwell, Oxford, pp 269–337
Kligman LHL, Takats S (1975) An actinomycin-sensitive endonuclease associated with DNA synthesis in *Tradescantia* nuclei. Exp Cell Res 95:176–190
Kornberg A (1974) DNA synthesis. Freeman, San Francisco
Kornberg A (1980) DNA replication. Freeman, San Francisco
Kornberg RD (1977) Structure of chromatin. Annu Rev Biochem 46:931–954
Kovacs CJ, Van't Hof J (1970) Synchronization of a proliferative population in a cultured plant tissue. Kinetic evidence for a G1/S population. J Cell Biol 47:536–539
Libbenga KR, van Iren F, Bogers RJ, Schraag-Lamers MF (1973) The role of hormones and gradients in the initiation of cortex proliferation and nodule formation in *Pisum sativum* L. Planta 114:29–39
Matsumoto Y-L, Yasuda H, Mita S, Marunouchi T, Yamada M-A (1980) Evidence for the involvement of Hl histone phosphorylation in chromosome condensation. Nature (London) 284:181–183
McLennan AG, Keir HM (1975a) Deoxyribonucleic acid polymerases of *Euglena gracilis*. Purification and properties of two distinct deoxyribonucleic acid polymerases of high molecular weight. Biochem J 151:227–238
McLennan AG, Keir HM (1975b) Deoxyribonucleic acid polymerases of *Euglena gracilis*. Primer-template utilization of, and enzyme activities associated with the two deoxyribonucleic acid polymerases of high molecular weight. Biochem J 151:239–247
McLennan AG, Keir HM (1975c) DNA polymerases of *Euglena gracilis:* heterogeneity of molecular weight and sub-unit structure. Nucl Acids Res 2:223–238
McLennan AG, Keir HM (1975d) Subcellular location and growth stage dependence of the DNA polymerases of *Euglena gracilis*. Biochim Biophys Acta 407:253–262
McLennan AG, Keir HM (1977) The deoxyribonucleic acid polymerases of nonvertebrate eukaryotes. Biochem Soc Symp 42:55–73
Minocha SC (1979) Abscisic acid promotion of cell division and DNA synthesis in Jerusalem artichoke tuber tissue cultured *in vitro*. Z Pflanzenphysiol 92:327–339
Mitchison JM (1971) The biology of the cell cycle. Univ Press, Cambridge
Mory YY, Chen D, Sarid S (1972) Onset of DNA synthesis in germinating wheat embryos. Plant Physiol 49:20–23
Mory YY, Chen D, Sarid S (1974) Deoxyribonucleic acid polymerase from wheat embryos. Plant Physiol 53:377–381
Mory YY, Chen D, Sarid S (1975) *De novo* biosynthesis of deoxyribonucleic acid polymerase during wheat embryo germination. Plant Physiol 55:437–442
Nasmyth KA (1977) Temperature-sensitive lethal mutants in the structural gene for DNA ligase in the yeast *Schizosaccharomyces pombe*. Cell 12:1109–1120
Nasmyth K, Nurse P, Fraser RSS (1979) The effect of cell mass on the cell cycle timing and duration of S-phase in fission yeast. J Cell Sci 39:215–233
Nurse P, Thuriaux P (1977) Controls over the timing of DNA replication during the cell cycle of fission yeast. Exp Cell Res 107:365–375
Nurse P, Thuriaux P, Nasmyth KA (1976) Genetic control of the cell division cycle in the fission yeast, *Schizosaccharomyces pombe*. Molec Gen Genet 146:167–178
Olszewska MJ, Kononowicz AK (1979) Activities of DNA polymerases and RNA polymerases detected *in situ* in growing and differentiating cells of root cortex. Histochemistry 59:311–324
Padilla GM, Cook JR (1964) The development of techniques for synchronizing flagellates. In: Zeuthen E (ed) Synchrony in cell division and growth. Intersci, New York, pp 521–535
Roberts K, Northcote DH (1970) The structure of sycamore callus cells during division in a partially synchronised suspension culture. J Cell Sci 6:299–321
Robinson NE, Bryant JA (1975) Development of chromatin-bound and soluble DNA polymerase activities during germination of *Pisum sativum* L. Planta 127:69–75
Ross CA, Harris WJ (1978a) DNA polymerases from *Chlamydomonas reinhardii*. Purification and properties. Biochem J 171:231–240

Ross CA, Harris WJ (1978b) DNA polymerases from *Chlamydomonas reinhardii*. Further characterization, action of inhibitors and associated nuclease activities. Biochem J 171:241–249
Sala F, Amileni AR, Parisi B, Spadari S (1980) A γ-like DNA polymerase in spinach chloroplasts. Eur J Biochem 112:211–217
Sala F, Galli MG, Levi M, Burroni D, Parisi B, Pedrali-Noy G, Spadari S (1981) Functional roles of the plant α-like and γ-like DNA polymerases. FEBS Lett 124:112–118
Sawai Y, Tsukada K (1977) Change of ribonuclease-H activity in dividing and regenerating rat liver. Biochim Biophys Acta 479:126–133
Sawai Y, Sugano N, Tsukada K (1978) Ribonuclease-H activity in cultured plant cells. Biochim Biophys Acta 518:181–185
Schimpff G, Müller H, Follmann H (1978) Age-dependent DNA labelling and deoxyribonucleotide synthesis in wheat seeds. Biochim Biophys Acta 520:70–81
Schmidt RR (1069) Control of enzyme synthesis during the cell cycle of *Chlorella* In: Padilla GM, Whitson GL, Cameron IL (eds) The cell cycle: gene-enzyme interactions. Academic Press, New York London, pp 159–177
Schönherr OT, Wanka F, Kuyper CMA (1970) Periodic change of deoxyribonuclease activity in synchronous cultures of *Chlorella*. Biochim Biophys Acta 224:74–79
Schwimmer S (1966) DNA polymerase activity of mung bean seedlings. Phytochemistry 5:791–794
Setterfield G (1963) Growth regulation in excised slices of Jerusalem artichoke tissue. Symp Soc Exp Biol 17:98–126
Shen S R-C, Schmidt RR (1966) Enzymic control of nucleic acid synthesis during synchronous growth of *Chlorella pyrenoidosa*. Arch Biochem Biophys 115:13–20
Söderhall S, Lindahl T (1976) DNA ligeases of eukaryotes. FEBS Lett 67:1–8
Srivastava BIS (1974) A 7S DNA polymerase in the cytoplasmic fraction from higher plants. Life Sci 14:1947–1954
Stevens C, Bryant JA (1978) Partial purification and characterization of the soluble DNA polymerase (polymerase-α) from seedlings of *Pisum sativum* L. Planta 138:127–132
Stevens C, Bryant JA, Wyvill PC (1978) Chromatin-bound DNA polymerase from higher plants: A DNA polymerase-β-like enzyme. Planta 143:113–120
Stewart GR, Smith H (1972) Effects of abscisic acid on nucleic acid synthesis and the induction of nitrate reductase in *Lemna polyrhiza*. J Exp Bot 23:875–885
Strasburger G (1880) Zellbildung und Zelltheilung. Gustav Fisher, Jena
Stout ER, Arens MQ (1970) DNA polymerase from maize seedlings. Biochim Biophys Acta 213:90–100
Sutcliffe JF, Bryant JA (1977) Biochemistry of germination and seedling growth. In: Sutcliffe JF, Pate JS (eds) The physiology of the garden pea. Academic Press, London New York, pp 45–82
Swift HH (1950) The desoxyribose nucleic acid content of animal nuclei. Physiol Zool 23:169–198
Tamiya H (1966) Synchronous cultures of algae. Annu Rev Plant Physiol 17:1–26
Terry OW, Edmunds LN (1970) Phasing of cell division by temperature cycles in *Euglena* cultured autotrophically under continuous illumination. Planta 93:106–127
Tseng BY, Erickson JM, Goulian M (1979) Initiator RNA of nascent DNA from animal cells. J Mol Biol 129:531–545
Tsukada K, Nishi A (1971) Polynucleotide ligase from cultured plant cells. J Biochem 70:541–542
Tymonko JM, Dunham VL (1977) Evidence for DNA polymerase-α and -β activity in sugar beet. Physiol Plant 40:27–30
Van Overbeek J, Loeffler JE, Masson MJR (1967) Dormin (abscisin II): inhibitor of plant DNA synthesis? Science 156:1497–1499
Van't Hof J (1968) Experimental procedures for measuring cell population kinetics parameters in plant root meristems. In: Prescott D (ed) Methods in cell physiology Vol 3. Academic Press, New York London, pp 95–124
Van't Hof J (1975) DNA fiber replication in chromosomes of a higher plant (*Pisum sativum*). Exp Cell Res 93:95–104

Van't Hof J (1976a) DNA fiber replication of chromosomes of pea root cells terminating S. Exp Cell Res 99:47–56

Van't Hof J (1976b) Replicon size and rate of fork movement in early S of higher plant cells (*Pisum sativum*). Exp Cell Res 103:395–403

Van't Hof J, Bjerknes CA (1977) 18 µm replication units of chromosomal DNA fibres of differentiated cells of pea (*Pisum sativum*). Chromosoma 64:287–294

Van't Hof J, Bjerknes CA (1979) Chromosomal DNA replication in higher plants. Bioscience 29:18–22

Van't Hof J, Kovacs CJ (1972) Mitotic cycle regulation in the meristems of cultured roots: the principal control point hypothesis. Adv Exp Med Biol 18:15–30

Van't Hof J, Bjerknes CA, Clinton JH (1978a) Replication properties of chromosomal DNA fibers and the duration of DNA synthesis of sunflower root-tip meristem cells at different temperatures. Chromosoma 66:161–171

Van't Hof J, Kuniyuki A, Bjerknes CA (1978b) The size and number of replicon families of chromosomal DNA of *Arabidopsis thaliana*. Chromosoma 68:269–285

Walker PM, Yates HB (1952) Nuclear components of dividing cells. Proc R Soc London Ser B 140:274–299

Walther WG, Edmunds LN (1970) Periodic increase in deoxyribonuclease activity during the cell cycle in synchronised *Euglena*. J Cell Biol 46:613–617

Waqar MA, Huberman JA (1975) Covalent linkage between RNA and nascent DNA in the slime mold, *Physarum polycephalum*. Biochim Biophys Acta 383:410–420

Weintraub H (1972) A possible role for histone in the synthesis of DNA. Nature (London) 240:449–453

Weissbach A (1977) Eukaryotic DNA polymerases. Annu Rev Biochem 46:25–47

Wever G, Takats ST (1970) DNA polymerase activities in *Tradescantia palidosa* pollen grain nuclei. Biochim Biophys Acta 199:8–17

Wimber DE (1966) Duration of the nuclear cycle in *Tradescantia* root tips at three temperatures as measured with H^3-thymidine. Am J Bot 53:21–24

Wintersberger E (1974) Deoxyribonucleic acid polymerases from yeast. Further purification and characterisation of DNA-dependent DNA polymerases A and B. Eur J Biochem 50:41–47

Wintersberger U (1974) Absence of low molecular-weight DNA polymerase from nuclei of the yeast *Saccharomyces cerevisiae*. Eur J Biochem 50:197–202

Wintersberger U, Wintersberger E (1970) Studies on deoxyribonucleic acid polymerases from mitochondria-free cell extracts. Eur J Biochem 13:11–19

Woodard J, Rasch E, Swift H (1961) Nucleic acid and protein metabolism during the mitotic cycle in *Vicia faba*. J Biophys Biochem Cytol 9:445–462

Worcel A, Han S, Wang ML (1978) Assembly of newly replicated chromatin. Cell 15:969–977

Wyers F, Sentenac A, Fromageot P (1973) Role of DNA-RNA hybrids in eukaryotes. Ribonuclease H in yeast. Eur J Biochem 35:270–281

Yarranton GT, Banks GR (1977) A DNA polymerase from *Ustilago maydis*. Evidence of proof-reading by the associated 3′→5′ deoxyribonuclease activity. Eur J Biochem 77:521–527

Yeoman MM, Aitchison PA (1973) Growth patterns in tissue (callus) cultures. In: Street HE (ed) Plant tissue and cell culture. Blackwell, Oxford, pp 240–268

Yeoman MM, Aitchison PA (1976) Molecular events of the cell cycle: a preparation for division. In: Yeoman MM (ed) Cell division in higher plants. Academic Press, London New York, pp 109–133

4 DNA Endoreduplication and Differential Replication

W. Nagl

1 Introduction

For a long time it has been thought that nuclear DNA is an extremely stable constant, and that the genetic information stored in the nucleus is identical in all cells of an individual, because of identical DNA replication during the S phase and identical chromosome segregation during mitosis. This includes the idea that during development the expression of specific genes is brought about by the evocation of certain gene activities from this constant source by environmental, cytoplasmic and hormonal stimuli. The dogma runs: The result of differential gene expression is differentiation (Kafatos 1979). In reality, however, cell differentiation and morphogenesis are not yet understood in terms of control and causality (Nagl 1979b, Wright 1979). To avoid uncritical satisfaction, regulation systems should be sought for, which may help to explain the realization of the *Bauplan* (phenotypic organization).

Since Geitler (1939) detected the process of endomitosis, we have learned that the nucleus and its DNA exhibit a dynamic organization rather than a constant one, and that most mature cells of most organisms do have, in addition to their diploid genome, quantitatively and/or qualitatively different DNA fractions in their nuclei.

In this chapter, I shall discuss the origin and occurrence of such nuclei, the mechanisms which lead to the modified genomic state, and its possible significance for cell physiology and differentiation. For more extensive reviews the reader is referred to D'Amato (1977) and Nagl (1978a).

2 Somatic Polyploidization Cycles

The term "somatic polyploidization" covers various kinds of cell cycles which are curtailed before cytokinesis (cell division). The result is increase in nuclear DNA content and the number of chromosome complements in a geometrical order. The evolution and regulation of such cell cycles in eukaryotes can be best understood if they are seen as steps of an evolutionary strategy which leads to cell-specific DNA increase by step-wise reduction and curtailment of the mitotic cell cycle and the complex mitotic machinery (Nagl 1978a; Figs. 1 and 2). The first step lies in the omission of cytokinesis only, leading to a multinucleate or polyenergid cell. Such a cell is functionally polyploid. The last step is the omission of all stages except DNA replication of a specific DNA sequence, e.g., a single gene (=DNA amplification, Sect. 3.2).

2.1 Polyenergid Cells

Polyenergid organization is frequently found in endosperm, at least during early differentiation (MAHESHWARI 1963). Binucleate and multinucleate cells occur regularly in the anther tapetum of many plants (CARNIEL 1952, 1963, GREILHUBER 1974). In several legumes, the embryo suspensor is made up of multinucleate chambers, e.g., in *Pisum* and *Lathyrus* (NAGL 1962, NAGL and FUHRMANN in press). Some taxa among algae and fungi exhibit polyenergid organization throughout their soma. While lysis of the cell wall and subsequent cell fusion, whose result is also a multinucleate tissue, occurs only rarely in plants (e.g., in galls caused by nematodes; HESSE 1970), cell fusion is the common way of synctium formation in mammals (e.g., syncytiotrophoblast formation in humans). Binucleate cells are, for example, a characteristic feature of liver (BRODSKII and URYVAEVA 1977).

The next rigorous short cuts of the mitotic cell cycle affect mitotic stages and hence directly lead to somatic polyploidy: the restitution cycle, the endomitotic cycle, and the endoreduplication cycle.

2.2 Nuclear Restitution Cycles

Nuclei in the anther tapetum, the endosperm, the suspensor, and galls of many angiosperms often undergo restitution, that is, the chromosomes enter mitosis but do not complete anaphase, and re-enter an interphase state within one and the same nucleus. Hence the chromosomes undergo a condensation – decondensation cycle, but karyokinesis is omitted, mostly due to failure of the spindle

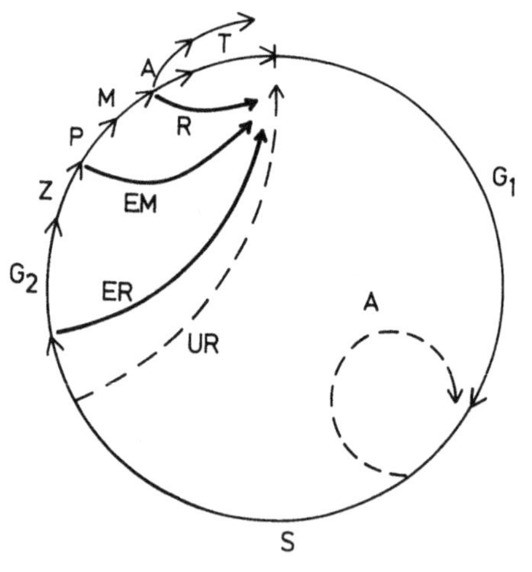

Fig. 1. Diagram illustrating the course of the mitotic cell cycle and the curtailed cell cycles: G_1 presynthetic or postmitotic period of the interphase; *S* DNA synthesis period: G_2 postsynthetic or premitotic "gap"; *Z* Z phase (dispersion period); *P* prophase; *M* metaphase; *A* anaphase; *T* telophase and cytokinesis; *R* nuclear restitution; *EM* andomitosis of angiosperm type; *ER* endoreduplication; *UR* endocycle with DNA underreplication; *A* DNA amplification (extra synthesis). (Modified from NAGL 1976a)

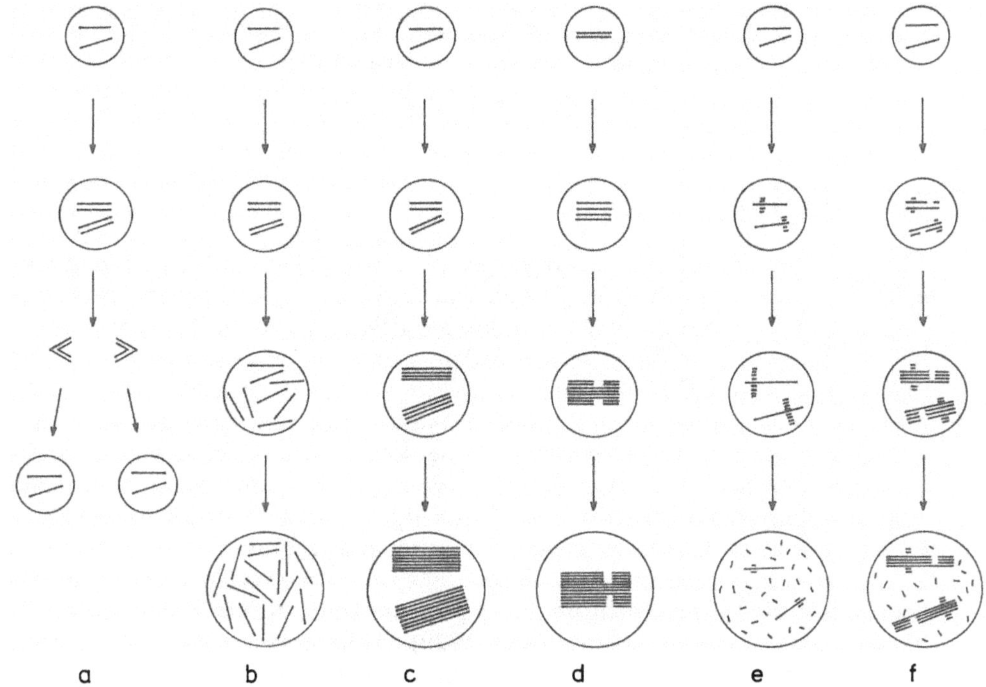

Fig. 2 a–f. The result of various cell cycles occurring in higher organisms. (Amended from NAGL 1978a). Each *long bar* indicates one haploid genome, each *short bar* a certain gene (or DNA sequence): **a** Mitotic cycle: the result is two nuclei which are genetically identical to the ancestor nucleus. **b** Endomitotic cycle: the result is a polyploid nucleus; the genomes are duplicated during each cycle. **c** Endoreduplication or polytenization cycle: although the same number of genomes are present as in the nuclei shown under **b**, the endochromosomes do not become separated, because no condensation phase occurs. **d** Underreplication cycle: in this case underreplication of a certain DNA sequence is shown in an endoreduplication cycle. **e** Amplification cycle: a certain gene or regulatory sequence is repeatedly additionally replicated. The amplified DNA becomes detached from the chromosome and is later degraded. **f** Combined endoreduplication–underreplication–amplification cycle, as suggested to occur, e.g., in the embryo-suspensor of *Phaseolus coccineus* (LIMA-DE-FARIA et al. 1975)

mechanism. Restitution, however, may take place as early as prophase. This shows especially clearly the continual transition of the mitotic cell cycle to an endo-cycle. Moreover, it demonstrates the independency in regulation of chromosome condensation and decondensation, break-down and reconstruction of the nuclear envelope, and formation and function of the spindle apparatus. Nuclear restitution also takes place after poisoning of the spindle by the tubulin-binding agent, colchicine.

The irregular shape, a typical characteristic of restitution nuclei, is the result of close attachment of the nuclear envelope around the randomly distributed chromosomes, and the minimal motion of the chromatin domains after reconstruction of the interphase state. Such nuclei, particularly dumb-bell-shaped anaphase restitution nuclei, have been often misinterpreted as amitotic figures.

Nuclear restitution is the common mechanism of somatic polyploidization in mammals (reviewed by BRODSKII and URYVAEVA 1977). In mouse liver, all or nearly all cells are polyploid up to 16 C (DIGERNES and BOLUND 1979).

In angiosperms, as in insects, endo-cycles predominate.

2.3 Endo-Cycles

The term "endo-cycle" was introduced by NAGL (1976a, 1978a) to designate a DNA replication cycle within the nuclear envelope and without spindle formation, i.e., the endomitotic and the endoreduplication cycle. Both cycles irrevocably lead to endopolyploidy and are evidently under strict genetic control. Endomitosis (GEITLER 1939) and endoreduplication (LEVAN and HAUSCHKA 1953) differ from each other in the following manner. In an endoreduplication cycle no mitosis-like structural changes can be seen in the nucleus, and G and S phases follow each other. On the other hand, as the term suggests, structural changes comparable with those seen in mitosis occur in a nucleus during an endomitotic cycle. Chromosome condensation and separation during endomitosis is, however, visible in certain insect taxa only. In angiosperms a structural change is visible if heterochromatin is present. There occurs a stage called "Z phase" (*Zerstäubungsstadium;* HEITZ 1929) or dispersion stage, during which the heterochromatin undergoes decondensation (Fig. 3). The Z stage can also be observed in the mitotic cycle, immediately before onset of prophasic chromosome coiling. This dispersion stage is the only structural feature by which an endomitotic cycle can be recognized in plants (GEITLER 1941, TSCHERMAK-WOESS and HASITSCHKA 1953, NAGL 1968, 1972a, TSCHERMAK-WOESS 1971). After considering the differences between the insect-type and angiosperm-type of endomitosis, one can interpret the endomitotic cell cycle in angiosperms as a cycle in which mitosis was curtailed by the suppression of events later than Z phase (NAGL 1970a, 1972a).

Because Z phase cannot be seen in those species which lack sufficient masses of heterochromatin, many authors refer all kinds of endo-cycles in plants to DNA endoreduplication (e.g., D'AMATO 1964, LIBBENGA and TORREY 1973; for discussion of the pro and contra see also NAGL 1977, 1978a). Moreover, a controversy exists as to whether the Z phase coincides with the late S period (time of asynchronous replication of hetero-

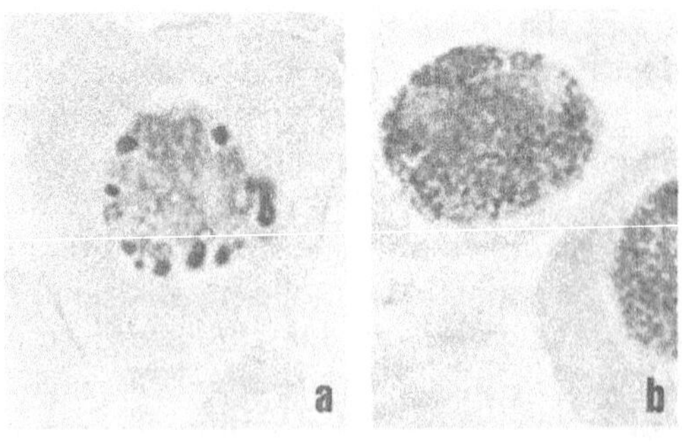

Fig. 3. Nuclei of *Allium carinatum* in respectively G_1 (**a**) and Z phase (**b**) of an endomitotic cycle. (×1,200)

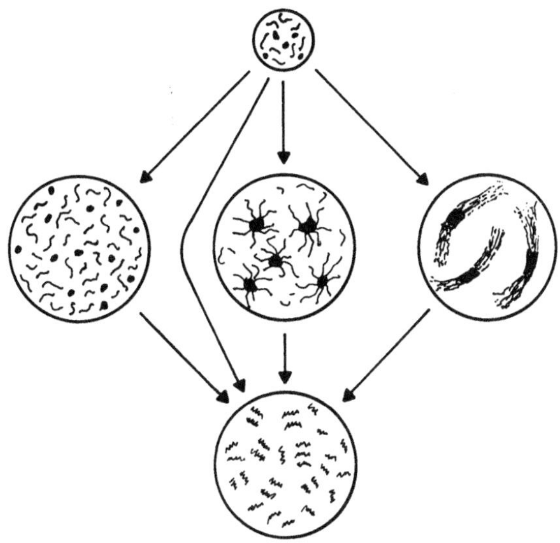

Fig. 4. Development of chromatin structures, specific for endopolyploid nuclei, from the species-specific structure (note that the proportion of chromatin remains the same if the total genome is endoreduplicated). The structure may be unchanged in the endopolyploid nucleus, or the chromocenters may form endochromocenters by fusion the heterochromatic regions of the sister chromosomes, or both the eu- and heterochromatin may be unseparated, thus forming polytene or giant chromosomes. In some instances, the endochromosomes coil up in a prophasic manner. This inhibits the formation of endochromocenters and polytene chromosomes, and disintegrates already existing ones. (Amended from NAGL 1978a)

chromatin; see BARLOW 1976, 1977, NAGL 1977, 1978a, for discussion). A strong argument against this view may be seen in the failure of Z phase in the endoreduplication cycle, during which DNA replication evidently occurs in condensed chromatin (e.g., in bands of giant chromosomes).

Nuclei which have passed through endo-cycles are called endopolyploid. Their structure may either be similar to the diploid ancestor nuclei, or different (Fig. 4). As the sister chromatids do not undergo any coiling in an endoreduplication cycle, they normally remain close together, thus forming bundles of sister chromatids (or endochromosomes). The endoreduplication cycle (or polytenization cycle), therefore, consequently leads to polytene nuclei. Z phase in the plant endomitotic cycle causes some loosening of the chromatid bundles, so that banded polytene (or giant) chromosomes are only visible after extreme condensation of the bands due to experimental inactivation of RNA synthesis (Fig. 5; NAGL 1969, 1970b).

3 Differential DNA Replication

During phylogenesis cell cycles evolved in which curtailment already occurred during the DNA replication period (S phase). Such cell cycles result in incomplete, or differential, DNA replication, and disproportionate increase of the nuclear DNA content (for reviews see NAGL 1976a, 1978a, 1979a, BUIATTI 1977). In several plants and animals, polyploidization takes place only in the euchromatin, while the heterochromatin (representing the site of highly reiterated DNA) is not at all or less often, replicated. Examples of heterochromatin underreplication are the salivary gland chromosomes of *Drosophila* (and many

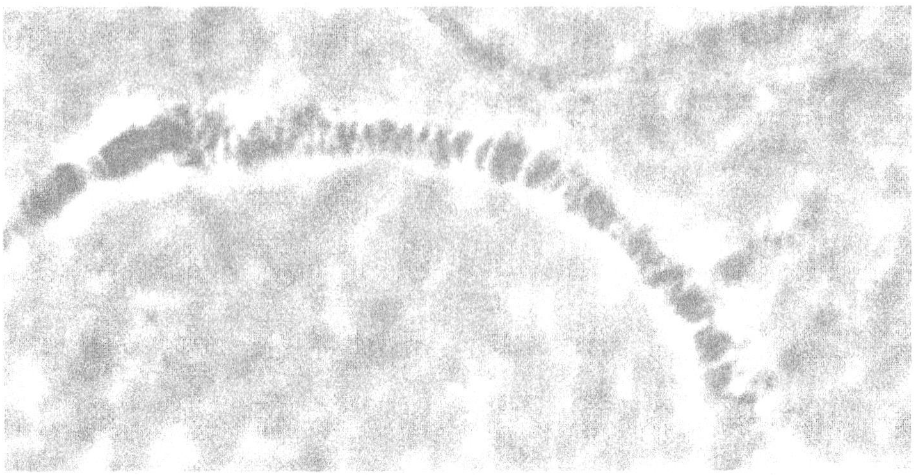

Fig. 5. Polytene chromosome in a suspensor giant cell in *Phaseolus vulgaris* (the banded appearance was induced by chilling the plants, which have developed seed pods, see NAGL 1969, 1970b). (Phase contrast, ×2,500)

other tissues of *Drosophila*) and various other insects, the embryo suspensor of *Tropaeolum* (NAGL 1976a, 1978b), the epicotyl cells of *Pisum* (VAN OOSTVELDT and VAN PARIJS 1972, 1976, BROEKAERT et al. 1979), leaves and fruits of *Cucumis* species (PERSON et al. 1974), and probably adult phase shoots of the ivy, *Hedera helix* (KESSLER and RECHES 1977, SCHÄFFNER and NAGL 1979). Differential DNA replication may include under-replication of the one, and extra replication of another portion of the genome. This was suggested to occur in the polytene nuclei of the *Phaseolus* suspensor (LIMA-DE-FARIA et al. 1975, FORINO et al. 1979).

At the end of the evolutionary strategem of curtailed cell cycles one can place DNA amplification, i.e., the replication of only one gene or regulatory DNA sequence. Most common is extra replication of the ribosomal genes in oocytes.

However, while meiotic DNA amplification is well established, the cases of somatic DNA amplification are somewhat unclear, due to methodical limitations (for a critical review see NAGL 1979a). Nevertheless, there is increasing evidence for a role of DNA amplification in cytodifferentiation of animals (e.g., STROM and DORFMAN 1976, SHARMA 1977, STROM et al. 1978, SCHMALENBERGER and NAGL 1979) and plants. For instance, the differentiation of several cell types in the orchid *Cymbidium* includes amplification of an AT-rich DNA fraction, located within the heterochromatin (certain regions of some chromocenters: Fig. 6; NAGL 1972b, NAGL and RÜCKER 1976, SCHWEIZER and NAGL 1976). DNA amplification seems to be a frequent event in tissue cultures, where it is clearly related with differentiation steps induced by phytohormone treatment (PARENTI et al. 1973, SACRISTAN and DOBRIGKEIT 1973, SCHÄFER et al. 1978, reviewed by BUIATTI 1977). Floral induction may also include the extra

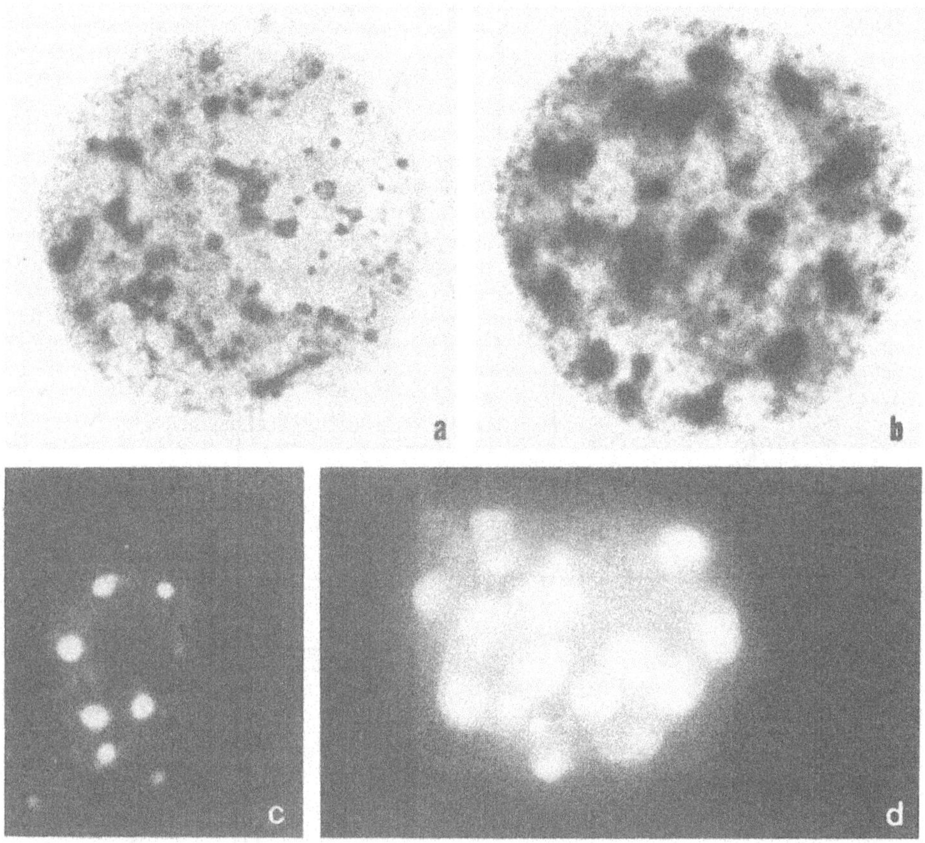

Fig. 6 a–d. Nuclei from protocorms of the orchid *Cymbidium* which have either undergone endo-cycles alone, or endo-cycles and additional transitory heterochromatin amplification. **a, b** Giemsa-stained (C-banding method) endopolyploid nuclei, in **b** with amplified heterochromatin (×1,100, from SCHWEIZER and NAGL 1976). **c, d** A similar pair of nuclei, stained with the AT-specific fluorochrome quinacrine. (×1,000; courtesy of D. SCHWEIZER, Vienna)

synthesis of a "floral DNA" which acts as switch from the vegetative to the generative pattern of gene expression of growth (WARDELL and SKOOG 1973, WARDELL 1976, FRÖLICH and NAGL 1979). Further examples of somatic DNA amplification in plants are under study in several laboratories and include salt-induced cell growth (I. ESSIGMANN personal communication) and crystal cell differentiation (H.T. HORNER and A. KAUSCH personal communication). The dynamic nature of plant ribosomal RNA genes was discussed by INGLE (1979).

Figure 7 shows examples of differential DNA replication in plants as evidenced by different techniques.

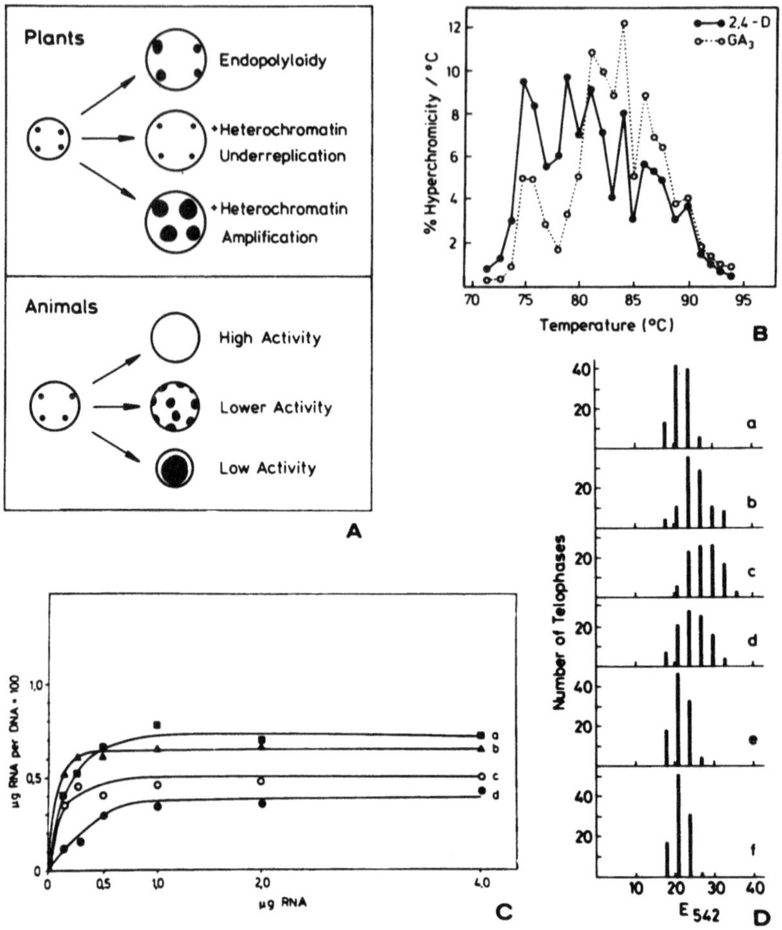

Fig. 7 A–D. Examples of differential DNA replication in plants as evidenced by various methods. **A** Structural evidence, as the proportion of condensed chromatin in cell nuclei of plants does not change during differentiation and function as it does in mammals. (NAGL 1979c); good examples are the *Tropaeolum* suspensor for underreplication, and *Cymbidium* protocorms for amplification. **B** Evidence obtained by derivative melting profiles of DNA; protocorm cultures were enriched with "normal" nuclei by GA_3, and with amplifying nuclei by 2,4-D, respectively, and the DNA isolated. The DNA from amplifying nuclei exhibits more AT-rich fractions than that of the "normal" nuclei. (NAGL and RÜCKER 1976). **C** Evidence obtained by saturation hybridization of ribosomal RNA to the DNA of various organs of *Phaseolus coccineus: a* shoot; *b* integument; *c* root; *d* suspensor. (Redrawn from LIMA-DE-FARIA et al. 1975). **D** Evidence obtained by scanning cytophotometry of Feulgen-stained nuclei in *Rhoeo discolor. a–f* are subsequent stages of flower formation from the tiny bud stage to the opening flower. Note the transitory increase of the DNA content (shift of histograms to the right) during early differentiation. (FRÖLICH and NAGL 1979). It must be emphasized that several methods have to be employed in order to obtain good evidence, because all possible techniques include sources of possible errors. (Reviewed by NAGL 1979a). There are more methods available than shown here, such as density gradient centrifugation, DNA reassociation (cot curves), in situ hybridization and other autoradiographic techniques

4 Physiological Significance of Somatic DNA Increase

As shown in Fig. 2, any cell cycle including DNA endoreduplication and/or differential DNA replication leads to excess DNA in addition to the diploid genome. The genetic effects of such cell cycles can be estimated from the DNA composition in the mature cell. Giant cells with polyploid and polytene nuclei dispose of a great potentiality of gene activity, because they are supplied with multiple templates (up to 500,000 per nucleus). It is well known from the developmental-specific and cell-specific puffing patterns in dipters that qualitatively and quantitatively different patterns of gene activity are induced in such nuclei, and that this allows a wide range of specialization in cell function. Therefore, high degrees of somatic polyploidy are found in cells and organs which rapidly differentiate and which reach high functional activity, but which often produce only a few gene products in abundance. Well-known examples are the salivary gland and silk gland cells of insect larvae, the nurse cells of oocytes in meroistic insect ovaries, suspensor haustoria of plant embryos, antipodal cells, endosperm nuclei, giant cells in the trophoblast of rodents, and other specialized, highly active, but normally short-lived cells. Lower degrees of polyploidy are, however, found in many cells of nearly all organisms, normally in a tissue-specific pattern of certain degrees of polyploidy (reviewed by NAGL 1978a). In angiosperms, estimates of the proportion of polyploid cells have been made in *Beta vulgaris* (BUTTERFASS 1966) and *Scilla decidua* (FRISCH and NAGL 1979), and a percentage of 70%–80% could be established.

The advantage of multiple copies of each gene may be only one side of endopolyploidy. The other side can be seen in the fact that somatic polyploidy is an important tool in tissue and organ modeling, as the size of a cell increases proportionately with the level of polyploidy. Actually, cell size depends on nuclear DNA content. This has been shown to be the case in species differing in the basic 2C DNA content, as well as in endopolyploid cells of an in individual (e.g., BENNETT 1973, PRICE et al. 1973, CAVALIER-SMITH 1978, NAGL 1978a). A clear example of tissue modeling by variation of cellular volume by polyploidy is, for instance, the pattern of differently sized scales in a butterfly wing being a consequence of the pattern of different degrees of polyploidy in the nuclei of the scale cells (KÜHN 1965; further examples are listed in NAGL 1978a).

The functional significance of DNA underreplication may be found in the economy of cell metabolism. The process occurs in cells with short life span, in cells which are end-differentiated, and which will never divide again. Therefore, DNA sequences and chromatin domains, which may be involved in mitotic spindle nucleation and in meiotic functions, are omitted from replication. However, one cannot yet exclude an effect on the regulatory machinery of the cell itself.

The functional significance of gene amplification is evident. Only the one gene is replicated that is actually needed by a cell (e.g., the ribosomal genes in oocytes, the dihydrofolate reductase gene in methotrexate-resistent murine cells: ALT et al. 1978). In other cases, a regulatory sequence (e.g., a repetitive sequence, or heterochromatin) may be amplifiied (NAGL 1972b, STROM et al.

1978). Such an amplified noncoding sequence may influence the development of an organ by altering the growth parameters (nucleotypic effect), or by acting as a trigger (switch) of inactivation and activation of certain gene batteries. However, details of such a mechanism are not yet understood.

5 An Evolutionary Perspective

DNA endoreduplication and differential replication can be understood as evolutionary strategies (NAGL 1976b). The close relationship between phylogenesis and ontogenesis was already suggested by HAECKEL (1866), and more recently, in terms of molecular biology, by FLICKINGER (1975), GALAU et al. (1976); reviewed by AYALA (1976). The ideas culminated in the "DNA optimization model" (NAGL 1978a), in which the ontogenetic changes in nuclear DNA are represented as predefined programs, depending on the phylogenetic state of a given species. In other words, the large-scale variation in somatic DNA amounts per cell may be explained on the basis of the large-scale variation of 2C values. This idea was proposed for the following reasons. Endopolyploidy is a phenomenon found very frequently in plants and several animal taxa. Critical examination of the literature reveals characteristic clusters of high frequency in certain systematic groups, and very stable tissue-specific patterns of the degrees of endopolyploidy (see the tables in NAGL 1978a). It has been suggested that the trend to develop endopolyploid tissues depends on a low 2C DNA content and thus on the phylogenetic state of the genome (NAGL 1976a). The DNA amount required by a species in differentiated cells may be achieved either during phylogenesis (by tandem increase of genome length), or during ontogenesis (by lateral multiplication of chromosome in endo-cycles; Fig. 8). Similar suggestions have been made by CAVALIER-SMITH (1978) and BRASCH (1980). The relationships assumed might by covered in some species due to

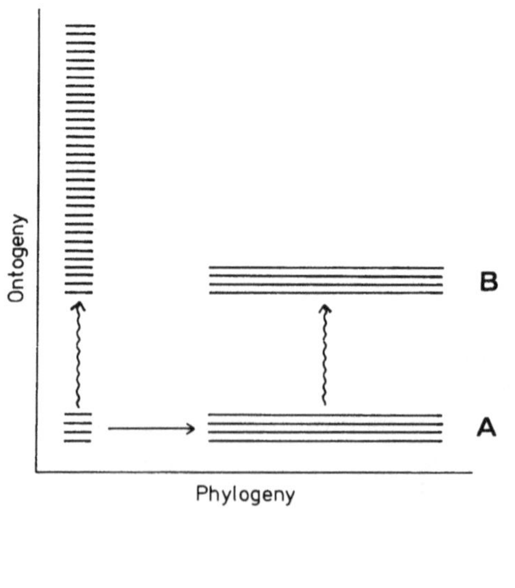

Fig. 8. Diagram to illustrate the possible strategem which underlies the occurrence and degree of endopolyploidy (exemplified as polyteny). *A* The *bars* represent the genome (chromosome complement) of a species with little DNA (at *left*) and a species whose genome increased during evolution by tandem addition of DNA. *B* It is assumed that highly active and specialized cells of eukaryotes need a defined mass of DNA of both types, coding and regulatory sequences. In the DNA-poor species this DNA amount is synthesized by endo-cycles; the DNA-rich species does not need "lateral" multiplication of the genome. (Amended from NAGL 1978a)

changes in connection with speciation and other cladogenetic events. It can be hoped that the biological significance of DNA endoreplication and differential replication will be more understandable, when genome organization in eukaryotes (i.e., repetitive DNA clusters, interspersion patterns, noncoding single-copy sequences, introns, palindromes) is better understood.

6 Conclusions

a) Most differentiated cells of nearly all angiosperms (and other plant and animal taxa) own more DNA than corresponds to the diploid genome.
b) DNA increase is achieved by curtailment of the mitotic cell cycle at a certain stage of mitosis or prior to mitosis leading to somatic polyploidy. In some cases, differential DNA replication, i.e., the selective replication of either a large, but incomplete part of the genome, or a small portion (often only one gene), leads to disproportionate DNA increase.
c) There is increasing evidence that somatic polyploidization and differential DNA replication play some role in cell differentiation, cell function, and general morphogenesis (organ and tissue modeling).
d) DNA endoreduplication and differential DNA replication may be understood as evolutionary trends toward economic synthesis of that macromolecule which may have selective advantage. In addition, the somatic DNA variation seems to be a process which is dependent on the phylogenetic state of the species' genome.
e) The species-specific diploid DNA content and its somatic variation can be understood as an optimization process with respect to various nucleo-cytoplasmic interactions. Future experiments should be designed to study the role of the cell's position in various gradients, the thermodynamic and electrodynamic constraints, and the photon-storage capacity of DNA in this optimization process.

References

Alt FW, Kellems RE, Bertino JR, Schminke RT (1978) Selective multiplication of dihydrofolate reductase genes in methotrexate-resistant variants of cultured murine cells. J Biol Chem 253:1357–1370
Ayala FJ (ed) (1976) Molecular evolution. Sinauer, Sunderland
Barlow PW (1976) The relationship of the dispersion phase of chromocentric nuclei in the mitotic cycle to DNA synthesis. Protoplasma 90:381–391
Barlow PW (1977) Changes in chromatin structure during the mitotic cycle. Protoplasma 91:207–211
Bennett MD (1973) Nuclear characters in plants. Brookhaven Symp Biol 25:344–366
Brasch K (1980) Endopolyploidy in vertebrate liver: an evolutinary perspective. Cell Biol Int Rep 4:217–226
Brodskii V Ya, Uryvaeva IV (1977) Cell polyploidy: its relation to growth and differentiation. Academic Press, London New York

Broekaert D, Van Oostveldt P, Van Parijs R (1979) Differential DNA replication in *Pisum sativum* L seedlings at the onset of germination. Biochem Physiol Pflanz 174:629–640

Buiatti M (1977) Gene amplification and tissue cultures. In: Reinert J, Bajaj S (eds) Plant cell, tissue and organ culture. Springer, Berlin Heidelberg New York, pp 358–374

Butterfass T (1966) Neue Aspekte der Polyploidieforschung und -züchtung. Mitt Max-Planck-Ges 1:47–58

Carniel K (1952) Das Verhalten der Kerne im Tapetum der Angiospermen mit besonderer Berücksichtigung von Endomitosen und sogenannten Endomitosen. Oesterr Bot Z 99:318–362

Carniel K (1963) Das Antherentapetum. Oesterr Bot Z 110:145–176

Cavalier-Smith T (1978) Nuclear volume control by nucleosceletal DNA, selection for cell volume and nucleosceletal DNA, selection for cell volume and cell growth rate, and the solution of the DNA C-value paradox. J Cell Sci 34:247–278

D'Amato F (1964) Endopolyploidy as a factor in plant tissue development. Caryologia 17:41–52

D'Amato F (1977) Nuclear cytology in relation to development. Cambridge Univ Pres, New York

Digernes V, Bolund L (1979) The ploidy classes of adult mouse liver cells. Virchows Arch B 32:1–10

Flickinger R (1975) Relation of an evolutionary mechanism to differentiation. Differentiation 3:155–159

Forino LMC, Tagliasacchi AM, Avanzi S (1979) Different structure of polytene chromosomes of *Phaseolus coccineus* suspensors during early embryogenesis. Protoplasma 101:231–246

Frisch B, Nagl W (1979) Patterns of endopolyploidy and 2C nuclear DNA content (Feulgen) in *Scilla* (Liliaceae). Plant Syst Evol 131:261–276

Frölich E, Nagl W (1979) Transitory increase in chromosomal DNA (Feulgen) during floral differentiation in *Rhoeo discolor*. Cell Differ 8:11–18

Galau GA, Chamberlin ME, Hough BR, Britten RJ, Davidson EH (1976) Evolution of repetitive and non-repetitive DNA. In: Ayala JF (ed) Molecular evolution. Sinauer, Sunderland, pp 200–224

Geitler L (1939) Die Entstehung der polyploiden Somakerne der Heteropteren durch Chromosomenteilung ohne Kernteilung. Chromosoma 1:1–22

Geitler L (1941) Das Wachstum des Zellkerns in tierischen und pflanzlichen Geweben. Ergeb Biol 18:1–54

Greilhuber J (1974) Uncommon karyological features in the tapetum of some Pedicularia-ceae (Scrophulariaceae). Caryologia 27:169–182

Haeckel E (1866) Generelle Morphologie der Organismen. Reimer, Berlin

Heitz E (1929) Heterochromatin, Chromozentren, Chromomeren. Ber Dtsch Bot Ges 47:274–284

Hesse M (1970) Cytologische Untersuchungen an Nematodengallen. Oesterr Bot Z 118:517–541

Ingle J (1979) Plant ribosomal RNA genes – a dynamic system? In: Rubinstein I, Phillips RL, Green CE, Gegenbach BG (eds) Molecular biology of plants. Academic Press, London New York, pp 139–164

Kafatos FC (1979) The insect chorion: programmed expression of specific genes during differentiation. In: Meints RH, Davies E (eds) Control mechanisms in development. Plenum, New York, pp 103–121

Kessler B, Reches S (1977) Structural and functional changes of chromosomal DNA during aging and phase change in plants. Chromosomes Today 6:237–246

Kühn A (1965) Vorlesungen über Entwicklungsphysiologie. Springer, Berlin, Heidelberg, New York

Levan A, Hauschka TS (1953) Endomitotic reduplication mechanisms in ascites tumors of the mouse. J Natl Cancer Inst 14:1–43

Libbenga KR, Torrey JG (1973) Hormone-induced endoreduplication prior to mitosis in cultured pea root cortex cells. Am J Bot 60:293–299

Lima-de-Faria A, Pero R, Avanzi S, Durante M, Ståhle U, D'Amato F, Granström H (1975) Relation between rRNA genes and the DNA satellites of *Phaseolus coccineus*. Hereditas 79:5–20

Maheshwari P (1963) Recent advances in the embryology of angiosperms. Univ Delhi Press, Delhi

Nagl W (1962) Über Endopolyploidie, Restitutionskernbildung und Kernstrukturen im Suspensor von Angiospermen und einer Gymnosperme. Oesterr Bot Z 109:431–494

Nagl W (1968) Der mitotische und endomitotische Kernzyklus bei *Allium carinatum*. I. Struktur, Volumen und DNS-Gehalt der Kerne. Oesterr Bot Z 115:322–353

Nagl W (1969) Banded polytene chromosomes in the legume *Phaseolus vulgaris*. Nature (London) 221:70–71

Nagl W (1970a) The mitotic and endomitotic nuclear cycle in *Allium carinatum*. II. Relations between DNA replication and chromatin structure. Caryologia 23:71–78

Nagl W (1970b) Temperature-dependent functional structures in the polytene chromosomes of *Phaseolus*, with special reference to the nucleolus organizers. J Cell Sci 6:87–107

Nagl W (1972a) Molecular and structural aspects of the endomitotic chromosome cycle in angiosperms. Chromosomes Today 3:17–23

Nagl W (1972b) Evidence of DNA amplification in the orchid *Cymbidium* in vitro. Cytobios 5:145–154

Nagl W (1976a) Zellkern und Zellzyklen. Ulmer, Stuttgart

Nagl W (1976b) Endopolyploidy and polyteny understood as evolutionary strategies. Nature (London) 261:614–615

Nagl W (1977) Nuclear structures during cell cycles. In: Rost TL, Gifford EM Jr (eds) Mechanisms and control of cell division. Dowden, Hutchison and Ross, Stroudsburg, pp 147–193

Nagl W (1978a) Endopolyploidy and polyteny in differentiation and evolution. Elsevier North Holland, Amsterdam

Nagl W (1978b) Biochemical evidence for differential DNA replication during polyploidization of the embryo-suspensor of *Tropaeolum*. Cell Chrom Newslett 1:14–17

Nagl W (1979a) Differential DNA replication in plants: a critical review. Z Pflanzenphysiol 95:283–314

Nagl W (1979b) Search for the molecular basis of diversification in phylogenesis and ontogenesis. Plant Syst Evol Suppl 2:3–25

Nagl W (1979c) Condensed interphase chromatin in plant and animal cell nuclei: fundamental differences. Plant Syst Evol Suppl 2:247–260

Nagl W, Fuhrmann B (in press) The embryo-suspensor and trophoblast as model systems of differentiation. Springer, Berlin Heidelberg New York

Nagl W, Rücker W (1976) Effects of phytohormones on thermal denaturation profiles of *Cymbidium* DNA: Indication of differential DNA replication. Nucl Acids Res 3:2033–2039

Parenti R, Guillé E, Grisvard J, Durante M, Giorgi L, Buiatti M (1973) Transient DNA satellite in dedifferentiating pith tissue. Nature New Biol 246:237–239

Pearson GG, Timmis JN, Ingle J (1974) The differential replication of DNA during development. Chromosoma 45:281–294

Price HJ, Sparrow AH, Nauman AF (1973) Correlations between nuclear volume and DNA content in meristematic cells of herbaceous angiosperms. Experientia 29:1028–2029

Sacristan MD, Dobrigkeit I (1973) Variable content of a (GC) rich satellite DNA in tumorous and normal cultures of *Crepis capillaris*. Z Naturforsch 28c:564–567

Schäfer A, Blaschke JR, Neumann KH (1978) On DNA metabolism of carrot tissue cultures. Planta 139:97–101

Schäffner K-H, Nagl W (1979) Differential DNA replication involved in transition from juvenile to adult phase in *Hedera helix*. Plant Syst Evol Suppl 2:105–110

Schmalenberger B, Nagl W (1979) Different DNA content, chromatin condensation, and transcription activity in retina cell nuclei of the Guinea pig. Plant Syst Evol Suppl 2:119–125

Schweizer D, Nagl W (1976) Heterochromatin diversity in *Cymbidium*, and its relationship to differential DNA replication. Exp Cell Res 98:411–423

Sharma AK (1977) Evidence of dynamism as a basis of chromosomal control of genetic reactions. Nucleus 20:4–10

Strom CM, Dorfman A (1976) Amplification of moderately repetitive DNA sequences during chick cartilage differentiation. Proc Natl Acad Sci USA 73:3428–3432

Strom CM, Moscona M, Dorfman A (1978) Amplification of DNA sequences during chicken cartilage and neural retina differentiation. Proc Natl Acad Sci USA 75:4451–4454

Tschermak-Woess E (1971) Endomitose. In: Altmann H-W (ed) Handbuch der Allgemeinen Pathologie Vol 2/2/1. Springer, Berlin Heidelberg New York, pp 569–625

Tschermak-Woess E, Hasitschka G (1953) Veränderungen der Kernstruktur während der Endomitose, rhythmisches Kernwachstum und verschiedenes Heterochromatin bei Angiospermen. Chromosoma 5:574–614

Van Oostveldt P, Van Parijs R (1972) Reiterated nucleotide sequences in DNA of dividing and elongating plant cells. Arch Int Physiol Biochim 80:416–418

Van Oostveldt P, Van Parijs R (1976) Underreplication of repetitive DNA in polyploid cells of *Pisum sativum*. Exp Cell Res 98:210–221

Wardell WL (1976) Floral activity in solutions of DNA extracted from tobacco stems. Plant Physiol 57:855–861

Wardell WL, Skoog F (1973) Flower formation in excised tobacco stem segments. III. DNA content in stem tissue of vegetative and flowering tobacco plants. Plant Physiol 52:215–220

Wright BE (1979) Causality in biological systems. Trends Biochem Sci 4:N 110–N 111

5 RNA Polymerase and Regulation of Transcription

R. WOLLGIEHN

1 Introduction

The DNA dependent RNA polymerase (EC 2.7.7.6) responsible for the synthesis of RNA catalyzes the formation of phosphodiester bonds using the four ribonucleoside triphosphates as substrate and DNA as template in the presence of a divalent metal ion. The RNA polymerase is a key enzyme implicated in the first step of gene expression and the control of transcription is of central importance during differentiation processes.

In order to understand the regulation of transcription, it is not only necessary to analyze the structure and function of the RNA-synthesizing enzymes, but also, in eukaryotes, to understand the structure and function of chromatin. Many data have been accumulated on the regulation of transcription in prokaryotes, but the specific mechanisms involved in the regulation of transcription in eukaryotic cells are widely unknown. However, during the last few years great progress has been made in analyzing the structure and function of eukaryotic nuclear RNA polymerases and chromatin (cf. NAGL, Chap. 1, this Vol.). In eukaryotes the situation is further complicated by the presence of specific mitochondrial and plastid RNA polymerases. Therefore the differential transcription of nuclear and organellar genes is an additional aspect of the control of differentiation and development of the eukaryotic cell.

Several reviews have appeared on RNA polymerases from prokaryotes (BURGESS 1971; BAUTZ 1972, CHAMBERLIN 1974, 1976, ZILLIG 1976) and eukaryotes, i.e., animal systems (JACOB 1973, CHAMBON 1975, BISWAS et al. 1975, ROEDER 1976), lower eukaryotes (HAGER et al. 1976, SENTENAC et al. 1976), and plants (DUDA 1976, BECKER 1979). This review will attempt to summarize our current knowledge of the plant RNA polymerases and the mechanisms for regulation of transcription. A brief review of the findings with prokaryotic and eukaryotic RNA polymerases will be given to facilitate comparisons with the plant enzymes.

2 RNA Polymerases from Prokaryotes and Eukaryotes

2.1 Structure and Function

It is relatively easy to solubilize and purify RNA polymerases from bacteria and blue-green algae, and great progress has been made in the elucidation of the structure and function of prokaryotic RNA polymerase. The bacterial transcription cycle can be subdivided into several steps: promoter *recognition* and *binding* of the enzyme to the template; RNA chain *initiation* which yields a tight complex of RNA polymerase with the promoter and in which the enzyme catalyzes the coupling of ATP or GTP to a second ribonucleoside triphosphate to generate a dinucleoside tetraphosphate of the structure pppPupX; *elongation,* in which successive monophosphate residues from substrate nucleoside triphosphates are added at the 3′-OH terminus of the initial dinucleoside tetraphosphate; chain *termination* with release of the newly formed RNA chain and of the enzyme from the DNA (for details see CHAMBERLIN 1976, KRAKOW et al. 1976).

The *Escherichia coli* enzyme consists of two α subunits of mol. wts. 40,000 each, one β subunit (150,000), one β' subunit (160,000) and one dissociable sigma (σ) subunit (86,000) (CHAMERLIN 1976). Two active forms of the enzyme are known. The holoenzyme ($\alpha_2\beta\beta'\sigma$) functions in initiation of transcription, the core enzyme lacks the σ subunit and functions in elongation. The σ subunit is necessary for specific initiation of RNA synthesis, i.e., for promoter selection. This subunit seems also to interact with regulatory proteins modifying the rate of transcription. After initiation, sigma is released, leaving the core enzyme responsible for further elongation of the RNA chain. Sigma is now free to function again as the initiation factor in a new transcription cycle. The β subunit contains at least part of the catalytic center of the enzyme, it participates in binding of substrate, it contributes to the interaction of the core enzyme with sigma and appears to be involved in translocation. The β' subunit appears to be involved in template binding of the enzyme. The α subunit is the major structural part of the enzyme. The function of this subunit is unknown, but seems to be involved in the specificity of transcription, possibly by interaction of RNA polymerase with promoter regions (ZILLIG 1976). Different mechanisms seem to exist for selective termination of RNA chains, the direct termination of growing RNA chains by interacting of the RNA polymerase with terminator sites, and the rho-factor-dependent termination of growing RNA chains at terminator sites on the DNA, which requires the presence of this factor in the reaction (ROBERTS 1976).

In vitro studies of the transcription process in eukaryotic cells began with the discovery of an enzyme capable of synthesizing RNA from four nucleoside triphosphates as substrates in isolated rat liver nuclei (WEISS 1960). Solubilization of stable RNA polymerase from mammalian nuclei, free of template, was first obtained by RAMUSZ et al. (1965) and JACOB et al. (1968). Multiple forms of RNA polymerase were first detected by ROEDER and RUTTER (1969) from rat liver and *Xenopus laevis*. Three polymerases could be separated by chromatography of the solubilized enzyme preparations on DEAE-Sephadex (ROEDER and RUTTER 1969) or DEAE-cellulose (KEDINGER et al. 1970) with linear gradients of ammonium sulfate or potassium chloride and were designated RNA polymerases I, II, and III (or A, B, and C) in order of their elution from DEAE-Spehadex (ROEDER and RUTTER 1969, ROEDER 1976, BLATTI et al. 1970). The major characteristics of the three eukaryotic RNA polymerases are summarized in Table 1. They differ in ionic strength optima, relative responses to Mn^{2+}, sensitivity to the fungal toxin α-amanitin, localization within the nucleus, in the kind of RNA which they synthesize, and in their subunit composition.

Polymerase I is localized within the nucleolus (ROEDER and RUTTER 1969, JACOB et al. 1970), transcribes ribosomal DNA cistrons and is refractory to α-amanitin (LINDELL et al. 1970, JACOB et al. 1970). RNA polymerase II is local-

Table 1. Properties of eukaryotic nuclear RNA polymerases I, II and III. (Most data from ROEDER 1976)

Characteristics	Enzyme		
	I	II	III
Elution from DEAE-Sephadex Ammonium sulfate (M)	~0.1	~0.2	0.2–0.3
Catalytic properties			
Ammonium sulfate optima (M)	~0.05	~0.10	0.05–0.20
Mn^{2+}/Mg^{2+} activity ratio	1–2	5–10	2
α-Amanitin sensitivities (μg/ml for 50% inhibition)			
Animal and plant cells	Insensitive	0.01–0.05	10–25
Insects	Insensitive	0.03–0.06	Insensitive
Yeast	300–600	~1	Insensitive
Localization within the nucleus	Nucleolus	Nucleoplasm	Nucleoplasm
Product of transcription	Precursors of the two high molecular weight ribosomal RNA's, 5.5S rRNA	mRNA precursors	5S rRNA, tRNA precursors
Subunit composition in mouse plastocytoma cells (mol. wt. $\times 10^{-3}$)	195, 117 61[a], 49 29, 19	205[b], 140 41, 29 27, 22 19.5, 19 16.5	155, 138 89, 70 53, 49 41, 32[c] 29, 19
Subunit composition of plant polymerases see Table 3			

[a] Two forms of I were found (I_A, I_B). I_B lacks the subunit with 61 kd
[b] Three forms of II were found (II_O, II_A, II_B). M_r in the largest subunit of II_O has 240 kd, II_A 205 kd, II_B 170 kd
[c] Two forms of III were found (III_A,III_B). In III_B this subunit has 33 kd

ized within the nucleoplasm (ROEDER and RUTTER 1969, JACOB et al. 1970), synthesizes heterogenous RNA (i.e., in part precursors of the cytoplasmic polysomal messenger RNA (ZYLBER and PENMAN 1971) and is inhibited by 10^{-8} to 10^{-9} M α-amanitin (JACOB et al. 1970, LINDELL et al. 1970). RNA polymerase III is also a nucleoplasmic enzyme (ROEDER and RUTTER 1969, SCHWARZ et al. 1974), synthesizes ribosomal 5S RNA and transfer RNA (WEINMANN and ROEDER 1974), and is inhibited by 10^{-3} to 10^{-4} M α-amanitin in higher animals (SCHWARTZ et al. 1974, WEINMANN and ROEDER 1974, SEIFART and BENECKE 1975) but refractory in lower eukaryotes (yeast), where enzyme I is sensitive to high α-amanitin concentrations (HAGER et al. 1976).

In various enzyme preparations, subspecies of the three polymerases have been detected when the purified enzyme preparations were further analyzed by ion-exchange chromatography and by polyacrylamide gel electrophoresis under nondenaturing conditions. Generally, two electrophoretic forms of polymerase I, two or three forms of polymerase II and two distinct forms of polymerase III have been determined (for refs. see ROEDER 1976), their biological role has not yet been established (see also Sect. 4.2.2).

All eukaryotic RNA polymerases are multisubunit enzymes with complexities greater than the bacterial enzymes. The mol.wts. of the whole enzymes are in the range of 500,000. The subunit composition of the enzymes is strikingly similar in many eukaryotic systems with always two large and several smaller polypeptides (see CHAMBON 1975, ROEDER 1976, HAGER et al. 1976, SENTENAC et al. 1976), though the molecular weights of the subunits and stoichiometry of subunit composition evidently vary. As an example the results from analysis of mouse plasmacytoma cell polymerases are given in Table 1 (ROEDER 1976).

2.2 Regulation

We are far from understanding all details of the function of the RNA polymerase subunits, of the factors participating in transcription, and of the possibilities of transcription regulation. It is assumed that in prokaryotes, beside the well-known subunits, additional positive and negative control factors are also involved in specific transcription. Alterations in transcription specificity may be possible by means of mechanisms including modifications of the enzyme structure, de novo synthesis of new RNA polymerases (e.g., new enzymes coded for by bacteriophage or virus genomes; BAUTZ 1976), or synthesis of regulatory proteins interacting with the DNA template to alter utilization of promoter or terminator signals by RNA polymerases (GILBERT 1976, LOSICK and PERO 1976).

From much experimental data obtained particularly with animal systems and lower eukaryotes one can postulate several possibilities for the regulation of eukaryotic nuclear polymerase activity and selectivity (for details see ROEDER 1976). Under several physiological conditions (e.g., growth, hormonal response), altered levels of individual polymerases have been correlated with altered rates of synthesis of specific classes of RNA. Here the concentration of a specific enzyme is correlated only in a quantitative way with the rate of transcription of specific genes. In other cases (e.g., early embryonic development), changes in gene activity and intracellular RNA polymerase activity are not associated with changes in the amounts of the enzymes, therefore the activity or selectivity of the polymerases must be altered by other mechanisms, including enzyme modification resulting in quantitative changes in activity, alteration in the availability of specific genes, or the involvement of regulatory factors influencing the interaction between a specific enzyme and specific genes. One type of factor is the binding proteins for several hormones (e.g., steroid hormones). The hormone-receptor complex combines with the genome leading to modulation of nuclear RNA transcription. From several systems other protein factors have been isolated, which have stimulatory effects on in vitro RNA synthesis by purified RNA polymerases, but in most cases only unspecific, quantitative changes in polymerase activity were observed. Most of these protein factors seem to act primarily by stimulating chain elongation (LENTFER and LEZIUS 1972, SEIFART et al. 1973). Stimulating factors can either be bound to the polymerase (REVIE and DAHMUS 1979) or to the DNA template (LEGRAVEREND and GLAZER 1980). Two protein factors were isolated from calf thymus which dissociated from the rest of the polymerase II during the early stages of enzyme purification. SF_1 was identified as an initiation factor, and SF_2, as elongation factor (SPINDLER 1979). Using polymerase II from different animals, selective and accurate transcription initiation at the major late adenovirus promoter was shown to depend on the addition of a crude protein fraction obtained from KB cells. Without this initiation factor transcription starts randomly on both DNA strands (WEIL et al. 1979a). In a similar way specific initiation of transcription of cloned conalbumin and ovalbumin genes has been obtained in the presence of purified calf thymus RNA polymerase II and a crude protein fraction (WASYLIK et al. 1980). It was also shown that components in the cell-free extracts from human KB cells are active in directing the accurate transcription by RNA polymerase III (WEIL et al. 1979b). A factor which is responsible for the selective initiation of the 5S rRNA genes in *Xenopus* by RNA polymerase III was also detected (ENGELKE et al. 1980).

3 Plant RNA Polymerases

3.1 Nuclear RNA Polymerases

Very soon after methods for the isolation and fractionation of animal polymerases had been developed, the first reports on the isolation and characterization of RNA polymerases from nuclei and chloroplasts of maize leaves (STOUT and MANS 1967, BOTTOMLEY et al. 1971, STRAIN et al. 1971), wheat leaves (POLYA and JAGENDORF 1971a, b), and from coconut nuclei (MONDAL et al. 1970) appeared.

Since then, nuclear RNA polymerases have been isolated and characterized from a variety of plants. The materials most intensively investigated are soybean hypocotyls (HARDIN and CHERRY 1972, HORGEN and KEY 1973, RIZZO and CHERRY 1975, RIZZO et al. 1974, LIN et al. 1974, 1975, 1976, GUILFOYLE and HANSON 1973, 1974, GUILFOYLE and JENDRISAK 1978, GUILFOYLE and KEY 1977a, b, GUILFOYLE and MALCOLM 1980, GUILFOYLE et al. 1975, 1976, 1980, GURLEY et al. 1976, JENDRISAK and GUILFOYLE 1978), cauliflower inflorescences (SASAKI et al. 1974, 1976, 1978, FUKASAWA and MORI 1974, MIZUOCHI and FUKASAWA 1975, GUILFOYLE 1976, GOTO et al. 1978, JENDRISAK and GUILFOYLE 1978), wheat germs (POLYA and JAGENDORF 1971a, b, POLYA 1973, JENDRISAK and BECKER 1973, 1974, JENDRISAK and BURGESS 1975, 1977, JENDRISAK and GUILFOYLE 1978, JENDRISAK 1980, MAZUS and BRODNIEWICZ-PROBA 1976, PETRANYI et al. 1977, HODO and BLATTI 1977, TEISSERE et al. 1977, 1979, KIDD et al. 1979, HAHN and SERVOS 1980), maize (STOUT and MANS 1967, STRAIN et al. 1971, MULLINIX et al. 1973, JENDRISAK and GUILFOYLE 1978, KIDD et al. 1979), and parsley cell cultures (LINK and RICHTER 1975, 1977, LINK et al. 1978, KIDD et al. 1979, GROSSMANN and SEITZ 1979, GROSSMANN et al. 1980). The polymerases from pea seedlings (GLICKLICH et al. 1974, SASAKI et al. 1973, 1979a, b), sugar beet roots (DUNHAM and CHERRY 1973), lentil roots (TEISSERE et al. 1973), Jerusalem artichoke tubes (GORE and INGLE 1974), rye embryos (FABISZ-KIJOWSKA et al. 1975, JENDRISAK and GUILFOYLE 1978), coconut endosperms (MONDAL et al. 1970,

1972a, c), and the unicellular flagellate *Euglena gracilis* (BRANDT and WIESNER 1977, FALCHUK et al. 1976, 1977, 1978) have also been investigated.

Plant cells possess three nuclear RNA polymerase forms I, II, and III according to the generally accepted nomenclature, similar to the enzymes of animal nuclei (see Sect. 2). These three enzymes can be separated by DEAE-Sephadex chromatography using increasing salt gradient (ROEDER and RUTTER 1969). The three classes were also distinguished according to the inhibitory effect of α-amanitin, structural differences, and their function and localization within the nucleus.

3.1.1 Isolation and Separation

3.1.1.1 Extraction and Solubilization

Various methods have been employed for the extraction of RNA polymerases from plant sources The starting material is mostly whole plant tissue, in several cases chromatin (e.g., RIZZO et al. 1974, DUNHAM and CHERRY 1973, GORE and INGLE 1974, TEISSERE et al. 1973, GUILFOYLE and KEY 1977b, GUILFOYLE et al. 1976) and in a few cases nuclei or nucleoli, although intact nuclei are relatively difficult to prepare in large amounts (GUILFOYLE 1976, LIN et al. 1975, HAHN and SERVOS 1980). Isolation of chromatin under low salt conditions has been done by different modifications of the method of HUANG and BONNER (1962). The polymerases are then easily solubilized and separated from DNA in the presence of high salt conditions; but this method has to be used with caution. Chromatin prepared at pH 8 contains mainly polymerase I, most of the polymerase II activity is soluble and remains in the high speed supernatant (LIN et al. 1974, GURLEY et al. 1976, GUILFOYLE and KEY 1977b). If an isolation medium of pH 6 is used for chromatin isolation, a several-fold greater recovery of chromatin-bound polymerase II is attained (LIN et al 1974). Methodical details for the isolation of chromatin and the solubilization of RNA polymerase from chromatin have been described, e.g., for sugar beet tissue (DUNHAM and CHERRY 1973), Jerusalem artichoke tissue (GORE and INGLE 1974) and soybean hypocotyls (LIN et al. 1974, GUILFOYLE et al. 1976). Direct isolation and solubilization of RNA polymerase from different plant tissue are possible after homogenization of the tissues in suitable media. This is possible since a portion of the RNA polymerase activity is not tightly associated with nuclei or chromatin and can be readily extracted from the tissues. Details of the procedures are given in e.g., for wheat germ (JENDRISAK and BURGESS 1975, TEISSERE et al. 1977, JENDRISAK 1980, HODO and BLATTI 1977), soybean hypocotyls (GUILFOYLE and JENDRISAK 1978), cauliflower inflorescences (FUKASAWA and MORI 1974, GOTO et al. 1978), parsley cell cultures (LINK and RICHTER 1975), maize leaves (STRAIN et al. 1971, MULLINIX et al. 1973), rye embryos (FABISZ-KIJOWSKA et al. 1975) and *Euglena gracilis* (BRANDT and WIESNER 1977, FACHUK et al. 1976, 1977).

3.1.1.2 Separation of Polymerases I, II, and III

The purification and separation of the multiple forms of RNA polymerase require a multistep procedure. Chromatography on DEAE-Sephadex (ROEDER and RUTTER 1969) is the most effective means for the resolution of the three enzymes in a single chromatographic step. The enzymes are eluted from the column in the order I (0.1 M ammonium sulfate), II (0.2 M), III (0.3 M) (JENDRISAK 1980). However, the separation of enzymes II and III is not always satisfactory. DEAE cellulose chromatography is another method of polymerase separation, but enzyme III has the unique property of co-eluting with enzyme I at low salt on DEAE cellulose. These difficulties explain in part the failure of many investigations to detect RNA polymerase III. A combination of the two separation methods makes it possible to separate enzymes I plus III from II on DEAE cellulose and then I from III by rechromatography on DEAE-Sephadex (GUILFOYLE 1976).

Additional chromatographic steps and other purification procedures are necessary to obtain highly purified (chromatographic homogeneous) enzymes: ammonium sulfate precipitation, sucrose or glycerol gradient centrifugation, polyethyleneimine (Polyamin P) precipitation (JENDRISAK and BURGESS 1975, HODO and BLATTI 1977, GOTO et al. 1978, GUILFOYLE and KEY 1977b, JENDRISAK 1980), phase partition chromatography in polyethylen glycol/dextran (MULLINIX et al. 1973), chromatography on Sepharose 4B and 6B or on phosphocellulose (JENDRISAK and BECKER 1974, JENDRISAK and BURGESS 1975, GUILFOYLE 1976, GUILFOYLE and KEY 1977b, GUILFOYLE and JENDRISAK 1978, GOTO et al. 1978, JENDRISAK 1980), DNA-cellulose (MULLINIX et al. 1973), DNA-Agarose (LINK and RICHTER 1975, GUILFOYLE and JENDRISAK 1978), Agarose (GUILFOYLE 1976, GUILFOYLE and KEY 1977b), Heparin-Sepharose (TEISSERE et al. 1977, 1979, HAHN and SERVOS 1980, JENDRISAK 1980), and Aminobutyl-Sepharose (FABISZ-KIJOWSKA et al. 1975, TEISSERE et al. 1977, 1979).

Several representative procedures for the isolation and purification of the individual enzymes are briefly summarized in Table 2. For details of the procedures the reader is referred to some original papers: for polymerase I: GUILFOYLE et al. 1976, GUILFOYLE and KEY 1977b (soybean), GROSSMANN et al. 1980 (parsley cell cultures); for polymerase II: MULINIX et al. 1973 (maize leaves); JENDRISAK and BURGESS 1975 (wheat germ); LINK and RICHTER 1975, 1977 (parsley cell cultures); GOTO et al. 1978 (cauliflower); GUILFOYLE and KEY 1977b (soybean hypocotyls); for polymerase III: TEISSERE et al. 1977 (wheat embryos); SASAKI et al. 1978 (cauliflower).

The three plant RNA polymerases can also be separated simultaneously in a more or less purified form (MONDAL et al. 1972a, HORGEN and KEY 1973, TEISSERE et al. 1973, 1979, FUKASAWA and MORI 1974, FABISZ-KIJOWSKA et al. 1975, GUILFOYLE 1976, HAHN and SERVOS 1980, BRANDT and WIESNER 1977, JENDRISAK 1980). TEISSERE et al. (1979, and personal communication) and JENDRISAK (1980) have developed methods for the isolation of the three polymerases from wheat germ. TEISSERE's method involves solubilization in ammonium sulfate, ammonium sulfate fractionation, affinity chromatography on Heparin-Sepharose, hydrophobic chromatography on ω-aminobutyl-Sepharose which yields two peaks of activity A and B, concentration of the peaks on a Heparin-Sepharose column and on a Sepharose 4B column, chromatography on DEAE Sephadex A-25, which allows the separation of enzymes I and III from fraction A and enzyme II from B, and finally purification of enzyme I by chromatography on CM-Sephadex and denatured DNA-Sepharose. On the other hand, JENDRISAK's method starts with the polyethyleneimine precipitation of the poly-

Table 2. Procedures for purification of the individual nuclear RNA polymerases from plant sources

Reference	GUILFOYLE et al. (1976) GUILFOYLE and KEY (1977b)	GOTO et al. (1978)	TEISSERE et al. (1977)
Source	Soybean hypocotyl	Cauliflower inflorescences	Wheat embryos
Enzyme	I	II[a]	III
Crude extract preparation	Homogenization of hypocotyls Preparation of pH 8 chromatin Solubilization of RNA polymerase Centrif. at 50,000 rpm Amm. sulfate precip. of the supernatant Solubilization of the enzyme	Homogen. in mortar Filtration Centrifugation Precipitation of polymerase with polyethyleneimine Extraction of the enzyme with 0.5 M ammon. sulfate Precip. of polymer. with ammon. sulfate	Grinding with Waring blender Filtration Solubilization of enzyme by ammon. sulfate addition Centrifugation Precipitation of the enzyme with ammon. sulfate Solubilization of enzyme Centrifugation
Purification procedures	Agarose A-1.5 m gel filtration DEAE cellulose chromatography CM Sephadex chromatography Phosphocellulose chromatography Sucrose density gradient centrifugation	DEAE Sephadex A-25 chromatography Phosphocellulose chromatography Glycerol gradient centrifugation	1st. Heparin Sepharose chromatography DEAE Sephadex chromatography 2nd. Heparin Sepharose chromatography δ-Aminobutyl-Sepharose chromatography
Yield	14%	20%	–
Purification	300-fold from chromatin; 20,000-fold from total tissue protein	2,560-fold	60,000-fold relative to the crude extract
Specific activity	200–300 nmol UMP incorp./mg protein/ 30 min./28 °C, using denat. DNA as template	231 nmol UMP incorp./mg protein/ 10 min./30 °C	125 nmol UMP incorp./mg protein/ 30 min./35 °C, using denat. DNA
Analytical assay	Polyacrylamide gel electrophoresis	Polyacrylamide gel electrophoresis	Polyacrylamide gel electrophoresis

[a] This procedure is also suitable for preparation of polymerase III. (SASAKI et al. 1978)

merase from the crude extract, followed by ammonium sulfate precipitation, DEAE-Sepharose CL-6B chromatography, and Heparin-Agarose chromatography. The three polymerases were separated by DEAE-Sephadex chromatography and further purified by phosphocellulose chromatography (enzymes I and II) or on DEAE-cellulose (enzyme III). Using this method approximately 3, 30, and 2 mg of RNA polymerase I, II, and III, respectively, were isolated from 1 kg of wheat embryos each at a yield of 30%–50% (JENDRISAK 1980).

3.1.2 Subunit Structure

Analysis of the subunits using polyacrylamide gels under denaturing conditions requires that sufficient amounts of highly purified enzymes are available. Nevertheless, during the last few years the subunit composition of the three highly purified polymerases from several plants has been relatively well characterized. The results, some of them summarized in Table 3, show a great similarity in the plant enzymes to the cognate enzymes of other eukaryotes (for subunit composition of animal enzymes see Table 1, of yeast polymerases see HAGER et al. 1976, and SENTENAC et al. 1976). All polymerases contain two high mol. wt. polypeptides of about 120,000–220,000 associated with several smaller polypeptides.

Only few data are available on RNA polymerase I and the molecular weights of the subunits are not in good agreement (Table 3). They contain the typical two high molecular weight subunits and, with the exception of the *Euglena* enzyme, only a few smaller polypeptides. It remains uncertain whether all the subunits are resolved on the gels (GUILFOYLE 1976). The cauliflower enzyme IB contains additionally a third high molecular weight polypeptide (168,000), but GUILFOYLE (1976) suggested that this may be due to contamination by enzyme IA, since the purified enzyme I could be separated in two bands by electrophoresis under nondenaturing conditions.

More data have been accumulated on polymerase II, because higher plant tissues contain high levels of the soluble enzyme II which can be relatively easily purified to homogeneity by routine procedures. The similarities in the subunit composition (number, mol. wts. and molar ratios of the polypeptides) of enzyme II from different plants show a general subunit structure for all plant RNA polymerases II (Table 3). All enzymes have two high mol. wt. subunits of approximately 220,000 and 140,000 (polymerase A) or 180,000 and 140,000 (polymerase B), one or two subunits of about 40,000, one or two of 25,000–27,000 and several smaller polypeptides between 14,000 and 22,000 mol. wt. In parsley cell cultures, three different RNA polymerases II were found and classified as IIO, IIA, and IIB corresponding to the three types of polymerase II in animal cells (LINK et al. 1978, KIDD et al. 1979). Several reports have shown that polymerase IIA is found in dormant (ungerminated) embryonic plant tissues and polymerase IIB in actively growing (germinated) tissues (GUILFOYLE and KEY 1977a, b, LINK et al. 1978, JENDRISAK and GUILFOYLE 1978, GUILFOYLE and JENDRISAK 1978, KIDD et al. 1979, JENDRISAK 1980, GUILFOYLE et al. 1980). The occurrence of proteolytic activity in metabolically active tissue (soybean hypocotyls, cauliflower inflorescences) but not in quiescent tissues

Table 3. Subunit structures of RNA polymerases I, II, and III from several plants

References	4, 5	6	3	2	1	6, 7	6, 7	6, 7	8, 9, 10	6, 7	8	11	6	12
Source	Soybean hypo-cotyl	Wheat germ	Cauliflo-wer inflo-rescenses	Parsley cell cultures	*Euglena gracilis*	Soybean hypo-cotyl	Wheat germ	Cauliflo-wer inflo-rescences	Parsley cell cultures	Maize meso-cotyl	Maize leaves	Pea seed-lings	Wheat germ	Wheat germ
Polymerase	I	I	I B	I	I	II A B	II A B	II	II A B C	II	II	II	III	III
Subunit molecular weights $\times 10^{-3}$	183	200	185	185		215, 180	220, 180	180	220, 200, 180	185	180			
			168		162						160	170		
					148								150	155
	136	125	135	130		138	140	140	140	145		140	130	132
													100	91
			70	70	74									70
											66			66
											64			
	50		56								51		55	53
	46		45	47	46									
						42	42 + 40		43		43	43		
	40							40		40	41			
		38		37	36								38	37
	33				34									33
				30										31
	28		28		28						28		28	28
						27	27 + 25			27 + 26				
									26			26		26
								25	25				25	
		24			24									
						22		22		22	22	22,5		
							21					21,5		
		20					20			20			20	
						19		19				18,5		
		17,8			18	17,6	17,8	17,5		17,8		17,5	17,8	
		17				17	17	17		17,5			17	
						16,2	16,3	16,2		16,3				
						16,1	16		16	16,1				16
						14	14	14		14				
					8									

References: 1. FALCHUK et al. (1977), 2. GROSSMANN et al. (1980), 3. GUILFOYLE (1976), 4. GUILFOYLE et al. (1976), 5. GUILFOYLE and KEY (1977b), 6. JENDRISAK (1980), 7. JENDRISAK and GUILFOYLE (1978) (see also HAHN and SERVOS (1980), 8. KIDD et al. (1979), 9. LINK and RICHTER (1975), 10. LINK et al. (1978), 11. SASAKI et al. (1979a), 12. TEISSERE et al (1977) (soybean embryonic axis, wheat germs) converting the 220,000 to the 180,000 mol. wt. polypeptide in vitro (GUILFOYLE and JENDRISAK 1978) and the similarities of the peptide maps of the two largest subunits of type II enzymes within one species (soybean, wheat, maize, parsley) (GUILFOYLE and JENDRISAK 1978, LINK et al. 1978, JENDRISAK 1980) favor the possibility that the 220,000 subunit is the precursor of the 180,000 subunit (see also Sect. 4.2.2).

In addition to the structural similarities of the polymerases II from different plants it was shown that they also have common antigenic determinants. Antibodies directed against wheat polymerases II react with the polymerases II from other monocotyledonous (rye) and also from dicotyledonous species (cauliflower, soybean), but they do not cross-react with yeast and calf thymus polymerase II and *E. coli* RNA polymerase. This shows that the basic structure is common to and is therefore highly conserved in both monocotyledonous and dicotyledonous plants (HODO and BLATTI 1977, JENDRISAK and GUILFOYLE 1978). On the other hand, the similarities between polymerases II of different plants are limited by the fact that two-dimensional tryptic maps of the large subnits of polymerase II from wheat, parsley, and maize have more differences than similarities (KIDD et al. 1979). Therefore, the extent of similarity among corresponding polypeptides of different plants remains uncertain.

Polymerase III subunit structure has been investigated only from wheat germ, but shows the same complexity as polymerase II (Table 3). The two subunits of 70,000 and 66,000 mol.wt. in Teissere's analysis could be contaminants, for their stoichiometry is variable during the steps of purification (TEISSERE et al. 1977).

From the limited number of data one can conclude that the enzymes within one class from different plants have more structural similarities than the polymerases I, II, and III from a single cell type, although the general structures (two large and several small subunits) are similar. This is not surprising as the classes of enzymes have distinct properties and functions (see Sect. 3.1.4). The functions of individual subunits of the polymerases and the significance of their heterogeneity are not understood. It may be possible that shared components participate in common functions and other subunits are responsible for the transcriptional properties of a particular enzyme or for independent regulation of the various enzymes by interaction with other cellular components.

3.1.3 General Properties

The requirement for RNA synthesis by plant polymerases in vitro are the same as those for other eukaryotic and prokaryotic enzymes: DNA, four ribonucleoside triphosphates, metal ions (Mg^{2+} or Mn^{2+}). The enzyme activity is also dependent on sulfhydryl components and the ionic strength of the medium. In general, all plant RNA polymerases appear to function optimally at pH of approximately 8.0 and an incubation temperature in the range of 25 to 30 °C.

3.1.3.1 Ion Strength and Metal Ion Optima

As with other eukaryotic polymerases, the three purified plant nuclear enzymes differ with respect to their divalent cation requirements, ionic strength and relative activities with different templates. It is, however, not possible to give a unified view of the catalytic properties of the three enzymes, because the data, determined in different laboratories with different plants, vary considerably. Catalytic properties have only rarely been determined under the same conditions in different laboratories, and this information should not be considered as invariant characteristics of the enzymes. In particular, ionic strength optima and the relative stimulating effects of Mg^{2+} and Mn^{2+} are difficult to compare because of their interdependence. On the other hand, both properties are highly dependent on the nature of the DNA (natural or synthetic), the state of the DNA (native or denaturated) and its concentration in the reaction. These problems have been discussed in more detail in DUDA's review (1976).

Despite these difficulties some generalization can be made. Enzyme I prefers a low ionic strength, enzyme II a higher ionic strength (JENDRISAK and BECKER 1973, GLICKLICH et al. 1974, RIZZO et al. 1974, GUILFOYLE 1976, GUILFOYLE et al. 1976, SASAKI et al. 1978, TEISSERE et al. 1979, GROSSMANN et al. 1980). The activation curve of polymerase III shows a broad plateau (GUILFOYLE 1976) or a biphasic profile (SASAKI et al. 1978) which was also found for other eukaryotic polymerases III (ROEDER 1976).

All three enzymes are activated by Mn^{2+} and Mg^{2+}, but the ion concentration for optimal activity is lower for Mn^{2+} (1–2 mM) than for Mg^{2+} (usually 5–10 mM). Whereas the concentration profile for Mn^{2+} is sharp, that obtained for Mg^{2+} is very broad (TEISSERE et al. 1973, FABISZ-KIJOWSKA et al. 1975, LINK and RICHTER 1975, SASAKI et al. 1978, GUILFOYLE and JENDRISAK 1978, GROSSMANN et al. 1980, HAHN and SERVOS 1980). Generally the polymerases are more active with Mn^{2+} than with Mg^{2+} when assayed at their respective metal ion optima and in some cases the Mn^{2+}/Mg^{2+} activity ratios were found to be greater for enzyme II than for enzyme I (TEISSERE et al. 1973, GLICKLICH et al. 1974, LINK and RICHTER 1975, GUILFOYLE et al. 1976).

Although essential for polymerase activity, Mn^{2+} and Mg^{2+} are not bound to the polymerase II from wheat germs or to polymerase I and II from *Euglena gracilis* (less than 0.1 gram-atom per mole of enzyme). On the other hand, these polymerases contain tightly bound zinc, the wheat enzyme 7 gram-atoms, the *Euglena* enzymes 2.2 gram-atoms zinc per mole enzyme (FALCHUK et al. 1976, 1977, PETRANYI et al. 1977). The function of the zinc is not presently known (see also Sect. 5.2.2), but it seems to be a catalytically essential component of the enzyme.

3.1.3.2 Inhibitors

Activity of plant RNA polymerases is completely inhibited by actinomycin D by the same mechanism as transcription is inhibited in all other organisms. Class I, II, and III polymerases from plants, as from other eukaryotes, are distinguished by their sensitivities to α-amanitin (Fig. 1). Polymerase I is resistent to α-amanitin (STRAIN et al. 1971, JENDRISAK and BECKER 1973, TEISSERE et al. 1973, GLICKLICH et al. 1974, RIZZO et al. 1974, FABISZ-KIJOWSKA et al. 1975, GUILFOYLE 1976, FALCHUK et al. 1977, GUILFOYLE and KEY 1977b). Polymerase II is sensitive to the drug and 50% inhibited at 0.012–0.05 μg/ml or nearly completely at 4–20 μg/ml α-amanitin (JENDRISAK and BECKER 1973, GLICKLICH et al. 1974, FABISZ-KIJOWSKA et al. 1975, LINK and RICHTER 1975, FALCHUK

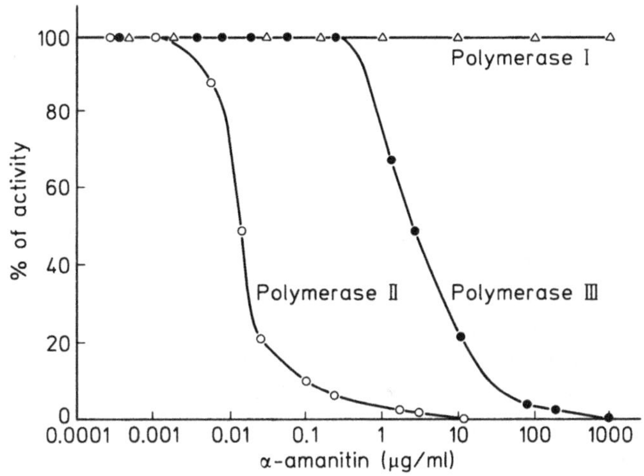

Fig. 1. Sensitivity of RNA polymerases from rye embryos to α-amanitin. The three enzymes were tested with calf thymus DNA as template. (FABISZ-KIJOWSKA et al. 1975)

et al. 1976, GUILFOYLE 1976, JENDRISAK and GUILFOYLE 1978, SASAKI et al. 1978, JENDRISAK 1980). Polymerase III is sensitive to high concentrations of α-amanitin. The cauliflower enzyme is inhibited by 50% at 200 μg/ml (SASAKI et al. 1978) or 1–2 mg/ml (GUILFOYLE 1976), the rye and wheat enzymes by 50% at about 50 μg/ml and completely inhibited at 500 μg/ml (FABISZ-KIJOWSKA et al. 1975, JENDRISAK 1980). Rifampicin or rifamycin SV have no inhibitory effect on plant nuclear polymerases (TEISSERE et al. 1973, FABISZ-KIJOWSKA et al. 1975, GUILFOYLE et al. 1976, GUILFOYLE and KEY 1977b, STRAIN et al. 1971).

3.1.3.3 Template Requirements

RNA polymerases transcribe homologous and heterologous DNA but also synthetic polynucleotide templates. Some enzymes have shown a preference for a homologous DNA template rather than for the commonly used calf thymus DNA (STRAIN et al. 1971, POLYA and JAGENDORF 1971a, b, MONDAL et al. 1972a, FUKASAWA and MORI 1974). In general, RNA polymerase I tends to prefer native to denatured DNA, whereas enzyme II shows higher activity with denatured DNA as template. Since a crude fraction of maize polymerases prefers native DNA and the purified enzyme II denatured DNA, it was concluded that this experiment might reflect the loss of a specific factor necessary for reading native DNA (MULLINIX et al. 1973). All three enzymes also show remarkable activities with synthetic polynucleotide templates. For example, poly(dC), poly(dT) and poly(dA,dT) appear to be much more efficient templates for the polymerases from different plants than native or denatured DNA (SASAKI et al. 1976, 1978, GUILFOYLE 1976, GUILFOYLE and KEY 1977b).

All the experiments with highly active natural DNA or synthetic polynucleotide templates appear to represent in large part nonspecific initiation by the polymerases. Native DNA preparations usually contain single-strand breaks (nicks) as well as denatured regions, and initiation of RNA chains occurs artifactually at nicks, gaps, or loose ends. The question whether the purified polymerases can initiate RNA synthesis can only be answered using native and intact virus DNA as templates, which are relatively easy to prepare. It was first shown by GUILFOYLE (1976) that cauliflower polymerase I and II are able to transcribe cauliflower mosaic virus (CaMV) DNA, a circular double-stranded DNA molecule of 4.5×16^6 mol.wt. at a low rate. The polymerases purified from wheat embryos transcribe this DNA (with Mn^{2+} as divalent cation) with an efficiency of 50%, 20% and 90% for polymerase I, II, and III in comparison with commercial (i.e., not intact) calf thymus DNA. With Mg^{2+} only enzymes II and III were able to transcribe (30% and 100%), whereas enzyme I was unable to initiate RNA synthesis (TEISSERE et al. 1979). The ability of isolated enzyme III to transcribe fully intact duplex DNA seems to be an intrinsic property (HOSSENLOPP et al. 1975, SKLAR et al. 1976, JAEHNING et al. 1976).

These results show that purified plant RNA polymerases (especially enzyme III) are principally able to initiate transcription and to transcribe native DNA. However, it remains uncertain whether they are also able to transcribe intact native DNA's in vitro, since the natural templates of these enzymes are never naked double-stranded DNA's, but rather very complex chromatin structures. Alterations of the DNA structure within the chromatin or the binding of some regulatory factors could be of fundamental importance for the function of the polymerases under natural conditions (for further discussion see Sect. 2.2.3, and CHAMBON 1975, ROEDER 1976).

3.1.4 Localization and Function

The localization of the plant RNA polymerases I, II, and III within the nucleus corresponds to the localization of the enzymes in other eukaryotic cells. In several cases enzyme activity (specially III) was also found in cytoplasmic fractions; however, the possibility of nuclear leakage was not ruled out.

All three enzymes could be solubilized from chromatin and isolated nuclei (MONDAL et al. 1972a, TEISSERE et al. 1973, GORE and INGLE 1974, LIN et al. 1974, RIZZO et al. 1974, GUILFOYLE 1976, GUILFOYLE and KEY 1977b). Isolated nucleoli contain only polymerase I (LIN et al. 1975, GUILFOYLE and KEY 1977b), which shows that the two other enzymes are localized in the nucleoplasm.

As shown in Table 1, one can assume that the three polymerases have the same function in all eukaryotes (see reviews CHAMBON 1975, ROEDER 1976) in synthesizing ribosomal RNA (enzyme I), mRNA (II) and low molecular weight 5S RNA and tRNA (III). It was first shown by MONDAL et al. (1972c) with hybridization experiments that isolated coconut nuclear polymerases I and II (CII and CI according to their nomenclature), after addition of regulating protein factors, appear to synthesize ribosomal RNA or nonribosomal RNA, respectively. In isolated nuclei or chromatin, rRNA is the main transcription product of the α-amanitin-insensitive polymerase I (GURLEY et al. 1976, GUILFOYLE and KEY 1977b, LUTHE and QUATRANO 1980), whereas the RNA whose synthesis is inhibited by α-amanitin (polymerase II) sediments at 6–10 S and may represent nonribosomal RNA (LUTHE and QUATRANO 1980). Nucleoli isolated from mung bean synthesize mainly ribosomal RNA, since hybridization to mung bean DNA of the RNA synthesized by nucleoli in vitro is reduced by 60–70% with unlabeled rRNA as a competitor (GRIERSON et al. 1980). Poly(A) RNA could not be detected in the in vitro product of maize nuclei (SLATER et al. 1978). Thus, there is no direct proof that plant polymerase II synthesizes precursors to messenger RNA and that polymerase III synthesizes precursors of tRNA or 5S RNA.

3.2 Chloroplast RNA Polymerase

Chloroplasts are genetically semiautonomous organelles; they synthesize DNA, RNA's, and proteins. It was shown that isolated chloroplasts are able to synthesize distinct classes of RNA (WOLLGIEHN and MUNSCHE 1972, CARRITT and EISENSTADT 1973, HARTLEY and ELLIS 1973, BOHNERT et al. 1977, WOLLGIEHN and PARTHIER 1979). The genes coding for ribosomal RNA's (23S, 16S, 5S), all species of transfer RNA, and several messenger RNA's (for the large subunit of ribulose 1,5-bisphosphate carboxylase, the P-32,000 membrane protein and the α and β subunits of the ATP synthetase complex) were localized within the circular chloroplast DNA of different origins (for references see BOHNERT et al., Chap. 14, this Vol.).

3.2.1 Isolation

Although marked progress has made during the last years, our knowledge about the DNA-dependent RNA polymerase from chloroplasts is still incomplete. The enzyme was shown to be firmly bound with DNA to the thylakoid membranes (for refs. see WOLLGIEHN and PARTHIER 1980). The solubilization efficiency depends on many factors including species specificity; however, removal of magnesium seems to be the most important one (BOTTOMLEY et al. 1971). Recently, a transcription complex was isolated from *Euglena* (HALLICK et al. 1976, SCHIEMANN et al. 1977) spinach (BRIAT et al. 1979) and *Chlamydomonas* (DRON et al. 1979) chloroplasts by means of 1% Triton X-100 in the absence of magnesium ions. This complex consists of DNA, RNA polymerase, and several unspecified proteins, and is highly active in RNA synthesis. Free plastid RNA polymerase, completely dependent on exogenous DNA was solubilized from maize with a Mg^{2+}-free, EDTA-containing medium (BOTTOMLEY et al. 1971, SMITH and BOGORAD 1974), from wheat by the use of high salt concentration (POLYA and JAGENDORF 1971a, b), from pea with Triton X-100 (JOUSSAUME 1973) or by simple osmotic shock in water containing mercaptoethanol (BENNETT and ELLIS 1973), from spinach by DNAse treatment at high ionic strength of the DNA–RNA polymerase complex (BRIAT and MACHE 1980) or by DNAase treatment of a high-speed supernatant of osmotically shocked chloroplasts (BRIAT et al. and LERBS et al., in preparation) and from *Euglena gracilis* (BRANDT and WIESSNER 1977). Further purification includes the same chromatographic and gradient centrifugation steps as used for the purification of nuclear polymerases (see Sect. 4.1.1).

With one exception, no evidence for multiple forms of chloroplast RNA polymerases has been provided. Only JOUSSAUME (1973) reported the presence of two RNA polymerases in pea chloroplasts, differing in localization (one in the stroma, the other membrane-bound), pH optima, DNA preference, and G+C content of the reaction product.

3.2.2 Properties of the Enzyme

Maximal activity of the solubilized cRNA polymerase requires DNA, the four nucleoside triphosphates, low ionic strength, pH between 7.8 and 8.5, and 10–20 mM Mg^{2+}. Manganese ions are much less active, in contrast to the effect of this cation on nuclear polymerases. The temperature optimum is in the range of 25–30 °C for wheat, pea, and *Euglena* enzymes, but 37–40 °C for the spinach enzyme (30 °C for the spinach DNA-RNA polymerase complex), and about 48 °C for the maize enzyme. Chloroplast RNA polymerase shows a preference for denatured DNA over native DNA (BOTTOMLEY et al. 1971, BENNETT and ELLIS 1973, BRIAT and MACHE 1980) and for homologous over heterologous DNA (BOTTOMLEY et al. 1971, JOUSSAUME 1973, POLYA and JAGENDORF 1971a, b, BRIAT and MACHE 1980).

The polymerase reaction of the solubilized enzyme is inhibited by actinomycin D and pyrophosphate, but not inhibited by α-amanitin or rifampicin, which is a selective inhibitor of prokaryotic RNA polymerase initiation reaction (POLYA and JAGENDORF 1971a, b, HALLICK et al. 1976, BRANDT and WIESSNER 1977, BRIAT et al. 1979, BRIAT and MACHE 1980). RNA synthesis in isolated chloroplasts is also refractory to rifampicin, whereas the drug reduces chloroplast RNA synthesis in intact cells of green algae, tobacco leaves, and *Acetabularia*, but no effect in vivo has been reported so far (for refs. see WOLLGIEHN and PARTHIER 1980). This controvery in the action of rifampicin in vivo obtained in various laboratories seems to be due to the different treatment conditions used, and perhaps also the species-specific uptake and transport peculiarities of the drug. Since a partial inhibition of enzyme activity was observed also with the solubilized crude plastid polymerase (BOTTOMLEY et al. 1971, BOGORAD et al. 1973) or with a reconstituted purified enzyme system (SURZYCKI and SHELLENBARGER 1976), it seems to be possible that the loss of initiation factors during the enzyme purification and consequently a loss in initiation specificity by the purified enzymes may be the reason for lack of rifampicin sensitivity.

3.2.3 Subunit Composition

The polypeptide subunit composition has been determined for chloroplast RNA polymerases from maize and spinach with different results. The maize enzyme has a molecular mass in excess of 500,000, consisting of polypeptides of mol. wts. (in thousands) of 180, 140, 120, 110, 100, 95, 85, 75, 70, 55, 42, 40, 38, 27 (SMITH and BOGORAD 1974, KIDD and BOGORAD 1980). The maize nuclear polymerase II also possesses subunits of 180,000, 140,000 and 40,000 mol. wts., but despite their similar molecular weights, the corresponding subunits are unrelated in primary structure, as shown by comparison of their proteolytic fragments (KIDD and BOGORAD 1979).

The RNA polymerase solubilized from the DNA-RNA polymerase complex from spinach chloroplasts was found to consist of five subunits with mol. wts.

of 69,000, 60,000, 58,000, 34,000 and 15,000. Additional subunits of 80,000, 50,000 and 40,000 mol. wts. were found to be present only in very small amounts. This enzyme preparation showed only little activity (BRIAT and MACHE 1980). With a newly developed method a highly active and stable polymerase was isolated from a high-speed supernatant of osmotically shocked spinach chloroplasts. This enzyme consists of seven subunits with mol. wts. in the region of 155,000, 105,000, 96,000, 75,000, 71,000, 39,000 and 26,000. Two of these subunits (75,000, 71,000) possess DNA binding activity (LERBS, BRIAT and MACHE, in preparation).

3.2.4 In Vitro Products

Free chloroplast RNA polymerase, as well as the transcriptionally acitve DNA-RNA polymerase complex, are able to initiate and elongate RNA molecules in vitro. The product of the free pea polymerase was determined to be heterogeneous in size, consisting of RNA species with molecular weights less than 500,000 (BENNETT and ELLIS 1973). The products of the DNA–RNA polymerase complex from spinach have mol. wts. between 0.07 and 2×10^6. Seventy five percent of this RNA is hybridizable with chloroplast DNA and 40% of these products are ribosomal RNA, showing that ribosomal DNA is preferentially transcribed in vitro (BRIAT et al. 1979). The in vitro product of the corresponding transcriptionally active complex from *Euglena,* as well as the RNA extracted from purified *Euglena* chloroplasts (i.e., the in vivo product), hybridize to 20–23% of the chloroplast DNA, which shows that the extent of transcription of chloroplast DNA is essentially the same as in vivo. Both types of RNA contain the same nucleotide sequences (HALLICK et al. 1976). The most abundant in vitro transcripts hybridize to restriction endonuclease fragments of chloroplast DNA coding for 23S, 16S, and 5S ribosomal RNA's. Non-rDNA sequences of chloroplast DNA are transcribed to a much lower level (RUSHLOW et al. 1980).

4 Regulation of Transcription

4.1 RNA Synthesis During Development

Many developmental changes in plants have been shown to be accompanied by changes in gene transcription, although control of gene expression in eukaryotic cells including plant cells is not necessarily at the transcriptional level (KAMALEY and GOLDBERG 1980, GOLDBERG 1980). RNA synthesis increases markedly in number of situations, including seed germination, hormonal responses, and photomorphogenesis.

4.1.1 Seed Germination

During seed germination metabolic processes are initiated immediately after the beginning of water uptake by the air-dry seeds. Simultaneously protein synthesis in the embryo also rises rapidly, mediated by preformed long-lived messenger RNA's transcribed already during seed formation (for review see BECKER 1979, PAYNE 1976, HECKER 1978, MÜNTZ 1982, BEWLEY 1982). Earlier reports have shown that at least in seeds of some plant species, RNA synthesis seems to be activated only after a lag phase of several hours of imbibition or germination (CHEN et al. 1968, CHEN and OSBORNE 1970; WALBOT 1971, HALLAM et al. 1972, SIELIWANOWICZ and CHMIELEWSKA 1973, BHAT and PADAYATTI 1974). According to more recent results obtained by methods using labeled precursors of a high specific radioactivity, synthesis of RNA is resumed immediately or shortly after exposure of the seed to favorable germination conditions. Many studies have been done on the sequence of synthesis of mRNA, rRNA, and tRNA during germination; the results vary widely according to the plants examined or even within a given species. Some authors have found a predominant or an exclusive synthesis of transfer RNA and ribosomal RNA during early germination, followed by messenger RNA synthesis (CHEN et al. 1971, BHAT and PADAYATTI 1975, CLAY et al. 1975), whereas other reports indicated an early transcription of heterologous nuclear RNA and messenger RNA, followed by ribosomal RNA and transfer RNA (DOBRZANSKA et al. 1973, REJMAN and BUCHOWITZ 1973, VAN DE WALLE et al. 1976). In contrast to these findings recent investigations have shown that at least in the seeds of wheat (DOSHCHANOW et al. 1975, SPIEGEL et al. 1975), rye (SEN et al. 1975, PAYNE 1977), *Agrostemma githago* (HECKER et al. 1977), radish (DELSENY et al. 1977) and *Vaccaria pyramidata* (HECKER and KÖHLER 1979) synthesis of mRNA, rRNA, and tRNA are initiated simultaneously immediately after the start of imbibition (Fig. 2). Less is known about the nature and function of the preformed and newly synthesized messenger RNA's during early germination.

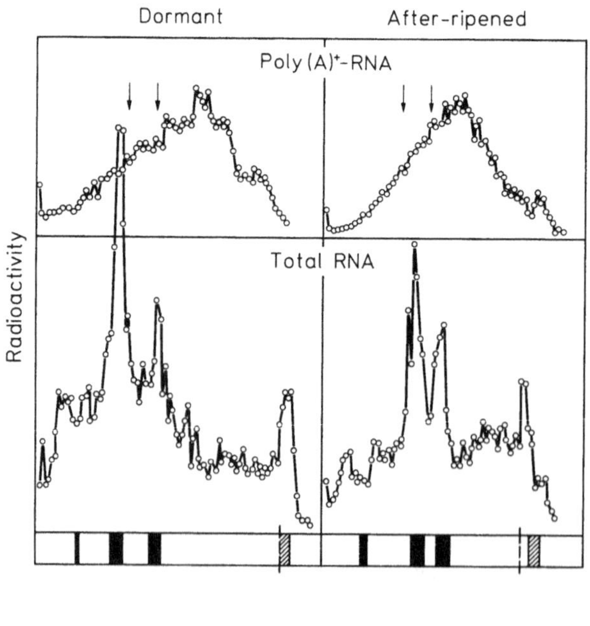

Fig. 2. Poly(A)-containing RNA and total RNA synthesized in dormant and after-ripened embryos of *Agrostemma githago*. Embryos were isolated from the dry seeds and incubated for 3 h with ^{3}H-uridine. Poly(A)-containing RNA was prepared by affinity chromatography on poly(U)-Sepharose. Total RNA and Poly(A)$^{+}$-RNA were analyzed by polyacrylamide gel electrophoresis. Before slicing the gels RNA fractions were stained with methylene blue: (DNA, 25 and 18S-rRNA, 5/4S-RNA (*from the left*). The scale for poly(A)$^{+}$-RNA is expanded (the positions of 25S and 18S-RNA are indicated by *arrows*). (HECKER et al. 1977)

CAERS et al. (1979) have found that polypeptides translated from mRNA extracted from both dry and 6 h imbibed wheat embryos were qualitatively identical and only minor quantitative differences were observed. On the other hand, CUMING et al. (1979), working also with wheat embryos, found gross quantitative differences between the translational capacities of poly(A)-RNA from dry and imbibing (2–24 h) embryos, the latter directing a greater proportion of high molecular weight polypeptides than did poly(A)$^+$-RNA from dry embryos. During germination of maize some new proteins were synthesized which were not translated from mRNA extracted from nongerminated embryos (VAN DE WALLE et al. 1979). In dry castor bean seeds (ROBERTS and LORD 1979) and unimbibed cucumber cotyledons (WEIR et al. 1978) only low levels of preformed mRNA can be detected, but during the first 4 days of germination the translational capacity of the mRNA isolated from the endosperm cells or cotyledons increases dramatically. Analysis of the in vitro translation product from cucumber mRNA revealed a changing pattern of labeled polypeptides during germination. The increase of the specific messengers for isocitrate lyase and malate synthase could be deduced from the results from immunoprecipitation of the polypeptides. This observation correlates with the in vivo increase of the activities of the two enzymes during the first days of germination (WEIR et al. 1980).

4.1.2 Hormonal Response

Another field of intensive research is the regulation of transcription by plant hormones. It has been shown that in many different plants and plant tissues auxins, gibberellin, and cytokinins are able to stimulate synthesis of ribosomal RNA, transfer RNA, and messenger RNA, and also to induce synthesis of several enzyme proteins. In some cases differential effects on the individual RNA species were observed. These results will not be discussed here since they have been summarized in several review articles (KEY and VANDERHOEF 1973, KULAEVA 1973, HALL 1973, JACOBSON 1977). Here it will only be mentioned that cell-free translation techniques were used to demonstrate that hormones are able to increase the level of specific mRNA's. It was first shown by VERMA et al. (1975) that the auxin-promoted synthesis of cellulase in pea cotyledons is preceded by an increase in the level of the cellulase mRNA. In response to gibberellic acid, the aleurone layers of barley synthesize the enzyme α-amylase (Fig. 3). The enhanced level of the amylase messenger after gibberellic acid treatment shows that the de novo synthesis of α-amylase is the result of a control at the level of transcription (HIGGINS et al. 1976, MUTHUKRISHNAN et al. 1979, HO 1980). From the data available it is not justified to decide between primary and secondary effects of the hormones on transcription.

4.1.3 Photomorphogenesis

Illumination of dark-grown seedlings or certain algae (*Euglena gracilis*) initiates a complex sequence of changes, whereby etioplasts or proplastids develop into mature chloroplasts. In etiolated seedlings the nucleus synthesizes all types of cytoplasmic RNA, but also within the developing etioplasts rRNA, tRNA, and

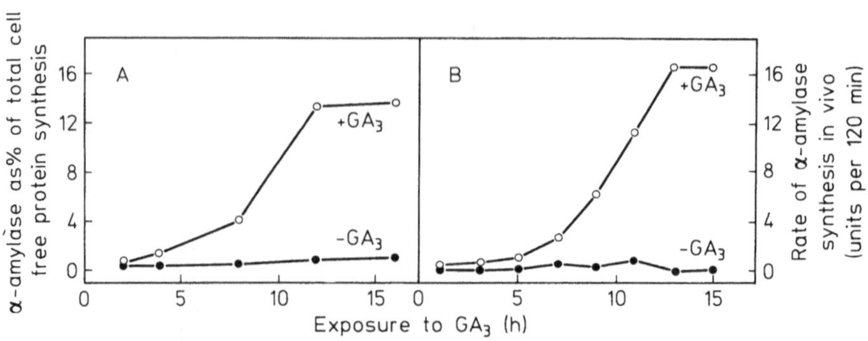

Fig. 3A, B. The increase with time in the level of translatable messenger RNA for α-amylase (**A**) and the increase in the rate of synthesis of the enzyme in vivo (B) in response to GA_3 treatment of barley aleurone layers. α-Amylase synthesized in the cell-free system from wheat embryos was estimated from autoradiographs of gels. (HIGGINS et al. 1976)

mRNA are transcribed. On the other hand, the proplastids of bleached algae contain only very small amounts of RNA. The mechanism of the light-induced conversion of etiolated plastids to chloroplasts requires not only synthesis of proteins, but also of several RNA species. Therefore synthesis of RNA is one of the early events in chloroplast differentiation. Illumination increases not only synthesis of chloroplast RNA but also the transcriptional activity of the nucleus, resulting in an increase in cytoplasmic RNA content. We have recently reviewed the details of these processes elsewhere (WOLLGIEHN and PARTHIER 1980) and therefore we will mention only briefly some results demonstrating the induction of specific mRNA's under the influence of light.

In several plant species, after illumination quantitative and qualitative changes occur in the patterns of mRNA transcribed as well in the nucleus (VERDIER et al. 1973, VERDIER 1975, GRIERSON and COVEY 1975, CURTIS and RAWSON 1977, APEL and KLOPPSTECH 1978) as in the chloroplast (RAWSON and BOERMA 1976, 1979, CHELM and HALLICK 1976, HAFF and BOGORAD 1976, BEDBROOK et al. 1978, CHELM et al. 1979, VERDIER 1979a, b). For *Euglena* chloroplast differentiation, a temporal program for RNA transcription from defined regions of the chloroplast genome was described (CHELM et al. 1979). Some direct evidence has been provided that light-stimulated individual mRNA's are transcribed on nuclear DNA or plastid DNA. Thus synthesis of at least two plastid DNA-coded mRNA's is controlled by light. In *Spirodela* (EDELMAN and REISFELD 1978, 1980), maize (BEDBROOK et al. 1978) and spinach chloroplasts (GRIERSON 1979) an RNA of 0.5×10^6 mol. wt. was accumulated during illumination representing the messenger for a thylakoid membrane protein called P-32,000 (Fig. 4). In the same way the synthesis of the chloroplast-coded mRNA for the large subunit of ribulose 1,5-bisphosphate carboxylase in *Chlamydomonas* (HOWELL 1978) and cucumber (WALDEN and LEAVER 1981) was also stimulated by light as well as the synthesis of several nuclear DNA-coded, individual mRNA's: the mRNA for phenylalanin ammonia lyase (Fig. 5), flavanone synthease, and UDP-apiose synthase (enzymes of flavonoid biosynthesis, SCHRÖDER et al. 1979, HAHLBROCK et al. 1980), and for two chloroplast proteins, the small

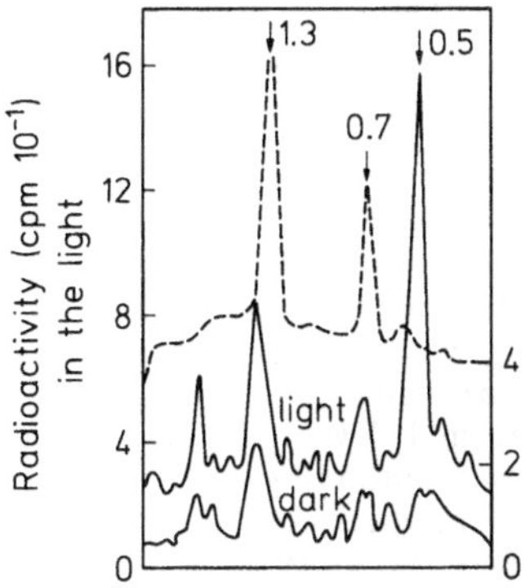

4

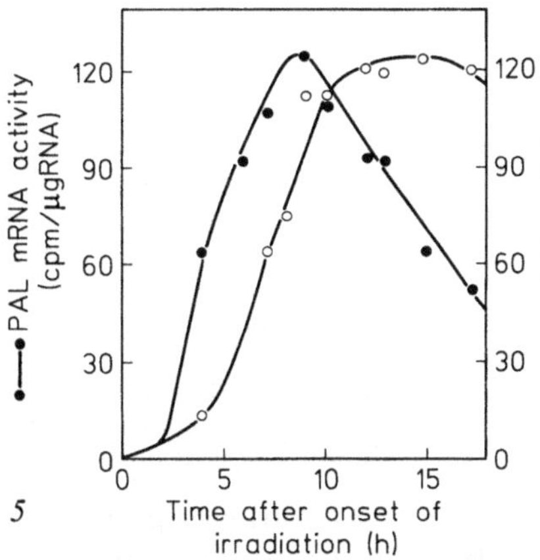

5

Fig. 4. The effect of light on the rate of synthesis of 0.5×10^6 mol. wt. RNA, the messenger for the P-32,000 chloroplast membrane protein, in *Spirodela* plants. Dark-grown plants and plants 3.5 h after transfer to light were labeled with [5-^{3}H]uridine for 45 min. Whole cell RNA was fractionated on polyacrylamide gels. (After ROSNER et al. 1977)

Fig. 5. Changes in phenylalanine ammonialyase (PAL) and PAL-mRNA activities in continuously irradiated cell suspension cultures from parsley. Crude extracts from the cells were used for measuring PAL activity. The corresponding mRNA activity was determined with RNA preparations from the same cell using the reticulocyte lysate assay. The product was isolated by the combined immunoprecipitation procedure and subsequent gel electrophoretic analysis. (After SCHRÖDER et al. 1979)

subunit of ribulose 1,5-bisphosphate carboxylase in cucumber (WALDEN and LEAVER 1981) and *Lemna* (TOBIN 1978, TOBIN and SUTTIE 1980), and the apoprotein of the light-harvesting chlorophyll a/b protein in barley (APEL and KLOPPSTECH 1978, APEL 1979), peas (CUMING and BENNETT 1981) and *Lemna* (TOBIN 1981). Another type of control by light at the level of mRNA transcription was found by APEL (1981) and SANTEL and APEL (1981). The NADPH:protochlorophyllide oxidoreductase is one of the major plastid membrane polypeptides of dark-grown barley plants which decreases rapidly during illumination. The enzyme is synthesized on cytoplasmic ribosomes. Upon illumination a rapid decline of the translatable mRNA coding for the enzyme occurs which is triggered by phytochrome.

4.2 Mechanisms of Control of Transcription

The rates of synthesis of all RNA classes (rRNA, mRNA, tRNA) change during different physiological states, but little is known about the regulatory mechanisms. It was observed in several plants or plant tissues that alterations in the in vivo RNA synthesis under quite different physiological conditions were correlated with RNA synthesis measured in vitro in the nuclei or in isolated chromatin.

Such correlations were found during wheat germination (Dobrzanska et al. 1973, Spiegel et al. 1975, Yoshida and Sazaki 1977), in developing pea cotyledons (Cullis 1976, 1978), during the development of potato tubers (Wechselberger et al. 1979), in isolated, normal, or gibberellic acid-treated potato tuber tissues (Kahl and Wechselberger 1977, Wielgat and Kahl 1979a, b), in isolated beet root slices (Duda and Cherry 1971), in isolated Jerusalem artichoke tuber tissue (Gore and Ingle 1974), in normal growing and in auxin-treated soybean hypocotyls (Hou and Pillay 1975, Guilfoyle and Key 1977b), after treatment of lentil roots (Miassod et al. 1970, Teissere et al. 1973) or of mung bean plants (Grierson et al. 1980) with auxin, in expanding internodes from dwarf pea plants after gibberellic acid treatment (Broughton 1968, McComb et al. 1970), in the gibberellic acid-treated and normal developing embryonic axis of hazel (Jarvis et al. 1968), in cytokinin-treated excised barley leaves (Romanko et al. 1978), during the diurnal rhythm of transcription in the duckweed (Nakashima 1978), and after illumination of dark-grown pea seedlings (Bottomley 1970).

These changes in transcriptional activity can be explained by different possibilities. First by altered concentrations of the different RNA polymerases within the cell (or nucleus), second by qualitative changes in existing RNA polymerases (e.g., alteration in the molecule structure, influence of regulatory factors), and third by modification of the template availability, i.e., availability of a greater number or other kinds of genetic loci by reversible binding of repressor proteins (histones and nonhistone proteins) to particular DNA sequences. Many attempts have been made to distinguish between these possibilities, but the results are still contradictory, due to methodical difficulties (1) in quantitative measurements of polymerase levels and analysis of the enzyme structure which requires extensive purification of the enzymes and of factors influencing the enzyme activity at different steps of plant development and (2) in the analysis of chromatin structure and the identification of specific repressor proteins (see also Roeder 1976, who gives several examples for regulation of transcription in animal systems).

In the following I will try to show which mechanisms for the regulation of the transcription may occur in plants.

4.2.1 Selective Gene Recognition

The existence of multiple RNA polymerases suggests that transcription might be regulated in part via enzymes which could specifically transcribe different genes or gene classes. This would presume that each enzyme recognizes specific DNA nucleotide sequences that are common to all genes which this enzyme transcribes. But since polymerase II transcribes a heterogeneous class of genes and also polymerase III transcribes different types of RNA (5S RNA and tRNA), other factors seem to be necessary to recognize specific promoter sites, e.g., several protein factors are responsible for specific gene initiation. Those proteins could be a part of the enzyme within the nucleus but lost during polymerase preparation (Spindler 1979). Therefore it must be evaluated critically whether results from experiments with isolated DNA or synthetic polynucleotides and different polymerases without addition of hitherto unknown factors are able to give information on the mechanisms of selective gene transcription (see also Sect. 3.1.3). The use of nuclei from different tissues of the same plant in which the chromatin, including regulatory proteins, is maintained in the native state seems to be another way of investigating possibilities of selective gene recognition (BOUMAN et al. 1979). These problems have been discussed in detail in the reviews of CHAMBON (1975) and ROEDER (1976).

Only few preliminary studies have been done with isolated plant RNA polymerases to investigate the possibility of selective gene transcription. BROOKS and MANS (1973) have analyzed the RNA synthesized in vitro by maize RNA polymerase II with maize DNA and calf thymus DNA as templates. DNA-RNA hybridization has shown that this RNA arises in part from repeated DNA sequences, and that three times as much of this RNA type is transcribed from homologous maize DNA than from calf thymus DNA. As already discussed above (Sect. 3.1.3) such experiments with double-stranded nuclear DNA as template have to be evaluated carefully since it is nearly impossible to obtain intact DNA from eukaryotes. Therefore SASAKI et al. (1976, 1978) used synthetic polynucleotides of known sequences as model DNA and found that cauliflower polymerases I, II, and III had different specificities to several synthetic templates in vitro. Although differences in the rates of chain initiation, elongation, and termination have not been distinguished, the experiments seem to show that each enzyme is capable of recognizing specific sequences and initiation of transcription from these regions. More detailed experiments are necessary, however, to prove the authors' conclusion that "each enzyme seems to play a role in the regulation of gene expression by discriminating different DNA sequences".

As mentioned already, TEISSERE et al. (1979) have investigated the in vitro transcription of the cauliflower mosaic virus genome by wheat polymerase I, II, and III. This intact DNA can be transcribed in the presence of Mn^{2+} by all three polymerases, although with different efficiency. Only enzyme III transcribes this template with the same activity as denatured calf thymus DNA. These studies demonstrate that the three polymerases are different, but they allow no conclusion concerning the ability for selective transcription of the enzyme in vivo. SEIDMAN et al. (1979) studied the kinetics of formation of stable binary complexes between wheat RNA polymerase II and adenovirus-2 DNA. They found that 5 of the 9 to 11 high-affinity binding sites corresponded to the in vivo promoters of virus transcription. Binding experiments with restriction endonuclease fragments of chloroplast DNA and chloroplast and *E. coli* RNA polymerases have shown that the chloroplast enzyme binds to sites located near the 5′ end of both the gene for 16S RNA and the gene for the large subunit of ribulose 1,5-bisphosphate carboxylase. The *E. coli* enzyme, on the other hand, binds to different sites near the 5′ end of the carboxylase gene (MCINTOSH et al. 1979) see also Sect. 2.2.5).

A more detailed analysis of a specific gene product was possible after transcription of yeast DNA by purified RNA polymerase I from yeast (VAN KEULEN et al. 1975, HAGER et al. 1976). Polymerase I but not enzyme II synthesized preferentially ribosomal RNA from total DNA. The polymerase I in vitro transcript contained a 20-fold higher concentration of ribosomal sequences than predicted for random transcription of the DNA (HAGER et al. 1976).

These results show that DNA-dependent RNA polymerases are not only passive elements of transcription but that they participate in the specificity of transcription recognizing specific sites of initiation and termination on the DNA (for factors influencing transcription specificity see Sect. 4.2.4).

4.2.2 Alterations in the Level of RNA Polymerases and Modulation of Polymerase Activity

4.2.2.1 Soluble RNA Polymerase Activity During Changes of Physiological States

Several reports describe changes in the activities of solubilized RNA polymerases I and II during different physiological transitions. These alterations are correlated with in vivo RNA synthesis activity and RNA synthesis measured in isolated nuclei or chromatin.

Dry seeds (wheat, rye, soybean) were shown to contain all three RNA polymerases measured directly by estimating the activity of the solubilized enzymes (MAZUS and BUCHOWICZ 1972, MAZUS and BRODNIEWICZ-PROBA 1976, JENDRISAK and BECKER 1973, 1974, TEISSERE et al. 1977, 1979, JENDRISAK and GUILFOYLE 1978, GUILFOYLE et al. 1980, HAHN and SERVOS 1980), or indirectly by determining the enzyme activity in isolated chromatin or nuclei (BARKER and RIEBER 1967, GUILFOYLE et al. 1980). Little is known about changes in RNA polymerase activity during the earliest stages of germination. During the first 48 h of wheat germination soluble RNA polymerase activity increases moderately (MAZUS and BUCHOWICZ 1972). Whereas polymerase II activity was found to be relatively high in the quiescent embryo, it was constant during the first 24 h of germination and increased sharply thereafter. The activity of polymerase I was very low in the resting embryo and, after a short lag period, increased gradually over 36 h of the experimental period (MAZUS and BRODNIEWICZ-PROBA 1976). On the other hand, the solubilized RNA polymerase II, as well as I, activity from soybean was found to be constant during the first 24 h of germination, although the polymerase activity in the isolated nuclei increased several-fold during that time period, the α-amanitin-resistent activity (polymerase I and III) increased more than the α-amanitin-sensitive activity (enzyme II) (GUILFOYLE et al. 1980, GUILFOYLE and MALCOLM 1980).

Changes in RNA polymerase activity were also found in soybean seedlings at later stages of postgerminative development (RIZZO and CHERRY 1975, LIN et al. 1976, GUILFOYLE and KEY 1977b). Nuclei isolated from the apical meristematic region of the etiolated soybean hypocotyl (with a high RNA synthesis in vivo) have a much higher ratio of polymerase I to II activity than the nuclei isolated from the mature tissue in the lower hypocotyl with a low level of in vivo RNA synthesis (the polymerase I to II activity ratio changed from 1.3 to 0.35). In the solubilized enzymes the relative activities of polymerase I to II are similar. Thus, whereas polymerase II activity remained relatively constant, the decrease in polymerase I activity during growth transition from the meristematic to the mature state showed a positive correlation with the change in rRNA synthesis in vivo.

When 3-day-old etiolated soybean seedlings were treated with an auxin (2,4-dichlorophenoxyacetic acid) the quiescent cells in the hypocotyl proliferated abnormally; this was associated with a massive synthesis of ribosomal RNA and ribosomes (Fig. 6). Polymerase II activity in nuclei or chromatin, isolated from the hypocotyls, remained nearly constant, but polymerase I activity increased severalfold after auxin treatment (reviewed by GUILFOYLE and KEY 1977b). First analysis seemed to show that after solubilization of the enzymes only polymerase I, but not polymerase II, activities increased during auxin-induced growth (GUILFOYLE et al. 1975, LIN et al. 1976, GUILFOYLE and KEY 1977b). However, in a more recent report it was shown (Fig. 6) that within 48 h after 2,4-D application solubilized polymerase I and II activities increased about tenfold and sixfold, respectively, when equivalent amounts of tissue are analyzed (GUILFOYLE et al. 1980).

TEISSERE et al. have shown that during short incubation periods (less than 2 h) auxin stimulates the synthesis of heterogenous nuclear RNA's and polysomal messenger RNA in lentil roots, whereas during longer incubation periods (15 h) the synthesis of ribosomal RNA is stimulated (reviewed by TEISSERE et al. 1973). During the short treatment of the tissue with auxin, the solubilized nuclear RNA polymerase activity was not influenced, and the hormonal effect was explained by an enhanced template availability. On the other hand, after longer incubation with auxin, the activity of polymerase I increased, but not the activities of enzymes II and III, which showed a hormonal control of the polymerase activity during this period (TEISSERE et al. 1973).

Enhanced activity of solubilized RNA polymerase was also shown in 2,4-D treated maize seedlings (ARENS and STOUT 1972) and isolated Jerusalem artichoke tuber tissue (GORE and INGLE 1974), as well as after illumination of dark-grown tissues (STOUT and MANS 1967, STOUT et al. 1967, BOTTOMLEY 1970, POULSON and BEEVERS 1970, APEL and BOGORAD 1976, SASAKI et al. 1979b).

Only a few cases allowed a decision on whether changes in polymerase activity during physiological transitions were due to changes in the amounts of enzyme proteins or due to modulations of enzyme activity. Such examples will be given in the following paragraphs.

4.2.2.2 Alterations in the Polymerase Content

Quantitative changes in the RNA polymerase level have been determined during developmental processes of plants in only very few cases. MAZUS and BRODNIEWICZ-PROBA (1976) have observed that RNA polymerase I activity was practically absent in resting wheat embryos but appeared several hours after the beginning of imbibition. The de novo synthesis was shown by labeling experiments with ^{14}C-valine. On the other hand, polymerase II was present in the resting embryo and its activity did not increase until later stages of development. However, HAHN and SERVOS (1980) isolated all three RNA polymerases also from nuclei of quiescent wheat embryos, but they gave no data concerning possible changes in the enzyme level during germination. During auxin-induced cellular proliferation in mature soybean hypocotyls (see above), RNA polymerase I and II contents increase about tenfold and sixfold, respectively (Fig. 6). Labeling with ^{35}S-methionine indicated enhanced de novo synthesis of the two enzymes after auxin application (GUILFOYLE et al. 1980).

4.2.2.3 Modulation of RNA Polymerase Activity

In contrast to the above results, the constancy in amount of RNA polymerase during other physiological states indicates that the rates of cellular RNA synthesis are not always correlated with amounts of RNA polymerases. Constancy in the amounts of the enzymes was found during increased RNA synthesis after illumination of etiolated pea seedlings (SASAKI et al. 1979b), after long auxin treatment of lentil roots (enzyme I, TEISSERE et al. 1973), in germinating wheat (enzyme II, JENDRISAK 1980), and during the early soybean germination (GUILFOYLE et al. 1980, GUILFOYLE and MALCOLM 1980). In the soybean axis

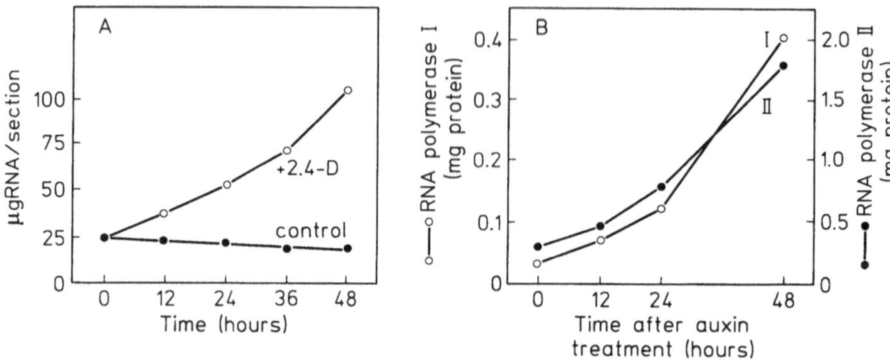

Fig. 6A, B. RNA and RNA polymerase accumulation in auxin (2,4-dichlorophenoxyacetic acid)-treated soybean embryonic axis. **A** In vivo rate of RNA accumulation; **B** Changes in the level of RNA polymerase I and II in embryonic axis after application of 2,4-D to etiolated 4-day (**A**) and 3-day (**B**) old soybean seedlings. (After GUILFOYLE and KEY 1977b, GUILFOYLE et al. 1980). The amounts of the two enzymes do not change in the control tissues not treated with auxin. (GUILFOYLE and MALCOLM 1980, see also Fig. 7)

(Fig. 7) a 25-fold increase in RNA polymerase activity was measured in isolated nuclei during the first 24 h of germination, but the amounts of RNA polymerases I and II solubilized from the whole tissue remained fairly constant (GUILFOYLE and MALCOLM 1980).

Several possibilities have been discussed to explain modulations in RNA polymerase activity (for changes in template availability see below):

1. Although the multiple forms of RNA polymerases are localized within the nucleus, it was found, for example in soybean hypocotyls (LIN et al. 1974, GUILFOYLE and KEY 1977a, b, GUILFOYLE et al. 1980), wheat germ (JENDRISAK and BURGESS 1975), and Jerusalem artichoke tissue (GORE and INGLE 1974) that after homogenization of the tissues the polymerases (particularly enzyme II) are not only associated with the chromatin but also found in the post-chromatin supernatant. The proportions between "bound" and "soluble" enzyme activities seem to differ in relation to the physiological activity of the cell, and the chromatin-bound activity is correlated with RNA accumulation (GORE and INGLE 1974). It can be suggested that RNA polymerases are dissociated from the chromatin in "resting" tissues and in this way are inactivated until bound to the chromatin again during germination or other physiological activation of the tissue (GORE and INGLE 1974, GUILFOYLE et al. 1980).

2. Activity and specificity of RNA polymerases can be regulated by protein factors interacting with the template or with the polymerase molecule itself. These factors include initiation, elongation, and termination factors, but also those proteins which are mediated in the enhancement of the RNA synthesizing activity by hormones. This subject will be discussed in detail below (see Sect. 4.2.4).

3. It seems possible that alteration in RNA polymerase II subunit structure during plant development may produce different states of the enzyme which might regulate the rate of transcription. Two forms of polymerase II (A and

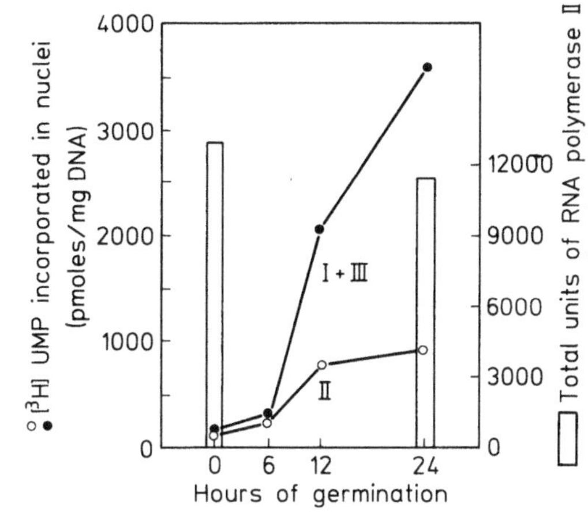

Fig. 7. RNA polymerase activities associated with nuclei, isolated from soybean embryonic axis excised at various stages during the first 24 h of seed germination, and RNA polymerase content of the cotyledons. Nuclei were isolated and assayed for α-amanitin resistant (polymerase I + III) and α-amanitin sensitive (polymerase II) activities. RNA polymerase was isolated from unimbibed and 24-h-germinated soybean axis and purified to homogeneity. (After GUILFOYLE and MALCOLM 1980, and GUILFOYLE et al. 1980)

B) were isolated from soybean hypocotyls (GUILFOYLE and KEY 1977b, GUILFOYLE and JENDRISAK 1978, GUILFOYLE et al. 1980, GUILFOYLE and MALCOLM 1980), wheat germs (JENDRISAK 1980) and parsley cell cultures (LINK and RICHTER 1975, LINK et al. 1978, KIDD et al. 1979), which differ in the molecular weight of the largest subunit (for details see Sect. 3.1.2).

Ungerminated soybean axis contain RNA polymerase IIA with a largest subunit of 215,000 mol. wt. The enzyme purified from the axis at different steps of germination shows a gradual convertion of the 215,000 subunit to a 180,000 mol. wt. polypeptide (in enzyme IIB), whereas no differences in charge or molecular weight of the other subunits were observed at any stage of axis growth (GUILFOYLE and JENDRISAK 1978, GUILFOYLE et al. 1980, GUILFOYLE and MALCOLM 1980). During the first 36 h of wheat germination the nuclear RNA polymerase activity increased 13-fold compared to the activity in nuclei isolated from ungerminated embryos, although the amount in RNA polymerase II in the embryo remained constant over this period. However, the 220,000 mol. wt. subunit was converted to a 180,000 mol. wt. subunit. In this way half of the RNA polymerase II has been altered 24 h after imbibition. In addition, the quantity of the 27,000 mol. wt. subunit increased during germination at the expense of the 25,000 mol. wt. subunit while maintaining the additive stoichiometry of 2.0 (JENDRISAK 1980).

These results lead to the hypothesis that RNA polymerase IIA may be a storage or precursor enzyme which is activated for transcription by conversion to the enzyme IIB (GUILFOYLE and JENDRISAK 1978, JENDRISAK 1980, GUILFOYLE et al. 1980).

Wheat RNA polymerase III was also found in two fractions during DEAE Sephadex chromatography. The physical basis for this heterogeneity is still unknown (JENDRISAK 1980).

4. Several studies have shown that transcription in eukaryotes may be regulated by phosphorylation of nuclear RNA polymerases (JUNGMANN and KRANIAS 1977). However, no apparent alteration was observed during imbibi-

tion, germination, and growth of soybean axes (GUILFOYLE and MALCOLM 1980) or during wheat germination (JENDRISAK 1980).

5. Another regulatory mechanism could result from the dependency of RNA polymerase activity on divalent cations (Mn^{2+} or Mg^{2+}) and nucleoside triphosphate (NTP) concentrations. With kinetic studies GROSSMANN and SEITZ (1979) have shown that RNA polymerase I from parsley cell cultures and soybean hypocotyls are allosterically regulated enzymes. Divalent cations are essential activators of polymerase I, whereas NTP's not complexed by divalent cations act as allosteric inhibitors. However, with nucleoside disphosphates and inorganic phosphate exceeding a ratio of 1:1 between divalent cations and NTP's, the Michaelis-Menten kinetics of the polymerase was maintained, in contrast to the results obtained with free NTP's. This shows that an enzymatic splitting of nucleoside triphosphate into nucleoside diphosphate and P_i by a nucleoside triphosphatase, which was shown to be localized within the nucleus, can prevent inhibition of RNA polymerase activity by free NTP molecules. In contrast to RNA polymerase, the NTPase was inhibited by Mg^{2+}, whereas free ATP acted as an activator. Therefore it seems possible that nucleoside triphosphatase controls the nuclear NTP pool in relation to divalent cations and thus regulates the RNA polymerase I activity by modifying the kinetics from an allosteric to a Michaelis-Menten behavior.

An additional effect of Mg^{2+} was observed in duckweed cells. As will be shown in Sect. 4.2.4, RNA polymerase I activity in duckweed exhibits a diurnal rhythm, which seems to be regulated by a movement of regulatory proteins between nuclei and cytoplasm (NAKASHIMA 1979a, b). Experiments with isolated nuclei have shown that these stimulatory factors are bound to nuclei at optimal Mg^{2+} concentration but are detached in a low Mg^{2+} medium. Since the Mg^{2+} uptake from the culture medium also changes diurnally (KONDO and TSUDZUKI 1978), it was suggested that changes in the Mg^{2+} concentration within the cell or nucleus could be responsible for the movement of stimulatory factors and finally for the diurnal rhythm in RNA synthesis (NAKASHIMA 1979b).

4.2.3 Template Availability

Only a few reports have shown changes in the levels of chromatin during physiological transitions which are coupled with alterations in RNA synthesis in vivo. In most experiments template availability of chromatin was measured to exogenously added bacterial RNA polymerase, perhaps an unreliable parameter, but nevertheless an indicator of changes in chromatin structure.

Chromatin from cucumber seedlings treated in vivo with auxin or gibberellin synthesizes more RNA than chromatin from control plants. During the first hours of hormone treatment the level or activity of the polymerase increased with no measurable increase in template activity. But after longer periods there was an increase in template availability as measured under bacterial polymerase saturated conditions (JOHNSON and PURVES 1970). On the other hand, chromatin isolated from gibberellin-treated hazel embryos (JARVIS et al. 1968) and auxin-treated lentil roots (TEISSERE et al. 1973) showed increased template activity after short hormone treatment of the tissues. After longer incubation periods

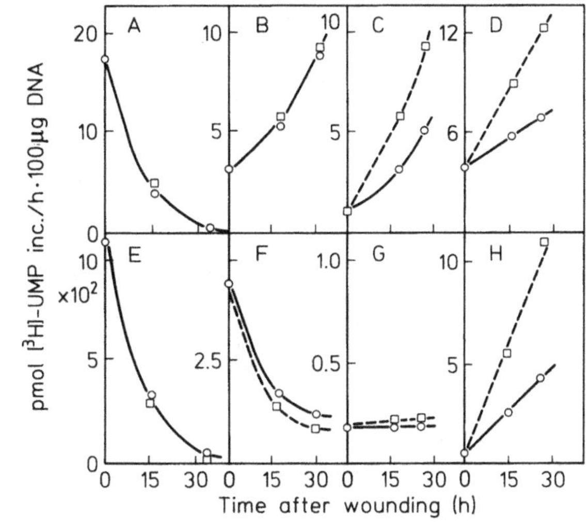

Fig. 8. Potato tuber tissue age and responsiveness of transcription (**A–D**) and template availability (**E–H**) toward wounding and gibberellic acid treatment. White potato tuber tissue from different developmental stages was wounded (○–○) or additionally treated with 10^{-7} mol l^{-1} GA_3 (□—□). Chromatin was isolated and purified and chromatin-associated RNA polymerase was measured (**A–D**) and template accessibility determined with *E. coli* RNA polymerase (**E–H**). Material: **A** and **E** small tubers (3–5 cm ∅) in a state of rapid growth (harvested in July); **B** and F large tubers (>7 cm ∅) at the time of growth completion (late August); **C** and **G** large tubers in the state of dormancy (November); **D** and **H** large tubers at the beginning of spronting after 5-month storages at +7 °C (January). (WIELGAT et al. 1979)

in the presence of hormones, the RNA polymerase activity increased in hazel embryos and lentil roots. In the latter an activation was observed of polymerase I only, but not of enzymes II and III (TEISSERE et al. 1973).

If storage tissues such as potato tubers or sugar beet roots are sliced and washed, an increase in metabolic activity ensues. These wound reactions are accompanied by dedifferentiation of the storage cell to a mitotic active cell and de novo synthesis of ribosomal RNA, transfer RNA, messenger RNA, and proteins (DUDA and CHERRY 1971, KAHL 1971, KAHL and WIELGAT 1976). In sugar beet root slices chromatin-bound RNA polymerase activity increases sevenfold but template availability (measured by saturated levels of *E. coli* RNA polymerase) about three times during 25 h after slicing (DUDA and CHERRY 1971). The reaction of potato tuber tissue upon wounding was found to be strictly dependent on the tuber age (Fig. 8). Wounding of young, rapidly growing tubers results in decrease of both DNA-dependent RNA polymerase activity and template availability, and the tissue is not responsive toward gibberellic acid. However, at the onset of dormancy of the tubers, the activity of the chromatin-bound RNA polymerase activity was enhanced after wounding. On the other hand, the tissue does not change its state of template after wounding (or hormone application) from the beginning of dormancy. Only older but still dormant tubers increase the template availability after injury (KAHL and WECHSELBERGER 1977, WIELGAT et al. 1979).

Gibberellic acid and auxin had no effect on in vivo RNA synthesis in sugar beet tissue, nevertheless the hormones increased the template availability of the chromatin (83% or 35%, respectively). The chromatin-bound polymerase activity was only slightly affected (DUDA and CHERRY 1971). Different effects of GA_3 were found after wounding of dormant potato tuber tissue (Fig. 8). GA_3 enhanced the synthesis of all kinds of RNA in vivo (KAHL and WIELGAT 1976, WIELGAT and KAHL 1979a) and increased the activity of the chromatin-bound RNA polymerase about 300% over that of the control tissue. Polymerase I and hence rRNA synthesis was preferentially stimulated, but polymerase II activity was only slightly affected by GA_3 (WIELGAT and KAHL 1979b, WIELGAT et al. 1979). The template availability of the chromatin was stimulated by the hormone only in tissues taken from old dormant tubers and not from tubers in the early period of dormancy (WIELGAT et al. 1979).

From these results one might conclude that plant hormones can affect both template availability and polymerase activity and that the effect may differ from tissue to tissue. Some experimental data have shown that regulatory factors mediate the hormone effect on RNA synthesis in general, but also on template availability. MATTHYSSE and PHILLIPS (1969), for example, isolated a protein from tobacco and soybean nuclei which interacts with auxin and enhances template availability in isolated nuclei and chromatin. A factor with similar function was isolated from coconut endosperm nuclei (ROY and BISWAS 1977).

In several other plant organs, including developing wheat embryos (YOSHIDA and SASAKI 1977), in cotyledons of developing pea seeds (MILLERD and SPENCER 1974, CULLIS 1976, 1978), and in germinating soybean axis (GUILFOYLE and MALCOLM 1980) changes in RNA synthesis are at least partially explained by changes in template availability, which in wheat embryos and potato tubers were found to be coupled with alterations in the patterns of nonhistone proteins of the chromatin (YOSHIDA and SASAKI 1977, KAHL et al. 1979).

4.2.4 Factors Influencing Polymerase Activity and Specificity

The initiation specificity of prokaryotic RNA polymerase is known to depend on the presence of the sigma protein, which binds reversibly to the core enzyme. One can assume that nuclear RNA polymerases also contain a core enzyme catalyzing the polymerization reaction and additionally loosely attached components (factors), presumably exerting regulatory functions during transcription (HUET et al. 1975). The detection of several factors responsible for initiation and elongation of transcription in animals and in lower eukaryotes was already mentioned (Sect. 2.2).

It is difficult to define the term "protein factor". Generally all proteins present in a cell that stimulate or inhibit RNA synthesis in vitro could be designated as "factors". Here we will omit most of those proteins that interact with a DNA template or with chromatin, thus altering the DNA availability for transcription (histones and nonhistone chromosomal proteins). We focus our attention on proteins which may interact directly with the polymerase molecules. It seems too early to discuss whether or not these factors or at least some of them belong to the subunits of the polymerase molecule.

One may assume that factors are regulatory elements necessary for specific transcription (e.g., specific initiation) in vivo, in contrast to the subunits as obligatory structural elements. Factors are only loosely bound to the core enzyme and are lost during early steps of enzyme purification. The determination of their function is difficult since they are not involved in the basic template-directed polymerization reaction assayed routinely in vitro. Especially earlier reports concerning RNA synthesis stimulating factors from plant tissues need critical avaluation. Not all methods of isolation, purification, characterization, and the determination of possible functions of the factors (including the quality or heterologous kind of the DNA templates used for the experiments) satisfy modern standards.

Two main groups of factors will be discussed here. The first group includes factors which directly influence RNA polymerase activity (initiation, elongation) and the second group consists of hormone-binding proteins which influence transcriptional activity within the cell.

Several factors influencing RNA polymerase activity in vitro have been isolated from coconut chromatin endosperm (MONDAL et al. 1970, 1972c, GANGULY et al. 1973). Factor B (mol. wt. 76,000) acts as an initiation factor. It stimulates the activity of the three coconut polymerases only with native eukaryotic DNA by binding to the enzyme or enzyme-DNA complex. Factor C was described as acting as a termination factor. Another protein factor was separated during the purification of RNA polymerase II by phosphocellulose chromatography from *Zea mays* (HARDIN et al. 1975) and from cell cultures of parsley (LINK and RICHTER 1977). It enhances RNA synthesis in the presence of native homologous DNA. The parsley factor consists of several small polypeptides (mol. wts. 14,000–30,000) and changes the metal ion requirement and ionic strength for optimal activity of the enzyme. Since RNA molecules of greater mean chain length were produced in the presence of the factor, chain elongation appears to be facilitated.

At least two initiation factors which are without effect on the elongation process were isolated from lentil roots (TEISSERE et al. 1975, 1976). Some characteristics of these factors are summarized in Table 4. It is of special interest that the level of factor γ in the cell is controlled by auxin, whereas that of factor δ is not. The authors discussed the possibility that the factor γ could modulate the activity of the polymerase I by giving the enzyme the capacity of recognizing new promoters on the DNA, resulting in a massive synthesis of ribosomal RNA.

RNA synthesis-stimulating factors have also been found in duckweed plants. The capacity of RNA synthesis in cells, isolated nuclei, and chloroplasts from duckweed alternated diurnally due to rhythmic changes in the activity of polymerase I but not of polymerase II (NAKASHIMA 1978, 1979a). Since the nuclei contain factors for stimulating RNA synthesis which seem to move diurnally between nuclei and cytoplasm, one may conclude that these factors cause the diurnal rhythm in the activity of polymerase I (NAKASHIMA 1979b).

Several hormone-binding proteins were isolated influencing transcription in vitro. MATTHYSSE and ABRAMS (1970) isolated a cytokinin-reactive heat-stable protein from pea chromatin. When added to a nucleoside triphosphate-incorporating system consisting of chromatin as a template and *E. coli* RNA polymerase, it caused an increase in the rate of nucleoside triphosphate incorporation in the presence of kinetin. Since also homologous (pea) DNA could be used as template, it was concluded that protein and hormone were able to influence

Table 4. Characterization of two transcription factors from lentil roots. The protein factors were isolated from non-histon chromosomal protein and separated on a carboxymethyl Sephadex C-25 column. (After TEISSERE et al. 1975)

	Factor γ	Factor δ
Stimulation effect on transcription in % over control	90	700
Polymerase which is stimulated	Ib	Ib, II
Function	Initiation	Initiation
Stability at 90 °C in %	90	10
DNA preferred	Double-stranded	Double-stranded
Increase in content after auxin treatment of the tissue	100	0

polymerase binding or initiation of RNA transcription. However, it would be of interest to reinvestigate this effect by methods which allow identification of specific transcription products.

A protein factor obtained from soybean cotyledons promoted the activity of chromatin-bound polymerase from control soybean hypocotyls, but not the polymerase from auxin-treated tissue (HARDIN et al. 1970). The authors postulated an action mechanism by which the factor modifies the RNA polymerase resulting in specific gene transcription. Another protein factor was isolated by affinity chromatography on a column of 2,4-D-substituted agarose from extracts of pea and corn shoots. This factor stimulated RNA synthesis up to 100% by *E. coli* RNA polymerase on purified DNA (VENIS 1971). RIZZO et al. (1977) used the same procedure (2,4-D-substituted Sepharose) in order to isolate a similar but more active transcription factor from soybean hypocotyls. This factor stimulated RNA synthesis two- to sevenfold when using *E. coli* polymerase and native calf thymus DNA. It also stimulated solubilized soybean RNA polymerase I by 25–80% after 2,4-D was included into the assay mixture. RNA polymerases II and III were not affected.

Other auxin-binding factors seem to interact with the chromatin, resulting in changes of the template availability. MATTHYSSE and PHILLIPS (1969) isolated such a protein factor from tobacco and soybean nuclei, which could interact with auxin and enhanced template activity of chromatin in the presence of saturating amounts of *E. coli* RNA polymerase. Auxin-binding proteins were also isolated from coconut endosperm nuclei (MONDAL et al. 1972b, ROY and BISWAS 1977). The complex from auxin and the nucleoplasmic receptor protein stimulated transcription of isolated chromatin from coconut endosperm saturated with *E. coli* RNA polymerase. In particular, the synthesis of a heterodisperse RNA fraction (9S–12S) was stimulated (ROY and BISWAS 1977).

A nonprotein factor, stimulating RNA polymerase II activity in vitro, has been isolated from soybean hypocotyls. This factor can be extracted from plasma membranes with ethanol (CLARK et al. 1976), but was also released by incubating the isolated plasma membranes with auxin. Therefore this factor was postulated as part of the mechanism of auxin action at the transcriptional level (HARDIN et al. 1972). It was also found that transcription in isolated nuclei or chromatin from different plant species could be directly stimulated by auxin (MATTHYSSE and PHILLIPS 1969, MAHESHWARI et al. 1966, SALOMON and MASCARENHAS 1972), gibberellin (JOHRI and VARNER 1968, DUDA and CHERRY 1971) and cytokinin (MAHESHWARI et al. 1966, MATTHYSSE and ABRAMS 1970, DUDA and CHERRY 1971, SELIVANKINA et al. 1979), when the hormones were added to the incubation assay or were present in the media during the preparation of nuclei or chromatin. Other experimentalists failed to observe the direct stimulation of in vitro RNA synthesis by hormones without addition of hormone-binding proteins (O'BRIEN et al. 1968, JOHNSON and PURVES 1970, ROY and BISWAS 1977, MENNES et al. 1978, WIELGAT and KAHL 1979b, GRIERSON et al. 1980). It is possible that the hormone-binding proteins were either absent in the chromatin (MONDAL et al. 1972b, ROY and BISWAS 1977), or if present in nuclei, that they were already saturated with hormone, so that the addition of external hormone was without effect on the RNA synthesis (MENNES et al. 1978).

Evidence available indicates that a hormone is first bound by specific receptor proteins in the target cell. The actual target of the hormone-receptor complex is a matter of controversy. The complex may react with distinct parts of the chromatin to mark specific initiation sites for RNA polymerase, or the complex may modify a specific RNA polymerase, resulting in altered gene recognition or simply increasing transcription activity of the enzyme.

All these observations demonstrate the existence of many cellular components that can effect changes in RNA polymerase activity and DNA transcription. However, much more information is needed to understand the mechanism of the transcription and the role of factors involved in its regulation and possibly even in selective gene recognition.

4.2.5 Chloroplast RNA Polymerase

Illumination of dark-grown seedlings or unicellular algae initiates a sequence of changes which result in the development of mature, functional chloroplasts from etioplasts or proplastids. One of the primary responses to light is the rapid increase in synthesis of all types of chloroplast RNA (see review by WOLLGIEHN and PARTHIER 1980) and in the activity of chloroplast RNA polymerase (STOUT et al. 1967, BOTTOMLEY 1970, APEL and BOGORAD 1976). It is widely unknown whether the reason for the increase in polymerase activity is an increase in the amount of enzyme or an activation of already-existing enzyme molecules.

Dark-grown *Euglena* cells synthesize only very small amounts of plastid RNA. Illumination results in a gradual increase in RNA synthesis after a short lag period (HEIZMANN et al. 1975, COHEN and SCHIFF 1976). This suggests that plastid RNA polymerase de novo synthesis in the cytoplasm is a prerequisite for intensive plastid RNA synthesis. The low level of RNA formation during the first phase of illumination may be catalyzed by enzyme molecules already present in proplastids. Another explanation of the lag period is that the enzyme is present in proplastids in sufficient amounts but not in the active form. On the other hand, in dark-grown higher plants the etioplasts are able to synthesize relatively large amounts of RNA, although RNA synthesis of plastids is strongly stimulated during illumination (see WOLLGIEHN and PARTHIER 1980). APEL and BOGORAD (1976) measured a fourfold increase of maize plastid polymerase activity after 16 h of illumination. But this change was neither due to a quantitative enhancement of the enzyme content nor was it the result of qualitative alteration of the purified enzyme. It was suggested that other light-inducible factors may be responsible for light-induced enhancement of RNA polymerase activity.

Factors influencing the activity of the chloroplast RNA polymerase core enzyme were isolated from *Chlamydomonas* and maize chloroplasts. The two factors from *Chlamydomonas* exhibit sigma-like activity (SURZYCKY and SHELLENBARGER 1976). Factor 2 (51,000 mol. wt.) was shown to be responsible for the initiation of transcription by interaction with homologous (*Chlamydomonas* chloroplast) or heterologous (*E. coli*) core enzyme, though less with the *E. coli* enzyme. The activity of nuclear RNA polymerases from *Chlamydomonas* was not affected by factor 2 even when chloroplast DNA was used as a template.

Another, the transcription accelerating factor was isolated from maize chloroplasts (JOLLY and BOGORAD 1980). This, 27,500 mol. wt. polypeptide, designated S-factor, has no effect on transcription by *E. coli* RNA polymerase or nuclear polymerase II. An interesting effect was found when maize chloroplast DNA sequences incorporated in cloned chimeric bacterial plasmids (pZmc 134 DNA) were transcribed by maize chloroplast RNA polymerase. With supercoiled pZmc 134 DNA as template, both vehicle DNA and chloroplast DNA fragments were transcribed in the absence of S, however, chloroplast DNA was preferentially transcribed in the presence of the S-factor. When circular (not supercoiled) pZmc 134 DNA was used, no effect of S on the specificity of transcription was observed.

Further work will show whether other factors increasing transcription specificity exist in the chloroplasts and whether the state of the DNA (supercoiled or relaxed) plays a role in the regulation of chloroplast genome transcription.

Finally it should be noted that transcription of both plastid and cytoplasmic RNA are under the photocontrol of the phytochrome system (SCOTT et al. 1971, THIEN and SCHOPFER 1975, APEL 1979, 1981, LINK 1981) and that it is also influenced by phytohormones, especially cytokinins (PARTHIER 1979).

5 Conclusions

The three plant nuclear DNA-dependent RNA polymerases have essentially the same structure, functions and catalytic properties as all other eukaryotic nuclear RNA polymerases. Compared with the nuclear enzymes, chloroplast RNA polymerase is much less studied because isolation of an active enzyme from the organelles is more difficult. The few data available indicate that the structure of the chloroplast enzyme is less complex than the nuclear polymerases, but further experiments have to show whether or not this enzyme corresponds in subunit structure and function to prokaryotic bacterial RNA polymerase. No information is hitherto available about plant mitochondrial RNA polymerase.

The regulation of transcription is a very complex phenomenon. It involves not only the regulation of polymerase activity but also changes in chromatin structure (template availability) and the influence of different internal and external regulating factors. This chapter is restricted mainly to those aspects which are related to the regulation of the activity of the different RNA polymerases.

At present one can assume that transcription in plants is regulated similarly to the ways known for animal or fungal systems. Numerous different possibilities are realized to regulate the polymerase or template activity and thus to regulate the synthesis of different types of RNA. The details of these mechanisms are still widely unknown. Since quantitative changes in transcription or induction of transcription of specific genes can be observed during different physiological transitions, such as seed germination, the influence of phytohormones, or photomorphogenesis, it is to hope that in the near future further details of the hitherto described mechanisms of transcription regulation in plants will be elucidated. These mechanisms include changes in template availability, RNA polymerase modification, the involvement of regulating proteins, the role of phytohormones and the mechanisms triggered by light. The results of further work will show the specific role of regulation of transcription for different developmental processes in plants.

References

Apel K (1979) Phytochrome-induced appearance of mRNA activity for the apoprotein of the light-harvesting chlorophyll a/b protein of barley (*Hordeum vulgare*). Eur J Biochem 97:183–188

Apel K (1981) The protochlorophyllide holochrome of barley (*Hordeum vulgare* L.). Phytochrome-induced decrease of translatable mRNA coding for the NADPH:protochlorophyllide oxidoreductase. Eur J Biochem 120:89–93

Apel K, Bogorad L (1976) Light-induced increase in the activity of maize plastid DNA-dependent RNA polymerase. Eur J Biochem 67:615–620

Apel K, Kloppstech K (1978) The plastid membranes of barley (*Hordeum vulgare*). Light-induced appearance on mRNA coding for the apoprotein of the light-harvesting chlorophyll a/b protein. Eur J Biochem 85:581–588

Arens MQ, Stout ER (1972) Enhanced activity of the soluble ribonucleic acid polymerase from 2,4-dichlorophenoxyacetic acid-treated maize seedlings. Plant Physiol 50:640–641

Barker GR, Rieber M (1967) The development of polysomes in the seeds of *Pisum arvense*. Biochem J 105:1195–1201

Bautz EKF (1972) Regulation of RNA synthesis. Prog Nucleic Acid Res Mol Biol 12:129–160

Bautz EKF (1976) Bacteriophage-induced DNA-dependent RNA polymerases. In: Losick R, Chamberlin M (eds) RNA polymerase. Cold Spring Harbor Lab, New York, pp 273–284

Becker WM (1979) RNA polymerases in plants. In: Hall TC, Davies J (eds) Nucleic acids in plants. CRC Press, Boca Raton, pp 111–141

Bedbrook JR, Link G, Coen DM, Bogorad L, Rich A (1978) Maize plastid gene expressed during photoregulated development. Proc Natl Acad Sci USA 75:3060–3064

Bennet J, Ellis RJ (1973) Solubilization of the membrane-bound deoxyribonucleic acid-dependent ribonucleic acid polymerase of pea chloroplasts. Biochem Soc Trans 1:892–894

Bewley JD (1982) Germination. In: Boulter D, Parthier B (eds) Nucleic acids and proteins in plants I. Encyclopedia of plant physiology new ser. Vol 14A. Springer, Berlin, Heidelberg, New York pp 559–591

Bhat SP, Padayatty JD (1974) Presence of conserved messenger RNA in rice embryos. Indian J Biochem Biophys 11:47–50
Bhat SP, Padayatty JD (1975) Transcriptional event during early phase of germination in rice embryos. Nature (London) 256:227–228
Biswas BB, Ganguly A, Das A (1975) Eukaryotic RNA polymerases and factors that control them. Prog Nucleic Acid Res Mol Biol 15:145–184
Blatti SP, Ingles CJ, Lindell TJ, Morris PW, Weaver RI. Weinberg F, Rutter WJ (1970) Structure and regulatory properties of eukaryotic RNA polymerases. Cold Spring Harbor Symp Quant Biol 35:649–658
Bogorad L, Mets LJ, Mullinix KP, Smith HJ, Strain GC (1973) Possibilities for intracellular integration: the ribonucleic acid polymerases of chloroplasts and nuclei, and genes specifying chloroplast ribosomal proteins. Biochem Soc Symp 38:17–41
Bohnert HJ, Driesel AJ, Herrmann RG (1977) Transcription and processing of transcripts in isolated unbroken chloroplasts. In: Bogorad L, Weil JH (eds) Acides nucléiques et synthése des proteins chez les végétaux. Ed CNRS, Paris, pp 213–218
Bottomley W (1970) Deoxyribonucleic acid-dependent ribonucleic acid polymerase activity of nuclei and plastids from etiolated peas and their response to red and far red light in vivo. Plant Physiol 45:608–611
Bottomley W, Smith HJ, Bogorad L (1971) RNA polymerase of maize: partial purification and properties of the chloroplast enzyme. Proc Natl Acad Sci USA 68:2412–2416
Bouman H, Mennes AM, Libbenga KR (1979) Transcription of nuclei isolated from tobacco tissue. FEBS Lett 101:369–372
Brandt P, Wiessner W (1977) Unterschiedliche Temperaturoptima der DNA-abhängigen RNA Polymerase von *Euglena gracilis* Stamm und ihre Bedeutung für die experimentelle Erzeugung der permanenten Apochlorose durch höhere Temperaturen. Z Pflanzenphysiol 85:53–60
Briat JF, Mache R (1980) Properties and characterization of a spinach chloroplast RNA polymerase isolated from a transcriptionally active DNA-protein complex. Eur J Biochem 111:503–509
Briat JF, Laulhere JP, Mache R (1979) Transcription activity of a DNA-protein complex isolated from spinach plastids. Eur J Biochem 98:285–292
Broughton WJ (1968) Influence of gibberellic acid on nucleic acid synthesis in dwarf pea internodes. Biochim Biophys Acta 155:308–310
Brooks RR, Mans RJ (1973) Selection of repeated sequences of homologous and heterologous DNA during in vitro transcription by maize RNA polymerase. Biochem Biophys Res Commun 52:608–613
Burgess RR (1971) RNA polymerase. Annu Rev Biochem 40:711–740
Caers LI, Penmans WJ, Carlier AR (1979) Preformed and newly synthesized messenger RNA in germinating wheat embryos. Planta 144:491–496
Carritt B, Eisenstadt JM (1973) RNA synthesis in isolated chloroplasts: characterization of the newly synthesized RNA. FEBS Lett 36:116–120
Chamberlin MJ (1974) Bacterial DNA-dependent RNA polymerase. In: Boyer PD (ed) The enzymes Vol X. 3rd edn. Academic Press, London, New York, pp 17–67
Chamberlin MJ (1976) RNA polymerase – an overview. In: Losick R, Chamberlin M (eds) RNA polymerase. Cold Spring Harbor Lab, New York, pp 17–67
Chambon P (1975) Eukaryotic nuclear RNA polymerases. Annu Rev Biochem 44:613–633
Chen D, Osborne DJ (1970) Hormones in the translational control of early germination in wheat embryos. Nature (London) 226:1157–1160
Chen D, Sarid S, Katchalski E (1968) Studies on the nature of messenger RNA in germinating wheat embryos. Proc Natl Acad Sci USA 60:902–909
Chen D, Schulz G, Katchalski E (1971) Early ribosomal RNA transcription and appearance of cytoplasmic ribosomes during germination of the wheat embryo. Nature New Biol 231:69–72
Chelm B, Hallick R (1976) Changes in the expression of the chloroplast genome of *Euglena gracilis* during chloroplast development. Biochemistry 15:593–599

Chelm BK, Hallick RB, Gray PW (1979) Transcription program of the chloroplast genome of *Euglena gracilis* during chloroplast development. Proc Natl Acad Sci USA 76:2258–2262

Clark JE, Morre DJ, Cherry JH, Yunghans WN (1976) Enhancement of RNA polymerase activity by non-protein components from plasma membranes of soybean hypocotyls. Plant Sci Lett 7:233–238

Clay WF, Hattermann FRH, Hammett JR (1975) Nucleic acid metabolism during germination of pima cotton (*Gossypium barbadense*). Plant Physiol 55:231–236

Cohen D, Schiff JA (1976) Events surrounding the early development of *Euglena* chloroplasts. Photoregulation of the transcription of chloroplast and cytoplasmic ribosomal RNA. Arch Biochim Biophys 177:201–216

Cullis CA (1976) Chromatin bound DNA-dependent RNA polymerase in developing pea cotyledons. Planta 131:293–298

Cullis CA (1978) Chromatin-bound DNA-dependent RNA polymerase in developing pea cotyledons. II. Polymerase activity and template availability under different growth conditions. Planta 144:57–62

Cuming AC, Bennett J (1981) Biosynthesis of the light-harvesting chlorophyll a/b protein. Control of messenger RNA activity by light. Eur J Biochem 118:71–80

Cuming AC, Kennedy TD, Lane BG (1979)Wheat embryo ribonucleates. XIV. Mass isolation of mRNA from wheat germ and comparison of its translational capacity with that of mRNA from imbibing wheat embryos. Can J Biochem 57:1170–1175

Curtis SE, Rawson JRY (1979) Measurement of the transcription of nuclear single-copy deoxyribonucleic acid during chloroplast development in *Euglena*. Biochemistry 18:1170–1175

Delseny M, Aspart L, Got A, Cooke R, Guitton Y (1977) Early synthesis of polyadenylic acid, polyadenylated and ribosomal nucleic acids in germinating radish embryo axes. Physiol Veg 15:413–428

Dobrzanska M, Tomaszewski M, Grzelczak Z, Rejman E, Buchowicz J (1973) Cascade activation of genome transcription in wheat. Nature (London) 244:507–509

Doshchanow KH, Ajtkhozhin M, Darkanbaev T (1975) Synthesis of cytoplasmic RNAs at the early stages of germination of wheat embryos. Physiol Rast 22:368–375

Dron M, Robreau S, Legal Y (1979) Isolation of a chromoid from the chloroplast of *Chlamydomonas reinhardi*. Exp Cell Res 119:301–305

Duda CT (1976) Plant RNA polymerases. Annu Rev Plant Physiol 27:119–132

Duda CT, Cherry JH (1971) Chromatin- and nuclei-directed ribonucleic acid synthesis in sugar beet root. Plant Physiol 47:262–268

Dunham VL, Cherry JH (1973) Solubilization and characterization of RNA polymerase from a higher plant. Phytochemistry 12:1897–1902

Edelman M, Reisfeld A (1978) Identification, in vitro translation, and control of the main chloroplast membrane protein synthesized in *Spirodela*. In: Akoyunoglou G, Argyroudi-Akoyunoglou H (eds) Chloroplast differentiation. Elsevier North-Holland, Amsterdam, pp 641–652

Edelman M, Reisfeld A (1980) Synthesis, processing and functional probing of p-32000, the major membrane protein translated within the chloroplast. In: Leaver CJ (ed) Genome organization and expression in plants. Plenum, New York, London, pp 353–362

Engelke DR, Ng SY, Shastry BS, Roeder RG (1980) Specific interaction of a purified transcription factor with an internal control region of 5S RNA genes. Cell 19:717–728

Fabisz-Kijowska A, Dullin P, Walerych W (1975) Isolation and purification of RNA polymerases from rye embryos. Biochim Biophys Acta 390:105–116

Falchuk KH, Mazus B, Ulpino L, Vallee BL (1976) *Euglena gracilis* DNA dependent RNA polymerase II: A zinc metalloenzyme. Biochemistry 15:4468–4475

Falchuk KH, Ulpino L, Mazus B, Vallee BL (1977) *E. gracilis* RNA polymerase I: A zinc metalloenzyme. Biochem Biophys Res Commun 74:1206–1212

Falchuk KH, Hardy C, Ulpino L, Vallee BL (1978) RNA metabolism, manganese and RNA polymerases of zinc-sufficient and zinc-deficient *Euglena gracilis*. Proc Natl Acad Sci USA 75:4175–4179

Fukasawa H, Mori K (1974) Cauliflower RNA polymerase: partial purification and preliminary characterization of DNA-dependent enzymes. Plant Sci Lett 2:391–396
Ganguly A, Das A, Mondal H, Mandal RK, Biswas BB (1973) Molecular weight and subunit structure of RNA polymerase I and initiation factor from chromation of plant cell nuclei. FEBS Lett 34:27–30
Gilbert W (1976) Starting and stopping sequences for the RNA polymerase. In: Losick R, Chamberlin M (eds) RNA polymerase, Cold Spring Harbor Lab, New York, pp 193–205
Glicklich D, Jendrisak JJ, Becker WM (1974) Separation and partial characterization of two ribonucleic acid polymerases from pea seedlings. Plant Physiol 54:356–359
Goldberg RB (1980) Structural gene expression in tobacco. In: Leaver CJ (ed) Genome organization and expression in plants. Plenum, New York, London, pp 117–126
Gore JR, Ingle J (1974) Ribonucleic acid polymerase activities in Jerusalem artichoke tissue. Biochem J 143:107–113
Goto H, Sasaki Y, Kamikubo T (1978) Large-scale purification and subunit structure of DNA-dependent RNA polymerase II from cauliflower inflorescence. Biochem Biophys Acta 517:195–204
Grierson D (1979) Light-stimulated accumulation of 14S RNA during chloroplast development in spinach cotyledons. Z Pflanzenphysiol 95:171–177
Grierson D, Covery S (1975) Changes in the amount of ribosomal RNA and poly(A)-containing RNA during leaf development. Planta 127:77–86
Grierson D, Thompson JR, Garcia-Mora R, Slater RJ (1980) Ribonucleic acid synthesis by isolated mung bean nucleoli. Z Pflanzenphysiol 97:437–447
Grossmann K, Seitz U (1979) RNA polymerase I from higher plants. Evidence for allosteric regulation and interaction with a nuclear phosphatase activity controlled NTP pool. Nucl Acids Res 7:2015–2029
Grossmann K, Friedrich H, Seitz U (1980) Purification and characterization of chromatin-bound DNA-dependent RNA polymerase I from parsley. Influence of nucleoside triphosphates. Biochem. J 191:165–171
Guilfoyle TJ (1976) Purification and characterization of DNA-dependent RNA polymerases from cauliflower nuclei. Plant Physiol 58:453–458
Guilfoyle TJ, Hanson JB (1973) Increased activity of chromatin-bound ribonucleic acid polymerase from soybean hypocotyl with spermidine and high ionic strength. Plant Physiol 51:1022–1025
Guilfoyle TJ, Hanson JB (1974) Greater length of ribonucleic acid synthesized by chromatin-bound polymerase from auxin-treated soybean hypocotyls. Plant Physiol 53:110–113
Guilfoyle TJ, Jendrisak JJ (1978) Plant DNA-dependent RNA polymerases: Subunit structures and enzymatic properties of the class II enzymes from quiescent and proliferating tissues. Biochemistry 17:1860–1866
Guilfoyle TJ, Key JL (1977a) The subunit structure of soluble and chromatin-bound RNA polymerase II from soybean. Biochem Biophys Res Commun 74:308–313
Guilfoyle TJ, Key JL (1977b) Purification and characterization of soybean DNA-dependent RNA polymerases and the modulation of their activities during development. In: Bogorad L, Weil J (eds) Nucleic acids and protein synthesis in plants. Plenum, New York, London, pp 37–63
Guilfoyle TJ, Malcolm S (1980) The amounts, subunit structures, and template-engaged activities of RNA polymerases in germinating soybean axis. Dev Biol 78:113–125
Guilfoyle TJ, Lin CY, Chen YM, Nago RT, Key JL (1975) Enhancement of soybean RNA polymerase I by auxin. Proc Natl Acad Sci USA 72:69–72
Guilfoyle TJ, Lin CY, Chen YM, Key JL (1976) Purification and characterization of RNA polymerase I from a higher plant. Biochim Biophys Acta 418:344–357
Guilfoyle T, Olszewski N, Zurfluh L (1980) RNA polymerase during developmental transitions in soybean. In: Leaver CJ (ed) Genome organization and expression in plants. Plenum, New York, London, pp 93–104
Gurley WB, Lin CY, Guilfoyle TJ, Nagato RT, Key JL (1976) Analysis of plant RNA polymerase I transcript in chromatin and nuclei. Biochim Biophys Acta 425:168–174

Haff LA, Bogorad L (1976) Poly(adenylic acid)-containing RNA from plastids of maize. Biochemistry 15:4110–4115

Hager G, Holland M, Valenzuela P, Weinberg F, Rutter WJ (1976) RNA polymerases and transcriptive specificity in *Saccharomyces cerevisiae*. In: Losick R, Chamberlin M (eds) RNA polymerase. Cold Spring Harbor Lab, New York, pp 745–762

Hahlbrock K, Gardiner SE, Matern U, Ragg H, Schröder J (1980) The role of light in the induction of mRNAs for phenylalanine ammonia-lyase and related enzymes in plant cultures. In: Leaver CJ (ed) Genome organization and expression in plants. Plenum, New York, London, pp 187–194

Hahn H, Servos D (1980) The isolation of multiple forms of DNA dependent RNA polymerases from nuclei of quiescent wheat embryos by affinity chromotography. Z Pflanzenphysiol 97:43–57

Hall RH (1973) Cytokinins as a probe of developmental processes. Annu Rev Plant Physiol 24:415–444

Hallam ND, Roberts BE, Osborne DJ (1972) Embryogenesis and germination in rye (*Secale cereale* L.). Planta 105:293–309

Hallick RB, Lipper C, Richards O, Rutter W (1976) Isolation of a transcriptionally active chromosome from chloroplasts of *Euglena gracilis*. Biochemistry 15:3039–3045

Hardin JW, Cherry JH (1972) Solubilization and partial characterization of soybean chromatin-bound RNA polymerase. Biochem Biophys Res Commun 48:299–306

Hardin JW, O'Brien TJ, Cherry JH (1970) Stimulation of chromatin-bound RNA polymerase activity by a soluble factor. Biochim Biophys Acta 224:667–670

Hardin JW, Cherry JH, Morré DJ, Lembi CA (1972) Enhancement of RNA polymerase activity by a factor released by auxin from plasma membrane. Proc Natl Acad Sci USA 69:3146–3150

Hardin JW, Apel K, Smith J, Bogorad L (1975) In: Markert CL (ed) Isozymes Vol I. Academic Press, London, New York, pp 55–68

Hartley MR, Ellis RJ (1973) Ribonucleic acid synthesis in chloroplasts. Biochem J 134:249–262

Hecker M (1978) Nucleinsäuren und Proteinbiosynthesen in quellenden pflanzlichen Embryonen und ihre Bedeutung für den Aktivitätswechsel von Samen. Biol Rundsch 16:137–154

Hecker M, Köhler K-H (1979) Protein and nucleic acid synthesis during imbibition of dormant and after-ripened *Vaccaria pyramidata* seeds. Dev Biol 69:270–280

Hecker M, Köhler K, Wiedmann M (1977) Reactivation of RNA synthesis during early germination of *Agrostemma githago* embryos. Biochem Physiol Pflanz 171:401–408

Heizmann P, Verdier G, Nigon V (1975) Light induction of RNA synthesis in *Euglena*. In: Les cycles cellulaires et leur blocage chez plusieurs protistes. CNRS, Paris, pp 247–254

Higgins TJV, Zwar JA, Jacobsen JV (1976) Gibberellic acid enhances the level of translatable mRNA for α-amylase in barley aleurone layers. Nature (London) 260:166–169

Ho TD (1980) Hormonal and genetic regulation of α-amylase synthesis in barley. In: Leaver CJ (ed) Genome organization and expression in plants. Plenum, New York, London, pp 147–157

Hodo HG, Blatti SP (1977) Purification using polyethyleneimine precipitation and low molecular weight subunit analysis of calf thymus and wheat germ DNA-dependent RNA polymerase II. Biochemistry 16:2334–2342

Horgen PA, Key JL (1973) The DNA-directed RNA polymerases of soybean. Biochim Biophys Acta 294:227–235

Hossenlopp P, Wells D, Chambon P (1975) Animal DNA-dependent RNA polymerases. Partial purification and properties of three classes of RNA polymerases from uninfected and adenovirus-infected *HeLa* cells. Eur J Biochem 58:237–251

Hou GC, Pillay DTN (1975) Chromatin RNA polymerase activity from soybean hypocotyls treated with gibberellic acid and AMO-1618. Phytochemistry 14:403–407

Howell SH (1978) Regulation of mRNAs coding for major chloroplast components in *Chlamydomonas reinhardii*. In: Akoyunoglou G, Argyroudi-Akoyunoglou JH (eds) Chloroplast differentiation. Elsevier North-Holland, Amsterdam, pp 679–686

Huang RC, Bonner J (1962) Histone, a suppressor of chromosomal RNA synthesis. Proc Natl Acad Sci USA 48:1216–1222
Huet J, Buhler JH, Sentenac A, Fromageot P (1975) Dissociation of two polypeptide chains from yeast RNA polymerase A. Proc Natl Acad Sci USA 72:3034–3038
Jacob ST (1973) Mammalian RNA polymerases. Prog Nucleic Acid Res Mol Biol 13:93–126
Jacob ST, Sajdel EM, Munro HN (1968) Altered characteristics of mammalian RNA polymerase following solubilization from nuclei. Biochem Biophys Res Commun 32:831–838
Jacob ST, Sajdel EM, Munro HN (1970) Different responses of soluble whole nuclear RNA polymerase and soluble nucleolar RNA polymerase to divalent cations and to inhibition by α-amanitin. Biochem Biophys Res Commun 38:765–770
Jacobson JV (1977) Regulation of ribonucleic acid metabolism by plant hormones. Annu Rev Plant Physiol 28:537–564
Jaehning JA, Weinman R, Brendler TG, Raskas HJ, Roeder RG (1976) Function and Regulation of RNA polymerase II and III in adenovirus-infected KB cells. In: Losick R, Chamberlin M (eds) RNA polymerase. Cold Spring Harbor Lab, New York, pp 819–834
Jarvis BC, Frankland B, Cherry JH (1968) Increased DNA template and RNA polymerase accociated with the breaking of seed dormancy. Plant Physiol 43:1734–1736
Jendrisak J (1980) Purification, structures and functions of the nuclear RNA polymerases from higher plants. In: Leaver CJ (ed) Genome organization and expression in plants, Plenum, New York, pp 77–92
Jendrisak J, Becker WM (1973) Isolation, purification, and characterization of RNA polymerases from wheat germ. Biochim Biophys Acta 319:48–54
Jendrisak JJ, Becker WM (1974) Purification and subunit analysis of wheat germ ribonucleic acid polymerase II. Biochem J 139:771–777
Jendrisak JJ, Burgess RR (1975) A new method for the large-scale purification of wheat germ DNA-dependent RNA polymerase II. Biochemistry 14:4639–4645
Jendrisak JJ, Burgess RR (1977) Studies on the subunit structure of wheat germ ribonucleic acid polymerase II. Biochemistry 16:1959–1964
Jendrisak J, Guilfoyle TJ (1978) Eukaryotic RNA polymerases: comparative subunit structures, immunological properties, and α-amanitin sensitivities of the class II enzymes from higher plants. Biochemistry 17:1322–1327
Johnson KD, Purves WK (1970) Ribonucleic synthesis by cucumber chromatin. Developmental and hormone-induced changes. Plant Physiol 46:581–585
Johri MM, Varner JE (1968) Enhancement of RNA synthesis in isolated pea nuclei by gibberellic acid. Proc Natl Acad Sci USA 59:269–276
Jolly SO, Bogorad L (1980) Preferential transcription of cloned maize chloroplast DNA sequences by maize chloroplast RNA polymerases. Proc Natl Acad Sci USA 77:822–826
Joussaume M (1973) Mise en évidence de deux formes de RNA polymérase dépendente du DNA dans les chloroplastes isolés de feuilles de pois. Physiol Veg 11:69–82
Jungmann RA, Kranias EG (1977) Nuclear phosphoprotein kinases and the regulation of gene transcription. Int J Biochem 8:819–830
Kahl G (1971) Synthesis of rRNA, tRNA and other RNA species concomitant with polyribosome formation in aging tuber slices. Z Naturforsch 26b:1058–1064
Kahl G, Wechselberger M (1977) Activation of chromatin-bound DNA-dependent RNA polymerase (E.C.2.7.7.6) in plant storage tissue slices. Z Naturforsch 32c:229–235
Kahl G, Wielgat B (1976) Regulation of transcriptional activity in wounded potato tuber tissues. Physiol Veg 14:725–738
Kahl G, Schäfer W, Wechselberg M (1979) Changes in non-histone chromosomal proteins during the development of potato tubers; their involvement in wound- and hormone-induced processes. Plant Cell Physiol 20:1217–1228
Kamalay JC, Goldberg RB (1980) Regulation of structural gene expression in tobacco. Cell 19:935–946
Kedinger C, Gniazdowski M Mandel JL, Gissinger F, Chambon P (1970) α-Amanitin:

A specific inhibitor of one of two DNA-dependent RNA polymerase activities from calf thymus. Biochem Biophys Res Commun 38:165–171
van Keulen H, Planta RJ, Retél J (1975) Structure and transcription specificity of yeast RNA polymerase. A Biochim Biophys Acta 395:179–190
Key JL, Vanderhoef LN (1973) Plant hormones and developmental regulation: Role of transcription and translation. In: Coward SJ (ed) Aspects of cell differentiation. Academic Press, London, New York, pp 49–83
Kidd GH, Bogorad L (1979) Peptide maps comparing subunits of maize chloroplast and type II nuclear DNA-dependent RNA polymerases. Proc Natl Acad Sci USA 76:4890–4892
Kidd GH, Bogorad L (1980) A facile procedure for purifying maize chloroplast RNA polymerase from whole cell homogenates. Biochim Biophys Acta 609:14–30
Kidd GH, Link G, Bogorad L (1979) Comparison of large subunits of type II DNA-dependent RNA polymerases from higher plants. Plant Physiol 64:671–673
Kondo T, Tsudzuki T (1978) Rhythm in potassium uptake by a duckweed, *Lemna gibba* G3. Plant Cell Physiol 19:1465–1473
Krakow JS, Rhodes G, Jovin TM (1976) RNA polymerase: Catalytic mechanisms and inhibitors. In: Losick R, Chamberlin M (eds) RNA polymerase. Cold Spring Harbor Lab, New York, pp 127–157
Kulaeva ON (1973) Cytokinins, their structures and functions (in Russian). Nauka, Moscow
Legraverend M, Glazer RI (1980) Characterization of a nonhistone chromosomal protein which stimulates RNA polymerase II. Biochim Biophys Acta 607:92–101
Lentfer D, Lezius AG (1972) Mouse-myeloma RNA polymerase B. Template specificities and the role of a transcription-stimulating factor. Eur J Biochem 30:278–284
Lin CY, Guilfoyle TJ, Chen MM, Nagao RT, Key JL (1974) The separation of RNA polymerases I and II achieved by fractionation of plant chromatin. Biochem Biophys Res Commun 60:498–506
Lin CY, Guilfoyle TJ, Chen YM, Key JL (1975) Isolation of nucleoli and localization of ribonucleic acid polymerase I from soybean hypocotyl. Plant Physiol 56:850–852
Lin CY, Chen T, Guilfoyle TJ, Key JL (1976) Selective modulation of RNA polymerase I activity during growth transitions in soybean seedlings. Plant Physiol 58:614–617
Lindell TJ, Weinberg F, Morris PW, Roeder RG, Rutter WJ (1970) Specific inhibition of nuclear RNA polymerase II by α-amanitin. Science 170:447–449
Link G (1981) Enhanced expression of a distinct plastid DNA region in mustard seedlings by continuous far-red light. Planta 152:379–380
Link G, Richter G (1975) Properties and subunit composition of RNA polymerase II from plant cell cultures. Biochim Biophys Acta 395:337–346
Link G, Richter G (1977) Characterization of a protein factor stimulating RNA synthesis by DNA-dependent RNA polymerase II from plant cell cultures. Eur J Biochem 76:119–128
Link G, Kidd GH, Richter G, Bogorad L (1978) Structural relationship among the multiple forms of DNA-dependent RNA polymerase II from cultured parsley cells. Eur J Biochem 91:363–368
Losick R, Pero J (1976) Regulatory subunits of RNA polymerase. In: Losick R, Chamberlin M (eds) RNA polymerase. Cold Spring Harbor Lab, New York, pp 227–246
Luthe DS, Quatrano RS (1980) Transcription in isolated wheat nuclei. II. Characterization of RNA synthesized in vitro. Plant Physiol 65:309–313
Maheshwari SC, Guha S, Gupta S (1966) The effect of indoleacetic acid on the incorporation of (^{32}P)orthophosphate and (^{14}C)adenine into plant nuclei in vitro. Biochim Biophys Acta 117:470–472
Matthysse AG, Abrams M (1970) A factor mediating interaction of kinetin with the genetic material. Biochim Biophys Acta 199:511–518
Matthysse AG, Phillips C (1969) A protein intermediary in the interaction of a hormon with the genome. Proc Natl Acad Sci USA 63:897–907
Mazus B, Brodniewicz-Proba T (1976) RNA polymerases I and II in germinating wheat embryo. Acta Biochem Pol 23:261–266

Mazus B, Buchowicz J (1972) RNA polymerase activity in resting and germinating wheat seeds. Phytochemistry 11:2443–2446

McComb AJ, McComb JA, Duda CT (1970) Increased ribonucleic acid polymerase activity associated with chromatin from internodes of dwarf pea plants. Plant Physiol 46:221–223

McIntosh L, Jolly SO, Bogorad L (1979) Chloroplast RNA polymerase binding sites near three maize plastid genes. In: Genome organization and expression in plants. Conference Edinburgh, July 1979, Abstract B32

Mennes AM, Bouman H, van der Burg MPM, Libbenga KR (1978) RNA synthesis in isolated tobacco callus nuclei, and the influence of phytohormones. Plant Sci Lett 13:329–339

Miassod R, Penon P, Teissere M, Ricard J, Cecchini J-P (1970) Contribution à l'étude d'acides ribonucleiques a marquage rapide de la racine de lentille. Isolement et caractérisation, action d'une hormone sur leur synthèse et sur celle de peroxydases. Biochim Biophys Acta 224:423–440

Millerd A, Spencer D (1974) Changes in RNA synthesizing activity and template activity in nuclei from cotyledons of developing pea seeds. Aust J Plant Physiol 1:331–341

Mizuochi T, Fukasawa H (1975) Characterization of an α-amanitin insensitive RNA polymerase from cauliflower. Phytochemistry 14:1707–1710

Mondal H, Mandal RK, Biswas BB (1970) Factors and rifampicin influencing RNA polymerase isolated from chromatin of eukaryotic cell. Biochem Biophys Res Commun 40:1194–1200

Mondal H, Mandal RK, Biswas BB (1972a) RNA polymerase from eukaryotic cells. Isolation and purification of enzymes and factors from chromatin of coconut nuclei. Eur J Biochem 25:463–470

Mondal H, Mandal RK, Biswas BB (1972b) RNA stimulated by indole acetic acid. Nature New Biol 240:111–113

Mondal H, Ganguly A, Das A, Mandal RK, Biswas BB (1972c) Ribonucleic acid polymerase from eukaryotic cells. Effects of factors and rifampicin on the activity of RNA polymerase from chromatin of coconut nuclei. Eur J Biochem 28:143–150

Mullinix KP, Strain GC, Bogorad L (1973) RNA polymerases of maize. Purification and molecular structure of DNA-dependent RNA polymerase II. Proc Natl Acad Sci USA 70:2386–2390

Müntz K (1982) Seed development. In: Boulter D, Parthier B (eds) Nucleic acids and proteins in plants. Encyclopedia of plant physiology new ser vol 14A. Springer, Berlin, Heidelberg, New York, pp 505–558

Muthukrishnan S, Maxwell ES, Chandra GR (1979) Messenger RNA for α-amylase in aleurone layers of barley seeds treated with gibberellic acid. In: Genome organization and expression in plants. Conference Edinburgh, July 1979, Abstract A 38

Nakashima H (1978) Further studies on the diurnal rhythm of uridine incorporation into RNA in the long-day duckweed, *Lemna gibba* G 3. Plant Cell Physiol 19: 375–383

Nakashima H (1979a) Diurnal rhythm of nuclear RNA polymerase I activity in a duckweed, *Lemna gibba* G 3, under continuous light conditions. Plant Cell Physiol 20:165–176

Nakashima H (1979b) Effects of incubation of nuclei from a longday duckweed, *Lemna gibba* G 3, under different ionic conditions on the activity of RNA synthesis. Plant Cell Physiol 20:715–724

O'Brien TJ, Jarvis BC, Cherry JH, Hanson JB (1968) Enhancement by 2,4-dichlorophenoxyacetic acid of RNA polymerase in soybean hypocotyl tissue. Biochim Biophys Acta 169:35–43

Parthier B (1979) The role of phytohormones (cytokinins) in chloroplast development. Biochem Physiol Pflanz 174:173–214

Payne PI (1976) The long lived messenger ribonucleic acid of flowering plant seeds. Biol Rev 51:329–363

Payne PI (1977) Synthesis of poly(A)-rich RNA in embryos of rye during imbibition and early germination. Phytochemistry 16:431–434

Petranyi P, Jendrisak JJ, Burgess RR (1977) RNA polymerase II from wheat germ contains tightly bound zinc. Biochem Biophys Res Commun 74:1031–1038

Polya GM (1973) Wheat leaf RNA polymerases III. Purification of the soluble RNA polymerase. Arch Biochem Biophys 155:125–135

Polya GM, Jagendorf AT (1971a) Wheat leaf RNA polymerases. I Partial purification and characterization of nuclear, chloroplast and soluble DNA-dependent enzymes. Arch Biochem Biophys 146:635–648

Polya GM, Jagendorf AT (1971b) Wheat leaf RNA polymerases. II. Kinetic characterization and template specificities of nuclear, chloroplast and soluble enzymes. Arch Biochem Biophys 146:649–657

Poulson R, Beevers L (1970) Nucleic acid metabolism during greening and unrolling of barley leaf segments. Plant Physiol 46:315–319

Ramusz M, Doly J, Mandel P, Chambon P (1965) A soluble DNA-dependent RNA polymerase in nuclei of non-dividing animal cells. Biochem Biophys Res Commun 19:114–120

Rawson J, Boerma C (1976) A measurement of the fraction of chlorplast DNA transcribed during chloroplast development in *Euglena gracilis*. Biochemistry 15:588–592

Rawson JRY, Boerma CL (1979) Hybridization of EcoRI chloroplast DNA fragments of *Euglena* to pulse labeled RNA from different stages of chloroplast development. Biochem Biophys Res Commun 89:743–749

Rejman E, Buchowicz J (1973) RNA synthesis during the germination of wheat seeds. Phytochemistry 12:271–276

Revie D, Dahmus ME (1979) Purification and partial characterization of a stimulatory factor for lamb thymus RNA polymerase II. Biochemistry 18:1813–1820

Rizzo PJ, Cherry JH (1975) Developmental changes in multiple forms of deoxyribonucleic acid dependent ribonucleic acid polymerase in soybean hypocotyl. Plant Physiol 55:574–577

Rizzo PJ, Cherry JH, Pederson K, Dunham VL (1974) Separation and partial characterization of multiple ribonucleic acid polymerases from soybean hypocotyl. Plant Physiol 54:349–355

Rizzo PJ, Pedersen K, Cherry JH (1977) Stimulation of transcription by a soluble factor isolated from soybean hypocotyl by 2,4-D affinity chromatography. Plant Sci Lett 8:205–211

Roberts JW (1976) Transcription termination and its control in *E. coli*. In: Losick R, Chamberlin M (eds) RNA polymerase. Cold Spring Harbor Lab, New York, pp 247–271

Roberts LM, Lord JM (1979) Developmental changes in the activity of messenger RNA isolated from germinating castor bean endosperm. Plant Physiol 64:630–634

Roeder RG (1976) Eukaryotic nuclear RNA polymerases. In: Losick R, Chamberlin M (eds) RNA polymerase. Cold Spring Harbor Lab, New York, pp 285–329

Roeder RG, Rutter WJ (1969) Multiple forms of DNA-dependent RNA polymerase in eukaryotic organisms. Nature (London) 224:234–237

Romanko EG, Selivankina SY, Kureyedow VA (1978) Effect of cytokinin on RNA synthesis and on the activity of chromatin-bound RNA polymerase in excised barley leaves. Physiol Rast 25:1199–1205

Rosner A, Reisfeld A, Jacob KM, Gressel J, Edelman M (1977) Shifts in the RNA and protein metabolism of *Spirodela* (duckweed) In: Bogorad L, Weil JH (eds) Acides nucléiques et synthèse des proteines chez les végétaux. CNRS, Paris, pp 561–568

Roy P, Biswas BB (1977) A receptor protein for indoleacetic acid from plant chromatin and its role in transcription. Biochem Biophys Res Commun 74:1597–1606

Rushlow KE, Orozco EM, Lipper C, Hallick RB (1980) Selective in vitro transcription of *Euglena* chloroplast ribosomal RNA genes by a transcriptionally active chromosome. J Biol Chem 255:3786–3792

Santel H-J, Apel K (1981) The protochlorophyllide holochrome of barley (*Hordeum vulgare* L.). The effect of light on the NADPH:protochlorophyllide oxidoreductase. Eur J Biochem 120:95–103

Salomon D, Mascarenhas JP (1972) Auxin and cyclic 3′,5′-adenosine monophosphate during the isolation of chromatin from *Avena* coleoptiles: Effects on cell free RNA synthesis. Biochem Biophys Res Commun 47:134–141

Sasaki Y, Sasaki R, Hashizume T, Yamada Y (1973) The solubilization and partial characterization of pea RNA polymerases. Biochem Biophys Res Commun 50:785–792

Sasaki Y, Goto H, Wake T, Sasaki R (1974) Purine ribonucleotide homopolymer formation activity of RNA polymerase from cauliflower. Biochim Biophys Acta 366:443–453

Sasaki Y, Goto H, Ohta H, Kamikubo T (1976) Template activity of synthetic deoxyribonucleotide polymers in the eukaryotic DNA-dependent RNA polymerase reaction. Eur J Biochem 70:369–375

Sasaki Y, Goto H, Toni H, Kamikubo T (1978) DNA-dependent RNA polymerase III from cauliflower. Characterization and template specificity. Biochim Biophys Acta 517:205–215

Sasaki Y, Ishiye M, Goto H, Kamikubo T (1979a) Purification and subunit structure of RNA polymerase II from the pea. Biochim Biophys Acta 564:437–447

Sasaki Y, Tomi H, Kamikubo T (1979b) Effect of light on the solubilized RNA polymerase level in pea buds. Plant Sci Lett 14:355–364

Schiemann J, Wollgiehn R, Parthier B (1977) Isolation of a transcription-active RNA polymerase-DNA complex from *Euglena* chloroplasts. Biochem Physiol Pflanz 171:474–478

Schröder J, Kreuzaler F, Schäfer E, Hahlbrock K (1979) Concomitant induction of phenylalanine ammonia-lyase and flavone synthase mRNAs in irradiated plant cells. J Biol Chem 254:57–65

Schwartz LB, Sklar VEF, Jaehning JA, Weinman R, Roeder RG (1974) Isolation and partial characterization of the multiple forms of deoxyribonucleic acid-dependent ribonucleic acid polymerase in the mouse myoloma MOPC 315. J Biol Chem 249:5889–5897

Scott NS, Nair H, Smillie RM (1971) The effect of red irradiation on plastid ribosomal RNA synthesis in dark-grown pea seedlings. Plant Physiol 47:385–388

Seidman S, Witney F, Surzycki S (1979) In vitro transcription of Ad 2 DNA by eukaryotic RNA polymerase II. I. Kinetics of formation of stable binary complexes at specific sites. Miami Winter Symp 1979, 16 (From Gene Protein: Inf. Transfer Norm. Abnorm. Cells), p 626 (After: Chem Abstracs, Biochem Sect, 92:89943e (1980)

Seifart KH, Benecke BJ (1975) DNA dependent RNA polymerase C. Occurence and localization in various animal cells. Eur J Biochem 53:293–300

Seifart KH, Jahasz PP, Benecke BJ (1973) A protein factor from rat-liver tissue enhancing the transcription of native templates by homologous RNA polymerase B. Eur J Biochem 33:181–191

Selivankina SYu, Romanko EG, Kuroyedov VA, Kulaeva ON (1979) Enhancement of the activity of chromatin-bound RNA polymerase with 6-benzylaminopurine added in the course of chromatin isolation. Fisiol Rast 26:41–47

Sen S, Payne PI, Osborne DJ (1975) Early ribonucleic acid synthesis during the germination of rye (*Secale cereale*) embryos and the relationship to early protein synthesis. Biochem J 148:381–387

Sentenac A, Dezélée S, Iborra F, Buhler J-M, Huet J, Wyers F, Ruet A, Fromageot P (1976) Yeast RNA polymerases. In: Losick R, Chamberlin M (eds) RNA polymerase. Cold Spring Harbor Lab, New York, 763–778

Sieliwanowicz B, Chmielewska I (1973) Studies on the initiation of protein synthesis in the course of germination of pea seeds. Bull Acad Polon Sci Ser Sci Biol 21:399–404

Sklar VEF, Yamamoto M, Roeder RG (1976) Molecular structures of eukaryotic class III RNA polymerases. In: Losick R, Chamberlin M (eds) RNA polymerase. Cold Spring Harbor Lab, New York, pp 803–817

Slater RJ, Venis MA, Grierson D (1978) Characterization of ribonucleic acid synthesis by nuclei isolated from *Zea mays*. Planta 144:89–93

Smith HJ, Bogorad L (1974) The polypeptide subunit structure of the DNA-dependent RNA polymerase of *Zea mays* chloroplasts. Proc Natl Acad Sci USA 71:4839–4842

Spiegel S, Obendorf RL, Marcus A (1975) Transcription of ribosomal and messenger RNAs in early wheat embryo germination. Plant Physiol 56:502–507

Spindler SR (1979) Deoxyribonucleic acid dependent ribonucleic acid polymerase II. Specific initiation and elongation factors from calf thymus. Biochemistry 18:4042–4048

Stout ER, Mans RJ (1967) Partial purification and properties of RNA polymerase from maize. Biochim Biophys Acta 134:327–336

Stout ER, Parenti R, Mans RJ (1967) An increase in RNA polymerase activity after illumination of dark-grown maize seedlings. Biochem Biophys Res Commun 29:322–326

Strain GC, Mullinix KP, Bogorad L (1971) RNA polymerase of maize. Nuclear RNA polymerases. Proc Natl Acad Sci USA 68:2647–2651

Surzycki SJ, Shellenbarger DL (1976) Purification and characterization of a putative sigma factor from *Chlamydomonas reinhardii*. Proc Natl Acad Sci USA 73:3961–3965

Teissere M, Penon P, Ricard J (1973) Hormonal control of chromatin availability and of the activity of purified RNA polymerases in higher plants. FEBS Lett 30:65–70

Teissere M, Penon P, v Huystee RB, Azon Y, Ricard J (1975) Hormonal control of transcription in higher plants. Biochim Biophys Acta 402:391–402

Teissere M, Penon P, Azou Y, Ricard J (1976) On the mode of action of transcriptional factors in higher plants. Plant Sci Lett 6:49–55

Teissere M, Penon P, Azou Y, Ricard J (1977) RNA polymerase III from wheat embryos. Purification by affinity and hydrophobic chromatographies. Characterization and molecular properties. FEBS Lett 82:77–81

Teissere M, Durant R, Ricard J, Croke R, Penon P (1979) Transcription in vitro of cauliflower mosaic virus DNA by RNA polymerase I, II and III purified from wheat embryos. Biochem Biophys Res Commun 89:526–533

Thien W, Schopfer P (1975) Control by phytochrome of cytoplasmic and plastid rRNA accumulation in cotyledons of mustard seedlings in the absence of photosynthesis. Plant Physiol 56:660–664

Tobin EM (1978) Light regulation of specific mRNA species in *Lemna gibba* L. G-3. Proc Natl Acad Sci USA 75:4749–4753

Tobin EM (1979) Light regulation of mRNAs coding for the small subunit of ribulose-1,5 bisphosphate carboxylase and chlorophyll a/b binding protein in *Lemna gibba*. In: Genome organization and expression in plants. Conference Edinburgh/Scotland, July 1979. Abstract B 36

Tobin EM (1981) White light effects on the mRNA for the light-harvesting chlorophyll a/b-protein in *Lemna gibba* L. G-3. Plant Physiol 67:1078–1083

Venis MA (1971) Stimulation of RNA transcription from pea and corn DNA by protein retained on Sepharose coupled to 2,4-dichlorophenoxyacetic acid. Proc Natl Acad Sci USA 68:1824–1827

Verdier G (1975) Synthesis and translation of light-induced mRNAs in etiolated *Euglena gracilis*. Biochim Biophys Acta 407:91–98

Verdier G (1979a) Poly(adenylic acid)-containing RNA of *Euglena gracilis* during chloroplast development. 1. Analysis of their complexity by hybridization to complementary DNA. Eur J Biochem 93:573–580

Verdier G (1979b) Poly(adenylic acid)-containing RNA of *Euglena gracilis* during chloroplast development. 2. Transcriptional origin of the different RNA. Eur J Biochem 93:581–600

Verdier G, Trabuchet G, Heizmann P, Nigon V (1973) Effet de l'éclairement sur les synthèses de RNA et de séquences polyadényliques dans des cultures d'*Euglena gracilis* étiolées. Biochem Biophys Acta 312:528–539

Verma DPS, MacLachlan GA, Byrne H, Ewings D (1975) Regulation and in vitro translation of messenger ribonucleic acid for cellulase from auxin-treated pea cotyledons. J Biol Chem 250:1019–1026

Walbot V (1971) RNA metabolism during embryo development and germination of *Phaseolus vulgaris*. Dev Biol 26:369–379

Walden R, Leaver CJ (1981) Synthesis of chloroplast proteins during germination and early development of cucumber. Plant Physiol 67:1090–1096

Walle van de C, Bernier G, Deltour R, Bronchart R (1976) Sequence of reactivation of ribunucleic acid synthesis during early germination of the maize embryo. Plant Physiol 157:632–639

Walle van der C, Neuray J, Dommes J (1979) Translation of newly synthesized messenger RNA during germination of maize. In: Genome organization and expression in plants. Conference Edinburgh/Scotland, July 1979, Abstract A 24.

Wasylyk B, Kedinger C, Corden J, Brison D, Chambon P (1980) Specific in vitro initiation of transcription on conalbumin and ovalbumin genes and comparison with adenovirus-2 early and late genes. Nature (London) 285:367–373

Wechselberger M, Wielgat B, Kahl G (1979) Rhythmic changes in transcriptional activity during the development of potato tubers. Planta 147:199–204

Weil PA, Luse DS, Segall J, Roeder RG (1979a) Selective and accurate initiation of transcription at the Ad 2 major late promoter in a soluble system dependent on purified RNA polymerase II and DNA. Cell 18:469–484

Weil PA, Segall J, Harris B, Ng S-Y, Roeder RG (1979b) Faithful transcription of eukaryotic genes by RNA polymerase III in systems reconstituted with purified DNA templates. J Biol Chem 254:6163–6173

Weinmann R, Roeder RG (1974) Role of DNA-dependent RNA polymerase III in the transcription of the tRNA and 5S RNA genes. Proc Natl Acad Sci USA 71:1790–1794

Weir EM, Riezman H, Grienenberger JM, Becker WM, Leaver CJ (1980) Regulation of glyoxysomal enzymes during germination of cucumber. Temporal changes in translatable mRNAs for isocitrate lyase and malate synthase. Eur J Biochem 112:469–477

Weiss SB (1960) Enzymatic incorporation of ribonucleoside triphosphates into the interpolynucleotide linkages of ribonucleic acid. Proc Natl Acad Sci USA 46:1020–1030

Wielgat B, Kahl G (1979a) Enhancement of polyribosome formation and RNA synthesis of gibberellic acid in wounded potato tuber tissue. Plant Physiol 64:863–866

Wielgat B, Kahl G (1979b) Gibberellic acid activates chromatin-bound DNA-dependent RNA polymerase in wounded potato tuber tissue. Plant Physiol 64:867–871

Wielgat B, Wechselberger M, Kahl G (1979) Age-dependent variations in transcriptional response to wounding and gibberellic acid in a higher plant. Planta 147:205–209

Wollgiehn R, Munsche D (1972) RNS-Synthese in isolierten Chloroplasten von *Nicotiana rustica*. Biochem Physiol Pflanz 163:137–155

Wollgiehn R, Parthier B (1979) RNA synthesis in isolated chloroplasts of *Euglena gracilis*. Plant Sci Lett 16:203–210

Wollgiehn R, Parthier B (1980) RNA and protein synthesis in plastid differentiation. In: Reinert J (ed) Chloroplast differentiation. Springer, Berlin, Heidelberg, New York, pp 97–145

Yoshida K, Sasaki K (1977) Changes in template activity and proteins of chromatin during wheat germination. Plant Physiol 59:497–501

Zillig W (1976) Function and reassembly of subunits of DNA-dependent RNA polymerase. In: Losick R, Chamberlin M (eds) RNA polymerase. Cold Spring Harbor Lab, New York, pp 101–125

Zylber EA, Penman S (1971) Products of RNA polymerases in *HeLa* cell nuclei. Proc Natl Acad Sci USA 68:2861–2865

6 RNA Sequences

T.A. DYER

1 Introduction

All RNA molecules have a primary structure which consists of a backbone of alternating ribose and phosphate residues with a purine (guanine or adenine) or pyrimidine (uracil or cytosine) base attached to each ribose (Fig. 1). Although these four bases (abbreviated as G, A, U and C respectively) predominate, some bases may be modified in the formation of the mature molecule. Also there is 2′-O-methylation of a few ribose residues. A comprehensive list of the structures of these modified residues and of the symbols used to denote them has been published (DUNN and HALL 1975).

The sequence of bases in an RNA molecule represents one of its most basic characteristics and it is ultimately the sequence which determines its biological properties. From the sequence one can

- predict the way in which the RNA may be folded;
- identify the sites at which it may interact with protein or other nucleic acids;
- determine the relationship between gene and gene product in order to find out, for instance, whether there is a precursor to the mature molecule and also to help define the coding sequence;
- compare homologous molecules from different organisms to determine what features they have in common and are therefore important for them to function;
- determine the identity of the RNA and its specific function such as, for example, the coding properties (and codon usage) of a messenger RNA molecule or the amino acid specificity of a transfer RNA molecule.

Furthermore, the sequence of most RNA molecules has only been changed very infrequently and usually at a fairly constant rate during the evolution of an organism. Therefore the sequence provides a good indication of the evolu-

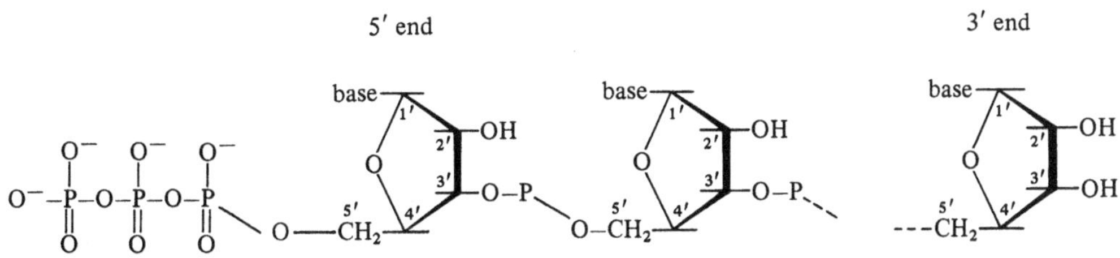

Fig. 1. Structure of RNA showing the 5′ and 3′ ends found in primary transcripts

Abbreviations: $tRNA_i^{Met}$, initiator tRNA.

tionary origin of the RNA and of the organism of which it is part. Thus there are a number of compelling reasons to sequence RNA.

2 Conventions for the Graphical Representation of RNA Sequences

An RNA molecule has a polarity. At one end (the 5′ end) the carbon in the 5′ position of the ribose usually has one, two or three phosphate residues attached to it and the carbon in the 3′ position of this ribose is in a phosphodiester bond with the next ribose (Fig. 1). At the other end of the molecule (the 3′ end) the carbon in the 5′ position of the terminal ribose is in a phosphodiester bond with the previous ribose residue and the 3′ carbon is usually unphosphorylated.

In representing the sequence of nucleotides in an RNA molecule in a linear fashion, the 5′ end is usually shown to the left and the 3′ end to the right of the diagram. Mono-, di- and triphosphates at the 5′ end are written as pN, ppN and pppN respectively, N being the symbol used to denote an unidentified nucleoside and p a phosphate residue. A phosphate residue at the 3′ end is shown to the right of the nucleoside symbol (Np). Symbols for internal phosphate residues are not usually shown. A hydroxyl group on the 5′ carbon at the 5′ end of the molecule is not shown but that on the 3′ carbon at the 3′ end of the molecule often is (N_{OH}), particularly when it is necessary to emphasize that it is not phosphorylated in this position.

Conventions in Denoting tRNA Sequences. The numbering of residues as in yeast $tRNA^{Phe}$ is used following the rules proposed by the participants of the 1978 Cold Spring Harbor Meeting on tRNA (see GAUSS and SPRINZL 1981). This numbering permits comparisons with the three-dimensional structure which has been determined for the $tRNA^{Phe}$ of yeast. When the sequence is written in linear form, the regions which are probably in secondary structure (Watson-Crick-type base pairs) may be indicated by specific underlining. For additional information concerning the representation of sequences in this way the reader is referred to the tRNA sequence compilation of GAUSS and SPRINZL (1981).

3 Structure of tRNA

Transfer RNA's contain, in some molecules, as few as 73 nucleotides and in others, as many as 90 residues. The three nucleotides of the anticodon are approximately in the middle of the chain and the 3′ end always terminates with the sequence $-CCA_{OH}$. The tRNA is charged by the attachment of an amino acid through its carboxyl group to the 2′ or the 3′ carbon of the ribose of the terminal adenosine residue.

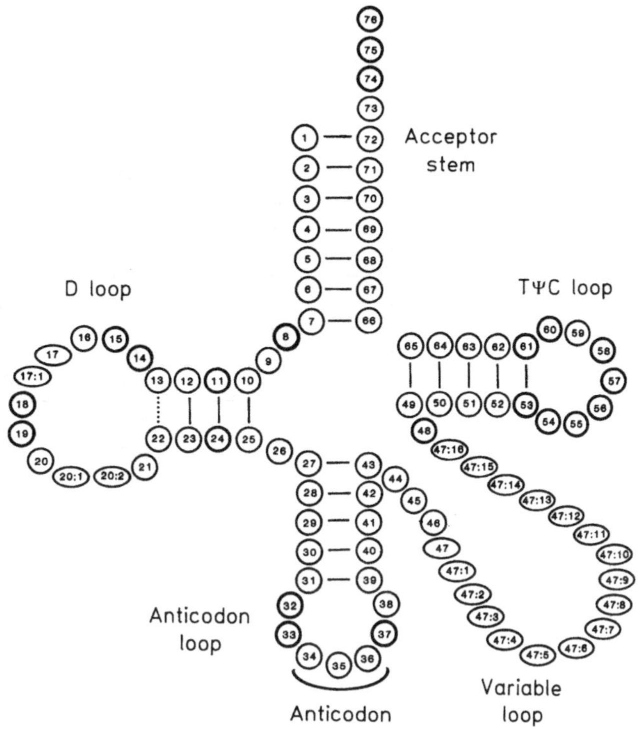

Fig. 2. Cloverleaf representation of tRNA structure showing possible Watson-Crick base pairing between bases. The *numbering of the residues* in the tRNA is according to that for phenyl-alanine tRNA of yeast. Residues in excess of those in the yeast RNA^{Phe} are given by a *colon followed by a further number. Thick-lined circles* denote residues which are invariant or semi-invariant

A basic feature of tRNA structure is the folding back of the chain upon itself in a highly ordered fashion. In yeast $tRNA^{Phe}$ the molecule is roughly L-shaped with the amino acid acceptor site at one extremity of a single-stranded segment protruding from one end of this "L", with the nucleotides of the anticodon in a relatively exposed position at the other end (see RICH and KIM 1978). This structure is stabilized by a combination of different forces. Just over half the bases form hydrogen-bonded base pairs of the type found in DNA (Watson-Crick base pairs), resulting in the formation of three helical regions. However, the largest contribution to the stability of the molecule is due to extensive base stacking which occurs as a result of the orientation of most of the bases, so that there is interaction between their hydrophobic flat faces.

The nucleotide sequence of a tRNA molecule is frequently drawn to show Watson-Crick base pairing with the result that the diagram has a cloverleaf appearance. The terms used to refer to the different parts of a tRNA molecule are derived from a description of such a diagram (Fig. 2). Regions in which there is base pairing are called stems and the single-stranded segments which they subtend are referred to as loops. However, many of the bases shown in "loops" also interact with each other through tertiary hydrogen bonding. One noteworthy feature of the tRNA molecule is the relatively large number of modified bases present. These may occur in or adjacent to the anticodon, while others predominate in the loop regions. The full structures of most of the modi-

Table 1. The genetic code

First position (5′ end)	Second position				Third position (3′ end)
	U	C	A	G	
U	Phe	Ser	Tyr	Cys	U
	Phe	Ser	Tyr	Cys	C
	Leu	Ser	Term	Term	A
	Leu	Ser	Term	Trp	G
C	Leu	Pro	His	Arg	U
	Leu	Pro	His	Arg	C
	Leu	Pro	Gln	Arg	A
	Leu	Pro	Gln	Arg	G
A	Ileu	Thr	Asp	Ser	U
	Ileu	Thr	Asp	Ser	C
	Ileu	Thr	Lys	Arg	A
	Met	Thr	Lys	Arg	G
G	Val	Ala	Asp	Gly	U
	Val	Ala	Asp	Gly	C
	Val	Ala	Glu	Gly	A
	Val	Ala	Glu	Gly	G

fied bases which have been identified are given by DUNN and HALL (1975). One of the most interesting types are the N^6 derivatives of adenine which occur next to the third position of the anticodon which have pronounced cytokinin activity (BURROWS 1975).

Different tRNA's for the same amino acid (isoacceptors) exist in all organisms. Those identified in preparations from plants have been listed by WEIL (1979; WEIL and PARTHIER, this Series, Chap. 2, Vol. 14A). Of the 64 triplet codons (see Table 1), three usually signal termination of transcription and of the remaining 61, the maximum likely to occur is 54 as A is not found in the first position of the anticodon (JUKES 1977). Twenty amino acids commonly occur in proteins, so at least one tRNA species must exist for each of these and protein synthesis is started by a specific initiator methionine tRNA. Various suggestions have been made as to what is the minimum number of tRNA species necessary to unambiguously read all the codons and this could be as few as 24 (LAGERKVIST 1978). The number of species in the cytosol seems to be close to the maximum, but in chloroplast and mitochondria there appear to be fewer species.

3.1 Specific Plant tRNA Sequences

3.1.1 Sequences of Cytosolic tRNA's

Several plant cytosolic tRNA's have now been sequenced, mainly from wheat germ (Table 2). Their sequences are typical of those found in the cytosol of eukaryotes. For example the plant $tRNA^{Phe}$ differs in composition from that of yeast $tRNA^{Phe}$ in only 13 out of 76 positions and all but one of these differences are in base paired regions. Nearly all the alterations in bases have occurred so as to preserve base pairing. Except for the initiator tRNA, the Tψ C loop of seven bases contains the sequence TψCG. However, some of the glycine and threonine isoacceptors and at least one tyrosine tRNA have U rather than T in this sequence (MARCU et al. 1978). The anticodon loop also contains seven bases with a purine next to the 3′ end and pyrimidine next to the 5′ end of the anticodon. The most unusual feature of the initiator $tRNA_i^{Met}$ of wheat is that it contains Aψ instead of Tψ in the TψC loop. The structure of this loop Aψ (or U)CGm^{1}AAA has been preserved in all the cytosol initiator tRNA's so far examined.

3.1.2 Sequences of Organelle tRNA's

Chloroplast tRNA's have been extensively studied (DRIESEL et al. 1979) and a number of species sequenced (Table 2). The main feature of interest is that they resemble the tRNA's of prokaryotes rather than the cytosolic components of eukaryotes. For example the initiator $tRNA_i^{Met}$ of bean chloroplasts has the typical TψCAAAU sequence of prokaryote initiator tRNA's in the TψC loop and also the first base cannot pair, another prokaryotic feature (CANADAY et al. 1980a). Furthermore the modified bases in chloroplast tRNA's resemble those of prokaryotes (GAUSS and SPRINZL 1981).

No mitochondrial tRNA's from plants have been sequenced yet. However, judging by what has been found in *Neurospora* they are likely to have some prokaryotic features but might be unique in other respects (HECKMAN et al. 1979). Mammalian mitochondrial tRNA's are even more individualistic than those of fungi (EPERON et al. 1980). Recent results have also shown that the genetic code in mitochondria is different to that found universally elsewhere and to accommodate this there are mitochondrial tRNA species which read codons differently from the usual (see HALL 1979).

4 mRNA Structure

The mRNA's of eukaryotes contain between 400 and 4000 nucleotides. In addition to the protein coding sequence, they usually contain a non-coding region at the 5′ end of the molecule. Apart from this common feature, the cytosolic and organelle mRNA's seem to differ in several fundamental respects from one another.

4.1 Cytosolic mRNA

A typical mRNA (Fig. 3) from the cytosol of a eukaryote has a modified base (cap structure) at its 5′ end followed by the segment of non-coding nucleotides (5′ non-coding sequence, leader sequence, 5′ untranslated region). The coding

Table 2. Known sequences of plant tRNA's

	Aminoacyl stem		D stem	D loop	D stem		Anticodon stem	Anticodon loop	Anticodon stem
	1 2 3 4 5 6 7	8 9	10 11 12 13	14 15 16 17 17:1 18 19 20 20:1 21	22 23 24 25	26	27 28 29 30 31	32 33 34 35 36 37 38	39 40 41 42 43
Glycine									
10	G C A Cm C A G	U m^1G	G U C Ψ	A G D G G U A	G A A U	A	G U A C C	C U G C C A m^5G	G U A C C
Leucine									
10	G G G G A U A	U G	G C G A	A A U U Gm G D A G	A C G C	Ψ	A C G G A	C U U* A A i^6A A	Ψ C C G U
11	G G C U U G A	U G	G U G A	A A U U Gm G D A G	A C A C	G	C G A G A	C U Cm A A A* A	U C U C G
2	G C C G C U A	U G	G U G A	A A U U Gm G D A G	A C A C	G	C U G C U	C U U A m^7G m^1G A	A G C A G
20	G C C G C U A	U G	G U G A	A A U U Gm G D A G	A C A C	G	C U G C U	C U U A m^7G m^1G A	A G C A G
Methionine									
0	A C C U A C U	U A	A C U C	A G C Gm G D D A	G A G Ψ	A	Ψ Ψ G C Ψ	U U C A U A C	G G C G G
Methionine initiator									
0	C G C G G A G	U A	G A G C	A A C U U Gm G D A	G C U C	G	C A A G G	C U C A U A A	C C U U G
1	A U C A G A G	U m^1G	m^2G C G C	A G C G G A A	G C G U	m^2G	G U G G G	C C C A U t^6A A	C C C A C
20	A U C A G A G	U m^1G	m^2G C G C	A G C G G A A	G C G U	m^2G	G Ψ G G G	C C C A U t^6A A	C C C A C
30	A G C U G A G	U m^1G	m^2G C G C	A G D G G A A	G C G Ψ	m^2G	A Ψ G G G	C U C A U t^6A A	C C C A U
30	C G C G G G G	U A	G A G C	A G U U U Gm G D A	G C U C	G	C A A G G	C U C A U A A	C C U U G
Phenylalanine									
0	G U C G G G A	U A	G C U C	A G U D Gm G D A	G A G C	A	G A G G A	C U G A A A* A	Ψ C C U C
20	G C U G G G A	U A	G C U C	A G D U G Gm U/D A	G A G C	G	G A G G A	C U G A A A* A	Ψ C C U U
1	G C C G A C U	U A	m^2G C U C	m^6A G D D G G G A	G A G C	m_2^2G	Ψ Ψ A G A	Cm U Gm A A yW A	Ψ C U A A
0	G C G G G G A	U A	m^2G C U C	A G D D G G G A	G A G C	m_2^2G	Ψ C A G A	Cm U Gm A A O_2yW A	Ψ C U G A
0	G U C G G G A	U A	G C U C	A G C U Gm G D A	G A G C	A	G A G G A	C U G A A A* A	U C C U C
Threonine									
0	G C C C C U U	U A	A C U C	A G U Gm G D A	G A G U	A	A C G. C C	A U G G U mt^6A A	G G C G U
Tryptophan									
0	G C G C U C U	U A	G U U C	A G U C Gm G D A	G A A C	m^2G	Ψ G G G Ψ	C U C C A A* A	A C C C G

0710 MARCU et al. (1977)
1010, 1011, 1012 OSORIO-ALMEIDA et al. (1980)
1020 CANADAY et al. (1980b)
1210 PIRTLE et al. (1981)
1310, 1311 CANADAY et al. (1980a)
1320 GHOSH et al. (1978)
1330 OLINS and JONES (1980)
1340 CALAGAN et al. (1980)
1410 GUILLEMAUT and KEITH (1977)
1420 CHANG et al. (1976)
1421 CHANG et al. (1978)
1430 DUDOCK and KATZ (1969); EVERETT and MADISON (1976), RAFALSKI et al. (1977); JANOWICZ et al. (1979)

sequence which follows has an initiator codon (AUG) at its 5′ end and a terminator codon (UAG, UAA or UGA) at its 3′ end. Then there is a 3′ non-coding segment (3′ non-coding sequence, 3′ untranslated region) and finally often a poly(A) tract [poly(A) tail] which contains up to 200 residues.

Variable loop 44 45 46	Variable loop 47 47:1 47:2 47:3 47:4 47:5 47:6 47:7 47:8 47:9 47:10 47:11	48	TΨC stem 49 50 51 52 53	TΨC loop 54 55 56 57 58 59 60	TΨC stem 61 62 63 64 65	Aminoacyl stem 66 67 68 69 70 71 72	73 74 75 76		
A G A		m^{5}C	m^{5}C m^{5}C G G G	U Ψ C G m^{1}A U U	C C C G G	C U G G U G C	A C C A	Cytosol	Wheat
C G A	C U U A A U A A A U C A	U	G A G G G	T Ψ C A A G U	C C U C C	U A U C C C C	A C C A	Chloroplast 1	Bean
U G C	U A A A G A G C G	U	G G A G G	T Ψ C G A G U	C C U C U	U C A A G U C	A C C A	Chloroplast 2	
U G C	U A G A G C A	U	C U C G G	T Ψ C G A G U	C C G A G	U A G C G G C	A C C A	Chloroplast 3	
U G C	G A G A G C A	U	C U C G G	T Ψ C G A G U	C C G A G	U A G C G G C	A C C A	Chloroplast 3	Spinach
G A m^{7}G	N	C	A U U G G	T Ψ C A A A U	C C A A U	A G U A G G U	A C C A	Chloroplast	Spinach
A A m^{7}G	U*	U	A C G G G	T Ψ C A A A U	C C C G U	C U C C G C A	A C C A	Chloroplast	Bean
A G m^{7}G	D	m^{5}C	C C A G G	A Ψ C G m^{1}A A A	C C U Gm G	C U C U G A U	A C C A	Cytosol	Bean
A G m^{7}G	D	m^{5}C	m^{5}C C A G G	A Ψ C G m^{1}A A A	C C U G G*	C U C U G A U	A C C A	Cytosol	Wheat
A G m^{7}G	D	m^{5}C	A C A G G	A U C G m^{1}A A A	C C U Gm U	C U C A G C U	A C C A	Cytosol	*Scenedesm obliquus*
A G m^{7}G	U	C	A C G G G	T Ψ C A A A U	C C U G U	C U C C G C A	A C C A	Chloroplast	Spinach
G U m^{7}G	U*	C	A C C A G	T Ψ C A A A U	C U G G U	U C C U G G C	A C C A	Chloroplast	Bean
A U m^{7}G	U*	C	A C C A G	T Ψ C A A A U	C U G G U	U C C U A G C	A C C A	Chloroplast	*Euglena*
A G m^{7}G	U*	C	C C U G G	T Ψ C G m^{1}A U C	C C G G G	A G Ψ C G G C	A C C A	Cytosol	*Euglena*
A G m^{7}G	D	C	G* C G U G	T Ψ C G m^{1}A U C	C A C G C*	U C A C C G C	A C C A	Cytosol	Wheat, pe lupin, barl
G U m^{7}G	U*	C	A C C A G	T Ψ C A A U	C U G G U	U C C U G G C	A C C A	Chloroplast	Spinach
A A m^{7}G	D	C	A U C G G	T Ψ C A A A U	C C G A U	A A G G G G C	U C C A	Chloroplast	Spinach
A U G	N	C	G U A G G	T Ψ C A A G U	C C A U C	A G A G C G U	G C C A	Chloroplast	Spinach

1440 CANADAY et al. (1980b)
1710 KASHDAN et al. (1980)
1810 CANADAY et al. (1981)
1010/34 Unknown derivative of uridine
1011/37 A derivative, probably i^{6}A, ms^{2}i^{6}A, zeatin or ms^2 zeatin
1210/47, 1810/47 N unidentified modified nucleotide
1310/47, 1410/47, 1420/47, 1440/47 3-N-(3-amino-3-carboxpropyl)uridine
1220/65 Modified derivative of guanosine
1410/37, 1420/37, 1440/37 ms^{2}i^{6}A
1421/47 Probably a derivative of uridine
1430/49 In lupin mainly A
1430/65 In lupin mainly U
1810/37 i^{6}A or ms^{2}i^{6}A

Eukaryote cytosolic mRNA's are functionally monocistronic and although they may contain more than one initiation site on a single mRNA, only the site nearest the 5′ end is active in initiation. Any internal site is inactive or cryptic (ROSENBERG and PATTERSON 1979).

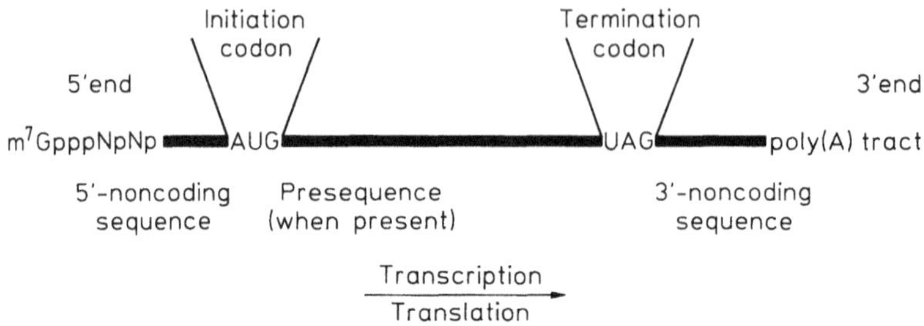

Fig. 3. Structure of messenger RNA found in the cytosol of eukaryotes

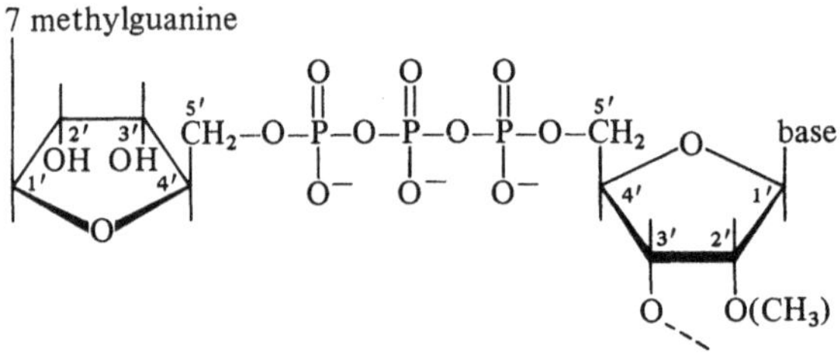

Fig. 4. "Cap" structure found at the 5′ end of many cytosolic messenger RNA's of eukaryotes

The 5′-terminal Cap. The 5′ terminal "cap" structure consists of 7-methylguanosine linked through its 5′-hydroxyl group to a triphosphate bridge. This connects it to the 5′ carbon of the penultimate nucleotide (Fig. 4). Cap structures are found in most eukaryote and viral mRNA's (reviews of SHATKIN 1976, FILIPOWICZ 1978, c.f. also HIRTH, Chap. 10, this Vol.; VAN VLOTEN-DOTING, Chap. 11, this Vol.) and they have been described in plants as well (SAINI and LANE 1977, HAUGLAND and CLINE 1978, NICHOLS 1979).

The 5′-Non-Coding Sequence. In eukaryotic cells initiation of translation occurs at the AUG triplet closest to the 5′ end of the molecule. The number of residues between this and the 5′ end of the molecule is highly variable. It may be as few as 9 (in brome mosaic virus 4) or over 200 (for example in the SV40 major-capsid-protein mRNA). Furthermore there is considerable sequence heterogeneity in this region in the different mRNA's which have been studied and, in some cases, even within a single species of mRNA.

The Coding Region. The linear sequence of nucleotides in the coding segment of mRNA is translated into the amino acid sequence of protein. The nucleotides are "read" in groups of three, referred to as codons, each codon specifying a particular amino acid, with the order of amino acids in the polypeptide being very important in determining the structure and function of the resulting protein. There are 64 different combinations of the four nucleotides and of these 61 have definite amino acid assignments and three signal termination of a polypeptide being synthesized (Table 1). The code is said to be degenerate, as many of the 20 different types of amino acid are selected for by more than one codon. This degeneracy is characterised by frequent third-place equivalence of cytosine and uracil and of guanine and adenine. The code appears to be identical in the nucleus-cytosol of plants and animals as plant mRNA's are apparently translated with complete fidelity in vitro by reticulocyte lysates, and animal mRNA's are correctly translated in wheat germ cell-free systems.

Synthesis of polypeptides starts with a methionine residue coded for by an AUG triplet. As this codon is the first one, it exactly specifies the amino acid composition and sequence of the polypeptide being synthesized as the rest of the coding sequence is then read in register. During translation, the N-terminal amino acid of the nascent polypeptide is specified by the initiator codon. Synthesis then continues without interruption to the carboxyl terminus where the polypeptide is terminated.

Termination of the completed polypeptide chain is dependent upon the presence of a specific termination codon, a protein release factor and GTP. The three termination codons, UAA, UAG and UGA are not recognized by tRNA's but are recognized instead by the specific release factor when they enter the ribosome.

The primary translation product specified by the coding sequence of the mRNA is not necessarily identical to that found in the mature protein. For instance polypeptides which are destined for transport across or insertion into membranes seem all to have a "signal sequence" which may be cleaved from the molecule either as it is discharged into or across the membrane (see BLOBEL and DOBBERSTEIN 1975, STEINER 1979); the signal sequences of cytosolic-synthesized chloroplast proteins, however, are cleaved in the stroma of the chloroplasts (SMITH and ELLIS 1979).

The signal sequences are usually found at the N-terminus of the nascent polypeptide and are comprised of 20 to 40 hydrophobic amino acids. The mRNA's for such polypeptides therefore contain a set of codons ("presequences") located just after the initiator codon specifying these signal sequences. In plants the storage proteins of legumes (HIGGINS and SPENCER 1980) and cereals (BURR et al. 1978, WEINAND and FEIX 1978, LARKINS and HURKMAN 1978) which are located within membrane-bound structures, all seem to be coded for by mRNA's which contain segments coding for such signal sequences.

The 3′-Non-Coding Sequence. Non-coding nucleotides are present towards the 3′ end of cytosolic mRNA's; over 600 residues may be present in such sequences. In any one family of mRNA's this segment seems to be highly conserved in sequence and one hexanucleotide AAUAAA has been found in the 3′-non-coding region of almost every mRNA sequenced so far.

The 3′-Poly(A) Tract. Between 30 and 40% of the cytosolic mRNA's of eukaryotes, including those of plants, contain a poly(A) tract at their 3′ terminus. The number of residues which are present varies from 50 to 200 which results in heterogeneity in size even in one type of mRNA molecule. The definition of polyadenylated mRNA is, however, a practical one based upon affinity chromatography which only selects molecules containing tracts of more than 10 to 20 residues.

4.2 Specific Cytosolic mRNA Sequences

The sequence of one plant cytosolic mRNA is now almost completely known (BEDBROOK et al. 1980). It is that of the mRNA for the small subunit of ribulose-

Table 3. Partial sequence of the messenger RNA coding for the small subunit of ribulose-1,5-bisphosphate carboxylase

←				Presequence				→				Start of mature molecule	↓			
Asn	Thr	Asp	Ile	Thr	Ser	Asn	Gly	Glu	Arg	Val	Lys	Cys	Met	Gln	Val	Trp
AAC	ACU	GAC	AUU	ACA	AGC	AAU	GGU	GAA	AGA	GUA	AAG	UGC	AUG	CAG	GUG	UGG
			10			20			30				40			50
Pro	Pro	Ile	Gly	Lys	Lys	Lys	Phe	Glu	Thr	Leu	Ser	Tyr	Leu	Pro	Pro	Leu
CCU	CCA	AUU	GGA	AAG	AAG	AAG	UUU	GAG	ACU	CUU	UCC	UAU	UUG	CCA	CCA	UUG
		60				70			80			90				100
Thr	Arg	Asp	Gln	Leu	Leu	Lys	Glu	Val	Glu	Tyr	Leu	Leu	Arg	Lys	Gly	Trp
ACG	AGA	GAU	CAA	UUG	UUG	AAA	GAA	GUU	GAA	UAC	CUU	CUG	AGG	AAG	GGA	UGG
		110			120				130			140			150	
Val	Pro	Cys	Leu	Glu	Phe	Glu	Leu	Leu	Lys	Gly	Phe	Val	Tyr	Gly	Glu	His
GUU	CCA	UGC	UUG	GAA	UUU	GAG	UUG	CUC	AAA	GGA	UUU	GUG	UAC	GGU	GAG	CAC
		160			170			180				190			200	
Asn	Lys	Ser	Pro	Arg	Tyr	Tyr	Asp	Gly	Arg	Tyr	Trp	Thr	Met	Trp	Lys	Leu
AAC	AAG	UCA	CCA	AGA	UAC	UAU	GAU	GGA	AGA	UAC	UGG	UCA	AUG	UGG	AAG	CUU
	210				220			230			240				250	
Pro	Met	Phe	Gly	Thr	Thr	Asp	Pro	Ala	Gln	Val	Val	Lys	Glu	Leu	Asp	Glu
CCU	AUG	UUU	GGC	ACC	ACU	GAU	CCU	GCU	CAA	GUC	GUG	AAG	GAG	GUU	GAU	GAA
	260			270				280			290			300		
Val	Val	Ala	Ala	Tyr	Pro	Glu	Ala	Phe	Val	Arg	Val	Ile	Gly	Phe	Asn	Asn
GUU	GUU	GCC	GCU	UAC	CCC	GAA	GCU	UUC	GUU	CGU	GUC	AUC	GGU	UUC	AAC	AAC
	310			320			330				340			350		
Val	Arg	Gln	Val	Gln	Cys	Ile	Ser	Phe	Ile	Ala	His	Thr	Pro	Glu	Ser	Tyr
GUU	CGU	CAA	GUU	CAA	UGC	AUC	AGU	UUC	AUU	GCA	CAC	ACA	CCA	GAA	UCC	UAC
360				370			380			390				400		

Term
UAA GUUCACUGCAUUGGAGUUCCUAUUUAUAUGUUAUGCUUUUAAGUUCCUUUUG
410 420 430 440 450 460

UUGUGUAUUUUUAUAAUUUCUGUUUUUGGAUUUCCAAAUUGCAAAUGGGAUGUGUG
470 480 490 500 510

UAAGAGUUAAUGAAUGAUAUGGUUAACUUUAUUCCCAAGUUUACUUGGCGGUUUGU
20 530 540 550 560 570

ACUGUGUGGCUUUCGUUGUUCAGUGUUCGACAUCUGUUGGUCCAAGUGCUAACACU
580 590 600 610 620 630

GGCACACGGUACGUAUUGGUCGGUGUUUGUGUAUCCU-PolyA
640 650 660 668

1,5-bisphosphate carboxylase of pea which is a nuclear encoded chloroplast protein. The intact messenger contains between 900 and 1000 residues of which about 100 are in the poly(A) tract and 260 in the 3′ non-coding region (Table 3). The mature protein contains 123 amino acids which are coded for by 369 residues. In the precursor protein there is a signal sequence of about 50 amino acids and the coding sequence of the last 13 of these has been determined but not that for the other 37 amino acids which occur near the beginning of the protein or that of the 5′ non-coding region. One notable feature of the molecule is that it lacks the AAUAAA sequence found in most cytosolic mRNA's near the poly(A) tract.

About 40% of the sequence of the mRNA coding for the major storage protein (phaseolin Gl-globulin) of the French bean is now also known (SUN et al. 1981). A comparison of this sequence with the gene sequence for this protein shows that the primary transcription product probably contains intervening segments which are excised during the maturation of the mRNA.

4.3 Organelle mRNA's

No chloroplast mRNA's have been sequenced yet. However, the sequence of the maize gene coding for the large subunit of ribulose-1,5-biphosphate carboxylase is known and from this the sequence of the protein-coding part of it can be inferred (see BOHNERT et al., Chap. 14, this Vol.). A feature of note is that just before this protein-coding part there is a sequence which would give GGAGG in the mRNA and this is complementary to the CCUCC sequence which occurs near the 3′ end of the chloroplast 16S ribosomal RNA (see Section 5). Such complementarity is thought to be important in initiation of protein synthesis (SHINE and DALGARNO 1974).

There have been reports recently that some chloroplast RNA's contain poly(A) tracts (SANO et al. 1979, VERDIER 1979a, b, BARTOLF and PRICE 1979). However, the poly(A) containing RNA has not yet been shown to be mRNA. In fact, it seems unlikely that most chloroplast mRNA molecules do contain tracts of this sort (WHEELER and HARTLEY 1975).

Nothing is known yet about plant mitochondrial mRNA sequences. In mammalian mitochondria some of the mRNA's contain poly(A) tracts (ALMALRIC et al. 1978) but there are few if any of these in the mRNA's in mitochondria of yeast (MOORMAN et al. 1978). The genetic code as used in mitochondria is slightly changed from that found elsewhere and therefore there are different codon assignments for some amino acids (see HALL 1979).

5 Types of Ribosomal RNA

The different types of ribosomal RNA which occur in plants are listed in Table 4. The large subunit of cytosolic ribosomes contains three RNA's: a high molecular weight species, 25S RNA (LOENING 1968), and two much smaller molecules, 5.8S (PAYNE and DYER 1972) and 5S RNA (PAYNE and DYER 1971, DYER and ZALIK 1979). The small subunit contains just one structural RNA component, 18S RNA (LOENING 1968).

The outstanding feature of the 16S and 23S RNA of the small and large subunits, respectively, of chloroplast ribosomes (LOENING and INGLE 1967) is that they are substantially smaller than the equivalent 18S and 25S RNA of the plant cytosolic ribosomes and about the same size as those of prokaryotes. As well as the 23S RNA, the large subunit of a chloroplast ribosome also contains a 5S (DYER and LEECH 1968, DYER and BOWMAN 1979) and, in higher

Table 4. Estimated size of the ribosomal RNA's of plants

Source of ribosomal RNA		Sedimentation coefficients (S)	Size parameters mol.wt.	No. nucleotides
Cytosol ribosomes	Large subunit	25S	1.3×10^{6} [a]	3,580[b]
		5.8S	5.07×10^{4}	157
		5S	3.84×10^{4}	120
	Small subunit	18S	0.7×10^{6}	1,926[b]
Chloroplast ribosomes	Large subunit	23S	1.05×10^{6} [a]	2,890[b]
		5S	3.94×10^{4}	122
		4.5S	$2.1–3.3 \times 10^{4}$	65–103
	Small subunit	16S	0.56×10^{6} [a]	1,541[b]
Mitochondrial ribosomes	Large subunit	24S	$1.12–1.26 \times 10^{6}$ [a]	3,082–3,470[b]
		5S	3.88×10^{4}	120
	Small subunit	18.5S	$0.69–0.78 \times 10^{6}$ [a]	1,800–2,146[b]

[a] Determined from their electrophoretic mobility relative to the high molecular weight RNA's of *E. coli* which were assumed to have molecular weights of 1.1 and 0.56×10^{6} (KURLAND 1960). These values are presumably for the salts of these polymers (VAN HOLDE and HILL 1974)

[b] Calculated from their molecular weights relative to that of *E. coli* 16S rRNA which contains 1541 nucleotides (BROSIUS et al. 1978, CARBON et al. 1978)

plants, a 4.5S RNA component too (WHITFELD et al. 1978, BOWMAN and DYER 1979). Two small RNA molecules in addition to 5S RNA have also been found in the large subunit of *Chlamydomonas reinhardii* (ROCHAIX and MALNOË 1978) but neither is equivalent to the 4.5S RNA of higher plants.

In the mitochondria of different plants the homologous high molecular weight ribosomal RNA's are not exactly the same size (PRING 1974, LEAVER 1975). Furthermore, they are substantially larger than the equivalent ribosomal RNA's of mammalian (12S and 16S) and yeast mitochondria (15S and 21S) and resemble more the plant cytosolic ribosomal RNA's in size. Also, in contrast to the ribosomes of all other mitochondria, those of plants also contain a 5S RNA molecule (LEAVER and HARMEY 1976, CUNNINGHAM et al. 1976).

5.1 High Molecular Weight Ribosomal RNA Sequences

There is only limited sequence data available concerning any of the plant high molecular weight ribosomal RNA's. This is mostly in the form of catalogues of the sequences of oligonucleotides formed by their digestion with T_1 ribonuclease. Such data is available for the chloroplast 16S RNA's of *Euglena gracilis* (ZABLEN et al. 1975), the red alga *Porphyridium* (BONEN and DOOLITTLE 1975, 1976) and the higher plant *Lemna minor* (WOESE and FOX 1977, WOESE and DYER unpublished results) and for the 18.5S rRNA of wheat mitochondria

(Bonen et al. 1977). The catalogues indicate a greater similarity between the rRNA of the organelles and those of prokaryotes than with the cytosolic ribosomal RNA of eukaryotes. The evolutionary implications of these findings are discussed fully elsewhere (Phillips and Carr 1977, Woese and Fox 1977).

Although none of the plant high molecular weight ribosomal RNA's has been sequenced yet, the complete sequence of the gene coding for maize chloroplast 16S RNA is known (Schwarz and Kössel 1980) and from this the sequence of the RNA can be inferred. The main findings of this study are that there is 76% homology between the 16S RNA of maize and *E. coli* and that there are regions of up to 53 consecutive bases identical in the two. The data indicates that the maize 16S RNA is 50 nucleotides shorter than the *E. coli* 16S RNA, possibly due to deletions. Because of these are fewer possibilities for the formation of hairpin loops in the chloroplast RNA.

The sequences which occur near the 3′ end of several small ribosomal subunit RNA's of plants have been determined (Table 5). It has been suggested that in prokaryotes, but not in eukaryotes, there is a sequence in this region which pairs with a complementary sequence in the 5′ non-coding region of the messenger RNA during initiation of protein synthesis (Shine and Dalgarno 1974). Maize and duckweed chloroplast 16S RNA's contain this sequence (CCUCC) but the wheat cytosolic 18S RNA does not, nor surprisingly does the 16S RNA of *Euglena*.

From what is known about ribosomal RNA sequences, it can be inferred that these molecules contain numerous hairpin loops. They appear to have appreciable secondary and tertiary structure, with the result that regions remote from one another in the linear molecule are associated when it is folded (Wollenzein et al. 1979).

Table 5. Sequences which occur at the 3′ end of the RNA from the small ribosomal subunit

Organism	Sequences	References
Maize chloroplast 16S[a]	– A A[b] GGUGCGGCUGGAUCACCUCCUU	Schwarz and Kössel (1980)
Duckweed chloroplast 16S	$-GAUCACCUCCU_{OH}$	Woese and Dyer, unpublished results
Euglena 16S	$-GAACAACUCN_{OH}$	Zablen et al. 1975
Wheat cytosol 18S	$-m^6_2Am^6_2ACCUGCGGAAGGAUCA*****UUG_{OH}$	Hagenbüchle et al. 1978; Darzynkiewicz et al. 1980

[a] Deduced from the sequence of the gene coding for the maize 16S RNA

[b] this adenosine residue may be methylated

5S ribosomal RNA

```
Cytosol
                                          10                  20                  30                  40                  50
                                        (AC)                                                  *
A1 rye[a]                                       * *
2  duckweed[b]      *       G                                       G
3  broad bean[b]    pA      *                                         U                                               C
4  dwarf bean[b]    *       *                                         U                                               C
5  sunflower[b]     *       *                                         U                                               C
6  tomato[b]        *                                                 C
7  spinach          *       G                                         U                                               C
8  wheat            p   G G A U G C G A U C A U A C C A G C A C U A A A G C A C C G G A U C C C A U C A G A A C U C C G A A
9  Chlorella        ppp A U G C U A C G U U C A U A C A C C A C G A A A G C A C C C G A U C C C A U C A G A A C U C G G A A

Chloroplast
B1 broad bean[b]                            C                                     A
2  dwarf bean[b]
3  tomato[b]
4  tobacco
5  duckweed         p   U A U U C U G G U G U C C U A G G C G U A G A G G A A C C A C A C C A A U C C A U C C C G A A C U U

Cytosol
                                          60                  70                  80                  90                  100
A1 rye[a]
2  duckweed[b]                                                  C
3  broad bean[b]
4  dwarf bean[b]
5  sunflower[b]                                                                                     C
6  tomato[b]                                                                                        C
7  spinach
8  wheat                G U U A A G C G U G C U U G G G C G A G A G U A G U A C U A G G A U G G G U G A C C U C C U G G G A
9  Chlorella            G U U A A A C G U G G U U G G G C U C G A C U A G U A C U G G G U U G G A G G A U U A C C U G A G U

Chloroplast
B1 broad bean[b]                              A                           A                                       C
2  dwarf bean[b]                                          *
3  tomato[b]                                              *
4  tobacco
5  duckweed             G G U G G U U A A A C U C U A C U G C G G U G A C G A U A C U G U A G G G G A G G U C C U G C G G A

Cytosol
                                          110                 120
A1 rye[a]                                                           *     U_OH
2  duckweed[b]                                                    C       *
3  broad bean[b]                    U                                     U_OH
4  dwarf bean[b]                                                          U N_OH
5  sunflower[b]                                                           *
6  tomato[b]                                                              *
7  spinach                                                      * C       U_OH
8  wheat                A G U C C U C G U G U U G C A U U C C C_OH
9  Chlorella            G G G A A C C C C G A C G U A G U G U_OH

Chloroplast
B1 broad bean[b]                              G
2  dwarf bean[b]                              G
3  tomato[b]
4  tobacco                                                              G
5  duckweed             A A A A U A G C U C G A C G C C A G A U U_OH
```

4·5S ribosomal RNA

```
                       10                  20                  30                  40                  50
C1  HOU                                                                                                     duckweed^b
2   HOU           A G                                     A * * * * * * *                                   maize
3   HOG A A G G U C A C G G C G A G A C G A G C C G U U A U C A U U A C G A U A G G U G U C A A G U G G A   tobacco

                       60                  70                  80                  90                 100
C1                                      A G                                           A G       U           duckweed^b
2                                       A G                               *     A     A A C     U           maize
3     A G U G C A G U G A U G U A U G C(G A)C U G A G G C A U C C U A A C A G A C C G G U A G A C U U G A   tobacco

      102
C1            duckweed^b
2             maize
3     A C_OH  tobacco
```

5·8S ribosomal RNA

```
                       10                  20                  30                  40                  50
D1  p(C)A C   C                                           C             C                                   wheat
2   p A G A A U G A C U C U C G G C A A C G G A Ψ A U C U A G G C U C U U G C A U C G A U G A A G A_m A C G   broad bean

                       60                  70                  80                  90                 100
D1                                      C                                           C                       wheat
2     U A G C G A A A U G C G A U A C U U G G U G U G A A U Ψ G_m C A G A A U C C C G U G A A C C A U C G A   broad bean

                      110                 120                 130                 140                 150
D1                                        (G)         *         C   C   * C C                   C           wheat
2     G U C U U U G A A C G C A A G U U G C∨C C C G A U G C C A U U A G G U U G A G G G C A C G U C U G C   broad bean

                      160
D1              C           G C_OH                                                                          wheat
2     C U G G G U G U C A C A U_OH                                                                          broad bean
```

[a] The inconsistencies between the rye and wheat sequences at positions 9–10 and 13–14 and at 36 are probably due to errors in the published rye sequence

[b] Sequences deduced by alignment of oligonucleotides produced by RNAase digestion

* Gaps placed in sequences to obtain maximal homology. Where there is appreciable homology with the main (first published) sequence, only base differences are shown

A1, 3–6 Payne and Dyer (1976)
A2 Dyer et al. (1977)
A7 Delihas et al. (1981)
A8 MacKay et al. (1980)
A8 Jordan et al. (1974)
B1–3, 5 Dyer and Bowman (1979)
B4 Takaiwa and Sugiura (1980b)
C1–2 Dyer and Bedbrook (unpublished results)
C3 Takaiwa and Sugiura (1980a)
D1 MacKay et al. (1980)
D2 Tanaka et al. (1980)

5.2 Low Molecular Weight Ribosomal RNA Sequences

Because of their small size and the relative ease with which they may be prepared, the low molecular weight ribosomal RNA's of plants have been studied in some detail. The sequence of examples of all but one of them has been established, the only exception being the mitochondrial 5S RNA for which there is a catalogue of the oligonucleotides produced by ribonuclease digestion (CUNNINGHAM et al. 1976).

5S Ribosomal RNA. All plant 5S RNA's contain about 120 residues (Table 6). Those from chloroplasts differ appreciably in sequence from their cytosol counterparts but bear a strong resemblance to the 5S RNA species of prokaryotes. In the higher plants the sequence of the cytosolic 5S RNA is highly conserved and there is even less variation in the sequences of the chloroplast 5S RNA. The difference between the cytosolic 5S RNA from higher plants and that of the green alga *Chlorella* is as great as that between these and mammalian and fungal cytosolic 5S RNA. This indicates that there was a very early evolutionary divergence between higher plants and green algae.

Despite the high degree of sequence conservation within a particular group the sequence of 5S RNA, as such, does not appear to be critical. Only the sequence CCGAAC between residues 40 and 50 is common to all the prokaryote and chloroplast 5S rRNA studied so far (see ERDMANN 1981). It is thought that the GAAC part of this might interact with the GTψC sequence of tRNA during protein synthesis (ERDMANN 1976). In cytosolic 5S RNA's there is a similar sequence which in the plants studied so far is AGAAC.

Physical and chemical studies show that 5S RNA has a high degree of secondary structure. From the sequences, a universal model for the folding of this molecule has been proposed (FOX and WOESE 1975a, b) and both the chloroplast (DYER and BOWMAN 1979) and cytosol 5S RNA (PAYNE and DYER 1976) conform to this model. The two ends of the molecule appear to be base paired and there are probably at least two other regions of base pairing. These results indicate that widely differing sequences may give rise to a molecule of similar shape. This shape may be essential for it to function and probably only those base substitutions that do not change this are tolerated.

There is frequently a di- or triphosphate residue at the 5′ end of cytosolic 5S RNA's, suggesting that they are a primary product of transcription. In contrast only monophosphate residues occur at the 5′ ends of chloroplast 5S RNA which is consistent with it having been excised from a larger precursor molecule.

Cytosol 5.8S RNA. The sequence of 5.8S RNA of broad bean has recently been determined (Table 6). It contains 163 nucleotides of which several are modified: ψ occurs at positions 22 and 78 and there are two 2′-O-methylated residues. The molecule contains regions of appreciable homology with human and yeast 5.8S RNA. In common with these it contains a G_{46} AAC_{49} sequence and it has been suggested that this, rather than a region in the 5S RNA, interacts with the TψC sequence in tRNA during protein synthesis in the cytosol of eukaryotes (WREDE and ERDMANN 1977).

The 5.8S RNA molecule appears to have a considerable amount of secondary structure. Models based on sequence data have been proposed for the way in which the human and yeast molecules can be folded (see LUOMA and MARSHALL 1978).

Chloroplast 4.5 S RNA. The sequence of chloroplast 4.5S RNA is now known (Table 6). It seems to bear some resemblance to the 3′ end of *E. coli* 23S RNA (MACKAY 1981) and may therefore be homologous to this region of the bacterial molecule. It is very unusual in having no 5′ phosphate residue and is probably cleaved from the precursor to the 23S RNA during the maturation of the latter (HARTLEY 1979).

6 Prospects

There can be little doubt that sequence data can be of immense value when evaluating the biological potential of an RNA molecule. Also, such data can be decisive in determining the homologies between RNA's or in establishing their evolutionary derivation. For these reasons there will be considerable incentive for further RNA's to be sequenced. The methodology for doing this is improving so rapidly that it should not be long before the sequence of all the major species of plant RNA is known. When this is achieved we will be in a good position to resolve many of the uncertainties referred to in this chapter. However, the tremendous amount of data generated will require sophisticated data-handling techniques to make it comprehensible. Furthermore, we have as yet hardly started looking at the minor RNA species which abound in cells and which are likely to be no less important than the major ones in determining how an organism functions.

References

Almalric F, Merkel C, Gefland R, Attardi G (1978) Fractionation of mitochondrial RNA from HeLa cells by high-resolution electrophoresis under strongly denaturing conditions. J Mol Biol 118:1–25

Bartolf M, Price CA (1979) Synthesis of poly(A)-containing RNA by isolated spinach chloroplasts. Biochemistry 18:1677–1680

Bedbrook JR, Smith SM, Ellis RJ (1980) Molecular cloning and sequencing of cDNA encoding the precursor to the small subunit of the chloroplast enzyme ribulose-1,5-bisphosphate carboxylase. Nature (London) 287:692–697

Blobel G, Dobberstein B (1975) Transfer of proteins across membranes 1. Presence of proteolytically processed and unprocessed nascent immunoglobulin light chains on membrane-bound ribosomes of murine myeloma. J Cell Biol 67:835–851

Bonen L, Doolittle WF (1975) On the prokaryotic nature of red algal chloroplasts. Proc Natl Acad Sci USA 72:2310–2314

Bonen L, Doolittle WF (1976) Partial sequences of 16S RNA and the phylogeny of blue-green algae and chloroplasts. Nature (London) 261:669–673

Bonen L, Cunningham RS, Gray MW, Doolittle WF (1977) Wheat embryo mitochondrial 18S ribosomal RNA: evidence for its prokaryotic nature. Nucl Acids Res 4:663–671
Bowman CM, Dyer TA (1979) 4.5 S ribonucleic acid, a novel ribosome component in the chloroplasts of flowering plants. Biochem J 183:605–613
Brosius J, Palmer ML, Kennedy PJ, Noller HF (1978) Complete nucleotide sequence of a 16S ribosomal gene from *Escherichia coli*. Proc Natl Acad Sci USA 75:4801–4805
Burr B, Burr FA, Rubenstein I, Simon MN (1978) Purification and translation of zein messenger RNA from maize endosperm protein bodies. Proc Natl Acad Sci USA 75:696–700
Burrows WJ (1975) Mechanisms of action of cytokinins. Curr Adv Plant Sci 7:837–847
Calagan JL, Pirtle RM, Pirtle IL, Kashdan MA, Vreman HJ, Dudock BS (1980) Homology between chloroplast and prokaryotic initiator tRNA. Nucleotide sequence of spinach chloroplast methionine initiator tRNA. J Biol Chem 255:9981–9984
Canaday J, Guillemaut P, Weil JH (1980a) The nucleotide sequences of the initiator transfer RNAs from bean cytoplasm and chloroplasts. Nucl Acids Res 8:999–1008
Canaday J, Guillemaut P, Gloeckler R, Weil JH (1980b) Comparison of the nucleotide sequences of chloroplast $tRNAs^{Phe}$ and $tRNAs_3^{Leu}$ from spinach and bean. Plant Sci Lett 20:57–62
Canaday J, Guillemaut P, Gloeckler R, Weil JH (1981) The nucleotide sequence of spinach chloroplast tryptophan transfer RNA. Nucl Acids Res 9:47–53
Carbon P, Ehresmann C, Ehresmann B, Ebel JP (1978) The sequence of *Escherichia coli* ribosomal 16S RNA determined by new rapid gel sequencing methods. FEBS Lett 94:152–156
Chang SH, Hecker M, Silberklang M, Brum CK, Barnett WE, RajBhandary UL (1976) The first nucleotide sequence of an organelle transfer RNA: Chloroplastic $tRNA^{Phe}$. Cell 9:717–724
Chang SH, Brum CK, Schnabel JJ, Heckman JH, RajBhandary UL (1978) Similarities in nucleotide sequence between *Euglena gracilis* and mammalian cytoplasmic phenylalanine tRNAs. Fed Proc 37:1768
Cunningham RS, Bonen L, Doolittle WF, Gray MW (1976) Unique species of 5S, 18S and 26S ribosomal RNA in wheat mitochondria. FEBS Lett 69:116–122
Darzynkiewicz E, Nakashima K, Shatkin AJ (1980) Base pairing in the conserved 3′ end of 18S rRNA as determined by psoralen photoreaction and RNase sensitivity. J Biol Chem 255:4973–4975
Delihas N, Andersen J, Sprouse HM, Kashdan M, Dudock BS (1981) The nucleotide sequence of spinach cytoplasmic 5S RNA. J Biol Chem 256:7515–7517
Driesel AJ, Crouse EJ, Gordon K, Bohnert HJ, Herrmann RG, Steinmetz A, Mubumbila M, Keller M, Burkard G, Weil JH (1979) Fractionation and identification of spinach chloroplast transfer RNAs and mapping of their genes on the restriction map of chloroplast DNA. Gene 6:285–306
Dudock BS, Katz G (1969) Large oligonucleotide sequences in wheat germ phenylalanine transfer ribonucleic acid. Derivation of total primary structure. J Biol Chem 244:3069–3074
Dunn DB, Hall RH (1975) Purines, pyrimidines, nucleosides and nucleotides: physical constants and spectral properties. In: Fasman GD (ed) Handbook of biochemistry and molecular biology Vol I. CRC Press, Cleveland, pp 65–215
Dyer TA, Bowman CM (1979) Nucleotide sequences of chloroplast 5S ribosomal ribonucleic acid in flowering plants. Biochem J 183:595–604
Dyer TA, Leech RM (1968) Chloroplast and cytoplasmic low-molecular-weight ribonucleic acid components of the leaf of *Vicia faba* L. Biochem J 106:689–698
Dyer TA, Zalik S (1979) Analysis of a 5S RNA-protein complex isolated from ribosomes of rye embryos. Can J Biochem 57:1400–1406
Dyer TA, Bowman CM, Payne PI (1977) The low-molecular-weight RNAs of plant ribosomes: their structure, function and evolution. In: Bogorad L, Weil JH (eds) Nucleic acids and protein synthesis in plants. Plenum, New York, pp 121–133
Eperon IC, Anderson S, Nierlich DP (1980) Distinctive sequence of human mitochondrial ribosomal RNA genes. Nature (London) 286:460–466

Erdmann VA (1976) Structure and function of 5S and 5.8S RNA. In: Cohn WE (ed) Progress in nucleic acid research and molecular biology. Academic Press, London New York, Vol 18, pp 45–90

Erdmann VA (1981) Collection of published 5S and 5.8S RNA sequences and their precursors. Nucl Acids Res 9:r25–r42

Everett GA, Madison JT (1976) Nucleotide sequence of phenylalanine transfer ribonucleic acid from pea (*Pisum sativum,* Alaska). Biochemistry 15:1016–1021

Filipowitz W (1978) Function of the 5′-terminal m^7G cap in eukaryotic mRNA. FEBS Lett 96:1–11

Fox GE, Woese CR (1975a) 5S RNA secondary structure. Nature (London) 256:505–507

Fox GE, Woese CR (1975b) The architecture of 5S rRNA and its relation to function. J Mol Evol 6:61–76

Gauss DH, Sprinzl M (1981) Compilation of tRNA sequences. Nucl Acids Res 9:r1–r23

Ghosh HP, Ghosh K, Simsek M, RajBhandary UL (1978) Spring Harbor Meeting on tRNA, Abstracts p6

Guillemaut P, Keith G (1977) Primary structure on bean chloroplastic $tRNA^{Phe}$. Comparison with *Euglena* chloroplastic $tRNA^{Phe}$. FEBS Lett 84:351–356

Hagenbüchle O, Santer M, Steitz JA, Mans RJ (1978) Conservation of the primary structure at the 3′ end of 18S rRNA from eukaryotic cells. Cell 13:551–563

Hall BD (1979) Mitochondria spring surprises. Nature (London) 282:129–130

Hartley MR (1979) The synthesis and origin of chloroplast low molecular weight ribosomal ribonucleic acid in spinach. Eur J Biochem 96:311–320

Haugland RA, Cline MG (1978) Capping structures at the 5′-terminus of polyadenylated ribonucleic acid in Avena coleoptiles. Plant Physiol 62:838–840

Heckman JE, Alzner-Deweerd B, RajBhandary UL (1979) Interesting and unusual features in the sequence of *Neurospora crassa* mitochondrial tyrosine transfer RNA. Proc Natl Acad Sci USA 76:717–721

Higgins TJV, Spencer D (1980) Biosynthesis of pea seed proteins: evidence for precursor forms from in vivo and in vitro studies. In: Leaver CJ (1980) Genome organization and expression in plants. Plenum, New York, pp 245–258

Janowicz Z, Wower JM, Augustyniak J (1979) Primary structure of barley embryo $tRNA^{Phe}$ and its identity with wheat germ $tRNA^{Phe}$. Plant Sci Lett 14:177–183

Jordan BR, Galling G, Jourdan R (1974) Sequence and conformation of 5S RNA from *Chlorella* cytoplasmic ribosomes: comparison with other 5S RNA molecules. J Mol Biol 87:205–225

Jukes TH (1977) How many anticodons? Science 198:319–320

Kashdan MA, Pirtle RM, Pirtle IL, Calagan JL, Vreman HJ, Dudock BS (1980) Nucleotide sequence of a spinach chloroplast threonine tRNA. J Biol Chem 255:8831–8835

Kurland CG (1960) Molecular characterization of ribonucleic acid from *Escherichia coli* ribosomes. I. Isolation and molecular weight. J Mol Biol 2:83–91

Lagerkvist U (1978) "Two out of three": an alternative method for codon reading. Proc Natl Acad Sci USA 75:1759–1762

Larkins BA, Hurkman WJ (1978) Synthesis and deposition of zein in protein bodies of maize endosperm. Plant Physiol 62:256–263

Leaver CJ (1975) The biogenesis of plant mitochondria, In: Harborne JB, van Sumere CF (eds) The chemistry and biochemistry of plant proteins. Academic Press, London New York, pp 137–166

Leaver CJ, Harmey MA (1976) Higher-plant mitochondria contain a 5S ribosomal ribonucleic acid component. Biochem J 157:275–277

Loening UE (1968) Molecular weight of ribosomal RNA in relation to evolution. J Mol Biol 38:355–365

Loening UE, Ingle J (1967) Diversity of RNA components in green plant tissues. Nature (London) 215:363–367

Luoma GA, Marshall AG (1978) Laser Raman evidence for new cloverleaf secondary structures for eukaryotic 5.8S RNA and prokaryotic 5S RNA. Proc Natl Acad Sci USA 75:4901–4905

Mackay RM (1981) The origin of plant chloroplast 4.5S ribsosomal RNA. FEBS Lett 123:17–18
Mackay RM, Spencer DF, Doolittle WF, Gray WM (1980) Nucleotide sequences of wheat embryo cytosol 5S and 5.8S ribosomal ribonucleic acids. Eur J Biochem 112:561–576
Marcu KB, Mignery RE, Dudock B (1977) Complete nucleotide sequence and properties of the major species of glycine transfer RNA from wheat germ. Biochemistry 16:797–806
Marcu K, Marcu D, Dudock B (1978) Wheat germ rRNAs containing uridine in place of ribothymidine: a characterization of an unusual class of eukaryotic tRNAs. Nucl Acids Res 5:1075–1092
Moorman AFM, Van Ommen GJB, Grivell LA (1978) Transcription in yeast mitochondria: isolation and physical mapping of messenger RNAs for subunits of cytochrome C oxidase and ATPase. Mol Gen Genet 160:13–24
Nichols JL (1979) N^6-methyladenosine in maize poly(A)-containing RNA. Plant Sci Lett 15:357–361
Olins PO, Jones DS (1980) Nucleotide sequence of *Scenedesmus obliquus* cytoplasmic initiator tRNA. Nucl Acids Res 8:715–729
Osorio-Almeida ML, Guillemaut P, Keith G, Canaday J, Weil JH (1980) Primary structure of three leucine transfer RNAs from bean chloroplast. Biochem Biophys Res Commun 92:102–108
Payne PI, Dyer TA (1971) Characterization of cytoplasmic and chloroplast 5S ribosomal ribonucleic acid from broad-bean leaves. Biochem J 124:83–89
Payne PI, Dyer TA (1972) Plant 5.8S RNA is a components of 80S but not 70S ribosomes. Nature New Biology 235:145–147
Payne PI, Dyer TA (1976) Evidence for the sequence of 5-S rRNA from the flowering plant *Secale cereale* (Rye). Eur J Biochem 71:33–38
Phillips DO, Carr NG (1977) Nucleic acid analysis and the endosymbiont hypothesis. Taxon 26:3–42
Pirtle RM, Pirtle IL, Kashdan MA, Vreman HJ, Dudock BS (1981) The nucleotide sequence of spinach chloroplast methionine elongator tRNA. Nucl Acids Res 9:183–188
Pring DR (1974) Maize mitochondria: purification and characterization of ribosomes and ribosomal ribonucleic acid. Plant Physiol 53:677–683
Rafalski AJ, Barciszewski J, Gulewicz K, Twardowski T, Keith G (1977) Nucleotide sequence of $tRNA^{Phe}$ from the seeds of lupin (*Lupinus luteus*). Comparison of the major species with wheat germ $tRNA^{Phe}$. Acta Biochem Pol 24:301–318
Rich A, Kim SH (1978) The three dimensional structure of transfer RNA. Sci Am 238:52–62
Rochaix JD, Malnoë P (1978) Anatomy of the chloroplast ribosomal DNA of *Chlamydomonas reinhardii*. Cell 15:661–670
Rosenberg M, Patterson BM (1979) Efficient cap-dependent translation of polycistronic prokaryotic mRNAs is restricted to the first gene of the operon. Nature (London) 279:696–701
Saini MS, Lane BG (1977) Wheat embryo ribonucleates VIII The presence of 7-methylguanosine 'cap structures' in the RNA of imbibing wheat embryos. Can J Biochem 55:819–824
Sano H, Spaeth E, Burton WG (1979) Messenger RNA of the large subunit of ribulose-1,5-bisphosphate carboxylase from *Chlamydomonas reinhardii*. Eur J Biochem 93:173–180
Schwarz Zs, Kössel H (1980) The primary structure of 16S rDNA from *Zea mays* chloroplasts is homologous to E coli 16S rRNA. Nature (London) 283:739–742
Shatkin AJ (1976) Capping of eukaryotic mRNAs. Cell 9:645–653
Shine J, Dalgarno L (1974) The 3′-terminal sequence of *Escherichia coli* 16S ribosomal RNA: complementary to nonsense triplets and ribosome binding sites. Proc Natl Acad Sci USA 71:1342–1346

Smith SM, Ellis RJ (1979) Processing of small subunit precursor of ribulose bisphosphate carboxylase and its assembly into whole enzyme are stromal events. Nature (London) 278:662–664

Steiner DF (1979) Processing of protein precursors. Nature (London) 279:674–675

Sun SM, Slightom JL, Hall TC (1981) Intervening sequences in a plant gene – comparison of the partial sequence of cDNA and genomic DNA of French bean phaseolin. Nature (London) 289:37–41

Takaiwa F, Sugiura M (1980a) The nucleotide sequence of 4.5S ribosomal RNA from tobacco chloroplasts. Nucl Acids Res 8:4125–4129

Takaiwa F, Sugiura M (1980b) Nucleotide sequences of the 4.5S and 5S ribosomal RNA genes from tobacco chloroplasts. Mol Gen Genet 180:1–4

Tanaka Y, Dyer TA, Brownlee GG (1980) An improved direct RNA sequence method; its application to *Vicia faba* 5.8S ribosomal RNA. Nucl Acids Res 86:1259–1272

Van Holde KE, Hill WE (1974) General physical properties of ribosomes. In: Nomura M, Tissiéres A, Lengyel P (eds) Ribosomes. Cold Spring Harbor Lab, New York, pp 53–91

Verdier G (1979a) Poly(adenylic acid)-containing RNA of *Euglena gracilis* during chloroplast development. I Analysis of their complexity by hybridisation to complementary DNA. Eur J Biochem 93:573–580

Verdier G (1979b) Poly(adenylic acid)-containing RNA of *Euglena gracilis* during chloroplast development. 2 Transcriptional origin of the different RNA. Eur J Biochem 93:581–586

Weil JH (1979) Cytoplasmic and organellar tRNAs in plants. In: Hall TC, Davies J (eds) Nucleic acids in plants Vol I. CRC Press, Boca Raton, pp 143–192

Weinand U, Feix G (1978) Electrophoretic fractionation and translation in vitro of poly(rA)-containing RNA from maize endosperm. Eur J Biochem 92:605–611

Wheeler AM, Hartley MR (1975) Spinach chloroplast messenger RNA does not contain poly(A). Nature (London) 257:66–67

Whitfeld PR, Leaver CJ, Bottomley W, Atchison BA (1978) Low-molecular-weight (4.5S) ribonucleic acid in higher plant-chloroplast ribosomes. Biochem J 175:1103–1112

Woese CR, Fox GE (1977) Phylogenetic structure of the prokaryotic domain: the primary kingdoms. Proc Natl Acad Sci USA 74:5088–5090

Wollenzein P, Hearst JE, Thammana P, Cantor CR (1979) Base-pairing between distant regions of the *Escherichia coli* 16S ribosomal RNA in solution. J Mol Biol 135:255–269

Wrede P, Erdmann VA (1977) *Escherichia coli* 5S RNA binding proteins L18 and L25 interact with 5.8S RNA but not with 5S RNA from yeast ribosomes. Proc Natl Acad Sci USA 74:2706–2709

Zablen LB, Kissil MS, Woese CR, Buetow DE (1975) The phylogenetic origin of the chloroplasts and prokaryotic nature of its ribosomal RNA. Proc Natl Acad Sci USA 72:2418–2422

7 RNA Processing and Other Post-Transcriptional Modifications

D. GRIERSON

1 Introduction

RNA molecules play a central part in cell metabolism, either by directing the synthesis of proteins or functioning as essential components of the protein-synthesising machinery. A detailed knowledge of their structure and metabolism is, therefore, an essential pre-requisite for an understanding of the control of protein synthesis. Since most RNA molecules arise as direct products of DNA transcription, such studies may also help to unravel the problems related to the differential control of gene expression during development.

A variety of studies over the last 20 years has shown that almost all polyribonucleotide sequences undergo extensive modification after transcription. Such modifications include (1) nucleolytic cleavage of precursor RNA molecules, leading to the production of smaller, "mature", RNA's, (2) terminal modifications such as the addition of CCA to the 3′ end of tRNA or the attachment of "cap" structures or poly(A) sequences to the 5′ and 3′ ends of mRNA molecules, (3) internal modifications such as the methylation of bases or ribose residues or the transformation of uridine to pseudouridine and (4) excision and splicing of coding sequences transcribed from split genes. Such post-transcriptional modifications are included here under the general heading of RNA "processing" or "maturation".

It is difficult to give a simple answer to the question: what is the significance of RNA processing? In view of the range of RNA types that occur, each with a different cell location and function, and the variety of processing reactions known, a simple explanation would probably be incomplete. Nevertheless, a few rules appear to be emerging. It seems probable that extra nucleotide sequences, present in precursor molecules and removed during the formation of mature RNA's, may function in a transient manner to induce the formation of a specific secondary or tertiary structure after synthesis. Internal methylation, or other modifications may provide binding sites or promote or preclude certain interactions important for the secondary or tertiary structure of an RNA or its association with other cell components. It is also suspected that other types of processing reaction are involved in the transport of RNA from nucleus to cytoplasm. Furthermore, specific nucleotide sequences may be involved in determining whether an RNA sequence is destroyed after transcription or accumulated. Finally, although the significance of split genes is not clear, the processing

---

Abbreviations: p16S RNA, precursor molecule of 16S rRNA; p23S RNA, precursor molecule of 16S rRNA; p5S RNA, precursor molecule of 5S RNA, hnRNA; heterogeneous nuclear RNA

of their transcription products is absolutely essential in order to produce an mRNA molecule with nucleotide triplets colinear with the polypeptide product.

Experience suggests that the general principles of RNA processing are very similar in all types of cell. For this reason, and because much more is known about processing in microorganisms and animals, selected aspects of RNA processing in bacteria and mammalian cells will be outlined in order to provide a framework against which the results of work from plants can be discussed. Further information on RNA precursors and RNA processing in microorganisms and animals can be obtained from reviews by MADEN (1971), PACE (1973), SMITH (1976), PERRY (1976), RUNGGER and CRIPPA (1977), HADJIOLOV and NIKOLAEV (1976), ABELSON (1979).

2 Methods of Studying RNA Processing

Evidence for RNA processing was first obtained by studying the synthesis of RNA molecules in vivo. Cells, tissue segments or whole organisms were fed radioactive precursors of RNA, such as [^{32}P]-phosphate or [^{3}H]-uridine, and the pulse-labelled RNA extracted from whole cells or cell fractions and analysed, generally by a method which separates RNA on the basis of size. In various organisms such studies have revealed the existence of rapidly-synthesised, metabolically unstable, RNA's present in small quantities. Generally speaking, such RNA's are larger than the more stable molecules of rRNA, tRNA and mRNA. The disappearance of radioactivity from these unstable molecules during chase incubations coincides with the accumulation of label in rRNA, tRNA and mRNA, suggesting that the larger molecules are converted to mature RNA's after transcription. Further, careful, labelling and pulse-chase experiments in vivo showed that the size reduction (i.e., processing) often occurs in discrete steps and is frequently accompanied by methylation or other chemical modification. Evidence for a direct relationship between sequence content of precursors, processing intermediates and mature, processed, products has been obtained by a variety of methods, such as analysis of the nucleotide composition of various RNA fractions, DNA–RNA hybridisation and competition hybridisation studies and determination of the frequency and distribution of methylation sites in the various molecules. In this way a general idea of the processing pathway for the production of a particular RNA has been built up.

The importance of in vivo studies is that they provide direct evidence for a metabolic relationship between various precursors and putative intermediates in a processing pathway. However, it is important to bear in mind that using this approach some processing steps may be missed, either because they occur very rapidly or because their effects are not easily detected. For example, there is an important distinction between the primary transcription product – which contains a 5′ triphosphate – and the first stable product which can be detected. Additional processing steps have sometimes been discovered by studying mutants, such as the ribonuclease III-deficient strains of *E. coli,* or by carrying out in vitro experiments with isolated nuclei, nucleoli or chloroplasts. However, an understanding of the mechanism of the various processing reactions can only come from knowledge of the sequence and secondary structure of the RNA being processed and by purifying the individual processing enzymes and studying how they interact with their substrates in vitro.

3 Synthesis and Processing of rRNA

3.1 Processing of rRNA Transcripts in Bacteria

The rRNA genes of *Escherichia coli* are arranged in one transcription unit comprising rRNA-coding sequences interspersed with spacer regions which contain tRNA sequences. The genes are transcribed in the order 16S→tRNA→23S→5S→tRNA. There are from five to ten such transcription units per chromosome (Fig. 1). Initiation begins before the 5′ end of the 16S RNA and transcription continues beyond the 3′ end of the tRNA sequences. Processing of the transcript normally occurs before synthesis is completed and the first stable products of transcription that can be detected by analysis of RNA from pulse-labelled cells are individual precursors to the stable rRNA's (p23S, p16S and p5S in Fig. 1; HECHT and WOESE 1968, ADESNIK and LEVINTHAL 1969). However, studies on a mutant strain of *E. coli* deficient in ribonuclease III have revealed a 30S RNA precursor ($M_r = 2.1 \times 10^6$) which is a complete transcript containing 16S, 23S, 5S, tRNA and spacer sequences (GINSBURG and STEITZ 1975).

The 30S RNA precursor contains pppACUG as the major 5′ terminus, suggesting that it is a primary transcription product (GINSBURG and STEITZ 1975). Furthermore the rRNA promotor, to the 5′ side of the 16S RNA sequence, is required for transcription of all the RNA sequences shown in Fig. 1 including the tRNA genes to the 3′ side of the 5S RNA sequence (MORGAN et al. 1978). Processing of the 30S RNA precursor occurs in a number of separate steps, involving several different enzymes. Ribonuclease III, which is specific for double-stranded regions in RNA, is thought to be responsible for the initial endonucleolytic cleavage of the precursor. This enzyme cleaves the 30S RNA in vitro into two large sequences ($M_r = 0.65$ and 1.2×10^6, which are slightly larger than the p16S and p23S RNA's shown in Fig. 1) plus several smaller fragments, including a 300-nucleotide fragment, containing the 5S RNA sequence (GINSBURG and STEITZ 1975).

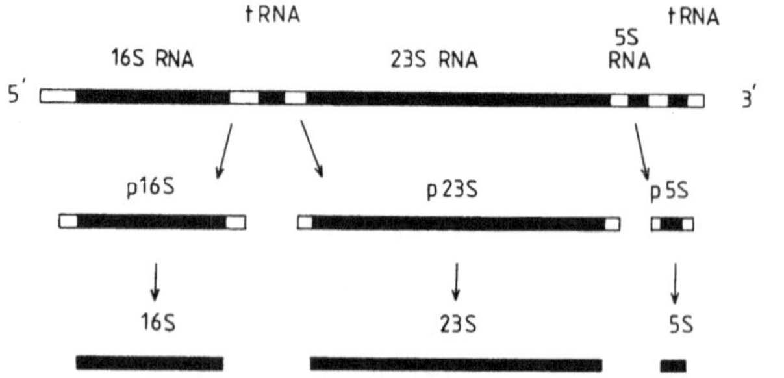

Fig. 1. An outline of the processing pathway for the 30S primary transcription product of the *E. coli* rRNA genes. Coding regions are shown in *black*, transcribed spacer regions in *white*

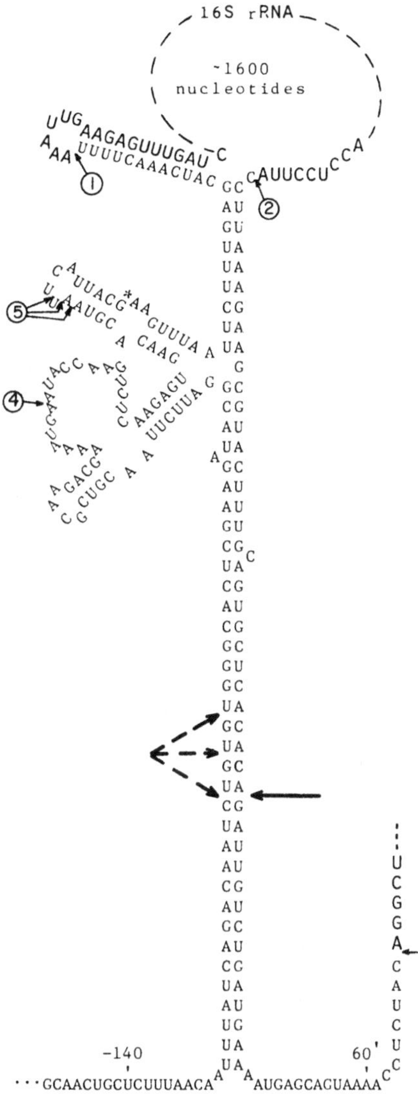

Fig. 2. Probable secondary structure of the rRNA precursor from *E. coli* involving the regions on either side of the 16S rRNA sequence. From YOUNG and STEITZ (1978). The *arrows* in the centre of the main stem show the ribonuclease III cleavage site (there is some uncertainty, indicated by the *dotted arrows*). The mature 5′ end of 16S RNA is at site *1*, cleaved by ribonuclease M16 (DAHLBERG et al. 1978). Cleavage at site *2* produces the mature 16S 3′ end (HAYES and VASSEUR 1976). Site *3* contains the 5′ terminus of tRNA^{ileu}. Site *4* contains the 5′ end of p16S RNA, found in 30S ribosomes from cells unable to cleave at site *1* (DAHLBERG et al. 1978). Site *5* is the cleavage site for a further processing stage. (LUND and DAHLBERG 1977)

There must, therefore, be several ribonuclease III recognition sites at intervals along the 30S RNA. One such site, for the excision of the 16S RNA precursor is shown in Fig. 2. The 5′ and 3′ spacer regions flanking the 16S RNA sequences are capable of forming a 26 base-pair stem, with the entire 16S RNA sequence looped out (YOUNG and STEITZ 1978). The base-paired stem contains a recognition site for ribonuclease III (Fig. 2), plus several recognition sites for other enzymes, discussed below. The spacer sequences on either side of the 23S RNA form a similar base-paired stem (see ABELSON 1979).

The processing scheme shown in Fig. 1 is oversimplified and the sequence of events shown in Fig. 3 represents the processing pathway more accurately (see HADJIOLOV and NIKOLAEV 1976):

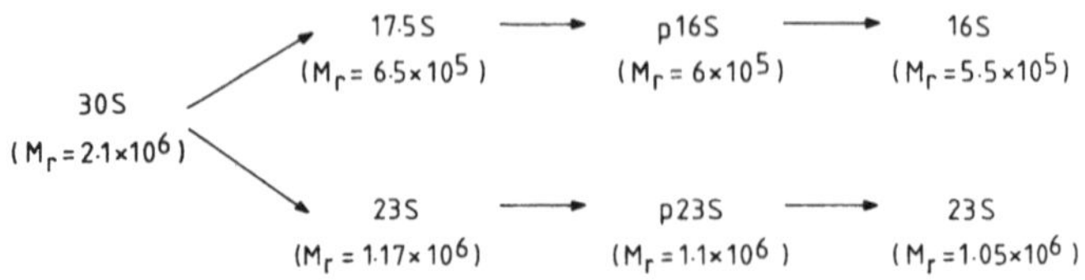

Fig. 3. Additional stages in the processing of the rRNA sequences from the 30S *E. coli* rRNA precursor. (After HADJIOLOV and NIKOLAEV 1976)

After the initial excision steps, the 16S and 23S precursor RNA's become associated with ribosomal proteins and undergo further processing by separate enzymes in pre-ribosomal particles (DAHLBERG et al. 1978). Ribonuclease M16, which is inactive against naked RNA, hydrolyses the 16S precursor at site 4 in Fig. 2, to generate a further processing product and mature 16S RNA is produced by nucleolytic cleavage at sites 1 and 2 in Fig. 2. Analogous modifications are made to the p23S RNA during processing.

Evidence for the cleavage of tRNA-containing sequences by a different processing enzyme, ribonuclease P, has been obtained from studies with certain ribonuclease III-deficient mutants of *E. coli* (GEGENHEIMER and APIRION 1978). A putative site for excision of the 5′ end of the tRNA-containing sequence located between the 16S and 23S RNA is shown in Fig. 2.

Cleavage of the 30S precursor in vitro by ribonuclease III yields a 300-nucleotide fragment which contains the 5S RNA sequence internally (GINSBURG and STEITZ 1975). Details of the processing of this RNA are not clear, but processing of a smaller fragment has been studied in *Bacillus subtilis*. In this bacterium two distinct precursors to 5S RNA have been discovered, p5A and p5B. These precursors are of different lengths and are transcribed from different coding sequences. SOGIN et al. (1977) have partially purified an endonuclease from *B. subtilis* with processing activity and STAHL et al. (1979) have studied the conversion of p5A (179 nucleotides) to mature 5S RNA (116 nucleotides) by this enzyme, called ribonuclease M5. The p5A molecule contains a complementary nucleotide sequence in the excess RNA flanking the 5′ and 3′ ends of the 5S sequence, thus forming a double-stranded stem with the 5S RNA looped out (Fig. 4). Ribonuclease M5 cuts the p5A molecule at two sites within this stem, liberating mature 5S RNA. The excised sequences, 21 nucleotides from the 5′ end and 42 nucleotides from the 3′ end are probably degraded by exonuclease scavenging enzymes. STAHL et al. (1979) have suggested that some of the extra nucleotides at the 5′ end of the precursor govern the folding of the molecule and induce the correct conformation of the 5S RNA sequence before processing. Support for this idea is provided by experiments carried out in vitro, where the 5′ terminal eight or nine nucleotides were removed, inducing an altered conformation in the remainder of the molecule. The function of the remaining nucleotides in the precursor, other than those which provide the M5 processing site, is not clear, although STAHL et al. (1979) point out that the 3′ terminal U^6G_{OH} resembles a transcribed termination sequence.

There is evidence that some of the RNA processing steps are intimately associated with the formation of pre-ribosomal particles and that the excess RNA in the precursors may induce conformations more favorable for ribosome assembly than those adopted by mature 16S and 23S RNA. For example, the final processing stages are slowed down if the supply of ribosomal protein is inhibited by chloramphenicol (see also the Sect. 3.3). Furthermore, in reconstitution experiments, purified p30S RNA binds some ribosomal proteins more

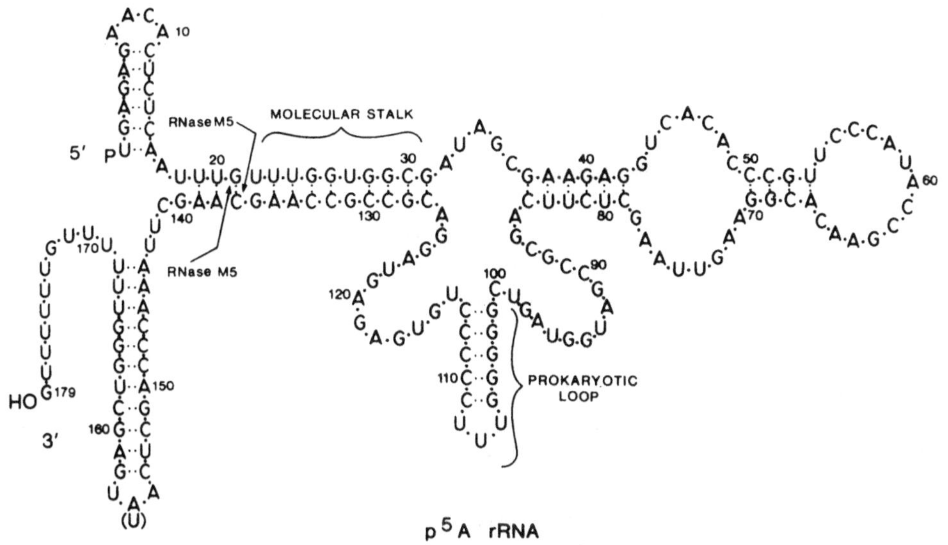

Fig. 4. Nucleotide sequence of the 5S rRNA precursor from *B. subtilis*. (STAHL et al. 1979). The cleavage sites for ribonuclease M5 are indicated by *arrows*

readily than either 16S or 23S RNA (NIKOLAEV et al. 1975) and p16S RNA is more efficient than 16S RNA in combining with ribosomal proteins to produce 30S ribosomal subunits (MANGIAROTTI et al. 1975).

Modification of nucleosides occurs at specific sites in 23S and 16S rRNA after synthesis. Most of the modifications involve methylation of the bases and are carried out by RNA methylases which use S-adenosyl methionine as methyl donor. *E. coli* 16S RNA has 10 methylated nucleotides, with 14 methyl groups and 23S RNA has 14 methylated nucleotides with 16 methyl groups. All of the methylated sites present in mature 23S RNA are found in the p30S RNA, whereas there is little methylation of the 16S RNA sequence until p16S is excised (see PERRY 1976, HADJIOLOV and NIKOLAEV 1976).

3.2 Blue-Green Algae

Ribosomal RNA precursors have been found in the unicellular blue-green alga *Anacystis nidulans* (SZALAY et al. 1972, DOOLITTLE 1972, SEITZ and SEITZ 1973) and in the filamentous species *Tolypothrix distorta* (GRIERSON and SMITH 1973). These precursors seem to be analogous to p23S and p16S in bacteria (see Fig. 1) since they migrate slightly more slowly than the mature rRNA's during electrophoresis in polyacrylamide gels. In *Tolypothrix distorta* a precursor with $M_r = 1.22 \times 10^6$ is converted to 23S rRNA ($M_r = 1.1 \times 10^6$) and the sequence of events leading to 16S rRNA production appears to be $0.76 \times 10^6 \rightarrow 0.68 \times 10^6 \rightarrow 0.56 \times 10^6$ (16S) (GRIERSON and SMITH 1973). Although the processing pathway has not been studied in great detail there are a number of similarities with rRNA biosynthesis in *E. coli*. The rRNA's are methylated (SEITZ and SEITZ 1973) and accumulate in cells treated with chloramphenicol (DOOLITTLE 1972). However, there is little information on the organisation of the genes coding for these rRNA's and it is not clear how far the similarity with *E. coli* extends.

3.3 Chloroplasts and Mitochondria

Chloroplast ribosomes contain four types of rRNA (cf. Dyer, Chap. 6, this Vol., Table 4). Each molecule is coded by the chloroplast DNA, which is a closed circle. The precise molecular weight of chloroplast DNA varies between species but in higher plants an average value is 90×10^6 or approximately 145,000 base pairs. Chloroplast DNA circles from some legume contain one copy of each rRNA gene, there are two copies in spinach (*Spinacea oleracea*), lettuce (*Lactuca sativa*), *Chlamydomonas reinhardii* and *Zea mays,* while *Euglena gracilis* chloroplast DNA contains three sets of each gene (see BOHNERT et al., Chap. 14, this Vol.).

In maize and spinach, where the two sets of genes are located with opposite polarity on different regions of the circle, the order of the coding sequence is: 16S—tRNA—23S—4.5S—5S (see BOHNERT et al., Chap. 14, this Vol.). There is a possibility that there are two 4.5S and 5S RNA sequences in this region (HARTLEY 1979). This general arrangement is very similar to the organisation of the rRNA genes in *E. coli* (Fig. 1), including the location of tRNA's in the spacer region between the 16S and 23S RNA's. In addition, there is strong sequence homology between the chloroplast and bacterial rRNA's (see DYER, Chap. 6, this Vol.).

Several authors have reported the synthesis of immediate precursors to 23S and 16S chloroplast rRNA in algae and higher plants (HARTLEY and ELLIS 1973, MUNSCHE and WOLLGIEHN 1973, ROSNER et al. 1973, DETCHON and POSSINGHAM 1973, GRIERSON and LOENING 1974, HEIZMANN 1974, MILLER and MCMAHON 1974, BOHNERT et al. 1974, POSNER and ROSNER 1975, SCOTT 1976, BOHNERT et al. 1977). These precursors have approximate molecular weights of 1.2 and 0.65×10^6 and seem analogous to the p23S and p16S RNA's of *E. coli* (Fig. 5). Kinetic experiments suggest that they are precursors to mature 23S and 16S RNA and competition hybridisation experiments have shown that the $M_r = 1.2 \times 10^6$ RNA contains the 23S rNA sequence and the $M_r = 0.65 \times 10^6$ RNA contains the 16S RNA sequence (HARTLEY et al. 1977, BOHNERT et al. 1977, HARTLEY and HEAD 1979). In addition to the 23s RNA sequence, the $M_r = 1.2 \times 10^6$ RNA also contains the 4.5S RNA sequence (HARTLEY 1979, cf. DYER, Chap. 6, this Vol.).

There is also evidence for a polycistronic chloroplast rRNA precursor, analogous to the 30S precursor RNA in *E. coli* (Fig. 5). A rapidly labelled RNA with a molecular weight in the range $2.7–2.9 \times 10^6$ has been detected in chloroplasts of spinach (HARTLEY and ELLIS 1973) and mung bean (GRIERSON and LOENING 1974). However, this large precursor does not always accumulate to any great extent (see, for example, DETCHON and POSSINGHAM 1973, SCOTT 1976), probably because under some growth conditions it is processed very rapidly, perhaps even before transcription is complete, as in *E. coli.*

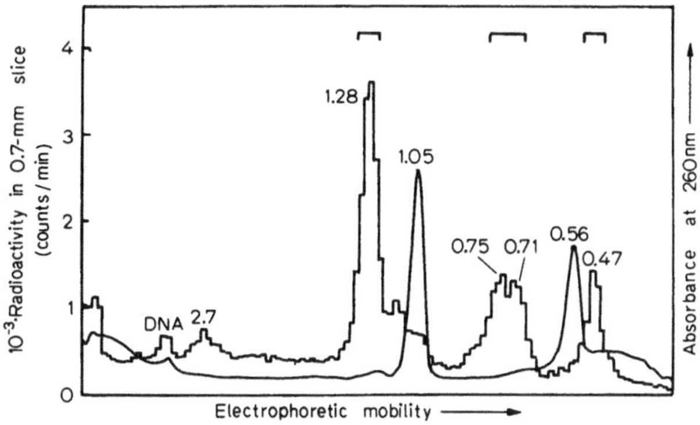

Fig. 5. Polyacrylamide gel electrophoresis of chloroplast RNA showing rRNA precursors and processing intermediates. (HARTLEY and HEAD 1979). Spinach plants were treated with chloramphenicol and the leaves labelled with ^{32}P-phosphate. RNA was prepared from chloroplasts isolated after a 6-h labelling period. The *smooth curve* shows the 23S and 16S rRNA's, the *histogram* shows the radioactivity

The first detectable transcription product of the rRNA genes in spinach chloroplasts has a molecular weight of approximately 2.7×10^6 (Fig. 5). This molecule has been shown to be synthesised in vivo, since it is present in RNA extracted from chloroplasts isolated from pulse-labelled leaves (HARTLEY and ELLIS 1973). It also accumulates in isolated chloroplasts carrying out light-driven RNA synthesis, thus confirming that it is a true transcription product of chloroplast DNA (HARTLEY and ELLIS 1973, HARTLEY et al. 1977, BOHNERT et al. 1974, 1977, HARTLEY and HEAD 1979). The molecular weight estimate for this RNA (2.7×10^6) can only be regarded as approximate because it has not been carried out under fully denaturing conditions (see the discussion relating to nucleolar RNA synthesis). However, the molecule remains intact during electrophoresis in 8 M urea at 37 °C (HARTLEY and HEAD 1979) and is therefore probably not an aggregate. It is not known whether the $M_r = 2.7 \times 10^6$ RNA has a 5′-triphosphate and there is no direct evidence that it is a primary transcription product, other than the fact that it accumulates rapidly. Nevertheless, it is interesting to note that an RNA of this size would correspond to approximately 8,000 nucleotides, which is sufficient to accommodate the complete sequence established for the rRNA-coding region of spinach chloroplast DNA mentioned above (see BOHNERT et al., Chap. 14, this Vol.).

Competition hybridisation experiments have demonstrated that one copy of each of the 23S and 16S rRNA sequences is present in this precursor (HARTLEY et al. 1977, BOHNERT et al. 1977, HARTLEY and HEAD 1979). From the pattern of labelling in vitro the 16S sequence appears to be closer to the 5′ end than the 23S sequence (HARTLEY and HEAD 1979). BOHNERT et al. (1977) have shown that the $M_r = 2.7 \times 10^6$ precursor RNA also contains low-molecular weight rRNA sequences, but it is not clear from their results whether these correspond to 5S RNA, 4.5S RNA or both. Results obtained by HARTLEY (1979), however, suggest that the 4.5S RNA may be part of the precursor and that either the 5S sequence is excised from the precursor as a very early event, or that it is transcribed separately. Support for the suggestion that the 5S RNA may be transcribed separately comes from a study of the DNA sequence of the rRNA-coding region in *Zea mays,* where there is a putative termination site following the 4.5S RNA region, followed by a putative initiation site up-

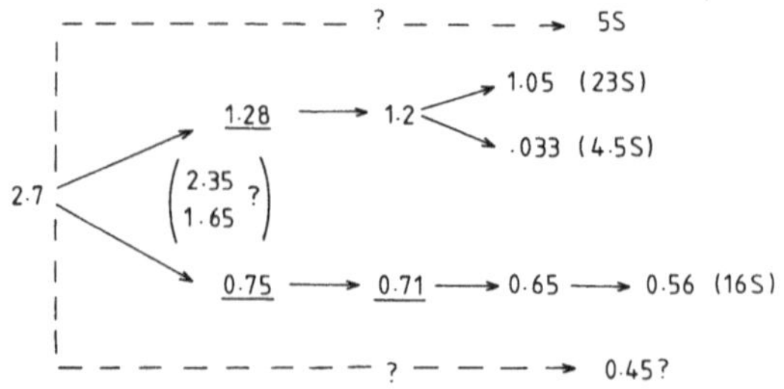

Fig. 6. Possible processing scheme for spinach chloroplast rRNA precursor. Based on data discussed in the text

stream from the 5S sequence (see BOHNERT et al., Chap. 14, this Vol.). There is no direct evidence for the presence of tRNA sequences in the precursor although these would be expected from the DNA map.

The complete processing pathway for the chloroplast rRNA precursor is not known. When isolated spinach chloroplasts carry out light-driven RNA synthesis in vitro in tricine-$MgCl_2$-KCl medium the major product is the $M_r = 2.7 \times 10^6$ precursor (BOHNERT et al. 1977, HARTLEY et al. 1977, HARTLEY and HEAD 1979) and there is very little processing of the precursor. However, when the chloroplasts are incubated in the presence of sorbitol-tricine medium several smaller RNA molecules are also produced. These are presumably processing products and have molecular weights of 2.35, 1.65, 1.2 and 0.7×10^6 (BOHNERT et al. 1977; HARTLEY et al. 1977, HARTLEY and HEAD 1979). Complete processing to produce mature 23S and 16S rRNA does not occur in vitro, possibly because some factor of cytoplasmic origin, such as nuclear-coded ribosomal proteins, are required for RNA maturation. Processing can also be inhibited in vivo by administering D(−)-*threo*-chloramphenicol to leaves (HARTLEY and HEAD 1979). Under these conditions the $M_r = 2.7 \times 10^6$ precursor continues to be processed but radioactivity accumulates in molecules with $M_r = 1.28$, 0.75 and 0.71×10^6 (Fig. 5). An additional RNA with $M_r = 0.45 \times 10^6$ also accumulates in the presence of chloramphenicol. It has been suggested that it is cleaved from the large precursor but its role is not clear. The inhibition of some processing reactions by chloramphenicol is analogous to the situation in *E. coli*. This suggests that a continuous supply of certain proteins synthesised in the chloroplasts, perhaps ribosomal proteins or labile processing enzymes, are necessary for the final maturation steps.

A tentative scheme for the processing of the rRNA precursor in spinach chloroplast is shown in Fig. 6. It should be stressed that this is a speculative scheme and the role of the various processing intermediates needs to be confirmed by sequence studies.

An interesting variation on the rRNA gene arrangement has been found in *Chlamydomonas reinhardii* chloroplasts. In this organism the chloroplast DNA carries genes for rRNA in two clusters. The linear arrangement of the coding sequences in each cluster is: 16S—7S—3S—23S—5S (compare with other chloroplast rRNA's, Table 4, Chap. 6, this Vol.) (ROCHAIX and MALNOË 1978). The most interesting feature is that the 23S-coding region is interrupted by an intervening sequence which occurs 270 bases from the 3′ end of the 23S RNA. The intervening sequence is approximately 900 bases long and does not appear in mature 23S rRNA (ROCHAIX and MALNOË 1978, ALLET and ROCHAIX 1979). Since both 23S rRNA genes contain this intervening sequence it seems probable that it is transcribed as part of the rRNA precursor and subsequently removed by a splicing reaction analogous to that which occurs during the processing of yeast tRNA and mRNA precursors (see Sects. 4.2 and 5). Repeated nucleotide sequences are present close to each end of the intervening sequence and these may be involved in recognition by splicing enzymes (ALLET and ROCHAIX 1979).

Much less work has been done on rRNA synthesis in mitochondria but the general pattern seems to be similar to that found for chloroplasts. In *Neurospora crassa,* for example, the first detectable transcript has a molecular weight of 2.4×10^6 and is processed via intermediates of 1.6×10^6 and 0.9×10^6 to yield mature mitochondrial rRNA's ($M_r = 1.28$ and 0.72×10^6) (KURIYAMA and LUCK 1973).

Biosynthesis of *N. crassa* mitochondrial rRNA must involve splicing (see Sects. 4 and 5). Recently, the mitochondrial rRNA genes have been shown to contain an insert which is not present in the 32S rRNA precursor. In temperature-sensitive mutants of *N. crassa*, grown at the non-permissive temperature, there is an altered ratio of 25S/19S rRNA and a 35S RNA precursor builds up. This latter molecule contains the 25S rRNA sequence, plus most or all of the transcribed insert (MANNELLA et al. 1979). It presumably represents an intermediate in the processing pathway. Splicing of the rRNA transcript must also occur in mitochondria of *S. cerevisiae* since the only 23S rRNA gene on the mitochondrial chromosome contains a 1,160 base-pair insert (BOS et al. 1978, FAYE et al. 1979).

3.4 Processing of Cytoplasmic rRNA in Nucleoli of Eukaryotes

The transcription and processing of cytoplasmic rRNA takes place in specialised organelles called nucleoli. These are dense nuclear organelles without a limiting membrane. They arise in association with one or a few specific regions of chromosomes, called nucleolar organisers, when cells are synthesising ribosomes. Nucleoli consist of a central region, comprising rRNA genes being actively transcribed, surrounded by regions containing fibrils and granules of ribonucleoprotein. These latter structures probably represent rRNA transcripts being processed and combined with proteins to produce ribosomes (see BUSCH and SMETANA 1970, FAKAN and BERNHARD 1971, HADJIOLOV and NIKOLAEV 1976). Tandem repeats of the genes coding for 18S, 5.8S and 25S (28S in mammals) rRNA, up to several thousand copies in some higher plants (INGLE 1973), are clustered together in nucleoli, where they are organised in polycistronic transcription units. The nucleolar rRNA genes are transcribed by RNA polymerase I, whereas the 5S RNA genes, which are located on a separate region of the chromosomes, except in lower eukaryotes such as *Dictyostelium* and *Saccharomyces* (PARDUE et al., 1973, WIMBER and STEFFENSEN 1970, WIMBER et al. 1974, DURANTE et al. 1977, HEMLEBEN and GRIERSON 1978), are transcribed by RNA polymerase III (see WOLLGIEHN, Chap. 5, this Vol.). Very little is known about the processing of 5S rRNA in animals and plants.

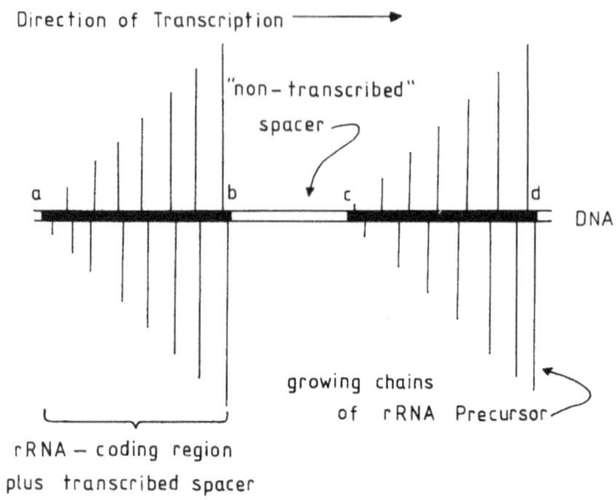

Fig. 7. Diagram of the arrangement of rRNA genes from nucleoli of eukaryotes as visualized by electron microscopy

In most organisms, including higher plants (FRIEDRICH et al. 1979), the nucleolar DNA sequence is:

$$5'\text{–}_{\text{spacer}}^{\text{transcribed}}\text{–}18S\text{–}_{\text{spacer}}^{\text{transcribed}}\text{–}5.8S\text{–}_{\text{spacer}}^{\text{transcribed}}\text{–}\underset{(28S)}{25S}\text{–}3'.$$

The first detectable rRNA precursor generally comprises a transcript of this sequence. Repeats of the basic DNA sequence are separated by so-called "non-transcribed spacer" regions (Fig. 7). The length of the precursor and the rRNA-coding regions vary in different organisms. The largest rRNA in higher plants (25S) has a molecular weight of 1.25×10^6, whereas the counterpart in mammals (28S) is 1.76×10^6. The "transcribed spacer" regions are much more variable, both in length and composition, between species, and the length of the "non-transcribed" region is also very variable.

Transcription of the rRNA genes has been studied in both animals and plants by visualising the spread nucleolar contents by electron microscopy (MILLER and BEATTY 1969, SCHEER et al. 1973, TRENDELENBURG et al. 1974, SPRING et al. 1974, 1975, BERGER and SCHWEIGER 1975) and by analysing pulse-labelled RNA by biochemical and biophysical techniques. Although polycistronic precursors to cytoplasmic rRNA have been reported in scores of eukaryotic species (some examples from work with plants are given in Table 1) it is not clear whether these are primary transcription products. The reason for this is that the location of the initiation and termination sites for RNA synthesis is unknown. Electron microscopy of spread nucleolar DNA frequently shows that there are growing chains of rRNA precursors molecules attached to each coding region (Fig. 7). The precursors appear to stop elongation when the RNA polymerase molecules complete transcription of the coding region. The unsolved question is whether there is an initiation and termination site for each separate coding region or whether molecules of RNA polymerase I, after completing transcription of DNA sequence a–b in Fig. 7, "read through" the "non-transcribed" spacer region into the adjacent coding region (c–d in Fig. 7). There is some evidence for this latter proposal. Firstly, biochemical analysis of the

Table 1. Molecular weight of precursors to cytoplasmic rRNA in various plants

Species	Pre-rRNA ($M_r \times 10^6$)	References
Algae		
Chlamydomonas reinhardii	2.7	Woodcock et al. (1975)
	2.4	Miller and McMahon (1974)
Acetabularia mediterranea	>2.08[a]	Spring et al. (1976)
Euglena gracilis	3.5	Brown and Haselkorn (1971)
	4.2+2.4	Scott (1976)
Fungi		
Saccharomyces cerevisiae	2.5	Udem and Warner (1972)
	2.8[a]	Nikolaev et al. (1979)
Schizosaccharomyces pombe	2.6+2.3	Grierson et al. (1970)
Higher Plants		
Pea (*Pisum sativum*)	2.3	Rogers et al. (1970)
Artichoke (*Helianthus tuberosus*)	2.3	Rogers et al. (1970)
Carrot (*Daucus carota*)	2.8+2.2	Leaver and Key (1970)
	2.1[a]	Seitz and Seitz (1979)
Mung bean (*Phaseolus aureus*)	2.7−2.5	Grierson and Loening (1974)
Sycamore (*Acer pseudoplatanus*)	3.4+2.4	Cox and Turnock (1973)
	3.4+2.5	Grierson and Covey (1976)
	2.1[a]	Seitz and Seitz (1979)
	2.9+2.4	Cecchini and Miassod (1976)
		Cecchini et al. (1972)
Parsley (*Petroselinum crispum*)	2.9+2.4	Richter (1973)
	2.3	Seitz and Seitz (1972)
Tobacco (*Nicotiana tabacum*)	2.7	Perry et al. (1970)
Duckweed *Lemna perpusilla*	2.3	Rosner et al. (1973)
Spirodela oligorrhiza		Posner and Rosner (1975)
Rye (*Secale cereale*)	2.3	Sen et al. (1975)
Stock (*Matthiola incana*)	2.5	Grierson and Hemleben (1977)
	2.1[a]	Grierson and Hemleben (1977)
Radish (*Raphanus sativus*)	2.3	Aspart et al. (1978)

[a] Molecular weight estimates made after electrophoresis in polyacrylamide gels under denaturing conditions

first stable transcription products in *Xenopus* has shown that the rRNA precursor (M_r = approx 2.6×10^6) does not contain a 5′ triphosphate, indicating that it is not a primary transcription product (SLACK and LOENING 1974). Secondly, evidence for transcriptional activity in the form of short RNA fibrils emanating from the "non-transcribed" regions can occasionally be seen with the electron microscope (see RUNGGER and CRIPPA 1977, HADJIOLOV and NIKOLAEV 1976). Thirdly, several investigators have detected giant transcripts of rRNA sequences in mammalian cells in vitro which may represent complete transcripts of the region a–d in Figure 7 (GRUMMT and LINDIGKEIT 1973, GRUMMT 1975).

The demonstration by NIKOLAEV et al. (1979) that the 37S precursor to cytoplasmic rRNA in *Saccharomyces cerevisiae* has a 5′ triphosphate does not resolve this problem, since the arrangement of transcription units may be different in higher eukaryotes. It seems possible, therefore, that the stable transcription products readily detectable in most cells, and with which most work has been done, represent partially processed products.

The molecular weights of precursors to cytoplasmic rRNA vary quite a lot in different species. One of the smallest precursors is found in *Acetabularia* ($M_r = 2.08 \times 10^6$) and one of the largest in *Homo sapiens* (HeLa cells: $M_r = 4.7 \times 10^6$). Comparisons between organisms gave rise to the suggestion that the size of the rRNA precursor has increased during evolution (PERRY et al. 1970, GRIERSON et al. 1970). This correlates with the general tendency for the largest rRNA to increase in size from, for example, 1.3×10^6 in plants to 1.56×10^6 in frogs to 1.76×10^6 in man, although the size of the 18S rRNA remains fairly constant (LOENING 1968). The increase in size of the precursor and the largest stable rRNA has been interpreted to indicate an increase in complexity, perhaps related to the control of ribosome formation or protein synthesis. This idea remains to be tested. What is clear is that the greatest differences exist between birds and mammals compared to most other organisms. However, the size of the rRNA precursor cannot be regarded as a good taxonomic character, particularly when comparing plants, amphibia and fish (see GRIERSON et al. 1970, RUNGGER and CRIPPA 1977).

The first detectable transcription product in HeLa cells, which has been most intensively studied, is 45S RNA (M_r = approx. 4.7×10^6). An outline of the processing pathway is given in Fig. 8 (see WEINBERG and PENMAN 1970, MADEN et al. 1974, HADJIOLOV and NIKOLAEV 1976). The 45S precursor is methylated in the rRNA regions either during, or shortly after synthesis. It contains the majority of methyl groups present in rRNA (70 in 28S, 40 in 18S and 1 in 5.8S RNA). These are mainly 2′-O-ribose methylations, although a few base methylations are present. The methyl groups in the precursor are conserved during processing, which involves a series of endonucleolytic cleavages. The initial event is removal of a 5′ non-rRNA sequence to liberate the 41S RNA (Fig. 8, see HADJIOLOV and NIKOLAEV 1976). This latter molecule gives rise to a 20S RNA, which is the immediate precursor to 18S rRNA plus 32S RNA, which contains the 5.8S and 28S rRNA sequences. During the final stages of processing of the 20S precursor six further methyl groups are added to the 18S rRNA. Processing of the 32S precursor involves the removal of the transcribed sequence joining 5.8S and 28S rRNA, although the two molecules remain hydrogen-bonded together by complementary base-pairing (Fig. 8). It is probable that all the processing events occur in the nucleolus (see HADJIOLOV and NIKOLAEV 1976), although 18S rRNA is sometimes difficult to detect in the nucleolus because it enters the cytoplasm rapidly.

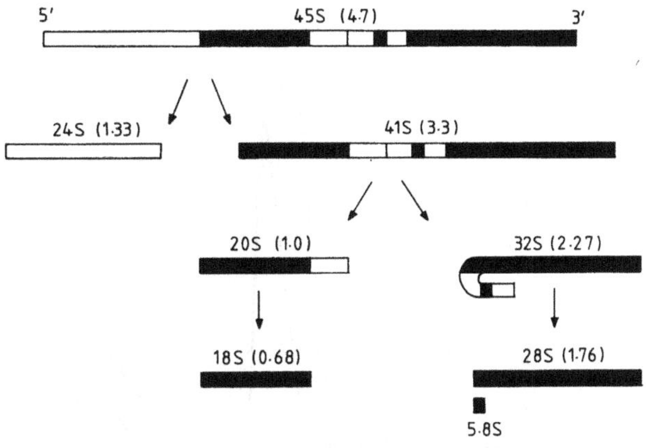

Fig. 8. Outline of the processing pathway of the 45S rRNA precursor in HeLa cells. (After HADJIOLOV and NIKOLAEV 1976). Non-conserved regions are shown in *white,* rRNA sequences are *black.* The approximate molecular weights in millions of the various RNA's are shown in *brackets*

Some variations in precursor processing have been found in different mammalian cells. In mouse L cells, for example, although the initial step involves excision of the 5′ non-rRNA sequence, the second processing event directly liberates the 18S rRNA, without a 20S intermediate (cf. Fig. 8). Furthermore, alternative processing pathways appear to operate side by side in rat liver cells. It therefore seems that the sequence of processing events varies in closely related species and even in the same cells under different physiological conditions (see HADJIOLOV and NIKOLAEV 1976). The interested reader is referred to the work of DUDOV et al. (1978) on the permutations of the sequence of processing reactions in rat liver.

In some animals inserts have been found in the rRNA-coding regions of the DNA (GLOVER and HOGNESS 1977). A similar insert has been detected in the 23S rRNA genes of *Chlamydomonas* chloroplasts and the 23S mitochondrial rRNA genes in *N. crassa* and *S. cerevisiae* (see Sect. 3.3). No such intervening sequences have yet been detected in the nucleolar DNA of higher plants. In situations where these genes actually function, such as in *Chlamydomonas,* the insert is probably transcribed as part of the precursor and is subsequently removed by a splicing reaction (see Sects. 4 and 5).

During processing in the nucleolus the precursor becomes associated with a number of proteins to produce ribonucleoprotein particles. Some of these proteins are identical to those found in ribosomes while others are confined to the nucleus and presumably function during RNA maturation and ribosome assembly (see HADJIOLOV and NIKOLAEV 1976, for short review).

The rRNA precursors detected in plants have molecular weights which range from 4.2×10^6 for *Euglena* to 2.08×10^6 in *Acetabularia* (Table 1). However, in many cases the molecular weight estimates have been carried out after polyacrylamide gel electrophoresis under non-denaturing conditions, sometimes without appropriate molecular weight markers. Therefore, the values quoted may be subject to error, due to unusual effects of base composition, RNA conformation or gel conditions. It is not clear whether the differences reported by laboratories working on the same species are due to genuine differences between the strains used. It is interesting to note that where molecular weights of plant precursors RNA's have been estimated under denaturing conditions the values are generally lower (see Table 1). In view of the foregoing discussion

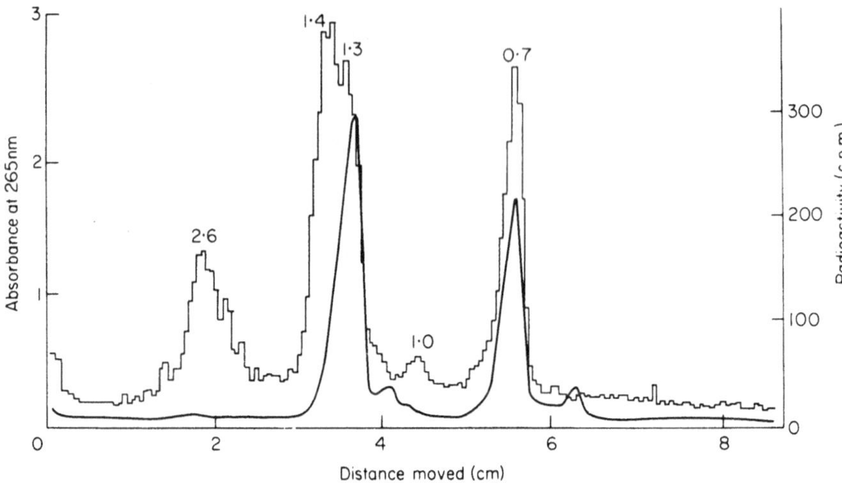

Fig. 9. Polyacrylamide gel electrophoresis of rRNA precursors and processing intermediates from roots of mung beans. (GRIERSON 1977). Roots were incubated for 1 h in ^{32}P-phosphate and RNA extracted and fractionated. The *smooth curve* shows the 25S and 18S rRNA's, the *histogram* shows the radioactivity. The approximate molecular weights of the RNA components are shown in millions

about the difficulty of identifying the primary product of a transcription unit, it is probably best to describe these precursors as the first stable products of transcription.

In some plants, such as carrot, sycamore, mung bean and duckweed, two precursors have been detected (LEAVER and KEY 1970, COX and TURNOCK 1973, GRIERSON and LOENING 1972, ROSNER et al. 1973). In mung bean leaves and in duckweed it has been shown that one precursor is in the chloroplast and another is in the nucleus (GRIERSON and LOENING 1974, POSNER and ROSNER 1975). In mung bean roots only the smaller precursor is found (Fig. 9). This RNA peak, which has a rather broad distribution in the molecular weight range $2.6–2.5 \times 10^6$ has been shown to be present in isolated nucleoli (GRIERSON et al. 1980). However, both precursors in carrot and sycamore appear to be in the nuclear fraction. Labelling kinetics do not support the view that the larger precursors in carrot and sycamore are cleaved during processing to produce the smaller ones (LEAVER and KEY 1970, COX and TURNOCK 1973), but see CECCHINI and MIASSOD (1976), who disagree. It has been suggested that where two precursors exist they might arise by variation in the location of the sites for initiation and termination of transcription in different genes, or by partial processing of some, but not all, transcripts before RNA synthesis is completed. It is not possible to distinguish unequivocally between these alternatives at present. However, SEITZ and SEITZ (1979) found only one precursor in carrot with $M_r = 2.3 \times 10^6$ (2.1×10^6 in formamide). Similar experiments with sycamore also gave a single rRNA precursor after electrophoresis in formamide, with M_r of about 2.1×10^6 (SEITZ and SEITZ 1979, GRIERSON, unpublished). These

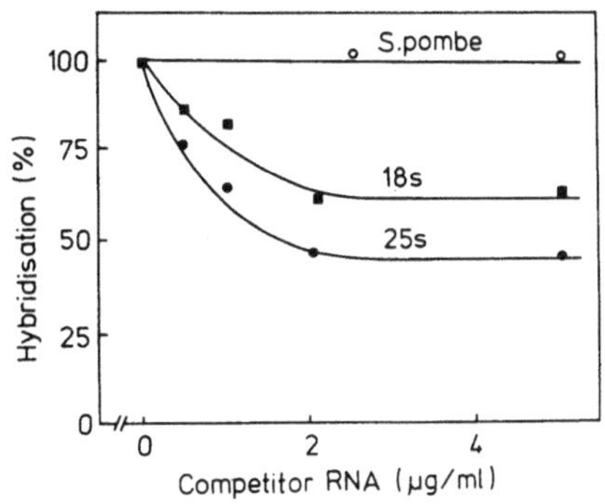

Fig. 10. Competition hybridisation between DNA, rRNA, precursor and 25S and 18S rRNA from mung beans. (GRIERSON and LOENING 1974). Mung bean rRNA precursor was prepared by gel electrophoresis and hybridised to DNA; 25S or 18S mung bean rRNA or RNA from *Schizosaccharomyces pombe* was used as competitor

results suggest that the larger precursors are particularly unstable and that they may contain hidden breaks in the polynucleotide chain.

Several criteria have been used to establish that the plant precursors are polycistronic transcription products containing rRNA sequences. (1) In pulse-chase experiments, radioactivity disappears from the putative precursors and subsequently appears in mature rRNA's (ROGERS et al. 1970, LEAVER and KEY 1970, COX and TURNOCK 1973, GRIERSON and LOENING 1974). (2) The nucleotide composition of the precursors is similar to that of mature rRNA (ROGERS et al. 1970, LEAVER and KEY 1970, GRIERSON and LOENING 1974). (3) In sycamore, the precursors are methylated, as are 25S and 18S rRNA (COX and TURNOCK 1973). (4) In mung bean, both unlabelled 25S and 18S cytoplasmic rRNA compete with labelled precursor for hybridisation to mung bean DNA (Fig. 10). (5) In rye, ribonuclease T_1 digestion of [^{32}P]-labelled precursor and 25S and 18S rRNA, followed by two-dimensional fractionation of the oligonucleotides, has shown that all large oligonucleotides present in 25S and 18S RNA are found in the precursor (SEN et al. 1975).

A scheme for plant nucleolar RNA processing is shown in Fig. 11. The uncertainty about the size of the initial transcription product, and the occasional observation of two large precursors, suggests that parts of the processing scheme have not always been detected and that the stable transcription products may represent intermediates in rRNA maturation. Processing generally involves the production of RNA's with molecular weights of $1.45–1.4 \times 10^6$ and $1.0–0.9 \times 10^6$, which are the immediate precursors of 25S and 18S rRNA. The $M_r = 1.45 \times 10^6$ RNA is also thought to contain the 5.8S RNA (originally called 7S RNA; HEPBURN and INGLE 1975).

There are some results which do not fit easily into the scheme in Fig. 11. However, there may be more than one sequence of processing steps, as found when different animal cells have been compared, and this would give rise to variation between plant species (GRIERSON 1977). In rye, no immediate precursor to 25S rRNA is found (SEN et al. 1975) and it seems that either the 25S rRNA is excised directly from the polycistronic precursor or that the immediate precursor to 25S rRNA is smaller than usual and difficult to distinguish from mature 25S rRNA. Similarly, the $M_r = 1.0–0.9 \times 10^6$ intermediate is not always detected. This is probably because it is processed very rapidly; it can be shown to accumu-

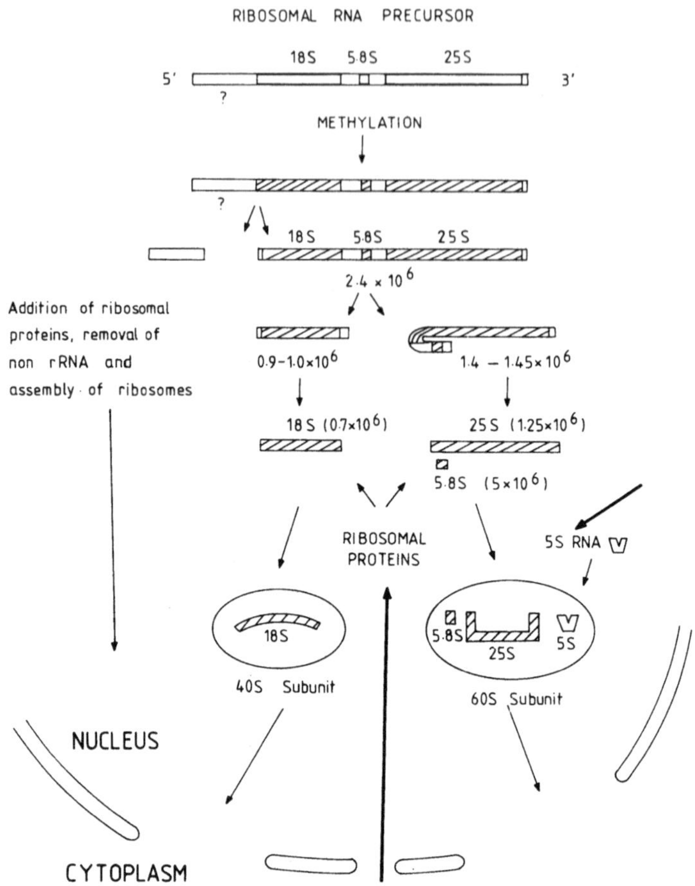

Fig. 11. An outline of the synthesis and processing of cytoplasmic rRNA precursors and the assembly of ribosomes in higher plants. The authenticity of the first processing cleavage has not been clearly established. The molecular weight of the stable polycistronic precursor varies in different plants (Table 1) as do some of the processing steps (see text for discussion)

late in cells treated with cycloheximide (LEAVER and KEY 1970, GEBAUER et al. 1975). Inhibition of protein synthesis by cycloheximide generally slows down or inhibits RNA processing (see also Sect. 4.1) and the protein synthesis inhibitor 2-(4-methyl-2,6 dinitroanalino)-N-methylpropionamide causes accumulation of the $M_r = 2.3 \times 10^6$ precursor in radish (ASPART et al. 1978). Other, previously unknown, processing intermediates may sometimes be detected after cycloheximide treatment (GEBAUER et al. 1975). It is probable that further processing reactions will be discovered by studies on rRNA synthesis by isolated nucleoli.

A number of factors have been shown to stimulate rRNA synthesis or processing including light (THIEN and SCHOPFER 1975, GRIERSON et al. 1976) and hormones (MELANSON and INGLE 1978) and β particles have been shown to inhibit processing (JACKSON and INGLE 1973). The molecular basis for these responses is not yet clear.

4 Processing of tRNA

There is considerable variation in the organisation of tRNA genes in different organisms and this is reflected in different patterns of transcription and processing of tRNA sequences. The one common feature seems to be that precursor RNA's are always involved.

4.1 Bacteria

In mutants of *E. coli* temperature-sensitive for ribonuclease P, tRNA's accumulate as precursors at the non-permissive temperature (SAKANO and SHIMURA 1975, IKEMURA et al. 1975, ILGEN et al. 1976). These precursors include monomers, dimers and trimers; the multiple tRNA precursors either consist of tandem repeats of a single tRNA sequence or contain several different tRNA's in one molecule (Fig. 13). They probably arise by processing of larger transcripts by the action of ribonuclease O (SAKANO and SHIMURA 1975). Some of the precursors are presumably released from the 30S rRNA transcript (Fig. 1).

Transcription of a single *E. coli* tRNA tyr gene, carried on a transducing phage, produces a precursor which contains an extra 41 nucleotides at the 5′ end, beginning with 5′pppG (ALTMAN and SMITH 1971, see Fig. 12). The extra 5′ sequence is removed by ribonuclease P (ROBERTSON et al. 1972). In phage T_4, seven different tRNA genes are clustered together in the following order, separated by short transcribed spacer regions:

$tRNA^{glu}$—$tRNA^{leu}$—$tRNA^{glyc}$—$tRNA^{pro}$—$tRNA^{ser}$—$tRNA^{thr}$—$tRNA^{ileu}$. A single promotor for transcription of these genes is about 1,000 base pairs upstream from $tRNA^{glu}$ and all seven sequences are part of a single transcription unit (see ABELSON 1979). Endonucleolytic cleavage by an enzyme other than ribonuclease P produces the following dimers: $tRNA^{glu}$—$tRNA^{leu}$; $tRNA^{pro}$—$tRNA^{ser}$; $tRNA^{thr}$—$tRNA^{ileu}$ and the $tRNA^{glu}$ precursor is released as a monomer. This initial processing may involve ribonuclease O, as suggested from studies on ribonuclease P-deficient *E. coli* (SAKANO and SHIMURA 1975). Further processing at the 5′ ends is carried out by ribonuclease P. There is also some endonuclease processing at the 3′ end, but the final maturation is thought to be carried out by an exonuclease variously named ribonuclease D, PIII or Q (ABELSON 1979). Processing occurs up to the 3′ CCA terminus, where this is present, but this sequence is not coded in all tRNA genes and is sometimes added later by a nucleotidyl transferase (DEUTSCHER 1973). The CCA terminus appears to be important for the correct processing of the tRNA precursors by ribonuclease P. During processing, extensive modification occurs to many of the bases in tRNA (MCGLOSKEY and NISHIMURA 1977).

Two points of wider interest have emerged from studies of tRNA processing in bacteria. Firstly, ribonuclease P, which generates mature 5′ ends, contains a tightly bound RNA component (STARK et al. 1978, ABELSON 1979) which is essential for processing activity. It has been suggested that this RNA may be involved in recognising the nucleotide sequence at the cleavage site in the

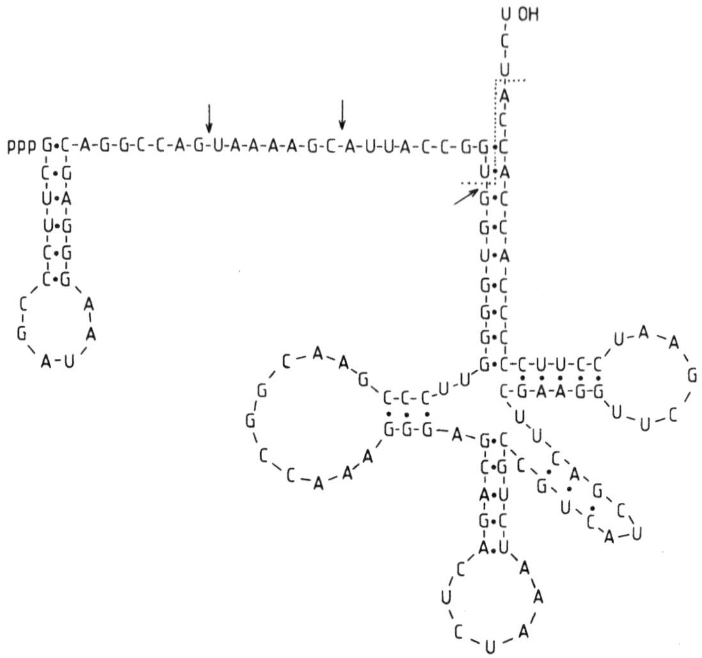

Fig. 12. Nucleotide sequence of precursor to $tRNA^{tyr}$. (After ALTMAN and SMITH (1971). *Arrows* show the ribonuclease cleavage sites; the mature tRNA is shown by the *dotted line.* From SZEKELY (1980), with permission from Macmillan, London and Basinkstoge

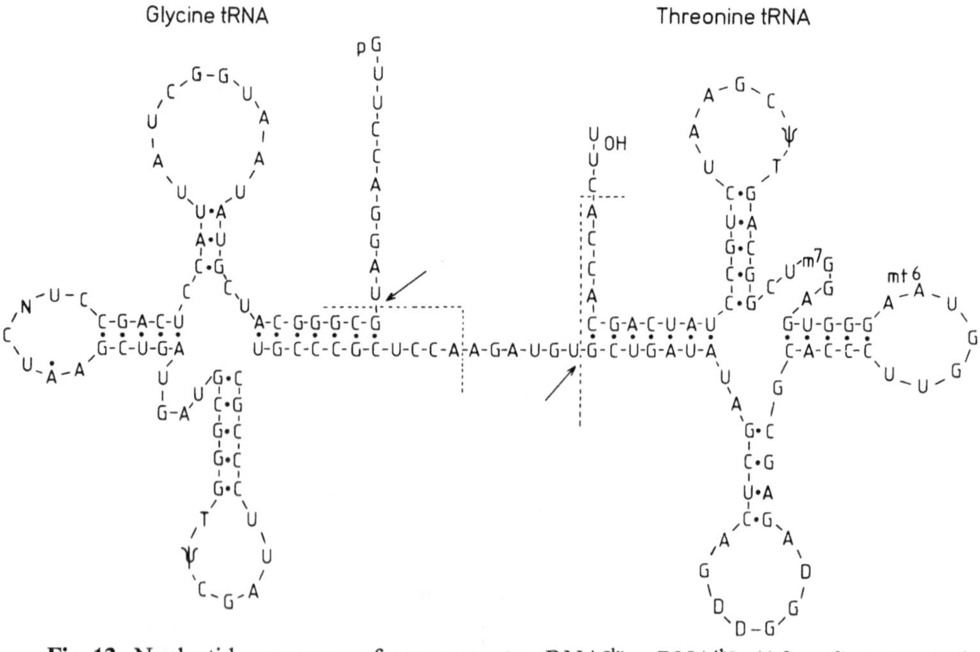

Fig. 13. Nucleotide sequence of precursor to $tRNA^{gly}$—$tRNA^{thr}$. (After CARBON et al. 1975). *Arrows* show the ribonuclease cleavage sites and *dotted lines* indicate the mature tRNA sequences. From SZEKELY (1980), with permission from Macmillan, London and Basingstoke

tRNA precursors. Secondly, the seven tRNA's present in the single transcription unit of phage T_4, mentioned above, do not all accumulate in the same proportions after processing. This suggests that some feature of the precursor sequence or secondary structure may affect the accumulation of individual mature tRNA's (ABELSON 1979).

4.2 Eukaryotes

Multiple precursors to tRNA have not been detected in eukaryotes. In some species relatively large spacer regions separate the coding sequences (CLARKSON et al. 1973) and it is possible that these spacers are either not transcribed or that they are rapidly eliminated by processing reactions. Nevertheless precursors approximately 20%–40% larger than the mature tRNA's are detected by pulse-labelling experiments with animal and plant cells.

Investigation of tRNA genes in *Saccharomyces cerevisiae* has shown that some contain inserts, ranging from 14–60 base pairs long, which do not appear in mature tRNA's. The gene for $tRNA^{tyr}$, for example, has a 14 base-pair intervening sequence adjacent to the anticodon (GOODMAN et al. 1977). KNAPP et al. (1978) studied the tRNA precursors which accumulated in a temperature-sensitive yeast mutant and showed that the RNA precursors contained the intervening sequence. Transcription and processing of the yeast $tRNA^{tyr}$ gene has been studied by injecting the DNA into frog oocytes (MELTON et al. 1980). The initial precursor, which contains about 108 nucleotides, has additional sequences at the 3′ and 5′ ends, plus the 14 nucleotide intervening sequence next to the anticodon. Several processing reactions occur in the nucleus before the tRNA enters the cytoplasm. These modifications include stepwise removal of 3′ and 5′ sequences, to reveal the mature termini, and excision of the intervening sequence and the joining (splicing) of the two halves of the mature tRNA. Addition of a 3′ terminal CCA and the modification of at least seven ribonucleotides also occurs before the tRNA appears in the cytoplasm (MELTON et al. 1980).

The tRNA-splicing reaction, which removes the intervening sequence, has been studied in vitro by PEEBLES et al. (1979) and KNAPP et al. (1979). The tRNA precursors which accumulate at the non-permissive temperature in the temperate-sensitive yeast mutants contain mature 3′ and 5′ termini but retain the intervening sequence. Incubation of these molecules with soluble extracts from normal yeast cells results in endonucleolytic cleavage, to liberate the intervening sequence plus two half-tRNA molecules carrying a 5′ OH and a 3′ phosphate. Ligation of these two half-tRNA's, to produce mature tRNA, is carried out by a soluble enzyme which requires ATP (KNAPP et al. 1979).

5 Processing of mRNA

5.1 General Features

Messenger-RNA from prokaryotes is particularly unstable. It is quite common for translation to begin before transcription is completed and for the mRNA to be degraded rapidly from the 5′ end shortly after synthesis. There is, therefore, not much evidence for processing of bacterial mRNA in the sense used in this article. However, DUNN and STUDIER (1975) have described processing of a large precursor to the "early" mRNA's of bacteriophage T7 and it is possible that further investigation may provide evidence for processing of mRNA coded by bacterial DNA.

Early experiments on mRNA synthesis in eukaryotic cells led to the detection of large RNA molecules in the nucleus which are heterogeneous in size (hnRNA or nuclear heterogeneous RNA; reviewed by LEWIN 1975 and PERRY et al. 1976). Much of this RNA, which is synthesised in the nucleoplasm by RNA polymerase II (cf. WOLLGIEHN, Chap. 5, this Vol.), is rapidly degraded after synthesis and does not enter the cytoplasm. Several investigators suggested that mRNA sequences are excised from hnRNA precursors by processing enzymes and selected for transport to the cytoplasm. A certain amount of controversy surrounds early work in this field. This was partly because the apparent size of hnRNA is affected by the experimental protocol and the occurrence of aggregates or degradation products of RNA can affect the interpretation of results. However, recent experiments, carried out with more modern techniques, have confirmed the existence of mRNA precursors. The size of the transcription units in eukaryotes ranges from about 2,000–20,000 nucleotides (DARNELL 1979).

It has been calculated that about 25% of hnRNA from tobacco leaf nuclei enters cytoplasmic ribosomes (GOLDBERG 1980). Processing involves (1) "capping" of the 5′ ends of mRNA precursors. Sometimes this is preceded by 5′ end trimming or may involve splicing. (2) Addition of poly(A) sequences at the 3′ end, which may also be preceded by trimming. (3) The excision of intervening sequences and the splicing together of mRNA coding-regions.

5.2 Capping

Most mRNA's from eukaryotic cells have no free 5′ phosphate or triphosphate; a special 5′ "cap" is added after synthesis. This consists of a 7-methyl G, joined via a 5′-5′ triphosphate linkage to the 5′ terminal nucleotide. This is an unusual linkage since all the other nucleotides are joined by 5′-3′ phosphate bonds. The general formula of the cap structure is $m^7GpppXpY$ (see DYER, Chap. 6, this Vol., Fig. 4). "X" may be the original 5′ terminal nucleotide used to initiate RNA synthesis, frequently pppG or pppA, or may be a 5′ mononucleotide produced by processing. Caps containing unmodified nucleotides in positions "X" and "Y" (called "cap zero") have been found in mRNA's of several plant viruses (SHATKIN 1976, cf. also HIRTH, Chap. 10, this Vol., VAN VLOTEN-DOTING, Chap. 11, this Vol.). Cap zero structures have also been found in polyadenylated RNA (presumed to be mRNA) of *Neurospora crassa* (SEIDEL and SOMBERG 1978), *Dictyostelium discoideum* (DOTTIN et al. 1976), *Saccharomyces cerevisiae* (SRIPATI et al. 1976, DE KLOET and ANDREAN 1976), *Avena sativa* (HAUGLAND and CLINE 1978) and *Triticum cereale* (KENNEDY and LANE 1979, SAINI and LANE 1977).

Many nuclear precursors to mRNA are capped and this modification therefore appears to occur as an early processing step. Capping of vaccinia virus mRNA (reviewed by SHATKIN 1976) occurs very rapidly after the initiation of mRNA synthesis in vitro. In the first step, a phosphate is removed from the initiating trinucleotide (pppG) by a phosphohydrolase to yield ppG. Guanosine monophosphate is then added in a 5′-5′ linkage, using GTP as substrate, and liberating pyrophosphate. At this stage the 5′ terminus is GpppG. The terminal G is then methylated at the N^7 position by a methylase, using S-adenosyl methionine as methyl donor.

In some instances the initiator nucleotide itself is not capped, but some nucleotides are removed from the 5′ end, exposing an internal nucleotide with a 5′ phosphate. A slightly different reaction mechanism would be necessary for capping of such a 5′ mononucleotide.

Messenger-RNA molecules do not have to be capped in order to be translated in vitro, but there is evidence that the cap increases translation efficiency (SHATKIN 1976, VAN VLOTEN-DOTING, Chap. 11, this Vol.). Recent evidence suggests that the stimulation of translation is due to a cap-binding protein which recognises m^7GpppN (SONENBERG et al. 1980).

5.3 Polyadenylation

Many eukaryotic mRNA's contain sequences of polyadenylic acid at their 3′ ends (LIM and CANELLAKIS 1970, DARNELL et al. 1971, LEE et al. 1971; EDMONDS et al. 1971, BRAWERMAN 1974). These vary in length up to about 250 A residues. Among the plant mRNA's shown to be polyadenylated are those coding for cellulase (VERMA et al. 1975), leghaemoglobin (VERMA et al. 1974), α-amylase (HIGGINS et al. 1976), phenylalanine ammonia lyase (RAGG et al. 1977), and the nuclear-coded chloroplast proteins: small subunit of ribulose bisphosphate carboxylase (HIGHFIELD and ELLIS 1978, CASHMORE 1979) and the light-harvesting protein of thylakoids (APEL and KLOPPSTECH 1978). Some of the hnRNA from animals and plants is also polyadenylated (SPOHR et al. 1976, GRIERSON and COVEY 1976, CHAPMAN and INGLE 1976). Labelling experiments suggest that the poly(A) is added to mRNA's or their precursors in the nucleoplasm after transcription. There do not appear to be regions of poly dA·dT in eukaryotic DNA of sufficient length to code for poly(A) (SHENKIN and BURDON 1974, GRIERSON 1975) but a poly(A) polymerase, which attaches poly(A) residues to the 3′ OH of primer RNA, is present in the nuclei of animals and plants (EDMONDS and ABRAMS 1962, HYATT 1967, NIESSING and SEKERIS 1972, WINTERS and EDMONDS 1973, BURKHARD and KELLER 1974, MANS and HUFF 1975). The in vitro synthesis of high molecular weight polydisperse RNA by RNA polymerase II has been demonstrated in maize nuclei (SLATER et al. 1978) but under the conditions used no poly(A) sequences were found in the RNA synthesised. JACOBSON et al. (1974) have suggested that in *Dictyostelium* about 25 A residues at the 3′ end of mRNA are coded by DNA and a further 100 are added post-transcriptionally.

Experiments by a number of different laboratories suggest that approximately half the mRNA sequences in higher plant cytoplasm contain poly(A) (KEY and SILFLOW 1975, COVEY and GRIERSON 1976, GRAY and CASHMORE 1976,

RAGG et al. 1977). This conclusion is based on the operational criterion of whether the mRNA binds to oligo(dT) cellulose or poly(U) sepharose. Messenger-RNA's with short regions of poly(A) (i.e., less than about 20 nucleotides) would normally be classified as poly(A)$^-$ unless special precautions are taken. In vitro translation of plant mRNA has shown that the poly(A)$^+$ and poly(A)$^-$ mRNA sequences often code for the same proteins (RAGG et al. 1977, GRAY and CASHMORE 1976). The most convincing explanation for these observations is that there is wide variation in the length of poly(A) sequences present at the 3′ ends of individual mRNA's. The length may vary with age of the mRNA or the physiological state of the cells and may be connected with mRNA turnover. For example, HUEZ et al. (1975) injected poly(A)$^+$ and poly(A)$^-$ globin mRNA into frog oocytes and showed that the mRNA was much more stable in vivo when it contained poly(A).

A poly(A) polymerase is also present in higher plant chloroplasts (BURKHARD and KELLER 1974). WHEELER and HARTLEY (1975) originally concluded that no poly(A) sequences longer than 20 nucleotides are present in spinach chloroplast mRNA; subsequently evidence was obtained for short sequences of oligo(A) (COVEY and HARTLEY, personal communication). The presence of poly(A)$^+$ RNA has also been demonstrated in maize chloroplasts (HAFF and BOGORAD 1976) and synthesis of poly(A)$^+$ RNA observed in isolated spinach chloroplasts (BARTOLF and PRICE 1979).

5.4 Splicing

There are two types of splicing which may occur during processing of mRNA precursors: the removal of non-conserved nucleotide sequences from *between* the mRNA coding regions and the addition of sequences at the *start* of the mRNA molecule. The latter case will be discussed first.

Sequencing studies have shown that mature mRNA's contain untranslated nucleotides at the 5′ and 3′ ends (see DYER, Chap. 6, this Vol.). These regions are transcribed from the DNA. However, they are not always contiguous with the mRNA-coding sequence in the chromosome, but can be attached to the mRNA after transcription by splicing of a mRNA precursor. Such a situation has been found for adenovirus, where different mRNA's are initially transcribed as a polycistronic precursor, which is subsequently processed in the nucleus to produce individual mRNA's each carrying the same 5′ "leader sequence" (Fig. 14). The leader sequence, which is about 200 nucleotides long, is some distance away from the mRNA sequences in the original RNA transcript. It is itself formed by the splicing of three separate regions of the precursor RNA (Fig. 14). This spliced, tripartite leader sequence is capped and then attached to the 5′ end of several different mRNA's. At least 13 adenovirus mRNA's, each with the same leader sequence, are produced by this type of reaction. Processing at the 5′ end or splicing of a mRNA leader sequence does not always occur, however. For example, in some adenovirus mRNA's the original 5′ end remains attached to the mRNA and is capped (ZIFF and EVANS 1978). Furthermore, although different mRNA's from *Dictyostelium discoideum* have the same

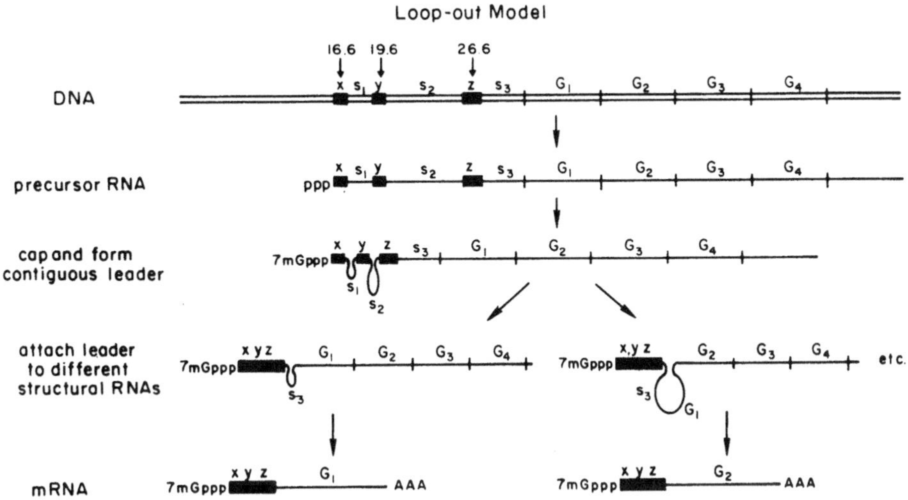

Fig. 14. Model showing the splicing of the leader sequence to different adenovirus mRNA's. The complete DNA sequence is copied into RNA. G_1, G_2, G_3 and G_4 represent different mRNA sequences; x, y and z represent the three segments of the leader sequence which are spliced together and then attached to individual mRNA's. Reproduced from SZEKELY (1980) with permission from Macmillan, London and Basingstoke, modified from KLESSIG (1977)

5′ end sequences these seem to be transcribed from the DNA immediately in front of the mRNA-coding sequence (KIMMEL and FIRTEL 1979) and no splicing reaction seems to be involved. Leader-sequences may be involved in regulation of RNA synthesis or translation. Furthermore, in some cases part of the 5′ "untranslated sequence" actually codes for amino acids present in protein precursors but removed after protein transport or assembly is completed.

Examples of internal splicing have also been described. DNA sequencing, together with hybridisation of mRNA, or cDNA copies of mRNA, to DNA, has shown that eukaryotic genes are often interrupted at intervals by nucleotide sequences which do not appear in mature mRNA. Such "split genes" have been found, for example, for globin (JEFFREYS and FLAVELL 1977), ovalbumin (BREATHNACH et al. 1977, DUGAICZYK et al. 1978, WEINSTOCK et al. 1978), immunoglobin (BRACK and TONEGAWA 1977) and a $M_r = 72 \times 10^3$ adenovirus DNA-binding protein (GOLDENBERG and RASKAS 1979). The intervening sequences are transcribed, with the coding regions, as part of the mRNA precursors. During processing the intervening sequences are removed by endonucleases and the ends of the mRNA coding-regions are joined together in the correct sequence by splicing reactions. This produces an mRNA with nucleotide triplets colinear with the amino sequence of the protein it specifies.

At least two putative precursors for globin mRNA have been detected, a 27S RNA (approximately 5,000 nucleotides) and a 15S RNA (approximately 1,500 nucleotides) (BASTOS and AVIV 1977, ROSS and KNECHT 1978, KWAN et al. 1977, TILGHMAN et al. 1978, CURTISS et al. 1977, CURTISS and WEISSMAN 1976). These precursors appear to contain the transcribed inserts present in the globin gene. The 27S precursor presumably has additional nucleotides at either the 5′ or 3′ ends (or both), which are removed during processing. The 15S precursor has the same terminal nucleotides as the 9S messenger and is already capped and carries a 3′ poly(A) tail (BASTOS and AVIV 1977, CURTISS et al. 1977, ROSS and KNECHT 1978). The transcribed inserts are excised from the 15S precursor as a late processing event and the three coding regions are spliced together to produce functional 9S mRNA. Splicing of some adenovirus mRNA also occurs after polyadenylation (NEVINS and DARNELL 1978).

The significance of splicing is not fully understood. It may be that it is an indication, at the molecular level, of the way in which specific proteins have evolved, i.e., not by single nucleotide changes but by the juxtaposition of comparatively long, and different, polynucleotide sequences, thereby producing new proteins with substantially different properties. There is evidence that splicing may be involved in subtle changes in gene expression. EARLY et al. (1980) have shown that two different mRNA's can be produced from a single immunoglobulin "gene" by alternative processing pathways. The μ Ig M antibody exists in two forms, as a membrane-bound protein (μm) and a secreted protein (μs). These two forms of the antibody differ in the C-terminal amino acids only. The mRNAs for both polypeptides are produced by splicing of similar regions of a mRNA precursor, transcribed from the same "gene". Different 3′ terminal sequences can be spliced onto the mRNA coding for μm and μs proteins, thereby altering the precise nature of the gene product synthesised.

Recently, SMITH and ELLIS (personal communication) have detected large precursors to mRNA for the nuclear-coded small subunit of the chloroplast enzyme ribulose-1,5-bisphosphate carboxylase. This is the first report of a nuclear precursor to an identified mRNA from a higher plant. At present there is no information as to whether the precursor contains inserts or whether it simply contains extra 3′ or 5′ nucleotides.

The enzymes responsible for mRNA splicing have not, as yet, been well characterised. Recently STEITZ and her colleagues have shown that the 5′ end of u1RNA (a small nuclear RNA) exhibits complementarity with the nucleotide sequences which occur across the splice junctions in mRNA precursors. It is, therefore, possible that this small RNA molecule may combine with processing enzymes and function in the recognition of sequences in mRNA precursors which require splicing (LERNER et al. 1980). This is an intriguing suggestion, especially when it is recalled that ribonuclease P, required for tRNA processing, contains a tightly bound RNA (see Sect. 4).

References

Abelson J (1979) RNA processing and the intervening sequence problem. Annu Rev Biochem 48:1035–1069

Adesnick M, Levinthal C (1969) Synthesis and maturation of ribosomal RNA in *Escherichia coli*. J Mol Biol 46:281–303

Allet B, Rochaix JD (1979) Structure analysis at the ends of the intervening DNA sequences in the chloroplast 23S ribosomal genes of *Chlamydomonas reinhardii*. Cell 18:55–60

Altman S, Smith JD (1971) Tyrosine tRNA precursor molecule polynucleotide sequence. Nature New Biol 233:35–39

Apel K, Kloppstech K (1978) The plastid membranes of barley (*Hordeum vulgare*) Light-induced appearance of mRNA coding for the apoprotein of the light-harvesting chlorophyll a/b protein. Eur J Biochem 85:581–588

Aspart L, Cooke R, Delseny M, Guitton Y (1978) Effect of the protein-synthesis-initiation inhibitor 2-(4-methyl-2,6-dinitroanilino)-N-methylpropionamide on ribonucleic acid synthesis in radish seedlings. Biochem J 176:607–611

Bartolf M, Price CA (1979) Synthesis of poly(A)-containing RNA by isolated spinach chloroplasts. Biochemistry 18:1677–1680

Bastos RN, Aviv H (1977) Globin RNA precursor molecules: biosynthesis and processing in erythroid cells. Cell 11:641–650

Berger S, Schweiger HG (1975) An apparent lack of non-transcribed spacers in rDNA of a green alga. Mol Gen Genet 139:269–275

Bohnert HJ, Schmitt JM, Hermann RG (1974) Structural and functional aspects of the plastome. III DNA- and RNA-synthesis in isolated chloroplasts. Portugal Acta Biol 14:71–90

Bohnert HJ, Driesel AJ, Herrmann RG (1977) Transcription and processing of transcripts in isolated unbroken chloroplasts. In: Bogorad L, Weil JH (eds) Acides nucléiques et synthèse des proteines chez les végétaux. CNRS, Paris, pp 213–217

Bos JL, Heyting C, Borst P, Arnberg AC, Van Bruggen EFJ (1978) An insert in the single gene for the large ribosomal RNA in yeast mitochondrial DNA. Nature (London) 275:336–338

Brack C, Tonegawa S (1977) Variable and constant parts of the immunoglobulin light chain gene of a mouse myeloma cell are 1250 nontranslated bases apart. Proc Natl Acad Sci USA 74:5652–5656

Brawerman G (1974) Eukaryotic messenger RNA. Annu Rev Biochem 43:621–642

Breathnach R, Mandel JL, Chambon P (1977) Ovalbumin gene is split in chicken DNA. Nature (London) 270:314–319

Brown RD, Haselkorn R (1971) Synthesis and maturation of cytoplasmic ribosomal RNA in *Euglena gracilis*. J Mol Biol 59:491–503

Burkhard G, Keller EB (1974) Poly (A) polymerase and poly (G) polymerase in wheat chloroplasts. Proc Natl Acad Sci USA 71:389–393

Busch H, Smetana K (1970) The nucleolus. Academic Press, London New York

Carbon J, Chang S, Kirk LI (1975) Clustered tRNA genes in *Escherichia coli*: transcription and processing. Brookhaven Symp Biol 26:23–36

Cashmore AR (1979) Reiteration frequency of the gene coding for the small subunit of ribulose-1,5-bisphosphate carboxylase. Cell 17:383–388

Cecchini JP, Miassod R (1976) Ribosomal cistrons in higher plant cells 1. A definitive scheme for the maturation pathway of the primary transcriptional product of ribosomal cistrons in *Acer pseudoplatanus* L. cells. Biochim Biophys Acta 418:104–116

Cecchini JP, Miassod R, Ricard J (1972) Processing of precursor ribosomal RNA in suspensions of higher plant cells. FEBS Lett 28:183–187

Chapman KSR, Ingle J (1976) The stability, polyadenylic acid content and ribonucleoprotein form of nuclear ribonucleic acid in artichoke. Biochem J 159:585–600

Clarkson SG, Birnstiel ML, Purdom IF (1973) Clustering of transfer RNA genes of *Xenopus laevis*. J Mol Biol 79:411–429

Covey SN, Grierson D (1976) Subcellular distribution and properties of poly(A)-containing RNA from cultured plant cells. Eur J Biochem 63:599–606

Cox BJ, Turnock G (1973) Synthesis and processing of ribosomal RNA in cultured plant cells. Eur J Biochem 37:367–376

Curtiss PJ, Weissman C (1976) Purification of globin messenger RNA from dimethylsulphoxide-induced friend cells and detection of a putative globin messenger RNA precursor. J Mol Bil 106:1061–1075

Curtiss PJ, Mantei N, Weissmann C (1977) Characterization and kinetics of synthesis of 15S β-globulin RNA, a putative precursor of β-globulin mRNA. Cold Spring Harbor Symp 42:971–984

Dahlberg AE, Dahlberg JE, Lund E, Tokimatsu H, Rabson AB, Calvert PC, Reynolds F, Zahalak M (1978) Processing of the 5′ end of *Escherichia coli* 16S ribosomal RNA. Proc Natl Acad Sci USA 75:3598–3602
Darnell JE Jr (1979) Transcription unit for mRNA production in eukaryotic cells and their DNA viruses. Prog Nucleic Acid Res Mol Biol 22:327–353
Darnell JE, Wall R, Tushinski RJ (1971) An adenylic acid-rich sequence in messenger RNA of HeLa cells and its possible relationship to reiterated sites in DNA. Proc Natl Acad Sci USA 68:1321–1325
Detchon P, Possingham JV (1973) Chloroplast ribosomal ribonucleic acid synthesis in cultured spinach leaf tissue. Biochem J 136:829–836
Deutscher MP (1973) Synthesis and functions of the C-C-A terminus of transfer RNA. Progr Nucleic Acid Res Mol Biol 13:51–92
Doolittle WF (1972) Ribosomal ribonucleic acid synthesis and maturation in the blue green alga *Anacystis*. J Bacteriol 111:316–324
Dottin RP, Weiner AM, Lodish HF (1976) 5′ terminal nucleotide sequences of the messenger RNAs of *Dictyostelium discoideum*. Cell 8:233–244
Dudov KP, Debeva MD, Hadjiolov AA (1978) Processing and migration of ribosomal ribonucleic acids in the nucleolus and nucleoplasm of rat liver nuclei. Biochem J 171:375–383
Dugaiczyk A, Woo SLC, Lai EC, Mace MJJr, McReynolds L, O'Malley BW (1978) The natural ovalbumin gene contains seven intervening sequences. Nature (London) 274:328–333
Dunn JJ, Studier FW (1975) Processing, transcription and translation of bacteriophage T7 messenger RNA. Brookhaven Symp Biol 26:267–276
Durante M, Cionini PG, Avanzi S, Cremoni R, D'Amato F (1977) Cytological localization of the genes for the four classes of ribosomal RNA (25S, 18S, 5.8S and 5S) in polytene chromosomes of *Phaseolus coccineus*. Chromosoma 60:269–282
Early P, Rogers J, Davis M, Calame K, Bond M, Wall R, Hood L (1980) Two mRNAs can be produced from a single immunoglobulin gene by alternative RNA processing pathways. Cell 20:313–319
Edmonds M, Abrams R (1962) Nature of a polynucleotide required for polyribonucleotide formation from adenosine triphosphate with an enzyme from thymus nuclei. J Biol Chem 237:2636; 2642
Edmonds M, Vaughan M Jr, Nakazato H (1971) Polyadenylic acid sequences in the heterogeneous nuclear RNA and rapidly-labelled polyribosomal RNA of HeLa cells: possible evidence for a precursor relationship. Proc Natl Acad Sci USA 68:1336–1340
Fakan S, Bernhard W (1971) Localisation of rapidly and slowly labelled nuclear RNA as visualised by high resolution autoradiography. Exp Cell Res 67:129–141
Faye G, Dennebouy N, Kujawa C, Jacq C (1979) Inserted sequence in the mitochondrial 23S ribosomal RNA gene of the yeast *Saccharomyces cerevisiae*. Mol Gen Genet 168:101–110
Friedrich J, Hemleben V, Meagher RB, Key JL (1979) Purification and restriction endonuclease mapping of soybean 18S and 25S ribosomal RNA genes. Planta 146:467–473
Gebauer HU, Seitz U, Seitz U (1975) Transport and processing of ribosomal RNA in plant cells after treatment with cycloheximide. Z Naturforsch 301:213–218
Gegenheimer P, Apirion D (1978) Processing of rRNA by RNAase P: spacer tRNAs are linked to 16S rRNA in an RNAase III mutant strain of *E. coli*. Cell 15:527–539
Ginsburg D, Steitz JA (1975) The 30S ribosomal precursor RNA from *Escherichia coli*. J Biol Chem 250:5647–5654
Glover DM, Hogness DS (1977) A novel arrangement of the 28S sequences in a repeating unit of *Drosophila melanogaster* rDNA. Cell 10:167–176
Goldberg RB (1980) Structural gene expression in tobacco. In: Leaver CJ (ed) Genome organization and expression in plants. Plenum, New York, London, pp 117–126
Goldenberg CJ, Raskas HJ (1979) Splicing patterns of nuclear precursors to the mRNA for adenovirus 2 DNA binding protein. Cell 16:131–138
Goodman HM, Olson MV, Hall BD (1977) Nucleotide sequence of a mutant eukaryotic

gene: the yeast tyrosine-inserting ochre suppressor SUP40. Proc Natl Acad Sci USA 74:5453–5457

Gray RE, Cashmore AR (1976) RNA synthesis in plant leaf tissue: the characterisation of messenger RNA species lacking and containing polyadenylic acid. J Mol Biol 108:595–608

Grierson D (1975) The hybridisation of [^{3}H] poly(uridylic) acid to DNA from *Phaseolus aureus* Roxb. Planta 127:87–91

Grierson D (1977) The nucleus and the organisation and transcription of nuclear DNA. In: Smith H (ed) The molecular biology of plant cells. Blackwell, Oxford, pp 213–255

Grierson D, Covey S (1976) The properties and function of rapidly-labelled nuclear RNA. Planta 130:317–321

Grierson D, Hemleben V (1977) Ribonucleic acid from the higher plant *Matthiola incana*. Molecular weight measurements and DNA-RNA hybridisation studies. Biochim Biophys Acta 475:424–436

Grierson D, Loening UE (1972) Distinct transcription products of the ribosomal genes in two different tissues. Nature New Biol 235:80–82

Grierson D, Loening UE (1974) Ribosomal RNA precursors and the synthesis of chloroplast and cytoplasmic ribosomal ribonucleic acid in leaves of *Phaseolus aureus*. Eur J Biochem 44:501–507

Grierson D, Smith H (1973) The synthesis and stability of ribosomal RNA in blue-green algae. Eur J Biochem 36:280–285

Grierson D, Rogers ME, Sartirana ML, Loening UE (1970) The synthesis of ribosomal RNA in different organisms: structure and evolution of the rRNA precursor. Cold Spring Harbor Symp 35:589–597

Grierson D, Covey SN Giles AB (1976) The effect of light on RNA metabolism in developing leaves. In: Smith H (ed) Light and plant development. Butterworth, London, pp 229–253

Grierson D, Thompson JR, Garcia-Mora R, Slater RJ (1980) Ribonucleic acid synthesis by isolated mung bean nucleoli. Z Pflanzenphysiol 97:437–447

Grummt I (1975) Synthesis of RNA molecules larger than 45S by isolated rat-liver nucleoli. Eur J Biochem 57:159–167

Grummt I, Lindigkeit R (1973) Pre-ribosomal RNA synthesis in isolated rat-liver nucleoli. Eur J Biochem 36:244–249

Hadjiolov AA, Nikolaev N (1976) Maturation of ribosomal ribonucleic acids and the biogenesis of ribosomes. Prog Biophys Mol Biol 31:95–144

Haff LA, Bogorad L (1976) Poly(adenylic acid)-containing RNA from plastids of maize. Biochemistry 15:4110–4115

Hartley MR (1979) The synthesis and origin of chloroplast low-molecular-weight ribosomal ribonucleic acid in spinach. Eur J Biochem 96:311–320

Hartley MR, Ellis RJ (1973) Ribonucleic acid synthesis in chloroplasts. Biochem J 134:249–262

Hartley MR, Head C (1979) The synthesis of chloroplast high-molecular-weight ribosomal ribonucleic acid in spinach. Eur J Biochem 96:301–309

Hartley MR, Head C, Gardiner J (1977) The synthesis of chloroplast RNA. In: Bogorad L, Weil JH (eds) Acides nucléiques et synthèse des proteines chez les végétaux, CNRS, Paris, pp 419–423

Haugland RA, Cline MG (1978) Capping structures at the 5′-terminus of polyadenylated ribonucleic acid in *Avena* coleoptiles. Plant Physiol 62:838–840

Hayes F, Vasseur M (1976) Processing of the 17-S *Escherichia coli* precursor RNA in the 27-S pre-ribosomal particle. Eur J Biochem 61:433–442

Hecht NB, Woese CR (1968) Separation of bacterial robosomal ribonucleic acid from its macromolecular precursors by polyacrylamide gel electrophoresis. J Bacteriol 95:986–990

Heizmann P (1974) Maturation of chloroplast rRNA in *Euglena gracilis*. Biochem Biophys Res Commun 56:112–118

Hemleben V, Grierson D (1978) Evidence that in higher plants the 25S and 18S rRNA genes are not interspersed with genes for 5S rRNA. Chromosoma 65:353–358

Hepburn AG, Ingle J (1975) Origin of 7S ribosomal RNA in artichoke. Phytochemistry 14:1157–1160
Higgins TJV, Zwar JA, Jacobsen JV (1976) Gibberellic acid enhances the level of translatable mRNA for α-amylase in barley aleurone layers. Nature (London) 260:166–168
Highfield PE, Ellis RJ (1978) Synthesis and transport of the small subunit of chloroplast ribulose bisphosphate carboxylase. Nature (London) 271:420–424
Huez G, Marbaix G, Hubert E, Clenter V, Leclerq M, Chantrenne H, Devos R, Soreq H, Nudel U, Littauer UZ (1975) Readenylation of polyadenylate-free globin m-RNA restores its stability in vitro. Eur J Biochem 59:589–592
Hyatt EA (1967) Polyriboadenylate synthesis by nuclei from developing sea urchin embryos. 1. Characterisation of the ATP polymerase reaction. Biochim Biophys Acta 142:246–253
Ikemura T, Shimura Y, Sakarno H, Ozeki H (1975) Precursor molecules of *Escherichia coli* transfer RNAs accumulated in a temperature-sensitive mutant. J Mol Biol 96:69–86
Ilgen C, Kirk LL, Carbon J (1976) Isolation and characterization of large transfer ribonucleic acid precursors from *Escherichia coli*. J Biol Chem 251:922–929
Ingle J (1973) The regulation of ribosomal RNA synthesis. In: Milborrow BV (ed) Biosynthesis and its control in plants. Academic Press, London, New York, pp 69–91
Jackson M, Ingle J (1973) The interpretation of studies on rapidly labeled ribonucleic acid in higher plants. Plant Physiol 51:412–414
Jacobsen A, Firtel RA, Lodish MF (1974) Synthesis of messenger and ribosomal RNA precursors in isolated nuclei of the cellular slime mould *Dictyostelium discoideum*. J Mol Biol 82:213–230
Jeffreys AJ, Flavell RA (1977) The rabbit β-globin gene contains a large insert in the coding sequence. Cell 12:1097–1108
Kennedy TD, Lane BG (1979) Wheat embryo ribonucleates XII. methyl-substituted nucleoside constituents and 5′-terminal dinucleotide sequences in bulk poly A-rich RNA from imbibing wheat embryos. Can J Biochem 57:927–931
Key JL, Silflow C (1975) The occurrence and distribution of poly(A)RNA in soybean. Plant Physiol 56:364–369
Kimmel AR, Firtel RA (1979) A family of short, interspersed repeat sequences at the 5′ end a set of *Dictyostelium* single-copy mRNAs. Cell 16:787–796
Klessig DF (1977) Two Adenovirus mRNAs have a common 5′ terminal leader sequence encoded at least 10 Kb upstream from their main coding regions. Cell 12:9–21
De Kloet SR, Andrean BAG (1976) Methylated nucleosides in polyadenylate-containing yeast messenger ribonucleic acid. Biochim Biophys Acta 425:401–408
Knapp G, Beckmann JA, Johnson PF, Fuhrman SA, Abelson J (1978) Transcription and processing of intervening sequences in yeast tRNA genes. Cell 14:221–236
Knapp G, Ogden RC, Peebles CL, Abelson (1979) Splicing of yeast tRNA precursors: structure of the reaction intermediates. Cell 18:37–45
Kuriyama Y, Luck DJL (1973) Ribosomal RNA synthesis in mitochondria of *Neurospora crassa*. J Mol Biol 73:425–437
Kwan S-P, Wood TG, Lingrel JB (1977) Purification of a putative precursor of globulin messenger RNA mouse nucleated erythroid cells. Proc Natl Acad Sci USA 74:178–182
Leaver CJ, Key JL (1970) Ribosomal RNA synthesis in plants. J Mol Biol 49:671–680
Lee SY, Mendecki J, Brawerman G (1971) A polynucleotide segment rich in adenylic acid in the rapidly-labelled polyribosomal RNA component of mouse sarcoma 180 ascites cells. Proc Natl Acad Sci USA 68:1331–1335
Lerner MR, Boyle JA, Mount SM, Wolin SL, Steitz JA (1980) Are snRNPs involved in splicing? Nature (London) 283:220–224
Lewin B (1975) Units of transcription and translation: the relationship between heterogeneous nuclear RNA and messenger RNA. Cell 4:11–20
Lim L, Canellakis ES (1970) Adenine-rich polymer associated with rabbit-reticulocyte messenger RNA. Nature (London) 227:710–712
Loening UE (1968) Molecular weights of ribosomal RNA in relation to evolution. J Mol Biol 38:355–365

Lund E, Dahlberg JE (1977) Spacer transfer RNAs in ribosomal RNA transcripts of E. coli: processing of 30S ribosomal RNA in vitro. Cell 11:247–262
McGloskey JA, Nishimura S (1977) Modified nucleosides in transfer RNA. Acc Chem Res 10:403–410
Maden BEH (1971) The structure and formation of ribosomes in animal cells. Prog Biophys Mol Biol 22:127–177
Maden BEH, Salim M, Robertson JS (1974) Progress in the structural analysis of mammalian 45S and ribosomal RNA. In: Nomura M, Tissieres A, Lengyl P (eds) Ribosomes. Cold Spring Harbor Lab, New York
Mangiarotti G, Turco E, Perlo C, Altruda F (1975) Role of precursor 16S RNA in assembly of *E. coli* 30S ribosomes. Nature (London) 253:569–571
Mannella CA, Collins RA, Green MR, Lambowitz AM (1979) Defective splicing of mitochondrial rRNA in cytochrome-deficient nuclear mutants of *Neurospora crassa*. Proc Natl Acad Sci USA 76:2635–2639
Mans RJ, Huff NJ (1975) Utilization of ribonucleic acid and deoxy oligomer primers for polyadenylic acid synthesis by adenosine triphosphate:polynucleotidyl exotransferase from maize. J Biol Chem 250:3672–3678
Melanson DL, Ingle J (1978) Regulation of ribosomal RNA accumulation by auxin in artichoke tissue. Plant Physiol 62:761–765
Melton DA, DeRobertis EM, Cortese R (1980) Order and intracellular location of the events involved in the maturation of a spliced tRNA. Nature (London) 284:143–148
Miller MJ, McMahon D (1974) Synthesis and maturation of chloroplast and cytoplasmic ribosomal RNA in *Chlamydomonas reinhardii*. Biochim Biophys Acta 266:35–44
Miller OL, Beatty BR (1969) Visualization of nucleolar genes. Science 164:955–957
Morgan EA, Ikemura T, Lindahl L, Fallon AM, Nomura M (1978) Some rRNA operons in *E. coli* have tRNA genes at their distal ends. Cell 13:335–344
Munsche D, Wollgiehn R (1973) Die Synthese von ribosomaler RNA in Chloroplasten von *Nicotiana rustica*. Biochim Biophys Acta 294:106–117
Nevins JR, Darnell JE (1978) Steps in the processing of AdZ mRNA: poly(A)$^+$ nuclear sequences are conserved and poly(A) addition precedes splicing. Cell 15:1477–1493
Niessing J, Sekeris CE (1972) A homoribopolynucleotide synthetase in rat liver nuclei associated with ribonucleoprotein particles containing DNA-like RNA. FEBS Lett 22:83–88
Nikolaev N, Birge CH, Gotoh S, Glazier K, Schlessinger D (1975) Primary processing of high molecular weight pre-ribosomal RNA in *Escherichia coli* and HeLa cells. Brookhaven Symp Biol 26:175–193
Nikolaev N, Georgiev OI, Venkov PV, Hadjiolov AA (1979) The 37S precursor to ribosomal RNA is the primary transcript of ribosomal RNA genes in *Saccharomyces cerevisiae*. J Mol Biol 127:297–308
Pace NR (1973) Structure and synthesis of the ribosomal ribonucleic acid of prokaryotes. Bacteriol Rev 37:562–603
Pardue ML, Brown DD, Birnstiel ML (1973) Location of the genes for 5S ribosomal RNA in *Xenopus laevis*. Chromosoma 42:191–203
Peebles CL, Ogden RC, Knapp G, Abelson J (1979) Splicing of yeast tRNA precursors: a two-stage reaction. Cell 18:27–35
Perry RP (1976) Processing of RNA. Annu Rev Biochem 45:605–629
Perry RP, Cheng T-Y, Freed JJ, Greenberg JR, Kelly DE, Tartof KD (1970) Evolution of the transcription unit of ribosomal RNA. Proc Natl Acad Sci USA 65:609–616
Perry RP, Bard E, Hames BD, Kelley DE, Schibler U (1976) The relationship between hnRNA and mRNA. Prog Nucleic Acid Res Mol Biol 19:275–292
Posner HB, Rosner A (1975) Effects of chloramphenicol on RNA synthesis in *Spirodela* chloroplasts. Plant Cell Physiol 16:361–365
Ragg H, Schroder J, Hahlbrock K (1977) Translation of poly(A)-containing and poly(A)-free messenger RNA for phenylanine ammonia-lyase, a plant specific protein, in a reticulocyte lysate. Biochim Biophys Acta 474:226–233
Richter G (1973) Vorstufen der ribosomalen RNA in frei suspendierten Calluszellen der Petersilie (*Petroselium sativum*). Planta 113:79–96

Robertson HD, Altman S, Smith JD (1972) Purification and properties of a specific *Escherichia coli* ribonuclease which cleaves a tyrosine transfer ribonucleic acid precursor. J Biol Chem 247:5243–5251
Rochaix JD, Malnoë P (1978) Anatomy of the chloroplast ribosomal DNA of *Chlamydomonas reinhardii*. Cell 15:661–670
Rogers ME, Loening UE, Fraser RSS (1970) Ribosomal RNA precursors in plants. J Mol Biol 49:681–692
Rosner A, Posner HB, Gressel J (1973) Synthesis and processing of RNA in *Lemna* characterization by gel electrophoresis. Plant Cell Physiol 14:555–564
Ross J, Knecht DA (1978) Precursors of alpha and beta globin messenger RNAs. J Mol Biol 119:1–20
Rungger D, Crippa M (1977) The primary ribosomal DNA transcript in eukaryotes. Prog Biophys Mol Biol 31:247–269
Saini MS, Lane BG (1977) Wheat embryo ribonucleates VIII. The presence of 7 methylguanosine 'cap structure' in the RNA of imbibing wheat embryos. Can J Biochem 55:819–824
Sakano H, Shimura Y (1975) Sequential processing of precursor tRNA molecules in *Escherichia coli*. Proc Natl Acad Sci USA 72:3369–3373
Scheer U, Trendelenburg MF, Franke WW (1973) Transcription of ribosomal RNA cistrons. Correlation of morphological and biochemical data. Exp Cell Res 80:175–190
Scott NS (1976) Precursors of chloroplast ribosomal RNA in *Euglena gracilis*. Phytochemistry 15:1207–1213
Seidel BL, Somberg EW (1978) Characterization of *Neurospora crassa* polyadenylated messenger ribonucleic acid: structure of the 5′ terminus. Arch Biochem Biophys 187:108–112
Seitz U, Seitz U (1972) Kern-Plasma-Transport neusynthetisierter rRNS in Zellen einer Suspensionskultur von *Petroselinum sativum*. Planta 106:141–148
Seitz U, Seitz U (1973) Biosynthese der ribosomalen RNS bei der blaugrünen Alge *Anacystis nidulans*. Arch Mikrobiol 90:213–222
Seitz U, Seitz U (1979) The molecular weight of rRNA precursor molecules and their processing in higher plant cells. Z Naturforsch 34c:253–258
Sen S, Payne PI, Osborne DJ (1975) Early ribonucleic acid synthesis during germination of rye (*Secale cereale*) embryos and the relationship to early protein synthesis. Biochem J 148:381–387
Shatkin AJ (1976) Capping of eucaryotic mRNAs. Cell 9:645–653
Shenkin A, Burdon RH (1974) Deoxyadenylate-rich and deoxyguanylate-rich regions in mammalian DNA. J Mol Biol 85:19–40
Slack JMW, Loening UE (1974) 5′-ends of ribosomal and ribosomal precursor RNAs from *Xenopus laevis*. Eur J Biochem 43:59–67
Slater RJ, Venis MA, Grierson D (1978) Characterisation of ribonucleic acid synthesis by nuclei isolated from *Zea mays*. Planta 144:89–93
Smith JD (1976) Transcription and processing of transfer RNA precursors. Prog Nucleic Acid Res Mol Biol 16:25–73
Sogin ML, Pace B, Pace NR (1977) Partial purification and properties of a ribosomal RNA maturation endonuclease from *Bacillus subtilis*. J Biol Chem 252:1350–1357
Sonenberg N, Trachsel H, Hecht S, Shatkin AJ (1980) Differential stimulation of capped mRNA translation in vitro by cap binding protein. Nature (London) 285:331–333
Spohr G, Dettori G, Manzari V (1976) Globin mRNA sequences in polyadenylated and nonpolyadenylated nuclear precursor-messenger RNA from avian erythroblasts. Cell 8:505–512
Spring H, Trendelenburg MF, Scheer U, Franke WW, Herth W (1974) Structural and biochemical studies of the primary nucleus of two green algal species, *Acetabularia mediterranea* and *Acetabularia major*. Cytobiologie 10:1–65
Spring H, Scheer U, Franke WW, Trendelenburg MF (1975) Lambrush-type chromosomes in the primary nucleus of the green alga *Acetabularia mediterranea*. Chromosoma 50:25–43

Spring H, Krohne G, Franke WW, Scheer U, Trendelenburg MF (1976) Homogeneity and heterogeneity of sizes of transcriptional units and spacer regions in nucleolar genes of *Acetabularia*. J Microsc Biol Cell 25:107–116

Sripati CE, Groner Y, Warner JR (1976) Methylated, blocked 5′ termini of yeast mRNA. J Biol Chem 251:2898–2904

Stahl DA, Walker TA, Meyhack B, Pace NR (1979) Precursor-specific nucleotide sequences can govern RNA folding. Cell 18:1133–1143

Stark BC, Kole R, Bowman EJ, Altman S (1978) Ribonuclease P: an enzyme with an essential RNA component. Proc Natl Acad Sci USA 75:3717–3721

Szekely M (1980) From DNA to protein: the transfer of genetic information. MacMillan, London

Szalay A, Munsche D, Wollgiehn R, Parthier B (1972) Ribosomal ribonucleic acid in *Anacystis nidulans*. Biochem J 129:135–140

Thien W, Schopfer P (1975) Stimulations of precursor rRNA synthesis in the cotyledons of mustard seedlings by phytochrome. Planta 124:215–217

Tilghman SM, Curtis PJ, Tiemeier DC, Leder P, Weissmann C (1978) The intervening sequence of a mouse *β*-globin gene is transcribed within the 15S *β*-globin RNA precursor. Proc Natl Acad Sci USA 75:1309–1313

Trendelenburg MF, Spring H, Scheer U, Franke WW (1974) Morphology of nucleolar cistrons in a plant cell, *Acetabularia mediterranea*. Proc Natl Acad Sci USA 71:3626–3630

Udem SA, Warner JR (1972) Ribosomal RNA synthesis in *Saccharomyces cerevisiae*. J Mol Biol 65:227–242

Verma DPS, Nash DT, Schulman H (1974) Isolation and in vitro translation of soybean leghaemoglobin mRNA. Nature New Biol 251:74–77

Verma DPS, Maclachlan GA, Byrne H, Ewings D (1975) Regulation and translation in vitro of mRNA for cellulase from auxin treated pea epicotyls. J Biol Chem 250:1019–1026

Weinberg RA, Penman S (1970) Processing of 45S nucleolar RNA. J Mol Biol 47:169–178

Weinstock R, Sweet R, Weiss M, Cedar H, Axel R (1978) Intragenic DNA spacers interrupt the ovalbumin gene. Proc Natl Acad Sci USA 75:1299–1303

Wheeler AM, Hartley MR (1975) Major mRNA species from spinach chloroplasts do not contain poly(A). Nature (London) 257:66–67

Wimber DE, Steffensen DM (1970) Localisation of 5S RNA genes on *Drosophila* chromosomes by DNA-RNA hybridisation. Science 170:639–641

Wimber DE, Duffey PA, Steffensen DM, Prensky W (1974) Localisation of the 5S RNA genes in *Zea mays* by RNA-DNA hybridisation in situ. Chromosoma 47:353–359

Winters MA, Edmonds M (1973) A poly(A) polymerase from calf thymus. J Biol Chem 248:4756–4762

Woodcock CLF, Stanchfield JE, Gould RR (1975) Morphology and size of ribosomal cistrons in two plant species: *Acetabularia mediterranea* and *Chlamydomonas reinhardii*. Plant Sci Lett 4:17–23

Young RA, Steitz JA (1978) Complementary sequences 1700 nucleotides apart form a ribonuclease III cleavage site in *Escherichia coli* ribosomal precursor RNA. Proc Natl Acad Sci USA 75:3593–3597

Ziff EB, Evans RM (1978) Coincidence of the promotor and capped 5′ terminus of RNA from the Adenovirus 2 major late transcription unit. Cell 15:1463–1475

8 Ribonucleases and Ribonucleic Acid Breakdown

G.L. FARKAS

1 RNA-Splitting Enzymes

1.1 Definitions, Terminology, and Classification

The RNA molecules are stable under physiological conditions, and can be broken down in vivo only by enzymatic hydrolysis. All RNA molecules are subject to hydrolytic breakdown; this makes their turnover possible. The hydrolytic breakdown of RNA, in addition to the regulation of RNA synthesis at the transcriptional level, plays a major role in the selection of those molecular

Table 1. Classification of RNA-splitting enzymes

Mode of action	Exonucleolytic			
Substrate attacked	RNA		RNA as well as DNA	
Products	5′-P-mono-nucleotides; 5′-P-mono-ester oligo-nucleotides	3′-P-mono nucleotides	5′-P-mono-nucleotides	3′-P-mono-nucleotides
Enzyme nomenclature classification	3.1.13	3.1.14	3.1.15	3.1.16
Typical examples ("prototypes")	Ribonuclease II 3.1.13.1 (NOSSAL and SINGER 1968)	Yeast ribonuclease 3.1.14.1 (OHTAKA et al. 1963)	Venom exonuclease 3.1.15.1 (LASKOWSKI 1966)	Spleen exonuclease 3.1.16.1 (BERNARDI and BERNARDI 1971)
	Exoribonuclease H 3.1.13.2 (STAVRIANOPOULOS and CHARGAFF 1973)			

Abbreviations: ABA, abscisic acid; ds, double-stranded; ss DNA, single-stranded DNA

species of RNA which are needed by the cell at a given moment of its development. Therefore, a thorough knowledge of the RNA-splitting enzymes is pivotal for the understanding of a number of regulatory and developmental processes in which RNA molecules are involved.

A high number of enzymes degrade RNA. The RNA-splitting enzymes belong to the large group of hydrolases, enzymes that catalyze the hydrolytic cleavage of C-O, C-N and C-C bonds (Enzyme Commission code number: 3). The actual linkages split by the nucleolytic enzymes are ester bonds; the RNA splitting enzymes are, therefore, esterases (EC code: 3.1.).

The more specific aspects of the mode of action of the RNA-splitting enzymes vary a great deal, and it has always been difficult to propose a correct and logical nomenclature. The 1979 edition of *Enzyme Nomenclature* revised the classification of nucleolytic enzymes and established a number of main types which now serve as "models" to which other nucleic acid splitting enzymes can and should be compared. Part of the system, which does not cover the DNases, is shown in Table 1. As may be seen, we distinguish the nucleolytic enzymes on the basis of three properties. The exo- and the endonucleolytic

Endonucleolytic			
RNA		RNA as well as DNA	
5′-P-mono nucleotides; 5′-P-mono-ester oligo-nucleotides	3′-P-mono- and oligo-nucleotides; 2,3′-P-cyclic P intermediates	5′-P-mono- and oligo-nucleotides	3′-P-mono- and oligo-nucleotides
3.1.26	3.1.27	3.1.30	3.1.31
Ribonuclease (*Physarum polycephalum*) 3.1.26.1. (HIRAMARU et al. 1969) Ribonuclease II 3.1.26.3 (ROBERTSON et.al. 1968) Endoribo-nuclease H (calf thymus) 3.1.26.4 (STAVRIANO-POULOS and CHARGAFF 1973)	Ribo-nuclease T_2 3.1.27.1 (HEPPEL 1966) Ribonuclease T 3.1.27.3 (KASAI et al. 1969)	Endonuc-lease S_1 (*Aspergillus*) 3.1.30.1 (VOGT 1973)	Micrococcal endonuclease 3.1.31.1 (ANFINSEN et al. 1971)

enzymes attack the terminal or the internal diester bonds of the polynucleotide chain, respectively. Some enzymes of both groups attack only RNA while others attack RNA as well as DNA. The enzymes produce, as end products, either 5′-phosphomonoesters or products "other than" 5′-phosphomonoesters (e.g., 3′-P-mono- and oligonucleotides and/or 2′,3′-cyclic P intermediates). These alternatives yield eight categories into which most, if not all, RNA-splitting enzymes can be placed. It should be stressed that these categories are necessary for the sake of clarity only. The sole purpose of the system is that the individual enzymes should not be mixed up. The different categories do not reveal much about the function of the enzyme. Enzymes with distinctly different functions may fall into the same category. For example: category 3.1.26 ("endonucleases producing 5′-phosphomonoesters") includes such different enzymes as ribonuclease III (specific for double-stranded RNA), ribonuclease H (acting on RNA–DNA hybrids) and ribonuclease P (specific for tRNA precursors).

RNA-splitting enzymes have been investigated in the past from three main points of view: (a) enzymological properties, including mode of action, (b) possible use in studies of nucleic acid structure, and (c) physiological role. Only points (a) and (c) will be dealt with in detail. In spite of some early efforts to find suitable enzymes, to the knowledge of the author, only one plant nuclease is being used routinely for the elucidation of nucleic acid structure. This enzyme, the single-strand specific mung bean nuclease (KOWALSKI et al. 1976), has recently become commercially available.

One would expect a close correlation between the mode of action of a nucleolytic enzyme and its physiological function. Strangely enough, the large amount of work devoted to the isolation and characterization of plant nucleases is not very illuminating in this respect. The section devoted to the description of RNA-splitting enzymes occurring in plants may seem to be little more than a catalog of proteins with well-defined in vitro capacities, the in vivo utilization of which is almost entirely speculation. In spite of this, a critical survey of the RNA-splitting enzymes found in plants is necessary, since the only hope of "finding some function" for the individual nucleases is to establish a clear-cut correlation between the presence and/or activity of a well-defined nuclease and a well-defined metabolic process. At present, we are not very close to success in such an approach. A prerequisite of a breakthrough, among other things, is a better characterization of the enzymes described. A comparison of different enzymes is sometimes difficult because of the different methods and/or criteria used in enzyme purification and characterization. Therefore, we shall first deal with the problems of purification and identification of nucleolytic enzymes and then try to reach a conclusion as to the types of RNA-splitting enzymes actually identified in plants.

1.2 Problems of Purification and Identification

As of this writing, the number of papers "describing" nucleolytic enzymes from plants is approaching 100. Do plants indeed contain that many different RNA-splitting enzymes? In the opinion of the author this is certainly not the case, not even if we allow for hitherto unidentified RNA-splitting enzymes, e.g., RNases involved in RNA processing. The discovery of a new enzyme is often implied by the description of enzyme properties different from those already reported, without drawing attention to the following problems: (a) possibility of isolation artifact formation, (b) differences due to different degrees of enzyme purity, and (c) the selection of arbitrary properties to "characterize" the enzyme with complete disregard of previous work. This noncritical approach favors publication and leads to the accumulation of a large number of papers difficult to evaluate. The differences described often do not seem to be significant enough to make either the confirmation or the rejection of the data attractive. This is why, in spite of the 100-odd papers published, we are still in doubt about the actual number of RNA-splitting enzymes occurring in plants.

In the following, we shall deal briefly with the most commonly neglected aspects of enzyme purification and characterization.

1.2.1 Formation of Artifacts by Oxido-Reductive Processes

Most plant extracts are rich in phenolic substrates and phenol-oxidizing enzymes. Oxidized phenols readily bond covalently to protein molecules. This may not inactivate enzymes but can significantly alter their properties. Apparent molecular weights may be increased and specificities changed. Noncovalent protein–protein associations may also lead to changes in enzymological properties. For example, bovine pancreatic ribonuclease A is not capable of significantly degrading double-stranded RNA, but acquires this ability when artificially dimerized (LIBONATI 1971). Seminal ribonuclease, treated with glutathione, forms a monomeric species exhibiting twice the specific activity of the native dimer (SMITH and SCHAFFER 1979). Although no similar phenomena have, as yet, been reported for plant nucleolytic enzymes, the probability of their occurrence is high.

Aggregate formation is an even more serious problem for molecular weight determination. Nucleolytic enzyme activity is well preserved in large molecular weight protein aggregates. Therefore, unusually high values for the molecular weight of an enzyme, e.g., 155,000 for sugar beet RNase (GUERRIERO and DUNHAM 1979), should be treated with caution.

1.2.2 Formation of Artifacts Due to Proteolytic Effects

A number of enzymes, including some mammalian and microbial RNases, retain their activity after partial proteolysis. Therefore, unusually low molecular weights, e.g., 9,000 for wheat germ RNase reported by TORTI et al. (1973) and 5,000 for potato leaf lysosomal RNase reported by PITT (1975) may well be due to proteolytic artifact formation. In such cases, "control" extraction and purification of the proteins in the presence of protease inhibitors would be necessary. Also, it should be tested whether or not the molecular weight of the purified enzyme is affected by the length of preincubation (auto-digestion) of the crude extract, before purification.

1.2.3 Dependence of the pH Optima on a Variety of Factors

Small differences in pH optima should not be regarded as significant. Since the pH optimum of an enzyme depends on the buffer, substrate, ionic strength, and composition of the milieu, claims for "different" pH optima should be supported by comparative data on a well-known (standard) enzyme investigated under identical conditions.

1.2.4 The Use of Homopolymers for the Assay of Base Specificity

Since the (relative) base specificity is one of the most important characteristics of an RNA-splitting enzyme, a large number of data have accumulated on the base specificity of plant RNases. Since it is easy to compare the rate of hydrolysis of different homopolymers as substrates, this approach starts to replace the more tedious work of assaying the relative rates of release of the four mononucleotides and/or the determination of the bases at either end of the oligonucleotides liberated during the early stage of substrate hydrolysis. The most common mistake is the use of different homopolymers as substrates, without regard to their respective chain lengths. Under otherwide identical conditions, an assay in which short-chain substrate is used tends to give higher values than the same assay with longer chain substrate, both with endo- and exonucleolytic enzymes, if the acid precipitation method is used. Another, often overlooked problem is that the homopolymers having a secondary structure stabilized by adequate ionic strength are practically resistant to digestion by a number of RNases. Since in the same ionic milieu the different homopolymers have different secondary structure, a simple comparison of hydrolytic rates may be misleading.

1.2.5 Electrophoretic Variants

Since gels can be stained for nuclease activity after electrophoresis, it appears to be an easy task to separate different enzymes by this method. However, the electrophoretic mobility of a protein, due to a number of factors mentioned above (protein–protein aggregation, oxido-reductive changes, proteolytic degradation etc.), is subject to nonspecific alterations, and even a clear-cut gel pattern is no guarantee for the absence of artifact formation. Specific controls are necessary to prove that an electrophoretic variant detected in a crude extract is an individual enzyme (or "isoenzyme"). At least partial purification of the enzymes corresponding to the "isoenzyme" bands and, if possible, a genetic analysis of the isozyme pattern provide adequate proof (cf. WILSON 1971, 1978). With no attempt to verify the protein-chemically distinct nature of an enzyme activity appearing as a band, isoenzyme studies in relation to physiological changes (e.g., hormone treatment, disease, development etc.) can, at best, suggest further studies.

1.3 Types of RNA-Splitting Enzymes in Plants

The problems associated with the characterization of plant nucleolytic enzymes were, of course, known to the earlier reviewers. This is clear from the last review on plant nucleases (WILSON 1975) which gives an almost complete survey of the nucleolytic enzymes from plants, up to 1974. Wilson's approach was to try to establish the *main types* of nucleolytic enzymes present in plant tissues. His attempt was marvellously successful, in spite of the kaleidoscope-like literature. The picture emerged that at least three, reasonably well-characterized "enzyme families", rather than individual enzymes, occur widely in plants. As discussed earlier, it is difficult to ascertain whether the differences described for various plants and enzyme preparations reflect inherent differences in enzyme properties or are due to uncertainties and errors of enzyme characterization. Perhaps the application of more refined immunological techniques will provide a clue (HAWTHORNE et al. 1980).

The main properties of these three enzyme families are shown in Table 2. The enzymes may briefly be described as follows: (a) Soluble RNase with endonucleolytic mode of action and marked preference for the diester bonds adjacent to guanine; the final products of substrate cleavage are 3′-nucleotides but 2,3,-cyclic nucleotides are the first products. (b) Particle-bond endonuclease, splitting both RNA and DNA [preferentially ssDNA (single-stranded DNA)], with a marked preference for phosphodiester bonds adjacent to adenine; the final products are 5′-nucleotides. (c) Soluble exonuclease, which splits both RNA and DNA (preferentially ssDNA) in an exonucleolytic fashion, releasing 5′-nucleotides.

Examples of these three types of enzymes have been described from a large variety of plant material. The RNase has been isolated from pea (MARKHAM and STROMINGER 1956), tobacco (JERVIS 1974), ryegrass (SHUSTER et al. 1959), spinach (TUVE and ANFINSEN 1960), soybean (MEROLA and DAVIS 1962), corn (WILSON 1968a, b), oat (WYEN et al. 1969), mung bean (WALTERS and LORING 1966), sugarcane (TANG and MARETZKI 1970), barley (LANTERO and KLOSTERMAN 1973), wheat (TORTI et al. 1973), *Chlorella* (BROWN and MARSHALL 1977) and barley (PIETRZAK et al. 1980). The endonuclease has been reported from mung bean (JOHNSON and LASKOWSKI 1968), potato (SUNO et al. 1973), maize (WILSON 1968a, b), oats (WYEN et al. 1971), wheat (KROEKER et al. 1975), and tobacco (OLESON 1976). The exonuclease has been isolated from barley (HOLBROOK et al. 1966), carrot (HARVEY et al. 1970), oats (UDVARDY et al. 1970), sugar beet (LERCH and WOLF 1972), barley (PIETRZAK et al. 1980), and *Petunia* (PLISCHKE and HESS 1980).

It is a matter of definition, and taste, as to what differences should be regarded great enough for the creation of one or more subclasses. The best candidate for such a subgroup is the enzyme described by WILSON (1968a, b, 1971) as "corn RNase II". The major properties of this enzyme are very similar to those of the large group of enzymes listed above as soluble RNase (see also Table 2). This enzyme differs from the "major" plant RNase by (a) the "microsomal" localization, (b) the tendency to hydrolyze both the purine and pyrimidine cyclic phosphates, (c) a somewhat higher pH optimum, and

Table 2. Properties of the main RNA-splitting enzymes isolated from plant cells

Characteristics	RNase	Endonuclease	Exonuclease
Substrate specificity	RNA	RNA, DNA	RNA, DNA
Preference for strandedness	ss	ss	ss
Mode of action	endo-	endo-	exo-
Localization	Soluble	Particle-bound	Soluble
Intermediate products	2′,3′-cyclic- -phosphates	–	–
End-products	3′-nucleotides	5′-nucleotides	5′-nucleotides
Relative base specificity	G>A=U>C	A>U(T), G, C	None
pH optimum	5.0–5.5	5.0–6.5	7.0–9.0
Molecular weight	20–25,000	31–35,000	100,000
EDTA sensitivity (ion requirement)	Low	High	High
EC No.	3.1.27.1	3.1.30.1	3.1.15.1

(d) a lower molecular weight. The best evidence for the existence of a distinct maize "RNase II" is, however, not biochemical but genetic in nature. Maize "RNase II" has several electrophoretic variants ("isozymes") which are genetically controlled. The inbred parents have one (or more) "isoenzymes" each, whereas the hybrids always have a combination of the isozymes of the parents (Wilson 1971, 1978). A particle-bound RNase, similar to corn "RNase II", has also been reported from tobacco (Reddi and Mauser 1965, Jervis 1974).

There are additional nucleolytic enzymes, of which but a single description is available, and the existence of which has not yet been confirmed. The more important of these are the following:

A soluble endo-ribonuclease from cucumber with a preference to split bonds adjacent to cytidine (Kado 1968).

A pyrimidine specific RNase, from *Vicia faba* (Beopoulos et al. 1978).

A soluble endoribonuclease from rye germ with an alkaline pH optimum and an ability to hydrolyze pyrimidine, but not purine, 2′,3′-cyclic phosphates (Kuligowska et al. 1976).

Two cytosolic endoribonucleases and a nuclease from wheat leaves; all of them with a relative U specificity (Chevrier and Sarhan 1980).

A cytosolic endoribonuclease from rye germ with an acidic pH optimum and a relative adenine specificity (Kuligowska et al. 1980).

"Factors" supposed to be responsible for the specific "nicking" of leucyl- (Babcock and Morris 1973) and chloroplast tyrosyl- (Locy and Cherry 1976) tRNAs.

Ribonuclease H, which specifically degrades the RNA strand of DNA–RNA hybrids, from carrot cell cultures (Sawai et al. 1978).

An endoribonuclease strongly bound to fenugreek and soybean ribosomes, which, however, can be released by high ionic strength buffers; on the basis of the preferential splitting of UpA, the RNase may be an U-specific enzyme (Rijven 1978).

A nuclease from *Tradescantia* leaves with a relative guanosine specificity (CLAPHAM 1980).

One wonders why the work on some of these highly interesting enzymes has not been followed up.

1.4 Subcellular Localization

1.4.1 Soluble Enzymes

Although there are large individual differences, the major part (up to 80–90%) of nucleolytic activity in crude extracts can usually be recovered in the soluble fraction. This activity is often referred to as "cytosolic", although enzymes localized in easily disruptible compartments, like vacuoles (lysosomes) and chloroplasts, certainly represent major sources of the "cytosolic" enzyme fraction. It is reasonable to suppose that at least part of this enzyme activity is cytoplasmic. However, the author is not aware of any method which would prove unequivocally the localization of a nucleolytic enzyme in the cytoplasm of a plant cell.

Most nucleolytic enzymes can be solubilized even if they are attached to or contained by cell organelles. Thus the term "soluble enzyme" is misleading. We do not know which fraction of the "soluble" enzymes described in Section 1.4. is actually localized in cell compartments. Much refinement of the available methods is necessary to prove the cytoplasmic localization of a nucleolytic enzyme.

1.4.2 Particle-Bond Enyzmes

The total nucleolytic enzyme activity of plant tissue extracts has often been separated into particulate and soluble fractions. A significant portion of the activity is usually recovered in the particulate fraction. The major interest, from a physiological point of view, is the nature of cell particles involved and the specificity of enzyme localization. Unfortunately, the majority of published work is not conclusive. Most papers are not quantitative, i.e., they do not pay attention to more than one or two fractions or enzymes, minimizing thereby the value of their conclusion as to the specificity of localization. Furthermore, the fractions are most often not characterized biochemically by marker properties.

Nucleus. The nucleus certainly does contain RNA-splitting enzymes. This follows a priori from the fact that various RNA's are processed in this cell organelle. Still, our knowledge about the nucleolytic enzymes of plant cell nuclei is scanty.

LYNDON (1966) found only 4% of the total RNase activity in the nuclei of pea root cells. Part of even this low activity was removed by washing the nuclei in 1.8 M sucrose. In contrast, BORUCKA-MANKIEWICZ and SZARKOWSKI (1977) found significant nuclease activity in isolated rye germ nuclei. The purified enzyme showed a number of characteristics similar to those of the main plant endonuclease (see Table 2) excepting that it appears to split dsDNA preferentially. The "nuclear fraction" from which the enzyme was isolated had not been characterized and the occurrence of the same enzymatic activity in other cell fractions was not studied.

Another nuclease in cell nuclei, specifically associated with barley leaf chromatin, has been described from a single laboratory. The properties of the enzyme, which has been purified to a small extent, have been claimed to be characteristically different from those of other RNases (SRIVASTAVA et al. 1971). In fact, however, the basic properties of the enzyme (except for a slight difference in the pH optimum) are similar to those of the nucleases belonging to the main plant endonuclease group (cf. Table 2). The chromatin nature of the preparation has been satisfactorily characterized. The arguments for the specific association of the enzyme with chromatin, because they are dissociated only by high ionic strength, are less convincing (SRIVASTAVA et al. 1971). In view of the common occurrence of the "major plant endonuclease" in cereals (see Sect. 1.3.), the absence of this enzyme from barley leaves or its exclusive association with chromatin would be surprising. The problem of the nature and function of chromatin-associated nuclease(s) is, however, most interesting and the confirmation and extension of this work is eagerly looked for.

Chloroplasts. The chloroplasts certainly contain nucleolytic enzymes. They do process chloroplast mRNA and rRNA. In addition, the wide-spread nicking of chloroplast rRNA, a post-maturation process, is likely to be due to an RNase effect in vivo. The reports on chloroplast nucleases are, however, few in number and erratic in their contents. This is probably due to the permeability of the chloroplast envelope and the loss of enzymes during chloroplast isolation. The first detailed work on the occurrence of RNase activity in the chloroplasts took advantage of the application of nonaqueous isolation media (ODINTSOVA et al. 1963). HADZIYEV et al. (1969) succeeded in isolating high RNase chloroplasts from wheat leaves both in nonaqueous and in aqueous media. By choosing a suitable buffer, the leakage of RNase from the chloroplasts was greatly reduced. Indirect evidence for the localization of RNase(s) in chloroplasts is provided by the detection of "RNA degrading activity" associated with the ribosomes isolated from spinach leaf chloroplasts (HOWE and URSINO 1972). Clearly, the chloroplasts appear to be more amenable to a search for specific chloroplast nucleases than previously believed. The use of isolated protoplasts as a starting material for the isolation of chloroplasts is highly recommended. In view of the phylogenetic implications of the problem, it would be interesting to see if an RNA-splitting enzyme possessing the characteristics of the RNase isolated recently from *Anacystis nidulans*, a cyanobacterium (LEHMANN et al. 1979), could be found in the chloroplasts of the higher plants.

Ribosomes. Ribosomal preparations usually contain nucleolytic enzymes. Low (WYEN and FARKAS 1971), medium (HSIAO 1968) and high (SIWECKA and SZARKOWSKI 1971) RNase levels have been reported for various ribosomal preparations. Reference is not being made to several other papers because the purity of the preparations described is doubtful. Therefore, not even a tentative estimate can be made as to how much RNase might be attached to a "typical" plant ribosome. It is certain, however, that even thoroughly purified and well-characterized ribosomes, like those isolated from soybean seedlings by MATSUSHITA et al. (1966) or maize by HSIAO (1968), do contain significant nucleolytic activity.

With some preparations, the activity has been found to be partially latent in ionic conditions which maintain the structural integrity of the ribosomes. In these cases the RNase was activated by agents known to disrupt ribosome structure (WYEN and FARKAS 1971, DYER and PAYNE 1974). High pH or high ionic strength milieu help to liberate both the inactive and active ribosome-bound enzymes (HSIAO 1968, DYER and PAYNE 1974, RIJVEN 1978), although tenaciously bound RNase activity has also been reported. Clearly, different nucleolytic enzymes are bound with different firmness to even the same ribosomes, as shown by RIJVEN (1978). Thus, the tenaciously bound RNase of ACTON (1974) could be easily solubilized with compounds such as streptomycin, which have a high affinity for ribosome-binding sites. It has been shown that the enzyme released can often be readsorbed to the ribosomes, e.g., by changing the pH (HSIAO 1968). The soluble enzyme of one species, broad bean, was adsorbed to the low-RNase ribosomes of rye embryos (DYER and PAYNE 1974). In the only case, *Avena*, where it has been tested in some detail, the ribosome-bound enzyme has been found to be similar to, if not identical with, the major plant RNase of the soluble fraction (WYEN and FARKAS 1971).

All these observations suggest that in a majority of cases most of the ribosome-bound RNase activity may be an artifact. However, the possibility that some RNase with a specific function may be bound to ribosomes in vivo cannot be entirely discounted. The phytochrome-mediated increase in RNase activity in the ribosomal fraction of etiolated lupin hypocotyls has been claimed to be physiologically important (ACTON 1974).

However, the ribosome-bound enzymes would be expected to exhibit unique properties if they had special functions. HOWE and URSINO (1972) described an RNA-degrading "activity" attached to chloroplast ribosomes in wheat. The enzyme has been claimed to be an EDTA-insensitive, partly latent exonuclease which produces 3′-nucleotides. The data are, however, not convincing. RIJVEN (1978) characterized the specificity of two nonpurified enzymatic activities released from fenugreek and soybean ribosomes at low and at high ionic strength, respectively. The easily removable activity is probably identical with or similar to the ubiquitous, relative purine (guanine) specific endoribonuclease on the basis of its preferential splitting of GpA and ApA. The strongly bound enzyme seems to be different, since it has shown a preference for UpA.

In conclusion, in spite of these preliminary results, the existence in plants of ribosome-bound enzymes with unique properties has not been proven convincingly.

Microsomes. The microsomes are, by definition, crude ribosomal preparations containing membranes and membrane fragments (endoplasmic reticulum) to which part of the ribosomes may be attached. Strangely enough, this fraction, the properties of which are difficult to define, yielded the best-characterized particle-bound nucleolytic enzyme, "RNase II" of maize (WILSON 1968a, b), the main properties of which are described in Section 1.3. The existence of a distinct RNase, termed RNase II, is supported by the isolation and purification of this enzyme from another plant, tobacco, by an entirely different procedure (JERVIS 1974). WILSON and SHANNON (1963) took advantage of the microsomal localization of this enzyme during its isolation and paid special attention to ionic strength, chemical composition, and pH of the extraction medium to obtain a high degree of attachment of "RNase II" to the microsomes and an enrichment of the enzyme in this particular fraction. This work, as well as that of HIRAI and ASAHI (1973) on pea, clearly show the paramount importance of well-defined conditions under which a particular enzyme can be preferentially recovered from (or adsorbed to ?!) a certain cell particle. In view of these uncertainties, we shall not deal with the fairly extensive literature which describes microsomal or particle-bound nucleolytic enzyme activities in plants, without detailed analysis of the systems. Still, one general conclusion can be drawn: particle-bound enzymes, even after solubilization, tend to exhibit a higher pH optimum than the typical soluble ones, and many of them are inhibited by EDTA. The reason for this "rule" is not known. The bound enzymes may be mostly endonuclease or RNase II.

Mitochondria. The mitochondria have not been shown conclusively to contain nucleolytic enzymes. However, as with "RNase II", the study of a particular, not highly purified cell fraction, defined as "mitochondrial", enabled WILSON (1968a, 1971) to isolate and characterize the endonuclease of maize. The same type of endonuclease has been found, also in particle-bound form, in oat leaf extracts by WYEN et al. (1971).

1.4.3 Lysosomal Localization

The lysosomes are, by definition, subcellular organelles which are surrounded by a double membrane and contain a variety of hydrolytic enzymes in a sedimentable and often latent form. The term, originally coined for an organelle of the animal cell, has been used in a broader sense with plant cells to include the vacuoles, aleurone grains, and spherosomes. MATILE (1978) believes that it would be more appropriate to speak about the "lytic compartment" of the plant cells, instead of lysosomes, because of the ontogenetic relationship between these organelles, all of which ultimately derive probably from the endoplasmic reticulum. In an operational sense, however, the term "lysosome" can be kept.

In view of the technical problems involved in the isolation of intact (and reasonably pure) lysosomes, and vacuoles, it is not surprising that the early literature on the lysosomal localization of nucleolytic enzymes in plant cells is not very convincing. However, a recent review by MATILE (1978) already listed a larger number of reliable reports on the lysosomal (vacuolar) localization of RNase in a wide variety of plants and tissues. The use of new techniques, such as immunochemistry, hitherto not applied in this field, corroborated the earlier findings (BAUMGARTNER and MATILE 1976). The refinement of the classical fractionation techniques, especially the use of isolated protoplasts as a starting material, beautifully demonstrated the presence of RNase(s), along with other hydrolytic enzymes, in the mature vacuole of higher plant cells (BOLLER and KENDE 1979), and confirmed its lysosomal nature as postulated by MATILE. Most importantly, the concept could be extended to the vacuoles in storage organs such as the endosperm tissues of castor bean (NISHIMURA and BEEVERS 1978) and the cotyledons of mung bean (VAN DER WILDEN et al. 1980). Another important aspect of enzyme localization is that the RNase activity found in a number of particulate preparations is actually due to the presence of either endoplasmic reticulum or of its derivatives (HIRAI and ASAHI 1975), which appear to be secretory sites involved in the synthesis and intracellular transport of hydrolytic enzymes (MATILE 1978).

Are the lysosomal (vacuolar) nucleolytic enzymes identical with the major RNA-splitting enzymes of the plant cell or are they endowed with some other type of specificity? The fact that a significant portion, up to 80%, of the RNA-splitting activity may be localized in carefully isolated vacuoles (BOLLER and KENDE 1979) suggests that the major lysosomal-vacuolar RNase(s) are identical with the soluble major plant endoribonuclease(s). However, this idea has not yet been confirmed. The lysosomal localization of some RNA-splitting enzymes makes their identification difficult. Since the lysosomes (vacuoles) are rich in proteases (MATILE 1978), and the enzymes within the lysosome are apparently not compartmentalized, proteolytic breakdown or modification of the lysosomal nucleases is likely. Indeed, PITT (1975) reported the occurrence of a low molecular weight (5,000) RNase in potato leaf lysosomes, which might well be an enzymatically active degradation product.

1.5 RNA-Splitting Enzymes in Relation to Development

1.5.1 Seed Germination

The reserve tissues of seeds, especially the cotyledons of the legumes, contain considerable amounts of stored RNA which is usually broken down during germination. The breakdown products may be used up, after hydrolysis, for the synthesis of new nucleic acids in the rapidly growing organs (BARKER and HOLLINSHEAD 1964, BEEVERS and GUERNSEY 1966). Since the RNase activity rapidly increases during germination in the cotyledons of legume seeds (BEEVERS and SPLITTSTOESSER 1968, BARKER et al. 1974), as well as in the storage tissues of cereal grains (INGLE and HAGEMAN 1965, VOLD and SYPHERD 1968, SUTCLIFFE and BASET 1973), the idea has often been entertained that the RNA breakdown and the increase in RNase level in the storage organs are causally correlated. The idea is certainly not correct for the endosperm of the cereals but may not be correct even for the legume seeds. The endosperm of cereal seeds usually contains comparatively little RNA and at the same time so much RNase (INGLE and HAGEMAN 1965) that the overall RNase level can hardly be a controlling factor in RNA breakdown. The same is true for the nuclease activity in the cotyledons of *Pisum sativum* (BRYANT et al. 1976a). Most importantly, the kinetics of increase in RNase activity often differs from that of the decrease in RNA both in monocots (VOLD and SYPHERD 1968, SUTCLIFFE and BASET 1973) and dicots (BRYANT et al. 1976a). As we shall see in the forthcoming sections, early hopes of finding a correlation between RNase activity and RNA level have in most cases not been fulfilled. High RNase activity is, more often than not, associated with a high rate of RNA accumulation in the cell. Thus, in the germinating seed, RNase activity may increase simultaneously with an increase in RNA content in the embryo axis tissues (INGLE and HAGEMAN 1965, TAKAIWA and TANIFUJI 1978). Still, it would be premature to reject completely the idea of the participation of RNases in RNA breakdown in the cotyledons of germinating seeds. The reserve materials of most cotyledons become highly depleted during germination. The cotyledonary tissues start to senesce and eventually may perish. Perhaps in the more advanced stages of senescence, associated with the breakdown of cellular compartmentation, the RNases do have a role in the degradation of RNA in the cotyledons.

No final conclusions can be drawn as to whether the ubiquitous increase in RNase activity (amount?) during seed germination is due to de novo enzyme synthesis or to enzyme activation. Density labeling experiments suggested that the RNase is synthetized de novo in the aleurone cells of germinating barley seeds (BENNETT and CHRISPEELS 1972). On the other hand, BARKER et al. (1974), using three independent methods, application of inhibitors, [^{14}C]-labeling, and density labeling, found no evidence for the de novo synthesis of the RNase in the cotyledons of *Pisum arvense* seeds, in the early phase of increase in RNase activity.

Developmental processes are often associated with preferential changes in the amount of one nucleolytic enzyme only (see Sect. 1.5.4. on senescence and injury). This might be the case with seed germination as well. BRYANT et al. (1976b) separated the RNase of the cotyledons of germinating pea seeds into three chromatographic fractions. The increase in RNase activity during germination was due, at least in part, to the appearance of a "new" isozyme. The enzyme fractions, however, have not been characterized.

1.5.2 Seed Maturation

RNase activity usually increases during seed development but may decrease to some extent in its later stages (INGLE et al. 1965, DALBY and CAGAMPANG 1970, JOHARI et al. 1977). With the developing seed, as a rule, there is an even poorer correlation between increase in RNase activity and decrease in RNA level than with the storage tissues of germinating seeds. If there is any correlation, it is usually a positive one between increase in RNase activity and RNA content.

A marked exception is the endosperm of maize seed in which the RNA content dramatically decreases at maturity, together with a concomitant increase in RNase activity (INGLE et al. 1965, MEHTA et al. 1972).

It is difficult to see the biochemical "sense" of a positive correlation between increase in RNA level and increase in RNase activity. From the very fact that the rate of increase and the maximum level of RNase activity is, at least in the endosperm of maize seeds, genetically controlled (DALBY and CAGAMPANG 1970, WILSON 1971, 1973), one would be tempted to postulate an important role for the enzyme in seed development. The idea of a direct participation of RNase activity in the control of zein biosynthesis, an attractive hypothesis, had to be rejected since not all mutations affecting the zein level lead to alterations in RNase activity (DALBY and CAGAMPANG 1970, WILSON 1973). Still, it is becoming increasingly evident that changes in RNase level are in some way linked to developmental patterns in corn endosperms, as shown recently by WILSON (1980).

The maturing endosperm of corn contains mostly the major endoribonuclease of plants, an enzyme also widespread in other tissues (WILSON 1971). Preliminary results suggest that the preponderance of the same endoribonuclease in mature *Sorghum* seeds is due to a progressive shift between the relative amounts of various RNases during grain development (JOHARI et al. 1977). Such a shift in the relative proportion of different nucleolytic enzymes should, however, be analyzed in more detail before we can draw any conclusions as to the frequency of occurrence, let alone the biological significance, of this phenomenon.

1.5.3 Root Growth and Differentiation

RNase activity shows a clear-cut correlation with the growth and differentiation of root cells. RNase activity is usually low in the young, meristematic cells and gradually increases with cell age, i.e., with cell elongation and maturity (ROBINSON and CARTWRIGHT 1958, PILET and BRAUN 1970, TREBAL et al. 1979). The extreme tip of the root (root cap) may, however, exhibit exceptionally high RNase activity (PILET and BRAUN 1970). The change in enzyme level, as a function of cell growth and differentiation, may be associated with differences in the cellular localization of nucleolytic enzymes. In the young cells of pea roots a very high percentage of RNase activity seems to be associated with the endoplasmic reticulum, whereas during cell maturation the enzyme activity starts to increase both in the soluble and particle bound (ribosomal?) fractions (HIRAI et al. 1975). Electrophoretic analysis of isozymes (SAHULKA 1971) and the study of chromatographic enzyme fractions (TREBAL et al. 1979) has not revealed a major shift in the ratio of various RNA splitting enzymes during root cell growth and maturation. The corn root tip cells, however, appear to have a nuclease specific for this zone of the root (WILSON 1968a). Clearly, the data are too scanty and do not allow a final conclusion as to the significance of the observations. Thus, in spite of the well-documented correlation between nucleolytic enzyme activity and cell growth or differentiation in roots, the physiological significance of this correlation remains to be elucidated.

1.5.4 Senescence

From the vast literature, one might get the overall impression that leaf senescence is generally associated with an increase in the activity of RNA-splitting enzymes. However, it is questionable whether or not the work done on excised leaves or isolated leaf discs is relevant to normal senescence. Since cell injury per se induces a dramatic increase in RNase activity, the excised leaf might not be the best system to study senescence-associated phenomena, unless rigorous controls are applied. Although the early work by LEWINGTON et al. (1967)

has directed our attention to the differences in the behavior of attached and detached leaves, the seriousness of this problems has not always been appreciated.

Whereas the RNase activity almost invariably increases in detached leaves, great variations have been found in the drifts of RNase level in aging, attached leaves. These differences are partly due to the poor definition of the experimental system, first of all to that of the poor definition of leaf age. One should pay much more attention to the evergreen problem: "how old is old?" (FARKAS 1978). To the knowledge of the author, the plastochron index, introduced by ERICKSON and MICHELINI (1957), has in no case been used to define leaf age in relation to RNase level. In addition, RNase activity is expressed by different authors on such different bases as total leaf area, unit of leaf area, fresh weight, dry weight, protein content, DNA content etc. Consequently, one cannot draw final conclusions from the kaleidoscope-like material; rather, one has only "impressions". These impressions are the following: with progressing leaf growth, maturation and senescence, the RNase activity of the leaf tissues may (a) increase, then decrease (beans: PHILLIPS et al. 1969, NAITO et al. 1979), (b) steadily increase (tobacco: LÁZÁR and FARKAS 1970), sometimes to a plateau (tomato: MCHALE and DOVE 1968, morning glory: BAUMGARTNER et al. 1975), or (c) decrease (barley: LÁZÁR and FARKAS 1970, LONTAI et al. 1972). All this variation is probably not due exclusively to the varying experimental conditions and/or different ways of calculation. It seems likely that all kinds of patterns are possible. This rules out a causal relationship between RNase level and senescence in that sense that the overall increase in RNase level would be a major factor triggering the onset of senescence. On the contrary, the deteriorative processes associated with senescence seem to trigger the increase in RNase activity. It is difficult to see clearly this temporal sequence with the sluggish, foliar leaf system. The beautiful *Ipomoea tricolor* system, with its rapidly senescing corolla, demonstrates the sequence convincingly: the onset of senescence precedes the increase in RNase level (KENDE and BAUMGARTNER 1974, BAUMGARTNER et al. 1975).

In the detached leaf the RNase activity invariably increases. Results pertaining to seven species are listed in the review by DOVE (1973) and will not be dealt with in detail.

The very fact that the increase in RNase activity in the isolated leaf tissue, if plotted as a function of time, is often biphasic, with a rapid increase early after isolation followed by a slight decrease and a second increase in enzyme level (DOVE 1971, PITT and GALPIN 1971), suggests that different control systems and/or enzymes may be involved in the change early after leaf detachment and in the later periods, respectively. We have demonstrated in a series of papers that this is indeed the case. By working out semi-quantitative methods to determine the amounts of the major nucleolytic enzymes in *Avena* leaf tissues, we have found that in the attached leaf aging is associated with the steady and specific increase of the sugar nonspecific endonuclease. This enzyme is a minor component in the spectrum of nucleolytic enzyme activities in the young tissues but becomes a dominant one during senescence. The level of the major nucleolytic enzyme of the young leaves, that of the relatively guanine-specific endoribonuclease, steadily decreases during senescence. In the detached leaf, the level of the endoribonuclease rapidly increases upon detachment and the amount of the sugar nonspecific nuclease starts to increase only when the visible signs of senescence (yellowing) appear (WYEN et al. 1971). The accumulation of the sugar nonspecific nuclease is stimulated in the *Avena* leaf tissues if premature senescence is triggered in young leaves by the application of abscisic acid (UDVARDY and FARKAS 1972).

In addition to the relatively adenine-specific endonuclease (WYEN et al. 1971), another "specific" nuclease has been claimed to be associated with or involved in senescence, a chromatin-bound nuclease described by SRIVASTAVA et al. (1971) from barley leaves. As pointed out earlier, the enzyme is poorly characterized and convincing evidence that it is basically different from the aging-specific nuclease of the *Avena* leaf, which is also particle-bound in crude homogenates, is missing.

The change in RNase isoenzyme patterns in senescing morning glory corolla (BAUMGARTNER and MATILE 1977), as well as the change in the EDTA-sensitivity of total RNase activity during leaf senescence in wheat (SODEK and WRIGHT 1969), support the idea that the specific change in the relative amount of various enzymes during senescence is real and widespread. However, much has to be done before the physiological significance of this intriguing correlation can be understood. In any case, senescence is correlated with increases of one or more specific nucleases, rather than with that of the overall nucleolytic activity. The ubiquitous increase in RNase level found in isolated leaf and other tissues is, however, due to a great extent to injury, a problem which will be dealt with in Section 1.6.1.

1.6 RNA-Splitting Enzymes and the Environment

1.6.1 Effect of Cellular Injury

Perhaps the most uniform and predictable response of a plant cell to an "environmental" effect is the increase in RNA-splitting enzyme activity upon cellular injury. The term "stress" has often been applied to describe this phenomenon. In the opinion of the author, however, most "stress" effects, e.g., water stress, heat effects, rubbing of the leaf surface, osmotic shock, leaf detachment, punching out of leaf discs, parasitic attack, can be traced back to mechanical and/or chemical injury. Therefore, we shall deal first with the effect of mechanical wounding.

Wounding (rubbing or cutting) of leaf (BAGI and FARKAS 1967, PITT and GALPIN 1971), storage (DE LEO and SACHER 1970, SACHER et al. 1975), and fruit tissues (DE LEO and SACHER 1971) induces a rapid increase in RNA-splitting enzyme activity, in some cases within 1–2 h after treatment. The response is ubiquitous and does not seem to be confined to any particular taxonomic group of the plant kingdom. Equally widespread is the increase in RNase activity in excised leaves, a phenomenon described for at least seven plant species from more than ten laboratories (for a review see DOVE 1973). The short-term response of RNase activity to leaf detachment is apparently due to mechanical injury. This phase of increase in RNase level does not seem to be an early "biochemical marker" of the onset of senescence, since the pattern of nucleolytic enzymes, once established in the damaged tissues, remains unchanged for a considerable time and does not resemble the typical nucleolytic enzyme spectrum of the aging leaves (UDVARDY et al. 1969, WYEN et al. 1971, UDVARDY and

FARKAS 1972). Although damage to the plasma ultrastructure, due to the rapid uptake of water or to the excision-induced disconnection of the water-continuum in the entire plant, cannot be discounted as a primary event leading to the increase in RNase activity in detached leaves (DOVE 1971), other factors also appear to be involved. The extent of the increase in RNase activity in excised leaf discs is correlated with the number of cells actually damaged, and the stimulus is translocated from the directly affected cells to zones not directly affected (BAGI and FARKAS 1967). The most striking example for a translocated stimulus is the high and rapid increase in RNase activity in the leaves of *Avena* seedlings simply pulled out from the soil and placed immediately in water. This effect appears to be due to the rupture of root hairs and to a substance translocated to the leaves from the site of damage (ERDEI and FARKAS, unpublished).

In spite of repeated efforts, the factor(s) inducing the rapid increase in RNase activity in injured and adjacent cells have not been identified. Plant hormones may be involved, especially ethylene and/or abscisic acid, which are known to be synthesized in injured cells. Abscisic acid was found, indeed, to stimulate the early rise of RNase activity in different systems (DE LEO and SACHER 1970, 1971, WYEN et al. 1972b, UDVARDY and FARKAS 1972) and the stimulatory effect of ethylene on the induction of wound-RNase has also been reported (SACHER et al. 1979). The role of abscisic acid in the regulation of RNase level is indirectly confirmed by numerous data showing the effect of cytokinins, well-known antagonists of abscisic acid, on the RNase level in wounded and isolated tissues (BAGI and FARKAS 1967, SRIVASTAVA 1968, SODEK and WRIGHT 1969, WYEN et al. 1972b, HODGE and SACHER 1975). Although abscisic acid and auxin are not such clear-cut antagonists of each other's action as abscisic acid and kinetin, it is noteworthy that auxin also tends to decrease the injury-induced rise in RNA-splitting enzyme activity (HODGE and SACHER 1975, VARIAN and SACHER 1978).

Much effort has been expended to find out whether or not the damage-induced increase in RNase activity is due to de novo synthesis or enzyme activation. An unequivocal demonstration of de novo synthesis of an enzyme protein is not easy. Most often the data are only "suggestive" of de novo protein synthesis but do not prove it. This is the case with the wound-induced RNase activity. Taking into account the limitations of the density-labeling technique, the only full proof for de novo enzyme synthesis would be the isolation of the enzyme, after induction of enzyme "synthesis" in the presence of a suitable isotope, and fingerprinting of the pure enzyme preparation to show that the whole protein molecule is labeled. Such data are not available for the wound-induced RNase. Still, the fact that actinomycin D, if added at the instant of wounding, as well as various inhibitors of protein synthesis, inhibit the development of wound RNase activity (BAGI and FARKAS 1967, SACHER et al. 1975, VARIAN and SACHER 1978), together with the results of density labeling experiments (SACHER and DAVIES 1974), strongly suggest that the wound RNase is synthesized de novo in injured *Rhoeo* leaf sections and bean pod tissue. On the basis of immunochemical and incorporation studies, including preliminary experiments by the use of density labeling, PITT (1971, 1974, 1975) has come to the conclusion that only a small portion of (the lysosomal) RNase is synthesized de novo in damaged potato leaves and tuber. However, negative results obtained with isotopes are rarely conclusive, especially if the enzyme preparations studied are not sufficiently pure. The immunochemical techniques also have their limitations.

No sound hypothesis can be advanced as to the physiological significance of wound-RNase production. The enzyme(s), probably soon after their synthesis, become sequestrated in the lysosomes (PITT 1975). This mechanism provides an effective separation of the enzymes and their substrates until, in a later phase, due to autophagy and total cell collapse, the RNA-degrading enzymes may digest a major part of the cellular RNA (MATILE 1978). As recently stressed by the author (FARKAS 1978) "there is no evident reason why the plant needs such a large reservoir of hydrolytic enzymes. However, there must be some reason. We only know that the amount of lysosomal enzymes is greatly increased during ageing and probably during disease. Perhaps, in plant disease and senescence, lysosomes temporarily play a protective role by compensating for a regulatory system that went wrong because of a defect in disease and/or senescence, and overproduced hydrolytic enzymes". These ideas, albeit entirely speculative, apply to the injury-induced increase in RNase activity. Maybe we do not have to look for a role of wound RNase.

1.6.2 Nucleolytic Enzymes in the Diseased Plant

Ever since the early work of REDDI (1959) on virus-infected tobacco leaves, evidence is accumulating which shows that the RNase activity is invariably higher in a number of diseased plant tissues than in the comparable controls (for reviews see DOVE 1973, WILSON 1975). RNase was one of the enzymes, the parasitically induced alterations of which served as a basis for the recognition of a common pattern of enzymatic changes in detached leaves and tissues attacked by parasites (FARKAS et al. 1964). The underlying causes of the common pattern are clear from the foregoing discussion. Since a wide variety of diseased tissues are physically or chemically damaged and show signs of premature senescence (for a review see FARKAS 1978), there is little doubt that a major part of increase in nucleolytic activity in diseased plant tissues can be traced back to these two major factors. However, it is tempting to postulate a parasitically induced synthesis of specific RNA-splitting enzymes, and to claim that the changes in nucleolytic enzyme activity have a bearing on the altered RNA transcription patterns in the diseased plant tissue (CHAKRAVORTY et al. 1974a, b, CHAKRAVORTY and SCOTT 1979a, b).

Most of the work suggesting the formation of specific enzymes upon infection, possibly important for the outcome of host–parasite relationship, has been done on tissues attacked by biotrophic parasites like rusts and powdery mildew. These pathogens elicit a biphasic increase in RNase activity in the infected tissues. The second peak is induced in susceptible host–parasite combinations only (CHAKRAVORTY et al. 1974a, b, CHAKRAVORTY and SCOTT 1979a). The separation of host and parasite functions is not easy in host–parasite systems but the results suggest that nucleolytic enzymes in the infected tissues differ in some of their properties from those found in the corresponding control plants, e.g., pH optimum, temperature stability, cleavage of ribohomopolymers, effects of ions, EDTA, urea and last but not least polysomal mRNA and rRNA (CHAKRAVORTY et al. 1980). Since most enzyme preparations were fairly crude, some of the interesting details might need confirmation. The observation that the barley RNase (soluble at pH 5.0), which is the major RNase component of the tissues, does not degrade polysome bound mRNA, whereas it does degrade polysome-bound rRNA may prove to be of great importance, if confirmed (SIMPSON et al. 1979, 1980). At present, the result seems strange in view of the well-known RNase sensitivity of polysomal mRNA. One should recall that in the hands of DAVIES and LARKINS (1974) the rRNA in polysomes was measurably affected only at levels of RNase at least 1,000 times as high as those needed for polysomal mRNA hydrolysis.

Another observation worthy of reinvestigation is that of REDDI (1966) on a "new" RNase in cells transformed by *Agrobacterium tumefaciens.*

With virus diseases, the analysis of the properties of "new" enzymes, if any, is more easy because the viruses of higher plants do not contain nucleolytic enzymes. The RNase(s) "induced" in the virus-infected cells have been identified in one case only. In a tobacco/TMV local lesion combination, the early increase in RNase activity was found to be due to an increase in the amount of the relatively guanine-specific endoribonuclease, the activity of which also increases upon wounding the tissue (WYEN et al. 1972a). It may be worth mentioning that in some cases, but not in all, the development of RNase activity in virus-infected leaves has also been found to be bi-phasic (DIENER 1961). This observation, and an electrophoretic analysis of RNase isozymes in Chinese cabbage systemically infected with turnip yellow mosaic virus (RANDLES 1968), suggest that the second peak of the bi-phasic RNase activity curve may be due to senescence, but this aspect of the problem has not been studied in detail.

Other efforts to detect *specific* changes in nucleolytic enzymes in connection with plant disease have also failed. The treatment of maize roots, the nucleolytic enzymes of which are well characterized, with the specific toxin of *Helminthosporium maydis,* race T, induced an increase in all RNases together, but only when growth was drastically reduced, a phenomenon suggesting a secondary reaction (WILSON and APEL 1975). This work, in spite of the negative results, shows clearly the advantage of using well-characterized systems. Unfortunately, a number of phytopathological laboratories have paid no attention to the presence in plants of several different enzymes with different well established properties, and did not even try to fit their results into our general knowledge on plant nucleolytic enzymes. Thus, in the opinion of the author, the fulfillment of the hope to find specific nucleolytic reactions involved in pathogenesis is still remote.

1.6.3 Light Effects

Light in chlorophyll-containing tissues naturally promotes protein and enzyme synthesis. Thus, it is not surprising that the development of wound-RNase activity is stimulated by illumination (BAGI and FARKAS 1967, SODEK and WRIGHT 1969). The light effect in itself, and also the observation that dichloromethylurea (DCMU) inhibits the increase in wound-RNase level (FARKAS unpublished), supports the view that energy-dependent enzyme protein synthesis is involved.

In addition to the overall light-effect, more specific actions of light on RNase level have been described. We have found that the illumination of etiolated oat leaves results in a drastic decrease in the amount of the sugar nonspecific endonuclease, which is a major component of both etiolated and senescent oat leaves. The amount of the relatively guanine-specific endoribonuclease was found to increase simultaneously (WYEN et al. 1971). It is not known whether the shift in the ratio of the major nucleolytic enzymes during greening is a general phenomenon or a special attribute of the *Avena* leaf.

Another specific response to light is the increase in the level of an RNase attached to the ribosomes of lupin hypocotyls (ACTON and SCHOPFER 1974). The light-induced increase in enzyme level, which on the basis of density labeling experiments is due to de novo synthesis (ACTON and SCHOPFER 1974), seems to be controlled by phytochrome (ACTON 1974). Phytochrome-mediated control of the polysome level has also been reported (ACTON 1978); the effect is, however, not necessarily connected with corresponding changes in RNase activity (FOURCROY et al. 1979).

1.6.4 Water Stress

One of the most regular responses of a plant cell to water shortage is an increase in RNase activity (DIENER 1961, DOVE 1967, 1971, ARAD et al. 1973, YI and TODD 1979). In order to obtain a large response, the water loss must be slow (DIENER 1961, DOVE 1973). The increase in RNase activity parallels the increase in water deficit and, at least in part, seems to be due to an increase in the concentration of abscisic acid (ARAD and RICHMOND 1976). Actual water loss and shrinkage of the cell (protoplast) volume is necessary to induce an increase in RNase activity in plant cells. The decrease in water potential in the cell per se is not sufficient. This was shown by the comparison of the effect of rapidly permeating, slowly permeating, and nonpermeating plasmolytica, respectively, on the RNase level in isolated tobacco leaf tissues. Only the nonpermeating plasmolytica induced an increase in RNase level (PREMECZ et al. 1977).

Isolated protoplasts, in spite of the watery medium in which they are suspended, are also exposed to water stress resulting in an increased RNase activity (LÁZÁR et al. 1973, FUCHS and GALSTON 1976). The protoplast suspension was found to be a suitable system to show that in addition to the mechanical and/or chemical injury the protoplasts suffer during isolation, the osmotic stress (shrinking of the protoplasts) induced by high osmotic media is a major factor inducing the increase in RNase activity. Experiments carried out with various inhibitors suggest that the increase in RNase level is probably due to enzyme protein synthesis (PREMECZ et al. 1977). No evidence was found for the formation of new types of RNA-splitting enzymes in the osmotically stressed cells as compared to the enzyme spectrum in the control. The RNase accumulating during osmotic stress in tobacco tissues has been identified as the ubiquitous, relatively guanine-specific plant endoribonuclease (LÁZÁR et al. 1973). On the other hand, the appearance of a new electrophoretic RNase component has been observed in wheat leaves exposed to dry air. The RNases have been purified 200-fold from both control and stressed leaves. They were found to differ in their molecular weights and specific activities. The results were interpreted as a reversible dimer$\rightleftharpoons$monomer transition of the enzyme upon water loss and rehydration, respectively, the monomeric form being more active (BLEKHMAN 1977a, b).

Since plants are often exposed to water stress, which reduces protein synthesis, the nature of metabolic changes involved in the impairment of protein synthesis is of a major theoretical and practical importance. It seems logical that the water-stress-induced increase in RNase activity causes a breakdown of polysome-bound mRNA, leading thereby to a decrease in the rate of protein synthesis. However, there is no good correlation between the increase in RNase activity and the decrease in polysome level during desiccation. The mRNA is conserved in desiccated moss tissues and remains available to support protein synthesis upon rehydration (DHINDSA and BEWLEY 1978). Similarly, in osmotically stressed isolated protoplasts, in which the protein synthesis is depressed (PREMECZ et al. 1978), the decrease in polysome population does not seem to be due to the increased RNase activity (RUZICSKA et al. 1979). These observations show that cellular compartmentation helps the plant to avoid the most serious consequences of a stress-induced increase in RNase level. Clearly, the plant cell has very efficient ways of keeping its nucleases under control. Normally, the nascent RNA chains in the cell "meet" only those nucleolytic enzymes which they are supposed to. The fate of exogeneously supplied RNA molecules is, however, completely different. Carrot protoplasts break down the alien RNA taken up by the mediation of liposomes (MATTHEWS et al. 1979).

Still, the idea that the increase in overall RNase activity can be the cause rather than the consequence of specific physiological changes is revived from time to time. Galston's group has found a good correlation between the viability of isolated oat leaf protoplasts and the RNase content, which increases upon protoplast isolation but can be kept at a low level by the addition of dibasic amino acids and polyamines (GALSTON et al. 1978). Although it cannot be excluded that in addition to the RNase(s) other unknown components of the system are also affected, the observation is worthy of serious attention because the protoplasts, especially right after isolation, are "leaky" cells in which the normal compartmentation is certainly disturbed.

1.7 Control of RNase Activity

1.7.1 Genetic Control

Very little work has been done on the genetics of plant nucleolytic enzymes. Early interest in the observation that the "opaque-2" mutation, which reduces zein synthesis in maize endosperm, is associated with an increased RNase level (DALBY and DAVIES 1967, WILSON and ALEXANDER 1967) was lost to a great extent when no causal correlation could be found between the increased RNase level and reduced zein synthesis (DALBY and CAGAMPANG 1970). Still, some properties of the enzyme affected by the mutation have been studied. It proved to be the "usual" plant endoribonuclease (WILSON 1968a, b); the gel-electrophoretic patterns of nucleases from wild-type and opaque-2 corn endosperms were identical (WILSON 1971). There is a great deal of variation in the RNase content of normal inbreds. The hybrid endosperms contain an intermediate level of RNase and a segregation of this character occurs in the F_2 generation (WILSON 1973).

Isozymes of nucleolytic enzymes in plants have been studied in a number of cases (LONTAI et al. 1972, OLESON et al. 1974, VAN LOON 1975, KOWALSKI et al. 1976). The data obtained have been used even for the identification of cultivars (ALMGARD and LANDERGREN 1974). However, if the genetic background of the material used is not known, these types of results cannot be interpreted in terms of genetics. Even the term "isozyme" is misleading. One should speak about electrophoretic variants, some of which might well be artifacts. True genetic variation of RNase isoenzymes, however, does occur in plants, as shown for conifer seed endosperm (MEJNARTOWICZ and BERGMANN 1977) and, more elegantly, for corn (WILSON 1978). As shown by gel electrophoresis, in inbred lines of corn, the root microsomal RNase ("RNase II" according to the nomenclature of WILSON) is present in the form of 1 to 3 isozymes. The hybrids contain all of the isoenzymes of the parents (WILSON 1978).

1.7.2 Hormonal Control

There are very few convincing data on the in vitro effect of plant hormones on the RNA-splitting enzymes. Usually little (LONTAI et al. 1972) or no effect was found. An interesting observation, which warrants further study, is that cytokinin ribosides inhibit the breakdown of ribohomopolymers by fenugreek RNases (RIJVEN 1978). However, one should not be over-optimistic because the effect has not been obtained with other RNases and the only result reported for a natural substrate was negative.

The scarcity of "positive" results obtained in vitro is in sharp contrast with the wide array of observations made on living material. Interestingly, however, little work has been done on intact leaves or other intact organs. Benzyladenine, a cytokinin, has been shown to increase the RNase level in intact leaves (FLETCHER 1969, NAITO et al. 1979). Still, due to the paucity of data, no final conclusions can be drawn as to the effect of growth regulators on the level of RNA-splitting enzymes in intact plants.

It should be born in mind that in *intact* plants complex interactions might obscure the "primary" effect of hormones on the RNase level. Thus, in intact leaves, ABA and kinetin affect the opening and closure of stomata. As shown by ARAD and RICHMOND (1976), treatment with ABA leads to a closure of stomata, thereby "saving" water for the plant under conditions which otherwise would lead to more serious water shortage. Kinetin has the opposite effect. The consequences of stoma closure are easy to predict: the water stress will be less and the RNase level will be low. On the contrary, kinetin treatment, leading to an enhanced water loss, will increase the RNase level. This has, indeed, been found by ARAD and RICHMOND (1976). However, if the stomata-regulating function of the hormones is excluded, e.g., in experiments at the cellular (protoplast) level, the effect of hormones becomes "normal", i.e., ABA will increase and kinetin will decrease the RNase activity. These strikingly opposite responses of the plant cell to the same hormone treatment, i.e., that the effect depends on the level of cellular organization, beautifully illustrate the difficulties involved in plant hormone research, and also indicate, with a new example, how plants manage to have specific hormonal regulation with comparatively few hormones. For experimental data on, and a more detailed discussion of, this problem see the work of PREMECZ et al. (1977). The same conclusion can be drawn from another work, which shows that the specific effect of ABA on the sugar nonspecific endonuclease in leaves (UDVARDY and FARKAS 1972) cannot be obtained with root tissue, even though the same enzyme is present in roots (SIVÓK et al. 1977).

Externally supplied cytokinins invariably decrease the injury-induced rise in RNase level in leaf tissues (BAGI and FARKAS 1967), including that occurring in detached leaves (SRIVASTAVA 1968, SODEK and WRIGHT 1969, DOVE 1971, HODGE and SACHER 1975). Abscisic acid increases the level of RNA-splitting enzymes in such diverse systems as isolated leaves (SRIVASTAVA 1968, WYEN et al. 1972), cut bean endocarp tissue (DE LEO and SACHER 1971) and isolated tobacco protoplasts (PREMECZ et al. 1977). The well-known antagonistic effects of ABA and cytokinins and/or auxin are characteristic for the wound-induced RNase as well (SRIVASTAVA 1968, DE LEO and SACHER 1971, HODGE and SACHER 1975). In this type of work, some variation of the results can be expected. The response is dependent on the endogenous hormone levels, light regime, etc. Most importantly, however, the hormones act on the different nucleolytic enzymes in different ways. The methods worked out for the semi-quantitative determination of the major nucleolytic enzymes in the *Avena* leaf have made it possible to detect which enzyme component of the overall enzyme activity responds, and to what extent, to treatments with different hormones. This work has also demonstrated that the effect of hormones is temporal, i.e., the short-term effect of the same hormone on the same tissue may be different from its long-term effect. ABA, in short-term treatment, has been shown to increase the level of the relatively guanine-specific endoribonuclease. In long-term treatment, however, ABA increased the amount of the relatively adenine-specific endonuclease in excised *Avena* leaves. The short-term effect of kinetin was almost entirely accounted for by a reduction in the level of the relatively purine-specific endoribonuclease (WYEN et al. 1972b, UDVARDY and FARKAS 1972).

It is largely unknown how the hormones regulate the level of RNases. Treatment with inhibitors like actinomycin D or puromycin leads not only to an inhibition of the hormone-induced increase in RNase activity but also to an actual decrease in the total wound-RNase level (BAGI and FARKAS 1967, SACHER and DE LEO 1977). This suggests that the enzyme has a short half-life. As far as can be judged from experiments applying actinomycin D at different times after wounding, the wound-RNase may be transcribed from a long-lived mRNA (SACHER and DE LEO 1977). Actinomycin D applied at the time of cutting inhibits the development of wound-RNase. This observation suggests that the regulation is at the level of transcription. Auxin, if added early, also inhibits the development of wound-RNase, i.e., it might also act at transcription. However, the auxin treatment accelerates the natural decay of the enzyme as well. Clearly, much has to be done to clarify, at the molecular level, the mode of action of growth regulators. Interestingly, the new results on the rapid effects of plant growth regulators on the proton (ion) pump of the cell membranes has not yet influenced our thinking about the hormonal regulation of nucleolytic enzymes.

2 Ribonucleic Acid Degradation

2.1 RNA "Level", "Breakdown", and "Turnover", Use and Mis-Use of the Terms and Methods

In contrast to the vast literature on RNA synthesis in plants, the comparable literature on RNA degradation, let alone our knowledge of it, is surprisingly limited. In addition, most of the data available deserve criticism rather than publicity. Even the terminology used is dishearteningly vague. What has actually been measured in most cases is RNA level, although authors write about RNA synthesis and breakdown. Of course, a decrease in the amount of the RNA can only be observed if RNA breakdown has taken place. However, a factor which induces an overall decrease in RNA level may either increase the rate of RNA breakdown or decrease the rate of RNA synthesis.

Isotope techniques have been used extensively to study RNA synthesis, and, to some extent, RNA breakdown in vivo. The amount of incorporation of a precursor into RNA has often been equated with the rate of RNA synthesis, although especially if the labeling period is long, the amount of label incorporated into RNA depends on at least three factors: (a) rate of RNA synthesis, (b) rate of RNA breakdown and (c) size of the particular precursor pools. As aptly pointed out by TREWAVAS (1976), the knowledge of the turnover rate of macromolecules is of crucial importance in the correct assessment of their "rate of synthesis".

Any effect which stimulates the synthesis of two different macromolecules to the same extent gives an apparent higher value for the synthetic rate of the one with the longer half-life. Alternatively, a factor which affects the rate of breakdown of nucleic acid molecules also affects their apparent rates of synthesis. These very important conceptual and methodological aspects of the problem have been taken into account in very few cases only. Therefore, the relevant literature is restricted and, as compared to the section on RNases, the section on RNA breakdown will be, seemingly, disproportionately short.

2.2 RNA Breakdown During Specific Physiological Processes

2.2.1 Seed Germination

It has already been outlined in Sect. 1.5.1. that in the storage tissues of seeds the RNA level, with or without a transitory increase, decreases during germination. This applies to the cotyledons of pea (BEEVERS and GUERNSEY 1966), bean (WALBOT 1971), and peanut (CHERRY 1963), as well as to the endosperm of cereal seeds (INGLE et al. 1964). However, in the castor bean endosperm, in which gluconeogenetic enzymes are synthesized during germination, the decrease in RNA content is transitory and insignificant. This decrease is followed by a massive increase in the amount of both low molecular weight and rRNA, until a plateau is reached on the 3rd day of germination (ROBERTS and LORD 1979).

Although it has seldom been studied in detail, isotope data suggest that, whether the trend is a decrease or an increase in RNA level in the tissue, RNA breakdown and synthesis do occur simultaneously. The factors involved in the regulation of RNA level are being studied intensively. The current interest centers around the embryo, in which massive protein synthesis and accumulation are initiated during germination. Both preformed and newly synthesized mRNA-[poly(A)-rich RNA] appear to play a role in the regulation of this process (PAYNE 1977, CAERS et al. 1979).

Part of the stored poly(A), presumably attached to mRNA, may break down rapidly in germinating seeds. The breakdown appears to be functionally important since it parallels a decrease in in vivo protein synthesis, probably coded for by the stored mRNA, in the embryo axis (DELSENY et al. 1977). The degradation of stored mRNA is especially clear-cut in wheat embryos germinating in the presence of cordycepin, an inhibitor of mRNA synthesis (CAERS et al. 1979).

An explanation, at the enzymological level, of this instability of stored mRNA is eagerly awaited, especially because in some cases the newly synthesized poly(A)-rich RNA appears to be stable for several hours (PAYNE 1977). Even this relative stability of the newly synthesised mRNA may not be general. In other cases, a fairly rapid turnover of the new mRNA (half-life 2 h) has been reported (CAERS et al. 1979).

Considerably less data are available for the endosperm. Instability of the poly(A)-rich RNA, especially that of the cytoplasmic fraction, has been demonstrated for the castor bean endosperm (ROBERTS and LORD 1979).

The role of mRNA breakdown and/or turnover in the regulation of transcriptional pattern in relation to development is evident. Much less clear is the function, if any, of "stable" RNA breakdown during germination. There has been much speculation as to whether the RNA breakdown products are reutilized for RNA synthesis in other organs. This probably does occur in most cases. Its significance is, however, not evident because of the extensive de novo nucleotide and polynucleotide synthesis in the embryonic axis and in the first leaves during germination. In the embryonic axis the "scavenging" pathways for nucleotide synthesis do not appear to be more important than the de novo routes (ROBINSON and BRYANT 1975; see also Chap. 9, this Vol.).

Even if little is known about the details of the regulation of breakdown, turnover, and processing of RNA in germinating seeds, it is clear that these events are under stringent control in the viable seed. The loss of seed viability in rye and pea is associated with the cleavage of rRNA (BRAY and CHOW 1976) as well as with the impairment of normal processing of precursor rRNAs (SEN and OSBORNE 1977). The appreciation that the synthetic processes cannot be studied without taking the breakdown processes into consideration (TREWAVAS 1976) will certainly catalyze the investigation of RNA breakdown in general and in seed germination in particular.

2.2.2 Senescence

Senescence in both intact and detached leaf tissues is associated with a decrease in their RNA content. The reports are especially numerous for detached leaves, often used as a model of "accelerated senescence", and need not be described in detail. It is sufficient to refer to a few "classical" papers (WOLLGIEHN 1961, OSBORNE 1962). It is to be regretted that the literature on intact leaves is considerably smaller (TAKEGAMI 1975, NAITO et al. 1979), because the regulatory aspects of the two systems, as already discussed in connection with the RNase level, are certainly different.

The major problems of senescence-associated RNA metabolism, studied mainly in excised tissues, are the following: (a) Are the different RNA species affected equally in the senescent tissue? (b) Is RNA synthesis or breakdown primarily affected? (c) What is the biochemical mechanism of the alterations? (d) What is the significance of the changes for the plant? These aspects of the problem will be discussed below.

a) The decrease in the amount of different RNA species in isolated leaves is not uniform, the level of tRNA being more stable than that of the rRNA

(CHROBOCZEK and CHERRY 1966, TESTER 1977). The breakdown of chloroplast rRNA is especially rapid and extensive (TAKEGAMI 1975). The level of DNA is quite stable, probably due to its low turnover rate. The special behavior of chloroplast rRNA during senescence reflects the generally high sensitivity of the chloroplasts to practically any deleterious effect. Not surprisingly, the chloroplast rRNA breaks down in isolated chloroplasts; the breakdown does not require the co-operation of the cytoplasm (PANIGRAHI and BISWAL 1979).

Little is known about the less evident, and possibly more specific, differences in the behavior of nucleic acids during senescence, although cotyledonary senescence in soybean has been reported to be associated with shift in the relative amount of some specific RNA's (BICK et al. 1970). Similar results have been reported for the aging tomato fruit (METTLER and ROMANI 1976). Changes in the tRNA spectrum of senescing tissues may be widespread. These changes make sense, because the translational regulation of protein synthesis is supposed to be an important factor in the control of developmental processes. The idea is only a nice hypothesis which has not yet been proven.

It would be even more interesting to detect specific changes in the mRNA spectrum of senescing leaves. This is, however, technically difficult. The amount of mRNA appears to be co-ordinately linked to total RNA production. This has been shown for ripening tomato fruits in which the poly(A) content gradually declines with aging (RATTANAPANONE et al. 1977). The general decline of poly(A) content, a rough estimate of the amount of mRNA, has also been shown for the aging tobacco leaf. The amount of mRNA decreased parallel with rRNA and tRNA (FRASER and GERWITZ 1980). Specific changes in the mRNA population have not been described.

Reports that actinomycin D promotes while cordycepin inhibits (TAKEGAMI and YOSHIDA 1975) the senescence of tobacco leaf discs are interesting, but difficult to interpret. The actinomycin D effect may be secondary, due to nonspecific damage induced by the inhibitor, which results in accelerated senescence. The cordycepin effect is intriguing, since it implies the direct involvement of mRNA in senescence – a logical idea if we regard senescence as a natural, programmed phase of plant development.

b) Very few data are relevant to the problem as to whether RNA synthesis or RNA breakdown is more affected in the senescing leaf. As already outlined, the technical problems are great, especially with excised tissues, where the effect of excision, per se, on isotope uptake and precursor pool sizes adds to the difficulties. No elegant and detailed work, comparable to that of TREWAVAS (1970) on intact *Lemna* fronds, has been done on excised leaves. The results on *Lemna,* a unique experimental system, show that the cytoplasmic rRNA's have a half-life of 4 days, whereas the half life of the chloroplast rRNA's is as long as 15 days. The rates of both synthesis and breakdown are greater for the cytoplasmic rRNA's. Starvation ("step down" to a nutritionally deficient medium or water) results in a dramatic decrease in the rate of rRNA synthesis and in an even greater increase in the rate of rRNA breakdown. Although the nutritional stepdown cannot be fully equated with normal or excision-induced leaf senescence, the results obtained with *Lemna* certainly have a bearing on the problem of nucleic acid metabolism in leaf senescence. In fact, the data obtained with *Lemna* are in line with part of the incorporation data obtained with leaves, suggesting a decrease in the rate of RNA synthesis in detached, senescing leaf tissues (OSBORNE 1962). It remains to be elucidated whether the rate of rRNA breakdown increases in detached leaves. The author is not aware of any work which would be conclusive in this respect. The fast decrease of chloroplast RNA in detached leaves may depend on an accelerated decay of chloroplast ribosomes, since the rRNA turnover appears to be regulated via the turnover of the whole ribosome (TREWAVAS 1970). This conclusion is compatible with the already mentioned fragility of the chloroplast structure, upon which the integrity of chloroplast ribosomes might depend.

c) The biochemical mechanism of the changes leading to RNA breakdown in the senescent leaf is not understood. We can safely say that, in spite of numerous claims, an increase in the RNase level cannot by itself be responsible for an increase in the rate of RNA breakdown.

One attractive possibility is that senescence-associated changes in hormone levels govern the RNA turnover in the senescing leaf. RNA breakdown can, indeed, effectively be inhibited by cytokinins (WOLLGIEHN 1961, OSBORNE 1962, TAKEGAMI 1975). This does not prove, however, that the endogenous regulation is also hormonal in nature. Cytokinins tend to normalize the effect of a wide variety of adverse conditions, and they might be regarded as veritable anti-stress factors, the action of which certainly cannot be limited to senescence phenomena and/or rejuvenation. The well-known "antagonistic" hormone pairs may exert their antagonistic effects in different ways. In the best-studied case, *Lemna*, benzyladenine increased both the synthetic and degradative rates of RNA metabolism. Abscisic acid, however, reduced the rate of RNA synthesis but left the rate of RNA degradation unaltered (TREWAVAS 1970). The effect of hormones on the rate of RNA synthesis and degradation may well be indirect. Although it is fashionable to explain hormonally regulated phenomena "at the transcriptional and translational levels", the extremely widespread anti-stress effect of kinetin suggests that the primary target of this hormone may not necessarily be the protein-synthesizing apparatus. Maintenance of the structural integrity of cell organelles (membranes etc.) would affect indirectly the synthetic rates in the cell compartment involved.

d) The significance and/or benefit of leaf senescence for the entire plant cannot be assessed properly. The competition among the developing organs is, however, a good case to state. The growing points, the developing young leaves and the fruits, are vitally important for survival and act as sinks for the nitrogenous constituents exported from the senescent leaves. Probably the RNA breakdown products are also reutilized in other tissues. This is possible because during RNA turnover they are not recycled into the same RNA precursor pool (TREWAVAS 1970).

2.2.3 Pathological Processes

Since genetic factors, localized both in the host and the parasite, play a part in the regulation of host–parasite relationships, it is not surprising that great attention has been paid to the nucleic acid metabolism of various host–parasite complexes. The parameters usually measured were total RNA level or RNA "synthesis", as judged from isotope incorporation data. The turnover of RNA in diseased tissues has not been studied. This, of course, would be very difficult to do since, at least with fungal and bacterial parasites, the extra precursor pools and extra compartmentation of synthetic and degradative sites of the pathogen should also be considered. In spite of these difficulties, a few general statements can be made.

a) In tissues attacked by biotrophic parasites, like rusts and powdery mildews, there is intensive RNA synthesis and accumulation in and close to the infection centers. Part of this may be due to the contribution of the cells of the parasite but a wounding effect inducing increased RNA synthesis in the host cells is also important.

In a certain distance from the infection centers, the leaf tissues often turn yellow as an indication of premature senescence. In these areas of the leaf, the chloroplasts show early signs of deterioration. The chloroplast rRNAs start to incorporate less label and their amount decreases preferentially (TANI et al. 1973). Clearly, the temporal sequence of rRNA breakdown in these cells is similar to that observed in healthy, aging tissues.

b) Infection with necrotrophic parasites, especially with pathogens producing toxins, leads to premature yellowing of the leaves, reminiscent of "normal" aging or senescence. RNA breakdown has been observed in such systems (LOVREKOVICH et al. 1964). These effects are not necessarily host and/or pathogen-specific. They are due to the metabolic load and mechanical or chemical injury associated with almost any plant disease.

c) Plant hormones, especially the cytokinins and auxins, may be indirectly involved in disease-induced senescence and in senescence-associated RNA breakdown. The infection sites of biotrophic parasites represent metabolic sinks that, like young organs, rob the surrounding tissues of low molecular weight metabolites, thereby accelerating their aging and senescence. The local accumulation of growth substances, especially of cytokinins, appears to be responsible for the directed transport of metabolites to the infection sites and/or retention of compounds synthesized in situ. These processes result in the formation on the infected leaves of "green islands" of high synthetic activity, which are often surrounded by senescent tissues.

d) RNA breakdown induced by pathogens or their toxins can be antagonized by the addition of kinetin (LOVREKOVICH et al. 1964). This does not mean, however, that a decrease of cytokinin concentration is the cause of RNA breakdown. RNA breakdown in such cases is a secondary phenomenon and the protection provided by kinetin is the manifestation of its general anti-stress effect rather than that of its specific action. For a more detailed discussion of points (a)–(d) see FARKAS (1978).

2.3 Regulation of RNA Breakdown

2.3.1 Hormonal Regulation

Enormous effort has been expended to prove, or disprove, that various facets of RNA metabolism are under hormonal control in plants. The gross aspects are clear. Hormones do affect the accumulation and/or breakdown of RNA. Cytokinins, gibberellins, and auxins usually favor the synthetic processes (RNA accumulation), whereas abscisic acid acts antagonistically. The details are, however, hotly debated. Most of the propositions, especially those which concern the primary action of hormones, have been challenged by later work.

A critical discussion of RNA breakdown as affected by hormones, is difficult because most laboratories took a "positive" approach in studying the hormonal regulation and attributed much greater importance to the controls of RNA synthesis than to those of RNA breakdown. Thus, their findings have, at best, an indirect bearing on the RNA breakdown problem.

A major concern of most workers has been whether the hormones exert their effect on RNA (and protein) metabolism at the transcriptional or translational level. A detailed survey of the literature, and also a recent review (JACOBSEN and HIGGINS 1978a, b), shows that there is fairly good evidence, with practically all plant hormones, for both transcriptional and translational regulation but convincing proof is generally lacking.

The regulation of transcription by hormones, whether quantitative or qualitative, may have at least an indirect effect on RNA breakdown. As to the quantitative regulation, it is clear that a change in the amount of primary transcripts can influence the apparent rate of breakdown of both mRNA and stable RNA. There is fairly good evidence that, in some cases, auxin does increase the amount of polysomes and hence (?!) that of mRNA (VERMA et al. 1975, DAVIES 1976). The same applies to the cytokinins. Similarly, one can expect that the amount of primary transcripts increases or decreases whenever phytohormones (kinetin, abscisic acid) increase or decrease the rate of rRNA synthesis. Through the regulation of rRNA synthesis, indirectly, the apparent rate of the degradation of rRNA is also affected. When, however, the rate of breakdown of a stable RNA, like that of the rRNA of *Lemna,* is stimulated by hormone treatment (TREWAVAS 1970), more direct post-transcriptional control must be operating. The nature of this post-transcriptional control is unknown. In the case of transcriptional control, the effect of the hormone on the RNA polymerase might come into consideration (BEX 1972).

Hormone-induced, qualitative changes in the transcriptional pattern may lead to alterations in the degradative rates if the transcription products (different mRNA's) have different stabilities. Although small differences are difficult to detect, at least two major stability components, one with an average half-life of 0.6 h and another with that of approximately 30 h have recently been detected in soybean suspension culture cells (SILFLOW and KEY 1979).

Plant hormones can have a differential effect on the transcription (and processing) of "related" molecules, such as the precursors of cytoplasmic and chloroplastic rRNA's. In the same cell, depending on the target organelle (nucleus versus chloroplast) in which the transcription takes place, the plant hormones affect the transcription (and processing) of the two kinds of rRNA's differently. Since the rate of synthesis of the final (processed) products was assayed (TREWAVAS 1970), the experiments do not differentiate between effects on transcription and processing.

Somewhat surprisingly, the plant hormones appear to regulate the *total* RNA level, including the stable RNA species and the rapidly labeled, heterodisperse RNA (PARANJOTHY and WAREING 1971). There is little evidence for selective effects. It remains to be seen whether this "overall" effectiveness of the hormones is only apparent, being due to the lack of sufficiently sensitive techniques. Some evidence for the specific stimulation of the transcription of single mRNA's is already available. The finding that auxin stimulates the preferential synthesis of an mRNA coding for cellulase in pea epicotyls (VERMA et al. 1975) and that gibberellic acid promotes the synthesis of α-amylase mRNA in barley aleurone cells (HIGGINS et al. 1976) gives hope that more specific effects will be soon discovered. At present, the number of good examples is still negligible.

In addition to the work which revealed the interesting hormone effects on *Lemna,* other studies also pinpoint the chloroplasts as special target organelles for plant hormones. The breakdown of chloroplast rRNA has been found to be preferentially arrested by kinetin in isolated tobacco leaf discs (TAKEGAMI 1975). The presence of chloroplasts in the target tissue (DYER and OSBORNE 1971) as well as chloroplast division (GRIERSON et al. 1977) seem to be important for obtaining optimal kinetin effects. These experiments, however, do not rule out the possibility that the cytokinins preserve the chloroplast rRNA "specifically", not because of any effect on transcription per se but because they do preserve or "improve" the structure of chloroplasts by an entirely unknown mechanism.

2.3.2 Light Effects

Light has at least two basically different effects on the RNA level in plants. First, as the ultimate energy source, light is necessary for the maintenance of anabolic metabolism in plants. In addition, light can regulate, via the phytochrome system, the RNA metabolism in a more specific way.

The effect of light on RNA synthesis has been investigated in detail. We know much less about RNA breakdown. Only a few conclusions can be drawn.

Light inhibits the breakdown of RNA in isolated leaf tissues. In tobacco, where the phenomenon has been analyzed in detail, the early decrease in RNA level in the dark was accounted for by the breakdown of chloroplast-rRNA. This breakdown was reduced by illumination. The level of cytoplasmic rRNA remained steady in the dark and was actually increased in the light, in spite of leaf tissue excision. Benzyladenine further reduced the decrease in chloroplast rRNA in the dark but not in the light, and had no effect whatsoever on the cytoplasmic rRNA (TAKEGAMI 1975). These results show once again the special sensitivity of the chloroplast system to environmental effects.

In contrast to the isolated foliar leaf tissues, isolated cotyledons, at least in the short run, do not necessarily lose RNA. This system is, therefore, not directly comparable with the isolated foliar leaf. Still, it is noteworthy that in excised, etiolated cucumber cotyledons light also tends to increase the level of chloroplast rRNA rather than that of the cytoplasmic species (KINOSHITA et al. 1979). In the cotyledons of intact radish seedlings, the preferential stimulatory effect of light on chloroplast rRNA synthesis was very strong (INGLE 1968).

Light affects the RNA metabolism in plants not only as an energy donor. Although the mechanism is not clear, illumination usually increases the relative amount of polysomes in plant cells (POULSON and BEEVERS 1970), a phenomenon that might or might not be connected with a light-induced increase in RNA polymerase activity and a correspondingly higher mRNA level. The bearing of these observations, out of which the light-stimulated increase in polysome level appears to be the most widespread, on the general problem of RNA breakdown is not clear.

A fascinating field is represented by the phytochrome-controlled processes. In addition to the general (nonspecific) light effects, phytochrome-mediated assembly of polysomes (SMITH 1976) and the involvement of phytochrome in light-stimulated rRNA synthesis have also been described (THIEN and SCHOPFER 1975). Undoubtedly, these processes do play a role, even if secondarily, in the regulation of RNA turnover or breakdown, but their exact function cannot be properly assessed at present.

Acknowledgment. I am grateful to Dr. C.M. WILSON of the Agricultural Research, Science and Education Administration, USDA, Urbana, Illinois, for his critical comments concerning this review.

References

Acton GJ (1974) Phytochrome controlled acid RNase: an "attached" protein of ribosomes. Phytochemistry 13:1303–1310

Acton GJ (1978) Phytochrome-mediated loss of polyribosomes in etiolated mustard hypocotyls. Photochem Photobiol 28:361–367

Acton GJ, Schopfer P (1974) Phytochrome induced synthesis of ribonuclease de novo in lupin hypocotyl sections. Biochem J 142:449–455

Almgard G, Landergren U (1974) Isoenzymatic variation used for the identification of barley cultivars. Z Pflanzenzuechtg 72:63–73

Anfinsen CB, Cuatrecasas P, Taniuchi H (1971) Staphylococcal nuclease, chemical properties and catalysis. In: Boyer PD, Lardy H, Myrbäck K (eds) The enzymes Vol 4, 3rd edn. Academic Press, London, New York, pp 177–204

Arad S, Richmond AE (1976) Leaf cell water and ezyme activity. Plant Physiol 57:656–658

Arad S, Mizrahi Y, Richmond AE (1973) Leaf water content and hormone effects on ribonuclease activity. Plant Physiol 51:817–824

Babcock DF, Morris RO (1973) Specific degradation of a plant leucyl transfer ribonucleic acid by a factor in the homologous synthetase preparations. Plant Physiol 52:292–297

Bagi G, Farkas GL (1967) On the nature of increase in ribonuclease activity in mechanically damaged tobacco leaf tissue. Phytochemistry 6:161–169

Barker GR, Hollinshead JA (1964) Nucleotide metabolism in germinating seeds. The ribonucleic acid of *Pisum arvense*. Biochem J 93:78–83

Barker GR, Bray CM, Walter TJ (1974) The development of ribonuclease and acid phosphatase during germination of *Pisum arvense*. Biochem J 142:211–219

Baumgartner B, Matile P (1976) Immunochemical localization of acid ribonuclease in morning glory flower tissue. Biochem Physiol Pflanz 170:279–285

Baumgartner B, Matile P (1977) Isoenzymes of RNase in senescing morning glory petals. Z Pflanzenphysiol 82:371–374

Baumgartner B, Kende H, Matile P (1975) Ribonuclease in senescing morning glory. Purification and demonstration of *de novo* synthesis. Plant Physiol 55:734–737

Beevers L, Guernsey FS (1966) Changes in some nitrogenous components during the germination of pea seeds. Plant Physiol 41:1455–1458

Beevers L, Splittstoesser WE (1968) Protein and nucleic acid metabolism in germinating peas. J Exp Bot 19:698–771

Bennett PA, Chrispeels MJ (1972) *De novo* synthesis of ribonuclease and β-1,3-glucanase by aleurone cells of barley. Plant Physiol 49:445–447

Beopoulos N, Esnault R, Buri JF (1978) Study on plant RNases. Isolation and properties of several activities from *Vicia faba* root cells. Biochim Biophys Acta 517:216–227
Bernardi A, Bernardi G (1971) Spleen acid exonuclease. In: Boyer PD, Lardy H, Myrbäck K (eds) The enzymes Vol 4, 3rd edn. Academic Press, London, New York, pp 329–336
Bex JHN (1972) Effects of abscisic acid on the soluble RNA polymerase activity in maize coleoptiles. Planta 103:11–17
Bick MD, Liebke A, Cherry JH, Strehler BL (1970) Changes in leucyl- and tyrosyl-tRNA of soybean cotyledons during plant growth. Biochim Biophys Acta 204:175–182
Blekhman GI (1977a) Quaternary structure of a cytoplasmic protein with ribonuclease activity from wheat seedling leaves: kinetic manifestation of its native, dissociated and reassociated forms. (in Russian) Biokhimija 42:815–822
Blekhman GI (1977b) Possible mechanism of changes in cytoplasmic ribonuclease activity in leaves of wheat seedlings during dehydration and rehydration. (in Russian) Fiziol Rast 24:507–512
Boller T, Kende H (1979) Hydrolytic enzymes in the central vacuole of plant cells. Plant Physiol 63:1123–1132
Borucka-Mankiewicz M, Szarkowski JW (1977) Purification and some properties of a nuclease from rye germ nuclei. Acta Biochim Pol 24:289–299
Bray CM, Chow TY (1976) Lesion in the ribosomes of non-viable pea (*Pisum arvense*) embryonic axis tissue. Biochim Biophys Acta 442:14–23
Brown EG, Marshall AJ (1977) Properties and subcellular distribution of ribonuclease activity in *Chlorella*. Phytochemistry 16:435–441
Bryant JA, Greenway SC, West GA (1976a) Development of nuclease activity in cotyledons of *Pisum sativum* L. Planta 130:137–140
Bryant JA, Greenway SC, West GA (1976b) Iso-enzymes of acid ribonuclease in cotyledons of *Pisum sativum* L. Planta 130:141–144
Caers LI, Penmans WJ, Carlier AR (1979) Preformed and newly synthesized messenger RNA in germinating wheat embryos. Planta 144:491–496
Chakravorty AK, Scott KJ (1979a) Changes in two barley leaf ribonuclease fractions during infection by the powdery mildew fungus. Physiol Plant Pathol 14:85–97
Chakravorty AK, Scott KJ (1979b) Changes in barley leaf ribonucleases during early stages of infection by *Erysiphe graminis* f. sp. hordei. Phytopathology 69:369–371
Chakravorty AK, Shaw M, Scrubb LA (1974a) Ribonuclease activity of wheat leaves and rust infection. Nature (London) 247:577–580
Chakravorty AK, Shaw M, Scrubb LA (1974b) Changes in ribonuclease activity during rust infection. II. Purification and properties of ribonuclease from healthy and infected flax cotyledons. Physiol Plant Pathol 4:355–358
Chakravorty AK, Simpson RS, Scott KJ (1980) Messenger and ribosomal RNA hydrolysis by ribonucleases. II. Changes in ribonuclease activities and ribosomes of barley leaves during the early stages of powdery mildew infection. Plant Cell Physiol 21:425–432
Cherry JH (1963) Nucleic acid, mitochondria, and enzyme changes in cotyledons of peanut seed during germination. Plant Physiol 38:440–446
Chevrier N, Sarhan F (1980) Partial purification and characterization of two RNAases and one nuclease from wheat leaves. Plant Sci Lett 19:21–31
Chroboczek H, Cherry JH (1966) Characterization of nucleic acids in peanut cotyledons. J Mol Biol 19:28–37
Clapham D (1980) Properties of plant nuclease 1 purified from *Tradescantia* leaves by binding to Concanavalin A-Sepharose. Hereditas 93:137–146
Dalby A, Cagampang GB (1970) Ribonuclease activity in normal, opaque-2, and floury-2 maize endosperm during development. Plant Physiol 46:142–144
Dalby A, Davies I ab I (1967) Ribonuclease activity in the developing seeds of normal and opaque-2 maize. Science 155:1573–1575
Davies E (1976) Polyribosomes from peas. VI. Auxin-stimulated recruitment of free monosomes into membrane-bound polysomes. Plant Physiol 57:516–518
Davies E, Larkins BA (1974) Polyribosome degradation as a sensitive assay for endolytic messenger-ribonuclease activity. Anal Biochem 61:155–164

De Leo P, Sacher JA (1970) Control of ribonuclease and acid phosphatase by auxin and abscisic acid during senescence of *Rhoeo* leaf sections. Plant Physiol 46:806–811
De Leo P, Sacher JA (1971) Effect of abscisic acid and auxin on ribonuclease during ageing of bean endocarp tissue sections. Plant Cell Physiol 12:791–796
Delseny M, Aspart L, Guitton Y (1977) Disappearance of stored polyadenylic acid and mRNA during early germination of radish (*Raphanus sativus* L.) embryo axes. Planta 135:125–128
Dhindsa RR, Bewley JD (1978) Messenger RNA is conserved during drying of the drought-tolerant moss *Tortula ruralis*. Proc Natl Acad Sci USA 75:842–846
Diener TO (1961) Virus infection and other factors affecting ribonuclease activity of plant leaves. Virology 14:177–189
Dove LD (1967) Ribonuclease activity of stressed tomato leaflets. Plant Physiol 42:1176–1178
Dove LD (1971) Short term response and chemical control of ribonuclease activity in tomato leaflets. New Phytol 70:397–401
Dove LD (1973) Ribonucleases in vascular plants: cellular distribution and changes during development. Phytochemistry 12:2561–2570
Dyer TA, Osborne DJ (1971) Leaf nucleic acids. II. Metabolism during senescence and the effect of kinetin. J Exp Bot 22:552–560
Dyer TA, Payne PI (1974) Catabolism of plant cytoplasmic ribosomes: A study of the interaction between ribosomes and ribonuclease. Planta 117:259–268
Enzyme Nomenclature (1979) Recommendations (1978) of the Nomenclature Committee of the International Union of Biochemistry, Academic Press, New York, London
Erickson RO, Michelini FJ (1957) The plastochron index. Am J Bot 44:297–305
Farkas GL (1978) Senescence and plant disease. In: Horsfall JG, Cowling EB (eds) Plant disease Vol III. 2nd edn. Academic Press, New York, London, pp 391–409
Farkas GL, Dézsi L, Horváth M, Kisbán E, Udvardy J (1964) Common pattern of enzymatic changes in detached leaves and tissues attacked by parasites. Phytopathol Z 49:343–354
Fletcher RA (1969) Retardation of leaf senescence by benzyladenine in intact bean plants. Planta 98:1–8
Fourcroy P, Lambert C, Rollin P (1979) Far-red mediated poly-ribosome formation in radish cotyledons. Effect of endogenous ribonucleases on polyribosome recovery. Planta 147:1–5
Fraser RSS, Gerwitz A (1980) Tobacco mosaic virus infection does not alter the polyadenylated messenger RNA content of tobacco leaves. J Gen Virol 46:139–148
Fuchs Y, Galston AW (1976) Macromolecular synthesis in oat leaf protoplasts. Plant Cell Physiol 17:475–482
Galston AW, Altman A, Kaur-Sawhney R (1978) Polyamines, ribonuclease and the improvement of oat leaf protoplasts. Plant Sci Lett 11:69–79
Grierson D, Chambers SE, Pennket LP (1977) Nucleic acid and protein synthesis in disks cut from mature leaves of *Nicotiana tabacum* L. and cultured on nutrient agar with and without kinetin. Planta 134:29–34
Guerriero VJr, Dunham VL (1979) The presence of a ribonuclease of high molecular weight in sugar beet storage tissue. Plant Physiol 47:15–18
Hadziyev D, Mehta SL, Zalik S (1969) Nucleic acids and ribonucleases of wheat leaves and chloroplasts. Can J Biochem 47:273–282
Harvey CL, Olson KC, Wright R (1970) Further purification and properties of phosphodiesterase from carrot. Biochemistry 9:921–925
Hawthorne SB, Hillam RP, Kenefick DG (1980) Serological characterization of a ribonuclease from *Hordeum vulgare*. Phytochemistry 19:1587–1591
Heppel LA (1966) Pig liver nuclei ribonuclease. In: Cantoni GL, Davies DR (eds) Procedures in nucleic acid research. Harper and Row, New York, pp 31–36
Higgins TJV, Zwar JA, Jacobsen JV (1976) Gibberellic acid enhances the level of translatable mRNA for α-amylase in barley aleurone layers. Nature (London) 260:166–169
Hirai M, Asahi T (1973) Membranes carrying acid hydrolases in pea seedling roots. Plant Cell Physiol 14:1019–1029

Hirai M, Asahi T (1975) Purification and properties of two ribonucleases in different intracellular compartments in pea root tissue. J Biochem 78:485–492

Hirai M, Saito S, Asaki T (1975) Activities of RNases in different cell compartments in different regions of pea root. Plant Cell Physiol 16:119–126

Hiramaru M, Uchida T, Egami F (1969) Studies on two nucleases and a ribonuclease from *Physarum polycephalum*. Purification and mode of action. J Biochem 5:701–708

Hodge ET, Sacher JA (1975) Effect of kinetin, auxin and abscisic acid on ribonuclease and acid phosphatase during senescence of leaf tissue. Biochem Physiol Pflanz 168:433–441

Holbrook J, Ortanderl F, Pfleiderer G (1966) Reinigung und Eigenschaften einer Exophosphodiesterase aus Malzkeimen. Biochem Z 345:427–439

Howe RC, Ursino DJ (1972) Ribonucleic acid degrading activity associated with ribosomes from spinach leaf chloroplasts. Can J Bot 50:691–695

Hsiao TC (1968) Ribonuclease associated with ribosomes of *Zea mays*. Plant Physiol 43:1355–1361

Ingle J (1968) The effect of light and inhibitors on chloroplast and cytoplasmic RNA synthesis. Plant Physiol 48:1850–1854

Ingle J, Hageman RH (1965) Metabolic changes associated with the germination of corn. II. Nucleic acid metabolism. Plant Physiol 40:48–53

Ingle J, Beevers L, Hageman RH (1964) Metabolic changes associated with the germination of corn I. Changes in weight and metabolites and their redistribution in the embryo axis, scutellum and endosperm. Plant Physiol 39:735–740

Ingle J, Beitz D, Hageman RH (1965) Changes in composition during development and maturation of maize seeds. Plant Physiol 40:835–839

Jacobsen JV, Higgins TJV (1978a) The influence of phytohormones on replication and transcription. In: Letham DS, Goodwin PB, Higgins TJV (eds) Phytohormones and related compounds: a comprehensive treatise Vol I. Elsevier, North-Holland, Amsterdam, New York, pp 515–582

Jacobsen JV, Higgins TJV (1978b) Post-transcriptional translational and post-translational effects of plant hormones. In: Letham SD, Goodwin PB, Higgins TJV (eds) Phytohormones and and related compounds: A comprehensive treatise Vol I. Elsevier, North-Holland, Amsterdam, New York, pp 583–621

Jervis L (1974) Partial purification and characterization of two *Nicotiana tabacum* ribonucleases. Phytochemistry 13:709–714

Johari RR, Mehta SL, Naik MS (1977) Protein synthesis and changes in nucleic acids during grain development of *Sorghum*. Phytochemistry 16:19–24

Johnson PH, Laskowski MSr (1968) Sugar un-specific mung bean nuclease I. J Biol Chem 243:3421–3424

Kado CJ (1968) Purification and properties of a ribonuclease isolated from etiolated cucumber seedlings. Arch Biochem Biophys 125:86–93

Kasai K, Uchida T, Egami F, Yoshida K, Nomoto M (1969) Purification and crystallization of ribonuclease N_1 from *Neurospora crassa*. J Biochem 66:389–396

Kende H, Baumgartner B (1974) Regulation of ageing in flowers of *Ipomoea tricolor* by ethylene. Planta 116:279–289

Kinoshita I, Katagiri K, Tsuyi H (1979) Effects of benzyladenine and light on changes in various RNA species in etiolated cucumber cotyledons. Plant Cell Physiol 20:707–713

Kowalski D, Kroeker WD, Laskowski MSr (1976) Mung bean nuclease I. Physical, chemical and catalytic properties. Biochemistry 15:4457–4463

Kroeker WD, Hanson DM, Fairley JL (1975) Activity of wheat seedling nuclease toward single-stranded nucleic acids. J Biol Chem 250:3767–3772

Kuligowska E, Klarkowska D, Szarkowski JW (1976) Alkaline ribonuclease from rye germ cytosol. Acta Biochim Pol 23:115–126

Kuligowska E, Klarkowska D, Szarkowski JW (1980) An acid ribonuclease from rye germ cytosol. Phytochemistry 19:31–35

Lantero OJ, Klosterman HJ (1973) Purification and properties of barley leaf ribonuclease. Phytochemistry 12:775–784

Laskowski MSr (1966) Pancreatic deoxyribonuclease. In: Cantoni GL, Davies RD (eds) Procedures in nucleic acid research. Harper and Row, New York, pp 85–101
Lázár G, Farkas GL (1970) Patterns of enzyme changes during leaf senescence. Acta Biol Acad Sci Hung 21:389–396
Lázár G, Borbély G, Udvardy J, Premecz G, Farkas GL (1973) Osmotic shock triggers an increase in ribonuclease level in protoplasts isolated from tobacco leaves. Plant Sci Lett 1:53–57
Lehmann J, Völkl W, Udvardy J, Borbély G, Sivók B, Farkas GL (1979) Characterization of a ribonuclease from *Anacystis nidulans* infected with cyanophage AS-1. Phytochemistry 18:541–544
Lerch B, Wolf G (1972) Isolation of phosphodiesterase from sugar beet leaves. Biochim Biophys Acta 258:206–218
Lewington PJ, Talbot M, Simon EW (1967) The yellowing of attached and detached cucumber cotyledons. J Exp Bot 18:526–534
Libonati M (1971) Degradation of poly A and double-stranded RNA by aggregates of pancreatic ribonuclease. Biochim Biophys Acta 228:440–445
Locy RD, Cherry JH (1976) Evidence for a chloroplast specific tyrosyl tRNA degarding activity. Biochem Biophys Res Commun 72:15–23
Lontai I, van Loon LC, Bruinsma J (1972) Effects of auxin on the activity of RNA-hydrolysing enzymes from senescing and ageing barley leaves. Z Pflanzenphysiol 67:146–154
Van Loon LC (1975) Polynucleotide-acrylamide gel electrophoresis of soluble nucleases from tobacco leaves. FEBS Lett 51:266–269
Lovrekovich L, Klement Z, Farkas GL (1964) Toxic effect of *Pseudomonas tabaci* on RNA metabolism in tobacco and its counteraction by kinetin. Science 145:165
Lyndon RF (1966) Intracellular distribution of ribonuclease activity in pea roots. Biochim Biophys Acta 113:110–119
Markham R, Strominger JL (1956) The action of leaf ribonuclease. Biochem J 64:46P
Matile P (1978) Biochemistry and function of vacuoles. Annu Rev Plant Physiol 29:193–213
Matsushita S, Mori T, Hata T (1966) Enzyme activities associtated with ribosomes from soybean seedlings. Plant Cell Physiol 7:535–545
Matthews B, Dray S, Widholm J, Ostro M (1979) Liposome-mediated transfer of bacterial RNA into carrot protoplasts. Planta 145:37–44
McHale JS, Dove LD (1968) Ribonuclease activity in tomato leaves as related to development and senescence. New Phytol 67:305–313
Mehta SL, Srivastava KN, Mali PC, Naik MS (1972) Changes in the nucleic acid and protein fractions in opaque-2 maize kernels during development. Phytochemistry 11:937–942
Mejnartowicz L, Bergmann F (1977) Variation and genetics of ribonucleases and phosphodiesterases in conifer seeds. Can J Bot 55:711–717
Merola AJ, Davis FF (1962) Preparation and properties of soybean ribonuclease. Biochim Biophys Acta 55:431–439
Mettler LJ, Romani RJ (1976) Quantitative changes in tRNA during ethylene induced ripening (ageing) of tomato fruits. Phytochemistry 15:25–28
Naito K, Iida A, Suzuki H, Tsui H (1979) The effect of benzyladenine on changes in nuclease and protease activities in intact bean leaves during ageing. Physiol Plant 46:50–53
Nishimura M, Beevers H (1978) Hydrolases in vacuoles from castor bean endosperm. Plant Physiol 62:44–48
Nossal NG, Singer MF (1968) The processive degradation of individual polyribonucleotide chains. I. *Escherichia coli* ribonuclease II. J Biol Chem 243:913–922
Odintsova MS, Rakovskaya MV, Sissakian NM (1963) Activity of some enzymes of phosphorus metabolism in chloroplasts isolated from a non-aqueous medium. (in Russian) Biokhimiya 28:616–621
Ohtaka Y, Uchida K, Sakai T (1963) Purification and properties of ribonuclease from yeast. J Biochem 54:322–327

Oleson AE (1976) Inhibition of plant nuclease I by ATP and other nucleotides. Phytochemistry 15:1203–1205

Oleson AE, Janski AM, Clark ET (1974) An extracellular nuclease from suspension cultures of tobacco. Biochim Biophys Acta 366:89–100

Osborne DJ (1962) Effect of kinetin on protein and nucleic acid metabolism in *Xanthium* leaves during senescence. Plant Physiol 37:595–602

Panigrahi PK, Biswal VC (1979) Ageing of chloroplasts *in vitro* I. Quantitative analysis of the degradation of pigments, protein and nucleic acids. Plant Cell Physiol 20:775–779

Paranjothy K, Wareing PF (1971) The effects of abscisic acid, kinetin and 5-fluorouracil on ribonucleic acid and protein synthesis in senescing radish leaf disks. Planta 99:112–119

Payne PI (1977) Synthesis of polyA-rich RNA in embryos of rye during inbibition and early germination. Phytochemistry 16:431–434

Phillips DR, Horton RF, Fletcher RA (1969) Ribonuclease and chlorophyllase activity in senescing leaves. Physiol Plant 22:1050–1054

Pietrzak M, Cudny H, Maluszynski M (1980) Purification and properties of two ribonucleases and a nuclease from barley seeds. Biochim Biophys Acta 614:102–112

Pilet PE, Braun R (1970) Ribonuclease activity and auxin effects in the *Lens* root. Physiol Plant 23:245–250

Pitt D (1971) Purification of a ribonuclease from potato tubers and its use as an antigen in the immunochemical assay of this protein following tuber damage. Planta 101:33–35

Pitt D (1974) Activation and *de novo* synthesis of ribonuclease following mechanical damage to leaves of *Solanum tuberosum* L. Planta 117:43–55

Pitt D (1975) Changes in activity of lysosomal ribonuclease following mechanical damage to leaves of *Solanum tuberosum* L. Planta 123:125–136

Pitt D, Galpin M (1971) Increase in ribonuclease activity following mechanical damage to leaf and tuber tissues of *Solanum tuberosum* L. Planta 101:317–332

Plischke W, Hess D (1980) Nucleases from *Petunia hybrida*: A sugar-unspecific nuclease isolated from leaves. Biochem Physiol Pflanz 175:629–636

Poulson R, Beevers L (1970) Nucleic acid metabolism during greening and unrolling of barley leaf segments. Plant Physiol 46:315–319

Premecz G, Oláh T, Gulyás A, Nyitrai A, Pálfi G, Farkas GL (1977) Is the increase in ribonuclease level in isolated tobacco protoplasts due to osmotic stress? Plant Sci Lett 9:195–200

Premecz G, Ruzicska P, Oláh T, Farkas GL (1978) Effect of "osmotic stress" on protein and nucleic acid synthesis in isolated tobacco protoplasts. Planta 141:33–36

Randles JW (1968) Ribonuclease isozymes in Chinese cabbage systemically infected with turnip yellow mosaic virus. Virology 36:556–563

Rattanapanone N, Grierson D, Stein H (1977) Ribonucleic acid metabolism during the development and ripening of tomato fruits. Phytochemistry 16:629–633

Reddi KK (1959) Tobacco leaf ribonuclease. III. Its role in the synthesis of tobacco mosaic virus nucleic acid. Biochim Biophys Acta 33:164–167

Reddi KK (1966) Ribonuclease induction in cells transformed by *Agrobacterium tumefaciens*. Proc Natl Acad Sci USA 56:1207–1214

Reddi KK, Mauser LJ (1965) Studies on the formation of tobacco mosaic virus ribonucleic acid. VI. Mode of degradation of host ribonucleic acid to ribonucleosides and their conversion to ribonucleoside 5′-phosphates. Proc Natl Acad Sci USA 53:607–613

Rijven AHGC (1978) Ribosomal wash ribonucleases from fenugreek (*Trigonella foenum graecum* L.) and soybean (*Glycine max* (L) Merr.) cotyledons and their interactions with poly(A) and some modified nucleosides. Plant Sci Lett 11:293–303

Roberts LM, Lord JM (1979) Ribonucleic acid synthesis in germinating castor bean endosperm. J Exp Bot 117:739–749

Robertson HD, Webster RE, Zinder ND (1968) Purification and properties of ribonuclease III from *Escherichia coli*. J Biol Chem 243:82–91

Robinson E, Cartwright PM (1958) Nucleolytic enzymes in growing root cells. J Exp Bot 9:430–435
Robinson NE, Bryant JA (1975) Onset of nucleic acid synthesis during germination of *Pisum sativum* L. Planta 127:63–68
Ruzicska P, Mettrie R, Dorokhov YuL, Premecz G, Oláh T, Farkas GL (1979) Polyribosomes in protoplasts isolated from tobacco leaves. Planta 145:199–203
Sacher JA, Davies DD (1974) Demonstration of *de novo* synthesis of RNase in *Rhoeo* leaf sections by deuterium oxide labelling. Plant Cell Physiol 15:157–161
Sacher JA, De Leo P (1977) Wound-induced RNase in senescing bean pod tissue: post-transcriptional regulation of RNase. Plant Cell Physiol 18:161–171
Sacher JA, Morgan EJ, De La Rosa D (1975) Paradoxical effect of actinomycin D. Regulation of synthesis of wound RNase at translation in turnip tissue. Plant Physiol 56:442–449
Sacher JA, Engstrom D, Broomfield D (1979) Ethylene regulation of wound-induced ribonuclease in turnip root tissue. Planta 144:413–418
Sahulka J (1971) Electrophoretic investigation of ribonuclease in roots and leaves of *Vicia faba* L. Biol Plant 13:243–248
Sawai Y, Sugano N, Tsukada K (1978) Ribonuclease H activity in cultured plant cells. Biochim Biophys Acta 518:181–185
Sen S, Osborne DJ (1977) Decline in ribonucleic acid and protein synthesis with loss of viability during early hours of inhibition of rye (*Secale cereale* L) embryos. Biochem J 166:33–38
Shuster L, Khorana HG, Heppel LA (1959) The mode of action of ryegrass ribonuclease. Biochim Biophys Acta 33:452–461
Silflow CD, Key JL (1979) Stability of polysome-associated polyadenylated RNA from soybean suspension culture cells. Biochemistry 18:1013–1018
Simpson RS, Chakravorty AK, Scott KJ (1979) Selective hydrolysis of barley leaf polysomal messenger RNA during the early stages of powdery mildew infection. Physiol Plant Pathol 14:245–258
Simpson RS, Chakravorty AK, Scott KJ (1980) Messenger and ribosomal RNA hydrolysis by ribonucleases. I. The action of two barley leaf ribonucleases on the messenger and ribosomal RNA of isolated polysomes. Plant Cell Physiol 21:413–423
Sivók B, Udvardy J, Balogh A, Farkas GL (1977) Nuclease I in wheat root. Z Pflanzenphysiol 83:327–333
Siwecka MA, Szarkowski JW (1971) The distribution of nucleolytic activity in organelles of rye embryos and seedlings (*Secale cereale*). Bull Acad Sci Pol 19:445–449
Smith GS, Schaffer SW (1979) Selective reduction of seminal ribonuclease by glutathione. Arch Biochem Biophys 196:102–108
Smith H (1976) Phytochrome-mediated assembly of polyribosomes in etiolated bean leaves. Evidence for post-transcriptional regulation of development. Eur J Biochem 65:161–170
Sodek L, Wright STC (1969) The effect of kinetin on ribonuclease, acid phosphatase, lipase and esterase levels in detached wheat leaves. Phytochemistry 8:1629–1640
Srivastava BIS (1968) Acceleration of senescence and of the increase of chromatin-associated nuclease in excised barley leaves by abscisin II and its reversal by kinetin. Biochim Biophys Acta 169:534–536
Srivastava BIS, Matsumoto H, Chadha KC (1971) Studies on chromatin-associated nuclease from barley leaves. Plant Cell Physiol 12:609–618
Stavrianopoulos JG, Chargaff E (1973) Purification and properties of ribonuclease H of calf thymus. Proc Natl Acad Sci USA 70:1959–1963
Suno M, Nomura A, Mizumo Y (1973) Studies on 3′-nucleotidase-nuclease from potato tubers. II. Further studies on substrate specificity and mode of action. J Biochem 73:1291–1297
Sutcliffe JF, Baset QA (1973) Control of hydrolysis of reserve materials in the endosperm of germinating oat (*Avena sativa*) grains. Plant Sci Lett 1:15–20
Takaiwa J, Tanifuji S (1978) Development of RNase activity in embryonic axes of germinating pea seeds. Plant Cell Physiol 19:1507–1518

Takegami T (1975) A study on senescence in tobacco leaf disks. II. Chloroplast and cytoplasmic rRNAs. Plant Cell Physiol 16:417–425

Takegami T, Yoshida K (1975) Remarkable retardation of the senescence of tobacco leaf disks by cordycepin, an inhibitor of RNA polyadenylation. Plant Cell Physiol 16:1163–1166

Tang WJ, Maretzki A (1970) Purification and properties of leaf ribonuclease from sugar cane. Biochim Biophys Acta 212:300–307

Tani T, Yoshikawa M, Naito N (1973) Effect of rust infection of oat leaves on cytoplasmic and chloroplast ribosomal nucleic acids. Phytopathology 63:491–494

Tester CF (1977) Nucleic acid metabolism in the developing primary leaf and senescing coleoptile of germinating oats. Physiol Plant 41:305–312

Thien W, Schopfer P (1975) Control by phytochrome of cytoplasmic and chloroplastic rRNA accumulation in cotyledons of mustard seedlings in the absence of photosynthesis. Plant Physiol 56:660–664

Torti G, Mapelli S, Soave C (1973) Acid ribonuclease from wheat germ: purification, properties and specificity. Biochim Biophys Acta 324:254–266

Trebal JP, Beopoulos N, Esnault R (1979) Ribonuclease activities in bean roots. Phytochemistry 19:1635–1637

Trewavas A (1970) The turnover of nucleic acids in *Lemna minor*. Plant Physiol 45:742–751

Trewavas A (1976) Plant growth substances. In: Bryant JA (ed) Molecular aspects of gene expression in plants. Academic Press, London New York, pp 249–298

Tuve TW, Anfinsen CB (1960) Preparation and properties of spinach ribonuclease. J Biol Chem 235:3437–3441

Udvardy J, Farkas GL (1972) Abscisic acid stimulation of the formation of an ageing specific nuclease in *Avena* leaves. J Exp Bot 23:914–920

Udvardy J, Farkas GL, Marrè E (1969) On RNase and other hydrolytic enzymes in excised *Avena* leaf tissues. Plant Cell Physiol 10:375–396

Udvardy J, Marré E, Farkas GL (1970) Purification and properties of a phosphodiesterase from *Avena* leaf tissues. Biochim Biophys Acta 206:392–403

Varian A, Sacher JA (1978) Wound-induced RNase in bean pod tissue. II. Auxin regulation of RNase synthesis at transcription. Plant Cell Physiol 19:1185–1193

Verma DPS, McLachlan GA, Byrne H, Ewings D (1975) Regulation and *in vitro* translation of messenger RNA for cellulase from auxin-treated pea epicotyls. J Biol Chem 250:1019–1026

Vogt VM (1973) Purification and further properties of single-strand specific nuclease from *Aspergillus oryzae*. Eur J Biochem 33:192–200

Vold BS, Sypherd PS (1968) Changes in soluble RNA and ribonuclease activity during germination of wheat. Plant Physiol 43:1221–1226

Walbot V (1971) RNA metabolism during embryo development and germination of *Phaseolus vulgaris*. Dev Biol 26:369–379

Walters TL, Loring HS (1966) Enzymes of nucleic acid metabolism from mung bean sprouts. I. Fractionation and concentration of phosphomonoesterase, ribonuclease M_1 and M_3, 3′-nucleotidase, and deoxyribonuclease. J Biol Chem 241:2870–2875

van der Wilden W, Herman EM, Chrispeels MJ (1980) Protein bodies of mung bean cotyledons as autophagic organelles. Proc Natl Acad Sci USA 77:428–432

Wilson CM (1968a) Plant nucleases. I. Separation and purification of two ribonucleases and one nuclease from corn. Plant Physiol 43:1332–1338

Wilson CM (1968b) Plant nucleases. II. Properties of corn ribonucleases I and II and corn nuclease I. Plant Physiol 43:1339–1346

Wilson CM (1971) Plant nucleases. III. Polyacrylamide gel electrophoresis of corn ribonuclease isoenzymes. Plant Physiol 48:64–68

Wilson CM (1973) Plant nucleases. IV. Genetic control of ribonuclease activity in corn endosperm. Biochem Genet 9:53–62

Wilson CM (1975) Plant nucleases. Annu Rev Plant Physiol 26:187–208

Wilson CM (1978) Plant nucleases. V. Survey of corn ribonuclease II isoenzymes. Plant Physiol 61:861–863

Wilson CM (1980) Plant nucleases. VI. Genetic and developmental variability in ribonuclease activity in inbred and hybrid corn endosperms. Plant Physiol 66:119–125

Wilson CM, Shannon JC (1963) The distribution of ribonucleases in corn, cucumber and soybean seedlings. Effects of isolation media. Biochim Biophys Acta 68:311–313

Wilson CM, Alexander DE (1967) Ribonuclease activity in normal and opaque-2 mutant endosperm of maize. Science 155:1575–1576

Wilson CM, Apel GA (1975) Effect of *Helminthosporium maydis,* race T, pathotoxin on growth and ribonuclease levels of corn roots. Crop Sci 15:385–389

Wollgiehn R (1961) Untersuchungen über den Einfluß des Kinetins auf den Nucleinsäure- und Proteinstoffwechsel isolierter Blätter. Flora 151:411–437

Wyen NV, Farkas GL (1971) On the ribonuclease associated with cytoplasmic ribosomes isolated from *Avena* leaves. Biochem Physiol Pflanz 162:220–224

Wyen NV, Udvardy J, Solymosy F, Marrè E, Farkas GL (1969) Purification and properties of a ribonuclease from *Avena* leaf tissues. Biochim Biophys Acta 191:588–597

Wyen NV, Erdei S, Farkas GL (1971) Isolation from *Avena* leaf tissues of a nuclease with the same type of specificity towards RNA and DNA. Accumulation of the enzyme during leaf senescence. Biochim Biophys Acta 232:472–483

Wyen NV, Udvardy J, Erdei S, Farkas GL (1972a) The level of a relatively purine specific ribonuclease increases in virus-infected hypersensitive or mechanically injured tobacco leaves. Virology 48:337–341

Wyen NV, Erdei S, Udvardy J, Bagi G, Farkas GL (1972b) Hormonal control of nuclease level in excised *Avena* leaf tissue. J Exp Bot 23:37–44

Yi C, Todd GW (1979) Changes in ribonuclease activity of wheat plants during water stress. Physiol Plant 46:13–18

9 Metabolism of Pyrimidines and Purines

C. WASTERNACK

1 Introduction

As in bacterial and animal systems, the purine and pyrimidine nucleotides in plants as well as their derivatives are operative as constituents of nucleic acids and coenzymes as well as in regulatory acting compounds. They are involved in the synthesis of thiamine, riboflavine, folic acid pteridines, histidine, or are constituents of cytokinins, purine alkaloids, and further unusual N-compounds representing one way of N-accumulation or N-excretion.

The functions of nucleotides and their derivatives are linked to the synthesis, transfer, and utilization of stored energy in energy metabolism, with the group-transfer reactions by their functions in nucleotide coenzymes, with a structural and functional role in nucleic acids or vitamins, with physiological effects of cyclic AMP or cytokinins, as well as with the allosteric regulations of enzymes and with a role in the energy charge of the cell (HENDERSON and PATERSON 1973).

The present article is focused on synthesis, interconversions, and degradation of pyrimidine and purine nucleotides in plants. This will be discussed in relation to nucleic acid synthesis. Several reviews on this subject have dealt with microbial and animal systems (GOTS 1970, MURRAY 1971, MURRAY et al. 1970, O'DONNOVAN and NEUHARD 1970, HARTMAN 1970, HENDERSON 1972, HENDERSON and PATERSON 1973). However, purine and pyrimidine metabolism in plants has not been discussed in detail before.

2 Occurrence of Pyrimidines and Purines in Plants

Pyrimidine and purine compounds are detectable in the acid-soluble fraction of plant material as bases, nucleosides, ribonucleotides, and deoxyribonucleotides. Figure 2 shows acid-soluble nucleotides fractionated from bean leaves (Fig. 2A) (WEINSTEIN et al. 1969) and *Euglena gracilis* (Fig. 2B) (KRAUSS and REINBOTHE 1977). The adenine nucleotides, NAD as well as nucleotide sugars, represent the main part of nucleotides in many different plants, while deoxyri-

---

Abbreviations: ATC, aspartate transcarbamylase; CPS, carbamoylphosphate synthetase; dNDP, deoxynucleoside diphosphate; dNTP, deoxynucleoside triphosphate; 5-FdUMP, 5-fluorodeoxyuridine-5′-monophosphate; IMP, inosine-5′-monophosphate; OMP, orotidine-5′-monophosphate; PRPP, 5′-phosphoribosyl-l-pyrophosphate; SAH, S-adenosylhomocysteine; XMP, xanthosine-5′-monophosphate

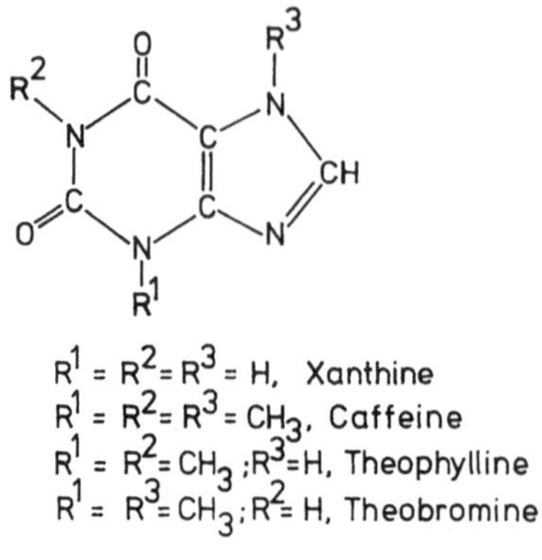

Fig. 1. Structures of methylated purines

bonucleotides, as well as cytidine nucleotides, are present in small or negligible percentage. Adenine, guanine, uracil, and their respective nucleosides were detectable in mature wheat grains (GRZELCZAK and BUCHOWICZ 1975) related to the RNA synthesis. UDP-glucose, ADP-glucose, and GDP-glucose are detectable in young leaves of *Phaseolus vulgaris* (WEINSTEIN et al. 1969). They are related to the sucrose, starch, and cellulose syntheses respectively. On the other hand, GDP-mannose functions in the glycolipid synthesis.

Some purines and pyrimidines are secondary products, e.g., methylated derivatives, such as caffeine, theobromine, and theophylline, as well as pyrimidine-containing products such as lathyrine, willardiine, isowillardiine, vicine, and convicine. Only some aspects of these will be discussed here.

Pyrimidinyl Amino Acids. (See reviews by REINBOTHE et al. 1981 and LAMBEIN et al. 1976). Pyrimidinyl amino acids have been detected in legume seeds in remarkable amounts. Lathyrine (β-[2-amino-pyrimidine]-4-yl alanine) was found in *Lathyrus tingitanus* (BELL 1961). Willardiine (β-[2,4-dihydroxy-pyrimidine]-1-yl alanine) as well as isowillardiine (β-[2,4-dihydroxy-pyrimidine]-3-yl alanine) were detected in germinating pea seedlings (cf. LAMBEIN et al. 1976) and other legumes (KRAUSS and REINBOTHE 1973). Biosynthesis of lathyrine, willardiine, and isowillardiine were discussed in terms of two independent routes: (1) Their formation from a preformed heterocyclic compound such as uracil; (2) The cyclization of γ-hydroxyhomoarginine (ASHWORTH et al. 1972, BROWN and AL-BALDAWI 1977). More recently, willardiine and isowillardiine have been synthesized in vitro by use of uracil, O-acetyl-L-serine, and an enzyme extract from *Pisum* and other sources, this evidence favoring the former pathway (MURAKOSHI et al. 1978). Different synthases for both compounds were suggested. The occurrence of pyrimidinyl amino acids is of chemotaxonomic significance (LAMBEIN et al. 1976).

Methylated Purines. Methylated purines, such as caffeine, theophylline, and theobromine (Fig. 1), are widely distributed in different higher plants (cf. SUZUKI and TAKAHASHI 1977). Caffeine is synthesized in the pericarp and is accumulated in leaves or seeds during fruit formation (BAUMANN and WANNER 1972). On

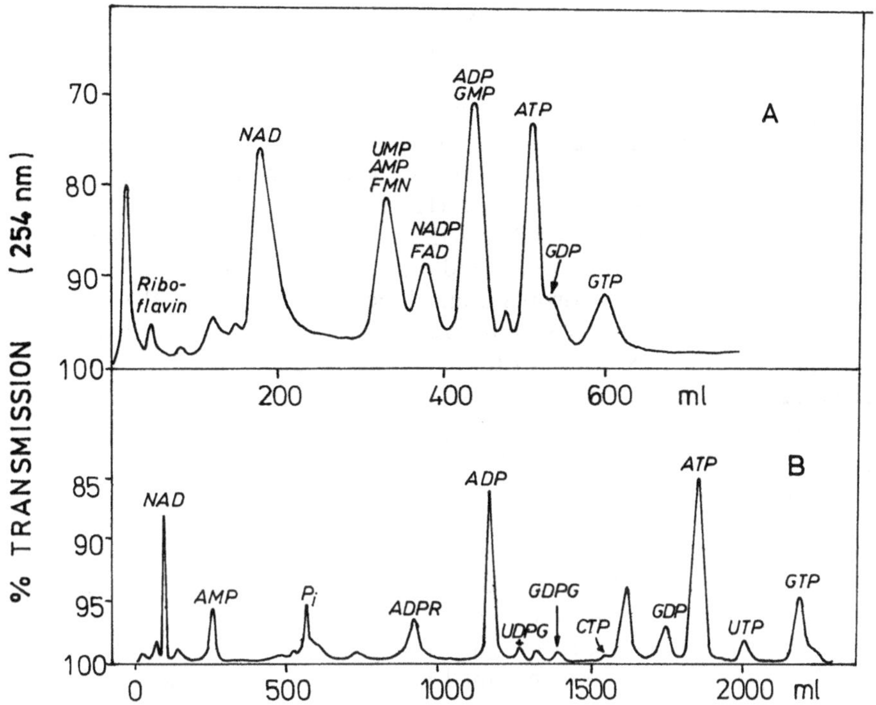

Fig. 2. Acid-soluble nucleotides of *Euglena gracilis Z* (**A**) and pinto bean leaves (*Phaseolus vulgaris L.*) (**B**). [Redrawn from KRAUSS and REINBOTHE 1977 (**A**) and WEINSTEIN et al. 1969 (**B**)]

the basis of tracer studies, their biosynthesis is suggested along the two following pathways: (1) Nucleotide pools → 7-methylated purine nucleotides → 7-methylxanthosine → 7-methylxanthine → 3,7-dimethylxanthine → 1,3,7-trimethylxanthine (SUZUKI and TAKAHASHI 1976b, BAUMANN et al. 1978). (2) Nucleic acid methylation → 1-methyl-AMP → 1-methylxanthine → 1,3-dimethylxanthine (SUZUKI and TAKAHASHI 1976a, 1977). S-Adenosylmethionine is clearly determined as the source of the methyl group in both pathways (SUZUKI and TAKAHASHI 1976c, SUZUKI 1972). The degradation of methylated xanthines takes place by demethylation and subsequent oxidation by the oxidative purine degradation pathway (Fig. 10).

The functions of methylated xanthines are not established in higher plants (SUZUKI and TAKAHASHI 1977). Their inhibitory effect on cyclic nucleotide phosphodiesterase has been repeatedly determined (BOLLIG et al. 1978, HARTFIEL and AMRHEIN 1976). Thus, the action of methylated xanthines can be mediated by cyclic AMP levels, but generalizations are speculative for different systems (HARTFIEL and AMRHEIN 1976) (see Sect. 2).

Cyclic AMP. Cyclic AMP exists in higher plants, but its function is still highly controversial (cf. lit. until 1977 in AMRHEIN 1977); only some comments are possible. It seems that cyclic AMP can be identified without doubt from some species of algae and higher plants (AMRHEIN 1977). In axenic rye grass endosperm cultures the extreme low levels of 2–12 pmol cAMP·g^{-1} fresh weight was determined (ASHTON and POLYA 1978). But the existence of 3,5-cAMP phosphodiesterase, an adenylate cyclase, is questionable for higher plants. Adenylate cyclase was not detectable in soybean hypocotyls or onion meristem (YUNGHANS and MORREE 1977) or *Zea mays* and other higher plants when different sensitive assays were used (HINTERMANN and PARISH 1979). Contrary to this, adenylate cyclase was detected in *Phaseolus vulgaris* (BROWN et al. 1979). More recently, 3,5-cAMP phosphodiesterase was purified from spinach leaves and localized in the chloroplast envelope (BROWN et al. 1980), although its function is not clear. To summarize, cyclic AMP is detected in some plant systems but evidently does not function as a second messenger in higher plants as far as is known. Furthermore, the participation of cyclic AMP, cAMP-specific phosphodiesterase, and adenylate cyclase in the circadian clock cycle as proposed as a theoretical model by CUMMINGS (1975) is highly controversial (BOLLIG et al. 1978).

In diverse fungi changed endogenous levels of cAMP were shown to correlate with morphological and developmental processes. Correlation between cAMP levels and carbon catabolite repression in yeasts on the one hand (MAHLER and LIN 1978), and N-starvation in Cyanophyceae on the other hand (HOOD et al. 1979) has been observed. This is not true for other fungi, however, where cAMP levels seem to be determined by the plasma membrane potential (PALL 1977). Thus, depolarization by membrane-active antibiotics, as well as by uncouplers of the oxidative phosphorylation, lead to a rapid transient increase in endogenous cAMP in different fungi (TREVILLYAN and PALL 1979). As suggested, elevated cAMP levels would function in cell wall synthesis resulting in altered membrane permeability (TREVILLYAN and PALL 1979).

Among phytoplanktonic algae such as *Anabaena flos-aquae, Microcystis aeroginosa, Scenedesmus communis, Chlorella pyrenoidosa, Pediastrum biradiatum* and *Ochromonas malhamensis*, cAMP is produced and released into the medium (FRANCKO and WETZEL 1980). Depending on species, cellular cAMP of 92–394 pmol·g^{-1} and extracellular cAMP of 8–440 pmol·l^{-1} were determined.

3 Formation of Pyrimidines and Purines

3.1 Pyrimidines

3.1.1 Pathway Reactions

Pyrimidines are synthesized de novo by a sequence of six reactions called the "orotic acid pathway". Starting with the elementary precursors such as CO_2, the amide group of glutamine as well as aspartic acid, the synthesis of the pyrimidine ring is finished before its N-glycosidic linkage with PRPP. UMP is the common precursor of all pyrimidine nucleotides (Fig. 3). Using 2 ATP,

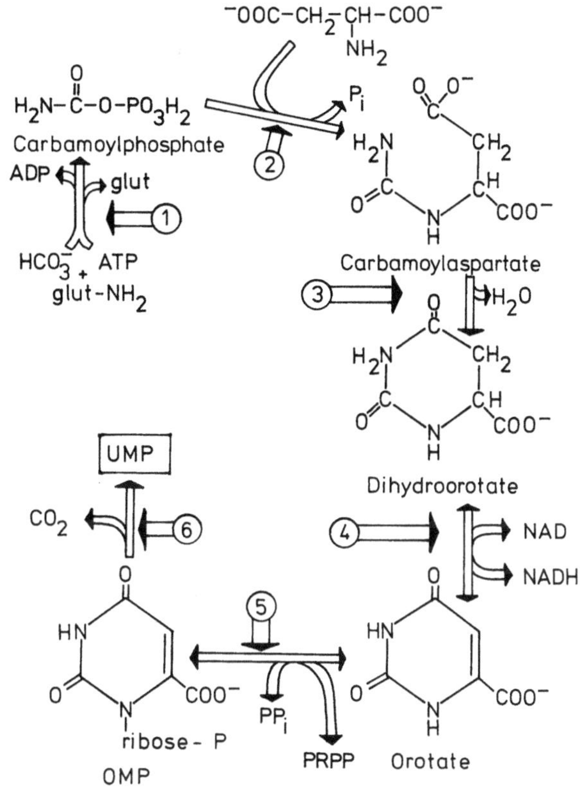

Fig. 3. Pathway of pyrimidine synthesis de novo. Enzymes involved: *1* carbamoyl phosphate synthetase I/II (EC 6.3.4.15/6.3.5.5); *2* aspartate transcarbamylase (EC 2.1.3.2); **3** dihydroorotase (EC 3.5.2.3); *4* dihydroorotate dehydrogenase (EC 1.3.1.14/15); *5* orotidylate phosphoribosyltransferase (EC 2.4.2.10); *6* orotidylate decarboxylase (EC 4.1.1.23)

HCO_3^-, and glutamine, carbamoyl phosphate is synthesized by the carbamoyl phosphate synthetase II (CPS II) which is widely distributed in plants, microorganisms, and animal systems, and is located in the cytoplasma. It differs from the mitochondrial carbamoyl phosphate synthetase I (CPS I) of animal systems in that the N-requirement and N-acetylglutamate do not affect the enzyme activity (cf. review of MAKOFF and RADFORD 1978). The energy-rich carbamoyl phosphate is linked to aspartate by aspartate transcarbamylase (ATC). Under dehydratation, the heterocyclic ring is closed by dihydroorotate dehydrogenase. Using orotate and PRPP, the first pyrimidine nucleotide OMP is formed by OMP phosphoribosyltransferase and is irreversibly decarboxylated to UMP by OMP decarboxylase. This orotic acid pathway (Fig. 3) is the main pathway in microorganisms, animals, and plants. It is likewise well established in fungi such as *Neurospora crassa* and *Saccharomyces cerevisiae*. This will be discussed in greater detail in connection with the metabolic compartmentation of carbamoyl phosphate.

Carbamoyl phosphate is the common intermediate of both pyrimidine and arginine synthesis. Therefore, structural and functional relationships between enzymes and metabolites involved in the beginning of these pathways are of special interest.

Higher plants incorporate $NaHCO_3$, aspartate, and intermediates of the orotic acid pathway into pyrimidines (e.g., WANG 1967, KING et al. 1965, LOVATT et al. 1979), and the enzymes responsible have been detected in cell-free extracts of a variety of plant species. Hence it is generally accepted that the orotic acid pathway serves as the de novo source of pyrimidine nucleotides in higher plants.

The suggestions of BUCHOWICZ and REIFER (1961) that wheat leaves converted orotic acid to uridine by direct orotidylate decarboxylation were based on poor experimental evidence (BUCHOWICZ and LESNIEWSKA 1970). No evidence for this suggestion has come from other authors (WOLCOTT and ROSS 1967, ASHIHARA 1978). Similar suggestions that pyrimidines can be formed by reverse reactions of reductive uracil degradation were not confirmed (BUCHOWICZ et al. 1963).

3.1.2 Enzymes

Carbamoyl Phosphate Synthetase. The properties of CPS have been studied using the enzyme purified from pea seedlings (O'NEAL and NAYLOR 1968, 1969, 1976). The enzyme appears to be the only one catalyzing carbamoyl phosphate synthesis in both pyrimidine and arginine biosynthesis. In a glutamine-dependent reaction ($K_m = 0.12$ mM), UMP acts as a strong competitive inhibitor ($K_i <$ 2 μM!), while ornithine and inosine-5′-monophosphate can activate several fold, partially by increasing the affinity of the enzyme to ATP (O'NEAL and NAYLOR 1976). Similar properties were detected for CPS from *Phaseolus aureus* (ONG and JACKSON 1972a). Again, glutamine was a better substrate than ammonium, ($K_m[NH_4]$ 45 times that of glutamine), N-acetyl-glutamine was without effect, and the glutamine analog, azaserine, inhibited reactions in presence of glutamine as well as NH_4^+, suggesting only one binding-site for glutamine (ONG and JACKSON 1972a). Thus like CPS of bacterial systems, e.g., *Escherichia coli* (MEISTER and POWERS 1978) the CPS of higher plants seems to be a class I enzyme (MAKOFF and RADFORD 1978) providing a common carbamoyl phosphate pool for both the arginine and pyrimidine synthesis. Most of the available carbamoyl phosphate (in low concentration) is used in the pyrimidine biosynthesis, since the ratio of aspartate transcarbamylase and ornithine transcarbamylase was found to be 1:3 as determined at near saturating kinetics (ONG and JACKSON 1972a). The carbamoyl phosphate content is sensitively regulated by UMP (O'NEAL and NAYLOR 1976). Carbamoyl phosphate could be limiting for arginine synthesis, if CPS activity were decreased too much (Fig. 4). Under these conditions, increasing ornithine activates CPS antagonizing the feedback-inhibition of UMP. The observed activation of CPS by purine nucleotides such as inosine-5′-monophosphate and guanosine-5′-monophosphate might balance the ratio of purine and pyrimidine nucleotides for nucleic acid synthesis (O'NEAL and NAYLOR 1968, 1976). Besides CPS, ATC is the second regulatory point in the pyrimidine biosynthesis of higher plants.

Aspartate Transcarbamylase. This enzyme was purified from mung bean (ONG and JACKSON 1972a, b, ACHAR et al. 1974, SAVITHRI et al. 1978 in purer form)

Fig. 4. Regulations of the pyrimidine synthesis in plants. Pathway reactions ⟹, feedback-inhibition →, antagonism to feedback-inhibition ......▸

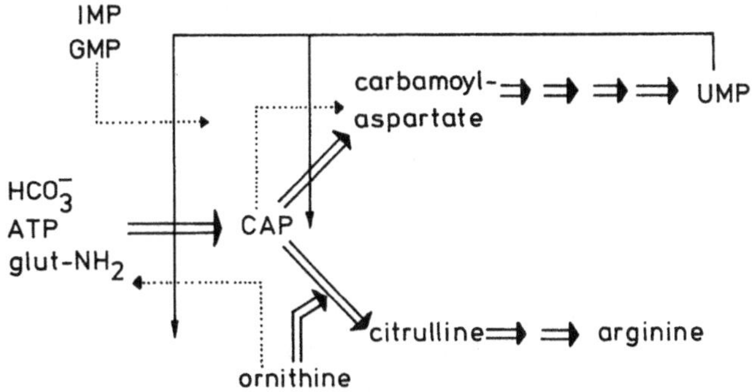

and from wheat germ (YON 1972, 1973). The determined size of the enzyme was similar to the catalytic subunit of the ATC of *Escherichia coli*. UMP acts as a competitive-acting allosteric inhibitor (Hill coefficient n = 2) with respect to carbamoyl phosphate. The enzyme exhibits sigmoidal kinetics with carbamoyl phosphate, hyperbolic kinetics with aspartate (ACHAR et al. 1974) (opposed to ONG and JACKSON 1972b), and is desensitized for UMP inhibition, suggesting an allosteric nature. For ATC of *Phaseolus aureus* a Bi Bi mechanism was suggested with sequence carbamoyl phosphate-binding (irreversible), aspartate-binding, carbamoyl aspartate-release, P_i-release (ONG and JACKSON 1972b). Interestingly, maximal catalytic activity and UMP sensitivity occur at pH 10.5 for both enzymes prepared from wheat and mung bean, suggesting unknown activating events under cellular conditions.

With the wheat germ ATC, the different protection to inactivation by UMP and carbamoyl phosphate has been interpreted as due to two conformational states reversibly accessible to the enzyme (YON 1973). The relevance of these conformational states is unclear (YON 1973) with regard to kinetic properties which were partially explained by an allosteric mechanism (YON 1972).

To summarize, pyrimidine synthesis seems to be controlled effectively in the ATC-step together with the regulatory properties of CPS. Both CPS and ATC were feedback-inhibited by UMP (Fig. 4). With regard to low K_m-value for carbamoyl phosphate (ATC) and the irreversible binding of carbamoyl phosphate to ATC, this substrate should be channeled into the pyrimidine pathway (ONG and JACKSON 1972a) (Fig. 4). However, the in vivo situation is quite unexpected. Apart from an incomplete study of the subcellular localization of UMP-synthesizing enzymes from *Vinca rosea* (KANAMORI et al. 1980), the location of enzymes and substrates has not been established in higher plants.

Turnover of ATC might represent a second regulatory level in pyrimidine synthesis, as suggested by its changed concentration throughout the cell cycle of *Chlorella* (DUNN et al. 1977).

In view of the enzymatic studies, earlier assumptions of an alternative carbamoyl phosphate synthesis from citrulline suggested by tracer experiments in wheat seedlings (RYBICKA et al. 1967) have been ruled out.

Dihydroorotase, Dihydroorotate Dehydrogenase, Orotidylate Phosphoribosyltransferase, Orotidylate Decarboxylase. Dihydroorotase was purified from pea seedlings (MAZUS and BUCHOWICZ 1968). It catalyzed the reversible conversion of carbamoyl aspartate to dihydroorotate, did not require any cofactors, and was highly stable unlike the enzyme of bacterial systems. There were no indications as to regulatory function of this enzyme in higher plants. Further enzymes in the orotic acid pathway were detected and purified from wheat embryos (KAPOOR and WAYGOOD 1965), bean leaves (WOLCOTT and ROSS 1967), blackgram seedlings (ASHIHARA 1978), and more recently from *Euglena gracilis* (WALTHER et al. 1981). OMP phosphoribosyltransferase and OMP decarboxylase prepared from blackgram and *Euglena gracilis* appeared to exist as a complex with noncooperative properties (ASHIHARA 1978, WALTHER et al. 1981), and were localized as soluble enzymes (ASHIHARA 1978, WALTHER et al. 1980). The observed inhibition by UMP, XMP, and other nucleotides had no physiological significance (WALTHER et al. 1981).

Enzymatic studies have been done on plant PRPP synthesis (ASHIHARA 1977a, b, ASHIHARA and KOMAMINE 1974). The role of PRPP synthesis in pyrimidine synthesis seems to be questionable on the basis of experiments done by KANAMORI et al. (1979, 1980). The interesting point of the significance of PRPP availability in pyrimidine synthesis suggested from mammalian systems (cf. BECKER et al. 1979) must be controlled for plants in coordinated terms of enzymes, actual metabolite concentrations and their cellular compartmentation.

OMP phosphoribosyltransferase as well as OMP decarboxylase seems to be missing in regulatory functions for pyrimidine synthesis, regarding their kinetic properties (WALTHER et al. 1981). The predominant role of CPS and ATC and their inhibition by micromolar amounts of UMP (O'NEAL and NAYLOR 1976, YON 1973) agree with in vivo experiments on the regulation of pyrimidine synthesis of intact cells of *Cucurbita pepo* (LOVATT et al. 1979).

The operation of de novo pyrimidine synthesis is widespread in plants, but the occurrence of repression/derepression and feedback-inhibition is of varying significance. As discussed above, feedback-inhibition by pyrimidine nucleotides is ultimatively a property of ATC and/or CPS in higher plants, fungi as well as green alga (DUNN et al. 1977) and *Cyanophyceae* (CURRIER and WOLK 1978). Due to the lack of repression in *Cyanophyceae* (CURRIER and WOLK 1978) or the limited state of repression in yeasts (LACROUTE 1968), a more comprehensive discussion is of interest.

In *Saccharomyces cerevisiae* the genes for the enzymes of pyrimidine synthesis are scattered in the genome (LACROUTE 1968). The pyrimidine-specific glutamine-dependent CPS and the ATC are both encoded by the uracil-2 locus which contains two cistrons (DENIS-DUPHIL and KAPLAN 1976). The cistrons are transcribed into a single polycistronic message in the order CPS → ATC (DENIS-DUPHIL and KAPLAN 1976). Therefore, nonsense mutations in the CPS-region results in a complete loss of ATC activity through a total polar effect suggesting only one protein initiation site typical for eukaryotic mRNA's. Reappearance of ATC activity in these mutants is explained by the creation of a reinitiation site by mutation inside the uracil-2 locus (EXINGER and LACROUTE 1979).

As demonstrated by mutational loss of the pyrimidine and/or the arginine-specific CPS, carbamoyl phosphate produced in one pathway is available in the other, suggesting a common pool of carbamoyl phosphate (LACROUTE et al. 1965). The pyrimidine-specific CPS and the ATC were purified as a multienzyme complex consisting of four subunits of $\alpha_2\beta_2$-type separable after heat treatment (LUE and KAPLAN 1969, 1971, AITKEN et al. 1975). The powerful feedback-inhibition of UTP which acts on both enzymes is lost in the ATC subunit. The CPS activity is present only in the aggregated state, while ATC is fully active as subunits (LUE and KAPLAN 1971). As demonstrated by tracer experiments in both separate as well as aggregate state of CPS and ATC, carbamoyl phosphate produced by the pyrimidine-specific CPS is channeled in this multienzyme complex without "dilution" by external carbamoyl phosphate (LUE and KAPLAN 1970). Similarly, ornithine transcarbamylase cannot compete with ATC for carbamoyl phosphate formed by the aggregate (LUE and KAPLAN 1970). In view of the carbamoyl phosphate exchange between the pyrimidine and the arginine pathway suggested by mutants (LACROUTE et al. 1965), the physiological role of the carbamoyl phosphate channeling in the wild type is not clear. The following useful integration of enzyme regulation by compartmentation seems to exist in *S. cerevisiae:* Carbamoyl phosphate is synthesized for the pyrimidine and arginine biosynthesis by pathway-specific glutamine-dependent synthetases, differing in their gene loci, enzyme properties, repression, and feedback-inhibition. But carbamoyl phosphate is exchangeable. Thus, feedback-inhibition of the pyrimidine-specific CPS by UTP will be reasonable only by the simultaneous feedback-inhibition by UTP on ATC. On the other hand, carbamoyl phosphate will be channeled only in the CPS-ATC multienzyme, if arginine-specific CPS is strongly repressed in arginine-rich medium. Regarding the extramitochondrial location of the arginine-specific CPS, the ornithine transcarbamylase, and the arginase (JANNIAUX et al. 1978), the difference in their location between fermentative and aerobic yeasts (URRESTARAZU et al. 1977) and the sensitive "epi-arginastic regulation", the availability of arginine-specific carbamoyl phosphate for the pyrimidine pathway seems to be restricted.

Further enzymes of pyrimidine synthesis in yeasts are sequentially induced by ureidosuccinate and dihydroorotate (LACROUTE 1968, KORCH et al. 1974). But there is some evidence that feedback-inhibition is the key process in regulation of this pathway and organism (LACROUTE 1968). Regulatory interactions between the pyrimidine and the purine synthesis seem to exist in *S. cerevisiae* (KORCH et al. 1974).

Using yeast OMP phosphoribosyltransferase VICTOR et al. (1979a, b) have shown divalent metal ion activation of the enzyme as one prerequisite of its Bi Bi ping-pong mechanism. Contrary to the complex detected in various mammalian systems, the OMP phosphoribosyl transferase and OMP decarboxylase from yeasts are separable. Synthesis of the transferase seems to be regulated on the transcriptional level (BACH et al. 1979) regarding the coordinate variation of uracil-3 mRNA and the enzyme activity.

Neurospora crassa, similarly to *S. cerevisiae,* possesses two pathway-specific glutamine-dependent CPS determined by separate loci. They are separable by gel filtration, differ in properties and in their cellular location. The pyrimidine-specific CPS is associated with ATC in a multienzyme complex of 650,000 which is encoded from the pyrimidine-3 gene through a unique mRNA starting in the CPS part. Purification studies on the CPS–ATC complex were of limited success due to the presence of an endogenous protease (MAKOFF and RADFORD 1977) and the instability of CPS (WILLIAMS et al. 1971). Using a genetic approach a complex of two ATC trimers were proposed by MAKOFF et al. (1978), linked by three CPS dimers which were composed in two domains of six homologous subunits. However, the important question about one bifunctional chain or separate polypeptide chains remains open. Carbamoyl phosphate channeling in the CPS–ATC complex is in accordance with this subunit model of MAKOFF et al. (1978). The arginine-2 gene coding for glutamine utilization in the arginine-specific CPS is extremely close to the pyrimidine-3 gene coding for CPS–ATC. However, the suggested functional relationship through a common glutaminase activity could not be found (MAKOFF and RADFORD 1976b). The CPS–ATC complex was localized histochemically in the nucleus, while the highly unstable arginine-specific CPS ($t_{1/2}=35$ min at 25 °C) (DAVIS et al. 1980) and ornithine transcarbamylase were localized in the mitochondria (BERNHARDT and DAVIS 1972). UTP binding as well as glutamine binding takes place only in the CPS part of the multienzyme complex when tested by the genetic approach (MAKOFF and RADFORD 1976a, b) and by in vitro studies (WILLIAMS and DAVIS 1970). This is reasonable regarding the enzyme and metabolite compartmentation in *N. crassa.* Carbamoyl phosphate is synthesized separately in the arginine and the pyrimidine pathway without mixing. Both pathways differ in their location. Mutants which are blocked in the transcarbamylase reaction or carbamoyl phosphate synthetase reaction of arginine or pyrimidine synthesis possess nutritional requirements for uridine or arginine. Carbamoyl phosphate synthesized by one pathway cannot be used in the other. However, under conditions of blocked transcarbamylases in the parallel pathway, carbamoyl phosphate can overflow and this is of nutritional significance. In the wild type, pyrimidine-produced carbamoyl phosphate is channeled along the CPS–ATC complex. This complex is present in two to five times higher molar concentrations than carbamoyl phosphate. Thus, one prerequisite for metabolite channeling is given. Carbamoyl phosphate can overflow, if concentrations are elevated in mutants lacking ATC activity (WILLIAMS et al. 1971). Thus, feedback-inhibition by UTP on the pyrimidine-specific CPS can be complete for the pyrimidine pathway contrary to the situation in yeasts. The carbamoyl phosphate channeling in the CPS–ATC complex and its nuclear location may reflect an advantage in regard to a near contact to the feedback-sensitive CPS and the end product UTP which is used most rapidly (DAVIS 1972).

The next enzyme in UMP synthesis, the constitutive biosynthetic dihydroorotate dehydrogenase, was localized in the mitochondrial membrane linked with the respiratory chain through ubiquinone (MILLER 1971). Enzyme release from membrane particles resulted in a solubilized protein complex containing bound lipids and inactive hydrophobic proteins (MILLER 1975). The catalytically active enzyme contained a small content of bound flavine mononucleotide which acted with a high turnover rate (MILLER 1975). The intracellular transport pathway of dihydroorotate and orotate, suggested by the nuclear location of CPS–ATC and the mitochondrial location of dihydroorotate dehydrogenase, remains unestablished. Regulation of enzyme synthesis was observed for ATC and dihydroorotate dehydrogenase (CAROLINE 1969). The first enzyme was regulated by derepression by end-product depletion, and the second enzyme was substrate-induced.

3.2 Purines

As in animal and bacterial systems, purines are synthesized in plants by a stepwise addition of small molecules or their constituents to preformed ribose. As shown in Fig. 5, N and C are derived from different sources. The sequence is started by the synthesis of PRPP from ribose-5-phosphate and ATP catalyzed by the PRPP synthetase. Due to the fact that PRPP functions as a phosphoribosyl donor in the synthesis on histidine, tryptophan, NAD coenzymes, pyrimidines and other phosphoribosyltransfer reactions (BECKER et al. 1979), the availability of PRPP is highly important for regulation. PRPP synthesis is inhibited competitively by ADP (with respect to Mg-ATP as substrate), by PRPP and 2,3-diphosphoglycerate (both with respect to ribose-5-phosphate as substrate), and it is inhibited noncompetitively by many di- and trinucleotides with respect to both Mg-ATP and ribose-5-phosphate as substrates (WYNGAARDEN 1976, ASHIHARA and KOMAMINE 1974). Contrary to the situation in spinach leaves (ASHIHARA 1977b), P_i is an allosteric activator of the PRPP synthetase in most systems studied. The plant enzyme seems to be influenced by the energy charge (ASHIHARA 1977a). Using the amide-N of glutamine or NH_4^+, PRPP is transformed into the unstable phosphoribosylamine (PRA) by the PRPP amidotransferase which is highly allosterically regulated by AMP but not by pyrimidine nucleotides (SATYANARAYANA and KAPLAN 1971). This amidotransferase reaction seems to be the most important regulatory step in purine synthesis de novo. The enzyme exists in a small (active) and a large (inactive) form. It is transferred between the two by PRPP (activation) and purine nucleotides

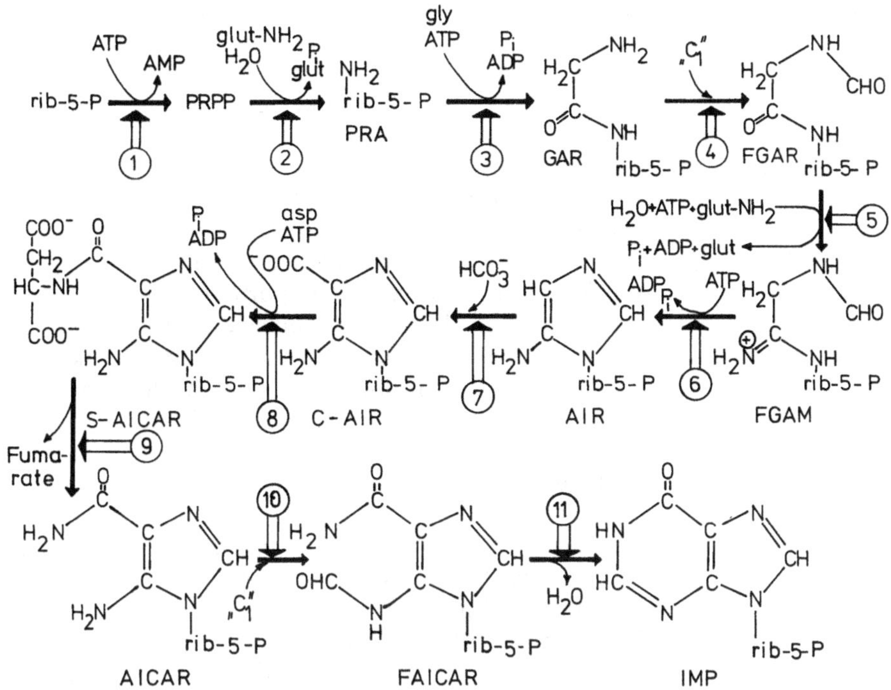

Fig. 5. Pathway of purine synthesis de novo. Enzymes involved: *1* PRPP synthetase (EC 2.7.6.1); *2* PRPP amidotransferase (EC 2.4.2.14); *3* GAR synthetase (EC 6.3.1.3); *4* GAR transformylase (EC 2.1.2.2); *5* FGAR amidotransferase (EC 6.3.5.3); *6* AIR synthetase (EC 6.3.3.1); *7* AIR carboxylase (EC 4.1.1.21); *8* s-AICAR synthetase (EC 6.3.2.6); *9* adenylosuccinate lyase (EC 4.3.2.1); *10* AICAR transformylase (EC 2.1.2.3); *11* IMP cyclohydrolase (EC 3.5.4.10)

(inactivation) (WYNGAARDEN 1976). The control of the amidotransferase reaction is possible by means of three different molecular mechanisms: (1) Changes of the *substrate levels* of PRPP or glutamine by metabolic events. (2) Changes in the *amount* or *intrinsic activity* of amidotransferase. (3) Changes in the *amount of nucleotides* as the inhibitors (WYNGAARDEN 1976).

Glycinamide ribotide (GAR) is formed from phosphoribosylamine and glycine catalyzed by a GAR synthase. The C-8 atom of the purine ring is introduced from N^5-N^{10}-methenyl-tetrahydrofolate by GAR transformylase giving formylglycinamide ribotide (FGAR). A second amide group of glutamine is used in the subsequent synthesis of formylglycinamidine ribotide (FGAM) which is catalyzed by FGAR amidotransferase. The next steps are an ATP-dependent cyclization to aminoimidazole ribotide (AIR) by AIR synthetase and a reversible carboxylation to aminoimidazole carboxylate ribotide (c-AIR) by AIR carboxylase. The N-2 of the purine ring is derived from aspartate by a succino-AICAR synthetase forming succinocarboxamide aminoimidazole ribotide (s-AICAR) and by an adenylosuccinate lyase forming AICAR. AICAR is formylated by N^{10}-formyl-tetrahydrofolate in the AICAR transformylase reaction. The terminal cyclization to IMP is catalyzed by IMP-cyclohydrolase.

IMP is the common precursor for all purine nucleotides. It is metabolized into AMP by the two-step reaction using the amino group of aspartate. The two enzymes, adenylosuccinate synthase and adenylosuccinate lyase, were purified from wheat seeds and pea seeds (HATCH 1966) and were identical in most essential properties with those enzymes detected for animal and microbial systems. AMP synthesis and s-AICAR cleavage seem to be catalyzed by the same enzyme (HATCH 1966). GMP which is formed by IMP dehydrogenase as well as XMP aminase, and AMP can be reutilized for IMP formation by AMP deaminase and GMP reductase (Fig. 7). Besides purine salvage reactions, such cycling of purines in the mononucleotide level is highly important for regulation with respect to the balance of purine nucleotides for nucleic acid synthesis as well as for the energy charge. Thus, AMP deaminase purified from baker's yeast (YOSHINO et al. 1979) seems to function in the stabilization of the energy charge. The enzyme shows cooperative binding of AMP, and ATP acts as a positive effector, P_i as a negative one (YOSHINO et al. 1979). The enzyme is distributed among eukaryotes, but its function is fulfilled by AMP nucleosidases in prokaryotes.

Contrary to such cycling, [^{14}C]-glycine was incorporated 10–26 times as much into ATP as GTP during 24 h in germinating soybean embryonic axis, and [^{14}C]-hypoxanthine and [^{14}C]-inosine were converted about five times as much into GTP as ATP (ANDERSON 1979). Thus, different triphosphate synthesis seems to be exist along salvage and de novo reactions without recycling of the purine label. These results have been discussed on the basis of different IMP pools (ANDERSON 1979).

Most of the information concerning purine synthesis in higher plants comes from tracer studies (SUZUKI and TAKAHASHI 1977). Regulatory and enzymatic studies on purine nucleotide synthesis and its interconversions in higher plants are poorly understood compared with microbial and animal systems.

4 Salvage Reactions of Pyrimidines and Purines

4.1 Pyrimidines

Naturally occurring nucleosides and bases formed by nucleotide breakdown or derived from exogenous sources can be reutilized in nucleotide synthesis. Reactions forming nucleotides from nucleosides and bases have been designated "salvage" pathways. Salvage reactions represent careful use of pyrimidine and purine ring systems by the cell and different routes are known. Firstly, a one-step formation of monophosphates (Fig. 6) by a uracil phosphoribosyltransferase operating in cell cultures of *Vinca rosea* (KANAMORI et al. 1980). The enzyme is characterized from pea seedlings, and is detected in different plants (BRESSAN et al. 1978). Secondly, a two-step formation of monophosphates by the sequence uridine phosphorylase → uridine kinase (Fig. 6). The phosphoribosyltransferase reaction is well established for purines and pyrimidines in microbial and animal systems (HOCHSTADT 1974).

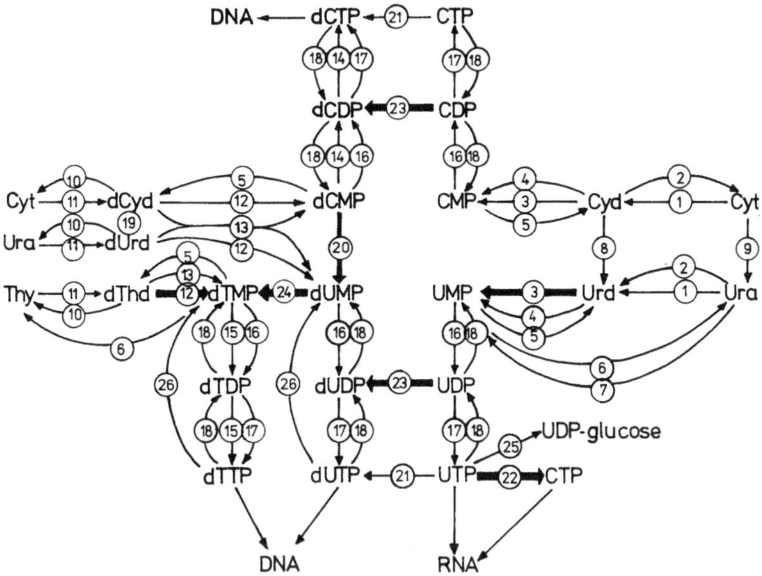

Fig. 6. Pyrimidine salvage reactions, pyrimidine nucleotide interconversions and the respective enzymes. *1* uridine phosphorylase (EC 2.4.2.2); *2* uridine nucleosidase (EC 3.2.2.3); *3* uridine kinase (EC 2.7.1.48); *4* nucleoside phosphotransferase (EC 2.7.1.77); *5* 5′-nucleotidase (EC 3.1.3.5); *6* pyrimidine-5′-nucleotide nucleosidase (EC 3.2.2.10); *7* uracil phosphoribosyltransferase (EC 2.4.2.9); *8* cytidine deaminase (EC 3.5.4.5); *9* cytosine deaminase (EC 3.5.4.1); *10* deoxyuridine nucleosidase (EC 3.2.2.3); *11* deoxyuridine phosphorylase (EC 2.4.2.23); *12* thymidine kinase (EC 2.7.1.75); *13* thymidine AMP-phosphotransferase (EC 2.7.1.77); *14* dCMP kinase (EC 2.7.4.14); *15* dTMP kinase (EC 2.7.4.9); *16* nucleoside monophosphate kinase (EC 2.7.4.4); *17* nucleoside diphosphate kinase (EC 2.7.4.6); *18* nucleoside diphosphatase (EC 3.6.1.6); *19* deoxycytidine deaminase (EC 3.5.4.14); *20* dCMP deaminase (EC 3.5.4.12); *21* nucleoside triphosphate reductase (EC 1.17.4.2); *22* CTP synthetase (EC 6.3.4.2); *23* ribonucleotide diphosphate reductase (EC 1.17.4.1); *24* thymidylate synthetase (EC 2.1.1.6); *25* glucose-1-phosphate uridyltransferase (EC 2.7.7.9); *26* nucleoside triphosphate diphosphatase (EC 3.6.1.19)

Two classes of nucleoside phosphorylases exist (Fig. 6) which operate specifically with ribose-1-phosphate or deoxyribose-1-phosphate without discrimination with respect to the base component. Similarly, nucleoside hydrolases only possess a strict specificity with regard to the sugar moiety (Dewey and Kidder 1973). The uridine hydrolase purified from mung bean (Achar and Vaidyanathan 1967) may function catabolically by its irreversible acting to prevent the accumulation of uridine nucleotides which interfere with the regulation of the de novo synthesis. On the other hand, hydrolytic nucleoside cleavage seems to be of limited function in yeast with high levels of ribose-5-phosphate, since pentose phosphate metabolites act as inhibitors (Magni et al. 1976). Catabolic cleavage of pyrimidine-5′-monophosphates is catalyzed by unspecific 5′-nucleotidases and by pyrimidine-5′-mononucleotide nucleosidases, or by specifically acting dTMP phosphatase. Interconversions may be performed by nucleoside ribosyltransferase and nucleoside deoxyribosyltransferase, while nucleoside phosphotransferase functions anabolically like kinases (see Sect. 5). Phosphorylation by ATP-dependent kinases (Fig. 6) is the main important reaction in pyrimidine salvage similar to microbial and animal systems. Two enzymes are also observed in plants which operate with thymidine or with uridine as well as cytidine.

Thymidine Kinase. Earlier reports on so-called thymidine kinase in grain (WANKA et al. 1964), peanuts (SCHWARZ and FITES 1970) and *Vicia faba* have to be revised, if they do not discriminate this enzyme from nucleoside phosphotransferase (DENG and IVES 1972). Only nonspecific nucleoside phosphotransferase was demonstrated in different plants, as suggested by the lack of end-product control in plant extracts (DENG and IVES 1972). Discrimination between phosphotransferase and thymidine kinase is possible (GRIVELL and JACKSON 1976) using inhibition of nucleoside phosphotransferase from wheat by cyclic AMP (POLYA and ASHTON 1973). The phosphotransferase of *Asplenium* and *Helianthus* which is closely associated with an AMP phosphohydrolase does not seem to serve in salvage functions (GRIVELL and JACKSON 1976). This is suggested by its late occurrence in maturation after the highly unstable thymidine kinase (BRUNORI et al. 1974). In *Acetabularia* a true ATP-dependent thymidine kinase of low K_m value was determined which was dTTP-feedback-sensitive, and was localized outside the chloroplasts and mitochondria (BANNWARTH et al. 1977b). Moreover, cytoplasmic nucleoside phosphotransferase was found, suggesting that earlier reports on thymidine phosphorylation may be caused by detection of this enzyme only (BANNWARTH et al. 1977a). Using rye chloroplasts, GOLASZEWSKI et al. (1975) demonstrated parallel occurrence of thymidine kinase and nucleoside phosphotransferase with a ratio of approximately 4:1. Thymidine kinase, localized mainly in the chloroplast matrix, was found to be concerned primarily with DNA synthesis in chloroplasts of *Chlamydomonas* (CHIANG et al. 1975). Interestingly, in the partially purified enzyme the final product was dTTP, suggesting a concerted action of several enzymes or of a single multifunctional enzyme (SWINTON and CHIANG 1979). In *Acetabularia,* in contrast to *Chlamydomonas,* the thymidine kinase seems to be encoded by the chloroplast DNA, translated by 70S chloroplast ribosomes and integrated into cytoplasmic membranes as well as participating in the nuclear DNA synthesis (BANNWARTH et al. 1977b). In view of the genetically distinct thymidine kinases in mitochondrial and nuclear DNA synthesis (e.g., BOGENHAGEN and CLAYTON 1976), it is possible that there are different thymidine kinases in plant systems.

In *Physarum polycephalum,* the temporal expression of a single thymidine kinase which undergoes post-translational modifications into multiple enzyme variants (GRÖBNER and SACHSENMAIER 1976), and their modulations through enzyme degradation depending upon temperature (WRIGHT and TOLLON 1979) suggest a more complicated relationship of this enzyme with DNA synthesis. Thus, thymidine incorporation as the measure of DNA synthesis should be carefully verified.

Uridine Kinase. This enzyme was partially purified from ungerminated *Zea mays* (DENG and IVES 1975), and catalyzes UMP synthesis in the ordered Bi Bi kinetic

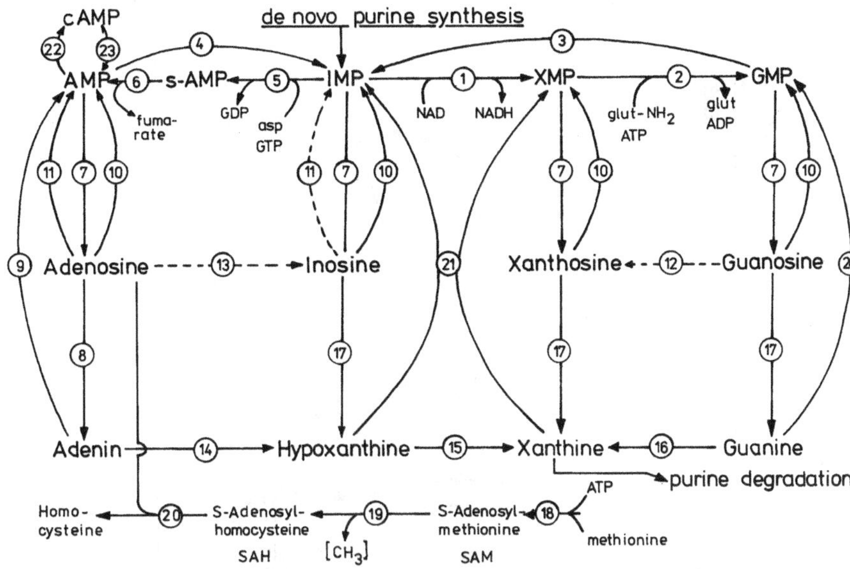

Fig. 7. Purine salvage reactions and the respective enzymes. *1* IMP dehydrogenase (EC 1.2.1.14); *2* GMP synthetase (EC 6.3.4.1); *3* GMP reductase (EC 1.6.6.8); *4* AMP aminohydrolase (EC 3.5.4.6); *5*-adenylosuccinate synthetase (EC 6.3.4.4); *6* adenylosuccinate lyase (EC 4.3.2.2); *7* 5′-nucleotidase (EC 3.1.3.5); *8* adenosine nucleosidase (EC 3.2.2.7); *9* adenine phosphoribosyltransferase (EC 2.4.2.7); *10* nucleoside phosphotransferase (EC 2.7.1.77); *11* adenosine kinase (EC 2.7.1.20); *12* guanosine deaminase (EC 3.5.4.15); *13* adenosine deaminase (EC 3.5.4.4); *14* adenine deaminase (EC 3.5.4.2); *15* xanthine dehydrogenase (EC 1.2.1.37); *16* guanine deaminase (EC 3.5.4.3); *17* purine nucleoside phosphorylase (EC 2.4.2.1); *18* S-adenosylmethionine synthetase; *19* S-adenosylmethionine: acceptor methyltransferase; *20* S-adenosyl homocysteinase (EC 3.3.1.1); *21* hypoxanthine phosphoribosyltransferase (EC 2.4.2.8); *22* adenylate cyclase (EC 4.6.1.1); *23* 3′,5′-cyclic nucleotide phosphodiesterase (EC 3.1.4.17); *24* guanine phosphoribosyltransferase (EC 2.4.2.8)

pattern. CTP and UTP compete with the phosphate-donor site suggesting nonallosteric regulation. Reversal of UTP/CTP inhibition by GTP can balance the ratio of pyrimidine and purine nucleotides (DENG and IVES 1975). Uridine kinase is found in 0- to 1-day-old maize seedlings as followed by concerted action of ATPase and a nonspecific nucleoside phosphotransferase (DENG and IVES 1972). Thus, the salvage function may be carried out preferentially by the former enzyme (see Sect. 4.2).

4.2 Purines

Like pyrimidine salvage, different reactions participate in the purine salvage of higher plants (Fig. 7). Purine nucleotides may be formed by adenosine kinase and nucleoside phosphotransferase, as well as by guanine-hypoxanthine phosphoribosyltransferase studied from *Lupinus luteus* cotyledons (GURANOWSKI and BARANKIEWICZ 1979, GURANOWSKI 1979a, b) and other sources. On the other

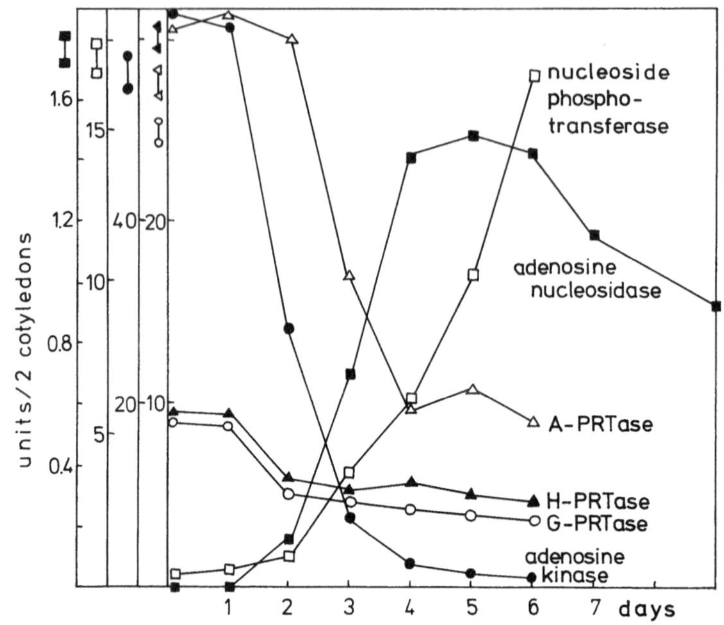

Fig. 8. Changes in activities of purine salvage enzymes in cotyledons of lupin seeds during germination. (Redrawn from GURANOWSKI and BARANKIEWICZ 1979; GURANOWSKI and PAWELKIEWICZ 1978)

hand, purine nucleosides are formed by 5′-nucleotidases, and in the case of adenosine, by SAH hydrolase. The latter is irreversibly hydrolyzed by adenosine nucleosidase. SAH hydrolase and adenosine nucleosidase were purified and characterized from *Lupinus* seeds, spinach, sugar beet, and barley leaves (GURANOWSKI and PAWELKIEWICZ 1977, POULTON and BUTT 1976a, b, GURANOWSKI and SCHNEIDER 1977).

Deamination of adenosine and guanosine is not widespread in higher plants (BARANKIEWICZ and PASZKOWSKI 1980). In contrast to this, deamination and phosphorolysis are the major routes of purine nucleoside cleavage in mammalian and microbial systems.

The activities of all these purine salvage enzymes undergo distinct changes during the germinating of *Lupinus* seeds (Fig. 8). The purine phosphoribosyl transferases were found to have the highest activity in the first two days, while purine nucleoside phosphorylase was absent. Thus, bases were utilized mainly via the one-step salvage reaction (GURANOWSKI and BARANKIEWICZ 1979). The adenine phosphoribosyltransferase as well as the adenosine kinase decreased during germination (Fig. 8). They were the main salvage enzymes in the early stages of RNA synthesis during germination. Inversely, nucleoside phosphotransferase activity was increased during germination, and is thus the most important enzyme of purine salvage during the later period of seedling development (GURANOWSKI and BARANKIEWICZ 1979). A similar relationship between kinases and nucleoside phosphotransferases has been observed for the pyrimidines (see Sect. 4.1.). On the other hand, adenosine nucleosidase is absent from an anabolizing tissue such as cotyledons of maturating seeds, but its activity increased drastically on the 4th day of germination (Fig. 8) (GURANOWSKI and PAWELKIEWICZ 1978). Thus, the nucleosidase acts as a typical lytic enzyme which mobilizes adenine from cotyledons for reutilization in other organs of seedlings (GURANOWSKI and PAWELKIEWICZ 1979). Despite this remarkable nucleosidase activity, adenine accumulation is not observed. This is possibly also caused by adenine phosphoribosyltransferase action (GURANOWSKI and BARANKIEWICZ 1979). In mesophyll protoplasts of tobacco, the purine salvage takes place only by the phosphoribosyltransferase reaction (BARANKIEWICZ and PASZKOWSKI 1980).

In addition, adenosine nucleosidase may function in the SAH hydrolysis in higher plants (Fig. 7). S-Adenosylmethionine (SAM) formed from ATP and methionine acts as a methyl donor in O,S,N, and C-methylation which are catalyzed by specific methyltransferases (POULTON and BUTT 1975). The resulting SAH acts as an inhibitor of the methylation process, and the SAM/SAH ratio could determine the degree of methylation (POULTON and BUTT 1975). Efficient SAH hydrolysis (POULTON and BUTT 1976a) takes place only by removing both products (Fig. 7). Thus adenosine nucleosidase could relieve the induction inhibition of caffeic acid methylation by SAH through facilitation of SAH hydrolysis (POULTON and BUTT 1975, 1976b). Purine salvage is quite important in the first hours of imbibition. As detected by exogenous addition of adenosine or adenine in germinating soybean embryonic axes, purine and pyrimidine nucleotides can be enhanced with preference to ATP (ANDERSON 1977a). Accordingly, the content of purine and pyrimidine bases as well as nucleosides was increased after 12 h in germinating wheat seeds (GRZELCZAK and BUCHOWICZ 1975). On the other hand, decreased synthesis of ATP from adenine and adenosine in deteriotated soybean seeds reflected the reduced anabolic capacity of the seeds (ANDERSON 1977b).

Like the purine metabolism of germinating seeds, adenine is preferentially anabolized by phosphoribosyltransferase in both healthy and *Phytophthora*-infected potato leaves, which is important for the predisposing of uninfected host leaves as well as the rate of the pathogenic action (CLARK et al. 1978). But adenosine is split by an adenosine-specific nucleosidase. Thus, adenine is available for anabolic reactions (CLARK et al. 1972).

To summarize, purine salvage takes place by a partially concerted pattern of the activities of adenosine kinase, adenine phosphoribosyltransferase, nucleoside phosphotransferase, and nucleosidase. Adenine salvage seems to serve mainly in the restoration of the adenylate level in higher plants (ANDERSON 1977b). The cellular location of the purine salvage pathway has not been studied.

5 Interconversions of Nucleotides

As illustrated in Figs. 6 and 7, many interconversions of nucleotides exist. Interestingly, reactions which change the base component of the sugar moiety are limited, but highly important for regulation. The thymidylate synthetase, the dCMP deaminase and the ribonucleotide reductase are key reactions for DNA synthesis. Beside thymidine salvage along the thymidine kinase or thymidine phosphotransferase (see Sect. 4.1.), dTMP is formed by thymidylate synthetase. Caused by insensitivity of assay procedures and the low activity of normal plant tissue, the enzyme has so far been detected only in *Chlamydomonas* cells, and cultured tissues of *Nicotiana, Pinus,* and *Daucus* (VANDIVER and FITES 1979), and has not been purified and characterized from them. Similarly, the dCMP deaminase which is regulated allosterically in bacterial and mammalian systems by dTTP is detected in *Chlorella* (SHEN and SCHMIDT 1966) but unestablished in higher plants. As in eukaryotic systems, CTP is formed by glutamine-dependent CTP synthetase in plants.

Ribonucleotide Reductase. A highly important branch point in nucleotide metabolism is the formation of deoxyribonucleotides. There are two reducing systems. Almost exclusively among bacteria, a monomeric ribonucleotide reductase requiring 5′-deoxyadenosylcobalamine and catalyzing reduction of all common ribonucleotide-5′-triphosphates is present in cyanophytes, *Clostridia,* and further prokaryotic systems except *Escherichia coli* (GLEASON and WOOD 1976). The enzyme of *Anabaena* recently purified (GLEASON and FRICK 1980) is a monomeric small molecule (mol.wt. 72,000) with a strict requirement for divalent cations and with a simple pattern of allosteric control by deoxyribonucleotides. On the other hand, in mammalian systems, *E. coli,* green algae, yeasts, and higher plants, a ribonucleotide reductase containing NADPH-dependent thioredoxin reductase reduces ribonucleotide-5′-diphosphates highly allosterically regulated by dATP. In spite of the detailed knowledge of this enzyme in microbial and animal systems (SINGH et al. 1977, THELANDER and REICHARD 1979, FOLLMANN 1974), plant ribonucleotide reductase is understood but poorly. The enzyme function may be correlated with high levels of DNA synthesis, but demonstration is difficult in higher plants. Ribonucleotide reductase activity was observed only in germinating wheat embryos (MÜLLER et al. 1973, SCHIMPFF et al. 1978), root tips of broad bean, and cultured soybean cells (HOVEMANN and FOLLMANN 1979). In wheat lacking deoxyribonucleotides in dry embryos, occurrence of ribonucleotide reduction *after* onset of DNA synthesis and a dinucleotide polyphosphate-like inhibitor of reduction were observed (SCHIMPFF et al. 1978). On the other hand, ribonucleotide reductase is detectable in germinating broad beans only during a limited period of growth (30–32 h after imbibition of seeds), or in fast-growing cultures of soybean only in the 2nd and 3rd day after inoculation, both preceding maximum of DNA synthesis (HOVEMANN and FOLLMANN 1979). Due to continuous presence of DNA polymerase, ribonucleotide reductase seems to be a larger limiting process than the replicating enzymes for the onset of DNA synthesis (HOVEMANN and FOLLMANN 1979).

In synchronously cultured algae such as *Scenedesmus obliquus* and *Chlorella pyrenoidosa,* ribonucleotide reductase activity coincides with DNA synthesis (FELLER and FOLLMANN 1976), and thioredoxin has been found (WAGNER and FOLLMANN 1977). In synchronous cultures of *S. cerevisiae* ribonucleotide reductase was distinctly periodically correlated with DNA synthesis. Maximum de novo synthesis of the enzyme is present during the onset of DNA synthesis (LOWDON and VITOLS 1973). In each case of coordinate changes of ribonucleotide reductase activity and DNA synthesis, pools of deoxyribonucleotides were extremely low.

More recently, FELLER et al. (1980) have demonstrated a precise coincidence between the maximum of the cycloheximide-sensitive ribonucleotide reductase and concentrations of dATP, dGTP, dCTP, dTTP, and the midpoint of DNA synthesis in synchronous *Scenedesmus obliquus.* This is the first demonstration of the dNDP → dNTP → DNA pathway in plant cells. Moreover, a depletion of the dTTP pool and a cessation of DNA synthesis in this algae, treated with 5-fluorodeoxyuridine (inhibition of thymidylate synthetase by 5-FdUMP) or methotrexate (inhibition of dihydrofolate reductase), is accompanied with overproduction of dATP, a continuous increase of ribonucleotide synthesis and a lack of expansion of the dGTP and dCTP pools. This suggest an allosteric regulation of reductase activity as well as a correlation of its synthesis and the dTTP content.

The properties of ribonucleotide reductase and thioredoxin reductase from green algae and higher plants, respectively, strongly suggest ribonucleotide reduction in plant organismus as in mammalian systems. On the other hand, in *Euglena gracilis,* NADPH-dependent thioredoxin reductase of unusual molecular weight was found (MUNAVALLI et al. 1975), and B_{12}-dependent ribonucleotide reductase was purified which differed from that of other systems (HAMILTON 1974). This implies a phylogenetic difference between *Euglena* and green algae as well as higher plants (GLEASON and WOOD 1976).

For mammalian and microbial systems, allosteric inhibition of ribonucleotide reductase by dATP and dGTP is being extensively studied. This inhibition causes numerous toxic effects of deoxyadenosine as well as deoxyguanosine in vivo (cf. review of HENDERSON et al. 1980). Such an inhibitory activity of deoxyadenosine was also determined for plant systems (e.g. FERNANDEZ-GOMEZ et al. 1970), suggesting similarity in ribonucleotide reductase regulation.

Contrary to limited changes of base- and sugar moiety in nucleotide interconversions, levels of phosphorylations of nucleotides are changed by largely different reactions. Anabolic functions can be performed by unspecific nucleoside monophosphate as well as by nucleoside diphosphate kinases acting with ribonucleotides or with deoxyribonucleotides. For dTMP, (d)CMP, GMP, and AMP as substrates specific kinases were described. Their distribution in plants is poorly understood. In synchronous *Chlorella* cells dTMP kinase activity rises with DNA synthesis (JOHNSON and SCHMIDT 1966).

Catabolic reactions in nucleotide metabolism can be performed by unspecific nucleoside diphosphatase, nucleoside triphosphatase, nucleoside triphosphate pyrophosphatase, nucleotide pyrophosphatase, 5′-nucleotidase, 3′-nucleotidase, nucleoside-5′-monophosphate nucleosidase as well as by the unspecific alkaline and acidic phosphatase (Fig. 6). For dTMP, ATP, dUTP, dCTP as substrates phosphatases are involved, and for IMP, AMP, NAD, NADP as substrates nucleosidases were detected acting specifically.

Their distribution and function in plants is not known satisfactorily.

A nucleoside triphosphate diphosphatase was detected from germinated pea cotyledons, as a chromatin-associated inhibitor of RNA synthesis and a kind of nonhistone protein (HIRASAWA et al. 1979a). The enzyme was heat-stabilized by association with chromatin. It catalyzes phosphohydrolysis of nucleoside triphosphates or diphosphates to monophosphates and P_i (HIRASAWA et al. 1979a, b).

In dry and germinated seeds of *Triticum*, nucleotide pyrophosphatase activity was localized cytochemically (GAHAN et al. 1979). It was found preferentially in those cells in which the contents were hydrolyzed and mobilized, e.g., endosperm (particularly in some cells adjacent to the aleurone layer) and xylem and phloem, indicating association with transport processes. In addition, ATPase is associated with plasma membrane of transfer cells of the phloem, suggesting a function in the transport of photosynthetic products (BENTWOOD and CRONSHAW 1978).

A preferential role of dTMP kinase (SOHAWA and HASE 1968) and dCMP deaminase (SHEN and SCHMIDT 1966) in DNA synthesis was observed in *Chlorella*. Induction of DNA synthesis by slicing of a potato tuber is accompanied by slight increase of nucleoside phosphotransferase and a more rapid increase of dTMP kinase (WATANABE and IMASEKI 1977).

Specific functions possess nucleotides in eukaryotic cilia and flagella. They are used by energy-transducing proteins such as dynein, and the binding of GTP to tubulin as prerequisite for the assembly of microtubules. These processes were catalyzed by different enzymes in *Chlamydomonas:* A heterogenous Mg^{2+} or Ca^{2+}-dependent ATPase, a Ca^{2+}-specific ATPase, two species of adenylate kinases and two species of nucleoside diphosphokinases (WATANABE and FLAVIN 1976).

Nucleotidases represent a large group of enzymes which are also potential regulators of nucleotide metabolism indicated by the existence of a mechanism for the regulation of their activity (FRITZSON 1978). Acting with 3′ or 5′-nucleotides, these enzymes are important determinants in the phosphorylation state of nucleotides and their interconversions, e.g., in the UMP cycle and the purine-nucleotide cycle (Figs. 6 and 7). They were mostly localized on the microsomal and plasma membranes with highly inhibitory properties of nucleoside di- and triphosphates. Retardation of DNA synthesis is related to the 5′-nucleotidase, while 3′-nucleotidase may function in the reutilization of nuclear material by the forming of nucleosides from nucleic acid breakdown products (FRITZSON 1978). This agrees with the action of 3′-nucleotidase as well as of IMP nucleosidase during germination of wheat embryos (PRICE and MURRAY 1969).

6 Free Nucleotides in Relation to Nucleic Acid Synthesis

The function of nucleotides is related to their regulatory effects as well as to the synthesis of nucleic acids. Changes in the pool sizes of nucleotides should be important for nucleic acid synthesis. Starting from such a schematic statement, many investigations have been performed on the relationship between free nucleotides and nucleic acid synthesis (e.g., KEYS 1968, BROWN and SHORT 1969, BROWN 1965), however the relation of nucleotide pools to the nucleic acid synthesis is a complicated one. The function of different nucleotide pools and their cellular location will now be considered with regard to the better investigated microbial and mammalian systems.

1. *Different precursor pools* have been determined for mRNA as well as rRNA of HeLa cells (WIEGERS et al. 1976), suggesting two *different local concentrations* of UTP in the nucleus. On the other hand, a rapid equilibrium between a nuclear and a cytoplasmic nucleotide pool seems to vary in different mammalian systems (KHYM et al. 1978).

2. In DNA synthesis, a *small precursor pool* of *high turnover* is used which is separated from the content of cellular nucleotides (KUEBBING and WERNER 1975). Such a replication pool can be produced by a multienzyme complex as observed for T4-infected *E. coli* resulting in a channeling of nucleotides (REDDY and MATHEWS 1978), and is also observed for mammalian DNA replication (REDDY and MATHEWS 1980). Thus, the question of a limited function for nucleotide pools in nucleic acid synthesis is only focused on a distinct part of the nucleotides available in the whole cell.

3. Such a *limitation* of the *nucleotide availability* can be observed in *E. coli* (PATO 1979), and can be produced in mouse Ehrlich ascites cells by amino acid starvation (GRUMMT and GRUMMT 1976). Thus, the *actual nucleotide triphosphate level* may be control the rate of pre-rRNA synthesis, but the exact experimental evidence for this is restricted to a few systems.

In plants, deoxycytidine-induced bleaching of mesophyll cells of *Lemna* (FRICK 1978) was followed by a significant growth-independent decrease in plastid replication (FRICK 1979). The experiments suggest functional compartmentation of pyrimidine nucleotides. Most studies are from older reports on free nucleotide content, neglecting those aspects of compartmentations mentioned above. Therefore only some recent reports will be discussed here.

In most acid-soluble nucleotide compositions studied, a small or even negligible content of deoxyribonucleotides is determined (NYGAARD 1972, JENNER 1968, BROWN 1965). Dry wheat embryos do not contain deoxyribonucleotides, but these compounds were synthesized before starting ribonucleotide reduction during germination, possibly from preformed deoxyribonucleosides (SCHIMPFF et al. 1978). On the other hand, in synchronous *Scenedesmus obliquus* the intracellular dTTP pool reaches its maximum with 4–5 pmol and the other deoxyribonucleotide pools with 2–3 pmol/10^6 cells being precisely coincident with ribonucleotide reductase activity and DNA synthesis (FELLER et al. 1980). In *Euglena gracilis* the B_{12}-dependent ribonucleoside triphosphate reductase activity (see Sect. 5) was increased up to 20-fold if cells progress into vitamin B_{12} deficiency, suggesting accumulation of its apoenzyme (CARELL and SEEGER 1980). But the dNTP pools were undetectable under B_{12} deficiency, appeared rapidly on replenishment with the vitamin up to about six times of normal exponentially growing cells, and were modulated by cell division (GOETZ and CARELL 1978). In addition to a rate-limiting function of ribonucleotide reduction, a low level of deoxyribonucleotides is highly important for regulation in the onset of DNA synthesis in plants.

Among the ribonucleotides adenine nucleotides and UDP-glucose represent the main fractions in different plant systems. A similar relation exists for different developmental stages (KEYS 1968, BROWN 1965, WEINSTEIN et al. 1969, JENNER 1968, NYGAARD 1973). The ATP pool is predominantly determined by photophosphorylation (BOTTOMLEY and STEWART 1976) and other energy-transducing systems, undergoing transient changes under light, dark, ammonia assimilation and so on (BOTTOMLEY and STEWART 1976, OHMORI and HATTORI 1978, LARSSON et al. 1978). However, the steady-state levels of adenylate pools were the same under different conditions (LARSSON et al. 1978) according to the energy charge concept.

Changes in adenylate pools during the onset of nucleic acid synthesis are preferentially related to their function in energy supply. Thus, the ATP pool increases ten-fold in the first hour of the germination of wheat embryos but is constant in the later phases (OBENDORF and MARCUS 1974).

UDP-glucose, one of the most important glycosyl donors, increases in the starch synthesis of developing wheat grains (JENNER 1968) or tube growth of germinating pine pollen (NYGAARD 1973). Its content is related to the function in sucrose synthesis as suggested by the properties of the sucrose phosphate synthetase (FEKETE 1971), and to germination (BHATIA and UPPAL 1979). The large increase of UDP-glucose during the log phase of growth of sycamore cell cultures suggests its important function in cell wall synthesis (BROWN and SHORT 1969). Accordingly, inhibition of cotton fiber growth induced by an insufficiency of boron is accompanied with lower UDP-glucose synthesis (WAINWRIGHT et al. 1980) as well as a reduced incorporation of UDP-glucose into cell wall material (DUGGER and PALMER 1980).

Metabolic changes during the different developmental stages of plants may doubtless be reflected in the acid-soluble nucleotides. This has been established for many germinating and developing seeds (cf. WEINSTEIN et al. 1969). However, due to the compartmentation mentioned above, as well as the diverse functions of nucleotides, the acid-soluble nucleotide content is of limited informational use when attempting to understand nucleic acid synthesis. Thus, a close functional relationship between changing nucleotide pattern and nucleic acid synthesis must be carefully established using kinetic methods.

7 Degradation of Pyrimidines and Purines

7.1 Pyrimidines

Pyrimidines can be degraded along an oxidative or a reductive pathway as well as by oxidative demethylation and reductive orotic acid degradation. Of these pathways, reductive degradation is realized in a great variety of organisms, as reviewed for microbial systems by VOGELS and VAN DER DRIFT (1976) as well as for microbial, animal, and plant systems by WASTERNACK (1978). As in animal systems, reductive degradation in plants represents the irreversible exhaustion of pyrimidine bases, competing their use in nucleic acid synthesis. In some plants and plant-like organisms pyrimidines act as a N-source (VOGELS and VAN DER DRIFT 1976), and are involved in the synthesis of uncommon amino acids containing pyrimidines (see Sect. 2), or in fact pyrimidine degradation represents one form of N-excretion (REINBOTHE et al. 1981) (Fig. 9).

Reductive degradation of pyrimidines is studied in plants only by use of radioactive tracers (cf. WASTERNACK 1978). In germinating pea seeds uracil and thymine were degraded reductively without correlation to the germination process (ROSS et al. 1971), while uridine and uracil degradation was increased during germination for the first 48 h in *Phaseolus mungo* (ASHIHARA 1977c) which degraded 80% of the portion taken up. The anabolic and catabolic uracil transformations is inverse close correlated in animal systems (WEBER et al. 1971), and is one example of the so-called molecular correlation concept of WEBER (1974). This seems to be invalid for plants.

Characterization of pyrimidine-degrading enzymes was studied with *Euglena gracilis* only. Here, β-ureidopropionase was purified; it showed Michaelis-Menten characteristics but lacked regulatory properties (WASTERNACK et al. 1979). Its function was not correlated with RNA synthesis during light-induced chloroplast development and the enzyme was localized in the cytosolic fraction (WALTHER et al. 1980).

β-Alanine, one end-product of uracil degradation, is a substrate of pantothenic acid synthesis (cf. WASTERNACK 1978). But an alternative pathway from malonic acid semialdehyde by γ-aminobutyrate transaminase observed for *Aspergillus nidulans* (ARST 1978) and suggested for higher plants, must also be considered.

Cytosine deaminase seems to be absent in higher plants. Thus, cytosine reutilization or degradation occur in the sequence cytosine→cytidine→uridine (ROSS and COLE 1968). A cytidine deaminase was partially characterized from extracts of *Lolium perenne* (FRISCH and CHARLES 1966).

The methyl group of thymine is degraded and eliminated by an uncommon but well-studied pathway of oxidative demethylation in *N. crassa* (ABBOTT and UDENFRIED 1974). Catalyzed by thymine-7-hydroxylase, a mixed functional oxygenase, thymine is converted into 5-hydroxymethyluracil, 5-formyl-uracil and

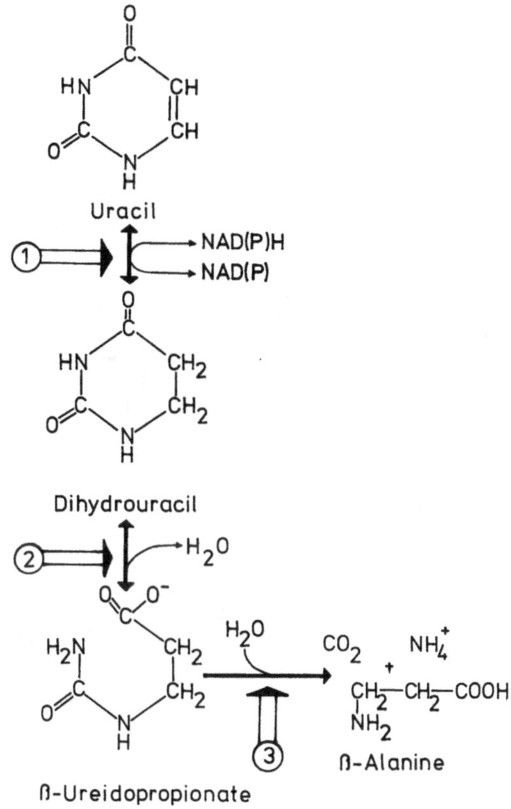

Fig. 9. Pathway of reductive pyrimidine degradation. Enzymes involved: *1* uracil reductase/dihydrouracil dehydrogenase (EC 1.3.1.1/2); *2* dihydropyrimidinase (EC 3.5.2.2); *3* β-ureidopropionase (EC 3.5.1.6)

uracil-5-carboxylic acid decarboxylase. A similar mixed-functional oxygenation takes place with thymidine by the pyrimidine deoxyribonucleoside-2′-hydroxylase. This enzyme forms thymine riboside which is split by a hydrolase. Thus, *N. crassa* can grow on thymidine, deoxyuridine, and thymine, although thymidine kinase is absent (GRIVELL and JACKSON 1968). Both oxygenations are dependent on Fe^{2+}, molecular oxygen, ascorbate, and α-ketoglutarate, resulting in a stoichiometric production of 5-hydroxymethyluracil and thymine riboside, or succinate and CO_2. Transformation of thymine into uracil by oxydative demethylation suggests careful use of thymine as a radioactive tracer in nucleic acid-labeling.

7.2 Purines

Purines are degraded in plants by an oxidative pathway (Fig. 10) which is also widely distributed in animal and microbial organisms. Starting with deamination of adenine or guanine, the purine ring is oxidized to uric acid via the step of hypoxanthine and xanthine by xanthine dehydrogenase which seems to be dependent on NADH as well as on NADPH in plants (NGUYEN and FEIERABEND

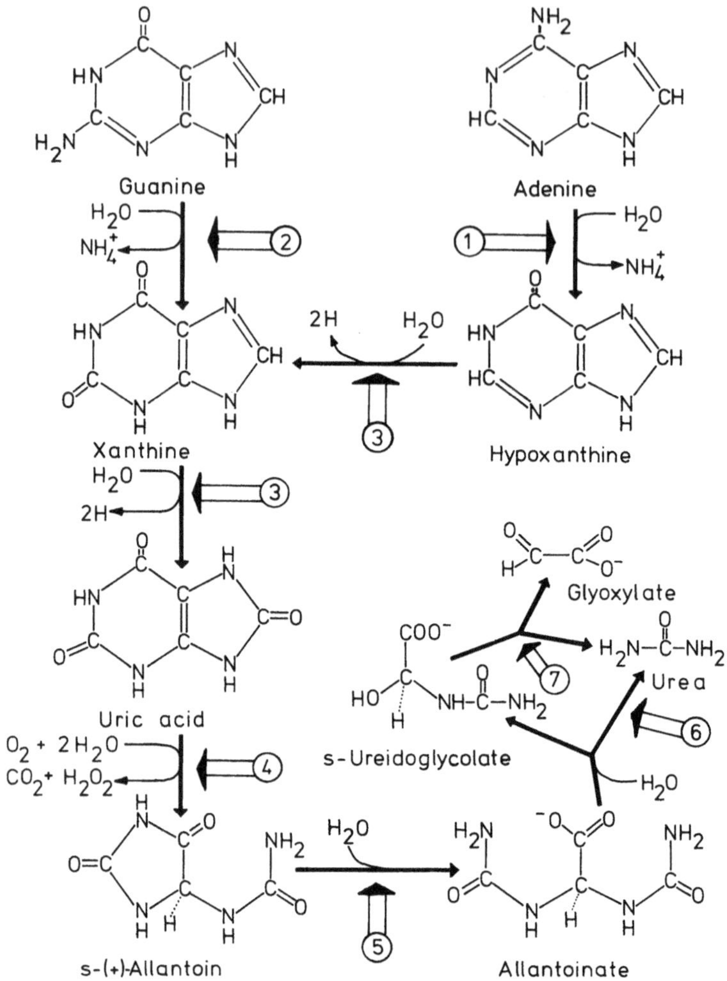

Fig. 10. Pathway of purine degradation. Enzymes involved: *1* Adenine deaminase (EC 3.5.4.2); *2* guanine deaminase (EC 3.5.4.3); *3* xanthine dehydrogenase (EC 1.2.1.37); *4* uricase (EC 1.7.3.3); *5* allantoinase (EC 3.5.2.5); *6* allantoicase (EC 3.5.3.4); *7* ureidoglycolase (EC 4.3.2.3)

1978). In *Aspergillus nidulans* an additional enzyme, purine hydroxylase II was observed which acts with hypoxanthine and nicotinate (SEALY-LEWIS et al. 1978). Uric acid is transformed by O_2-consumption into S(+)-allantoin, CO_2 and H_2O_2 by uricase involving presumably (−)2-oxo-4-hydroxy-4-carbhydroxy-5-ureido-imidazoline as a levorotatory intermediate as well as an oxidative, decarboxylative, and hydrolytic step of a high degree of positional and stereochemical specificity (BONGAERTS and VOGELS 1979). Further hydrolytic cleavage takes place by allantoinase and allantoicase forming ureidoglycolic acid which is cleaved by ureidoglycolase into glyoxylic acid and urea.

In higher plants incomplete oxidative purine degradation is widely distributed, resulting in allantoin or allantoate (cf. REINBOTHE and MOTHES 1962, REINBOTHE et al. 1981). In algae, fungi, and yeasts growth on purines is a widespread property using purine-nitrogen by degradation. This has been excellently established by enzymatic and genetic studies for *Chlorella* (e.g., AMMAN and LYNCH 1964), *N. crassa* (e.g., REINERT and MARZLUF 1975), *A. nidulans* (e.g., SCAZZOCHIO and DARLINGTON 1968), and *S. cereviseae* (COOPER and LAWTHER 1973) and has been reviewed by VOGELS and van der DRIFT (1976). From higher plants only some aspects of purine degradation will be discussed here in more detail.

Allantoin and allantoic acid are the major soluble nitrogen compounds in a wide variety of plants (REINBOTHE and MOTHES 1962). Their amounts vary greatly at different developmental stages and are thus insignificant for taxonomic studies (HOFMANN et al. 1969). Both ureides function as the main form of N-transport, N-accumulation, and N-excretion, as well as N-reutilization in some legumes (HERRIDGE et al. 1978). They were synthesized preferentially in the nodules as shown by [^{15}N]-studies (MATSUMOTO et al. 1977b, HERRIDGE et al. 1978), by grafting experiments (MATSUMOTO et al. 1978) as well as the increasing activities of purine oxidizing enzymes during nodule formation (ATKINS et al. 1980). They were accumulated predominantly in the stem internodes. Clearly dependent on nodule formation (MATSUMOTO et al. 1978), ureide synthesis decreased by exogenous addition of soluble nitrogen compounds (MATSUMOTO et al. 1977c). The rapid decrease of allantoin accumulation after denodulation can be retarded by ammonia, urea, or nitrate MATSUMOTO et al. 1978). Allantoin produced in the bacteroids is transported in roots, stems, and leaves, and is the N-source for vegetative growth (MATSUMOTO et al. 1977a). Allantoin is used effectively for seed formation.

Allantoin accumulation produced by inoculation with *Rhizobium* can be replaced by a supply of exogenous nitrogen (MATSUMOTO et al. 1978). Accordingly, uricase activity is increased in roots of leaf cuttings by N-supplementation (THEIMER and HEIDINGER 1974) and is present in roots of nodulating and non-nodulating varieties of *Glycine max* (TAJIMA and YAMAMOTO 1977). Uricase from nodules differs from that of a non-nodulating variety (TAJIMA and YAMAMOTO 1975). In non-nodulating *Glycine max* regulation of uricase activities takes place through the continuous presence of the apoprotein and a varying amount of a heat-stable low-molecular cofactors (TAJIMA and YAMAMOTO 1977). Bacteroids were proposed as the site of ureide synthesis (TAJIMA and YAMAMOTO 1975), but due to the lack in xanthine-oxidizing enzymes in bacteroid fractions more recently NAD-dependent xanthine dehydrogenase, uricase and allantoinase were localized in the cytosol of the bacteroid-containing cells (ATKINS et al. 1980, TRIPLETT et al. 1980). The NAD-dependent xanthine dehydrogenase was determined as playing a critical and energy-conserving role in the ureide synthesis and there were similar rates of xanthine oxidation and N_2 fixation by bacteroids (TRIPLETT et al. 1980). This and further recent evidence suggest ureide formation of N_2-fixing nodules via de novo purine synthesis (ATKINS et al. 1980, TRIPLETT et al. 1980).

In glyoxysomes of *Ricinus* endosperm (THEIMER and BEEVERS 1971) or peroxisomes from maize root tips (PARISH 1972) the uricase was located thus almost operating near to the H_2O_2-consuming catalase. On the other hand, NADH-dependent xanthine dehydrogenase is located in the supernatant (NGUYEN and FEIERABEND 1978). Contrary to the findings of BROWN (1965) the enzyme is clearly detected in higher plants, but there is a little knowledge of this enzyme. On the basis of selected mutants formed in callus culture of *Nicotiana tabacum* it appears likely that xanthine dehydrogenase and nitrate reductase use a common Mo-cofactor (MENDEL and MÜLLER 1976).

The capacity to degrade purines seems to be related to the photosynthetic phosphorylation (NGUYEN 1979, HARTMANN and GEISSLER 1973). The xanthine dehydrogenase activity is decreased if *Pharbitis nil* is placed under increasing light conditions, and is increased either when DCMU is given or when CO_2 is removed (NGUYEN et al. 1980).

Intracellular compartmentation seems to function in purine metabolism. Vacuolar sequestration is also observed with the purine compound S-adenosylmethionine (SVIHLA et al. 1963). It is actively transported into vacuoles (SCHWENCKE and ROBICHON-SZULMAJSTER 1976). On the other hand, in isolated vacuoles only guanosine and inosine were taken up by a saturable transport system (K_m 0.63 and 0.15 mM), and adenosine transport is strongly dependent on the growth phase of the cell culture (NAGY 1979), while SAH is the only purine compound isolated from the whole cells grown in the absence of SAH and S-adenosylmethionine. Intracellular purine storage, depending on its use in catabolic and anabolic reactions, was observed for *Chlorella fusca* and was influenced only by ammonia supply (PETTERSEN 1975). To summarize, purine degradation and the utilization of ureides are expressed with different capacities in various higher plants. They are related to nitrogen metabolism.

8 Concluding Remarks

Purine and pyrimidine metabolism in plants follow the known pathways generally accepted for microbial and animal systems. A large body of results on higher plants has been discussed in overall terms, but more recently attention has been focused on regulatory and developmental aspects. In this respect, the functional relationship between precursor metabolism and nucleic acid synthesis can be identified as one of the most important questions. Due to the rate of nucleic acid synthesis during different developmental stages, methodological difficulties restrict experimental evidence from higher plants compared with microbial and animal systems.

We need more knowledge about the relationship between enzymes and metabolites in purine and pyrimidine metabolism and their function in nucleic acid metabolism. Thus, regulatory aspects and compartmentation are only partially understood as yet.

Note Added in Proof

After preparation of this manuscript some reviews were published concerning problems mentioned: (1) Occurrence and function of cyclic AMP in fungi (PALL 1981), (2) cyclic AMP in higher plants (BROWN and NEWTON 1981), (3) cellular nucleotide measurements for microbial ecology (KARL 1980), (4) biosynthesis of nucleotides in plants (ROSS 1981), (5) purine degradation related to N-metabolism in fungi (MARZLUF 1981), (6) purine degradation and ureide metabolism in higher plants (THOMAS and SCHRADER 1981a), (7) nucleotide compartmentation and its function in nucleic acid synthesis (WASTERNACK and BENNDORF 1983). The problem of allantoin degradation and compartmentation in *S. cerevisiae* exhausted from this section was reviewed by Cooper (1980). Important new results were published in the following fields:

Pyrimidine Synthesis. The appearance of only one CPS catalyzing carbamoyl phosphate synthesis in both arginine and pyrimidine biosynthesis is recently shown by reciprocal regulation of both pathways. Using phaseolotoxin, a specific inhibitor of the ornithine transcarbamylase, and 5-fluorouracil-sensitive and -resistent cell lines of carrot cultures a reciprocal use of carbamoyl phosphate in both pathways was shown (JACQUES and SUNG 1981). With respect to the plastidal location of CPS in *Pisum sativum* (TAYLOR and STEWART 1981) CPS seems to be different in properties *and* location between higher plants and fungi (cf. Sect. 3.1.2.). As discussed in Sect. 3.1.2. CPS is a regulatory enzyme in pyrimidine synthesis which is true also for non-dividing storage tissues like pea cotyledons (KOLLÖFFEL and VERKERK 1981). Regulation of pyrimidine synthesis in the ATC step by UMP (cf. Sect. 3.1.2.) or UTP (GUERN and HERVE 1980) was shown for UMP in synchronously dividing cells of *Helianthus tuberosus* (PARKER and JACKSON 1981) and tomato cell suspension cultures (KOCH et al. unpublished). But, question on UMP compartmentation arises by the parallel increase in the activity of ATC and the total amount of UMP during the cell cycle (PARKER and JACKSON 1981). As shown for *Euglena gracilis* (WASTERNACK et al. 1982) and tomato cell suspension cultures (WASTERNACK unpublished), a UMP pool was separable far from the cytosolic OMP phosphoribosyltransferase/OMP decarboxylase and ATC by sucrose density gradient centrifugation. This suggests a separate location of the ATC and its feedback-inhibitor UMP in plant cells argueing against our general ignorance of compartmental aspects of enzymes, metabolites and effectors in discussing regulation.

Nucleotides. In carrot cell suspension cultures there is a triphosphate-cleaving acid phosphatase which is inhibited by ATP and is released into the medium (CIARROCHI et al. 1981). The chromatin-bound nucleoside triphosphatase of suspended callus cells of *Petroselinum crispum* modulates the nuclear ribonucleotide triphosphate pools thus regulating RNA polymerase I (GROSSMANN et al. 1981) (cf. Chapter 5, pp. 151 ff.). A Mg^{2+}-dependent ATP-sensitive uridine diphosphatase purified from soybean (HUBER and PHARR 1981) may be function in removing UDP which inhibits sucrose-phosphate synthetase in the cytoplasm of mesophyll cells.

Contents of the adenylates were widely used to describe the energy status of a cellular system by the energy charge (EC). The AMP deaminase seems to be the main enzyme in stabilization of the EC as observed for yeasts (YOSHINO and MURAKAMI 1981) and spinach leaves (YOSHINO and MURAKAMI 1980), whereas the cytosolic 5′-nucleotidase functions predominantly in removing excess of IMP (ITOH 1981). Independent of the use of the EC for a distinct enzymatic reaction, there are some different arguments from higher plants against a general use of the EC concept of ATKINSON (1977): (1) Stabilization of the EC between 0.8 and 0.3 depending on O_2-limitation was observed for maize root tips (SAGLIO et al. 1980), lettuce seeds (RAYMOND and PRADET 1980, HOURMANT and PRADET 1981) and rice seedlings (MOCQUOT et al. 1981). This indicates that the EC is only an indicator of the energy status, if ATP is formed glycolytically by anoxia (RAYMOND and PRADET 1980). (2) With respect to the different amount of P_i in *Anacystis nidulans* exposed to dark, light and N_2 (BORNEFELD and WEISS 1981) and its sequestration in vacuolar polyphosphates (e.g. *Neurospora crassa*, CRAMER et al. 1980), the phosphorylation potential is favoured as a more convenient parameter. (3) The EC values determined by total extraction of cellular adenylates seems to be invalid, if we regard the recently observed transient change of adenylates from mitochondria into chloroplasts after the onset of light-induced chloroplast development (HAMPP et al. 1982). Furthermore, the near-equilibrium kinetics of cellular metabolism being the basis of Atkinson's concept (ATKINSON 1977) is often not fulfilled for ATP reactions (GIERSCH et al. 1980).

Purine Salvage. A 5′-methylthioadenosine nucleosidase was purified to homogenity from *Lupinus luteus* (GURANOWSKI et al. 1981). It is probably the only enzyme releasing adenine in the first hours of germinating to form the methylthio-group of methionine (GURANOWSKI et al. 1981). In *E. gracilis* 5-methylthioadenosine is cleaved phosphorolytically being the only source of free adenine (GURANOWSKI and WASTERNACK 1982). Adenine and adenosine were preferentially used in anabolic reactions independent of heterotrophic or autotrophic growth. In *E. gracilis* the pattern of adenine and adenosine metabolizing enzymes differs from that of animal and plant cells by the existence of adenine deaminase and the apparent lack of adenosine-cleaving activities (GURANOWSKI and WASTERNACK 1982).

Purine Degradation. Degradation of purines was extensively studied in relation to ureide formation during N_2-fixation of nodulated legumes (cf. Sect. 7.2.). The complete sequence of oxidative purine breakdown is present in a cell-free system of *Vigna unguiculata* (WOO et al. 1980), and its capacity is related to the rate of N_2-fixation (ATKINS 1981). The uricase seems to be the main control element (WOO et al. 1981, RAINBIRD and ATKINS 1981). Its O_2-demand in comparison to the O_2-sensitivity of the nitrogenase arizes the question of enzyme compartmentation. The following scheme seems to be highly probable (SCHUBERT 1981, HANKS et al. 1981): NH_4^+ formed by nitrogenase of bacteroids is assimilated through the proplastidal glutamine synthetase and glutamate synthase (HANKS et al. 1981) in amino acids which are used for the cytoplasmic-located de novo purine synthesis. Starting from hypoxanthine purine breakdown takes place by the cytosolic xanthine dehydrogenase. Uric acid is oxidized by the peroxisomal uricase to allantoin and H_2O_2 which is degraded by the catalase. Allantoin can be degraded to allantoic acid by allantoinase located in the endoplasmic reticulum or can be directly transported into the xylem sap. This sequence of assimilation and storage of fixed nitrogen through the purine synthesis and breakdown represents a apparent detour, but in terms of energy cost and of the high N/C ratio of allantoin and allantoic acid it is more efficient for the plant cell than asparagine and glutamine (SCHUBERT 1981). Allantoin and allantoic acid are the major forms of N-transport from nodules to shoots in soybeans, and the allantoinase activity increases during developing fruits (THOMAS and SCHRADER 1981b) (for further details cf. review of THOMAS and SCHRADER 1981a).

References

Abbott MT, Udenfried S (1974) α-Ketoglutarate-complexed dioxygenases. In: Hayaishi O (ed) Molecular mechanism of oxygen activation. Academic Press, London, New York, pp 167–214

Achar BS, Vaidyanathan CS (1967) Purification and properties of uridine hydrolase from mung-bean (*Phaseolus vulgaris*) seedlings. Arch Biochem Biophys 119:356–362

Achar BS, Savithri HS, Vaidyanathan CS, Appaji Rao N (1974) Studies on plant aspartate transcarbamylase. Purification and properties of the enzyme from mung-bean (*Phaseolus aureus*) seedlings. Eur J Biochem 47:15–22

Aitken DM, Lue PF, Kaplan JG (1975) Kinetic and reaction mechanism of the carbamoyl phosphate synthetase of a multienzyme aggregate from yeast. Can J Biochem 53:721–730

Amman ECB, Lynch VH (1964) Purine metabolism by unicellular algae. II. Adenine, hypoxanthine, xanthine degradation by *Chlorella pyrenoidosa*. Biochim Biophys Acta 87:370–379

Amrhein N (1977) The current status of cyclic AMP in higher plants. Annu Rev Plant Physiol 28:123–132

Anderson JD (1977a) Responses of adenine nucleotides in germinating soybean embryonic axes to exogenous applied adenine and adenosine. Plant Physiol 60:689–692

Anderson JD (1977b) Adenylate metabolism of embryonic axes from deteriorated soybean seeds. Plant Physiol 59:610–614

Anderson JD (1979) Purine nucleotide metabolism of germinating soybean embryonic axes. Plant Physiol 63:100–104

Arst HN Jr (1978) GABA transaminase provides as alternative route of β-alanine synthesis in *Aspergillus nidulans*. Mol Gen Genet 163:23–27

Ashihara H (1977a) Regulation of the activity of spinach phosphoribosylpyrophosphate synthetase by "energy charge" and end products. Z Pflanzenphys 85:383–392

Ashihara H (1977b) Characterization of phosphoribosylpyrophosphate synthetase from spinach leaves. Z Pflanzenphys 83:379–392

Ashihara H (1977c) Changes in activities of the de novo and salvage pathway of pyrimidine nucleotide biosynthesis during germination of black gram (*Phaseolus mungo*) seeds. Z Pflanzenphys 81:199–211

Ashihara H (1978) Orotate phosphoribosyltransferase and orotidine-5′-monophosphate decarboxylase of black gram (*Phaseolus mungo*) seedlings. Z Pflanzenphys 87:225–241

Ashihara H, Komamine A (1974) Regulatory properties of a plant phosphoribosylpyrophosphate synthetase. Plant Sci Lett 2:119–123

Ashton AR, Polya GM (1978) Cyclic adenosine-3′,5′-monophosphate in axenic reye grass endosperm cell cultures. Plant Physiol 61:718–722

Ashworth TS, Brown EG, Roberts FM (1972) Biosynthesis of willardiine and isowillardiine in germinating pea seeds and seedlings. Biochem J 129:897–905

Atkins CA (1981) Metabolism of purine nucleotides to form ureides in nitrogen-fixating nodules of cowpea (*Vigna unguiculata* L. Walp.). FEBS Lett 125:89–93

Atkins CA, Rainbird R, Pate JS (1980) Evidence for a purine pathway of ureide synthesis in N_2-fixing nodules of cowpea [*Vigna unguiculata* (L.) Walp]. Z Pflanzenphys 97:249–260

Atkinson E (1977) Cellular energy metabolism and its regulation. Academic Press, London, New York

Bach ML, Lacroute F, Botstein D (1979) Evidence for transcriptional regulation of orotidine-5′-phosphate decarboxylase in yeast by hybridization of mRNA to the yeast structural gene cloned in *Escherichia coli*. Proc Natl Acad Sci USA 76:386–390

Bannwarth H, Ikehara N, Schweiger HG (1977a) Thymidine phosphorylating enzymes in *Acetabularia*: Evidence for the occurence of a thymidine kinase. Proc R Soc London Ser B 198:155–176

Bannwarth H, Ikehara N, Schweiger HG (1977b) Nucleo-cytoplasmic interactions in the regulation of thymidine phosphorylation in *Acetabularia*. Proc R Soc London Ser B 198:177–190

Barankiewicz J, Paszkowski J (1980) Purine metabolism in mesophyll protoplasts of tobacco (*Nicotiana tabacum*) leaves. Biochem J 186:343–350

Baumann TW, Wanner H (1972) Untersuchungen über den Transport von Koffein in der Kaffeepflanze (*Coffea arabica*). Planta 108:11–20

Baumann TW, Dupont-Looser E, Wanner H (1978) 7-Methylxanthosine – an intermediate of caffeine biosynthesis. Phytochemistry 17:2075–2076

Becker MA, Raivio KO, Seegmüller JE (1979) Synthesis of PRPP in mammalian cells. Adv Enzymol 49:281–306

Bell EA (1961) Isolation of a new amino acid from *Lathyrus tingitanus*. Biochim Biophys Acta 47:602–603

Bentwood BJ, Cronshaw J (1978) Cytochemical localization of adenosine triphosphatase in the phloem of *Pisum sativum* and its relation to the function of transfer cells. Planta 140:111–120

Bernhardt SA, Davis RH (1972) Carbomoyl phosphate compartmentation in *Neurospora*: Histochemical localization of aspartate and ornithine transcarbamylases. Proc Natl Acad Sci USA 69:1868–1872

Bhatia IS, Uppal D (1979) Free and sugar nucleotides in seedlings of *Sorghum vulgare*. Plant Sci Lett 16:59–66

Bogenhagen D, Clayton DA (1976) Thymidylate nucleotide supply for mitochondrial DNA synthesis in mouse L-cells. Effect of 5-FUdR and methotrexate in thymidine kinase plus and thymidine kinase minus cells. J Biol Chem 251:2938–2944

Bollig J, Mayer K, Mayer WE, Engelmann W (1978) Effects of cAMP, theophylline, imidazole and 4-(3,4)-dimethoxybenzyl-2-imidazolidone on the leaf movement rhythm of *Trifolium repens* – a test of the cAMP hypothesis of circadian rhythms. Planta 141:225–230

Bongaerts GPA, Vogels GD (1979) Mechanism of uricase action. Biochim Biophys Acta 567:295–308

Bornefeld T, Weiss U (1981) Adenylate energy charge and phosphorylation potential in the blue-green bacterium *Anacystis nidulans*. Biochem Physiol Pflanz 176:71–82

Bottomley PJ, Stewart WDP (1976) ATP pools and transients in the blue-green alga *Anabaena cylindrica*. Arch Microbiol 108:249–258

Bressan RA, Murray MG, Gale JM, Ross CW (1978) Properties of pea seedling UPRTase and its distribution in other plants. Plant Physiol 61:442–446

Brown EG (1965) Changes in the free nucleotide and nucleoside pattern of pea seeds in relation to germination. Biochem J 95:509–514

Brown EG, Al-Baldawi NF (1977) Biosynthesis of the pyrimidinyl amino acid lathyrine by *Lathyrus tingitanus L.* Biochem J 164:589–594
Brown EG, Newton RP (1981) Cyclic AMP and higher plants. Phytochemistry 20:2453–2463
Brown EG, Short KC (1969) The changing nucleotide pattern of sycamore cells during culture in suspension. Phytochemistry 8:1365–1372
Brown EG, Al-Najafi T, Newton RP (1979) Adenosine-3′,5′-cyclic monophosphate, adenylate cyclase and a cyclic AMP binding-protein in *Phaseolus vulgaris.* Phytochemistry 19:23–30
Brown EG, Edwards MJ, Newton RP, Smith CJ (1980) The cyclic nucleotide phosphodiesterases of spinach chloroplasts and microsomes. Phytochemistry 19:23–30
Brunori A, Hotta Y, Stern H (1974) Instability in thymidine kinase in plant extracts. Plant Cell Physiol 15:421–427
Buchowicz J, Lesniewska A (1970) Conversion of orotate to uridine by an acetone powder from wheat seedlings. Phytochemistry 9:1083–1085
Buchowicz J, Reifer I (1961) The conversion of orotic acid to pyrimidine derivatives in plant material. Acta Biochem Pol 8:25–34
Buchowicz J, Reifer I, Geric I (1963) ^{14}C-Carbamoyl-β-alanine as precursor of pyrimidines in higher plants. Acta Biochem Pol 10:157–162
Carell EF, Seeger JW Jr (1980) Ribonucleotide reductase activity in vitamin B_{12}-deficient *Euglena gracilis.* Biochem J 188:573–576
Caroline DF (1969) Pyrimidine synthesis in *Neurospora crassa:* Gene-enzyme relationship. J Bacteriol 100:1371–1377
Chiang KS, Eves E, Swinton DC (1975) Variation of thymidine incorporation patterns in the alternating vegetative and sexual life cycles of *Chlamydomonas reinhardtii.* Dev Biol 42:53–63
Ciarrocchi G, Cella R, Nielsen E (1981) Release of nucleotide-cleaving acid phosphatase from carrot cells grown in suspension culture. Physiol Plant 53:375–377
Clark MC, Page OT, Fisher MG (1972) Purification and properties of N-ribosyl adenine ribohydrolase from potato leaves. Phytochemistry 11:3413–3419
Clark MG, Melanson DL, Page OT (1978) Purine metabolism and differential inhibition of spore germination in *Phytophthora infestans.* Can J Microbiol 24:1032–1038
Cooper TG (1980) Selective gene expression and intracellular compartmentation: two means of regulating nitrogen metabolism in yeast. TIBS 5:332–334
Cooper TG, Lawther RP (1973) Induction of allantoin degradative enzymes in *Saccharomyces cerevisiae* by the last intermediate of the pathway. Proc Natl Acad Sci USA 70:2340–2344
Cramer CL, Vaughn LE, Davis RH (1980) Basic amino acids and inorganic polyphosphates in *Neurospora crassa.* Independent regulation of vacuolar pools. J Bacteriol 142:945–952
Cummings FW (1975) A biochemical model for the circadian clock. J Theor Biol 55:455–470
Currier TC, Wolk CP (1978) Regulation of uridylic acid biosynthesis in the cyanobacterium *Anabaena variabilis.* J Bacteriol 136:682–687
Davis RH (1972) Metabolite distribution in cells. Science 178:835–840
Davis RH, Ristow JL, Hanson BA (1980) Carbamoyl phosphate synthetase A of *Neurospora crassa.* J Bacteriol 141:144–155
Deng Q, Ives DH (1972) Modes of nucleoside phosphorylation in plants: Studies on the apparent thymidine kinase and the true uridine kinase of seedlings. Biochim Biophys Acta 277:235–244
Deng Q, Ives DH (1975) Non-allosteric regulation of the uridine kinase from seeds of *Zea mays.* Biochim Biophys Acta 377:84–94
Denis-Duphil M, Kaplan JG (1976) Fine structure of the ura-2 locus in *Saccharomyces cerevisiae,* II. Meiotic and mitotic mapping studies. Mol Gen Genet 145:259–271
Dewey VC, Kidder GW (1973) Partial purification and properties of a nucleoside hydrolase from *Crithidia.* Arch Biochem Biophys 157:380–387

Dugger WM, Palmer RL (1980) Effect of boron on the incorporation of glucose from UDP-glucose into cotton fibers grown in vitro. Plant Physiol 65:266–273

Dunn JH, Jerris HH, Wilkins JH, Meredith MJ, Smith KT, Flora JB, Schmidt RR (1977) Coordinate and non-coordinate accumulation of aspartate transcarbamylase and dihydroorotase in synchronous *Chlorella* cells growing on different nitrogen sources. Biochim Biophys Acta 485:301–313

Exinger F, Lacroute F (1979) Genetic evidence for the creation of a reinitiation site by mutation inside the yeast ura-2 gene. Mol Gen Genet 173:109–113

Fekete MAR (1971) The regulative properties of UDP glucose: D-fructose-6-phosphate 2-glucosyltransferase (sucrose phosphate synthetase) from *Vicia faba* cotyledons. Eur J Biochem 19:73–80

Feller W, Follmann H (1976) Ribonucleotide reductase activity in green algae. Biochem Biophys Res Commun 70:752–758

Feller W, Schimpff-Weiland G, Follmann H (1980) Deoxyribonucleotide biosynthesis in synchronous algae cells. Eur J Biochem 110:85–92

Fernandez-Gomez ME, Stockert JC, Gonzalez-Fernandez A, Lopez-Saez JF (1970) Delays in prophase induced by adenosine-2-deoxyriboside and their relation with DNA synthesis. Chromosoma 29:1–11

Follmann H (1974) Die enzymatische Ribonucleotid-Reduktion: Biosyntheseweg der Desoxyribonucleotide. Angew Chem 86:624–634

Francko DA, Wetzel RG (1980) Cyclic adenosine-3′:5′-monophosphate: production and extracellular release from green and blue-green algae. Physiol Plant 49:65–67

Frick H (1978) Pyrimidine metabolism in *Lemna minor*. I. Functional compartmentation of chloroplast pyrimidine metabolism in a higher plant. Plant Physiol 61:989–992

Frick H (1979) Pyrimidine metabolism in *Lemna minor*. II. Specific inhibition of plastid replication in a higher plant by cytidine deoxyriboside. Plant Physiol 63:979–983

Frisch DM, Charles MA (1966) Deamination of 4-amino-pyrimidine nucleosides by extracts of rye grass (*Lolium perenne*). Plant Physiol 41:475–478

Fritzson (1978) Regulation of nucleotidase activities in animal tissues. Adv Enzyme Regul 16:43–61

Gahan PB, Sierakowska H, Dawson AL (1979) Nucleotide pyrophosphatase activity in dry and germinated seeds of *Triticum* and its relationship to general acid phosphatase activity. Planta 145:159–166

Giersch C, Heber U, Koayashi Y, Inoue Y, Shibata K, Heldt HW (1980) Energy charge, phosphorylation potential and proton motive force in chloroplasts. Biochim Biophys Acta 590:59–73

Gleason FK, Frick TD (1980) Adenosylcobalamine-dependent ribonucleotide reductase from the blue-green alga *Anabaena* sp. J Biol Chem 255:7728–7733

Gleason FK, Wood JM (1976) Ribonucleotide reductase in blue-green algae: Dependence on adenosylcobalamine. Science 192:1343–1344

Goetz GH, Carell EF (1978) Recovery from vitamin B_{12}-induced unbalanced growth. The shortened cell cycle and the deoxyribonucleoside triphosphate pools. Biochem J 170:631–636

Golaszewski T, Rytel M, Rogozinski J, Szarkowski JW (1975) Thymidine kinase activity in rye chloroplasts. FEBS Lett 58:370–373

Gots JS (1970) Regulation of purine and pyrimidine metabolism. In: Greenberg DM (ed) Metabolic pathways Vol V, 3nd edn. Academic Press, London, New York, pp 225–255

Grivell AR, Jackson JF (1968) Thymidine kinase: Evidence for its absence from *Neurospora crassa* and some other microorganisms, and the relevance of this to the specific labelling of deoxyribonucleic acids. J Gen Microbiol 54:307–317

Grivell AR, Jackson JF (1976) Thymidine phosphotransferase and nucleotide phosphohydrolase of the fern *Asplenium nidus*. General properties and inhibition by adenosine 3′,5′-cyclic monophosphate. Biochem J 155:571–581

Gröbner P, Sachsenmaier W (1976) Thymidine kinase enzyme variants in *Physarum polycephalum*; Change of pattern during the synchronous mitotic cycle. FEBS Lett 71:181–184

Grossmann K, Haschke H-P, Seitz HU (1981) Regulation of RNA synthesis in higher plant cells by the action of a nucleoside triphosphatase. Planta 152:457–460

Grummt I, Grummt F (1976) Control of nucleolar RNA synthesis by the intracellular pool sizes of ATP and GTP. Cell 7:447–453

Grzelczak Z, Buchowicz J (1975) Purine and pyrimidine bases and nucleosides of germinating *Triticum aestivum* seeds. Phytochemistry 14:329–331

Guern M, Herve G (1980) Polyploidy and aspartate transcarbamylase activity in *Hippocrepis comosa* L. Planta 149:27–33

Guranowski A (1979a) Plant adenosine kinase: purification and some properties of the enzyme from *Lupinus luteus* seeds. Arch Biochem Biophys 196:220–226

Guranowski A (1979b) Nucleoside phosphotransferase from yellow lupin seedling cotyledons. Biochim Biophys Acta 569:13–22

Guranowski A, Barankiewicz J (1979) Purine salvage in cotyledons of germinating lupin seeds. FEBS Lett 104:95–98

Guranowski A, Pawelkiewicz J (1977) Adenosylhomocysteinase from yellow lupin seeds; purification and properties. Eur J Biochem 80:517–523

Guranowski A, Pawelkiewicz J (1978) Adenosylhomocysteinase and adenosine nucleosidase activities in *Lupinus luteus* cotyledons during seed formation and germination. Planta 139:245–247

Guranowski A, Schneider Z (1977) Purification and characterization of adenosine nucleosidase from barley leaves. Biochim Biophys Acta 482:145–158

Guranowski A, Wasternack C (1982) Adenine and adenosine metabolizing enzymes in cell-free extracts of *Euglena gracilis*. Comp Biochem Physiol 71B:483–488

Guranowski AB, Chiang PK, Cantoni GL (1981) 5′-Methylthioadenosine nucleosidase. Purification and characterization of the enzyme from *Lupinus luteus* seeds. Eur J Biochem 114:293–299

Hamilton FD (1974) Ribonucleotide reductase from *Euglena gracilis*. A 5′-deoxyadenosylcobalamine-dependent enzyme. J Biol Chem 249:4428–4434

Hampp R, Goller M, Ziegler H (1982) Adenylate levels, energy charge and phosphorylation potential during dark-light and light-dark transition in chloroplasts, mitochondria and cytosol of mesophyll protoplasts from *Avena sativa* L. Plant Physiol 69:448–455

Hanks JF, Tolbert NE, Schubert KR (1981) Localization of enzymes of ureide biosynthesis in peroxisomes and microsomes of nodules. Plant Physiol 68:65–69

Hartfiel G, Amrhein N (1976) The action of methylxanthines on motility and growth of *Chlamydomonas reinhardtii* and other flagellated algae. Is cyclic AMP involved? Biochem Physiol Pflanz 169:531–556

Hartman SG (1970) Purines and pyrimidines. In: Greenberg DM (ed) Metabolic pathways Vol III, 3nd edn. Academic Press, London, New York, pp 1–68

Hartmann E, Geissler G (1973) Über den Ureidstoffwechsel beim Laubmoosprotonema von *Funaria hygrometrica L.* (Sibth.). I. Der Einfluß von verschiedenfarbigem Licht. Biochem Physiol Pflanz 164:614–622

Hatch MD (1966) Adenylosuccinate synthetase and adenylosuccinate lyase from plant tissues. Biochem J 98:198–203

Henderson JF (1972) Regulation of purine biosynthesis. ACS monograph 170, Am Chem Soc, Washington DC

Henderson JF, Paterson ARP (1973) Nucleotide metabolism. Academic Press, London, New York

Henderson JF, Scott FW, Lowe JK (1980) Toxicity of naturally occuring purine deoxyribonucleosides. Pharm Ther 8:573–604

Herridge DF, Atkins CA, Pate JS, Rainbird RM (1978) Allantoin and allantoic acid in the nitrogen economy of the Cowpea (*Vigna unguiculata* L. Walp.). Plant Physiol 62:495–498

Hintermann R, Parish RW (1979) Determination of adenylate cyclase activity in a variety of organisms: evidence against occurrence of the enzyme in higher plants. Planta 146:459–461

Hirasawa E, Takahashi E, Matsumoto H (1979a) Association of nucleotide phosphohydrolyzing activity with chromatin in germinated pea cotyledons. Plant Cell Physiol 20:219–224

Hirasawa E, Matsumoto H, Takahashi E (1979b) Nucleoside triphosphate diphosphatase from germinated pea cotyledon chromatin. Plant Cell Physiol 20:1041–1046

Hochstadt (1974) The role of the membrane in the utilization of nucleic acid precursors. CRC Crit Rev Biochem 2:259–310

Hofmann E, Schlee D, Reinbothe H (1969) Zum Vorkommen und zur Verbreitung von Allantoin in Boraginaceae. Flora (Jena) Abt A 159:510–518

Hood EE, Armour S, Ownby JD, Handa AK, Bressan RA (1979) Effect of nitrogen starvation on the level of adenosine 3′, 5′-monophosphate in *Anabaena variabilis*. Biochim Biophys Acta 588:193–200

Hourmant A, Pradet A (1981) Oxidative phosphorylation in germinating lettuce seeds (*Lactuca sativa*) during the first hours of imbibition. Plant Physiol 68:631–635

Hovemann B, Follmann H (1979) Deoxyribonucleotide synthesis and DNA polymerase activity in plant cells (*Vicia faba* and *Glycine max*). Biochim Biophys Acta 561:42–52

Huber SC, Pharr DM (1981) Occurrence and regulatory properties of uridine diphosphatase in fully expanded leaves of soybean (*Glycine max* Merr.) and other species. Plant Physiol 68:1294–1298

Itoh R (1981) Regulation of cytosol 5′-nucleotidase by adenylate energy charge. Biochim Biophys Acta 659:31–37

Jacques S, Sung ZR (1981) Regulation of pyrimidine and arginine biosynthesis investigated by the use of phaseolotoxin and 5-fluorouracil. Plant Physiol 67:287–291

Janniaux JC, Urrestarazu LA, Wiame JM (1978) Arginine metabolism in *Saccharomyces cerevisiae*. Subcellular localization of the enzymes. J Bacteriol 133:1096–1107

Jenner CF (1968) The composition of soluble nucleotides in the developing wheat grain. Plant Physiol 43:41–49

Johnson RA, Schmidt RR (1966) Enzymatic control of nucleic acid synthesis during synchronous growth of *Chlorella pyrenoidosa*. I. Deoxythymidine monophosphate kinase. Biochim Biophys Acta 129:140–144

Kanamori I, Ashihara H, Komamine A (1979) Changes in the activities of the pentose phosphate pathway and pyrimidine nucleotide biosynthesis during growth of *Vinca rosea* cells in suspension culture. Z Pflanzenphys 93:437–448

Kanamori I, Ashihara H, Komamine A (1980) Subcellular distribution and activity of enzymes involved in uridine-5-monophosphate synthesis in *Vinca rosea* cells. Z Pflanzenphys 96:7–16

Kapoor M, Waygood ER (1965) Orotidine-5-phosphate pyrophosphorylase of wheat embryos. Can J Biochem 43:143–151

Karl DM (1980) Cellular nucleotide measurements and applications in microbial ecology. Microbiol Rev 44:739–796

Keys AJ (1968) The intracellular distribution of free nucleotides in the tobacco leaf. Biochem J 108:1–8

Khym JX, Jones MH, Lee WH, Regan JD, Volkin E (1978) On the question of compartmentation of the nucleotide pool. J Biol Chem 253:8741–8746

King J, Wang D, Waygood ER (1965) Biosynthesis of nucleotides in wheat. II. Pyrimidines from ^{14}C labelled compounds. Can J Biochem 43:237–244

Kollöffel C, Verkerk BC (1982) Carbamoyl phosphate synthetase activity from cotyledons of developing and germinating pea seeds. Plant Physiol 69:143–145

Korch CT, Lacroute F, Exinger F (1974) A regulatory interaction between pyrimidine and purine biosynthesis via ureidosuccinic acid. Mol Gen Genet 133:63–75

Krauss GJ, Reinbothe H (1973) Die freien Aminosäuren in Samen von Mimosaceae. Phytochemistry 12:125–142

Krauss G, Reinbothe H (1977) Purification and fractionation of free nucleotides from *Euglena gracilis* by a combined procedure of ligand-exchange and anion-exchange gel chromatography. Anal Biochem 78:1–8

Kuebbing D, Werner R (1975) A model for compartmentation of de novo and salvage thymidine nucleotide pools in mammalian cells. Proc Natl Acad Sci USA 72:3333–3336

Lacroute F (1968) Regulation of pyrimidine biosynthesis in *Saccharomyces cerevisiae*. J Bacteriol 95:824–832
Lacroute F, Pierard A, Grenson M, Wiame JM (1965) The biosynthesis of carbamoyl phosphate in *Saccharomyces cerevisiae*. J Gen Microbiol 40:127–142
Lambein F, Kuo YH, van Parijs R (1976) Isoxalin-5-ones. Chemistry and biology of a new class of plant products. Heterocycles 4:567–593
Larsson CM, Tillberg JE, Hallman G (1978) Light-induced phosphate binding in relation to photophosphorylation and levels of ATP, ADP, and AMP in green algae *Scenedesmus*. Physiol Plant 44:115–121
Lovatt CJ, Albert LS, Tremblay GC (1979) Regulation of pyrimidine biosynthesis in intact cells of *Cucurbita pepo*. Plant Physiol 64:562–569
Lowdon M, Vitols E (1973) Ribonucleotide reductase activity during the cell cycle of *Saccharomyces cerevisiae*. Arch Biochem Biophys 158:177–184
Lue PF, Kaplan JG (1969) The aspartate transcarbamylase and carbamoyl phosphate synthetase of yeast: a multifunctional enzyme complex. Biochem Biophys Res Commun 34:426–433
Lue PF, Kaplan JG (1970) Metabolic compartmentation at the molecular level: the function of a multienzyme aggregate in the pyrimidine pathway of yeast. Biochim Biophys Acta 220:365–372
Lue PF, Kaplan JG (1971) Aggregation states of a regulatory enzyme complex catalyzing the early steps of pyrimidine biosynthesis in baker's yeast. Can J Biochem 49:403–411
Magni G, Natalini BP, Ruggieri S, Vita A (1976) Baker's yeast uridine nucleosidase is a regulatory copper containing protein. Biochem Biophys Res Commun 69:724–730
Mahler HR, Lin CC (1978) Exogenous adenosine-3′, 5′-monophosphate can release yeast from catabolite repression. Biochem Biophys Res Commung 83:1039–1047
Makoff AJ, Radford A (1976a) The location of the feedback-specific region within the pyrimidine-3 locus of *Neurospora crassa*. Mol Gen Genet 146:247–251
Makoff AJ, Radford A (1976b) Glutamine utilization in both the arginine-specific and pyrimidine-specific carbamoyl phosphate synthetase enzymes of *Neurospora crassa*. Mol Gen Genet 149:175–178
Makoff AJ, Radford A (1977) Dissociation of an enzyme complex from *Neurospora crassa* – evidence for a protease. Biochim Biophys Acta 485:314–329
Makoff AJ, Radford A (1978) Genetics and biochemistry of carbamoylphosphate biosynthesis and its utilization in the pyrimidine biosynthesis pathway. Bacteriol Rev 42:307–328
Makoff AJ, Buxton FP, Radford A (1978) A possible model for the structure of the *Neurospora* carbamoyl phosphate synthetase-aspartate transcarbamylase complex enzyme. Mol Gen Genet 161:297–304
Marzluf GA (1981) Regulation of nitrogen metabolism and gene expression in fungi. Microbiol Rev 45:437–461
Matsumoto M, Yatazawa M, Yamamoto Y (1977a) Distribution and change in the contents of allantoin and allantoic acid in developing nodulating and non-nodulating soybean plants. Plant Cell Physiol 18:353–359
Matsumoto M, Yatazawa M, Yamamoto Y (1977b) Incubation of ^{15}N into allantoin in nodulated soybean plants supplied with $^{15}N_2$. Plant Cell Physiol 18:459–462
Matsumoto M, Yatazawa M, Yamamoto Y (1977c) Effects of exogenous nitrogen-compounds on the concentration of allantoin and various constituents in several organs of soybean plants. Plant Cell Physiol 18:613–624
Matsumoto T, Yatazawa M, Yamamoto Y (1978) Allantoin metabolism in soybean plants as influenced by grafts, a delayed inoculation with *Rhizobium* and a late supply of nitrogen compounds. Plant Cell Physiol 19:1161–1168
Mazus B, Buchowicz J (1968) Purification and properties of dihydroorotase from pea plants. Acta Biochem Pol 15:317–325
Meister A, Powers SG (1978) Glutamine-dependent carbamoyl phosphate synthetase: catalysis and regulation. Adv Enzyme Regul 16:289–315
Mendel RR, Müller AJ (1976) A common genetic determinant of xanthine dehydrogenase and nitrate reductase in *Nicotiana tabacum*. Biochem Physiol Pflanz 170:538–541

Miller RW (1971) Dihydroorotate-quinone reductase of *Neurospora crassa* mitochondria. Arch Biochem Biophys 146:256–270
Miller RW (1975) A high molecular weight dihydroorotate dehydrogenase of *Neurospora crassa*. Purification and properties of the enzymes. Biochem J 53:1288–1300
Mocquot B, Prat C, Mouches C, Pradet A (1981) Effect of anoxia of energy charge and protein synthesis in rice embryo. Plant Physiol 68:636–640
Müller H, Wahl R, Kuntz I, Follmann H (1973) Enzymatic ribonucleotide reduction in wheat. Hoppe-Seyler's Z Physiol Chem 354:1299–1303
Munavalli S, Parker DV, Hamilton FD (1975) Identification of NADPH-thioredoxin reductase system in *Euglena gracilis*. Proc Natl Acad Sci USA 72:4233–4237
Murakoshi I, Ikegami F, Ookawa N, Ariki T, Haginiwa J, Kuo YH, Lambein F (1978) Biosynthesis of uracilylalanines willardiine and isowillardiine in higher plants. Phytochemistry 17:1571–1576
Murray AW (1971) The biological significance of purine salvage. Annu Rev Biochem 40:811–826
Murray AW, Elliott DC, Atkinson MR (1970) The nucleotide biosynthesis from preformed purines in mammalian cells: Regulatory mechanisms and biological significance. Prog Nucleic Acid Res Mol Biol 10:87–119
Nagy M (1979) Studies on purine transport and on purine content in vacuoles isolated from *Saccharomyces cerevisiae*. Biochim Biphys Acta 558:221–232
Nguyen J (1979) Effect of light on deamination and oxidation of adenylic compounds in cotyledons in *Pharbitis nil*. Physiol Plant 46:255–259
Nguyen J, Feierabend J (1978) Some properties and subcellular localization of xanthine dehydrogenase in pea leaves. Plant Sci Lett 13:125–132
Nguyen J, Cornic G, Imhoff C, Louason G (1980) Effect of carbon dioxide, oxygen, and dichlorophenyldimethylurea on the light-sensitive activity of xanthine dehydrogenase in cotyledons of *Pharbitis nil*. Can J Bot 58:1160–1164
Nygaard P (1972) Deoxyribonucleotide pools in plant tissue culture. Physiol Plant 26:29–33
Nygaard P (1973) Nucleotide metabolism during pine pollen germination. Physiol Plant 28:361–371
Obendorf RL, Marcus A (1974) Rapid increase of ATP during early wheat embryo germination. Plant Physiol 53:779–781
O'Donnovan GA, Neuhard J (1970) Pyrimidine metabolism in microorganisms. Bacteriol Rev 34:278–343
Ohmori M, Hattori A (1978) Transient change in the ATP pool of *Anabaena cylindrica* associated with ammonia assimilation. Arch Microbiol 117:17–20
O'Neal TD, Naylor AW (1968) Purine and pyrimidine nucleotide inhibition of carbamoyl phosphate synthetase from pea seedlings. Biochem Biophys Res Commun 31:322–327
O'Neal TD, Naylor AW (1969) Partial purification of carbamoyl phosphate synthetase of Alaska pea (*Pisum sativum L. cv.* Alaska). Biochem J 113:271–279
O'Neal TD, Naylor AW (1976) Some properties of pea leaf carbamoyl phosphate synthetase. Plant Physiol 57:23–28
Ong BL, Jackson JF (1972a) Pyrimidine nucleotide biosynthesis in *Phaseolus aureus*. Enzymatic aspects of the control of carbamoyl phosphate synthesis and utilization. Biochem J 129:583–593
Ong BL, Jackson JF (1972b) Aspartate transcarbamylase from *Phaseolus aureus*. Partial purification and properties. Biochem J 129:571–581
Pall ML (1977) Cyclic AMP and the plasma membrane potential in *Neurospora crassa*. J Biol Chem 252:7146–7150
Pall ML (1981) Adenosine 3′,5′-phosphate in fungi. Microbiol Rev 45:462–480
Parish RW (1972) Urate oxidase in peroxisomes from maize root tips. Planta 104:247–251
Parker NF, Jackson JF (1981) Control of pyrimidine biosynthesis in synchronously dividing cells of *Helianthus tuberosus*. Plant Physiol 67:363–366
Pato ML (1979) Alterations of deoxyribonucleotide triphosphate pools in *Escherichia coli*: Effects in DNA replication and evidence for compartmentation. J Bacteriol 140:518–524

Petterson R (1975) Control of ammonium of intercompartmental guanine transport in *Chlorella*. Z Pflanzenphys 76:213–223

Polya GM, Ashton AR (1973) Inhibition of wheat seedling 5′ (3′)-ribonucleotide phosphohydrolase by adenosine-3′, 5′-cyclic monophosphate. Plant Sci Lett 1:349–357

Poulton JE, Butt VS (1975) Purification and properties of S-adenosyl-L-methionine: caffeic acid O-methyltransferase from leaves of spinach beet (*Beta vulgaris L.*). Biochim Biophys Acta 403:301–314

Poulton JE, Butt VS (1976a) Purification and properties of S-adenosyl-L-homocysteine hydrolase from leaves of spinach beet. Arch Biochem Biophys 172:135–142

Poulton JE, Butt VS (1976b) Partial purification and properties of adenosine nucleosidase from leaves of spinach beet (*Beta vulgaris L.*). Planta 131:179–185

Price CE, Murray AW (1969) Purine metabolism in germinating wheat embryos. Biochem J 115:129–133

Rainbird RM, Atkins CA (1981) Purification and some properties of urate oxidase from nitrogen-fixing nodules of cowpea. Biochim Biophys Acta 659:132–140

Reddy GPV, Mathews CK (1978) Functional compartmentation of DNA precursors in T 4-phage infected bacteria. J Biol Chem 253:3461–3467

Reddy GPV, Matthews CK (1980) Multienzyme complex for metabolic channeling in mammalian DNA replication. Proc Natl Acad Sci USA 77:3312–3316

Reinbothe H, Mothes K (1962) Urea, ureides and guanidines in plants. Annu Rev Plant Physiol 16:271–282

Reinbothe H, Miersch J, Mothes K (1981) Special problems of nitrogen metabolism in plants. In: Neuberger A, van Deenen LLM (eds) Comprehensive biochemistry Vol 19A, Elsevier/North-Holland Biomedical Press, Amsterdam New York, pp 51–163

Raymond P, Pradet A (1980) Stabilization of adenine nucleotide ratios at various values by an oxygen limitation of respiration in germinating lettuce (*Lactuca sativa*) seeds. Biochem J 190:39–44

Reinert WR, Marzluf GA (1975) Genetic and metabolic control of purine catabolic enzymes of *Neurospora crassa*. Mol Gen Genet 139:39–55

Ross CW (1981) Biosynthesis of nucleotides. In: Marcus A (ed) Biochemistry of Plants, Vol 6, Academic Press, London, New York, pp 169–205

Ross C, Cole C (1968) Metabolism of cytidine and uridine in bean leaves. Plant Physiol 43:1227–1231

Ross C, Coddington RL, Murray MG, Bledson CS (1971) Pyrimidine metabolism in cotyledons of germinating Alaska peas. Plant Physiol 47:71–75

Rybicka H, Buchowicz J, Reifer I (1967) Carbamoyl-^{14}C-citrulline in pyrimidine synthesis in wheat plant seedlings. Acta Biochem Pol 14:249–254

Saglio PH, Raymond P, Pradet A (1980) Metabolic activity and energy charge of excised maize root tips under anoxia. Control of soluble sugars. Plant Physiol 66:1053–1057

Satyanarayana T, Kaplan JG (1971) Regulation of the purine pathway in baker's yeast: Activation and feedback inhibition of phosphoribosyl pyrophosphate amidotransferase. Arch Biochem Biophys 142:40–47

Savithri HS, Vaidyanathan CS, Appaji Rao N (1978) Studies on plant aspartate transcarbamylase: Regulatory properties of the enzyme from mung bean (*Phaseolus aureus*) seedlings. Proc Indian Acad Sci 87B:67–79

Scazzochio C, Darlington AJ (1968) The induction and repression of the enzymes of purine breakdown in *Aspergillus nidulans*. Biochim Biophys Acta 166:557–568

Schimpff G, Müller H, Follmann H (1978) Age-dependent DNA labeling and deoxyribonucleotide synthesis in wheat seeds. Biochim Biophys Acta 520:70–81

Schubert KR (1981) Enzymes of purine biosynthesis and catabolism in *Glycine max*. I. Comparison of activities with N_2-fixation and composition of xylem exudate during nodule development. Plant Physiol 68:1115–1122

Schwarz OJ, Fites RC (1970) The appearence of pyrimidine nucleoside and deoxynucleoside kinase activities in peanut axes. Phytochemistry 9:1899–1905

Schwencke J, de Robichon-Szulmajster H (1976) The transport of S-adenosyl-L-methionine in isolated yeast vacuoles and spheroplasts. Eur J Biochem 65:49–60

Sealy-Lewis HM, Scazzochio C, Lee S (1978) A mutation defective in the xanthine alternative pathway of *Aspergillus nidulans*. Mol Gen Genet 164:303–308

Shen SRC, Schmidt RR (1966) Enzymatic control of nucleic acid synthesis during synchronous growth of *Chlorella pyrenoidosa*. I. Deoxycytidine monophosphate deaminase. Arch Biochem Biophys 115:13–20

Singh D, Tamao Y, Blakeley RL (1977) Allosterism, regulation and cooperativity: The case of ribonucleotide reductase of *Lactobacillus leichmannii*. Adv Enzyme Regul 15:81–100

Sohawa Y, Hase E (1968) Suppressive effect of light on the formation of DNA and the increase of deoxythymidine monophosphate kinase activity in *Chlorella protothecoides*. Plant Cell Physiol 9:461–466

Suzuki T (1972) The participation of S-adenosylmethionine in the biosynthesis of caffeine in the tea plants. FEBS Lett 24:18–20

Suzuki T, Takahashi E (1976a) Further investigation of the biosynthesis of caffeine in tea plants (*Camellia sinensis L.*). Methylation of tRNA by tea leaf extracts. Biochem J 160:181–184

Suzuki T, Takahashi E (1976b) Caffeine biosynthesis in *Camellia sinensis* L. Phytochemistry 15:1235–1239

Suzuki T, Takahashi E (1976c) Metabolism of methionine and biosynthesis of caffeine in the tea plant (*Camellia sinensis* L.). Biochem J 160:171–179

Suzuki T, Takahashi E (1977) Biosynthesis of purine nucleotides and methylated purines in higher plants. Drug Metab Rev 6:213–242

Svihla G, Dainko L, Schlenk F (1963) Ultraviolet microscopy of purine compounds in the yeast vacuole. J Bacteriol 85:399–409

Swinton DC, Chiang KS (1979) Characterization of thymidine kinase and phosphorylation of deoxyribonucleosides in *Chlamydomonas reinhardtii*. Mol Gen Genet 176:399–409

Tajima S, Yamamoto Y (1975) Enzymes of purine catabolism in soybean plants. Plant Cell Physiol 16:271–282

Tajima S, Yamamoto Y (1977) Regulation of uricase activity in developing roots of *Glycine max*, non-nodulating variety A 62-2. Plant Cell Physiol 18:247–253

Taylor A, Stewart GR (1981) Tissue and subcellular localization of enzymes of arginine metabolism in *Pisum sativum*. Biochem Biophys Res Commun 101:1281–1289

Theimer RR, Beevers H (1971) Uricase and allantoinase in glyoxysomes. Plant Physiol 47:246–251

Theimer RR, Heidinger P (1974) Control of particulate urate oxidase activity in bean roots by external nitrogen supply. Z Pflanzenphys 73:360–370

Thelander L, Reichard P (1979) Reduction of ribonucleotides. Annu Rev Biochem 48:133–158

Thomas RJ, Schrader LE (1981a) Ureide metabolism in higher plants. Phytochemistry 20:361–371

Thomas RJ, Schrader LE (1981b) The assimilation of ureides in shoot tissues of soybeans. 1. Changes in allantoinase activity and ureide contents of leaves and fruits. Plant Physiol 67:973–976

Trevillyan JM, Pall ML (1979) Control of cyclic adenosine-3′, 5′-monophosphate levels by depolarizing agents in fungi. J Bacteriol 138:397–403

Triplett EW, Blevins DG, Randall DG (1980) Allantoic acid synthesis in soybean root nodule cytosol via xanthine dehydrogenase. Plant Physiol 65:1203–1206

Urrestarazu LA, Vissers S, Wiame LM (1977) Changes in location of ornithine carbamoyltransferase and carbamoyl phosphate synthetase among yeasts in relation to the arginine/ornithine transferase regulatory complex and the energy status of the cell. Eur J Biochem 79:473–481

Vandiver VV, Fites RC (1979) Thymidylate synthetase activity from *Chlamydomonas* cells and cultured tissues of *Nicotiana, Pinus* and *Daucus*. Plant Physiol 64:668–670

Victor J, Greenberg LB, Sloan DL (1979a) Studies of the kinetic mechanism of orotate phosphoribosyltransferase from yeast. J Biol Chem 254:2647–2655

Victor J, Leo-Mensah A, Sloan DL (1979b) Divalent metal ion activation of the yeast orotate phosphoribosyltransferase catalyzed reaction. Biochemistry 18:3597–3604

Vogels GD, Drift van der C (1976) Degradation of purines and pyrimidines in microorganisms. Bacteriol Rev 40:403–468

Wagner W, Follmann H (1977) A thioredoxin from green algae. Biochem Biophys Res Commun 77:1044–1051

Wainwright IM, Palmer RL, Dugger WM (1980) Pyrimidine pathway in boron-deficient cotton fiber. Plant Physiol 65:893–896

Walther R, Wasternack C, Helbing D, Lippmann G (1980) Pyrimidine metabolizing enzymes in *Euglena gracilis* – Synthesis and localization of OPRTase, ODCase and β-ureidopropionase. Biochem Physiol Pflanz 175:764–771

Walther R, Krauss GJ, Reinbothe H (1981) Purification and properties of orotidine-5-phosphate phosphoribosyltransferase and orotidine-5-phosphate decarboxylase of *Euglena gracilis*. Biochem Physiol Pflanz 176:116–128

Wang D (1967) Biosynthesis of nucleotides in wheat. IV. Metabolism of specifically labelled ^{14}C-orotic acid. Can J Biochem 45:721–728

Wanka F, Vasil IK, Stern H (1964) Thymidine kinase: The dissociability and its bearing on the enzyme activity in plant materials. Biochim Biophys Acta 85:50–59

Wasternack C (1978) Degradation of pyrimidines – enzymes, localization and role in metabolism. Biochem Physiol Pflanz 173:467–499

Wasternack C (1980) Degradation of pyrimidines and pyrimidine analogs – pathways and mutual influences. Pharmac Ther 8:629–651

Wasternack C, Benndorf R (1983) Nucleotide compartmentation and its consequence for nucleic acid synthesis. Biol Zentralblatt **102** No 1 (in press)

Wasternack C, Lippmann G, Reinbothe H (1979) Pyrimidine degrading enzymes. Purification and properties of β-ureidopropionase of *Euglena gracilis*. Biochim Biophys Acta 570:341–351

Wasternack C, Walther R, Glund K, Reinbothe H (1982) Pyrimidine metabolizing enzymes in plants – Selected problems from *Euglena gracilis* and cell suspension cultures of *Lycopersicon esculentum* cv. Lukullus. In: Molecular and Cellular Regulation of Enzyme Activity, Proc. of a Symp. held in Halle, August 10–16 (1981) G. Fischer Verlag (Jena) (in press)

Watanabe T, Flavin M (1976) Nucleotide-metabolizing enzymes in *Chlamydomonas* flagella. J Biol Chem 251:182–192

Watanabe A, Imaseki H (1977) Studies on enzymes involved in DNA synthesis and thymine nucleotide formation in potato tube slices. Plant Cell Physiol 18:859–868

Weber G (1974) Molecular Correlation Concept. In: Busch H (ed) Molecular biology of cancer. Academic Press, London, New York, pp 437–521

Weber G, Queener SF, Ferdinandus JA (1971) Control of gene expression in carbohydrate, pyrimidine and DNA synthesis. Adv Enzyme Regul 9:63–95

Weinstein LH, McCune DC, Mancini JF, van Lenken P (1969) Acid soluble nucleotides in pinto bean leaves at different stages of development. Plant Physiol 44:1499–1510

Whright M, Tollon Y (1979) *Physarum* thymidine kinase. A step or a peak enzyme depending upon temperature of growth. Eur J Biochem 96:177–181

Wiegers U, Kramer G, Klapproth K, Hilz H (1976) Separate pyrimidine-nucleotide pools for messenger RNA and ribosomal RNA synthesis in HeLa S3 cells. Eur J Biochem 64:535–540

Williams LG, Davis RH (1970) Pyrimidine-specific carbamoyl phosphate synthetase in *Neurospora crassa*. J Bacteriol 103:335–341

Williams LG, Bernhardt SA, Davis RH (1971) Evidence for two discrete carbamoyl phosphate pools in *Neurospora crassa*. J Biol Chem 246:973–978

Wolcott JH, Ross C (1967) Orotidine-5′-phosphate decarboxylase and pyrophosphorylase of bean leaves. Plant Physiol 42:275–279

Woo KC, Atkins CA, Pate JS (1980) Biosynthesis of ureides from purines in cell-free system from nodule extracts of cowpea (*Vinga unguiculata* (L.) Walp.). Plant Physiol 66:735–739

Woo KC, Atkins CA, Pate JS (1981) Ureide synthesis in a cell-free system from cowpea (*Vigna unguiculata* (L.) Walp.) nodules. Plant Physiol 67:1156–1160

Wyngaarden JB (1976) Regulation of purine biosynthesis and turnover. Adv Enzyme Regul 14:25–42

Yon RJ (1972) Wheat-germ aspartate transcarbamylase. Kinetic behaviour suggesting an allosteric mechanism of regulation. Biochem J 128:311–320

Yon RJ (1973) Wheat-germ aspartate transcarbamylase. The effect of ligands on the inactivation of the enzyme by trypsin and denaturating agents. Biochem J 131:699–706

Yoshino M, Murakami K (1980) AMP deaminase from spinach leaves. Purification and some regulatory properties. Z Pflanzenphys 99:331–338

Yoshino M, Murakami K (1981) In situ studies on AMP deaminase as a control system of the adenylate energy charge in yeasts. Biochim Biophys Acta 672:16–20

Yoshino M, Murakami K, Tshushima K (1979) AMP deaminase from baker's yeast. Purification and some regulatory properties. Biochim Biophys Acta 570:157–166

Yunghans WN, Morree DJ (1977) Adenylate cyclase not found in soybean hypocotyl and onion meristem. Plant Physiol 60:144–149

10 Structure of Plant Viral Genomes

L. Hirth

1 Introduction

Until two years ago, the subject matter of an article such as this, would have been restricted to plant viral RNA's. Indeed nothing or very little was known about the organization of plant viral DNA genomes. With the recent determination of the complete sequence of cauliflower mosaic virus (CaMV) DNA, the situation was altered, and since these data represent the first determination of the structure of genes related to higher plants, a number of important questions can be raised for the first time. This justifies the division of this section into two distinct parts. In the first, I shall discuss the recent data concerning the organization and the expression of the genome of cauliflower mosaic virus. The second part will be devoted to the organization of the genome of plant RNA viruses and to their various modes of expression. In this latter case, several excellent recent reviews (Zaitlin 1979, Lane 1979, Davies 1979) have already been published and I shall emphasize new data related to the general concept of the in vitro and in vivo expression of eukaryotic genes.

2 DNA Viruses

2.1 Organization of the Genome of Cauliflower Mosaic Virus (CaMV)

CaMV was the first to be identified as a DNA plant virus (Shepherd et al. 1968). Since then about seven other viruses have been reported to be similar to CaMV (see Matthews: *Classification and Nomenclature of Viruses,* 1979). However, very little is known about them and I shall restrict my review to CaMV.

2.1.1 General Properties of the Virus

These have been described in detail by Hull in a recent review (1979). Various purification procedures are also described; but whatever the procedure used the virus yield is very low, compared to that obtained in the case of many RNA plant viruses. It is not clear whether the low amount of purified virus obtained from an infected plant reflects the low concentration of virus in the

A list of abbreviations of plant viruses can be found on p. XVIII

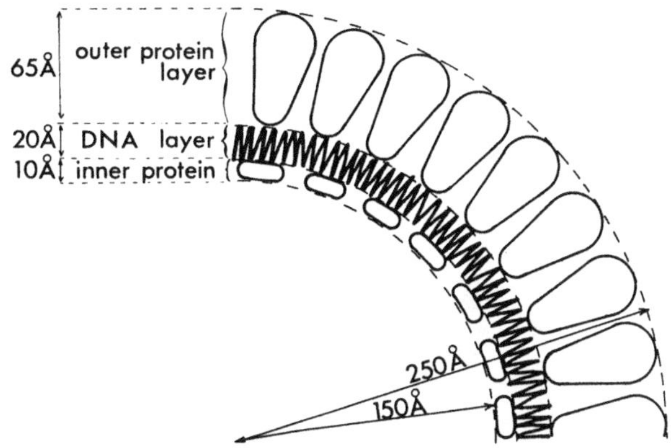

Fig. 1. Structure of CaMV obtained by neutron diffraction methods

host plant or whether it is due to inefficient release of virus from specific stable inclusion bodies. It seems very difficult to extract efficiently the totality of the virions from the infected plant cells.

Recent experiments performed mainly in my laboratory have established the following characteristics of the CaMV.

2.1.2 Structure of the CaMV Capsid

In contrast with earlier reports (TEZUKA and TANIGUCHI 1972a, BRUNT et al. 1975), it is now established that the CaMV capsid consists of a single protein subunit with a molecular weight of about 42,000 (AL ANI et al. 1979b). The arrangement of the subunits is not exactly known, but seems to be icosahedral and of the type T=7. The existence of a central core arising from the association between viral RNA and some protein, as claimed by others (TEZUKA and TANIGUCHI 1972b, HULL and SHEPHERD 1976), has been recently ruled out by neutron diffraction studies (CHAUVIN et al. 1979) and Fig. 1 shows how the DNA interacts with the protein subunits. The inner protein layer shown in the scheme is of undetermined origin. This inner shell could explain the presence of a small amount of two specific proteins discovered in addition to the major 42,000 and found in all virus preparations.

2.1.3 DNA Interruptions

The DNA of the virus, which is circular and possesses three interruptions (VOLOVITCH et al. 1978), is folded in such a manner that the majority of its nucleotides interact with the inner part of the polypeptide chain of each protein subunit. Thus the negative charges of the phosphate residues are neutralized by the positive charges of the basic amino acids of the protein subunits.

2.1.4 Interaction DNA – Coat Protein

At pH 11.25, the virus swells and part of the DNA protrudes out of the capsid (Fig. 2); at a pH higher than 12, the virions collapse, but by back-dialysis

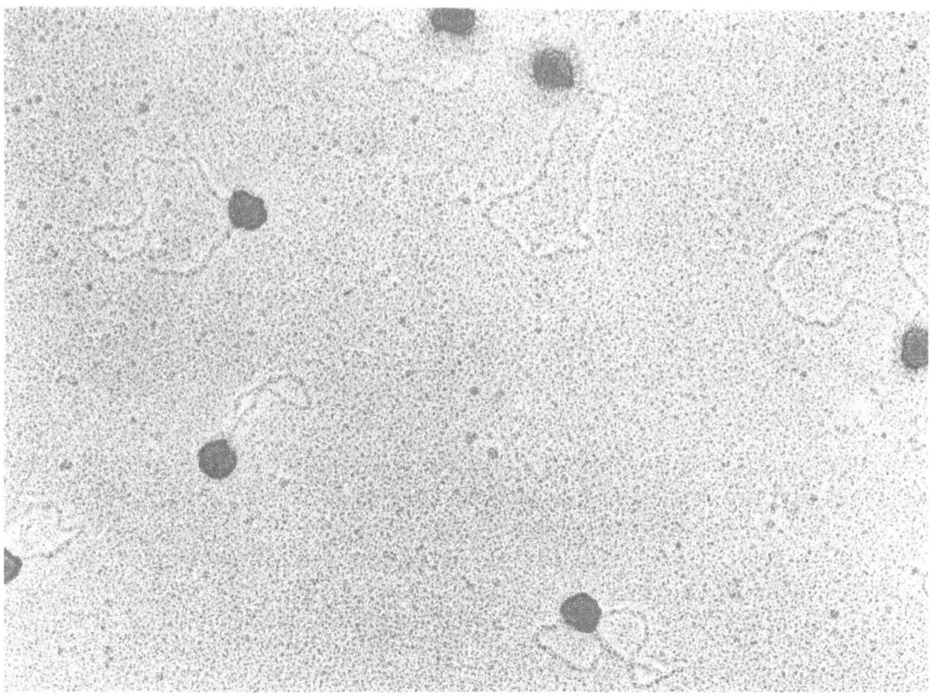

Fig. 2. Action of high pH on CaMV: Part of the DNA protrudes from the capsid

to pH 7, the DNA reinserts itself into the capsid and the "reconstructed virions" appear identical to the native ones, at least from the point of view of the protection of the DNA (AL ANI et al. 1979b).

2.2 Structure of the Genome

2.2.1 Viral DNA

The DNA of CaMV (Cabb-S strains) has a molecular weight of 5.1×10^6 and is composed of 8024 base pairs. In the electron microscope and also in agarose gel electrophoresis, two kinds of molecules may be identified: linear and circular. The proportion of both types of molecules varies greatly in different preparations but in general, linear molecules are less numerous than circular ones (LEBEURIER et al. 1980). An extensive discussion concerning the various properties of CaMV DNA and its electron microscopy and electrophoretic behaviour have been recently published by HULL (1979). Several of these conclusions should be reconsidered in view of the recent determination of the complete sequence of the DNA of the CaMV "Strasbourg strain".

VOLOVITCH et al. (1977) had reported for the first time that heat or alkali treatment of CaMV DNA results in the denaturation of the DNA and that agarose gel electrophoresis of the products reveal the existence of three bands.

The longest corresponds to the single-stranded linear form of the whole molecule and represents one strand with one interruption. The two other bands arise from the complementary strand which includes two interruptions: the sum of the lengths of the two bands fit well with the length of the whole genome. The location, structure, and meaning of these three interruptions will be discussed below. Apparently, the majority of the molecules of CaMV DNA have three interruptions but recent observations by RICHARDS et al. (personal communication) in my laboratory seem to suggest that a small number of molecules have only two nicks and some may even have only one. Apparently, no molecule lacking all three nicks is found in DNA extracted from virus preparations.

2.2.2 Restriction Map of the CaMV DNA

Several restriction maps of CaMV DNA have been established from various strains. Figure 3 gives the most recent restriction map obtained with Cabb S strain. The results are the same with CaMV DNA cloned in pBR322 and amplified in *E. coli* (HOHN et al. 1980). This was confirmed by the recent determination of the DNA sequence of the genome of CaMV (FRANCK et al. 1980).

2.2.3 Sequence of CaMV DNA

The determination of the DNA sequence of the Cabb S strain (an isolate of the laboratory of Virology of Strasbourg) has been established from noncloned DNA by means of adaptation of the classical method of MAXAM and GILBERT (1977). The short discontinuities present in both strands are referred to as gaps. The gap 1 is in strand α (which is the transcribed strand) and it was chosen as the zero point of our restriction map. The gaps 2 and 3 are located at 20 and 53 maps units, respectively (see Fig. 4). The positions of the three gaps are conserved in all CaMV strains examined to date (HULL 1979) with the exception of CM_4-184, which has undergone a small deletion in the region of gap 3 (HULL et al. 1979).

The complete established sequence tells us much about the organization of the CaMV genome, although a straightforward extrapolation from DNA sequence to the properties of the final gene products is made difficult by the complexity of maturation in eukaryotes (splicing). Nevertheless, analysis of the coding capacity of the nucleotide sequence reveals the general outlines of CaMV genetic organization.

The analysis of the distribution of UGA, UAG, and UAA termination codons in the three possible coding frames of the theoretical α strand transcript (i.e., sequence identical to β strand) has shown that α strand transcript can encode long proteins; the transcript of the β strand has no potential coding capacity. This fits well with the results obtained earlier by means of hybridization methods by HULL et al. (1979) and AL ANI et al. (1980).

The β strand sequence, apart from a region of about 1,000 nucleotides in the vicinity of gap 1, is free of nonsense codons for considerable distances in one or another of the three reading frames. The longest such potential coding sequence, Region V, is 2082 nucleotides long. It is noteworthy that, with the

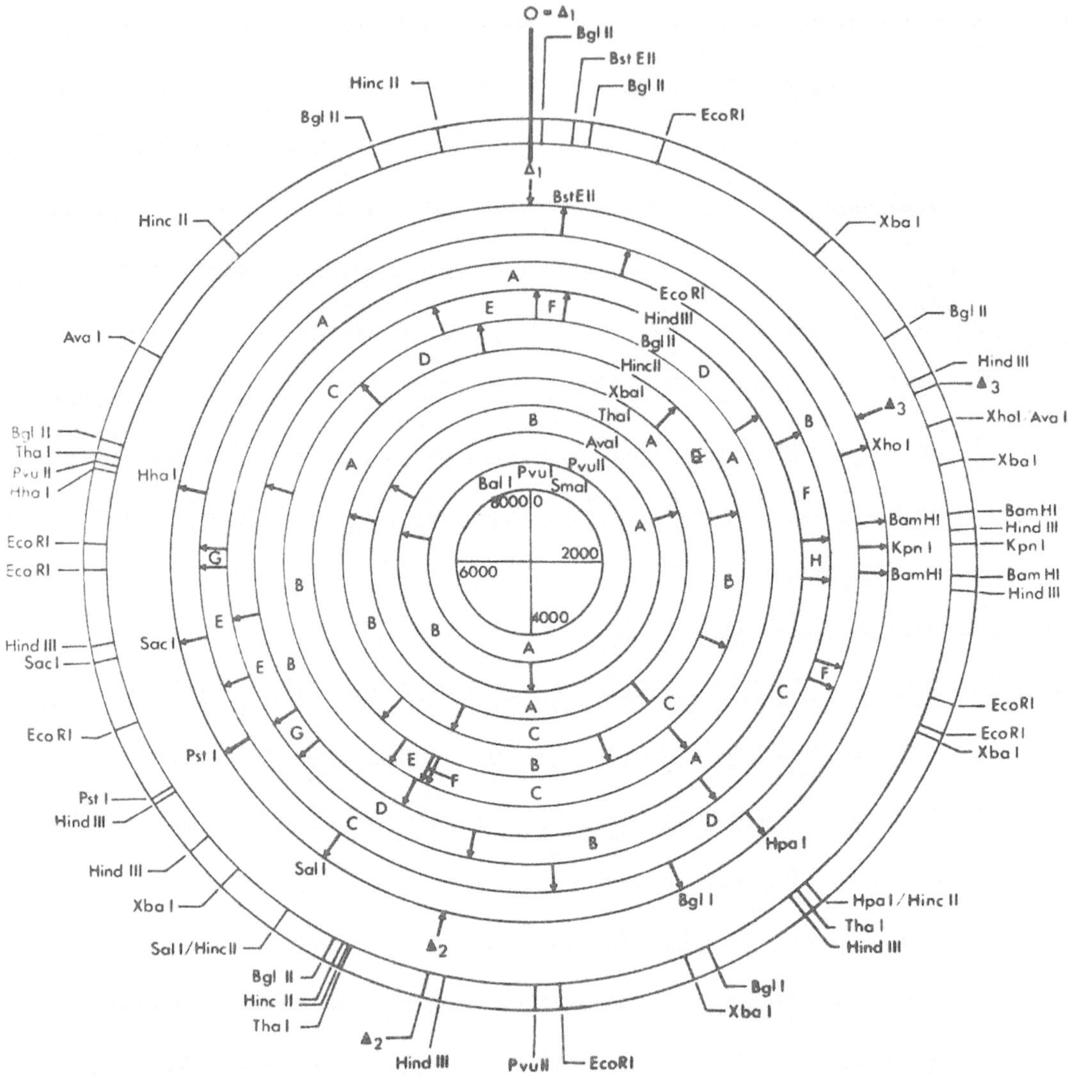

Fig. 3. Restriction map of CaMV DNA (Strasbourg strain)

exception of about 120 nucleotides at the junction between Region IV and V, there is little overlap between successive coding regions. Figure 4 shows the lengths of the polypeptides for which the six regions may code (if, as is in general the case for eukaryotes, translation starts with the first in-phase AUG in each region and there is no read-through of termination codons).

However, till now only two proteins corresponding to these two putative genes have been identified: the viroplasm (inclusions bodies) gene and that of the precursor of the major coat protein subunit. The first of these two genes was located by using the hybrid-arrested translation (HART) technique (ODELL

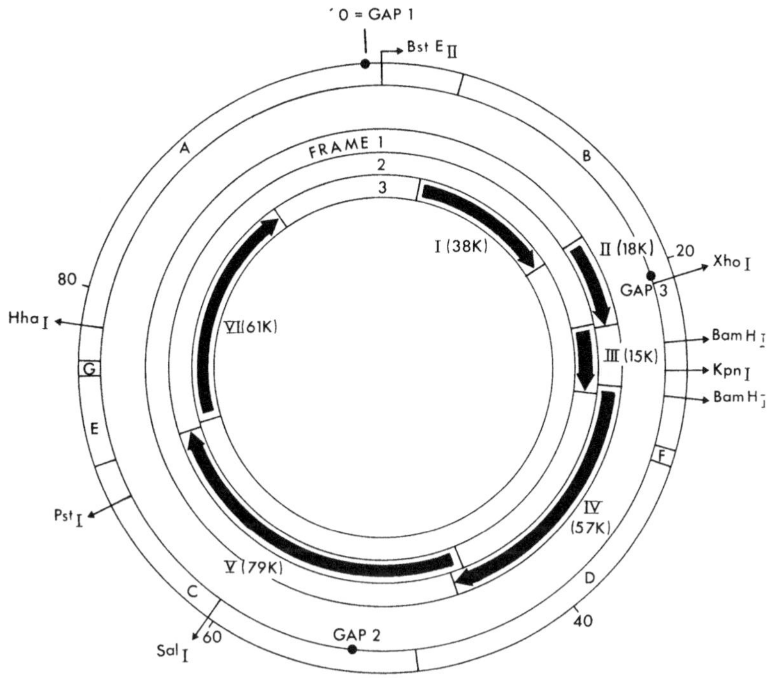

Fig. 4. Genetic map of CaMV DNA: *Arrows* indicate the six putative genes and direction of transcription. *Numbers in brackets* indicate the molecular weights of the six putative proteins. *Letters on the outer circle* indicate the seven EcoRI restriction fragments. The positions of the three gaps are also shown

et al. 1980, AL ANI et al. 1980) and the molecular weight of this putative gene (62,000) fits well with that of the viroplasm protein isolated from infected leaves.

The second is the gene coding for the major coat protein subunit. This coat protein is unusually lysine-rich (about 18%). The examination of the coding capacity of each of the six putative coding regions has shown that only Region IV has a potential to code for a protein approaching this degree of richness in lysine. Figure 5 shows the sequence of such protein with a molecular weight of 55,000, which does not fit with the molecular weight of 42,000 of the major protein subunit. However, if this protein is taken to be a precursor, the cleavage of the N and C terminal ends, which are very acidic, gives rise to a protein having the exact amino acid composition of the protein subunit as established by BRUNT et al. (1975). The boundaries of the putative coat protein polypeptide were chosen to optimize the fit.

The 42,000 protein is never found in an in vitro system stimulated by mRNA extracted from infected plants. However, a virus-specific polypeptide with an estimated molecular weight of about 55,000 was discovered and might correspond to the 55,000 protein arising from the translation of the Region IV. Thus viral coat protein would then be synthesized in the form of a precursor. The sequence of the amino acids of the coat protein subunits shows some

```
1               5                   10                  15                  20                  25                 30
fmet Ala Glu Ser Ile Leu Asp Arg Thr Ile Asn Arg Phe Trp Tyr Asn Leu Gly Glu Asp Cys Leu Ser Glu Ser Gln Phe Asp Leu Met
Ile Arg Leu Met Glu Glu Ser Leu Asp Gly Asp Gln Ile Ile Asp Leu Thr Ser Leu Pro Ser Asp Asn Leu Gln Val Glu Gln Val Met
Thr Thr Thr Glu Asp Ser Ile Ser Glu Glu Glu Ser Glu Phe Leu Leu Ala Ile Gly Glu Thr Ser Glu Glu Glu Ser Asp Ser Gly Glu
Glu Pro Glu Phe Glu Gln Val Arg Met Asp Arg Thr Gly Gly Thr Glu Ile Pro Lys Glu Glu▼Asp Gly Glu Gly Pro Ser Arg Tyr Asn
Glu Arg Lys Arg Lys Thr Pro Glu Asp Arg Tyr Phe Pro Thr Gln Pro Lys Thr Ile Pro Gly Gln Lys Gln Thr Ser Met Gly Met Leu
Asn Ile Asp Cys Gln Thr Asn Arg Arg Thr Leu Ile Asp Asp Trp Ala Ala Glu Ile Gly Leu Ile Val Lys Thr Asn Arg Glu Asp Tyr
Leu Asp Pro Glu Thr Ile Leu Leu Leu Met Glu His Lys Thr Ser Gly Ile Ala Lys Glu Leu Ile Arg Asn Thr Arg Trp Asn Arg Thr
Thr Gly Asp Ile Ile Glu Gln Val Ile Asp Ala Met Tyr Thr Met Phe Leu Gly Leu Asn Tyr Ser Asp Asn Lys Val Ala Glu Lys Ile
Asp Glu Gln Glu Lys Ala Lys Ile Arg Met Thr Lys Leu Gln Leu Cys Asp Ile Cys Tyr Leu Glu Glu Phe Thr Cys Asp Tyr Glu Lys
Asn Met Tyr Lys Thr Glu Leu Ala Asp Phe Pro Gly Tyr Ile Asn Gln Tyr Leu Ser Lys Ile Pro Ile Ile Gly Glu Lys Ala Leu Thr
Arg Phe Arg His Glu Ala Asn Gly Thr Ser Ile Tyr Ser Leu Gly Phe Ala Ala Lys Ile Val Lys Glu Glu Leu Ser Lys Ile Cys Asp
Leu Ser Lys Lys Gln Lys Lys Leu Lys Lys Phe Asn Lys Lys Cys Cys Ser Ile Gly Glu Ala Ser Thr Glu Tyr Gly Cys Lys Lys Thr
Ser Thr Lys Lys Tyr His Lys Lys Arg Tyr Lys Lys Lys Tyr Lys Ala Tyr Lys Pro Tyr Lys Lys Lys Lys Lys Phe Arg Ser Gly Lys
Tyr Phe Lys Pro Lys Glu Lys Lys Gly Ser Lys Gln Lys Tyr Cys Pro Lys Gly Lys Lys Asp Cys Arg Cys Trp Ile Cys Asn Ile Glu
Gly His Tyr Ala Asn Glu Cys Pro Asn Arg Gln Ser Ser Glu Lys Ala His Ile Leu Gln Gln Ala Glu Lys Leu Gly Leu Gln Pro Ile
Glu Glu Pro Tyr Glu Gly Val Gln Glu Val Phe Ile Leu Glu Tyr Lys▼Glu Glu Glu Glu Glu Thr Ser Thr Glu Glu Ser Asp Gly Ser
Ser Thr Ser Glu Asp Ser Asp Ser Asp Umber
```

Fig. 5. Sequence of the putative precursor of the coat protein subunit. *Triangles* indicate the possible cleavage sites of this precursor

similarity with H1 histone and with respect to this observation, morphogenesis of this virus should be a very interesting problem to solve.

From the above experiments, it is uncertain whether the maturation process would involve the joining together (splicing) of discontinuous regions of the mRNA precursor, as observed for many eukaryotic mRNA's. Negative evidence comes from several arguments. For example, electron microscopic studies of RNA–DNA heteroduplexes between CaMV DNA and poly(A)$^+$-RNA's extracted from infected plants have failed to reveal a single-stranded DNA loop. Furthermore, the fact that the reading frame in the coding region jumps abruptly from one phase to another so that successive open regions often overlap slightly does not suggest the existence of intervening sequences (introns). But further studies are necessary to be sure that no splicing exists at the level of leader sequences as it is the case for adenoviruses.

It was surprising to ascertain that the putative gaps were in fact nicks, as no nucleotides are missing but are instead regions of sequence overlap: short sequences of about 18 nucleotides for the three gaps at one terminus of each discontinuity, probably the 5′ terminus is displaced from the double helix by an identical sequence at the other boundary of the discontinuity. Figure 6 gives the nucleotide sequence of the regions around the three gaps. The role of the gaps is unknown but they may be start/stop points for replication of viral DNA as it is evident that redundant terminal sequences arise if a round of DNA replication proceeds for a short way beyond the original starting point. From recent experiments (Lebeurier et al. 1980, Howell et al. 1980), CaMV DNA cloned in plasmid pBR322 and amplified has no gaps and is highly infectious. Thus gaps are not necessary for the multiplication of the viral DNA.

On the other hand, linear and circular DNA are both infectious and plants seem to be able to repair DNA molecules efficiently (Lebeurier et al. 1980).

From the above considerations it may be deduced that CaMV DNA should be a good tool for genetic engineering in higher plants.

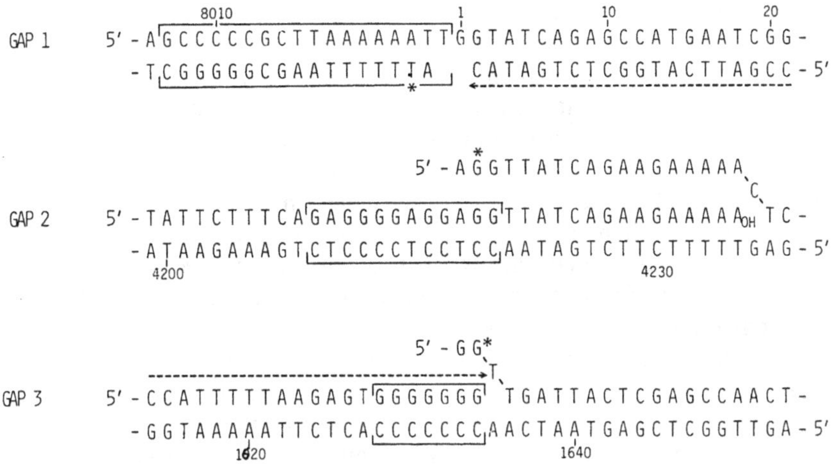

Fig. 6. Sequence around the three gaps. *Stars* indicate possibly modified nucleotides

3 RNA Viruses

RNA viruses constitute the largest group of plant viruses. This is in contrast with animal viruses, many of which are DNA viruses. The reasons for this difference between the two kingdoms of the eukaryotes regarding the genome chemistry of their viruses is not known. The origin of RNA viruses presents a difficult and not yet solved problem.

If we consider that the replication of a viral RNA requires some specific enzyme or at least one or more specific factors, it is clear that several genes are needed to ensure the replication of the RNA as well as its coating. In general the RNA's of plant viruses are plus-strand RNA's, that is to say, RNA's that play the role of mRNA's. For this reason plant viral RNA's are a good tool for studying how messenger RNA's from higher plants are translated and how this translation is regulated. In the case of the simplest plant viruses the whole genetic information is contained in one piece of RNA which is polycistronic; this contrasts with the eukaryotic mRNA's which in general seem to be monocistronic. In spite of the difference between viral and cellular mRNA, the former have been used as a model to propose a tentative hypothesis for understanding some critical steps of the translation of eukaryotic mRNA's.

This first section of the review will be divided in to three parts. The first will be devoted to the general organization of the viral genome and to the distribution of the various genes necessary for the expression of pathogenicity. In the second part the structure of the regions of viral RNA's playing a strategic role in the expression of genes and/or in the regulation of their expression will be described and discussed. In the third part the various hypotheses regarding the mechanism of mRNA's reading by eukaryotic ribosomes will be discussed.

3.1 Distribution of the Genes

The evolution of this problem is conceptually very interesting. After the discovery by GIERER and SCHRAMM (1956) that the pathogenic power of tobacco mosaic virus (TMV) is entirely contained in a unique molecule of RNA and that a single break in this molecule of 2×10^6 mol.wt. abolishes the infectivity, it was believed that linkage of the various genes in a single RNA molecule was a general rule. When a RNA preparation obtained from alfalfa mosaic virus (AMV) was found to contain RNA's of different molecular weights (GILLASPIE and BANCROFT 1965), it was thought that only the largest RNA was infectious and that the others were either degradation products of the large one or RNA from cellular origin. In 1970 VAN VLOTEN-DOTING et al. established that the three largest nucleocapsids present in AMV preparations were necessary for the expression of pathogenicity. This discovery was followed by several other reports which demonstrated that viral genes can be distributed in several ways.

This phenomenon is not specific for plant viruses since many animal RNA viruses also have divided genomes. However, in contrast with plant viruses, these animal viruses belong in general to the "minus strand" RNA group: such RNA's are not infectious and must be first transcribed to a complementary copy which has the properties of a mRNA. This is the case for the orthomyxoviruses (Influenza virus) (see review by BISHOP 1977: *Comprehensive Virology*), the bunya-viruses (BISHOP and SHOPE 1979) and the arenaviruses (RAWLES and LEUNG 1979). It is worth noting that only few plant virus particles have been found to contain negative RNA's. This is well established for tomato spotted wilt virus (VAN DEN HURK et al. 1977), the particles of which contain three molecules of RNA which are not infectious. It is also suspected that some plant viruses closely resemble the animal rhabdoviruses, e.g., rabies or vesicular stomatitis viruses. In this case too, the RNA which is in one piece is a negative strand. It is also necessary to mention that in the case of some viruses such as wound tumor virus, the virions contain double-stranded RNA's (12 components: REDDY et al. 1974). However, these viruses will not be discussed in this review.

3.1.1 The RNA's of Monopartite Plant Viruses

A large number of plant viruses belong to this group. The organization of their RNA extremities is very different from one virus to another. Two viruses have been extensively studied with respect to the organization of their RNA: tobacco mosaic virus (TMV) and turnip yellow mosaic virus (TYMV).

Tobacco Mosaic Virus is a rod-shaped virus containing one piece of RNA of about 2×10^6 mol.wt. (genomic RNA) (see review by HIRTH and RICHARDS 1980) protected by 2,200 identical protein subunits, having a molecular weight of 17,500. When put in an in vitro system (wheat germ or reticulocytes), TMV RNA induces the synthesis of two proteins with molecular weights of 130,000 and 165,000, respectively (KNOWLAND et al. 1975).

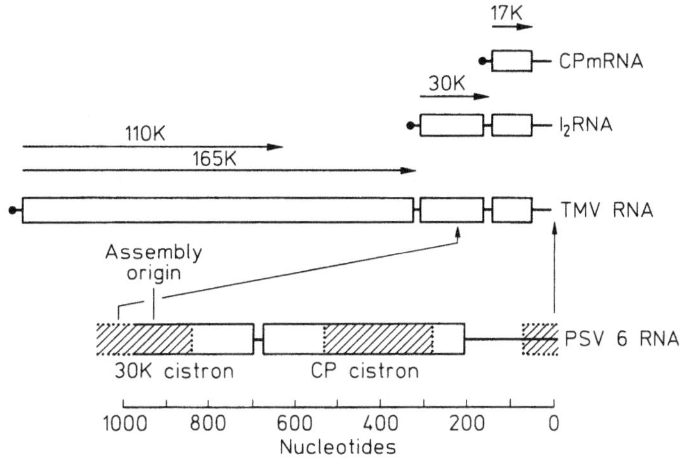

Fig. 7. Structure of TMV RNA and of the two corresponding subgenomic RNA's. The translated sequences are shown by *white rectangles*. The bottom of the figure represents part of 3′ end of the TMV RNA sequence which is presently known. • represents the cap: – represents the nontranslated part of the various RNA's (leader and intercistronic sequences). *CPmRNA* is the subgenomic RNA for coat protein. I_2 is the subgenomic RNA from which the 30,000 mol.wt. protein is translated

The sequences of these two proteins overlap to a considerable extent at their N-terminal extremities (PELHAM 1978). Coat protein is not made from the whole RNA but only from a subgenomic RNA having a low molecular weight (LMC) (HIGGINS et al. 1976, BEACHY and ZAITLIN 1977). A 30,000 protein is also made from another subgenomic RNA (I_2) (BRUENING et al. 1976). Figure 7 shows the various RNA's formed from the genomic RNA of TMV. It is clear that in this case the only gene which is translated (from the 3 RNA's) is that near the 5′ end of the viral RNA. The other genes are "silent genes". It is worth noting that in contrast with the 30,000 mol.wt. protein gene, the coat protein gene of the common strain of TMV is not coated but is found as a naked RNA in infected plants (this is not the case for the cowpea strain of TMV) (SIEGEL et al. 1976). The mode of formation of the subgenomic RNA's is still unknown. It must be emphasized that the inoculation of the sole genomic RNA to plants is sufficient to induce infection.

Turnip Yellow Mosaic Virus (TYMV). In contrast with TMV, TYMV is a virus with icosahedral symmetry. Preparations of TYMV contain several types of particles with various densities (MATTHEWS 1970). The particles with the lowest density are empty shells (without RNA), those with the highest density are nucleocapsids (top and bottom components respectively). Particles with intermediate densities between top and bottom components are often present in TYMV preparation (PLEIJ et al. 1976). They contain only part of the genomic RNA and/or subgenomic RNA molecules. When the whole RNA of a purified preparation of virions is extracted and fractionated by gel filtration, it is possible to isolate, in addition to the genomic RNA, one subgenomic RNA (SRNA) which codes for the coat protein (KLEIN et al. 1976).

In contrast with the LMC of TMV this subgenomic RNA is coated. It is located partly in the bottom component and partly in the intermediate components (KLEIN 1976). If the existence of subgenomic RNA in nucleocapsids is a general rule for tymoviruses, its location varies greatly from one type of virus to another; for example in eggplant mosaic virus (EMV) the SRNA is located in the top component and apparently not in the bottom one (SZYBIAK et al. 1978); in okra mosaic virus (OMV), on the other hand, both top and bottom components contain the SRNA (GIVORD, unpublished results). Although other RNA monopartite plant viruses are known to exist, in general their mechanism of multiplication has not been studied in detail. This is the case for luteoviruses (e.g., barley yellow dwarf virus, tomato bushy stunt virus, southern bean mosaic virus, etc.). As will be shown below, tobacco necrosis virus (TNV) and carnation mottle virus seem to escape the general rule according to which the genes contained in the genomic RNA are expressed via a subgenomic RNA.

3.1.2 RNA's of Multipartite Plant Viruses

In this case, several RNA's carry the various genes necessary for the expression of pathogenicity. The RNA can be coated in separate capsids as in the case of tobraviruses [tobacco rattle virus (TRV); see review by HIRTH 1976] or some of the RNA's can be present in the same capsid (bromoviruses, cucumoviruses, etc.; see review by LANE 1979). In general, the number of RNA pieces varies from one virus to another but it usually is between two and four. Table 1 gives some examples of multipartite RNA viruses. It is worth noting that multipartite viruses can possess either icosahedral or helical symmetry.

In general, each RNA species expresses only one gene. If it contains the information for two genes, the gene close to the 3′ end of the RNA needs the formation of the corresponding subgenomic RNA for its expression (MOHIER et al. 1976, GERLINGER et al. 1977). However, in the case of nepoviruses and comoviruses, each RNA behaves as a monocistronic message, although the synthesized polypeptide chain is cleaved to several functional proteins by a protease (DAVIES et al. 1977, FRITSCH et al. 1980). This enzyme could be a cleavage product of the polypeptide chain coded by one of the viral RNA's (FRITSCH et al. 1980). These viruses are very similar to animal viruses such as picornaviruses (poliovirus) as far as their mode of gene expression is concerned (HUANG and BALTIMORE 1977).

3.1.3 Satellite Viruses

In some cases, monopartite and multipartite plant virus preparations contain one additional RNA which is enclosed in a coat of protein subunits coded by itself or by a helper virus. Indeed, this RNA is unable to replicate by itself and requires another virus for its multiplication. In general this association between helper and satellite is specific. Although a relatively great number of satellites have been described, only three of them will be briefly evoked.

Table 1. Classification of the various groups of divided genome plant viruses

		Group of viruses	Characteristics of RNA	Mol.wt. of protein subunits	Examples
Virus with helical symmetry		Tobraviruses	4 RNA's Mol.wt. 2.4, 1.3, 1.15, 0.85×10^6	22,000	Tobacco rattle virus Pea early browning virus
		Hordeiviruses	4 RNA's ranging from mol.wt. 1 to 1.5×10^6	21,000	Barley stripe mosaic virus Lychnis ringspot virus
		Beet necrotic yellow vein virus	4 RNA's Mol.wt. 2.3, 1.8, 0.7, 0.6×10^6	17,500	Soil-borne wheat mosaic virus
Viruses with icosahedral symmetry	Bipartite RNA's	Nepoviruses	2 RNA's Mol.wt. 2.4 and 1.4×10^6	55 to 60,000	Tobacco ringspot virus Tomato black ring virus
		Comoviruses	2 RNA's Mol.wt. 2.0 and 1.4×10^6	2 capsid proteins 25 and 44,000	Cowpea mosaic virus Squash mosaic virus
		Pea enation mosaic virus[a]	2 RNA's Mol.wt. 1.7 and 1.3×10^6	22,000	Pea enation virus
	Multipartite RNA's	Bromoviruses[a]	3 RNA's Mol.wt. 1.1, 1.0, and 0.7×10^6	20,000	Brome mosaic virus Broad bean mottle virus
		Cucumoviruses[a]	3 RNA's Mol.wt. 1.27, 1.13 and 0.82×10^6	24,000	Cucumber mosaic virus Tomato aspermy virus
		Ilarviruses[a]	3 RNA's Mol.wt. 1.1, 0.9, and 0.7×10^6	25,000	Tobacco streak virus Tulare apple mosaic virus
	Multipartite RNA's (Bacilliform morphology)	Alfalfa mosaic virus[a]	3 RNA's Mol.wt. 1.2, 0.9, 0.7×10^6	24,300	

[a] From the preparations of this virus a fourth RNA can be extracted. In general its molecular weight is comprised between 0.2 and 0.3×10^6. This small RNA is in fact a subgenomic RNA of one of the three main RNA's. Further details are given in MATTHEWS, 1979

Satellite Tobacco Necrosis Virus (STNV). This is a very small virus of icosahedral symmetry containing a single molecule of 400,000 mol.wt. This RNA contains the information for the coat protein only (LEUNG et al. 1976) and depends apparently on tobacco necrosis virus (TNV) for its multiplication (KASSANIS 1972). When both viruses are present in the same plant, STNV multiplies very efficiently and reduces the yield of TNV. The molecular mechanism of this competition is unknown. The relationship between "helper" and satellite is specific as some strains of STNV are activated only by specific strains of TNV (KASSANIS and PHILIPPS 1970).

Cucumber Mosaic Virus Associated RNA 5 (Carna 5). This RNA is encapsidated by CMV coat protein. It has been purified and found not to multiply when inoculated alone in susceptible plants (KAPER et al. 1976, KAPER and TOUSIGNANT 1977). It possesses all the properties of a true satellite of CMV. Its presence in infected plants can considerably change the severity of the disease. The presence of Carna 5 is expressed either as a new disease syndrome (which can be lethal for tomato, for example) or by the attenuation of viral infection (WATERWORTH et al. 1979). As for STNV, no significant sequence homology was detected between Carna 5 and its helper (SHOULDER et al. 1974, LOT et al. 1977).

The origin of Carna 5 is unknown but it was recently suggested (WATERWORTH et al. 1979) that it might have originated from CMV host plants and that it could play a regulatory role in the multiplication of the helper virus.

Whatever the origin and the role of satellites in disease expression, it is clear that they are usually able to multiply very actively. This is apparently also the case for a third satellite RNA: the *tomato black ring virus satellite* which is very efficiently translated in vitro (FRITSCH et al. 1980). Some authors have claimed that the system "helper-satellite" has some similarities with the competitive relationship of defective interfering particles (DI particles) and their helper virus. However, the DI particles found in the case of infection of animal cells by picornaviruses or rhabdoviruses (HUANG and BALTIMORE 1977) contain RNA which are directly related to the RNA of the helper. In the case of poliovirus, the RNA contained in DI particles is a part of the 5′ OH region of the genomic RNA. The production of DI particles might play a role in the epidemiology of animal diseases and in their latency. DI particles have not yet been described for plant viruses but the fact that satellites are found with increasing frequency in plant virus infections suggests that they may play a similar role in plant virus epidemiology in spite of their different origin. In the third part of this chapter I shall describe the structure of satellite RNA's.

3.2 Structure of RNA of Plant Viruses

Although the elucidation of complete sequences of the RNA's of mono and multipartite plant viruses is likely to make an important contribution to our understanding of viral replication, the progress in this field has been slow and complete sequences are known only for three subgenomic RNA's: LMC of TMV (GUILLEY et al. 1979), TYMV (GUILLEY and BRIAND 1978) and RNA

4 of AMV (BREDERODE et al. 1980). But the fact that virus multiplication in higher plants has in common several initial events such as ribosome recognition, protein synthesis initiation and termination followed by RNA replication suggests that the strategy of viral gene expression may be the same in different viruses.

On the other hand, the knowledge of the structure of these regions might throw light on the general problem of the mechanism of translation of eukaryotic mRNA's. This is the rationale behind much of the work recently done in my own and several other laboratories, which is directed at determining the primary structure of the 5′ and 3′ end of various plant virus RNA's. The most surprising result of these studies was the discovery that a great variety of RNA terminal structures in plant viruses are capable of playing the same role, for instance, the recognition in vivo and in vitro of the same type of ribosomes. The reason for such a variability is still an open question. Thus, with respect to their 5′ and 3′ termini, RNA plant viruses fall in several main categories.

3.2.1 Categories of 5′ and 3′ Termini of Plant Virus RNA's

a) Prokaryote-Like Plant Virus RNA's. Prokaryotic mRNA's are characterized by a 5′ terminus of the type PPPY and a 3′ end, the structure of which varies greatly from one messenger to another, but which have no apparent specific function except in the case of RNA phages where it recognizes the replicase system. This situation is found in the case of STNV RNA (LEUNG et al. 1979). Figure 8a gives the structure of the 5′ end of STNV RNA and shows that the 5′ terminal nucleotide is (p)ppA. The 42 5′ terminal nucleotides may be folded in a hairpin which might play the role of a signal for recognition of the RNA by the 40S subunit of eukaryotic ribosomes. Recently the sequence of the 41 3′ terminal nucleotides of STNV has been determined and is given in Fig. 8b (MERREGAERT et al. 1979).

This sequence contains a termination codon UGA, but it is not certain whether the preceding region has any coding capacity. This terminal sequence cannot be folded, even partly, into a tRNA-like structure. Moreover no binding of any amino acid has yet been reported. It should be interesting to know whether tobacco necrosis virus (TNV), which is probably replicated by the same replicase as STNV, has the same 3′ terminal sequence. It has been reported earlier (WIMMER et al. 1968) that the 5′ terminal sequence of TNV and STNV are identical but this is only true for the three first 5′ nucleotides: ppApGpUp..].

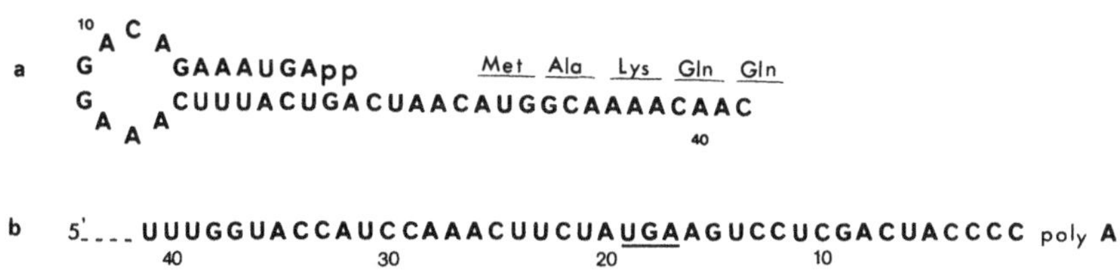

Fig. 8. Structures of **a** the 5′ end (LEUNG et al. 1979) and **b** the 3′ end of STNV (MERREGAERT et al. 1979)

It is worth noting that this type of "5′ prokaryotic terminus" is not necessarily characteristic of all satellites. Figure 11 which gives the complete sequence of Carna 5 satellite (see below) shows that the 5′ end is capped[1] (RICHARDS et al. 1978).

b) Capped 5′ Terminus and 3′ End with Various Termini. It was discovered several years ago that most eukaryotic mRNA's are capped, that is to say have a $m^7G^{5'}ppp^{5'}XpY$ at their 5′ end (in general X and Y represent two nucleotides which are methylated on their ribose moiety, but this is not the case for plant virus RNA's).

With respect to their 3′ terminus the capped plant virus RNA's may be divided into three main groups of unequal importance: those having a 3′ terminus which is unable to bind any amino acids, those having a t-RNA like structure and those having a poly(A) at their 3′ end.

c) 5′ Capped with 3′ "Free" End. Only two viruses and a satellite were identified as belonging to this group. Alfalfa mosaic virus (AMV), tobacco rattle virus (TRV) and Carna 5 satellite of CMV.

Extensive studies have been performed with AMV and I shall report here the most significant and recent results on the subject (for a general review on AMV, see VAN VLOTEN-DOTING and JASPARS (1977).

5′ Termini of AMV. In 1975 it was demonstrated that the four viral RNA's have a cap of the type m^7 GpppGp. (For structure see DYER, Chap. 6, this Vol.). This cap (like the cap of other viral and eukaryotic mRNA's) can be removed by using a nucleotide pyrophosphatase isolated from tobacco cell culture or from potato (FILIPOWICZ et al. 1978). This removal allows [^{32}P]-labeling of the free 5′ end of viral RNA by polynucleotide kinase and thus establishing the sequences of the nucleotides of the 5′ end of RNA 3, and those of the 3′ terminus of the four RNA's. Figure 9 (a, b, c) shows the sequence of the nontranslated part of the 5′ ends of RNA 2–3 of S strain (PINCK et al., unpublished results) and 4 of strain 425 (BREDERODE et al. 1980). It is clear that these sequences have nothing in common except for the cap. It is also remarkable that the leader sequence for RNA 3 is rather long: the first AUG is present only after 257 nucleotides (it is not yet firmly established whether this AUG is the initiation codon for the 35,000 mol.wt. protein coded by the 5′ cistron of RNA 3). The leader sequence of RNA 1 is unknown. In contrast, RNA 4 which codes for coat protein and is, in fact, a subgenomic RNA corresponding to RNA 3, has a relatively short leader sequence which is given in Fig. 10c. This sequence is completely devoid of G except for the G following the cap.

3′ Termini of AMV. Figure 10 gives the structures of the 3′ termini of the three genomic RNA's of AMV. They are identical over a relatively short length

1 It has been reported (ABOU HAIDAR and HIRTH 1977) that the long RNA's of TRV were not capped, but recent data (C. FRITSCH, personal communication) suggest that this is not the case, but that for unknown reasons only a small number of molecules seem to be capped

10 50

m^7GpppGUUUUUAUCUUUUCGCGAUUGAAAAGAUAAGUUUCAGAGUCUAAUCUUUUCAAUAUG

60 100

UUCACUCUUUUGAGAUGUCUUGGAUUUGGCGUCAACGACCUUACCGAUAU...

a RNA 2

10 50

m^7GpppGUUUUCAUCUUACACACGCUUGUGCAAGAUAGUUAAUCAUUCCAAUUCAACUCAAUUAA

100

CGUUUUUUACAGUGUAAUUCGUACUUUUCGUAAGUAAGUUUCUGUAAAAGCGUUUCUUGUUUUAA

RNA 3'

130 150

UUUGGCCUAACACGUAAUUCGUACUCUUCGUGAGUAAGUUGUGUUAGCCAUACCUAUCCUUUAAA

200 250

UUCCUGUCAAUUUAAAAAGAAAAUCAUUCCCAUUCGCGUAAUUCGUACUCUUCGUGAGUAAGUUG

(Met Ser Lys Leu

300

CAAAUGGAGAAUAGAAAAACAAAUGCCUCGAGUUCUGGAAUGUUUUCUUCCUCUAGCUUUUCAG

Gln) Met Glu Asn Thr Lys Thr Asn Gly Ser Ser Ser Gly Met Phe Ser Ser Ser Ser Phe Ser

b RNA 3

10

m^7GpppGUUUUUAUUUUUAAUUUUCUUUCAAAUACUUCCAU

36 70

CAUGAGUUCUUCACAAAAGAAAGCUGGUGGGAAAGCUGG

90

UAAACCUACUAAACG

c RNA 4

Fig. 9. a Structure of the 106 nucleotides from the 5′ end AMV–RNA 2. **b** Sequence of the 318 nucleotides from the 5′ of AMV–RNA 3. *Underlined regions* are protected by the ribosomes in vitro. *Arrow* indicates the 5′ terminus of RNA 3′ found in RNA preparations of top b component (GUG is a potential prokaryotic initiation codon: It is unlikely that it plays this role). **c** Sequence of the 5′ end of AMV–RNA 4 (for both strains S and 425)

```
                 -110                                                      -60
RNA 3  AGGCCUAUGAAGUCCUAUUCGCUCCUGAUAGGAAGGACUUCAUAUUGCUUAUAUAUGUG
RNA 2                    ——————————————————GAAAUUCU———————————————————
RNA 1      GUAUAGAAC—————————————————————AAGAUUCU—————————————————CA—

              -50                                          -10
RNA 3  CUGGUGCACAUAUAUAAAUGCUCAUGCAAAACUGCAUGAAUGCCCCUAAGGGAUGC(OH)
RNA 2  —UA——————————————————————————————U—————————————————————.
RNA 1  —UA————UG———————————————U————U————————————————————————
```

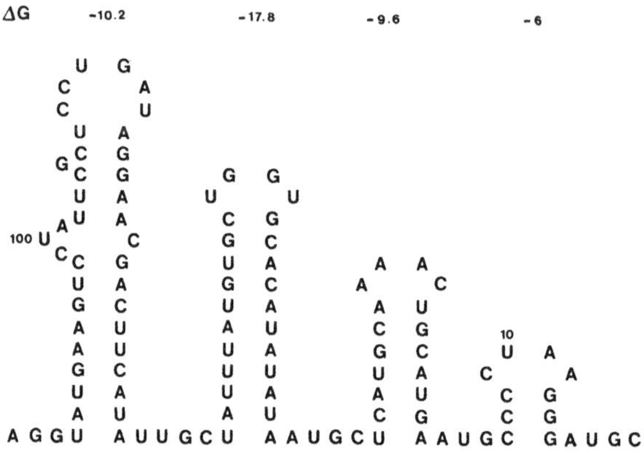

Fig. 10. The top of the figure are the 3′ terminal sequences of the three major AMV–RNA's (*lines* indicate identical sequences). At the bottom the possible secondary structure of the 3′ end of RNA 3 is shown

(26 nucleotides) and in the regions where relatively large differences in primary structure are found, the same type of stable secondary structure is present at least through 100 nucleotides (PINCK and PINCK 1979). In spite of the observed differences a certain degree of sequence homology is apparent between the 3′ terminal sequences of AMV. This is in good agreement with the result reported earlier by mean of hybridization studies (BOL et al. 1975, GOULD and SYMONS 1978). Although these terminal sequences are unable to bind any amino acid, the four RNA's have a great affinity for the coat protein subunits. HOUWING and JASPARS (1978) have presented strong evidence for the location of sites with a high affinity for coat protein near the 3′ terminus of RNA 4 and it is probable that these sites are entirely located in the extracistronic part of RNA 4. But as the four RNA's have a similar structure at their 3′ terminus, they all bind coat protein at the same position. If we remember that coat protein is needed to render the three large RNA's of AMV infectious (VAN VLOTEN-DOTING and JASPARS 1977), it might be speculated that interactions between coat protein and viral nucleic acid induce some conformational change which is required for the recognition of the 3′ end by the replicase.

It seems that *tobacco streak virus*, which is a multipartite virus in which the nucleic acids need coat protein for the expression of their infectivity, shows great similarity to AMV. It is not known if it is capped but its 3′ end seems to be pXOH and cannot be charged.

met glu asn cys ala glu gly leu tyr leu arg
m^7GpppGUUUUGUUUG AUG GAG AAU UGC GCA GAG GGG UUA UAU CUG CGU

glu asp leu ser leu gly gly val gly tyr leu pro ala lys ala
GAG GAU CUG UCA CUC GGC GGU GUG GGA UAC CUC CCU GCU AAG GCG

gly met ser
GGU UGAGUGAUGUUCCCUCGGACUGGGGACCGCUGGCUUGCGAGCU AUG UCC

ala thr leu ser thr thr leu ser phe glu pro pro leu ser leu
GCU ACU CUC AGU ACU ACA CUC UCA UUU GAG CCC CCG CUC AGU UUG

leu ala glu pro gly thr trp phe ala asp thr met asp phe leu
CUA GCA GAA CCC GGC ACA UGG UUC GCC GAU ACU AUG GAU UUU CUA

lys lys his ser val arg trp tyr glu ser
AAG AAA CAC UCU GUU AGG UGG UAU GAG UCA UGACGCACGCAGGGAGAGGCU

AAGGCUUAUGCUAUGCUGAUCUCCGUGAAUGUCUAUCAUUCCUCUGCAGGACCC$_{OH}$

Fig. 11. The complete sequence of satellite RNA of CMV (RNA 5). The possible initiation sites and the sequence of the corresponding peptides are also indicated

Tobacco Rattle Virus (TRV), which is a multipartite virus with helical symmetry, has its various RNA's (four probably) "capped". Nothing is known concerning their leader sequences or their 3′ termini. However, aminoacylation of the 3′ terminus of the RNA has not been reported. The sequence homology found between the largest RNA and the smallest one suggests a situation similar to that of AMV where the 3′ termini of the various RNA's are identical (DARBY and MINSON 1973). No specific affinity of the coat protein for TRV RNA has been found.

Carna 5: Satellite of CMV. Figure 11 shows the complete sequence of this satellite (RICHARDS et al. 1978). It is capped, and has a short potential leader sequence and a 3′ terminus which is never aminoacylated. This satellite (385 nucleotides) contains two termination codons, UGA, and the sequences which precede these codons code for two putative small peptides. It was claimed that in vitro synthesis of two corresponding peptides has been obtained (OWENS and KAPER 1977) but in my laboratory, assays of translation of Carna 5 in both wheat germ and reticulocyte cell-free systems were unsuccesful (MOHIER, personal communication). It is worth noting that the putative leader sequence is poor in G and that the whole RNA contains 41.5% of GC and 58.5% of AU. This suggests that this RNA possesses stable secondary structure in solution.

d) 5′ Capped with a 3′ tRNA-Like Structure. It was reported several years ago that TYMV was able to bind valine in conditions very similar to those in which a true tRNA is charged (PINCK et al. 1970). This observation was extended later to several other plant viruses.

Table 2 gives the list of viral RNA's for which aminoacylation has been demonstrated.

Table 2. Binding of amino acids by various virus plant RNA's

	Viral RNA's	Amino acid fixed
Viruses with monopartite genomes	TMV (*vulgare*) genomic and subgenomic RNA's	Histidine
	TMV (cowpea strain) genomic and subgenomic RNA's	Valine
	TYMV genomic and subgenomic RNA's	Valine
	Eggplant mosaic virus and probably other members of tymovirus group	Valine
Viruses with multipartite genomes	Brome mosaic virus (the four RNA's)	Tyrosine
	Cowpea chlorotic mottle virus (the four RNA's)	Tyrosine
	Cucumber mosaic virus (the four RNA's)	Tyrosine

The 5′ and 3′ ends of several of these viruses have been extensively studied and their nucleotide sequence established.

e) 5′ Terminal Sequence of Monopartite Genomic RNA's with a tRNA-Like Structure

TMV. Figure 12 gives the primary structure of the leader sequence of the common strain of TMV (RICHARDS et al. 1978) and those of two other strains of TMV (KUKLA et al. 1979). It is clear that whatever the strain, this leader sequence has several unusual features. Among these features is the complete absence of G (except the G following the cap), the tract of 27 nucleotides (17–43) composed exclusively of A and C in which the triplet AAC is repeated seven times, and the tetranucleotide UUAC six times. It was also clearly demonstrated that the first AUG is a functional initiation codon for protein synthesis. The sequence following the first AUG corresponds to the common 5′ terminus of the 110,000 and 140,000 mol.wt. proteins translated in vitro from genomic TMV RNA and found also in infected tobacco protoplasts and plants (JONARD et al. 1978). The lack of G suggests that the TMV RNA leader sequence has a weak secondary structure. Figure 12 also shows that mutations affect not only the coat protein genes (and perhaps also the other coded proteins having unknown function), but also the leader sequences (KUKLA et al. 1979). Although changes and deletions appear in the sequence, G is never found (KUKLA et al. 1979). The complete sequence of the subgenomic RNA coding for the coat protein of TMV strain *vulgare* has been determined. Except for the G following the cap, its leader sequence contains no G and is very short (only nine nucleotides).

TYMV. The nucleotide sequence of the leader sequence of the genomic and subgenomic RNA's of the common strain of this virus (BRIAND et al. 1977, GUILLEY and BRIAND 1978) shows that they have nothing in common with

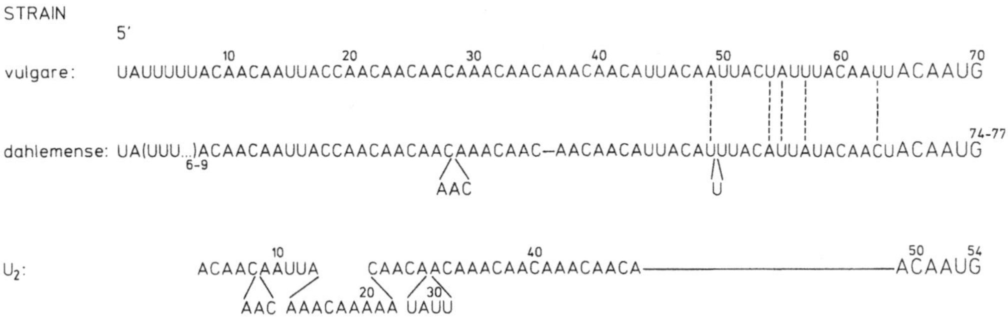

Fig. 12. Leader sequences of three strains of TMV

those found in TMV RNA. However, if they are not completely devoid of G residues their G content is very low.

An adenine residue and not a guanine follows the cap for the TYMV subgenomic RNA. In the genomic RNA the first potential initiation codon comes at position 88–90 and is closely followed by another at position 95–97. Although no conclusive evidence exists as yet, the study of binding of reticulocyte ribosomes to TYMV RNA and the identification of the dipeptide obtained in a cell-free system stimulated by TYMV genomic RNA in the presence of sparsomycin, are consistent with initiation occurring at AUG codon 95–97. A surprising feature of the leader sequence of TYMV genomic RNA is its homology with the 5′ terminus of rabbit β globin mRNA (Briand et al. 1977).

Brome Mosaic Virus (BMV). Recently, the leader sequence of the 5′ terminus of RNA 3 of this virus has been established (Fig. 13) (Ahlquist et al. 1979). This sequence has nothing in common with the other plant viral RNA leader sequences, except for its relatively low content in G residues. This sequence is 70 or 90 nucleotides long depending on whether GUG or AUG is taken as initiation codons (it is unlikely that GUG is an initiation codon in eucaryotic mRNA). The 73 nucleotides coding for the N terminus of the 35,000 mol.wt. protein have also been determined. Apparently, the amino acid sequence shown is the only one possible whatever the initiation codon chosen. As will be discussed in the third part of this review, this leader sequence may be folded in a possible secondary structure which is not a very stable one. The leader sequence of RNA 4 determined previously was found to be a very short one (Dasgupta et al. 1975). The shortness of the leader sequence seems to be a common characteristic of subgenomic viral RNA's.

f) 3′ tRNA-Like Structure

TYMV. The 3′OH end of its RNA has been described as serving as a substrate for a number of tRNA-specific enzymes, including tRNA synthetase, tRNA nucleotidyltransferase, RNAse P (a tRNA maturation endonuclease), and elongation factor EF_1. Of these functions, the only one that has been firmly established and extensively studied is the ability to recognize valyl tRNA synthetase

pGm^7
p
pGUAAAAUACCAACUAAUUCUCGUUCGAUUCCGGCGAACAUUCUAUUUUACCAACA

UCGGUUUUUUCAGUAGUGAUACUGUUUUUGUUCCCGAUGUCUAACAUAGUUUCU
(Met Ile Leu Phe Leu Phe Pro) Met Ser Asn Ile Val Ser
CCCUUCAGUGGUUCCUCACGAACUACGUCUGACGUUGGCAAGCAAGCGGGAGGUA
Pro Phe Ser Gly Ser Ser Arg Thr Thr Ser Asp Val Gly Lys Gln Ala Gly Gly

Fig. 13. Leader sequence of BMV RNA 3 (AHLQUIST et al. 1979). The *underlined triplets* are termination codons which are in phase

from yeast or *E. coli*. All these observations suggest the existence of a tRNA-like structure at the 3′ end of this viral genome (GIEGÉ et al. 1978). It was of course of interest to compare the structure of the 3′ end of TYMV with that of true eukaryotic valyl-tRNA's. The sequence of the 159 nucleotides from the 3′ end of TYMV RNA which, isolated, binds valine very efficiently (BRIAND et al. 1977) has been established. This sequence contains five termination codons located after the 111th nucleotide, and it has been shown that the sequence at the 5′ end of this fragment codes for the terminal part of the coat protein (BRIAND et al. 1977). The nontranslated part located immediately after the first UAA (108 nucleotides) can be folded into a clover-leaf structure as shown in Fig. 14. Surprisingly, very little correspondence with constant tRNA features can be observed. There are, for instance, no modified bases in the viral RNA's; also missing is the sequence GUUCR which is the normal base equivalent of the pentanucleotide GTψCR found in virtually all tRNA's. On the positive side the 3′ terminal tetranucleotide ACCCA^{OH} is the same in this case as in all known species of tRNA^{val}. In addition, arm II of the viral clover-leaf has a remarkable number of points in common with the anticodon arm of major species of tRNA^{val} of several types of yeasts. The CAC valine anticodon is properly positioned.

In conclusion, the TYMV 3′ end RNA can be folded in a clover-leaf structure which is much larger than that of a true tRNA^{val} but which has several points in common with it. We could of course propose other types of folding but all these clover-leaf structures could be conceivably accomodated in an L-shaped tertiary structure of the type proposed for tRNA^{phe}. Furthermore it has been demonstrated that the affinity of valyl tRNA synthetase for TYMV RNA is of the same order as for a true tRNA^{val}. This suggests that the recognition between synthetase and tRNA is probably due to a tridimensional structure in which a small number of nucleotides have a specific position in space, the terminal CCA playing a crucial role. This point of view is reinforced by the primary structure of the 3′ end of TMV and BMV RNA's. Figure 15 shows a possible secondary structure for the TMV 3′ terminus (KUKLA et al. 1979). No real analogy exists between this structure and a true tRNA, except the terminal CCA and the UUCG corresponding to the TψCG of Tψ loop of all tRNA. However, this isolated part of TMV RNA binds histidine very efficiently (GIEGÉ, personal communication). No histidine anticodon is located at the expected place in the putative anticodon arm.

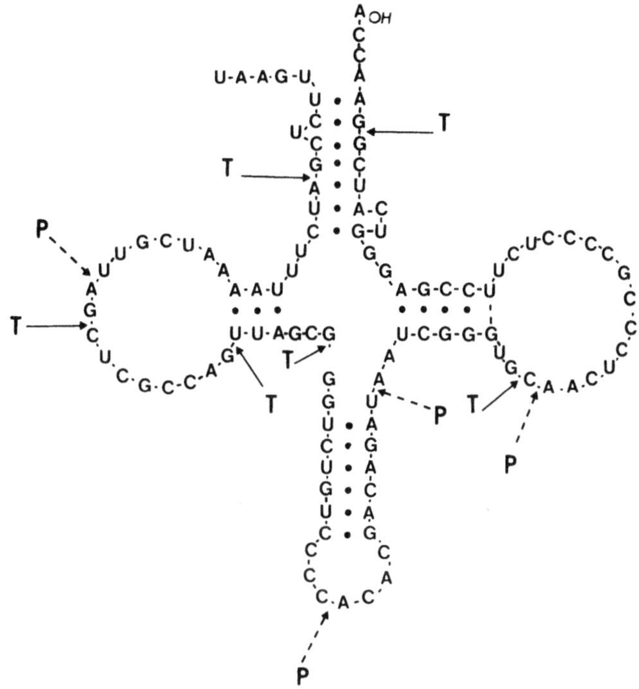

Fig. 14. A possible secondary structure for the 108 untranslated nucleotides at the 3′ end of TYMV RNA. *Letters* indicate the cleavage sites for T_1 and pancreatic RNases

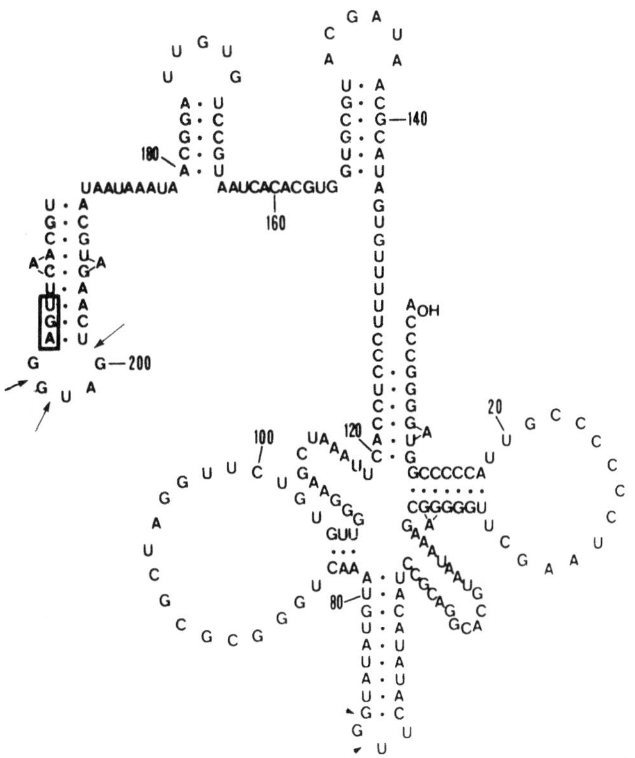

Fig. 15. A possible secondary structure for the 214 nucleotides at the 3′ end of TMV RNA. The *boxed UGA* could be a termination codon for the coat protein

BMV. The same type of observation can be made for BMV RNA's (HALL et al. 1972, KOHL and HALL 1974), the 3′ ends of which bind tyrosine. In fact four RNA's of BMV have a T1 cleavage site situated 161 bases from their 3′ terminus. These fragments can be aminoacylated. The last 161 nucleotides in each of the four RNA's have nearly the same sequence (DASGUPTA and KAESBERG 1977). This similarity in sequence is also found in the case of LMC of TMV and subgenomic RNA of TYMV with the 3′ end of the corresponding genomic RNA. The 3′ terminus of BMV RNA 4 has a possible secondary structure resembling as much as possible a clover-leaf structure. In fact, this structure is very far from a tyrosyl tRNA: only the eight bases adjacent to the 3′ terminus of the RNA 4 fragment (which was the only sequenced fragment) are identical in sequence to the eight terminal bases of tyrosine tRNA from *Torula utilis* and eleven interior bases are identical in sequence to eleven bases encompassing the anticodon region of tyrosine tRNA from *Saccharomyces cerevisiae, T. utilis,* and *E. coli*. In the case of BMV, too, tyrosine seems to be bound in conditions which do not mimic the mischarging of a tRNA by an heterologous amino acid.

g) 5′ Terminus-Linked Protein. It was shown several years ago that picornaviruses (poliovirus, mengovirus) have a 5′ terminus covalently blocked by a protein (see Sect. 3.1.2) linked to the 5′ terminal nucleotide probably by a phosphodiester bond between the N-terminal part of the protein and the residue at the 5′ of the uridine. It was also demonstrated that the RNA-linked protein was not required for the infection of cells by free RNA: indeed poliovirus mRNA extracted from polysomes lacks the 5′-linked protein. In this case the protein does not play a role in translation, but is probably involved in RNA replication (HUANG and BALTIMORE 1977, DASGUPTA et al. 1980). With respect to the 3′ end of the genome-linked protein, several situations exist in the case of RNA plant viruses.

The 3′ End is pXOH. This is the case of southern bean mosaic virus (M. ROSSMANN, personal communication) but very little is known concerning the structure of the RNA of this virus although much information is available regarding the three-dimensional structure of the protein subunit.

The 3′ End as a Poly(A)Tail. This is the case of several viruses. Most of them have an icosahedral symmetry.

The viruses with *icosahedral symmetry* belong to the group of como- and nepoviruses. In 1977 it was demonstrated that the 5′ ends of the RNA's of cowpea mosaic virus (a comovirus) have neither cap nor mono, di or triphosphate (KLOOTWIJK et al. 1977). Later it was found that a protein of a molecular weight of about 5,000 was bound covalently by an amino acid to the first 5′ nucleotide (STANLEY et al. 1978). It was also demonstrated that, as in the case of poliovirus, viral RNA is infectious even without its 5′-linked protein: however, the RNA's of the progeny have the linked protein which is then coded by the viral genome (STANLEY and VAN KAMMEN 1979). The situation is slightly different for tomato black ring virus (TBRV), a nepovirus which is often accompanied by a satellite RNA. The two main infectious RNA's of this virus have a 5′-linked protein which can be removed by proteinase K but, in contrast with CMV, the infectivity of the RNA's disappears. The genome-linked protein seems in this case necessary for infectivity. The reason for this is still unknown (HARRISON and BARKER 1978). How the protein is bound to the RNA is also unknown.

It is worth noting that the satellite RNA has 5′ and 3′ termini similar to that of the "helper virus": this raises the interesting question of the origin of the satellites: in this case, although no sequence RNA homology has been found with "helper RNA's", some similarities exist and it is difficult to accept that this satellite arose from cellular RNA as was claimed for Carna 5 of CMV.

Poly(A)Tail. Many viruses with a 5′ terminal-linked protein have poly(A)tails. This was first discovered by EL MANNA and BRUENING in 1973 for the 3′ end of cowpea mosaic virus (CMV) RNA's. Evidence has been presented that the approximately 200 adenosine residues detected are located at the 3′ end and not internally. The same type of observation was reported later with tomato black ring virus (TBRV) as well as for its satellite. Apparently, the length of the poly(A)tail varies from one preparation to another and also from one RNA to another. It is not clear whether these differences correspond to artifacts of preparation.

In the case of beet necrotic yellow vein virus RNA's (PUTZ, personal communication) the four RNA's are polyadenylated but three sets of fragments can be distinguished with a mean length of 65, 120, and 140 nucleotides, respectively. The reproducibility of these results suggests that each family of poly(A) corresponds to a viral RNA.

Barley stripe mosaic virus (BSMV) is a particular case of polyadenylated RNA virus. Depending on the strain of BSMV, the number of RNA's varies from 2 to 4 and apparently all these RNA's have a poly(A)tail (AGRANOWSKY et al. 1978), the length of which was not established. However, using poly(U)sepharose or oligo-(dT)columns, it is possible to separate polyadenylated from nonpolyadenylated RNA's. It was clear that whole preparations of BSMV RNA contain two types of molecules, poly(A)$^-$ and poly(A)$^+$. Both types of molecules seem to be infectious in contrast with the observations made by SPECTOR and BALTIMORE (1975) with poliovirus: they have demonstrated that poly(A)tails are required for infectivity. The origin and the role of the poly(A) tail is unknown in the case of plant viruses. It could be added after replication or transcribed, as for poliovirus, from a minus-strand having a poly(U)tail. However, RNA of nonpolyadenylated RNA viruses acting as messenger in the infected cells were never isolated in a polyadenylated form.

It is possible that poly(A) sequences protect the viral RNA's against the attack of cellular exonuclease. When the size of the RNA tail of poliovirus is reduced experimentally, the viral RNA is rapidly degraded.

Tobacco etch virus also has a poly(A)tail but the nature of its 5′ terminus is unknown (HARI et al. 1979). It should be emphasized that plant viruses known as having polyadenylated RNA's are relatively rare.

3.2.2 Considerations on the Role of 5′ and 3′ Ends of Plant Virus RNA's

From the above consideration, it is clear that the 3′ and 5′ ends are very different, depending on which virus is considered. It might be tempting to relate this variability *to host specificity*. In fact it is not the case either for plants or for animal viruses. Indeed, the same plant can be infected by viruses differing in their 5′ and 3′ termini. On the other hand and in view of the origin of RNA viruses, the existence of viruses with a 5′-linked genome protein or a 3′ tRNA-like structure does not fit well with the concept of RNA viruses as arising from the stabilization of mRNA's originating from the transcription of cellular DNA fragments isolated during the evolution process.

In addition to host specificity, three main events related to the behavior of the RNA are common to any virus. There are translation (and its regulation), replication, and coating. Also in these various cases it is difficult to relate the various terminal structure to any specific function.

Translation. Concerning the cap it is clear that it is not necessary for translation of viral RNA's. For example, STNV, which is uncapped, is very efficiently translated. The minor RNA 3 of AMV (see Sect. 2), which is not capped, is very well translated. However, the viral RNA's treated by decapping enzymes have lost a significant part of their stimulating properties of an in vitro proteosynthesis system. But the role of the cap is extensively discussed by VAN VLOTEN-DOTING (Chap. 11, this Vol.).

Concerning translation, both poly(A) and protein-linked genomes are apparently unnecessary except if we consider, in the latter case, the stability of the viral mRNA in the in vitro system. The same observation has been made for tRNA-like structure.

Replication of RNA (for review, see ZAITLIN 1979). It was assumed that "replicase" (or replication system) recognizes the 3′ end of viral RNA's but, in the case of plants, no firm evidence is available. It was suggested that the tRNA-like structure, which is a template for several enzymes (see above) might be a signal for "replication system" when it is charged. Indeed, some recent experiments seem to indicate that TYMV RNA could be charged in vivo (see VAN VLOTEN-DOTING, Chap. 11, this Vol.). However, since the discovery of the tRNA-like structure 10 years ago, nothing is in fact known on the biological role of this structure. The most surprising results from studies of such structures are the big differences between them and authentic tRNA's; this observation opens the way for cross-linking experiments between tRNA synthetases and their specific tRNA and the corresponding viral tRNA-like fragment, allowing an original approach of the still unsolved problem of recognition of a tRNA by its tRNA synthetase.

Have some RNA terminal structures some role in coating? This is not likely, at least in view of the very limited number of examples we have. In the case of TMV, an internal nucleotide sequence is recognized by a double disc of protein subunits (for review, see HIRTH and RICHARDS, 1980) and the 3′ terminal tRNA-like structure binding histidine does not apparently play any role. Concerning isometric viruses such as BMV, CCMV, and AMV, the relative lack of specificity of recognition between protein subunits and any viral RNA does not favor the hypothesis of a participation of a terminal structure of the RNA to viral morphogenesis. In fact, the meaning of the variability of the 3′ and 5′ termini of viral RNA is still an open question.

3.3 Plant Viral RNA's and Binding to Eukaryotic Ribosomes

Various aspects of translation of RNA viruses are extensively discussed in VAN VLOTEN-DOTING, Chapter 11, and I shall evoke now only the mechanism of recognition of viral RNA by eukaryotic ribosomes and the consequences of this mechanism on the expression of genes in the case of polycistronic RNA's. From several observations concerning animal and plant viruses, it was deduced by KOZAK (1978) that only the 5′ proximal cistron of each such RNA was translatable. The following cistrons need the formation of subgenomic RNA's as was described in the preceding section. The general rule put forth by KOZAK is that eukaryotic ribosomes are unable to recognize internal cistrons and that 40S ribosome subunits run from the cap along the leader sequence till the first AUG which, in these conditions, is necessarily the first initiation codon. Several observations are in agreement with this concept, but others do not fit well with it.

I shall briefly report some recent experiments done in my laboratory and which, together with experiments from other laboratories, may contribute to clarify the important question of the recognition of eukaryotic mRNA by homologous ribosomes.

First of all I want to emphasize that no RNA–RNA interactions between the 3′ end of *E. coli* 16S RNA and a purine-rich region located in several nucleotides upstream of the initiation codons for prokaryotic mRNA have been unambiguously discovered in the case of eukaryotes (except perhaps in the case of Rous sarcoma virus). However, in the case of TYMV RNA six consecutive base pairs can be drawn between the 3′ terminal sequence $_{HO}$UUUACU of 18S ribosomal RNA and position 93–98 of this noncoding region of turnip yellow mosaic virus RNA (BRIAND et al. 1977). Observations of the same type have been made for β globin mRNA (BARALLE and BROWNLEE 1978), for tobacco mosaic virus, and alfalfa mosaic virus. However, the very low stability of the base pairing schemes proposed makes unlikely that such interactions have a role in the recognition process between mRNA and 18S ribosomal RNA.

PINCK et al. (1979) have recently tested the in vitro formation of polysomes by AMV RNA 3 (which is bicistronic) and wheat germ ribosomes. RNA 4, which corresponds to the 3′ terminal part of RNA 3, was used as control. They found that RNA 4 associates with one ribosome but RNA 3 with several ribosomes: in fact in sucrose gradient centrifugation, disomes are found in addition to aggregates containing more than two ribosomes. The experiments performed with uniformly labeled viral RNA's 3 and 4 in the presence of sparsomycin (an inhibitor of elongation) have made it possible to identify the sequences of the RNA protected by the ribosomes. Figure 16 shows the sequences of

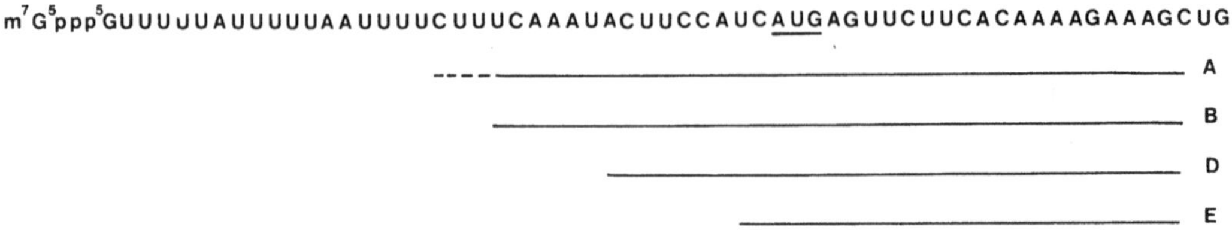

Fig. 16. Binding of ribosomes to AMV RNA 4. The *lines A, B, D, E* indicate the various fragments of RNA 4 isolated after action of the pancreatic ribonuclease and protected by the ribosomes

protected RNA 4 obtained by both T1 and pancreatic RNase digests. The lack of G in the leader sequence of RNA 4 does not allow the localization of the 5′ terminus of this protected sequence. However, partial RNAse digest shows that about 40 nucleotides are protected; the AUG, which is surely the initiation codon, taking up a central position. The same experiment performed with RNA 3 has shown that, in addition to sequences belonging to the 5′ terminus of this capped RNA, a part of the leader sequence of RNA 4 (which, in this case, is included at the 3′ end of RNA 3) is protected, but that only a small number of ribosomes reacts with it. On the other hand, the protected fragment does not include the part of the sequence on the right of the AUG, suggesting a wrong positioning of the ribosomes. This could suggest that inner initiation site is recognized. But experiments performed with RNA 3 suggest that the RNA molecules could have hidden breaks which play the role of a free 5′OH end and that the putative inside initiation site is probably an artifact.

This point of view was reinforced by recent experiments performed in my laboratory (PINCK et al. 1981). They first established the sequence of the first 318 nucleotides of RNA 3 and that of 79 nucleotides preceding the leader sequence of RNA 4 when included in RNA 3. The complete sequence of RNA 4 has been established by BREDERODE et al. (1980). In addition to this sequence determination, they have tested the ribosome binding capacity of both RNA 3 and 4 and of some of their fragments. They also discovered in all preparation of AMV RNA the existence of an additional RNA 3 (RNA 3′) lacking the 155 5′-terminal nucleotides of the corresponding leader sequence and devoid of cap (see Fig. 9b). Figure 17 shows gradient sucrose analysis of binding of viral RNA's 3 and 3′, 4, and some of their fragments to ribosomes in the presence and absence of sparsomycin. It is clear that RNA 3 binds several ribosomes. In addition to a small peak of monosomes, a peak corresponding to an aggregate of neighbouring ribosomes was observed. RNA 3′ seems to bind only one ribosome with perhaps a very small quantity of disomes. RNA 4 binds only one ribosome. Analysis of the sequences protected by RNA shows that for RNA 3, sequences from 19 to 61, from 130 to 165, and from 226 to 279, are protected (see Fig. 9b). Some other fragments in small quantity are present in isolated RNA's. This suggests that three ribosomes can bind to this long sequence of RNA 3. In contrast, RNA 3′ binds only one ribosome. In addition, binding experiments have been performed with fragments arising from isolation of part of RNA 3 preceding the leader sequence of the coat

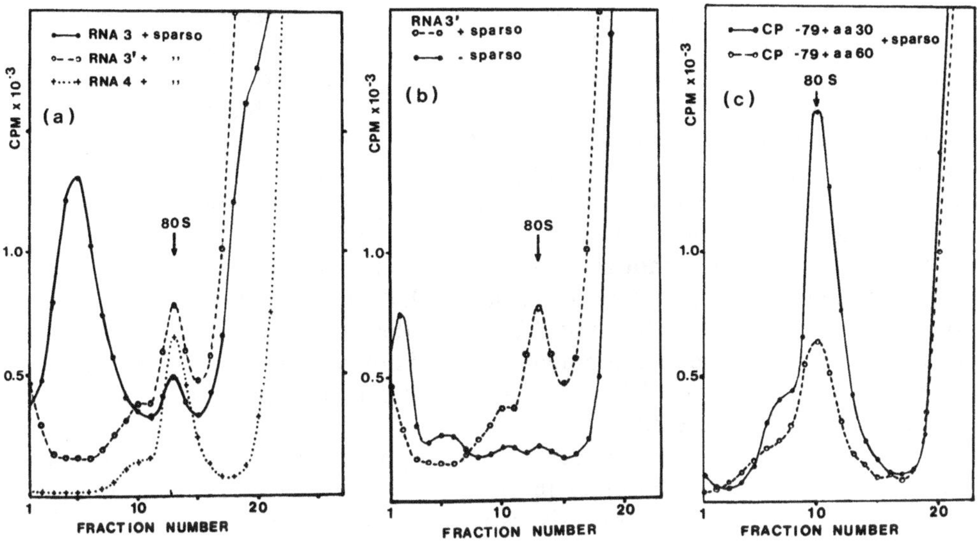

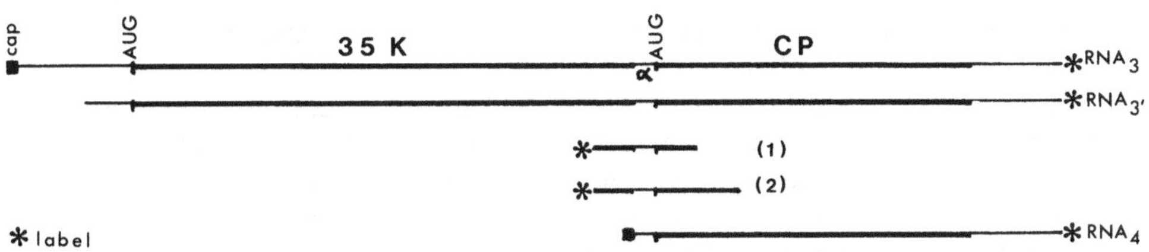

Fig. 17a–c. Sucrose gradient analysis of RNA 3, 3′, 4, and fragments including intercistronic region between the end of the first cistron of RNA 3 and the beginning of the coat protein cistron. α leader sequence of RNA 4. *1* and *2* fragments containing a part of the 5′ end of RNA 4 plus 79 nucleotides of the end of the putative preceding cistron (combined fragments). **a** Sucrose gradient analysis of the binding of RNA's 3, 3′ and 4 to ribosomes of wheat germ extract in the presence of sparsomycin. **b** Binding of RNA 3′ to ribosomes in the presence and the absence of sparsomycin. **c** Binding of combined fragments 1 and 2 to ribosomes in the presence of sparsomycin

protein cistron, as indicated in Fig. 17. The sequence of these fragments is known and includes a more or less long part of the coat protein cistron. In any case the isolated fragments contained a sequence of 115 nucleotides preceding the first AUG. Figure 17 shows that the sucrose gradient analysis gives the same patterns as those obtained with RNA 4. The artificially lengthened leader sequence of RNA 4 is thus unable to bind more than one ribosome. This observation and the fact that there is no relation between the number of ribosomes bound and the length of leader sequence suggests strongly that some specific "entry site" for ribosome exists (AHLQUIST et al. 1979). Indeed, it was clearly demonstrated that for TMV the 69 nucleotide-long leader sequence binds two ribosomes (FILIPOWICZ and HAENNI 1979). The same has been demonstrated for BMV (91 nucleotides: AHLQUIST et al. 1979) and was suggested for TYMV (92 nucleotides: FILIPOWICZ and HAENNI 1979). But RNA 3′ having a 103 nucleotide-long leader sequence binds only one ribosome. The reported experiments seem to indicate that some specific structure on mRNA is recognized by eukaryotic ribosomes. The observations reported with AMV are in relative agreement with those of AHLQUIST et al. (1979) obtained with BMV. It is also clear that only one "entry site" seems to exist on the fragments of RNA 3 containing the leader sequence of the coat protein gene, but the recent nucleotide sequence determination of the fragment shows the presence of a AUG not very far from its 5′ end which could be a putative initiation site for the fragment, although it is followed very soon by nonsense codons. This experiment does not prove (or disprove) that ribosomes (40S or 80S) move from the binding site to the first AUG. On the other hand, the role of the cap appears as not essential although it is not known from the reported experiments whether uncapped RNA's or fragments can be capped in the in vitro proteosynthesis system used.

It is of course tempting to suggest nonrecognition specificity of eukaryotic mRNA by eukaryotic ribosomes in view of the preceding considerations on the absence of common characteristics in leader sequences of these mRNA's, but I wish to emphasize the fact that, in several cases, and especially in that of the tRNA-like structure of the 3′ end of plant viral RNA's, very different sequences seem to be recognized specifically by the same enzyme. Similar spatial configuration could probably arise from different sequences.

In spite of the preliminary nature of the preceding considerations, they in general confirm and precise the hypothesis of KOZAK (1978) the main characteristic of which is that the first AUG after the 5′ terminus is the initiation codon and consequently that eukaryotic mRNA's, even when they carry several genes, behave as a monocistronic RNA. However, there are several examples which do not fit with Kozak's scheme.

This is the case for carnation mottle virus RNA which carries three genes which translated in vitro yield three unrelated proteins (SALOMON et al. 1978).

Together these three proteins represent the total genetic information present in the RNA. Disappearance of the large protein is related to limited phosphorolysis of the 3′ terminus of the RNA which seems to play a regulatory role. It seems that this virus deserves more work. In vitro translation of tobacco necrosis virus RNA yields coat protein and two other large proteins (SALVATO and FRAENKEL-CONRAT 1977), suggesting a translation of a prokaryotic type, but not enough is known concerning the chemical relations between the three proteins and they might arise from "read-through" of termination codons (see VAN VLOTEN-DOTING, Chap. 11). I will now describe the results of experiments performed very recently in my laboratory with beet necrotic yellow vein virus RNA's (BNYVV). This virus has four RNA's (see above). Each of them was carefully purified two times in denaturating conditions. Labeling of the 3′ end of each RNA by pCp and RNA ligase indicates that each preparation is homogeneous. However, the in vitro translation of each RNA yields, in addition to a protein of a relatively large molecular weight, coat protein. In the case of RNA 3, which codes apparently only for this coat protein, various treatments and especially proteinase K treatment do not qualitatively or quantitatively change the translation products. This might be due to the fact that the coat protein gene is present in every viral RNA and can be translated from each of them (each behaving like a polycistronic prokaryotic mRNA) or that RNA is linked to RNA's 1, 2, and 4 in such a manner that disruption of noncovalent interactions or protein-linkage is inefficient. The behaviour of this RNA deserves further research.

In conclusion, plant viral RNA's have been very useful for understanding how eukaryotic mRNA works and it seems that they will be a good tool, in the near future, to elucidate the mechanism of regulation of the expression of eukaryotic genes at the level of translation.

Note Added in Proof. Since this paper was written new data are available concerning both DNA and RNA plant viruses. We indicate below only the very important points which are worthy of note.

RNA Viruses. The complete sequences of BMV RNA 3 and RNA 4 are known (AHLQUIST et al., 1981) and show that for RNA 3 a noncoding sequence of 250 nucleotides joining the 3a gene and the coat protein gene.

In the case of the Bromoviruses and Cucumoviruses, sequence data show a marked intraviral homology at the 3′ termini which is reflected in a high degree of secondary structure homology over at least 190 nucleotides (AHLQUIST et al., 1981). A sequence homology has also been reported concerning the 3′ ends of various strains of AMV. The conservation of the 3′ secondary structure of various RNAs of plant viruses is probably related to recognition signals for the replicase.

Concerning cowpea mosaic virus it was recently reported that Middle RNA of this virus codes for two overlapping proteins which are precursors of the two viral coat proteins and are matured by proteases coded by Bottom RNA. The overlapping of the two precursors suggests the existence of two initiation sites for protein synthesis and appears as an exception to the rule of Kozak concerning the recognition of eucaryotic mRNAs by ribosomes (FRANSSEN et al., 1982).

DNA Viruses: Recently the in vivo (COVEY and HULL, 1981; DUDLEY et al., 1982; GUILLEY et al., 1982) and *in vitro* GUILLEY et al., 1982) transcription products of various CaMV strains have been described. There are two long transcripts corresponding to the total length of the α-strand, a 19 S transcript corresponding to gene VI, and an 8 S RNA containing the nonsense region of the α-strand. The mode of expression of the genes other than gene VI is at the present time an unsolved problem. The beginning and end of each viral transcript has been localized and it has been demonstrated that the longest species probably arises by transcription of a supercoiled (non gapped) DNA form present in infected plants. This form is never found in virions (MÉNISSIER et al., 1982). Finally, total sequences for two other CaMV strains (GARDNER et al., 1981; BALÀZS et al., 1982) have been established and have shown very few differences in sequence with Cabb-S strain.

For the first time proof has been obtained at the molecular level that plants are able to carry out recombination and ligation of DNA HOWELL et al., 1981; LEBEURIER et al., 1982); it has been demonstrated that short foreign DNA sequences inserted at specific places in the viral DNA are replicated as part of the viral genome.

References

Abou Haidar M, Hirth L (1977) 5′terminal structure of tobacco rattle virus RNA: evidence for polarity of reconstitution. Virology 76:173–185

Agranowsky AA, Dolja VV, Kavsan VM, Atabekow JG (1978) Detection of polyadenylate sequences in RNA components of barley stripe mosaic virus. Virology 91:95–105

Ahlquist P, Dasgupta R, Shi D, Zimmern D, Kaesberg P (1979) Two step binding of eukaryotic ribosomes to brome mosaic virus RNA 3. Nature (London) 281:277–282

Ahlquist P, Luckow V, Kaesberg P (1981a) Complete sequence of BMV RNA 3. J Mol Biol 153:23–38

Ahlquist P, Dasgupta R, Kaesberg P (1981b) Near identity of 3′ RNA secondary structure in bromoviruses and cucumber mosaic virus. Cell 23:183–189

Al Ani R, Pfeiffer P, Lebeurier G, Hirth L (1979a) The structure of cauliflower mosaic virus. I. pH-induced structural changes. Virology 93:175–187

Al Ani R, Pfeiffer P, Lebeurier G (1979b) The structure of cauliflower mosaic virus. II. Identity and location of the viral polypeptides. Virology 93:188–197

Al Ani R, Pfeiffer P, Whitechurch O, Lesot A, Lebeurier G, Hirth L (1980) A virus specified protein produced upon infection by cauliflower mosaic virus (CaMV). Ann Virol (Inst Pasteur) 131E:33–53

Balàzs E, Guilley H, Jonard G, Richards KE (1982) Nucleotide sequence of cauliflower mosaic virus DNA, isolate D/H. (in press)

Baralle FE, Brownlee GG (1978) AUG is the only recognisable signal sequence in the 5′ non coding regions of eucaryotic mRNA. Nature (London) 274:84–87

Beachy RN, Zaitlin M (1977) Characterization and in vitro translation of the RNAs from less – than full length – virus related nucleoprotein rods present in tobacco mosaic virus preparations. Virology 81:160–169

Bishop DH (1977) Virion polymerases. In: Fraenkel-Conrat H, Wagner RA (eds) Comprehensive virology Vol X. Plenum, New York, pp 117–278

Bishop DHL, Shope RE (1979) Bunyaviridae. In: Fraenkel-Conrat H, Wagner RA (eds) Comprehensive virology Vol XIV. Plenum, New York, pp 1–132

Bol JF, Brederode FT, Janze GC, Rauh DC (1975) Studies on sequence homology between the RNAs of alfalfa mosaic virus. Virology 65:1–15

Brederode FT, Koper-Zwarthoff EC, Bol JF (1980) Complete nucleotide sequence of alfalfa mosaic virus RNA 4. Nucl Acids Res 8:2213–2223

Briand JP, Jonard G, Guilley H, Richards KE, Hirth L (1977) Nucleotide sequence (n = 159) of the amino acid accepting 3′ OH extremity of turnip yellow mosaic virus RNA and the last portion of its coat protein cistron. Eur J Biochem 72:453–463

Bruening G, Beachy RN, Scalla R, Zaitlin M (1976) In vitro and in vivo translation of the ribonucleic acid of cowpea strain of tobacco mosaic virus. Virology 71:498–517

Brunt AA, Barton RJ, Tremaine JH, Stace-Smith R (1975) The composition of cauliflower mosaic virus protein. J Gen Virol 27:101–106

Chauvin C, Jacrot B, Lebeurier G, Hirth L (1979) The structure of cauliflower mosaic virus: a neutron diffraction study. Virology 96:640–641

Covey SN, Hull R (1981) Transcription of cauliflower mosaic virus DNA. Detection of transcripts, properties and location of the gene encoding the virus inclusion body protein. Virology 111:463–474

Darby G, Minson AC (1973) The structure of tobacco rattle virus ribonucleic acids: common nucleotide sequences in the RNA species. J Gen Virol 21:285–295

Dasgupta A, Zabel P, Baltimore D (1980) Dependence of the activity of the poliovirus replicase on a host cell protein. Cell 19:423–429

Dasgupta R, Kaesberg P (1977) Sequence of an oligonucleotide derived from the 3′ end of each of the four brome mosaic viral RNAs. Proc Natl Acad Sci USA 74:4900–4904

Dasgupta R, Shih DS, Saris C, Kaesberg P (1975) Nucleotide sequence of a viral RNA fragment that binds to eukaryotic ribosomes. Nature (London) 256:624–628

Davies JW (1979) The translation of plant viral RNA. In: Hall TC, Davies JW (eds) Nucleic acids in plants Vol II. CRC Press, Boca Raton, pp 111–149

Davies JW, Aalbers AMJ, Stuck EJ, Kammen Van A (1977) Translation of cowpea mosaic virus RNA in a cell free extract from wheat germ. FEBS Lett 77:265–269

Dudley RK, Odell JT, Howell SH (1982) Structure and 5′ termini of the large and 19 S RNA transcripts encoded by the cauliflower mosaic virus genome. Virology 117:19–28

El Manna MM, Bruening G (1973) Polyadenylate sequences in the ribonucleic acids of cowpea mosaic virus. Virology 56:198–206

Filipowicz W, Haenni AL (1979) Binding of ribosomes to 5′ terminal leader sequences of eukaryotic messenger RNAs. Proc Natl Acad Sci USA 76:3111–3115

Filipowicz W, Szczesna E, Wodmar-Filipowicz A (1978) Role of 5′ terminal 7-methylguanosine in mRNA function: correlation of the effect of enzymatic mRNA decapping with effect of cap analogs on in vitro translation of TMV RNA and globin mRNA. Biochimie 60:569

Franck A, Guilley H, Jonard G, Richards K, Hirth L (1980) Nucleotide sequence of cauliflower mosaic virus DNA. Cell 21:285–294

Franssen H, Golbach R, Broekhuysen MM, Kammen Van A (1982) Expression of Middle component RNA of cowpea virus: In vitro generation of a precursor to both capsid proteins by Bottom-component RNA encoded protease from infected cells. J Virol 41:8–17

Fritsch C, Mayo MA, Murant AF (1978) Translation of the satellite RNA of tomato black ring virus in vitro and in tobacco protoplasts. J Gen Virol 40:587–593

Fritsch C, Mayo MA, Murant AF (1980) Translation products of genome and satellite RNAs of tomato black ring virus. J Gen Virol 46:381–389

Gardner RC, Howarth AJ, Hahn P, Brown-Luedi M, Shepherd RJ, Messing J (1981) The complete nucleotide sequence of an infectious clone of cauliflower mosaic by M13mp7 shotgun sequencing. Nucleic Acids Res 9:2871–2888

Gerlinger P, Mohier E, Le Meur MA, Hirth L (1977) Monocistronic translation of alfalfa mosaic virus RNAs. Nucl Acids Res 4:813–826

Giegé R, Briand JP, Mengual R, Ebel JP, Hirth L (1978) Valylation of the two RNA components of TYMV and specificity of the RNA aminoacylation reaction. Eur J Biochem 84:251–256

Gierer A, Schramm G (1956) Infectivity of RNA from tobacco mosaic virus. Nature (London) 177:702–703

Gillaspie AG, Bancroft JB (1965) Properties of ribonucleic acid from alfalfa mosaic virus and related component. Virology 27:391–397

Gould A, Symons RH (1978) Alfalfa mosaic virus. Determination of the sequence homology between the four RNA species and comparison with the four RNA species of cucumber mosaic virus. Eur J Biochem 91:269–278

Guilley H, Briand JP (1978) Nucleotide sequence of turnip yellow mosaic virus coat protein messenger RNA. Cell 15:113–122

Guilley H, Jonard G, Kukla B, Richards KE (1979) Sequence of 1000 nucleotides at the 3′ end of tobacco mosaic virus RNA. Nucl Acids Res 6:1287–1308

Guilley H, Dudley RK, Jonard G, Balàzs E, Richards KE (1982) Transcription of cauliflower mosaic virus DNA: Detection of promotor sequences by in vitro transcription and characterisation of the polyadenylated transcripts found in virus infected plants. Cell (in press)

Hall TC, Shi DS, Kaesberg P (1972) Enzyme mediated binding of tyrosine to brome mosaic virus ribonucleic acid. Biochem J 129:969–976

Hari V, Siegel A, Rozek C, Timberlake WE (1979) The RNA of tobacco etch virus contains poly (A). Virology 92:568–571

Harrison BD, Barker H (1978) Protease sensitive structure needed for infectivity of nepovirus RNA J Gen Virol 40:711–715

Higgins TJV, Goodwin PB, Whitfeld PR (1976) Occurrence of short particles in beans infected by the cowpea strain of TMV. II. Evidence that short particles contain the cistron for the coat protein. Virology 72:486–493

Hirth L (1976) Reconstitution of helical viruses other than TMV. In: Fraenkel-Conrat H, Wagner RA (eds) Comprehensive virology Vol XI. Plenum, New York, pp 39–63

Hirth L, Richards K (1981) Tobacco mosaic virus: Model for structure and function of a simple virus. In: Lauffer MA, Bang FB, Maramorosch K, Smith KM (eds) Advances in virus research, Vol XXVI. Academic Press, London New York, pp 145–199

Hohn T, Hohn B, Lesot A, Lebeurier G (1980) Restriction map of native and cloned cauliflower mosaic virus DNA. Gene 11:23–32

Houwing CJ, Jaspars EMJ (1978) Coat protein binds to 3′ terminal part of RNA 4 of alfalfa mosaic virus. Biochemistry 17:2927–2933

Howell SH, Walker LL, Dudley RK (1980) Cloned cauliflower mosaic virus DNA infects turnip (*Brassica rapa*). Science 208:1265–1269

Howell HS, Walker LL, Walden RM (1981) Rescue of in vitro generated mutants of cloned cauliflower mosaic virus genome in infected plants. Nature 293:483–486

Huang AS, Baltimore D (1977) Defective interfering animal viruses. In: Fraenkel-Conrat H, Wagner RA (eds) Comprehensive virology Vol X. Plenum, New York, pp 73–116

Hull R (1979) The DNA of plant DNA viruses. In: Hall TC, Davies JW (eds) Nucleic acids in plants Vol I CRC Press, Boca Raton, pp 3–29

Hull R, Shepherd RJ (1976) The coat protein of cauliflower mosaic virus. Virology 70:217–220

Hull R, Covey SN, Stanley J, Davies JW (1979) The polarity of the cauliflower virus genome. Nucl Acids Res 7:669–677

Jonard G, Richards KE, Mohier E, Gerlinger P (1978) Nucleotide sequence at the 5′ extremity of tobacco mosaic virus RNA. II. The coding region (N=69–236). Eur J Biochem 84:521–531

Hurk Van den J, Tas PWL, Peters D (1977) The ribonucleic acid of tomato spotted wilt virus. J Gen Virol 36:81–91

Kaper JM, Tousignant ME (1977) Cucumber mosaic virus associated RNA 5. I. Role of host plant and helper strain in determining amount of associated RNA 5 with virions. Virology 80:186–195

Kaper JM, Tousignant ME, Lot H (1976) A low molecular weight replicating RNA associated with a divided genome plant virus, defective or satellite RNA. Biochem Biophys Res Commun 72:1237–1243

Kassanis B (1972) Properties and behaviour of a virus depending for its multiplication on another. J Gen Microbiol 27:477

Kassanis B, Phillips MP (1970) Serological relationship of strains of tobacco necrosis virus and their ability to activate strains of satellite virus. J Gen Virol 9:119–126

Klein C (1976) Variété structurale du RNA de deux Tymovirus. Thesis, Univ Strasbourg

Klein C, Fritsch C, Briand JP, Richards KE, Jonard G, Hirth L (1976) Physical and functional heterogeneity of TYMV RNA: evidence for the existence of an independent messenger coding for coat protein. Nucl Acids Res 3:3043–3062

Klootwijk J, Klein I, Zabel P, Kammen Van A (1977) Cowpea mosaic virus RNAs have neither m^7 GpppN... nor mono-, di- or triphosphates at their 5′ ends. Cell 11:73–82

Knowland J, Hunter T, Hunt T, Zimmern D (1975) Translation of tobacco mosaic virus RNA and isolation of the messenger for TMV coat protein. In: Haenni AL, Beaud G (eds) In vitro transcription and translation of viral genomes. Proc INSERM-EMBO colloquium, Paris, pp 211–216

Kohl RJ, Hall TC (9174) Aminocylation of RNA from several: amino acid specificity and differential activity of plant, yeast, and bacterial synthetases. J Gen Virol 25:257–261

Koper-Zwarthoff EC, Brederode FT, Walstra P, Bol JF (1979) Nucleotide sequence of the 3′-noncoding region of alfalfa mosaic virus RNA 4 and its homology with the genomic RNAs. Nucl Acids Res 7:1887–1900

Kozak M (1978) How do eukaryotic ribosomes select initiation regions in messenger RNA. Cell 15:1109–1123

Kukla B, Guilley H, Jonard G, Richards KE, Mundry RW (1979) Characterization of long guanosine-free RNA sequences from the Dahlemense and U2 strains of tobacco mosaic virus. Eur J Biochem 98:61–66

Lane LC (1979) The nucleic acids of multipartite, defective and satellite plant viruses. In: Hall TC, Davies JW (eds) Nucleic acids in plants Vol II. CRC Press, Boca Raton, pp 65–110

Lebeurier G, Hirth L, Hohn T, Hohn B (1980) Infectivity of native and cloned DNA of cauliflower mosaic virus. Gene 12:139–146
Lebeurier G, Hirth L, Hohn B, Hohn T (1982) In vivo recombination of cauliflower mosaic virus DNA. Proc Natl Acad Sci USA 79:2933–2936
Leung DW, Gilbert CW, Smith RE, Sasavag NL, Clark JM (1976) Translation of satellite tobacco necrosis virus ribonucleic acid by an in vitro system from wheat germ. Biochemistry 15:4943–4950
Leung DW, Browning KS, Heckman JE, RajBhandary UL, Clark JM (1979) Nucleotide sequence of the 5′ terminus of satellite tobacco necrosis virus ribonucleic acid. Biochemistry 18:1361–1366
Lot H, Jonard G, Richards K (1977) Cucumber mosaic virus RNA 5. Partial characterization and evidence for no large sequence homologies with genomic RNAs. FEBS Lett 80:395–400
Matthews REF (1970) Turnip yellow mosaic virus. CAB/AAB Descriptions of plant viruses Set 1, sheet 2
Matthews REF (1979) Classification and nomenclature of viruses. Intervirology 12:129–296
Maxam AM, Gilbert W (1977) A new method for sequencing DNA. Proc Natl Acad Sci USA 74:560–564
Ménissier J, Lebeurier G, Hirth L (1982) Free cauliflower mosaic virus supercoiled DNA in infected plants. Virology 117:322–328
Merregaert J, Van Emmeld J, Fiers W (1979) 3′ terminal nucleotide sequence (N=41) of satellite tobacco necrosis virus RNA. Virology 98:182–190
Mohier E, Hirth L, Le Meur MA, Gerlinger P (1976) Analysis of alfalfa mosaic virus 17 S RNA translation products. Virology 71:615–618
Odell JT, Howell SH (1980) The identification, mapping and characterization of mRNA for P66, a cauliflower mosaic virus-coded protein. Virology 102:349–359
Owens RA, Kaper JM (1977) Cucumber mosaic virus-associated RNA 5. II. In vitro translation in a wheat germ protein-synthesis system. Virology 80:196–203
Pelham HRB (1978) Leaky UAG termination codon in tobacco mosaic virus RNA. Nature (London) 272:469–471
Pinck L, Pinck M (1979) Sequence homology at the 3′-ends of alfalfa mosaic virus RNAs. FEBS Lett 107:61–65
Pinck L, Franck A, Fritsch C (1979) Formation of ribosome-RNA initiation complexes with alfalfa mosaic virus RNA 4 and 5. Nucl Acids Res 7:151–166
Pinck M, Yot P, Chapeville F, Duranton H (1970) Enzymatic binding of valine to the 3′ end of TYMV RNA. Nature (London) 226:954–956
Pinck M, Fritsch C, Ravelonandro M, Thivent C, Pinck L (1981) Binding of ribosomes to the 5′ leader sequence of RNA 3 from alfalfa mosaic virus. Nucl Acids Res 9:1087–1099
Pleij CWA, Mellama JR, Noort A, Bosch L (1976) The occurrence of the coat protein messenger RNA in the minor components of turnip yellow mosaic virus. FEBS Lett 80:19–22
Rawles WE, Leung WC (1979) Arenaviridae. In: Fraenkel-Conrat H, Wagner RA (eds) Comprehensive virology Vol XIV. Plenum, New York, pp 157–185
Reddy DVR, Kimura I, Black LM (1974) Coelectrophoresis of an RNA from wound tumor and rice dwarf viruses. Virology 60:293–296
Richards KE, Guilley H, Jonard G, Hirth L (1978) Nucleotide sequence at the 5′ extremity of tobacco mosaic virus RNA. I. The non coding region (Nucleotides 1–68) Eur J Biochem 84:513–519
Richards KE, Jonard G, Jacquemond M, Lot H (1978) Nucleotide sequence of cucumber mosaic virus-associated RNA 5. Virology 89:395–408
Salomon R, Bar-Joseph M, Soreq H, Gozes I, Littauer UZ (1978) Translation in vitro of carnation mottle virus RNA. Regulatory function of the 3′-region. Virology 90:288–298
Salvato MS, Fraenkel-Conrat H (1977) Translation of tobacco necrosis virus and its satellite in a cell free wheat germ system. Proc Natl Acad Sci USA 74:2288–2292

Shepherd RJ, Wakeman RJ, Romanko RR (1968) DNA in cauliflower mosaic virus. Virology 36:150–152

Shoulder A, Darby G, Minson T (1974) RNA-RNA hybridation using ^{125}I-labeled RNA from tobacco necrosis virus and its satellite. Nature (London) 251:733–734

Siegel A, Hari V, Montgomery I, Kolacz R (1976) A messenger for capsid protein isolated from tobacco mosaic virus infected tissue. Virology 73:363–371

Spector DH, Baltimore D (1975) Poly (A) on mengovirus. RNA. J Virol 16:1081

Stanley J, Kammen Van A (1979) Nucleotide sequences adjacent to the proteins covalently linked to the cowpea mosaic virus genome sequence determination after in vitro labelling using RNA ligase. Eur J Biochem 101:45–49

Stanley J, Rottier P, Davies JW, Zabel P, Kammen Van A (1978) A protein linked at the 5′ termini of both RNA components of the cowpea mosaic virus genome. Nucl Acids Res 5:4505–4522

Szybiak U, Bouley JP, Fritsch C (1978) Evidence for the existence of a coat protein messenger RNA associated with the top component of three tymoviruses. Nucl Acids Res 5:1821–1831

Tezuka N, Taniguchi T (1972a) Stepwise degradation of cauliflower mosaic virus by pronase. Virology 48:142–146

Tezuka N, Taniguchi T (1972b) Structural protein of cauliflower mosaic virus. Virology 48:297–299

Vloten-Doting Van L, Jaspars EMJ (1977) Plant covirus systems: Three components systems. In: Fraenkel-Conrat H, Wagner RA (eds) Comprehensive virology, Vol XI. Plenum, New York, pp 1–44

Vloten-Doting Van L, Dingjan-Versteegh A, Jaspars EMJ (1970) The nucleoprotein components of alfalfa mosaic virus necessary for infectivity. Virology 40:419–430

Volovitch M, Drugeon G, Yot P (1978) Studies on the single-stranded discontinuity of the cauliflower mosaic virus genome. Nucl Acids Res 8:2913–2925

Volovitch M, Dumas JP, Drugeon G, Yot P (1977) Single stranded interruptions in cauliflower mosaic virus DNA. In: Bogorad L, Weil JH (eds) Acides nucléiques et synthèse des protéines chez les végétaux. CNRS, Paris, pp 635–639

Waterworth HE, Kaper JM, Toussignant ME (1979) Carna 5, the small cucumber mosaic virus-dependent replicating RNA, regulates disease expression. Science 204:845–847

Wimmer E, Chang AY, Clark JM, Reichmann ME (1968) Sequence studies of satellite tobacco necrosis virus RNA. Isolation and characterisation of a 5′ terminal trinucleotide. J Mol Biol 38:59–73

Zaitlin M (1979) The RNAs of monopartite plant viruses. In: Hall TC, Davies JW (eds) Nucleic acids in plants, Vol II, CRC Press, Boca Raton, pp 31–64

11 Translation of Plant Virus RNA's

L. VAN VLOTEN-DOTING and L. NEELEMAN

1 Introduction

In vitro translation of plant virus RNA's yields, besides valuable information about the proteins encoded by these RNA's, information about the protein-synthesizing machinery of the host.

In this chapter we will confine ourselves to results obtained with plant viruses with a genome consisting of single-stranded RNA(s) with the plus polarity (genomic RNA has the same polarity as messenger RNA). In this type of viruses the genetic information is often not located in one RNA molecule, but distributed over two or three different RNA's (for reviews see BRUENING 1977, VAN VLOTEN-DOTING and JASPARS 1977). Besides the genomic RNA(s), virus preparations sometimes contain subgenomic mRNA(s) (mostly coding for the coat protein). With some viruses these subgenomic RNA(s) are not encapsidated and they can only be found in the infected cell. The different RNA's (genomic as well as subgenomic) of one particular virus are numbered in order of decreasing molecular weights. Up to now there are no indications for splicing of plant virus RNA's.

Recently two review articles on translation of plant virus RNA's have been published (DAVIES 1979, ATABEKOV and MOROZOV 1979). In both papers the data are discussed per virus, and an extensive survey of the size of the proteins directed by the different RNA's is given. For this type of information the reader is referred to the papers cited above. Information about plant virus RNA's may be found in LANE (1979), ZAITLIN (1979) and HIRTH, Chapter 10, this Volume.

In the present contribution we have concentrated on the possible functions of the structural features of the RNA, and on the common modes of expression of the genetic information employed by different viruses.

In Section 2 the viruses are discussed according to the type of structure present at the 5′ and 3′ terminus of their RNA(s), respectively. It is remarkable that for all plant viruses studied so far, all RNA's (genomic as well as subgenomic) of one particular virus have always the same type of structure at their termini. Section 3 deals with the fidelity of translation in vitro. The various strategies employed by plant viruses to express their genetic information are outlined in Section 4. Followed by two sections on regulation of translation; virus versus host (Sect. 5) and regulation of the virus information (Sect. 6). In Section 7 the possible functions of the virus-coded proteins are discussed.

A list of abbreviations of plant viruses can be found on p. XVIII

Table 1. Terminal structures of plant virus RNA's

Type of genome	Virus group	Members	Virus	Terminal structure 5′	(Ref)	3′	(Ref)
Monopartite	tobamo	10	TMV	Cap	(1, 2)	tRNA his	(3)
			CcTMV	–		tRNA val	(4)
	tymo	16	TYMV	Cap	(5)	tRNA val	(6)
			EMV	–		tRNA val	(6)
	potex	8	PMV	Cap	(7)	–	
			PVX	Cap	(8)	pX_{OH}[a]	(8)
	poty	34	TEV	Protein	(9)	poly(A)	(10)
			SBMV	Protein	(11)	pX_{OH}	(11)
			TRosV	Protein	(12)	pX_{OH}	(12)
			CarMV	–		pX_{OH}[a]	(13)
			TNV	ppX	(14)	–	
Bipartite	como	10	CPMV	Protein	(15, 16)	poly(A)	(17)
			BPMV	–		poly(A)	(18)
			RCMV	–		poly(A)	(19)
			EAMV	Protein	(20)	–	
			SqMV	Protein	(20)	–	
	nepo	10	TBRV	Protein	(21)	poly(A)	(22)
			TRSV	Protein	(21, 23)	poly(A)	(22)
			RRV	–		poly(A)	(22)
			Tomato ringspot virus	–		poly(A)	(22)
			Strawberry latent ringspot virus	–		poly(A)	(22)
	tobra	2	TRV	Cap	(24)	pX_{OH}[a]	(25)
?[b]	hordei	3	BSMV	Cap	(26)	tRNA tyr	(27)
Tripartite	bromo	3	BMV	Cap	(28)	tRNA tyr	(29)
			CCMV	–		tRNA tyr	(30)
	cucumo	3	CMV	Cap	(31)	tRNA tyr	(30)
	Ilar	14	AMV[c]	Cap	(32)	pX_{OH}	(33)
			TSV	–		pX_{OH}	(33)
Satellite			STNV	ppX	(13)	pX_{OH}[a]	(34)
			CARNA[d]	Cap	(35)	pX_{OH}	(36)

Grouping of viruses according to MATTHEWS (1979), except grouping of AMV

[a] Indicates that the 3′ terminus is pX_{OH}, and that there are no reports about amino acylation

[b] Number of genome parts of hordei viruses is 2 or 3 (see, e.g., AGRANOVSKY et al. 1979, 1981 and BOYKOV et al. 1981 and references there in).

[c] Officially AMV is classified separately, however, there is increasing evidence that AMV belongs to the ilar group (HALK and FULTON 1978 and references therein)

[d] CARNA is also called SAT RNA (MOSSOP and FRANCKI 1979)

2 Virus RNA Structures

2.1 Structure at the 5′ Terminus

Three different types of structures have been found at the 5′ termini of plant virus RNA's: "cap", genome-linked protein or (p)ppX (cf. HIRTH, Chap. 10, this Vol.).

2.1.1 $m^7G^{5'}ppp^{5'}X^{(m)}pY^{(m)}p\ldots = $ "cap"

Like most eukaryotic mRNA's a number of plant virus RNA's (Table 1) have been shown to be capped. It is remarkable that all caps found to date in plant viruses are devoid of the methyl groups at the ribose moiety of nucleosides X and Y. The role of cap structures in the biosynthesis of proteins has been studied extensively (for review see FILIPOWICZ 1978, see also SONENBERG et al. 1979, FRESNO and VASQUEZ 1980). A summary of the present state of knowledge is given here.

The presence of the cap facilitates the formation of the initiation complex, probably by enhancing the affinity of the RNA for one of the initiation factors. In polycistronic mRNA's mainly the 5′ proximal cistron seems to profit from the presence of the cap (PATERSON and ROSENBERG 1979, VAN STEEG et al. 1979). In in vitro systems the cap protects the RNA from digestion by exonucleases. The presence of the cap is not obligatory for the initiation of protein biosynthesis. The role played by the cap during initiation of protein biosynthesis is more important for some RNA's than for others. Furthermore the presence of the cap is more important at low than at high mRNA concentration, at low than at high salt concentration. Cap analogs inhibit the translation of capped as well as of uncapped mRNA's. Under some conditions uncapped mRNA's can outcompete capped mRNA's (HERSON et al. 1979).

Initially it was found that the rabbit reticulocyte lysate was less sensitive to inhibition by cap analogs than the wheat germ system (LODISH and ROSE 1977, ASSELBERGS et al. 1978a). Later it was shown that the two systems behave rather similarly provided that the endogeneous globin mRNA is degraded by nuclease, and that the experiments are performed at the appropriate ionic conditions (WU et al. 1977, ASSELBERGS et al. 1978b, CHU and RHOADS 1978, 1980,

---

References. 1 KEITH and FRAENKEL-CONRAT 1974; 2 ZIMMERN 1975; 3 OBERG and PHILIPSON 1972; 4 BEACHY et al. 1976; 5 KLEIN et al. 1976; 6 PINCK et al. 1972; 7 ABOU HAIDAR and BANCROFT 1978; 8 SONENBERG et al. 1978; 9 HARI 1981; 10 HARI et al. 1979; 11 GHOSH et al. 1980; 12 MORRIS-KRSINICH 1980; 13 SALOMON et al. 1978; 14 LESNAW and REICHMAN 1970; 15 DAUBERT et al. 1978; 16 STANLEY et al. 1978; 17 EL MANNA and BRUENING 1973; 18 SEMANCIK 1974; 19 OXELFELT 1976; 20 DAUBERT and BRUENING 1979; 21 HARRISON and BARKER 1978; 22 MAYO et al. 1979a; 23 MAYO et al. 1979b; 24 ABOU HAIDAR and HIRTH 1977; 25 MINSON and DARBY 1973; 26 AGRANOWSKY et al. 1979; 27 AGRANOWSKY et al. 1981; 28 DASGUPTA et al. 1976; 29 HALL et al. 1972; 30 KOHL and HALL 1974; 31 SYMONS 1975; 32 PINCK 1975; 33 HALL 1979; 34 HORST et al. 1971; 35 LOT et al. 1977; 36 RICHARDS et al. 1978b

Table 2. Effect of proteinase K treatment on the infectivity and translatibility of protein-linked plant virus RNAs[a]

Proteinase K-treated RNA from	Infectivity[b]	Translatibility in[b]	
		Wheat germ	Reticulocytes
CPMV (1)	+	+	
TRSV (2, 3)	−	+	+
SBMV (4, 5)	−	−	+

[a] The results obtained with CPMV and TRSV seem to be representative for the comovirus and nepovirus group, respectively
[b] Activity unaltered is indicated by +, activity lost is indicated by −

References. 1 STANLEY et al. 1978; 2 MAYO et al. 1979b, C. FRITSCH, personal communication; 3 CHU 1980; 4 VEERISETTY and SEHGAL 1980; 5 SALERNO-RIFE 1980

BERGMANN and LODISH 1979). Apparently the total mRNA concentration is a strong determinant in the sensitivity of mRNA translation to inhibition by cap analogs. Perhaps because the endogeneous RNA, the exogeneous RNA, and the cap analog all compete for the same factor.

It is unknown if the role played by the cap in vitro is similar to the role in vivo. The fact that cap analog had no effect on mRNA translation in *Xenopus* oocytes (ASSELBERGS et al. 1978c) might be an indication that there can be severe differences between the in vivo and the in vitro situation.

Summarizing, one could state that the cap structure is important for translation of eukaryotic mRNA's, but we have no precise knowledge of its function.

2.1.2 Genome-Linked Protein

The 5′ terminus of the RNAs from several como- and nepoviruses and also of SBMV have been shown to be linked to a protein (Table 1). The function of this protein is as yet unknown. For some viruses the RNA linked protein is required for infectivity, for other viruses not (Table 2). Treatment of the RNA with proteinase K mostly has no effect on the messenger function of the RNA. The exception is SBMV–RNA, the messenger function of this RNA is severely diminished when assayed in wheat germ extracts (Table 2). From the fact that protein-linked SBMV–RNA is a better messenger in wheat germ extracts than proteinase K-treated SBMV–RNA, we have to conclude that plant ribosomes are able to translate protein-linked RNA, or alternatively plant extracts can remove the intact protein but not the residue left by proteinase K.

2.1.3 (p)ppX

To date only two plant virus RNA's have been found which are neither "capped" nor "protein-coupled" at their 5′ terminus (Table 1). Although it cannot be ruled out that in vivo these RNA's become "capped", it has been shown that STNV–RNA is not "capped" prior to its translation in vitro (LEUNG et al. 1976, ROMAN et al. 1976).

2.2 Structure at the 3′ Terminus

Three different types of structures have been found at the 3′ termini of plant virus RNA's: poly(A)tail, tRNA-like or X_{OH} (cf. HIRTH, Chap. 10, this Vol.).

2.2.1 Poly(A)Tail

In contrast to most eukaryotic mRNA's only a few plant virus RNA's have been shown to carry a poly(A)tail (Table 1). The function of the 3′ terminal poly(A) of eukaryotic mRNA's is still uncertain. It has been reported (MARBAIX et al. 1975) that the functional life-time of globin mRNA is much longer when at least 30 A residues are present. This is in agreement with the finding that polyadenylated histon mRNA's were much longer functioning as messengers in oocytes than nonpolyadenylated histon mRNA's (HUEZ et al. 1978). It has been suggested that the poly(A)track protects the coding sequences from exonuclease digestion. This is contradicted by the fact that for both α_{2u}-globine mRNA (DESHPANDE et al. 1979) and for AMV–RNA 4 (MARBAIX et al. 1979) the functional life-time was independent of the length of the 3′ terminal poly(A) (which in the case of AMV–RNA 4 was added in vitro). Alternatively the proteins associated with the poly(A)tail could be involved in the transport of mRNA through the nuclear membrane (CHISICK et al. 1979), or in the processing of the mRNA (BINA et al. 1980).

It is not yet known if the poly(A)track of the plant virus RNA's is virus-coded [the minus strand carries a poly(U)track] or if it is added to the viral RNA after replication by host or virus coded enzymes. For CPMV the nucleotide sequence adjacent to the poly(A) has been determined (DAVIES et al. 1979), this sequence did not contain the AAUAAA sequence found in most eukaryote polyadenylated mRNA's (PROUDFOOT and BROWNLEE 1976). Recently it was suggested that the polyadenylation signal for VSV (an animal virus) mRNA's was not AAUAAA but a sequence rich in U residues (CLERX-VAN HAASTER et al. 1980). It has been found for both BSMV (AGRANOVSKY et al. 1978) and TEV (HARI et al. 1979) that the infectivity of poly(A)$^+$RNA is about equal to that of poly(A)$^-$RNA. Unfortunately the progeny of the poly(A)$^-$RNA has not been analyzed.

Theoretically it is possible that the encapsidated virus RNA is not, while the virus RNA functioning in vivo as messenger is polyadenylated. However, at least for AMV, it has been shown that the virus RNA's present in polysomes of infected tobacco leaves do not bind to oligo(dT)columns (BOL et al. 1976). Thus if the viral mRNA is polyadenylated at all, the tracks are extremely short.

Virus RNA's, in contrast to cellular mRNA's, have to serve as messenger for protein synthesis *and* as template for RNA replication. It is conceivable that the poly(A)tail is a disadvantage in the latter process.

2.2.2 "tRNA-Like" Structure

A feature found up to now only for virus mRNA's (from both animal and plant viruses) is the presence of a "tRNA-like" structure, which can be charged with an amino acid at the 3′ terminus (for review see HALL 1979; HIRTH Chap. 10, this Vol.).

The RNA from EMC (an animal virus) contains an internal tRNA-like structure (LINDLEY and STEBBING 1977). At present it cannot be excluded that plant virus RNA's terminating with poly(A) or pX_{OH} contain such internal tRNA-like structures.

The tRNA-like structure could be the precursor of a virus-coded tRNA, which could play a role in the translation of virus RNA's during the infection cycle. It has been shown by JOSHI et al. (1978) that in oocytes the tRNA-like structure of TYMV–RNA is liberated and charged with valine. During the in vitro protein synthesis the valine bound to the viral RNA can be donated to the growing peptide chain (HAENNI et al. 1973). The latter result is in contrast to the results obtained with valylated EMV, a close relative of TYMV (HALL et al. 1979). Furthermore CHEN and HALL (1973) showed that neither the intact tyrosylated BMV–RNA nor the tyrosylated 3′ fragment of 160 nucleotides was capable of transferring tyrosine to peptide material during protein synthesis. One could speculate that the tRNA-like structures of BMV and EMV have to be further maturated to become active, for instance by removal of an intervening sequence. It was found recently that several eukaryotic tRNA precursors contain intervening sequences (VENEGAS et al. 1979 and references therein), which are removed during maturation.

It is also possible that the 3′ terminal tRNA-like structure is not involved in translation, but does play a role during the replication of virus RNA (HALL and WEPPRICH 1976). The aminoacylated BMV–RNA's have a high affinity for EF-1. The replicase of the bacteriophage $Q\beta$ contains the complex EF.Tu. EF.Ts (BLUMENTHAL and CARMICHAEL 1979 and references therein). The binding of EF-1 to aminoacylated BMV–RNA could point to a homology in subunit structure between the replicase of a bacteriophage and a plant virus (BASTIN and HALL 1976).

2.2.3 pX_{OH}

It has been shown that the RNA's from several viruses, which do not terminate with a poly(A)tail, cannot be charged with an amino acid (Table 1). For AMV and TSV (HALL 1979) this was rather surprising since these viruses have a lot in common with BMV and CMV, the latter two can both be charged with tyrosine.

In contrast to the RNA's from BMV and CMV, the RNA's from AMV and several ilar viruses have sites with a high affinity for the viral coat protein (GONSALVES and FULTON 1977, VAN VLOTEN-DOTING and JASPARS 1977). For AMV–RNA 4 it is known (HOUWING and JASPARS 1978, KOPER-ZWARTHOFF and BOL 1979) that these high affinity sites are located in the extracistronic 3′ terminal part of the RNA. Considering the homology between the 3′ terminal region of all four AMV–RNA's, sites with high affinity for the viral coat protein on the three genomic RNA's will also be located at the 3′ termini (KOPER-ZWARTHOFF et al. 1979). This does not mean that other parts of the RNA's cannot be involved in the binding of viral coat protein to high affinity sites.

The presence of coat protein is probably not important for translation, since the presence of coat protein during in vitro translation did not influence the products directed by the AMV–RNA's in wheat germ extracts (RUTGERS 1977) or in reticulocyte lysates (in the latter case only the products directed by RNA 1 and 4 were investigated, VAN TOL unpublished results).

From Table 1 it is evident that there is no correlation between the type of structure present at the 5′ terminus with the type of structure present at the 3′ terminus. Nor is there a correlation between one of these terminal structures and the type of genome. It is remarkable that although all species of RNA of one particular virus always have the same type of structures at either terminus, this does not hold for satellite RNA's. For the two satellite RNA's with known 5′ termini, it was found that this structure was identical to that of the helper virus, however, in contrast to the RNA's of its helper virus CARNA could not bind tyrosine to its 3′ terminus.

3 Fidelity of Translation

3.1 Comparison of in Vitro Products with in Vivo Products

Most in vitro translation studies have been performed with cell-free systems derived from wheat embryo or wheat germ (MARCUS, Chap. 3, Vol. 14A) or with a lysate of rabbit reticulocytes treated with micrococcus nuclease to destroy the endogeneous messenger RNA (PELHAM and JACKSON 1976). In general this latter system will give "clearer" patterns. Part of this difference might be due to a difference in nuclease activity in the two systems. Recently the preparation of active cell-free extracts from yeast has been reported (GASIOR et al. 1979, SZCZESNA and FILIPOWICZ 1980, TUITE et al. 1980). The yeast system may become especially useful for the study of regulation of translation: e.g., suppression of termination (see Sect. 6.2).

In vitro translation of plant virus RNA's only gives useful information about the proteins directed by these RNA's in vivo when the in vitro translation occurs with fidelity, and vice versa faithful translation of a messenger RNA suggest that the results obtained with the in vitro system are important for understanding the in vivo principles. It has been shown for several viruses that the coat protein directed by the subgenomic coat protein messenger (in several cell-free systems) is identical to the coat protein formed in vivo (see review by DAVIES 1979, ATABEKOV and MOROZOV 1979). Thus in vitro the translation of these small messengers seems to occur with fidelity. A more difficult question to answer is whether the larger RNA's are also translated with fidelity, especially whether the proteins are terminated at the correct position (Sect. 3.2).

For TMV the noncoat protein of molecular weight 130,000 produced in vitro as well as in vivo has been analyzed (SCALLA et al. 1978). The peptide patterns obtained after cyanogen bromide cleavage of the two proteins were very similar, suggesting that these proteins are similar or even identical. These results prove that in vivo TMV–RNA is also translated into a 130,000 protein, which is apparently not extensively modified after translation. However, they do not prove that this large protein is the functionally active protein.

The potyviruses are the only group of plant viruses which are known to induce large amounts of virus-coded nonstructural proteins in infected plants. DOUGHERTY and HIEBERT (1980b, c) isolated three different inclusion proteins from infected plants. With the help of antisera raised against these proteins it could be shown that these inclusion proteins are virus-coded. Analysis of V8 protease digestion products showed that the in vitro and in vivo produced proteins are very similar or even identical.

For several viruses the proteins induced after infection of protoplasts have been studied. The size of these proteins has been compared to the size of the proteins directed by the viral RNA's in vitro. However, the results should be considered with care, since it has not been proven that the proteins seen are indeed virus-coded and not virus-induced host proteins. Based on size comparison, it has been suggested that the proteins induced in vivo by TMV–RNA (SAKAI and TAKEBE 1972), satellite RNA of TBRV (FRITSCH et al. 1978), all four RNA's of CCMV (SAKAI et al. 1977) and BMV (SAKAI et al. 1979, OKUNO and FURUSAWA 1979) and the two RNA's of CPMV (STUIK 1979, ROTTIER 1980) are identical to the largest products directed by these RNA's in vitro. However, one should be aware of the fact that to be able to see differences in protein patterns between virus-infected and mock-infected protoplasts it is mostly necessary to inhibit host protein synthesis by treatment with actinomycin D or irradiation with UV light. Since both treatments have also a severe effect on the amount of infectious virus produced (TAKEBE 1977, ALBLAS and BOL 1977), the possibility that the virus-coded proteins produced in treated protoplasts differ (e.g., in processing) from the proteins present during the normal infection cycle may not be overlooked.

3.2 Comparison of Products Formed in Different Cell-Free Systems

Another approach to the question of fidelity is the comparison of the products directed by the same RNA in different cell-free systems: wheat germ, wheat embryo (MARCUS this Chap. 3, Vol. 14A); or reticulocyte lysate (PELHAM and JACKSON 1976) or in amphibian oocytes (GURDON et al. 1971). The latter system can be considered as an intermediate between an in vivo and an in vitro system. It has been shown with some animal virus RNA's that both oocytes and the reticulocyte lysate are capable of processing the primary translation product into proteins similar or identical to the products found in vivo (GHYSDAEL et al. 1977, SHIH et al. 1978). A disadvantage of the oocyte system is the very high background of products made under the direction of endogenous mRNA's. To permit a detailed analysis of the virus-coded proteins, specific antisera against these proteins (or their precursors) are required. Unfortunately such antisera are not available for most of the proteins coded for by plant virus RNA's.

The main difference between products synthesized in the different eukaryotic systems seems to be the point of termination. This difference might be due to a difference in nuclease activity or in a difference in the capacity of the

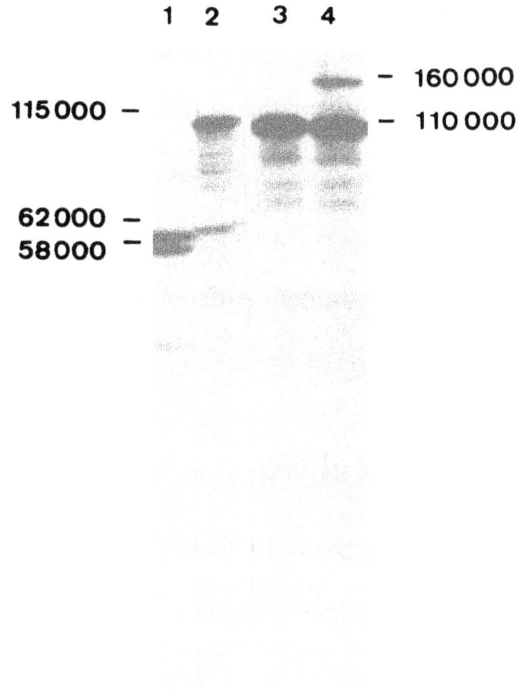

Fig. 1. Effect of paromomycine on the translation of AMV–RNA 1 and TMV–RNA in the mRNA-dependent rabbit reticulocyte cell-free system. Products directed by 100 μg/ml AMV–RNA 1 (lanes *1,2*) and 100 μg/ml TMV–RNA (lanes *3,4*) in the absence (lanes *1,3*) and the presence (lanes *2,4*) of 10 μg/ml paromomycine. (For details of translation experiments see VAN TOL et al. 1980)

systems used to suppress termination (see Sect. 6.2). The formation of smaller products can be due to degradation of the messenger, but also to premature termination at certain "difficult" points on the RNA. From the results obtained with paromomycine, an antibiotic able to suppress, in vivo as well as in vitro, several missense and nonsense mutations in yeast (PALMER et al. 1979; SINGH et al. 1979), it is evident that mismatching enhances the possibility to proceed along these points (Fig. 1, VAN VLOTEN-DOTING and NEELEMAN 1980), suggesting that the difficulty is due to the translation of certain codons and not to the secondary structure of the mRNA. From the fact that under the direction of several plant virus RNA's the largest product made in the presence of paromomycine is the same size as the largest product made in the absence of paromomycine, it follows that the termination at this point is not suppressed. This could mean that not all three stop codons are suppressed in the presence of paromomycine in reticulocyte lysates, *or* that at the end of the cistron several termination codons succeed each other, *or* that the termination codon at the true end of the cistron is embedded in an RNA sequence (or structure) which enhances the efficiency of termination, making read-through impossible.

It is rather remarkable that even in the presence of paromomycine AMV–RNA 3 in reticulocyte lysates does not direct a significant amount of a "read-through" product of 65,000. It has been shown previously (VAN VLOTEN-DOTING 1976, RUTGERS 1977) that in wheat germ extracts AMV–RNA 3 can direct the synthesis of a small amount of a protein of about 65,000 mol.wt. This protein is thought to be a read-through product since it can be precipitated with antiserum raised against AMV particles and it contains tryptic peptides which are also present in the major product (35,000 mol.wt.) directed by this RNA.

A difference in the point of initiation of protein synthesis has only been found upon translation of AMV–RNA 4 in a prokaryotic cell, free system (derived from *E. coli*). Besides initiation on the normal AUG codon initiation took also place on a UUU codon located close to the cap (Castel et al. 1977, 1979, Koper-Zwarthoff et al. 1977). Initiation on this latter point resulted in the formation of a protein which was identical to the coat protein but did contain 12 additional amino acids at the N-terminus (Castel et al. 1977, 1979, van Vloten-Doting et al. 1977).

4 Strategy of Expression of the Information

In vitro translation of one purified plant virus RNA often yields more than one protein. In several cases it has been demonstrated that the products represent overlapping chains (Davies 1979 and references therein, Atabekov and Morozov 1979 and references therein). Most of the numerous bands found in the wheat germ system are probably artifacts, since variation in the incubation conditions may lead to the formation of products with a higher molecular weight. However, also in the reticulocyte lysate one purified RNA often directs the synthesis of two or three main products. The sum of these proteins greatly exceeds the coding capacity of the messenger.

For a number of virus RNA's (TYMV, Bénicourt et al. 1978, TMV and TRV, Pelham 1978, 1979a, AMV, van Tol and van Vloten-Doting 1979) it has been shown that these proteins represent overlapping chains with the same N-terminus. Since it is rather difficult to find in vivo virus-coded proteins other than the coat protein, it is hard to determine which of the products found in vitro represent the "real" protein(s). At present we cannot decide whether these partial translates have functional significance or are merely artifacts of the in vitro system (see also Sect. 6.2).

For some virus RNA's (CPMV, Pelham 1979b, Rezelman et al. 1980, TYMV, Morch and Bénicourt 1980a) it has been shown that the primary translation product is cleaved in vitro. This might indicate that the primary product has to be processed to become functional, or alternatively that this product (at least in vitro) is not very stable. To decide between these possibilities we have to know which of these proteins are present *and* functional in vivo.

Theoretically the three-letter code offers the possibility of coding three chemically different proteins on one nucleotide sequence. For some bacteriophages double or even triple use of the same nucleotide has been reported (Shaw et al. 1978 and references therein). To date there is no evidence that plant virus RNA's are read in more than one reading frame.

Recently, it has been shown that in vitro *E. coli* ribosomes can sometimes, determined by the tRNA population, change their reading frame (Atkins et al.

1979). This phenomenon leads to the formation of a family of proteins with identical N-terminal parts and varying C-terminal parts. It is unknown if a similar phenomenon can also take place in eukaryotic systems, either in vivo or in vitro.

The different RNA's from one particular virus do not necessarily employ the same strategy of expression. For instance the two largest RNA's of the viruses with a tri-partite genome can, at least in vitro, be translated into full-length products, which might be processed later, while in the case of the smallest genomic RNA only the 5′ half is translated directly, the 3′ half being translated from a subgenomic messenger (DAVIES 1979 and references therein, ATABEKOV and MOROZOV 1979 and references therein).

4.1 Functionally Monocistronic: Expression of the Internal Cistron is Mediated by a Subgenomic mRNA

Like most eukaryotic mRNA's, the majority of the plant virus RNA's studied so far are functionally monocistronic (only one open initiation site per RNA molecule). A model for the initiation of protein biosynthesis in eukaryotes, the so-called scanning hypothesis, has been published by KOZAK (1978).

If the information for a second (or a third) cistron is present, that part of the RNA is amplified in vivo (DAVIES 1979, ATABEKOV and MOROZOV 1979). In the majority of plant viruses which produce subgenomic RNA's to generate several proteins from one genomic RNA, the information for the coat protein is located proximal to the 3′ terminus. The only possible exceptions at present are TRV and BSMV. RNA 2 from the CAM strain of TRV and RNA 2 from the Norwich strain of BSMV both direct in vitro only the synthesis of coat protein (MAYO et al. 1976, DOLJA et al. 1979).

BISARO and SIEGEL (1980) found in leaves infected with TRV CAM a subgenomic RNA, which contained sequences present in RNA 2. This RNA directed in vitro the synthesis of 30 K protein. Preparations of the Norwich strain of BSMV contain an RNA 2 (TALIANSKY et al. 1979) which direct the synthesis of a 75,000 mol.wt. protein. These results could suggest that for TRV and BSMV the cistron coding for the coat protein is located proximal to the 5′ terminus.

It is not yet known if these subgenomic mRNA's arise by partial transcription or by processing of the genomic RNA. GONDA and SYMONS (1979) used hybridization to cDNA to follow the synthesis of CMV–RNA's in protoplasts (not treated with UV light or actinomycine D). They found that the ratio of sequences coding for the coat protein (present in both RNA 3 and 4) to sequences unique for RNA 3 was close to 1. From these results the authors concluded that RNA 4 arises by specific cleavage of RNA 3. However, these results are not conclusive since the same ratio was found at 0 h post-infection when it should have been 2:1.

The sequence at the 5′ terminus of the subgenomic mRNA's of plant viruses is different from that of the corresponding genomic RNA (KOPER-ZWARTHOFF et al. 1977, RICHARDS et al. 1978a, GUILLEY and BRIAND 1978, BRIAND et al. 1978, GUILLEY et al. 1979, PINCK et al. 1979). This is in contrast to the situation found in several animal viruses where the same leader sequence is spliced onto several messengers (CRICK 1979).

Although there is no doubt that the main translation product is derived from the cistron located near the 5′ terminus, small amounts of translation products from the internal cistrons are often found. In most cases this has been explained by the assumption that the genomic RNA is slightly contaminated with subgenomic mRNA's or that the RNA is processed in vitro. However, the question whether the internal cistron is absolutely not accessible for translation or is only expressed at a much lower efficiency remains open.

4.2 Monocistronic: The Primary Product is a "Polyprotein"

In vitro the two RNA's from CPMV both direct the synthesis of a large product corresponding approximately to the total of genetic information present in the RNA's (DAVIES et al. 1977). Neither of the two products reacted with antiserum against CPMV capsids (ROTTIER 1980). No products corresponding in size to one or both of the viral capsid proteins were formed. Up to now no subgenomic messengers for the coat proteins have been found (STUIK 1979, ROTTIER 1980). These results suggest that the primary products have to be processed into functional units. PELHAM (1979b) reported that the 200,000 mol.wt. protein directed by RNA 1 is cleaved into a 32,000 and a 170,000 protein, and that one of these two proteins is a protease which is responsible for the cleavage of the product directed by RNA 2. REZELMAN et al. (1980) also observed a 30,000 and a 170,000 mol.wt. protein directed by RNA 1 both in vitro and in vivo (protoplasts), in vivo the 170,000 protein is cleaved into several smaller proteins (Fig. 2) (none of these corresponding to capsid proteins), in vitro no processing of the products directed by RNA 2 was observed (in the absence or in the presence of the products directed by RNA 1).

The translation strategy of the nepoviruses seems to be similar to that of the comoviruses. In vitro the two RNA's of TBRV both direct very large proteins. The product directed by RNA 2 did react with antiserum to TBRV particles, showing that this protein contains sequences similar or identical to those in the capsid protein, and could represent a precursor for the capsid protein (FRITSCH et al. 1980).

HELLMANN et al. (1980) reported the synthesis of a 300,000 protein under the direction of RNA from TVMV (a potyvirus). A protein of this size could correspond to the total of genetic information present in this RNA, functional units will have to be formed by processing. However, DOUGHERTY and HIEBERT (1980c) and KOZIEL et al. (1980) did not find indications for the formation of a polyprotein (see also Sect. 4.3).

Posttranslational cleavage has also been reported for the largest protein directed by TYMV–RNA. It is remarkable that the smaller protein which is also directed by this RNA and does contain the regions that are processed in the larger (overlapping) protein is not cleaved in vitro.

One should be aware of the fact that in none of the three cases of processing of large proteins reported above can we distinguish between the formation of

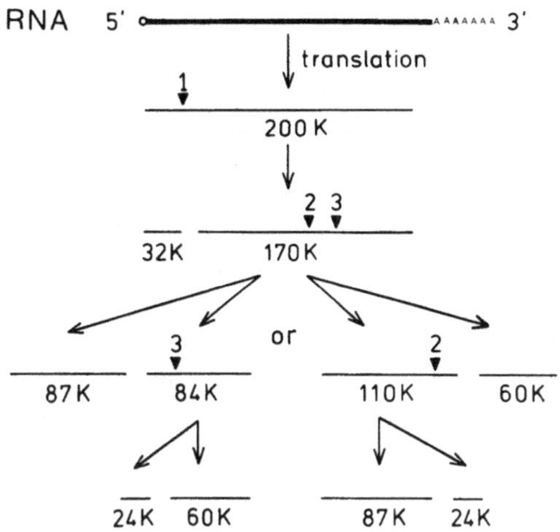

Fig. 2. Model for the expression of CPMV–RNA 1, as proposed by REZELMAN et al. (1980). The position of cleavage point 1 is based on the in vitro translation experiments of PELHAM (1979). Since the 110K, 87K, 84K, and 60K proteins are apparently simultaneously produced, the order of cleavage at point 2 and 3 is not fixed. (Figure by courtesy of Dr. R. GOLDBACH)

active smaller proteins from a precursor protein or the inactivation by stepwise degradation of the functional protein.

The posttranslational cleavage of polio virus polyprotein into functional units is well documented (COOPER 1977). It has been noticed that the comoviruses and nepoviruses (which probably employ the same mechanism) resemble poliovirus in having a protein linked to the 5′ terminus and a poly(A)tail to the 3′ terminus of the genomic RNA's.

The same might hold for potyvirus RNA's. These RNA's are known to be polyadenylated, the structure at the 5′ terminus has not been identified but the presence of a cap seems to be unlikely (HELLMANN et al. 1980). The latter statement is based on the fact that cap analogs stimulate rather than inhibit the translation of potyvirus RNA's, furthermore treatment with $NaIO_4$ and $NaB[^3H]_4$ failed to label cap structures.

However, there is not necessarily a correlation between structural features and the mode of translation. SBMV–RNA and TRosV–RNA are both linked to a protein, but virus preparations do contain a subgenomic messenger for the coat protein (RUTGERS et al. 1980, MORRIS-KRSINICH 1980).

4.3 Di- or Polycistronic

At present it is often suggested that initiation of translation by eukaryotic ribosomes is completely restricted to the AUG codon close to the 5′ terminus (KOZAK 1978). However, it might be an oversimplification to confine the translation of all eukaryotic mRNA's to one model. A problem in nature is often solved in several different ways. We have already seen above that at the 3′ or the 5′ terminus of a mRNA at least three different structures can be found.

The polycistronic messenger has proven very successful in prokaryotes, and eukaryotic cell-free systems are capable of translating polycistronic prokaryotic messengers (LEGON et al. 1977, DAVIES 1979). Thus it is quite conceivable that although the majority of eukaryotic messengers are functionally monocistronic, polycistronic messengers will be found. Indeed there are already several reports suggesting that such polycistronic eukaryotic mRNA's do exist. (See for animal viruses, e.g., KNAUERT and EHRENFELD 1979, VAN STEEG et al. 1979, WESTAWAY 1977).

Ribosome-binding studies have shown that more than one ribosome can be bound to some plant virus RNA's under initiation conditions. However, in itself this is not sufficient to conclude that such an RNA has two open initiation sites. It has been shown that the two ribosomes present on TMV–RNA (FILIPOWICZ and HAENNI 1979) as well as on BMV–RNA 3 (AHLQUIST et al. 1979) are both present in the leader sequence. The same has been suggested for TYMV–RNA (FILIPOWICZ and HAENNI 1979). However, the two ribosomes on AMV–RNA 3 (VAN VLOTEN-DOTING et al. 1977, RUTGERS 1977, NEELEMAN and VAN VLOTEN-DOTING 1979, PINCK et al. 1979) protect two different pieces of RNA against digestion with RNAases (PINCK et al. 1979). One of these is identical to the 5′ terminus of the subgenomic mRNA for the coat protein. Furthermore in vitro translation of this RNA suggests that the coat protein cistron, which is located internally, is translated at a very low efficiency (NEELEMAN et al. 1977, RUTGERS 1977). Taken together these results are suggestive, however, one should be aware of the possibility that the ribosomes did not bind to intact RNA 3 but to degradation products, and that the coat protein was directed by a "pseudo" RNA 4 formed in vitro. It has been shown for several other RNA's that multiple initiation sites are available on fragmented RNA (PELHAM 1979c). However, it is also possible that the low rate of protein synthesis initiation at the internal cistron is reminiscent of the evolutionary older (before the appearance of the cap?) situation of a dicistronic messenger.

In vitro translation of TNV–RNA (purified under denaturing conditions) yields mainly coat protein and two larger proteins (SALVATO and FRAENKEL-CONRAT 1977). At present it has not yet been shown that these larger products are unrelated to coat protein, so the possibility that they represent read-through products (see also Sect. 6.2) cannot be ruled out.

In vitro translation of the intact CarMV–RNA (purified under denaturing conditions) yields three unrelated proteins (SALOMON et al. 1978). Together these three proteins could represent the total genetic information present in this RNA. It is remarkable that limited phosphorolysis of the 3′ terminus with polynucleotide phosphorylase caused selective disappearance of the largest protein. From these results it was concluded that the 3′ terminus might have a regulatory function. RUTGERS (1977) and HOUWING and JASPARS (1978) found no indications for a regulatory role (in vitro nor in vivo) of the 3′ terminal sequence of the monocistronic AMV–RNA 4.

The results obtained with in vitro translation of RNA 2 preparations from BSMV differ per strain (DOLJA et al. 1979). RNA 2 preparations from the Russian strain direct in wheat embryo cell-free systems the synthesis of two proteins: coat protein and a 85,000 mol.wt. protein. Translation of RNA 2 preparations from the Norwich strain yields only coat protein, while a subgenomic RNA (homologous to RNA 2, TALIANSKY et al. 1979) directs the synthesis of a 75,000 protein. The RNA 2 from the Russian strain might be translated as a polycistronic message. However, in our opinion it is less likely that two strains from one virus would employ different strategies for the expression of the genetic information. It is quite possible that preparations of RNA 2 from the Russian strain consist of a mixture of two nearly indentical RNA's, the largest of these directs the synthesis of the 5′ proximal cistron (probably coat protein, see Sect. 4.1) while the "internal" protein is directed by an RNA which is slightly smaller (missing the AUG at the 5′ terminus?).

It has been reported that RNA 2 from TRV strain PRN directs the synthesis of two different proteins (FRITSCH et al. 1977). However, careful analysis of preparation of RNA 2 from this strain revealed the presence of two RNA's each coding for one protein (PELHAM 1979a).

The translation strategy of potyvirus RNA's is still unclear, conflicting results have been published by different groups (Fig. 3). DOUGHERTY and HIEBERT (1980a, b) found that in vitro translation of TEV–RNA and PeMV–RNA in mRNA-dependent rabbit reticulocyte lysates resulted in the formation of several proteins. Four of these proteins could by immune precipitation be identified as the capsid protein and three inclusion proteins. Besides these four proteins another major protein (87,000 mol.wt.) is formed in vitro, which does not react with any of the antisera. Since these proteins are serologically not related it is unlikely that they represent overlapping chains. Time course experiments did not reveal a precursor for the different proteins, and all proteins were already present after 15 min incubation, which is too short a time to complete the synthesis of a precursor protein of about 300,000 mol.wt. From these results it was concluded that the potyvirus RNA's are translated as a polycistronic messenger (DOUGHERTY and HIEBERT 1980a). However, in a later paper (DOUGHERTY and HIEBERT 1980c) it was concluded that the intact 39S RNA is probably translated as a monocistronic RNA, directing only the synthesis of the 87,000 protein and several read-through products, and that slightly degraded RNA directed the synthesis of the smaller proteins.

One of the problems with the interpretation of the results from DOUGHERTY and HIEBERT is that the situation around the cistron for the nuclear inclusion protein is very complicated (Fig. 3). In order to explain the immune precipitation data the authors assume that ribosomes which initiate at the beginning of the 54,000 mol.wt. protein will terminate at the end of this cistron or at the end of the capsid cistron, while ribosomes which initiate in phase upstream will terminate somewhere in the capsid cistron. In our opinion the conclusions drawn by DOUGHERTY and HIEBERT (1980c) about the translation strategy of potyvirus RNA's is not in accordance with the results presented. Even the "best" RNA preparations always direct the synthesis of several proteins, and in the tentative genetic map these proteins do not result from the same initiation site.

Completely different results were obtained by KOZIEL et al. (1980), they found that TEV–RNA in wheat germ embryo extracts directed only the synthesis of one protein of 40,000. Since HELLMANN et al. (1980) found that TVMV–RNA directed the same proteins in reticulocyte lysates as in wheat germ extracts it is unlikely that the differences between the results obtained by KOZIEL et al. (1980) and DOUGHERTY and HIEBERT (1980a, b, c) are due to the fact that they used different cell-free systems.

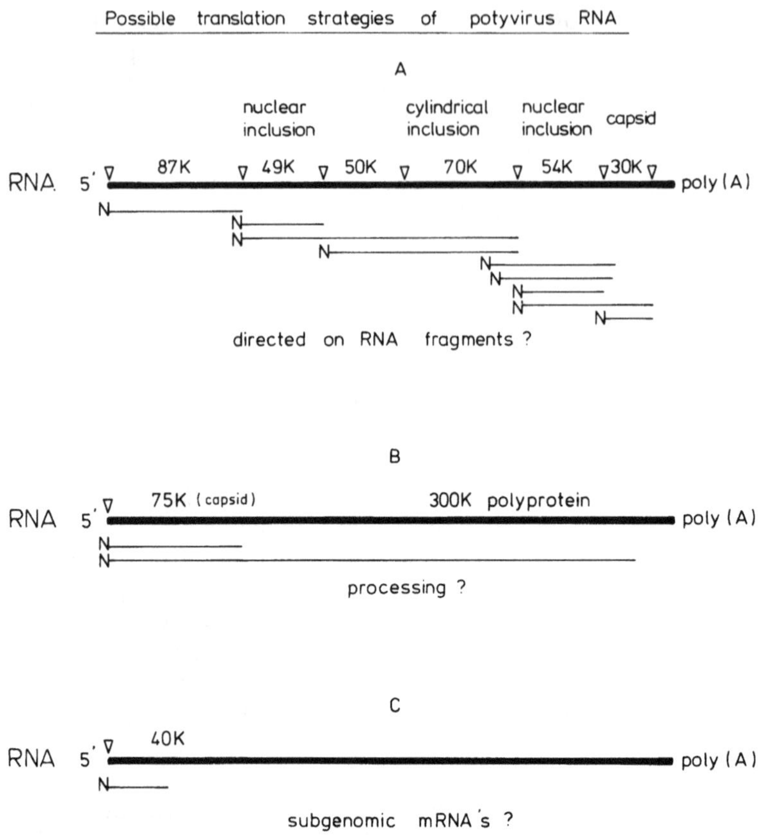

Fig. 3. Translation strategy of potyvirus RNA according to the results of **A** DOUGHERTY and HIEBERT (1980a, 1980c); **B** HELLMANN et al. (1980); **C** KOZIEL et al. (1980). (For further explanation see text)

HELLMANN et al. (1980) (studying TVMV–RNA translation in rabbit reticulocyte lysates and in wheat germ extracts) suggest that potyvirus RNA translation starts only at the 5′ terminus while termination of translation occurs at a number of sites, leading to a nested set of polypeptides all with common N termini. Also in this case there are indications that some products could be directed by RNA fragments. At high ionic strength there is one dominant termination site, the main product under these conditions is about 75,000. This latter product contains, probably at its C terminus, the sequences of the coat protein. Time course experiments suggest that TVMV–RNA can be translated into a product of about 300,000 corresponding to the coding capacity of the RNA.

The main differences between the results of DOUGHERTY and HIEBERT (1980c) and those of HELLMANN et al. (1980) are that according to the former investigators there is no polyprotein of 300,000, the different products originate of different initiation sites and the information for the coat protein is located close to the 3′ terminus of the RNA, while according to HELLMANN et al. (1980)

the RNA can be translated into a full length product and the information for the coat protein is located close to the 5′ terminus of the RNA.

5 Competition Between Host and Viral mRNA's

One intriguing question is how a virus manipulates its host cell so that this cell starts to manufacture new virus particles at its own expense. In intact plants the virus production seems to be limited in some way. FÖGLEIN et al. (1975) isolated protoplasts from one half of a tobacco leaf which had been inoculated with TMV 8–10 days previously. Replication of viral RNA was measured in these protoplasts as well as in the control half-leaf. In the isolated protoplasts a burst of TMV–RNA synthesis occurred, whereas little viral RNA was synthesized in the control half-leaf (while uptake of ^{32}P in both systems was about equal as judged by the level of incorporation in ribosomal RNA). These results indicate that in leaf tissue the virus replication is controled. This control is apparently released in isolated protoplasts. Whether this control operates at the level of RNA or protein synthesis is unknown.

Plant viruses do not show the phenomenon of "shut-off" of the host protein synthesis. Nevertheless they are often produced in massive amounts. FRASER and GERWITZ (1980) found that the amount of poly(A)$^+$-RNA (assumed to be representative for host mRNA) did not decrease during the infection of tobacco plants with TMV, while the amount of host protein synthesis decreased considerably. The authors concluded that regulation did not take place at the level of transcription or mRNA turnover. To prove that regulation does take place at the level of translation it has to be shown that the poly(A)$^+$-RNA isolated from infected plants is as active in in vitro protein synthesis as is the poly(A)$^+$ RNA isolated from healthy plants.

HERSON et al. (1979) showed that the poly(A)$^+$-RNA isolated from wheat germ is in wheat germ extracts a less efficient messenger than several (subgenomic) plant virus RNA's. To explain this and several other results it has been postulated (KAEMPFER et al. 1978, HERSON et al. 1979) that one of the initiation factors recognizes specific sequences in the mRNA other than the cap or the AUG, and that different initiation sequences differ in their affinity for this factor.

CHROBOCZEK et al. (1980a) have proposed that the seed transmissibility of plant viruses could be modulated by competition between viral and cellular messengers. This idea is sustained by the fact that the genomic RNA's of both BMV and TMV are "outcompeted" by the mRNA's for storage proteins of *Vicia faba* L. Although this is an attractive idea, to make plausible that this is the basis of exclusion of viruses during seed formation it has to be shown that the storage protein messenger RNA's are either stronger competitors or present in more abundance than normal host mRNA's.

Interpretation of competition experiments is severely hampered by the fact that we cannot exclude the possibility that the preferential translation of an

RNA species is due to the in vitro lability of a factor required for the translation of the competing RNA species.

6 Regulation of Expression of Virus Information

Although the function of most virus-coded proteins is unknown (see Sect. 7), it is unlikely that all products have to be made in equal abundance. The fact that the majority of virus-coded proteins is hardly detectable in vivo, while coat protein is present in large amounts, is in accordance with the idea that the amount of the different gene products is regulated. For viruses with a multipartite genome and/or subgenomic messengers the abundance of the different messengers can be regulated on the level of the RNA. Besides (or in addition to) the regulation on the level of transcription two different modes of regulation on the level of translation can be envisaged: preferential initiation and read-through of leaky termination codons. An additional possibility of regulation may be found in differences in the stability of the various gene products. All types of regulation may be found "on top of each other" in one virus.

6.1 Regulation by Preferential Initiation

The plant viruses studied so far all contain at least two different initiation sites, and regulation of initiation may play a role in all plant viruses. Up to now no studies on preferential initiation have been performed with the presumed polycistronic mRNA's. In studies on messenger RNA competition with different RNA's from the same virus the subgenomic coat protein messenger is mostly preferentially translated (Davies 1979 and references therein, Bénicourt and Haenni 1978, Zagorski 1978, Herson et al. 1979, Chroboczek et al. 1980b). Several lines of evidence suggest that this phenomenon is due to preferential initiation (Pyne and Hall 1979, Herson et al. 1979, Chroboczek et al. 1980b).

Wheat germ extracts contain a protein kinase (Rychlik and Zagorski 1980) which, at least in vitro, is involved in the regulation of the translation of the BMV–RNA's (Rychlik et al. 1980).

From competition experiments with the AMV–RNA's it was concluded that, at least in rabbit reticulocyte lysates, RNA 1 and RNA 3 require a (initiation) factor which is not required for the translation of AMV–RNA 2 (Van Tol and Van Vloten-Doting 1979).

6.2 Regulation by the Use of Leaky Termination Codons

Recently it has been found that the formation of two or three overlapping chains under the direction of one purified RNA species may be due to the presence of "suppressable" stop codons in the sequence coding for the largest product. At present "internal" UAG codons have been found in TMV–RNA (PELHAM 1978) and TYMV–RNA (BÉNICOURT and MORCH 1979). In both cases the addition of amber suppressor tRNA from yeast enhanced the formation of the largest product (160,000 and 205,000 mol.wt. respectively). Besides these UAG codons "internal" stop codons are present in TRV–RNA 1 (PELHAM 1979a), TYMV–RNA (unpublished results cited in MORCH and BENICOURT 1980b), AMV–RNA 1 (VAN TOL and VAN VLOTEN-DOTING 1979, VAN TOL et al. 1980) in potyvirus RNA's (DOUGHERTY and HIEBERT 1980c, HELLMANN et al. 1980) and probably in PVX–RNA (WODNAR FILIPOWICZ et al. 1980). Although none of these stop codons has been positively identified by the use of well-defined suppressor tRNA's, there are indications that at least some of them could represent UGA codons.

The suppressable stop codons in AMV–RNA 1 have been investigated in some detail. In Fig. 4 the incorporation of amino acids under the direction of increasing concentrations of AMV–RNA 1 and AMV–RNA 2 is shown. The curve for AMV–RNA 1 shows an optimum followed by a sharp decrease. Product analysis shows that at low mRNA concentration a product (115,000), corresponding to the total of genetic information present, is formed. At high mRNA concentration two smaller products (58,000 and 62,000) are formed. These three proteins represent a nested set of proteins with the same N-terminus (VAN TOL and VAN VLOTEN-DOTING 1979). This peculiar behavior of AMV–RNA 1 is not due to some general deficiency of the in vitro system, since the products directed by other RNA's are not influenced by the mRNA concentration (Fig. 4).

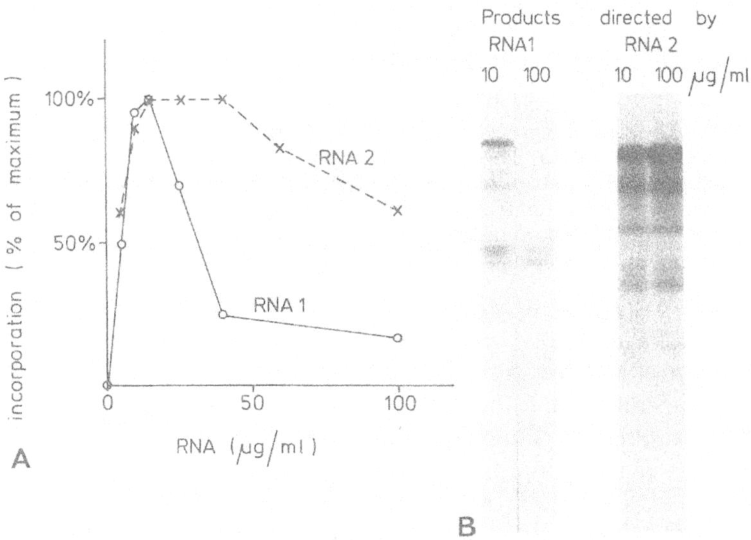

Fig. 4. Effect of the concentration of AMV–RNA 1 and AMV–RNA 2 on **A** the incorporation of amino acids, **B** the products directed in the mRNA dependent rabbit reticulocyte lysate. (For details of translation experiments see VAN TOL et al. 1980)

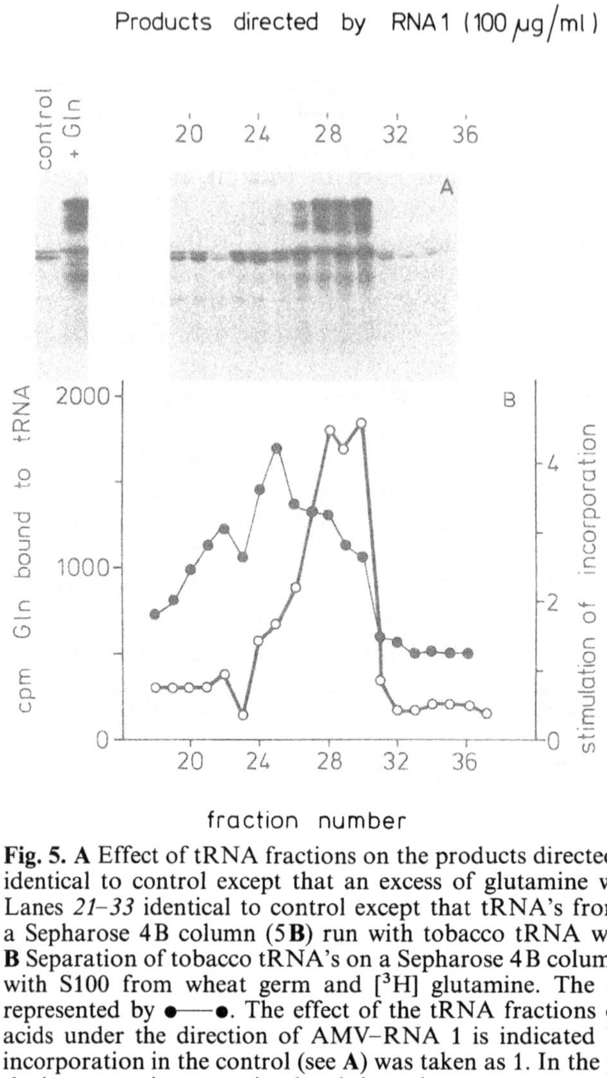

Fig. 5. A Effect of tRNA fractions on the products directed by AMV–RNA 1. Lane + *Gln* identical to control except that an excess of glutamine was present during translation. Lanes *21–33* identical to control except that tRNA's from the indicated fractions from a Sepharose 4B column (5**B**) run with tobacco tRNA were present during translation. **B** Separation of tobacco tRNA's on a Sepharose 4B column. All fractions were incubated with S100 from wheat germ and [^{3}H] glutamine. The glutamine bound to tRNA is represented by •—•. The effect of the tRNA fractions on the incorporation of amino acids under the direction of AMV–RNA 1 is indicated by o—o. The amount of the incorporation in the control (see **A**) was taken as 1. In the presence of excess of glutamine the incorporation was stimulated three times

The formation of partial translation products can be suppressed by the addition of an excess of glutamine or wheat germ- or tobacco tRNA's. Further analysis showed that the suppression of termination was due to one particular tRNA coding for glutamine (Fig. 5; VAN TOL et al. 1980, VAN TOL 1981). This tRNA is not required for the translation of AMV–RNA 4, an mRNA known to contain both code words for glutamine. These results might suggest that this particular $tRNA^{Gln}$ can read one of the termination codons, and there are indications that it can read UGA (VAN TOL et al. 1980, VAN TOL 1981).

Based on these results we assume that AMV–RNA 1 does contain in its coding sequence three suppressible stop codons: one closely succeeding the initiation codon (this one is responsible for the sharp decrease in incorporation of amino acids under nonsuppression conditions) and two located C terminal of the 58,000 and 62,000 mol.wt. proteins.

No effect of the RNA concentration on the ratio of products directed by TRV–RNA has been reported.

The ratio of the products directed by TYMV–RNA was not influenced by the concentration of mRNA, tRNA or amino acids (A.L. HAENNI and S. JOSHI personal communication), this might suggest that the "internal" stop codons in AMV–RNA 1 are different from those in TYMV–RNA (one UAG and one unidentified stop codon).

The read-through of the UAG codon apparently depends on a certain amount of mismatching which takes place on this codon depending on the surrounding structure of the RNA (PELHAM 1978), on the presence of spermidine (MORCH and BÉNICOURT 1980b), or of high concentration of magnesium ions (PELHAM 1978, MORCH and BÉNICOURT 1980b).

In eukaryotes (yeast mitochondria, FOX 1979, rabbit reticulocytes, GELLER and RICH 1980) a tRNA corresponding to the UGA codon is present, although sometimes in very low amount. So in the strict sense this codon is not leaky, and read-through of UGA depends mainly on the balance between mRNA and this suppressor tRNA. That the balance between mRNA and tRNA is very important is clearly illustrated by the results obtained with AMV–RNA 1. Low RNA 1 concentration will result in the formation (by read-through) of large protein, high RNA 1 concentration results (by stop at the presumed stop codons) in a mixture of two smaller products, high RNA 1 concentration in the presence of "suppressor" tRNA will again lead to the formation of one large product (Fig. 5; VAN TOL and VAN VLOTEN-DOTING 1979, VAN TOL et al. 1980). Also in this case the RNA around the termination codon (VAN TOL et al. 1980) and the concentration of magnesium ions (PELHAM 1979a, VAN TOL et al. 1980) but not the concentration of spermidine (MORCH and BÉNICOURT 1980b, VAN TOL et al. 1980) influences the efficiency of termination.

Why are there "leaky" stop codons present in plant virus RNA's? It is possible that the read through found in vitro is only an artifact. This does not seem very likely since the phenomenon "stop and/or read-through" is found in at least seven viruses belonging to five different groups (cf. Table 1).

HELLMANN et al. (1980) found that the frequency of termination of protein synthesis after the 75,000 protein, which is probably a precursor for the capsid protein, is influenced by the translation conditions. If a virus RNA is translated into a polyprotein, which is processed in functional units, all products are made in equal amounts. The "leaky" termination site directly after the information for the coat protein might be involved in the regulation of the relative synthesis of coat protein and nonstructural proteins. Another indication that the presence of "leaky" termination codons might be functional is the fact that the tRNA required for the synthesis of full length products under the direction of AMV–RNA 1 is present (in low amounts) in the host (VAN TOL 1981). Studies on the population of tRNA's from tobacco cell organelles showed that the "suppressor" $tRNA^{Gln}$ was principally present in the cytoplasm. However, tRNA preparations isolated from chloroplasts as well as from mitochondria did contain some "suppressor" tRNA, which could be due to cross-contamination (unpublished results). If the mechanism of "stop and/or read-through" does play a role in vivo, the "suppressor" tRNA should be present in the compartment where the translation of the AMV–RNA's takes place: the cytoplasm.

The use of "suppressible" stop codons makes it possible to generate two or three overlapping proteins from one RNA strand *and* to regulate the ratio of the different products. For instance for AMV–RNA 1 one could envisage the possibility that early in infection (when there is only a low amount of RNA) the large protein is required, while lateron (after amplification of the RNA) the two smaller products have to be made. The (presumed) stop codon succeeding the initiation codon, may function analogous to the attenuator during RNA synthesis. Although RNA 1 has a "strong" initiation site (necessary to supply enough protein early in infection), only small amounts of protein are formed late in infection, because under non-suppression conditions most ribosomes will terminate very quickly.

It has been suggested that the in-phase overlapping organisation of genes may constitute a strategy to facilitate the mutual interaction of two or three proteins through their common structural domains (SHAW and MURIALDO 1980).

Alternatively the sequence common to the two (or three) proteins may be required for a common function; e.g., interaction with membranes or templates. Till it has been shown that the different gene products are present and functional in vivo this is pure speculation. For the bacteriophage Qβ it has been shown that the read-through protein A1, which arises by suppression of a stop codon, is an essential protein (HOFSTETTER et al. 1974).

7 Function of Virus-Coded Proteins

At present the only virus-coded proteins for which a possible function can be grasped are the coat protein and for some viruses the genome-linked protein.

It has been assumed that RNA viruses have to code for their own replicase (subunit). However, it is well documented that healthy plants do contain an RNA polymerase (IKEGAMI and FRAENKEL-CONRAT 1979 and references there in, CHIFFLOT et al. (1980) and references therein). To date the only evidence for the involvement of virus-coded protein(s) in the RNA replicase is the preference of the replicases induced by CPMV (ZABEL et al. 1979) and BMV (HARDY et al. 1979) for their own RNA as template. KUMARASAMY and SYMONS (1979) showed that in extensive purified preparations of CMV-induced RNA replicase a protein of 100,000 was present which was absent from a preparation purified in a comparable way from healthy plants. As yet it has not been proven that this protein is a virus-coded functional unit of the enzyme. That virus-coded proteins do play a role during the process of RNA replication is evident from the fact that two ts mutants of TMV have been described in which the RNA replication is thermosensitive (DAWSON and WHITE 1978, 1979).

Besides in the RNA replication, virus-coded proteins other than coat protein are thought to be involved in the transport of virus through the host (DINGJAN-VERSTEEGH et al. 1972, NISHIGUCHI et al. 1978).

From in vitro translation studies of plant virus RNA's we have learned the size of the products which might be encoded by these RNA's. Furthermore we have obtained some idea about the possible modes of the expression and the regulation of expression of the genetic information. To correlate these data with the size and the function of the products found in vivo, studies of (thermosensitive) mutants of plant viruses (DAWSON and WHITE 1978, 1979, NISHIGUCHI et al. 1978, VAN VLOTEN-DOTING et al. 1980), preferentially in protoplasts (TAKEBE 1977) or in synchronised infections of whole plants (DAWSON et al. 1975) are indispensible.

Acknowledgments. We thank Dr. E.M.J. JASPARS for critical reading of the manuscript. We are grateful to Drs. J.W. DAVIES; R. GOLDBACH; A.-L. HAENNI; T.C. HALL; E. HIEBERT; P. KAESBERG; M. MAYO; S.A. RUTGERS and W. ZAGORSKI, for sending us information or manuscripts prior to publication.

References

Abou Haidar M, Bancroft JB (1978) The structure of the 5′-terminus of papaya mosaic virus RNA. J Gen Virol 39:559–563

Abou Haidar M, Hirth L (1977) 5′-Terminal structure of tobacco rattle virus RNA: Evidence for polarity of reconstitution. Virology 76:173–185

Agranovsky AA, Dolja VV, Kavsan VM, Atabekov JG (1978) Detection of polyadenylate sequences in RNA of barley stripe mosaic virus. Virology 91:95–105

Agranovsky AA, Dolja VV, Gorbulev VG, Kozlov YV, Atabekov JG (1981) Aminoacylation of barley stripe mosaic virus RNA: polyadenylate-containing RNA has a 3′-terminal tyrosine-accepting structure. Virology 113:174–187

Agranovsky AA, Dolja VV, Kagramanova VK, Atabekov JG (1979) The presence of a cap structure at the 5′-end of barley stripe mosaic virus RNA. Virology 95:208–210

Ahlquist P, Dasgupta R, Shih DS, Zimmern D, Kaesberg P (1979) Two-step binding of eukaryotic ribosomes to brome mosaic virus RNA 3. Nature (London) 281:277–282

Alblas F, Bol JF (1977) Factors influencing the infection of cowpea mesophyll protoplasts by alfalfa mosaic virus. J Gen Virol 36:175–185

Asselbergs FAM, Peters WHM, Venrooij Van WJ, Bloemendal H (1978a) Diminished sensitivity of re-initiation of translation to inhibition by cap analogues in reticulocyte lysates. Eur J Biochem 88:483–488

Asselbergs FAM, Peters WHM, Venrooij Van WJ, Bloemendal H (1978b) Inhibition of translation of lens mRNAs in a messenger dependent reticulocyte lysate by cap analogues. Biochim Biophys Acta 520:577–587

Asselbergs FAM, Peters WHM, Venrooij Van WJ, Bloemendal H (1978c) Cap analogues do not inhibit mRNA translation in *Xenopus laevis* oocytes. FEBS Lett 94:195–198

Atabekov JG, Morozov S Yu (1979) Translation of plant virus messenger RNAs. Adv Virus Res 25:1–91

Atkins JF, Gesteland RF, Reid BR, Anderson CW (1979) Normal tRNAs promote ribosomal frameshifting. Cell 18:1119–1131

Bastin M, Hall TC(1976) Interaction of elongation factor 1 with aminoacylated brome mosaic virus and tRNAs. J Virol 20:117–122

Beachy RN, Zaitlin M, Bruening G, Israel HW(1976) A genetic map for the cowpea strain of TMV. Virology 73:498–507

Bénicourt C, Haenni A-L (1978) Differential translation of turnip yellow mosaic virus mRNAs in vitro. Biochem Biophys Res Commun 84:831–839

Bénicourt C, Morch MD (1979) Organization of the genome of TYMV deduced from its messenger properties. Biochem Trans 7:1095–1097

Bénicourt C, Péré J, Haenni A (1978) Translation of TYMV RNA into high molecular weight proteins. FEBS Lett 86:268–272

Bergmann JE, Lodish HF (1979) Translation of capped and uncapped vesicular stomatitis virus and reovirus mRNAs: Sensitivity to m^7 GpppAm and ionic conditions. J Biol Chem 254:459–468

Bina M, Feldmann RJ, Deeley RG (1980) Could poly(A) align the splicing sites of messenger RNA precursors. Proc Natl Acad Sci USA 77:1278–1282

Bisaro DM, Siegel A (1980) A new viral RNA species in tobacco rattle virus-infected tissue. Virology 107:194–201

Blumenthal T, Carmichael GG (1979) RNA replication: function and structure of $Q\beta$ replicase. Annu Rev Biochem 48:525–548

Bol JF, Bakhuizen CEGC, Rutgers T (1976) Composition and biosynthetic activity of polyribosomes associated with alfalfa mosaic virus infections. Virology 75:1–17

Briand JP, Keith G, Guilley H (1978) Nucleotide sequence at the 5′ extremity of turnip yellow mosaic virus genome RNA. Proc Natl Acad Sci USA 75:3168–3172

Boykov SV, Taliansky ME, Malyshenko SI, Kozlov YV, Atabekov JG (1981) Homology in the 3′-terminal regions of different barley stripe mosaic virus RNA species. Virology 113:168–173

Bruening G (1977) Plant covirus systems: Two component systems. In: Fraenkel-Conrat H, Wagner RR (eds) Comprehensive virology, Vol XI. Plenum, New York, pp 55–142

Castel A, Kraal B, Kerklaan PRM, Klok J, Bosch L (1977) Initiation of polypeptide synthesis with various NH_2-blocked aminoacyl-tRNAs under the direction of alfalfa mosaic virus RNA 4. Proc Natl Acad Sci USA 74:5509–5513

Castel A, Kraal B, Konieczny A, Bosch L (1979) Translation by *Escherichia coli* ribosomes of alfalfa mosaic virus RNA 4 can be initiated at two sites on the monocistronic message. Eur J Biochem 101:123–133

Chen JM, Hall TC (1973) Comparison of tyrosyl transfer ribonucleic acid and brome mosaic virus tyrosyl ribonucleic acid as amino acid as amino acid donors in protein synthesis. Biochemistry 12:4570–4574

Chifflot S, Sommer P, Hartmann D, Stussi-Garaud C, Hirth L (1980) Replication of alfalfa mosaic virus RNA: Evidence for a soluble replicase in healthy and infected tobacco leaves. Virology 100:91–100

Chisick MC, Brennessel BA, Biswas DK (1979) Poly(A)-protein interactions and transport of mRNA in isolated nuclei. Biochem Biophys Res Commun 91:1109–1116

Chroboczek J, Bassüner R, Ostrowka K, Witt M, Zagorski W (1980a) Seed transmissibility of plant viruses may be modulated by competition between viral and cellular messengers. A proposal. Plant Science Lett 19:263–270

Chroboczek J, Puchkovka L, Zagorski W (1980b) Regulation of brome mosaic virus genes expression by restriction of initiation of protein synthesis. J Virol 34:330–335

Chu L-Y, Rhoads RE (1978) Translational recognition of the 5′-terminal 7-methylguanosine of globin messenger RNA as a function of ionic strength. Biochemistry 17:2450–2455

Chu L-Y, Rhoads RE (1980) Inhibition of cell-free messenger ribonucleic acid translation by 7-methylguanosine 5′-triphosphate: effect of messenger ribonucleic acid concentration. Biochemistry 19:184–191

Chu PWG (1980) Ph D Thesis, Univ Adelaide

Clerx-Van Haaster CM, Clewley JP, Bishop DHL (1980) Oligonucleotide sequence analyses indicate that vesicular stomatitis virus large defective interfering virus particle RNA is made by internal deletion: Evidence for similar transcription polyadenylation signals for the synthesis of all vesicular stomatitis virus in RNA species. Virology 33:807–817

Cooper PD (1977) Genetics of picornaviruses. In: Fraenkel-Conrat H, Wagner RR (eds) Comprehensive virology, Vol IX. Plenum, New York, pp 133–207

Crick F (1979) Split genes and RNA splicing. Science 204:264–271
Dasgupta R, Harada F, Kaesberg P (1976) Blocked 5′ termini in brome mosaic virus RNA. J Virol 18:260–267
Daubert S, Bruening G (1979) Genome-associated proteins of comoviruses. Virology 98:246–250
Daubert S, Bruening G, Najarian R (1978) Protein bound to the genome RNAs of cowpea mosaic virus. Eur J Biochem 92:45–52
Davies JW (1979) Translation of plant virus ribonucleic acids in extracts from eukaryotic cells. In: Hall TC, Davies JW (eds) Nucleic acids in plants, Vol II, Chap 10. CRC Press, Boca Raton
Davies JW, Aalbers AMJ, Stuik EJ, Kammen Van A (1977) Translation of cowpea mosaic virus RNA in a cell-free extract from wheat germ. FEBS Lett 77:265–269
Davies JW, Stanley J, Kammen Van A (1979) Sequence homology adjacent to the 3′ terminal poly(A) of cowpea mosaic virus RNAs. Nucl Acids Res 7:493–500
Dawson WO, White JL (1978) Characterisation of a temperature sensitive mutant of tobacco mosaic virus deficient in synthesis of all RNA species. Virology 90:209–213
Dawson WO, White JL(1979) A temperature sensitive mutant of tobacco mosaic virus deficient in synthesis of single-stranded RNA. Virology 93:104–110
Dawson WO, Schlegel DE, Lung MCY (1975) Synthesis of tobacco mosaic virus in intact tobacco leaves systemically inoculated by differential temperature treatment. Virology 65:565–573
Deshpande AK, Chatterjee B, Roy AK (1979) Translation and stability of rat liver messenger RNA for α_{2u}-globulin in Xenopus oocyte. J Biol Chem 254:8937–8942
Dingjan-Versteegh A, Vloten-Doting Van L, Jaspars EMJ (1972) Alfalfa mosaic virus hybrids constructed by exchanging nucleoprotein components. Virology 49:716–722
Dolja VV, Sokolova NA, Tjulkina LG, Atabekov JG (1979) A study of barley stripe mosaic virus (BSMV) genome. II. Translation of individual RNA species of two BSMV strains in homologous cell-free systems. Mol Gen Genet 175:93–97
Dougherty WG, Hiebert E (1980a) Translation of potyvirus RNA in a rabbit reticulocyte lysate: Reaction conditions and identification of capsid protein as one of the products of in vitro translation of tobacco etch and pepper mottle virus RNAs. Virology 101:446–474
Dougherty WG, Hiebert E (1980b) Translation of potyvirus RNA in a rabbit reticulocyte lysate: Identification of nuclear inclusion proteins as products of tobacco etch virus RNA translation and cylindrical inclusion protein as a product of the potyvirus genome. Virology 104:174–182
Dougherty WG, Hiebert E (1980c) Translation of potyvirus RNA in a rabbit reticulocyte lysate: Cell-free translation strategy and a genetic map of the poty viral genome. Virology 104:183–194
El Manna MM, Bruening G (1973) Polyadenylate sequences in the ribonucleic acids of cowpea mosaic virus. Virology 56:198–206
Filipowicz W (1978) Functions of the 5′-terminal m^7G cap in eukaryotic mRNA. FEBS Lett 96:1–11
Filipowicz W, Haenni A-L (1979) Binding of ribosomes to 5′-terminal leader sequences of eukaryotic messenger RNAs. Proc Natl Acad Sci USA 76:3111–3115
Föglein FJ, Kalpagam C, Bates DC, Premecz G, Nyitrai A, Farkas GL (1975) Viral RNA synthesis is renewed in protoplasts isolated from TMV infected xanthi tobacco leaves in an advanced stage of virus infection. Virology 67:74–79
Fox TD (1979) Five TGA "stop" codons occur within the translated sequence of the yeast mitochondrial gene for cytochrome c oxidase subunit II. Proc Natl Acad Sci USA 76:6534–6538
Fraser RSS, Gerwitz A (1980) Tobacco mosaic virus does not alter the polyadenylated messenger RNA content of tobacco leaves. J Gen Virol 46:139–148
Fresno M, Vasquez D (1980) Inhibitory effects of "cap" analogues on globin mRNA and encephalomyocarditis RNA translation in a reticulocyte cell-free system. Eur J Biochem 103:125–132

Fritsch C, Mayo MA, Hirth L (1977) Further studies on the translation products of tobacco rattle virus RNA in vitro. Virology 77:722–732

Fritsch C, Mayo MA, Murant AF (1978) Translation of the satellite of tomato black ring virus in vitro and in tobacco protoplasts. J Gen Virol 40:587–593

Fritsch C, Mayo MA, Murant AF (1980) Translation products of genome and satellite RNAs of tomato black ring virus. J Gen Virol 46:301–390

Gasior E, Herrera F, Sadnik I, McLaughlin CS, Moldave K (1979) The preparation and characterization of a cell-free system from *Saccharomyces cerevisiae* that translates natural messenger ribonucleic acid. J Biol Chem 254:3965–3969

Geller AI, Rich A (1980) A UGA termination suppression $tRNA^{Trp}$ active in rabbit reticulocytes. Nature (London) 283:41–46

Ghosh A, Dasgupta R, Salerno-Rife T, Rutgers T, Kaesberg P (1980) Southern bean mosaic viral RNA has a 5′-linked protein but lacks terminal poly(A). Nucl Acids Res 7:2137–2146

Ghysdael J, Hubert E, Travnicek M, Bolognesi DP, Burny A, Cleuter Y, Huez G, Kettmann R, Marbaix G, Portetelle D, Chantrenne H (1977) Frog oocytes synthesize and completely process the precursor polypeptide to virion structural proteins after microinjection of avian myoblastosis virus RNA. Proc Natl Acad Sci USA 74:3230–3234

Goldstein NO, Pardoe IU, Burness ATH (1976) Requirement of an adenylic acid-rich segment for the infectivity of encephalomyocarditis virus. J Gen Virol 31:271–276

Gonda T, Symons RH (1979) Cucumber mosaic virus replication in cowpea protoplasts: Time course of virus, coat protein and RNA synthesis. J Gen Virol 45:723–736

Gonsalves D, Fulton RW (1977) Activation of *Prunus* necrotic ringspot virus and rose mosaic virus by RNA 4 components of some ilarviruses. Virology 81:398–407

Guilley H, Briand JP (1978) Nucleotide sequence of turnip yellow mosaic virus coat protein mRNA. Cell 15:113–122

Guilley H, Jonard G, Kukla B, Richards KE (1979) Sequence of 1000 nucleotides at the 3′ end of tobacco mosaic virus RNA. Nucl Acids Res 6:1287–1308

Gurdon JB, Lane CD, Woodland HR, Marbaix G (1971) Use of frog eggs and oocytes for the study of messenger RNA and its translation in living cells. Nature (London) 233:177–182

Haenni A-L, Prochiantz A, Bernard O, Chapeville F (1973) TYMV valyl-RNA as an amino-acid donor in protein biosynthesis. Nature (London) New Biol 241:166–168

Halk EL, Fulton RW (1978) Stabilization and particle morphology of prune dwarf virus. Virology 91:434–443

Hall TC (1979) Transfer RNA-like structures in viral genomes. Int Rev Cytol 60:1–26

Hall TC, Wepprich RK (1976) Functional possibilities for aminoacylation of viral RNA in transcription and translation. Ann Microbiol (Inst Pasteur) 127A:143–152

Hall TC, Shih DS, Kaesberg P (1972) Enzyme-mediated binding of tyrosine to brome mosaic virus ribonucleic acid. Biochem J 129:969–976

Hall TC, Pinck M, Ma Y, Duranton HM, German TL (1979) Aminoacylation and messenger functions of eggplant mosaic virus RNA. Virology 97:354–365

Hardy SF, German TL, Loesch-Fries LS, Hall TC (1979) Highly active template-specific RNA-dependent RNA polymerase from barley leaves infected with brome mosaic virus. Proc Natl Acad Sci USA 76:4956–4960

Hari V (1981) The RNA of tobacco etch virus: Further characterization and detection of protein linked to RNA. Virology 112:391–399

Hari V, Siegel A, Rozek C, Timberlake WE (1979) The RNA of tobacco etch virus contains poly(A). Virology 92:568–571

Harrison BD, Barker H (1978) Protease-sensitive structure needed for infectivity of nepovirus RNA. J Gen Virol 40:711–715

Hellmann GM, Shaw JG, Lesnaw JA, Chu L-Y, Pirone TP, Rhoads RE (1980) Cell-free translation of tobacco vein mottling virus RNA. Virology 106:207–216

Herson D, Schmidt A, Seal S, Marcus A, Van Vloten-Doting L (1979) Competitive mRNA translation in an in vitro system from wheat germ. J Biol Chem 254:8245–8249

Hofstetter H, Monstein HJ, Weissmann C (1974) The read-through protein A_1 is essential for the formation of viable $Q\beta$ particles. Biochem Biophys Acta 374:238–251
Horst J, Fraenkel-Conrat H, Mandeles S (1971) Terminal heterogeneity at both ends of the satellite tobacco necrosis virus ribonucleic acid. Biochemistry 10:4748–4752
Houwing CJ, Jaspars EMJ (1978) Coat protein binds to the 3′-terminal part of RNA 4 of alfalfa mosaic virus. Biochemistry 17:2927–2933
Huez G, Marbaix G, Gallwitz D, Weinberg E, Devos R, Hubert E, Cleuter Y(1978) Functional stabilisation of HeLa cell histon messenger RNAs injected into *Xenopus* oocytes by 3′-OH polyadenylation. Nature (London) 271:572–573
Ikegami M, Fraenkel-Conrat H (1979) Characterization of the RNA-dependent RNA polymerase of tobacco leaves. J Biol Chem 254:149–154
Joshi S, Haenni A-L, Hubert E, Huez G, Marbaix G (1978) In vivo aminoacylation and "processing" of turnip yellow mosaic virus RNA in *Xenopus laevis* oocytes. Nature (London) 275:339–341
Kaempfer R, Rosen H, Israeli R (1978) Translational control: Recognition of the methylated 5′ end and an internal sequence in eukaryotic mRNA by the initiation factor that binds methionyl-$tRNA_f^{Met}$. Proc Natl Acad Sci USA 75:650–654
Keith J, Fraenkel-Conrat H (1975) Tobacco mosaic virus RNA carries 5′-terminal triphosphorylated guanosine blocked by 5′-linked 7-methylguanosine. FEBS Lett 57:31–33
Klein C, Fritsch C, Briand J-P, Richards KE, Jonard G, Hirth L (1976) Physical and functional heterogeneity in TYMV-RNA: Evidence for the existence of an independent messenger coding for the coat protein. Nucl Acids Res 3:3043–3061
Knauert F, Ehrenfeld E (1979) Translation of poliovirus RNA in vitro: Studies on n-formylmethionine-labeled polypeptides initiated in cell-free extracts prepared from poliovirus infected HeLa cells. Virology 93:537–546
Kohl RJ, Hall TC (1974) Aminoacylation of RNA from several viruses: Amino acid specificity and differential activity of plant, yeast and bacterial synthetases. J Gen Virol 25:257–261
Koper-Zwarthoff EC, Bol JF (1979) 3′ Terminal nucleotide sequence of alfalfa mosaic virus RNA 4. Proc Natl Acad Sci USA 76:1114–1117
Koper-Zwarthoff EC, Lockard RE, Alzner-De Weerd B, RajBhandary UL, Bol JF (1977) Nucleotide sequence of 5′ terminus of alfalfa mosaic virus RNA 4 leading into coat protein cistron. Proc Natl Acad Sci USA 74:5504–5508
Koper-Zwarthoff EC, Brederode FTh, Walstra P, Bol JF (1979) Nucleotide sequence of the 3′ noncoding region of alfalfa mosaic virus RNA 4 and its homology with the genomic RNAs. Nucl Acids Res 7:1887–1900
Kozak M (1978) How do eukaryotic ribosomes select initiation regions in messenger RNA. Cell 15:1109–1123
Koziel MG, Hari V, Siegel A (1980) In vitro translation of tobacco etch virus RNA. Virology 106:177–179
Kumarasamy R, Symons RH (1979) Extensive purification of the cucumber mosaic virus induced RNA replicase. Virology 96:622–632
Lane LC (1979) The nucleic acids of multipartite, defective and satellite plant viruses. In: Hall TC, Davies JW (eds) Nucleic acids in plants, Vol II, Chap 3. CRC Press, Boca Raton, pp 65–110
Legon S, Model P, Robertson HD (1977) Interaction of rabbit reticulocyte ribosomes with bacteriophage f1 mRNA and of *Escherichia coli* ribosomes with rabbit globin mRNA. Proc Natl Acad Sci USA 74:2692–2696
Lesnaw JA, Reichmann ME (1970) Identitiy of the 5′-terminal RNA nucleotide sequence of the satellite necrosis virus and its helper virus: Possible role of the 5′-terminus in the recognition by virus-specific RNA replicase. Proc Natl Acad Sci USA 66:140–145
Leung DW, Gilbert CW, Smith RE, Sasavage NL, Clark JM Jr (1976) Translation of satellite tobacco necrosis virus ribonucleic acid by an in vitro system from wheat germ. Biochemistry 15:4943–4950

Lindley IJD, Stebbing N (1977) Aminoacylation of encephalomyocarditis virus RNA. J Gen Virol 34:177–182
Lodish HF, Rose JK (1977) Relative importance of 7-methylguanosine in ribosome binding and translation of vesicular stomatitis virus mRNA in wheat germ and reticulocyte cell-free systems. J Biol Chem 252:1181–1188
Lot H, Jonard G, Richards K (1977) Cucumber mosaic virus RNAs. Partial characterization and evidence for no large sequence homologies with genomic RNAs. FEBS Lett 80:395–400
Marbaix G, Huez G, Burny A, Cleuter Y, Hubert E, Leclerq M, Chantrenne H, Soreq H, Nudel U, Littauer UZ (1975) Absence of polyadenylate segment in globine messenger RNA accelerates its degradation in *Xenopus* oocytes. Proc Natl Acad Sci USA 72:3065–3067
Marbaix G, Huez G, Soreq H, Gallwitz D, Weinberg E, Devos R, Hubert E, Cleuter Y (1979) Role of the polyadenylate segment in the stability of eukaryotic messenger RNAs. In: Rosenthal S (ed) Gene functions. FEBS Vol 51. Pergamon, New York
Matthews REF (1979) Third report of the international committee on taxonomy of viruses. Intervirology 12:126–296
Mayo MA, Fritsch C, Hirth L (1976) Translation of tobacco rattle virus RNA in vitro using wheat germ extracts. Virology 69:408–415
Mayo MA, Barker H, Harrison BD (1979a) Polyadenylate in the RNA of five nepoviruses. J Gen Virol 43:603–610
Mayo MA, Barker H, Harrison BD (1979b) Evidence for a protein covalently linked to tobacco ringspot virus RNA. J Gen Virol 43:735–740
Minson T, Darby G (1973) 3′-Terminal oligonucleotide fragments of tobacco rattle virus ribonucleic acids. J Mol Biol 77:337–340
Morch M-D, Bénicourt C (1980a) Post-translational proteolytic cleavage of in vitro synthesized turnip yellow mosaic virus RNA-coded high molecular weight protein. J Virol 34:85–94
Morch M-D, Bénicourt C (1980b) Polyamines stimulate in vitro suppression of amber termination by normal tRNAs. Eur J Biochem 105:445–451
Morris-Krsinich BAM (1980) In vitro translation of turnip rosette virus RNA in rabbit reticulocyte lysates. John Innes Annu Rep 1979, 100
Mossop DW, Francki RIB (1979) The stability of satellite viral RNAs in vivo and in vitro. Virology 94:243–253
Neeleman L, Vloten-Doting Van L (1979) Determination of the number of ribosomal binding sites on the RNAs of eukaryotic viruses. Methods Enzymol LX:410–417
Neeleman L, Rutgers T, Vloten-Doting Van L (1977) Internal initiation of protein synthesis on RNA of eukaryotic virus? In: Legocki AB (ed) Translation of natural and synthetic polynucleotides. Univ Agric Poznan, pp 292–298
Nishiguchi M, Motoyoshi F, Oshima N (1978) Behaviour of a temperature sensitive strain of tobacco mosaic virus in tomato leaves and protoplasts. J Gen Virol 39:53–61
Nomoto A, Kitamura N, Golini F, Wimmer F (1977) The 5′-terminal structures of poliovirion RNA and poliovirus mRNA differ only in the genome-linked protein VPg. Proc Natl Acad Sci USA 74:5345–5349
Oberg B, Philipson L (1972) Binding of histidine to tobacco mosaic virus RNA. Biochem Biophys Res Commun 48:927–932
Okuno T, Furusawa I (1979) RNA polymerase activity and protein synthesis in brome mosaic virus-infected protoplasts. Virology 99:218–225
Oxelfelt P (1976) Biological and physicochemical characteristics of three strains of red clover mottle virus. Virology 74:73–80
Palmer E, Wilhelm JM, Sherman F (1979) Phenotypic suppression of nonsense mutants in yeast by aminoglycoside antibiotics. Nature (London) 277:148–150
Paterson BM, Rosenberg M (1979) Efficient cap-dependent translation of polycistronic prokaryotic mRNAs is restricted to the first gene in the operon. Nature (London) 279:696–701
Pelham HRB (1978) Leaky UAG termination codon in tobacco mosaic virus RNA. Nature (London) 272:469–471

Pelham HRB (1979a) Translation of tobacco rattle virus RNA in vitro: Four proteins from three RNAs. Virology 97:256–265
Pelham HRB (1979b) Synthesis and proteolytic processing of cowpea mosaic virus proteins in reticulocyte lysates. Virology 96:463–477
Pelham HRB (1979c) Translation of fragmented viral RNA in vitro: Initiation at multiple sites. FEBS Lett 100:195–199
Pelham HRB, Jackson RJ (1976) An efficient mRNA-dependent translation system from reticulocyte lysates. Eur J Biochem 67:247–256
Pinck L (1975) The 5′-end groups of alfalfa mosaic virus RNAs are $^{7}mG^{5'}$ $ppp^{5'}Gp$. FEBS Lett 59:24–28
Pinck L, Franck A, Fritsch C (1979) Formation of ribosome-RNA initiation complexes with alfalfa mosaic virus RNA 4 and RNA 3. Nucl Acids Res 7:151–166
Pinck M, Chan S-K, Genevaux M, Hirth L, Duranton H (1972) Valine specific tRNA-like structure in two viruses of turnip yellow mosaic virus group. Biochemistry 54:1093–1094
Proudfoot NJ, Brownlee GG (1976) 3′ Non-coding region sequences in eukaryotic messenger RNA. Nature (London) 263:211–214
Pyne JW, Hall TC (1979) Efficient ribosome binding of brome mosaic virus (BMV) RNA 4 contributes to its ability to outcompete the other BMV RNAs for translation. Intervirology 11:23–29
Rezelman G, Goldbach R, Kammen Van A (1980) Expression of bottom component RNA of cowpea mosaic virus in cowpea protoplasts. J Virol 36:366–373
Richards KE, Guilley H, Jonard G, Hirth L (1978a) Nucleotide sequence at the 5′ extremity of tobacco mosaic virus RNA 1. The noncoding region (nucleotides 1–68). Eur J Biochem 84:513–519
Richards KE, Jonard G, Jacquemond M, Lot H (1978b) Nucleotide sequence of cucumber mosaic virus-associated RNA 5. Virology 89:395–408
Roman R, Brooker JD, Seal SN, Marcus A (1976) Inhibition of the translation of a 40S ribosome-Met-t-RNA_i^{Met} complex to an 80S ribosome-Met-$tRNA_i^{Met}$ complex by 7-methylguanosine 5′-phosphate. Nature (London) 260:359–360
Rottier PMJ (1980) Viral protein synthesis in cowpea mosaic virus infected protoplasts. Ph D Thesis, Wageningen
Rutgers AS (1977) In vitro and in vivo translation of the RNAs of alfalfa mosaic virus. Ph D Thesis, Univ Leiden
Rutgers T, Salerno-Rife T, Kaesberg P (1980) Messenger RNA for the coat protein of southern bean mosaic virus. Virology, 104:506–509
Rychlik W, Zagorski W (1980) Purification and characterisation of cyclic AMP-independent protein kinase from wheat germ, differentiating translation of natural templates. Eur J Biochem 106:653–659
Rychlik W, Kupidlowska E, Nowak E, Zagorski W (1980) Wheat germ protein kinase affects the translation of brome mosaic virus ribonucleic acid in vitro. Biochemistry 19:5249–5255
Sakai F, Takeba I (1972) A non-coat protein synthesized in tobacco mesophyll protoplasts infected by tobacco mosaic virus. Mol Gen Genet 118:93–96
Sakai F, Watts JW, Dawson JRO, Bancroft JB (1977) Synthesis of proteins in tobacco protoplasts infected with cowpea chlorotic mottle virus. J Gen Virol 34:285–293
Sakai F, Dawson JRO, Watts JW (1979) Synthesis of proteins in tobacco protoplasts infected with brome mosaic virus. J Gen Virol 42:323–328
Salerno-Rife T (1980) Ph D Thesis, Univ Madison
Salomon R, Bar-Joseph M, Soreq H, Gozes I, Littauer UZ (1978) Translation in vitro of carnation mottle virus RNA: Regulatory function of the 3′ region. Virology 90:288–298
Salvato MS, Fraenkel-Conrat H (1977) Translation of tobacco necrosis virus and its satellite in a cell-free wheat germ system. Proc Natl Acad Sci USA 74:2288–2292
Scalla R, Romaine P, Asselin A, Rigaud J, Zaitlin M (1978) An in vivo study of a non-structural polypeptide synthesized upon TMV infection and its identification with a polypeptide synthesized in vitro from TMV-RNA. Virology 91:182–193

Semancik JS (1974) Detection of polyadenylic acid sequences in plant pathogenic RNAs. Virology 62:288–291

Shaw DC, Walker JE, Northrop FD, Barrell BG, Godson GN, Fiddes JC (1978) Gene K, a new overlapping gene in bacteriophage G4. Nature (London) 272:510–515

Shaw J, Murialdo H (1980) Morphogenetic genes C and Nu3 overlap in bacteriophage. Nature (London) 283:30–35

Shih DS, Shih CL, Kew O, Pallansch M, Rueckert R, Kaesberg P (1978) Cell-free synthesis and processing of the proteins of poliovirus. Proc Natl Acad Sci USA 75:5807–5811

Singh A, Ursic D, Davies J (1979) Phenotypic suppression and misreading in *Saccharomyces cerevisiae*. Nature (London) 277:146–148

Sonenberg N, Shatkin AJ, Ricciardi RP, Rubin M, Goodman R (1978) Analysis of terminal structures of RNA from potato virus X Nucl Acids Res 5:2501–2512

Sonenberg N, Morgan MA, Testa D, Colonno RJ, Shatkin AJ (1979) Interaction of a limited set of proteins with different mRNAs and protection of 5′-caps against pyrophosphatase digestion in initiation complexes. Nucl Acids Res 7:15–29

Stanley J, Rottier P, Davies JW, Zabel P, Kammen Van A (1978) A protein linked to the 5′ termini of both RNA components of the cowpea mosaic virus genome. Nucl Acids Res 5:4505–4522

Stanley J, Goldbach R, Kammen Van A (1980) The genome linked protein of cowpea mosaic virus is coded by RNA from bottom component. Virology 106:180–182

Steeg Van H, Pranger MH, Zeijst Van der BAM, Benne R, Voorma HO (1979) In vitro translation of semliki forest virus 42S RNA initiation at two different sites. FEBS Lett 108:292–298

Stuik E (1979) Protein synthesis directed by cowpea mosaic virus RNAs. Ph D Thesis, Wageningen

Symons RH (1975) Cucumber mosaic virus RNA contains 7-methylguanosine at the 5′ terminus of all four RNA species. Mol Biol Rep 2:277–285

Szczesna E, Filipowicz W (1980) Faithfull and efficient translation of viral and cellular eukaryotic mRNAs in a cell-free S27 extract of *Saccharomyces cerevisiae*. Biochem Biophys Res Commun 92:563–569

Takebe I (1977) Protoplasts in the study of plant virus replication. In: Fraenkel-Conrat H, Wagner RR (eds) Comprehensive virology, Vol XI. Plenum, New York London, pp 237–283

Taliansky ME, Boykov SV, Malyshenko SI, Kavsan VM, Atabekov JG (1979) A study of barley stripe mosaic virus (BSMV) genome. I. Determination of sequence homology between BSMV RNA species. Mol Gen Genet 175:89–92

Tol Van RGL (1981) Ph D Thesis, Univ Leiden

Tol Van RGL, Vloten-Doting Van L (1979) Translation of alfalfa mosaic virus RNA 1 in the mRNA-dependent translation system from rabbit reticulocyte lysates. Eur J Biochem 93:461–468

Tol Van RGL, Gemeren Van R, Vloten-Doting Van L (1980) Two leaky termination codons in AMV-RNA 1 (suppressor tRNA in wheat germ). FEBS Lett 118:67–71

Tuite MF, Plesset J, Moldave K, McLaughlin CS (1980) Faithful and efficient translation of homologous and heterologous mRNAs in an mRNA-dependent cell-free system from *Saccharomyces cerevisiae*. J Biol Chem 255:8761–8766

Veerisetty V, Sehgal OP (1980) Proteinase K-sensitive factor essential for the infectivity of southern bean mosaic virus ribonucleic acid Phytopathology 70:282–284

Venegas A, Quiroga M, Zaldivar J, Rutter WJ, Valenzuela P (1979) Isolation of yeast $tRNA^{Leu}$ genes. J Biol Chem 254:12306–12309

Vloten-Doting Van L (1976) Similarities and differences between viruses with a tripartite genome. Ann Microbiol (Inst Pasteur) 127A:119–129

Vloten-Doting Van L, Jaspars EMJ (1977) Plant covirus systems: Three-component systems. In: Fraenkel-Conrat H, Wagner RR (eds) Comprehensive virology, Vol XI. Plenum, New York London, pp 1–53

Vloten-Doting Van L, Neeleman L (1980) Translation of plant virus RNAs. In: Leaver CJ (ed) Genome organisation and expression in plants. Plenum New York London, pp 511–527

Vloten-Doting Van L, Bol J, Neeleman L, Rutgers T, Dahlen Van D, Castel A, Bosch L, Marbaix G, Huez G, Hubert E, Cleuter Y (1977) In vivo and in vitro translation of the RNAs of alfalfa mosaic virus. In: Bogorad L, Weil JH (eds) Nucleic acids and protein synthesis in plants. Plenum, New York London, pp 387–411

Vloten-Doting Van L, Hasrat JA, Oosterwijk E, Van 't Sant P, Schoen MA, Roosien J (1980) Description and complementation analysis of 13 temperature sensitive mutants of alfalfa mosaic virus. J Gen Virol 46:415–426

Westaway EG (1977) Strategy of flavivirus genome: Evidence for multiple internal initiation of translation of proteins specified by kunjin virus in mammalian cells. Virology 80:320–335

Wodnar-Filipowicz A, Krzeczkowski LJ, Filipowicz W (1980) Translation of potato virus X RNA into high molecular weight proteins. FEBS Lett, 109:151–155

Wu JM, Cheung CP, Suhadolnik RJ (1977) Differential inhibition with partially purified and endogenous rabbit reticulocyte globin mRNA by 7-methylguanosine 5′-monophosphate. Biochem Biophys Res Commun 78:1079–1086

Yogo Y, Wimmer E (1975) Sequence studies of poliovirus RNA. III. Polyuridylic acid and polyadenylic acid as components of the purified poliovirus replicative intermediate. J Mol Biol 92:467–477

Zabel P, Jongen-Neven I, Van Kammen A (1979) In vitro replication of cowpea mosaic virus RNA III Template recognition by cowpea mosaic virus RNA replicase. J Virol 29:21–33

Zagorski W (1978) Translational regulation of expression of the brome mosaic virus RNA genome in vitro. Eur J Biochem 86:465–472

Zaitlin M (1979) The RNAs of monopartite plant viruses. In: Hall TC, Davies JW (eds) Nucleic acids in plants, Vol II, Chap 2. CRC Press, Boca Raton, pp 31–64

Zimmern D (1975) 5′-End group of tobacco mosaic virus RNA is $m^7G^{5'}$ $ppp^{5'}$ Gp. Nucl Acids Res 2:1189–1201

12 Biology, Structure, Functions and Possible Origin of Viroids[1]

H.L. SÄNGER

1 Introduction

The great interest viroids have attracted ever since their discovery is reflected in numerous review articles (DIENER 1972a, b, 1974, SEMANCIK 1976, DIENER and HADIDI 1977, DIENER et al. 1977, DIENER 1979a, DICKSON 1979, SÄNGER 1979a, b, SEMANCIK 1979, SÄNGER 1980a, b, GROSS and RIESNER 1980, KLEINSCHMIDT et al. 1981). Thus, one might get the impression that the viroid field is now in some way overreviewed especially after an entire book on viroids has appeared (DIENER 1979b). However, the rapid progress in viroid research still justifies up-dated treatments which are, nevertheless, obsolete in many parts at the time of their appearance. The same will apply for this contribution which surveys the data on the biology, structure, function and possible origin of viroids as available early in 1982.

Viroids are not only the most unusual RNA molecules occurring in plants but at the same time they are, together with the so-called "prions" or "virinos" (see Section 7), the smallest disease agents presently known. Viroids represent coat-protein-free RNA molecules with a unit length of their chain between about 240 to 380 nucleotides depending on the viroid "species". Thus, their genome is less than one tenth the size of that of the smallest bacterial virus. All viroids presently known are exclusively pathogens of higher plants, where they cause economically important diseases and considerable losses in different crop plants.

The discovery of viroids in 1970/72 is an interesting example of serendipity in modern biology, because they were found although not actually sought. In fact, they finally proved to be completely different than expected when the first efforts were made to characterize the elusive and presumed virus-like agents of the corresponding plant diseases. After early scepticism and even rejection, viroid research has now developed into a well-accepted field with interesting perspectives for both applied and basic science.

The term "viroid" has been proposed by DIENER (1971b, 1973a) in order to differentiate between the novel class of "naked" infectious RNA molecules of low molecular weight and the conventional viruses with their encapsidated genomes. Although the names "meta-virus" (SINGH and BAGNALL 1968, SINGH and CLARK 1971) and "low molecular weight pathogenic RNA" or "pathogene" (SEMANCIK and WEATHERS 1972a, SEMANCIK et al. 1973) had been proposed, the term "viroid" became generally accepted.

1 Abbreviations used for the viroids are listed in Table 1, p. 370

Viroids are of great importance for agriculture and horticulture because of their potential threat to crop plants and the economically important losses they cause. Therefore, not only improved methods for their detection and elimination are needed, but also an improved understanding of their mechanisms of pathogenicity. Because of their relative stability, their small size, and the complete knowledge of their structure, viroids are at the same time excellent model systems for biochemical and biophysical investigations. Viroids have evolved and replicate in an intimate adaptation with the nucleic acid-synthesizing machinery of plant cells. Therefore, it can finally be expected that the elucidation of the molecular basis of their unusual mode of replication, pathogenicity and interference with central functions of plant cell metabolism will contribute to a better understanding of normal cellular processes and their possible control by small RNA molecules.

2 The Biology of Viroids

2.1 Viroid Diseases

The eleven viroid-incited plant diseases presently known are listed in Table 1. The names of these diseases reflect the characteristic disease symptoms caused by the corresponding viroids. There is little doubt that additional viroid diseases will be detected, so that their number will gradually increase with time. From the established nucleotide sequence the five viroids PSTV, CEV, CSV (Gross et al. 1978, 1982, Haseloff and Symons 1981, Visvader et al. 1982), ASBV and CCCV (Symons 1981) may be considered as distinct viroid "species" because they differ from each other in more than 30–40% of their primary sequence. It is possible, on the other hand, that some of the remaining diseases, especially of tomato, could be causes by "strains" of one and the same viroid (PSTV?) which might differ from the "type strain" by a few nucleotide exchanges only, as has been shown for a pathogenic and a mild isolate of PSTV (Gross et al. 1981) and different "isolates" of CSV and CEV (Haseloff and Symons 1981, Gross et al. 1982, Visvader et al. 1982).

In susceptible host plants viroids may cause symptoms which range from malformations and discolorations of leaves (CEV, CCMV) and fruits (PSTV, CPFV) to retardation of general plant growth (CSV, BSV, HSV) (Fig. 1) and even death of the whole plant may occur (CCCV) (Fig. 2). The viroid aetiology of all these diseases has been established by demonstrating the extra band of viroid RNA by polyacrylamide gel electrophoresis of nucleic acid preparations directly extracted from diseased plant tissue with the aid of phenol. This band is absent in RNA samples from healthy control plants (Fig. 3). In addition, the viroid nature of the extra RNA band has been demonstrated by assaying its infectivity, wherever possible. For this purpose the gel is cut into slices and the buffer homogenates individually prepared from them are bioassayed (Fig. 4).

Table 1. The viroid diseases presently known and the abbreviations of the corresponding viroids

Viroid disease	Viroid	References[a]
1. Potato spindle tuber	PSTV	1, 2
2. Citrus exocortis	CEV	3, 4
3. Chrysanthemum stunt	CSV	5, 6
4. Chrysanthemum chlorotic mottle	CCMV	7, 8
5. Cucumber pale fruit	CPFV	9, 10
6. Coconut "cadang-cadang"	CCCV	11, 12
7. Hop stunt	HSV	13, 14
8. Avocado sunblotch	ASBV	15
9. Tomato bunchy top	TPTV	16, 17
10. Tomato "planta macho"	TPMV	18
11. Burdock stunt	BSV	19

[a] *References:* 1. DIENER 1971b; 2. SINGH and CLARK 1971; 3. SÄNGER 1972; 4. SEMANCIK and WEATHERS 1972b; 5. HOLLINGS and STONE 1973; 6. DIENER and LAWSON 1973; 7. ROMAINE and HORST 1975; 8. HORST and ROMAINE 1975; 9. VAN DORST and PETERS 1974; 10. SÄNGER et al. 1976; 11. RANDLES 1975; 12. RANDLES et al. 1976; 13. SASAKI and SHIKATA 1977a; 14. SASAKI and SHIKATA 1977b; 15. THOMAS and MOHAMED 1979; 16. MC CLEAN 1931; 17. WALTER 1981; 18. GALINDO et al. 1982; 19. CHEN et al. 1982.

Fig. 1. Symptoms of growth retardation and epinasty as caused by PSTV in tomato plants (cv. Rutgers). *Left* healthy control plants

The cadang-cadang disease of coconut differs from all other viroid diseases in that two RNA's are associated with it (RANDLES and PALUKAITIS 1979), which in addition occur as fast and slow electrophoretic variants so that altogether four extra viroid bands are found. The fast forms are found in early

Fig. 2. Coconut plantation in the Philippines with palm trees killed by the viroid of the cadang-cadang disease (CCCV) and subsequently decapitated by a hurricane

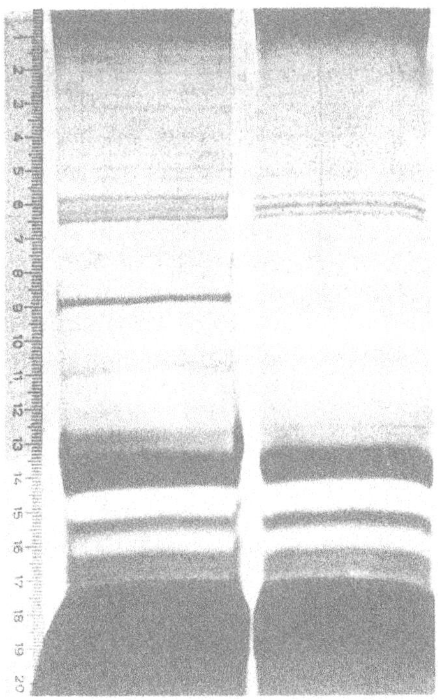

Fig. 3. Separation of 25 mg 2M LiCl-soluble RNA from healthy (*right lane*) and from viroid (PSTV-infected) (*left lane*) tomato leaf tissue on a non-denaturing 5% polyacrylamide slab gel (6 mm thick). The RNA bands are made visible by staining with methylene blue. Note the extra band of viroid RNA in the left lane which usually comprises less than 0.001% of the total cellular nucleic acids from which purification is started

stages of the disease while only the slow form is found at late stages (Imperial et al. 1981).

Polyacrylamide gel analysis may be used for routine diagnosis of viroid infestation in crop plants (Morris and Smith 1977; Morris and Wright 1975,

Fig. 4. Results of an infectivity test to show the presence of viroid (PSTV) RNA in a polyacrylamide gel after electrophoresis. Young plants were inoculated with the buffer homogenate from individual gel slices and for each row of plants one slice was used. The viroid-infected malformed plants in the central three rows showed the first symptoms 3–4 weeks after inoculation and substantiated the presence of the infectious PSTV RNA in the corresponding slices. In order to improve the symptoms all plants were decapitated 4 weeks after inoculation. The newly growing axillary shoots usually develop severe symptoms of epinasty and leaf malformations which greatly facilitates the evaluation of the bioassay. Only the relevant eight rows out of approximately 100 altogether are shown. It should be noted that all early work on viroids was essentially based on such laborious time- and greenhouse space-consuming methods

PFANNENSTIEL et al. 1980, MACQUAIR et al. 1981). But the absence of a visible viroid band cannot be considered as a reliable proof for the absence of the agent. Since viroid RNA may represent only a minute fraction (usually less than 0.001%) of the total nucleic acids directly extracted from diseased plants, its concentration could be beyond the level of detection after staining the gel with methylene blue, toluidine blue, or ethidium bromide. Although the recently developed and highly sensitive silver-staining (SAMMONS et al. 1981) allows to detect down to 1 ng (0.03 ng mm^{-2}) of nucleic acids in polyacrylamide gels (SOMMERVILLE and WANG 1981, GOLDMAN and MERRIL 1982) gel analysis should be complemented by appropriate bioassays if critical tests are required.

A completely new approach for assaying for viroids even under practical conditions is presently developed by applying recombinant DNA techniques and molecular hybridization. The required viroid-complementary cDNA can be synthetised with the aid of the reverse transcriptase using viroid RNA as template (OWENS 1978, PALUKAITIS and SYMONS 1978, PALUKAITIS et al. 1981). The use of tailor-made sequence-specific primers has allowed to obtain viroid-specific cDNA which corresponds to the full length of the corresponding viroid (ROHDE et al. 1981a, b). In the case of PSTV the corresponding cDNA has already been subjected to molecular cloning in order to obtain PSTV-specific hybridization probes (OWENS and CRESS 1980). This new technique allows the practically unlimited amplification and a very efficient radioactive labelling of the viroid-specific cDNA thus allowing in the future simplified techniques of molecular hybridization to screen for the presence of viroids. One of these highly specific and sensitive techniques is the so-called "dot-spot test".

It may even be used under practical conditions to routinely assay large numbers of samples as required in institutions involved, for example, in plant breeding and plant introduction (OWENS and DIENER 1981).

2.2 Economic Importance

Most of the viroid diseases known at present (Table 1) are of economic importance and the effect of a viroid infection may be quite dramatic as in the case of cadang-cadang, which causes disastrous losses in coconut palms in the Philippines (Fig. 2). It has been estimated that probably more than 20 million trees were killed from 1926 to 1978, which makes the disease the main threat to coconut production in the Philippines (RANDLES 1975). The chrysanthemum stunt disease has caused a serious epidemic in cultivated cut- and pot-flower chrysanthemums in the U.S.A. during 1945–1947 (HOLLINGS and STONE 1970). It had been spread inadvertently over the whole country with infested but symptomless cuttings, which were distributed mainly by one propagator (KELLER 1951, 1953). When it was realized that the disease was easily transmissible by contaminated tools and hands, pertinent control measures were rapidly adopted and only indexed plants kept in isolation were used for futher propagation. Considerable losses have also been reported for PSTV-infected potato (SINGH et al. 1971) and CEV-infected citrus species and varieties (CALAVAN et al. 1968).

2.3 Experimental Transmission

All viroids described so far have been experimentally transmitted with mechanical means using crude tissue homogenates. The success of transmission is considerably increased if measures are taken to eliminate the action of nucleases, which inactivate viroid infectivity quite rapidly by degrading the nuclease-sensitive viroid RNA molecule. Correspondingly, tissue homogenates are usually freshly prepared with cold and slightly alkaline buffers in the presence of the nuclease-absorbing clay mineral bentonite and immediately used for inoculation. Under these conditions standard inoculation techniques, as used with conventional plant viruses or their RNA's, may be applied. Thus, rubbing of the inoculum onto carborundum-dusted leaves is used experimentally in many viroid–host combinations (tomato-PSTV, CEV, CPFV). In certain host plants stem-puncturing or stem-slashing with razor blades or scalpels (gynura–CEV, chrysanthemum–CSV, cucumber–CPFV) is more efficient in viroid transmission. Grafting has played an important role in the early experimental transmission of the citrus exocortis disease.

Quantitative experimental work, based on the infectivity of viroids as determined by bioassays, is somewhat complicated, because infectivity is generally measured by dilution end point titrations using the all-or-none type response of a generalized (systemic) viroid infection. Since most experiments with viroids result in large effects the differences in titers can be determined with fair accuracy by this type of bioassay. There are two viroid-host systems for which a local lesion response upon infection has been reported. Thus, CSV produces chlorotic spots about a month after inoculation on leaves of florist's cineraria (*Senecio cruentus*). These lesions can be made visible already after 2 weeks by the so-called starch lesion test (LAWSON 1968b). PSTV, on the other hand, has been reported to produce necrotic local lesions on leaves of the solanaceous species *Scopolia sinensis* 1 to 2 weeks after inoculation (SINGH 1973). But in both cases, the local lesion development is strongly influenced by environmental conditions, and there is considerable variation in the response among plants and even between leaves of the same plant. Therefore, these local lesion hosts have not been commonly used for bioassays and quantitative work.

2.4 Experimental Host Range

The experimental host range of different viroids differs considerably. PSTV is able to systemically infect numerous plant species of several families (SINGH 1973), whereas ChCMV seems to infect only chrysanthemum varieties. CCCV has, so far, only been transmitted successfully to a few other palm species and ASBV seems to infect avocado (*Persea americana*) and cinnamon (*Cinnamonum zelanicum*), which are both members of the family Lauraceae. The host range of the other viroids is limited to certain species of a few specific plant families. Thus CEV can be transmitted from citrus to the herbaceous composite *Gynura aurantiaca* and to different solanaceous hosts including tomato, potato, *Petunia* and *Scopolia*. CPFV infects many species of the cucurbit family, but it may also be transmitted to certain solanaceous hosts.

An interesting problem in this context is the host-specificity of viroids, which can be studied on the molecular basis by comparing the primary sequence of different viroids after adaptation to certain hosts. First attempts to clarify this point were made by comparing the RNase T_1 ^{125}I fingerprints of PSTV and CEV, which are characteristically different from each other after propagation in the two host plants tomato and *Gynura* (DICKSON et al. 1978). It was found that both viroids displayed the most essential features of their characteristic fingerprints irrespective of the host plant in which they were propagated. Some variations were observed, the nature and origin of which has not yet been elucidated. These results clearly show that the infecting viroid must serve as a template from which the progeny viroid is copied and that the host plant has no pronounced influence on viroid structure.

2.5 Transmission Under Natural Conditions

The viroids of the spindle tuber disease of potato (PSTV) and the stunt disease of chrysanthemum (CSV) are known to spread rapidly to neighbouring plants by the combination of wounding and contact of foliage as occurring during culturing measures or handling in the field and in greenhouses. Vertical transmission through seed and pollen has been reported for PSTV in certain solanaceous host species, which causes serious problems in obtaining healthy plant material by breeding. Early work indicated that under field conditions the potato spindle tuber disease could also be accidentally transmitted at low frequencies by biting insects and by aphids (SCHULTZ and FOLSOM 1923a, b, c, GOSS 1928, 1930, 1931). Since at that time the disease was thought to be of viral origin, the possibility must be considered that its successful transmission can be related to insect-transmissible potato viruses which might also have been present in the diseased plants in question. Corresponding reinvestigations have recently shown that the two aphid species *Macrosiphum euphorbiae* (DE BOKX and PIRON 1981) and *Myzus persicae* (LEE and SÄNGER 1982) are capable of transmitting true PSTV from infected to healthy potato and tomato plants respectively at frequencies below 6%.

Most viroid diseases, however, are inadvertently spread by man via the vegetative propagation of viroid-infested but symptomless plants through tubers (potato), cuttings (chrysanthemum) or grafting (citrus). Thus, CEV has been disseminated inadvertently into all citrus-growing areas of the world by distributing and grafting citrus varieties, which were symptomless carriers of the agent. In addition, CEV may also be transmitted to neighbouring trees within a citrus plantation by contaminated tools during pruning and cutting. The mode of natural transmission of the cadang-cadang disease of the coconut palm, however, which is propagated only by seeds, is still unknown.

After it has now become possible to carry out reliable routine screening tests for the presence of viroids, epidemiological studies could clarify the involvement of wild plant species as potential reservoir plants, from which these agents could probably be transferred accidentally to susceptible cultivated plant species, where they become pathogenic and produce disease symptoms. Studies of the experimental host range have shown that viroids are often latent in wild plant species, whereas they are generally pathogenic in cultivated plants (SINGH 1973). In fact, the wild potato species *Solanum phureja* and *S. stenotonum* were found to harbor viroids which are capable of inducing potato spindle tuber disease in cultivated potato. Also the seemingly healthy ornamental plant *Columnea erythrophae* appeared to be viroid-infected (OWENS et al. 1978). Many observations lead to the suspicion that viroid diseases of cultivated plants are of recent origin and that their world-wide spreading by the agricultural and horticultural activities of man is strongly favoured by the monoculture of crop plants and ornamentals.

2.6 Expression of Symptoms

In herbaceous host plants such as potato, tomato, *Gynura,* chrysanthemum and cucumber, the first symptoms of a generalized viroid infection usually occur in newly developing leaves 2–8 weeks after inoculation with PSTV, CEV, CSV, ChCMV and CPFV, respectively. On the other hand, symptoms may take several months and even years to develop, as in the case of CEV in certain citrus

Fig. 5. Tomato plants (cv Rutgers) 8 weeks after infection with PSTV isolated which differ in their pathogenicity. *From left to right* plant infected with a lethal, severe, intermediate and mild PSTV, uninfected control plant

varieties, CCCV in coconut palms and ASBV in avocado. However, it should be noted here that especially in many wild host plant species, viroid infection and replication does not produce any recognizable disease symptoms. Thus, viroids may often be latent in certain hosts (O'Brien and Raymer 1964) and only cause disease when transmitted to cultivated plant and species varieties.

The disease symptoms are of course specified by both the host plant and the viroid. For example, three different tomato cultivars may respond to an infection with the common strain of PSTV with severe, mild or no symptoms (Mühlbach and Sänger 1977, Mühlbach et al. 1977). On the other hand, viroid strains producing severe, intermediate and mild symptoms in a given host have been reported for PSTV (Fernow 1967), CEV (Weathers and Calavan 1961, Calavan et al. 1964) and ChCMV (Horst 1975). The availability of such mild and severe strains of PSTV (Fig. 5) has initiated relating these obvious differences in pathogenicity to possible differences in the primary sequence of the corresponding viroid strains. Fingerprint analysis has shown that severe and mild strains are uniquely characterized by the presence and absence of particular spots, from which it was estimated that they differ in 2–10 of their nucleotides (Dickson et al. 1979). The comparison of the nucleotide sequences of the pathogenic PSTV "type strain" with mild and severe PSTV isolates has established (Fig. 23) that difference in pathogenicity can be related to a few nucleotide exchanges only (Gross et al. 1981).

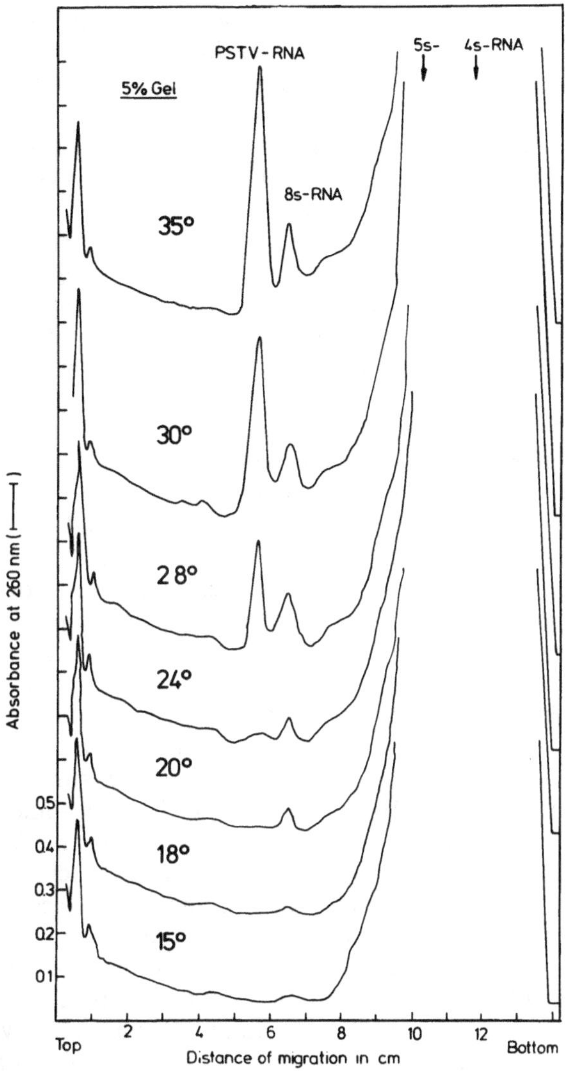

Fig. 6. Electrophoretic analysis on 5% polyacrylamide gels of "2M LiCl-soluble RNA" from PSTV-infected tomatoes grown for 4 weeks at different temperatures. On each gel the RNA-equivalent from 50 g leaf tissue was run. The increase of the concentration of PSTV-RNA with increasing growth temperature becomes evident from the height of the corresponding peak

If symptoms are produced, their expression is critically dependent on a threshold concentration of viroid and on environmental conditions. High temperatures, high light intensities and inoculation with highly infectious inocula generally increase the severity of symptoms and the rapidity of their appearance. This phenomenon can be directly related to the inordinate temperature dependence of viroid replication, which differs strikingly from the replication of conventional plant viruses. At 18–20 °C, where most plant viruses are replicating well, the replication of PSTV, CEV, CSV and CCMV in tomato, *Gynura* and chrysanthemum, respectively, is usually so low that no stainable viroid band can be detected in gel analysis. Rising the growth temperature of infected plants to 32–35 °C may cause a more than 500-fold increase in viroid yields (SÄNGER and RAMM 1975) (Fig. 6). Under these conditions, severe symptoms usually appear 10–14 days after inoculation, whereas it may take 2 to 3 months at 18–20 °C. The influence of a threshold viroid concentration for symptom expression can be observed after inoculation with highly dilute inocula as in end point dilution experiments. In this case clear-cut symptoms may only be recognized when the apparently healthy-looking test plants tomato, *Gynura* and chrysanthemum are decapitated 4 weeks later. As a direct consequence of the viroid concentration build-up in the meantime, the growing axillary buds of infected plants exhibit pronounced symptoms of epinasty and growth retardation.

The symptoms of viroid infection suggest that they might be based on a hormonal malfunction resulting in abnormal cell development. The corresponding analysis of endogenous plant growth substances indicated a significant decrease in the concentration of gibberellins in CEV-infected *Gynura* tissue whereas no change was observed in the amounts of abscisic acid and indoleacetic acid. The reduction in gibberellic acid concentration observed at 30 and 60 days after inoculation could be correlated with the severe stunting response of the CEV-infected host plant (RODRIGUEZ et al. 1979).

Studies of the response of healthy and CEV-infected tissue to exogenously applied phytohormones have shown that both tissue types respond similarly to cytokinins and gibberellic acid. However, a dramatic inhibition of root formation is observed when either leaf discs, stems or callus cultures containing CEV are exposed to a variety of auxins (DURAN-VILLA and SEMANCIK 1982). The response appears to be independent of the level of differentiation. This suggests that viroid infection could affect the auxin metabolism and the auxin/cytokinen balance. Also a morphological alteration of the cells affecting their ability to respond to exogenous auxin could be assumed as the viroid-induced induction of symptoms. If viroid pathogenicity is in some way be exerted via the metabolism of the endogenous plant hormones, an additional indirect way of interference of viroid replication with the nucleic acid metabolism of the host is conceivable: The viroid-induced hormonal changes could themselves regulate or modify the nucleic acid metabolism of the host as has been postulated for example for gibberellic acid (JACOBSEN 1977). The alteration in the phytohormone content of plants after viroid-infection is also reflected in the impaired rooting ability of cuttings from CCMV- and CSV-infected chrysanthemum (HORST et al. 1977) and from CEV-infected *Gynura aurantiaca* (FLORES and RODRIGUEZ 1981).

2.7 Cytopathic Effects of Viroid Infection

In susceptible plant species viroids produce conspicuous symptoms of mosaic, veinal necrosis, blistering and epinasty of leaves, which is generally combined with a severe stunting of the malformed whole plant (Fig. 1). Studies of the cytopathological changes in *Gynura aurantiaca* leaf tissue after CEV infection did not reveal any significant changes in the fine structure of nuclei, mitochondria, chloroplasts, ribosomes and the intraplasmatic membrane system. The only cytopathic change observed in this viroid-host system are malformations of the cell wall and of the cell wall associated so-called "paramural bodies" or "plasmalemmasomes". They are protrusions of the cellular membrane of

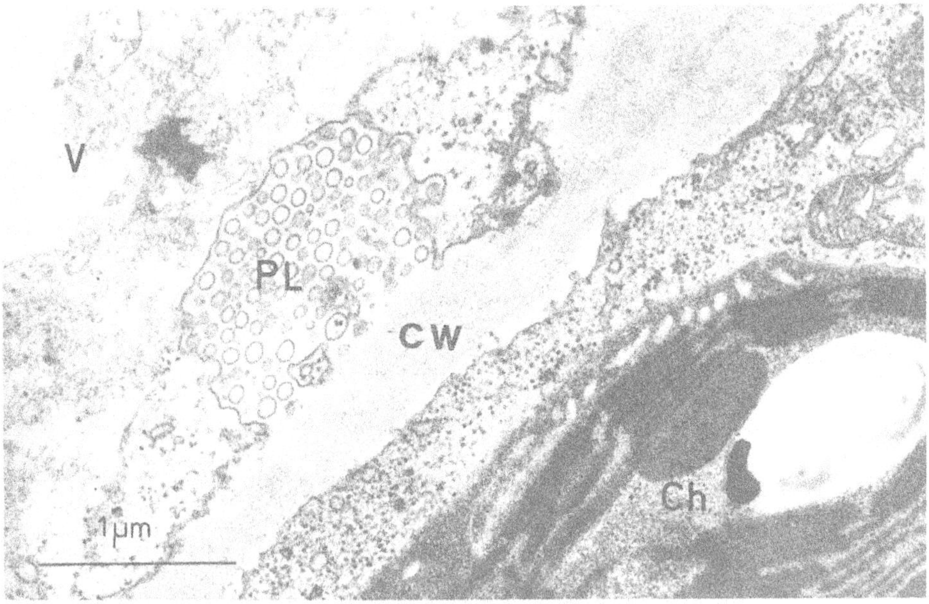

Fig. 7. Vesicular plasmalemmasome (*PL*) in tissue of healthy *Gynura aurantiaca*. *CW* cell wall; *Ch* chloroplast; *V* vacuole

variable size, shape and internal structure and their origin and function has not yet been clarified.

In individual healthy *Gynura* plants plasmalemmasomes with either vesicular (Fig. 7) or with tubular (Fig. 8) internal structures are found. In regions of cell wall thickening, plasmalemmasomes in the process of spreading between cell wall and plasmalemma are quite frequently observed. In CEV-infected *Gynura*, however, both types of plasmalemmasomes exhibit pronounced malformations in their size, shape and internal structures (Figs. 9 and 10), and they were never found in the process of "spreading" between cell wall and plasmalemma, which is characteristic for plasmalemmasomes in healthy tissue. Moreover, malformed plasmalemmasomes are occurring exclusively in association with the irregular cell walls typical for CEV-infected and symptom-bearing *Gynura* leaf tissue (Fig. 11). It should be emphasized that plasmalemmasomes are found at the same frequency in healthy as well as in CEV-infected *Gynura* (WAHN et al. 1980a, b). These results are at variance to the findings of SEMANCIK and VANDERWOUDE (1976) that plasmalemmasomes are only present in CEV- and PSTV-infected tissue, respectively, and that they are the primary cytopathic effect of viroid infection.

In the chlorotic spots characteristic for CSV-infected leaf tissue of "Mistletoe" chrysanthemum, the number of chloroplasts was found to be reduced to about one third as compared to the adjacent green tissue or to tissue from healthy plants. The electron microscope revealed that the thylakoid membrane system of these chloroplasts is irregular and filled with osmiophilic material.

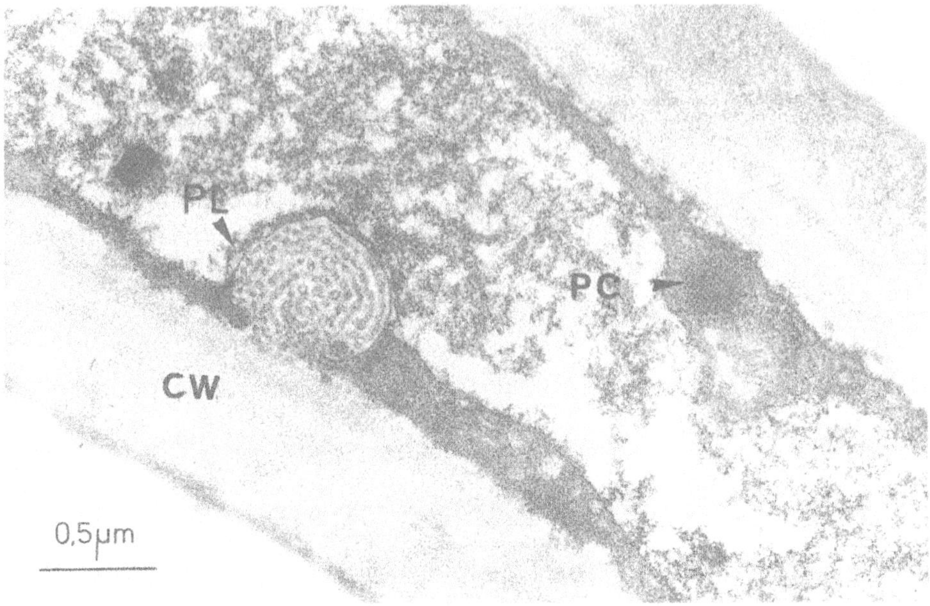

Fig. 8. Tubular plasmalemmasome (*PL*) in tissue of healthy *Gynura aurantiaca.* Some of the tubules are sectioned longitudinally. *CW* cell wall; *PC* protein crystal

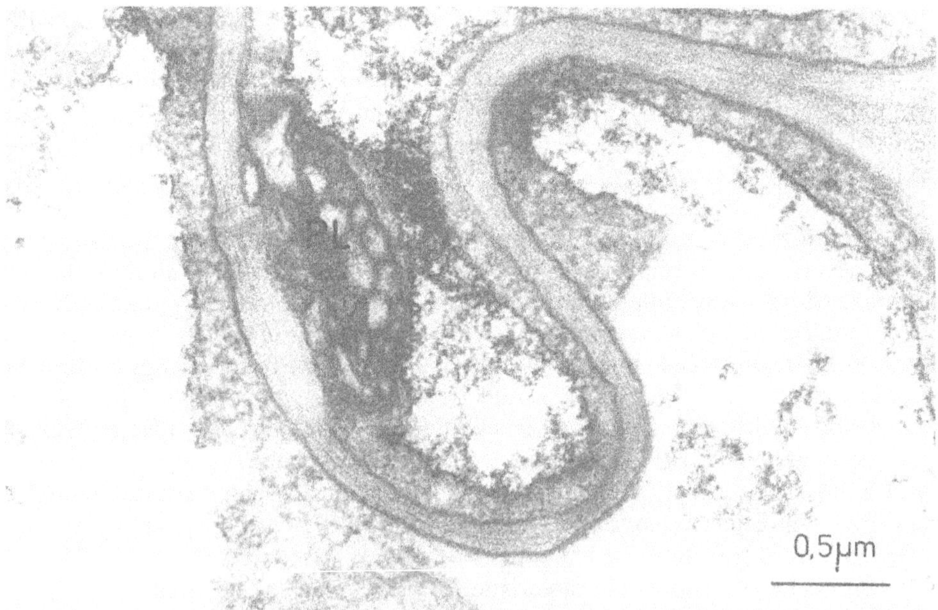

Fig. 9. Vesicular plasmalemmasome (PL) in CEV-infected *Gynura aurantiaca.* Note its irregular internal structures and its association with malformed cell walls

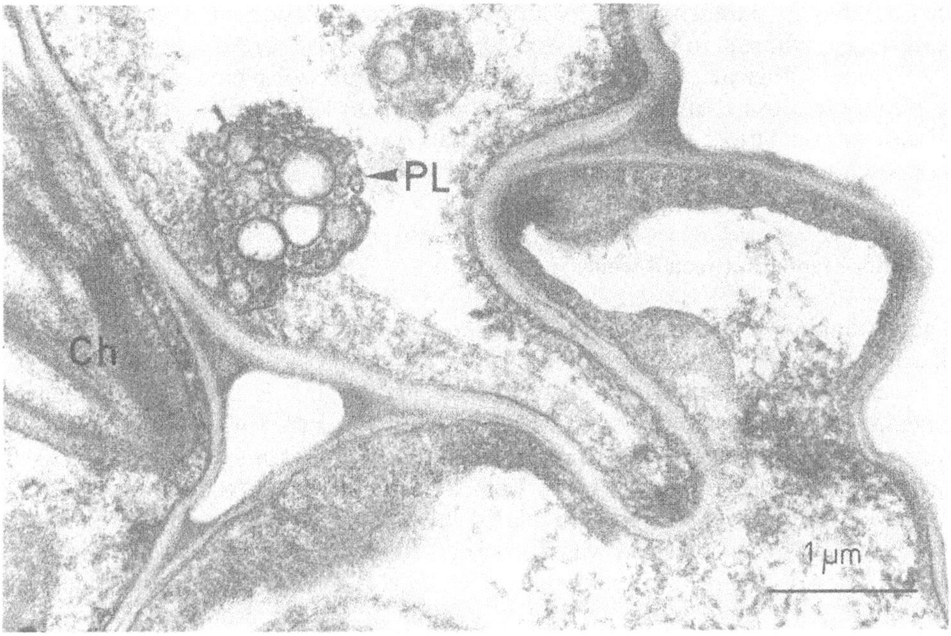

Fig. 10. Tubular plasmalemmasome (*PL*) in CEV-infected *Gynura aurantiaca*. Note its heavily contrasted irregular internal structures and its association with a malformed cell wall. *Ch* chloroplast

Fig. 11. *Gynura aurantiaca* DC (velvet plant) infected with CEV (*right*) as compared to healthy plant

In addition, in parenchyma cells an extreme accumulation of microfilaments is observed, whereas in healthy tissue microfilaments are rarely found.

Cytopathological changes induced by viroid infections have also been reported for leaf tissue of PSTV-infected tomato (HARI 1980) and *Scopolia sinensis* (PALIWAL and SINGH 1981) and for avocado leaves infected with the avocado sunblotch viroid (DA GRAÇA and MARTIN 1981).

2.8 Interference Between Viroids

The availability of severe and mild isolates of one and the same viroid and of common hosts for several different viroids allowed "cross-protection tests" between them. In such tests a viroid or viroid "strain" is used for challenge inoculation on a previously viroid-infected plant to probe for possible interference in symptom expression. Inoculations of tomato plants with mild PSTV isolates from field-grown potatoes were able to prevent the development of the symptoms of severe isolates inoculated subsequently (FERNOW 1967, FERNOW et al. 1969). Cross-protection was also observed in the chrysanthemum cultivar "Deep Ridge" where a latent strain of CCMV (CCMV-NS) was detected in different clones of this cultivar. The expression of the symptoms of the severe CCMV strain was totally prevented when the plants had been inoculated with CCMV(-NS) 8 days before the subsequent challenge inoculation with the severe CCMV (HORST 1975). In a systematic study with different viroid "species" it was found that CEV, CSV and a severe and a mild strain of PSTV exhibit cross-protection in a variety of viroid–host combinations. Interestingly, ChCMV did not protect against any of these viroids in chrysanthemum. Alternatively, PSTV did not protect against ChCMV in this host (NIBLETT et al. 1978). Although the actual mechanism of this phenomenon is not known, one must assume that the interfering viroids affect a common biological process in symptom expression although they differ significantly from one another in their fingerprints and hence their nucleotide sequence. On the other hand, the possibility cannot be ruled out that a threshold concentration of the challenging viroid is required for symptom expression. This concentration might not be reached in the presence of the protecting viroid as a result of partial inhibition of replication. Therefore, the level of replication by the individual viroids in doubly infected cross-protected plant has to be considered when the molecular basis for cross-protection is determined.

2.9 Control Measures

The dissemination of viroid diseases may be controlled efficiently by prompt removal of all apparently diseased plants and by the selection of healthy plant material, which is then used for further propagation. In all cases, where the disease agent is readily transmitted by mechanically or contaminated tools, their sterilization with flaming or with chemicals like sodium hypochloride or sodium hydroxide plus formalin has proven successful in preventing further spreading of the disease (ROISTACHER et al. 1969). In addition, indexed viroid-free plant material to be used as propagating stock should be maintained in isolation und be treated with great caution to prevent recontamination. If the pertinent control measures are strictly followed, the incidence and spreading of viroid diseases can be efficiently reduced.

In the case of PSTV-infected potatoes (STACE-SMITH and MELLOR 1970) and DSV-infected chrysanthemums (HOLLINGS and STONE 1970) a combination of heat treatment and meristem tip culture or axillary bud culture has resulted in the recovery of 2 to 4% of viroid-free regenerated plants. In view of the preferentially enhanced viroid replication at elevated temperatures, meristem tip cultures of potatoes after "cold-treatment" has been successfully applied and about half of the regenerated plants were found to be viroid-free (LIZARAGA et al. 1980).

Thus far, no chemical treatment has resulted in curing a viroid disease. The antibiotic tetracycline and penicillin which are effective against prokaryotic pathogens failed to cure the cadang-cadang disease of coconut (RANDLES et al. 1977). An emulsion containing 1% piperonyl butoxide, 0.1% Triton X-100 and water acts only as a protectant of PSTV infection in potato and *Scopolia sinensis* if viroid inoculation of the sprayed leaves is carried out within 4 days. A similar treatment kills tomato plants (SINGH et al. 1975, SINGH 1977).

3 The Structure of Viroids

For many years the molecular structure of viroids remained unknown because physically recognizable entities were lacking even in highly infectious viroid preparations. Accordingly, the structural properties had to be inferred indirectly from different experimental approaches utilizing viroid infectivity as the sole parameter for its characterization. All of this early work was carried out with PSTV and CEV.

The interesting course of attempts to unravel the properties of these unusual agents has been reviewed in detail (DIENER 1972b, 1974; DIENER and HADIDI 1977, DIENER 1979, SEMANCIK 1979). In summary, it turned out that in some early experiments viroids behaved like double-stranded RNA, whereas others indicated that they were single-stranded molecules. These ambiguous properties led to the assumption that viroids may be a single-stranded molecule with some sort of hairpin structure and an extensive base-pairing or a double-stranded but incompletely base-paired molecule. From electron microscopy it was concluded that PSTV is a single-stranded RNA with some kind of a hairpin-like structure and with a molecular weight in the range of $8.0–9.0 \times 10^4$ (SOGO et al. 1973). From runs in different gel systems the molecular weight estimates for PSTV and CEV ranged from 0.5 to 1.25×10^5. From the following discussion it will become apparent that this unusual behavior of the elusive agent in these early studies can now be explained on the basis of its unique structure for which no precedent was known.

3.1 Viroid Purification and Properties of Purified Viroids

Viroid RNA is present in infected plant tissue in extremely low amount and usually 0.1–0.5 μg can be recovered per g tissue. Under optimal conditions yields up to 1.5 $\mu g\ g^{-1}$ can be obtained. Therefore, viroid isolation and purification poses formidable problems as compared to the isolation of viral RNA's. Their association with virus particles and their protection by the viral protein coat and envelope first allows selective isolation of these virions and then isolation of the viral RNA from those. In the case of the "naked" viroid RNA molecules, the total cellular nucleic acids have to be isolated from the infected tissue from which the minute amount of viroid RNA has then to be separated and concentrated.

Various methods for viroid isolation have been reported and reviewed (DIENER and HADIDI 1977, DIENER 1979, DICKSON 1979) and the following generalizations can be drawn: Total cellular nucleic acids are extracted from fresh or frozen young leaves by homogenization in the presence of a buffer with pH range between 7 to 9.5, a phenol-chloroform mixture, a detergent like 1% sodium dodecylsulphate and a nuclease inhibitor like the clay mineral bentonite at a final concentration of 1 $mg\ ml^{-1}$. After centrifugation the upper aqueous phase is recovered, reextracted with phenol and centrifuged again. The ribosomal RNA's are then precipitated from the aqueous phase by a final concentration of 2 M NaCl or LiCl and pelleted by centrifugation. To the resulting supernatant $2^1/_2$ volumes of absolute ethanol are added and after storage overnight at −20 °C the precipitated nucleic acids consisting mainly of DNA, tRNA, 5S-RNA and the traces of viroid RNA are recovered as pellet by centrifugation and subsequent drying in a vacuum desiccator. The DNA may be removed from the resuspended nucleic acid preparation by DNase treatment or as cetyltrimethylammonium complex. The final DNA-free RNA preparation may be subjected to a molecular sieve (permeation) chromatography step (SINGH and SÄNGER 1976) in order to remove the bulk of the tRNA and 5S-RNA. The resulting RNA preparation is thus enriched in viroid RNA and may be used for preparative electrophoresis on 5% polyacrylamide gels. After the run the gels are stained in 20 $\mu g\ ml^{-1}$ ethidium bromide and the nucleic acid bands can be visualized under UV light. The extra band of viroid RNA can be readily detected when RNA preparations from healthy plant tissue are run side by side for comparison. Viroid RNA may finally be recovered from the cut-out gel band by phenol extraction or electroelution followed by ethanol precipitation and chromatography on hydroxyapatite or CF 11 cellulose. In order to achieve high purity of viroid RNA, certain steps should be repeated, prefereably under adverse conditions of separation by which certain contaminating cellular nucleic acid or their fragments are removed with certainty. The purity of the final viroid RNA sample should always be evaluated by analytical gel runs and/or fingerprint analysis after in vitro iodination or 5′-endgroup labelling of the RNase A and T_1 oligonucleotides.

It should be noted that certain plant species pose special problems with respect to nucleic acid extraction and purification because of their high content of phenolic substances, polysaccharides and other interfering constituents. Therefore different modifica-

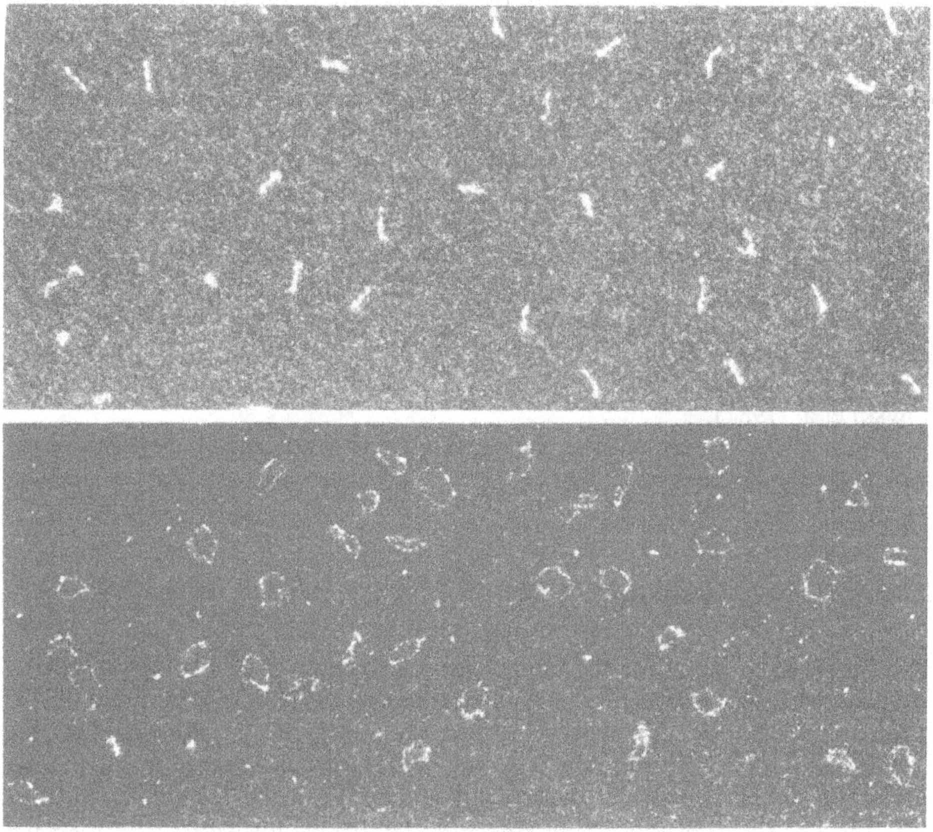

Fig. 12. *Top* Unselected dark-field electron micrograph of undenatured viroid (PSTV) molecules. The length of the native double-stranded rod-like structure is 37 ± 6 nm. *Bottom* Unselected dark-field micrograph of denatured PSTV. The contour length of the completely unfolded covalently closed circular viroid molecules is 100 ± 6 nm

tions of the general procedure have been introduced (RANDLES 1975, PALUKAITIS and SYMONS 1980, NIBLETT et al. 1980, OHNO et al. 1982) which greatly improved the yield of nucleic acids and of the minute amount of viroid RNA present in it.

Considerable progress was made when highly purified preparations became available in amounts sufficient to carry out direct biophysical and biochemical analysis. Then it was unabiguously demonstrated with thermodynamic methods that viroids exist in both their native and denatured form as one single RNA strand with a molecular weight of about 120,000, which corresponds to a chain length of about 360 nucleotides (SÄNGER et al. 1976). From electron microscopic and hydrodynamic analysis it was concluded that the native viroid molecules are double-stranded, rod-like structures with an axial ratio of about 20:1 (Fig. 12, top). Electron micrographs of denatured viroids finally demonstrated that they are actually single-stranded RNA circles (Fig. 12, bottom), which due to intramolecular base-pairing exist in their native state as double-stranded, rod-like structures. A small fraction (usually less than 1%) of the molecules were linears, which were considered to be generated by nicking of circles during the isolation procedure. The unexpected circularity of viroids was further substantiated biochemically by attempted end-group labelling, which failed to show the existence of 3′- and 5′-end groups in viroid molecules (SÄNGER et al. 1976). The circularity of viroids was later confirmed by electron microscopy (MCCLEMENTS and KAESBERG 1977), but the different viroid preparations investigated contained variable amounts of linear molecules. Their possible origin and their properties are still a matter of controversy, which will be discussed separately. Thus viroids became the first example of a single-stranded, circular RNA in nature. It should be mentioned in this context that in the meantime other circular forms of RNA have been found in the cytoplasm of eukaryotic cells (HSU and COCO-PRADOS 1979). Moreover, intron-derived circular RNA's have been detected in mitochondria of yeast and in isolated nuclei of *Tetrahymena* (GRABOWSKI et al. 1981).

Today, viroids represent the biochemically and biophysically best characterized nucleid acids next to tRNA.

3.2 The Primary Structure of PSTV

The unequivocal proof for the unique structural features of viroids was finally provided when the complete primary sequence established the structure of PSTV (GROSS et al. 1978). This work was severely hampered by the fact that the rate of viroid replication is rather low and viroid molecules cannot be sufficiently labelled with ^{32}P in vivo for subsequent sequencing work. Therefore, purified viroid RNA had to be labelled in vitro after controlled nuclease digestion using [γ-^{32}P ATP] and T4 phage-induced 5′-polynucleotide kinase from *E. coli*. The in vitro post-labelling technique can only be applied when viroid preparations free from any contaminating cellular RNA are available, and when the kinase is free from any traces of nucleases. The suitability of this approach was first proven by in vitro labelling of the complete RNase A and T1 digests of different viroids (CEV, PSTV and CSV), and by their comparison on the basis of their fingerprint patterns (GROSS et al. 1977), which established the existence of different viroid species (Fig. 13). The high resolution achieved in these fingerprints allowed the separation of the complete RNase T1 and RNase A fragments of PSTV and their sequence analysis by a combination of the conventional and the newly developed rapid gel-sequencing techniques. For establishing the total sequence, overlapping sequencing is needed. These were obtained by controlled digestion of PSTV with a set of different RNases by which longer viroid fragments could be generated (see legend of Fig. 14). These overlapping fragments were also 5′-end-group labelled in vitro and subject to sequence analysis. The complete sequence of the 359 nucleotides of PSTV could be established with certainty by the rather complex "puzzle" of arranging the large set of individual sequences in the appropriate overlapping array (Fig. 14). The resulting complete nucleotide sequence clearly substantiated that PSTV is a true single-stranded circular molecule.

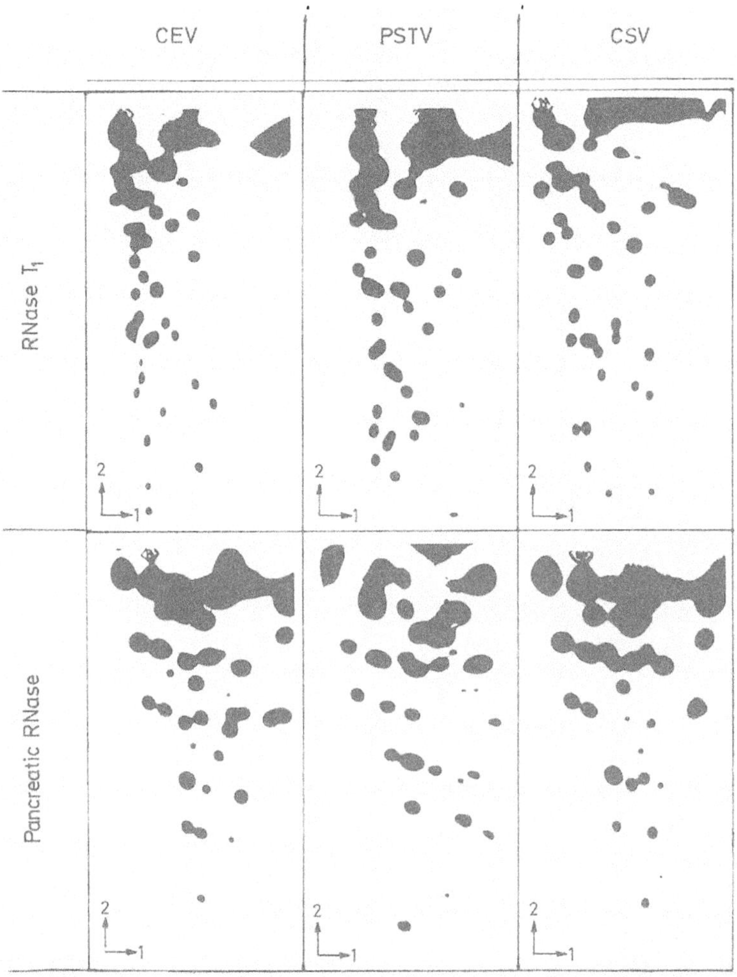

Fig. 13. Oligonucleotide fingerprints after complete digestion of three viroids with RNase T1 and pancreatic RNase, *CEV* citrus exocortis viroid from *Gynura aurantiaca; PSTV* potato spindle tuber viroid from tomato; *ChSV* chrysanthemum stunt viroid from *Cineraria.* The lower half of each pattern represents large oligonucleotides which are specific for each viroid "species"

3.3 The Secondary Structure of PSTV

When the primary nucleotide sequence of the PSTV molecule was arranged in such a way that a maximum number of intramolecular base pairs occurred, a largely double-stranded, rod-like structure was obtained (Fig. 15). It is characterized by the serial arrangement of 26 double-stranded stretches, which are separated by 25 segments of unpaired bases in single-stranded loops or bulge loops. In addition, there is one loop on each end of the rod-like molecule. The validity of this unique secondary structure was ascertained by considering

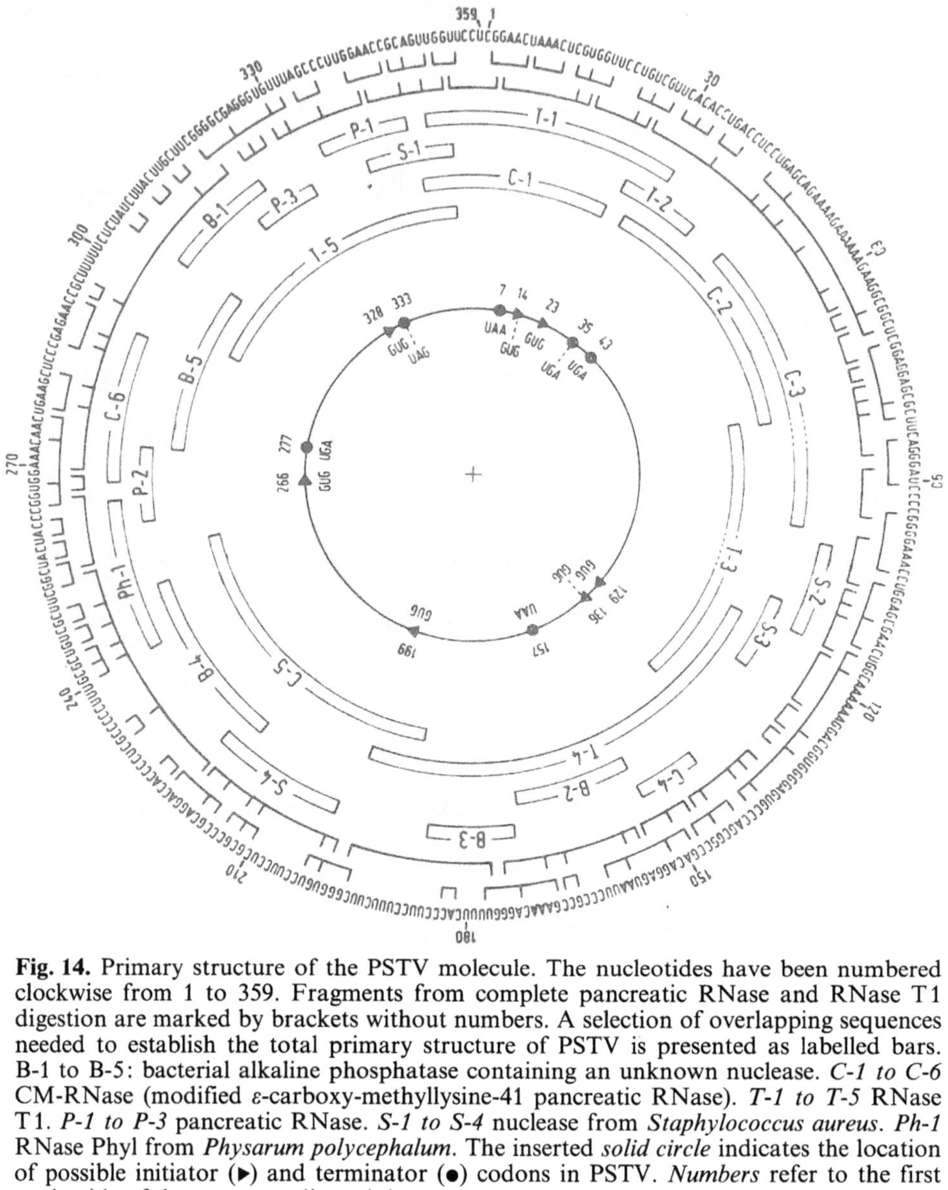

Fig. 14. Primary structure of the PSTV molecule. The nucleotides have been numbered clockwise from 1 to 359. Fragments from complete pancreatic RNase and RNase T1 digestion are marked by brackets without numbers. A selection of overlapping sequences needed to establish the total primary structure of PSTV is presented as labelled bars. B-1 to B-5: bacterial alkaline phosphatase containing an unknown nuclease. *C-1 to C-6* CM-RNase (modified ε-carboxy-methyllysine-41 pancreatic RNase). *T-1 to T-5* RNase T1. *P-1 to P-3* pancreatic RNase. *S-1 to S-4* nuclease from *Staphylococcus aureus*. *Ph-1* RNase Phyl from *Physarum polycephalum*. The inserted *solid circle* indicates the location of possible initiator (▶) and terminator (●) codons in PSTV. *Numbers* refer to the first nucleotide of the corresponding triplet

the location of the enzymatic cleavage sites and the results from chemical modification of cytidines to uridines in single-stranded loops by bisulphite (DOMDEY et al. 1978). In detail, PSTV consists of 73 × AMP (=20.3%), 77 × UMP (=21.4%), 101 × GMP (=28.1%) and 108 × CMP (=30.1%). The two partially complementary halves of the PSTV circle have virtually the same length namely 179 and 180 nucleotides, respectively. There are 73 G:C (=59.8%), 38 A:U

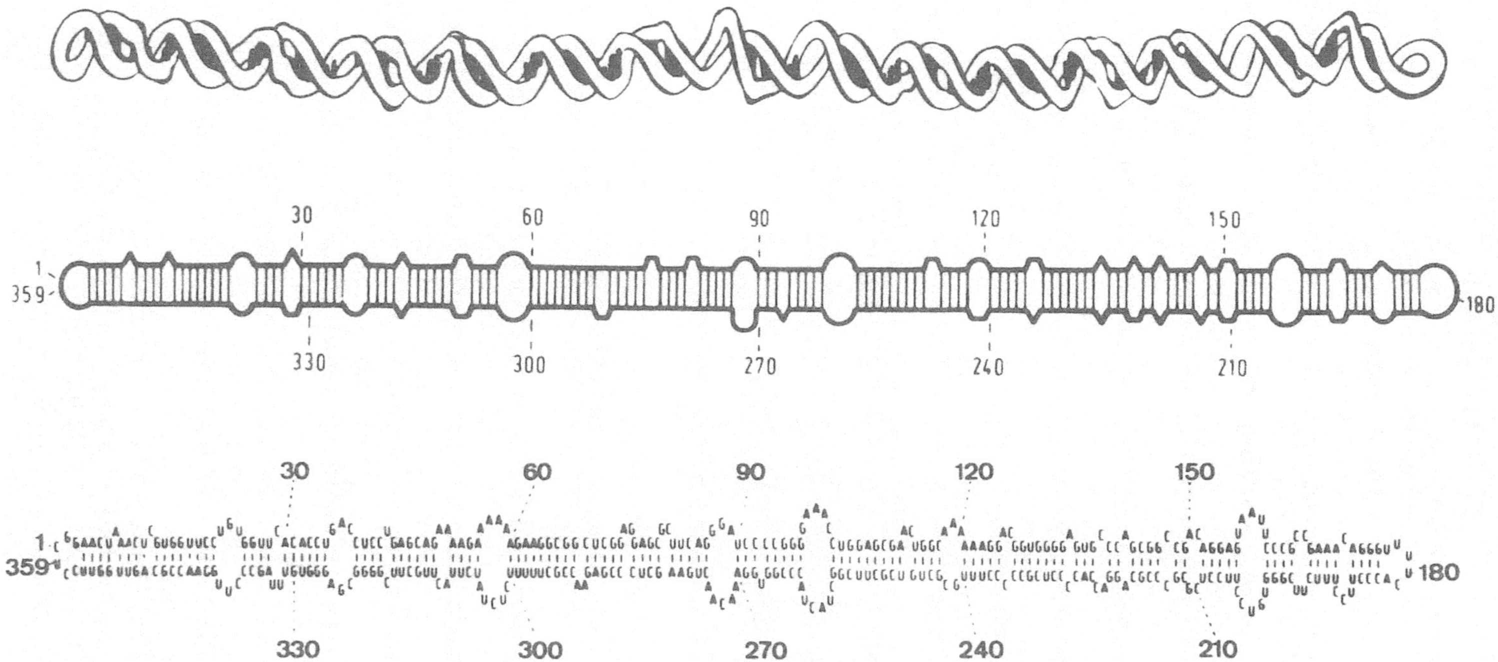

Fig. 15. The nucleotide sequence (primary structure) and the resulting secondary structure of PSTV as shown in a two-dimensional base-pairing scheme (*bottom*), in the corresponding diagrammatic representation (*middle*) and in a three-dimensional manner (*top*). This structure is based on maximal intramolecular base-pairing taking into consideration the observed nuclease and bisulphite sensitivity of PSTV and the data obtained from detailed physicochemical investigations including binding studies with dyes and specific tRNA's

(=31.2%) and 11 G:U (=9.0%) pairs in the native secondary structure resulting in a total of 122 base pairs (=67.7%) out of the 179 base pairs potentially possible if complete intramolecular complementarity would exist. This rod-like secondary structure of PSTV agrees well with its thermal stability, the high cooperativity of the thermal stability, the high cooperativity of the thermal denaturation process and the axial ratio of 20 as derived from hydrodynamic studies (SÄNGER et al. 1976).

Exactly the same secondary structure model was obtained by an independent theoretical approach based on thermodynamics. With the established primary nucleotide sequence at hand the free energy of helix formation was systemically minimized for all 359 nucleotides of the PSTV molecule (RIESNER et al. 1979). For this purpose the matrix procedure of TINOCCO et al. (1971) was applied with a slight modification so that it became applicable to circular molecules. From the corresponding graphical representation of all possible double-stranded stretches, the thermodynamically most stable secondary structure of PSTV under physiological conditions emerged and is called the native structure. This structure is in complete agreement with the secondary structure model proposed earlier on the basis of maximal base-pairing and biochemical data and arguments (GROSS et al. 1978). It should be noted that all other secondary structure alternatives which can be constructed and which lead to a more globular shape of the molecule are not in accordance with the biochemical results and the thermodynamic data and calculations. Theoretical evaluations finally show that the specific features of viroids, such as the number and type of base pairs, the high cooperativity in melting and the formation of very stable hair-pins are improbable in random sequences (RIESNER et al. 1979).

3.4 Absence of Tertiary Structure Folding in PSTV

The base-pairing scheme of PSTV does not preclude interactions between loops of distant parts of the molecule, which would fold the extended structure to a more globular shape. To probe for a possible tertiary structure folding, dye binding to the PSTV molecule was studied under a variety of ionic conditions. In case these alterations of ionic strength induce conformational changes, a change in dye binding should also be expected, because the accessibility of the dye-binding sites would be impaired. Ethidium bromide, which is well known as an intercalating dye without preference for special base pairs of defined sequences, was used to probe for base-paired segments. The dye H 8208 2-(4′-aminophenyl)-5-(4′-methylpiperazin-1″yl)benzimidazole on the other hand, binds specifically to A:U base pairs. It was found that these two dyes are bound nearly quantitatively to the PSTV molecule up to complete saturation under different ionic conditions. This shows that changes of ionic strength do not induce any tertiary structure folding in the PSTV molecule. In fact, one is tempted to correlate the 22 ethidium bromide molecules intercalating in the native PSTV structure with its 26 double-stranded segments. On the other hand, the 35 molecules of the dye H 8208 binding to the PSTV molecule could be related to the 38 A:U base pairs actually present in its native secondary structure (RIESNER et al. 1979).

Further evidence for the absence of tertiary structure comes from binding studies of specific tRNA's to PSTV (WILD et al. 1980). Transfer RNA molecules are well suited for such studies, because their anticodon loops are highly ordered by tertiary structure folding. The specific anticodon region allows probing of certain internal loops in the viroid molecule. It was found that five out of the seven tRNA species tested are able to bind specifically to those single-stranded PSTV loops which are characterized by sequence complementarity to the tRNA corresponding anticodon region (Fig. 16, Table 2). This finding is corroborated by the observation that tRNA molecules for which an anticodon-complementary loop is lacking ($tRNA_f^{Met}$ and $tRNA_1^{Val}$ from *E. coli*) do not bind to PSTV (see Table 2). The free accessibility of the anticodon binding loops distributed over the total viroid molecule clearly substantiates its completely extended structure and the absence of any tertiary structure folding. This conclusion is strengthened by the fact that tRNA's are rather voluminous conformational probes and that their binding would not only be hindered sterically by tertiary structure folding in a limited area but also by folding at more distant regions. From these results the question arises if such complexes between viroids and cellular RNA's might also exist in vivo. Although the concentration of viroid molecules and tRNA might be sufficiently high in the cell for complex formation, it is not possible at present to decide whether or not this feature is of any direct functional relevance. The binding studies with specific dyes and tRNAs corroborate the previous evidence derived from electron microscopy, sequence analysis and thermodynamic studies that native PSTV molecules are extended rod-like structures without any tertiary structure folding.

3.5 Properties of the RNA Molecule Complementary to PSTV

The established sequence of PSTV allows construction of a hypothetical molecule of complementary sequence. Since the sequenced PSTV is arbitrarily called plus PSTV, the complementary RNA molecule is the minus PSTV strand. Considering the experimental demonstration of the existence of the minus PSTV RNA molecules, one should keep in mind that they might be present in extremely low concentrations like the intermediate replicative forms in the case of conventional viruses. On the basis of the established sequence of PSTV, the properties of the circular minus PSTV were evaluated theoretically (RIESNER et al. 1979). Despite the relatively high self-complementarity of the plus PSTV molecule, the secondary structure of the minus PSTV RNA cannot be the same, because only G:C and A:U pairs but not G:U pairs would be copied into the same type of base pairs. There are 16 G:U base pairs present in PSTV which will not lead to the same pairs in the minus PSTV, whereas six new G:U pairs would be generated from six opposite A:C positions in the plus PSTV molecule. Thermodynamic optimization of the basepairing scheme of the minus strand leads to a PSTV-like secondary structure, which contains 117 base pairs instead of the 125 of the plus PSTV molecule. Its melting temperature is about 12 °C

Table 2. Binding of specific tRNA's to PSTV

tRNA	Origin	Anticodon		Molecules bound per PSTV molecule	K $[M^{-1}]$
Glu 2	*E. coli*	s^2U^+UC	(m^2A)	1	5.3×10^5
Asp	Yeast	GUC	(m^1G)	1	Not determined
Phe	Yeast	G_m-AA	(Y)	2	7.5×10^5
Ser 1+2	Yeast	IGA	(i^6A)	2	Not determined
Lys	*E. coli*	s^2U^+UU	(t^6A)	3	7.0×10^4
fMet	*E. coli*	CAU	(A)	No binding	–
Val 1	*E. coli*	VAC	(m^6A)	No binding	–

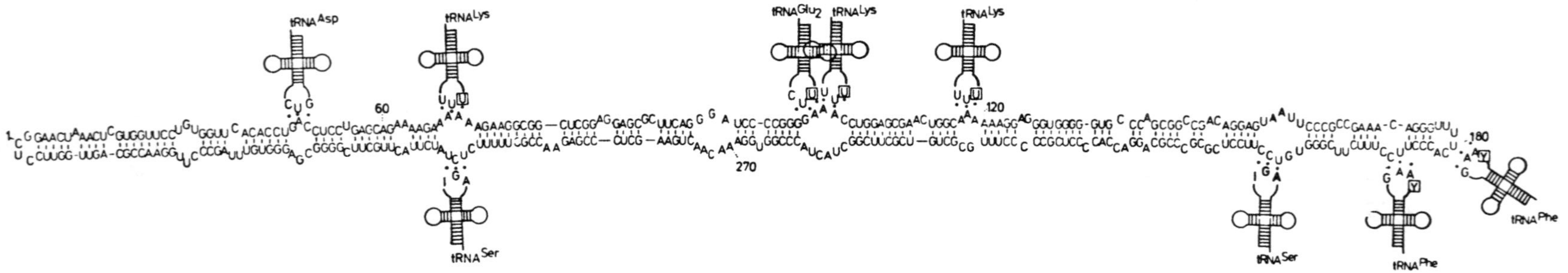

Fig. 16. Binding of specific tRNA's to the PSTV molecule. The tRNA's are depicted smaller than their actual size as compared to the size of the PSTV RNA

lower and the cooperativity of its melting is slightly decreased. From their rather similar structural features one can conclude that plus and minus viroid molecules of unit length would most probably migrate together in native as well as in fully denaturing gels. The vast excess of the plus PSTV may render the preparative separation of the minus PSTV a rather difficult experimental problem. But if successfully separated it can be clearly identified by RNase T_1 and RNase A fingerprints. On the basis of the complementary PSTV sequence and according to the previously established correlation between the gross nucleotide composition of any RNase T_1 or RNase A oligonucleotide and its position after homochromatography and electrophoresis (DOMDEY et al. 1978) the minus PSTV fingerprints must be completely different from the plus PSTV fingerprints.

The unique structural features of both the plus and the minus viroid molecules clearly indicate that viroid complementary sequences can only be unambigiously demonstrated by molecular hybridization when certain experimental prerequisites are fulfilled which take the circularity and the specific secondary structure of the probes into account. Thus, the extensive intramolecular complementarity of both molecules may result in the hybridization of ^{125}I-labelled plus PSTV with unlabelled circular and linear molecules of the same polarity under high salt/low temperature conditions (ROHDE and SÄNGER 1981). Consequently, a combination of defined and homogeneous hybridization probes and stringent hybridization conditions are an essential requirement for molecular hybridization, especially in viroid research.

3.6 Origin and Properties of Linear Viroid Molecules

A controversy over circularity as the intrinsic structural feature of viroids arose when it was reported that PSTV preparations may contain up to 80% linear viroid molecules, only the rest being circles (MCCLEMENTS and KAESBERG 1977). The two forms were later separated by gel electrophoresis in the presence of high concentrations of formamide and urea, and the bioassay of the corresponding gel fractions indicated that both molecular species are infectious (OWENS et al. 1977). From incorporation of ^{32}P into PSTV, it was finally concluded that the observed linear forms are a major product of viroid replication and that they are not arising in vitro by nicking of circles with nucleases during isolation and purification (HADIDI and DIENER 1978).

A plausible explanation for the occurrence of these unusually high concentrations of linear forms emerged when it was found that Mg^{2+} ions may produce linear molecules by nicking of viroid circles. Mg^{2+} ions have been used at slightly alkaline pH to cleave specifically individual tRNA species (WINTERMEYER and ZACHAU 1973) a reaction which generates 5′-hydroxyl endgroups and 2′- and 3′-phosphates. When this Mg^{2+}-catalyzed phosphodiester cleavage was applied to five different viroid species, it was found that a single nick occurred under a wide range of controlled conditions producing linear viroid molecules of the same size as the covalently closed viroid circles from which they arose (SÄNGER et al. 1979). The detailed analysis of the cleavage kinetics

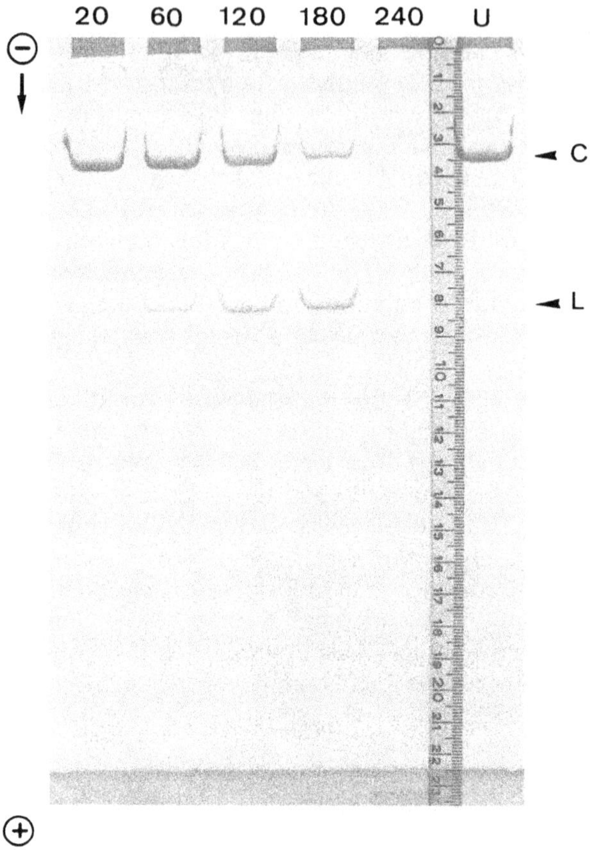

Fig. 17. Electrophoretic analysis of the kinetics of Mg^{2+}-catalyzed nicking of circular viroid (PSTV) molecules in a fully denaturing 5% polyacrylamide gel. The number above the slots give the min of treatment with 5 mM Mg^{2+} at 37 °C and pH 9.0. With increasing time of incubation linearization of viroid circles increases. After 180 min about 50% of the viroid circles are converted into linear forms. *U* untreated control sample; *C* circular molecules; *L* linear molecules

showed that 50% of the viroid circles are converted into linear forms after about 25, 50, 150 and 350 min at 60°, 50°, 37° and 25 °C, respectively. All five viroid species tested (PSTV, CPFV, CSV, CEV and CCMV) showed practically the same susceptibility to Mg^{2++} ions. Fig. 17 shows the kinetics of the Mg^{2+}-catalysed cleavage of PSTV circles on a fully denaturing 5% polyacrylamide gel.

The detailed analysis of the migration of circular and linear molecules under non-denaturing conditions (i.e., in the presence of buffer of high ionic strength and at a gel temperature of 12–15 °C) showed that both forms comigrate in 5% gels (Fig. 18). Under these conditions the size and the secondary structure of both forms must be very similar, if not identical. This finding is not surprising, because a single nick would certainly not change the rather stable rod-like secondary structure of the native viroid molecule. Consequently, if viroids are isolated by preparative electrophoresis in 5% polyacrylamide gels under non-denaturing conditions the possibility of selecting for viroid circles inadvertantly losing linear viroid molecules can be excluded.

The end group analysis of the linearized molecules revealed that the ratio of 5′-terminal pA, pC, pU and pG was 38:17:25:20. This shows that in contrast to tRNA the Mg^{2+}-catalyzed cleavage of viroids is of rather low specificity.

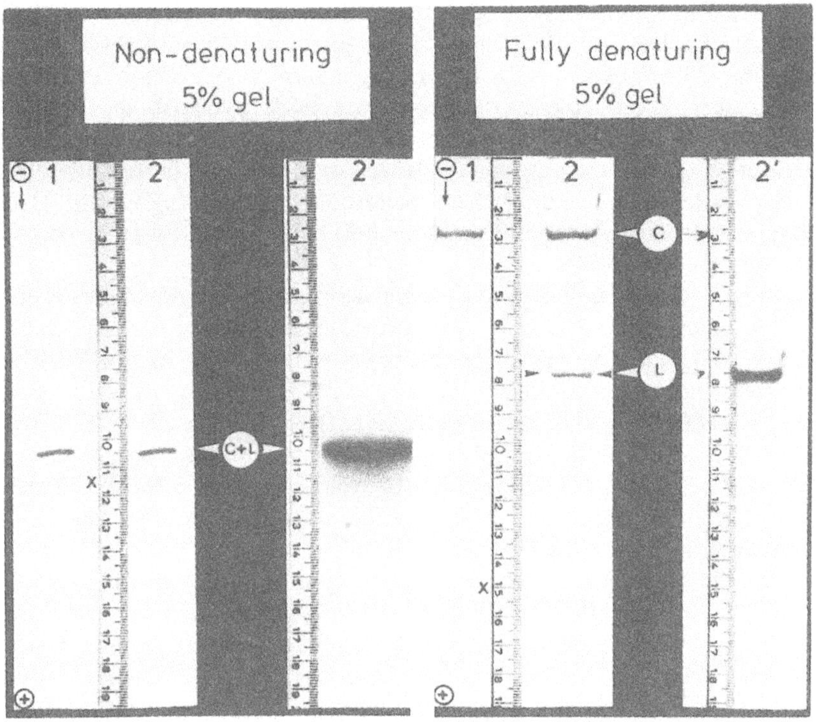

Fig. 18. Electrophoretic analysis of a viroid (PSTV) preparation after Mg^{2+}-catalyzed nicking followed by 5'-^{32}P end-group labelling and autoradiography: *C* circular viroid molecules; *L* linear viroid molecules; *X* xylene cyanol marker dye. Slots *1* untreated control samples; slots *2* Mg^{2+}-treated viroid RNA; slots *2'* autoradiographs corresponding to slots 2 of the stained gels. Comparison between the pattern of the stained gel and the corresponding autoradiograph shows that under non-denaturing conditions (*left*) the viroid circles and the Mg^{2+}-nicked and 5'-^{32}P end-group labelled viroid molecules co-migrate. In contrast, under fully denaturing conditions (*right*) the Mg^{2+}-nicked 5'-^{32}P-labelled linearized molecules migrate almost three times as fast as the unfolded viroid circles, which are not susceptible to 5'-end-group labelling

The nicking of viroids may also be detected in thermodynamic experiments. The midpoint-temperature and the cooperativity of the main denaturation process of the linear forms are significantly lower than those of the circular form. These results confirm the earlier theoretical prediction (LANGOWSKI et al. 1978) and support the interpretation of two melting peaks which appeared after repetitive heating in our previous calorimetric studies (KLUMP et al. 1978). The considerable amount of melting even at 48 °C indicates that a number of different transitions arising from heterogeneously nicked molecules are superimposed. This result is in good agreement with the end group analysis.

The relatively high and varying percentage (30–80%) of linear viroid molecules observed in certain preparations (MCCLEMENTS and KAESBERG 1977) can be attributed to Mg^{2+}-catalyzed cleavage of viroid circles, because the isolation and purification had been carried out in the presence of Mg^{2+}-ions. In fact, according to the procedure reportedly used to purify PSTV (DIENER et al. 1977, HADIDI and DIENER 1977, DIENER and HADIDI 1977), all the relevant samples must have been exposed for longer periods of time to 3 mM Mg^{2+} in alkaline buffers of pH 9.0 at elevated temperatures. Most probably Mg^{2+} ions were added to stabilize viroid structure and the alkaline pH is obviously a precaution to inhibit the action of possible traces of nucleases. Unfortunately, however, it is inevitable that under such conditions Mg^{2+} may exert the completely adverse effect of linearizing viroid circles by the catalysis of the cleavage of phosphodiester bonds.

Mg^{2+}-nicked and purified linear forms were found to be about 1,000 to 10,000 times less infectious than previously reported for the isolated viroid circles (SÄNGER et al. 1976). The very low infectivity of our purified samples of linear molecules could be due to a contamination with viroid circles of less than 0.1–0.001%. It is interesting to note that viroid circles were obtained again when inocula from individual symptom-bearing plants from the end-point dilution of the bioassay of isolated linear molecules were used for further propagation. The finding that linear PSTV molecules are non-infectious has been substantiated by MORRIS (1979).

On the other hand, there are reports that the linear forms of PSTV (OWENS et al. 1977) and CSV (PALUKAITIS and SYMONS 1980) are nearly as infectious as the corresponding circular molecules. The reinvestigation of this problem has shown that the relative concentration of linear PSTV molecules may greatly differ in different preparations even when EDTA is used to avoid inadvertant nicking. When isolated linear molecules of such preparations were bioassayed in dilution endpoint experiments it turned out that they had about the same specific infectivity as the corresponding circles (SÄNGER and RAMM, in preparation).

In conclusion, the nature of the linear forms as possible intermediates of viroid replication is still unresolved. In future experiments, reliable conclusions on this problem can only be drawn if viroid extraction and purification is carried out in the presence of an excess of EDTA in order to complex any polyvalent cations. Moreover, the first analytical (and preparative) electrophoretic runs should be carried out in 5% gels under non-denaturing conditions. The resulting viroid band will then contain circular and linear forms which may subsequently analyzed (or isolated) by polymerizing the band into a fully denaturing gel where both forms can be separated.

3.7 Structure Formation, Conformers and Multiple Forms of PSTV

Nucleic acids may exert their biological functions through conformational changes. Such structural transformations can be studied in vitro by the analysis of their thermal denaturation process. The resulting curve representing the structural transitions is the "melting curve", which on the basis of thermodynamic theory allows certain conclusions on nucleic acid structure. In Fig. 19 the differentiated melting curves for three structurally different nucleic acids are shown together with the corresponding structural changes occurring during the process

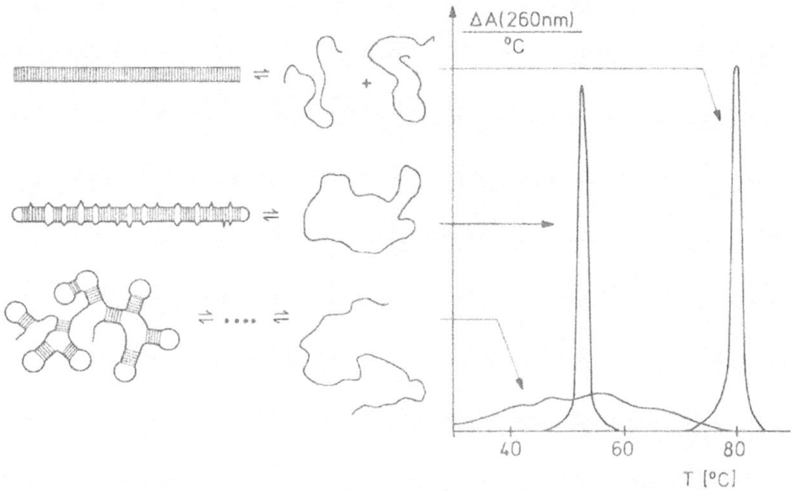

Fig. 19. Diagrammatic presentation of the differentiated melting curves as related to the secondary structure of different RNA molecules

of melting. The melting curve shows that the melting of viroid molecules is highly cooperative, and that it occurs at a relatively low temperature as compared to the melting of a homogeneous double-stranded RNA. In fact, it combines the narrow range of transition of a RNA molecule consisting of two individual strands which are totally base-paired in a homogeneous way with the low temperature of transition of a branched single-stranded RNA species. It becomes evident that these characteristic features are due to the unique structure of the viroid molecule. The serial arrangement of double-stranded segments and single-stranded loops is responsible for the low temperature of transition, whereas the completely unbranched structure leads to the high cooperativity of the melting process. From this unique combination and the circularity of the molecules, a general model for the secondary structure of viroids had already been proposed at a time before a single nucleotide sequence was known (LANGOWSKI et al. 1978, KLUMP et al. 1978). It turned out to be in perfect agreement with the detailed secondary structure model (Fig. 15), which was constructed after the first nucleotide sequence had been established (GROSS et al. 1978, RIESNER et al. 1979).

Investigations applying high-resolution melting together with fast and low temperature-jump techniques allowed a detailed analysis of the process of structure formation of PSTV on the basis of its established primary and secondary structure (HENCO et al. 1979). They revealed (Fig. 20) that in addition to the highly cooperative main transition (T_m between 56.5 and 49 °C for different viroid species) all viroids show at higher temperatures an intermediate transition ($T_m = 53$ °C) and a high temperature transition ($T_m = 68$ °C). The amplitudes of both transitions comprise only about 1% of that of the main transition, which represents the dissociation of 78 to 94 base pairs depending on the viroid species. During the intermediate transition two hairpins with 5–10 base pairs

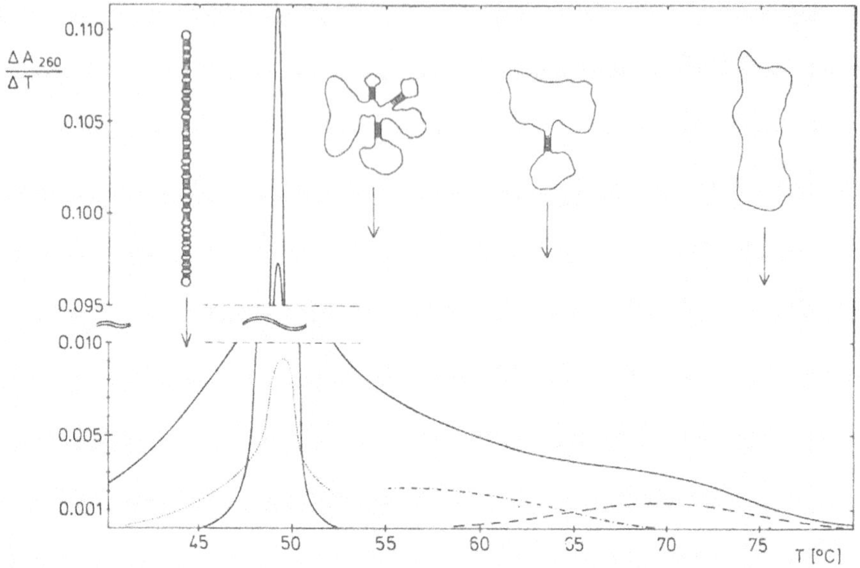

Fig. 20. Composite melting curve of the structural transitions of the PSTV molecule

and 10–20 nucleotides in the loops are dissociating. The high temperature transition corresponds to a hairpin of 9 G:C pairs and 1 A:U pair and more than 40 bases in the loop. It can be shown that these stable hairpins are not part of the native structure but are formed during the main melting process. From these results the mechanism of structure formation in PSTV can not only be described in detail on the basis of its primary sequence as shown in Fig. 21 (left) (HENCO et al. 1979) but the corresponding structures can also be found in electron micrographs of denatured PSTV molecules (Fig. 21, right).

According to this scheme, after the opening of one of the starting segments, the left half of the PSTV molecule opens up at about 50 °C with segments 1–13 in one cooperative step. Segments 14 and 15, however, are particularly stable and increase the T_m value of the dissociation of the right half of the molecule by about 10 °C as compared to the left half. The dissociation of the right half allows the formation of three new hairpins between distant complementary regions (see curved arrows in Fig. 21, left). Their high stability lowers the T_m value of the complete right half. This long-range interaction prevents a biphasic melting and allows the dissociation of the whole native structure in one co-operative transition with short-lived structural intermediates. The high temperature transition with only one very stable base-paired segment is of special interest, because it explains the nature and origin of the "tennis rackets" and "ping pong paddles" in electron micrographs (Fig. 21, right) of denatured viroid molecules (SÄNGER et al. 1976, KLOTZ and SÄNGER 1981). Thus, the melting of viroids is not a mere dissociation of base pairs but involves the concerted opening of 125 base pairs and the formation of 28 new base pairs. This process may be described as a switching from the native rod-like structure to a hairpin-containing circle. This process is unique as compared to other nucleic acids

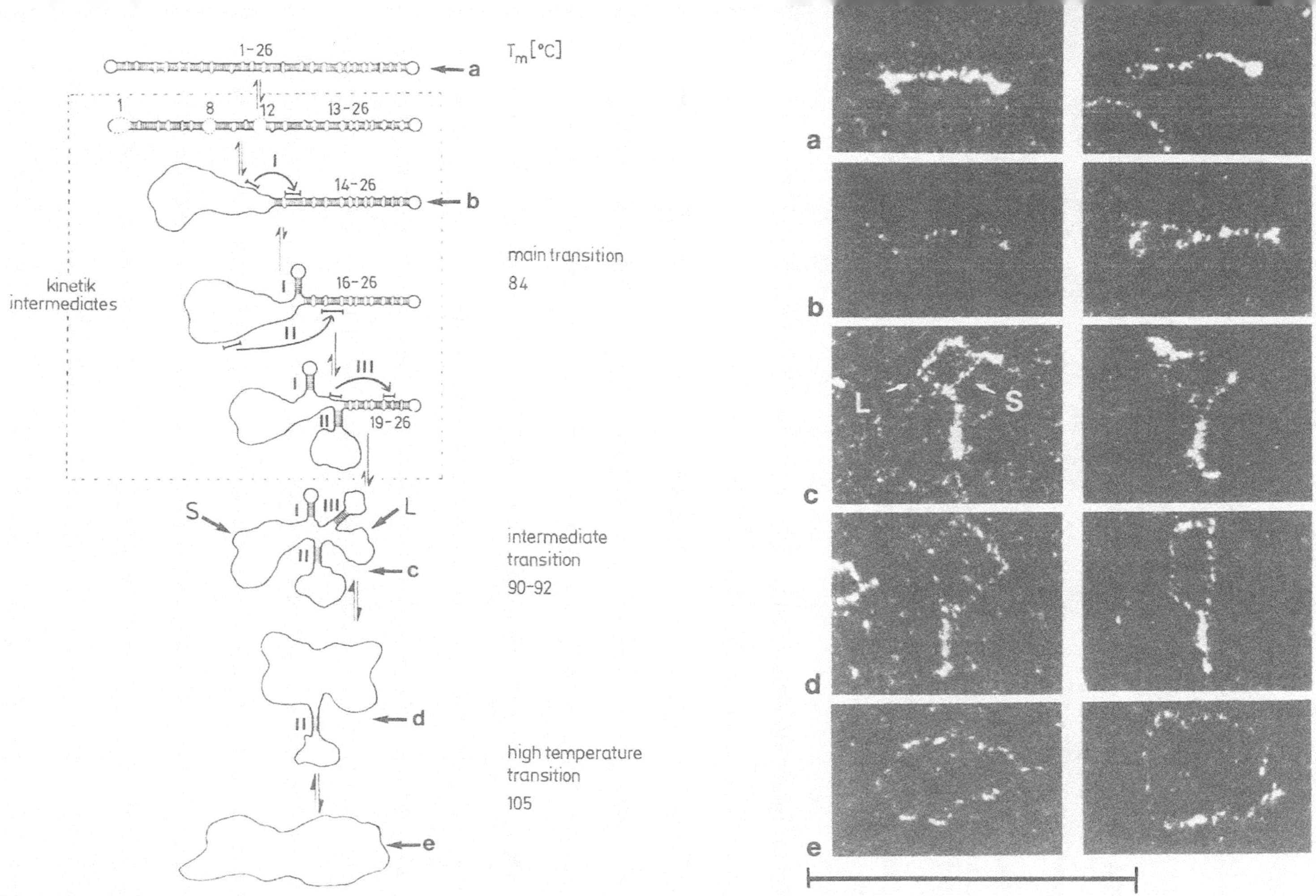

Fig. 21. *Left* Mechanism of denaturation and renaturation of the PSTV molecule with different structural intermediates. The double-stranded segments of the molecule are numbered 1–26 *from left to right*. The T_m values are given for 1 M NaCl. *Right* Selected PSTV molecules from dark-field electron micrographs obtained after spreading of denatured PSTV RNA. The different structural intermediates a to e in the micrographs correspond to the intermediates a to e in the denaturation scheme. S and L denote the small and large single-stranded

and remarkably similar in all viroid species. From the thermodynamic point of view, the combination of high cooperativity and low denaturation temperature may be described as an optimal combination between stability and flexibility. Although viroids are the largest RNA molecules for which so many details of structure and structure formation are known, the actual functional implications of their structural properties remain to be elucidated.

The finding of metastable viroid conformers would explain the previous observation that viroids exist in multiple forms and that viroid infectivity is found in well separated bands after chromatographic and electrophoretic separation (SINGH et al. 1974, DICKSON et al. 1975, RANDLES 1975, RANDLES et al. 1976). In fact, purification of viroids involves several denaturation steps such as phenolization, treatment at elevated temperature and ethanol precipitation. Electrophoretic analysis has shown that such structural intermediates may be obtained after redissolving ethanol precipitates in aqueous solution and after rapid cooling of denatured viroid preparations. Therefore, the reported appearance of viroid infectivity in well-separated fractions does not imply molecular heterogeneity, but is most probably due to conformational heterogeneity.

3.8 Structural Differences Between the Pathogenic PSTV "Type Strain" and a "Mild" PSTV Isolate

In 1967 FERNOW showed that from field-grown potatoes both mild and severe isolates of the potato spindle tuber "virus" could be recovered and that these biological properties remained stable during further propagation of the agent. Also a screening test in Eastern Canada revealed differences in the pathogenicity of PSTV isolates. Ninety-two percent of the PSTV-infected field potatoes contained mild "strains" and 8% contained severe "strains". When the mild "strains" were studied more closely it was found that different isolates differed from each other in severity and some produced such mild symptoms that they were referred to a latent "strain" (SINGH et al. 1970). Mild, intermediate, severe and lethal symptoms as caused by corresponding PSTV isolates in tomato are shown in Fig. 5. As mild isolates have also been described for CEV (SÄNGER 1972, ROISTACHER et al. 1969) and CCMV (HORST 1975), one can expect that strains with different pathogenicity exist for all viroids.

Since RNA is the only known component of viroids, the differences in viroid-pathogenicity must be reflected in differences in the nucleotide sequence of the corresponding viroid RNA's. The first comparison of fingerprints of PSTV isolates with different pathogenicity was carried out by DICKSON et al. (1979) after labelling viroid RNA in vitro with ^{125}I. The severe and mild strains were uniquely characterized by the presence and absence of particular spots in their fingerprints whereas the intermediate "type strain" exhibited no such singular features. Although the interpretation of ^{125}I-fingerprints is difficult because RNase T_1 oligonucleotides without a cytidine and pancreatic fragments with a 3′-terminal uridine will not become labelled, it could be concluded that only minor differences exist between the severe, intermediate and mild PSTV strains. Thus, the marked contrast between severe and mild pathogenicity must be explained on the basis of relatively few specific oligonucleotide sequence differences. Interestingly these mild PSTV strains cross-protect not only against severe PSTV strains but also against CEV and CSV (NIBLETT et al. 1978) which differ extensively from each other and from PSTV in their primary sequences (GROSS et al. 1982).

The elucidation of the structural differences between the sequenced pathogenic PSTV "type strain" (GROSS et al. 1978, DOMDEY et al. 1978) and a "mild" PSTV isolate became possible after the complete sequence of the latter had also been established (GROSS et al. 1981). By comparison, it turned out that the RNase T_1 fingerprints of both strains (Fig. 22A and C) differ in the location of oligonucleotides nos. 30 and 39. The nucleotide sequence of no. 30 is CAAAAAAG in the severe strain and was established to be CAAUAAG in the mild isolate. The sequence of no. 39 is known to be CUUUUUCuC-UAUCUUACUUG in the severe strain and turned out as CUUUUUCUCUA-UCUUUCUUUG in the mild PSTV.

The RNase A fingerprints of both strains (Fig. 22, B and D) differ in that oligonucleotide no. 31 (AAAAAAGGAC) of the intermediate strain is replaced by oligonucleotides AAU and AAGGAC which are now comigrating with nos. 9 and 23/24, respectively, in the mild isolate. These findings are summarized in Fig. 22, which shows the previously determined primary and secondary structure of the intermediate PSTV (top) together with that of the mild PSTV strain as newly established by sequence analysis. Thus it becomes evident that three nucleotide changes involving the introduction of three uridines by one insertion and by two replacements of adenosines, respectively, may change a pathogenic PSTV strain to a virtually non-pathogenic isolate. Interestingly these changes are such that the total number of the 359 ribonucleotides of the PSTV molecule is not altered.

Regarding the secondary-structure model of PSTV with respect to these sequence changes, it is interesting to note that they occur in or opposite to such oligopurine sequences (i.e., nucleotides 118–122 and 307–313) which have been found to exist in all viroids studied so far. Only one of the observed sequence differences, namely that in the left half of the molecule, seems to alter the secondary structure to some degree, in that the mild PSTV strain has two more base pairs in this region than the intermediate strain. Therefore, the question arises how the number of base pairs, i.e., the secondary structure around nucleotides nos. 50 and 310, could in some way determine the pathogenic response of the host plant during PSTV replication in tomato plants. Our recent sequence analysis of a severe PSTV isolate has revealed comparable nucleotide exchanges in the left half of the molecule. This may indicate that pathogenicity of PSTV is determined in some as yet unknown way by a structural domain around nucleotides 40 to 50 and 310 to 320.

Considering the dynamic aspects of PSTV structure, it is of interest that all the observed mutations are located outside those regions which have been shown to be directly involved in the formation of branched intermediated, i.e., in stable hairpins which are newly formed during the thermal denaturation of the PSTV molecule (HENCO et al. 1977, RIESNER et al. 1979). The existence of these regions and the formation of the corresponding hairpins in several

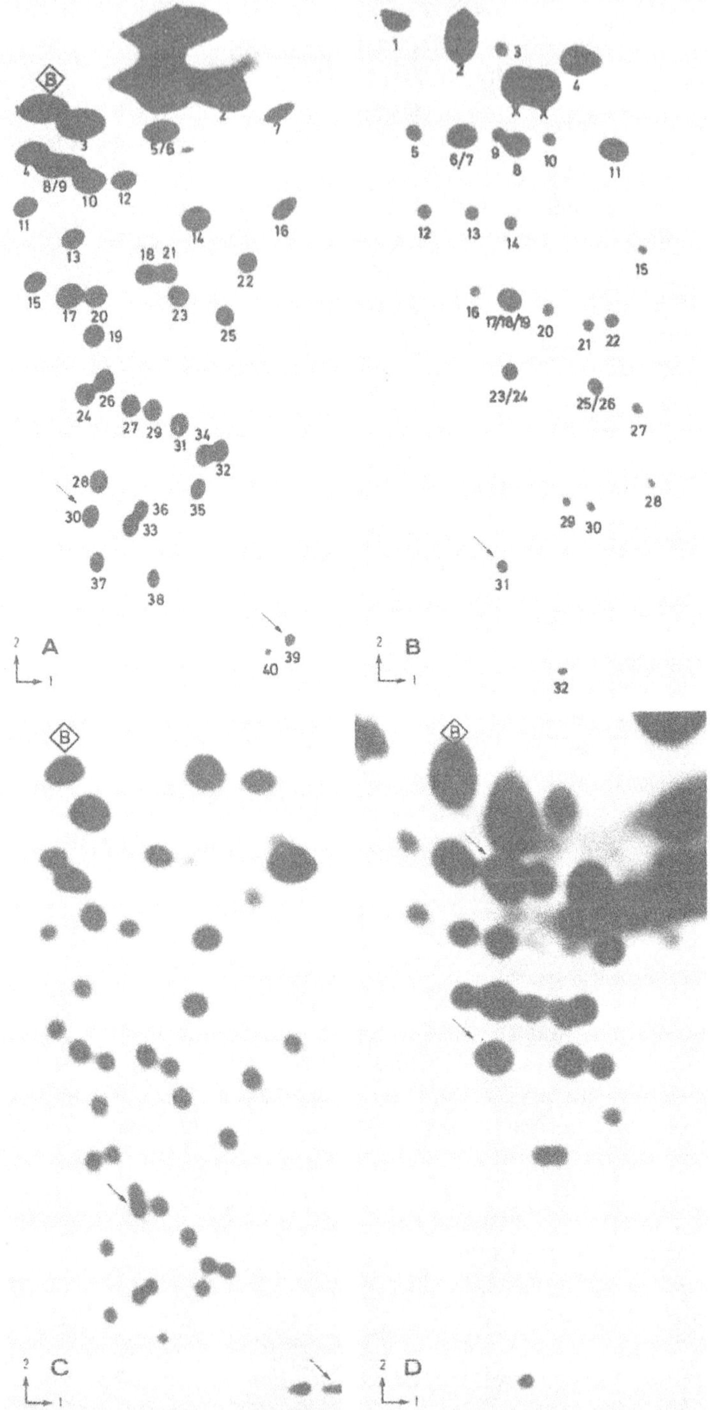

Fig. 22. Fingerprints of PSTV strains causing intermediate (**A, B**) and mild (**C, D**) symptoms in tomato. **A, C** RNase T_1 fingerprints; **B, D** RNase A fingerprints. Fingerprints were produced by electrophoresis and homochromatography. All oligonucleotides were quantified and sequenced. Those differing in the two PSTV strains are marked with *arrows*

viroid species suggests that they are of functional importance. Although their relevance for certain biological properties is not yet understood, there are indications that they could act as specific signals during viroid transcription (RACKWITZ et al. 1981). With respect to the theoretical translation potential of viroids, it should be noted that the described mutations would change the reading frame, in such a way that the infecting (+) RNA molecule could now code for peptides or proteins, and the replacement of adenosine in position 120 by uridine would create a possible UAA terminator codon.

The structural details of more PSTV isolates are needed before any definite correlation between the molecular structure and the biological functions of PSTV can be made.

3.9 The Problem of Different Viroid "Species"

In view of the different viroid diseases and the host specificity of the corresponding agents, the question of structural similarities and differences between these different viroids arises. Fingerprint analysis carried out with purified viroids after in vitro labelling with ^{125}I (DICKSON et al. 1975) and after in vitro 5′-phosphorylation with [α-^{32}P-ATP] and T4 phage-induced polynucleotide kinase (GROSS et al. 1977) showed that PSTV from tomato, CSV from cineraria and CEV from *Gynura* differ significantly from each other in their oligonucleotide patterns and hence their primary sequence (Fig. 13). From the analysis of their fingerprints of PSTV and CEV DICKSON et al. (1975) estimated that the differences in nucleotide sequence between these species is 13.3% or greater. They concluded that at least 50 out of the approximately 360 nucleotides should differ, and that these differences are distributed throughout the entire molecule. These findings supported the concept of individual viroid "species" and invalidated the previous assumption (SEMANCIK and WEATHERS 1972b, SINGH and CLARK 1973, SEMANCIK et al. 1973) that for example PSTV and CEV are independent isolates of the same pathogen, because they produce similar symptoms in certain hosts. In addition, fingerprint analysis showed that each of the three viroid preparations investigated consisted of a single molecular species of RNA and not of a mixture of several RNAs of about the same length, but of different nucleotide sequence (Fig. 13).

An interesting problem in this context is the influence of the host on the specificity of viroids as reflected in their fingerprint pattern and their primary sequence. To clarify this point the fingerprints of PSTV and CEV after propagation in tomato and *Gynura aurantiaca,* respectively, were compared (DICKSON et al. 1978). It was found that both viroids displayed the essential features of their characteristic fingerprints irrespective of the host plant, in which they were propagated. Some minor variations were observed, but their nature and origin has not yet been elucidated. These results clearly showed that viroids "breed true", i.e. the infecting viroid must serve as a template, from which the progeny viroid is copied and that the host plant has no pronounced influence on the primary sequence of the viroid replicating in it.

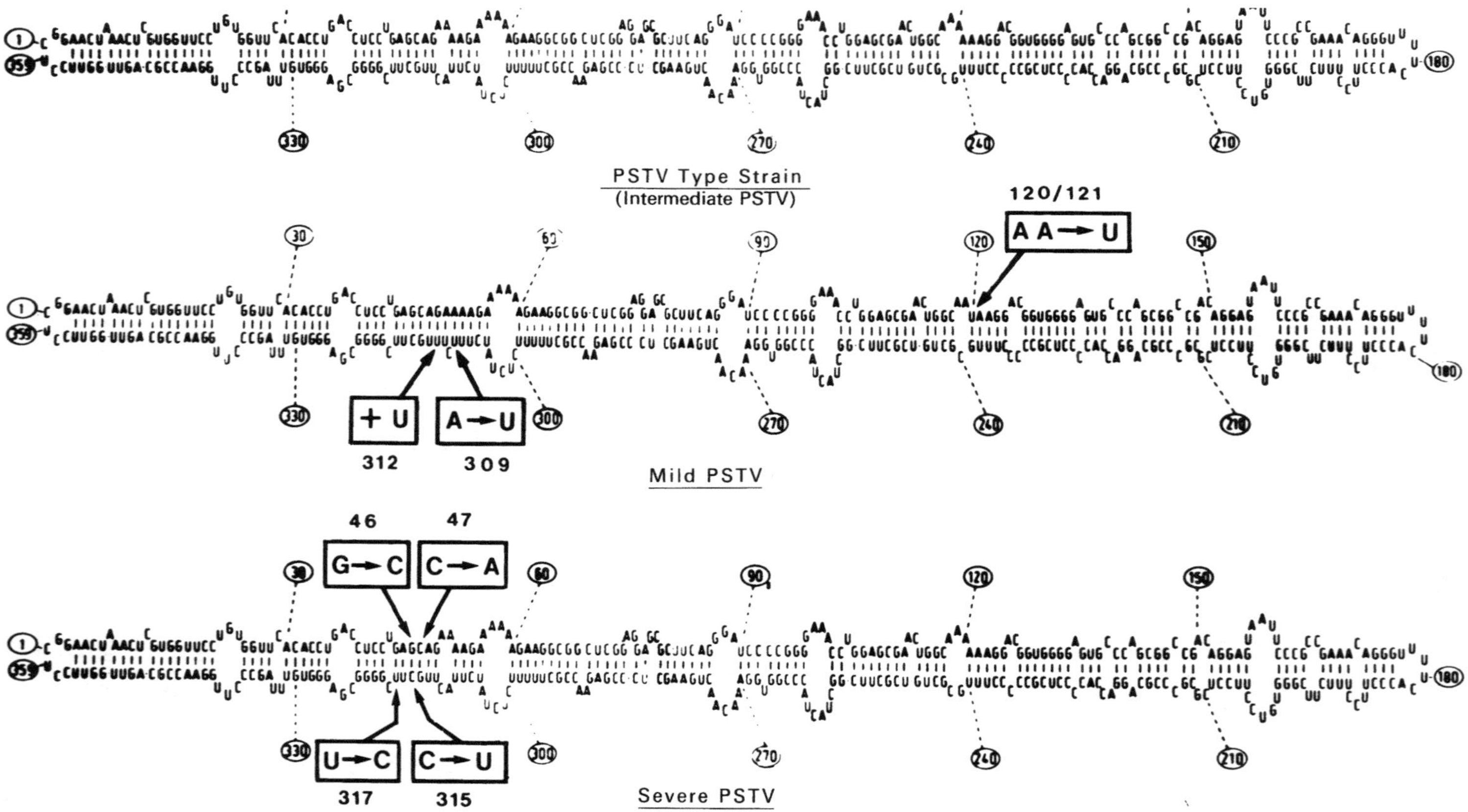

Fig. 23. Primary and secondary structure of PSTV strains causing intermediate (*top*), mild (*middle*) and severe (*bottom*) symptoms in tomato as shown in Fig. 5. Around nucleotide no. 120, two adenosines in the severe strain have been replaced by one uridine in the mild strain, whereas in the neighbourhood of nucleotide no. 310 an adenosine of the severe strain has been replaced by a uridine, and another uridine has been inserted, in the mild strain. There are no further differences between these strains, since the mild PSTV strain was sequenced as completely as the severe strain. Note that the two strains differ slightly in their secondary structures, the mild strain having two additional base pairs. It should be noted that the nucleotide exchanges in the severe strain are not yet incorporated into the secondary structure of the type strain but rather are indicated specifically by the boxes. These results suggest that pathogenicity of PSTV may be determined by a structural domain in the left part of the viroid molecule

The final proof for the existence of different viroid "species" as the agents of different viroid diseases came from the elucidation of the complete nucleotide sequence of different viroids and viroid isolates as in the case of CEV and Australian isolate (VISVADER et al. 1982) and a Californian isolate (GROSS et al. 1982). Moreover, the primary structure of CSV was established for an Australian isolate (HASELOFF and SYMONS 1981) and an English isolate (GROSS et al. 1982). In addition the sequence analysis was completed for ASBV (SYMONS 1981) and for CCCV with its four disease-associated RNA species (HASELOFF et al. 1982). The chain length of these viroids and viroid isolates and their sequence homology with the PSTV "prototype viroid" are listed in Table 3 together with the data available for two of the four known encapsidated viroid-like RNA's, the so-called virusoids. The corresponding primary sequences and optimized secondary structures are compared in Fig. 24.

Considering their overall secondary structure as derived upon on the basis of maximum base-pairing, all these viroids and virusoids represent well-defined rod-like RNA molecules of impressive similarity. Although their chain length varies between about 246 and 574 nucleotides, all of them are characterized by extensive intramolecular base-pairing leading to short helical segments which are separated by short single-stranded regions forming loops or bulge loops. But before any detailed comparison between the different viroids can be made, the complexity of the CCCV RNA species needs to be discussed.

3.10 The Complexity of the CCCV System

While all other viroid diseases are characterized by the presence in viroid-infected tissue of a single infectious RNA of unit length which predominates in its circular form, four variant RNA species are found to be associated with the cadang-cadang disease of coconut and named according to their electrophoretic mobility (IMPERIAL et al. 1981). Two of these four RNA's, called CCCV RNA 1 fast (246 nucleotides) and CCCV RNA 2 fast (492 nucleotides), are present in infected palms at early stages of the disease (IMPERIAL et al. 1981). As infection progresses over a period of years, two additional RNA's, called CCCV RNA 1 slow (287 nucleotides) and CCCV RNA 2 slow (574 nucleotides), respectively, appear and predominate (MOHAMED et al. 1982)

Sequence analysis has shown that four of the six isolates of CCCV RNA 1 fast consist of two populations of molecules 246 and 247 nucleotides long and differing in the presence or absence of a C at nucleotide 198. The relative proportions of the two CCCV RNA 1 fast subspecies vary between different isolates, but the various sequence differences between the isolates do not appear to correlate with differences in geographic location. Since RNA 1 fast is essentially identical in different isolates it can be considered as the unit length molecule of the CCCV RNA system (see Fig. 24). Its three other RNA species can be derived from RNA 1 fast by duplication of sequences (HASELOFF et al. 1982) (Fig. 25).

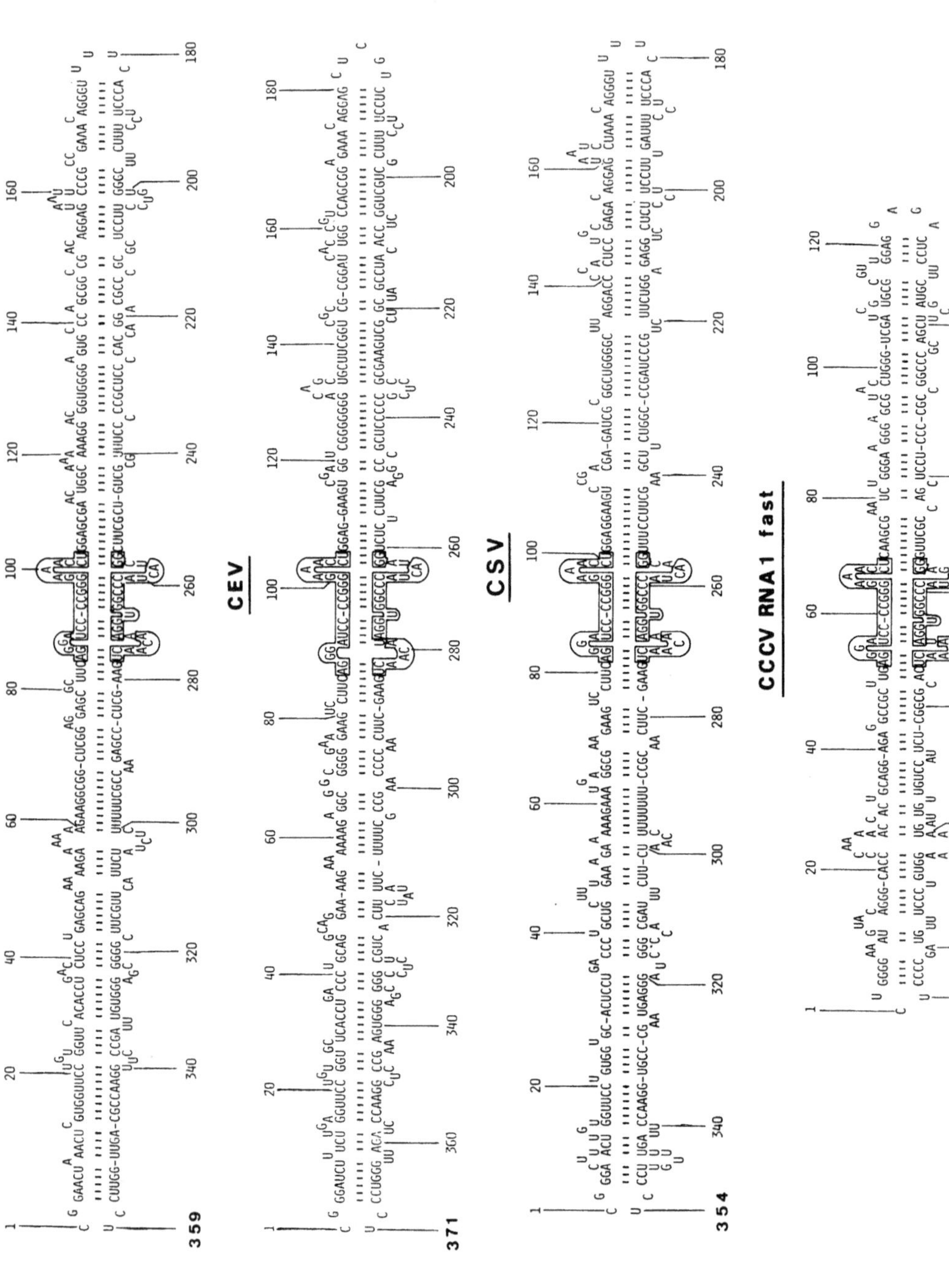
CEV
CSV
CCCV RNA1 fast

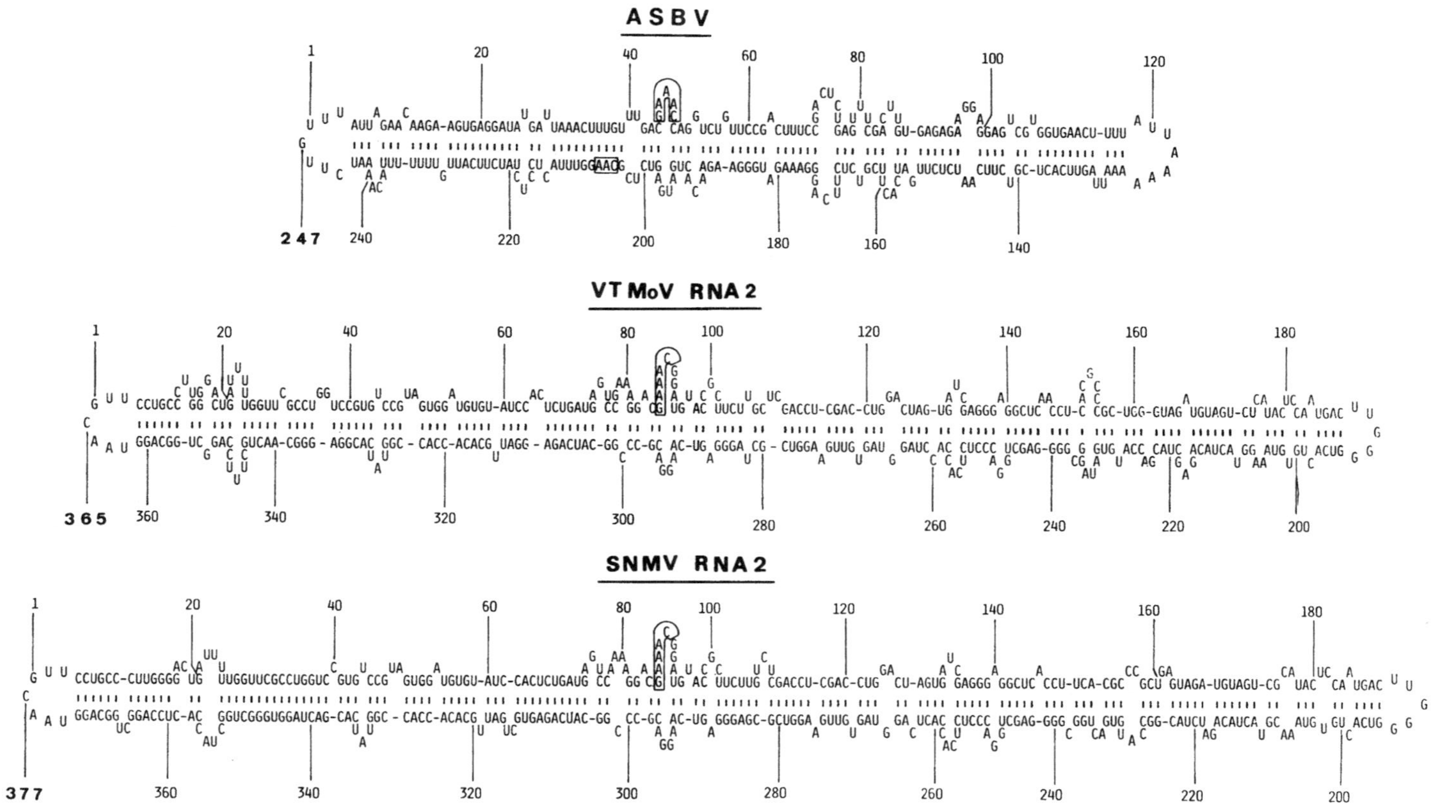

Fig. 24. The nucleotide sequence and the resulting secondary structure of the unit length molecules (monomers) of five viroid "species" (PSTV, CEV, CSV, CCCV and ASBV) and two "virusoids" (VTMo RNA 2 and SNMV RNA 2). The boxed-in sequences of PSTV, CEV, CSV and CCCV are common to these four viroids. The "virusoids" share only a hexanucleotide and ASBV a pentanucleotide with them. For references to the corresponding sequencing work see Table 3

Table 3. Chain length of different viroids, viroid isolates and "virusoids" and their sequence homology with PSTV

Agents		Number of ribonucleotides	Sequence homology with PSTV (in %)	References
Viroids				
PSTV	Pathogenic prototype	359	(100)	GROSS et al. (1978)
	Mild isolate	359	99	GROSS et al. (1981)
CEV	Californian isolate	371	73	GROSS et al. (1982)
	Australian isolate	371	59	VISVADER et al. (1981)
CSV	English isolate	354	73	GROSS et al. (1982)
	Australian isolate	356	69	HASELOFF and SYMONS (1981)
CCCV	RNA 1 fast[a]	246	11	HASELOFF et al. (1982)
	RNA 1 slow[b]	287	11	HASELOFF et al. (1982)
	RNA 2 fast	492	11	HASELOFF et al. (1982)
	RNA 2 slow	574	11	HASELOFF et al. (1982)
ASBV		247	18	SYMONS (1982)
Virusoids				
VTMoV	RNA 2	365	1.5	HASELOFF and SYMONS (1982)
		366	1.5	HASELOFF and SYMONS (1982)
SNMV	RNA 2	377	1.5	HASELOFF and SYMONS (1982)

[a] Certain isolates consisted of a mixture of two species of 246 or 247 nucleotides
[b] From nine isolates sequenced one consisted of 287 nucleotides, one of 297 nucleotides, six consisted of 296 nucleotides and one of 301 nucleotides, so that the duplicated sequences of RNA 1 fast by which RNA 1 slow can be produced consist of 41, 50 or 55 nucleotides

Thus, the sequence analysis of nine isolates of RNA 1 slow from single palms from different localities in the Philippines revealed that they contain the entire sequence and structure of the smaller CCCV RNA 1 fast plus an additional duplicated sequence and structure of 41, 50 or 55 nucleotides, respectively. They are derived from the right hand region of RNA 1 fast between nucleotides 95 to 150 and added to the right hand end of the native molecule of CCCV RNA 1 fast between nucleotides 123 and 124 to produce the corresponding species of RNA 1 slow (see Fig. 26) and corresponding molecules with chain lengths of 287, 296, 297 and 301 nucleotides have been found.

The CCCV RNAs 2 fast are 492 and 494 nucleotides long and represent perfect dimers of the RNA 1 fast with chain length of 246 and 247 nucleotides. The same applies for the sequenced RNA 2 slow of isolate Baao54 which proved to be a dimer of the corresponding RNA 1 slow with its 287 nucleotides. Fig-

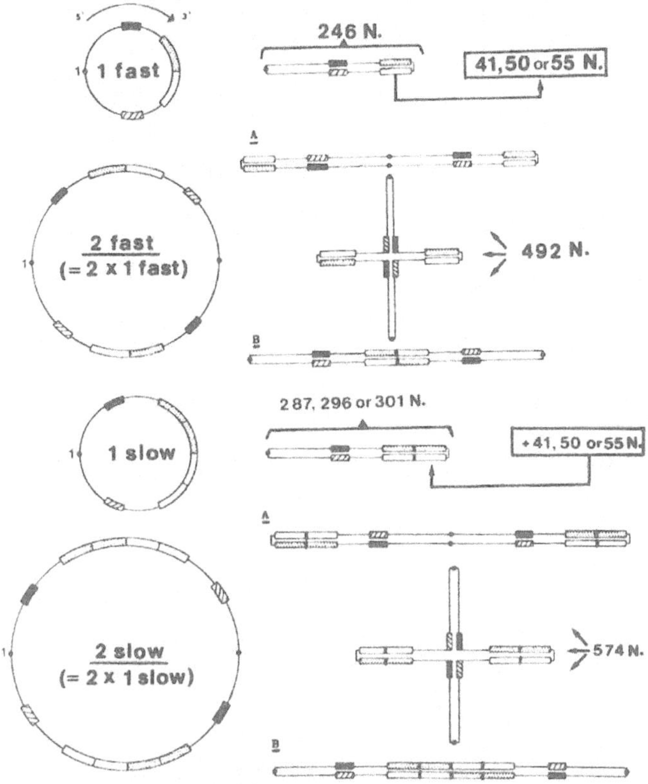

Fig. 25. Structural relationship between the four CCCV RNA's according to HASELOFF et al. 1972. RNA 1 fast with its 246 nucleotides represents the unit length molecule of CCCV from which RNA 2 fast is generated by duplication so that a molecule with 492 nucleotides is formed. Three of several structural alternatives are shown for RNA 2 fast. RNA 1 slow is derived from RNA 1 fast by duplication of a region of 41, 50 or 55 nucleotides, respectively, which originate from the right hand end of 1 fast as shown in detail in Fig. 26. RNA 2 slow is again created by duplication of RNA 1 slow and 3 of several structural alternatives are shown

ure 26 shows that despite these sequence duplications the native structure of the RNA's is still maintained rod-like. While the monomeric forms (RNA 1 fast and RNA 1 slow) exist as a single rod-like conformer, their corresponding duplexes (RNA 2 fast and RNA 2 slow) are potentially able to form two rod-like conformers as well as a large number of intermediate cruciform-shaped structures.

All these data would indicate that the CCCV RNA slow forms are generated from the RNA fast forms by single, rare sequence duplication events occurring separately in each infected palm. If so, all slow RNA's would originate from single parent molecules and accumulate in preference to the fast RNA's because of a competitive advantage in replication.

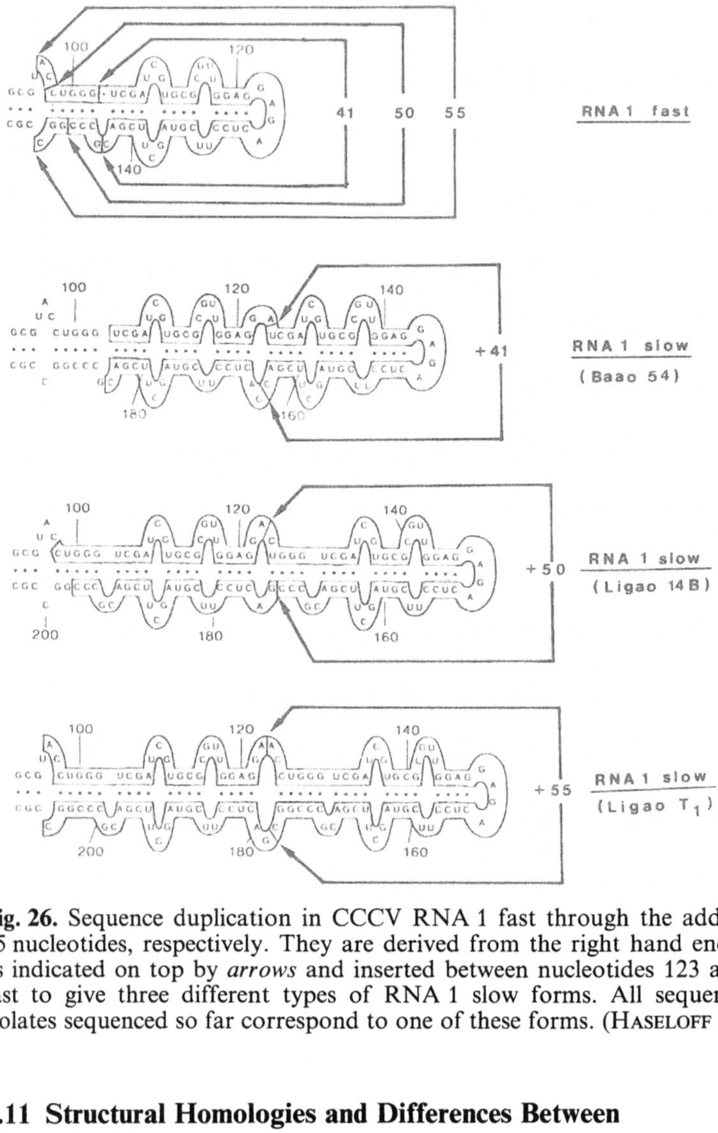

Fig. 26. Sequence duplication in CCCV RNA 1 fast through the addition of 41, 50 or 55 nucleotides, respectively. They are derived from the right hand end of this molecule as indicated on top by *arrows* and inserted between nucleotides 123 and 124 of RNA 1 fast to give three different types of RNA 1 slow forms. All sequenced RNA 1 slow isolates sequenced so far correspond to one of these forms. (HASELOFF et al. 1982)

3.11 Structural Homologies and Differences Between the Viroid "Species" PSTV, CSV, CEV, CCCV and ASBV

Regarding the structural features of the different viroids in more detail on the basis of their primary sequence the following generalizations can be made: As compared to CCCV and ASBV the three viroids PSTV, CEV and CSV share about 60%–70% sequence homology (cf. Figs. 27 and 28) and thus represent a separate group of viroids which may be derived from a common ancestral viroid. The comparison of CSV isolate from England (GROSS et al. 1982) (Figs. 24 and 27) and from Australia (HASELOFF and SYMONS 1981) (not shown) reveals that in the latter the two additional nucleotides, UA, are inserted (between residues 45 and 46 of the English isolate) so that its RNA consists of

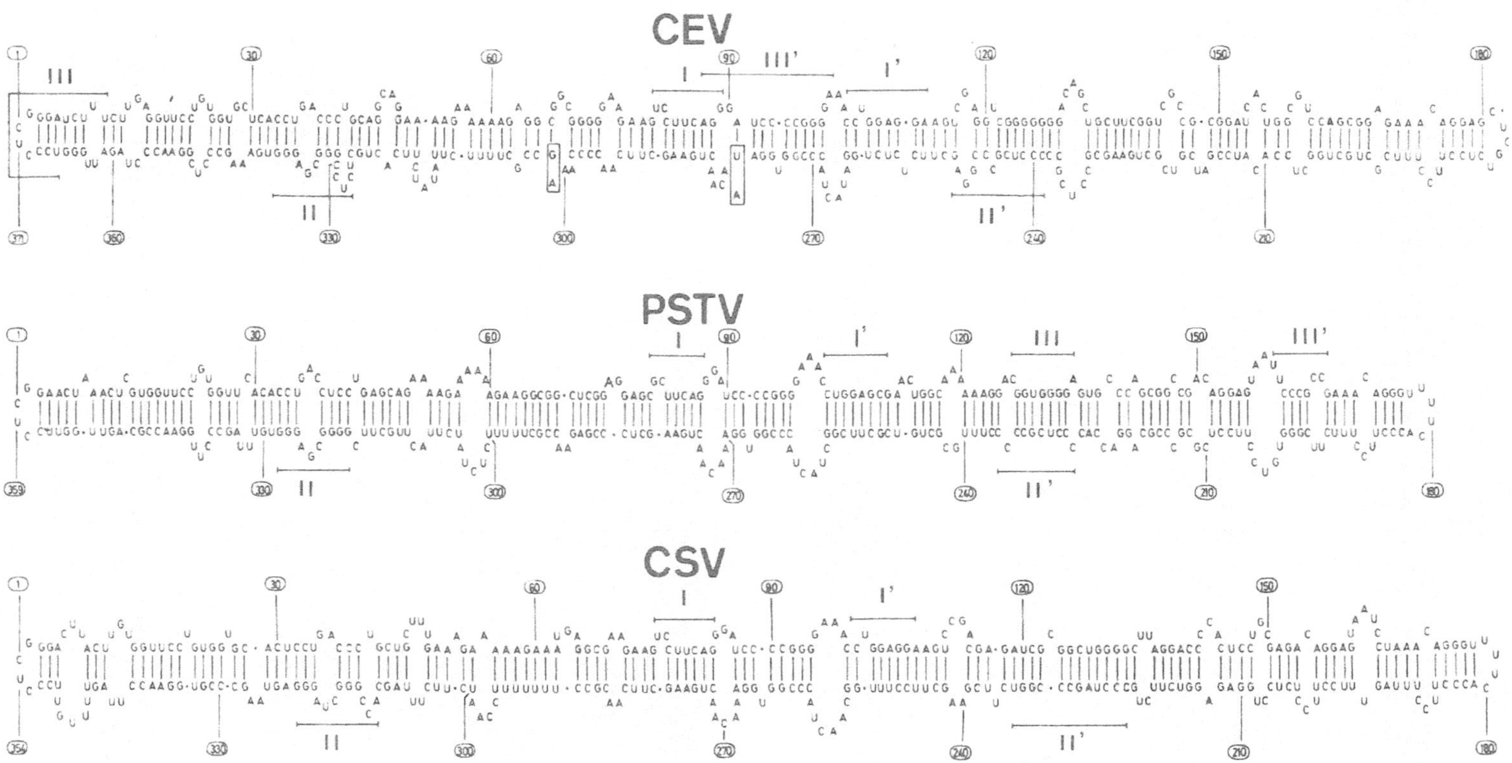

Fig. 27. Comparison of secondary structures of CEV, PSTV and CSV. The sequences are arranged to obtain a maximum number of base pairs and taking in consideration the sensitivity of specific sites to controlled enzymatic digestions. This results in the typical rod-like structure. Arrangements with minor differences are possible for several sections. The first possible A and U or G and C pairs in the end loops are considered as non-hydrogen-bonded on the basis of their reactivity. The segments which could form hairpin structures upon thermal denaturation are indicated by lines, indexed I, I′, II, II′ and III, III′. Segments I and I′ form hairpin I etc. The hairpins are shown in Fig. 29. Evidently hairpins I and III in CEV can not exist simultaneously; the low stability of hairpin III eliminates the possibility of its occurrence

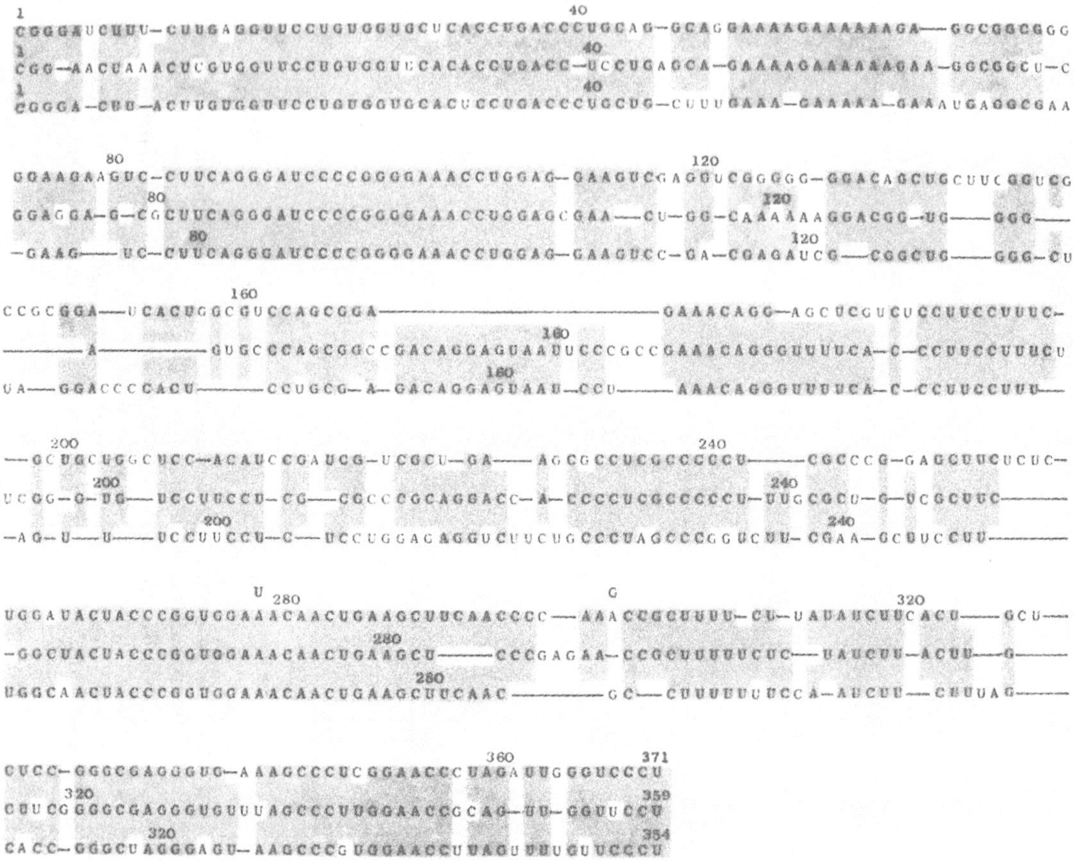

Fig. 28. Primary structures of CEV, PSTV and CSV (*top to bottom*). Sequences are aligned for maximum homology; numbering is as in Fig. 24. Homologous sequences are indicated by shaded areas. A considerable homology between the different viroids is found: CEV-PSTV, 73%, CSV-PSTV, 73%, CEV-CSV, 67%, CEV-CSV-PSTV, 59%

altogether 356 nucleotides. In addition, ten nucleotide exchanges were found in different positions of the Australian CSV (not shown) as compared to the English CSV (Figs. 24 and 27). The Australian (not shown) and Californian isolate of CEV (Figs. 24 and 27) on the other hand, consist both of 371 nucleotides and differ from each other only by four nucleotide exchanges (VISVADER et al. 1981, GROSS et al. 1982). In the depicted PSTV, CSV and CEV the G:C, A:U and purine: pyrimidine ratios are around one; there are almost twice as many G·C as A:U pairs; 9–13% of the base pairs are G·U pairs, and the overall number of base pairs (120–124) is remarkably constant although the total number of nucleotides differs slightly among these three viroid species (Table 4). All of them contain a long polypurine sequence around nucleotides 50–65. The detailed sequence homology is presented in Fig. 28, where shaded areas identify homologous structures. These homologous sequences, however,

Table 4. Comparison of nucleotide composition, and number and type of base pairs of CEV, PSTV and CSV

Viroid	Total number of nucleotides	Number of nucleotides and proportion of total								Ratio		Number of base pairs			
		A		U		C		G		A/U	G/C	total	G·C	A·U	G·U
		No.	(%)	No.	(%)	No.	(%)	No.	(%)						
CEV	371	71	19.1	77	20.8	112	30.2	111	29.9	0.92	0.99	122	74	37	11
PSTV	359	73	20.3	77	21.4	108	30.1	101	28.1	0.95	0.94	124	72	36	16
CSV	354	74	20.9	93	26.3	97	27.4	90	25.4	0.80	0.93	120	67	41	12

are not evenly distributed along the molecules. Rather surprisingly they are accumulated around the first and the last approximately 110 nucleotides, i.e., there is less homology in the middle of the RNA's. The ends and the central region of the linearized sequences (Fig. 28), however, comprise the left and the right halves, respectively, of the circular viroid secondary structures shown in Fig. 27. Consequently, these viroids in their native state contain two domains: a longer one of significantly more homology on the left as compared to a shorter one, with less homology, to the right. Interestingly, the secondary structure of the left halves appears to be less stable than that of the right halves. The question arises here whether these observed homologies are simply the reflection of a phylogenetic relationship among these viroids, or whether they are the result of a convergent development towards a common functional requirement. An answer to this question is not possible as yet.

The CCCV RNAs share only about 11% of sequence homology with PSTV, CSV and CEV and therefore represent members of a second group of viroids. The sequence homology is essentially confined to two regions each about 20 nucleotides long. Most intriguingly, they correspond to the two highly conserved regions in the predicted secondary structure of the native PSTV, CEV and CSV molecules (see Figs. 24 and 30). It should be noted that these regions were postulated to be possibly involved in base-pairing of viroid complementary RNA with U1a RNA (DIENER 1981, GROSS et al. 1982) and in the formation of a stabilizing stem-loop structure in the viroid (−)RNA (GROSS et al. 1982). Moreover, this structure and possible interaction were assumed to reflect the origin of viroids as "escaped introns" and one of the possible mechanisms of viroid pathogenesis (DIENER 1981, GROSS et al. 1982) as will be discussed in detail in sections 4.8 and 5.

Although ASBV shows a slightly higher sequence homology (ca. 18%) with PSTV, CEV and CSV than the CCCV RNAs, the regions of homology are limited and scattered. The similarity between ASBV and all other viroids is reduced to the hexanucleotide, GAAACC, which forms an unpaired loop in the central region of the secondary structure of ASBV. Again, this sequence is part of the highly conserved central region in all other viroids (see Figs. 24 and 30). Because of this lack of homology ASBV should be allocated to a third viroid group. The relevance of this limited homology for the possible aberrant origin and mechanism of pathogenesis of ASBV as compared to the other viroids will be discussed in section 4.8 and 5.

In order to stress the similarities rather than the differences, the highly conserved region may be aligned vertically with the corresponding sequences of all viroids. Then additional but rather short regions of homology can be observed in similar positions. From these similarities it becomes conceivable that ASBV could have also arisen from an ancestral similar to PSTV, CSV of CEV by elimination of about 90 nucleotides from the left hand end of the molecule (see Fig. 24) together with substantial sequence divergence in the rest of the molecule relative to these members of the first viroid group (SYMONS 1981).

In spite of the overall similarity in secondary structure there is 1.5% or less sequence homology or complementarity between the virusoids and the

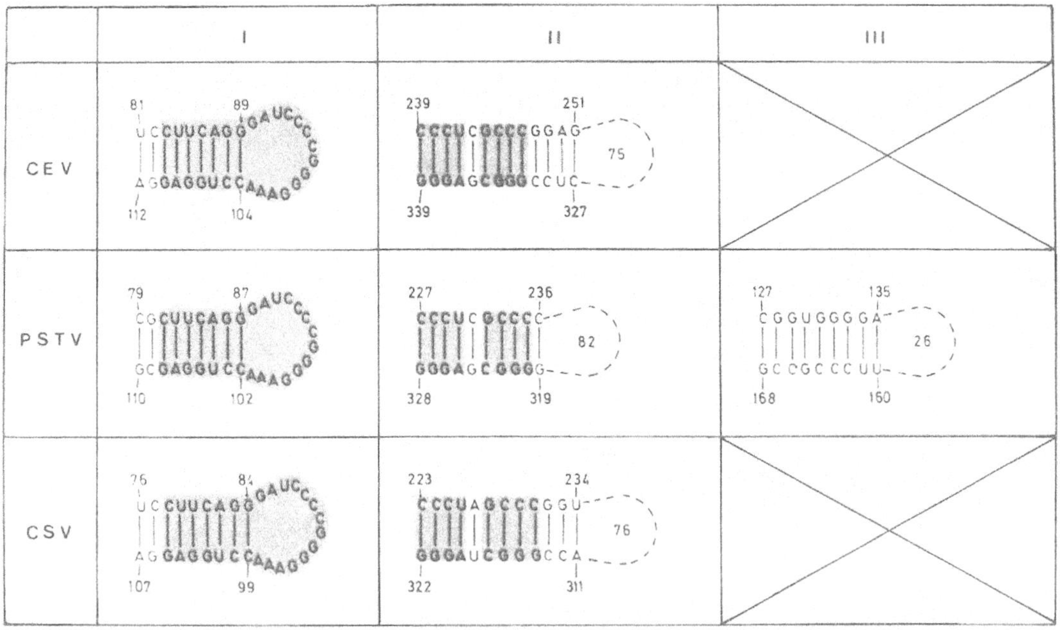

Fig. 29. Hairpins occurring in intermediate structures during thermal denaturation of viroids. Sequence identity of hairpin stems (*I* and *II*) and loops (*I*) is indicated by *shaded background*. Numbers in the loops specify the number of nucleotides in the corresponding loop

viroids. Nevertheless, VTMoV and SNMV RNA 2 contain the sequence GAAAC (nucleotides 86–90 in both RNAs) which is present in a single-strand region (see Fig. 30) in the center of the secondary structure (HASELOFF and SYMONS 1982). Again, also this pentanucleotide is part of the highly conserved region in all viroids. Therefore, it seems very likely that this common structures may reflect some function common to all these agents and it is well conceivable that they play an important role in the replication of these molecules by the corresponding host enzymes.

Considering now their secondary structure in detail, PSTV, CSV and CEV can be arranged on the basis of maximum base-pairing into well-defined secondary structures of impressive similarity. There is biochemical and physicochemical evidence that these rod-like structures represent isolated viroids in their native state in vitro. Other possible secondary structures with significantly less base pairs and occurring as branched molecules may exist in vivo and become stabilized by their interaction with components of the host cell. In fact, such metastable structural intermediates have been observed after heating and rapid cooling of viroid RNA (RIESNER et al. 1979). There is even more similarity between the three viroids than suggested by the overall rod-like secondary structures alone. It has been found for PSTV and it had been postulated for all viroids studied to date that the structural transitions of viroids during thermal denaturation is of well defined complexity and highly cooperative. Melting of the left

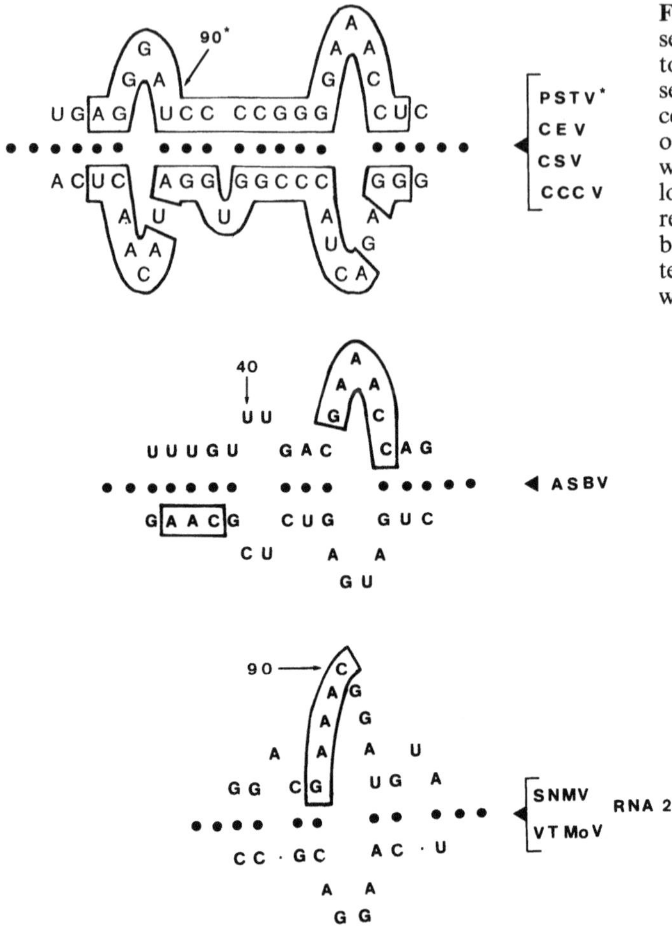

Fig. 30. The highly conserved sequences common to all viroids and virusoids sequenced so far. Only the central region of the secondary structure models where these sequences are located is shown. The corresponding nucleotides are boxed in. For details see text and compare also with Fig. 24

half of the viroid molecule permits the formation of new ("secondary") and more stable hairpin structures between denatured and yet double-stranded regions. These interactions across the molecule destabilize the corresponding double-stranded domain and lead in the main transition to a branched intermediate which, in case of PSTV, contains three "secondary" hairpins (RIESNER et al. 1979, HENCO et al. 1979). Such complementary sequences with the potential of forming new "secondary" hairpins are also found in the native structures of CEV and CSV. These sequences are marked with bars and Roman numerals in Figs. 27 and 21 and are presented as hairpins in Fig. 28.

From the fact that the "secondary" hairpin I and the stem of hairpin II have been strictly conserved one may postulate that they have vital functions for viroids in vivo. One could visualize that these hairpins are formed during viroid replication, where they would guarantee the circularization of linear precursor molecules by providing the close contact neccessary for ligation of their ends. A similar mechanism has been proposed for the circularization of the single-stranded DNA of bacteriophage fd (SCHALLER et al. 1969). The constant regions, including the characteristic polypurine sequence, however, could also be signals for initiation, elongation or termination of viroid replication. The exact knowledge of more viroid primary and secondary structures will finally enable us to understand and to explain, on the molecular level, the finding that viroids directly interact with host RNA polymerase II in vivo (MÜHLBACH and SÄNGER 1979) and in vitro (RACKWITZ et al. 1981) and are transcribed by this enzyme into complementary RNA.

This section on viroid structure should not be concluded without mentioning that there are several recent reviews which are dealing in great detail with different aspects of the unique structural features of viroids (SÄNGER 1979a, b, 1980a, b, GROSS and RIESNER 1980, KLEINSCHMIDT et al. 1981, RIESNER et al. 1981). The interested reader is referred to them.

4 The Functions of Viroids

4.1 Translation Properties of Viroids

The small size, the correspondingly limited genetic information content and the unique structure of the viroid molecule raise many questions related to viroid replication and function and the main question is whether viroids code for peptides or proteins. Previous attempts to translate PSTV and CEV into proteins in several in vitro systems (from wheat germ, *Bacillus stearothermophilus, E. coli* and *Pseudomonas aeruginosa*) where brome mosaic virus mRNA's and Qβ RNA stimulate translation, have failed. Moreover viroid RNA did not interfere with the in vitro translation of potent messenger RNA species in these systems (DAVIES et al. 1974, HALL et al. 1974, SEMANCIK et al. 1977). Aminoacylation reactions carried out with CEV showed that amino acids are not accepted by this viroid (HALL et al. 1974) which would also preclude any tRNA-like functions for viroids.

From the established primary sequence it appears that the circular arrangement of the uneven number of 359 nucleotides makes three rounds of translation potentially possible in PSTV. But the common AUG initiator triplet and several other codons are missing. There are seven possible GUG initiator and six possible terminator triplets (see inner circle in Fig. 14). The shortest possible translation product would be a tetrapeptide, whereas the longest possible protein could result from more than two rounds of translation, provided that in this case the weak terminator codon UGA were translated. The lack of ribosome binding sites at the 5′-side of the possible GUG initiator codons, the absence in viroids of the "cap" structure common to most eukaryotic mRNA's and several other aspects of the primary and the very stable secondary structure of PSTV seem to preclude its translation into peptides or proteins (GROSS et al. 1978). Especially the circularity of the viroid molecule would also argue against its

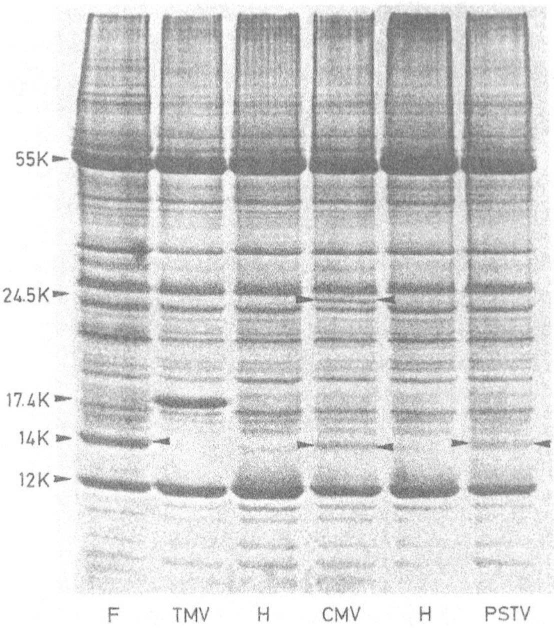

Fig. 31. Pattern of phenol-extractable proteins from tomato leaf tissue (cv. Rutgers) systematically infected with different pathogens. *F Cladosporium fulvum*; *TMV* tobacco mosaic virus; *DMV* cucumber mosaic virus; *CPFV* cucumber pale fruit viroid; *CEV* citrus exocortis viroid; *H* healthy control. Note the marked increase of the 14k protein band present in healthy tissue (H) after fungus = *Cladosporum fulvum* (F), virus (CMV and TMV) and viroid (CPFV) infection which produce necrotic foliar symptoms. TMV and CEV infection cause only mild symptoms of leaf malformation. At 17.4k and 24.5k the viral coat proteins are visible

translation after it was shown that the circularization of potent mRNA's prevents their binding to ribosomes and their translation into proteins (KOZAK 1979, KONARSKA et al. 1981).

On the basis of the PSTV sequence, the hypothesis has been put forward that viroids might exert their functions like the so-called "negative-strand viruses", namely via a complementary viroid molecule which would then be the coding strand (MATTHEWS 1978). In the corresponding PSTV minus strand four GUG initiator codons and nine possible terminators (4 × UAG, 2 × UAA and 3 × UGA) exist, from which at least four proteins composed of 28, 43, 79 or 108 amino acids, respectively, could result. The two larger of these hypothetical viroid-specific polypeptides would have molecular weights of about 8,500 and 11,300. Since the largest would contain 11% arginine and 5% lysine it should have histone-like properties. Although the presence of PSTV minus strands in PSTV-infected tissue has been repeatedly demonstrated on the basis of molecular hybridization, the putative viroid-specific proteins have not yet been found. The two 15,000 and 18,000 molecular weight proteins accumulating in viroid-(CEV)-infected *Gynura* tissue (CONEJERO and SEMANCIK 1977, FLORES et al. 1978, CONEJERO et al. 1979) are not viroid-specific and the 14,000 molecular weight protein in PSTV-infected tomato tissue accumulates as a general pathophysiological response of the host plant also after the infection with tobacco mosaic virus, cucumber mosaic virus and the leaf mould *Cladosporium fulvum* (Fig. 31). Interestingly, the accumulation of this protein can be related to the severity of the necrotic symptoms induced by these pathogens (CAMACHO and SÄNGER 1982a, b).

4.2 The Problems of Viroid Replication

The mechanism of viroid replication and pathogenesis is still largely a matter of conjecture. The lack of information in this area is mainly based on system-inherent experimental limitations. Thus, higher plants and the corresponding cell systems are less well suited to study the dynamic aspects of cell metabolism after viral infection than bacterial and animal cells.

Unlike certain viruses, viroids cannot carry their own replicase with them, because they are "naked" RNA molecules and not encapsidated in a protein coat and envelope. Furthermore, it is obvious that the potential genetic information of the viroid molecule with its about 360 nucleotides is insufficient to code for a complete viroid-specific replicase, even if over-lapping reading frames and three rounds of translation are assumed. Because of this limited coding capacity, one must assume that viroid replication largely depends on the nucleic acid-synthesizing machinery pre-existing in the host cell.

Although all attempts have failed so far to translate viroid RNA in vitro into proteins or peptides (see previous paragraph), it is nevertheless conceivable that the plus viroid molecule or its complementary strand could be translated in vivo into a small polypeptide which is capable of converting a host enzyme into a viroid-specific replicase. This would resemble a strategy similar to the one used by the RNA bacteriophage Qβ, in which case a phage-coded 65,000 molecular-weight protein is combined with three host proteins (ribosomal protein S1 and the protein synthesis elongation factors EF–Tu and EF–Ts) to create the Qβ-specific replicase with its four subunits (see KAMEN 1975). Also overlapping reading frames could be used to code for entirely different proteins, as realized by the DNA bacteriophage ΦX 174 (SANGER et al. 1977). Finally, the possibility exists that the corresponding viroid molecule might require specific processing and or modification before its translation proceeds.

Regarding the possible pathways of viroid replication by nucleic acid-synthesizing enzymes preexisting in the host, we have to consider that healthy plant cells are known to contain two classes of RNA-synthetising enzymes, namely the DNA-dependent RNA polymerases I, II and III (see DUDA 1976, WOLLGIEHN, Chap. 5, this Vol.), and the RNA-dependent RNA polymerase (see FRAENKEL-CONRAT 1979). Theoretically, viroids could therefore either be replicated through a complementary RNA, as are the RNA plant viruses, in which case the RNA-dependent RNA polymerase of the host could be utilized. The corresponding in vitro transcription experiments which showed that this enzyme from healthy tomato leaf tissue is capable of synthesizing PSTV copies of apparently full length when PSTV RNA is offered as template are described in detail in Section 4.6.

Viroid replication could also follow the strategy of the RNA tumour viruses, the genome of which is first transcribed by the reverse transcriptase into DNA, which then serves as template for the final synthesis of the retrovirus progeny RNA. Since no reverse transcriptase activity has been detected in higher plants so far, one would be forced to assume that upon viroid infection the viroid-specific DNA template is newly synthesized in a way as yet unknown.

On the other hand, the viroid-specific DNA complement could be already present in the host cell as a part of the host genome but in a repressed state. In this case one could visualize that the infecting viroid RNA acts as a regulatory molecule by triggering the de-repression of the transcription of the presumed viroid complementary sequences present in the host genome. If so, the viroid-specific DNA could serve as template for one of the DNA-dependent RNA polymerases from which the viroid progeny RNA could be synthesized. Unequivocal evidence for any of these routes would be the presence of viroid-specific DNA in viroid-infected tissue. The previously published experimental evidence for this and the corresponding refutal are described in Section 4.7.

One of the unique features of viroid RNA is its similarity to a DNA molecule of the same size and G:G content as far as the overall shape, the thermal stability and the cooperativity of denaturation is concerned (RIESNER et al. 1979). It was conceivable therefore, that viroids might be efficient templates for one of the DNA-dependent RNA polymerases of the host. This notion gained additional support from the finding that RNA polymerase from *E. coli* is able to select from a random mixtures of copolymers a RNA molecule with a chain length between 100 and 150 nucleotides which is capable of multiplying indefinitely (BIEBRICHER and ORGEL 1973). The results from in vivo and in vitro experiments which showed that the DNA-dependent RNA polymerase II of the host plant is involved in viroid replication and accepts viroid RNA as efficient template are discussed in Sections 4.3 and 4.5 respectively.

4.3 Replication of Viroids in Protoplasts and Cell Cultures

The first studies on viroid replication at the cellular level were carried out with protoplasts isolated from leaf tissue of the tomato cultivar "Hilda 72" (MÜHLBACH et al. 1977). The general metabolic capacity of these protoplasts was substantiated by their ability to regenerate new cell walls after 3 days, to divide within a week and to produce complete tabacco mosaic virus (TMV) particles after infection with TMV RNA in vitro (MÜHLBACH et al. 1977). Viroid replication can be followed in these protoplast by incorporation of ^3H uridine after infection in vitro (MÜHLBACH and SÄNGER 1977). To discriminate unambiguously between the different RNA polymerases possibly involved in viroid replication the mushroom toxin α-amanitin was applied. As summarized in Table 5, in eukaryotic cells α-amanitin is known to inhibit selectively the DNA-dependent RNA polymerases II and III at low and high concentrations, respectively, whereas RNA polymerase I is not inhibited at all (LINDELL et al. 1970, ZYLBER and PENMAN 1971, WEINMANN and ROEDER 1974, WIELAND and FAULSTICH 1978).

It turned out that α-amanitin has no appreciable effect on the synthesis of cellular and viroid RNA, if it is present at 10 μg per ml culture medium during the period of 48–72 h after inoculation, when viroid replication is detectably increased. At 50 μg per ml culture medium, however, α-amanitin inhibits viroid replication to about 75% (Fig. 32), whereas the biosynthesis of the prominent cellular RNA species tRNA, 5S RNA, 7S RNA and ribosomal RNA is again not appreciably affected (Table 6). It should be mentioned in this context

Table 5. Function and sensitivity to α-amanitin of DNA-dependent RNA polymerases

RNA polymerase	Catalyzed product	Sensitivity to α-amanitin
I :	Ribosomal RNA	Insensitive
II :	Messenger RNA	Sensitive to 10^{-8} M
III:	Transfer RNA	Sensitive to 10^{-5} M

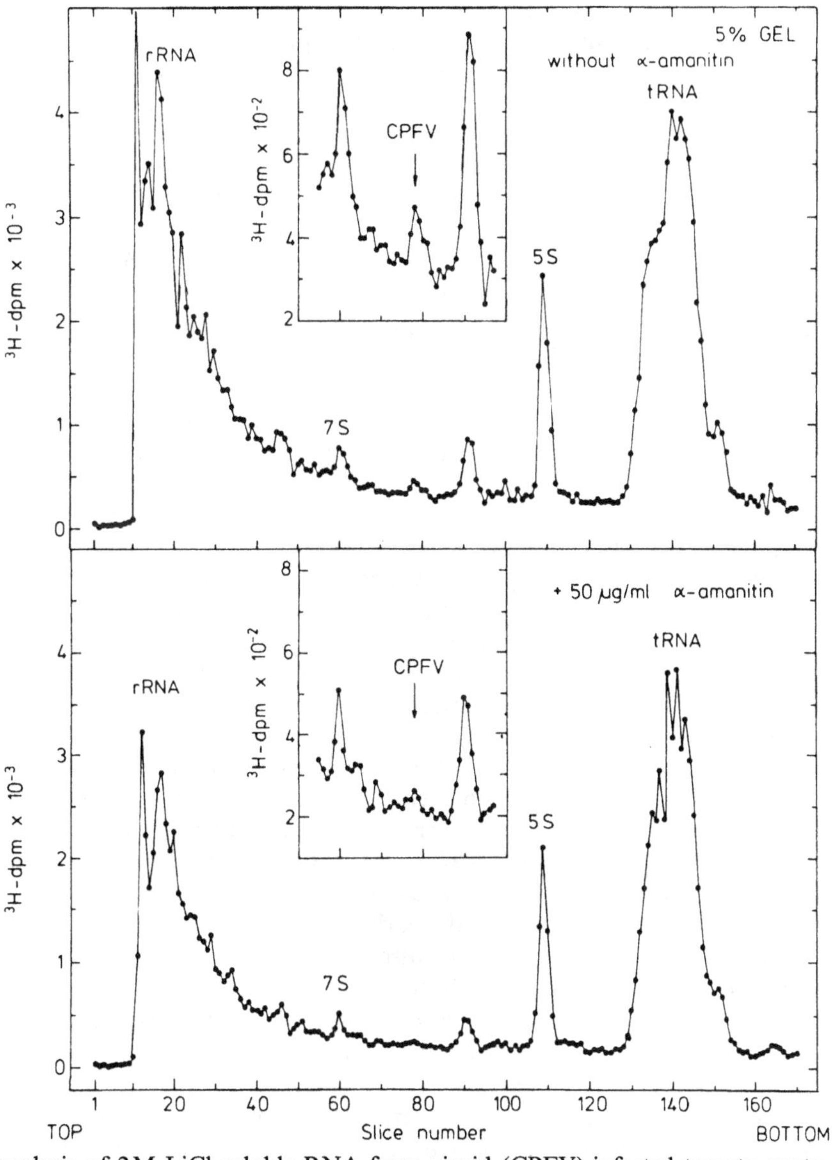

Fig. 32. Analysis of 2M LiCl-soluble RNA from viroid (CPFV)-infected tomato protoplasts in 5% gels. RNA pattern obtained in the absence (*top*) and in the presence of 50 μg ml^{-1} α-amanitin

that only the applied short time exposure (between 48 and 72 h) gives meaningful results because protoplast are killed if 50 μg α-amanitin per ml culture medium is present throughout the total time of culture. Control experiments with tomato protoplasts inoculated with tobacco mosaic virus (TMV) RNA showed that the replication and accumulation of TMV was not affected by α-amanitin under these conditions, where TMV replication is known to increase logarithmically. Therefore, the marked inhibition of viroid replication by α-amanitin is unlikely to be due to a secondary effect of this drug on cell metabolism in general (MÜHLBACH and SÄNGER 1979).

In view of the concentration-dependent specificity of the α-amanitin inhibition the interpretation of such in vivo experiments depends on knowing the intracellular concentration of this inhibitor, which was therefore determined by using the ^{3}H-labelled toxin. When tomato protoplasts were exposed to culture medium containing 10^{-5} (50 μg ml^{-1}) and 10^{-6} M (5 μg ml^{-1}) ^{3}H- α-amanitin for 20 h an intracellular concentration of 2×10^{-8} M and 2×10^{-9} M was already reached within the first 10 h of incubation. This clearly demonstrated that under the conditions where viroid replication is specifically inhibited the intracellular concentration of the inhibitor was in the range of 10^{-8} M. In vitro experiments with DNA-dependent RNA polymerases from different sources (WIELAND and FAULSTICH 1978) including higher plants (DUDA 1976) have shown (see Table 5) that RNA polymerase II is specifically inhibited at a concentration of 10^{-8} M α-amanitin whereas inhibition of RNA polymerase III requires a 1,000-fold higher concentration (WEINMAN and ROEDER 1974) RNA polymerase I is not affected at all by α-amanitin.

Studies with animal cell (ROTT and SCHOLTISSEK 1970, MAHY et al. 1972) and plant cell (SEITZ and SEITZ 1971, BRÄNDLE and ZETSCHE 1973) systems have shown that under comparable experimental conditions RNA polymerase II is also inhibited in vivo. As the intracellular concentration of α-amanitin which inhibits ^{3}H-uridine incorporation into viroid RNA in protoplasts is in the range where the nuclear DNA-dependent RNA polymerase II is specifically inhibited, it was postulated that this enzyme system is involved in viroid replication (MÜHLBACH and SÄNGER 1979).

RNA polymerase II is located in the nucleoplasm and it is responsible for the synthesis of heterogeneous nuclear RNA, i.e., precursors to messenger RNA which are transcribed from the corresponding sections of the DNA of the host genome (ZYLBER and PENMAN 1971; WEINAMN and ROEDER 1974). The assumption that the nuclear RNA polymerase II is involved in viroid replication would provide a plausible explanation the earlier findings (DIENER 1971, SÄNGER 1972) that the bulk of viroid infectivity also is found within the nuclei and closely associated with the chromatin. It should be mentioned here that these results are at variance with the findings of SEMANCIK et al. (1976) in CEV-infected *Gynura* tissue where a viroid-membrane complex rather than a viroid-nucleus complex has been postulated.

It should be emphasized, however, that the observed α-amanitin inhibition cannot be considered as definitive proof of direct replication of viroid RNA by the nuclear DNA-dependent RNA polymerase II. There are several possible mechanisms, by which this enzyme system could be indirectly involved as a

Table 6. Effect of α-amanitin, actinomycin D, and cycloheximide on RNA biosynthesis in viroid (CPFV)-infected tomato protoplasts

$\mu g\ ml^{-1}$		% Inhibition					
		Total RNA	Ribosomal RNA	7S RNA	5S RNA	Transfer RNA	Viroid RNA
α-Amanitin	0	–	–	–	–	–	–
	10	0	0	18.3	5.7	8.5	16.8
	50	10.3	1.9	19.2	0	5.5	75.6
Actinomycin D	20	80.5	80.3	97.1	86.4	82.5	84.5
Cycloheximide	1	24.8	20.4	24.2	28.9	56.2	58.0
	50	66.4	52.4	100	87	91.7	100

result of its central role in the formation of mRNA's in the host cell. There is also the remote possibility that viroid replication may proceed through a yet unknown pathway which is α-amanitin-sensitive.

For comparison, the effect of actinomycin D on the RNA metabolism has also been studied. The detailed analysis in the protoplast system showed that not only viroid replication but also cellular RNA synthesis is inhibited to about 85% by this drug (Table 6). From comparable results with leaf discs (DIENER and SMITH 1975) and isolated nuclei (TAKAHASHI and DIENER 1975) and infiltrated foliar tissue or sprouts (GRILL and SEMANCIK 1980), the involvement of host DNA during viroid replication was assumed. But in view of the inhibition of the biosynthesis of all RNA species, including viroid RNA, a non-specific secondary effect of this drug cannot be ruled out. Therefore, it is only justified to exclude the involvement of pre-existing RNA-dependent RNA replicases during viroid replication, which are known to be actinomycin D-resistant in several host plant systems infected with conventional plant viruses (SÄNGER and KNIGHT 1963, TAKEBE and OTSUKI 1969, WHITE and MURAKISHI 1977).

In order to test the possible requirement of newly synthesized proteins for viroid replication, cycloheximide was included in the protoplast experiments because it is known to interfere specifically only with protein biosynthesis and not with nucleic acid transcription itself. It was found, however, that the biosynthesis of cellular and viroid RNA is strongly inhibited by cycloheximide at concentrations between 1 to 50 $\mu g\ ml^{-1}$ (Table 6). In view of this non-specific and rather general disturbance of cell metabolism, these results do not allow the conclusion that specific and newly synthesized proteins are a prerequisite for viroid replication.

Further progress in the study of viroid replication at the cellular level can be expected from permanent plant cell cultures which are continuously synthesizing viroid RNA. In callus cultures from a PSTV-infected wild-type potato (*Solanum demissum* L.) and tomato (*Lycopersicon peruvianum* L. Mill.) and in cell suspension cultures derived from potato (*Solanum tuberosum* L.) protoplasts inoculated in vitro with PSTV, viroid replication was found to persist for now over 2 years through more than 25 subcultures. The continuous de novo viroid synthesis was substantiated by the incorporation of ^{3}H-uridine and of ^{32}P-orthophosphate into PSTV RNA and its subsequent analysis by gel electrophoresis and autoradiography (MÜHLBACH and SÄNGER 1981). Similar observations of sustained long-term viroid replication were reported for cell suspensions prepared from the tomato cultivar "Rutgers" (ZELCER et al. 1981). The maximum rates of ^{32}P incorporation into PSTV RNA varied between 0.4% of that into soluble RNA (ZELCER et al. 1981) and 2–3% of that into tRNA (MÜHLBACH and SÄNGER 1981), which might be due to the different plant species used in these experiments. Previously reported attempts to culture cells from tumors induced by *Agrobacterium tumefaciens* on tomato plants infected with citrus exocortis viroid could on principle lead to similar systems, but de novo synthesis of viroid RNA has not been shown in these neoplastic cell cultures (SEMANCIK et al. 1978).

It is interesting to note that the continuous viroid (PSTV) replication in plant tissue culture through numerous cell divisions seems to differ from the experience with conventional plant viruses which are often eliminated from continuously dividing meristematic tissue and cells (HOLLINGS 1965, INGRAM 1976). This would provide additional evidence of fundamental differences between the mechanisms of replication of plant viruses and of viroids. The reader interested in the use and specific problems of protoplasts and cell cultures in viroid and plant virus research is referred to a recent review by MÜHLBACH (1982).

4.4 The Presumed DNA-Dependence of Viroid Replication

The DNA dependence of viroid RNA replication has previously been postulated from inhibition studies with actinomycin D, which is known to inhibit DNA-dependent RNA-transcription when this antibiotic was applied to leaf strips (DIENER and SMITH 1975) and nuclei prepared from PSTV-infected tomato leaves. However, as actinomycin inhibits the biosynthesis of all RNA species including viroid RNA, a non-specific effect of this inhibitor cannot be ruled out completely. Therefore, these results would only allow exclusion of involvement of pre-existing RNA-dependent RNA replicases during viroid relication, because these are actinomycin D-resistant as demonstrated in several systems infected with conventional RNA plant viruses (SÄNGER and KNIGHT 1963, TAKEBE and OTSUKI 1969, WHITE and MURAKISHI 1977).

Additional evidence for the DNA dependence seemed to come from molecular hybridization experiments with ^{125}I-labelled viroid RNA, from which the existence of cellular DNA sequences complementary to viroid RNA not only in viroid-infected (SEMANCIK and GEELEN 1975, HADIDI et al. 1976) but also in healthy host and non-host plant species (HADIDI et al. 1976) was inferred. Furthermore, from the observed lack of any significant differences in the properties of DNA-dependent RNA polymerase fractions from healthy and from CEV-infected *Gynura* leaves it was concluded that probably no de novo synthesis of polymerase is required for CEV replication and that it is accomplished in vivo via a DNA template (GEELEN et al. 1976). The reinvestigation of the presumed existences of this viroid-specific DNA under more rigorous conditions of molecular hybridization revealed that no such DNA is present in infected or healthy plant tissue (ZAITLIN et al. 1980, ROHDE and SÄNGER 1981, HADIDI et al. 1981). It was also established that the earlier results had been caused by cellular RNA's, especially by breakdown products of ribosomal RNA contaminating the viroid probes used for hybridization (BRANCH et al. 1981). Thus, the originally presumed DNA dependence of viroid replication has lost its experimental basis.

It should be mentioned in this context that conventional molecular hybridization techniques may not provide conclusive evidence for the claims which are derived from them. The circularity and high degree of self-complementarity of the viroid molecule and its complementary form (SÄNGER et al. 1976, GROSS et al. 1978) require very specific experimental conditions to obtain reliable hybridization data (ROHDE and SÄNGER 1981). In addition, the absence of any trace of contaminating cellular RNA species in the viroid probes used for hybridization must be rigorously demonstrated by appropriate controls and by fingerprint analysis (GROSS et al. 1977).

4.5 In Vitro Transcription of Viroid RNA by DNA-Dependent RNA Polymerase II of Plant Origin

From the specific inhibition of viroid replication in protoplasts by an intracellular concentration of 10^{-8} M α-amanitin (MÜHLBACH and SÄNGER 1979; see Sect. 4.3), it was assumed that the DNA-dependent RNA polymerase II of the host cell might be involved in viroid replication. This assumption was tested in in vitro experiments (RACKWITZ et al. 1981) for which this enzyme had been purified from callus or green leaf tissue of the wild tomato *Lycopersicon peruvianum* (L.) Mill. and from wheat germ. The purification followed essentially the procedure developed by JENDRISAK and BURGESS (1975) for wheat germ. It was found with both enzymes that standard single-stranded DNA templates are transcribed less efficiently than double-stranded DNA templates and that the transcription of single-stranded DNA templates is considerably improved in the presence of oligo-RNA primers.

Synthetic and natural RNA's also served as templates for transcription, thus confirming previous results on the same enzyme from pea and cauliflower (SASAKI et al. 1974a, b, 1976). Tobacco mosaic virus (TMV) RNA and soluble RNA (5S-RNA and tRNA) from yeast were relatively poor templates, whereas the synthetic template/primer poly(rA)/oligo(rU) was transcribed with high efficiency. Most surprisingly, however, of all natural RNA templates tested, viroid RNA was transcribed with the highest efficiency by DNA-dependent RNA polymerase II from tomato and wheat germ. When the three viroid "species" PSTV, CEV and CSV were compared with respect to their template efficiency, PSTV RNA was transcribed best by the tomato polymerase II. PSTV was propagated in and isolated from tomato (*Lycopersicon esculentum* Mill. cv. "Rentita") whereas for CEV *Gynura aurantiaca* DC and for CSV *Chrysanthemum morifolium* Ramat. cv. "Yellow Delaware" were used as host plants. Although overall

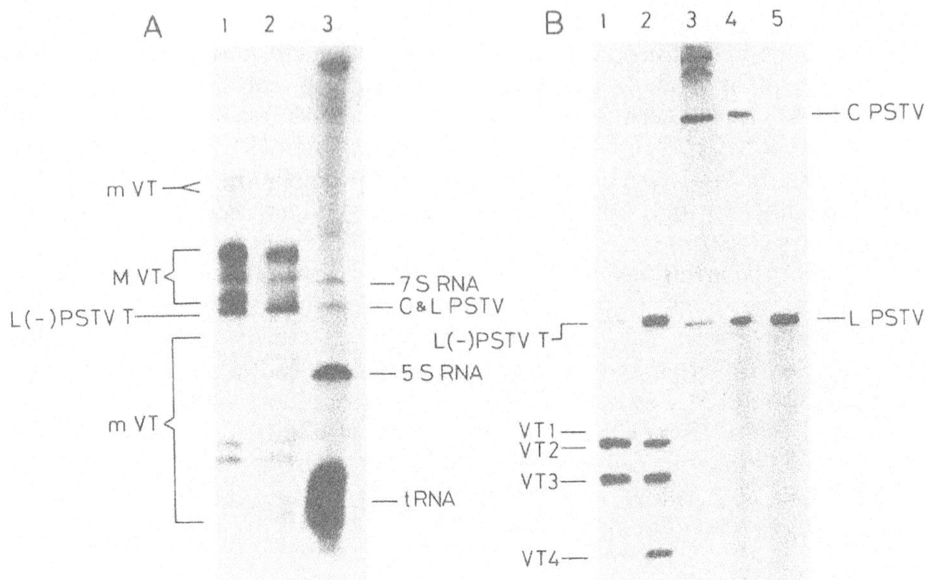

Fig. 33 A, B. Analysis by polyacrylamide gel electrophoresis (PAGE) and autoradiography of ^{32}P-labelled RNA's transcribed from viroid (PSTV) RNA by DNA-dependent RNA polymerase II from tomato tissue and wheat germ. Marker RNA's were ^{32}P-labelled, LiCl-soluble RNA from PSTV-producing potato callus and ^{125}I-labelled PSTV. **A** PAGE on a 5% polyacrylamide slab gel (140 × 190 × 1.5 mm) at 10 °C and in the presence of high salt (89) mM) TEB buffer and 8 M urea. *1* Transcriptional products of wheat-germ RNA polymerase II. *2* Transcriptional products of tomato RNA polymerase II. *3* ^{32}P-labelled cellular RNA from PSTV-infected tomato callus. *MVT* major viroid (PSTV) transcripts; *mVT* minor viroid (PSTV) transcripts. *L* (−) *PSTV T* linear viroid (PSTV) transcript. **B** PAGE on a 5% polyacrylamide gel at 40 °C in the presence of low salt (22.3 mM) TEB butter and 8 M urea. *1* Transcriptional products of wheat-germ enzyme. *2* Transcriptional products of tomato enzyme. *3* ^{32}P-labelled cellular RNA. *4* Circular (CPSTV) and linear (LPSTV) ^{125}I-labelled PSTV RNA. *5* Linear ^{125}I-labelled PSTV RNA. VT1–VT4, viroid-(PSTV) RNA-directed RNA transcripts

transcription was about the same when circular and linear viroid PSTV RNA molecules were used as templates, the pattern of product formation varied slightly between different template preparations.

The viroid RNA-directed ^{32}P-labelled transcription products were analyzed on 5% polyacrylamide gels containing 8 M urea. When the electrophoresis was carried out in high-salt TEB buffer (89 mM Tris/EDTA/borate, pH 8.3) and at low gel temperature (10 °C), both enzymes gave a very similar RNA product pattern consisting of several distinct bands (Fig. 33 A, lanes 1, 2). Surprisingly, transcripts are visible which are apparently larger than the corresponding viroid template itself (Fig. 33 A, MTV). However, no reliable conclusions about the actual size of the transcripts can be drawn from this pattern, because the gel provides denaturing conditions only for conventional RNA's, whereas viroid molecules, the corresponding transcripts and their presumptive hybrid complexes with the template remain largely undenatured. If the same products were analyzed in low-salt (22 mM) TEB buffer and at high gel temperature (40 °C),

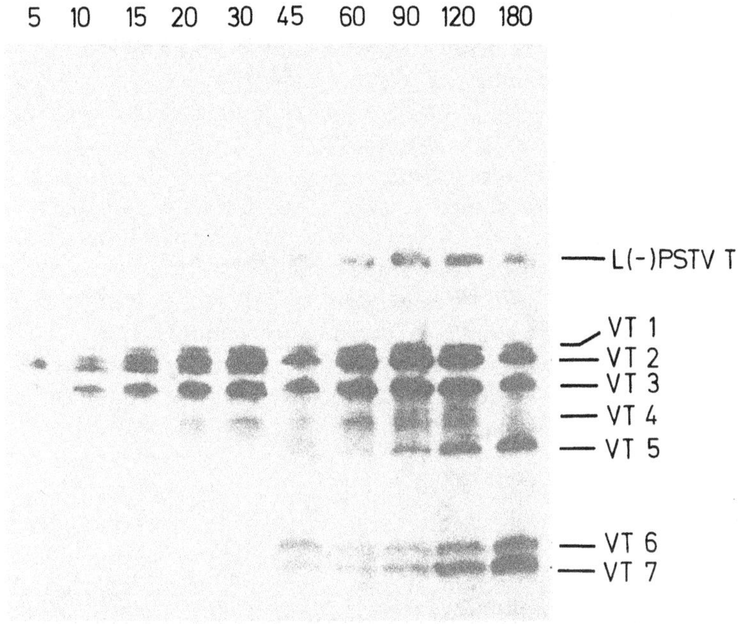

Fig. 34. Inhibition of viroid (PSTV)-directed RNA transcription by α-amanitin as analyzed by PAGE. Assay conditions were as in Fig. 33 in the absence (lane c) and presence of various concentrations of α-amanitin as indicated on top of the lanes, where the numbers correspond to the negative $\log_{10}$ of the molar concentrations of α-amanitin present in the assay. The assay mixture was incubated for 2 h and 10-μl aliquots were analyzed on the gel as described in legend to Fig. 1b. *C* control without α-amanitin; *M* circular and linear ^{125}I-labelled PSTV marker RNA; *VT1–VT9* viroid (PSTV) RNA-directed RNA transcripts

where viroid RNA is fully denatured, circular (C PSTV) and linear (L PSTV) viroid molecules are clearly separated (Fig. 33B, lanes 3–5), and it became evident that the largest transcripts from both polymerases had the same migrating properties and hence the same size as the linear viroid molecule. That this RNA is a genuine transcript and not due to terminal labelling of linear viroid molecules was demonstrated by the absence of transcription on omitting one of the labelled triphosphates, and by molecular hybridization. In addition to the full-length copies, several smaller distinct viroid transcription products were found.

The detailed analysis of the transcription kinetics showed that the different transcriptional intermediates are not produced in a strictly sequential order so that they could be artifacts of the in vitro system. The highly specific interaction between viroid RNA and polymerase II was substantiated by the inhibition of transcription at 10^{-8} α-amanitin (Fig. 34) and by binding studies between ^{125}I-labelled PSTV RNA and tomato polymerase II using the cellulose filter technique. Viroid RNA was found to form a binary complex with polymerase II, to compete with DNA for the template binding site on the enzyme and to strongly inhibit DNA-directed RNA-synthesis. Additional evidence for the specific viroid transcription came from appropriate molecular hybridization experiments using the Northern blot technique according to ALWINE et al. (1977). Viroid-relevant conditions were applied in order to be able to discriminate between (+) and (−) viroid RNA and to prevent self-annealing during hybridization (see ROHDE and SÄNGER 1981).

Together with the previous in vivo inhibition of viroid replication by α-amanitin, all these data suggested that the first step in the viroid replication process – the copying of the infecting viroid RNA molecule – could proceed with the aid of the DNA-dependent RNA polymerase II of the host plant. The finding that a DNA-dependent polymerase can quite efficiently accept and transcribe viroid (+) RNA into its complementary (−) RNA form may seem surprising. However, in view of the previously described DNA-like features of the viroid RNA, it is intelligible from a mechanistic point of view that the viroid molecule is capable of directing DNA-dependent RNA polymerase for its replication.

From the appearance of the distinct viroid-specific RNA transcripts during synthesis of the full-length viroid copies, one may conclude that the secondary structure of the viroid molecule causes the enzymes to pause at specific sites analogous to the transcription pattern of $Q\beta$ midivariant MDV-1 RNA by $Q\beta$ replicase and that specific recognition sites must exist in the viroid RNA molecule for the RNA polymerase. It is still unclear why the polymerase stops after one round of in vitro transcription, whereas transcription products larger than the viroid molecule itself are detectable in vivo. Further experiments are needed to decide whether the smaller transcripts which accumulate later in transcription are caused by a site-specific nuclease activity possibly present in the polymerase preparation, or whether they are transcripts of specifically nicked viroid molecules, in which case the transcription would stop at the nick. The transcription products should finally be analyzed in detail by biochemical techniques, preferably by sequence analysis which would hopefully allow determination of the signals responsible for initiation, pausing and termination of viroid transcription. It will also be of interest to analyze the transcription of viroid RNA by the DNA-dependent RNA polymerases I and III.

The previous observation that viroids replicate and accumulate in the nucleus (DIENER 1971, SÄNGER 1972, TAKAHASHI and DIENER 1975) and most efficiently in rapidly dividing plant tissue (MÜHLBACH and SÄNGER 1981) may now be related to the fact that, due to its normal function of synthesizing genomic messenger RNA's, polymerase II is highly active in the nuclei of dividing cells, an it would hence be available there for intensive viroid replication. Viroid pathogenesis could also be explained in these terms. As the infecting viroid molecule directs the nuclear DNA-dependent RNA polymerase II to perform its "selfish" replication the synthesis of genomic messenger RNA's of the host cell may become inhibited or repressed. Thus, normal differentiation would be disturbed, leading to the expression of the typical symptoms of leaf malformation and retardation of growth.

It should be mentioned here that viroid RNA is also accepted as template by different polymerases from prokaryotes in vitro. Thus, PSTV RNA is recognized by bacteriophage $Q\beta$ replicase in the presence of either Mg^{++} or Mn^{++}

ions. Although the in vitro reaction products were heterogeneous, the largest ones had an apparent molecular weight of about 100,000 dalton (OWEN and DIENER 1977). DNA-dependent RNA polymerase from *E. coli* is able to transcribe PSTV RNA into RNA copies of apparently full length provided that a viroid-specific, tailor-made DNA primer consisting of 15 nucleotides is used in transcription (ROHDE et al. 1982). When DNA-dependent DNA polymerase from *E. coli* is used a viroid-specific cDNA can be obtained (ROHDE et al. 1982) but transcription is less efficient than with reverse transcriptase (ROHDE et al. 1981a, b).

4.6 In Vitro Transcription of Viroid RNA by RNA-Dependent RNA Polymerase Purified from Healthy Host Tissue

Although the presently available in vivo and in vitro data strongly suggest that DNA-dependent RNA polymerase II is involved in viroid replication, the possible participation of other RNA-synthesizing host enzymes cannot be excluded. One such candidate is the RNA-dependent RNA polymerase of plants. Such enzyme activities have been detected in several healthy plant species (DUDA et al. 1973, DUDA 1976, IKEGAMI and FRAENKEL-CONRAT 1978, 1979b, ROMAINE and ZAITLIN 1978, LAZAR et al. 1979, TAKANAMI and FRAENKEL-CONRAT 1982) including tomato, a well-known host plant for many viroids (BOEGE and SÄNGER 1980). Therefore, one can safely assume that this enzyme is generally present in tissue of all higher plants. Interestingly, an RNA-dependent RNA-polymerase activity is increased about two-to six fold after infection by various RNA plant viruses (FRAENKEL-CONRAT 1979). Especially cucumber mosaic virus has been reported to elicit particularily high enzyme activity in cucumber cotelydons (GILLILAND and SYMONS 1968, MAY et al. 1969, 1970, MAY and SYMONS 1971). It thus appears that the upon virus infection these enzymes are taken over for the replication of the viral RNA, a conjecture which is still controversial. In analogy to the well-characterized RNA replicase from *Escherichia coli* infected with RNA bacteriophage $Q\beta$ (KAMEN 1975), the RNA-dependent RNA polymerase in virus-infected plants has therefore also been designated "RNA-replicase". Although the catalytical and biochemical properties of the RNA replicase seem to be very similar to those of the RNA-dependent RNA polymerase pre-existing in the uninfected plant cell, the actual relationship between these two enzyme activites is still unresolved. Therefore, the enzyme from healthy plants should be called RNA-dependent RNA polymerase. Its biological role in plant cell metabolism is still unknown. It has been suggested that it may serve to synthesize double-stranded RNA's possiblfy with regulatory functions (IKEGAMI and FRAENKEL-CONRAT 1979b).

In a recent study it was found that RNA-dependent RNA polymerase from virus-free healthy tomato leaf tissue accepts PSTV RNA as template and produces apparently full-length PSTV copies in vitro as judged from their electrophoretic mobility on denaturing 5% polyacrylamide gels and from molecular hybridization on Northern blots under stringent conditions (BOEGE et al. 1982). Viroid transcription requires the presence of Mn^{2+} and/or Mg^{2+} ions and is not inhibited by concentration of 10^{-5} M α-amanitin. The percentage of viroid template copying under optimal conditions was found to vary around 0.1%. Of special interest is the appearance of high molecular weight products which barely enter the 2.5% stacking gel or the 5% resolving gel. With other RNA templates, these enzymes were found to produce extremely heterodisperse transcription products of low molecular weight irrespective of the nature and size of the template and the source from which they were isolated (ROMAINE and ZAITLIN 1978, DUDA 1976, CHIFFLOT et al. 1980). Therefore the full-length viroid copy is the first well-defined and homogeneous product reported to be synthesized in vitro by an RNA-dependent RNA polymerase from healthy plants. A comparable synthesis of distinct transcription products has so far only been reported for the RNA-dependent RNA replicase from barley leaves infected with brome mosaic virus (HARDY et al. 1979).

Regarding the relation between the structure of the RNA templates and their transcription efficiency, it was previously found that RNA-dependent RNA polymerase transcribes single-stranded RNA more efficiently than double-stranded RNA (FRAENKEL-CONRAT 1979, IKEGAMI and FRAENKEL-CONRAT 1978, 1979a, b, BOEGE and SÄNGER 1980). In view of these results, the observed viroid transcription into copies of full length is not too surprising, because viroid RNA displays unique structural properties which might specifically predispose this molecule as an efficient template. Viroid RNA exhibits features of both double-stranded and single-stranded RNA's, which may be considered as an optimal compromise between structural stability and functional flexibility. Thus a high degree of self-protection is combined with properties allowing efficient and complete transcription by RNA-dependent RNA polymerase.

The synthesis of full-length viroid RNA copies might indicate that viroid transcription is initiated and terminated at specific sites which are most probably loop regions. One may anticipate that full-length transcription is also favoured by the comparatively small size of the viroid RNA molecule with a chain length of 359 nucleotides only and by the relatively high nuclease resistance of the largely double-stranded template as well as its RNA copy. The elucidation of the details of this process and of the nature of the transcripts larger than unit length requires further biochemical analysis.

4.7 Properties of RNA Intermediates of Viroid Replication

The first attempts to demonstrate complementary viroid RNA's as intermediates of viroid replication in CEV-infected *Gynura* leaves (SEMANCIK and GEELEN 1975) or in PSTV-infected tomato leaves (HADIDI et al. 1976) by molecular hybridization have failed and indicated the presence of viroid complementary DNA. Therefore, it was assumed the viroids replicate via DNA intermediates. However, more rigorous hybridization experiments have shown that no DNA complementary to PSTV is present in infected or uninfected tomato tissue (ZAITLIN et al. 1980, BRANCH and DICKSON 1980, HADIDI et al. 1981). Thus, the involvement of complementary RNA molecules in viroid replication as postulated from its α-amanitin sensitivity (MÜHLBACH and SÄNGER 1979) and from in vitro transcription experiments (RACKWITZ et al. 1981) became finally accepted. The corresponding RNA's have been detected by molecular hybridization in plants infected with either CEV (GRILL and SEMANCIK 1978, GRILL et al. 1980) or PSTV (OWENS and CRESS 1980, ROHDE and SÄNGER 1981, BRANCH et al. 1981), whereas in extracts from healthy plants no viroid complementary RNA was found. From characterization studies it was concluded that the PSTV complementary RNA strands are heterogenous in size, forming either seven (ROHDE and SÄNGER 1981) or four (BRANCH et al. 1981) discrete bands in polyacrylamide gels. The latter four bands were shown to contain RNA molecules approximately 700, 1,050, 1,500 and 1,800 nucleotides long, thus representing multimers of the PSTV unit length RNA with 350 nucleotides. They seem to be present in complexes containing extensive double-stranded regions, but after treatment with RNases under conditions favouring digestion of single-stranded regions, the high molecular weight unit-length minus strands appear (BRANCH et al. 1981). Fingerprint analysis of hybridized ^{125}I-labelled PSTV following recovery from the hybrids demonstrated that all regions of the PSTV molecules are represented in the complementary RNA strands (ZELCER et al. 1982). Further characterization of the double-stranded structures indicated that monomeric circular of linear PSTV strands are complexed with multimeric complementary RNA strands and the synthesis of the double-stranded complexes increases simultaneous with to synthesis of the single-stranded PSTV-RNA (OWENS and DIENER 1982). From all these data it appears that these complexes reflect the structure of viroid replication intermediates in which the plus strand is present at unit length and the minus strands are composed of approximately viroid-length tandem repeats. Interestingly, circular double-stranded RNA has been detected in PSTV-infected tomatoes by electron microscopy (FRENCH et al. 1982) but their origin and nature is still uncertain.

All these data would suggest that the mechanism for the production of long minus viroid strands is based on the rolling-circle model previously proposed to explain the replication of certain viral RNA's (BROWN and MARTIN 1965) in particular the plus strand synthesis of the DNA bacteriophage ΦX 174 (GILBERT and DRESSLER 1968). The specific α-amanitin sensitivity of viroid synthesis in vivo (MÜHLBACH and SÄNGER 1979) and the efficient in vitro transcription of viroid RNA suggest that the multisubunit DNA-dependent RNA polymerase II enzyme system is involved in the viroid replication process (RACKWITZ et al. 1981). The efficient in vitro viroid (PSTV) transcription by RNA-dependent RNA polymerase from healthy leaves suggests that this enzyme might also take part in viroid replication (BOEGE et al. 1982).

4.8 Possible Mechanisms of Viroid Pathogenesis

Viroids are composed entirely of RNA and their small genomes (240–380 nucleotides) are characterized by the absence of the protein synthesis initiation triplet AUG from both the plus and minus strand sequences. Moreover, PSTV and CEV have failed to stimulate protein synthesis in vitro and no viroid-specific proteins could so far be detected in vivo. All these data suggest that viroids

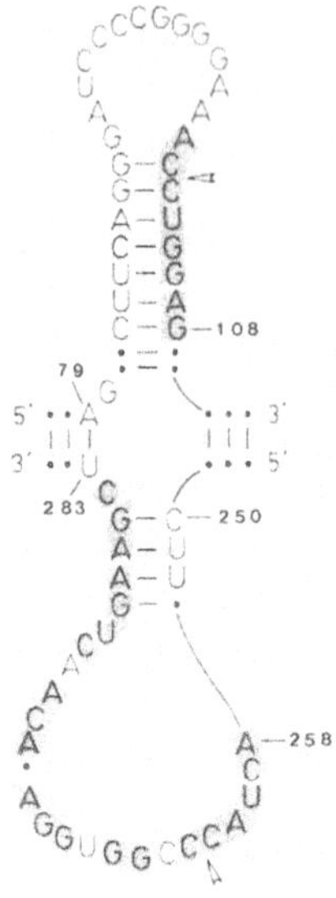

Fig. 35. Possible palindromic arrangements of a highly conserved region in the center of viroid secondary structure. *Black dots* indicate different nucleotides in homologous positions, whereas a *line* in the nucleotide sequence indicates major differences. Nucleotides identical in PSTV, CEV and CSV are presented in a palindrome-like structure. Nucleotide numbers refer to PSTV. Shaded nucleotides are homologous with the 5′ end of U1a RNA from chicken shown in detail for nucleotides 258–282 only, in Fig. 36. The stem region of the hairpin pointing downwards is longer and more stable for the three individual viroids than that shown above for the nucleotides identical in the three viroids

may not code for any protein products, that they become replicated by the nucleic acid synthetizing machinery pre-existing in the host cell and that viroid pathogenesis may be based on the direct interference of the viroid RNA molecules themselves with the nucleic acid metabolism of the host cell. In fact, it has been speculated by numerous authors (SEMANCIK and WEATHERS 1972a, ROBERTSON and DICKSON 1974, REANNY 1975, DICKSON and ROBERTSON 1976, ROBERT 1978, DIENER 1978, DICKSON 1979, CRICK 1979, DIENER 1979a, b, ROBERTS 1980) that viroids might be some escaped regulatory RNA that became replicable. This notion has gained support from the comparison of the established viroid sequences with the sequence of the small stable nuclear U1a RNA (DIENER 1981a, b, 1982, DICKSON 1981, GROSS et al. 1982) which is assumed to play a key role in the splicing of eukaryotic mRNA (LERNER et al. 1980). U1a RNA is 165 nucleotides long (REDDY et al. 1974) and has been strictly conserved in its primary sequence during evolution (BRALANT et al. 1980). Its proposed function in the splicing process is based on the considerable sequence complementarity of a large part of its 5′ end to the consensus sequence at the 5′- and 3′-proximal regions of the intervening sequences (introns) of mammalian genes. Therefore, it has been assumed that due to this complementarity a U1a RNA molecule could simultaneously form base pairs with both ends of the intron so that the two coding regions of the mRNA precursor are brought so close together that splicing could occur (LERNER et al. 1980, ROGERS and WALL 1980).

Most interestingly, the A-C-C-U-G sequences of U1a RNA which might be involved in the recognition of pre-mRNA splice junctions and similar sequences (e.g., A-C-C-C-G) occur in several locations of all viroids sequenced so far with the exception of ASBV. One of these homologous regions (nucleotides 101–108 in PSTV) is part of the stem region of the "secondary" hairpin I (Figs. 21, 29) of a metastable viroid conformation. A more extended homology,

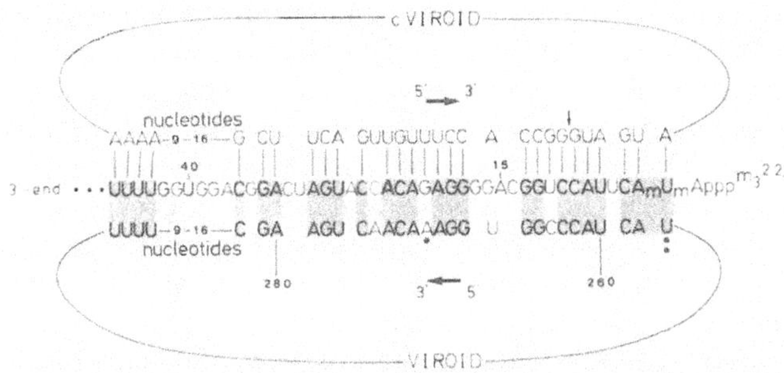

Fig. 36. Homology of a highly conserved viroid sequence with, and complementarity of the corresponding complementary viroid (cVIROID) to, the 5′-terminal sequence of U1a RNA from chicken. Nucleotide numbers in the viroid sequence (*below*) refer to PSTV. Homology and complementarity to U1a RNA are presented in detail for nucleotides 258–282 only. For a possible significance of the uridine stretch (nucleotides 43–46) of U1a RNA see text. *Shaded* nucleotides indicate homology between the three viroids and the 5′ end of U1a RNA. Possible base pairs between a complementary viroid and the 5′ end of U1a RNA are indicated by *vertical bars*. *Vertical arrow* points to a hypothetical splice junction *A** indicates that a CEV mutant contains U in this position; *U*** indicates that CSV contains A in this position

however, is found from nucleotides 258 to 282 (PSTV) in the opposite region of nucleotide 90 in of the viroid secondary structure (Figs. 27, 30). This latter region, homologous to the 5′ end of U1a RNA, is part of a very stable double-stranded segment of the in vitro secondary structures of all these viroids. Secondary structure in vivo, however, may be different from that shown in Fig. 27. If interaction with cellular components of the host plant favored and stabilized the highly conserved "secondary" hairpin I, the opposite region in Fig. 27 could also form a hairpin, thus giving rise to a palindrome-like structure (Fig. 35). In this case, the extended sequence homologous to U1a RNA (nucleotides 258–282 in PSTV) becomes mainly single-stranded (Fig. 35). Both stretches of shaded sequences in the palindrome (Fig. 36) are homologous to the region of U1a RNA believed to the directly involved in RNA splicing.

Since viroids replicate via complementary RNA molecules, these replicative intermediates would also be able to form a corresponding palindrome-like structure which would be able to form base pairs with U1a RNA. Fig. 36 summarizes (a) the extensive sequence homology between the 5′ end of U1a RNA and a conserved region of the three viroids, and (b) the extensive base-pairing possible between U1a RNA and the corresponding sequence of a complementary viroids. The vertical arrow points to a hypothetical splicing site in the complementary viroid molecule. Moreover, viroids contain a conserved oligo(uridine) sequence which corresponds to an oligo(uridine) stretch in U1a RNA. Consequently, complementary viroids contain an oligo(adenosine) sequence, also occurring frequently in introns, 50–80 nucleotides upstream of the splice site of eukaryotic pre-mRNA.

All these considerations reveal the possibility of interference between viroids and mRNA synthesis and processing which would explain the pathogenic action of viroids. First, the in vitro transcription experiments (Rackwitz et al. 1981) have shown that plus (and minus) viroids RNA molecules can act as potent competitors for the genomic DNA templates of the host by occupying the DNA-dependent RNA polymerase II and directing it towards viroid replication, so that mRNA synthesis is disturbed. Second, pathogenicity could also be exerted at the subsequent stages of mRNA processing. The viroid RNA's could act as competitors for U1a RNA and mRNA precursors, respectively, in the splicing process, i.e., the correct recognition and alignement of intron-exon junctions, the removal of introns and the ligation of the corresponding exons to give functional mRNA molecules.

From the existence in the host cell of linear and circular viroid molecules it may be concluded that the two characteristic steps in splicing namely nucleolytic cleavage and subsequent ligation are also essential steps in viroid maturation. The insufficient genetic information content of viroids would indicate that also for this process the machinery pre-existing in the host cell is utilized. The recent finding that RNA ligase can be isolated from wheat germ, which is able to circularize linear RNA by the formation of a 2′-phosphomonoester, 3′,5′-phosphodiester linkage (KONARSKA et al. 1981) strongly suggests that other higher plants, including viroid hosts, might also contain RNA ligases which could produce viroid circles from the corresponding linear precursors molecules. The wheat germ ligase, in fact, is capable of transforming linear viroid RNA into viroid circles in vitro (BRANCH et al. 1982, KIKUCHI et al. 1982).

In conclusion one may postulate that viroids are replicated and exert their pathogenic functions by imitating the genomic DNA and the splice sites of the host. The differences in viroid pathogenicity could be explained by sequence variations which might result in affinity changes for different classes of mRNA precursors leading to the observed differences in symptom expression.

A note of caution should be added here in order to avoid too far reaching conclusions. Although intervening sequences of plant genes with splice junctions similar to those of animal genes have been detected (SUN et al. 1981, JENSEN et al. 1981) the U1 RNA from plants has not yet been found. Moreover, none of the above mentioned U1 RNA sequences is present in the plus or minus strand of ASBV (SYMONS 1981) so that this viroid must be assumed to be of different origin and to exert its pathogenic function in a different way. Finally, the possible involvement of U1 RNA in splicing is, in itself, still a matter of conjecture (TRAPNELL et al. 1980, SHARP 1981, ZIEVE 1981).

Approaching the problem from a more biological standpoint, it must be emphasized that pathogenicity is a complex biological property and that the expression of disease symptoms is not only specified by the genome of the viroid isolate but also by the genome of the host plant. Thus not only do severe, intermediate, and mild PSTV strains exist, but different tomato cultivars may respond with severe (cv. Rutgers), mild (cv. Rentita), or practically no symptoms (cv. Hilda 72) upon infection with one and the same PSTV isolate, although replication and accumulation is about the same in all three host plants (MÜHLBACH and SÄNGER 1977). Moreover, symptom development is dependent on a threshold concentration of the viroid, as is demonstrated after inoculation with highly dilute inocula. In this case pronounced symptoms may be recognized only after an extended period or in the newly growing axillary buds after decapitation, which can be directly related to the increased viroid concentration in the corresponding tissue. Finally, environmental factors such as the nature, intensity, and length of illumination and the temperature during plant growth may greatly influence the response of the host plant to viroid infection and also the intensity of viroid replication itself (SÄNGER and RAMM 1975). It can be assumed, therefore, that the interaction between viroid and host cell and the mechanism of pathogenicity will turn out to be much more complex than it appears at first sight at present.

5 The Possible Origin of Viroids

One of the most interesting questions arising during discussions on viroids centers around their possible origin. However, the lack of solid facts along these lines leads to the simple answer that the origin of viroids is still unknown. It is not surprising, on the other hand, that this uncertain situation has evoked many speculations.

Before the unique structure of viroids had been established, the possibility was considered that viroids might have originated from conventional viruses by degeneration (DIENER 1974). Alternatively they were regarded as a kind of "primitive viruses that had not yet developed the genetic sophistication to code for one or more capsid proteins capable of assembling into a protective capsid" (DIENER 1979a, b). The erroneous finding that sequences complementary to PSTV RNA exist in the DNA from several uninfected host species (HADIDI et al. 1976) suggested that PSTV originated from host genetic material. Since the affinity between PSTV and the plant DNA seemed to be higher in several solanaceous species than in plants of more distant phylogenetic relationship the idea arose that PSTV originated from genes normally present in certain solanaceous plant species (HADIDI et al. 1976, DIENER 1978). When the absence of any viroid-specific DNA in healthy or viroid-infected was unequivocally established (BRANCH and DICKSON 1980, ZAITLIN et al. 1980, ROHDE and SÄNGER 1981) this idea was, of course, no more tenable.

In light of present knowledge five additional hypotheses on viroid origin have been discussed:

1. Viroids might have specifically originated from low molecular weight nuclear RNA's which are believed to perform normal regulatory functions within the cell nucleus. Mutation of these RNA's to allow their replication or/and by chance introduction into a foreign plant species, especially into cultivated ones, could have changed these normal RNA's into pathogenic ones (DIENER 1974, 1979a, b).

2. Viroids may have originated during RNA splicing processes from spliced out and circularized introns, which aquired the ability to escape degradation and to become efficiently replicated by the nucleic acid synthetizing machinery of the host cell (ROBERTS 1978, CRICK 1979, DIENER 1979a, b, 1981a, b, GROSS et al. 1982). This hypothesis gained some support from the finding that a surprising sequence homology between U1 RNA and PSTV RNA exists. U1 RNA is a small nuclear RNA (REDDY et al. 1974) which has been highly conserved during evolution (BRANLANT et al. 1980). It is believed that it is involved in

mRNA processing because its 5′ end is complementary to the intron-exon junctions of pre-messenger RNA's. Hence, the postulated function of U1 RNA is based on their potential ability to form base pairs with these junctions which are thus brought into close and correct contact so that the subsequent splicing out of the intron and the furter processing of mRNA can proceed (LERNER et al. 1980, ROGERS and WALL 1980, OSHIMA et al. 1981). From this sequence homology between the (+) PSTV RNA and the 5′-end of U1 RNA it follows that the PSTV (−) RNA must contain sequences which are homologous to the intron-exon junctions. These sequences may therefore be considered as relics of the hypothetical origin of viroids as "escaped introns".

One testable corollary of this hypothetical model would be that the sequences of all viroids should have in common sequences homologous to the 5′-end of the U1 RNA. Most interestingly such a sequence is present in PSTV (nucleotides 258 to 282) CSV (nucleotides 257 to 280) CEV (nucleotides 263 to 290) and CCCMV (in RNA 1 fast nucleotides 175 to 194 and in RNA 1 slow nucleotides 216 to 232).

The general principles of this homology and complementarity have been discussed in Section 4.8 and need not be repeated here. It should only be mentioned again that in addition, all viroids contain a conserved oligo(uridine) sequence which corresponds to an oligo(uridine) stretch in U1a RNA. Consequently, complementary viroids contain an oligo(adenosine) sequence, also occurring frequently in introns, 50–80 nucleotides upstream of the splice site of eukaryotic pre-mRNA.

Although the "escaped-intron" hypothesis gained some support from the finding that intron-derived circular RNA exists in mitochondria of yeast (HALBREICH et al. 1980) and in nuclei of *Tetrahymena thermophila* (GRABOWSKI et al. 1981) a note of caution should be added here. None of the U1 RNA sequences is present in the plus or minus strand of ASBV (SYMONS 1981) so that it must be assumed that this viroid is of different origin. Moreover, there is no published evidence as yet that U1 RNA exists in plants and the involvement of this RNA in splicing is still largely a matter of conjecture. However, intervening sequences with splice functions similar to those of animal origin have been detected in plants in the genes for bean phaseolin (SUN et al. 1981) and for soybean leghaemoglobin a and b (JENSEN et al. 1981). Thus, future work will have to show whether or not this very suggestive "escaped intron" hypothesis of the origin of viroids is applicable.

3. Viroids may have originated from the small circular RNA components of certain recently discovered plant viruses with multipartite genomes (see Sect. 6). However, these viroid-like RNA's (which are also called "virusoids") have only their overall secondary structure, size and circularity in common with the true viroids. The detailed analysis of several thermodynamic and kinetic properties has revealed that the "virusoids" are clearly distinct from viroids with regard to the Gibbs free energy of structure formation, the mean T_m value cooperativity and the hydrodynamic stiffness. Since they are much more similar in all these parameters to calculated random sequences than the unique viroid RNA molecules this hypothesis becomes less probable (RIESNER et al. in preparation).

4. According to a hypothesis which has been put forward recently, viroids and RNA viruses might form a novel class of RNA elements that exchange genetic information between eukaryotic cells (ZIMMERN 1982). This speculation is based on the previously discussed conjectures (see preceding paragraph) which connect the biosynthesis and structure of viroids with the splicing machinery of the eukaryotic cells and the existence of intron-derived circular RNA's in mitochondrial of yeast and nuclei of *Tetrahymena*.

The hypothetical novel cellular elements would consist of two interacting classes of RNA molecules which are called "signal RNA's" and "antenna RNA's". The "signal RNA's" can be excised and/or inserted directly into another RNA and they can also be transported across membranes, between intracellular compartments and between cells. The "antenna RNA's" are the target of the "signal RNA's" and the fused product of both is called a "fuson". Based primarily on inductive arguments this unorthodox system has been advanced to provide a mechanism for cellular regulation by which the cellular genetic control via RNA splicing and the spatial arrangement of cells in an organism could be linked. This mechanism would have advantage that once the signal and antenna RNA's have been transcribed the system could operate independently of genomic DNA. Moreover it could be a means by which the homogeneity of dispersed repeated genes is maintained.

Regarding the origin of viroids, according to this highly speculative hypothesis they are considered to represent aberrant "signal RNA's" or small extensively deleted "fusons" which have retained little more than an origin of replication and a splice junction and that are no longer under the control of copy number.

5. Viroids might have evolved as RNA pathogens through the infection of higher plants by prokaryotes. This hypothesis on viroid origin is based on the finding that not only eukaryotic but also prokaryotic DNA-dependent RNA polymerase and DNA-dependent DNA polymerase from *E. coli* are able to transcribe viroid RNA quite efficiently in vitro into viroid-complementary RNA and DNA copies, respectively (ROHDE et al. 1982). It is conceivable, therefore, that viroids may have originated in prokaryotes either as DNA or RNA and that viroid host plants originally acquired viroid-specific sequences through infection by prokaryotes which are the most common causal agents of numerous plant diseases.

The transfer of nucleic acid sequences from a plant pathogen to its host is, for example, well established in the case of infections by virulent strains of *Agrobacterium tumefaciens*. They are known to contain T-plasmids which induce so-called "Crown gall"-tumors in dicotyledoneous plants (DRUMMOND 1979). Part of the tumorigenic T-DNA becomes stably integrated into the plant cell genome (VAN MONTAGU and SCHELL 1982) and the expression of these sequence is under control of the DNA-dependent RNA-polymerase of the host (WILLMITZER et al. 1981). Secondly, the prokaryotic plasmid pBR 322 contains "TATA" box-like sequences that promote specific initiation of in vitro transcription by eukaryotic DNA-dependent RNA polymerase II (SASSONE-CORSI et al. 1981) the enzyme which is evidently also involved in viroid replication (MÜHLBACH and SÄNGER 1979, RACKWITZ et al. 1981). Thus, one could postulate that certain plants may have aquired specific DNA sequences from prokaryotic plant pathogens either by integration or in the form of extrachromosomal and autonomously replicating elements. Alternatively, viroid RNA might have already existed as such in a bacterial or fungal plant pathogen prior to the uptake by the susceptible plant. The finding that in the presence of an appropriate primer viroid-complementary DNA synthesis is catalyzed in vitro by *E. coli* DNA polymerase I might be taken as an indication that viroid-specific or viroid-like nucleic acids might be found in non-plant systems provided that such elements would become replicated and escape the regulation through the corresponding hosts once they are introduced.

Without any doubt all these hypotheses will catalyze new experimental approaches not only in viroid research but also in the field of mRNA processing and regulation. Even in case that they will not succeed in clarifying the actual origin of viroids they will certainly contribute to the further elucidation of their enigmatic functions. As long as speculations are partly rooted in established grounds they, in addition, can provide a gratifying intellectual stimulus and delight.

6 Viroid-Like RNA's Encapsidated in Virions ("Virusoids")

In Australia four previously undescribed viruses have recently been found in wild plants which are characterized by their unusual RNA complement. Using the traditional method of employing the vernacular name of host in which they were first found, together with the description of the disease they cause, the first virus was named *velvet tobacco mottle virus* (VtMoV). Its polyhedral particles about 30 nm in diameter contain five single-stranded RNA components. They are designated RNA's 1, 1a, 1b, 2 and 3 in decreasing order of apparent molecular weight: 1.5×10^6, 0.63×10^6, 0.16×10^6, and 0.12×10^6, respectively. The relative proportion of each RNA encapsidated is 7%, 4%, 7%, 47% and 35%, respectively, for RNA's 1–3 (RANDLES et al. 1981). It appeared that RNA's 1a and 1b are breakdown products of RNA 1 which, like RNA 3, represent linear molecules. Most interestingly, RNA 2 is a circular molecule with about the same length as those of linear RNA 3. Complementary DNA (cDNA) analysis indicated that the circular and linear forms of these viroid-like RNA's 2 and 3 contain undistinguishable nucleotide sequences and are unique RNA species with no sequence homology with any of the other encapsidated RNA's (GOULD 1981). In order to discriminate them from viroids they are called "virusoids".

The second example of an encapsidated viroid-like RNA is found in the *Solanum nodiflorum mottle virus* (SNMV) which is serologically related with VTMoV and contains two unique RNA species as shown by cDNA analysis. They are RNA 1, a long linear single-stranded molecule with a molecular weight of 1.5×10^6 and RNA 2 and RNA 3 which are circular and linear forms of the same molecule, respectively. Electron microscopy and thermal denaturation studies indicate a viroid-like RNA for RNA 2 of SNMV. Electrophoretic analysis indicated that it has a slightly larger molecular weight (0.13×10^6) than the viroid-like RNA 2 of VTMoV (0.12×10^6). cDNA analysis revealed that SNMV RNA 2 contains the entire nucleotide sequence of VTMoV RNA 2 plus an additional 40–50 nucleotides not present in the latter. Comparative sequence homology studies between the large RNA's encapsidated in SNMV and VTMoV (both of molecular weight 1.5×10^6) revealed only 20% homology (GOULD and HATTA 1981).

The third and fourth example of an encapsidated viroid-like RNA is realized in the *lucerne transient streak virus* (LTSV) (TIEN et al. 1981) and the subterranean clover mottle virus (SCMV) (RANDLES et al. 1981, GOULD and HATTA, 1981, TIEN et al. 1981).

The rapid progress in RNA sequencing and the availability of the corresponding RNA's as isolated from purified virions has allowed the elucidation of the primary sequence and secondary structure of the viroid-like RNA's of VTMV and SNMV. RNA 2 of SNMV consists of 377 residues whereas that of VTMV consists of two approximately equimolar species, one of 366 residues and the other, with a single nucleotide deletion, of 365 residues. There is 92–95% sequence homology between the RNA 2 species of the two viruses. The predicted secondary structures possess extensive intramolecular base pairing to give rod-like structures similar to those of viroids. The structural similarities between the viroid-like RNA's 2 of both viruses may reflect functional similarities.

As has been discussed in Sect. 3.9 there is a central region in the rod-like structures of the viroids PSTV (GROSS et al. 1978), CSV (HASELOFF and SYMONS 1981, GROSS et al. 1982), CEV (GROSS et al. 1982, VISVADER et al. 1982) and CCCV (HASELOFF et al. 1982), which is highly conserved in both sequence and structure. ASBV does not share this common structure (SYMONS 1981) except for the residues GAAACC (residues 45–50) which, as in the other viroids, are present on a single-stranded loop in the central region of the native molecule. Interestingly, VTMV and SNMV RNA 2 also contain the sequence GAAAC (residues 86–90 in both molecules) which is also present in a single strand region in the center of the proposed secondary structure. However, there is no extensive sequence homology or complementarity between the viroids PSTV, CSV, CEV, ASBV and CCCV and the RNA 2 of VTMV and SNMV. It will be of considerable interest to determine the functional implications of these unique structural features and especially, the role of the pentanucleotide GAAAC in all these nucleic acids.

The analysis of the functional relationship between the RNA components of these viruses by GOULD et al. (1981) led to the following rather surprising results: neither the virus-like RNA 1 nor the viroid-like RNA 2 of VTMoV is capable of independent replication in *Nicotiana clevelandii.* The function of RNA 2 for the RNA 1 replication could not be fulfilled by conventional viroids such as those of CSV, CEV or the viroid-like RNA's associated with avocado sunblotch and cadang-cadang disease of coconut. Similarly, RNA 2 failed to replicate when co-inoculated with RNA's of conventional viruses such as those of southern bean mosaic, sowbean mosaic, *Galinsoga* mosaic, red clover mosaic, and carnation mottle, all of which possess genomes with molecular weights similar to those of VTMoV RNA 1. The interdependence of the two VTMoV RNA's for biological activity was so highly specific that the function of VTMoV

RNA 2 for the replication of RNA 1 could not be replaced by the viroid-like RNA 2 from the closely related SNMV. Similarly SNMV RNA 2 would not be replaced by VTMoV RNA 2 for the replication of SNMV RNA 1. These data support the view that RNA 2 cannot be a satellite RNA or an independent viroid.

The appearance of a viroid-like RNA encapsidated by virus particles forces critical investigation of all circular RNA's discovered in diseased plants in the future, because they may also require the interaction of a latent or specific viral species in their replication cycle. It is not a prerequisite of this proposed system that the circular RNA's be encapsidated by the helper virus, nor need it necessarily involve the appearance of viral particles in the infected host.

Considering the origin of viruses such as VTMoV, SCMU, SNMV and LTSV the possibility that their RNA 2 is a satellite must be rejected because RNA's 1 and 2 are obligatory for infection. Similarily the suggestion that they represent an association of a virus and a conventional viroid, the RNA's of which are both encapsidated in the coat protein of the virus, is untenable. It seems possible, however, that these viruses may have originated from such associations and that during their evolution both RNA's lost functions necessary for their separate autonomous replication. These functions could be fulfilled by their respective attendant RNA's. One could also speculate that these viruses represent stages in the evolution of viroids from viruses. Further studies on the replication of these viruses will undoubtedly yield some interesting clues as to the origin of viroids and their interaction with the nucleic acid-synthesizing machinery of the host cell.

7 Viroids, "Prions" and "Virinos"

One of the most interesting questions in viroid research centers around the possible existence of viroids and viroid-like pathogens as causative agents of certain unconventional diseases of man and animal. It is not surprising, therefore, that parallels have been drawn between certain properties of the newly discovered viroids and the elusive and enigmatic agents of certain neurological diseases of man and animal in particular with the agent of the scrapie disease of sheep and goat (DIENER 1972, 1973b). These parallels were based on similar target size values of viroids (PSTV and CEV) and the scrapie agent which gave molecular weight estimates ranging from 64,000 to 150,000 as well as on other properties of both types of agents. Especially their high resistance to heating and proteases, their inability to elicit an immune response, and the absence of virus-like particles in infected tissue were compatible with the hypothesis that both agents could have a similar molecular structure. As has been shown in the preceding chapters, not only the complete molecular structure of several viroids has been established since then, but is has also been shown that no proteins are associated with viroids.

Regarding the scrapie agent, on the other hand, all efforts to verify viroid-like properties have been fruitless so far. Thus, no infectious nucleic acids could

Table 7. Comparison of the stabilities of a viroid (PSTV) and a "prion" (scrapie agent) after treatment with different chemicals. (DIENER et al. 1982)

Treatment	Concentration	PSTV	Scrapie agent
RNase A	0.1–100 $\mu g\ ml^{-1}$	+	–
DNase	100 $\mu g\ ml^{-1}$	–	–
Proteinase K	100 $\mu g\ ml^{-1}$	–	+
Trypsin	100 $\mu g\ ml^{-1}$	–	+
Et_2PC[a]	10–20 mM	[–]	+
NH_2OH	0.1–0.5 M	+	–
Psoralen (AMT)[b]	10–500 $\mu g\ ml^{-1}$	+	–
Phenol	H_2O-saturated	–	+
SDS[c]	1–10%	–	+
Zn^{2+}	2 mM	+	–
Urea	3–8 M	–	+
Alkali	pH 10	[–]	+
KSCN	1 M	–	+

[a] Diethylpyrocarbonate; [b] 4′-Aminomethyl-4,5′, 8-trimethylpsoralen hydrochloride; [c] Sodiumdodecyl sulfate. –, No change; [–], small change; +, inactivation

be isolated from brain homogenates of scrapie-infected animals and claims that a low molecular weight DNA component is essential for scrapie infectivity have not been confirmed (for a detailed discussion see PRUSINER 1982). In contrast, recent progress in the purification and characterization of the scrapie agent has provided evidence that it contains a hydrophobic protein which is essential for the expression of infectivity. Moreover, even in highly enriched preparations of the scrapie agent no nucleic acid could be demonstrated (see PRUSINER 1982).

The results of a recent comparison of the properties of PSTV with those of the scrapie agent (DIENER et al. 1982) is shown in Table 7. It becomes evident that PSTV is inactivated by RNase digestion, psoralen photoadduct formation, Zn^{2+}-catalyzed hydrolysis and chemical modification with NH_2OH, all procedures which modify nucleic acids. The scrapie agent, on the other hand, is inactivated by proteinase K, trypsin digestion, chemical modification with diethylpyrocarbonate and by exposure to phenol, sodium dodecylsulfate, KSCN and urea, all procedures which modify proteins and by which PSTV RNA is not inactivated. The results of these recent investigations clearly differentiate between both entities and place the scrapie agent (together with comparable agents causing neurotropic diseases in other animals and the Kuru, Creutzfeldt-Jakob and Alzheimers disease in man) outside the realm of viroids. These fundamental differences have led to the introduction of the new operational term "prion" for the agents of scrapie and related diseases. "Prions" are small proteinaceous infectious particles that resist inactivation by most procedures that modify nucleic acids. Their smallest form seems to have a molecular weight of 50,000 or less and their unusual features distinguish them from viruses, plasmids and viroids (PRUSINER 1982). The basic properties of viroids and "prions" are compared in Table 8.

It should be noted here that the assumption that "prions" are infectious, i.e. replicable proteinaceous agents devoid of any nucleic acid moiety is an

Table 8. Properties of viroids and "prions". (DIENER et al. 1982)

Properties	Viroids	"Prions"
Host range	Higher plants	Mammals
Nucleic acid	RNA	Not demonstrated conclusively to date
Protein	−	+
Molecular weight (monomer)	$\leqq$130,000	$\leqq$50,000
Sedimentation coefficient	6.7 S	$\leqq$2 S[a]
D_{37}[b] at 254 nm, $J\ m^{-2}$	5,000	42,000

[a] The S value was calculated with the assumption that the particle density is $>1.05\ g\ cm^{-3}$. Because of the hydrophobicity of the agent and the unknown amount of detergent bound to it the observed value could be in error if the particles would float

[b] Dose of UV-light that permits 37% (1/e) survival of the infectious particle

exiting and heretical challenge to the central dogma of molecular biology. It raises all the problems how a non-nucleic acid-containing particle can replicate by itself or become replicated by the host. Consequently, many more hard facts have to be assembled from rigorous experimentation before the concept of "prions" as infectious proteinaceous agents can be accepted without any reservation.

In fact, a much simpler working hypothesis has been proposed which fits both established facts and the more recent data and does not require hypotheses outside the current frame-work of molecular biology (KIMBERLIN 1982): It is based on two assumptions only: First, that there is indeed a scrapie-specific nucleic acid as indicated by the replicability, strain variation and mutation of the scrapie agent, and second, that this scrapie nucleic acid, like the viroid nucleic acid, is not translated. The scrapie nucleic acid could be very small, become replicated by host enzymes, and could act as, or code for, some kind of regulatory nucleic acid which would produce the disease. Alternatively, the scrapie-specific nucleic acid could bind a functionally important host protein and thus induce a pathological condition. The apparent absence of specific antigenicity in the scrapie disease could be simply explained by the host origin of this protein. If the scrapie-specific nucleic acid is covered with host protein it would be hard to purify and to become recognized. An agent of this kind would fit into the niche between viruses which specify some of their own proteins, and viroids which need no proteins at all. For this kind of agents the neologism, "virino", has already been proposed (DICKINSON and FRAZER 1979) which may, indeed be more appropriate than the term "prion". By analogy with neutrinos the term "virino" only implies that the agents are small, immunological neutral particles with high penetration properties which need special criteria to detect their presence and unlike the term "prion" it does not emphasize any macromolecular species responsible for infectivity. It should be remembered in this context that also tobacco mosaic was considered a proteinaceous agent in the early phase of its molecular characterization (STANLEY 1935). It will be interesting to await the purification of the scrapie agent to homogeneity and the results of the subsequent identification of its infectious moiety.

8 Concluding Remarks

Viroids are the first pathogens the complete molecular structure of which is known in detail. After the early years of scepticism and even rejection, their well-established existence as a rather unusual reality has removed a number of psychological blocks in our thinking. Viroids have extended our concept of disease agents into the region of small RNA molecules. Along the way several unwritten dogmas have broken down. Infectious nucleic acids were previously considered to only exist and survive in an encapsidated form, the virus particle. Viroids showed, however, that also "naked" infectious nucleic acid molecules can exist, because they developed a unique structure, which confers the necessary stability to the molecule. Furthermore, the minimal size of infectious replicating nucleic acids is no longer restricted to molecules of 10^6 molecular weight, which corresponds to a chain length of about 3,600 nucleotides. Viroids are only one tenth of that size and yet infectious and replicating.

Infectious viral nucleic acids exert some of their pathogenic functions through the virus-specific proteins they code for. Their lacking in vitro translation potentials together with the obvious absence of viroid-specific proteins in infected plants indicates that also this concept does not apply for viroid pathogenesis. Viroid RNA has the ability to act in vitro as an efficient template for DNA-dependent RNA polymerase II, to form a binary complex with this polymerase, to compete with DNA for the template binding site on the enzyme and to strongly inhibit DNA-directed RNA synthesis. If viroid RNA is, in fact, capable of directing DNA-dependent RNA polymerase II to perform its "selfish" replication in vivo then viroid pathogenicity could result from the direct interference of viroid replication with the messenger RNA synthesis of the host. The presence in different viroids of strictly conserved sequences which exhibit a surprising homology and complementarity with the known consensus sequences of exon-intron junctions and the 5′-end of the nuclear U1 RNA furthermore suggests that viroid RNA might also interfere with messenger RNA maturation, in particular with the mechanisms of splicing and ligation. Thus, viroids could function like some of the tentative regulatory RNA's which have been repeatedly postulated in the past.

Viroids differ from all presently known nucleic acids by the combination of circularity, unusual secondary structure and the specific dynamics of highly cooperative structural transitions. In addition viroids represent RNA molecules with thermodynamic and kinetic properties of a DNA of the same size and G:C content. However, these features do not result from a DNA-like secondary structure but are the consequence of the serial arrangement of double-helical segments and internal loops which are only guaranteed by a very specific nucleotide sequence. Random sequences of the same number nucleotides and composition do not yield the high degree of base-pairing and the extraordinary cooperativity observed for viroids. There is little doubt that the combination of these unique features is of functional importance for replication, ligation, pathogenicity and "survival". After the elucidation of viroid structure the first light has been shed on the possible mechanisms of viroid replication and pathogenesis.

One can safely assume already that both mechanisms must differ fundamentally from the corresponding ones of conventional viruses. The study of the more dynamic aspects of the viroid–host plant interaction on the cellular level will progress rapidly because several technical problems inherent in plant systems have been overcome and suitable plant cells and protoplasts are now available for such investigations. Nevertheless, one should remember that in higher plants certain pathogenic responses are based on functions of the intact tissue or even of the entire plant.

One of the most interesting questions centres around the possible existence of viroids or viroid-like pathogens as causative agents of certain unconventional diseases of man and animal. It must be emphasized that so far viroids have been found only in higher plants and that there is no direct evidence yet that they are also existing in other organisms. Experience has shown, however, that viruses, mycoplasms, bacteria and fungi are found as pathogens in all organisms. Therefore, it would be a rather unique situation, if viroids were confined to higher plants only. Despite the recent developments in the elucidation of the unusual nature of the scrapie agent and its fundamental differences from viroids (Sect. 7), it is well justified to continue the search for viroids or viroid-like molecules in all those transmissible diseases where the involvement of conventional viruses has been ruled out by the classical methods of virology.

In conclusion, viroids represent a completely new class of pathogenic RNA's in plants with unique structural features and a still enigmatic mechanism of replication and pathogenesis. Regarding the relation between structure and function, viroids represent an optimal compromise between structural stability and functional flexibility, which combines maximal self-protection with efficient replication.

With respect to their interaction with the host plant, viroids may be considered as replicating competitive inhibitors of messenger RNA synthesis and maturation. Their stable yet flexible DNA-like structure confers to them a selective advantage over other cellular nucleic acids by guaranteeing their replication by host enzymes and their "survival".

Because of their small size, their known structure, their relative stability and their intimate relationship to central functions of the host cell metabolism, viroids are not only an excellent system to investigate the relationship between molecular structure and biological function, but also the molecular biology and biochemistry of higher plants in general. All these features render viroids fascinating models for future investigations by various scientific disciplines. The recent discovery that small viroid-like RNA's, the so-called "virusoids", exist as an integral and functional part of the multipartite genome of certain plant viruses opens unexpected new perspectives in this field.

Acknowledgements. I wish to acknowledge the continual collaboration and the excellent efforts of my associates identified in the respective publications. I also thank all my colleages who made available their unpublished data. The generous support of our work by the Deutsche Forschungsgemeinschaft through Sonderforschungsbereich 47 (Virologie), personal grants and stipends is gratefully acknowledged.

References

Alwine JC, Kemp DJ, Stark GR (1977) Method for detection of specific RNAs in agarose gels by transfer to diazobenzyloxymethyl-paper and hybridization with DNA probes. Proc Natl Acad Sci USA 74:5350–5354

Biebricher CK, Orgel LE (1973) An RNA that multiplies indefinitely with DNA-dependent RNA polymerase: Selection from a random copolymer, Proc Natl Acad Sci USA 70:934–938

Boege F, Sänger HL (1980) RNA-dependent RNA polymerase from healthy tomato leaf tissue. FEBS Lett 121:91–96

Boege F, Rohde W, Sänger HL (1982) In vitro transcription of viroid RNA into full length copies by RNA-dependent RNA polymerase from healthy tomato leaf tissue. Biosci Rep 2:185–194

Brändle E, Zetsche K (1973) Zur Lokalisation der α-Amanitin sensitiven RNA-Polymerase in Zellkernen von *Acetabularia*. Planta 111:209–217

Branlant C, Krol A, Ebel J-P, Lazar E, Gallinaro H, Jacob M, Sri-Widada J, Jeanteur P (1980) Nucleotide sequences of nuclear U1A RNAs from chicken, rat and man. Nucl Acids Res 8:4143–4154

Branch AD, Dickson E (1980) Tomato DNA contains no detectable regions complementary to potato spindle tuber viroid as assayed by southern hybridization. Virology 104:10–26

Branch AD, Robertson HD, Dickson E (1981) Longer-than-unit-length viroid minus strands are present in RNA from infected plants. Proc Natl Acad Sci USA 78:6381–6385

Branch AD, Robertson AD, Greer C, Gegenheimer P, Peebles C, Abelson J (1982) Cell-free circularization of viroid progeny RNA by an RNA ligase from wheat germ. Science 217:1147–1149

Brown F, Martin SJ (1965) A new model for virus ribonucleic acid replication. Nature (London) 208:861–863

Busch H, Reddy R, Rothblum L, Choi VC (1982) SnRNAs, SnRNPs, and RNA processing. Annu Rev Biochem 51:617–654

Calavan EC, Frolich EF, Carpenter JB, Roistacher CN, Christiansen DW (1964) Rapid indexing for exocortis of citrus. Phytopathology 54:1359–1362

Calavan EC, Weathers LG, Christiansen DW (1968) Effect of exocortis on production and growth of Valencia orange trees on trifoliate orange rootstock. In: Childs JFL (ed) Proc 4th Conf Int Org Citrus Virologists. Univ Florida Press, Gainessville

Camacho Henriquez A, Sänger HL (1982a) Analysis of acid-extractable tomato leaf proteins after infection with a viroid, two viruses and a fungus and partial purification of the "pathogenesis-related" protein p14. Arch Virol (in press)

Camacho Henriquez A, Sänger HL (1982b) Gelelectrophoretic analysis of phenol-extractable leaf proteins from different viroid/host combinations. Arch Virol (in press)

Chen W, Tien P, Zhu YX, Liu Y (1982) The viroid-like RNAs associated with burdock stunt disease. (submitted)

Chifflot S, Sommer P, Hartmann D, Stussi-Garaud C, Hirth L (1980) Replication of alfalfa mosaic virus RNA: Evidence for a soluble replicase in healthy and infected tobacco leaves. Virology 100:91–100

Conjero V, Semancik JS (1977) Exocortis viroid: Alteration in the proteins of *Gynura aurantiaca* accompanying viroid infection. Virology: 77:221–232

Conejero V, Picazo I, Segado P (1979) Citrus exocortis viroid (CEV): Protein alterations in different hosts following viroid infection. Virology 97:454–456

Crick F, (1979) Split genes and RNA splicing. Science 204:264–271

Da Graça JV, Martin MM (1981) Ultrastructural changes in avocado leaf tissue infected with avocado sunblotch. Phytopathol Z 102:185–194

Dale JL, Allen RN (1979) Avocado affected by sunblotch disease contains low molecular weight ribonucleic acid. Aust Plant Pathol 8:3–4

Davies JW, Kaesberg P, Diener TO (1974) Potato spindle tuber viroid. XII. An investigation of viroid RNA as a messenger for protein synthesis. Virology: 61:281–286

De Bokx JA, Piron PGM (1981) Transmission of potato spindle tuber viroid by aphids. Neth J Plant Pathol 87:31–34
Dickson E (1979) Viroids: Infections RNA in plants. In: Hall TC, Davies JW (eds) Nucleic acids in plant Vol II. CRC Press, Boca Raton
Dickinson AG, Frazer GW (1979) The scrapie replication-site hypothesis and its implications for pathogenesis. In: Prusiner SB, Hadlow WJ (eds) Slow transmissible diseases of the nervous system Vol 2. Academic Press, London New York
Dickson E (1981) A model for the involvement of viroids in RNA splicing. Virology 115:216–221
Dickson E, Robertson HD (1976) Potential regulatory roles for RNA in cellular development. Cancer Res 36:3387–3393
Dickson E, Prensky W, Robertson HD (1975) Comparative studies of two viroids: Analysis of potato spindle tuber and citrus exocortis viroids by RNA fingerprinting and polyacrylamide-gel electrophoresis. Virology 68:309–316
Dickson E, Diener TO, Robertson HD (1978) Potato spindle tuber and citrus exocortis viroid undergo no major sequence changes during replication in two different hosts. Proc Natl Acad Sci USA 75:951–954
Dickson E, Robertson HD, Niblett CL, Horst RK, Zaitlin M (1979) Minor differences between nucleotide sequences of mild and severe strains of potato spindle tuber viroid. Nature (London) 277: 60–62
Diener TO (1971a) Potato spindle tuber virus: A plant virus with properties of a free nucleic acid. III. Subcellular location of PSTV-RNA and the question of whether virions exist in extracts or in situ. Virology 43:75–89
Diener TO (1971b) Potato spindle tuber "virus". IV. A replicating, low molecular weight RNA. Virology 45:411–428
Diener TO (1972a) Viroids. Adv Virus Res 17:295–313
Diener TO (1972b) Is the scrapie agent a viroid? Nature (New Biol) 235:218–219
Diener TO (1973a) Virus terminology and the viroid: a rebuttal. Phytopathology 63:1328–1329
Diener TO (1973b) Similarities between the scrapie agent and the agent of potato spindle tuber disease. Ann Clin Res 5:268–278
Diener TO (1974) Viroids: The smallest known agents of infetious disease. Ann Rev Microbiol 28:23–39
Diener TO (1978) Are viroids autoinducing regulatory RNAs? In: Persistent Viruses, Academic Press, London New York
Diener TO (1979a) Viroids: Structure and function. Science 205:859–866
Diener TO (1979b) Viroids and viroid diseases. Wiley and Sons, New York
Diener TO (1981a) Are viroids escaped introns? Proc Natl Acad Sci USA 78:5014–5015
Diener TO (1981b) Viroids: abnormal products of plant-metabolism. Annu Rev Plant Physiol 32:313–325
Diener TO (1982) Viroids – minimal biological-systems. Bioscience 32:38–44
Diener TO, Hadidi A (1977) Viroids. In: Fraenkel-Conrat H, Wagner RR (eds) Comprehensive virology vol XI. Plenum, New York
Diener TO, Lawson RH (1973) Chrysanthemum stunt: A viroid disease. Virology 51:94–101
Diener TO, Smith DR (1975) Potato spindle tuber viroid. XIII. Inhibition of replication by actinomycin D. Virology 63:421–427
Diener TO, Hadidi A, Owens RA (1977) Methods for studying viroids. In: Koprowski H, Maramosch K (eds), Vol 6. Academic Press, London New York
Diener TO, McKinley MP, Prusiner SB (1982) Viroids and prions. Proc Natl Acad Sci USA 79:5220–5224
Domdey H, Jank P, Sänger HL, Gross HJ (1978) Studies on the primary and secondary structure of potato spindle tuber viroid: Products of digestion with ribonuclease A and ribonuclease T_1 and modification with bisulfite. Nucl Acids Res 5:1221–1236
Drummond M (1979) Crown gall disease. Nature (London) 281:343–347
Duda CT (1976) Plant RNA Polymerases. Annu Rev Plant Physiol 27:119–132
Duda CT, Zaitlin M, Siegel A (1973) In vitro synthesis of double-stranded RNA by an enzyme system isolated from tobacco leaves. Biochim Biophys Acta 319:62–71

Duran-Villa N, Semancik JS (1982) Differential response of tomato tissue infected with citrus exocortis viroid to exogenous auxin. Phytopathology (in press)

Fernow KH (1967) Tomato as a test plant for detecting mild strains of potato spindle tuber virus. Phytopathology 57:1347–1352

Fernow KH, Peterson LC, Plaisted RL (1969) The tomato test for eliminating spindle tuber from potato planting stock. Am Potato J 46:424–429

Flores R, Chroboczek J, Semancik JS (1978) Some properties of the CEV-P_1 protein from citrus exocortis viroid-infected *Gynura aurantiaca* D.C. Physiol Plant Pathol 13:193–201

Flores R, Rodriguez JL (1981) Altered pattern of root-formation on cuttings of *Gynura aurantiaca* infected by citrus exocortis viroid. Phytopathology 71:964–966

Fraenkel-Conrat H (1979) RNA-dependent RNA polymerases of Plants. Trends Biochem Sci 4:184–186

French RC, Price MA, Derrick KS (1982) Circular double-stranded-RNA in potato spindle tuber viroid-infected tomatoes. Nature (London) 295:259–260

Galindo J, Smith DR, Diener TO (1982) Etiology of planta macho, a viroid disease of tomato. Phytopathology 72:49–54

Geelen JLMC, Weathers LG, Semancik JS (1976) Properties of RNA polymerases of healthy and citrus exocortis viroid-infected *Gynura aurantiaca* DC. Virology 69:539–546

Gilbert W, Dressler D (1968) DNA-replication: The rolling circle model. Cold Spring Harbor Symp Quant Biol 33:473–484

Gilliland JM, Symons RH (1968) Properties of a plant virus-induced RNA polymerase in cucumbers infected with cucumber mosaic virus. Virology 36:232–240

Goldmann D, Merril CR (1982) Silver staining of DNA in polyacrylamide gels: Linearity and effect of fragment size. Electrophoresis 3:24–26

Goss RW (1928) Transmission of potato spindle-tuber by grasshoppers (Locustidae). Phytopathology 18:445–448

Goss RW (1930) Insect transmission of potato-virus diseases. Phytopathology 20:136

Goss RW (1931) Infection experiments with spindle tuber and unmottled curly dwarf of the potato. Univ Nebraska Agric Exp Stan Res Bull 53:36

Gould AR (1981) Studies on encapsidated viroid-like RNA II. Purification and characterization of a viroid-like RNA associated with velvet tobacco mottle virus (VTMoV). Virology 108:123–133

Gould AR, Hatta T (1981) Studies on encapsidated viroid-like RNA. III. Comparative studies on RNAs isolated from velvet tobacco mottle virus and *Solanum nodiflorum* mottle virus. Virology 109:137–147

Gould AR, Francki RIV, Randles JW (1981) Studies on encapsidated viroid-like RNA IV. Requirement for infectivity and specificity of 2 RNA components from velvet tobacco mottle virus. Virology 110:420–426

Grabowski PJ, Zang AJ, Cech TR (1981) The intervening sequence of the ribosomal RNA precursor is converted to a circular RNA in isolated nuclei of *Tetrahymena.* Cell 23:447–476

Grill LK, Semancik JS (1978) RNA sequences complementary to citrus exocortis viroid in nucleic acid preparations from infected *Gynura aurantiaca.* Proc Natl Acad Sci USA 75:896–900

Grill LK, Semancik JS (1980) Viroid synthesis: the question of inhibition by actinomycin D. Nature (London) 283:399–400

Grill LK, Negruk VI, Semancik JS (1980) Properties of the complementary RNA sequences associated with infection by the citrus exocortis viroid. Virology 107:24–33

Gross HJ, Riesner D (1980) Viroids: A class of subviral pathogens. Angew Chem Int Ed Engl 19:231–243

Gross HJ, Domdey H, Sänger HL (1977) Comparative oligonucleotide fingerprints of three plant viroids. Nucl Acids Res 4:2021–2028

Gross HJ, Domdey H, Lossow C, Jank P, Raba M, Alberty H, Sänger HL (1978) Nucleotide sequence and secondary structure of potato spindle tuber viroid. Nature (London) 273:203–208

Gross HJ, Liebl U, Alberty H, Krupp G, Domdey H, Ramm K, Sänger HL (1981)

A severe and a mild potato spindle tuber viroid isolate differ in 3 nucleotide exchanges only. Biosci Rep 1:235–241

Gross HJ, Krupp G, Domdey H, Raba M, Alberty H, Lossow CH, Ramm K, Sänger HL (1982) Nucleotide sequence and secondary structure of citrus exocortis and chrysanthemum stunt viroid. Eur J Biochem 121:249–257

Hadidi A, Diener TO (1977) De novo synthesis of potato spindle tuber viroid as measured by incorporation of ^{32}P. Virology 78:99–107

Hadidi A, Diener TO (1978) In vivo synthesis of potato spindle tuber viroid: kinetic relationship between the circular and linear forms. Virology 86:57–65

Hadidi A, Jones DM, Gillespie DH, Wong-Staal F, Diener TO (1976) Hybridization of potato spindle tuber viroid to cellular DNA of normal plants. Proc Natl Acad Sci USA 73:2453–2457

Hadidi A, Cress DE, Diener TO (1981) Nuclear DNA from uninfected or potato spindle tuber viroid-infected tomato plants contains no detectable sequences complementary to cloned double-stranded viroid cDNA. Proc Nat Acad Sci USA 78:6932–6935

Hall TC, Wepprich RK, Davies JW, Weathers LG, Semancik JS (1974) Functional distinctions between the ribonucleic acids form citrus exocortis viroid and plant viruses: Cell-free translation and aminoacylation reactions. Virology 61:486–492

Hardy SF, German TL, Loesch-Fries LS, Hall TC (1979) Highly active template-specific RNA-dependent RNA polymerase from barley leaves infected with brome mosaic virus. Proc Natl Acad Sci USA 76:4956–4960

Hari V (1980) Ultrastructure of potato spindle tuber viroid-infected tomato leaf tissue. Phytopathology 70:385–387

Haseloff J, Symons RH (1981) Chrysanthemum stunt viroid-primary sequence and secondary structure. Nucl Acids Res 9:2741–2752

Haseloff J, Symons R (1982) Comparative sequence and structure of viroid-like RNAs of two plant viruses. Nucl Acids Res 10:3681–3691

Haseloff J, Mohamed NA, Symons RH (1982) Viroid RNAs of the cadang-cadang disease of coconuts. Nature (London) 299:316–322

Henco K, Riesner D, Sänger HL (1977) Conformation of viroids. Nucl Acids Res 4:177–194

Henco K, Sänger HL, Riesner D (1979) Fine structure melting of viroids as studied by kinetic methods. Nucl Acids Res 6:3041–3059

Hollings M (1965) Disease control though virus-free stock. Annu Rev Phytopathol 3:367–396

Hollings M, Stone OM (1970) Attempts to eliminate chrysanthemum stunt from chrysanthemum by meristem-tip culture after heat-treatment. Ann Appl Biol 65:311–315

Hollings M, Stone OM (1973) Some properties of chrysanthemum stunt, a virus with the characteristics of an uncoated ribonucleic acid. Ann Appl Biol 74:333–348

Horst RK (1975) Detection of a latent infectious agent that protects against infection by chrysanthemum chlorotic mottle viroid. Phytopathology 65:1000–1003

Horst RK, Romaine CP (1975) Chrysanthemum chlorotic mottle: A viroid disease. NY Food Life Sci Q 8:11–14

Horst RK, Langhans RW, Smith SH (1977) Effects of chrysanthemum stunt, chlorotic mottle, aspermy and mosaic on flowering and rooting of chrysanthemums. Phytopathology 67:9–14

Hsu M, Coco-Prados M (1979) Electron microscopic evidence for the circular form of RNA in the cytoplasm of eukaryotic cells. Nature (London) 280:339–340

Ikegami M, Fraenkel-Conrat H (1978) RNA-dependent RNA polymerase of tobacco plants. Proc Natl Acad Sci USA 75:2122–2124

Ikegami M, Fraenkel-Conrat H (1979a) Characterization of the RNA-dependent RNA-polymerase of tobacco leaves. J Biol Chem 254:149–154

Ikegami M, Fraenkel-Conrat H (1979b) Characterization of double-stranded ribonucleic acid in tobacco leaves. Proc Natl Acad Sci USA 76:3637–3646

Imperial JS, Rodrigue JB, Randles JM (1981) Variation in the viroid-like RNA associated with cadang-cadang disease: evidence for an increase in molecular-weight with disease progress. J Gen Virol 56:77–85

Ingram DS (1976) Growth of biotrophic parasites in tissue culture. In: Heitefuss R,

Williams PH (eds) Physiological plant pathology. Encyclopedia of plant physiologie new ser Vol IV. Springer, Berlin Heidelberg New York, pp 743–759

Jacobsen JV (1977) Regulation of ribonucleic acid metabolism by plant hormones. Annu Rev Plant Physiol 28:537–564

Jendrisak JJ, Burgess RR (1975) A new method for the large-scale purification of wheat germ DNA-dependent RNA polymerase II. Biochemistry 14:4639–4645

Jenden EO, Poludan K, Hyldig-Nielson JJ, Jorgensen P, Marcker KA (1981) The structure of a chromosomal leg-haemoglobin gene from soybean. Nature (London) 291:677–679

Kamen RI (1975) Structure and function of the Q3 replicase. In: Zinder ND (ed) RNA Phages. Cold Spring Harbor Lab, Cold Spring Harbor 203–234

Keller JR (1951) Report on indicator plants for chrysanthemum stunt virus and on a previously unreported chrysanthemum virus. Phytopathology 41:947–949

Keller JR (1953) Investigations on chrysanthemum stunt virus and chrysanthemum virus Q. Cornell Univ Agric Exp Stan Memoir 324:40

KiKuchi Y, Tyc K, Filipowicz W, Sänger HL, Gross HJ (1982) Circularization of linear viroid RNA via 2′-phosphomonoester, 3′, 5′-phosphodiester bonds by a novel type of RNA ligase from wheat germ and *Chlamodomonas*. Nucl Acids Res (in press)

Kimberlin RH (1982) Scrapie agent: Prions or virinos? Nature (London) 297:107–108

Kleinschmidt AK, Klotz G, Seliger H (1981) Viroid structure. Annu Rev Biophys Bioeng 10:115–132

Klotz G, Sänger HL (1981) Electron microscopic evidence for viroid conformers. Eur J Cell Biol 25:5–7

Klump H, Riesner D, Sänger HL (1978) Calorimetric studies on viroids. Nucl Acids Res 5:1581–1587

Konarska M, Filipowica W, Domdey H, Gross HJ (1981) Formation of a 2′-phosphomonoester, 3′,5′-phosphoester linkage by a novel RNA ligase in wheat germ. Nature (London) 293:112–116

Kozak M (1979) Inability of circular nRNA to attach to eucaryotic ribosomes. Nature (London) 280:82–85

Langowski J, Henco K, Riesner D, Sänger HL (1978) Common structural features of different viroids: Serial arrangement of double helical sections and internal loops. Nucl Acids Res 5:1589–1610

Lawson RH (1968) Cineraria varieties as starch lesion test plants for chrysanthemum stunt virus. Phytopathology 58:690–695

Lazar E, Walter B, Stussi-Garaud C, Hirth L (1979) RNA dependent RNA polymerases from healthy and tobacco necrosis virus-infected *Phaseolus aureus*: Assay of localization in fractions of cellular homogenates. Virology 96:553–563

Lee J, Sänger HL (1982) Transmission of potato spindle tuber viroid (PSTV) by the peach aphid *Myzus persicae*. (in prep)

Lerner MR, Boyle JA, Mount SM, Wolin SL, Steitz JA (1980) Are snRNPs involved in splicing. Nature (London) 283:220–224

Lindell TJ, Weinberg F, Morris PW, Roeder RG, Rutter WJ (1970) Specific inhibition of nuclear RNA polymerase II by α-amanitin. Science 170:447–449

Lizaraga RE, Salazar LF, Roca WM, Schilde-Rentschler L (1980) Elimination of potato spindle tuber viroid by low temperature and meristem culture. Phytopathology 70:754–755

Macquair G Monsion M, Bachelie JC, Faydi C, Dunez J (1981) The slab gel-electrophoretic assay for detection and investigation of chrysanthemum chlorotic mottle viroid (CCHMV) in infected plants. Agron Trop 1:99–103

Mahy BWJ, Hastie ND, Armstrong SH (1972) Inhibition of influenza virus replication by α-amanitin: Mode of action. Proc Natl Acad Sci USA 69:1421–1424

Matthews REF (1978) Are viroids negative-strand viruses? Nature 276:850

May JT, Symons RH (1971) Specificity of the cucumber mosaic virus-induced RNA polymerase for RNA and polynucleotide templates. Virology 44:517–526

May JT, Gilliland JM, Symons RH (1969) Plant virus-induced RNA polymerase: Properties of the enzyme partly purified from cucumber cotyledons infected with cucumber mosaic cirus. Virology 39:54–65

May JT, Gilliland JM, Symons RH (1970) Properties of a plant virus-induced RNA polymerase in particulate fractions of cucumber infected with cucumber mosaic virus. Virology 41:653–664

McClean APD (1931) Bunchy top disease of tomato. S Af Dept Sci Bull 100:36

McClements WL, Kaesberg P (1977) Size and secondary structure of potato spindle tuber viroid. Virology 76:477–484

Morris TJ (1979) Evidence for a single infectious species of potato spindle tuber viroid. Intervirology 11:89–96

Morris TJ, Smith EM (1977) Potato spindle tuber disease: Procedures for the detection of viroid RNA and certification of disease-free potato tubers. Phytopathology 67:145–150

Morris TJ, Wright NS (1975) Detection on polyacrylamide gel of a diagnostic nucleic acid from tissue infected with potato spindle tuber viroid. Am Potato J 52:57–63

Mühlbach H-P (1982) Plant cell cultures and protoplasts in plant virus research. Curr Top Microbiol Immunol 99:81–129

Mühlbach H-P, Sänger HL (1977) Multiplication of cucumber pale fruit viroid in inoculated tomato leaf protoplasts. J Gen Virol 35:377–386

Mühlbach H-P, Sänger HL (1979) Viroid replication is inhibited by α-amanitin. Nature (London) 278:185–188

Mühlbach H-P, Sänger HL (1981) Continous replication of potato spindle tuber viroid (PSTV) in permanent cell-cultures of potato and tomato. Biosci Rep 1:79–87

Mühlbach H-P, Camacho-Henriquez A, Sänger HL (1977) Isolation and properties of protoplasts from leaves of healthy and viroid-infected tomato plants. Plant Sci Lett 8:183–189

Niblett CL, Dickson E, Fernow KH, Horst RK, Zaitlin M (1978) Cross protection among four viroids. Virology 91:198–203

Niblett CL, Dickson E, Horst RK, Romaine CP (1980) Additional hosts and an efficient purification procedure for four viroids. Phytopathology 70:610–615

O'Brien MJ, Raymer WB (1964) Symptomless hosts of the potato spindle tuber virus. Phytopathology: 54:1045–1047

Ohno T, Akiya J, Higuchi M, Okada Y, Yoshikawa N, Takahashi T, Hashimoto J (1982) Purification and characterization of hop stunt viroid. Virology 118:54–63

Ohshima Y, Itoh M, Okada N, Miyata T (1981) Novel models for RNA splicing that involve a small nuclear RNA. Proc Natl Acad Sci USA 78:4471–4474

Owens RA (1978) In vitro synthesis and characterization of DNA complementary to potato spindle tuber viroid. Virology 89:380–387

Owens RA, Cress DE (1980) Molecular cloning and characterization of potato spindle tuber viroid cDNA sequences. Proc Natl Acad Sci USA 77:5302–5306

Owens RA, Diener TO (1977) Synthesis of RNA complementary to potato spindle tuber viroid using Qβ replicase. Virology 79:109–120

Owens RA, Diener TO (1981) Sensitive and rapid diagnosis of potato spindle tuber viroid disease by nucleic-acid hybridization. Science 213:670–672

Owens RA, Diener TO (1982) RNA intermediates in potato spindle tuber viroid replication. Proc Natl Acad Sci USA 79:113–117

Owens RA, Erbe E, Hadidi A, Steere RL, Diener TO (1977) Separation and infectivity of circular and linear forms of potato spindle tuber viroid. Proc Natl Acad Sci USA 74:3859–3863

Owens RA, Smith DR, Diener TO (1978) Measurement of viroid sequence homology by hybridization with complementary DNA prepared in vitro. Virology 89:388–394

Paliwal YC, Singh RP (1981) Cytopathological changes induced by potato spindle tuber viroid in *Scopolia sinensis*. Can J Bot 59:677–682

Palukaitis P, Symons RH (1978) Synthesis and characterization of complementary DNA probe for chrysanthemum stund viroid. FEBS Lett 92:268–272

Palukaitis P, Symons RH (1980) Purification and characterization of the circular form of chrysanthemum stunt viroid. J Gen Virol 46:477–489

Palukaitis P, Halla T, Alexander D McE, Symons RH (1979) Characterization of a viroid associated with avocado sunblotch disease. Virology 99:145–151

Palukaitis P, Rakowski AG, Alexander DM, Symons RH (1981) Rapid indexing of

the sunblotch disease of avocados using a complementary-DNA probe to avocado sunblotch viroid. Ann Appl Biol 98:439–449
Pfannenstiel MA, Slack SA, Lane LC (1980) Detection of PSTV in field-grown potatoes by an improved electrophoretic assay. Phytopathology 70:1015–1018
Prusiner SB (1982) Novel proteinaceous infectious particles cause scrapie. Science 216:136–144
Rackwitz HR, Rohde W, Sänger HL (1981) DNA-dependent RNA polymerase-II of plant-origin transcribes viroid RNA into full-length copies. Nature (London) 291:297–301
Randles JW (1975) Association of two ribonucleic acid species with cadang-cadang disease of coconut palm. Phytopathology 65:163–167
Randles JW, Palukaitis P (1979) In vitro synthesis and characterization of DNA complementary to cadang-cadang-associated RNA. J Gen Virol 43:649–662
Randles JW, Rillo EP, Diener TO (1976) The viroidlike structure and cellular location of anomalous RNA associated with the cadang-cadang disease. Virology 74:128–139
Randles JW, Boccardo G, Retuerma ML, Rillo EP (1977) Transmission of the RNA species associated with cadang-cadang of coconut palm, and the insensitivity of the disease to antibiotics. Phytopathology 67:1211–1216
Randles JW, Davies C, Hatta T, Gould AR, Francki RIB (1981) Studies on encapsidated viroid-like RNA. I. Characterization of velvet tobacco mottle virus. Virology 108:111–122
Reanny DC (1975) A regulatory role for viral RNA in eucaryotes. J Theor Biol 49:461–492
Reddy R, Ro-Choi TS, Henning D, Busch H (1974) Primary sequence of U-1 nuclear ribonucleic acid of Novikoff hepatoma ascites cells. J Biol Chem 249:6486–6494
Riesner D, Henco K, Rokohl U, Klotz G, Kleinschmidt AK, Gross HJ, Domdey H, Sänger HL (1979) Structure and structure formation of viroids. J Mol Biol 133:85–115
Riesner D, Steger G, Schumacher J, Gross HJ, Sänger HL (1981) Structure and function of viroids. Biophys Struct Mech 7:240–241
Roberts RJ (1978) Intervening sequences excised in vitro. Nature (London) 274:530
Roberts R (1980) Small RNAs and splicing. Nature (London) 283:132–133
Robertson HD, Dickson E (1974) RNA processing and the control of gene expression. In: Process. Brookhaven Symp Biol 26:240–266
Rodriguez JL, Garcia-Martinez JL, Flores R (1979) The relationship between plant growth substances content and infection of *Gynura aurantiaca* D. C. by citrus exocortis viroid. Physiol Plant Pathol 13:355–363
Rogers J, Wall R (1980) A mechanism for RNA splicing. Proc Natl Acad Sci USA 77:1877–1879
Rohde W, Sänger HL (1981) Detection of complementary RNA intermediates of viroid replication by northern blot hybridization. Biosci Rep 1:327–336
Rohde W, Schnölzer M, Rackwitz HR, Haas B, Seliger H, Sänger HL (1981a) Specifically primed synthesis in vitro of full-length DNA complementary to potato-spindle-tuber viroid. Eur J Biochem 118:151–157
Rohde W, Schnölzer M, Sänger HL (1981b) Sequence-specific priming of the in vitro synthesis of DNA complementary to citrus exocortis viroid. FEBS Lett 130:208–212
Rohde W, Rackwitz H-R, Boege F, Sänger HL (1982) Viroid RNA is accepted as a template for in vitro transcription by DNA-dependent DNA polymerase I and RNA polymerase from *Escherichia coli*. Biosci Rep (in press)
Roistacher CN, Calavan EC, Blue RL (1969) Citrus exocortis virus – Chemical inactivation on tools, tolerance to heat and separation of isolates. Plant Dis Rep 53:333–336
Roistacher CN, Calavan EC, Blue RL, Navarro L, Gonzales R (1977) A new more sensitive citron indicator for detection of mold isolates of citrus exocortis viroid (CEV). Plant Dis Rep 61:135–139
Romaine CP, Horst RK (1975) Suggested viroid etiology for chrysanthemum chlorotic mottle disease. Virology 64:86–95
Romaine CP, Zaitlin M (1978) RNA-dependent RNA polymerases in uninfected and tobacco mosaic virus-infected tobacco leaves: Viral-induced stimulation of a host polymerase activity. Virology 86:241–253

Rott R, Scholtissek C (1970) Specific inhibition of influenza replication by α-amanitin. Nature (London) 228:56

Sänger HL (1972) An infectious and replicating RNA of low molecular weight: The agent of the exocortis disease of citrus. Adv Biosci 8:103–116

Sänger HL (1979a) Structure and function of viroids. In: Prusiner SB, Headlow WJ (eds) Slow transmissible diseases of the nervous system. Vol 2. Academic Press, London New York

Sänger HL (1979b) Viroide, eine neue Klasse molekularer Krankheitserreger. Mitt Ber Robert-Koch-Stiftung 1:71–99

Sänger HL (1980a) Structure and possible functions of viroids. Ann NY Acad Sci 354:251–278

Sänger HL (1980b) Viroids: Biology, structure and possible functions. In: Leaver CJ (ed) Genome organisation and expression in plants. Plenum, New York London

Sänger HL, Knight CA (1963) Action of actinomycin D on RNA synthesis in healthy and virus-infected tobacco leaves. Biochem Biophys Res Commun 13:455–461

Sänger HL, Ramm K (1975) Radioactive labelling of viroid-RNA. In: Markham R, Davies DR, Hopwood DA, Horne RW (eds) Modification of the information content of plant cells, North-Holland/American Elsevier, Amsterdam, pp 229–252

Sänger HL, Klotz G, Riesner D, Gross HJ, Kleinschmidt AK (1976) Viroids are single-stranded covalently closed circular RNA molecules existing as highly basepaired rod-like structures. Proc Natl Acad Sci USA 73:3852–3856

Sänger HL, Ramm K, Domdey H, Gross HJ, Henco K, Riesner D (1979) Conversion of circular viroid molecules to linear strands. FEBS Lett 99:117–122

Sammons DW, Adams LD, Nishizawa E (1981) Ultrasensitive silver-based color staining of polypeptides in polyacrylamide gels. Electrophoresis 2:135–141

Sanger F, Air GM, Barrell BG, Brown NL, Coulson AR, Fiddes JC, Hutchinson III CA, Slocombe PM, Smith M (1977) Nucleotide sequence of bacteriophage ΦX174 DNA. Nature (London) 265:687–695

Sasaki M, Shikata E (1977a) Studies on the host range of hop stunt disease in Japan. Proc Jpn Acad 53B:103–108

Sasaki M, Shikata E (1977b) On some properties of hop stunt disease agent, a viroid, Proc Jpn Acad 53B:109–112

Sasaki R, Goto H, Arima K, Sasaki Y (1974a) Effect of polyribonucleotides on eukaryotic DNA-dependent RNA polymerases. Biochim Biophys Acta 366:435–442

Sasaki Y, Goto H, Wake T, Sasaki R (1974b) Purine ribonucleotide homopolymer formation activity of RNA polymerase from cauliflower. Biochim Biophys Acta 366:443–453

Sasaki Y, Goto H, Ohta H, Kamikubo R (1976) Template activity of synthetic deoxyribonucleotide polymers in the eukaryotic DNA-dependent RNA polymerase reaction. Eur J Biochem 70:369–375

Sassone-Corsi P, Corden J, Kedinger C, Chambon P (1981) Promotion of specific in vitro transcription by excised "TATA" box sequences inserted in a foreign nucleotide environment. Nucl Acids Res 9:3941–3958

Schaller J, Voss H, Gucker S (1969) Structure of the DNA of bacteriophage fd. II. Isolation and characterisation of a DNA Fraction with double strand-like properties. J Mol Biol 44:445–458

Schultz ES, Folsom D (1923a) Transmission, variation, and control of certain degeneration diseases of Irish potatoes. J Agric Res 25:43–117

Schultz ES, Folsom D (1923b) Spindling-tuber and other degeneration diseases of Irish potatoes. Phytopathology 13:40

Schultz ES, Folsom DA (1923c) A "spindling-tuber disease" of Irish potatoes. Science 57:149

Seitz U, Seitz U (1971) Selektive Hemmung der Synthese der AMP-reichen RNS durch α-Amanitin in Zellen höherer Pflanzen. Planta 97:224–229

Semancik JS (1976) Structure and replication of plant viroids. In: Animal virology. Academic Press, London New York

Semancik JS (1979) Small pathogenic RNA in plants – the viroids. Annu Rev Phytopathol 17:461–484

Semancik JS, Desjardins PR (1980) Multiple small RNA species and the viroid hypothesis for the sunblotch disease of avocado. Virology 104:117–121

Semancik JS, Geelen JLMC (1975) Detection of DNA complementary to pathogenic viroid RNA in exocortis disease. Nature (London) 256:753–756
Semancik JS, Vanderwoude WJ (1976) Exocortis viroid: Cythopathic effects at the plasma membrane in association with pathogenic RNA. Virology 69:719–726
Semancik JS, Weathers LG (1972a) Exocortis disease: Evidence for a new species of "infectious" low molecular weight RNA in plants. Nature (New Biol) 237:242–244
Semancik JS, Weathers LG (1972b) Pathogenic 10 S RNA from exocortis disease recovered from tomato bunchy-top plants similar to potato spindle tuber virus infection. Virology 49:622–625
Semancik JS, Magnuson DS, Weathers LG (1973) Potato spindle tuber disease produced by pathogenic RNA from citrus exocortis disease: Evidence for the identity of the causal agent. Virology 52:292–294
Semancik JS, Tsuruda D, Zaner L, Geelen JLMC, Weathers JG (1976) Exocortis disease: Subcellular distribution of pathogenic (viroid) RNA. Virology 69:669–676
Semancik JS, Conjero V, Gerhart J (1977) Citrus exocortis viroid: Survey of protein synthesis in *Xenopus laevis* oocytes following addition of viroid RNA. Virology 80:218–221
Semancik JS, Grill LK, Civerolo EL (1978) Accumulation of viroid RNA in tumor cells after double infection by *Agrobacterium tumefaciens* and citrus exocortis viroid. Phytopathology 68:1288–1292
Sharp PA (1981) Speculations on RNA splicing. Cell 23:643–646
Singh A, Sänger HL (1976) Chromatographic behaviour of the viroids of the exocortis disease of citrus and of the spindle tuber disease of citrus. Phytopathol 87:143–160
Singh RP (1973) Experimental host range of the potato spindle tuber "virus". Am Potato J 50:111–123
Singh RP (1977) Piperonyl butoxyde as a protectant against potato spindle tuber viroid infection. Phytopathology 67:933–935
Singh RP, Bagnall RH (1968) Infectious nucleic acid from host tissues infected with potato spindle tuber virus. Phytopathology 58:696–699
Singh RP, Clark MC (1971) Infectious low-molecular-weight ribonucleic acid. Biochem Biophys Res Commun 44:1077–1082
Singh RP, Clark MC (1973) Similarity of host response to both potato spindle tuber and citrus exocortis viruses. FAO Plant Prot Bull 21:121–125
Singh RP, Finnie RE, Bagnall RH (1970) Relative prevalence of mild and severe strains of potato spindle tuber virus in Eastern Canada. Am Potato J 47:289–293
Singh RP, Finnie RE, Bagnall RH (1971) Losses due to the potato spindle tuber virus. Am Potato J 48:262–267
Singh RP, Michniewicz JJ, Narang SA (1974) Multiple forms of potato spindle-tuber metavirus ribonucleic acid. Can J Biochem 52:809–812
Singh RP, Michniewicz JJ, Narang SA (1975) Piperonyl butoxyde, a potent inhibitor of potato spindle tuber viroid in *Scopolia sinensis*. Can J Biochem 53:1130–1132
Sogo JM, Koller T, Diener TO (1973) Potato spindle tuber viroid. X. Visualization and size determination by electron microscopy. Virology 55:70–80
Somerville LL, Wang K (1981) The ultrasensitive silver "protein" stain also detects nanograms of nucleic acids. Biochem Biophys Res Commun 102:53–58
Stace-Smith R, Mellor FC (1970) Eradication of potato spindle tuber virus by thermotherapy and axillary bud culture. Phytopathology 60:1857–1858
Stanley WM (1935) Isolation of a crystalline protein prossessing the properties of tobacco-mosaic virus. Science 81:644–645
Sun SM, Slightom JL, Hall TC (1981) Intervening sequences in a plant gene – comparison of the partial sequence of cDNA and genomic DNA of French bean phaseolin. Nature (London) 289:37–41
Symons RH (1981) Avocado sunblotch viroid – primary sequence and proposed secondary structure. Nucl Acids Res 9:6527–6537
Takahashi T, Diener TO (1975) Potato spindle tuber viroid. XIV. Replication in nuclei isolated from infected leaves. Virology 64:106–114
Takanami Y, Fraenkel-Conrat H (1982) Noviral gene is able to elicit RNA-dependent RNA polymerase in cucumber mosaic virus-infected cucumber cotelydons. Virology 116:372–374

Takebe I, Otsuki Y (1969) Infection of tobacco mesophyll protoplasts by tobacco mosaic virus. Proc Natl Acad Sci USA 64:843–848

Thomas W, Mohamed NA (1979) Avocado sunblotch – a viroid disease? Aust Plant Path Soc Newslett 1–2

Tien P, Davies C, Hatta T, Francki RIB (1981) Viroid-like RNA encapsidated in lucerne transient streak virus. FEBS Lett 132:353–356

Tinocco I Jr, Uhlenbeck OC, Levine MD (1971) Estimation of secondary structure in ribonucleic acids. Nature (London) 230:362–367

Trapnell BC, Tolstoshev P, Crystal RG (1980) Secondary structures for splice junctions in eukaryotic and viral messenger RNA precursors. Nucl Acids Res 8:3659–3672

Van Dorst HJM, Peters D (1974) Some biological observations on pale fruit, a viroid-incited disease of cucumber. Neth J Plant Path 80:85–96

Van Montagu M, Schell J (1982) The Ti plasmids of *Agrobacterium. Curr* Top Microbiol Immunol 96:236–254

Visvader JE, Gould AR, Bruening GE, Symons RH (1982) Citrus exocortis viroid-nucleotide-sequence and secondary structure of an Australian isolate. FEBS Lett 137:288–292

Wahn K, Rosenberg de Gomez R, Sänger HL (1980a) Cytopathologie von viroid-infiziertem Pflanzengewebe. I. Veränderungen des Plasmalemmas und der Zellwand bei *Gynura aurantiaca* DC nach Infektion mit dem Viroid der Citrus Exocortis Krankheit (CEV). Phytopathol Z 98:1–18

Wahn K, Rosenberg de Gomez F, Sänger HL (1980b) Cytopathic changes in leaf tissue of *Gynura aurantiaca* infected with the viroid of the exocortis disease of citrus. J Virol 49:355–365

Walter B (1981) Un viroide de la Tomate en Afrique de l'Ouest: identité avec le viroide du "potato spindle tuber"? CR Acad Sci Paris 292 III:537–542

Weathers LG, Calavan EC (1961) Additional indicator plants for exocortis and evidence for strain differences in the virus. Phytopathology 51:262–264

Weinmann R, Roeder RG (1974) Role of DNA-dependent RNA polymerase III in the transcription of the tRNA and 5S RNA genes. Proc Natl Acad Sci USA 71:1790–1795

White JL, Murakishi HH (1977) In vitro replication of tobacco mosaic virus RNA in tobacco callus cultures: Solubilization of membrane-bound replicase and partial purification. J Virol 21:484–492

Wieland Th, Faulstich H (1978) Amatoxins, Phallotoxins, Phallolysin and Antamanide: The biologically active components of poisonous *Amanita* mushrooms. Crit Rev Biochem 5:185–260

Wild U, Ramm K, Sänger HL, Riesner D (1980) Loops in viroids. Eur J Biochem 103:227–235

Willmitzer L, Schmalenbach W, Schell J (1981) Transcription of T-DNA in octopine and nopaline crown gall tumours is inhibited by low concentrations of α-amanitin. Nucl Acids Res 9:4801–4812

Wintermeyer W, Zachau HG (1973) Mg^{2+}-katalysierte, spezifische Spaltung von tRNA. Biochim Biophys Acta 299:82–90

Zaitlin M, Niblett CL, Dickson E, Goldberg RB (1980) Tomato DNA contains no detectable regions complementary to potato spindle tuber viroid as assayed by solution and filter hybridization. Virology 104:1–9

Zelcer A, Vanadels J, Leonhard DA, Zaitlin M (1981) Plant-cell suspension cultures sustain longterm replication of potato spindle tuber viroid. Virology 109:314–322

Zelcer A, Zaitlin M, Robertson HD, Dickson E (1982) Potato spindle tuber viroid-infected tissues contain RNA complementary to the entire viroid. J Gen Virol 59:139–148

Zieve GW (1981) Two groups of small stable RNAs. Cell 35:296–297

Zimmern D (1982) Do viroids and RNA viruses derive from a system that exchange genetic information between eukaryotic cells. Trends Biochem Sci 205–207

Zylber EA, Penman S (1971) Products of RNA polymerases in HeLa cell nuclei. Proc Natl Acad Sci USA 68:2861–2865

13 The Ti-Plasmids of *Agrobacterium tumefaciens*

J. SCHELL

1 Introduction

Many species of higher plants, in fact most dicotyledonous plants, are susceptible to crown gall disease (DE CLEENE and DE LEY 1976). This disease is characterized by the formation of tumors, often at the crown separating stem from roots, as a result of infection of wounded sites by gram-negative soil bacteria (*Agrobacterium tumefaciens*). Although these bacteria are required for tumor induction and are mostly found associated with crown gall harboring plants in nature, they are not necessary for tumor maintenance and growth (BRAUN 1953). In fact, sterile crown gall tissues can readily be cultured indefinitely on simple media lacking any added growth hormones. Crown gall plant cells therefore are persistently altered cells with a capacity for autonomous growth in the absence of any outside stimulation by the inciting bacteria. The normal mechanisms controlling growth and differentiation of plant cells do not operate properly in these transformed cells and, as such, crown gall cells can be compared with animal tumor cells.

BRAUN and MANDLE (1948) were the first to conclude from their observations that a factor (tumor-inducing or TIP factor) is transmitted from the inciting bacteria to the host cells, resulting, in a relatively short time (about 36 h), in a stable transformation of the plant cells into crown gall tumor cells. As a result of the action of the TIP, the normal differentiation and growth control mechanisms are altered or inactivated. The cells acquire a capacity for autonomous growth, possibly because they persistently synthesize or otherwise become independent of the growth-regulating substances and metabolites that are externally required for cellular growth and division of normal, untransformed cells.

Because of its relation to animal cancers and its value as an ideal experimental model system to study the control mechanisms underlying cellular growth and differentiation, crown gall has been extensively and continuously studied ever since the beginnings of this century (SMITH 1916, JENSEN 1918, BRAUN and WHITE 1943, BRAUN 1980).

Most conceivable components of the bacterial cells, such as toxins, bacterially produced plant growth stimulators, endoantigens, bacterial DNA, bacteriophages, small RNA's, etc., were at one time or another implicated as being the hypothetical tumor-inducing principle. None of these claims was ultimately

---

Abbreviations. Onc, oncogenicity; Ti, tumor-inducing; TIP, tumor-inducing principle; T-DNA, DNA containing the Ti plasmid DNA fragment; T-region (of Ti plasmid), region transferred into host cell genome; Tra genes, genes controlling Ti plasmid transfer, (see also legends to Fig. 1, 2)

confirmed. Until around 1970, most of the crown gall research was centered on the study of the properties of transformed plant cells. It is largely because some of the attention shifted to the study of the biological and genetic properties of the inciting bacteria, that not only the nature of the tumor-inducing principle could be discovered and firmly established, but also that the biological significance of the crown gall phenomenon could be understood.

The observations of MOREL and his collaborators were one of the grounds for a more fundamental implication of the inciting bacteria: these authors (PETIT et al. 1970) demonstrated that there are at least two different types of crown gall-inducing bacteria, because they induce tumors with different properties. One type of *Agrobacterium* strains induced crown gall tumors in which N-α-(D-1-carboxyethyl)-L-arginine (octopine) was synthesized, whereas the other type induced tumors containing N-α-(1,3-dicarboxypropyl)-L-arginine (nopaline). The important fact here is that the type of arginine derivative synthesized in the tumor was found to be specified by the particular *Agrobacterium* strain used to incite the tumor and to be dependent of the host plant on which the tumors were induced.

It was furthermore demonstrated that *Agrobacterium* strains that induce the synthesis of octopine in crown gall can selectively use this product, but not nopaline, as sole energy, carbon and/or nitrogen source, whereas *Agrobacterium* strain that induce the synthesis of nopaline in crown gall can selectively use it, but not octopine, as energy, carbon, and nitrogen source. Thus, a genetic linkage between oncogenicity and "opine" metabolism was first established and the hypothesis proposed that genetic information could be somehow transferred from bacterium to plant.

The name "opine" was subsequently proposed (TEMPÉ and SCHELL 1977, TEMPÉ et al. 1978) to describe products, specifically synthesized by crown gall plant cells, that can be used by *Agrobacteria* as specific growth substances.

The identification of the nature and function of the elusive tumor-inducing principle resulted primarily from experiments designed to identify the genetic determinants in *Agrobacterium* responsible for its tumor-inducing capacity. Thus, it was first reported in 1973 (see SCHELL 1975), that the tumor-inducible principle of *Agrobacteria* was carried by extrachromosomal DNA elements of the plasmid type and not of the lysogenic type as had been previously reported.

The experimental demonstration of the involvement of Ti-plasmids (ZAENEN et al. 1974) in oncogenicity followed rapidly from work in different laboratories (VAN LAREBEKE et al. 1974, 1975, WATSON et al. 1975, ENGLER et al. 1975, BOMHOFF et al. 1976) and relied on the further analysis of some of the many contradictory observations previously published about the nature of the tumor-inducing principle. Thus, the observations by HAMILTON and FALL (1971) that certain select strains of crown gall bacteria, such as, for example, strain C58, irreversibly lost their tumor-inducing capacity when grown at 36 °C, were confirmed and shown to be due to the irreversible loss by the non-oncogenic bacteria of their large extrachromosomal Ti-plasmids. The remarkable observations of KERR (1969, 1971), who discovered that oncogenicity could be transferred from one strain of *Agrobacterium* to another by inoculating both strains, together or in succession, into the same susceptible plant, were also confirmed and shown to be due to the transmission of Ti-plasmids between *Agrobacterium* strains. Since the discovery of Ti-plasmids and their central role in the crown gall phenomenon, a number of important observations have allowed the formulation of a general concept describing and explaining this neoplasmic transformation of plant cells. Thus, the demonstration that the genes controlling opine catabolism in *Agrobacteria* and the genes determining opine synthesis in transformed plant cells are both localized on Ti-plasmids (BOMHOFF et al. 1976, KERR and ROBERTS 1976, GENETELLO et al. 1977, KERR et al. 1977, MONTOYA et al. 1977, HOOYKAAS et al. 1977, VAN LAREBEKE et al. 1977, SCHELL et al. 1977, HOLSTERS et al. 1978a, KLAPWIJK et al. 1978) explain the linkage between opine metabolism and oncogenicity and provide genetic evidence in favor of a model involving the Ti-plasmid in a mechanism of DNA transfer from bacterium to plant. The obvious way to explain the involvement of the Ti-plasmids both in oncogenicity and in opine synthesis control was to assume that part or the whole of the Ti-plasmid was somehow transferred to – and maintained and expressed in – the transformed plant cells. The experiments demonstrating that this assumption is correct were based on a detailed genetic analysis of the Ti-plasmid and on the demonstration, by means of DNA/DNA hybridizations, of the physical presence of a Ti-plasmid DNA fragment (the so-called T-DNA) in the nucleus of transformed plant cells. These observations will be described in more detail in the following sections and can be fitted in a general concept that was called "genetic colonization" (SCHELL et al. 1979). According to this concept, the Ti-plasmids are natural gene vectors for plant cells, evolved by and for the benefit of the bacteria that harbor Ti-plasmids. Crown gall cells as a direct result of genetic transformation by the Ti-plasmid proliferate to form tumors in which various substances (opines) are produced and released. Free-living *Agrobacteria* specifically use these opines as energy, carbon, and/or nitrogen sources. Bacteria, in general, are known to be able to conquer an ecological niche by acquiring the capacity to catabolize certain organic compounds not readily degradable by most other bacterial species. Several groups of soil bacteria, living in and around the rhizosphere of plants, are able to decompose organic compounds released by plants. With the advent of Ti-plasmids, *Agrobacterium* has carried this capacity one step further by genetically forcing plant cells – via a gene transfer event – to produce specific compounds (opines) which they are uniquely equipped to catabolize.

2 Ti-Plasmids Are Catabolic Plasmids and Natural Gene Vectors for Plants

All Ti-plasmids in different *Agrobacterium* strains are not identical. The DNA base sequence in different Ti-plasmids can vary considerably (CURRIER and NESTER 1976, GENETELLO et al. 1977, VAN LAREBEKE et al. 1977, SCHELL and VAN MONTAGU 1977a, SCIAKY et al. 1978, ENGLER et al., submitted). However, consistent classification of Ti-plasmids can be obtained when one groups the

Ti-plasmids according to the type of opine they specify (synthesis in crown galls and catabolism by bacteria). Thus far, the Ti-plasmids that have been studied fall into three classes: (1) "octopine" Ti-plasmids which code for octopine (and related opines such as octopinic acid, lysopine, and histopine, and also for agropine metabolism) (FIRMIN and FENWICK 1978, COXON et al. 1980); (2) "nopaline" Ti-plasmids which code for nopaline and ornaline metabolism (FIRMIN and FENWICK 1977, KEMP et al. 1979) and for a newly discovered opine, called agrocinopine (ELLIS, KERR, PETIT, TEMPÉ, unpublished data); and (3) "agropine" Ti-plasmids which are related to octopine plasmids and code for agropine metabolism only (GUYON et al. 1980, COXON et al. 1980). Octopine and nopaline plasmids have been studied most extensively. We shall first review our knowledge about these plasmids and subsequently compare them with other Ti-plasmids and with related plasmids from *A. rhizogenes*.

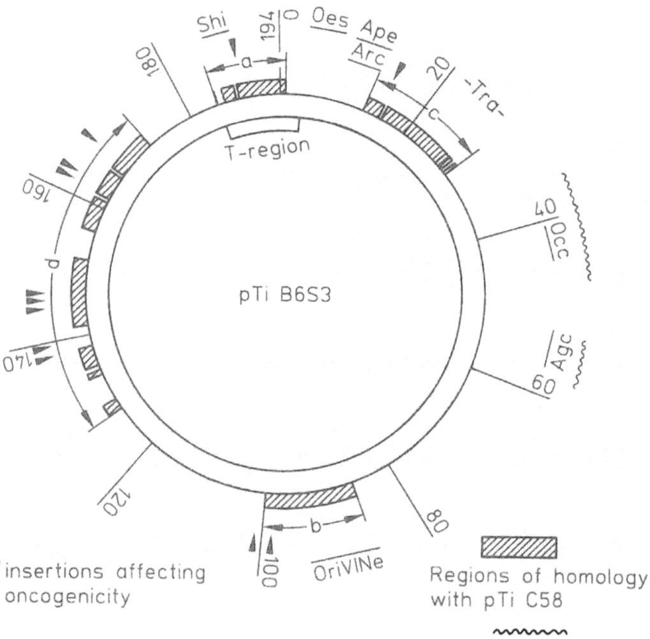

a

Fig. 1 a, b. Functional map of a standard octopine (Ach5) and nopaline (C58) Ti-plasmid. The four areas of homology, *a, b, c, d,* between these plasmids are indicated by *hatched boxes*. *Arrow heads* indicate the sites where nontumorogenic (Onc^-) mutants were localized. *Wavy lines* indicate the areas where functions involved in opine catabolism were mapped. The location of some further mutants such as mutants in incompatibility (*Inc*), shoot induction (*Shi*), transfer function (*Tra*), phage exclusion (*Ape*) as well as biosynthesis of nopaline (*Nos*), catabolism of nopaline (*Noc*), catabolism of octopine (*Occ*), catabolism of arginine (*Arc*), catabolism of agropine (*Agr*) and catabolism of phosphorylated sugars (*Psc*) are indicated

2.1 Genetic and Functional Organization of Octopine and Nopaline Ti-Plasmids

Ti-plasmid- encoded functions were defined by comparing the properties of a given *Agrobacterium* strain with and without a given Ti-plasmid. Ti-plasmids were thus found to be responsible for the following properties: (1) crown gall tumor induction; (2) specificity of opine synthesis in transformed plant cells; (3) catabolism of specific opines; (4) agrocin sensitivity; (5) conjugative transfer of Ti-plasmids; and (6) catabolism of arginine and ornithine (ZAENEN et al. 1974, VAN LAREBEKE et al. 1974, 1975, SCHELL 1975, WATSON et al. 1975, ENGLER et al. 1975, BOMHOFF et al. 1976, GENETELLO et al. 1977, KERR et al. 1977, PETIT et al. 1978a, 1978b, FIRMIN and FENWICK 1978, KLAPWIJK et al. 1978, ELLIS et al. 1979, GUYON et al. 1980).

In order to further establish that these properties are determined by plasmid genes, mutant plasmids were isolated by transposon-insertion mutagenesis (HERNALSTEENS et al. 1978, VAN MONTAGU and SCHELL 1979, HOLSTERS et al. 1980, OOMS et al. 1980, DE GREVE et al. 1981) and by deletion (KOEKMAN et al. 1979, HOLSTERS et al. 1980). Finally, the different mutations were localized on the physical maps established for nopaline (DEPICKER et al. 1980) and octopine Ti-plasmids (DE VOS et al. 1981), thus allowing the establishment of functional genetic maps for these plasmids. A summary of these results is diagrammed in Fig. 1.

A number of general conclusions can be derived from this genetic analysis of Ti-plasmids. Mutants of all the expected Ti-determined phenotypes have

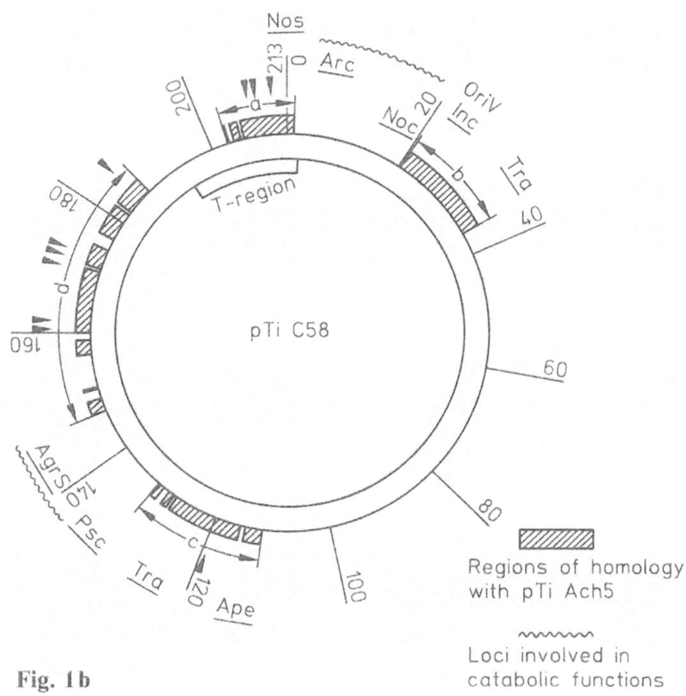

Fig. 1b

been found and assigned to specific regions of the Ti-plasmid maps. Several regions of Ti-plasmids must be directly or indirectly involved in determining the different steps leading to neoplastic transformation. Indeed, mutants affecting so-called Onc (oncogenicity) functions are distributed over about half of the Ti-plasmid map; the other half does not appear to carry any genes involved in transformation. No Onc^- mutants were observed in this region of various Ti-plasmids (right half of maps in Fig. 1) and large deletions, eliminating most of this region, were obtained that do not affect crown gall formation by these plasmids.

One class of mutations localized within the so-called T-region of Ti-plasmids is of particular interest, since this region is transferred and maintained in transformed plant cells (see below). The mutations thus far localized in this region are of four general types:

1. Mutant plasmids inducing crown gall tumors in which no nopaline synthesis occurs. These mutations demonstrate that opine synthesis and tumor induction and maintenance are coded for by different genes. Furthermore, since these opine synthesis-deficient Ti-plasmid mutants still allow *Agrobacterium* harboring them to catabolize opines, it must be concluded that genes involved in opine synthesis (after transfer to the plant cells) are not in any way involved in opine catabolism.
2. Mutants that totally or incompletely abolish tumor formation.
3. Mutants that strikingly influence the organogenic activity of transformed plant cells. Three types of such mutants have been isolated thus far, each localized in a well-defined segment of the T-region: (1) shoot formers, (2) root formers and (3) shoot *and* root formers. From this last class of mutants normal plants can be regenerated which synthesize opines in all their tissues (stems, leaves, and roots).
4. Mutants that do not visibly affect any of the known crown gall phenotypes.

On the basis of the localization of the various mutants, a functional map of the T-region can be drawn (Fig. 2). The most important conclusion from this genetic analysis is that the T-region must contain genes that play a direct, active, and specific role after transfer into plant cells. Some of these genes specify opine biosynthesis, whereas other genes directly or indirectly control plant cellular growth and differentiation.

Interestingly, some Onc mutants (both within and outside of the T-region) suggest a host-specific interaction. For example, some mutants affect functions which are dispensable for tumor formation on *Kalanchoe,* but not on tobacco or potatoes. There is no information about the products encoded by Onc regions outside of the T-region except that several of these DNA segments, when cloned in *E. coli* plasmids, synthesize proteins in minicells (DHAESE, unpublished data). Ti-plasmids, like most large bacterial plasmids, were found to be conjugative or autotransferable (KERR et al. 1977, GENETELLO et al. 1977, HOLSTERS et al. 1978b): i.e., they promote their own transfer from one bacterial host to another. It was therefore to be expected that Ti-plasmids would carry an elaborate set of genes (so-called Tra genes) to control this transfer. Tra^- plasmid mutants were obtained by deletion and insertion mutagenesis and mapped in at least two different regions of nopaline Ti-plasmids (HOLSTERS et al. 1980). It had

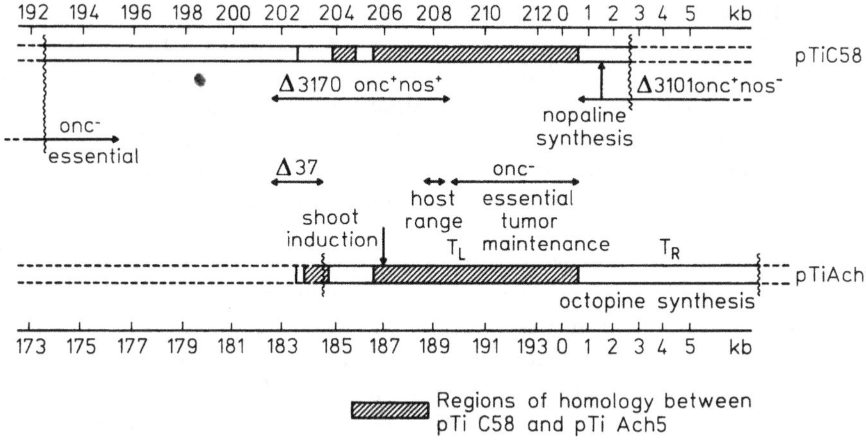

Fig. 2. The functional organization of the T-region of the nopaline plasmid, pTiC58, and the octopine plasmid, pTiAch5. The regions of homology between these plasmids are indicated by *hatched boxes*. The coordinates are given in kilobases (kb). *Horizontal arrows* mark the extent of some deletion mutants; certain characteristic properties such as tumorogenicity (*onc*) and the capacity to synthesize nopaline (*nos*) are indicated. It is important to note that although the T-regions of octopine and nopaline Ti-plasmids differ significantly in DNA sequence, they nevertheless have a similar organization. The central part is made up by a DNA stretch which is highly conserved in different Ti-plasmids (DEPICKER et al. 1978; CHILTON et al. 1978). It is in this conserved DNA that most of the mutations affecting tumor maintenance or tumor growth and differentiation are localized. To the right of this conserved region the genes involved in specifying opine synthesis are found. In nopaline tumors the T-DNA extends to the left of the conserved region. No phenotypes have thus far been associated with this region of the T-DNA

been suggested that some of the Tra functions essential for bacterial conjugation would also be involved in the transfer of plasmid DNA to plant cells (TEMPE et al. 1977). However, most of the Tra^- mutations obtained thus far are still oncogenic, indicating that the transfer of plasmid DNA to plants occurs normally, notwithstanding the fact that bacterial conjugation is impaired.

A very significant part of the T-region (more than half in the case of nopaline Ti-plasmids) is not essential for tumor induction or maintenance. It can therefore be postulated that most of this DNA codes for opine biosynthetic functions. If this assumption is correct, one might expect to find yet other and novel types of opines in crown gall plant cells. The catabolic functions of Ti-plasmids provide *Agrobacterium* with the capacity to use opines as sole energy, carbon and/or nitrogen source. These functions have been mapped (see Fig. 1) and the plasmid DNA fragments thus identified have been isolated and propagated by molecular cloning techniques. Their expression in Ti-plasmid-free *Agrobacteria* was studied (unpublished results from the author's laboratory) and points to the existence of a complex operon structure possibly containing both positive and negative controlling elements. Such an operon probably consists of genes for a permease (KLAPWIJK et al. 1977), the octopine or nopaline dehydrogenase and enzymes involved in the further degradation of arginine and ornithine (ELLIS et al. 1979).

Mutants defective in the catabolism of agropine map in a Ti-plasmid segment distinct from the segment involved in octopine catabolism. Likewise, the genes involved in the uptake and degradation of agrocinopine (ELLIS, KERR, PETIT and TEMPÉ, unpublished data) by nopaline-type Ti-plasmids map in a segment opposite the segment involved in nopaline catabolism. The degradation of the different opines is therefore controlled by a set of different genes and each Ti-plasmid may harbor catabolic functions for more than one type of opine. Plasmids that code for the degradation of a given opine but do not harbor genes specifying the synthesis of this opine in their T-region have been observed and can usually be regarded as naturally occurring mutant plasmids.

It is important to note that most of the mutants which have thus far been found to affect tumor formation are localized in Ti-plasmid sequences that are homologous between octopine and nopaline Ti-plasmids (see Fig. 1 and ENGLER et al. 1981). Interestingly, both the genes in the T-region involved in opine synthesis and the various genes outside the T-region, involved in opine and amino acid catabolism have different (nonhomologous) DNA sequences in different Ti-plasmid types. On the other hand, most of the Tra functions (involved in the conjugative transfer of Ti-plasmids) are localized in DNA sequences conserved amongst different types of Ti-plasmids. The Ti-plasmids studied thus far turn out to be repressed for autotransferability, i.e., the conjugative Tra functions are normally repressed but are induced by the presence of specific opines, thus further stressing the central role of opine synthesis and catabolism in the evolution of Ti-plasmids (PETIT et al. 1978a, KLAPWIJK et al. 1978).

2.2 Generality of the Opine and Genetic Colonization Concepts

In the introduction we have described the observations that led to the proposal that the biological rationale underlying the evolution of Ti-plasmids is that they are catabolic plasmids with the added capacity to transform plant cells in order to reprogram them for the biosynthesis of specific and highly selective organic nutrients (opines) for the benefit of the inciting bacteria. If these concepts are correct one would expect: (1) that most, if not all, crown gall lines should produce opines; (2) that opine synthesis and catabolism should be coupled; (3) that opines should have played a central role in the evolution and spreading of different Ti-plasmid types; (4) that other, functionally related, plasmids would exist that likewise function to promote host/parasite or host/symbiont interactions.

Evidence that fulfills these predictions is accumulating rapidly. GUYON et al. (1980) have recently demonstrated that crown galls, induced by "null type" Ti-plasmid-harboring bacteria (SCIAKY et al. 1978), do in fact produce an opine, identical to the agropine first observed as a second opine in octopine tumor lines (FIRMIN and FENWICK 1978). This opine is a condensation of the lactam form of glutamic acid and a hexitol (COXON et al. 1980). Except for some opine synthesis-defective Ti-plasmid mutants (experimental or natural) producing crown galls devoid of one or another opine, the prediction is therefore fulfilled that all crown gall lines produce one or more types of opines that can specifically be catabolized by bacteria containing the inciting type of Ti-plasmid.

The Ti-plasmids studied thus far turn out to be repressed for autotransferability. Conjugation is, however, induced by the presence of opines (PETIT et al. 1978a, KLAPWIJK et al. 1978, KLAPWIJK and SCHILPEROORT 1979). *Agrobacterium* strains will therefore spread their Ti-plasmids when in favorable presence of crown gall tissue and opines play the control role in this spreading mechanism.

The general value of these observations is perhaps most elegantly demonstrated by the recent work of ELLIS, KERR, MURPHY, PETIT and TEMPÉ. Most nopaline strains are sensitive to a nucleotide antibiotic called agrocin (TATE et al. 1979), a compound synthesized by some other *Agrobacterium* strains. Agrocin-producing *Agrobacterium* strains are able to compete efficiently with classical nopaline strains for territory (NEW and KERR 1972), and they have been put to good use to protect fruit trees against crown gall formation (ELLIS and KERR 1978). The sensitivity of nopaline strains to the action of agrocin was shown to be determined by Ti-plasmid functions (ENGLER et al. 1975), which probably encode an efficient uptake mechanism for agrocin. Recently, it was shown that a membrane protein was responsible for the high-affinity uptake of agrocin (MURPHY and ROBERTS 1979). One may wonder what beneficial function, important enough to be present in most nopaline strains, this uptake protein performs. Since agrocin contains a phosphorylated sugar moiety, a type of compound normally not taken up by bacteria, it was conceivable that some phosphorylated sugars, present in crown gall, could be substrates for the agrocin-sensitive strains and thus form a new class of opines. By studying crown gall compounds able to enhance the agrocin sensitivity of nopaline strains, a set of compounds, denoted agrocinopines, were found which belong to this opine class (ELLIS, KERR, PETIT and TEMPÉ, unpublished data). These agrocinopines were also found to induce the conjugative properties of nopaline Ti-plasmids (such as pTiC58), although nopaline itself cannot induce the autotransferability of this strain (ELLIS and MURPHY 1981). It is also important to note that only opines that are actually substrates for a given Ti-plasmid can induce the conjugative properties of this plasmid.

Another argument in favor of the generality of the opine concept stems from the study of the hairy root tumors induced by *Agrobacterium rhizogenes* strains. The disease is characterized by abundant proliferation of roots at the wound sites. Large Ti-like plasmids were shown to be responsible for the rhizogenicity (MOORE et al. 1979, WHITE and NESTER 1980a) because the transfer of *rhizogenes* plasmids to a Ti-plasmidless *Agrobacterium tumefaciens* strain resulted in receptor strains that induce hairy root disease.

These *rhizogenes* plasmids are compatible with Ti-plasmids (COSTANTINO et al. 1980) but show only a limited base sequence homology with Ti-plasmids (WHITE and NESTER 1980b). Especially the fact that there is only a weak homology with the T-region of Ti-plasmids (WHITE and NESTER 1980b) is of particular interest.

Notwithstanding this striking differences with Ti-plasmids, *A. rhizogenes* strains were shown to be able to utilize the opine agropine and TEMPÉ et al. (unpublished data) have shown that hairy roots in axenic culture produce agropine. Thus, it is shown that the opine concept is really at the basis of the plant cell transformations induced by these bacterial plasmid genes.

2.3 The Transfer of the T-Region to Plant Cells

To establish whether or not Ti-plasmid DNA is stably transferred and maintained in crown gall plant cells, DNA/DNA hybridizations were performed. The most detailed results were obtained by Southern blotting hybridizations involving DNA isolated from several independently isolated crown gall tissues. The DNA's were digested with various restriction enzymes, and radioactively labeled isolated fragments covering the whole of the Ti-plasmid were used as probes in the Southern analysis (DE BEUCKELEER et al. 1978, SCHELL et al. 1979, ZAMBRYSKI et al. 1980, THOMASHOW et al. 1980, YADAV et al. 1980, LEMMERS et al. 1980).

The results show that all crown gall tumors contain a DNA segment, called the T-DNA, which is homologous to DNA sequences in the Ti-plasmid used to induce the tumor line. In all cases, this T-DNA corresponds to – and is colinear with – a continuous stretch of Ti-plasmid DNA, which has therefore been called the T-region (see above).

In different tumors induced on tobacco by the nopaline strains C58 and T37, the T-DNA segment was found to be identical with a size of about 23 Kb. The T-DNA in octopine tobacco crown gall lines was found to be smaller (about 14–15 Kb) and somewhat variable in size, although always derived from the T-region. A detailed restriction map of the T-DNA of nopaline strains has been published (DEPICKER et al. 1980).

By preparing DNA from purified nuclei, chloroplasts and mitochondria isolated from crown gall tissues, it was shown that the T-DNA is located in the nucleus and not in chloroplast or mitochondria (WILLMITZER et al. 1980, CHILTON et al. 1980). When left- or right-hand border fragments of the T-region were used as probes in the Southern blot hybridization analysis, a small number of bands (usually 1 to 3) became apparent. The exact number and the sizes of these composite fragments, presumably consisting of T-DNA covalently linked to plant DNA, varied from tumor line to line (and sometimes within a given line) as a function of the restriction enzyme used. These observations indicated that more than one copy of T-DNA is integrated at more than one site in plant DNA.

In contrast, when internal fragments of the T-region were used as hybridization probes, only single bands were revealed, corresponding in each case to the full length of the T-region restriction fragment used as probe. The bands hybridizing to different internal fragments of the T-region demonstrated the same relative intensity.

Taken together, these data suggest that tumor cells contain a limited number (possibly up to 5) of complete copies of T-DNA associated with different plant DNA sequences. Further demonstration of the integration of T-DNA in plant DNA and an understanding of the structure of T-DNA was obtained when a complete *Eco*RI digest of T37 tobacco crown gall DNA was cloned in a phage λ vector, thus allowing the isolation and detailed study of T-DNA border sequences (ZAMBRYSKI et al. 1980). Two clones were studied in detail. One clone contains the left and right borders of the T-DNA linked together and the other clone contains the right end of the T-DNA linked to repetitive plant DNA sequences. These data therefore suggest that the T-DNA in nopaline tumors is organized as tandem repeats which are inserted into repetitive sequences of plant DNA.

It should be noted, however, that it has not yet been proven whether the T-DNA is a stable part of a plant chromosome. It is also possible that there may be two states of the T-DNA, one integrated and one not.

The sum of the evidence derived from comparative studies involving different tumor lines induced with the same Ti-plasmid; tumor lines induced with Ti-plasmid mutants carrying transposon-insertions in the T-region (LEMMERS et al. (1980); and also from a detailed study of the nucleotide sequences bordering the T-DNA segment in cloned T-DNA fragments (ZAMBRYSKI et al. 1980) suggest that the "ends" of the T-region are involved in the integration of the T-DNA. Either the T-region ends recombine before integration into the plant DNA or the entire T-region is excised and subsequently inserted.

The T-DNA therefore has one central property in common with bacterial transposons since it appears to be a discrete unit of DNA with the capacity to integrate into nonhomologous DNA.

3 Expression of T-DNA in Plant Cells

3.1 Transcription of T-DNA Sequences

The first positive evidence that the T-DNA is transcribed within the plant cell was reported by DRUMMOND et al. (1977), who demonstrated that total RNA isolated from crown gall tumors hybridized to a specific fragment of the Ti-plasmid assumed to be present within the plant cell. A more extensive study was performed by GURLEY et al. (1979), who studied transcription in three different octopine tumors and found that the right part of the T-DNA was transcribed most actively, whereas little transcription was detected from the left part. A serious drawback in both studies, however, was that the extent of the T-DNA in the tumor lines studied was not known at the time. Therefore, these studies did not allow the construction of a detailed picture of the transcribed parts of the T-DNA. Whereas the transcription studies by DRUMMOND et al. (1977) and GURLEY et al. (1979) were performed with total RNA, we decided to study in detail the transcription of the T-DNA in two octopine tumors with special emphasis on comparison between nuclear and polysomal RNA. For these studies, tumor lines with accurately mapped T-DNA's were used.

The following approach was used: (1) nuclear RNA was isolated from crown gall cells and labeled by polynucleotide kinase. Furthermore, radioactive cDNA was made from nuclear RNA using oligo (dT) as primer. (2) Polysomal RNA was isolated from crown gall cells. After separation into poly $(A)^+$ and poly

(A)$^-$ fractions by oligo (dT)-cellulose chromatography, the RNA's were labeled by polynucleotide kinase. (3) Highly purified and physiologically active nuclei were isolated from crown gall cells and incubated under conditions allowing RNA synthesis by endogenous RNA polymerase. Synthesized RNA was labeled by incorporation of [α-^{32}P]UTP.

To map transcribed regions, the labeled RNA was hybridized to Southern blots containing cloned parts of the T-DNA digested by various restriction enzymes. To find out which RNA polymerase is responsible for the transcription of the T-DNA, transcription was studied both in the presence and in the absence of various concentrations of α-amanitin.

These studies have indicated that the T-DNA is transcribed over its entire length. However, well-defined regions of the T-DNA are reproducibly actively transcribed, whereas other parts are only weakly transcribed. Surprisingly, no extensive differences were found between nuclear transcripts and polysomal transcripts. This might indicate that extensive processing steps do not occur in the formation of polysomal RNA. T-DNA derived transcripts were found both in the poly(A)$^+$ and the poly(A)$^-$ fraction of the polysomal RNA. Both fractions gave the same relative hybridization pattern of T-DNA fragments. By comparing the transcriptional pattern of the T-DNA (Fig. 3) with the functional organization of the T-region (Fig. 2), it can be seen that the areas that are most actively transcribed correspond to those that are genetically correlated with opine synthesis (right end of T-region). The results of the transcription studies are summarized in Fig. 3.

Finally, the transcription of the T-DNA in isolated nuclei was shown to be inhibited by low concentrations of α-amanitin (0.7 µg/ml). This concentration is known to inhibit specifically RNA polymerase II from plants. Therefore

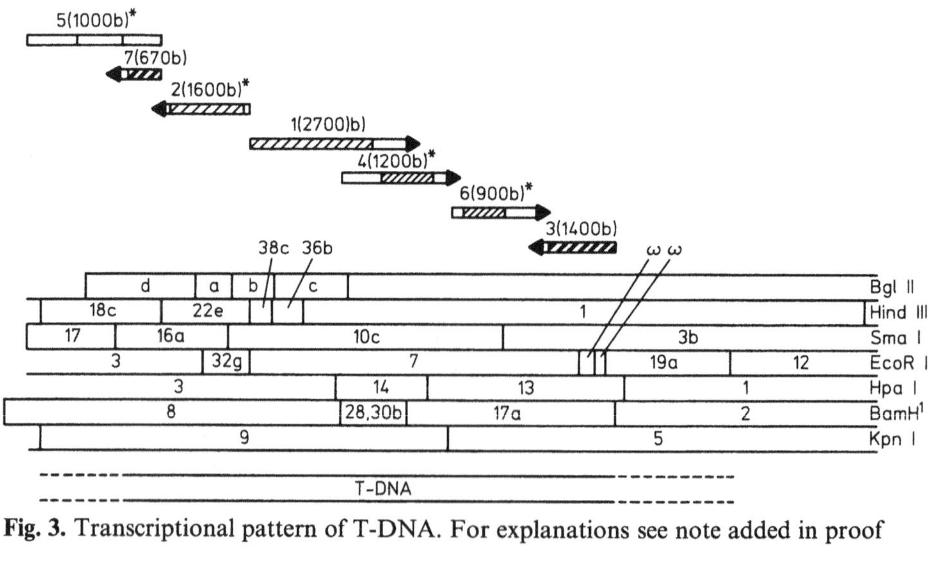

Fig. 3. Transcriptional pattern of T-DNA. For explanations see note added in proof

the transcription of the prokaryotic-derived T-DNA seems to be provided by the host-RNA polymerase II.

3.2 Translation of T-DNA-Derived mRNA

To demonstrate that the transcription of T-DNA results in the production of functional mRNA's, T-DNA-specific transcripts were isolated, translated in vitro, and the radioactive protein products identified.

Messenger RNA was isolated from transformed (B6, octopine) and from habituated tobacco tissue culture cells and tested for translational activity in two cell-free systems for protein synthesis. Polysomal RNA, as well as poly(A)-containing RNA, was efficiently translated in wheat germ extracts and in a rabbit reticulocyte lysate. Inhibition studies showed that 0.5 mM pm^{7}G suppressed activity by about 90%, suggesting that the majority of mRNA's were capped, as would be expected for eukaryotic mRNA. Gel electrophoretic analysis in the presence of dodecyl-sulfate revealed a broad distribution of radioactive proteins with molecular weights ranging from 10 to 90×10^3. In all of these experiments mRNA from transformed and habituated tissues showed no significant differences.

To detect T-DNA-specific mRNA's, poly(A)-containing RNA from both tissues was hybridized to two clones derived from the T-region, which covered the entire length of the T-DNA. After separation of the DNA-RNA hybrids from nonhybridized RNA by chromatography on agarose, the DNA-RNA hybrids were melted, and the RNA was translated in the wheat germ extract. Four distinct protein bands were observed above the background. These proteins are good candidates for T-DNA specific proteins, but do not represent major proteins; they were never detected as distinct bands of radioactive protein after translation of total mRNA from transformed or from habituated cells.

Experiments in progress will show which part(s) of the T-DNA code for these proteins. Of special interest is the protein with molecular weight$=41 \times 10^3$, since its size is close to the published values for lysopine dehydrogenase (mol. wt.$=38$–39×10^3). Experiments with specific antibodies will be necessary to demonstrate unequivocally lysopine dehydrogenase as a product of the T-DNA. (Lysopine dehydrogenase produces opines of the octopine series).

T-DNA-specific proteins have recently been described (McPherson et al. 1980). In these experiments, a different procedure was employed for DNA-RNA hybridizations, and the tissue used was a different tobacco line which had, however, also been transformed with B6 (octopine plasmid). The size of the T-DNA-specific proteins (mol. wt$=15$, 16.5, <30, and 30×10^3) was different from those found in the experiments described here (mol. wt.$=19.5$, 22, 25, and 41×10^3). These discrepancies could be due to inaccurate molecular weight determinations, since one would expect that T-DNA encoded proteins would be identical in different tissues. Other explanations are possible, however, and more experiments with different tissues would seem necessary to clarify these points.

4 The Development of the Ti-Plasmid as an Experimental Gene Vector

In the previous section we have summarized the evidence showing that Ti-plasmids in *Agrobacterium* are natural gene vectors which can transfer a number of genes into the plant nucleus such that these genes are stably maintained and expressed. The obvious question therefore arose (SCHELL and VAN MONTAGU 1977b) whether the Ti-plasmid could be developed as a general plant gene vector. In order to answer this question the following points had to be settled.

4.1 Are Genes, Inserted in the T-Region, Cotransferred to the Plant Nucleus?

In view of the observed involvement of the "ends" of the T-region in the integration of T-DNA, one could expect that any DNA segment inserted between these "ends" would be cotransferred, provided no function essential for T-DNA transfer and stable maintenance was inactivated by the insertion. The genetic analysis of the T-region by transposon-insertion provided us with Ti-plasmid mutants suited to test this hypothesis. A Tn7 insertion in the nopaline synthase locus (see Fig. 2), produced a Ti-plasmid able to initiate T-DNA transfer and tumor formation. Analysis of the DNA extracted from these tumors showed that the T-region containing the Tn7 segment had been transferred as a single 38×10^3 base pair segment without any major rearrangements (HERNALSTEENS et al. 1980).

4.2 Can Genes, Inserted via T-DNA into Plant Nuclei, Be Expressed?

The expression of the Tn7 insertion in the nopaline synthase locus was assayed as a model system for other genes inserted in this area. Tn7 codes for a dihydrofolate reductase which is resistant to methotrexate (TENNHAMMER-EKMAN and SKÖLD 1979). Suspension cultures of both untransformed and of crown gall tobacco tissue were found to be completely inhibited by 2 µg/ml of methotrexate. In contrast, cultures established with Tn7-containing crown gall cells grew well on media containing 2 µg/ml of methotrexate. In addition, nuclei isolated from these methotrexate tobacco lines were shown to synthesize Tn7 transcripts. Further experiments are in progress to determine whether transcription of the inserted Tn7 is initiated by a T-DNA promotor or by a Tn7 promotor.

4.3 Can Normal Plants Be Regenerated from T-DNA-Containing Plant Cells?

The ultimate aim of many gene transfer attempts in plants is to produce fertile cultivars harboring and transmitting new genetic properties. It was, therefore, essential to determine whether T-DNA transfer could be dissociated from neoplastic transformation and whether normal, fertile plants could be derived from T-DNA-transformed plant cells. In order to answer this question, a large set of insertion mutants were obtained with an octopine TiB6S3 plasmid (DE GREVE et al. 1981). One of the mutant plasmids (pGV2100) was clearly less oncogenic on tobacco and sunflower hypocotyls when compared to the wild-type plasmid. Tumors appeared only after prolonged incubation time. Furthermore, shoots proliferated from the greenish tumors, in contrast to the undifferentiating white tumors induced by strains harboring the wild-type TiB6S3 plasmid. The Tn7 insertion in pGV2100 was mapped and found to be located in the left arm of the common DNA of the T-region (Fig. 2).

As a test for transformation, tumor tissue and the shoots on tobacco were assayed for the presence of lysopine dehydrogenase (OTTEN and SCHILPEROORT 1978). The tumor tissue was found to be positive. Most of the shoots were negative but some of the proliferating shoots were positive. One such shoot was grown further on growth hormone-free media and found to develop roots and later to grow into a fully normal, flowering plant. Each part of this plant, leaves, stem, and roots was found to contain lysopine dehydrogenase activity and polysomal RNA was found to contain T-DNA transcripts homologous to the opine synthesis locus. No transcripts of the conserved segment of the T-region (Figs. 2 and 3) were observed.

These observations therefore demonstrate that normal plants can be obtained from plant cells transformed with Ti-plasmids genetically altered in specific segments of the T-region. Seeds obtained by self-fertilization of these plants produced new plants with active T-DNA linked genes, thus demonstrating that genes introduced in plant nuclei, via the Ti-plasmid, can be sexually inherited.

5 General Conclusions

The phenomenon of crown gall formation on plants is of fundamental importance in the study of the molecular biology of plant development. The abnormal growth pattern of crown gall plant cells has been shown to be the direct consequence of the presence and expression of specific "tumor" genes. These genes have been transferred into the plant nuclei by gram-negative bacteria, which contain natural gene vectors in the form of large Ti-plasmids. These Ti-plasmids carry the plant tumor-inducing genes and can transfer them to the plant nucleus. Because these tumor genes and a series of mutations have been isolated which affect directly the growth and morphogenetic properties of the transformed plant cells, the crown gall system represents a model system uniquely suited to the study of the genes and gene products involved in the control of cellular growth and differentiation. The Ti-plasmid, because of its gene vector properties, is also suited to the analysis, via reversed genetics, of the relation between structure and function of plant genes. Finally, the Ti-plasmid can be used to introduce selected and isolated genes into plants and may therefore open new avenues for plant breeding.

Note Added in Proof. Generality of the opine and genetic colonization concepts. A strong argument in favour of the general validity of the opine and a genetic colonization concept was recently provided by the case of the hairy roots induced by the *Agrobacterium rhizogenes*. The formation of hairy roots was shown to be due to the transfer and expression of part of the Ti-like plasmid present in rhizogenic *A. rhizogenes* strains to the hairy root plant tissues (CHILTON et al. 1982, WILLMITZER et al. 1982a).
Expression of T-DNA in Plant Cells. The analysis of T-DNA derived transcripts in octopine crown gall tumors by Northern analysis has given evidence that the T-DNA codes for at least seven genes which are located on the T-DNA as shown in figure 3 (WILLMITZER et al., 1982b). By mutagenizing one or a combination of these genes, data were obtained indicating that some of the T-DNA genes suppress the formation of shoots whereas others are involved in suppressing the formation of roots in crown gall tissues (LEEMANS et al. 1982).

Acknowledgements. The investigations reported here were supported by grants from the Kankerfonds van de A.S.L.K, from the Instituut tot aanmoediging van het Wetenschappelijk Onderzoek in Nijverheid and Landbouw (I.W.O.N.L.)(248/A), and from the Fonds voor Wetenschappelijk Geneeskundig Onderzoek (3.0052.78). The author wishes to thank all the members of the cooperating laboratories for their help and contributions.

The author wishes to acknowledge that this chapter was written with the help of Dr. M. VAN MONTAGU, Dr. M. HOLSTERS, Dr. A. DEPICKER, Dr. G. ENGLER, Dr. M. LEMMERS, Dr. P. DHAESE, and M. DE BEUCKELEER (Laboratorium voor Genetica, Rijksuniversiteit Gent); Dr. L. WILLMITZER, Dr. J. SCHRÖDER, and Dr. L. OTTEN (Max-Planck-Institut für Züchtungsforschung, Köln); and Dr. J.P. HERNALSTEENS, J. LEEMANS, and H. DE GREVE (Laboratorium voor Genetische Virologie, Vrije Universiteit Brussel).

References

Beuckeleer De M, Block De M, Greve De H, Depicker A, Vos De R, Vos De G, Wilde De M, Dhaese P, Dobbelaere MR, Engler G, Genetello C, Hernalsteens JP, Holsters M, Jacobs A, Schell J, Seurinck J, Silva B, Haute Van E, Montagu Van M, Vliet Van F, Villarroel R, Zaenen I (1978) The use of the Ti-plasmid as a vector for the introduction of foreign DNA into plants. Proc IVth Int Conf Plant Path Bacteriol, Angers, pp 115–126

Bomhoff G, Klapwijk PM, Kester HCM, Schilperoort RA, Hernalsteens JP, Schell J (1976) Octopine and nopaline synthesis and breakdown genetically controlled by a plasmid of *Agrobacterium tumefaciens*. Mol Gen Genet 145:177–181

Braun AC (1953) Bacterial and host factors concerned in determining tumor morphology in crown gall. Bot Gaz 114:363–371

Braun AC (1980) A history of the crown gall problem. In: Kahl G (ed) The molecular biology of plant tumors. Academic Press, London, New York

Braun AC, Mandle RJ (1948) Studies on the inactivation of the tumor-inducing principle in crown gall. Growth 12:255–269

Braun AC, White PR (1943) Bacteriological sterility of tissues derived from secundary crown gall tumors. Phytopathology 33:85–100

Chilton M-D, Drummond MH, Merlo DJ, Sciaky D (1978) Highly conserved DNA of Ti-plasmids overlaps T-DNA, maintained in plant tumors. Nature (London) 275:147–149

Chilton M-D, Saiki RK, Yadav N, Gordon MP, Quetier F (1980) T-DNA from *Agrobacterium* Ti plasmid is in the nuclear DNA fraction of crown gall tumor cells. Proc Natl Acad Sci USA 77:4060–4064

Cleene De M, Ley De J (1976) The host range of crown-gall. Bot Rev 42:389–466

Costantino P, Hooykaas PJJ, Dulk-Ras den H, Schilperoort RA (1980) Tumor formation and rhizogenicity of *Agrobacterium rhizogenes* carrying Ti-plasmids. Gene 11:79–87

Coxon DT, Davies AMC, Fenwick GR, Self R, Firmin JL, Lipkin D, Janes NF (1980) Agropine, a new amino acid derivative from crown gall tumours. Tetrahedron Lett 21:495–498

Currier TC, Nester EW (1976) Evidence for diverse types of large plasmids in tumor inducing strains of *Agrobacterium*. J Bacteriol 126:157–165

Depicker A, Montagu Van M, Schell J (1978) Homologous DNA sequences in different Ti-plasmids are essential for oncogenicity. Nature (London) 275:150–153

Depicker A, Wilde De M, Vos De G, Vos De R, Montagu Van M, Schell J (1980) Molecular cloning of overlapping segments of the nopaline Ti-plasmid pTiC58 as a means to restriction endonuclease mapping. Plasmid 3:193–211

Drummond MH, Gordon MP, Nester EW, Chilton M-D (1977) Foreign DNA of bacterial plasmid origin is transcribed in crown gall tumours. Nature (London) 269:535–536

Ellis JD, Kerr A (1978) Developing biological control agents for soil borne pathogens. Proc IVth Int Conf Plant Bacteriol, Angers, pp 245–250

Ellis JD, Murphy PJ (1981) Four new opines from crown gall tumours – their detection and properties. Mol Gen Genet 181:36–43

Ellis JD, Kerr A, Tempé J, Petit A (1979) Arginine catabolism: a new function of both octopine and nopaline Ti-plasmids of *Agrobacterium*. Mol Gen Genet 173:263–269

Engler G, Holsters M, Montagu Van M, Schell J, Hernalsteens JP, Schilperoort RA (1975) Agrocin 84 sensitivity: a plasmid determined property in *Agrobacterium tumefaciens*. Mol Gen Genet 138:345–349

Engler G, Depicker A, Maenhaut R, Villarroel-Mandiola R, Montagu Van M, Schell J (1981) Physical mapping of DNA base sequence homologies between an octopine and a nopaline Ti-plasmid of *Agrobacterium tumefaciens*. J Mol Biol 151:183–208

Frimin JL, Fenwick RG (1977) N^2(1-3 dicarboxypropyl) ornithine in crown gall tumours. Phytochemistry 16:761–762

Firmin JL, Fenwick GR (1978) Agropine – a major new plasmid-determined metabolite in crown gall tumours. Nature (London) 276:842–844

Genetello C, Larebeke Van N, Holsters M, Depicker A, Montagu Van M, Schell J (1977) Ti-plasmids of *Agrobacterium* as conjugative plasmids. Nature (London) 265:561–563

Greve De H, Decraemer H, Seurinck J, Montagu Van M, Schell J (1981) The functional organization of the octopine *Agrobacterium tumefaciens* plasmid pTiB6S3. Plasmid 6:235–248

Gurley WB, Kemp JD, Albert MJ, Sutton DW, Callis J (1979) Transcription of Ti plasmid-derived sequences in three octopine-type crown gall tumor lines. Proc Natl Acad Sci USA 76:2828–2832

Guyon P, Chilton M-D, Petit A, Tempé J (1980) Agropine in "null type" crown gall tumors: evidence for the generality of the opine concept. Proc Natl Acad Sci USA 77:2693–2697

Hamilton RH, Fall MZ (1971) The loss of tumor initiating ability in *Agrobacterium tumefaciens* by incubation at high temperature. Experientia 27:229–230

Hernalsteens JP, Greve De H, Montagu Van M, Schell J (1978) Mutagenesis by insertion of the drug resistance transposon Tn7 applied to the Ti plasmid of *Agrobacterium tumefaciens*. Plasmid 1:218–225

Hernalsteens JP, Vliet Van F, Beuckeleer De M, Depicker A, Engler G, Lemmers M, Holsters M, Montagu Van M, Schell J (1980) The *Agrobacterium tumefaciens* Ti plasmid as a host vector system for introducing foreign DNA in plant cells. Nature (London) 287:654–656

Holsters M, Waele De D, Depicker A, Messens E, Montagu Van M, Schell J (1978a) Transfection and transformation of *Agrobacterium tumefaciens* Mol Gen Genet 163:181–187

Holsters M, Silva B, Vliet Van F, Hernalsteens JP, Genetello C, Montagu Van M, Schell J (1978b) In vivo transfer of the Ti-plasmid of *Agrobacterium tumefaciens* to *Escherichia coli*. Mol Gen Genet 163:335–338

Holsters M, Silva B, Vliet Van F, Genetello C, Block De M, Dhaese P, Depicker A, Inzé D, Engler G, Villarroel R, Montagu Van M, Schell J (1980) The functional organization of the nopaline *A. tumefaciens* plasmid pTiC58. Plasmid 3:212–230

Hooykaas PJJ, Klapwijk PM, Nuti MP, Schilperoort RA, Rörsch A (1977) Transfer of the *Agrobacterium tumefaciens* Ti-plasmid to avirulent Agrobacteria and to *Rhizobium* ex planta. J Gen Microbiol 98:477–484

Jensen CO (1918) Undersøgelser vedrørende nogle svulstlignende dannelser hos planter. Vet Landbohøjsk Arsskr 1918:91–143

Kemp JD, Hack E, Sutton DW, El Wakio M (1978) Unusual amino acids and their relationship to tumorigenesis. Proc IVth Int Conf Plant Path Bacteriol, Angers, pp 183–188

Kerr A (1969) Transfer of virulence between isolates of *Agrobacterium*. Nature (London) 223:1175–1176

Kerr A (1971) Acquisition of virulence by non-pathogenic isolation of *Agrobacterium radiobacter*. Physiol Plant Pathol 1:241–246

Kerr A, Roberts WP (1976) *Agrobacterium*: Correlation between and transfer of pathogenicity, octopine and nopaline metabolism and bacteriocin 84 sensitivity. Physiol Plant Pathol 9:205–211

Kerr A, Manigault P, Tempé J (1977) Transfer of virulence in vivo and in vitro in *Agrobacterium*. Nature (London) 265:560–561

Klapwijk PM, Schilperoort RA (1979) Negative control of octopine degradation and transfer genes to octopine Ti-plasmids in *Agrobacterium tumefaciens*. J Bacteriol 139:424–431

Klapwijk PM, Oudshoorn M, Schilperoort RA (1977) Inducible permease involved in the uptake of octopine, lysopine and octopinic acid by *Agrobacterium tumefaciens* strains carrying virulence-associated plasmids. J Gen Microbiol 102:1–11

Klapwijk PM, Scheuldermon T, Schilperoort RA (1978) Coordinated regulation of octopine degradation and conjugative transfer of Ti-plasmids in *Agrobacterium tumefaciens:* evidence for a common regulatory gene and separate operons. J Bacteriol 136:775–785

Koekman BT, Ooms G, Klapwijk PM, Schilperoort RA (1979) Genetic map of an octopine Ti-plasmid. Plasmid 2:347–357

Larebeke Van N, Engler G, Holsters M, Elsacker Van den S, Zaenen I, Schilperoort RA, Schell J (1974) Large plasmid in *Agrobacterium tumefaciens* essential for crown gall-inducing ability. Nature (London) 252:169–170

Larebeke Van N, Genetello C, Schell J, Schilperoort RA, Hermans AK, Hernalsteens JP, Montagu Van M (1975) Acquisition of tumour-inducing ability by non-oncogenic agrobacteria as a result of plasmid transfer. Nature (London) 255:742–743

Larebeke Van N, Genetello C, Hernalsteens JP, Depicker A, Zaenen I, Messens E, Montagu Van M, Schell J (1977) Transfer of Ti-plasmids between *Agrobacterium* strains by mobilization with the conjugative plasmid RP4. Mol Gen Genet 152:119–124

Lemmers M, Beuckeleer De M, Holsters M, Zambryski P, Depicker A, Hernalsteens JP, Montagu Van M, Schell J (1980) Internal organization, boundaries and integration of Ti-plasmid DNA in nopaline crown gall tumours. J Mol Biol 144:355–378

McPherson JC, Nester EW, Gordon MP (1980) Proteins encoded by *Agrobacterium tumefaciens* Ti plasmid DNA (T-DNA) in crown gall tumors. Proc Natl Acad Sci USA 77:2666–2670

Montagu Van M, Schell J (1979) The plasmids of *Agrobacterium tumefaciens*. In: Timmis K, Pühler A (eds) Plasmids of medical, environmental and commercial importance. Elsevier, Amsterdam, pp 71–96

Montoya A, Chilton M-D, Gordon MP, Sciaky D, Nester EW (1977) Octopine and nopaline metabolism in *Agrobacterium tumefaciens* and crown-gall tumor cells: role plasmid genes. J Bacteriol 129:101–107

Moore L, Warren G, Strobel G (1979) Involvement of a plasmid in het hairy root disease of plants caused by *Agrobacterium rhizogenes*. Plasmid 2:617–626

Murphy PJ, Roberts WP (1979) A basis for agrocin 84 sensitivity in *Agrobacterium*. J Gen Microbiol 114:207–213

New PB, Kerr A (1972) Biological control of crown gall: field measurements and glasshouse experiments. J Appl Bacteriol 35:279–287

Ooms G, Klapwijk PM, Poulis JA, Schilperoort RA (1980) Characterization of Tn*904* insertions in octopine Ti plasmid mutants of *Agrobacterium tumefaciens*. J Bacteriol 144:82–91

Otten LA, Schilperoort RA (1978) A rapid microscale method for the detection of lysopine- and nopaline dehydrogenase activities. Biochim Biophys Acta 527:497–500

Petit A, Delhaye S, Tempé J, Morel G (1970) Recherches sur les guanidines des tissus de crown gall. Mise en évidence d'une relation biochimique spécifique entre les souches d' *Agrobacterium* et les tumeurs qu'elles induisent. Physiol Veg 8:205–213

Petit A, Tempé J, Kerr A, Holsters M, Montagu Van M, Schell J (1978a) Substrate induction of conjugative activity of *Agrobacterium tumefaciens*. Nature (London) 271:570–572

Petit A, Dessaux Y, Tempé J (1978b) The biological significance of opines. I. A study of opine catabolism by *Agrobacterium tumefaciens*. Proc IVth Int Conf Plant Path Bacteriol, Angers, pp 143–152

Schell J (1975) The role of plasmids in crown gall formation by *A. tumefaciens*. In: Ledoux L (ed) Genetic manipulations with plant materials. Plenum, New York, London, pp 163–181

Schell J, Montagu Van M (1977a) On the transfer, maintenance and expression of bacterial Ti-plasmid DNA in plant cells transformed with *A. tumefaciens*. Brookhaven Symp Biol 29:36–49

Schell J, Montagu Van M (1977b) The Ti plasmid of *Agrobacterium tumefaciens*, a natural vector for the introduction of *NIF* genes in plants? In: Hollaender A (ed) Genetic engineering for nitrogen fixation. Plenum, New York, London, pp 159–179

Schell J, Montagu Van M, Depicker A, Waele De D, Engler G, Genetello C, Hernalsteens JP, Holsters M, Messens E, Silva A, Elsacker Van den S, Larebeke Van N, Zaenen I (1977) *Agrobacterium tumefaciens:* what segment of the plasmid is responsible for the induction of crown gall tumors? In: Bogorad L, Weil JH (eds) Nucleic acids and protein synthesis in plants. Plenum, New York, London, pp 329–342

Schell J, Montagu Van M, Beuckeleer De M, Block De M, Depicker A, Wilde De M, Engler G, Genetello C, Hernalsteens JP, Holsters M, Seurinck J, Silva B, Vliet Van F, Villarroel R (1979) Interactions and DNA transfer between *Agrobacterium tumefaciens*, the Ti-plasmid and the plant host. Proc R Soc London Ser B 204:251–266

Sciaky D, Montoya AL, Chilton M-D (1978) Fingerprints of *Agrobacterium* Ti plasmids. Plasmid 1:238–253

Smith EF (1916) Studies on the crown gall of plants. Its relation to human cancer. J Cancer Res 1:231–309

Tate ME, Murphy PJ, Roberts WP, Kerr A (1979) Adenine N^6-substituent of agrocin 84 determines its bacteriocin-like specificity. Nature (London) 280:697–699

Tempé J, Schell J (1977) Is crown gall a natural instance of gene transfer? In: Legocki AB (ed) Translation of natural and synthetic polynucleotides. Univ Agric, Poznan, pp 416–420

Tempé J, Petit A, Holsters M, Montagu Van M, Schell J (1977) Thermosensitive step associated with transfer of the Ti-plasmid during conjugation: possible relation to transformation in crown gall. Proc Natl Acad Sci USA 74:2848–2849

Tempé J, Estrade C, Petit A (1978) The biological significance of opines. II. The conjugative activity of the Ti-plasmids of *Agrobacterium tumefaciens*. Proc IVth Int Conf Plant Path Bacteriol, Angers, pp 153–160

Tennhammer-Ekman B, Sköld O (1979) Trimethoprim resistance plasmids of different origin encode different drug-resistant dihydrofolate reductases. Plasmid 2:334–346

Thomashow MF, Nutter R, Montoya AL, Gordon MP, Nester EW (1980) Integration and organisation of Ti-plasmid sequences in crown gall tumors. Cell 19:729–739

Vos De G, Beuckeleer De M, Montagu Van M, Schell J (1981) Restriction endonuclease mapping of the octopine tumor inducing pTiAch5 of *Agrobacterium tumefaciens*. Plasmid 6:249–253

Watson B, Currier TC, Gordon MP, Chilton M-D, Nester EW (1975) Plasmid required for virulence of *Agrobacterium tumefaciens*. J Bacteriol 123:255–264

White FF, Nester EW (1980a) Hairy root: plasmid encodes virulence traits in *Agrobacterium rhizogenes*. J Bacteriol 141:1134–1141

White FF, Nester EW (1980b) Relationship of plasmids responsible for hairy root and crown gall tumorigenicity. J Bacteriol 144:710–720

Willmitzer L, Beuckeleer De M, Lemmers M, Montagu Van M, Schell J (1980) The Ti-plasmid derived T-DNA is present in the nucleus and absent from plastids of crown gall cells. Nature (London) 287:359–361

Yadav NS, Postle K, Saiki RK, Thomashow MF, Chilton M-D (1980) T-DNA of a crown gall teratoma is covalently joined to host plant DNA. Nature (London) 287:458–461

Zaenen I, Larebeke Van N, Teuchy H, Montagu Van M, Schell J (1974) Supercoiled circular DNA in crown gall inducing *Agrobacterium* strains. J Mol Biol 86:109–127

Zambryski P, Holsters M, Kruger K, Depicker A, Schell J, Montagu Van M, Goodman HM (1980) Tumor DNA structure in plant cells transformed by *A. tumefaciens*. Science 209:1385–1391

14 Organization and Expression of Plastid Genomes

H.J. Bohnert, E.J. Crouse, and J.M. Schmitt

1 Introduction

A green plant contains at least three types of organelles – nucleus, mitochondrion and plastid – which replicate, transcribe and express their genetic information in a coordinated way. The existence of DNA in plastids might have been inferred already from the genetic studies of Baur (1909) and Correns (1909), provided that the DNA-chromosome-gene concept had existed at that time. Evidence that plastids distribute genetic markers in a non-Mendelian mode of inheritance accumulated from the work performed with higher plants (Renner 1936, Rhoades 1946, Michaelis 1955, Schötz 1958, Stubbe 1959, Hagemann 1964, Röbbelen 1966, von Wettstein 1967, Tilney-Bassett 1970), *Chlamydomonas* (Sager 1954) and *Euglena* (Lyman et al. 1961).

The actual presence of DNA in chloroplasts (which since has become abbreviated as c-, chl-, chloro-, cl-, cp-, ct-, p-, pl- or ptDNA) was first demonstrated by Ris and Plaut (1962). Subsequent work over the following decade has brought the assignment of the correct cellular DNA component to plastids and the elucidation of the physicochemical parameters and structural aspects of chloroplast DNA.

Recent work on plastid DNA has profited much from the newly developed techniques to selectively isolate, clone and characterize specific parts of a genome. For the elucidation of its function this DNA may be used for in vitro transcription and translation experiments and the products may be identified. Finally, the primary structure can be determined by nucleic acid sequencing. These general techniques have been adapted and optimized to chloroplast research (Edelman et al. 1982). Consequently, the structure of the plastid chromosome from a constantly increasing number of species is known, allowing us to generalize different types of chromosome organization. Common features of plastid gene organization may be pointed out and, finally, it will become possible to trace plastid evolution and origin.

Abbreviations. DCMU = 3-(3,4-dichlorophenyl)-1,1-dimethylurea; LSU = large subunit protein of RuBPCase; CFI = coupling factor, F1-particle of ATP-synthetase; trnA, trnI = tRNA gene symbols; atpA, atpB, atpE = gene symbols for alpha, beta, and epsilon subunit proteins of the plastid ATP-synthetase F1-complex; rbcL = gene symbol for large subunit protein of the ribulose bisphosphate carboxylase; rbcS = gene symbol for small subunit protein of the ribulose-1,5-bisphosphate carboxylase; psbA = gene symbol for thylakoid membrane protein (mnemonic: 32,000 molecular weight protein, photogene); rrnA, B = rRNA operon(s); rrs = gene for 16S rRNA; rrl = gene for 23S rRNA; rrf = gene for 5S rRNA; rrg = gene for 4.5S rRNA; for gene symbols see (Bachman and Low 1980).

In contrast to our knowledge of plastid DNA, relatively little is known about the mechanisms of plastid gene transcription, control elements and regulation of gene expression in different plastid types or at different stages of plastid development. Characterization of these processes is essential for the understanding of the three genetic components of the eukaryotic plant cell.

Ultimately, the majority of organisms depend on the action of chloroplasts for reduction of carbon dioxide and the generation of oxygen. It will therefore be of importance to study and understand the processes of organellar coordination of gene expression on the genetic, biochemical and molecular levels. While plant productivity has been mainly increased by classical breeding, which will have to be continued, it is to be hoped that a better understanding of chloroplast gene structure and coordinated expression of genes among organelles will provide additional information necessary to engineer plants by recombinant DNA technologies.

By reference to selected papers, this article summarizes our knowledge on the structure and replication of plastid DNA, the expression of chloroplast genes and the control of their expression. Several review articles on plastid genome structure have been written (KUNG 1977, KIRK and TILNEY-BASSETT 1978, TEWARI 1979, BEDBROOK and KOLODNER 1979, HERRMANN and POSSINGHAM 1980, HAGEMANN and BÖRNER 1981, EDELMAN 1981, GROOT 1983, STUTZ 1983). Transcription and translation of plastid DNA and related topics have been covered by WHITFELD (1977), WOLLGIEHN and PARTHIER (1980), DYER and LEAVER (1981), GUILFOYLE (1981), STEINBACK (1981), ELLIS (1981), BOHNERT et al. (1983) and in Chapters 5, 15, 6, 7, this Volume, and in Chapter 2 Volume 14A, this Series.

2 Physicochemical Properties and Structural Aspects of Plastid DNAs

Genetic studies which had suggested organellar inheritance independent of the nucleus prompted attempts to isolate, identify and characterize the DNA component involved. RIS and PLAUT (1962) demonstrated DNA-like filaments in low electron density areas within chloroplasts from *Chlamydomonas*. Subsequently, DNA was also found in chloroplasts from other organisms and DNA has since then been located in proplastids (EDELMAN et al. 1964), etioplasts (HERRMANN and KOWALLIK 1970), chromoplasts (HERRMANN 1972) and leucoplasts (SIU et al. 1976). For some time, the assignment of the correct cellular DNA component to the chloroplast posed considerable problems until a consistent view had been established (for reviews see KIRK and TILNEY-BASSETT 1978, TEWARI 1979, HERRMANN and POSSINGHAM 1980). Highly purified, intact chloroplasts, which can be treated with DNAase prior to DNA isolation, are a reliable source of chloroplast DNA free from nuclear contamination (KOLODNER and TEWARI 1975c, d, HERRMANN et al. 1975, SCHMITT and HERRMANN 1977, EDELMAN et al. 1982). Modifications of the basic techniques according to needs have also been described (RHODES and KUNG 1981, BOHNERT and CROUSE 1981).

In some cases, chloroplast DNA can be isolated from total cellular DNA by preparative CsCl gradient centrifugation if the nuclear and plastid DNA differ sufficiently in buoyant density. This is often the case in algae, but not in most species of higher plants, since both types of DNA band at the same or relatively close density positions (for reviews see KIRK 1976, EDELMAN 1981). Alternatively, chloroplast DNA can be isolated from total DNA using column chromatography followed by CsCl-ethidium bromide density gradient centrifugation to recover the covalently closed circular DNA (KOLLER and DELIUS 1980) or from plastids isolated in nonaqueous media (BOWMAN and DYER 1982).

Table 1 lists physicochemical characteristics of plastid DNAs which were obtained by different methods such as determination of buoyant density, thermal melting, renaturation studies, chemical analysis of base composition, electron microscopy and digestion with restriction endonucleases (see EDELMAN 1981, and, for earlier studies, KIRK 1976).

2.1 Nucleotide Composition of Plastid DNAs

The buoyant densities of higher plant chloroplast DNAs in neutral CsCl are rather uniform, centering around 1.697 g cm^{-3}. This uniformity may be due to an evolutionary relatedness of the species, although with a larger number of species studied, plastid DNAs of different buoyant density have been discovered. In algal chloroplast DNAs, a more widespread range of buoyant densities between 1.685 g cm^{-3} (*Euglena gracilis*) and 1.706 g cm^{-3} (*Acetabularia cliftonii*) is found, reflecting most probably the evolutionary diversity of the different algal classes.

Buoyant densities may be used to calculate the average dG+dC content relative to a standard DNA (MANDEL et al. 1968). These values can then be compared to those calculated from thermal denaturation midpoints (MANDEL and MARMUR 1968) and with those from direct chemical analysis. The average dG+dC values from all three, or, more often, from two methods, coincide reasonably well. Thus plastid DNAs range in average dG+dC contents from the unusually low value of 24%–28% for *Euglena* to 47% for *Acetabularia* among the algae and span the relatively narrow range of 36% to 40% for higher plants (Table 1).

No modified bases such as methyl cytosine are present in plastid DNA (KIRK 1976) while, in contrast, nuclear DNA in plants is often composed up to 10% by this base. Absence of methyl cytosine was therefore sometimes taken as criterion of purity for chloroplast DNA (WHITFELD and SPENCER 1968). This generalization still appears to be valid, although it has been shown recently that plastid DNA may be methylated in *Chlamydomonas* during gametogenesis (BURTON et al. 1979, BOLEN et al. 1982). It has been suggested that this might be a mechanism underlying maternal inheritance in this organism by which non-methylated DNA from one parent is degraded via a restriction-modification system (BURTON et al. 1979, SAGER et al. 1981). The influence of methylation on the transmission of chloroplast markers in *Chlamydomonas* is, however, still questioned (BOLEN et al. 1982).

Table 1. Physicochemical properties of plastid DNAs[b]

Taxonomic position and species	Buoyant density	% dG+dC	Kinetic complexity	Size from electron microscopy	Restriction analysis	Ploidy	References
Phycophyta							
Euglenophyceae							
Euglena gracilis	1.685	24–28	1.8 (0.92)	40–44	125–143	59–72	EDELMAN et al. (1964), RAY and HANAWALT (1964), GRAY and HALLICK (1977), BRAWERMAN and EISENSTADT (1964), STUTZ and VANDREY (1971), MANNING et al. (1971), MANNING and RICHARDS (1972b), CROUSE et al. (1978), RAWSON et al. (1978), SLAVIK and HERSHBERGER (1975)
Xanthophyceae							
Olisthodiscus luteus	1.691		0.92	46	150	13–44	ALDRICH and CATTOLICO (1981), ERSLAND et al. (1981), CATTOLICO (1978), LÜTTKE (1980)
Phaeophyceae							
Pylaiella littoralis	1.693	34					DALMON and LOISEAUX (1981)
Sphacellaria sp.	1.695						DALMON and LOISEAUX (1981)
Uncertain relationship							
Cyanophora paradoxa[a]	1.692 1.716	33	1.17				HERDMAN and STANIER (1977), MUCKE et al. (1980), BOHNERT et al. (1982), EDELMAN et al. (1967), JAYNES et al. (1981)
Chlorophyceae							
Acetabularia cliftonii	1.706	47	15.2	Linear and minicircles		3	GREEN et al. (1977), PADMANABHAN and GREEN (1978)
Acetabularia acetabulum	1.703		15	Linear and minicircles			MAZZA et al. (1979, 1980)
Chlamydomonas reinhardii	1.695	36	1.9	62	196	26–52	BASTIA et al. (1971), HOWELL and WALKER (1976), BEHN and HERRMANN (1977), WELLS and SAGER (1971), ROCHAIX (1978), ROCHAIX and MALNOE (1978)

Chlorella pyrenoidosa	1.687		1.2–1.3			20	BAYEN and RODE (1973)
Codium fragile		37		27.3	85		HEDBERG et al. (1981)
Scenedesmus obliquus	1.694						PRYKE et al. (1979)
Spirogyra sp.	1.690	30					TAKAYA and SASAKI (1973)
Bryophyta							
Sphaerocarpos donellii	1.691	31		37			HERRMANN et al. (1980b)
Pteridophyta							
Asplenium nidus	1.697	38		44			HERRMANN et al. (1980b)
Pteris vittata	1.697	38		43			HERRMANN et al. (1980b)
Spermatophyta							
Angiospermae, monocotyledoneae							
Avena sativa (oats)	1.698	39		37			KOLODNER and TEWARI (1975d)
Narcissus pseudonarcissus (daffodil)	1.697			44	161		FALK et al. (1974), THOMPSON et al. (1981)
Pennisetum americanum (pearl millet)					125–130		RAWSON et al. (1981)
Secale cereale (rye)	1.699	39			136–144		HERRMANN and FEIERABEND (1980)
Spirodela oligorrhiza (duckweed)	1.698	37		54	182		VAN EE et al. (1980, 1981, 1982)
Triticum aestivum (wheat)	1.698				135		BOWMAN et al. (1981)
Tulipa gessneriana (tulip)				43.6			WUTTKE (1976)
Zea mays (maize)	1.698			43			MANNING et al. (1972), VEDEL et al. (1976)
	1.700	39		38	136		KOLODNER and TEWARI (1975c, d), BEDBROOK and BOGORAD (1976), BEDBROOK et al. (1977)
Angiospermae, dictoyledoneae							
Antirrhinum majus (snapdragon)	1.697	38		46			HERRMANN et al. (1975)
Atriplex triangularis (orach)					152		PALMER (1982)
Atropa belladonna (deadly nightshade)					158–161		FLUHR and EDELMAN (1981a)
Beta vulgaris (Swiss chard)	1.697	38		46			HERRMANN et al. (1975)
Cucumis sativus (cucumber)	1.701				155		PALMER (1982)
Lactuca sativa (lettuce)	1.697	38		44–46		18	WELLS and BIRNSTIEL (1969), MANNING et al. (1972), KOLODNER et al. (1976), KOLODNER and TEWARI (1975d)

Table 1 (continued)

Taxonomic position and species	Buoyant density	% dG+dC	Kinetic complexity	Size from electron microscopy	Restriction analyis	Ploidy	References
Nicotiana tabacum (tobacco)	1.700		1.14				Tewari and Wildman (1970),
Oenothera parviflora	1.697				160	30–90	Fluhr and Edelman (1981a)
(evening primrose)					160		Seyer et al. (1981)
	1.697	38		45	152		Herrmann et al. (1975), Gordon et al. (1981, 1982)
Petunia hybrida (petunia)	1.699	39		46	153		Bovenberg et al. (1981)
Phaseolus aureus (mung bean)					150		Palmer and Thompson (1981a)
Phaseolus sp.		39		40		14–50	Kolodner and Tewari (1975d), Bennett and Radcliffe (1975)
Pisum sativum (pea)	1.698	33–39	0.95	39	136	32–97	Kolodner and Tewari (1972, 1975d), Chu et al. (1981), Kolodner et al. (1976)
					120		Palmer and Thompson (1981a)
Sinapis alba (mustard)					158		Link et al. (1981)
Spinacia oleracea (spinach)	1.698	38		46	145	48–200	Whitfeld and Spencer (1968), Herrmann et al. (1975), Crouse et al. (1978), Driesel et al. (1979)
		37–39		40			Manning et al. (1972), Kolodner and Tewari (1975d), Scott and Possingham (1980)
Tropaeolum majus (nasturtium)				44	155		Thompson et al. (1981)
Vicia faba (broad bean)	1.696	37		39	121		Koller and Delius (1980)

[a] *Cyanophora paradoxa* is described as a flagellate containing endosymbiotic cyanobacteria, called cyanelles (see Trench 1979)

[b] Buoyant density as g cm^{-3}; kinetic complexity in 10^8 molecular weight; size from electron microscopy in 10^{-6} m; size from restriction endonuclease analysis in Kbp; ploidy (analytical complexity) is given in number of unit size genomes per organelle

A small number of ribonucleotides (12–18 depending on the species) have been identified in higher plant chloroplast DNAs by alkali treatment and RNAase digestion. These are covalently linked to the deoxyribonucleotides and have been found in both strands. The ribonucleotide sites in pea chloroplast DNA have been mapped by electron microscopy (KOLODNER et al. 1975, see also TEWARI 1979). It will be of importance to locate these sites on the physical maps constructed for pea plastid DNA, as they may be related to DNA replication (see BEDBROOK and KOLODNER 1979).

2.2 Kinetic Complexity

Heat- or alkali-denatured DNA renatures under controlled salt and temperature conditions at a rate which is inversely proportional to its complexity (WETMUR and DAVIDSON 1968). This can be monitored photometrically and compared with the renaturation rate of a DNA of known complexity under identical conditions. The kinetic complexities of chloroplast DNAs are in the range of approximately 1×10^8 molecular weight (for reviews see HERRMANN and POSSINGHAM 1980, EDELMAN 1981). The kinetic complexity of *Euglena* chloroplast DNA, which has a very low dG+dC content, cannot be estimated this way unless a controversial correction factor is introduced into the calculations. Without this factor a complexity of 1.8×10^8 molecular weight is observed, whereas the complexity which is derived from restriction endonuclease analysis of *Euglena* chloroplast DNA is 0.92×10^8 molecular weight (see Sect. 4). *Chlamydomonas* chloroplast DNA has a much higher complexity (1.9×10^8 mol.wt.) than that found for other plastid DNAs. The chloroplast DNAs from *Acetabularia cliftonii* (PADMANABHAN and GREEN 1978) and *A. mediterranea* (WELLS and SAGER 1971) are reported to have a complexity that is almost one order of magnitude higher (15×10^8 mol.wt.), which is comparable to the complexity for *E. coli* chromosomal DNA (25×10^8).

2.3 Size, Uniformity and Intramolecular Heterogeneity of Plastid DNAs

The size values obtained from kinetic complexities of chloroplast DNAs can be compared to those determined from length measurements of plastid DNAs in the electron microscope relative to a length standard in the same experiment. The factor used to convert length into molecular weight was often different in different reports, as were the conditions of preparing DNA for electron microscopy. A widely accepted factor of 2.07×10^6 molecular weight per μm length of relaxed DNA is now used (STÜBER and BUJARD 1977). The values given in Table 1 are original values. Examination of plastid DNAs from an ever-increasing number of species reveals circular molecules of uniform size which is species-specific (Table 1). The proportion of circular molecules increases, when yield is sacrificed for gentle preparation conditions (KOLODNER and TEWARI 1975b, d, HERRMANN et al. 1975, KOLLER and DELIUS 1980). Thus, there is reason to believe that these circular molecules represent the intact chloro-

Table 2. Morphology of circular plastid DNA molecules

Species	Size (Kbp)	Organization in segments (Kbp)[d]			
Euglena gracilis Z-Ha[a]	140	*5.7*	*5.7*	*5.7*	122
Euglena gracilis Z-S[b]					
Chlamydomonas reinhardii	195	*21*	76	*21*	77
Atriplex purpurea	152	*24.1*	nd	*24.1*	nd
Cucumis sativus	155	*>14*	24	*>14*	<104
Lactuca sativa	156	*24.4*	20	*24.4*	87
Narcissus pseudonarcissus[c]	161	*28.5*	16.6	*28.5*	87.8
Nicotiana tabacum	160	*20.4*	24	*20.4*	95
		22.6	nd	*22.6*	nd
		24	13	*24*	99
Oenothera parviflora	152	*23*	20	*23*	86
Pennisetum americanum	125–130	nd	12	nd	nd
Petunia hybrida	153	*>15*	20	*>15*	<102
Phaseolus aureus	150	*23*	21	*23*	83
Pisum sativum[b]	124				
	135				
Sinapis alba	150–158	*~22*	~30	*~22*	~84
Spinacia oleracea	145	*23*	19	*23*	80
Spirodela oligorriza	182	*27*	28	*27*	100
Triticum aestivum	135	*21*	12.8	*21*	80.2
Tropaeolum majus[c]	155	*27.1*	18.5	*27.1*	82.3
Vicia faba[b]	121				
Zea mays	136	*22.5*	12.6	*22.5*	78.5
		22.5	13	*22.5*	77

Repeat structures were determined by electron microscopy (E) or by restriction endonuclease analysis (R).

[a] *Euglena gracilis* Z-Ha contains in addition a fourth 16S rRNA gene within the single-copy DNA region

plast chromosome. The molecular weight calculated from length measurements coincides reasonably well with the values obtained by other methods (Tables 1, 2; see Sects. 2.1 and 3.1). Of the circular molecules revealed by electron microscopy, a variable proportion appears in a covalently closed circular conformation and a small number occur as unicircular dimers or two interlocked monomers (for reviews see Tewari 1979, Kolodner and Tewari 1979, Herrmann and Possingham 1980, Edelman 1981).

The sizes of plastid DNAs from dicotyledoneous plants are between 39 μm and 46 μm, and those of monocotyledoneous plants range from 37 μm to 55 μm. Algal plastid DNAs are more divergent in length, ranging from 27 μm in *Codium fragile* to 62 μm in *Chlamydomonas reinhardii.* The sizes of plastid DNAs from two ferns are similar to those of angiosperms, while the chloroplast DNA from the bryophyte *Sphaerocarpos donellii* is slightly smaller than most higher plant plastid DNAs (Table 1). The chloroplast DNA in several species of *Acetabularia* was recovered as a structurally mixed population consisting predominantly of a complex network of linear DNA molecules ranging in size up to 200 μm

Repeat structure	References
Tandem (R,E)	GRAY and HALLICK (1978), RAWSON et al. (1978), JENNI and STUTZ (1978), EL-GEWELY et al. (1981), KOLLER and DELIUS (1982a)
No repeat (R)	WURTZ and BUETOW (1981)
Inverted (R)	ROCHAIX (1978), ROCHAIX and MALNOE (1978)
Inverted (R)	PALMER (1982)
Inverted (R)	PALMER (1982)
Inverted (E)	KOLODNER and TEWARI (1979)
Inverted (R)	THOMPSON et al. (1981)
Inverted (R)	SEYER et al. (1981)
Inverted (R)	JURGENSON and BOURQUE (1980)
Inverted (R)	FLUHR and EDELMAN (1981a)
Inverted (R)	GORDON et al. (1981, 1982)
Inverted (R)	RAWSON et al. (1981)
Inverted (R,E)	BOVENBERG et al. (1981)
Inverted (R)	PALMER and THOMPSON (1981a)
No repeat (R)	PALMER and THOMPSON (1981a)
No repeat (R,E)	CHU et al. (1981)
Inverted (R)	LINK et al. (1981)
Inverted (R,E)	CROUSE et al. (1978), DRIESEL et al. (1979), HERRMANN et al. (1980a), KOLODNER and TEWARI (1979)
Inverted (R)	VAN EE et al. (1981, 1982)
Inverted (R,E)	BOWMAN et al. (1981)
Inverted (E)	THOMPSON et al. (1981)
No repeat (E,R)	KOLLER and DELIUS (1980)
Inverted (E)	KOLODNER and TEWARI (1979)
Inverted (R) (E)	BEDBROOK and BOGORAD (1976), BEDBROOK et al. (1977)

[b] Only one set of rRNA genes per chromosome
[c] Chromoplast DNA investigated
[d] The size of the repeat segments is given in italics

and a small amount of minicircles (GREEN 1976, PADMANABHAN and GREEN 1978, GREEN et al. 1977, MAZZA et al. 1979, 1980). The average size of the minicircular DNA in *A. cliftonii* is 4.2 μm. Its buoyant density of 1.712 g cm^{-3} is different from that of the plastid main band DNA (1.706 g cm^{-3}). Equal in buoyant density with the bulk DNA (1.703 g cm^{-3}) are the 0.6 μm minicircles from *A. acetabulum* which were found as replicating structures (rolling-circle mechanism) in high frequency (MAZZA et al. 1980). In the few species where DNA was isolated from chromoplasts (daffodil, tulip and nasturtium) circle sizes of about 44 μm have been found (Tables 1, 2).

One report concerning variation within a specific region of the chloroplast genome from *Euglena* has appeared. JENNI et al. (1981) mapped a variable DNA region of approximately 0.8 Kbp (0.6% of genome length) in *E. gracilis* Z. This variability is likely to be due to a different number of clustered small repeats.

Various techniques including CsCl gradient centrifugation, elution from hydroxyapatite, micrococcal nuclease digestion, restriction endonuclease digestion and denaturation mapping by electron microscopy have been used to study overall sequence organization in chloroplast DNAs. These methods, which have been listed in the approximate order of increasing resolution, give estimates of the distribution of dG·dC and dA·dT basepairs along the circular molecules. Intramolecular heterogeneity had first been found in algal chloroplast DNAs. The best-studied organism in this respect is *Euglena*. Sheared chloroplast DNA shows a bimodal density distribution in neutral CsCl gradients with the bulk of the DNA banding at 1.685 g cm^{-3} and a satellite at 1.703 g cm^{-3} (STUTZ and VANDREY 1971, SLAVIK and HERSHBERGER 1976, CROUSE et al. 1974, KOPECKA et al. 1977). The buoyant density of this satellite varies between experiments, depending on the initial molecular weight of the DNA. By hybridization experiments it was shown that this DNA was preferentially enriched for rRNA genes (see Sect. 4.1). Similarly, the elution profile of *Euglena* chloroplast DNA from hydroxyapatite columns shows a shoulder at low phosphate concentrations. This indicates that a large portion of the molecule differ by about 20% in dG+dC content (SCHMITT et al. 1981). No such gross heterogeneity has been found in chloroplast DNA from higher plants (for a review see EDELMAN 1981). When *Euglena* and spinach were compared using more powerful methods such as computer analysis of thermal melting, determination of buoyant densities for restriction fragments and micrococcal nuclease digestion, base compositional heterogeneity became apparent in the higher plant chloroplast DNA as well (SCHMITT et al. 1981). In *Euglena* chloroplast DNA stretches containing about 12% dG+dC comprised nearly 30% of the circular molecule, whereas about 10% of the DNA was higher than 60% dG+dC. In spinach, about 30% of the molecule was found to be as low as 22% in dG+dC, while 10% was higher than 60% dG+dC. In both organisms, the "heavy" and "light" stretches must be closely interspersed and smaller than about 1 to 1.5 Kbp since restriction fragments of this or larger size, when examined in the analytical ultracentrifuge, ranged only from 27% dG+dC to 48% dG+dC in *Euglena* (CROUSE et al. 1978, see also GRAY and HALLICK 1979b for comparable values) and from 34% dG+dC to 56% dG+dC in spinach (CROUSE et al. 1978, SCHMITT et al. 1981).

Base compositional heterogeneity has been found in chloroplast DNAs from other higher plants as well by thermal melting (TABIDZE and BERIDZE 1979) and electron microscopic denaturation mapping (KOLODNER and TEWARI 1979, KOLLER and DELIUS 1980). Earlier work on base compositional heterogeneity has been reviewed by EDELMAN (1981).

It appears that physicochemical measurements, such as buoyant density, dG+dC content and circle size, are often useful parameters in describing plastid DNA. Care should be taken not to overinterpret these data by drawing definite conclusions on plastid phylogeny.

2.4 Amount and Structural Arrangement of DNA Within Plastids

The amount of DNA per chloroplast (analytical complexity) is much higher than if only one circular DNA molecule existed per organelle. Values of up to about 200 times the unit size of the circular chromosome (Table 1) have been reported (SCOTT and POSSINGHAM 1980) by using a cloned chloroplast DNA fragment as a probe for quantification. Values from earlier measurements have to be regarded as approximations because the estimates of nuclear DNA contamination, DNA losses during purification, DNA remaining unextracted and non-DNA contamination are prone to errors. Furthermore, studies with higher plant chloroplasts were made using preparations which contained cells with a spectrum of young to mature organelles, which differ in DNA content, depending on the topological origin of the cell in still-expanding leaves. This has been demonstrated recently by SCOTT and POSSINGHAM (1980) and SELLDEN and LEECH (1981). Finally, in some studies it was not taken into account that the amount of plastid DNA is dependent on nuclear ploidy (BUTTERFASS 1979). The following statements can be made at present:

1. Chloroplast DNA is arranged into regions (nucleoids) which are located in the stroma. Each nucleoid contains several unit-size DNA molecules and is therefore polyploid. The organelle is polyenergidic because several nucleoids can be seen in one chloroplast by fluorescence microscopy of DNA stained with the fluorochrome DAPI (JAMES and JOPE 1978, COLEMAN 1979, SCOTT and POSSINGHAM 1980, SELLDEN and LEECH 1981). The size and number of DAPI staining areas is dependent on the developmental state of the chloroplast. Size, number and distribution of these areas within the plastid appear to be species-specific (SELLDEN and LEECH 1981, SCOTT and POSSINGHAM 1980). The electron microscopic studies on these nucleoids have been extensively discussed by HERRMANN and POSSINGHAM (1980).

Ploidy of plastid DNA and nucleoid number change during chloroplast development. SCOTT and POSSINGHAM (1980) found that the approximately 20 chloroplasts from young spinach leaf cells each contained about 200 DNA molecules of unit size (equivalent to 3.2×10^{-14} g). In contrast, the approximately 150 chloroplasts of older cells each contained about 30 chloroplast DNA molecules in 10–15 nucleoids. The authors concluded that the amount of chloroplast DNA per cell does not change much during the later stages of cellular development. This means that replication of chloroplast DNA takes place mainly in young dividing, rather than in expanding, leaf cells in spinach, as also reported for wheat (BOFFEY et al. 1979). The chloroplast number in pea cells increased during expansion of primary leaves but the amount of DNA per chloroplast decreased. This resulted in a rather constant amount of chloroplast DNA per cell (LAMPPA et al. 1980). The same behaviour has also been found in the unicellular alga *Olisthodiscus* (CATTOLICO 1978).

More than 50% of the chloroplasts in several species of *Acetabularia* (and the related *Batophora oerstedii*) have been found to lack DNA by cytochemical methods (WOODCOCK and BOGORAD 1970, COLEMAN 1979, LÜTTKE and BONOTTO 1981). An explanation of this result may be that in these plants, where the analytical complexity is 1–3 chromosomes per plastid (GREEN et al. 1977), plastid division is faster than DNA replication at certain stages of development. It might be interesting to know whether minicircles, which cannot be stained by DAPI, are present in these plastids.

2. Electron microscopic studies suggest that the chloroplast DNA of a nucleoid is attached to one or several membrane sites (HERRMANN and KOWALLIK

1970). DNA from spinach chloroplasts attached to supposedly proteinaceous material can be visualized in the electron microscope after mild lysis of chloroplasts and sucrose or CsCl gradient centrifugation of lysates (SCHMITT JM, BOHNERT HJ, HERRMANN RG, unpublished; see Fig. 3a in HERRMANN and POSSINGHAM 1980). Similar complexes from spinach chloroplasts have also been reported by YOSHIDA et al. (1978). The DNA in these complexes can be obtained in a covalently closed circular conformation. Length measurements of this circular DNA, when in a relaxed conformation, suggested the presence of several of the circular monomers per complex, indicating that each complex is equivalent to the DNA from one nucleoid. Caution should be taken concerning the interpretation of electron microscopic data on DNA/protein complexes. Electron microscopic pictures similar to those obtained from lysates of chloroplasts could be generated when purified circular chloroplast DNA was spread with electrophoretically pure Fraction-1-protein (BOHNERT HJ, SCHMITT JM, unpublished). Since the association of proteins with plastid DNA is expected to be of regulatory importance, the characterization of the protein moiety from such complexes by independent methods seems desirable.

It should be pointed out that data on plastid DNAs are available from only approximately 40 species, most of which are angiosperms. The angiosperms alone comprise approximately 270,000 species. Among the ferns, mosses and what is often indiscriminately called algae, only few species have been investigated with respect to plastid DNA. In several algal classes, as, for example, the red algae, no plastid DNA has been characterized.

3 Physical Maps of Plastid DNAs

Class II restriction endonucleases (ROBERTS 1982), which recognize short DNA sequences and cleave at or nearby the recognition site, have proven to be useful tools in dissectioning the plastid genome. The ordering of DNA fragments produced by restriction endonuclease cleavage may then be accomplished by a variety of experimental strategies (see EDELMAN et al. 1982 for a compilation of methods).

3.1 Gross Morphology of Plastid DNAs

After cleavage of plastid DNA with restriction endonucleases and separation of the DNA fragments by gel electrophoresis according to size, DNA fragment patterns are obtained. The sum of the molecular weights of the individual DNA fragments produced by each enzyme is in reasonable agreement with the complexity determined by other methods (Tables 1, 2; see Sect. 2). The consecutive order of the DNA fragments produced by two or more enzymes can be arranged as a circular map (Figs. 2, 4). The similarity in size and base composition found for many higher plant plastid DNAs (see above) is reflected by a similar number of cleavage sites for a given endonuclease, although the DNA fragment patterns are easily distinguishable between species. In most cases, depending on the choice of restriction endonucleases, multiple copies of a DNA fragment are found which originate from segments present more than once on one circular molecule. The presence of repeated regions in most of the plastid DNAs studied up to now has been confirmed by electron microscopy (Fig. 1a; Table 2). In many higher plants and in *Chlamydomonas*, these repeated regions are arranged in an inverted mode (Fig. 1a, Fig. 2, Table 2). The two inverted-repeat units are rather uniform in size for all the species studied, ranging from 20 to 24 Kbp. Only in the case of *Spirodela* chloroplast DNA has a size of 27 Kbp been reported. For chromoplast DNA, however, sizes of 28.5 Kbp and 28 Kbp have been determined in daffodil and nasturtium, respectively. The invertedly repeated segments are separated by a small and a large single-copy DNA region (Figs. 1a, 4). These single-copy DNA segments vary considerably in length among species, ranging from approximately 12 Kbp (maize) to 76 Kbp (*Chlamydomonas*) for the smaller region and from 78.5 Kbp (maize) to 100 Kbp (*Spirodela*) for the larger single-copy region (Table 2). Published exceptions from this structure are broad bean, pea and *E. gracilis* Z-S chloroplast DNAs which show no sequence repetition and *Euglena gracilis* Z-Ha chloroplast DNA which has a closely spaced triple tandem repeat, 5.6–5.8 Kbp in size (Figs. 1b, 2).

A different type of sequence inversion occurs in *Chlamydomonas reinhardii* and *Euglena gracilis* Z. Between 4% and 7% of the sheared, denatured chloroplast DNA renatured spontaneously as approximately 0.1–0.3 Kbp duplex fragments located in various parts of the genome (GELVIN and HOWELL 1979). Similarly, ROCHAIX (1978, 1981) found sequences also interspersed throughout the genome which are homologous to parts of a 1.5 Kbp DNA fragment from the inverted repeat in the region of the 5S rRNA gene (see Sect. 4). KOLLER and DELIUS (1982a, b) have mapped two sets of small inverted repeats (~125 nucleotides long) in *Euglena* chloroplast DNA. These repeats frame the extra 16S rRNA gene (see below, JENNI and STUTZ 1979). One set of these elements is within approximately 0.5 Kbp on one side and the other set within 4.8 Kbp to the other side of the 16S rRNA gene.

3.2 Insertions, Deletions and Rearrangements

If closely related species are examined by restriction endonuclease analysis, similarities in the fragment patterns are obvious. For example, chloroplast DNA from three species of *Nicotiana* were indistinguishable by Eco RI digestion (ATCHISON et al. 1976). In contrast, restriction pattern polymorphisms in *Nicotiana* chloroplast DNA have been described (SCOWCROFT 1979, KUNG et al. 1981, RHODES et al. 1981). Small differences in DNA fragment sizes have also been found in *Pelargonium* cultivars (METZLAFF et al. 1981), in different *Chlamydomonas* species (LEMIEUX et al. 1980, METS 1980) and in several plastid mutants of *Chlamydomonas* (MYERS et al. 1982). In the genus *Oenothera*, five genetically different plastid types yielded the same number of DNA restriction fragments although variation in size of some fragments was observed (GORDON et al. 1981,

a

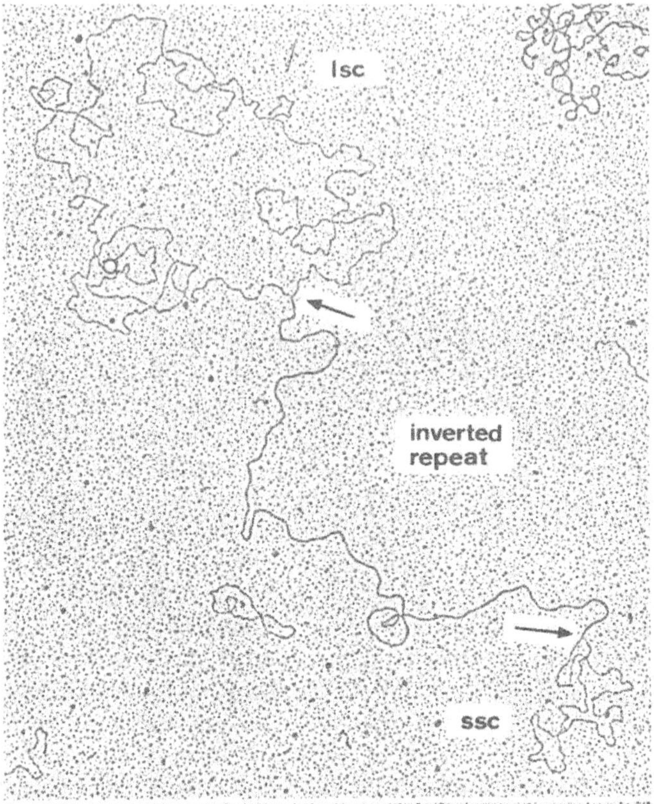

b

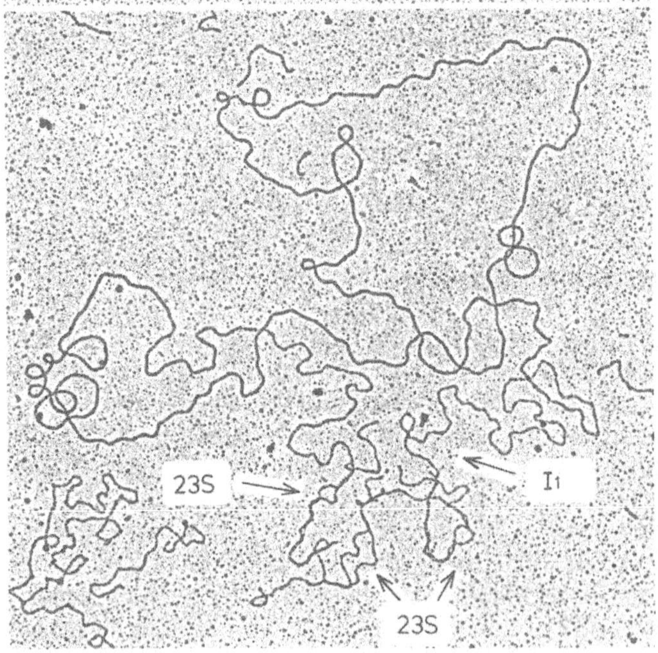

Fig. 2. Schematic presentation of different types of plastid DNA organization. *Arrows* repeat structures, if present, and direction of transcription of rrn operons. *Closed symbols* 16S rRNA genes; *Open symbols* 23S rRNA genes

1982). The RuBPCase-deficient *Oenothera* mutant I sigma (HALLIER et al. 1978) yields the same number of fragments, but again there are small differences in size from the wild type I plastid genome (GORDON et al. 1980). These results are an indication of small insertions/deletions (150–1,200 bp in the five *Oenothera* plastid genomes) which distinguish plastid DNAs from closely related species while the overall fragment sizes and restriction endonuclease cleavage sites are conserved. In the case of *E. gracilis* Z several sub-species have been distinguished meanwhile (WURTZ and BUETOW 1981). In addition, the results of JENNI et al. (1981), who found a region of the chloroplast DNA that varies in size in their strain Z, might be an indication that rearrangements in these fast-growing single-cell algae occur frequently.

The conserved gross organization of the majority of plastid DNAs studied (with inverted repeat and single-copy regions) cannot be seen by comparing restriction patterns, because single base pair changes may create or abolish cleavage sites. Several groups have tried to assess the degree of homology between plastid DNAs of different species by other methods. TABIDZE and BERIDZE (1979) have shown similarities in the fine structure of thermal melting curves of chloroplast DNAs from several species of higher plants. This indicates that not only the average dG+dC content (see Sect. 2.1), but also the mass fractions of individual subtransitions have been conserved. In contrast, on a wider evolutionary scale, spinach and *Euglena* exhibit strikingly different melting curves (SCHMITT et al. 1981).

◁

Fig. 1. Electron micrographs of plastid DNA. **a** Cytochrome spreading of single-stranded circular red bean (*P. vulgaria*) chloroplast DNA showing renaturation of the inverted repeat region. *lsc*/*ssc* large/small single-copy DNA region. **b** Heteroduplex between circular *Euglena gracilis* Z chloroplast DNA and the inserted DNA fragment in plasmid pBK 8 (BOROS et al. 1979) containing the *E. coli* rrnB operon. Three *E. coli* DNA fragments have annealed with the three plastid rrn operons.

Three 23S rDNAs ("23S") are melted (see DELIUS and KOLLER 1980) due to partial mismatches, whereas the 16S rDNA sequences are more conserved. The pBK 8 insert produces a long single-strand at the 3′-end of the plastid rrn operon. I_1 intramolecular renaturation of a short inverted repeat segment of *Euglena* chloroplast DNA (KOLLER and DELIUS 1982a). (By courtesy of Drs. B KOLLER, H DELIUS, EMBL, Heidelberg)

The order of homologous DNA regions in chloroplast DNA has been investigated by hybridization of radioactively labelled restriction fragments from one species to filter-immobilized DNA fragments of other species. The linear order of homologous sequences was similar between different species of the Solanaceae family, between tobacco and spinach and between *Atriplex* and spinach (FLUHR and EDELMAN 1981b, PALMER 1982). More extended rearrangements have been found between mung bean (a species with inverted-repeat units) and pea (no repeated regions) (PALMER and THOMPSON 1981a). BISARO and SIEGEL (1980) reported extensive sequence homologies between several higher plant plastid DNAs monitoring DNA/DNA reassociation in solution, although a considerable amount of mismatch was observed. The homology of pea chloroplast DNA to chloroplast DNAs from several other species was estimated to be approximately 30% with the resolution which can be obtained by reassociation experiments using sheared DNA (shorter than 1 Kbp). Heteroduplex analysis in maize and pea showed that small homologous regions were closely interspersed with heterologous DNA stretches (LAMPPA and BENDICH 1979a, 1981).

In summary, gross sequence changes in plastid DNAs have been shown by different techniques. The degree of heterology revealed between species from different families is largely dependent on the resolution of the methods. In Section 4.3 homologies among plastid DNAs with respect to specific genes are considered.

4 Gene Mapping

The present list of plastid genes (see BOTTOMLEY and BOHNERT Chap. 15, this Vol.) includes genes for rRNAs and tRNAs and approximately 15 proteins. About 10 ribosomal proteins are synthesized in chloroplasts (ENEAS-FILHO et al. 1981) and their genes will have to be added to this list. This statement is based on the assumption, as outlined by BOTTOMLEY and BOHNERT (Chap. 15, this Vol.), that RNAs found within the plastids are transcripts from plastid DNA and that proteins produced on the 70S translation system of chloroplasts are derived from these transcripts. This assumption appears reasonable because no proof exists for the import of mRNA into the chloroplast, although RNA exchange between organelles has been postulated on the basis of indirect evidence (BRADBEER et al. 1979).

4.1 Genes on Plastid DNAs

4.1.1 Genes for rRNAs

The 70S ribosomes of chloroplasts contain 3 (e.g. *Euglena*) or 4 (e.g. higher plants) RNA molecules which differ in size and nucleotide sequence (WHITFELD 1977, DYER Chap. 6, this Vol.). The 30S small subunit particle contains the 16S rRNA while the 23S and 5S rRNAs are located in the 50S ribosomal

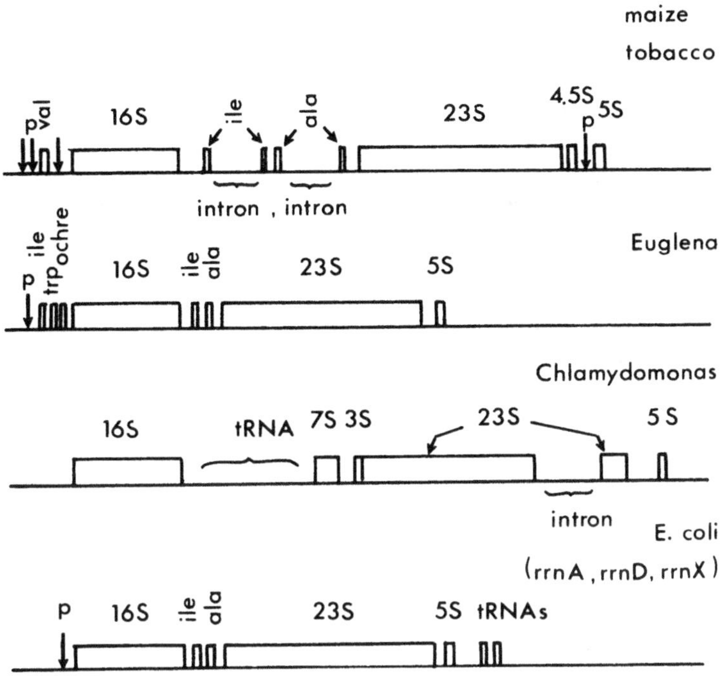

Fig. 3. Schematic presentation of plastid and *E. coli* rrn transcription units. *Arrows* point to promoters, or putative promoters, as judged by comparison with the *E. coli* "consensus sequence". (Redrawn from MORGAN et al. 1977, 1978, and from references in Sect. 4.1.1)

subunit. The large ribosomal subunit in flowering plants and in at least one fern (*Dryopteris acuminata*) contain an additional 4.5S rRNA (WHITFELD et al. 1978, BOWMAN and DYER 1979, DYER and LEAVER 1981, TAKAIWA et al. 1982). Some plastid DNAs contain multiple copies of the rRNA genes (STUTZ and VANDREY 1971, CROUSE et al. 1974, KOPECKA et al. 1977, THOMAS and TEWARI 1974). By hybridization of purified individual rRNAs to DNA fragments, it was shown that in plants where the plastid DNA has been mapped so far the rRNA genes are located close together as a set and probably constitute one operon. The genes for rRNAs are part of the inverted-repeat unit, wherever this structure is present (16 species reported), thus having a copy number per chromosome of two (Table 2). These rRNA genes and some tRNA genes (see Sect. 4.1.2) located in the repeat unit account for approximately 8 Kbp or 30% of the repeat size in higher plants and, most probably, for essentially the whole *Euglena* repeat (Table 2; WOLLGIEHN and PARTHIER 1979). The absence of repeated regions in pea, broad bean and *E. gracilis* Z-S plastid DNAs suggests that the additional copies of these genes in other species might not be functionally essential. It has been proposed that inverted repeat units may have a stabilizing effect, acting against recombinational losses or rearrangements of chromosomes (BEDBROOK and KOLODNER 1979). In *Euglena* either one or three rRNA operons has been found up to now (Tables 1, 2; Figs. 1b, 2). The three units are arranged as tandem repeats. A single 16S rRNA gene, physically separated from the three units in *Euglena* strain Z, is found in addition (JENNI and STUTZ 1979). For the pea and broad bean plastid genomes, which lack the inverted repeat unit, only one rRNA operon was demonstrated.

Detailed physical maps of the rRNA operons have been reported (Fig. 3). In *Euglena* (RAWSON et al. 1978, JENNI and STUTZ 1978, GRAY and HALLICK 1978, EL-GEWELY et al. 1981), the three operons are close together, being separated by not more than 0.8 Kbp (if the distance between the 5S and the following 16S rRNA gene are considered). Slight variations between strains exist in this region (EL-GEWELY et al. 1981). Because of the high buoyant density of the rRNA genes (47% dG+dC) subfractionation of the *Euglena* plastid chromosome is possible (see Sect. 2.3). In *Chlamydomonas* and most higher plant plastid DNAs, the rRNA genes are near to one end of the inverted repeat unit. The rRNA genes are clustered in the general order 5′–16S–spacer–23S–5S–3′ with some variations (Figs. 2, 3; Sect. 5.) which document differences among higher plants, *Chlamydomonas* and *Euglena* plastid DNAs, respectively. In some species the precise sizes of these rRNA's have been determined by sequencing rRNAs or rRNA genes. The maize and tobacco plastid 16S rRNA genes are 1,491 and 1,486 nucleotides long, respectively, with a sequence homology of 96% (SCHWARZ and KÖSSEL 1980, TOHDOH and SUGIURA 1982). Their homology with the *E. coli* 16S rRNA molecule (1,542 nucleotides) is 74%. For the 23S rRNA genes, sizes of 2,898 nucleotides (maize) and 2,804 nucleotides (tobacco) have been determined (EDWARDS and KÖSSEL 1981, TAKAIWA and SUGIURA 1982a). These plastid rRNA genes exhibited a sequence homology of 92% between themselves and homology of approximately 67% with the *E. coli* 23S rRNA (2,904 nucleotides). The 5S rRNAs from tobacco, maize, spinach and *Spirodela* contain 122 or 121 or 119 nucleotides (DYER and BOWMAN 1979, DELIHAS et al. 1981, TAKAIWA and SUGIURA 1980a, b, 1981).

The 4.5S rRNA molecules from various flowering plants and from the fern *Dryopteris acuminata* (TAKAIWA et al. 1982) are between 65 and 103 nucleotides in length, depending on the species (see DYER and LEAVER 1981 for a review). Variation of 4.5S rRNA size has even been found within *Spirodela* chloroplasts where related molecules of 65, 96 and 103 nucleotides have been seen (BOWMAN and DYER 1979). No molecule corresponding to the 4.5S rRNA has been found in two cyanobacterial species or in chloroplasts from algae, mosses and the fern *Adianthum*. The function of the 4.5S rRNA in the plastid ribosome is unknown. Several possibilities exist. It may be viewed as a breakdown product of 23S rRNA which is known to contain hidden nicks in vivo (WHITFELD 1977). It might also be an end product of 23S rRNA processing (HARTLEY and HEAD 1979, HARTLEY 1979) which is entrapped in the assembled ribosome. Sequence homologies between the 3′-end of the *E. coli* 23S rRNA, the 3′-end of eukaryotic 28S rRNAs and several sequenced 4.5S rRNAs or 4.5S rRNA genes (MACKAY 1981, EDWARDS et al. 1981, TAKAIWA and SUGIURA 1980b, MACHATT et al. 1981) suggest that the plastid 23S rRNA gene may have become interrupted by a short intron which is spliced out without the 23S rRNA becoming religated (TAKAIWA and SUGIURA 1982a). The 4.5S rRNA thus most probably represents the 3′-end of some plastid 23S rRNA genes. In support of this hypothesis, a short sequence at the 3′-end of the 4.5S rRNAs from both wheat and tobacco

(WILDEMAN and NAZAR 1980, TAKAIWA and SUGIURA 1980a) may form a double-stranded stem with the 5′-end of the *Euglena* 23S rRNA (OROZCO et al. 1980b). This secondary structure is also possible between the 3′-end of the 4.5S rRNA and the 5′-end of the 23S rRNA of maize chloroplasts (EDWARDS et al. 1981).

The 16S and 23S rRNA genes are separated by a spacer stretch of DNA (Fig. 3) whose length varies from 259 bp in *Euglena* (GRAF et al. 1980) to 2,408 bp or 2,080 bp in maize (KOCH et al. 1981) and tobacco (TAKAIWA and SUGIURA 1982b), respectively. The corresponding spacer size is 440 bp in the *E. coli* rrnB operon (BROSIUS et al. 1981) and 334 bp in *Bacillus subtilis* (LOUGHNEY et al. 1982) and approximately 540 bp in *Anacystis nidulans* (TOMIOKA et al. 1981). This spacer DNA in *E. coli*, in *B. subtilis* and in some chloroplast DNAs contains tRNA genes (see Sect. 4.1.2; Fig. 3). At least in the cases of *E. coli* and maize, this spacer is cotranscribed with the rRNA genes (see Sect. 5).

The 23S rRNA gene may also be interrupted by an intron sequence, as was found for *Chlamydomonas* chloroplast DNA (ALLET and ROCHAIX 1979, ROCHAIX 1981). This intron shares sequence homologies with introns found at corresponding positions in the gene for large ribosomal subunit RNA from yeast mitochondria and from the cytoplasm of *Tetrahymena* (ROCHAIX 1981). Again in *Chlamydomonas*, the 23S rRNA gene is preceded at its 5′-end by a 3S and a 7S RNA gene of yet unknown functions. Presumptive ribosomal RNAs of this size have only been found in *Chlamydomonas* up to now.

4.1.2 Genes for tRNAs

Plastid tRNAs, which are different from those of the cytosol and mitochondria (for reviews see BARNETT et al. 1978, WEIL 1979), are products of the plastid genome (reviewed in WEIL and PARTHIER, Vol. 14A, this Series; see also DYER Chap. 6, this Vol.). Although individual tRNAs for all amino acids have not yet been isolated and characterized, there is evidence that a complete set of tRNA genes, corresponding to all 20 amino acids, is present (MEEKER and TEWARI 1980). The actual number of tRNA genes exceeds the amount of 20, since isoaccepting species for individual amino acids are present. The number of plastid tRNA genes can be calculated from saturation hybridization experiments. *Euglena* plastids contain about 25 tRNA genes as indicated by plateau-hybridisation studies (GRUOL and HASELKORN 1976, MCCREA and HERSHBERGER 1976, SCHWARTZBACH et al. 1976). A similar number of plastid tRNA genes was reported for maize (HAFF and BOGORAD 1976a), whereas spinach and pea chloroplast DNAs might have about 40 tRNA genes (MEEKER and TEWARI 1980, 1982).

The plastid 4S RNAs from various plant species have been fractionated by two-dimensional gel electrophoresis and the individual tRNAs identified by in vitro aminoacylation using either chloroplast or *E. coli* aminoacyl-tRNA synthetases and [^{3}H] amino acids (Table 3). This has led to the identification of 27 spinach and 27 red bean (*Phaseolus vulgaris*) chloroplast tRNA species specific for 16 amino acids, 18 soybean (*Glycine max*) tRNAs specific for 14 amino acids, 25 maize tRNA species specific for 17 amino acids, 23 *Euglena*

Table 3. Plastid tRNA's identified and tRNA genes mapped

Amino acid	Spinach		Red bean		Soybean		Maize		Euglena		Cyanophora	
	a	b	a	b	a	b	a	b	a	b	a	b
Ala	1	1	1	1	1	nd	1	1	1	1	1	1
Arg	2	1	2	nd	nd	nd	1	1	2	1	2	1
Asn	1	1	2	2	nd	nd	1	1	1	1	1	nd
Asp	nd	nd	nd	nd	1	nd	1	1	nd	nd	1	1
Cys	nd	nd	nd	nd	nd	nd	nd	nd	nd	nd	nd	nd
Gln	nd	nd	nd	nd	nd	nd	nd	nd	1	nd	nd	nd
Glu	nd	nd	nd	nd	nd	nd	nd	nd	1	1	nd	nd
Gly	2	1	2	nd	1	nd	2	1	1	1	nd	nd
His	1	1	1	1	1	nd	1	1	1	nd	1	nd
Ile	2	2	2	2	2	nd	2	2	2	1	1	1
Leu	3	3	3	3	2	nd	3	3	2	2	3	nd
Lys	1	nd	1	1	1	nd	1	1	1	1	1	1
Met	3	3	2	2	2	nd	2	2	2	1	1	1
Phe	1	1	2	2	1	nd	1	1	1	1	1	nd
Pro	1	nd	1	2	nd	nd	1	1	1	nd	1	nd
Ser	3	3	3	2	1	nd	3	2	1	nd	3	2
Thr	2	1	1	nd	1	nd	2	2	2	1	1	1
Trp	1	1	1	nd	1	nd	1	1	1	1	1	nd
Tyr	1	1	1	1	2	nd	1	1	1	1	2	1
Val	2	1	2	1	1	nd	1	1	1	1	2	1
Total identified	27		27		18		25		23		23	
Total mapped		21		20				23		15		11

nd = not determined or not found;
a = number of isoaccepting tRNAs identified;
b = minimum number of different tRNA genes mapped on homologous DNA, multiple copies of a tRNA gene are not considered

Based on DRIESEL et al. (1979), MUBUMBILA et al. (1980), SWAMY and PILLAY (1982), KUNTZ et al. (1982), WEIL et al. (1982), EL-GEWELY et al. (1982); for tRNA or tRNA gene sequences see DYER, Chap. 7, this Vol. and SPRINZL and GAUSS (1982); for a list of additional identified tRNAs in other organisms see WEIL and PARTHIER, Vol. 14A, this Series; tRNA genes revealed from sequencing data are not included in this Table.

tRNA species specific for 18 amino acids and 23 cyanelle tRNA species specific for 16 amino acids (Table 3) (MUBUMBILA et al. 1980, KUNTZ et al. 1982, SWAMY and PILLAY 1982, WEIL et al. 1982). Genes for most of these identified and some yet unidentified tRNAs have been located on the physical maps of the plastid DNA from spinach red bean, maize and *Euglena* as well as on the cyanelle DNA from *Cyanophora paradoxa* (EL-GEWELY et al. 1982, DRIESEL et al. 1979, KUNTZ et al. 1982, WEIL et al. 1982) (Table 3). The identification of plastid tRNAs specific for some amino acids (e.g., asparagine, cysteine, glutamine and glutamic acid) has not been achieved.

Up to now, tRNA genes or pseudo-tRNA structures have been mapped and sequenced at the 5′-end of the 16S rRNA gene of spinach ($tRNA^{Val}$; BRIAT,

DRON, LOISEAUX, MACHE personal communication), of maize ($tRNA^{Val}$; SCHWARZ et al. 1981a), of *N. tabacum* (TOHDOH et al. 1981), of *E. gracilis* Z (pseudo-$tRNA^{Trp}$ and pseudo-$tRNA^{Ile}$; HALLICK et al. 1978) and of *E. gracilis* strain *bacillaris* (pseudo-$tRNA^{Trp}$, pseudo-$tRNA^{Ile}$ and a tRNA-like sequence containing the anticodon "ochre"; EL-GEWELY MR, HELLING RB, DIBBITS JGT personal communication). Recently, also in *Euglena gracilis*, it has been realized that this leader region of the rrn operon shows sequence homologies with the 16S–23S spacer region (MIYATA et al. 1982). It contains pseudo-tRNA genes and might have evolved from a duplicated spacer region. Although likely, it is not yet proven that these leader tRNA genes are cotranscribed with the rRNA genes (see Sect. 5). Two tRNA genes have also been sequenced in the spacer DNA separating the 16S and 23S rRNA genes in maize (KOCH et al. 1981), *Euglena* (GRAF et al. 1980, OROZCO et al. 1980a, b) and tobacco (TAKAIWA and SUGIURA 1982b). These spacer tRNAs, as in some *E. coli* and *B. subtilis* rRNA operons, are specific for the amino acids isoleucine and alanine. In *Chlamydomonas* an unidentified tRNA gene (Fig. 3) was mapped at the corresponding position (MALNOE and ROCHAIX 1978) and in spinach, up to now, a gene for a $tRNA^{Ile}$ isoacceptor is known (BOHNERT et al. 1979).

The size differences (1.8 to 2.5 Kbp) of the spacer DNAs among flowering plants appear at least in part to be due to intron sequences interrupting the tRNA coding sequences. The maize $tRNA^{Ile}$ gene is interrupted by a 949 bp intron and the $tRNA^{Ala}$ gene by a 806 bp intron (KOCH et al. 1981). The corresponding introns for tobacco are 707 bp and 710 bp (TAKAIWA and SUGIURA 1982b). The two tobacco introns appear to have originated from a common sequence, as they share extensive sequence homologies. No introns are found in the two *Euglena* spacer tRNA genes.

Besides in the rDNA spacer and preceding the rRNA operon, several other tRNA genes have been mapped in the inverted repeat-region of several flowering plants (e.g., spinach, Fig. 4). The majority of tRNA genes, however, are located in the larger of the two single-copy DNA regions (Fig. 4, DRIESEL et al. 1979, WEIL et al. 1982) from spinach, maize and red bean plastid DNAs. Also, the majority of the mapped 4S RNAs from *Chlamydomonas* and *Spirodela* plastid DNAs are positioned in a similar manner (MALNOE and ROCHAIX 1978, GROOT and VAN HARTEN-LOOSBROEK 1981). In *Euglena* chloroplast DNA, the tRNA genes, which are not a part of rDNA units, are distributed non-randomly over the whole genome (KUNTZ et al. 1982, OROZCO and HALLICK 1982a, b). Although several transcriptional units must exist for tRNA genes, some gene-clustering is found at specific loci (HALLICK et al. 1978, OROZCO and HALLICK 1982a, b). The clustering of tRNA genes in certain chromosomal regions appears to be found in all plastid genomes investigated in this regard.

4.1.3 Genes Coding for Proteins

The position of the gene for the large subunit of RuBPCase (rbcL) is located on maize, spinach, *Chlamydomonas*, wheat, pea, mung bean, *Atriplex*, and cucumber plastid DNAs (COEN et al. 1977, BEDBROOK et al. 1979, WHITFELD and

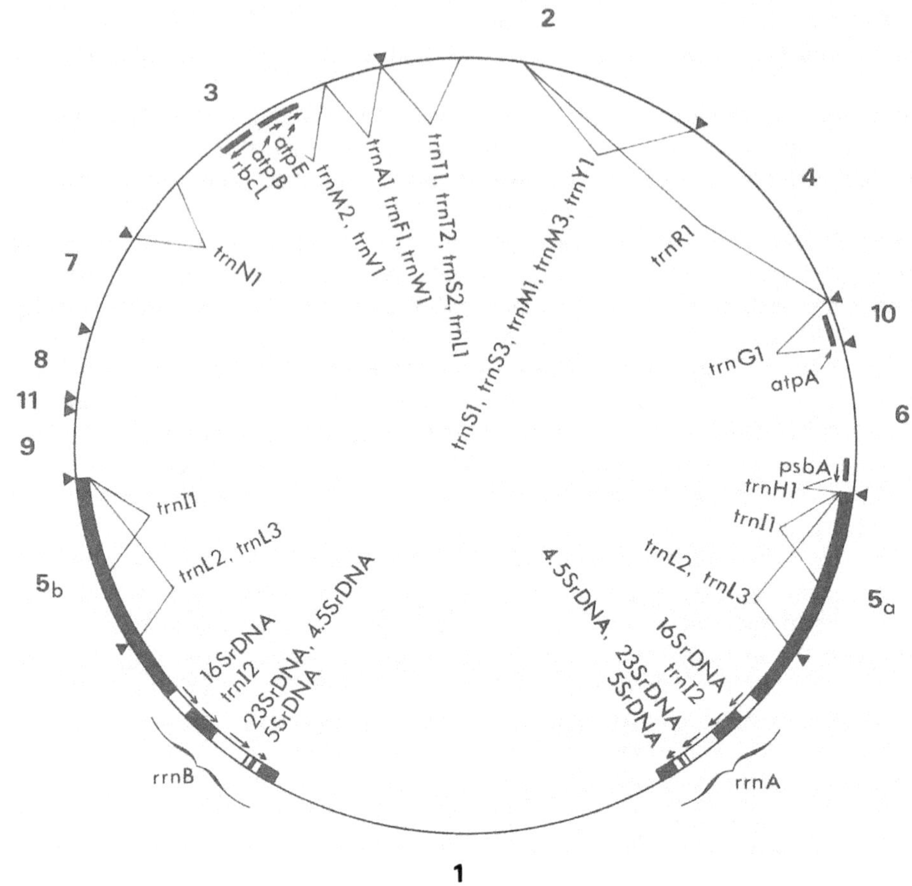

Fig. 4. Physical map and gene map of spinach chloroplast DNA. The positions of recognition sites of the restriction endonuclease Sal I are given by *large arrows*. The DNA fragments are numbered according to decreasing size. Lines extending from gene symbols enclose the DNA region within which a particular gene has been mapped. *Small arrows* give the direction of transcription of genes. (Redrawn from BOHNERT et al. 1979, DRIESEL et al. 1979, 1980, HERRMANN et al. 1980c, WESTHOFF et al. 1981)

BOTTOMLEY 1980, ERION et al. 1981, MALNOE et al. 1979, BOWMAN et al. 1981, KOLLER et al. 1982, PALMER and THOMPSON 1981a, PALMER 1982). Also, the map position of a thylakoid membrane protein gene (psbA) is known on the maize, spinach, pea, mung bean, *Atriplex,* cucumber and *Chlamydomonas* plastid chromosomes (BEDBROOK et al. 1978, DRIESEL et al. 1980, ROCHAIX 1981, PALMER and THOMPSON 1981a, PALMER 1982). This gene was initially named photogene since its transcription is regulated by light (BEDBROOK et al. 1978) or, according to the apparent molecular weight of the processed protein, 32,000 molecular weight protein gene (EDELMAN and REISFELD 1980). Finally, the genes for three subunits of the chloroplast ATP-synthetase F1-complex (atpA, atpB, and atpE) have been located in spinach (WESTHOFF et al. 1981). In maize plastid

DNA the positions of two as yet unidentified genes are also known. One of these, 2.2 Kbp in size, is located close to the rbcL gene, both transcribed in opposite directions (LINK and BOGORAD 1980, JOLLY et al. 1981). In analogy with the spinach genome structure (Fig. 4) this gene might contain the atpB or atpB and atpE coding sequences (WESTHOFF et al. 1981). The second gene, which is close to the psbA gene in maize, lies immediately inside the inverted-repeat segment of this DNA (SCHWARZ et al. 1981b) and overlaps partially with a $tRNA^{His}$ gene. A gene coding for an unidentified thylakoid membrane protein (MALNOE et al. 1979), presumably homologous to the psbA gene, is mapped on the *Chlamydomonas* plastid DNA. By using mRNA preparations from spinach chloroplasts (e.g., 14S RNA and 20S RNA, SPEIRS and GRIERSON 1978, DRIESEL et al. 1980) and probes from inside identified genes known from spinach chloroplast DNA, it was possible to localize small segments on *Spirodela* chloroplast DNA (VAN EE et al. 1982) which, in all probability, contain the corresponding genes. The approach of taking cloned sequences, which are derived from a particular gene whose position is known in one species, for the identification of the corresponding gene in plastid DNA from other plants, is now widely used (see Sect. 4.3).

4.2 Gene Structure

Bona fide transcription requires the recognition of specific nucleotide sequences at the start and, often, at the end of a gene or operon. In addition, sequences have been revealed, predominantly in bacterial systems, which act in the control of transcription or which facilitate the correct binding of the mRNAs to the ribosome (see ROSENBERG and COURT 1979, BUJARD 1980, CHAMBON 1975 for reviews).

Promoters. E. coli DNA-dependent RNA polymerase binds to stretches known as "−10" (Pribnow box) and "−35" regions (for a review see BUJARD 1980). A "consensus sequence" compiled by comparison of many such sites can be viewed as an ideal bacterial-type promoter. Plastid DNA from spinach has been probed with *E. coli* RNA polymerase under stringent conditions and found to contain several strong binding sites distributed over the unique-copy regions of the molecule (ZECH et al. 1981). The strongest binding of polymerase was observed at the site of the psbA gene. No comparably strong site was found in the region of the rRNA genes.

The extensive work on the characterization of chloroplast genes (see GROOT 1983, STUTZ 1983 for reviews) and the growing number of gene sequences now permits an examination of putative chloroplast promoters and a comparison to the *E. coli* consensus sequence (Table 4). It should, however, be kept in mind that sequence homology is no proof for a promoter function of the chloroplast sequence in question. Supporting evidence is required for this. The 5′-end of the transcript should, for instance, be located only a few basepairs downstream from the putative promoter site. This is the case for the spinach rbcL gene since its start codon is 10 or 11 nucleotides away from a sequence which shows strong homology with the *E. coli* promoter consensus sequence (ZUR-

Table 4. Putative promoter sequences on chloroplast DNAs

Organism	gene(s)		Sequence
Maize	tRNA^{Val}-16 SrRNA	F1	TAGGACCGGTTG ACAATTGAATCC AATTTTTCCCAT TATTTGACTGTC CAT −445 … −400
		F2	AGTGGCCTTGCG TTTCTCGCCCCT TTGCCTTAGGAT TCGTTAATTCTC TTTCTCGATGGG ACGGGGA-tRNA^{Val} −360 … −310
	16 SrRNA	F3	GGCTTGTGGGAT TGACGTGATAGG GTAGGGTTGGCT ATACTGCTGGTG GCGAACTCCAGG CTAATAATCTG −160 … −100
Tobacco	tRNA^{Val}-16 SrRNA		AGAATAGTGGCG TTGAGTTTCTCG ACCCTTTGACTT AGGATTAGTCAG TTCTATTTCTCG (12 bp)-tRNA^{Val} −370 … −300
Tobacco	tRNA^{Asn}		ATGAATGGAATT GGGATATGGATG GAATATAATGAA ATAGAGCCA −180 … −140
Maize	LSU		TTTAATGAGTTG ATATTAATTAAA TATCATTTTTTT TAGATTTTTGCA AAAGGTTTCTTT CGCCTAAT −130 … −70
Spinach	LSU		CGGTTACGGTTG GGTTGCGCCATA TATATGAAAGAG TATACAATAATG ATGTATTTG −220 … −170
Spinach photogene			TAGATATTGGTT GACACGGGCATA TATAAGGCATGT TATACTGTTGAA TAACAATCT −130 … −80
Consensus sequence			TGTTGACAATTT ←—(15–18 BP)—→ ATTTGT TATAATG (5–7 bp) CAT

DNA stretches protected against DNAase digestion by bound *E. coli* DNA-dependent RNA polymerase are marked by a line above the sequence. Sequences resembling promoters are underlined. Start nucleotides of mRNAs are marked by arrows. Numbers refer to distances in nucleotides from the first nucleotide in a gene or from the start of a translated sequence

AWSKI et al. 1981). Likewise, a similar position of a maize rbcL promoter relative to the start of transcription was revealed (MCINTOSH et al. 1980). The primary sequence of the spinach and *Nicotiana debneyi* psbA genes is now also available (ZURAWSKI et al. 1982). It shows that 6 out of 6 nucleotides ("−35" region) and 5 out of 6 nucleotides (Pribnow box) are identical with the consensus sequence of *E. coli*. In spinach, transcription starts 12 nucleotides downstream of the 5′-end of this putative promoter site.

In maize, several promoter-like sequences have been found by sequencing of tRNA genes. In two cases, these putative promoters are similar to the T7A2 promoter (STEINMETZ A, personal communication). In tobacco, a possible promoter site close to the 5′-end of a $tRNA^{Asn}$ gene was found (KATO et al. 1981) (Table 4). From sequencing the 5′-end of the *E. gracilis* strain *bacillaris* chloroplast 16S rDNA region, sequences are again found which match the consensus sequence of *E. coli* (HELLING RB personal communication). TOHDOH et al. (1981) reported that the rRNA genes in tobacco are preceded by a putative promoter site. Similar sequences are not that easily identified close to the chloroplast rDNA in maize. Since the start nucleotide for the primary transcript is not known, sequences upstream of the 5′-end of the 16S rRNA gene have to be analyzed. In maize chloroplast DNA, three positions around nucleotides −420 (F1), −330 (F2) and −130 (F3) from the start of the 16S rRNA gene have been found to bind *E. coli* RNA polymerase (SCHWARZ et al. 1981a), where upon the DNA is protected against nuclease attack. Sites F1 and F3 exhibit sequence homologies (Table 4) with *E. coli* rrn promoters (ROSENBERG and COURT 1979). A similar sequence has been found upstream from the 5′-end of the tobacco 16S rRNA gene (TOHDOH et al. 1981).

Terminators. In bacterial systems, one type of termination exists which requires a specific site on the template and is factor-independent. This termination sequence is located downstream relative to the protein termination codon and in a short hairpin structure. This structure appears to be recognized by the approaching RNA polymerase for termination, although a certain degree of readthrough occurs.

Such a possible terminator was found at the 3′-end of the rbcL (LSU) gene in spinach (ZURAWSKI et al. 1981) and was also deduced from the published sequence of the corresponding maize gene (ZURAWSKI et al. 1981). A terminator structure can also be seen adjacent to the 3′-end of the spinach psbA gene, between 38 and 96 nucleotides after the stop codon, and downstream the corresponding gene from *N. debneyi* between 30 and 105 nucleotides past the stop codon (ZURAWSKI et al. 1982) (Fig. 5).

Ribosome Binding Sites. Nucleotide sequences upstream of the translated part of a mRNA, close to the protein start codon, have been found to contain the sequence GGAGG (SHINE and DALGARNO 1974) which may basepair to the 3′-end of 16S rRNA (CCUCC). This structure is thought to represent a ribosome binding site for the initiation of protein synthesis. Sites complementary to the 3′-end of maize and tobacco 16S rRNAs, which have the CCUCC se-

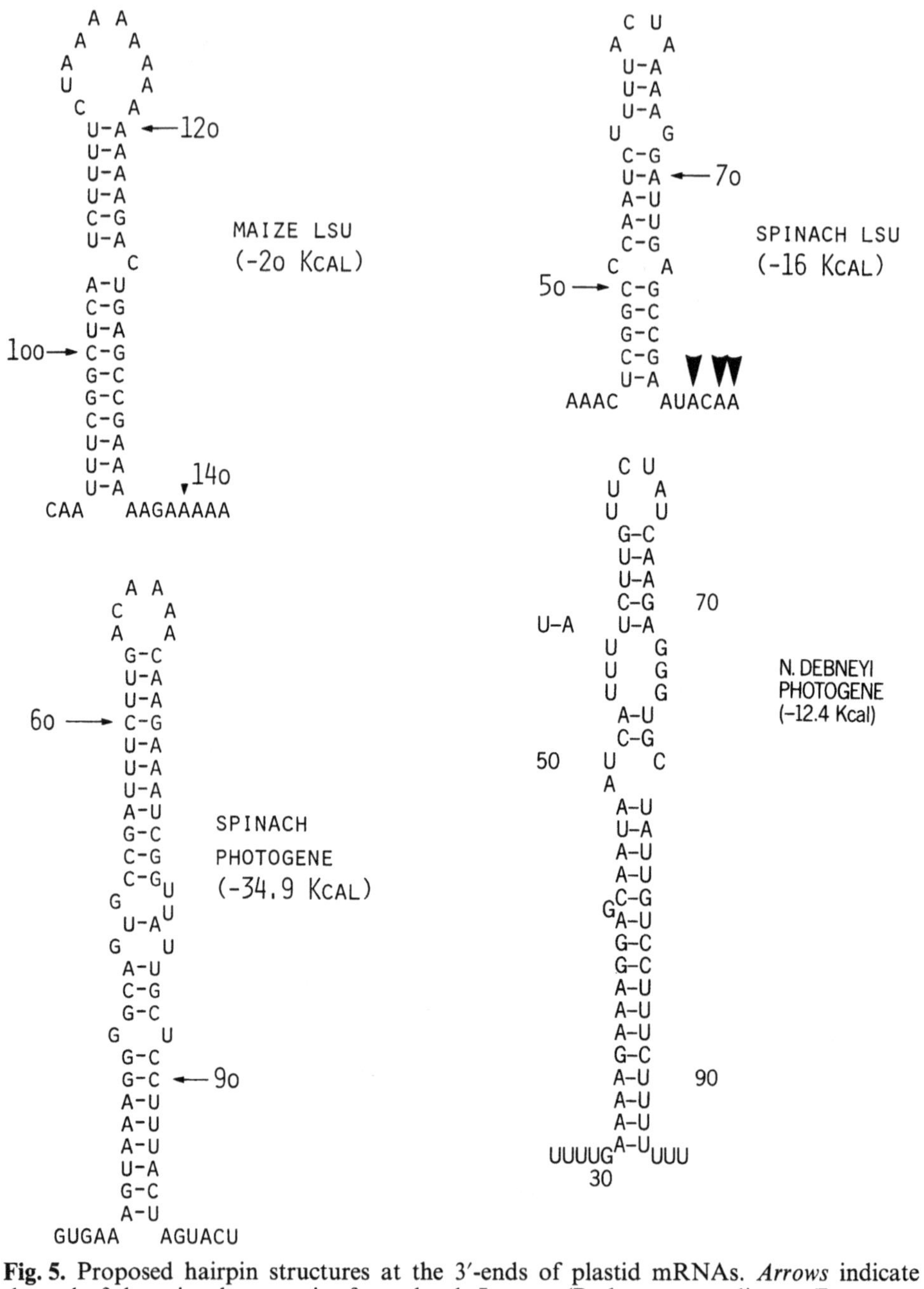

Fig. 5. Proposed hairpin structures at the 3′-ends of plastid mRNAs. *Arrows* indicate the end of the spinach transcript from the rbcL gene. (Redrawn according to ZURAWSKI et al. 1981, 1982)

Table 5. Sequences of nucleotides occurring at the 3′end of small-subunit rRNA and possible ribosome-binding sites of chloroplast mRNAs[b]

Organism	rRNA	Sequence[a]	References
Yeast cytoplasm	17 S	CUCGCGGAAGGAUCA__UUA$_{OH}$	DE JONGE et al. (1977)
Mouse cytoplasm	18 S	CCUGCGGAAGGAUCA__UUA$_{OH}$	HAGENBÜCHLE et al. (1978)
Wheat cytoplasm	18 S	CCUGCGGAAGGAUCA__UUG$_{OH}$	HAGENBÜCHLE et al. (1978)
Duckweed chloroplast	16 S	GAUCACCUCCU	WOESE and DYER, in DYER and LEAVER (1981)
Euglena chloroplast	16 S	UACUGGAAGGUGUGGCUGGAACAACUCC	GRAF et al. (1980)
Escherichia coli	16 S	UAGGGGAACCUGCGGUUGGAUCACCUCCUUA$_{OH}$	CARBON et al. (1978)
Maize chloroplast	16 S	UACUGGAAGGUGCGGCUGGAUCACCUCCUUU$_{OH}$	SCHWARZ and KÖSSEL (1980)
Spinach rbcL		$_{HO}$GUAUUCAGGGAGGGAUGUUGAAGUUUA	ZURAWSKI et al. (1981)
Maize rbcL		$_{HO}$GUAUUCAGGGAGGGAUGUUGAGUACUU	MCINTOSH et al. (1980)

[a] Underlined nucleotides differ from those of *E. coli*, the ribosome binding site sequence is shown by two lines, the arrow points to the protein initiation codon

[b] The spinach psbA gene (ZURAWSKI et al. 1982) shows no homology with this sequence close to the 1st AUG-codon, and only weak homology (-UGUGG-) at the 2nd AUG-codon. Its mRNA is only poorly translated in an *E. coli* transcription/translation system (BOTTOMLEY personal communication)

quence, have been found by sequencing the rbcL genes from maize and spinach (Table 5) (MCINTOSH et al. 1980, ZURAWSKI et al. 1981).

In support of the view presented here, possible transcription start and termination signals, as well as a putative ribosome-binding site, have recently been revealed on chloroplast DNA from *E. gracilis* Z (OROZCO and HALLICK 1982b) preceding an open reading frame in the chloroplast DNA fragment EcoG. The putative ribosome-binding site has the structure 5′-GGAGT-3′ (DNA) or 5′-CCUCA-3′ (RNA), which is complementary to the 3′-end of the *Euglena gracilis* Z chloroplast 16S rRNA (GRAF et al. 1980). For the spinach psbA gene, a similar site is not found close to the first methionine which begins the reading frame. Since the correct size of this protein is not yet known and since the spinach 16S rRNA gene has not been sequenced, it is at present difficult to determine whether such a ribosome binding site exists in this case.

In summary, it can be stated that for all plastid genes sequenced so far, at the start of transcription and beyond the translation stop codons, sequence

Table 6. Codon usage in the maize (a) and spinach (b) rbcL genes and in the spinach (c) psb A gene. Numbers, besides the termination codons, are normalized as frequencies per 1,000 nucleotide triplets. (According to McIntosh et al. 1980, Zurawski et al. 1981, 1982)

First letter	Second Letter U				C				A				G				Third letter
		a	b	c		a	b	c		a	b	c		a	b	c	
U	Phe	25	25	26	Ser	13	17	28	Tyr	19	29	20	Cys	13	10	6	U
	Phe	19	17	48	Ser	8	8	14	Tyr	17	10	14	Cys	10	8	0	C
	Leu	23	17	28	Ser	10	4	0	ochre	1	0	1	opal	0	0	0	A
	Leu	25	21	20	Ser	2	0	6	amber	0	1	0	Trp	17	17	28	G
C	Leu	10	21	17	Pro	13	25	26	His	23	13	12	Arg	23	25	23	U
	Leu	2	0	0	Pro	4	6	0	His	8	19	17	Arg	19	10	9	C
	Leu	15	23	23	Pro	19	8	17	Gln	23	19	17	Arg	6	13	3	A
	Leu	2	4	0	Pro	8	4	0	Gln	4	6	3	Arg	0	0	0	G
A	Ile	23	23	51	Thr	36	32	34	Asn	13	17	34	Ser	6	8	17	U
	Ile	19	17	28	Thr	15	21	9	Asn	15	17	28	Ser	0	0	12	C
	Ile	8	0	9	Thr	13	19	6	Lys	38	40	0	Arg	8	8	6	A
	Met	25	21	34	Thr	0	0	0	Lys	17	8	0	Arg	4	4	3	G
G	Val	25	31	28	Ala	38	42	68	Asp	50	44	17	Gly	44	46	68	U
	Val	2	2	0	Ala	13	8	6	Asp	13	10	3	Gly	2	10	6	C
	Val	29	34	34	Ala	38	29	17	Glu	42	53	43	Gly	27	29	17	A
	Val	6	4	0	Ala	10	8	6	Glu	17	15	14	Gly	19	13	3	G

homologies with eubacterial promoters and terminators exist. In addition, the presence of a sequence which in eubacteria is supposed to represent a ribosome binding site has been established.

Codon Usage. As complete nucleotide sequences of plastid genes become available, the use of codons in the translation process can be studied (Table 6). In the spinach and maize rbcL genes (McIntosh et al. 1980, Zurawski et al. 1981) and in the spinach psbA gene (Zurawski et al. 1982) all codons specifying amino acids occur with the exception of the codons CGG (arginine) and ACG (threonine). The termination codon "opal" (UGA) was also not detected in these genes. AUG is the first codon of long reading frames which are preceeded by promoter-like sequences. The UAA and UAG codons terminate these reading frames.

According to the wobble hypothesis (Crick 1966), approximately 32 tRNAs would be required for translation of the observed codons in chloroplast genes. In different plants, between 18 and 27 isoaccepting tRNA's have been identified (Table 3) and in all these plants several more 4S RNA's remain to be identified still.

Based on the two complete sequences available, a bias in the use of codons is apparent (Table 6), in which either A or U is preferred as the third base.

This occurs with a much higher frequency than would be expected from the average dG+dC content of the chromosome. More data are required in order to test the hypothesis that selective codon usage is a genomic mechanism for the modulation of gene expressivity (GRANTHAM et al. 1981).

Several open reading frames have been identified for which no proteins have yet been detected. These are found in the transcribed introns of the maize and tobacco rrn operons (KOCH et al. 1981, TAKAIWA and SUGIURA 1982b) and in *Euglena* chloroplast DNA (OROZCO and HALLICK 1982b). In maize chloroplasts these introns are stable transcripts after being cleaved from the rrn transcriptional units (KÖSSEL H personal communication). In these sequences, the codons CGG (arg), ACG (thr) and, as a termination codon, UGA (opal) are evident. It is thus reasonable to assume that plastids use the normal genetic code.

4.3 Interspecies Conservation of Gene Structure and Sequence Among Plastid DNAs

Sequence homologies between the plastid DNAs of different species have been reported (LAMPPA and BENDICH 1979a, 1981, BOHNERT et al. 1980, GORDON et al. 1981, FLUHR and EDELMAN 1981b, PALMER and THOMPSON 1981a, PALMER 1982) by using reassociation kinetics and heterologous DNA/DNA hybridizations (see also Sect. 3.2). A large sequence homology exists among the rRNA genes from eubacteria, chloroplasts and, at least some, mitochondrial ribosomal small subunit RNA genes (SCHWARZ and KÖSSEL 1980, BOHNERT et al. 1980, GRAF et al. 1980, EDWARDS and KÖSSEL 1981, BORBELY and SIMONCSITS 1981, and see ZWIEB et al. 1981, GLOTZ et al. 1981 for comparisons). Also, many tRNA gene sequences are highly conserved (see DYER Chap. 6, this Vol. and WEIL and PARTHIER Vol. 14A, this Series). This may be exploited in heterologous hybridization experiments (MUBUMBILA et al. 1980) although, because of the small size of tRNA molecules, a few base changes might be sufficient to eliminate formation of the hybrids. With respect to genes coding for proteins, a comparison of the sequences of the rbcL genes from maize and spinach (MCINTOSH et al. 1980, ZURAWSKI et al. 1981) reveals more than 90% conservation in the protein-coding part of the gene. Most of the changes which were revealed by sequencing occur in the third-base of codons and do not lead to a change in the amino acid. Homology to a much lower extent is found in the 5′-noncoding region and even less in the non-translated 3′-end of the gene. The same features have now been revealed when the sequences of the psbA genes from spinach and *N. debneyi* were compared (ZURAWSKI et al. 1982). Using heterologous hybridizations with pea, broad bean, red bean and *Euglena* plastid DNA fragments, it appears that the mapped and identified spinach plastid genes, or probes from inside the translated region of these cloned genes (BOTTOMLEY W, HARTLEY MR, WHITFELD PR personal communication; PALMER and THOMPSON 1981b and personal communication) may be used to specifically map corresponding genes on other, especially higher plant, plastid DNAs (BOHNERT HJ, CROUSE EJ unpublished results).

4.4 Comparison of Eubacterial and Plastid Genes

Plastid nucleoid structure, mode of plastid division, many biochemical characteristics and sensitivities to inhibitors of biosynthetic functions led to a concept that plastids may have evolved from a prokaryotically organized endosymbiont. Additional evidence can be given (see below) for an origin of extant plastids from ancient forms of present-day cyanobacteria, as the "endosymbiont hypothesis" may be phrased (see TAYLOR 1979, MARGULIS 1981, GRAY and DOOLITTLE 1982, for recent reviews). The structural and sequence similarities of plastid rrn operons with those from *E. coli* (LINDAHL et al. 1977, BOROS et al. 1979, BROSIUS et al. 1981) and also *Anacystis nidulans* (=*Synechococcus*) (TOMIOKA et al. 1981) chromosomal DNAs are striking. Chloroplast DNA fragments carrying these genes have been identified by heterologous hybridization (BOHNERT et al. 1980) and electron microscopy (DELIUS and KOLLER 1980). Between the maize or tobacco plastid rRNA genes and those of *E. coli* approximately 75% (16S rRNA), 67% (23S rRNA) and 55% (5S rRNA) sequence homologies have been detected (SCHWARZ and KÖSSEL 1980, EDWARDS and KÖSSEL 1981, TOHDOH and SUGIURA 1982, TAKAIWA and SUGIURA 1982a). Although small deletions or single-base exchanges were observed, the sequences, without doubt, must have evolved from a common ancestor. Hairpin structures and loops as well as large stretches of the sequences, which most likely are functionally essential regions of RNA/protein interaction, were highly conserved.

It may be predicted from this that ribosomal proteins or their genes should also be highly conserved. This is supported by cross-reactions of antibodies raised against *Chlamydomonas* chloroplast ribosomal subunits (SCHNEEMAN and SURZYCKI 1979) with *E. coli* ribosomal proteins, and by cross hybridization of DNA sequences containing the *E. coli* elongation factor Tu as well as ribosomal protein genes with *Chlamydomonas* chloroplast DNA (WATSON and SURZYCKI 1982). It is difficult to reconcile these facts (and the endosymbiont hypothesis) to the fact that, apart from approximately 10 genes (ENEAS-FILHO et al. 1981), most genes for chloroplast ribosomal proteins have been located in the nucleus by genetic investigations (see GILLHAM 1978 for a review). If these genes have been eliminated from plastid DNA, and have been incorporated into nuclear chromosomes, it might be worth probing nuclear DNA with cloned cyanobacterial ribosomal protein genes. Recently, we have used cloned DNA fragments, from within spinach genes coding for proteins, in heterologous hybridizations with chromosomal DNA from the cyanobacteria *Anacystis nidulans* and *Aphanocapsa* 6714 (BOHNERT HJ, BOTTOMLEY W and LOEFFELHARDT W unpublished results, ASTIER C and BOHNERT HJ unpublished results). With all genes tested (rbcL, psbA, atpA, atpB and atpE), strong hybridizations to one or two DNA fragments were obtained.

The study of the nucleic acids of photosynthetic organelles with cyanobacterial features is of obvious interest. *Cyanophora paradoxa,* a biflagellate alga, contains such organelles (called cyanelles) which were viewed as endosymbiotic cyanobacteria (TRENCH 1979). These organelles are surrounded by a peptidoglycan cell wall remnant (AITKEN and STANIER 1979) and contain cyanobacterial pigments. The DNA of these cyanelles is of the size and complexity of plastid

DNAs and contains a DNA segment 9 Kbp in size which is present twice and which harbors the rRNA genes (HERDMAN and STANIER 1977, MUCKE et al. 1980, LÖFFELHARDT et al. 1980, BOHNERT et al. 1982). When DNA from the spinach plastid genes rbcL, psbA, atpA, atpB, and atpE were hybridized to cyanelle DNA, as with the cyanobacterial DNA, strong hybridization with one or two DNA fragments was obtained (BOHNERT HJ, LÖFFELHARDT W unpublished results).

5 Transcription of Plastid Genes

RNA synthesis requires a DNA primer, the four ribonucleoside triphosphates, divalent cations and DNA-dependent RNA polymerase to catalyze the reaction. Bona fide transcription of genes requires certain protein factors that are at least transiently part of the DNA/RNA polymerase complex and that function in the recognition of specific nucleotide sequences at both ends of the gene. These promoter and terminator sequences are described in Sect. 4.2. The plastid RNA polymerase molecule is described by WOLLGIEHN (Chap. 5, this Vol.). Further reviews on prokaryotic or eukaryotic transcription (LOSICK and CHAMBERLIN 1976, ROSENBERG and COURT 1979, CHAMBON 1975, GUILFOYLE 1981, DYER and LEAVER 1981) are available.

The transcription of plastid DNAs has been studied on different levels of experimental complexity: in vivo, in organello (i.e., in isolated organelles) and in vitro. The basically prokaryotic characteristics of chloroplast transcription have already been discussed (WHITFELD 1977, WOLLGIEHN and PARTHIER 1980, BOHNERT et al. 1983). Many investigators have used the selective inhibition of either the cytosolic or the plastid functions. The limitations of this approach should be considered because several problems may occur (see GALLING Chap. 18, this Vol., BOHNERT et al. 1983).

5.1 Transcription of rRNA Genes

The transcripts of the rRNA genes are comparatively stable and amount to more than 90% of the RNA in plastids. The synthesis of 23S rRNA (1.08×10^6 molecular weight) and 16S rRNA (0.56×10^6 molecular weight) was demonstrated in vivo (see Table 7). The two large rRNAs appear after short-time labelling as precursor molecules with sizes of 1.25×10^6 molecular weight (p23S) and 0.65×10^6 molecular weight (p16S) (MUNSCHE and WOLLGIEHN 1973, HARTLEY and ELLIS 1973). These immediate precursors are derived from a still larger precursor molecule with the size of approximately 2.7×10^6 molecular weight (HARTLEY and ELLIS 1973, HARTLEY et al. 1977, BOHNERT et al. 1974, 1976, 1977, HARTLEY 1979, HARTLEY and HEAD 1979, RICHTER et al. 1977). The species of RNAs present in the primary transcript have been determined recently. Small DNA probes from this region of the maize chloroplast DNA

Table 7. Sizes and function of higher plant chloroplast RNAs

Type or source	Gene	Sedimentation coefficient (S)	Size (m.w. × 10^6)	Nucleotides	Species	Remarks	References
Ribosome[a] small subunit (30S)	rrs[b]	16	0.56[c]	1,542[d]	maize		SCHWARZ and KÖSSEL (1980)
p16		17–18	0.65		various	immediate precursor to 16S rRNA	HARTLEY and ELLIS (1973), MUNSCHE and WOLLGIEHN (1973), HEIZMANN (1974), POSNER and ROSNER (1975), SCOTT (1976)
RIBOSOME large subunit (50S)	rrl 1	23	1.08		maize		EDWARDS and KÖSSEL (1981)
	rr f	5	0.039	122	various		DYER and BOWMAN (1979),
	rr g	4.5	0.021–0.033	65–103	various		WILDEMAN and NAZAR (1980), TAKAIWA and SUIGIURA (1980a), GRAY and HALLICK (1979a)
p23		25	1.28		various	immediate precursor to 23S rRNA	HARTLEY and ELLIS (1973), MILLER and MCMAHON (1974), SCOTT (1976)
Spacer RNA from rDNA unit	tRNAs and introns	~14	0.42		maize spinach	intermediary and and stable processing products	KOCH et al. (1981), HARTLEY et al. (1977), BOHNERT et al. (1976, 1977)
4S RNA (tRNAs)	amino acid symbol	4		67–95	various	25–40 genes/chromosome	SPRINZL and GAUSS (1982), WEIL and PARTHIER Vol 14A, this series, DYER Chap. 6, this Vol.

mRNA	rbcL	17–19	0.6	1,690	spinach maize	translated sequence is 1,426 nucleotides	ZURAWSKI et al. (1981) MCINTOSH et al. (1980)
	psbA	13–14	0.45	1,230	spinach	translated sequence 1,068 nucleotides	ZURAWSKI et al. (1982)
	atpA atpB atpE				spinach spinach spinach	hybrid arrest, release, immunological detection of products	WESTHOFF et al. (1981)

[a] Differences exist for the genes in the rDNA region from the alga *Chlamydomonas* (MALNOE and ROCHAIX 1978) and in the rDNA spacer from *Euglena* (KELLER et al. 1980)

[b] The gene nomenclature follows, in part, the rules established for *E. coli* genes (BACHMAN and LOW 1980); rbcL – large subunit gene of the ribulose-bisphosphate carboxylase; psbA – thylakoid membrane protein gene, mnemonic: photogene (BEDBROOK et al. 1978), or peak D protein (EAGLESHAM and ELLIS 1974) or rapidly labelled membrane protein P32,000 (EDELMAN and REISFELD 1980); atpA, atpB, atpE – α, β and ε subunits of the CF1-complex (WESTHOFF et al. 1981)

[c] Determined from their electrophoretic mobility relative to the high-MW RNAs from ribosomes of *E. coli* (KURLAND 1960), cytosolic ribosomes from *Acanthamoeba castellanii* (LOENING 1968) and cytosolic ribosomes of mouse liver (see HAGENBÜCHLE et al. 1978)

[d] From DNA or RNA sequencing data

were hybridized to filter-immobilized chloroplast RNA. DNA probes covering portions of the rDNA region extending from the 5′-end of the 16S rRNA gene through the 16S/23S spacer, including the genes for $tRNA^{Ile}$ and $tRNA^{Ala}$ and their introns, the 23S rRNA gene and the 4.5S rRNA gene hybridized to the 2.7×10^6 molecular weight precursor. DNA probes containing the 5S rRNA gene failed to hybridize to the precursor RNA (ZENKE G, EDWARDS K, LANGRIDGE P, KÖSSEL H personal communication). The 5S rRNA gene in all probability represents an independent transcriptional unit, as has been shown by labelling kinetics (HARTLEY 1979, see BOHNERT et al. 1983) and sequencing (DYER and BEDBROOK 1980, TAKAIWA and SUGIURA 1980a, EDWARDS et al. 1981).

From *Euglena* chloroplasts a transcript of 1.8×10^6 molecular weight was isolated which contained both the 16S and 23S rRNA sequences (WOLLGIEHN and PARTHIER 1979). This may represent transcription of the whole repeat DNA unit (5.6–5.8 Kbp) in this organism which is equivalent to a 1.85×10^6 transcript from one strand. Finally, plastids of the brown alga *Pylaiella littoralis* were also shown to contain a 1.88×10^6 molecular weight precursor RNA molecule which contained nucleotide sequences of at least the two large rRNAs (LOISEAUX et al. 1980). Transcription and processing of plant rRNA's is reviewed by GRIERSON (Chap. 7, this Vol.) in more detail.

5.2 Transcription of tRNA Genes

Our knowledge concerning the genomic location of tRNA genes and the structure of tRNAs or tRNA genes (Table 3 and Sect. 4.1.2; for recent reviews see DYER Chap. 7, this Vol. and WEIL and PARTHIER Vol. 14A, this Series) contrasts sharply with our ignorance regarding tRNA synthesis. Studies on the synthesis of tRNAs or 4S RNAs, which constitute approximately 5% of the total plastid RNA, have been reported for *Euglena* (BARNETT et al. 1969, REGER et al. 1970, PARTHIER and KRAUSPE 1974). In vivo synthesis was also measured in higher plants (BURKARD et al. 1972, MERRICK and DURE 1972). In these cases a light-induced increase of chloroplast-specific tRNAs was observed. However, during cotton cotyledon development, increasing amounts of chloroplast tRNAs were observed even in the dark (BRANTNER and DURE 1975). Isolated intact chloroplasts from spinach are capable of light-driven synthesis of 4S RNAs (HARTLEY 1979, BOHNERT et al. 1977, ENEAS-FIHLO et al. 1981). The products have not been characterized further. No information is available about precursors of tRNAs or the possibility of polycistronic transcription of those tRNAs whose genes have been mapped close together (DRIESEL et al. 1979, WEIL et al. 1982, OROZCO and HALLICK 1982a, b).

5.3 Transcription of Protein Genes

Transcripts to be translated into proteins amount to only 2%–3% of the total chloroplast RNA. The genes for rRNAs and tRNAs occupy close to 10% of the coding capacity of plastid DNA. This calculation is based on the smallest mapped angiosperm chloroplast DNA, that of pea (120 Kbp, PALMER and THOMPSON 1981a; or 135 Kbp, CHU et al. 1981), on which one set of rRNA genes and approximately 40 tRNA genes were found (MEEKER and TEWARI 1982). Larger plastid DNAs might contain more and even different genes. Up to now there is, however, no experimental evidence in support of this hypothesis. Several groups have tried to measure the extent and complexity of chloroplast transcription by hybridization experiments using total plastid RNA or leaf RNA as probes.

In pea, the total amount of RNA which can be hybridized to filter-bound chloroplast DNA is equivalent to the complexity of one strand of the transcribed DNA (OISHI et al. 1981). Similarly, ZECH et al. (1981) could show that all the fragments generated by Sal I endonuclease cleavage of spinach chloroplast DNA hybridize with in vitro labelled high molecular weight RNA from chloroplasts, although with varying intensities between fragments. The same observation has been reported by MATSUDA and SURZYCKI (1980) for *Chlamydomonas* chloroplast DNA, with the exception of three small dA·dT-rich fragments to which no RNA sequences hybridized. HOWELL and WALKER (1976, 1977), by using saturation hybridization, showed that in *Chlamydomonas* between 40% and 60% (depending on developmental and cell cycle conditions) of the double-stranded plastid chromosome was transcribed. These results permit the conclusion that these plastid DNAs contain no major region which is transcriptionally silent.

In *Euglena,* depending on the developmental status of the plastids, 46%–52% of the DNA of one strand has been demonstrated to be transcribed. This corresponds to the transcription of about half the genome (CHELM et al. 1977a, RAWSON and BOERMA 1976a, MILNER et al. 1979). The limit of detection in these experiments was estimated to be 20 copies of a specific RNA per cell. Assuming approximately 10 chloroplasts per *Euglena* cell, 1–2 transcripts per chloroplast would have been detectable (CHELM and HALLICK 1976, CHELM et al. 1977a, b, 1978, 1979). The difference in the total amount of plastid DNA transcribed between pea and *Euglena* can partly be explained by the low average dG+dC content of *Euglena* plastid DNA (24%–28%), as compared to higher plant plastid DNA's (about 37%). In *Euglena* about 30% of the genome is as low as 12% dG+dC, which might be too low to contain genes (see Sect. 2.3).

It appears that most transcripts are present both in light-grown and dark-grown pea plants since only about 2%–3% difference in total hybridizable RNA was reported (OISHI et al. 1981). An alternative explanation might be that approximately equal amounts of both light-stimulated and light-repressed transcripts were found. The presence of as small an amount as 0.15%–0.2% of poly(A)$^+$RNA was found among these transcripts from pea chloroplasts (MEEKER and TEWARI 1980). The conclusion that plastid RNA contains little or no terminal poly(A) stretches had been reached by WHEELER and HARTLEY (1975). In contrast, HAFF and BOGORAD (1976b) reported 6% of total poly(A)$^+$-RNA from maize seedlings hybridized to maize chloroplast DNA, and MILNER et al. (1979) found that poly(A)$^+$RNA from *Euglena* hybridized to 2.8% of the chloroplast DNA. A value of 7%–10% of the total cellular poly(A)$^+$RNA has also been reported (VERDIER 1979) as being transcribed from *Euglena* chloroplast DNA. In this case, the presence of plastids stimulated the synthesis of nuclear poly(A)$^+$RNA, which subsequently became associated with plastids. BARTOLF and PRICE (1979) reported short (10–45 residues long) poly(A) stretches in RNA from spinach chloroplasts.

There is no proof that the RNAs which hybridize to various DNA fragments are mRNAs. This would require the demonstration that they can be translated into protein. The presence of several mRNAs in plastid RNA preparations has been shown via their translational products, either in isolated plastids (see BOTTOMLEY and BOHNERT Chap. 15, this Vol.) or in vitro in wheat germ, reticulocyte lysate or *E. coli* translation systems (SAGHER et al. 1976, HARTLEY et al. 1975, BEDBROOK et al. 1978, SANO et al. 1979, MALNOE et al. 1979, DRIESEL et al. 1980, WESTHOFF et al. 1981, see BOGORAD et al. 1978 for an outline of in vitro techniques, and EDELMAN et al. 1982 for a collection of methods). Selectively, specific mRNA may be isolated using the hybrid-selection method by which a mRNA is first annealed to a DNA fragment from which it was transcribed. The double-stranded hybrid can be removed from non-complementary mRNAs and the purified mRNA may be translated in vitro (DRIESEL et al. 1980, WESTHOFF et al. 1981). Alternatively, and elegantly, a coupled or linked DNA transcription/translation system has been used to both map and identify a product as being derived from a specific plastid DNA fragment (COEN et al. 1977, BOTTOMLEY and WHITFELD 1979). Among the mRNAs which have been used for localization of genes are the mRNAs of the rbcL, psbA, atpA, atpB, and atpE genes (see references above, and LINK et al. 1978, WHITFELD and BOTTOMLEY 1980, LINK 1981a). It is to be expected that mRNAs for other polypeptides will be isolated and characterized in the near future. Likely candidates are mRNAs for the proteins which are identified as products of protein synthesis in isolated organelles (see BOTTOMLEY and BOHNERT Chap. 15, this Vol.).

The mRNAs for the large subunit of RuBCase and for the approximately 32,000 molecular weight thylakoid membrane protein are the most abundant mRNA species in chloroplasts (BEDBROOK et al. 1978, HARTLEY et al. 1975, ROSNER et al. 1975, LINK et al. 1978, DRIESEL et al. 1980, LINK 1981b). Neither RNA is polyadenylated (ROSNER et al. 1975, WHEELER and HARTLEY 1975, SAGHER et al. 1976, HOWELL et al. 1977). The LSU mRNA sediments in non-denaturing sucrose gradients at a position of 18S–20S, corresponding to a molecular weight of approximately 0.7×10^6 or 2,100 nucleotides. The mRNA for the 32,000 protein sediments at 14S in sucrose gradients, corresponding to 0.5×10^6 molecular weight or 1,500 nucleotides (see Table 7) (SAGHER et al. 1976, SPEIRS and GRIERSON 1978, EDELMAN and REISFELD 1980, DRIESEL et al. 1980). Sequences coding for proteins in these genes on spinach chloroplast DNA are 1,425 nucleotides (rbcL) and 1,068 nucleotides (psbA) long (ZURAWSKI et al. 1981, 1982). The length of the transcribed region is 1,690 (rbcL) and 1,230 nucleotides (psbA), respectively.

5.4 Control of Transcription

Control of chloroplast gene expression is obvious from many observations which reveal biological changes in response to environmental or internal stimuli. Such stimuli include changes in hydration, temperature, light, and hormones, which can be both causing agents and responses to developmental programs (APEL and BOGORAD 1976, LINK et al. 1978, BEDBROOK et al. 1978, LINK 1981a, b, EDELMAN and REISFELD 1980). Despite extensive experimentation (for a review on chloroplast development see SCHIFF 1981), the regulatory connections between stimulus and transcriptional response have hardly been characterized in molecular terms. Articles which review some aspects, such as hormonal effects, and the use of cell and tissue cultures are available (BISWAS and ROY 1978, GALUN 1981, TRAN THANH VAN 1981).

Euglena gracilis has been used in many studies on chloroplast development. When this organism is kept in the dark, chloroplasts are converted to etioplasts. Upon illumination, these differentiate into chloroplasts again. This change is prevented by blocking mitochondrial functions. Since the photosynthetic apparatus is built up in the presence of DCMU, it was concluded that the light effect on development is not exerted on any of the components of this complex (SCHIFF 1981). In dark-grown *Euglena* cells, 53% of the plastid DNA of one strand was transcribed into RNA. This value, after illumination, dropped to 45% (RAWSON and BOERMA 1976a). In contrast, 32% of the chloroplast DNA was transcribed in rapidly growing cultures in the light (RAWSON 1975). Four different types of RNA populations were distinguished, according to their behaviour in a dark-light shift, in *Euglena* chloroplasts by CHELM et al. (1977a, b, 1978, 1979). A continuously synthesized RNA population, which consisted mainly of the rRNAs, could be discerned from either induced or repressed RNA populations during early chloroplast development and from RNAs induced at a later stage of development. Some of the differently regulated RNAs were located on specific chloroplast DNA fragments.

The development from proplastids to chloroplasts during germination of seeds is accompanied by an increase in DNA and RNA synthesis (SEN et al. 1975, MORY et al. 1972, MERRICK and DURE 1972). These studies have been extended to different intervals of a plant's lifetime, e.g., to developing leaflets, their expansion into a mature leaf and their senescence (PEARSON et al. 1978, NESS and WOOLHOUSE 1980a, b, PANIGRAHI and BISWAL 1979, SPEIRS and BRADY 1981, GRIERSON and COVEY 1975).

In *Spirodela,* old fronds retain their chlorophyll during prolonged darkness, while newly made fronds are yellowish. After transfer to the light, a rapid onset of RNA synthesis in chloroplasts was observed in these young fronds followed by a slower increase which consists mainly of newly made rRNA. The rapid synthesis is mainly due to the synthesis of a mRNA (GRESSEL 1978) which had a molecular weight of 0.5×10^6 and which could be translated in heterologous in vitro systems, yielding a peptide of molecular weight of 33,500 (ROSNER et al. 1975, EDELMAN and REISFELD 1980). The peptide is rapidly converted in vivo into an approximately 32,000 molecular weight protein that is found in the thylakoid membrane. This so-called photogene has since been found also in maize (BEDBROOK et al. 1978), spinach (DRIESEL et al. 1980), and mustard (LINK 1981a, b).

The protein from the corresponding psbA gene in spinach chloroplast DNA can be obtained by translation in a heterologous assay system (see Sect. 5.3).

Chloroplast mRNA from young plants contains mainly this mRNA, which is translatable in a reticulocyte lysate system (DRIESEL et al. 1980). When RNAs are prepared from etiolated spinach plants at different time intervals after the start of illumination and translated, it appears that the protein is not made with RNAs prepared during the first 6–8 h of illumination. This is in contrast to other proteins for which mRNAs appear to be present even in dark-grown plants, like the rbcL gene product, which was synthesized continuously (see HERRMANN et al. 1980a). This would seem to indicate a differential activation of individual genes although it is not yet excluded that some mRNA from the psbA gene is present in low amount in the dark and that it is not recognized in the heterologous system as well as others.

The light effect has also been studied in order to investigate coordination between the cytosolic and the chloroplast compartments (TOBIN 1981, SMITH and ELLIS 1981, SILVERTHORNE and ELLIS 1980, APEL and KLOPPSTECH 1980, BÖGE et al. 1981, CUMING and BENNETT 1981). SMITH and ELLIS (1981) who measured light and dark levels of the transcripts of the chloroplast DNA-coded rbcL gene and the nucleus-coded rbcS gene, reported that both transcripts were present in illuminated plants, but were not present or present only in minute amounts in poly(A)$^+$RNA (rbcS transcript) or chloroplast RNA (rbcL transcript) from dark-grown pea plants. In contrast, RNA from the flagellate *Chlorogonium elongatum,* grown in darkness with acetate as the carbon source, contain mRNAs for both the RuBPCase subunits, although in smaller amounts than autotrophically grown cells (WESTHOFF and ZETSCHE 1981).

It has been reported that phytochrome enhances the synthesis of chloroplast rRNA in mustard cotyledons in the absence of photosynthesis (THIEN and SCHOPFER 1975). By illumination with blue light, an enhancement of chloroplast rRNA synthesis has also been observed (RICHTER and DIRKS 1978). An enhanced expression of a distinct plastid DNA region in mustard dark-grown seedlings, initiated by irradiation with continuous far-red light, was recently reported by LINK (1981a, b). This phytochrome-controlled transcript is the mRNA for the gene psbA in mustard (LINK 1982).

A compartmentalization of chloroplast specific biochemical pathways is known for the C_4-plants, e.g., maize. Bundle-sheath chloroplasts are able to fix carbon dioxide by the photosynthetic carbon cycle, while mesophyll cell chloroplasts provide the four-carbon organic acids from which one carbon can be split. The biochemical difference between the two chloroplast types is also found on the transcriptional level. Bundle-sheath chloroplasts contain a functional RuBPCase and also the mRNA for the large subunit of this enzyme. Mesophyll cell chloroplasts have neither the enzyme nor this mRNA, within the limits of detection. It has been convincingly shown that the DNA from both types of chloroplasts contains the rbcL gene (LINK et al. 1978).

The control of transcription at the transition from, for example, chloroplasts to chromoplasts appears to be discernable from chloroplast senescence. While in ageing leaves chloroplast function in general decays (CALLOW 1974, NESS and WOOLHOUSE 1980a, b), young chromoplasts have been shown to follow a transcriptional program different from that found in chloroplasts. In *Tropaeolum majus* total chloroplast or chromoplast RNAs hybridize, in addition to

DNA fragments containing the rRNA genes, to different DNA fragments and with different relative intensities (THOMPSON JA and SITTE P personal communication).

Differential transcription of chloroplast DNA also occurs in *Chlamydomonas and Euglena* at specific stages of the cell cycle (LEDOIGT and CALVAYRAC 1974, HOWELL and WALKER 1977, MATSUDA and SURZYCKI 1980, ROCHAIX 1981, see also BUETOW et al. 1980). This will become an increasingly exploitable system for the study of transcriptional control mechanisms as probes for a variety of plastid genes are available.

6 Replication of Plastid DNAs

In the polyenergidically organized, polyploid chloroplasts the amount of DNA per organelle varies depending on developmental conditions. In pea, for example, a decrease of plastid analytical complexity from 50 in etioplasts to 14 in chloroplasts was observed (BENNETT and RADCLIFFE 1975). Both in angiosperms and *Chlamydomonas* the overall chloroplast DNA content is correlated with that of the nucleus, because the amount of plastid DNA depends on the nuclear ploidy (WHITEWAY and LEE 1977, BUTTERFASS 1979). The gene(s) coding for the chloroplast DNA polymerase(s) appear to be located in the nucleus (see HAGEMANN and BOERNER 1981, for a review). Inhibitors of chloroplast protein synthesis block nuclear DNA replication in *Chlamydomonas* (BLAMIRE et al. 1974), indicating in addition a possible role for a chloroplast protein in nuclear DNA replication. The validity of results obtained by using various inhibitors of either the eukaryotic or prokaryotic systems of biosynthesis has been reviewed by GALLING (Chap. 18, this Vol.).

In *Euglena,* an increase of plastid DNA per cell in response to light-induced development has been shown by several independent experimental techniques (RAWSON and BOERMA 1976b, CHELM et al. 1977a, b). In *Chlamydomonas,* chloroplast DNA synthesis proceeded throughout the light period, whereas nuclear DNA was synthesized after the onset of the dark period (TURMEL et al. 1980). GIBSON and HERSHBERGER (1975) found differing ratios of chloroplast DNA to nuclear DNA in *Euglena* depending on the growth conditions. In this organism, the ratio between chloroplast DNA and nuclear DNA is kept approximately constant with each cell cycle, plastid DNA being doubled following replication in the nucleus. Some overshooting of initial chloroplast DNA synthesis appears to be controlled by a degradative enzyme system which restores the initial ratio (SRINIVAS and LYMAN 1980). The total amount of plastid DNA per organelle is present in multiples of the chromosome size (see Table 1). No preferentially amplified fractions of plastid DNA have been demonstrated up to now. In higher plants, chloroplast DNA synthesis is found in tissues where cell divisions occur (LAMPPA and BENDICH 1979b, BOFFEY et al. 1979, ROSE et al. 1975, SCOTT and POSSINGHAM 1980, BENNETT and RADCLIFFE 1975).

Replication of chloroplast DNA is semi-conservative (CHIANG and SUEOKA 1967, MANNING and RICHARDS 1972a, RICHARDS and MANNING 1975). In pea and maize chloroplast DNAs, replication is initiated at two D-loops located on opposite strands (Kolodner and Tewari 1975a). The two sites are approximately 7 Kbp apart in both organisms. Replication proceeds bidirectionally until the two near forks fuse. Completion of the resulting "Cairns-type" replicative intermediate may be followed by a rolling-circle type of replication (Kolodner and Tewari 1975b) which leads to several copies of the genome from one initiation event. "Cairns-type" replication has also been shown for *Euglena* chloroplast DNA (Richards and Manning 1975) and appears to be initiated from one origin of replication. More data are needed to precisely locate the origin(s) of replication on the physical maps of plastid DNAs. Knowledge about these origins should be useful in constructing chloroplast-specific cloning vehicles or in the cloning of chloroplast genes in other systems. In addition, the location of the origin may serve as a zero point in describing plastid DNA molecules by coordinates.

7 Conclusions

Genes coding for the constituents of chloroplasts are dispersed among the nuclear, chloroplast and, possibly, mitochondrial (Bennoun 1981) genomes. Cooperation between these genetic systems is essential for the biogenesis and function of the plastid. Many, if not all, multimeric chloroplast structures or enzymes originate from products of both the nuclear and plastid genomes (see Ellis 1981 for a review). The control mechanisms involved in the coordinated expression of the three distinct plant genomic types are poorly understood. Their elucidation will require more information on the structure and organization of chloroplast and nuclear genes and on their expression. In this article, we focused on the chloroplast side of this problem, reviewing chloroplast genomic organization and gene expression. Work on other plastid types (Thompson et al. 1981) will widen the experimental approach. The size of the plastid chromosome varies over a wide range within plant families and even more so between phylae. This poses the question whether all plastids carry the same set of genes. It will be for the future to establish whether genes in *Chlamydomonas* (195 Kbp) are also encoded in *Codium* (57 Kbp). Up to now, it is generally accepted that plastids are similar with respect to number and types of genes. Although this may be true, the nuclear genomes of plants vary. It has to be expected that nuclear genomes may have developed different mechanisms to control plastid replication, development and basic functions (Stubbe 1971). This may be seen in the different ways by which plants of different phylogenetic categories respond to light and etiolation, or by the differential gene expression within different tissues of the same plant.

Present evidence suggests a complete transcription of plastid DNA. With pea as a model (120 Kbp), after subtracting genes for the rRNAs (8 Kbp) and for approximately 40 tRNA genes (10 Kbp), at least 100 Kbp may code for proteins. Assuming a size of 1.5 Kbp as an average which includes transcribed untranslated regions, a set of approximately 70 protein genes could be encoded.

The study of plastid genomic structure and transcription will contribute to this understanding. The following questions may be asked:

- How many genes are located on plastid DNAs and what is their function in organelle metabolism and in the regulation of expression of the plastid DNA?
- Do plastids from every species contain the same set of genes?
- Are there genes whose products are exported in order to function in the regulation of nuclear and/or mitochondrial gene expression?
- How closely are plastid genes spaced and to what extent do they overlap?
- How common are introns and what is their function?
- Are there structures equivalent to insertion sequences (IS elements) in plastid DNA, as might be suggested from recent work on *Euglena* plastid DNA (KOLLER and DELIUS 1982a)?
- Is there polycistronic transcription of genes which map closely on the genome, e.g., the atpB and atpE genes or several tRNA genes?
- Are all chloroplast promoters alike and can a "consensus sequence" be established, or are there families of promoters in order to switch on families of genes?
- Are there factors resembling bacterial repressors or inducers, or sigma-type RNA polymerase factors in plastids? If there are, where are they coded?
- If plastids evolved from endosymbionts, which lost part of their genes to the nucleus, are there cyanobacterial-type genes in the nucleus?

Acknowledgements. We wish to thank the colleagues who made available unpublished results and we are grateful for the comments made by Drs. W BOTTOMLEY, H DELIUS, R HELLING and B KOLLER. We are indepted to the technical staff and the computer group of the EMBL for their excellent support. Original results reported here (HJB) were funded by the Deutsche Forschungsgemeinschaft. HJB is recipient of a Heisenberg Stipendium.

References

Aitken A, Stanier RY (1979) Characterization of peptidoglycan from the cyanelles of *Cyanophora paradoxa*. J Gen Microbiol 112:219–223

Aldrich J, Cattolico RA (1981) Isolation and characterization of chloroplast DNA from the marine chromophyte, *Olisthodiscus luteus:* electron microscopic visualization of isomeric molecular forms. Plant Physiol 68:641–647

Allet B, Rochaix JD (1979) Structure analysis at the ends of the intervening DNA sequences in the chloroplast 23S ribosomal genes of *C. reinhardii*. Cell 18:55–60

Apel K, Bogorad L (1976) Light-induced increase in the activity of maize plastid DNA-dependent RNA polymerase. Eur J Biochem 67:615–620

Apel K, Kloppstech K (1980) The effect of light on the biosynthesis of the light-harvesting chlorophyll a/b protein. Evidence for the requirement of chlorophyll a for the stabilization of the apoprotein. Planta 150:426–430

Atchison BA, Whitfeld PR, Bottomley W (1976) Comparison of chloroplast DNAs by specific fragmentation with Eco RI endonuclease. Mol Gen Genet 148:263–269

Bachmann BJ, Low KB (1980) Linkage map of *Escherichia coli* K-12, Microbiol Rev 44:1–56

Barnett WE, Pennington CJ, Fairfield SA (1969) Induction of *Euglena* transfer RNAs by light. Proc Natl Acad Sci USA 63:1261–1268

Barnett WE, Schwartzbach SD, Hecker LI (1978) The transfer RNAs of eukaryotic organelles. Prog Nucleic Acid Res Mol Biol 21:143–179

Bartolf M, Price CA (1979) Synthesis of poly(A)-containing RNA by isolated spinach chloroplasts. Biochemistry 18:1677–1680

Bastia D, Chiang KS, Swift H, Siersma P (1971)Heterogeneity, complexity, and repetition of the chloroplast DNA of *Chlamydomonas reinhardtii*. Proc Natl Acad Sci USA 68:1157–1161

Baur E (1909) Das Wesen und die Erblichkeitsverhältnisse der „*varietates albomarginata*" von *Pelargonium zonale*. Z Vererbungsl 1:330–351

Bayen M, Rode A (1973) Heterogeneity and complexity of *Chlorella* chloroplastic DNA. Eur J Biochem 39:413–420

Bedbrook JR, Bogorad L (1976) Endonuclease recognition sites mapped on *Zea mays* chloroplast DNA. Proc Natl Acad Sci USA 73:4309–4313

Bedbrook JP, Kolodner R (1979) The structure of chloroplast DNA. Annu Rev Plant Physiol 30:593–620

Bedbrook JR, Kolodner R, Bogorad L (1977) *Zea mays* ribosomal RNA genes are part of a 22,000 base pair inverted repeat. Cell 11:739–749

Bedbrook JR, Link G, Coen DM, Bogorad L, Rich A (1978) Maize plastid gene expressed during photoregulated development. Proc Natl Acad Sci USA 75:3060–3064

Bedbrook JR, Coen DM, Beaton AR, Bogorad L, Rich A (1979) Location of the single gene for the large subunit of ribulosebisphosphate carboxylase on the maize chloroplast chromosome. J Biol Chem 254:905–910

Behn W, Herrmann RG (1977) Circular DNA in the beta-satellite DNA of *Chlamydomonas reinhardtii*. Mol Gen Genet 157:25–30

Bennett J, Radcliffe C (1975) Plastid DNA replication and plastid division in the garden pea. FEBS Lett 56:222–225

Bennoun P (1981) Does the chloroplast control mitochondrial functions? FEBS Lett 136:1–2

Bisaro D, Siegel A (1980) Sequence homology between chloroplast DNAs from several higher plants. Plant Physiol 65:234–237

Biswas BB, Roy P (1978) Plant growth substances as modulators of transcription. Subcell Biochem 5:187–219

Blamire J, Fletchner VR, Sager R (1974) Regulation of nuclear DNA replication by the chloroplast in *Chlamydomonas*. Proc Natl Acad Sci USA 71:2867–2871

Böge F, Westhoff P, Zimmermann K, Zetsche K (1981) Regulation of the synthesis of ribulose-1,5-bisphosphate carboxylase and its subunits in the flagellate *Chlorogonium elongatum*. I. The effect of light and acetate on the synthesis and the degradation of the enzyme. Eur J Biochem 113:581–586

Boffey SM, Ellis JR, Sellden G, Leech RM (1979) Chloroplast division and DNA synthesis in light-grown wheat leaves. Plant Physiol 64:502–505

Bogorad L, Bedbrook JR, Coen DM, Kolodner R, Link G (1978) Genes for chloroplast proteins and rRNAs. In: Akoyunoglou G, Argyroudi-Akoyunoglou JH (eds) Chloroplast development. Elsevier, Amsterdam, pp 541–551

Bohnert HJ, Crouse EJ (1981) A simple, inexpensive, general method for the isolation of high molecular weight chloroplast DNA. Plant Mol Biol Newslett 2:70–72

Bohnert HJ, Schmitt JM, Herrmann RG (1974) Structural and functional aspects of the plastome. III. DNA and RNA synthesis by isolated chloroplasts. Port Acta Biol 14:71–90

Bohnert HJ, Driesel AJ, Herrmann RG (1976) Characterization of the RNA compounds synthesized by isolated chloroplasts. In: Bücher T, Neupert W, Sebald W, Werner S (eds) Genetics and biogenesis of chloroplasts and mitochondria. North-Holland, Amsterdam New York, pp 629–636

Bohnert HJ, Driesel AJ, Herrmann RG (1977) Transcription and processing of transcripts in isolated unbroken chloroplasts. In: Bogorad L, Weil JH (eds) Acides nucléiques et synthèse des protéines chez les végétaux. Centre National de la Recherche Scientifique, Paris, pp 213–218

Bohnert HJ, Driesel AJ, Crouse EJ, Gordon K, Herrmann RG, Steinmetz A, Mubumbila M, Keller M, Burkard G, Weil JH (1979) Presence of a transfer RNA gene in the spacer sequence between the 16S and 23S rRNA genes of spinach chloroplast DNA. FEBS Lett 103:52–56

Bohnert HJ, Gordon KHJ, Crouse EJ (1980) Sequence homologies among chloroplast, mitochondrial and bacterial ribosomal RNA and messenger RNA. Mol Gen Genet 179:539–545

Bohnert HJ, Crouse EJ, Pouyet J, Mucke H, Löffelhardt W (1982) The subcellular localization of DNA components from *Cyanophora paradoxa*, a flagellate containing endosymbiotic cyanelles. Eur J Biochem 126:381–388

Bohnert HJ, Crouse EJ, Schmitt JM (1983) Chloroplast RNA synthesis. In: Ellis RJ (ed) Chloroplast biogenesis. Cambridge Univ Press, Cambridge, in press

Bolen PL, Grant DM, Swinton D, Boynton JE, Gillham NW (1982) Extensive methylation of chloroplast DNA by a nuclear gene mutation does not affect chloroplast gene transmission in *Chlamydomonas*. Cell 28:335–343

Borbely G, Simoncsits A (1981) 3′-terminal conserved loops of 16S rRNAs from the cyanobacterium *Synechococcus* AN PCC 6301 and maize chloroplast differ only in two bases. Biochem Biophys Res Commun 101:846–852

Boros I, Kiss A, Venetianer P (1979) Physical map of the seven ribosomal RNA genes of *Escherichia coli*. Nucleic Acids Res 6:1817–1830

Bottomley W, Whitfeld P (1979) Cell-free transcription and translation of total spinach chloroplast DNA. Eur J Biochem 93:31–39

Bovenberg WA, Kool AJ, Nijkamp HJJ (1981) Isolation, characterization and restriction endonuclease mapping of the *Petunia hybrida* chloroplast DNA. Nucleic Acids Res 9:503–517

Bowman CM, Dyer TA (1979) 4.5S ribonucleic acid, a novel ribonucleic acid, a novel ribosome component in the chloroplast of flowering plants. Biochem J 183:605–613

Bowman CM, Dyer TA (1982) Purification and analysis of DNA from wheat chloroplasts isolated in nonaqueous media. Anal Biochem 122:108–118

Bowman CM, Koller B, Delius H, Dyer TA (1981) A physical map of wheat chloroplast DNA showing the location of the structural genes for the ribosomal RNAs and the large subunit of ribulose 1,5-bisphosphate carboxylase. Mol Gen Genet 183:93–101

Bradbeer JW, Atkinson YE, Börner T, Hagemann R (1979) Cytoplasmic synthesis of plastid polypeptides may be controlled by plastid-synthesized RNA. Nature (London) 279:816–817

Brantner JH, Dure LS (1975) The developmental biochemistry of cotton seed embryogenesis and germination. VI. Levels of cytosol and chloroplast aminoacyl-tRNA synthetases during cotyledon development. Biochim Biophys Acta 414:99–114

Brawerman G, Eisenstadt JM (1964) Deoxyribonucleic acids from the chloroplasts of *Euglena gracilis*. Biochim Biophys Acta 91:477–485

Brosius J, Dull TJ, Sleeter DD, Noller H F (1981) Gene organization and primary structure of a ribosomal RNA operon from *Escherichia coli*. J Mol Biol 148:107–127

Buetow DE, Wurtz EA, Gallagher T (1980) Chloroplast biogenesis during the cell cycle. In: Whitson G (ed) Nuclear-cytoplasmic interactions in the cell cycle. Academic Press, London New York, pp 9–55

Bujard H (1980) The interactions of *E. coli* RNA polymerase with promoters. Trends Biochem Sci 5:274–278

Burkard G, Vaultier JP, Weil JH (1972) Differences in the level of plastid-specific tRNAs in chloroplasts and etioplasts of *Phaseolus vulgaris*. Phytochemistry 11:1351–1353

Burton WG, Grabowy CT, Sager R (1979) Role of methylation in the modification and restriction of chloroplast DNA in *Chlamydomonas*. Proc Natl Acad Sci USA 76:1390–1394

Butterfass T (1979) Patterns of chloroplast reproduction. Springer, Berlin Heidelberg New York

Callow JA (1974) Ribosomal RNA, fraction I protein synthesis and ribulose diphosphate

carboxylase activity in developing and senescing leaves of cucumber. New Phytol 73:13–20
Carbon P, Ehresmann C, Ehresmann B, Ebel JP (1978) The sequence of *Escherichia coli* ribosomal 16S RNA determined by new rapid gel methods. FEBS Lett 94:152–156
Cattolico RA (1978) Variation in plastid number. Effect on chloroplast and nuclear deoxyribonucleic acid complement in the unicellular alga *Olisthodiscus luteus*. Plant Physiol 62:558–562
Chambon P (1975) Eukaryotic nuclear RNA polymerases. Annu Rev Biochem 44:613–638
Chelm BK, Hallick RB (1976) Changes in the expression of the chloroplast genome of *Euglena gracilis* during chloroplast development. Biochemistry 15:593–599
Chelm BK, Hoben PJ, Hallick RB (1977a) Expression of the chloroplast ribosomal RNA genes of *Euglena gracilis* during chloroplast development. Biochemistry 16:776–781
Chelm BK, Hoben PJ, Hallick RB (1977b) Cellular content of chloroplast DNA and chloroplast ribosomal RNA genes in *Euglena gracilis* during chloroplast development. Biochemistry 16:782–786
Chelm BK, Gray PW, Hallick RB (1978) Mapping of transcribed regions of *Euglena gracilis* chloroplast DNA. Biochemistry 17:4239–4244
Chelm BK, Hallick RB, Gray PW (1979) Transcription program of the chloroplast genome of *Euglena gracilis* during chloroplast development. Proc Natl Acad Sci USA 76:2258–2262
Chiang KS, Sueoka N (1967) Replication of chloroplast DNA in *Chlamydomonas reinhardtii*: Its mode and regulation. Proc Natl Acad Sci USA 57:1506–1513
Chu NM, Oishi K, Tewari KK (1981) Physical mapping of the pea chloroplast DNA and localization of the ribosomal RNA genes. Plasmid 6:279–292
Coen DM, Bedbrook JR, Bogorad L, Rich A (1977) Maize chloroplast DNA fragment encoding the large subunit of ribulosebisphosphate carboxylase. Proc Natl Acad Sci USA 74:5487–5491
Coleman AW (1979) Use of the fluorochrome 4′6 diamino-2-phenylindole in genetic and developmental studies on chloroplast DNA. J Cell Biol 82:299–305
Correns C (1909) Vererbungsversuche mit blass(gelb)grünen und buntblättrigen Sippen bei *Mirabilis jalapa, Urtica pilulifera* und *Lunaria annua*. Z Vererbungsl 1:291–329
Crick FHC (1966) Codon-anticodon pairing: the wobble hypothesis. J Mol Biol 19:548–555
Crouse EJ, Vandrey JP, Stutz E (1974) Comparative analyses of chloroplast and mitochondrial DNAs from *E. gracilis* In: Avron M (ed) Proc 3rd Int Congr Photosynth. Elsevier, Amsterdam, pp 1775–1786
Crouse EJ, Schmitt JM, Bohnert HJ, Gordon K, Driesel AJ, Herrmann RG (1978) Intramolecular compositional heterogeneity of *Spinacia* and *Euglena* chloroplast DNAs. In: Akoyunoglou G, Argyroudi-Akoyunoglou JH (eds) Chloroplast development. Elsevier, Amsterdam, pp 565–572
Cuming AC, Bennett J (1981) Biosynthesis of the light-harvesting chlorophyll a/b protein. Control of messenger RNA activity by light. Eur J Biochem 118:71–80
Dalmon J, Loiseaux S (1981) The deoxyribonucleic acids of two brown algae: *Pylalella littoralis* (L.) Kjellm. and *Sphacellaria* sp. Plant Sci Lett 21:241–251
Delihas N, Andersen J, Sprouse HM, Dudock B (1981) The nucleotide sequence of the chloroplast 5S ribosomal RNA from spinach. Nucleic Acids Res 9:2801–2805
Delius H, Koller B (1980) Sequence homologies between *Escherichia coli* and chloroplast ribosomal DNA as seen by heteroduplex analysis. J Mol Biol 142:247–261
Driesel AJ, Crouse EJ, Gordon K, Bohnert HJ, Herrmann RG, Steinmetz A, Mubumbila M, Keller M, Burkard G, Weil JH (1979) Fractionation and identification of spinach chloroplast transfer RNAs and mapping of their genes on the restriction map of chloroplast DNA. Gene 6:285–306
Driesel AJ, Speirs J, Bohnert HJ (1980) Spinach chloroplast mRNA for a 32,000-dalton polypeptide: Size and localization on the physical map of the chloroplast DNA. Biochim Biophys Acta 610:297–310

Dyer T, Bedbrook JR (1980) The organization in higher plants of the genes coding for chloroplast ribosomal RNA. In: Leaver CJ (ed) Genome organization and expression in plants. Plenum Press, New York London, pp 305–311

Dyer TA, Bowman CM (1979) Nucleotide sequences of chloroplast 5S ribosomal ribonucleic acid in flowering plants. Biochem J 183:595–604

Dyer TA, Leaver CJ (1981) RNA: Structure and metabolism. In: Marcus A (ed) The biochemistry of plants, vol VI. Academic Press, London New York, pp 111–168

Eaglesham ARJ, Ellis RJ (1974) Protein synthesis in chloroplasts. II. Light-driven synthesis of membrane proteins by isolated pea chloroplasts. Biochim Biophys Acta 335:396–407

Edelman M (1981) Nucleic acids of chloroplasts and mitochondria. In: Marcus A (ed) The biochemistry of plants, vol VI. Academic Press, London New York, pp 249–301

Edelman M, Reisfeld A (1980) Synthesis, processing and functional probing of P-32,000, the major membrane protein translated within the chloroplast. In: Leaver CJ (ed) Genome organization and expression in plants. Plenum Press, New York London, pp 353–362

Edelman M, Cowan CA, Epstein HT, Schiff JA (1964) Studies of chloroplast development in *Euglena.* VIII. Chloroplast-associated DNA. Proc Natl Acad Sci USA 52:1214–1219

Edelman M, Swinton D, Schiff JA, Epstein HT, Zeldin B (1967) Deoxyribonucleic acid of the blue-green algae (*Cyanophyta*). Bacteriol Rev 31:315–331

Edelman M, Hallick R, Chua NH (eds) (1982) Methods in chloroplast molecular biology. Elsevier, Amsterdam, in press

Edwards K, Kössel H (1981) The rRNA operon from Zea mays chloroplasts: nucleotide sequence of 23S rDNA and its homology with *E. coli* 23S rDNA. Nucleic Acids Res 9:2853–2869

Edwards K, Bedbrook J, Dyer T, Kössel H (1981) 4.5S rRNA from *Zea mays* chloroplasts shows structural homology with the 3′-end of prokaryotic 23S rRNA. Biochem Int 2:533–538

Ee JH van, Man in't Veld W, Planta RJ (1980) Isolation and characterization of chloroplast DNA from the duckweed, *Spirodela oligorhiza.* Plant Physiol 66:572–575

Ee JH van, Vos YJ, Planta RJ (1981) Physical map of chloroplast DNA of *Spirodela oligorhiza:* Analysis by the restriction endonucleases Pst I, Xho I and Sac I. Gene 12:191–200

Ee JH van, Vos YJ, Bohnert HJ, Planta RJ (1982) Mapping of genes on the chloroplast DNA of *Spirodela oligorhiza.* Plant Mol Biol 1:117–131

El-Gewely MR, Lomax MI, Lau ET, Helling RB, Farmerie W, Barnett WE (1981) A map of specific cleavage sites and tRNA genes in the chloroplast genome of *Euglena gracilis* bacillaris. Mol Gen Genet 181:296–305

El-Gewely MH, Helling RB, Farmerie W, Barnett WE (1982) Location of a phenylanaline tRNA gene on the physical map of the *Euglena gracilis* chloroplast genome. Gene 17:337–339

Ellis RJ (1981) Chloroplasts proteins: Synthesis, transport and assembly. Annu Rev Plant Physiol 32:111–137

Eneas-Filho J, Hartley MR, Mache R (1981) Pea chloroplast ribosomal proteins: characterization and site of synthesis. Mol Gen Genet 184:484–488

Erion JL, Tarnowski J, Weissbach H, Brot N (1981) Cloning, mapping, and in vitro transcription-translation of the gene for the large subunit of ribulose-1,5-bisphosphate carboxylase from spinach chloroplasts. Proc Natl Acad Sci USA 78:3459–3463

Ersland DR, Aldrich J, Cattolico RA (1981) Kinetic complexity, homogeneity, and copy number of chloroplast DNA from the marine alga *Olisthodiscus luteus.* Plant Physiol 68:1468–1473

Falk H, Liedvogel B, Sitte P (1974) Circular DNA in isolated chromoplasts. Z Naturforsch 29c:541–544

Fluhr R, Edelman M (1981a) Physical mapping of *Nicotiana tabacum* chloroplast DNA. Mol Gen Genet 181:484–490

Fluhr R, Edelman M (1981b) Conservation of sequence arrangement among higher plant chloroplast DNAs: molecular cross hybridization among the *Solanaceae* and between *Nicotiana* and *Spinacia*. Nucleic Acids Res 9:6841–6853

Galun E (1981) Plant protoplasts as physiological tools. Annu Rev Plant Physiol 32:237–266

Gelvin S, Howell SH (1979) Small repeated sequences in the chloroplast genome of *Chlamydomonas reinhardi*. Mol Gen Genet 173:315–322

Gibson WH, Hershberger CL (1975) Changing proportions of DNA components in *Euglena gracilis*. Arch Biochem Biophys 168:8–14

Gillham NW (1978) Organelle heredity. Raven Press, New York

Glotz C, Zwieb C, Brimacombe R (1981) Secondary structure of the large subunit ribosomal RNA from *Escherichia coli, Zea mays* chloroplast, and human and mouse mitochondrial ribosomes. Nucleic Acids Res 9:3287–3306

Gordon KHJ, Hildebrandt JW, Bohnert HJ, Herrmann RG, Schmitt JM (1980) Analysis of the plastid DNA in an *Oenothera* plastome mutant deficient in ribulose bisphosphate carboxylase. Theor Appl Genet 57:203–207

Gordon KHJ, Crouse EJ, Bohnert HJ, Herrmann RG (1981) Restriction endonuclease cleavage site map of chloroplast DNA from *Oenothera parviflora* (*Euoenothera* plastome IV). Theor Appl Genet 59:281–296

Gordon KHJ, Crouse EJ, Bohnert HJ, Herrmann RG (1982) Physical mapping of differences in chloroplast DNA of the five wild-type plastomes in *Oenothera* subsection *Euoenothera*. Theor Appl Genet 61:373–384

Graf L, Kössel H, Stutz E (1980) Sequencing of 16S–23S spacer in a ribosomal RNA operon of *Euglena gracilis* chloroplast DNA reveals two tRNA genes. Nature (London) 286:908–910

Grantham R, Gautier C, Gouy M, Jacobzone M, Mercier R (1981) Codon catalog usage is a genome strategy modulated for gene expressivity. Nucleic Acids Res 9:r43–r74

Gray MW, Doolittle WF (1982) Has the endosymbiont hypothesis been proven? Microbiol Rev 46:1–42

Gray PW, Hallick RB (1977) Restriction endonuclease map of *Euglena gracilis* chloroplast DNA. Biochemistry 16:1665–1671

Gray PW, Hallick RB (1978) Physical mapping of the *Euglena gracilis* chloroplast DNA and ribosomal RNA gene region. Biochemistry 17:284–289

Gray PW, Hallick RB (1979a) Isolation of *Euglena gracilis* chloroplast 5S ribosomal RNA and mapping the 5S rRNA gene on chloroplast DNA. Biochemistry 18:1820–1825

Gray PW, Hallick RB (1979b) Base composition heterogeneity of *Euglena gracilis* chloroplast DNA. Biochim Biophys Acta 561:53–58

Green BR (1976) Covalently closed minicircular DNA associated with *Acetabularia* chloroplasts. Biochim Biophys Acta 447:156–166

Green BR, Muir BL, Padmanabhan U (1977) The *Acetabularia* chloroplast genome: small circles and large kinetic complexity. In: Woodcock CFL (ed) Progress in *Acetabularia* research. Academic Press, London New York, pp 107–122

Gressel J (1978) Light requirements for the enhanced synthesis of a plastid mRNA during *Spirodela* greening. Photochem Photobiol 27:167–169

Grierson D, Covey SN (1975) Changes in the amount of ribosomal RNA and poly(A) containing RNA during leaf development. Planta 127:77–86

Grierson D, Loening U (1974) Ribosomal RNA precursors and the synthesis of chloroplast and cytoplasmic ribosomal ribonucleic acid in leaves of *Phaseolus vulgaris*. Eur J Biochem 44:501–507

Groot GSP (1983) Higher plant chloroplast DNA. In: Ellis RJ (ed) Chloroplast biogenesis. Cambridge Univ Press, Cambridge, in press

Groot GSP, Harten-Loosbroek van N (1981) Physical mapping of 4S RNA genes on chloroplast DNA of *Spirodela oligorhiza*. Curr Genet 4:187–190

Gruol DJ, Haselkorn R (1976) Counting the genes for stable RNA in the nucleus and chloroplasts of *Euglena*. Biochim Biophys Acta 477:82–95

Guilfoyle TJ (1981) DNA and RNA polymerase. In: Marcus A (ed) The biochemistry of plants, vol VI. Academic Press, London New York, pp 207–247

Haff LA, Bogorad L (1976a) Hybridization of maize chloroplast DNA with transfer ribonucleic acids. Biochemistry 15:4105–4109
Haff LA, Bogorad L (1976b) Poly (adenylic acid)-containing RNA from plastids of maize. Biochemistry 15:4110–4115
Hagemann R (1964) Plasmatische Vererbung. Fischer, Jena
Hagemann R, Börner T (1981) Plastid genetics. In: Ellenberg H, Esser K, Kubitzki K, Schnepf E, Ziegler H (eds), Prog Bot 43:159–173
Hagenbüchle O, Santer M, Steitz JA, Mans RJ (1978) Conservation of the primary structure at the 3′-end of 18S rRNA from eucaryotic cells. Cell 13:551–563
Hallick RB, Gray PW, Chelm BK, Rushlow KE, Orozco EM (1978) *Euglena gracilis* chloroplast DNA structure, gene mapping, and RNA transcription. In: Akoyunoglou G, Argyroudi-Akoyunoglou JH (eds) Chloroplast development. Elsevier, Amsterdam, pp 619–622
Hallier UW, Schmitt JM, Heber U, Chaianova SS, Volodarsky AD (1978) Ribulose-1,5-bisphosphate carboxylase-deficient plastome mutants of *Oenothera*. Biochim Biophys Acta 504:67–83
Hartley MR (1979) The synthesis and origin of chloroplast low molecular weight ribosomal ribonucleic acid in spinach. Eur J Biochem 96:311–320
Hartley MR, Ellis RJ (1973) Ribonucleic acid synthesis in chloroplasts. Biochem J 134:249–262
Hartley MR, Head C (1979) The synthesis of chloroplast high molecular weight ribosomal ribonucleic acid in spinach. Eur J Biochem 96:301–309
Hartley MR, Wheeler A, Ellis RJ (1975) Protein synthesis in chloroplasts. V. Translation of messenger RNA for the large subunit of fraction I protein in a heterologous cell-free system. J Mol Biol 91:67–77
Hartley MR, Head CW, Gardiner J (1977) The synthesis of chloroplast RNA. In: Bogorad L, Weil JH (eds) Nucleic acids and protein synthesis in plants. Plenum Press, New York London, pp 419–423
Hedberg MF, Huang Y-S, Hommersand MH (1981) Size of the chloroplast genome in *Codium fragile*. Science 213:445–447
Heizmann P (1974) Maturation of chloroplast rRNA in *Euglena gracilis*. Biochem Biophys Res Commun 56:112–118
Herdman M, Stanier RY (1977) The cyanelle: chloroplast or endosymbiotic prokaryote. FEMS Microbiol Lett 1:7–12
Herrmann RG (1972) Do chromoplasts contain DNA? II. The isolation and characterization of DNA from chromoplasts, chloroplasts, mitochondria, and nuclei of *Narcissus*. Protoplasma 74:7–17
Herrmann RG, Feierabend J (1980) The presence of DNA in ribosome-deficient plastids of heat-bleached rye leaves. Eur J Biochem 104:603–609
Herrmann RG, Kowallik KV (1970) Multiple amounts of DNA related to the size of chloroplasts. Protoplasma 69:365–372
Herrmann RG, Possingham JV (1980) Plastid DNA-the plastome. In: Reinert J (ed) Results and problems in cell differentiation, vol X. Springer, Berlin Heidelberg New York, pp 45–96
Herrmann RG, Bohnert HJ, Kowallik KV, Schmitt JM (1975) Size, conformation and purity of chloroplast DNA of some higher plants. Biochim Biophys Acta 378:305–317
Herrmann RG, Seyer P, Schedel R, Gordon K, Bisanz C, Winter P, Hildebrandt JW, Wlaschek M, Alt J, Driesel AJ, Sears BB (1980a) The plastid chromosomes of several dicotyledons. In: Bücher T, Sebald W, Weiss H (eds) Biological chemistry of organelle formation. Springer, Berlin Heidelberg New York, pp 97–112
Herrmann RG, Palta HK, Kowallik KV (1980b) Chloroplast DNA from three archegoniates. Planta 148:319–327
Herrmann RG, Whitfeld PR, Bottomley M (1980c) Construction of a SalI/PstI restriction map of spinach chloroplast DNA using low-gelling-temperature- agarose electrophoresis. Gene 8:179–191
Howell SH, Walker LL (1976) Informational complexity of the nuclear and chloroplast genomes of *Chlamydomonas reinhardi*. Biochim Biophys Acta 418:249–256

Howell SH, Walker LL (1977) Transcription of the nuclear and chloroplast genomes during the vegetative cell cycle in *Chlamydomonas reinhardi*. Dev Biol 56:11–23

Howell SH, Heizmann P, Gelvin S, Walker LL (1977) Identification and properties of the messenger RNA activity in *Chlamydomonas reinhardii* coding for the large subunit of ribulose-1,5-bisphosphate carboxylase. Plant Physiol 59:464–470

James JW, Jope C (1978) Visualization by fluorescence of chloroplast DNA in higher plants by means of the DNA-specific probe 4′6-diphenyl-2-phenylindole. J Cell Biol 79:623–630

Jaynes JW, Vernon LP, Klein SM, Strobel GA (1981) Genome size of cyanelle DNA from *Cyanophora paradoxa*. Plant Sci Lett 21:345–356

Jenni B, Stutz E (1978) Physical mapping of the ribosomal DNA region of *Euglena gracilis* chloroplast DNA. Eur J Biochem 88:127–134

Jenni B, Stutz E (1979) Analysis of *Euglena gracilis* chloroplast DNA. Mapping of a DNA sequence complementary to 16S rRNA outside of the three rRNA gene sets. FEBS Lett 102:95–99

Jenni B, Fasnacht M, Stutz E (1981) The multiple copies of the *Euglena gracilis* chloroplast genome are not uniform in size. FEBS Lett 125:175–179

Jolly SO, Bogorad L (1980) Preferential transcription of cloned maize chloroplast DNA sequences by maize chloroplast RNA polymerase. Proc Natl Acad Sci USA 77:822–826

Jolly SO, McIntosh L, Link G, Bogorad L (1981) Differential transcription in vivo and in vitro of two adjacent maize chloroplast genes: the large subunit of ribulose bisphosphate carboxylase and the 2.2-kilobase gene. Proc Natl Acad Sci USA 78:6821–6825

Jonge de P, Klootwijk J, Planta RJ (1977) Sequence of the 3′-terminal 21 nucleotides of yeast 17S ribosomal RNA. Nucleic Acids Res 4:3655–3663

Jurgenson JE, Bourque DP (1980) Mapping of rRNA genes in an inverted repeat in *Nicotiana tabacum* chloroplast DNA. Nucleic Acids Res 8:3505–3516

Kato A, Shimada H, Kusada M, Sugiura M (1981) The nucleotide sequences of two $tRNA^{Asn}$ genes from tobacco chloroplasts. Nucleic Acids Res 9:5601–5607

Keller M, Burkard G, Bohnert HJ, Mubumbila M, Gordon K, Steinmetz A, Heiser D, Crouse EJ, Weil JH (1980) Transfer RNA genes associated with the 16S and 23S rRNA genes of *Euglena* chloroplast DNA. Biochem Biophys Res Commun 95:47–54

Kirk JTO (1976) Chloroplast nucleic acids. In: Fasman GD (ed) Handbook of biochemistry and molecular biology, vol II. CRC Press, Cleveland, pp 356–362

Kirk JTO, Tilney-Bassett RAE (1978) The plastids: their chemistry, structure, growth and inheritance. Elsevier/North Holland Biomedical Press, Amsterdam New York

Koch W, Edwards K, Kössel H (1981) Sequencing of the 16S–23S spacer in a ribosomal RNA operon of *Zea mays* chloroplast DNA. Cell 25:203–213

Koller B, Delius H (1980) *Vicia faba* chloroplast DNA has only one set of ribosomal RNA genes as shown by partial denaturation mapping and r-loop analysis. Mol Gen Genet 178:261–269

Koller B, Delius H (1982a) Electron microscopic analysis of the extra 16S rRNA gene and its neighbourhood in chloroplast DNA from *Euglena gracilis* strain Z. FEBS Lett 139:86–92

Koller B, Delius H (1982b) Parts of the sequence between the complete rRNA operons are repeated on either side of the extra 16S rRNA gene in chloroplast DNA of *Euglena gracilis* strain Z. FEBS Lett 140:198–202

Koller B, Delius H, Dyer T (1982) The organization of the chloroplast DNA in wheat and maize in the region containing the LS gene. Eur J Biochem 122:17–23

Kolodner R, Tewari KK (1972) Molecular size and conformation of chloroplast DNA from pea leaves. J Biol Chem 247:6355–6364

Kolodner R, Tewari KK (1975a) Presence of displacement loops in the covalently closed circular chloroplast deoxyribonucleic acid from higher plants. J Biol Chem 250:8840–8847

Kolodner R, Tewari KK (1975b) Chloroplast DNA from higher plants replicates by both the Cairns and the rolling circle mechanism. Nature (London) 256:708–711

Kolodner RD, Tewari KK (1975c) Denaturation mapping studies on the circular chloroplast deoxyribonucleic acid from pea leaves. J Biol Chem 250:4888–4895

Kolodner R, Tewari KK (1975d) The molecular size and conformation of the chloroplast DNA from higher plants. Biochim Biophys Acta 402:372–390

Kolodner R, Tewari KK (1979) Inverted repeats in chloroplast DNA from higher plants. Proc Natl Acad Sci USA 76:41–45

Kolodner R, Warner RC, Tewari KK (1975) The presence of covalently linked ribonucleotides in the closed circular deoxyribonucleic acid from higher plants. J Biol Chem 250:7020–7026

Kolodner R, Tewari KK, Warner RC (1976) Physical studies on the size and structure of the covalently closed circular chloroplast DNA from higher plants. Biochim Biophys Acta 447:144–155

Kopecka H, Crouse EJ, Stutz E (1977) The *Euglena* chloroplast genome: Analysis by restriction enzymes. Eur J Biochem 72:525–535

Kung S-D (1977) Expression of chloroplast genomes in higher plants. Annu Rev Plant Physiol 28:401–437

Kung S-D, Zhu YS, Chen K, Shen GF, Sisson VA (1981) *Nicotiana* chloroplast genome. II. Chloroplast DNA alteration. Mol Gen Genet 183:20–24

Kuntz M, Keller M, Crouse EJ, Burkard G, Weil JH (1982) Fractionation and identification of *Euglena gracilis* cytoplasmic and chloroplastic tRNAs and mapping of tRNA genes on chloroplast DNA. Curr Genet 6:63–69

Kurland CG (1960) Molecular characterization of ribonucleic acid from *Escherichia coli* ribosomes. J Mol Biol 2:83–91

Lagerkvist U (1978) "Two out of three": An alternative method for codon reading. Proc Natl Acad Sci USA 75:1759–1762

Lamppa GK, Bendich AJ (1979a) Chloroplast DNA sequence homologies among vascular plants. Plant Physiol 63:660–668

Lamppa GK, Bendich AJ (1979b) Changes in chloroplast DNA levels during development of pea (*Pisum sativum*). Plant Physiol 64:126–130

Lamppa GK, Bendich AJ (1981) Fine scale interspersion of conserved sequences in the pea and corn chloroplast genomes. Mol Gen Genet 182:310–320

Lamppa GK, Elliot LV, Bendich AJ (1980) Changes in chloroplast number during pea leaf development. An analysis of a protoplast population. Planta 148:437–443

Ledoigt G, Calvayrac R (1974) Synthèse d'acides ribonucléiques lourds au cours d'une génération céllulaire chez *Euglena gracilis*. Planta 121:181–191

Lemieux C, Turmel M, Lee RH (1980) Characterization of chloroplast DNA in *Chlamydomonas eugamentos* and *C. moewusii* and its inheritance in hybrid progeny. Curr Genet 2:139–147

Lindahl L, Yamamoto M, Nomura M, Kirschbaum JB, Allet B, Rochaix J-D (1977) Mapping of a cluster of genes for components of the transcriptional and translational machineries of *Escherichia coli*. J Mol Biol 109:23–47

Link G (1981a) Enhanced expression of a distinct plastid DNA region in mustard seedlings by continuous far-red light. Planta 152:379–380

Link G (1981b) Cloning and mapping of the chloroplast DNA sequences for two messenger RNAs from mustard (*Sinapis alba* L). Nucleic Acids Res 9:3681–3694

Link G (1982) Phytochrome control of plastid mRNA in mustard (*Sinapis alba* L.). Planta 154:81–86

Link G, Bogorad L (1980) Sizes, locations and directions of transcription of two genes on a cloned maize chloroplast DNA sequence. Proc Natl Acad Sci USA 77:1832–1836

Link G, Coen DM, Bogorad L (1978) Differential expression of the gene for the large subunit of ribulose bisphosphate carboxylase in maize cell types. Cell 15:725–731

Link G, Chambers SE, Thompson JA, Falk H (1981) Size and physical organization of chloroplast DNA from mustard (*Sinapis alba* L.). Mol Gen Genet 181:454–457

Löffelhardt W, Mucke H, Bohnert HJ (1980) Cyanelle DNA from *Cyanophora paradoxa*: Analogies to chloroplast DNA. In: Schwemmler W, Schenk HEA (eds) Endocytobiology. De Gruyter, Berlin New York, pp 523–530

Loening UE (1968) Molecular weights of ribosomal RNA in relation to evolution. J Mol Biol 38:355–365

Loiseaux S, Rozier C, Dalmon J (1980) Plastidal origin of a large ribosomal precursor molecule in the brown alga *Pylaiella littoralis* (L.) Kjellm. Plant Sci Lett 18:381–388
Losick R, Chamberlin M (1976) RNA Polymerase. Cold Spring Harbor Lab Press, Cold Spring Harbor
Loughney K, Lund E, Dahlberg JE (1982) tRNA genes are found between the 16S and 23S rRNA genes in *Bacillus subtilis*. Nucleic Acids Res 10:1607–1624
Lüttke A (1980) Relation between chloroplast replication and cell division in *Olisthodiscus luteus*. Plant Sci Lett 18:191–199
Lüttke A, Bonotto S (1981) Chloroplast DNA of *Acetabularia mediterranea*: Cell cycle related changes in distribution. Planta 153:536–542
Lyman H, Epstein HT, Schiff JA (1961) Studies of chloroplast development in *Euglena*. I. Inactivation of green colony formation by U.V. light. Biochim Biophys Acta 50:301–309
Machatt MA, Ebel JP, Branlant C (1981) The 3′-terminal region of bacterial 23S ribosomal RNA: structure and homology with the 3′-terminal region of eukaryotic 28S rRNA and with chloroplast 4.5S rRNA. Nucleic Acids Res 9:1533–1549
MacKay RM (1981) The origin of plant chloroplast 4.5S ribosomal RNA. FEBS Lett 123:17–18
Malnoe P, Rochaix JD (1978) Localization of 4S RNA genes on the chloroplast genome of *Chlamydomonas reinhardii*. Mol Gen Genet 166:269–275
Malnoe P, Rochaix JD, Chua NH, Spahr PF (1979) Characterization of the gene and messenger RNA of the large subunit of ribulose-1,5-diphosphate carboxylase in *Chlamydomonas reinhardii*. J Mol Biol 133:417–434
Mandel M, Marmur J (1968) Use of ultraviolet absorbance-temperature profile for determining the guanine plus cytosine content of DNA. Methods Enzymol 12b:195–206
Mandel M, Schildkraut CL, Marmur J (1968) Use of CsCl density gradient analysis for determining the guanine plus cytosine content of DNA. Methods Enzymol 12b:184–195
Manning JE, Richards OC (1972a) Synthesis and turnover of *Euglena gracilis* nuclear and chloroplast deoxyribonucleic acid. Biochemistry 11:2036–2043
Manning JE, Richards OC (1972b) Isolation and molecular weight of circular chloroplast DNA from *Euglena gracilis*. Biochim Biophys Acta 259:285–296
Manning JE, Wolstenholme DR, Ryan RS, Hunter JA, Richards OC (1971) Circular chloroplast DNA from *Euglena gracilis*. Proc Natl Acad Sci USA 68:1169–1173
Manning JE, Wolstenholme DR, Richards OC (1972) Circular DNA molecules associated with chloroplasts of spinach, *Spinacia oleracea*. J Cell Biol 53:594–601
Margulis L (1981) Symbiosis in cell evolution: Life and its environment on the early earth. WH Freeman, San Francisco
Matsuda Y, Surzycki SJ (1980) Chloroplast gene expression in *Chlamydomonas reinhardi*. Mol Gen Genet 180:463–474
Mazza A, Bonotto S, Felluga B, Casale A, Sassone-Corsi P (1979) Topological organization of the chloroplast genome of *Acetabularia acetabulum*. In: Bonotto S, Kefeli V, Puiseux-Dao S (eds) Developmental biology of *Acetabularia*. Elsevier/North-Holland, Amsterdam New York, pp 115–130
Mazza A, Casale A, Sassone-Corsi B, Bonotto S (1980) A minicircular component of *Acetabularia acetabulum* chloroplast DNA replicating by the rolling circle. Biochem Biophys Res Commun 93:668–674
McCrea JM, Hershberger CL (1976) Chloroplast DNA codes for transfer RNA. Nucleic Acids Res 3:2005–2018
McIntosh L, Poulsen C, Bogorad L (1980) Chloroplast gene sequence for the large subunit of ribulose bisphosphate carboxylase of maize. Nature (London) 288:556–560
Meeker R, Tewari KK (1980) Transfer ribonucleic acid genes in the chloroplast deoxyribonucleic acid of pea leaves. Biochemistry 19:5973–5981
Meeker R, Tewari KK (1982) Divergence of tRNA genes in chloroplast DNA of higher plants. Biochim Biophys Acta 696:66–75
Merrick WC, Dure LS (1972) The developmental biochemistry of cotton seed embryogenesis and germination. IV. Levels of cytoplasmic and chloroplastic transfer ribonucleic acid species. J Biol Chem 247:7988–7999

Mets L (1980) Uniparental inheritance of chloroplast DNA sequences in interspecific hybrids of *Chlamydomonas*. Curr Genet 2:131–138

Metzlaff M, Boerner T, Hagemann R (1981) Variations of chloroplast DNAs in the genus *Pelargonium* and their biparental inheritance. Theor Appl Genet 60:37–41

Michaelis P (1955) Modellversuche zur Plastiden- und Plasmavererbung. Zuechter 25:209–221

Miller MJ, McMahon D (1974) Synthesis and maturation of chloroplast and cytoplasmic ribosomal RNA in *Chlamydomonas reinhardii*. Biochim Biophys Acta 366:35–44

Milner JJ, Hershberger CL, Buetow DE (1979) *Euglena gracilis* chloroplast DNA codes for polyadenylated RNA. Plant Physiol 64:818–821

Miyata T, Kikuno R, Ohshima Y (1982) A pseudogene cluster in the leader region of the *Euglena* chloroplast 16S–23S rRNA genes. Nucleic Acids Res 10:1771–1780

Morgan EA, Ikemura T, Nomura M (1977) Identification of spacer tRNA genes in individual ribosomal RNA transcription units of *Escherichia coli*. Proc Natl Acad Sci USA 74:2710–2714

Morgan EA, Ikemura T, Lindahl L, Fallon AM, Nomura M (1978) Some rRNA operons in *E. coli* have tRNA genes at their distal ends. Cell 13:335–344

Mory YY, Chen D, Sarid S (1972) Onset of deoxyribonucleic acid synthesis in germinating wheat embryos. Plant Physiol 49:20–23

Mubumbila M, Burkard G, Keller M, Steinmetz A, Crouse EJ, Weil JH (1980) Hybridization of bean, spinach, maize and Euglena chloroplast tRNAs with homologous and heterologous chloroplast DNAs. An approach to the study of homology between chloroplast tRNAs from various species. Biochim Biophys Acta 609:31–39

Mucke H, Löffelhardt W, Bohnert HJ (1980) Partial characterization of the genome of the endosymbiotic cyanelles from *Cyanophora paradoxa*. FEBS Lett 111:347–352

Munsche D, Wollgiehn R (1973) Die Synthese von ribosomaler RNA in Chloroplasten von *Nicotiana rustica*. Biochim Biophys Acta 249:106–117

Myers M, Grant DM, Rabert DK, Harris EH, Boynton JE, Gillham JW (1982) Mutants of *Chlamydomonas reinhardtii* with physical alterations in their chloroplast DNA. Plasmid 7:133–151

Ness PJ, Woolhouse HW (1980a) RNA synthesis in *Phaseolus* chloroplasts. I. Ribonucleic acid synthesis in chloroplast preparations from *Phaseolus vulgaris* L. leaves and solubilization of the RNA polymerase. J Exp Bot 31:223–233

Ness PJ, Woolhouse HW (1980b) RNA synthesis in *Phaseolus* chloroplasts. II. Ribonucleic acid synthesis in chloroplasts from developing and senescing leaves. J Exp Bot 31:235–245

Oishi K, Sumnicht T, Tewari KK (1981) Messenger ribonucleic acid transcripts of pea chloroplast deoxyribonucleic acid. Biochemistry 20:5710–5717

Orozco EM, Hallick RB (1982a) *Euglena gracilis* chloroplast transfer RNA transcription units. I. Physical map of the transfer RNA gene loci. J Biol Chem 257:3258–3264

Orozco EM, Hallick RB (1982b) *Euglena gracilis* chloroplast transfer RNA transcription units. II. Nucleotide sequence analysis of a $tRNA^{Val}$ – $tRNA^{Arg}$ – $tRNA^{Leu}$ gene cluster. J Biol Chem 257:3265–3275

Orozco EM, Gray PW, Hallick RB (1980a) *Euglena gracilis* chloroplast ribosomal RNA transcription units. I. The location of transfer RNA, 5S, 16S, and 23S ribosomal RNA genes. J Biol Chem 255:10991–10996

Orozco EM, Rushlow KE, Dodd JB, Hallick RB (1980b) *Euglena gracilis* ribosomal RNA transcription units. II. Nucleotide sequence homology between the 16S–23S ribosomal RNA spacer and the 16S ribosomal RNA leader regions. J Biol Chem 255:10997–11003

Padmanabhan U, Green BR (1978) The kinetic complexity of *Acetabularia* chloroplast DNA. Biochim Biophys Acta 521:67–73

Palmer JD (1982) Physical and gene mapping of chloroplast DNA from *Atriplex triangularis* and *Cucumis sativa*. Nucleic Acids Res 10:1593–1605

Palmer JD, Thompson WF (1981a) Rearrangements in the chloroplast genomes of mung bean and pea. Proc Natl Acad Sci USA 78:5533–5537

Palmer JD, Thompson WF (1981b) Clone banks of the mung bean, pea and spinach chloroplast genomes. Gene 15:21–26

Panigrahi PK, Biswal UC (1979) Ageing of chloroplasts in vitro. I. Quantitative analysis of the degradation of pigments, proteins and nucleic acids. Plant Cell Physiol 20:775–779

Parthier B, Krauspe R (1974) Chloroplast and cytoplasmic transfer RNA of *Euglena gracilis*. Biochem Phys Pflanz 165:1–17

Pearson JA, Thomas K, Thomas H (1978) Nucleic acids from leaves of a yellowing and a non-yellowing variety of *Festuca pratensis* Huds. Planta 144:85–87

Posner HB, Rosner A (1975) Effects of chloramphenicol on RNA synthesis in *Spirodela* chloroplasts. Plant Cell Physiol 16:361–365

Pryke JA, Cranney MK, Jones DS (1979) Hybridisation studies on DNA isolated from *Scenedesmus obliquus*. Plant Sci Lett 16:125–128

Rawson JRY (1975) A Measurement of the fraction of chloroplast DNA transcribed in *Euglena*. Biochem Biophys Res Commun 62:539–545

Rawson RY, Boerma CL (1976a) A measurement of the fraction of chloroplast DNA transcribed during chloroplast development in *Euglena gracilis*. Biochemistry 15:588–592

Rawson RY, Boerma CL (1976b) Influence of growth conditions upon the number of chloroplast DNA molecules in *Euglena gracilis*. Proc Natl Acad Sci USA 73:2401–2404

Rawson JRY, Kushner SR, Vapnek D, Alton NK, Boerma CL (1978) Chloroplast ribosomal RNA genes in *Euglena gracilis* exist as three clustered tandem repeats. Gene 3:191–209

Rawson JRY, Clegg MT, Thomas K, Rinehart C, Wood B (1981) A restriction map of the ribosomal RNA genes and the short single-copy DNA sequence of the pearl millet chloroplast genome. Gene 16:11–19

Ray DS, Hanawalt PC (1964) Properties of the satellite DNA associated with the chloroplasts of *Euglena gracilis*. J Mol Biol 9:812–824

Reger BJ, Fairfield SA, Epler JL, Barnett WE (1970) Identification and origin of some chloroplast aminoacyl-tRNA synthetases and tRNA. Proc Natl Acad Sci USA 67:1207–1213

Renner O (1936) Zur Kenntnis der nicht-mendelnden Buntheit der Laubblätter. Flora (Jena) NF 30:218–290

Rhoades MM (1946) Plastid mutations. Cold Spring Harbor Symp Quant Biol 11:202–207

Rhodes PR, Kung SD (1981) Chloroplast deoxyribonucleic acid isolation: purity achieved without nuclease digestion. Can J Biochem 59:911–915

Rhodes PR, Zhu YS, Kung SD (1981) *Nicotiana* chloroplast genome. I. Chloroplast DNA diversity. Mol Gen Genet 182:106–111

Richards OC, Manning JE (1975) Replication of chloroplast DNA in *Euglena gracilis*. Les cycles cellulaires et leur blocage. Coll Int CNRS 240:213–221

Richter G, Dirks W (1978) Blue-light-induced development of chloroplasts in isolated seedling roots. Preferential synthesis of chloroplast ribosomal RNA species. Photochem Photobiol 27:155–160

Richter G, Misske R, Stempka R, Dirks W (1977) Mode of synthesis of ribosomal RNA in various plants. In: Bogorad L, Weil JH (eds) Nucleic acids and protein synthesis in plants. Plenum Press New York London, pp 85–96

Ris H, Plaut W (1962) The ultrastructure of DNA-containing areas in the chloroplasts of *Chlamydomonas*. J Cell Biol 13:383–391

Roberts RJ (1982) Restriction and modification enzymes and their recognition sequences. Nucleic Acids Res 10:r117–r144

Rochaix JD (1978) Restriction endonuclease map of the chloroplast DNA of *Chlamydomonas reinhardii*. J Mol Biol 126:597–617

Rochaix JD (1981) Organization, function and expression of the chloroplast DNA of *Chlamydomonas reinhardii*. Experientia 37:323–332

Rochaix JD, Malnoe P (1978) Anatomy of the chloroplast ribosomal DNA of *Chlamydomonas reinhardii*. Cell 15:661–670

Röbbelen G (1966) Chloroplasten-Differenzierung nach induzierter Plastommutation bei *Arabidopsis thaliana* (L.) Heynh. Z Pflanzenphys 55:387–403

Rose RJ, Cran DG, Possingham JV (1975) Changes in DNA synthesis during cell growth and chloroplast replication in greening spinach leaf disks. J Cell Sci 17:27–41

Rosenberg M, Court D (1979) Regulatory sequences involved in the promotion and termination of RNA transcription. Annu Rev Genetics 13:319–353

Rosner A, Jakob KM, Gressel J, Sagher D (1975) The early synthesis and possible function of a 0.5×10^6 Mr RNA after transfer of dark-grown *Spirodela* plants to light. Biochem Biophys Res Commun 67:383–391

Sager R (1954) Mendelian and non-Mendelian inheritance of streptomycin resistance in *Chlamydomonas reinhardi*. Proc Natl Acad Sci USA 40:356–363

Sager R, Grabowy C, Sano H (1981) The mat-1 gene in *Chlamydomonas* regulates DNA methylation during gametogenesis. Cell 24:41–47

Sagher D, Grosfeld H, Edelman M (1976) Large subunit ribulosebisphosphate carboxylase messenger RNA from *Euglena* chloroplasts. Proc Natl Acad Sci USA 73:722–726

Sano H, Spaeth E, Burton WG (1979) Messenger RNA of the large subunit of ribulose-1,5-bisphosphate carboxylase from *Chlamydomonas reinhardi*. Eur J Biochem 93:173–180

Schiff JA (1981) Development, inheritance and evolution of plastids and mitochondria. In: Tolbert NE (ed) The biochemistry of plants, vol I. Academic Press, London New York, pp 209–292

Schmitt JM, Herrmann RG (1977) Fractionation of cell organelles in silica sol gradients. Methods Cell Biol 15:177–200

Schmitt JM, Bohnert HJ, Gordon KHJ, Herrmann RG, Bernardi G, Crouse EJ (1981) Compositional heterogeneity of the chloroplast DNAs from *Euglena gracilis* and *Spinacia oleracea*. Eur J Biochem 117:375–382

Schneeman R, Surzycki S (1979) *E. coli* ribosomal proteins are cross reactive with antibody prepared against *Chlamydomonas reinhardi* chloroplast ribosomal subunit. Mol Gen Genet 176:95–104

Schötz F (1958) Beobachtungen zur Plastidenkonkurrenz bei *Oenothera* und Beiträge zum Problem der Plastiden-Vererbung. Planta 51:173–185

Schwarz Z, Kössel H (1980) The primary structure of 16S rDNA from *Zea mays* chloroplast is homologous to *E. coli* 16S rRNA. Nature (London) 283:739–742

Schwarz Z, Kössel H, Schwarz E, Bogorad L (1981a) A gene coding for $tRNA^{Val}$ is located near 5′-terminus of 16S rRNA gene in *Zea mays* chloroplast genome. Proc Natl Acad Sci USA 78:4748–4752

Schwarz Z, Jolly SO, Steinmetz A, Bogorad L (1981b) Overlapping divergent genes in the maize chloroplast chromosome and in vitro transcription of the gene for $tRNA^{His}$. Proc Natl Acad Sci USA 78:3423–3427

Schwartzbach SD, Hecker LI, Barnett WE (1976) Transcriptional origin of *Euglena* chloroplast tRNA. Proc Natl Acad Sci USA 73:1984–1988

Scott NS (1976) Precursors of chloroplast ribosomal RNAs in *Euglena gracilis*. Phytochemistry 15:1207–1213

Scott NS, Possingham JV (1980) Chloroplast DNA in expanding spinach leaves. J Exp Bot 31:1081–1092

Scowcroft WR (1979) Nucleotide polymorphism in chloroplast DNA of *Nicotiana debneyi*. Theor Appl Genet 55:133–137

Sellden G, Leech RM (1981) Localization of DNA in mature and young wheat chloroplasts using the fluorescent probe 4′-6-diamidino-2-phenylindole. Plant Physiol 68:731–734

Sen S, Payne PI, Osborne DJ (1975) Early RNA synthesis during the germination of rye (*Secale cereale*) embryos and the relationship to early protein synthesis. Biochem J 148:381–387

Seyer P, Kowallik KV, Herrmann RG (1981) A physical map of *Nicotiana tabacum* plastid DNA including the location of structural genes for ribosomal RNAs and the large subunit of ribulose bisphosphate carboxylase/oxygenase. Curr Genet 3:189–204

Shine J, Dalgarno L (1974) The 3′-terminal sequence of *Escherichia coli* 16S ribosomal RNA: Complementarity of nonsense triplets and ribosome binding sites. Proc Natl Acad Sci USA 71:1342–1346

Silverthorne J, Ellis JR (1980) Protein synthesis in chloroplasts. VIII. Differential synthesis of chloroplast proteins during spinach leaf development. Biochim Biophys Acta 607:319–330

Siu C-H, Swift H, Chiang K-S (1976) Characterization of cytoplasmic and nuclear genomes in the colorless alga *Polytoma*. II. General characterization of organelle nucleic acids. J Cell Biol 69:371–382

Slavik NS, Hershberger CL (1975) The kinetic complexity of *Euglena gracilis* chloroplast DNA. FEBS Lett 52:171–174

Slavik NS, Hershberger CL (1976) Internal structural organization of chloroplast DNA from *Euglena gracilis* Z. J Mol Biol 103:563–581

Smith SM, Ellis RJ (1981) Light-stimulated accumulation of transcripts of nuclear and chloroplast genes for ribulosebisphosphate carboxylase. J Mol Appl Gen 1:127–137

Speirs J, Brady CJ (1981) A coordinated decline in the synthesis of subunits of ribulosebisphosphate carboxylase in aging wheat leaves. II. Abundance of messenger RNA. Aust J Plant Physiol 8:603–618

Speirs J, Grierson D (1978) Isolation and characterization of 14S RNA from spinach chloroplasts. Biochim Biophys Acta 521:619–633

Sprinzl M, Gauss DH (1982) Compilation of tRNA sequences. Nucleic Acids Res 10:r1–r56

Srinivas U, Lyman H (1980) Photomorphogenic regulation of chloroplast replication in *Euglena*. Plant Physiol 66:295–301

Steinback KE (1981) Proteins of the chloroplast. In: Marcus A (ed) The biochemistry of plants, vol VI. Academic Press, London New York, pp 303–319

Stubbe W (1959) Genetische Analyse des Zusammenwirkens von Genom und Plastom bei *Oenothera*. Z Indukt Abstamm Vererbungsl 90:288–298

Stubbe W (1971) Origin and continuity of plastids. In: Reinert J, Ursprung H (eds) Origin and continuity of cell organelles, vol III. Springer, Berlin Heidelberg New York, pp 65–81

Stüber D, Bujard H (1977) Electron microscopy of DNA: Determination of absolute molecular weights and linear density. Mol Gen Genet 154:299–303

Stutz E (1983) Algal chloroplast DNA. In: Ellis RJ (ed) Chloroplast biogenesis. Cambridge Univ Press, Cambridge, in press

Stutz E, Vandrey JP (1971) Ribosomal DNA satellite of *Euglena gracilis* chloroplast DNA. FEBS Lett 17:277–280

Swamy GS, Pillay DTN (1982) Characterization of *Glycine max* cytoplasmic, chloroplastic and mitochondrial tRNAs and synthetases for phenylalanine, tryptophane and tyrosine. Plant Sci Lett 25:73–84

Tabidze V, Beridze T (1979) Fine melting curves of chloroplast DNA of higher plants. Plant Sci Lett 16:157–164

Takaiwa F, Sugiura M (1980a) Nucleotide sequences of the 4.5S and 5S ribosomal RNA genes from tobacco chloroplasts. Mol Gen Genet 180:1–4

Takaiwa F, Sugiura M (1980b) Cloning and characterization of 4.5S and 5S RNA genes in tobacco chloroplasts. Gene 10:95–103

Takaiwa F, Sugiura M (1981) Heterogeneity of 5S RNA species in tobacco chloroplasts. Mol Gen Genet 182:385–389

Takaiwa F, Sugiura M (1982a) The complete nucleotide sequence of a 23S rRNA gene from tobacco chloroplasts. Eur J Biochem 124:13–19

Takaiwa F, Sugiura M (1982b) Nucleotide sequence of the 16S–23S spacer region in an rRNA gene cluster from tobacco chloroplast DNA. Nucleic Acids Res 10:2665–2675

Takaiwa F, Kusuda M, Sugiura M (1982) The nucleotide sequence of chloroplast 4.5S rRNA from a fern, *Dryopteris acuminata*. Nucleic Acids Res 10:2257–2260

Takaya K, Sasaki K (1973) Biochemical properties of nucleic acids of chloroplasts in *Spirogyra*. Plant Cell Physiol 14:237–248

Taylor FJR (1979) Symbionticism revisited: a discussion of the evolutionary impact of intracellular symbioses. Proc R Soc London Ser B 204:267–286

Tewari KK (1979) Structure and replication of chloroplast DNA. In: Hall TC, Davies JW (eds) Nucleic acids in plants. Vol I. CRC Press, Boca Raton, pp 41–108

Tewari KK, Wildman SG (1970) Information content in the chloroplast DNA. In: Control of organelle development. Cambridge University Press. Symp Soc Exp Biol 24:147–179

Thien W, Schopfer P (1975) Control by phytochrome of cytoplasmic and plastid rRNA accumulation in cotyledons of mustard seedlings in the absence of photosynthesis. Plant Physiol 56:660–664

Thomas JR, Tewari KK (1974) Conservation of ribosomal RNA genes in the chloroplast DNAs of higher plants. Proc Natl Acad Sci USA 71:3147–3151

Thompson JA, Hansmann P, Knoth R, Link G, Falk H (1981) Electron microscopical localization of the 23S and 16S rRNA genes within an inverted repeat for two chromoplast DNAs. Curr Genet 4:25–28

Tilney-Bassett RAE (1970) Genetics and plastid physiology in *Pelargonium*. III. Effect of cultivar and plastids on fertilization and embryo survival. Heredity 25:89–103

Tobin EM (1981) Light regulation of the synthesis of two major nuclear-coded chloroplast polypeptides in *Lemna gibba*. In: Akoyunoglou G (ed) Photosynthesis V: chloroplast development. Balaban, Philadelphia, pp 949–959

Tohdoh N, Sugiura M (1982) The complete nucleotide sequence of a 16S ribosomal RNA gene from tobacco chloroplasts. Gene 17:213–218

Tohdoh N, Shinozaki K, Sugiura M (1981) Sequence of a putative promoter region for the rRNA genes of tobacco chloroplast DNA. Nucleic Acids Res 9:5399–5406

Tomioka N, Shinozaki K, Sugiura M (1981) Molecular cloning and characterization of ribosomal RNA genes from a blue-green alga, *Anacystis nidulans*. Mol Gen Genet 184:359–363

Tran Thanh Van KM (1981) Control of morphogenesis in in vitro cultures. Annu Rev Plant Physiol 32:291–311

Trench RK (1979) The cell biology of plant-animal symbiosis. Annu Rev Plant Physiol 30:485–531

Turmel M, Lemieux C, Lee RW (1980) Net synthesis of chloroplast DNA throughout the synchronized vegetative cell cycle of *Chlamydomonas*. Curr Genet 2:229–232

Vedel F, Quetier F, Bayen M (1976) Specific cleavage of chloroplast DNA by EcoRI restriction nuclease. Nature 236:440–442

Verdier G (1979) Poly(adenylic acid)-containing RNA of *Euglena gracilis* during chloroplast development. II. Transcriptional origin of the different RNA. Eur J Biochem 93:581–586

Watson JC, Surzycki SJ (1982) Extensive sequence homology in the DNA coding for elongation factor Tu from *Escherichia coli* and the *Chlamydomonas reinhardtii* chloroplast. Proc Natl Acad Sci USA 79:2264–2267

Weil JH (1979) Cytoplasmic and organellar tRNAs in plants. In: Hall TC, Davies JW (eds) Nucleic acids in plants. Vol I. CRC Press, Boca Raton, pp 143–192

Weil JH, Parthier B (1982) Transfer RNA and aminoacyl-tRNA synthetases in plants. In: Boulter D, Parthier B (eds) Encyclopedia of plant physiology, New Ser, vol 14A. Springer, Berlin Heidelberg New York

Weil JH, Mubumbila M, Kuntz M, Keller M, Steinmetz A, Crouse EJ, Burkard G, Guillemaut P, Selden R, McIntosh L, Bogorad L, Löffelhardt W, Mucke H, Bohnert HJ (1982) Gene mapping studies and sequence determination on chloroplast tRNAs from various photosynthetic organisms. In: Akoyunoglou G, Rentzepit P (eds) Proc Special FEBS meeting on Cell Function and Differentiation Liss, New York, in press

Wells R, Birnstiel M (1969) Kinetic complexity of chloroplastal deoxyribonucleic acid and mitochondrial deoxyribonucleic acid from higher plants. Biochem J 112:777–786

Wells R, Sager R (1971) Denaturation and renaturation kinetics of chloroplast DNA from *Chlamydomonas reinhardtii*. J Mol Biol 58:611–622

Westhoff P, Zetsche K (1981) Regulation of the synthesis of ribulose-1,5-bisphosphate carboxylase and its subunits in the flagellate *Chlorogonium elongatum*. Eur J Biochem 116:261–267

Westhoff P, Nelson N, Bünemann H, Herrmann RG (1981) Localization of genes for coupling factor subunits on the spinach plastid chromosome. Curr Genet 4:109–120

Wetmur JG, Davidson N (1968) Kinetics of renaturation of DNA. J Mol Biol 31:349–370

Wettstein von D (1967) Chloroplast structure and genetics. In: San Pietro A, Greer A, Army TJ (eds) Harvesting the sun. Academic Press, London New York, pp 153–190

Wheeler AM, Hartley MR (1975) Major mRNA species from spinach chloroplasts do not contain poly(A). Nature (London) 257:66–67

Whiteway MS, Lee RW (1977) Chloroplast DNA content increases with nuclear ploidy in *Chlamydomonas*. Mol Gen Genet 157:11–15

Whitfeld PR (1977) Chloroplast RNA. In: Stewart PR, Letham DS (eds) The ribonucleic acids, 2nd edn. Springer, Berlin Heidelberg New York, pp 297–332

Whitfeld PR, Bottomley W (1980) Mapping of the gene for the large subunit of ribulose bisphosphate carboxylase on spinach chloroplast DNA. Biochem Int 1:172–178

Whitfeld PR, Spencer D (1968) Buoyant density of tobacco and spinach chloroplast DNA. Biochim Biophys Acta 157:333–343

Whitfeld PR, Leaver CJ, Bottomley W, Atchison BA (1978) Low-molecular-weight (4.5S) ribonucleic acid in higher plant chloroplast ribosomes. Biochem J 175:1103–1112

Wildeman AG, Nazar RW (1980) Nucleotide sequence of wheat chloroplastid 4.5S ribonucleic acid. J Biol Chem 255:11896–11900

Wollgiehn R, Parthier B (1979) RNA synthesis in isolated chloroplasts of *Euglena gracilis*. Plant Sci Lett 16:203–210

Wollgiehn R, Parthier B (1980) RNA and protein synthesis in plastid differentiation In: Reinert J (ed) Results and problems in cell differentiation, vol X. Springer, Berlin Heidelberg New York, pp 97–145

Woodcock CLF, Bogorad L (1970) Evidence for variation in the quantity of DNA among plastids of *Acetabularia*. J Cell Biol 44:361–375

Wurtz EA, Buetow DE (1981) Intraspecific variation in the structural organization and redundancy of chloroplast ribosomal DNA cistrons in *Euglena gracilis*. Curr Genet 3:181–187

Wuttke H-G (1976) Circular DNA in chromoplasts of *Tulipa gesneriana*. Planta 132:317–319

Yoshida Y, Laulhere J-P, Rozier C, Mache R (1978) Visualization of folded chloroplast DNA from spinach. Biol Cell 32:187–190

Zech M, Hartley MR, Bohnert HJ (1981) Binding sites of *E. coli* DNA-dependent RNA polymerase on spinach chloroplast DNA. Curr Genet 4:37–46

Zurawski G, Perrot B, Bottomley W, Whitfeld PR (1981) The structure of the gene for the large subunit of ribulose 1,5-bisphosphate carboxylase from spinach chloroplast DNA. Nucleic Acids Res 9:3251–3270

Zurawski G, Bohnert HJ, Whitfeld PR, Bottomley W (1982) Nucleotide sequence of the gene for the 32,000-M_r thylakoid membrane protein from *Spinacia oleracea* and *Nicotiana debneyi* predicts a totally conserved primary translation product of M_r 38,800. Proc Natl Acad Sci USA (in press)

Zwieb C, Glotz C, Brimacombe R (1981) Secondary structure comparisons between small subunit ribosomal RNA molecules from six different species. Nucleic Acids Res 9:3621–3640

15 The Biosynthesis of Chloroplast Proteins

W. Bottomley and H. J. Bohnert

1 Introduction

Shortly after the rediscovery of Mendel's work which formed the basis of modern genetics, the first example of non-Mendelian inheritance was reported (Baur 1909, Correns 1909). This mode, which determined the inheritance of certain characteristic leaf colorations, subsequently became known as "extrachromosomal" or "cytoplasmic" inheritance. Although it was soon realized that the genetic determinants for these characteristics were associated with plastids, it took more than half a century to develop the experimental tools to study the processes underlying the phenomenon. These studies, which could be called "chloroplast molecular genetics", involve the investigation of the genes and gene products responsible for the development and functioning of chloroplasts within plant cells.

Chloroplast molecular genetics began in 1962 with the report that chloroplast preparations contained DNA (Ris and Plaut 1962) and that this DNA had physical properties which distinguished it from nuclear DNA (Kirk 1963). Also in 1962 Lyttleton described the first isolation of ribosomes from chloroplasts. The discovery of this distinct class of ribosomes was followed by the demonstration of the incorporation of radioactive amino acids into proteins by isolated chloroplasts from *Euglena* (Eisenstadt and Brawerman 1963) and later from higher plants (Spencer and Wildman 1964). These findings raised the question of whether or not these organelles were able to develop and function independently of the remainder of the cell. It has now been established that, although chloroplasts have the potential to be autonomous, they are, to a large extent, dependent on the products of cytosolic synthesis for their development. Although they have the enzymes to synthesize both DNA and RNA (Kirk and Tilney-Bassett 1978) in addition to protein synthetic capacity, many of these essential enzymes originate in the cytosol. In addition, it has become apparent that the genetic complexity of chloroplast DNA is insufficient to code for all the enzymes associated with the development and functioning of the organelles. DNA and protein synthesis alone require over 100 genes to code for DNA replicase, RNA polymerase, rRNA's and ribosomal proteins, tRNA's and aminoacyl-tRNA-synthetases, as well as soluble factors. When one adds to this approximately 40 enzymes for chlorophyll and carotenoid synthesis, as well as

---

Abbreviations. CF_0 and CF_1, coupling factors of ATPase; CP1, chlorophyll-protein complex of photosystem I; DCCD, dicyclohexyl carbodiimide; EF-G, elongation factor G; EF-Ts, elongation factor T sensitive; PS I and II, photosystems I and II; RuBP, ribulose 1,5-bisphosphate

over 40 enzymes involved in photosynthesis (Kirk and Tilney-Bassett 1978), then virtually all of the higher plant chloroplast DNA (about 150 kb, see, e.g., Bedbrook and Kolodner 1979, Tewari 1979) would be required for these proteins alone. Since chloroplasts contain in addition many of the enzymes for lipid synthesis, carbohydrate metabolism and nucleotide synthesis it is obvious that, even with extensive overlapping of genes, there is not enough DNA to code for the total chloroplast constituents.

It has been demonstrated that genes specifying chloroplast proteins in nuclear DNA compensate for this deficiency of coding capacity within the chloroplast DNA. In 1972 Kawashima and Wildman found that specific peptides from tryptic digests in the small subunit of RuBP carboxylase was inherited in a Mendelian manner in interspecific crosses of *Nicotiana* spp. This demonstrated that the genes coding for this chloroplast polypeptide were located in the nucleus. This method of studying the inheritance of biochemical variation in proteins has led to the discovery of a number of chloroplast constituents which are coded for by nuclear DNA (Kung et al. 1972, Gillham et al. 1977, Gray 1980).

The first efforts to locate the site of synthesis of chloroplast proteins exploited the known difference between the 80S-ribosomal cytosolic and the 70S-ribosomal chloroplast protein-synthesizing systems. By this means a substantial body of evidence has been accumulated, based mainly on inhibitors believed to specifically affect either the cytosolic or chloroplast system, which demonstrated the involvement of both in the synthesis of chloroplast proteins.

The next stage in chloroplast molecular genetics was the identification of the polypeptides synthesized in isolated organelles. This technique has so far resulted in the identification of about 13 polypeptides of known function as being synthesized on chloroplast ribosomes. These include the large subunit of RuBP carboxylase (Blair and Ellis 1973), six of the eight subunits of chloroplast ATPase (Mendiola-Morgenthaler et al. 1976, Doherty and Gray 1980, Nelson et al. 1980), cytochrome f (Doherty and Gray 1979), cytochrome b_{559} (Zielinski and Price 1980), two protein synthesis elongation factors (Ciferri et al. 1979) and two polypeptides of the P-700-chlorophyll a-protein complex (Geetha and Gnanam 1980).

The more difficult task of demonstrating conclusively that particular polypeptides are synthesized in the cytoplasm has been largely carried out by allowing cytosolic polysomes to complete the synthesis of polypeptides in vitro or by in vitro protein synthesis on poly-(A) containing mRNA's which are believed to come exclusively from the cytoplasm (Wheeler and Hartley 1975). By these approaches several chloroplast proteins which are translated in the cytosol have been identified, including the small subunit of RuBP carboxylase (Roy et al. 1976 1977, Cashmore et al. 1978), ferredoxin (Huisman et al. 1978), one polypeptide from the ATPase (Nelson et al. 1980) and a polypeptide of the light-harvesting chlorophyll a/b protein complex (Apel and Kloppstech 1978).

With the advent of the new techniques of molecular cloning (for a descriptive compilation of these methods see Wu, 1980) it has become possible to isolate and characterize genes from both chloroplast and nuclear genomes. The application of these methods has already enabled the gene coding for the large subunit of RuBP carboxylase to be precisely located on the chloroplast DNA of maize

(BEDBROOK et al. 1979), *Chlamydomonas* (MALNOË et al. 1979), spinach (WHITFELD and BOTTOMLEY 1980) and wheat (BOWMAN et al. 1981). In addition, the gene for a 32,000-M_r polypeptide has been mapped on the chloroplast DNA of maize (BEDBROOK et al. 1978) and spinach (DRIESEL et al. 1980). The more difficult task of cloning and identifying nuclear genes has so far yielded a clone containing the coding sequence for the small subunit of RuBP carboxylase (BEDBROOK et al. 1980).

The dispersal of the genes coding for chloroplast proteins into at least two compartments of the cell poses a number of problems. The transcription of the two genomes need to be coordinated by some means and the translation of the mRNA's at two sites also must be regulated. Finally, the cytosolic products have to be transported across the chloroplast membrane, often assembled into a complex and, in many cases, inserted into the thylakoid membranes.

In this review we will consider a number of aspects of the biosynthesis of chloroplast proteins. These include a description of the basic characteristics of the two protein-synthesizing systems, the nature of their products and also the transport, processing and assembly of those products into functional units.

2 The Protein-Synthesizing Systems of the Cytoplasm and the Chloroplast

Green plant tissue contains three discrete locations for the synthesis of proteins. These are the cytosol, the mitochondrion and the chloroplast. The composition of the cytosolic protein synthetic apparatus is closely related to that of other eukaryotic organisms (WOOL 1979, OCHOA and DE HARO 1979), while those of both mitochondria and chloroplasts have much in common with bacterial protein synthesis (HOOBER 1976, O'BRIEN 1976, LUCAS-LENARD and LIPMANN 1971).

Here, only a brief outline of the essential features of the components of the protein-synthesizing systems from the cytoplasm and chloroplasts will be given. For more extensive treatments of the various components see WEIL 1979, LEAVER 1979, HALL 1979, BARNETT et al. 1978, LEA and NORRIS, 1977, WHITFELD 1977 (cf. also WEIL and PARTHIER, Chap. 2, and MARCUS, Chap. 3, Vol. 14A, this Series.

2.1 The Ribosomes

The most readily detected difference between the protein-synthesizing apparatus of the cytoplasm and that of the chloroplasts is the size of the ribosomes and the ribosomal RNAs (see KIRK and TILNEY-BASSETT 1978, LEAVER 1979). The sedimentation constant of the cytoplasmic ribosomes is in the range of 80–87S, which is similar to that from most other eukaryotes (WOOL 1979), while that of chloroplasts corresponds to prokaryotic ribosomes and is in the range of

67–70S (WHITFELD 1977). The large subunits of the cytoplasmic ribosomes are approximately 60S and contain RNA of 25S (1.3×10^6 mol.wt.), 5.8S (50.4×10^3 mol.wt.) and 5S (40×10^3 mol.wt.), while the small subunits sediment at 40S and contain 18S (0.7×10^6 mol.wt.) RNA. The chloroplast large ribosomal subunits have sedimentation constants of approximately 50S and contain two small molecules of 5S and 4.5S ($25–33 \times 10^3$ mol.wt.) in addition to the large RNA of 23S (1.1×10^6 mol.wt.), while the small subunits sediment at 30S and contain 16S (0.56×10^6 mol.wt.) RNA.

In addition to the differences in the RNA species, the ribosomes from chloroplast and cytoplasm also have distinct ribosomal proteins (LYTTLETON 1962). Differences between the two sets of ribosomal proteins are obvious even when these are separated in one-dimensional gels and the differences become more pronounced when two-dimensional electrophoresis is used (JONES et al. 1972, YURINA and ODINTSOVA 1974, FREYSSINET and SCHIFF 1974, HANSON et al. 1974, METS and BOGORAD 1974, SPIESS 1977, FREYSSINET 1978). Although there is no positive evidence for proteins which are common to the two types of ribosomes, it should be noted that it was shown by STÖFFLER et al. (1974) that prokaryotic and eukaryotic ribosomal proteins carry common antigenic determinants. This suggests the possibility that there has been a high degree of conservation of the genes throughout evolution (WOOL 1979). From this it could be expected that there are some similarities between plant cytoplasmic and chloroplast ribosomal proteins. The prokaryotic nature of chloroplast ribosomes was recently illustrated when it was shown that antibodies prepared against the small subunit of *Chlamydomonas* chloroplast ribosomes cross-reacted with low molecular weight proteins from *E. coli* ribosomes (SCHNEEMAN and SURZYCKI 1979).

2.2 Transfer RNA

Initially it was demonstrated that pea leaves contain two sets of tRNA (ALIEV and FILIPOVICH 1968, BURKARD et al. 1969) and later it was found that $tRNA^{Met}$ from bean chloroplasts could be acylated and formylated by an enzyme isolated from the chloroplasts (BURKARD et al. 1969). This suggested that in the chloroplast, unlike the cytoplasm, peptide chain initiation used formyl-methionyl-tRNA. Initiation of peptide synthesis by $tRNA_f^{Met}$ has also been demonstrated in the chloroplasts of spinach (BIANCHETTI et al. 1971) and *Acetabularia* (BACHMAYER 1970), again emphasizing the prokaryotic nature of chloroplast protein synthesis.

In addition to both the cytoplasm and the chloroplasts containing distinct sets of tRNA's it is also known that isoaccepting species have been described from both sources for the majority of amino acids (LEA and NORRIS 1977, BARNETT et al. 1978; DRIESEL et al. 1979, WEIL 1979).

The prokaryotic nature of chloroplast protein synthesis is further exemplified by the fact that the chloroplast tRNAs can be charged by aminoacyl-tRNA-synthetases from chloroplasts, bacteria or blue-green algae (PARTHIER and KRAUSPE 1974, WEIL 1979). Cytoplasmic tRNAs, on the other hand, are charged by cytoplasmic enzymes but not by those of the chloroplasts, bacteria or mitochondria (GUILLEMAUT et al. 1975, GUILLEMAUT and WEIL 1975, JEANNIN et al. 1976, 1978, cf. also WEIL and PARTHER, Chap. 2, Vol. 14A, this Series).

2.3 Aminoacyl-tRNA-Synthetases

The activation of amino acids and their attachment to specific tRNAs is carried out by a series of aminoacyl-tRNA synthetases, each one of which is specific for a particular amino acid. In most cases in which the synthetases for a particular amino acid from the cytoplasm and chloroplast have been compared, they have been shown to be different. This has been carried out for some synthetases from peas (ALIEV and FILIPOVICH 1968, WRIGHT et al. 1974), *Euglena* (REGER et al. 1970, KRAUSPE and PARTHIER 1974, LEA and NORRIS 1977), *Phaseolus* (GUILLEMAUT and WEIL 1975, GUILLEMAUT et al. 1975, JEANNIN et al. 1976) and cotton (BRANTNER and DURE 1975, for details see WEIL and PARTHIER, Chap. 2, Vol. 14A, this Series.

2.4 Regulatory Factors

In addition to the components detailed above, both prokaryotic and eukaryotic protein synthesis require a variety of factors (LUCAS-LENARD and LIPMANN 1971, WEISSBACH and OCHOA 1976). While the basic mechanisms in both systems are similar, the additional factors are quite different. The greater variety of the eukaryotic factors probably reflects the greater complexity required for the regulation of protein synthesis in higher organisms.

While three initiation factors have been characterized in prokaryotes (LUCAS-LENARD and LIPMANN 1971) and at least six in animals (WEISSBACH and OCHOA 1976), their occurrence in plants is not as well studied. Several factors resembling those from animals have been found in wheat germ (SEAL et al. 1977, SPREMULLI et al. 1977), but so far none from chloroplasts has been characterized.

Two bacterial (FELICETTI and LIPMANN 1968) and two animal cytoplasmic (LIN et al. 1969) protein synthesis elongation factors have been described. One cytoplasmic and one chloroplast elongation factor have been partially purified from *Chlorella vulgaris* (CIFERRI and TIBONI 1973, 1976), while two factors from spinach chloroplasts have been described (TIBONI et al. 1978). One of the two cytoplasmic elongation factors has also been purified from wheat germ (GOLINSKA and LEGOCKI 1973).

2.5 Messenger RNA

Although mRNA is not, strictly speaking, a component of the chloroplast protein-synthesizing system but rather the means of programming it, it seems useful to compare the mRNA's from chloroplasts with those from the cytoplasm.

It is generally assumed that plant cytoplasmic mRNA's are "eukaryotic" in nature although not a great deal of specific evidence for this has been reported. The most significant features of eukaryotic mRNA have been recently described by HALL (1979). Many eukaryotic mRNA's have a "cap" of m^7 GpppN- at the 5′-end (cf. DYER, Chap. 6, this Vol.). Although such caps have been found in some plant viral mRNAs, it is not certain that such features are essential for the translation of such messengers. An important feature of eukaryotic mRNA is the existence of a poly(A) sequence at the 3′-end. This poly(A) tail, which is of variable length, may be used as a means of purifying mRNAs by hybridizing them via the poly(A) sequence to poly(U) or oligo(dT) linked to cellulose. This method has been applied to the isolation of mRNA's for the small subunit of RuBP carboxylase (DOBBERSTEIN et al. 1977, CHUA and SCHMIDT 1978b, SMITH and ELLIS 1979), ferredoxin (HUISMAN et al. 1978), the chlorophyll a/b apoprotein (APEL and KLOPPSTECH 1978) and a subunit of chloroplast ATPase (NELSON et al. 1980). Although the method is a very useful one for separation of mRNA from other RNA species (e.g., ribosomal RNA) it must be borne in mind that not all cytoplasmic mRNAs contain poly(A) tails (HALL 1979). The term poly(A)$^+$-RNA is often equated to cytoplasmic mRNA, as opposed to chloroplast mRNA, which has been reported to contain no poly(A) tail (WHEELER and HARTLEY 1975) or only very short stretches. GRAY and CASHMORE (1976) have examined the poly(A) content of mRNA's isolated from cytoplasmic polysomes. They found that only 36% of this fraction contained poly(A) as determined by its affinity for poly(U) cellulose under stringent conditions. They also reported that the major polypeptide products resulting from translation of the poly(A)$^+$ and poly(A)$^-$ fractions appeared to be identical.

The earlier report that the major species of chloroplast mRNA lack poly(A) sequences (WHEELER and HARTLEY 1975), was recently confirmed by MILNER et al. (1979) who found that none of the mRNA isolated from chloroplast polysomes of *Euglena* contained poly(A). They also reported that 7.7% of the total cell poly(A)$^+$-RNA hybridized to chloroplast DNA The amount of poly(A)$^+$-RNA hybridizing to chloroplast DNA was reduced to 2.8% on competition with chloroplast ribosomal RNA. HAFF and BOGORAD (1976) also reported that about 6% of total poly(A)$^+$-RNA from maize seedlings hybridized to maize chloroplast DNA but no attempt was made to determine if this RNA had messenger activity. The significance of these findings is not easy to evaluate, particularly as 3% of *E. coli* mRNA contains poly(A) sequences (NAKAZATO et al. 1975).

3 Techniques Used for the Study of the Biosynthesis of Chloroplast Polypeptides

Several experimental techniques have been developed and applied to the study of the biosynthesis of chloroplast proteins. To date, the most fruitful contributions have come from three experimental approaches:

1. In vivo labelling of proteins in mutants such as those deficient in plastid ribosomes or of wild-type plants in the presence of specific inhibitors of protein or RNA synthesis.
2. Examining the amount and nature of proteins synthesized in isolated organelles.
3. The in vitro synthesis of proteins in heterologous systems using chloroplast or cytosolic RNA or DNA as templates.

The application of recombinant DNA techniques combined with in vitro protein synthesis in systems using DNA as templates is already yielding detailed information concerning the coding sites for chloroplast proteins (COEN et al. 1977, MALNOË et al. 1979, WHITFELD and BOTTOMLEY 1980, DRIESEL et al. 1980). It is to be expected that sequencing will permit the organization of entire genomes to be studied. This technique, which is already being successfully applied in the mitochondrial field (NOBREGA and TZAGOLOFF 1980, BARRELL et al. 1980), will also greatly facilitate the elucidation of the factors and mechanisms involved in the regulation of gene expression as well as in the synthesis, transport, assembly and functioning of the chloroplast components. In this section, a brief outline will be given of the three above approaches as applied to the synthesis of chloroplast proteins.

3.1 Synthesis of Chloroplast Proteins in Vivo

The basis of this method is to follow the synthesis of specific proteins in vivo under conditions which remove, or substantially reduce, the contribution of either cytosol or chloroplast protein synthesis. Usually one of the systems available is blocked either by the application of a specific inhibitor of protein synthesis or by a mutation. The mutant approach is not widely used since it has to rely on well-characterized mutants with defined blocks in the synthetic capacity. This is seldom achieved with plastid mutants, many of which exhibit multiple biochemical defects for which the genetic lesion is unknown (cf. HENNINGSEN and STUMMANN, Chap. 15, this Vol.).

The inhibitor approach relies on the premise that prokaryotic (70S) and eukaryotic (80S) systems are distinguishable by their sensitivity to certain antibiotics and that the chloroplast protein-synthesizing system is indeed prokaryotic in nature.

The use of inhibitors in the study of protein synthesis has been reviewed (PESTKA 1971, 1976). The inhibitors most frequently used for the study of in vivo synthesis of chloroplast proteins are cycloheximide for the inhibition of cytosolic synthesis (HOOBER and BLOBEL 1969, AVADHANI and BUETOW 1972) and D-threo-chloramphenicol, lincomycin, spectinomycin or streptomycin to inhibit the chloroplast system (SPENCER and WILDMAN 1964; ELLIS 1969, ELLIS and HARTLEY 1971, MACHOLD and AURICH, 1972, CHUA et al. 1973, SCHWARTZBACH and SCHIFF 1974, cf. GALLING, Chap. 18, this Vol.).

Table 1 contains a list of the effects of inhibitors of chloroplast and cytoplasmic protein synthesis on a variety of chloroplast polypeptides. As can be seen, while many of the results agree with one another, there are a number of conflicting conclusions regarding the site of synthesis of some proteins.

Table 1. Sites of synthesis of chloroplast proteins as deduced from studies using inhibitors of protein synthesis

Component	Organism	Inhibitor	Effect	Site of synthesis	References
RuBP carboxylase	*Chlamydomonas reinhardii*	Spectinomycin	+	Chloropl. and cytosol	Armstrong et al. (1971)
		Cycloheximide	+		
		Chloramphenicol	+	Chloropl. and Cytosol	Margulies (1971)
		Spectinomycin	+		
		Cycloheximide	+		
	Phaseolus vulgaris	Chloramphenicol	+	Chloropl.	Ireland and Bradbeer (1971)
	Euglena gracilis	Streptomycin	+	Chloropl.	Schiff (1971)
		Chloramphenicol	+	Chloropl.	Schiff (1971)
		Cycloheximide	+	Cytosol	Davis and Merrett (1975)
	Pisum sativum	Lincomycin	+	Chloropl.	Ellis and Hartley (1971)
	Nicotiana tabacum	Chloramphenicol	+	Chloropl.	Beisenherz and Koth (1975)
		Cycloheximide	+	Cytosol	Beisenherz and Koth (1975)
	Hordeum vulgare	Chloramphenicol	+	Chloropl.	Criddle et al. (1970)
		Cycloheximide	+	Cytosol	Criddle et al. (1970)
Synthesis of the LSU	*Pisum sativum*	Chloramphenicol	+	Chloropl.	Cashmore (1976)
	Hordeum vulgare	Chloramphenicol	+	Chloropl.	Criddle et al. (1970)
Synthesis of the SSU	*Pisum sativum*	Cycloheximide	+	Cytosol	Cashmore (1976)
	Hordeum vulgare	Cycloheximide	+	Cytosol	Criddle et al. (1970)
Ribosomal proteins (some)	*Acetabularia*	Cycloheximide	+	Cytosol	Kloppstech and Schweiger (1974)
Aminoacyl-tRNA synthetases	*Euglena gracilis*	Cycloheximide	+	Cytosol	Hecker et al. (1974)
		Streptomycin	–	Cytosol	Hecker et al. (1974)
	Euglena gracilis	Cycloheximide	+	Cytosol	Parthier (1973)
RNA polymerase	*Pisum sativum*	Lincomycin	–	Cytosol	Ellis and Hartley (1971)
ALA synthetase ALA dehydrase porphobilinogenase chlorophyllase	*Nicotiana tabacum*	Chloramphenicol	–	Cytosol	Beisenherz and Schneider (1974)
		Cycloheximide	+	Cytosol	Beisenherz and Schneider (1974)
Fatty acid synthetase	*Euglena gracilis*	Cycloheximide	–	Chloropl.	Ernst-Fonberg et al. (1974)
	Chlamydomonas	Spectinomycin	+	Chloropl.	Sirevag and Levine (1972)

Phosphoglycerate kinase	*Euglena gracilis*	Streptomycin	–	Cytosol	SCHIFF (1971)
	Phaseolus vulgaris	Chloramphenicol	–	Cytosol	IRELAND and BRADBEER (1971)
	Pisum sativum	Lincomycin	–	Cystosol	ELLIS and HARTLEY (1971)
Triosephosphate -dehydrogenase (NADP)	*Phaseolus vulgaris*	Chloramphenicol	–	Cytosol	IRELAND and BRADBEER (1971)
	Pisum sativum	Lincomycin	–	Cytosol	ELLIS and HARTLEY (1971)
	Euglena gracilis	Streptomycin	–	Cytosol	SCHIFF 1971
Triosephosphate isomerase	*Phaseolus vulgaris*	Chloramphenicol	–	Cytosol	IRELAND and BRADBEER (1971)
FDP aldolase	*Phaseolus vulgaris*	Chloramphenicol	–	Cytosol	IRELAND and BRADBEER (1971)
Transketolase	*Phaseolus vulgaris*	Chloramphenicol	–	Cytosol	IRELAND and BRADBEER (1971)
Ribose-phosphate isomerase	*Phaseolus vulgaris*	Chloramphenicol	–	Cytosol	IRELAND and BRADBEER (1971)
	Pisum sativum	Lincomycin	–	Cytosol	ELLIS and HARTLEY (1971)
Phosphoribulokinase	*Phaseolus vulgaris*	Chloramphenicol	+	Chloropl.	IRELAND and BRADBEER (1971)
	Pisum sativum	Lincomycin	–	Cytosol	ELLIS and HARTLEY (1971)
	Chlamydomonas	Spectinomycin	–	Cytosol	ARMSTRONG et al. (1971)
Ferredoxin	*Chlamydomonas reinhardii*	Cycloheximide	+	Cytosol	ARMSTRONG et al. (1971)
		Spectinomycin	–	Cytosol	ARMSTRONG et al. (1971)
		Cycloheximide	+	Cytosol	ARMSTRONG et al. (1971)
	Euglena gracilis	Cycloheximide	+	Cytosol	MATSON and KIMURA (1976)
	Phaseolus vulgaris	Chloramphenicol	–	Cytosol	HASLETT et al. (1973)
		Lincomycin	–	Cytosol	HASLETT et al. (1973)
Ferredoxin-NADP- -reductase	*Chlamydomonas reinhardii*	Chloramphenicol	–	Cytosol	ARMSTRONG et al. (1971)
		Spectinomycin	–	Cytosol	ARMSTRONG et al. (1971)
Cytochrome f	*Phaseolus vulgaris*	Chloramphenicol	+	Chloropl.	GREGORY and BRADBEER (1973)
	Euglena gracilis	Chloramphenicol	+	Chloropl.	SMILLIE et al. (1971)
		Cycloheximide	+	Cytosol	SMILLIE et al. (1971)
	Chlamydomonas reinhardii	Chloramphenicol	+	Chloropl.	ARMSTRONG et al. (1971)
		Spectinomycin	+	Chloropl.	ARMSTRONG et al. (1971)
Cytochrome b_6 (b_{563})	*Chlamydomonas reinhardii*	Spectinomycin	+	Chloropl.	ARMSTRONG et al. (1971)
		Cycloheximide	+	Cytosol	ARMSTRONG et al. (1971)
	Phaseolus vulgaris	Chloramphenicol	+	Chloropl.	GREGORY and BRADBEER (1973)
	Euglena gracilis	Cycloheximide	+	Cytosol	WILDNER (1976)
		Chloramphenicol	–	Cytosol	WILDNER (1976)
Cytochrome b-559_{LP}	*Phaseolus vulgaris*	Chloramphenicol	+	Chloropl.	GREGORY and BRADBEER (1973)

Table 1 (continued)

Component	Organism	Inhibitor	Effect	Site of synthesis	References
Cytochrome b-552	*Euglena gracilis*	Streptomycin	+	Chloropl.	Schiff (1971)
	Euglena gracilis	Chloramphenicol	+	Chloropl.	Wildner (1976)
	Euglena gracilis	Cycloheximide	+	Chloropl.	Wildner (1976)
	Phaseolus vulgaris	Streptomycin	+	Chloropl.	Freyssinet et al. (1978, 1979)
Protein synthesis elongation factor EF-G_{chl}	*Chlorella vulgaris*	Cycloheximide	−	Chloropl.	Ciferri and Tiboni (1976)
		Chloramphenicol	+	Chloropl.	Ciferri and Tiboni (1976)
	Euglena gracilis	Cycloheximide	+	Cytoplasm	Breitenberger et al. (1979)
		Streptomycin	−	Cytoplasm	Breitenberger et al. (1979)
		Spectinomycin	−	Cytoplasm	Breitenberger et al. (1979)
Plastocyanin	*Phaseolus vulgaris*	Chloramphenicol	−	Cytosol	Haslett and Cammack (1974)
		Lincomycin	−	Cytosol	Haslett and Cammack (1974)
Phosphoglycollate phosphatase	*Euglena gracilis*	Cycloheximide	−	Chloropl.	Davis and Merrett (1975)
		Chloramphenicol	+	Chloropl.	Davis and Merrett (1975)
	Phaseolus vulgaris	Chloramphenicol	−	Cytosol	Murray et al. (1973)
Glyoxylate reductase (NADP)	*Phaseolus vulgaris*	Chloramphenicol	−	Cytosol	Murray et al. (1973)
ATPase (CFl)	*Phaseolus vulgaris*	Chloramphenicol	+	Chloropl.	Horak and Hill (1972)
		Cycloheximide	+	Cytosol	Horak and Hill (1972)
ATPase γ and δ subunits	*Pisum sativum*	Chloramphenicol	−	Cytoplasm	Bouthyette and Jagendorf (1978)
Chlorophyll-protein complex I	*Pisum sativum*	Lincomycin	+	Chloropl.	Ellis 1975
		Cycloheximide	+	Cytosol	Ellis 1975
	Vicia faba	Chloramphenicol	+	Chloropl.	Machold and Aurich (1972)
Chlorophyll-protein complex II	*Vicia faba*	Chloramphenicol	−	Cytosol	Machold and Aurich (1972)
		Cycloheximide	+	Cytosol	Machold and Aurich (1972)
	Pisum sativum	Lincomycin	−	Cytosol	Ellis (1975)
		Cycloheximide	+	Cytosol	Ellis (1975)
	Chlamydomonas reinhardii	Cycloheximide	+	Cytosol	Hoober (1970)

ELLIS (1975) measured the amounts of membrane proteins and RuBP carboxylase which accumulate in detached etiolated pea shoots that were caused to green in the presence of either the 70S ribosome-specific inhibitor lincomycin or of 2-(4-methyl-2,6-dinitroanilino)-N-methylpropionamide (MDMP) which inhibits the cytosolic 80S ribosome system. It was observed that the accumulation of protein belonging to photosystem I (PS I) was completely prevented by lincomycin but that PS II proteins were unaffected. However, proteins of PS I had not been observed as products of protein synthesis in isolated pea chloroplasts and proteins of both PS I and PS II were inhibited by cycloheximide. From this it was concluded that both PS I and PS II proteins are synthesized on cytosolic ribosomes, but that the insertion of PS I proteins into the chloroplast lamellae required a product of chloroplast protein synthesis. He also observed that MDMP inhibited the synthesis of both undissociated RuBP carboxylase and the large subunit of this protein. Inhibition of the large subunit was postulated to be due to a requirement for the presence of the small subunit as a prerequisite for its synthesis.

Similar results were obtained with green pea leaves by CASHMORE (1976). His results, however, differed in as much as he found that, in the presence of cycloheximide, small amounts of the large subunit of RuBP carboxylase were synthesized. In addition, he observed an inhibition of several chloroplast lamellar proteins in the presence of chloramphenicol and concluded that these are synthesized within the chloroplast.

CHUA and GILLHAM (1977) separated 33 chloroplast membrane polypeptides from *Chlamydomonas reinhardtii* on polyacrylamide gels and examined the sensitivity of their synthesis to a number of inhibitors. They concluded that at least nine of these were made on chloroplast ribosomes including two polypeptides from the PS I and PS II reaction centres.

In an attempt to demonstrate that two of the subunits (γ and δ) of the CF_1 part of chloroplast ATPase were made in the cytoplasm, BOUTHYETTE and JAGENDORF (1978) studied the synthesis of those subunits in excised pea shoots. They found that synthesis of the γ and δ subunits was inhibited by the presence of cycloheximide but not lincomycin, and concluded that these two subunits were indeed cytoplasmically synthesized. In contrast, NELSON et al. (1980) recently found the γ subunit in addition to α, β and ε (see Sect. 2.3.2) among the products of protein synthesis of isolated spinach chloroplasts and so have concluded that it is made on the chloroplast ribosomes.

Chlamydomonas mutant *ac-20*, which becomes deficient in chloroplast ribosomes when grown on a medium containing organic carbon, has been used by GIVAN (1979) to demonstrate tight coupling between the synthesis of the large and small subunits of RuBP carboxylase. When grown under conditions which result in the absence of chloroplast ribosomes, neither the large nor the small-subunits of RuBP carboxylase could be detected. When transferred to a medium which allowed the accumulation of chloroplast ribosomes, both subunits were synthesized. However, in the presence of chloramphenicol the recovery of the synthetic capacity for both subunits was inhibited.

FEIERABEND and WILDNER (1978), on the other hand, examined the synthesis of RuBP carboxylase in a rye variety which lacks chloroplast ribosomes when grown at a temperature in excess of 32 °C. They found that, under these conditions, there was an accumulation of free small subunits, suggesting that the coupling between the synthesis of the two subunits was not very tight. This again contrasts with other evidence from higher plants which suggests that the lack of large subunits has a repressive effect on the synthesis of the small subunit. HALLIER et al. (1978) found that a mutant of *Oenothera hookeri*, in which the synthesis of the large subunit of RuBP carboxylase is blocked by a mutation in the plastid DNA, failed to synthesize any small subunit.

It is obvious from the examples cited, that great care must be taken in interpreting the results of inhibitor studies in vivo (cf. GALLING, Chap. 18, this Vol). Many of the inhibitors used have been shown to affect other processes at the concentrations used (PESTKA 1976, and see ELLIS 1977a for a table of known activities affecting processes other than protein synthesis). A further drawback is the lack of knowledge of the inhibitor concentration at the site of action since the antibiotics vary in their ability to cross membranes as well as in their susceptibility to enzymatic, photolytic or oxidative destruction. In addition, it cannot be assumed that chloroplast components will behave in the same way as bacterial systems towards inhibitors. One striking example is the case with the inhibitor of the DNA-dependent RNA polymerase in *E. coli*, rifampicin. It has virtually no effect on maize chloroplast RNA polymerase even at high concentrations while it inhibits the bacterial RNA polymerase completely at concentrations as low as 2 μg/ml (BOTTOMLEY et al. 1971). At least in this instance the chloroplast does not react in a typical prokaryotic manner.

A great deal of work has been carried out using inhibitors to study variations in various chloroplast enzyme activities, particularly in studies on plant development. Because of the difficulty of distinguishing, in those reports, between the effects of the inhibitor on enzyme activation rather than enzyme synthesis, we will not attempt to summarize the work in this area.

3.2 Protein Synthesis in Isolated Chloroplasts

The most rigorous method, to date, of demonstrating the site of synthesis of chloroplast proteins has been the use of isolated organelles. If it can be shown that a polypeptide is synthesized by isolated chloroplasts in the absence of cytosolic contamination, then it can be assumed that the polypeptide is synthesized in the organelle in vivo.

One early problem associated with the use of isolated organelles was the possibility that proteins not related to chloroplast proteins could be synthesized by contaminating cytosolic inclusions or bacteria (APP and JAGENDORF 1964). However, the development of newer methods for the isolation of organelles has largely solved this problem (JENSEN and BASSHAM 1966, RAMIREZ et al. 1968, HEBER 1973, LARSSON and ALBERTSSON 1974, SALISBURY et al. 1975). These improved methods yield preparations of plastids with high proportions of the organelles which still have intact outer membranes and are almost completely devoid of contaminating cytosolic components, as judged either by microscopic examination or the measurement of biochemical activities. One cautionary note should be sounded, however. Under unfavourable conditions, usually when concentrated homogenates of young tissue are being prepared, it is possible to generate "multiorganelle complexes" containing other organelles, such as nuclei and mitochondria, in addition to chloroplasts embedded in a matrix of cytoplasm and apparently surrounded by a membrane (LARSSON and ALBERTSSON 1974, SCHMITT and HERRMANN 1977).

Although the identification of a particular protein among the products of polypeptide synthesis in isolated chloroplasts provides strong evidence for them being the site of synthesis in vivo, the failure to observe such synthesis cannot be used as evidence that a particular polypeptide is not synthesized within the chloroplast in vivo. It is possible that in a particular chloroplast preparation at the time of isolation, some regulatory mechanism might prevent the synthesis of a polypeptide. For example, isolated maize mesophyll chloroplasts do not synthesize any large subunit protein of RuBP carboxylase whereas the bundle-sheath plastids do (LINK et al. 1978). In *Euglena* chloroplasts (CHELM et al. 1978) as well as spinach chloroplasts (DRIESEL et al. 1980) the steady-state amounts of mRNAs detected by hybridization against chloroplast DNA fragments favours the concept of differential activity of the genes rather than that of a uniform transcription of the total DNA. Also, there is ample evidence that in chloroplasts changes in the amounts of mRNAs for specific proteins occur according to their stage of development (ROSNER et al. 1975, BEDBROOK et al. 1978, TOBIN 1978, APEL 1979, SILVERTHORNE and ELLIS 1980). Hence there may be many proteins whose synthesis is switched off or which are being made in isolated chloroplasts in amounts too low to be detected by present methods. In addition, it is possible that the synthesis of some peptides within the chloroplast depends on a continuous interaction with the cytosol and thus would not be observed in isolated chloroplasts. Finally it should also be mentioned that synthesis in isolated organelles might be mainly "run-off" synthesis of peptide chains already started at the time of isolation of the organelles. Measurement of the generation of N-formyl [^{35}S]-methionylpuromycin in the light in chloroplasts has, however led to the claim that initiation of new chains does occur in isolated organelles (HIGHFIELD and ELLIS 1976).

The highest number of soluble proteins which become labelled during synthesis in isolated chloroplasts was reported by ELLIS et al. (1978). After labelling isolated pea chloroplasts with [^{35}S]-methionine, up to 80 radioactive spots could be resolved by two-dimensional electrophoresis of soluble proteins released from these chloroplasts. When the total chloroplast proteins were separated on SDS-polyacrylamide-gradient gels only 37 radioactive bands could be observed. The radioactivity of the 80 spots did not correspond with the stained pattern of abundant soluble chloroplast proteins. This led the authors to speculate that these minor soluble products represent a different set of soluble proteins than the set normally seen. However, it is possible that these spots represent, in part, products of incomplete synthesis, degradation products, precursors to functional proteins, deaminated or otherwise modified proteins, or even proteins that are normally membrane-bound but which are soluble at some stage during synthesis. In addition to these soluble proteins the synthesis of at least 12 membrane-bound proteins has been detected in isolated spinach chloroplasts (BOTTOMLEY et al. 1974). The idea of a relatively large number of peptides synthesized in isolated chloroplasts is stimulating and should be followed in more detail.

The synthesis of a polypeptide within chloroplasts does not necessarily mean that the coding sequence for that polypeptide is located in the chloroplast DNA. Although in the three cases so far proven, i.e., the large subunit of RuBP carboxylase, a 32,000-M_r membrane component and cytochrome f (see below), the protein synthesized in the organelle has been shown to be coded for by chloroplast DNA, it remains possible for mRNA from the nucleus or even the mitochondrion to enter the chloroplast and be used in protein synthesis. It has already been proposed (BRADBEER et al. 1979) that RNA is transported from the chloroplast to the cytosol as a means of influencing transcription

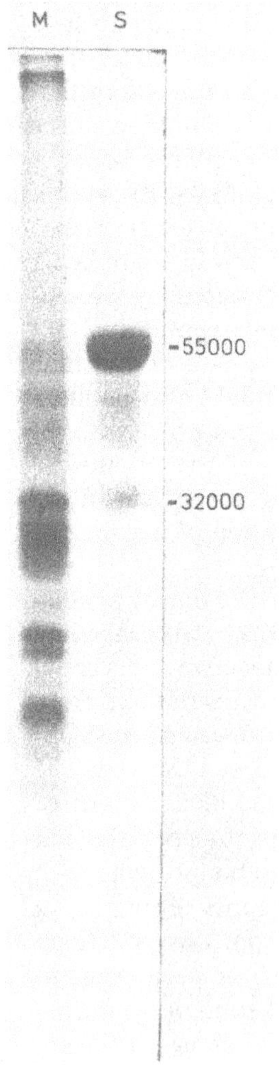

Fig. 1. Light-driven protein synthesis in isolated spinach chloroplasts. Incorporation of [^{35}S]-methionine was carried out as described by BOTTOMLEY et al. (1974). Chloroplasts were lysed in water and the membrane and supernatant fractions separated by centrifugation. The products were separated on a 15%–25% SDS polyacrylamide gel. *M*, membrane fraction; *S*, supernatant fraction. The position of the large subunit of RuBP carboxylase (55,000 mol.wt.) and the 32,000-M_r membrane protein are indicated

and translation in that compartment. An analogous case is the uptake of DNA molecules by Ca^{2+}-treated *E. coli* cells (MANDEL and HIGA 1970), which is now widely used to introduce DNA fragments into bacteria. It has recently been suggested (GRINIUS 1980) that this uptake of DNA is achieved by a transport process driven by an ion gradient such as is known to exist in chloroplasts. On the other hand, the existence of the cistrons coding for ribosomal RNAs as well as tRNAs within the chloroplast genome could indicate that RNA cannot be transported across the outer envelope so that all the RNAs required by the plastid must be synthesized within the organelle.

Isolated organelles have been used in two ways to study protein synthesis. One method is to isolate and label chloroplasts in isoosmotic solutions of sorbitol, sucrose or KCl, keeping the plastids as intact as possible. In this case light may be used as the energy source to drive the reaction via the ATP generated during photophosphorylation (BLAIR and ELLIS 1973, BOTTOMLEY et al. 1974) (Fig. 1). In the second method, isolated chloroplasts are resuspended in a medium without osmoticum in order to remove a permeability barrier to the introduction of substrates. The incorporation of amino acids into protein then requires ATP as the energy source and is largely unaffected by light (BLAIR and ELLIS 1973, BOTTOMLEY et al. 1974).

While both methods appear to yield similar products, the conditions for optimal synthesis vary (BOTTOMLEY et al. 1974). For example, light-driven protein synthesis in chloroplasts prepared in sorbitol is inhibited by concentrations of Mg^{2+} as low as 1 mM whereas those chloroplasts in 0.2 M KCl require around 7 mM Mg^{2+} for optimal synthesis. ATP-driven synthesis in the dark also has a requirement for added Mg^{2+}. While sodium ions are inhibitory to the light-driven system and potassium ions strongly stimulatory, the ATP-driven system is less sensitive to monovalent ions although it gives optimal

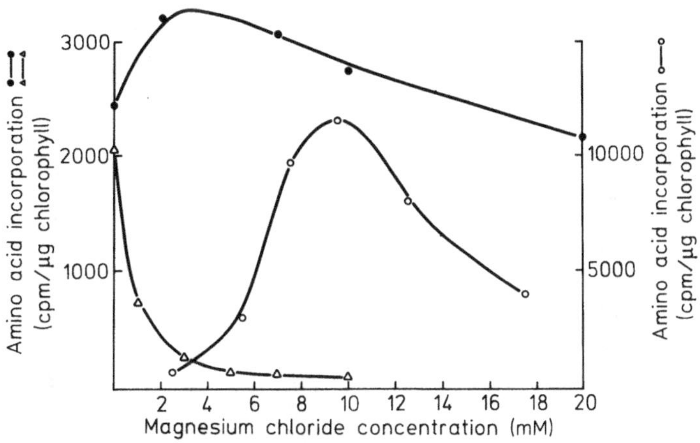

Fig. 2. Effect of $MgCl_2$ concentration on protein synthesis in isolated spinach chloroplasts; (△—△) light-driven synthesis in intact chloroplasts prepared in sorbitol medium; (●—●) light-driven intact chloroplasts prepared in KCl medium; (○—○) ATP-driven synthesis in broken chloroplasts. (BOTTOMLEY et al. 1974)

Table 2. Chloroplast proteins synthesized in isolated organelles

Polypeptide	Molecular weight	Source of chloroplasts	Method of[a] identification	Reference
Large subunit of RuBP carboxylase	55,000	Pea	Tryptic	BAIR and ELLIS (1973)
	55,000	Spinach	PAGE	BOTTOMLEY et al. (1974)
	55,000	*Euglena gracilis*	PAGE	VASCONCELOS (1976)
ATPase F_1 $\alpha, \beta, \varepsilon$	59,000 55,000 13,000	Spinach	PAGE	MENDIOLA-MORGENTHALER et al. (1976)
ATPase F_1 $\alpha, \beta, \gamma, \varepsilon$	59,000 55,000 37,500 13,000	Spinach	Immunol.	NELSON et al. (1980)
ATPase F_1 α, β	59,000 55,000	Maize	PPM	GREBANIER et al. (1978)
ATPase F_0 I	15,500	Spinach	Immunol.	NELSON et al. (1980)
ATPase F_0 DCCD binding protein	8,000	Pea	PAGE and DCCD binding	DOHERTY and GRAY (1980)
Cytochrome f	37,300	Pea	PPM	DOHERTY and GRAY (1979)
Cytochrome b_{559}	6,000	Spinach	PAGE	ZIELINSKI and PRICE (1980)
Elongation factors EF-G and EF-T	77,000 45,500	Spinach	2-D IEF/PAGE	CIFERRI et al. (1979)
P-700-chlorophyll a-proteins	70,000 58,000	Sorghum	PAGE	GEETHA and GNANAM (1980)
	68,000	Spinach	PAGE	ZIELINSKI and PRICE (1980)
32,000-M_r membrane protein	32,000	Spinach	PAGE	BOTTOMLEY et al. (1974)
	32,000	Pea	PAGE	EAGLESHAM and ELLIS (1974)
	34,500 (precursor)	Maize	PAGE & PPM	GREBANIER et al. (1978)
	31,000	*Euglena gracilis*	PAGE	VASCONCELOS (1976)

[a] Tryptic = Tryptic peptide mapping; PAGE = Polyacrylamide gel electrophoresis; Immunol. = Immunological precipitation with antibodies; 2D IEF/PAGE = Two-dimensional isoelectric focussing and electrophoresis; PPM = Partial proteolysis

synthesis in the presence of ammonium ions. Ammonium ions, which uncouple photophosphorylation, inhibit the light-driven protein synthesis almost completely.

Synthesis of a number of proteins with known function in isolated organelles has been demonstrated. This could mean that also their site of coding is inside the envelope, on chloroplast DNA (see Sects. 3 and 6). A list of these polypeptides is given in Table 2 together with the plants from which the chloroplasts were prepared. In addition, several other unidentified peptides, presumably constituents of membranes, have been reported (BOTTOMLEY et al. 1974). It has been demonstrated in some cases that the amount of protein produced by isolated chloroplasts reflects the amount of mRNA present in the organelles at the time of isolation (SILVERTHORNE and ELLIS 1980).

3.3 In Vitro Polypeptide Synthesis

One indisputable proof that the coding site of a structural gene is located at a given site is the demonstration that a protein can be synthesized by transcribing a specific DNA fragment and subsequently translating the RNA formed into the protein. The product may then be compared with the protein made in vivo. Substantial knowledge concerning both the site of synthesis of chloroplast proteins and the location of the genes coding for them has come from the use of in vitro synthesis systems. We shall briefly describe the three major approaches.

1. A relatively straightforward way of assaying the capacity to synthesize proteins is a ribosome "run-off" system, starting with polysomes in which, after isolation, the peptide chains already initiated are allowed to elongate and terminate. Although such systems have been developed for chloroplast ribosomes, they are much more widely used with cytosolic ribosomes.
2. Isolated mRNA can be translated in heterologous in vitro systems. One prerequisite is that out of many products, one or more particular protein can be isolated and identified. Several systems have been developed to perform in vitro translation, including those prepared from *E. coli* (ZUBAY et al. 1970), wheat germ (ROBERTS and PATERSON 1973, MARCU and DUDOCK 1973) and rabbit reticulocytes (PELHAM and JACKSON 1976). Hybrid arrested translation, which has been introduced recently (PATERSON et al. 1977), has made possible the extension of this method to the localization of the gene for a particular mRNA.
3. Coupled transcription-translation or linked transcription-translation systems are becoming increasingly important. In the former RNA and protein are made in one assay, whereas in the latter RNA synthesized on DNA is used to stimulate protein synthesis only after reoptimizing the conditions. Two forms of coupled systems (ZUBAY et al. 1970, FUCHS 1976, BOTTOMLEY 1982) derived from *E. coli* and one linked system using *E. coli* RNA polymerase and a rabbit reticulocyte translation system (ROBERTS et al. 1975, COEN et al. 1978) have been described.

3.3.1 Polysomal Run-Off System

The basis of this technique is the isolation of polysomes which are in the process of translating mRNA in vivo and causing them to complete polypeptide chain elongation in vitro by supplying them with the necessary substrates, cofactors and an energy source. Incorporation of radioactive amino acids during elongation facilitates the identification of the completed polypeptide. The different classes of ribosomes, cytosolic or organellar, can either be separated by physical means or cross contamination eliminated by the use of the same inhibitors which distinguish the protein synthesis systems in in vivo experiments (see Sect. 2.3.1). The use of inhibitors in vitro has the advantage that the concentration of the inhibitor at the site of action can be readily controlled.

The polysomal "run-off" technique has been applied successfully to the identification of the site of synthesis of the small subunit of RuBP carboxylase (GOODING et al. 1973, GRAY and KEKWICK 1973 1974, ROY et al. 1976 1977). Although difficulties were encountered because of the small proportion of the total radioactivity that was incorporated into the small subunit and the apparent lack of specificity when antibodies were used to select the polysomes, it was concluded that the 80S ribosome system did in fact synthesize the small subunit. It has been shown by DOBBERSTEIN et al. (1977) that the small subunit of RuBP carboxylase is made in vitro as a precursor about 3,500 mol.wt. larger than the mature polypeptide. DOBBERSTEIN et al. (1977) also examined the polysomal run-off products from *Chlamydomonas* and found only the processed mature small subunit. This was found to be due to the presence of a processing endoprotease in the polysome preparations. It had previously been noted by ROY et al. (1976) that at 20,000 mol.wt. peptide was a "contaminant" in their immunoprecipitation, while ROY et al. (1977) attributed a 20,000M_r band to an endogenous product of the wheat germ system, used as a source of protein synthesis factors for the polypeptide run-off. It is possible that this polypeptide was, in fact, the precursor of the small subunit and that the polysomal preparations contained little or no endoprotease.

While the run-off technique has provided valuable information concerning the synthesis of chloroplast proteins in the past, it is apparent that the use of in vitro synthesis, using partially purified mRNAs in heterologous systems, will be the method of choice in the future.

3.3.2 Heterologous in Vitro Synthesis of Chloroplast Proteins

The use of the term heterologous in vitro synthesis here is confined to the synthesis of polypeptides in partly purified cell-free systems from heterologous sources, using either RNA or DNA as the template.

Heterologous in vitro translation usually depends on the use of a crude cell-free supernatant obtained after breaking the cells of an organism. This supernatant fraction should contain all the components required for peptide initiation, elongation and termination, and also be dependent on the addition of exogenous mRNA for optimal polypeptide synthesis. The most commonly used sources of these supernatant fractions are *E. coli* (NIRENBERG and MATTHAEI 1961, NIRENBERG 1963, MODOLELL 1971), wheat germ (ROBERTS and PAT-

TERSON 1973, MARCU and DUDOCK 1974) and rabbit reticulocytes (PELHAM and JACKSON 1976).

If DNA is to be used as a template for transcription and translation in vitro then the supernatant fraction must also contain DNA-dependent RNA polymerase. This has been achieved either using the endogenous polymerase in *E. coli* cells (ZUBAY et al. 1970, BOTTOMLEY 1982) or by first transcribing the DNA with *E. coli* polymerase and then reoptimising the conditions and adding the supernatant fraction from rabbit reticulocytes (ROBERTS et al. 1975, COEN et al. 1978).

The products obtained from the coupled in vitro transcription-translation of chloroplast DNA from spinach, tobacco and *Oenothera hookeri* by the *E. coli* S-30 system are compared in Fig. 3 with those from transcription of total spinach chloroplast RNA.

The hybrid arrest method, which was first introduced by PATERSON et al. (1977), consists of translating a mRNA in a cell-free system and then repeating the assay, having first hybridized the mRNA to a particular DNA fragment. If, under these conditions, a particular polypeptide is eliminated from the products, it provides strong evidence that the DNA fragment contained the gene for that mRNA. Melting of the hybrid and again translating the mRNA should result in the reappearance of the polypeptide.

Translation of chloroplast mRNA has been carried out using protein-synthesizing systems from *E. coli* (HARTLEY et al. 1975, BOTTOMLEY et al. 1976) wheat germ (BOTTOMLEY et al. 1976, SAGHER et al. 1976) or rabbit reticulocytes (LINK et al. 1978, SILVERTHORNE and ELLIS 1980, DRIESEL et al. 1980), while the transcription-translation of DNA has used either *E. coli* (BOTTOMLEY and WHITFELD 1979, BOTTOMLEY 1982) or a linked *E. coli* polymerase/rabbit reticulocyte system (COEN et al. 1977).

The first identification of a chloroplast protein from the translation of chloroplast RNA was carried out by HARTLEY et al. (1975), who used total spinach chloroplast RNA to programme a cell-free protein synthesizing system from *E. coli*. Among the products was a polypeptide of 52,000 mol.wt. which was demonstrated to be the large subunit of RuBP carboxylase by comparison of the tryptic peptide map with that from authentic large subunit. In addition, they observed that a polypeptide of 35,000 mol.wt. was also synthesized. Both this peptide and the large subunit appeared to be of slightly larger molecular weight than the respective products synthesized in isolated chloroplasts.

SAGHER et al. (1976) used the wheat germ protein synthesizing system to translate the poly(A)-RNA fraction from *Euglena*. They found that the most prominent product was the large subunit of RuBP carboxylase, which they identified by two-dimensional electrofocussing and electrophoresis. The translation of poly(A)-RNA from *Chlamydomonas* (SANO et al. 1979) and *Spirodela* (REISFELD et al. 1978b) by the *E. coli* or the wheat germ system has also been demonstrated to yield the large subunit. The rabbit reticulocyte system has been used by LINK et al. (1978) to demonstrate that poly(A)-mRNA from bundle-sheath but not from mesophyll cells gives rise to the large subunit. This is consistent with the occurrence of Calvin cycle activity in the bundle-sheath and not in mesophyll cells (HATCH 1976).

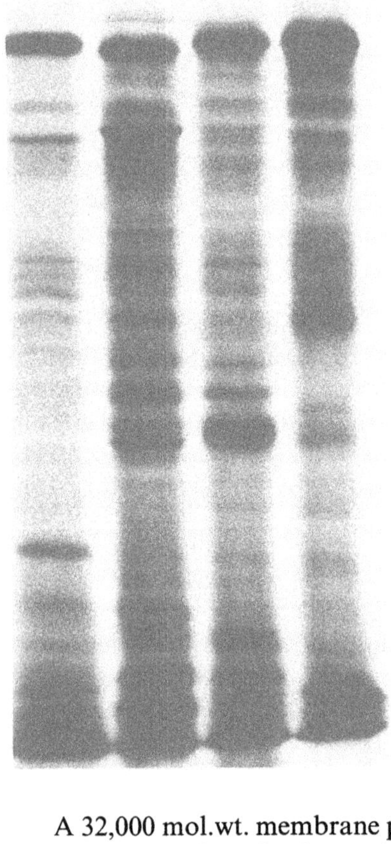

Fig. 3. Autoradiograph of a 15% SDS-polyacrylamide gel of the [^{35}S]-methionine-labelled products synthesized in the *E. coli* transcription-translation system programmed with: *L*eft to *R*ight (*Track 1*) chloroplast DNA from *Nicotiana tabacum;* (*Track 2*) chloroplast DNA from *Oenothera hookeri;* (*Track 3*) chloroplast DNA from spinach; (*Track 4*) chloroplast RNA from spinach. (BOTTOMLEY and WHITFELD 1979)

A 32,000 mol.wt. membrane polypeptide similar to that from spinach (HARTLEY et al. 1975) has also been observed in the translation products of *Spirodela* poly(A)-RNA using the wheat germ system (REISFELD et al. 1978a b) as well as from maize and spinach chloroplast RNA using the reticulocyte system (BEDBROOK et al. 1978, DRIESEL et al. 1980).

Finally, the small subunit of RuBP carboxylase has also been shown to be synthesized as a precursor during the translation of poly(A)$^+$-RNA in the wheat germ system (CASHMORE et al. 1978, DOBBERSTEIN et al. 1977, HIGHFIELD and ELLIS 1978).

Transcription-translation of a cloned chloroplast DNA fragment from maize has been carried out in a "linked" system consisting of a transcription assay

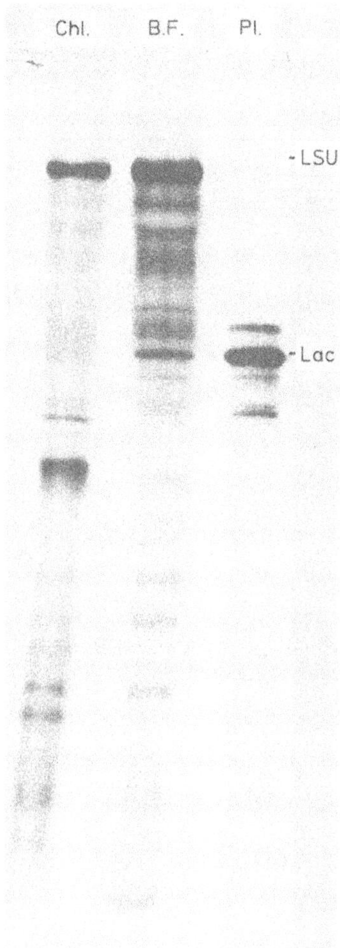

Fig. 4. Autoradiograph of a 15%–25% SDS-polyacrylamide gel of the products synthesized in the *E. coli* transcription-translation system programmed with: (*Chl*) total chloroplast DNA; (*BF*) chimeric cloned DNA containing the gene for the large subunit of RuBP carboxylase; (*Pl*) DNA from the plasmid vector alone. (*LSU*) Position of the 55,000-M_r large subunit of RuBP carboxylase; (*Lac*) position of the 27,000-M_r β-lactamase from the plasmid vector

with *E. coli* polymerase followed by a rabbit reticulocyte lysate translation system (COEN et al. 1977), while a coupled transcription-translation system from *E. coli* has been used for both total and cloned spinach chloroplast DNA (BOTTOMLEY and WHITFELD 1979, WHITFELD and BOTTOMLEY, 1980). In all cases the large subunit of RuBP carboxylase was identified among the products by comparison of the patterns of peptides obtained by partial proteolytic digestion with those from authentic large subunits.

Hybrid arrest translation has been used to demonstrate that the sequences homologous to those of the mRNA for a 32,000 mol.wt. light-induced protein are contained in the chloroplast DNA of maize (BEDBROOK et al. 1978), and that a gene for a protein of similar molecular weight is located on spinach chloroplast DNA (DRIESEL et al. 1980).

It must also be borne in mind that it is not necessary to have complete sequence homology between the mRNA and DNA to arrest the translation.

Table 3. Chloroplast proteins synthesized by in vitro systems

Polypeptide	Molecular weight	Precursor molecular weight	Organism
Large subunit of RuBP carboxylase	55,000	–	Spinach
	59,000	–	*Euglena gracilis*
	50,000–60,000	–	*Chlamydomonas reinhardii*
	52,000	–	*Spirodela oligorrhiza*
	55,000	–	Maize
	55,000	–	Spinach
	55,000	–	Spinach
	50,000–60,000	–	*Chlamydomonas reinhardii*
Small subunit of RuBP carboxylase	16,000	20,000	*Chlamydomonas reinhardii*
	14,000	20,000 18,000 16,000	Pea
	14,000	18,000	Spinach
Light-Harvesting chlorophyll a/b-protein	25,000	29,500	Barley
	28,000	33,000 32,000	Pea
	27,500 25,000 23,000		Sorghum
Ferredoxin	17,000	20,500	*Nicotiana* *Phaseolus* *Chlamydomonas*
P-700-chlorophyll a-protein	68,000	–	Sorghum
32,000-M_r membrane protein	32,000	–	Spinach
	32,000	34,500	Maize
	32,000	33,500	*Spirodela*
		34,000	*Chlamydomonas*

[a] Tryptic = Tryptic peptide mapping; PAGE = Polyacrylamide gel electrophoresis; Immunol. = Immunological precipitation with antibodies; 2D IEF/PAGE = 2 Dimensional isoelectric focussing and electrophoresis; PPM = Partial proteolysis

Template nucleic acid	Cell-free system	Method of identification[a]	Reference
chp RNA	*E. coli*	Peptide map	HARTLEY et al. (1975)
poly(A)$^-$ RNA	Wheat germ	2-D-IEF/ PAGE	SAGHER et al. (1976)
14S poly (A)$^-$ RNA	*E. coli*	Immunol. tryptic	HOWELL et al. (1977)
14S chloroplast RNA	*E. coli* and wheat germ	PPM	REISFELD et al. (1978b)
Cloned DNA fragment	*E. coli*/ rabbit retic.	PPM	COEN et al. (1977)
Total chp DNA	*E. coli*	PPM	BOTTOMLEY and WHITFELD (1979)
Cloned DNA fragment	*E. coli*	PPM	WHITFELD and BOTTOMLEY (1980)
Cloned DNA	*E. coli*	Immunol. tryptic	MALNOË et al. (1979)
poly (A)$^+$ RNA	Wheat germ	Immunol.	DOBBERSTEIN et al. (1977)
poly (A)$^+$ RNA	Wheat germ	Immunol. tryptic	CASHMORE et al. (1978)
poly (A)$^+$ RNA	Wheat germ	Immunol.	CHUA and SCHMIDT (1978a)
poly (A)$^+$ RNA	Wheat germ	Immunol.	APEL and KLOPPSTECH (1978)
poly (A)$^+$ RNA	Wheat germ and rabbit retic.	Immunol.	SCHMIDT et al. (1980)
total leaf RNA	Sorghum chloroplast	PAGE	GEETHA and GNANUM (1980)
poly (A)$^+$ RNA	Wheat germ	Immunol.	HUISMAN et al. 1978)
Total leaf RNA	Sorghum chloroplast	PAGE	GEETHA and GNANAM (1980)
chp RNA	*E. coli* Rabbit retic.	PAGE	HARTLEY et al. (1975) DRIESEL et al. (1980)
chp RNA	Rabbit retic.	PAGE	BEDBROOK et al. (1978)
chp RNA	Wheat germ	PPM	REISFELD et al. (1978b)
Cloned chp DNA	*E. coli*	PAGE	MALNOË et al. (1979)

Although the method does provide evidence for the location of the gene, it remains possible that coincidental homology could yield false results.

Table 3 lists the chloroplast proteins which have been reported to be products of various in vitro protein-synthesizing systems.

The identification of a particular protein among the products of in vitro translation of a mRNA fraction allows the conclusion that the mRNA contained the information for that protein. Information regarding the site of synthesis in vivo depends on the methods used for the isolation of the mRNA. When purified chloroplasts are the source of RNA, then it can be concluded that the chloroplast is the site of synthesis in vivo. However, since the observation by WHEELER and HARTLEY (1975) that spinach chloroplasts contain no poly(A)$^+$-RNA, it has become usual to equate poly(A) containing RNA with cytosolic RNA. While in most cases this appears to have been justified, the possibility remains that some chloroplast mRNA's contain poly(A) sequences. That this is indeed the case has already been demonstrated in maize chloroplasts by HAFF and BOGORAD (1976) and in *Euglena* chloroplasts by MILNER et al. (1979).

The absence of a protein from the translation products of a mRNA preparation does not exclude the possibility that the preparation contains that mRNA. The use of heterologous systems implies that there is a large degree of commonality between the mechanisms of protein synthesis of different organisms. That this is not necessarily so has already been reported by BOTTOMLEY et al. (1976) who found that, when crude RNA preparations were used, a cell-free system from *E. coli* appeared to translate preferentially chloroplast mRNAs whereas one from wheat germ preferred cytoplasmic messengers. REISFELD et al. (1978b) also found that, using chloroplast RNA from *Spirodela,* the *E. coli* system was most efficient for the synthesis of the large subunit of RuBP carboxylase, while a wheat germ translation system gave the 32,000-M_r membrane protein as the major product. Although it is tempting to interpret the differential translation of certain mRNA's in terms of them being either prokaryotic or eukaryotic in nature, further evidence is needed to determine whether these differences in translatability indicate significant biological differences or whether they have a more trivial explanation.

4 The Site of Synthesis of Chloroplast Proteins

We have estimated that the number of proteins required to carry out the various functions of chloroplasts is over 200 (see Sect. 1). However, since many of those are probably complex proteins made up of two or more polypeptides, the number of different polypeptides could be considerably greater. The greatest number of polypeptides so far separated as products of synthesis in isolated chloroplasts is 16 membrane-bound (ZIELINSKI and PRICE 1980) and 80 soluble polypeptides (ELLIS et al. 1978). The number of thylakoid proteins is certainly an underestimate because of the lack of resolution of the one-dimensional gels used, while the soluble polypeptides may be an overestimation caused by artefacts which often occur with isoelectric focussing, as well as those from such sources as the specific premature termination of synthesis. Of these polypeptides only very few (about 12) have been identified.

Although a similar estimate of the number of chloroplast polypeptides which are made in the cytoplasm has not been attempted, it is commonly believed to be significantly greater than the number made in the chloroplast. As yet only three or four such proteins have been identified.

While an experimental strategy for the identification of all cytoplasmically synthesized chloroplast proteins has not at present been developed, it may soon be possible to contemplate such a project for the polypeptides for which the genetic information resides in the chloroplast genome. With the advent of the techniques of recombinant DNA, a feasible approach may be to identify all the gene products of chloroplast DNA and to establish whether or not any interchange of mRNA takes place between the chloroplast and cytoplasm. One problem associated with such an approach arises from the fact that many polypeptides which are made in isolated chloroplasts do not appear to be identical with detectable chloroplast components. It is possible that these are made as precursors which are not processed by the isolated organelles, or that they may be turned over so rapidly in vivo that no significant amount of protein is accumulated. This would complicate the task of determining the identity and site of synthesis of these proteins.

4.1 RuBP Carboxylase

The first evidence that isolated chloroplasts could incorporate amino acids into RuBP carboxylase was provided in 1968 (MARGULIES and PARENTI 1968, MARGULIES 1970), when a coincidence of incorporated [^{14}C]-leucine with carboxylase activity in isolated bean chloroplasts was obtained. A similar coincidence of radioactivity with enzyme activity was found by HARRIS et al. (1973), using gel filtration to separate the products of amino acid incorporation into isolated *Euglena* chloroplasts. In 1968 RAMIREZ et al. isolated pea chloroplasts using 0.2 M KCl as the osmoticum and found that these chloroplasts exhibited a light-dependent incorporation of [^{14}C]-leucine into protein in the absence of added ATP. BLAIR and ELLIS (1973) repeated this work and obtained a single peak of radioactivity from the soluble products when they were separated by polyacrylamide gel electrophoresis. They showed this polypeptide to be identical with the large subunit of RuBP carboxylase by comparison of the tryptic peptid maps. Large subunit has been demonstrated to be a product of the translation of chloroplast RNA in in vitro protein synthesizing systems from *E. coli* (HARTLEY et al. 1975) as well as from wheat germ (SAGHER et al. 1976) and the rabbit reticulocyte system (LINK et al. 1978). It has also been shown to be a major product from the transcription and translation of total spinach chloroplast DNA (BOTTOMLEY and WHITFELD 1979) as well as cloned chloroplast DNA fragments from maize (COEN et al. 1977), *Chlamydomonas* (MALNOË et al. 1979) and spinach (WHITFELD and BOTTOMLEY 1980).

From this evidence it can be concluded that chloroplast DNA is the coding site of the large subunit of RuBP carboxylase, that it is transcribed in the chloroplast to form a mRNA which lacks poly(A) or has only short poly(A) sequences, and that this mRNA is translated by the chloroplast protein-synthesizing system to yield the final polypeptide.

The observation that the newly synthesized large subunit is partly assembled into the holoenzyme (BOTTOMLEY et al. 1974, MENDIOLA-MORGENTHALER et al. 1976), which was originally dismissed as a fortuitous comigration of the two bands (ELLIS 1977a, b), has recently been confirmed (BARRACLOUGH and ELLIS 1980). While this partial assembly takes place in chloroplasts prepared in sorbitol medium, it is absent in chloroplasts prepared in KCl medium. It has been observed that KCl chloroplasts appear to be deficient by other criteria in that they fail to carry out light-dependent oxygen evolution (BOTTOMLEY et al. 1974) and the granal stacks appear swollen when observed in the electron microscope (WHITFELD et al. 1973). Whether this incorporation of newly synthesized subunits into holoenzyme is the result of assembly with preexisting small subunits or merely exchange of radioactive large subunits with unlabelled subunits in assembled enzyme is yet to be determined.

It has recently been reported (LANGRIDGE 1981) that the large subunit is made as a slightly larger precursor molecule. This would make it analogous with the 32,000 mol.wt. membrane protein (see below) which is also synthesized in isolated chloroplasts as a precursor. Although up to now no satisfactory hypothesis exists for this observation, it is possible that the extra peptide sequences facilitate assembly into holoproteins in the case of the large subunit or insertion into the thylakoid membrane in the case of the 32,000-M_r polypeptide.

Since the small subunit of RuBP carboxylase is not found among the products of protein synthesis by isolated chloroplasts, it was postulated that it must be synthesized in the cytoplasm. Conclusive evidence for this postulation was difficult to achieve. Not only did the small subunit protein represent a very small proportion of the total proteins made on cytoplasmic ribosomes, but also the coprecipitation of other products during immunoprecipitation always raised the question of the specificity of the antibody (GOODING et al. 1973, GRAY and KEKWICK 1973 1974, ALSCHER et al. 1976, ROY et al. 1976 1977).

Probably the most conclusive experiment to determine the site of synthesis of the small subunit was reported in 1977 when DOBBERSTEIN et al. translated the poly(A)$^+$-RNA fraction from the alga *Chlamydomonas* (Fig. 4). On immunoprecipitation with small subunit antibodies they obtained only one protein with a molecular weight of approximately 20,000. This polypeptide had a mol.wt. 3,500 larger than the small subunit found in RuBP carboxylase and could be processed to the size of the mature small subunit by a specific soluble endoprotease found in *Chlamydomonas* polysomal supernatant fractions (Fig. 7). Hence, the larger polypeptide must be the precursor small subunit.

This work has been confirmed using peas (CASHMORE et al. 1978, CHUA and SCHMIDT 1978a, HIGHFIELD and ELLIS 1978) and spinach (CHUA and SCHMIDT 1978a).

The mRNA for the small subunit precursor has been partly purified from pea leaves (CASHMORE 1979). When this 11.5S RNA was translated in a wheat germ system the 20,000-M_r precursor was the predominant product. The kinetics of hybridization of this mRNA preparation to pea DNA suggest that there are only one or very few copies per haploid genome.

The nuclear gene which codes for the precursor of the small subunit in peas has recently been cloned and sequenced (BEDBROOK et al. 1980).

The small subunit is coded for by the nuclear DNA, synthesized in the cytoplasm as a precursor and transported into the chloroplast where it is processed and assembled into the holoenzyme. Since the location of the genes and the site of synthesis of both the subunits are now known, the problems remaining with the biosynthesis of the enzyme relate to the transport and assembly of the small subunit (Sect. 4) and finally, whether there are other factors involved in their regulation and coordination.

4.2 Proton-Translocating ATPase

The proton translocating ATPase from chloroplasts is a complex enzyme controlling a key step in energy transduction by linking proton translocation with ATP synthesis (BAKKER-GRUNWALD 1977, MCCARTY 1979). The enzyme complex, which will be referred to as ATPase, consists of two parts. The F_0-factor is the membrane-bound basal section of the complex while the other part, called CF_1, or coupling factor, is attached to the CF_0 but projects out from the membrane surface. CF_1 can be removed from the membrane by treatment with EDTA and reattached by removal of the EDTA in the presence of Mg^{2+} (RACKER 1967 1977).

Recent reports on the number of CF_0 subunits in spinach chloroplast ATPase vary. PICK and RACKER (1979) separated four subunits by two-dimensional gel electrophoresis. These they designated I (17,500 mol.wt.), II (15,000), III (13,000) and the DCCD binding protein (8,000). However, NELSON et al. (1980) resolved only three such subunits. The subunit which they designated I appears to be subunit II of PICK and RACKER, subunit II corresponds to subunit III of PICK and RACKER while their subunit III is the DCCD-binding protein.

In 1976 MENDIOLA-MORGENTHALER et al. demonstrated the presence of the α, β and ε subunits in the products of polypeptide synthesis by isolated spinach chloroplasts in response to light. Recently, NELSON et al. (1980), using immunoprecipitation of the products of protein synthesis in isolated spinach chloroplasts with monospecific antibodies to the subunits of CF_1, found that subunit γ is made in the chloroplast in addition to these subunits.

A report that chloramphenicol treated pea seedlings synthesize the γ-subunit of CF_1 and that it therefore is made in the cytoplasm (BOUTHYETTE and JAGENDORF 1978) is based on very low amounts of radioactivity, as well as the low resolution of gel-electrophoresis when using gel slices instead of autoradiography or fluorography to locate the radioactive peptides.

The CF_0 basal part of the ATPase has also been studied in isolated organelles. DOHERTY and GRAY (1980) found that the lipophilic DCCD binding subunit (mol.wt. 8,000) was made in isolated pea chloroplasts while NELSON et al. (1980) demonstrated that their subunit I is also made in isolated spinach chloro-

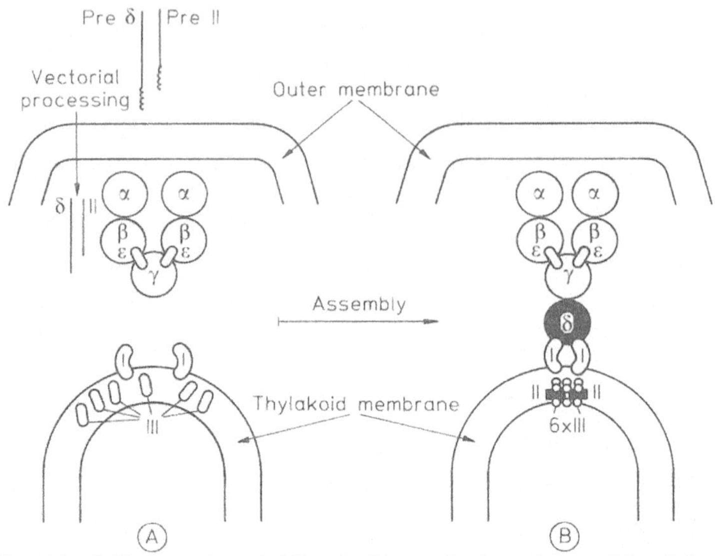

Fig. 5 A, B. Proposed model for the biosynthesis and assembly of the chloroplast proton-translocating ATPase. **A** Before assembly. **B** Assembled (NELSON et al. 1980, with permission)

plasts. Of the remaining subunits the CF_1 δ has been found among the products of translation of total spinach leaf RNA in a wheat germ system (NELSON et al. 1980). The fact that it is made as a precursor about 8,000 mol.wt. larger than the mature subunit led them to suggest that it is made in the cytoplasm and transported into the chloroplast by "vectorial transport". Since it is known that some polypeptides made in the chloroplast also are first synthesized as precursors (GREBANIER et al. 1978, ELLIS and BARRACLOUGH 1978, LANGRIDGE 1981), more positive evidence is needed before any conclusion can be made that this polypeptide is cytoplasmically synthesized.

It has been observed (MENDIOLA-MORGENTHALER et al. 1976, NELSON et al. 1980) that in addition to synthesizing some of the subunits, isolated organelles also assemble the holoenzyme. This indicates that a pool of the subunits which are presumed to be synthesized in the cytoplasm (CF_1 δ and one or two CF_0 subunits) exists within the chloroplasts. An alternative possibility is that the labelled subunits synthesized within the isolated chloroplasts exchanged with unlabelled subunits in existing assembled complexes to form labelled holoenzyme.

NELSON et al. (1980) have proposed a model for the assembly of the ATPase subunits (Fig. 5) which may in part serve as a more general model for the assembly of membrane-bound chloroplast proteins.

In contrast to the results obtained with spinach, GREBANIER et al. (1978) reported that in their experiments the α and β subunits were not correctly assembled into the thylakoids in isolated maize chloroplasts. They observed that extraction of the membrane with silicotungstic acid after amino acid incorporation removed virtually all of the unlabelled CF_1 but very little of the newly synthesized radiactive α and β subunits. Their results do not reveal how much of the newly synthesized subunits were removed with the soluble fraction during isolation of the membranes. It has been observed in this laboratory (BOTTOMLEY unpublished observations) that in isolated spinach chloroplasts, the newly synthesized α and β subunits largely remain in the soluble fractions. GREBANIER et al. (1978) suggested that the incorrect insertion into the membrane could be caused by a lack of a cytoplasmically synthesized factor in the isolated maize chloroplasts.

In summary, it appears that at least five of the nine subunits of the ATPase are synthesized in the chloroplast. The other subunits may be synthesized in the cytosol and transported into the organelle as a precursor. The holoenzyme is then assembled and inserted into the thylakoid membrane in the chloroplast. To date, no specific information on the sites of the genes coding for these subunits has been reported.

Although the subunit composition of the mitochondrial ATPase complex is very similar to that of chloroplasts, the sites of synthesis of the subunits from the two organelles are quite different. All five F_1 subunits of yeast mitochondrial ATPas are cytoplasmically synthesized, while the four F_0 subunits are all made within the mitochondria (SCHATZ and MASON 1974). In *Neurospora* (SEBALD et al. 1977) and rat liver (DE JONG et al. 1980) mitochondria, on the other hand, the DCCD binding F_0 polypeptide is synthesized in the cytosol.

4.3 Cytochromes

In 1973, GREGORY and BRADBEER treated greening bean leaves with (D-)-threo-chloramphenicol and found that the accumulation of cytochromes f, b-559_{LP} and b_{563} was inhibited, suggesting that these proteins are made within the chloroplast. It has been recently reported that the cytochrome b_{559} is indeed made also by isolated chloroplasts (ZIELINSKI and PRICE 1980) supporting this earlier finding. On the other hand, WILDNER (1976) found that the synthesis of cytochrome b_{563} in *Euglena* was not affected by chloramphenicol. Cytochrome c_{552} has been found to be inhibited by both cytoplasmic and chloroplast protein synthesis inhibitors (WILDNER 1976, FREYSSINET et al. 1979). Although cytochrome c_{552} is a small polypeptide (12,000 mol.wt.), its synthesis is complicated by the fact that it contains a haem group. The synthesis and assembly of the haem and polypeptide give ample opportunity for the involvement of both the chloroplast and cytosolic systems (FREYSSINET et al. 1979).

In 1977 HALLIER and HEBER reported that the synthesis of cytochrome f is under the control of the chloroplast genome in *Oenothera hookeri*. Later GRAY (1980) studied the inheritance of cytochrome f in reciprocal interspecific crosses between *Nicotiana tabaccum* and *N. glutinosa*. The pattern of inheritance of differences in the tryptic peptide maps of the proteins from the two species led him also to conclude that the genetic information was inherited maternally.

While it has been stated that cytochrome f was not synthesized by isolated pea chloroplasts (EAGLESHAM and ELLIS 1974); DOHERTY and GRAY (1979) used the solubility of cytochrome f in ethyl acetate/ethanol/ammonia to extract it specifically from chloroplasts prior to immunoprecipitation with the antibodies to cytochrome f. By this means they identified it among the products of polypeptide synthesis by isolated pea chloroplasts. It appears then, that this protein is coded for by the chloroplast DNA and that it is synthesized within the organelle.

4.4 Elongation Factors EF-G_{chl} and EF-T_{chl}

In 1976 CIFERRI and TIBONI reported that the chloroplast elongation factor EF-G_{chl} was synthesized in cells of *Chlorella vulgaris* which had been treated with cycloheximide but not in chloramphenicol treated cells. This led them to propose the site of synthesis of EF-G_{chl} to be on the 70S ribosomes of the chloroplasts. Later, CIFERRI et al. (1979) used two-dimensional (electrofocussing and electrophoresis) gels to show that the positions of both the elongation factors EF-G_{chl} and EF-T_{chl} of spinach chloroplasts coincide with minor spots of soluble proteins which become labelled when isolated chloroplasts are incubated in the light in the presence of [^{35}S] methionine. The molecular weights of these two proteins are approximately 75,000 and 46,000. The demonstration that another stromal protein, in addition to the large subunit of the RuBP carboxylase, is a product of synthesis within the chloroplast is an important observation and should be confirmed by data from different plants. Some of this importance lies in the fact that it may well be a first example of soluble proteins synthesized in the chloroplast which are not dependent on the cytosol system for the synthesis of a subunit.

In contrast, BREITENBERGER et al. (1979) found a 20-fold increase in EF-G_{chl} activity on transfer of dark-grown cells of *Euglena* to the light. The increase was prevented by cycloheximide but not by streptomycin or spectinomycin, which inhibit chloroplast protein synthesis. In addition, about 70% of this increase also occurred in the mutant strain W_3BUL for which no chloroplast DNA has been detected although membrane systems resembling chloroplasts are still present. From this, they concluded that in *Euglena,* EF-G_{chl} is nuclear coded and cytoplasmically synthesized. The chloroplast elongation factor EF-Ts from *Euglena* has recently been shown to be synthesized in a similar manner (FOX et al. 1980). It thus appears that in *Euglena,* the synthesis of the elongation factors can be compared to the synthesis of the corresponding factors in yeast mitochondria (PARISI and CELLA 1971, SCRAGG 1971, RICHTER 1971) or *Neurospora* mitochondria (BARATH and KÜNTZEL 1972) where these genes are also located in the nucleus and the proteins imported into the organelle after synthesis.

Confirmation that *Euglena* and higher plants have different sites of synthesis for analogous polypeptides and possibly also different locations of the genes, whould have interesting implications concerning the evolution of photosynthetic organisms.

4.5 32,000 M_r Membrane Protein

A membrane-bound polypeptide of an approximate mol.wt. of 32,000 has been observed among the products of protein synthesis in isolated organelles from peas (BLAIR and ELLIS 1973, SILVERTHORNE and ELLIS 1980), spinach (BOTTOMLEY et al. 1974) and maize (GREBANIER et al. 1978). In addition, a similar polypeptide has been found to be synthesized in fronds of *Spirodela* sp. (REISFELD et al. 1978a, b, WEINBAUM et al. 1979).

Isolated chloroplasts appear to incorporate [^{35}S]-methionine into this 32,000-M_r protein at rates comparable with those for the large subunit of RuBP carboxylase (SILVERTHORNE and ELLIS 1980). While the large subunit is the most abundant protein, constituting approximately 50% of the total soluble chloroplast protein, the 32,000-M_r polypeptide was not detected among the stained polypeptides. This implies that this polypeptide must have a very short half-life.

A 32,000-M_r polypeptide or its precursor has been reported to be a product of translation of chloroplast RNA in heterologous in vitro systems. In 1975, HARTLEY et al. found a protein of approximately this molecular weight among the products, when total spinach chloroplast RNA was translated in an *E. coli* system. *Spirodela* chloroplast RNA also yields a similar peptide in the wheat germ system but much less efficiently in the *E. coli* system (REISFELD et al. 1978a, b). The peak D from peas (ELLIS 1978, SILVERTHORNE and ELLIS 1980) and the product of a photogene of maize (BEDBROOK et al. 1978, GREBANIER et al. 1978) are of similar molecular weight and may also be related proteins.

A further characteristic of this polypeptide is that it is synthesized both in the organelle and from chloroplast RNA in vitro as a slightly larger precursor in maize (GREBANIER et al. 1978) and peas (ELLIS and BARRACLOUGH 1978). In *Spirodela* a similar precursor was shown to be the major product of short-time labelling in vivo (EDELMAN and REISFELD 1978). This precursor is processed within 1 h in *Spirodela* in vivo (EDELMANN and REISFELD 1978) and also in isolated pea chloroplasts (ELLIS 1978). In contrast, isolated maize chloroplasts did not process the 34,500-M_r precursor (GREBANIER et al. 1978). The synthesis and turnover of this protein during development will be dealt with in Section 7.

It seems probable that the 32,000 mol.wt. membrane proteins from the various plants are related and perform the same function. Although this function remains to be determined, it has been shown that severe depletion in *Spirodela* of this protein by chloramphenicol treatment has little or no effect on CO_2-fixation and hence a role in photosynthesis is excluded (EDELMAN and REISFELD 1978).

One remaining problem is the synthesis of a polypeptide within the chloroplast as a precursor. Since most precursors so far described are proteins which must be transported across a membrane, the observation of a precursor to a protein synthesized within the organelle was somewhat unexpected. It may be that the additional amino acids are needed to assure the correct insertion of the mature protein into the membrane.

4.6 Light-Harvesting Chlorophyll a/b Protein

The thylakoid membrane from chloroplasts may be solubilized by detergents to yield a variety of polypeptides which can be separated by polyacrylamide gel electrophoresis. Under these conditions there can be found between three and ten chlorophyll-containing bands depending on the conditions and techniques used (ANDERSON 1975, THORNBER 1975, MACHOLD et al. 1979, DELEPELAIRE and CHUA 1979). All of these complexes are aggregates of chlorophyll molecules and proteins, which are destroyed by treatments such as heating in detergents, to yield the apoprotein(s) of the complex and free chlorophyll. One of these proteins was termed light-harvesting chlorophyll a/b protein by THORNBER and HIGHKIN (1974) to describe its participation in a complex thought to be active in the harvesting of light for photosynthesis. Although reports from different laboratories and plants give slightly varying data, the molecular weights of the major proteins associated with this complex are about 25,000 and 26,000.

In 1972 KUNG et al. showed that the pattern of inheritance of this apoprotein in interspecific crosses of *Nicotiana* was Mendelian and that it therefore must be coded in the nucleus. MACHOLD and AURICH (1972) found that in *Vicia faba,* cycloheximide but not chloramphenicol inhibited the synthesis of the major apoprotein and concluded that it was synthesized on cytoplasmic ribosomes. Its synthesis is induced by light (APEL and KLOPPSTECH 1978) and these authors also reported that the protein was made from poly(A)$^+$-RNA in a wheat germ translation system as a precursor approximately 4,000 mol.wt. larger than the mature protein. Since the synthesis, not only of the apoprotein itself, but also of its mRNA, are induced by light, this induction must be exerted at the transcriptional level. The transcription of this mRNA is induced by the phytochrome system (APEL 1979). In addition, a second light reaction which induces the conversion of protochlorophyll to chlorophyll is required for the assembly of the chlorophyll-protein complex and its insertion into the thylakoid membrane (APEL 1979).

GEETHA and GNANAM (1980) used a preparation of chloroplasts from young sorghum seedlings which had been preincubated in the light until their capacity to synthesize proteins had been exhausted. They found that these preparations responded to exogenous RNA. When total leaf RNA was used in this "homologous cell-free system" they observed incorporation of label into three polypeptides from the light-harvesting chlorophyll a/b-protein complex (27,500, 25,000 and 23,000 mol.wts). Since these polypeptides were not labelled by freshly isolated chloroplasts in the absence of added RNA, they concluded that the mRNAs for all three were in the cytoplasm and that they were normally products of the cytosolic protein synthesis system.

4.7 P-700 Chlorophyll a-Complex Proteins

GEETHA and GNANAM (1980) labelled freshly isolated chloroplasts from the mesophyll of *Sorghum vulgare* and separated the P-700-chlorophyll a-protein complex (CP1). They found that two of the three polypeptides (M_r 70,000 and 58,000) were labelled. When they used total leaf RNA in preincubated chloroplasts (see Sect. 4.6), the third polypeptide (M_r 68,000) was found to have been synthesized in addition to the other two. They interpreted this as demonstrating that the mRNA of the two endogenously synthesized polypeptides originated in the chloroplast while the third was of cytosolic origin. ZIELINSKI and PRICE (1980) came to the same conclusion for a 68,000 protein from the CPI complex which they found among the products of synthesis in isolated spinach chloroplasts.

4.8 Ferredoxin

Ferredoxin is an iron-containing protein which acts as an electron carrier in the photosynthetic electron transport chain.

That the genetic information for the apoprotein is contained in the nuclear genome has been demonstrated by following the pattern of inheritance in *Nicotiana* interspecific hybrids (KWANYUEN and WILDMAN 1975, HUISMAN et al. 1977a, b), when ferredoxins from both parents were found to be transmitted to the progeny in a Mendelian manner.

In 1978, HUISMAN et al. isolated poly(A)$^+$-RNA from *Nicotiana tabacum*, *Phaseolus vulgaris* and *Chlamydomonas reinhardtii* and translated them in vitro in a wheat germ system. When the products were analyzed by immunoprecipitation with antibodies to ferredoxin, in each case a single polypeptide 3,000 to 4,000 mol.wt. larger than native ferredoxin precipitated. From this they concluded that ferredoxin was cytoplasmically synthesized. It was suggested that the precursor was transported into the chloroplast and processed by a mechanism similar to that of the small subunit of RuBP carboxylase. The incorporation of iron into the native protein was postulated to occur within the chloroplast where phytoferritine, the iron storage protein, is located (SECKBACK 1972).

4.9 Other Thylakoid Membrane Proteins

The above part of this section has dealt with chloroplast proteins which have been reasonably well characterized. That is, proteins of known function and one of unknown function (the 32,000 protein), which have been extensively studied in several plants. In addition to these proteins the sites of synthesis of a number of thylakoid membrane proteins have been reported. However, these proteins have been characterized as bands on gels and little or nothing is known about their function. It is likely that some of these unidentified polypeptides are identical with some of the proteins already discussed. For example, the various ATPase subunits as well as cytochrome f and chlorophyll a/b apoprotein are all membrane-bound and must be among the thylakoid polypeptides described.

The determination of the site of synthesis of membrane polypeptides relies heavily on the ability to resolve individual peptides. It has been reported that isolated thylakoids of *Chlamydomonas* can be resolved into 58 polypeptides on two-dimensional gels (BOSCHETTI et al. 1978). In spite of this even the precise number of polypeptides that are involved in the architecture of the thylakoid and envelope membranes remains uncertain.

The most extensive information on individual thylakoid membrane polypeptides comes from work on *Chlamydomonas reinhardtii* (GILLHAM et al. 1977) which will be used as a basis for the discussion of membrane proteins from other species.

Using short pulses of [^{14}C]acetate, which lead to well-defined products by preventing dispersal of the label into a variety of pools, CHUA and GILLHAM (1977) determined the labelling patterns in the presence and absence of inhibitors of either chloroplast or cytoplasmic protein synthesis. By separating the thylakoid proteins into chloroform/methanol-soluble and -insoluble fractions prior to electrophoresis they resolved 33 proteins of which 9 were apparently synthesized in the chloroplast. These results were confirmed by demonstrating that a mutant, whose chloroplast ribosomes were resistant to spectinomycin, gave the same pattern of labelling in the presence of both spectinomycin and cycloheximide as the wild-type *Chlamydomonas* did with cycloheximide alone.

In *Euglena* it was found that ten thylakoid polypeptides became labelled when chloroplasts were incubated with tritiated amino acids (BINGHAM and SCHIFF 1979). Of these, seven were absent from both dark-grown wild-type cells and cells from the mutant (W_3 *BUL*) which contains no detectable chloroplast DNA. Two other peptides, not detected in the products from isolated chloroplasts, were also absent from dark-grown W_3 *BUL* cells. This suggests that at least 7 to 12 thylakoid polypeptides are coded for by chloroplast DNA and/or synthesized in the chloroplast.

As described above, evidence from higher plants so far available points to at least eight membrane-associated peptides being made on chloroplast ribosomes. These include six subunits of the ATPase, cytochrome f and the 32,000 mol.wt. membrane protein. Whether or not these polypeptides are represented among the nine *Chlamydomonas* thylakoid proteins synthesized in the chloroplasts, is yet to be demonstrated. Recently, CHUA and his colleagues (DELEPELAIRE and CHUA 1979, CHUA and BLOMBERG 1979) have found that there is considerable immunological cross-reaction between antibodies raised against the algal thylakoid polypeptides and membrane polypeptides from pea and spinach chloroplasts. Although there are differences in some physical parameters such as molecular weight, this method should provide a useful tool for the determination of the function of the *Chlamydomonas* proteins and the search for the corresponding peptides in other species.

Current studies on chloroplast thylakoids are providing insights into the complex interactions between pigments, proteins and lipids which enable these membranes to carry out their unique functions (MACHOLD et al. 1979). With increasing knowledge of their structure and function it should be possible to link the biosynthetic and physiological evidence to give a picture of the synthesis, assembly and functioning of these membranes whose structure is of fundamental importance to the process of photosynthesis.

4.10 Chloroplast Ribosomal Proteins

The 70S chloroplast ribosomes consist of two subunits containing defined rRNA species and approximately 50 distinct ribosomal proteins. While the rRNAs are transcribed inside the chloroplast from genes on chloroplast DNA (BED-

BROOK and KOLODNER 1979), little is known about the site of synthesis of chloroplast ribosomal proteins, the regulation of their synthesis or the mechanism of their assembly. Most of the available information on chloroplast ribosomal proteins is derived from genetic studies on *Chlamydomonas* (see GILLHAM 1978 for a review). For this organism a variety of mutants are described which have defects in ribosome assembly or which are either resistant to or dependent on antibiotics affecting chloroplast ribosome function (BOGORAD et al. 1976, GILLHAM et al. 1976). While mutants with defective assembly processes might arise due to changes in RNA synthesis and processing it is generally thought that the effects of antibiotics are exerted on the ribosomal proteins. For these antibiotic effects approximately 20 genes have been described, seven or eight of which are involved in chloroplast ribosome assembly (HARRIS et al. 1974), while three confer antibiotic resistance on chloroplast ribosomes (DAVIDSON et al. 1974). All of these are nuclear mutants as shown by their pattern of inheritance. It has been reported that some of the other mutants isolated confer antibiotic resistance by alteration of at least seven loci on chloroplast DNA (SCHLANGER and SAGER 1974, BOYNTON et al. 1973 1976). In some cases, apparently a single chloroplast ribosomal protein was affected since its physical parameters had changed in the mutant (METS and BOGORAD 1972, DAVIDSON et al. 1974). While several methods for the isolation of chloroplast ribosomal proteins on two-dimensional gels have been described (BRÜGGER and BOSCHETTI 1975, METS and BOGORAD 1974, FREYSSINET and SCHIFF 1974), none of these proteins has, as yet, been identified among the products of protein synthesis of either the cytosol or the chloroplast.

KLOPPSTECH and SCHWEIGER (1974) followed the effect of transplantation of nuclei between different species of *Acetabularia*. They found that the pattern of proteins of the chloroplast large ribosomal subunit changed after a few weeks from that of the host species to that of the nucleus donor. From this they concluded that some of the chloroplast ribosomal proteins were nuclear-coded.

4.11 Aminoacyl-tRNA-Synthetases

Other important components of chloroplast protein synthesis are the aminoacyl-tRNA synthetases which charge the chloroplast-specific tRNAs with the appropriate amino acid (BARNETT et al. 1976, WEIL et al. 1976). The enzymes can be replaced in vitro by the corresponding enzymes from *E. coli* or enzymes from blue-green algae (PARTHIER 1973). The chloroplast synthetases are distinct from the enzymes found in the cytosol (REGER et al. 1970). The site of synthesis of chloroplast aminoacyl-tRNA synthetases has been studied mainly by following the induction of these enzymes in bleached greening *Euglena* cells in the presence of specific inhibitors of protein synthesis (PARTHIER 1973, HECKER et al. 1974). It was found that cycloheximide, but not streptomycin, inhibited the synthesis of the synthetases. In addition, the bleached mutant W_3 *BUL*, which lacks chloroplast DNA, was also found to contain these enzymes. From this, it could be concluded that the cytoplasm is the site of synthesis of the enzymes studied so far (cf. WEIL and PARTHIER, Chap. 2, Vol. 14A, this Series).

5 Transport of Cytoplasmically Synthesized Proteins into the Chloroplast

Since a significant proportion of chloroplast proteins is synthesized in the cytosol using mRNAs which probably originate from nuclear DNA, the elucidation of the mechanism of transport of these proteins into the chloroplast is one of the central questions in chloroplast molecular biology today. Some insights into the mechanism can be gained by analogy with mechanisms found in other trans-membrane processes such as protein secretion (DAVIS and TAI 1980) and the biogenesis of mitochondria (SCHATZ and MASON 1974, TZAGOLOFF et al. 1979). The transport of proteins is a complex process about which a number of questions can be asked e.g.: What is the site of synthesis of the proteins to be transported? How is the membrane site recognized? How are they recognized and selected for transport? By what means do they cross the membrane, and how is the whole process regulated? We are at present only able to provide partial answers to some of these questions. The transport mechanism must be sensitive enough to discriminate between proteins belonging to the chloroplast and those which must remain in the cytosol. They must also be distinguished from proteins which are to be stored in vesicles or secreted and also from those to be transported into mitochondria, lysosomes or other organelles.

It was originally thought that transport into chloroplasts was more complex than many other cases of membrane transport since it involved crossing the two distinct membranes of the outer envelope. More recently it has been observed that these two outer membranes are in a state of constant oscillation and that the two unit membranes often appear to join and act as a single membrane (DOUCE et al. 1973). This is analogous to the apparent continuous merging and separation which has been observed in mitochondrial membranes (BUTOW et al. 1975) and which has been suggested to be involved in transport into that organelle. There is evidence that there are 80S cytoplasmic ribosomes closely associated with the outer membrane of mitochondria (KELLEMS et al. 1974, BENNETT et al. 1976). Although such an association has been postulated for chloroplasts (GOODING et al. 1973, LAULHERE and DORNE 1977), no substantial evidence for a functional attachment of ribosomes to the chloroplast envelope in higher plants has been reported. Recently, GIBBS (1979) has reported evidence for a tight link between cytoplasmic ribosomes and chloroplasts in the eight classes of algae where the chloroplasts are enclosed in a cisterna of endoplasmic reticulum. It was proposed that these ribosomes are involved in the transport of cytoplasmically synthesized proteins into the chloroplasts.

BLOBEL and DOBBERSTEIN (1975a b) have proposed the "signal mechanism" for the transport of secretory proteins across the endoplasmic reticulum (ROTHMANN and LENARD 1977, WICKNER 1979, DAVIS and TAI 1980). Proteins to be secreted are made with a leader sequence of 15 to 30 amino acids at the N-terminus. The mRNA is translated on the membrane-bound ribosomes of the rough endoplasmic reticulum and transported through the membrane, where the signal sequence is cleaved off while the chain is still in the process of elongation.

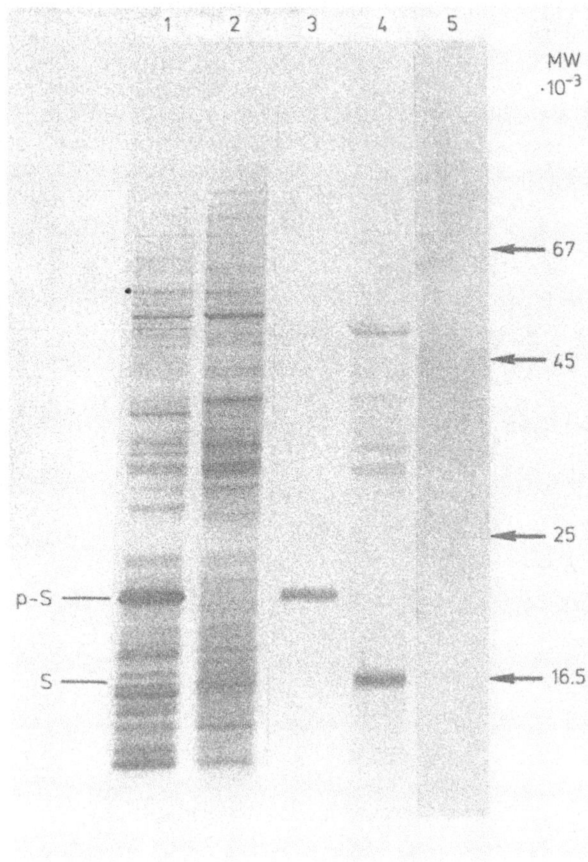

Fig. 6. SDS-polyacrylamide gels of the polypeptides synthesized by polysomes or poly-[A]$^+$-RNA of *Chlamydomonas reinhardtii* in a cell-free wheat germ system (*Track 1*) polypeptides synthesized on poly-[A]$^+$-RNA; (*Track 2*) products of polysomal run-off; (*Track 3*) immunoprecipitation of 1 with anti-SSU IgG fraction; (*Track 4*) immunoprecipitation of 2 with anti-SSU IgG fraction, (*Track 5*) immunoprecipitation of 1 with pre-immune IgG fraction. *p-S* position of SSU-precursor; *S* position of SSU (DOBBERSTEIN et al. 1977, with permission)

In 1977 DOBBERSTEIN et al. translated poly(A)$^+$ RNA from the alga *Chlamydomonas* in a wheat germ in vitro synthesis system and found that a polypeptide of molecular weight 20,000 was precipitated by the antibodies against purified small subunit of RuBP carboxylase (Fig. 6). The mature protein, which had a molecular weight of 15,500, was formed by cleavage of the precursor by a soluble endoprotease from postribosomal supernatants (Fig. 7).

Although the synthesis of a high molecular weight precursor is analogous to the mechanism proposed by BLOBEL and DOBBERSTEIN (1975a), the chloroplasts proteins do not appear to be transported in the same manner. Translation of the small subunit precursor is completed prior to transport and processing whereas, with the signal mechanism, transport and processing are co-translational. In addition, the leader sequence does not cause ribosomes to become attached to the chloroplast membrane as they do to the endoplasmic reticulum. DOBBERSTEIN et al. (1977) suggested that the chloroplast transport mechanism is more analogous to that of the transport of diphtheria toxin across animal plasma membranes (PAPPENHEIMER 1977). This toxin consists of two linked polypeptide chains, A and B. The B chain is believed to interact with the mem-

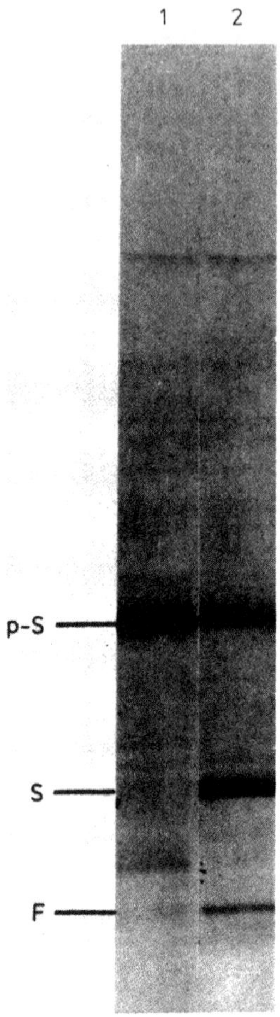

Fig. 7. SDS polyacrylamide gel of the products of endoproteolytic cleavage of the SSU precursor. The precursor was immunoprecipitated. (*Track 1*) incubated with heat inactivated postribosomal supernatant from *Chlamydomonas reinhardtii*; (*Track 2*) incubated with untreated postribosomal supernatant from *Chlamydomonas reinhardtii*; *p-S* position of SSU-precursor; *S* position of SSU. F a small fragment that presumably represents the difference between pS and S. (Dobberstein et al. 1977, with permission)

brane and facilitate the passage of the A chain across. After passage the chains are cleaved releasing the A chain, which contains the active site, into the cytoplasm. Schmidt et al. (1979) have proposed the term "transit sequence" for the leader sequences of transported chloroplast proteins.

That the mechanism demonstrated in *Chlamydomonas* is more generally applicable is indicated by the finding that spinach (Chua and Schmidt 1978b) and pea (Cashmore et al. 1978, Chua and Schmidt 1978b, Highfield and Ellis 1978) also synthesize and process precursor proteins which are 4,000 to 5,000 larger in mol. wt. than the mature small subunit.

Incubation of the supernatants from the in vitro synthesis of the precursor in *Chlamydomonas* or higher plants with intact chloroplasts leads to the appearance of mature processed small subunit in the assembled RuBP carboxylase molecules (Chua and Schmidt 1978b, Smith and Ellis 1979). Although this was first reported to be unaffected by light (Chua and Schmidt 1978b), when the experiments were repeated under conditions where ATP was limiting (Grossman et al. 1980), the uptake of proteins was found to be an energy-requiring process which was stimulated by either light or added ATP. In the light, in the absence of added ATP, uncouplers of photophosphorylation inhibited the transport process. COBB and WELLBURN (1976) had previously reported that the transport of proteins into the chloroplasts required light.

One interesting result was that, while the precursor from peas and spinach could be taken up and processed by chloroplasts from either plant, there was no such interchangeability between *Chlamydomonas* and the higher plant chloroplasts (CHUA and SCHMIDT 1978a).

The endoprotease, which was found in the soluble fraction from *Chlamydomonas* (DOBBERSTEIN et al. 1977), was first reported to be associated with the outer envelope of the chloroplasts of peas (HIGHFIELD and ELLIS 1978), however, on re-examination it was found that, as with *Chlamydomonoas*, the pea enzyme was also in the stromal fraction (SMITH and ELLIS 1979).

It has since been shown that other chloroplast proteins are apparently transported by the same mechanism. A protein which is associated with the light-harvesting chlorophyll a/b complex is made in vitro from barley poly(A)$^+$ RNA in a wheat germ system (APEL and KLOPPSTECH 1978) or from *Chlamydomonas* poly(A)$^+$ RNA in a wheat germ or a rabbit reticulocyte system (SCHMIDT et al. 1980) as a precursor of around 32,000 mol.wt. When the in vitro products were incubated with intact chloroplasts, the precursor was not only transported across the outer membrane and processed down to its mature size of 28,000 mol.wt., but it was also inserted into the thylakoid membrane. Since no precursor was found in the stromal fraction it was concluded that processing to the mature size occurred during transport across the outer membrane (SCHMIDT et al. 1980).

Chloroplast ferredoxin is also known to be encoded by nuclear DNA in *Nicotiana* (KWANYUEN and WILDMAN 1975, HUISMAN et al. 1977a b). In 1978, HUISMAN et al. found that translation of poly(A)$^+$ RNA from *Nicotiana tabaccum, Phaseolus vulgaris* or *Chlamydomonas reinhardtii* in a wheat germ system and immunoprecipitation of the products with antibodies to ferredoxin, gave a polypeptide of 20,500 mol.wt. which was 3,500 mol.wt. larger than mature ferredoxin. NELSON et al. (1980) also showed that the δ subunit of the spinach chloroplast coupling factor CF_1 is made from total leaf RNA in a reticulocyte system as a precursor 8,000 mol.wt. larger than the mature subunit. Although the processing of these precursors to the mature proteins was not studied, it seems reasonable to assume that they will also be found to be processed during, or immediately after, transport across the chloroplast envelope.

From the rather limited evidence so far available it is possible to suggest some generalizations regarding the mechanism of transport of cytoplasmically synthesized polypeptides to their sites in the chloroplast. Firstly, it appears that they are made as precursors of higher molecular weight and that this extra "transit sequence" (SCHMIDT et al. 1979) in some manner facilitates the transport of the precursor across the chloroplast envelope. Since pretreatment of the chloroplasts with a protease prevents this transport (CHUA and SCHMIDT 1978a). it seems probable that the "transit sequence" interacts with a protein on the outer surface of the membrane. Following passage through the membrane the "transit sequence" is rapidly removed by a soluble endoprotease in the stroma. The chloroplasts can then assemble these polypeptides into their final structures by combining them with other polypeptides of cytoplasmic or chloroplast origin, and, when necessary, inserting them into the thylakoid membranes.

It is by no means proven that this mechanism of transport of cytoplasmically synthesized proteins across the chloroplast envelope is the only one that operates. Other mechanisms may be involved. For instance, it is conceivable that the transit sequence could be an intrinsic part of the mature polypeptide, in which case it would need to be protected, by some mechanism, from cleavage by the processing enzyme.

6 Location of the Genes Specifying Chloroplast Polypeptides

Since the organization of the plant nuclear and chloroplast genomes are covered in detail in other chapters of these volumes, this section will be confined to some aspects of gene localization which are particularly relevant to the biosynthesis of chloroplast proteins. In addition to these articles a number of reviews on the structure and organization of chloroplast DNA and the location of chloroplast genes have appeared (KUNG 1977, BEDBROOK and KOLODNER 1979, TEWARI 1979, HERRMANN and POSSINGHAM 1980).

A substantial body of evidence concerning the location of the genes for chloroplast proteins comes from an approach which can be termed "exclusion genetics". This approach makes use of the many mutants, from higher plants as well as from green algae and *Euglena,* which are defective in chloroplast ribosome function. If the presence of a particular chloroplast protein can be demonstrated in such mutants, then it is almost certain that its gene is located on nuclear DNA. The concept assumes that the mutants do not contain even a small proportion of active chloroplast ribosomes and also that there is no export into the cytoplasm of mRNA transcribed from chloroplast DNA. On the other hand, if a protein is absent from these plastid mutants, it cannot be concluded that the gene is located in the chloroplast DNA. Many components of the chloroplast are complex proteins composed of subunits from both the chloroplast and the cytosol (cf. Sect. 3 and Table 3). Their assembly is often dependent on the coordinated synthesis of the various subunits. If the chloroplast system is shut off by a mutation, then it is quite possible that the synthesis of the cytosolic components will also be affected. An example of this is the *Oenothera* mutant *Iδ* in which a mutation in the chloroplast genome prevents the synthesis of the large subunit of RuBP carboxylase. The absence of this subunit apparently also results in the complete absence of the cytoplasmically synthesized small subunit (HALLIER et al. 1978).

A summary of the results obtained using higher plant mutants which are deficient in chloroplast ribosomes is given in Table 4. From this it can be seen that the chloroplast is dependent on the cytosol for the synthesis of a large variety of proteins.

While mutants which are deficient in ribosome activity have provided useful information, the complete removal of plastid protein synthesis results in such

Table 4. Components of chloroplasts synthesized in the cytosol as deduced from their presence in higher plants which are deficient in chloroplast ribosomes

Protein	Plant	Reference
DNA polymerase	*Zea* *Vicia* *Hordeum* *Pelargonium*	SHUMWAY and WEIER (1967) DYER et al. (1971) SPREY (1972) KNOTH et al. (1974)
RNA polymerase	*Secale*	FEIERABEND (1979)
Calvin-cycle enzymes (except RuBP carboxylase)	*Secale* *Triticum* *Pisum* *Avena* *Hordeum*	FEIERABEND and SCHRADER-REICHARDT (1976) FEIERABEND and MIKUS (1977) FEIERABEND and MIKUS (1977) FEIERABEND and MIKUS (1977) BRADBEER and BÖRNER (1978)
Enzymes of pigment synthesis	*Secale* *Pelargonium*	FEIERABEND (1977) HAGEMANN and BÖRNER (1978)
Electron transport chain proteins (except cytochromes f, b-559 and b-6	*Secale*	FEIERABEND (1979)
Enzymes of starch formation	*Pelargonium*	HAGEMANN and BÖRNER (1978)
Proteins of the chloroplast envelope	*Vicia* *Hordeum* *Pelargonium* *Secale*	DYER et al. (1971) SPREY (1972) BÖRNER et al. (1972) SCHÄFERS and FEIERABEND (1976)
Enzymes of PS I and PS II	*Pelargonium*	HAGEMANN and BÖRNER (1978)
Components of the prolamellar bodies	*Hordeum*	KNOTH and HAGEMANN (1977)

massive changes that it is often difficult to locate the precise position of the lesion. It would therefore be useful to have available a number of mutants in which the effects of the lesion are more restricted and so may be precisely characterized biochemically. At present, such mutants are largely restricted to *Chlamydomonas* (GILLHAM et al. 1977, GILLHAM 1978, SHEPHERD et al. 1979).

The chloroplast genome is small (average 150 kbp) in comparison to the nuclear genome and is therefore more amenable to biochemical manipulation. As a result of recent work the gross morphology of the genome, as well as the precise location of a number of genes are already known for several plants. DNA from maize (BEDBROOK et al. 1977), spinach (HERRMANN et al. 1980), wheat (BOWMAN et al. 1981), *Chlamydomonas* (ROCHAIX 1978) and *Oenothera* (GORDON et al. 1981) chloroplasts have a similar overall structure, consisting of large and small single copy regions separated by two inverted repeat regions containing the ribosomal RNA cistrons (BEDBROOK and KOLODNER 1979, (see BOHNERT et al., Chap. 14, this Vol. for details). The chloroplast DNA's of

Euglena (RAWSON et al. 1978) and *Vicia* (KOLLER and DELIUS 1980) were found to be different. In *Euglena* chloroplast DNA three copies of the rDNA are present close together in tandem rather than inverted arrangement, while in *Vicia* only a single copy of rDNA was found.

In addition to the rRNA genes, the genes for 27 specific tRNA genes together with 8 other 4S RNA genes have been mapped on spinach chloroplast DNA (DRIESEL et al. 1979) while only the distribution of the genes for unfractionated 4S RNA have been localized on *Chlamydomonas* (MALNOË and ROCHAIX 1978) and *Euglena* (HALLICK et al. 1978).

Of the genes so far localized on chloroplast DNA the one most extensively studied is that for the large subunit of RuBP carboxylase. This gene has been precisely mapped on the chloroplast DNA from maize (BEDBROOK et al. 1979) spinach (WHITFELD and BOTTOMLEY 1980) *Chlamydomonas* (MALNOË et al. 1979) and wheat (BOWMAN et al. 1981). In these plants it is located in the larger of the single copy regions although its position with respect to the orientation of the ribosomal genes in *Chlamydomonas* is opposite to that in the higher plants so far studied. Similarly, the gene coding for a 32,000-M_r membrane protein in both maize (BEDBROOK et al. 1978) and spinach (DRIESEL et al. 1980) has been shown to be close to one end of the large single copy region.

A number of studies have been carried out to determine the proportion of the total chloroplast DNA which is transcribed in vivo. Hybridization of total *Euglena* cell RNA with chloroplast DNA has given values of 47 to 53% (RAWSON and BOERMA 1976) and 26% (CHELM et al. 1977) of the genome transcribed. In *Chlamydomonas,* HOWELL and WALKER (1977) found that between 39% and 60% of the chloroplast DNA was transcribed at various stages of the cell cycle. When asynchronously grown cells were used as the source of RNA it hybridized to 60% of the single-stranded DNA, suggesting that in at least 10% of the DNA, both strands are transcribed (HOWELL and WALKER 1977).

If one third of the genome, or approximately 50,000 nucleotides of RNA, is transcribed, then, in addition to the two rRNA cistrons (16,000 nucleotides) and a set of tRNA genes (approximately 5,000 nucleotides, allowing for 35 to 40 genes and some transcribed non-genic RNA), there would be sufficient mRNA to code for about 40 polypeptides of an average molecular weight of 25,000. This calculation assumes that there are no large introns without any function. Since to date only the genes for two polypeptides have been definitely located on the chloroplast DNA and eight more have been demonstrated to be synthesized within the chloroplast (see Sect. 3) and hence are probably chloroplast DNA coded, there would be approximately 30 more genes for proteins yet to be identified.

7 Synthesis of Chloroplast Proteins During Development

There are several distinct but inter-related processes which must be clearly recognized when considering variations in protein synthesis during chloroplast development. Not only is there the normal ontogeny of the plant to be considered, but also the changes caused by the interaction of light with several photoreceptors which also influence the biosynthetic processes associated with organelle development. Not only do the photoreceptors involved in harvesting light energy for photosynthesis affect the synthesis of chloroplast proteins, but also photoreceptors outside the plastid have been shown to be involved (SCHWARTZBACH and SCHIFF 1979, APEL 1979).

A germinating seed normally contains a fully functioning protein synthesis system (WEEKS and MARCUS 1971) including ribosomes and polysomes (SPIEGEL and MARCUS 1975), stored mRNA (MAYER and SHAIN 1974, DURE and HARRIS 1977), tRNA's and aminoacyl-tRNA-synthetases (NORRIS et al. 1973). The growing organism uses these components, usually provided by the maternal parent, to carry out an initial development programme. Almost immediately after the beginning of germination, before the first cell division occurs, RNA synthesis commences (SEN et al. 1975) followed by DNA replication (MORY et al. 1972). At this stage proplastids begin to differentiate into chloroplasts in the light or into etioplasts if kept in the dark. Chlorophyll precursors accummulate in the dark and their light-dependent conversion to chlorophylls is accompanied by an increase in the synthesis of thylakoid proteins leading to the development of the chloroplasts into their mature form. In a number of plants, the incorporation of labelled precursors into proteins such as RuBP carboxylase ceases at about the time when leaf expansion ceases (BRADY and SCOTT 1976, DICKMANN and GORDON 1975, CALLOW 1974). The residual protein synthesis then seems to be sufficient to account only for protein turnover and no net changes in protein content take place. Finally, in the senescing leaf, both chloroplast and cytoplasmic protein-synthesizing systems slow down in such a way that degradation leads to a net loss of protein. In the case of the RuBP carboxylase it has been shown that the actual loss of the protein is preceded by a decrease in enzyme activity (CALLOW 1974).

In the course of this ontogenic development the plastids can undergo a variety of modifications, normally starting from a chloroplast, leading to forms which are often tissue-specific such as amyloplasts or chromoplasts (KIRK and TILNEY-BASSETT 1978).

Most interest has been centred on the investigation of the changes in protein synthesis during light-induced development. This can be carried out using bleached algal cells or mixotrophically grown *Euglena* (FREYSSINET et al. 1978 1979) and following protein synthesis during greening. Greening higher plants have also been subjected to extensive study, while some studies on chloroplast protein synthesis during normal development in the light have also yielded interesting results (BLACKWOOD and LEAVER 1977, EDELMAN and REISFELD 1978, SILVERTHORNE and ELLIS 1980).

The measurement of the responses of chloroplasts to light by monitoring the activity of chloroplast enzymes can often be interpreted as the result of two different effects. While the measured effects may be caused by variations in the synthesis of the enzyme, it is also possible that they are the result of variations in the activity of the enzyme being studied. For example, it has been found that the activity of chloroplast DNA-dependent RNA polymerase increases three- to four-fold following 16 h illumination of etiolated maize seedlings, whereas the amount of enzyme remained unaffected (Apel and Bogorad 1976). In this review we will confine ourselves to those studies which result in responses that can be attributed to the synthesis of the proteins themselves.

7.1 RuBP Carboxylase

The two subunits of RuBP carboxylase are made in different cell compartments and their genes are contained in different genomes (see Sect. 4.1). It can be expected therefore, that the study of the synthesis and assembly of the subunits, either during normal development or during the greening of dark-adapted tissue could provide information on the regulation and coordination of chloroplast and cytosolic protein synthesis. Because the chloroplast is the primary photoreceptor of the plant, the greening system should be expected to exhibit the greatest differential effects on the two sites of protein synthesis.

As early as 1956, Lyttleton observed that, when etiolated wheat leaves were exposed to 20 h of white light, there was a threefold increase in the amount of RuBP carboxylase (Fraction I) formed. Later Kupke (1962) showed that, when etiolated bean leaves were treated with light, there was a correlation between the amount of Fraction I protein formed and the level of chlorophyll in the plastid. Kleinkopf et al. (1970) found that greening barley leaves exhibited an increase in RuBP carboxylase activity and that this increase was correlated with an increase in the amount of enzyme itself.

In barley (Smith et al. 1974) the incorporation of labelled amino acids into the two subunits during greening was found to precede both the synthesis of chlorophyll and the increase in activity of the RuBP carboxylase. Chlorophyll synthesis had a lag of about 4 h and RuBP carboxylase activity of 6 h, whereas the synthesis of both subunits appeared to have no significant lag. They postulated that synthesis of the subunits is triggered early in greening and that assembly into the active enzyme occurs later.

In 1975 Siddel and Ellis examined the ability of chloroplasts isolated from pea apices at various stages of greening to incorporate [^{35}S]-methionine into specific proteins. They found that etioplasts could synthesis the large subunit of RuBP carboxylase when supplemented with ATP. The capacity to synthesize large subunits in response to either light or ATP increased for the first 48 h of greening. Between 48 and 96 h the isolated chloroplasts apparently completely lost their ability to make large subunit, even though the accumulation of this polypeptide, as judged by protein staining, continued.

Reisfeld et al. (1978a) followed the incorporation of [^{35}S]-methionine into the large subunit of RuBP carboxylase of *Spirodela* after illumination of dark-grown plants. They found a lag phase of 5 to 8 h during which the incorporation remained constant, followed by a gradual increase in the labelled carboxylase, which almost doubled after 25 h of light.

In *Lemna gibba* it was found (Tobin and Suttie 1980) that the [^{35}S]-methionine incorporation into the two subunits of RuBP carboxylase following illumination of dark-adapted plants appeared to be closely coordinated and to occur without any lag phase.

In *Chlamydomonas* the synthesis of the small and the large subunits appear to be very tightly coupled (IWANIJ et al. 1975). They are synthesized mainly during the light phase, and the ratio of the incorporation of labelled amino acids in the light to that in the dark is identical for both subunits. In addition, inhibition of both 70S protein synthesis and 80S protein synthesis abolished the in vivo incorporation of amino acids into the holoenzyme. This apparent close coordination of synthesis of the two subunits was also found in a mutant of *Chlamydomonas* (*ac-20*), which lacks chloroplast ribosomes when grown on organic carbon medium (GIVAN 1979). The recovery of the ability to synthesize either subunit of RuBP carboxylase when transferred to inorganic medium was inhibited by both chloramphenicol and cycloheximide. It had earlier been shown (GIVAN 1974) that cycloheximide stimulated the activity of RuBP carboxylase when the *ac-20* mutant was transferred from organic to inorganic media. This apparent lack of coordination between enzyme activity and synthesis indicates that caution should be exercised when interpreting changes in enzyme activity in terms of protein synthetic capacity.

7.2 32,000-M_r Membrane Protein

Since the early reports of the synthesis of a membrane bound polypeptide of M_r about 32,000 by isolated chloroplasts of pea (BLAIR and ELLIS 1973) and spinach (BOTTOMLEY et al. 1974), there have been a number of investigations into the synthesis and role of this protein (See Sect. 4.5). There have been no reports of significant differences between the proteins from various plants so we will assume that they are functionally related.

As described above, SIDDELL and ELLIS (1975) examined the ability of isolated pea chloroplasts to synthesize particular proteins using either light or ATP as an energy source. Under either condition the synthesis of the 32,000-M_r protein as well as the large subunit of RuBP carboxylase, increased up to 48 h greening. At 96 h there was virtually no protein synthesis in response to ATP while in the light-driven system there was an increase in the synthesis of the 32,000-M_r protein together with a decrease in the synthesis of all other proteins. The fact that between 48 and 96 h of greening there appeared to be an increase in protein accumulation, as judged by the density of stained proteins on their chromatograms, suggests that the amino acid incorporation in isolated plastids does not provide a true picture of the in vivo situation. A later suggestion (SILVERTHORNE and ELLIS 1980), that their 13-day-old pea plants were analogous to 40-day-old spinach, is of doubtful validity.

The synthesis of the 32,000-M_r protein during the change from dark heterotrophic growth to light, autotrophic growth in *Spirodela oligorrhiza* has been studied by REISFELD et al. (1978a). They found a sharp, transient increase in the synthesis of this polypeptide 3 h after the transfer to light, followed by a return to a steady state. This transient increase paralleled a previously described (ROSNER et al. 1975) brief increase in the synthesis of a 0.5×10^6 plastid mRNA as well as an increase in the appearance of the 32,000-M_r protein among the products of in vivo synthesis of polypeptides from total RNA translated in the wheat germ system. It is of interest that the data of Reisfeld et al. (1978a) also show the apparent transient formation of at least one other polypeptide of 25,000 mol.wt., at a later time during greening. Recently Weinbaum et al. (1979) found that the ability of mature fronds of *Spirodela* to synthesize the 32,000-M_r protein was much greater than those from younger plants. Pulse chase experiments also demonstrated that this protein is rapidly turned over in the plant.

In maize it has been shown that a 32,000-M_r protein is apparently made in isolated chloroplasts as a 34,500-M_r precursor (Grebanier et al. 1978). The mRNA which codes for a 34,500-M_r peptide has been found to be absent from etiolated maize seedlings but to occur in increasing amounts during greening (Bedbrook et al. 1978). A concomitant increase in the hybridization of plastid RNA to a restriction fragment of maize chloroplast DNA with greening suggested that the control of the synthesis of this protein is exercised at the transcriptional level.

7.3 Light-Harvesting Chlorophyll a/b Protein

In 1977, Armond et al. reported that illumination of etiolated barley caused the insertion of a chlorophyll a/b binding protein into the thylakoid membrane. Apel and Kloppstech (1978) later showed that a precursor to this protein is synthesized in the wheat germ system from poly(A)$^+$-mRNA of barley leaves. The precursor (29,500 mol.wt.) was identified by immunoprecipitation from the wheat germ translation products by antibodies to the chlorophyll a/b binding apoprotein. The level of this mRNA, which was undetectable in the etiolated tissue, increased during the first 6 h of illumination after which it remained steady. This light-induction of the mRNA was found to be a response to the phytochrome system (Apel 1979, Apel and Kloppstech 1978). However, following red-light treatment, the protein was not detected in chloroplast thylakoids. Since insertion into the membrane took place only during continuous illumination it was concluded that this insertion depended on the synthesis of chlorophyll. The appearance of the mature protein in the thylakoid membrane under the influence of white light occurred with a lag period of about 6 h. This lag period was reduced to about 3 h if the phytochrome system was activated by a previous red-light pulse. It was therefore suggested that, during the lag period, the formation of the light-harvesting chlorophyll a/b complex is limited by the transcription of the mRNA for the apoprotein.

The formation of this chlorophyll–protein complex appears to involve the phytochrome-induced transcription of a mRNA for the apoportein from the nuclear DNA which is then translated as a precursor on the cytoplasmic ribosomes. Following transport across the chloroplast envelope and removal of the transit sequence, the protein interacts with chlorophyll which has been synthesized in response to continuous light, and the complex is inserted into the thylakoid membrane.

The synthesis of this protein as a precursor (32,000 mol.wt.) has also been reported to occur when poly(A)$^+$-mRNA from *Lemna gibba* is translated by the wheat germ system (TOBIN 1978).

7.4 Cytochromes

FREYSSINET et al. (1979), using a sensitive immunological assay for cytochrome c_{552} found, that the synthesis of this protein in the chloroplasts of greening *Euglena* cells was very tightly coupled to the synthesis of chlorophyll. The lag phase in the synthesis of both pigments was the same and both were inhibited to the same extent by cycloheximide and streptomycin. In addition, neither could be detected in mutants which lacked DNA, suggesting that cytochrome c_{552} may be coded, at least in part, by chloroplast DNA. They suggested also that tight coupling of synthesis may be due to the constituents of the thylakoid membranes regulating each other's synthesis in order to ensure the correct sequence of membrane assembly. WILDNER (1976) also found that cytochrome c_{552} was inhibited by both 70S and 80S protein synthesis inhibitors.

The proposal that the syntheses of a number of membrane components are interdependent is also supported by the observation of GREGORY and BRADBEER (1973) that the light induction of the synthesis of cytochromes f, b_{559} and b_{563} follow similar kinetics.

8 Conclusions

The existence within the cells, of discrete organelles such as chloroplasts and mitochondria, which contain their own genomes and protein synthesis systems, raises a number of challenging questions. Outstanding among these are the problems of how these organelles arose and the nature of the mechanisms that regulate their development and function. In addition, the fact that the cytosolic and organellar protein-synthesizing systems each make significant contributions to these processes, raises the question of how the genes for the components became distributed between the genomes. BOGORAD (1977) has put forward two schemes to suggest how gene dispersal may have occurred. One proposal is that the organelle gene was duplicated and one copy integrated into the nuclear genome followed by loss of the gene from the organelle. The second suggestion is that the plastid gene mutated and made useless protein. If another protein from elsewhere in the cell could substitute for this protein, then the gene for this substitute protein would remain in the nucleus. Subsequent mutation could result in that particular gene serving the organelle alone. While such proposals are stimulating, they do not account for recent data concerning the role of transit sequences in protein transport. Any more appropriate scheme would need to account for the formation of such sequences as well as for the existence of processing enzymes and transport sites.

From an evolutionary standpoint, the current distribution of chloroplast genes between nuclear and chloroplast DNA could represent a stable balance between the two genomes. In this case the present cooperation would represent an endpoint of their competition for genes. A second possibility is that the present status of gene dispersal is a transitory one and that the chloroplast genome is either expanding or decreasing. If the coordination of the genomes was still progressing, it could be expected that the composition of chloroplast genomes would vary from plant to plant. The present knowledge of the structure of chloroplast genomes is not sufficient to make a definite decision. While it has been shown that there is a significant uniformity in the organization of maize, spinach and *Chlamydomonas* chloroplast DNA (Sect. 5), that of *Euglena,* which is remote from the others in an evolutionary sense, has some quite significant differences. In addition, it has recently been found that *Vicia faba* DNA has only one set of ribosomal RNA cistrons (KOLLER and DELIUS, 1980), whereas other higher plants have two. In some cases, it has been possible to use cloned genes from one species as probes for the position of the corresponding genes on chloroplast DNA from other species (WHITFELD and BOTTOMLEY 1980, BOWMAN et al. 1981). If this method proves to be generally applicable, it should become possible to compare the organization of various chloroplast genomes with relative ease.

A model for the interaction of chloroplast and cytosolic protein synthesis is shown in Fig. 8. It incorporates most of the features which are currently thought to be involved. It should be remembered that each arrow in the scheme may be subjected to a regulatory process.

The regulation of the synthesis of proteins in the two systems known to contribute to chloroplast components remains one of the most interesting problems in chloroplast molecular biology. While the involvement of both systems in the synthesis of such proteins as RuBP carboxylase and ATPase make it tempting to suggest that this cooperation is a general phenomenon, not sufficient evidence is yet available to allow such a generalization. The demonstration that the protein synthesis elongation factors are of chloroplast origin in *Chlorella* and higher plants, whereas they appear to be nuclear encoded in *Euglena* (see Sect. 4.4), raises the possibility that differences in organization and regulation exist in various organisms. Another possible example of such differences is the finding that low concentrations of cycloheximide actually stimulate protein synthesis in the chloroplasts of *Euglena* (SMILLIE et al. 1971, PRICE et al. 1978) and *Chlamydomonas* (GIVAN 1974, 1979). This suggests a negative regulatory effect of a cytoplasmic product on chloroplast protein synthesis, which has not been observed in higher plants. The finding that small subunit of RuBP carboxylase accumulates when chloroplast protein synthesis is inhibited by high temperature in rye (FEIERABEND 1978), while a blockage of large subunit synthesis by a chloroplast DNA mutation in *Oenothera* results in the complete absence of small subunit (HALLIER et al. 1978), also suggest that there may be differences in the mechanism of regulation. While evidence is emerging that the mechanism of transport of cytoplasmically synthesized polypeptides across the chloroplast membrane involves transit sequences (see Sect. 5), the recent suggestion by GIBBS (1979) that in some classes of algae, which appear to have rough endoplasmic

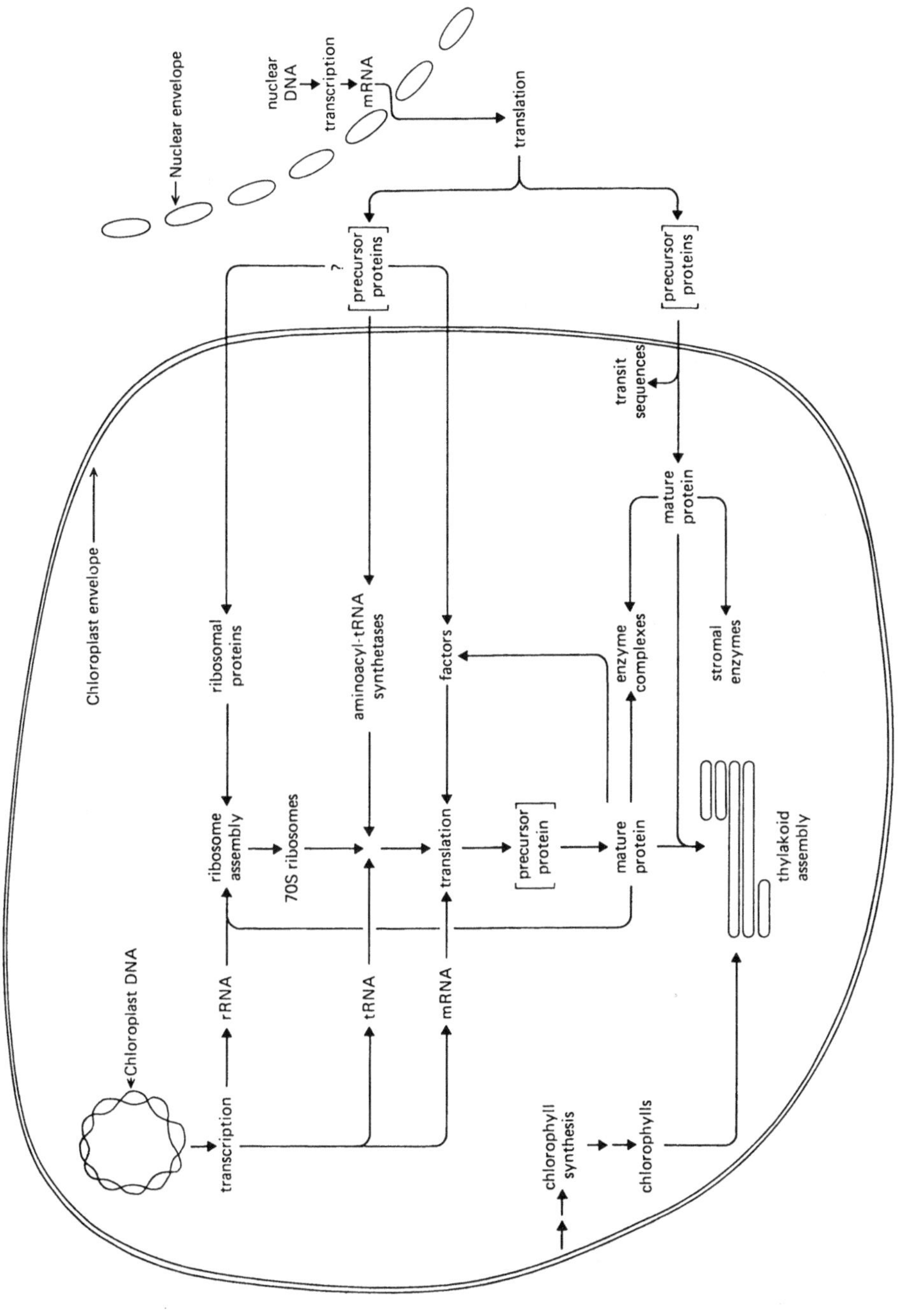

Fig. 8. A model for the interaction of the chloroplast and cytosolic systems in the synthesis of chloroplast proteins

reticulum associated with the chloroplasts, this structure may facilitate communication between the compartments, indicates that many plant families need to be investigated before any general transport and regulatory mechanisms can be postulated.

The suggestion that RNA is transported across chloroplast membranes in barley (BRADBEER et al. 1979) and conflicting reports of this in *Euglena* (MCCREA and HERSHBERGER 1978, SCHWARTZBACH et al. 1979), raise the problem of nucleic acid transport. Since a large number of chloroplast proteins are synthesized in the cytoplasm and transported into the organelle, there seems to be no a priori reason why nucleic acids should not also be transported. However, the presence of the full complement of chloroplast tRNA genes in chloroplast DNA in addition to the ribosomal RNA cistrons suggests that RNA molecules may be unable to be transported from the cytoplasm.

A number of schemes have been proposed to explain various aspects of chloroplast transport and assembly (GOODING et al. 1973, APEL and KLOPPSTECH 1978, EDELMAN and REISFELD 1978, ELLIS and BARRACLOUGH 1978, SCHIFF 1978, BRADBEER et al. 1979). However, it is apparent that a number of conflicting observations have been made and that there is not, as yet, sufficient information to propose a satisfactory model for the regulation of chloroplast protein synthesis.

The recent rapid advances in our knowledge of the organization of the chloroplast genome can be expected to be followed by detailed knowledge of the sequence of whole genes or eventually whole genomes. This should lay the foundation for a detailed investigation of the mechanism of regulation of chloroplast gene expression. Also, new probes for the isolation of chloroplast protein genes from the nucleus should result in similar knowledge becoming available for the nuclear encoded proteins. The identification of the products of these genes is being carried out largely with the aid of heterologous cell-free systems. Although the prokaryotic nature of the chloroplasts has been established, they do not always perform as prokaryotes and it may become necessary to develop homologous in vitro systems before the detailed mechanisms of regulation can be worked out. The isolation of a protein which apparently affects the specificity of the chloroplast RNA polymerase (JOLLY and BOGORAD 1980) as well as the insensitivity of this enzyme to the bacterial inhibitor rifampicin, suggests that chloroplasts may have their own and unique mechanisms of regulation. With the knowledge gained from the application of new techniques of molecular biology, there is no doubt that an understanding of the regulation of chloroplast development, and perhaps even of the mechanism of their evolution, will soon begin to emerge.

Acknowledgement. H.J.B. wishes to thank the Deutsche Forschungsgemeinschaft (Bo 573/4) for support.

Note Added in Proof. ENEAS-FILHO et al. (1981) have found that 11–15 chloroplast ribosomal proteins are synthesized in isolated chloroplasts. GREEN (1980) found the apoprotein of the P-700 chlorophyll-a complex among the products synthesized by isolated chloroplasts from *Acetabularia.*

The site of synthesis of 4 subunits of PS-I have been assigned on the basis of the response of *Spirodela* to inhibitors (NECHUSHTAI et al. 1981). Subunits I, V & VIb were found to be of chloroplast origin whereas subunit II was synthesized in the cytosol. The site of synthesis of subunits III, IV & VIa remain to be clarified. The NADPH-protochlorophyll oxidoreductase has been shown to be made in the cytoplasm (APEL 1981) as a precursor of 8000-M_r larger than the mature protein.

The 32,000-M_r protein has been shown to be responsible for herbicide binding to PS-II (STEINBACK et al. 1981; MATTOO et al. 1982). It has been reported to be rapidly synthesized in diverse species (HOFFMAN-FALK et al. 1982) while the amino-acid sequence from two species (spinach and *Nicotiana debneyi*), as derived from the DNA sequences have been shown to be identical (ZURAWSKI et al. 1982).

An alternative approach to the study of plastid genes has been introduced by GATENBY et al. (1981) who expressed the wheat and maize genes for large subunit of RuBPCase in *E. coli* and by GATENBY and CASTLETON (1982) who expressed these genes under the transcriptional control of a bacteriophage lambda promoter.

SCHMIDT et al. (1981) have followed the transport and assembly of two polypeptides of the light-harvesting chlorophyll *a*/*b*-protein complex in peas. These polypeptides are synthesized as precursors, the extra amino-acid chain conferring water solubility on the hydrophobic protein. These are imported, processed and inserted into the thylakoid where they bind chlorophyll.

The synthesis, transport and assembly of cytoplasmically-synthesized chloroplast proteins has been reviewed by ELLIS (1981a, b). For recent data on the location of genes for plastid proteins on chloroplast DNAs see BOHNERT et al. (Chap 14, this Vol.).

References

Aliev KA, Filipovich II (1968) Differences between transfer ribonucleic acids and aminoacyl-tRNA synthetases from the chloroplasts and from the cytoplasm of pea seedlings. Mol Biol 2:297–304 (translated from Molekulyarnaya Biologia 2:364–373)

Alscher R, Smith MA, Petersen LW, Huffaker RC, Criddle RS (1976) In vitro synthesis of the large subunit of ribulose diphosphate carboxylase on 70S ribosomes. Arch Biochem Biophys 174:216–225

Anderson JM (1975) The molecular organization of chloroplast thylakoids. Biochim Biophys Acta 416:191–235

Apel K (1979) Phytochrome-induced appearance of mRNA activity for the apoprotein of the light-harvesting chlorophyll a/b protein of barley (*Hordeum vulgare*). Eur J Biochem 97:183–188

Apel K (1981) The protochlorophyllide holochrome of barley (*Hordeum vulgare* L.). Phytochrome-induced decrease of translatable mRNA coding for the NADPH: protochlorophyllide oxido-reductase. Eur J Biochem 120:89–93

Apel K, Bogorad L (1976) Light-induced increase in the activity of maize plastid DNA-dependent RNA polymerase. Eur J Biochem 67:615–620

Apel K, Kloppstech K (1978) The plastid membranes of barley (*Hordeum vulgare*). Light-induced appearance of mRNA coding for the apoprotein of the light-harvesting chlorophyll a/b protein. Eur J Biochem 85:581–588

App AA, Jagendorf AT (1964) ^{14}C-Amino acid incorporation by spinach chloroplast preparations. Plant Physiol 39:772–776

Armond PA, Staehelin LA, Arntzen CJ (1977) Spatial relationship of photosystem I, photosystem II, and the light-harvesting complex in chloroplast membranes. J Cell Biol 73:400–418

Armstrong JJ, Surzycki SJ, Moll B, Levine RP (1971) Genetic transcription and translation specifying chloroplast components in *Chlamydomonas reinhardi*. Biochemistry 10:692–701

Avadhani NG, Buetow DE (1972) Isolation of active polyribosomes from the cytoplasm, mitochondria, and chloroplasts of *Euglena gracilis*. Biochem J 128:353–365

Bachmayer H (1970) Initiation of protein synthesis in intact cells and in isolated chloroplasts of *Acetabularia mediterranea*. Biochim Biophys Acta 209:584–586

Bakker-Grunwald T (1977) ATPase. In: Trebst A, Avron M (eds) Photosynthesis I. Encyclopedia of plant physiology new ser Vol. V Springer, Berlin Heidelberg New York, pp 369–373

Barath Z, Küntzel H (1972) Cooperation of mitochondrial and nuclear genes specifying the mitochondrial genetic apparatus in *Neurospora crassa*. Proc Natl Acad Sci USA 69:1371–1374

Barnett WE, Schwartzbach SD, Hecker LI (1976) The tRNAs and aminoacyl-tRNA synthetases of *Euglena* chloroplasts. In: Bücher Th et al. (eds) Genetics and biogenesis of chloroplasts and mitochondria. Elsevier North-Holland, Amsterdam, New York, pp 661–666

Barnett WE, Schwartzbach SD, Hecker LI (1978) The transfer RNAs of eukaryotic organelles. Prog Nucleic Acid Res Mol Biol 21:143–179

Barraclough R, Ellis RJ (1979) The biosynthesis of ribulose bisphosphate carboxylase. Uncoupling of the synthesis of the large and small subunits in isolated soybean leaf cells. Eur J Biochem 94:165–177

Barraclough R, Ellis RJ (1980) Protein synthesis in chloroplasts IX. Assembly of newly-synthesized large subunit into ribulose bisphosphate carboxylase in isolated intact pea chloroplasts. Biochim Biophys Acta 608:19–31

Barrell BG, Anderson S, Bankier AT, Bruijn de MHL, Chem E, Coulson AR, Drouin J, Eperon IC, Nierloch DP, Roe BA, Sanger F, Schreier PH, Smith AJH, Staden R, Young IC (1980) Different pattern of codon recognition by mammalian mitochondrial tRNAs. Proc Natl Acad Sci USA 77:3164–3166

Baur E (1909) Das Wesen und die Erblichkeitsverhältnisse der „varietates albomarginata-chort" von Pelargonium zonale. Z Vererbungsl 1:330–351

Beauchemin N, Larue B, Cedergren RJ (1973) The characterization of the tRNAs and aminoacyl-tRNA synthetases of the blue-green alga, *Anacystis nidulans*. Arch Biochem Biophys 156:17–25

Bedbrook JR, Kolodner R (1979) The structure of chloroplast DNA. Annu Rev Plant Physiol 30:593–620

Bedbrook JR, Kolodner R, Bogorad L (1977) *Zea mays* chloroplast ribosomal RNA genes are part of a 22,000 base pair inverted repeat. Cell 11:739–749

Bedbrook JR, Link G, Coen DM, Bogorad L, Rich A (1978) Maize plastid gene expressed during photoregulated development. Proc Natl Acad Sci USA 75:3060–3064

Bedbrook JR, Coen DM, Beaton AR, Bogorad L, Rich A (1979) Location of the single gene for the large subunit of ribulosebisphosphate carboxylase on the maize chloroplast chromosome. J Biol Chem 254:905–910

Bedbrook JR, Smith S, Ellis JR (1980) Cloning and nucleotide sequence of the structural gene for the ribulose bisphosphate carboxylase small subunit. Nature (London) 287:692–697

Beisenherz WW, Koth P (1975) Influence of CAP and CHI on synthesis of RuDPCO, NADP-GDH, and chlorophyll during leucoplast-chloroplast transformation in tissue cultures of *Nicotiana tabacum*. Z Pflanzenphysiol 75:201–210

Beisenherz WW, Schneider HjAW (1974) Syntheseort von Enzymen der Chlorophyllbiosynthese. Ber Dtsch Bot Ges 87:161–166

Bennett WF, Gutierrez-Hartmann A, Butow RA (1976) The role of mitochondria-bound 80S ribosomes in mitochondrial biogenesis. In: Bücher Th et al. (eds) Genetics and biogenesis of chloroplasts and mitochondria. Elsevier North-Holland, Amsterdam, New York, pp 801–806

Bianchetti R, Lucchini G, Sartirana ML (1971) Endogenous synthesis of formyl-methionine peptides in isolated mitochondria and chloroplasts. Biochem Biophys Res Commun 42:97–102

Bingham S, Schiff JA (1979) Events surrounding the early developments of *Euglena* chloroplasts. 15. Origin of plastid thylakoid polypeptides in wild-type and mutant cells. Biochim Biophys Acta 547:512–530

Blackwood GC, Leaver CJ (1977) The effect of light on protein synthesis in green leaves. In: Bogorad L, Weil JH (eds) Acides nucléiques et synthèse des protéines chez les végétaux. CNRS, Paris, pp 611–615

Blair GE, Ellis RJ (1973) Protein synthesis in chloroplasts. I. Light-driven synthesis of the large subunit of Fraction I protein by isolated pea chloroplasts. Biochim Biophys Acta 319:223–234

Blobel G, Dobberstein B (1975a) Transfer of proteins across membranes. I. Presence of proteolytically processed and unprocessed nascent immunoglobulin light chains on membrane bound ribosomes of murine myeloma. J Cell Biol 67:835–851

Blobel G, Dobberstein B (1975b) Transfer of proteins across membranes. II. Reconstitution of functional rough microsomes from heterologous components. J Cell Biol 67:852–862

Bogorad L (1977) Genes for chloroplast ribosomal RNAs and ribosomal proteins: gene dispersal in eucaryotic genomes. In: Brinkley BR, Porter KR (eds) International cell biology 1976–77. Rockefeller Univ Press, New York, pp 175–182

Bogorad L, Davidson JN, Hanson MR (1976) Genes affecting erythromycin resistance and sensitivity of *Chlamydomonas reinhardi* chloroplast ribosomes. In: Bücher Th et al (eds) Genetics and biogenesis of chloroplasts and mitochondria. Elsevier North-Holland, Amsterdam, New York, pp 61–68

Börner Th, Knoth R, Herrmann F, Hagemann R (1972) Struktur und Funktion der genetischen Information in den Plastiden. V. Das Fehlen von ribosomaler RNS in den Plastiden der Plastommutante "Mrs. Parker" von *Pelargonium zonale* Ait. Theor Appl Genet 42:3–11

Boschetti A, Souton-Heiniger E, Schaffner J-C, Eichenberger W (1978) A two-dimensional separation of proteins from chloroplast thylakoids and other membranes. Physiol Plant 44:134–140

Bottomley W (1980) Fraction 1 protein. In: Reinert J (ed) Chloroplasts. Results and problems in cell differentiation Vol X. Springer, Berlin, Heidelberg, New York, pp 180–199

Bottomley W (1982) Coupled transcription-translation of chloroplast DNA and the translation of chloroplast RNA using *E. coli* S-30 extracts in vitro. In: Edelman M, Hallick RB, Chua NH (eds) Methods in chloroplast molecular biology. Elsevier North-Holland, Amsterdam, New York, in press

Bottomley W, Whitfeld PR (1979) Cell-free transcription and translation of total spinach chloroplast DNA. Eur J Biochem 93:31–39

Bottomley W, Spencer D, Wheeler AM, Whitfeld PR (1971) The effect of a range of RNA polymerase inhibitors on RNA synthesis in higher plant chloroplasts and nuclei. Arch Biochem Biophys 143:269–275

Bottomley W, Spencer D, Whitfeld PR (1974) Protein synthesis in isolated spinach chloroplasts: Comparison of light-driven and ATP-driven synthesis. Arch Biochem Biophys 164:106–117

Bottomley W, Higgins TJV, Whitfeld PR (1976) Differential recognition of chloroplast and cytoplasmic messenger RNA by 70S and 80S ribosomal systems. FEBS Lett 63:120–124

Bouthyette P-Y, Jagendorf AT (1978) The site of synthesis of pea chloroplast coupling factor 1. Plant Cell Physiol 19:1169–1174

Bowman CM, Koller B, Delius H, Dyer TA (1981) A physical map of wheat chloroplast DNA showing the location of the structural genes for the ribosomal RNAs and the large subunit of ribulose 1,5-bisphosphate carboxylase. Mol Gen Genet 183:93–101

Boynton JE, Burton WG, Gillham NW, Harris EH (1973) Can a non-mendelian mutation affect both chloroplast and mitochondrial ribosomes? Proc Natl Acad Sci USA 70:3463–3467

Boynton JE, Gillham NW, Harris EH, Tingle CL, Winkle-Swift K, Adams GMW (1976) Transmission, segregation and recombination of chloroplast genes in *Chlamydomonas*. In: Bücher Th et al. (eds) Genetics and biogenesis of chloroplasts and mitochondria. Elsevier North-Holland, Amsterdam, New York, pp 313–322

Bradbeer JW, Börner Th (1978) Activities of glyceraldehyde-phosphate dehydrogenase ($NADP^+$) and phosphoribulokinase in two barley mutants deficient in chloroplast ribosomes. In: Akoyunoglou G, Argyroudi-Akoyunoglou JH (eds) Chloroplast development. Elsevier North-Holland, Amsterdam, New York, pp 727–732

Bradbeer JW, Atkinson YE, Börner T, Hagemann R (1979) Cytoplasmic synthesis of plastid polypeptides may be controlled by plastid-synthesized RNA. Nature (London) 279:816–817

Brady CJ, Scott NS (1976) The persistence of plastid polyribosomes and fraction 1 protein synthesis in ageing wheat leaves. In: Bogorad L, Weil JH (eds) Acides nucléiques et synthèse des protéines chez les végétaux. CNRS, Paris, pp 387–393

Brantner JH, Dure LS (1975) The developmental biochemistry of cotton seed embryogenesis and germination. VI. Levels of cytosol and chloroplast aminoacyl-tRNA synthetases during cotyledon development. Biochim Biophys Acta 414:99–114

Breitenberger CA, Graves MC, Spremulli LL (1979) Evidence for the nuclear location of the gene for chloroplast elongation factor G. Arch Biochem Biophys 194:265–270

Brügger M, Boschetti A (1975) Two-dimensional gel electrophoresis of ribosomal proteins from streptomycin-sensitive and streptomycin-resistant mutants of *Chlamydomonas reinhardi*. Eur J Biochem 58:603–610

Burkard G, Eclancher B, Weil JH (1969) Presence of N-formyl-methionyl transfer RNA in bean chloroplasts. FEBS Lett 4:285–287

Butow RA, Bennett WF, Finkelstein DB, Kellems RE (1975) Nuclear-cytoplasmic interactions in the biogenesis of mitochondria in yeast. In: Tzagoloff A (ed) Membrane biogenesis. Mitochondria, chloroplast, and bacteria. Plenum, New York, London, pp 155–199

Callow JA (1974) Ribosomal RNA, fraction I protein synthesis, and ribulose diphosphate carboxylase activity in developing and senescing leaves of cucumber. New Phytol 73:13–20

Cashmore AR (1976) Protein synthesis in plant leaf tissue. The sites of synthesis of the major proteins. J Biol Chem 251:2848–2853

Cashmore AR (1979) Reiteration frequency of the gene coding for the small subunit of ribulose-1,5-bisphosphate carboxylase. Cell 17:383–388

Cashmore AR, Broadhurst MK, Gray RE (1978) Cell-free synthesis of leaf protein: Identification of an apparent precursor of the small subunit of ribulose-1,5-bisphosphate carboxylase. Proc Natl Acad Sci USA 75:655–659

Chelm BK, Hoben PJ, Hallick RB (1977) Expression of the chloroplast ribosomal RNA genes of *Euglena gracilis* during chloroplast development. Biochemistry 16:776–781

Chelm BK, Gray PW, Hallick RB (1978) Mapping of transcribed regions of *Euglena gracilis* chloroplast DNA. Biochemistry 17:4239–4244

Chua N-H, Blomberg F (1979) Immunochemical studies of thylakoid membrane polypeptides from spinach and *Chlamydomonas reinhardtii*. A modified procedure for crossed immunoelectrophoresis of dodecyl sulfate protein complexes. J Biol Chem 254:215–223

Chua N-H, Gillham NW (1977) The sites of synthesis of the principal thylakoid membrane polypeptides in *Chlamydomonas reinhardtii* J Cell Biol 74:441–452

Chua N-H, Schmidt GW (1978a) In vitro synthesis, transport, and assembly of ribulose 1,5-bisphosphate carboxylase subunits. In: Siegelman HW, Hind G (eds) Photosynthetic carbon assimilation. Plenum, New York, London, pp 325–347

Chua N-H, Schmidt GW (1978b) Post-translational transport into intact chloroplasts of a precursor to the small subunit of ribulose-1,5-bisphosphate carboxylase. Proc Natl Acad Sci USA 12:6110–6114

Chua N-H, Blobel G, Siekevitz P (1973) Isolation of cytoplasmic and chloroplast ribosomes and their dissociation into active subunits from *Chlamydomonas reinhardtii*. J Cell Biol 57:798–814

Ciferri O, Tiboni O (1973) Elongation factors for chloroplast and mitochondrial protein synthesis in *Chlorella vulgaris*. Nature (London) New Biol 245:209–211

Ciferri O, Tiboni O (1976) Evidence for the synthesis in the chloroplast of elongation factor G. Plant Sci Lett 7:455–466

Ciferri O, Di Pasquale G, Tiboni O (1979) Chloroplast elongation factors are synthesized in the chloroplast. Eur J Biochem 102:331–335

Cobb AH, Wellburn AR (1976) Polypeptide binding to plastid envelopes during chloroplast development. Planta 129:127–131

Coen DM, Bedbrook JR, Bogorad L, Rich A (1977) Maize chloroplast DNA fragment encoding the large subunit of ribulosebisphosphate carboxylase. Proc Natl Acad Sci USA 74:5487–5491

Correns C (1909) (1) Vererbungsversuche mit blaß(gelb)grünen und buntblättrigen Sippen bei *Mirabilis jalapa*, *Urtica pilulifera* und *Lunaria annua*. Z Vererbungsl 1:291–329
Criddle RS, Dau B, Kleinkopf GE, Huffaker RC (1970) Differential synthesis of ribulose-diphosphate carboxylase subunits. Biochim Biophys Res Commun 41:621–627
Davidson JN, Hanson MR, Bogorad L (1974) An altered chloroplast ribosomal protein in *ery*-Ml mutants of *Chlamydomonas reinhardi*. Mol Gen Genet 132:119–129
Davis B, Merrett MJ (1975) The glycolate pathway and photosynthetic competence in *Euglena*. Plant Physiol 55:30–34
Davis BD, Tai PC (1980) The mechanism of protein secretion across membranes. Nature (London) 283:433–438
Delepelaire P, Chua N-H (1979) Lithium dodecyl sulfate/polyacrylamide gel electrophoresis of thylakoid membranes at 4° C: characterizations of two additional chlorophyll a-protein complexes. Proc Natl Acad Sci USA 76:111–115
Dickmann DI, Gordon JC (1975) Incorporation of ^{14}C-photosynthate into protein during leaf development in young *Populus* plants. Plant Physiol 56:23–27
Dobberstein B, Blobel G, Chua N-H (1977) In vitro synthesis and processing of a putative precursor for the small subunit of ribulose-1,5-bisphosphate carboxylase of *Chlamydomonas reinhardtii*. Proc Natl Acad Sci USA 74:1082–1085
Doherty A, Gray JC (1979) Synthesis of cytochrome f by isolated pea chloroplasts. Eur J Biochem 98:87–92
Doherty A, Gray JC (1980) Synthesis of a dicyclohexylcarbodiimide-binding proteolipid by isolated pea chloroplasts. Eur J Biochem 108:131–136
Douce R, Holtz RB, Benson AA (1973) Isolation and properties of the envelope of spinach chloroplasts. J Biol Chem 248:7215–7222
Driesel AJ, Crouse EJ, Gordon K, Bohnert HJ, Herrmann RG, Steinmetz A, Mubumbila M, Keller M, Burkhard J, Weil JH (1979) Fractionation and identification of spinach chloroplast transfer RNAs and mapping of their genes on the restriction map of chloroplast DNA. Gene 6:285–306
Driesel AJ, Speirs J, Bohnert HJ (1980) Spinach chloroplast mRNA for a 32 Kd polypeptide. Size and localization on the physical map of the chloroplast DNA. Biochem Biophys Acta 610:297–310
Dure III LS, Harris B (1977) Regulation of translation of stored mRNA of dicot seeds. In: Bogorad L, Weil JH (eds) Nucleic acids and protein synthesis in plants. Plenum, New York, London, pp 279–291
Dyer TA, Miller RH, Greenwood AD (1971) Leaf nucleic acids: I. Characteristics and role in the differentiation of plastids. J Exp Bot 22:125–136
Eaglesham ARJ, Ellis RJ (1974) Protein synthesis in chloroplasts. II. Light-driven synthesis of membrane proteins by isolated pea chloroplasts. Biochim Biophys Acta 335:396–407
Edelman M, Reisfeld A (1978) Characterization, translation and control of the 32,000 dalton chloroplast membrane protein in *Spirodela*. In: Akoyunoglou G, Argyroudi-Akoyunoglou JH (eds) Chloroplast development. Elsevier North-Holland, Amsterdam, New York, pp 641–652
Eisenstadt J, Brawerman G (1963) The incorporation of amino acids into the protein of chloroplasts and chloroplast ribosomes of *Euglena gracilis*. Biochim Biophys Acta 76:319–321
Ellis RJ (1969) Chloroplast ribosomes: Stereospecificity of inhibition by chloramphenicol. Science 163:477–478
Ellis RJ (1975) Inhibition of chloroplast protein synthesis by lincomycin and 2-(4-methyl-2,6-dinitroanilino)-N-methylpropionamide. Phytochemistry 14:89–93
Ellis RJ (1977a) Protein synthesis by isolated chloroplasts. Biochim Biophys Acta 463:185–215
Ellis RJ (1977b) The synthesis of chloroplast proteins. In: Bogorad L, Weil JH (eds) Nucleic acids and protein synthesis in plants. Plenum, New York, London, pp 195–212
Ellis RJ (1978) Chloroplast proteins and their synthesis. In: Norton G (ed) Plant proteins, Butterworth, London, pp 25–40
Ellis RJ (1981a) Protein transport across membranes: an introduction. Biochem Soc Symp 46:223–234

Ellis RJ (1981b) Chloroplast proteins: Synthesis, transport and assembly. Annu Rev Plant Physiol 32:111–137

Ellis RJ, Barraclough R (1978) Synthesis and transport of chloroplast proteins inside and outside the cell. In: Akoyunoglou G, Agyroudi-Akoyunoglou JH (eds) Chloroplast development. Elsevier North-Holland, Amsterdam, New York, pp 185–194

Ellis RJ, Hartley MR (1971) Sites of synthesis of chloroplast proteins. Nature (London) New Biol 233:193–196

Ellis RJ, Highfield PE, Silverthorne J (1978) The synthesis of chloroplast proteins by subcellular systems. In: Hall DO, Coombs J, Goodwin TW (eds) Photosynthesis 77. Biochem Soc, London, pp 497–506

Eneas-Filho J, Hartley MR, Mache R (1981) Pea chloroplast ribosomal proteins: characterization and site of synthesis. Mol Gen Genet 184:484–488

Ernst-Fonberg ML, Dubinskas F, Jonak ZL (1974) Comparison of two fatty acid synthetases from *Euglena gracilis* variety bacillaris. Arch Biochem Biophys 165:646–655

Feierabend J (1977) Capacity for chlorophyll synthesis in heat-bleached 70S ribosome-deficient rye leaves. Planta 135:83–88

Feierabend J (1978) Cooperation of cytoplasmic and plastidic protein synthesis in rye leaves. In: Akoyunoglou G, Argyroudi-Akoyunoglou JH (eds) Chloroplast development. Elsevier North-Holland, Amsterdam, New York, pp 207–213

Feierabend J (1979) Role of cytoplasmic protein synthesis and its coordination with plastidic protein synthesis in the biogenesis of chloroplasts. Ber Dtsch Bot Ges 92:553–574

Feierabend J, Mikus M (1977) Occurrence of a high temperature sensitivity of chloroplast ribosome formation in several higher plants. Plant Physiol 59:863–867

Feierabend J, Schrader-Reichardt U (1976) Biochemical differentiation of plastids and other organelles in rye leaves with a high-temperature-induced deficiency of plastid ribosomes. Planta 129:133–145

Feierabend J, Wildner G (1978) Formation of the small subunit in the absence of the large subunit of ribulose 1,5-bisphosphate carboxylase in 70S ribosome-deficient rye leaves. Arch Biochem Biophys 186:283–291

Felicetti L, Lipmann F (1968) Comparison of amino acid polymerization factors isolated from rat liver and rabbit reticulocytes. Arch Biochem Biophys 125:548–557

Fox L, Erion J, Tarnowski J, Spremulli L, Brot N, Weissbach H (1980) *Euglena gracilis* chloroplast EF-Ts. Evidence that it is a nuclear-coded product. J Biol Chem 255:6018–6019

Freyssinet G (1978) Determination of the site of synthesis of some *Euglena* cytoplasmic and chloroplast ribosomal proteins. Exp Cell Res 115:207–219

Freyssinet G, Schiff JA (1974) The chloroplast and cytoplasmic ribosomes of *Euglena*. II. Characterization of ribosomal proteins. Plant Physiol 53:543–554

Freyssinet G, Harris G, Schiff JA (1978) Synthesis of cytochrome c-552 in *Euglena gracilis*. In: Akoyunoglou G, Argyroudi-Akoyunoglou JH (eds) Chloroplast development. Elsevier North-Holland, Amsterdam, New York, pp 267–270

Freyssinet G, Harris GC, Nasatir M, Schiff JA (1979) Events surrounding the early development of Euglena chloroplasts. 14. Biosynthesis of cytochrome c-552 in wild type and mutant cells. Plant Physiol 63:908–915

Fuchs E (1976) The interdependence of magnesium with spermidine and phosphoenolpyruvate in an enzyme-synthesizing system in vitro. Eur J Biochem 63:15–22

Gatenby AA, Castleton JA (1982) Amplification of maize ribulose bisphosphate carboxylase large subunit synthesis in *E. coli* by transcriptional fusion with the lambda N operon. Mol Gen Genet 185:424–429

Gatenby AA, Castleton JA, Saul MW (1981) Expression in *E. coli* of maize and wheat chloroplast genes for large subunit of ribulose bisphosphate carboxylase. Nature (London) 291:117–121

Geetha V, Gnanam A (1980) Identification of P700-chlorophyll a-protein complex as a product of chloroplast protein synthesis. FEBS Lett 111:272–276

Gibbs SP (1979) The route of entry of cytoplasmically synthesized proteins into chloroplasts of algae possessing chloroplast ER. J Cell Sci 35:253–266

Gillham NW (1978) Organelle heredity. Raven, New York

Gillham NW, Boynton JE, Harris EH, Fox SB, Bolen PL (1976) Genetic control of chloroplast ribosome biogenesis in Chlamydomonas. In: Bücher Th, Neupert W, Sebald W, Werner S (eds) Genetics and biogenesis of chloroplast and mitochondria. North Holland Elsevier, Amsterdam, New York, pp 69–76

Gillham NW, Boynton JE, Chua N-H (1977) Genetic control of chloroplast proteins. In: Sanadi DR, Vernon LP (eds) Current topics in bioenergetics Vol VIII. Academic Press, London, New York, pp 211–260

Givan AL (1974) Ribulose disphosphate carboxylase synthesis in *Chlamydomonas reinhardii*: Inhibition by chloramphenicol and stimulation by cycloheximide. Planta 120:181–188

Givan AL (1979) Ribulose bisphosphate carboxylase from a mutant strain of *Chlamydomonas reinhardii* deficient in chloroplast ribosomes. The absence of both subunits and their pattern of synthesis during enzyme recovery. Planta 144:271–276

Golinska B, Legocki AB (1973) Purification and some properties of elongation factor I from wheat germ. Biochim Biophys Acta 324:156–170

Gooding LR, Roy H, Jagendorf AT (1973) Immunological identification of nascent subunits of wheat ribulose diphosphate carboxylase on ribosomes of both chloroplast and cytoplasmic origin. Arch Biochem Biophys 159:324–335

Gordon KHJ, Crouse EJ, Bohnert HJ, Herrmann RG (1981) Restriction endonuclease cleavage site map of chloroplast DNA from *Oenothera parviflora* (plastome iv). Theor Appl Genet, 59:281–296

Gray JC (1980) Maternal inheritance of cytochrome *f* in interspecific *Nicotiana* hybrids. Eur J Biochem 112:39–46

Gray JC, Kekwick RGO (1973) Synthesis of the small subunit of ribulose 1,5-diphosphate carboxylase on cytoplasmic ribosomes from greening bean leaves. FEBS Lett 38:67–69

Gray JC, Kekwick RGO (1974) The synthesis of the small subunit of ribulose 1,5-bisphosphate carboxylase in the french bean *Phaseolus vulgaris*. Eur J Biochem 44:491–500

Gray RE, Cashmore AR (1976) RNA synthesis in plant leaf tissue: the characterization of messenger RNA species lacking and containing polyadenylic acid. J Mol Biol 108:595–608

Grebanier AE, Coen DM, Rich A, Bogorad L (1978) Membrane proteins synthesized but not processed by isolated maize chloroplasts. J Cell Biol 78:734–746

Green BR (1980) Protein synthesis by isolated *Acetabularia* chloroplasts. In vitro synthesis of the apoprotein of the P-700-chlorophyll *a*-protein complex (CPI). Biochim Biophys Acta 609:107–120

Gregory P, Bradbeer JW (1973) Plastid development in primary leaves of *Phaseolus vulgaris*. The light-induced development of the chloroplast cytochromes. Planta 109:317–326

Grinius L (1980) Nucleic acid transport driven by ion gradient across cell membrane. FEBS Lett 113:1–10

Grossman A, Bartlett S, Chua N-H (1980) Energy-dependent uptake of cytoplasmically-synthesized polypeptides by chloroplasts. Nature (London) 285:625–628

Guillemaut P, Weil JH (1975) Aminoacylation of *Phaseolus vulgaris* cytoplasmic, chloroplastic and mitochondrial $tRNAs^{Met}$ and of *E. coli* $tRNAs^{Met}$ by homologous and heterologous enzymes. Biochim Biophys Acta 407:240–248

Guillemaut P, Steinmetz A, Burkard G, Weil JH (1975) Aminoacylation of $tRNA^{Leu}$ species from *E. coli* and from the cytoplasm, chloroplasts, and mitochondria of *Phaseolus vulgaris* by homologous and heterologous enzymes. Biochim Biophys Acta 378:64–72

Haff LA, Bogorad L (1976) Poly(adenylic acid)-containing RNA from plastids of maize. Biochemistry 15:4110–4115

Hagemann R, Börner Th (1978) Plastid ribosome-deficient mutants of higher plants as a tool in studying chloroplast biogenesis. In: Akoyunoglou G, Argyroudi-Akoyunoglou JH (eds) Chloroplast development. Elsevier North-Holland, Amsterdam, New York, pp 709–720

Hall TC (1979) Plant messenger RNA. In: Hall TC, Davies JW (eds) Nucleic acids in plants. CRC Press, Boca Raton, pp 217–251

Hallick RB, Gray PW, Chelm BK, Rushlow KE, Orozco EM (1978) *Euglena gracilis* chloroplast DNA structure, gene mapping, and RNA transcription. In: Akoyunoglou G, Argyroudi-Akoyunoglou JH (eds) Chloroplast development. Elsevier North-Holland, Amsterdam, New York, pp 619–622

Hallier UW, Heber UW (1977) Cytochrome *f* deficient plastome mutants of *Oenothera*. In: Miyachi S, Katoh S, Fujita Y, Shibata K (eds) Photosynthetic organelles. Structures and function. Jpn Soc Plant Physiol, Cent of Acad Publ, Tokyo, pp 257–273

Hallier UW, Schmitt JM, Heber U, Chaianova SS, Volodarsky AD (1978) Ribulose-1,5-bisphosphate carboxylase deficient plastome mutants of *Oenothera*. Biochim Biophys Acta 504:67–83

Hanson MR, Davidson JN, Mets LJ, Bogorad L (1974) Characterization of chloroplast and cytoplasmic ribosomal proteins of *Chlamydomonas reinhardi* by two-dimensional gel electrophoresis. Mol Gen Genet 132:104–118

Harris EH, Preston JF, Eisenstadt JM (1973) Amino acid incorporation and products of protein synthesis in isolated chloroplasts of *Euglena gracilis*. Biochemistry 12:1227–1234

Harris EH, Boynton JE, Gillham NW (1974) Chloroplast ribosome biogenesis in *Chlamydomonas*. Selection and characterization of mutants blocked in ribosome formation. J Cell Biol 63:160–179

Hartley MR, Wheeler A, Ellis RJ (1975) Protein synthesis in chloroplasts. V. Translation of messenger RNA for the large subunit of fraction I protein in a heterologous cell-free system. J Mol Biol 91:67–77

Haslett BG, Cammack R (1974) The development of plastocyanin in greening bean leaves. Biochem J 144:567–572

Haslett BG, Cammack R, Whatley FR (1973) Quantitative studies on ferredoxin in greening bean leaves. Biochem J 136:697–703

Hatch MD (1976) The C_4 pathway of photosynthesis: Mechanism and function. In: Burns RH, Black CC (eds) Carbon dioxide metabolism and plant productivity. Univ Park Press, Baltimore, pp 59–81

Heber U (1973) Stoichiometry of reduction and phosphorylation during illumination of intact chloroplasts. Biochim Biophys Acta 305:140–152

Hecker LI, Egan J, Reynolds RJ, Nix CE, Schiff JA, Barnett WE (1974) The sites of transcription and translation for *Euglena* chloroplastic aminoacyl-tRNA synthetases. Proc Natl Acad Sci US 71:1910–1914

Heizman P, Verdier G, Yonis H (1978) Transcription of nuclear and chloroplast genomes in *Euglena* during greening and in bleached mutants. In: Akoyunoglou G, Argyroudi-Akoyunoglou JH (eds) Chloroplast development. Elsevier North-Holland, Amsterdam, New York, pp 623–628

Herrmann RG, Possingham JV (1980) Plastid DNA-the plastome. In: Reinert J (ed) Chloroplasts. Results and problems in cell differentiation Vol X. Springer, Berlin, Heidelberg, New York, pp 45–96

Herrmann RG, Whitfeld PR, Bottomley W (1980) Construction of a SalI/Pst I restriction map of spinach chloroplast DNA using low-gelling-temperature-agarose electrophoresis. Gene 8:179–198

Highfield PE, Ellis JR (1976) Protein synthesis in chloroplasts. VII. Initiation of protein synthesis in isolated intact pea chloroplasts. Biochim Biophys Acta 447:20–27

Highfield PE, Ellis RJ (1978) Synthesis and transport of the small subunit of ribulose bisphosphate carboxylase. Nature (London) 271:420–424

Hoffman-Falk H, Mattoo AK, Marder JB, Edelman M, Ellis RJ (1982) General occurrence and structural similarity of the rapidly synthesized 32,000-dalton protein of the chloroplast membrane. J Biol Chem 257:4583–4587

Hoober JK (1970) Sites of synthesis of chloroplast membrane polypeptides in *Chlamydomonas reinhardi* y-1. J Biol Chem 245:4327–4334

Hoober JK (1976) Protein synthesis in chloroplasts In: McConkey EH (ed) Protein synthesis Vol II. Dekker, New York, pp 169–248

Hoober JK, Blobel G (1969) Characterization of the chloroplastic and cytoplasmic ribosomes of *Chlamydomonas reinhardi*. J Mol Biol 41:121–138

Horak A, Hill RD (1972) Adenosine triphosphatase of bean plastids. Its properties and site of formation. Plant Physiol 49:365–370

Howell SH, Walker LL (1976) Informational complexity of the nuclear and chloroplast genomes of *Chlamydomonas reinhardi*. Biochim Biophys Acta 418:249–256

Howell SH, Walker LL (1977) Transcription of the nuclear and chloroplast genomes during the vegetative cell cycle in *Chlamydomonas reinhardi*. Dev Biol 56:11–23

Howell SH, Heizmann P, Gelvin S, Walker LL (1977) Identification and properties of the messenger RNA activity in *Chlamydomonas reinhardi* coding for the large subunit of D-ribulose-1,5-bisphosphate carboxylase. Plant Physiol 59:464–470

Huisman JG, Bernards A, Liebregts P, Gebbink MGTh, Stegwee D (1977a) Qualitative and quantitative immunofluorescence studies of chloroplast ferredoxin. Planta 137:279–286

Huisman JG, Gebbink MGTh, Modderman P, Stegwee D (1977b) The coding site of chloroplast ferredoxin. Planta 137:97–105

Huisman JG, Moorman AFM, Verkley FN (1978) In vitro synthesis of chloroplast ferredoxin as a high molecular weight precursor in a cell-free protein synthesizing system from wheat germs. Biochem Biophys Res Commun 82:1121–1131

Ireland HMM, Bradbeer JW (1971) Plastid development in primary leaves of *Phaseolus vulgaris*. The effects of D-*threo* and L-*threo* chloramphenicol on the light-induced formation of enzymes of the photosynthetic carbon pathway. Planta 96:254–261

Iwanij V, Chua N-H, Siekevitz P (1975) Synthesis and turnover of ribulose biphosphate carboxylase and of its subunits during the cell cycle of *Chlamydomonas reinhardtii*. J Cell Biol 64:572–585

Jeannin G, Burkard G, Weil JH (1976) Aminoacylation of *Phaseolus vulgaris* cytoplasmic, chloroplastic and mitochondrial $tRNAs^{Pro}$ and $tRNAs^{Lys}$ by homologous and heterologous enzymes. Biochim Biophys Acta 442:24–31

Jeannin G, Burkard G, Weil JH (1978) Characterization of *Phaseolus vulgaris* cytoplasmic, chloroplastic and mitochondrial $tRNAs^{Phe}$; aminoacylation by homologous and heterologous enzymes. Plant Sci Lett 13:75–81

Jensen RG, Bassham JA (1966) Photosynthesis by isolated chloroplasts. Proc Natl Acad Sci USA 56:1095–1101

Jolly SO, Bogorad L (1980) Preferential transcription of cloned maize chloroplast DNA sequences by maize chloroplast RNA polymerase. Proc Natl Acad Sci USA 77:822–826

Jones BL, Nagabhushan N, Gulyas A, Zaliki S (1972) Two dimensional acrylamide gel electrophoresis of wheat leaf cytoplasmic and chloroplast ribosomal proteins. FEBS Lett 23:167–170

Jong De L, Holtrop M, Kroon AM (1980) The biogenesis of rat-liver mitochondrial ATPase. Evidence that the N,N′-dicyclohexyl carbodiimide-binding protein is synthesized outside the mitochondria. Biochim Biophys Acta 606:331–337

Joy KW, Ellis RJ (1975) Protein synthesis in chloroplasts. IV. Polypeptides of the chlorplast envelope. Biochim Biophys Acta 378:143–151

Kawashima N, Wildman SG (1972) Studies on fraction I protein. IV. Mode of inheritance of primary structure in relation to whether chloroplast or nuclear DNA contains the code for a chloroplast protein. Biochim Biophys Acta 262:42–49

Kellems RE, Allison VF, Butow RA (1974) Cytoplasmic type 80S ribosomes associated with yeast mitochondria. II. Evidence for the association of cytoplasmic ribosomes with the outer mitochondrial membrane in situ. J Biol Chem 249:3297–3303

Kirk JTO (1963) The deoxyribonucleic acid of broad bean chloroplasts. Biochim Biophys Acta 76:417–424

Kirk JTO, Tilney-Bassett (1978) The plastids. Their chemistry, structure, growth and inheritance, 2nd edn. Elsevier North-Holland, Amsterdam, New York

Kleinkopf GE, Huffaker RC, Matheson A (1970) Light-induced de novo synthesis of ribulose 1,5-diphosphate carboxylase in greening leaves of barley. Plant Physiol 46:416–418

Kloppstech K, Schweiger HG (1974) The site of synthesis of chloroplast ribosomal proteins. Plant Sci Lett 2:101–105
Knoth R, Hagemann R (1977) Struktur und Funktion der genetischen Information in den Plastiden. XVI. Die Feinstruktur der Plastiden und der elektronenmikroskopische Nachweis echter Mischzellen in Blättern der Plastommutation auslösenden Genmutante albostrians von *Hordeum vulgare* L. Biol Zentralbl 96:141–150
Knoth R, Herrmann FH, Böttger M, Börner Th (1974) Struktur und Funktion der genetischen Information in den Plastiden. XI. DNA in normalen und mutierten Plastiden der Sorte, „Mrs. Parker“ von *Pelargonium zonale*. Biochem Physiol Pflanz 166:129–148
Koller B, Delius H (1980) *Vicia faba* chloroplast DNA has only one set of ribosomal RNA genes as shown by partial denaturation mapping and R-loop analysis. Mol Gen Genet 178:261–269
Krauspe R, Parthier B (1974) Chloroplast and cytoplasmic aminoacyl-tRNA synthetases of *Euglena gracilis*. Biochem Physiol Pflanz 165:18–36
Kung SD (1977) Expression of chloroplast genomes in higher plants. Annu Rev Plant Physiol 28:401–437
Kung SD, Thornber JP, Wildman SG (1972) Nuclear DNA codes for photosystem II chlorophyll-protein of chloroplast membranes. FEBS Lett 24:185–188
Kupke DW (1962) Correlation of a soluble leaf protein with chlorophyll accumulation. J Biol Chem 237:3287–3291
Kwanyuen P, Wildman SG (1975) Nuclear DNA codes for *Nicotiana* ferredoxin. Biochim Biophys Acta 405:167–174
Langridge P (1981) Synthesis of the large subunit of ribulose bisphosphate carboxylase may involve a precursor. FEBS Lett 123:85–89
Larsson C, Albertsson PA (1974) Photosynthetic $^{14}CO_2$ fixation by chloroplast populations isolated by a polymer two-phase technique. Biochim Biophys Acta 357:412–419
Laulhere JP, Dorne AM (1977) Are cytoplasm ribosomes in chloroplast preparations functionally attached to the chloroplast membrane? Plant Sci Lett 8:251–256
Lea PJ, Norris RD (1977) tRNA and aminoacyl-tRNA synthetases from higher plants. Prog Phytochem 4:121–167
Leaver CJ (1979) Ribosomal RNA of plants. In: Hall TC, Davies JW (eds) Nucleic acids in plants Vol I. CRC Press, Boca Raton, pp 193–215
Lin S-Y, McKeehan WL, Culp W, Hardesty B (1969) Partial characterization of the enzymatic properties of the aminoacyl transfer ribonucleic acid binding enzymes. J Biol Chem 244:4340–4350
Link G, Coen DM, Bogorad L (1978) Differential expression of the gene for the large subunit of ribulose bisphosphate carboxylase in maize leaf cell types. Cell 15:725–731
Lucas-Lenard J, Lipmann F (1971) Protein biosynthesis. Annu Rev Biochem 40:409–448
Lyttleton JW (1956) Relationship between photosynthesis and a homogeneous protein component of plant cytoplasm. Nature (London) 177:283–284
Lyttleton JW (1962) Isolation of ribosomes from spinach chloroplasts. Exp Cell Res 26:312–317
Machold O, Aurich O (1972) Sites of synthesis of chloroplast lamellar proteins in *Vicia faba*. Biochim Biophys Acta 281:103–112
Machold O, Meister A, Adler K (1971) Spektroskopische Eigenschaften von elektrophoretisch getrennten Chlorophyll-Protein-Komplexen der Photosysteme I und II aus *Vica faba* und *Chlorella pyrenoidosa*. Photosynthetica 5:160–165
Machold O, Simpson DJ, Møller BL (1979) Chlorophyll-proteins of thylakoids from wild-type and mutants of barley (*Hordeum vulgare* L.). Carlsberg Res Commun 44:235–254
Malnoë P, Rochaix J-D (1978) Localization of 4S RNA genes on the chloroplast genome of *Chlamydomonas reinhardii*. Mol Gen Genet 166:269–275
Malnoë P, Rochaix J-D, Chua N-H, Spahr P-F (1979) Characterization of the gene and messenger RNA of the large subunit of ribulose 1,5-diphosphate carboxylase in *Chlamydomonas reinhardii*. J Mol Biol 133:417–434

Mandel H, Higa A (1970) Calcium dependent bacteriophage DNA infection. J Mol Biol 53:159–162

Marcu K, Dudock B (1974) Characterization of a highly efficient protein synthesizing system derived from commercial wheat germ. Nucl Acids Res 1:1385–1397

Margulies MM (1970) In vitro protein synthesis by plastids of *Phaseolus vulgaris*. V. Incorporation of ^{14}C-leucine into a protein fraction containing ribulose 1,5-diphosphate carboxylase. Plant Physiol 46:136–141

Margulies MM (1971) Concerning the site of synthesis of proteins of chloroplast ribosomes and Fraction I protein (ribulose-1,5-bisphosphate carboxylase). Biochem Biophys Res Commun 44:539–545

Margulies MM, Parenti F (1968) In vitro protein synthesis by plastids of *Phaseolus vulgaris*. III. Formation of lamellar and soluble chloroplast protein. Plant Physiol 43:504–514

Mattoo AK, Marder JB, Gressel J, Edelman M (1982) Presence of the rapidly-labelled 32,000-dalton chloroplast membrane protein in triazine resistant biotypes. FEBS Lett 140:36–40

Matson RS, Kimura T (1976) Ferredoxin biosynthesis in *Euglena gracilis*. Biochim Biophys Acta 442:76–87

Mayer AM, Shain Y (1974) Control of seed germination. Annu Rev Plant Physiol 25:167–193

McCarty R (1979) Roles of a coupling factor for photophosphorylation in chloroplasts. Annu Rev Plant Physiol 30:79–104

McCrea JM, Hershberger CL (1978) Chloroplast DNA codes for tRNA from cytoplasmic polyribosomes. Nature (London) 274:717–719

Mendiola-Morgenthaler LR, Morgenthaler JJ, Price CA (1976) Synthesis of coupling factor CF_1 protein by isolated spinach chloroplasts. FEBS Lett 62:96–100

Merrick WC, Dure III LS (1972) The developmental biochemistry of cotton seed embryogenesis and germination. IV. Levels of cytoplasmic and chloroplastic transfer ribonucleid acid species. J Biol Chem 247:7988–7999

Mets L, Bogorad L (1972) Altered chloroplast ribosomal proteins associated with erythromycin-resistant mutants in two genetic systems of *Chlamydomonas reinhardi*. Proc Natl Acad Sci USA 69:3779–3783

Mets LJ, Bogorad L (1974) Two-dimensional polyacrylamide gel electrophoresis: an improved method for ribosomal proteins. Anal Biochem 57:200–210

Milner JJ, Hershberger CL, Buetow DE (1979) *Euglena gracilis* chloroplast DNA codes for polyadenylated RNA. Plant Physiol 64:818–821

Modolell J (1971) The S-30 system from *Escherichia coli*. In: Last JA, Laskin AI (eds) Protein biosynthesis in bacterial systems. Dekker, New York, pp 1–64

Mory YY, Chen D, Sarid S (1972) Onset of deoxyribonucleic acid synthesis in germinating wheat embryos. Plant Physiol 49:20–23

Munoz-Calvo ML, Tiboni O, Ciferri O (1977) Immunological specificity of the mitochondrial elongation factor G of *Chlorella vulgaris*. Plant Sci Lett 8:345–353

Murray DR, Wara-Aswapati O, Ireland HMM, Bradbeer JW (1973) The development of activities of some enzymes concerned with glycollate metabolism in greening bean leaves. J Exp Bot 24:175–184

Nakazato H, Venkatesan S, Edmonds M (1975) Polyadenylic acid sequences in *E. coli* messenger RNA. Nature (London) 256:144–146

Nechushtai R, Nelson N, Mattoo AK, Edelman M (1981) Site of synthesis of subunits to photosystem I reaction centre and the proton-ATPase in *Spirodela*. FEBS Lett 125:115–119

Nelson N, Nelson H, Schatz G (1980) Biosynthesis and assembly of the proton-translocating adenosine triphosphatase complex from chloroplasts. Proc Natl Acad Sci USA 77:1361–1364

Nirenberg MW (1963) Cell-free protein synthesis directed by messenger RNA. Methods Enzymol 6:17–23

Nirenberg MW, Matthaei JH (1961) The dependence of cell-free protein synthesis in

E. coli upon naturally occurring or synthetic polyribonucleotides. Proc Natl Acad Sci USA 47:1588–1602

Nobrega FG, Tzagoloff A (1980) Assembly of the mitochondrial membrane system. Structure and location of the mitochondrial glutamic tRNA gene in *Saccharomyces cerevisiae*. FEBS Lett 113:52–54

Norris RD, Lea PJ, Fowden L (1973) Aminoacyl-tRNA synthetases in *Triticum aestivum* L. during development and germination. J Exp Bot 24:615–625

O'Brien T (1976) Mitochondrial protein synthesis. In: McConkey EH (ed) Protein synthesis Vol II. Dekker, New York, pp 249–307

Ochoa S, Haro de C (1979) Regulation of protein synthesis in eukaryotes. Annu Rev Biochem 48:549–580

Pappenheimer Jr AM (1977) Diphteria toxin. Annu Rev Biochem 46:69–94

Parisi B, Cella R (1971) Origin of the ribosome specific factors responsible for peptide chain elongation in yeast. FEBS Lett 14:209–213

Parthier B (1973) Cytoplasmic site of synthesis of chloroplast aminoacyl-tRNA synthetases in *Euglena gracilis*. FEBS Lett 38:70–74

Parthier B, Krauspe R (1974) Chloroplast and cytoplasmic transfer RNA of *Euglena gracilis*. Transfer RNA^{Leu} of blue-green algae as a substitute for chloroplast $tRNA^{Leu}$. Biochem Physiol Pflanz 165:1–17

Paterson BM, Roberts BE, Kuff EL (1977) Structural gene identification and mapping by DNA·mRNA hybrid-arrested cell-free translation. Proc Natl Acad Sci USA 74:4370–4374

Pelham HRB, Jackson RJ (1976) An efficient mRNA-dependent translation system from reticulocyte lysates. Eur J Biochem 67:247–256

Pestka S (1971) Inhibitors of ribosome functions. Annu Rev Biochem 40:697–710

Pestka S (1976) Insights into protein biosynthesis and ribosome function through inhibitors. Prog Nucleic Acid Res Mol Biol 17:217–245

Pick U, Racker E (1979) Purification and reconstitution of the N,N'-dicyclohexylcarbodiimide sensitive ATPase complex from spinach chloroplasts. J Biol Chem 254: 2793–2799

Price CA, Ortiz W, Gaynor JJ (1978) Regulation of protein synthesis in isolated chloroplasts of *Euglena gracilis*. In: Akoyunoglou G, Argyroudi-Akoyunoglou (eds) Chloroplast development. Elsevier North-Holland, Amsterdam, New York, pp 257–267

Racker E (1967) Resolution and reconstitution of the inner mitochondrial membrane. Fed Proc 26:1335–1340

Racker E (1977) Mechanisms of energy transformations. Annu Rev Biochem 46:1006–1014

Ramirez JM, Campo del FF, Arnon DI (1968) Photosynthetic phosphorylation as energy source for protein synthesis and carbon dioxide assimilation by chloroplasts. Proc Natl Acad Sci USA 59:606–612

Rawson JRY, Boerma CL (1976) A measurement of the fraction of chloroplast DNA transcribed during chloroplast development in *Euglena gracilis*. Biochemistry 15:588–592

Rawson JRY, Boerma CL (1979) Hybridization of EcoRI chloroplast DNA fragments of *Euglena* to pulse labeled RNA from different stages of chloroplast development. Biochem Biophys Res Commun 89:743–749

Rawson JRY, Kushner SR, Vapneck D, Alton NK, Boerma CL (1978) Chloroplast ribosomal RNA genes in *Euglena gracilis* exist as three clustered tandem repeats. Gene 3:191–209

Reger BJ, Fairfield SA, Epler JL, Barnett WE (1970) Identification and origin of some chloroplast aminoacyl-tRNA synthetases and tRNAs. Proc Natl Acad Sci US 67:1207–1213

Reisfeld A, Gressel J, Jakob KM, Edelman M (1978a) Characterization of the 32,000 dalton membrane protein I. Early synthesis during photoinduced plastid development of *Spirodela*. Photochem Photobiol 27:161–165

Reisfeld A, Jakob KM, Edelman M (1978b) Characterization of the 32,000 dalton chloroplast membrane protein. II. The molecular weight of chloroplast messenger RNAs translating the precursor to P-32,000 and full size RuDP carboxylase large subunit. In: Akoyunoglou G, Argyroudi-Akoyunoglou JH (eds) Chloroplast development. Elsevier North-Holland, Amsterdam, New York, pp 669–674

Richter D (1971) Production of mitochondrial peptide-chain elongation factors in yeast deficient in mitochondrial deoxyribonucleic acid. Biochemistry 10:4422–4425

Ris H, Plaut W (1962) The ultrastructure of DNA-containing areas in the chloroplasts of *Chlamydomonas*. J Cell Biol 13:383–391

Roberts BE, Paterson BM (1973) Efficient translation of tobacco mosaic virus RNA and rabbit globin 9S RNA in a cell-free system from commercial wheat germ. Proc Natl Acad Sci USA 70:2330–2334

Roberts BE, Gorecki M, Mulligan RC, Danna KJ, Rosenblatt S, Rich A (1975) Simian Virus 40 DNA directs synthesis of authentic viral polypeptides in a linked transcription-translation cell-free system. Proc Natl Acad Sci USA 72:1922–1926

Rochaix JD (1978) Restriction endonuclease map of the chloroplast DNA of *Chlamydomonas reinhardii*. J Mol Biol 126:597–617

Rosner A, Jakob KM, Gressel J, Sagher D (1975) The early synthesis and possible function of a 0.5×10^6 M_r RNA after transfer of dark-grown *Spirodela* plants to light. Biochem Biophys Res Commun 67:383–391

Rothman JE, Lenard J (1977) Membrane asymmetry. Science 195:743–753

Roy H, Gooding LR, Jagendorf AT (1973) Formation, release, and identification of peptidyl-[^{3}H]puromycin from wheat leaf ribosomes in vitro. Arch Biochem Biophys 159:312–323

Roy H, Patterson R, Jagendorf AT (1976) Identification of the small subunit of ribulose 1,5-bisphosphate carboxylase as a product of wheat leaf cytoplasmic ribosomes. Arch Biochem Biophys 172:64–73

Roy H, Terenna B, Cheong LC (1977) Synthesis of the small subunit of ribulose-1, 5-bisphosphate carboxylase by soluble fraction polyribosomes of pea leaves. Plant Physiol 60:532–537

Sagher D, Grosfeld H, Edelman M (1976) Large subunit ribulosebisphosphate carboxylase messenger RNA from *Euglena* chloroplasts. Proc Natl Acad Sci USA 73:722–726

Salisbury JL, Vasconcelos AC, Floyd GL (1975) Isolation of intact chloroplasts of *Euglena gracilis* by isopycnic sedimentation in gradients of silica. Plant Physiol 56:399–403

Sano H, Spaeth E, Burton WG (1979) Messenger RNA of the large subunit of ribulose-1,5-bisphosphate carboxylase from *Chlamydomonas reinhardi*. Isolation and properties. Eur J Biochem 93:173–180

Schäfer H-A, Feierabend J (1976) Ultrastructural differentiation of plastids and other organelles in rye leaves with a high-temperature-induced deficiency of plastid ribosomes. Cytobiol 14:75–90

Schatz G, Mason TL (1974) The biosynthesis of mitochondrial proteins. Annu Rev Bioch 43:51–87

Schiff JA (1971) Developmental interactions among cellular compartments in *Euglena*. In: Boardman NK, Linnane AW, Smillie RM (eds) Autonomy and biogenesis of mitochondria and chloroplasts. Elsevier North-Holland, Amsterdam, New York, pp 98–118

Schiff JA (1973) The development, inheritance, and origin of the plastid in *Euglena*. Adv Morphol 10:265–312

Schiff JA (1978) Photocontrol of chloroplast development in *Euglena*.In: Akoyunoglou G, Argyroudi-Akoyunoglou JH (eds) Chloroplast development. Elsevier North-Holland, Amsterdam, New York, pp 747–767

Schlanger G, Sager R (1974) Localization of five antibiotic resistances at the subunit level in chloroplast ribosomes of *Chlamydomonas*. Proc Natl Acad Sci USA 71:1715–1719

Schmidt GW, Devillers-Thiery A, Desruisseaux H, Blobel G, Chua N-H (1979) NH_2-terminal amino acid sequences of precursor and mature forms of the ribulose-1,5-bisphosphate carboxylase small subunit from *Chlamydomonas reinhardtii*. J Cell Biol 83:615–622

Schmidt GW, Bartlett S, Grossman AR, Cashmore AR, Chua N-H (1980) In vitro synthesis, transport, and assembly of the constituent polypeptides of the light-harvesting chlorophyll a/b protein complex. In: Leaver CJ (ed) Genome organization and expression in plants, Plenum, New York, London, pp 337–351

Schmidt GW, Bartlett SG, Grossman AR, Cashmore AR, Chua N-H (1981) Biosynthetic pathways of two polypeptide subunits of the light-harvesting chlorophyll *a/b*-protein complex. J Cell Biol 91:468–478

Schmitt JM, Herrmann RG (1977) Fractionation of cell organelles in silica sol gradients. Methods Cell Biol 15:177–200

Schneeman R, Surzycki S (1979) *E. coli* ribosomal proteins are cross reactive with antibody prepared against *Chlamydomonas reinhardi* chloroplast ribosomal subunit. Mol Gen Genet 176:95–104

Schwartzbach SD, Schiff JA (1974) Chloroplast and cytoplasmic ribosomes of *Euglena*: Selective binding of dihydrostreptomycin to chloroplast ribosomes. J Bacteriol 120:334–341

Schwartzbach SD, Schiff JA (1979) Events surrounding the early development of *Euglena* chloroplasts 13. Photocontrol of protein synthesis. Plant Cell Physiol 20:827–838

Schwartzbach SD, Barnett WE, Hecker LI (1979) Evidence that *Euglena* chloroplasts do not exports tRNAs. Nature (London) 280:86–87

Scragg AH (1971) Chain elongation factors of yeast mitochondria. FEBS Lett 17:111–114

Seal SN, Giesen M, Roman R, Marcus A (1977) Functional characterization of the initiation factors of wheat germ. In: Bogorad L, Weil JH (eds) Nucleic acid and protein synthesis in plants. Plenum, New York, London, pp 167–181

Sebald W, Sebald-Althaus M, Wächter E (1977) In: Bandlow W, Schweyen RH, Wolf K, Kaudewitz F (eds) Mitochondria 1977. Genetics and biogenesis of mitochondria. de Gruyter, Berlin, pp 433–440

Seckback J (1972) Electron microscopical observations of leaf ferritin from iron-treated *Xanthium* plants: Localization and diversity in the organelle. J Ultrastruct Res 39:65–76

Sen S, Payne PI, Osborne DJ (1975) Early RNA synthesis during the germination of rye (*Secale cereale*) embryos and the relationship to early protein synthesis. Biochem J 148:381–387

Shepherd HS, Boynton JE, Gillham NW (1979) Mutations in nine chloroplast loci of *Chlamydomonas* affecting different photosynthetic functions. Proc Natl Acad Sci USA 76:1353–1357

Shumway LK, Weier TE (1967) The chloroplast structure of iojap maize. Am J Bot 54:773–780

Sidell SG, Ellis RJ (1975) Protein synthesis in chloroplasts. Characteristics and products of protein synthesis in vitro in etioplasts and developing chloroplasts from pea leaves. Biochem J 146:675–685

Silverthorne J, Ellis RJ (1980) Protein synthesis in chloroplasts. VIII. Differential synthesis of chloroplast proteins during spinach leaf development. Biochim Biophys Acta 607:319–330

Sirevag R, Levine RP (1972) Fatty acid synthetase from *Chlamydomonas reinhardi*. Sites of transcription and translation. J Biol Chem 247:2586–2591

Smillie RM, Bishop DG, Gibbons GC, Graham D, Grieve AM, Raison JK, Reger BJ (1971) In: Boardman NK, Limnane AW, Smillie RM (eds) Autonomy and biogenesis of mitochondria and chloroplasts. Elsevier North-Holland, Amsterdam, New York, pp 422–433

Smith MA, Criddle RS, Peterson L, Huffaker RC (1974) Synthesis and assembly of ribulosebisphosphate carboxylase enzyme during greening of barley plants. Arch Biochem Biophys 165:494–504

Smith SM, Ellis RJ (1979) Processing of small subunit precursor of ribulose bisphosphate carboxylase and its assembly into whole enzyme are stromal events. Nature (London) 278:662–664

Spencer D, Wildman SG (1964) The incorporation of amino acids into protein by cell-free extracts from tobacco leaves. Biochemistry 3:954–959

Spiegel S, Marcus A (1975) Polyribosome formation in early wheat embryo germination independent of either transcription or polyadenylation. Nature (London) 256:228–230

Spiess H (1977) Analysis of the chloroplast ribosomal proteins from *Chlamydomonas reinhardii*, streptomycin-resistant and dependent mutants by two-dimensional gel electrophoresis. Plant Sci Lett 10:103–113

Spremulli LL, Walthall BJ, Lax SR, Ravel JM (1977) Purification and properties of a met-tRNA binding factor from wheat germ. Arch Biochem Biophys 178:565–575

Sprey B (1972) Ribosomale RNA und Thylakoidmembranen in Plastiden von Chlorophylldefektmutanten der Gerste. Z Pflanzenphysiol 67:223–243

Steinback KE, McIntosh L, Bogorad L, Arntzen CJ (1981) Identification of the triazine receptor protein as a chloroplast gene product. Proc Natl Acad Sci USA 78:7463–7467

Stöffler G, Wool IG, Lin A, Rak KH (1974) The identification of the eucaryotic ribosomal proteins homologous with *Escherichia coli* proteins L7 and L12. Proc Natl Acad Sci USA 71:4723–4726

Tewari KK (1979) Structure and replication of chloroplast DNA. In: Hall TC, Davies JW (eds) Nucleic acids in plants Vol I. CRC Press, Boca Raton, pp 41–108

Thornber JP (1975) Chlorophyll-proteins. Light-harvesting and reaction center components of plants. Annu Rev Plant Physiol 26:127–158

Thornber JP, Highkin HR (1974) Composition of the photosynthetic apparatus of normal barley leaves and a mutant lacking chlorophyll b. Eur J Biochem 41:109–116

Tiboni O, Di Pasquale G, Ciferri O (1978) Purification of the elongation factors present in spinach chloroplasts. Eur J Biochem 92:471–477

Tobin EM (1978) Light regulation of specific mRNA species in *Lemna gibba* L.G-3. Proc Natl Acad Sci USA 75:4749–4753

Tobin EM, Suttie JL (1980) Light effects on the synthesis of ribulose-1, 5-bisphosphate carboxylase in *Lemna gibba* L.G-3. Plant Physiol 65:641–647

Tzagoloff A, Macino G, Sebald W (1979) Mitochondrial genes and translation products. Annu Rev Biochem 48:419–441

Vasconcelos AC (1976) Synthesis of proteins by isolated *Euglena gracilis* chloroplasts. Plant Physiol 58:719–721

Weeks DP, Marcus A (1971) Preformed messenger of quiescent wheat embryos. Biochim Biophys Acta 232:671–684

Weil JH (1979) Cytoplasmic and organellar tRNAs in plants. In: Hall TC, Davies JW (eds) Nucleic acids in plants Vol I CRC Press, Boca Raton, pp 143–192

Weil JH, Burkard G, Guillemaut P, Jeannin G, Martin R, Steinmetz A (1976) tRNAs and aminoacyl-tRNA synthetases in plant organelles. In: Bücher Th, Neupert W, Sebald W, Werner S (eds) Genetics and biogenesis of chloroplasts and mitochondria. Elsevier North-Holland, Amsterdam, New York, pp 667–675

Weinbaum SA, Gressel J, Reisfeld A, Edelman M (1979) Characterization of the 32,000 dalton chloroplast membrane protein. III. Probing its biological function in *Spirodela*. Plant Physiol 64:828–832

Weissbach H, Ochoa S (1976) Soluble factors required for eukaryotic protein synthesis. Annu Rev Biochem 45:191–216

Wheeler AM, Hartley MR (1975) Major mRNA species from spinach chloroplasts do not contain poly(A). Nature (London) 257:66–67

Whitfeld PR (1977) Chloroplast RNA. In: Stewart PR, Letham DS (eds) The ribonucleic acids, 2nd edn. Springer, Berlin, Heidelberg, New York, pp 297–332

Whitfeld PR, Bottomley W (1980) Mapping of the gene for the large subunit of ribulose bisphosphate carboxylase on spinach chloroplast DNA. Biochem Int 1:172–178

Whitfeld PR, Spencer D, Bottomley W (1973) Products of chloroplast DNA-directed transcription and translation. In: Pollok JK, Lee WJ (eds) The biochemistry of gene expression in higher organisms. Australian and N Z Book Corp, Sydney, pp 504–522

Wickner W (1979) The assembly of proteins into biological membranes: The membrane trigger hypothesis. Annu Rev Biochem 48:23–45

Wildner GF (1976) The kinetics of appearance of chloroplast proteins and the effect of cycloheximide and chloramphenicol on their synthesis. Z Naturforsch 31(c):157–162

Wool IG (1979) The structure and function of eukaryotic ribosomes. Annu Rev Biochem 48:719–754

Wright RD, Kanabus J, Cherry JH (1974) Multiple leucyl-tRNA synthetases in pea seedlings. Plant Sci Lett 2:347–355

Wu R (ed) (1980) Methods in enzymology Vol. 68. Recombinant DNA. Academic Press, London , New York

Yurina NP, Odintsova MS (1974) Buoyant density of chloroplast ribosomes in CsCl. Plant Sci Lett 3:229–234

Zielinski RE, Price CA (1980) Synthesis of thylakoid membrane proteins by chloroplasts isolated from spinach. Cytochrome b_{559} and P700-chlorophyll a-protein. J Cell Biol 85:435–445

Zubay G, Chambers DA, Cheong LC (1970) Cell-Free studies on the regulation of the lac operon. In: Beckwith J, Siper D (eds) The lac operon. Cold Spring Harbor Labs, New York, pp 375–391

Zurawski G, Bohnert HJ, Whitfeld PR, Bottomley W (1982) Nucleotide sequence of the gene for the 32,000-M_r thylakoid membrane protein from *Spinacia oleracea* and *Nicotiana debeneyi* predicts a totally conserved primary translation product of M_r 38,950. Proc Natl Acad Sci USA (in press)

16 Use of Mutants in the Study of Chloroplast Biogenesis

K.W. HENNINGSEN and B.M. STUMMANN

1 Introduction

In order to gain information on the relation between structure, composition, and function of chloroplasts, an important approach is to undertake studies on the development of this organelle. The developmental approach can be divided into two lines of study: (1) By the use of wild-type plants in greening systems, where the sequential developmental steps can be resolved by means of synchronously occurring development of plastids during their light-induced development from etioplasts or plastids that have been dedifferentiated during a preceding growth in darkness. (2) By the use of mutants with lesions blocking chloroplast development at particular steps. A most powerful analytical system is obtained when the greening approach is combined by the use of mutants and the results subsequently compared with results from studies of the wild type.

The mutants might simply be used as a tool in biochemical or physiological studies to dissect particular steps in chloroplast development or to establish relations between structure, composition, and function. The mutants can also be used to identify genes with structural or regulatory functions in the synthesis of chloroplast-specific components. A further analysis will allow the genes to be located in either the nuclear or the chloroplast genome. As more genes with known function and location are identified, it will be possible to elucidate how the coordination between the nuclear and the chloroplast genomes functions to regulate the expression of genes with functions required in specific steps of the developmental sequence leading from proplastids to fully differentiated chloroplasts.

2 Induction, Detection, and Characterization of Mutants Defective in Plastid Development

The strategy for the work on mutant production strongly depends on the organisms of choice and the primary problem to be solved by the use of the resulting

Abbreviations. ALA, δ-aminolevulinic acid; Chl, chlorophyll; Chl-ide, chlorophyllide; CF_1, coupling factor 1 of ATPase; CP I and II, chlorophyll-protein complexes of reaction centers of photosystem I and II; DCIP, dichlorophenol indophenol; DCMU, 3-(3′,4′-dichlorophenyl) 1,1-dimethylurea; EF, endoplasmic fracture (of freeze-etched thylakoids); EPR, electron paramagnetic resonance, Pchl-ide, protochlorophyllide; PF, protoplasmic fracture; PF_s, protoplasmic fracture in stacked region; PS I and II, photosystem I and II; P-700, pigment absorbing at 700 nm (a chlorophyll a).

mutant material. In the following we are confining ourselves to the use made of mutants in green algae and higher plants where sexual recombination allows the location of genes to either the nuclear or chloroplast genome. However, the considerations and techniques used for photosynthetic bacteria (MARS et al. 1980, VAN BAALEN 1980) are rather similar to those used with green algae and *Euglena* (SCHIFF 1973, SCHMIDT and LYMAN 1976).

The induction of mutations can be performed either by irradiation or by chemical mutagens. Irradiation with either X-rays or γ-rays are convenient treatments for general purpose mutagenesis. Since a large fraction of the induced mutants results from chromosomal aberrations like deficiencies and translocations, irradiation is not the best choice for induction of single gene mutations. For both algae (GILLHAM 1965, LOPPES 1968) and higher plants (MILES 1980) the chemical mutagens ethyl methane sulfonate (EMS) and N-methyl-N′-nitro-N-nitrosoguanidine (NG) are very effective for induction of mutants. For higher plants chemical mutagens like sodium azide (NILAN et al. 1973), ethylene oxide, and ethylene imine (BLIXT 1972) have also been used. By the use of these and other chemical mutagens, point mutations, as well as small deletions and additions, can be induced with equal ease and alter single genes located in either the nuclear or the plastid genomes.

Mutations induced in nuclear genes can easily be detected in green algae since these organisms are haploid and a clone of mutant cells will form directly from a mutagenized cell. Higher plants are diploid and a mutation in one of the nuclear genomes can, if the mutant allele is a dominant, be detected in the M_1 generation or, if recessive, it can be revealed by the mutant phenotypes appearing in the M_2 generation. Though the induction of mutations in genes located in the plastid genome is similar to that for genes in the nucleus, the number of plastid genome copies and the way segregation occurs for the plastid DNA molecules (also termed plastid chromosomes) makes it difficult to obtain plants with only mutant plastids, each with a pure population of plastid chromosomes with the newly formed mutation. These problems can be reduced if cells of *Chlamydomonas* are pretreated with 5-fluoro-deoxyuridine that specifically inhibits DNA replication in the organelles (SHEPHERD et al. 1979). Cells treated this way have a reduced number of genome copies in the plastids. A mutation induced in a single plastid genome will have an increased probability of segregating from the normal genomes in a few divisions and thus allows expression of the mutant phenotype. Unfortunately, a similar technique to increase the yield of non-Mendelian mutants in higher plants has not yet been devised.

In the search for mutants blocked in plastid development, it is very important to realize that the spectrum of mutants finally obtained will be limited by the conditions used during the screening procedure as well as the sensitivity of the method used for detection of mutant phenotypes. Temperature and light conditions should be controlled as well as possible. This will help in defining the mutant phenotypes selected. The search should primarily be directed toward the mutant phenotypes exhibiting a complete block in plastid development. This causes a problem in maintaining the mutants, since a complete block, whether in pigment biosynthesis or in photosynthesis, will often result in mutant plants being lethal. In work with green algae and *Euglena,* lethal mutants can generally be maintained by the use of supplemented growth media in accordance with common techniques for the isolation of auxotrophic mutants in microbial genetics (LEVINE 1971, BISHOP 1971, HUDOCK and ROSEN 1976, SCHIFF et al. 1971, 1980). With higher plants the lethal mutant phenotypes are either impossible or too laborious to maintain by the use of techniques of supplemented growth. However, recessive lethal mutants of higher plants can be detected as segregants in the M_2 (F_2) generation from mutagenized plant material. The recessive lethal mutants are then maintained as heterozygous stocks by selfing or crossing of heterozygous plants and using the segregating homozygous recessive mutant phenotypes for experimental work at the seedling stage while growth is still supported by the seed reserve. This is a rather laborious procedure, but has been used with success for the analysis of recessive lethal mutants affecting plastid development in barley (VON WETTSTEIN 1957, HENNINGSEN et al. 1974), maize (MILES 1980), and pea (STUMMANN and HENNINGSEN 1980).

As pointed out earlier, the expression of some mutant phenotypes can be highly dependent on temperature and/or light conditions. This fact has been used in an attempt to isolate a type of conditional lethal mutant with plastid development that is sensitive to high temperature (SMILLIE et al. 1978). Mutant phenotypes can also result from the possession of a pigment system that is sensitive to high or moderate intensities of light. This bleaching type of mutant can in some cases be partially normalized when developed under low light intensity (VON WETTSTEIN et al. 1971, HENNINGSEN et al. 1973). A particular type of conditional lethal mutant is the *tigrina* phenotype observed in barley (VON WETTSTEIN et al. 1971, NIELSEN 1974, KANNANGARA et al. 1978). Developed under a light/dark cycle plants homozygous for the recessive mutant alleles at the *tigrina* loci will develop transverse segments of necrotic tissue alternating with normal leaf tissue and die as young seedlings. Grown under continuous light, the *tigrina* mutants do not develop necrotic bands and homozygous mutant plants are viable.

Pleiotropic effects on chloroplast development is a common phenomenon and can often make it extremely difficult to identify the primary lesion caused by a mutation. In such cases a genetic analysis in combination with biochemical and physiological studies of several developmental stages and under different growth conditions will be required in order to decide whether the mutant phenotype is caused by pleiotropic effects resulting from a single gene mutation or if the defective plastid development is a syndrome caused by a deficiency including several genes.

Two mutant phenotypes, pigment-deficient and photosynthesis-deficient, are major objects to select for and use in studies on chloroplast development. Mutants with changes in the content or type of the photosynthetic pigments can be detected by visual inspection. Mutants with the remaining pigments evenly distributed in all parts of the tissue have proved useful and are classified according to the level of pigments remaining: pale green, yellow-green, yellow, and albina. Screening procedures for detection of photosynthesis-deficient mutants with normal or close to normal content of pigments have been devised for green algae (LEVINE 1971, BISHOP 1971) and more resently for higher plants (MILES 1980). Incorporation of $^{14}CO_2$ as a measure of photosynthetic capacity has been used in green algae (LEVINE 1971). Chlorophyll fluorescence yield has been used to detect mutants in green algae and in higher plants (MILES 1980). Treatment of mutagenized algal cultures or plants with photodynamic inhibitors, including herbicides, have been used to select for photosynthesis mutants. Selection with sodium arsenate has been applied with *Euglena* (SHNEYOUR and AVRON 1975) and *Chlamydomonas* (TOGASAKI and HUDOCK 1972). Metronidazole has been used to select nonphotosynthetic mutants in *Chlamydomonas* (SCHMIDT et al. 1977). Bipyridylium herbicides have been used to screen or select for photosynthesis mutants in maize (MILES 1980). The use of these inhibitors for screening or selection of mutants is based on the observation that only cells with a functioning photosynthetic system will be killed during growth in the light. The nonphotosynthetic mutants can then be rescued and the lesion in photosynthesis be localized to either the primary photochemical reactions, the electron transport pathways, photophosphorylation, or the CO_2 fixation pathways. Numerous biochemical and biophysical techniques are employed in the final characterization of chloroplast mutants. The use of these techniques in the analysis of the primary lesion in the mutants is dealt with in the literature cited.

3 Structural Organization of Plastid Membranes

3.1 Developing Plastid Membranes

During dark development of angiosperm seedlings, proplastids develop into etioplasts with an internal membrane system and the chlorophyll (Chl) precursor, protochlorophyllide (Pchl-ide). The internal membranes in proplastids are restricted to a few vesicular membranes that appear to be generated by outgrowth of invaginations from the inner of the envelope membranes (VON WETTSTEIN 1958). As the proplastids differentiate into young etioplasts, lamellar layers (also termed primary thylakoids) form (cf. Fig. 1). The lamellar layers consist of double layers of membrane sheets with perforations. At the maximum development, etioplast membranes consist of a network of tubular membranes in the crystalline prolamellar body connected with the double membrane sheets of the lamellar layers (HENNINGSEN and BOYNTON 1969). In young etioplasts, the small amount of Pchl-ide with in vivo absorption maximum in the red region at 635–637 nm appears to be associated with the lamellar layers (KLEIN and SCHIFF 1972). Fully developed etioplasts have Pchl-ide associated with the prolamellar body membranes (KLEIN and POLJAKOFF-MAYBER 1961, KAHN 1968) and in vivo absorption maximum at 648–650 nm (HENNINGSEN and BOYNTON 1969, KLEIN and SCHIFF 1972). Etioplasts also contain substantial amounts of carotenoids. The carotenoids of etioplasts are the same as those found in chloroplasts (LICHTENTHALER 1969). Lutein is the predominating carotenoid in etioplasts. β-carotene and neoxanthine are present in smaller proportions and the contents of these carotenoids increase during the light-dependent development of etioplasts into chloroplasts. In mature chloroplasts β-carotene and xanthophylls are associated with both lamellar membranes and envelope membranes (SIEFERMANN-HARMS et al. 1978). Etioplasts also contain colorless lipids. Both polar lipids and quinones (LICHTENTHALER 1968) are present in etioplasts in amounts comparable to those found in chloroplasts. These pigments are also distributed between prolamellar body membranes, lamellar layers, and the envelope of etioplasts (BAHL 1977). Only the two major lipid classes of the chloroplast, mono- and digalactosyl diglycerides, are synthesized in larger quantities during greening (ROUGHAN and BOARDMAN 1972).

In addition to the pigments, etioplast membranes contain a majority of the polypeptides characteristic of fully developed chloroplast membranes (cf. Fig. 12 and Sect. 4.3.) as well as a number of etioplast-specific polypeptides. Some of the etioplast membrane polypeptides are constituents of the Pchl-ide holochrome complex. Etioplasts also contain a protein synthesizing system and soluble enzymes in the CO_2 fixation pathway. Many of the developmental changes in etioplasts are accelerated by activation of the phytochrome (SMILLIE et al. 1971), but further plastid development is dependent on light absorbed by Pchl-ide (VIRGIN et al. 1963).

Upon illumination of the etiolated leaves, Pchl-ide absorbing at 650 nm in vivo is photoconverted to chlorophyllide (Chl-ide), and this initiates the loss of the crystalline configuration of the prolamellar body (HENNINGSEN and BOYNTON 1969). Subsequently the absorption maximum of the Chl-ide shifts from 682 to 672 nm in vivo, the Chl-ide is esterified to give chlorophyll a (Chl a) and the prolamellar body membranes are dispersed into primary lamellar layers. These processes have similar time courses and are temperature-dependent in a similar manner, thus they appear interconnected (HENNINGSEN 1970, HENNINGSEN and THORNE 1974).

The dispersal of the prolamellar bodies involves a restructuring of the membraneous tubuli into membrane sheets with numerous perforations and protuberances and membrane components appear to be translocated (HENNINGSEN et al. 1974). During the dispersion process the profile length of the lamellar layers increases to reach a level that then remains constant during the early stages of greening (HENNINGSEN and BOYNTON 1974). Many two-disc grana of small diameter are formed before the phase of rapid Chl synthesis. During the lag in Chl synthesis and the phase of slow Chl synthesis, high ratios of Chl a to Chl b are observed (HENNINGSEN and BOARDMAN 1973). A Chl a to b ratio of 30–40 for leaves greened for 2 h decreases to about 4 after 4 h of greening. At this stage of development the plastid membrane system consists of unpaired lamellar layers and a few small grana each with maximum two discs. Photosystem I and II activities are high measured on a Chl basis (HENNINGSEN and BOARDMAN 1973). During the phase of rapid chlorophyll synthesis, the accumulating chlorophyll appears to become associated with previously assembled photosynthetic units. As the size of the photosynthetic units increases, the light-harvesting capacity increases and low light intensity is then sufficient to saturate the photosystems. After 6 h of illumination the Chl a to b ratio reaches 3 and then remains constant. At about 6 h of greening the photochemical activities measured on a Chl basis decline to the value observed with mature chloroplasts. Further pairing of the lamellar layers results in grana with three or more discs. During the phase of rapid Chl synthesis, paired regions of the lamellar layers increase proportional to the amount of Chl accumulated in the leaf. Though high photochemical activity can be detected with unpaired lamellar layers, it appears that an effective light-harvesting system occurs predominantly in connection with the paired regions of the lamellae in the multi-disc grana. Taken together, the observations strongly suggest that the prolamellar body membranes contain sufficient components to support the formation of the lamellar layers with capacity for photosynthetic electron transport. However, a reorganization of the components contained in the prolamellar body membranes is required before the photosynthetic units become active. It appears that the newly dispersed lamellar layers serve as a basis for the insertion of further membrane components synthesized in the light and required for the development of an effective light-harvesting system associated with the paired grana regions of the lamellar layers.

3.2 Mutants Affecting Structural Organization of Plastid Membranes in Relation to Photosynthetic Capacity

Recessive lethal mutants of the phenotypes *albina*, *xantha*, and *viridis* in barley have been studied (VON WETTSTEIN et al. 1971, HENNINGSEN et al. 1973, 1974) for their capacity to synthesize Pchl-ide in the dark and Chl in the light (Fig. 1). The pigment contents of the leaves have been related to the type of plastid membrane structures formed and the capacity to carry out photosynthetic reactions in leaves and with isolated plastids. The composition of the lamellar membranes has also been investigated with special reference to their protein constituents and attempts have been made to reveal the sites of transcription and translation of these plastid components.

3.2.1 Mutants Defective in Dark Development

Barley mutants at the *xantha* loci *xan-f*, *-g*, *-h*, *-l*, *-u*, and several *albina* loci are completely or partially blocked in the synthesis of Pchl-ide. Feeding of the leaves with δ-aminolevulinic acid (ALA), the first precursor specific to the porphyrin pathways, has been used to study the specific blocks in the biosynthesis of Pchl-ide. Leaves of the wild type convert externally supplied ALA to Pchl-ide. The completely blocked mutants at the gene loci *xan-f*, *-g*, and *-h* convert the supplied ALA to protoporphyrin IX. Protoporphyrin is also the major product accumulated in the incompletely blocked mutants at the loci *xan-f*, *-g*, *vir-y*, and *alb-e*. The incompletely blocked mutants *xan* $-l^{35}$ and *vir*-k^{23} convert externally supplied ALA predominantly into Mg-protoporphyrin, while the mutant *xan*-u^{21} accumulates uroporphyrin as well as protoporphyrin (cf. Sect. 4.1).

Barley mutants at the *xan-f*, *-g*, and *-h* loci with a complete block in the synthesis of Pchl-ide fail to form prolamellar bodies but primary lamellar layers are present (Fig. 2). These mutants also fail to form any Chl in the light and the amount of lamellar membranes is considerably reduced compared with plas-

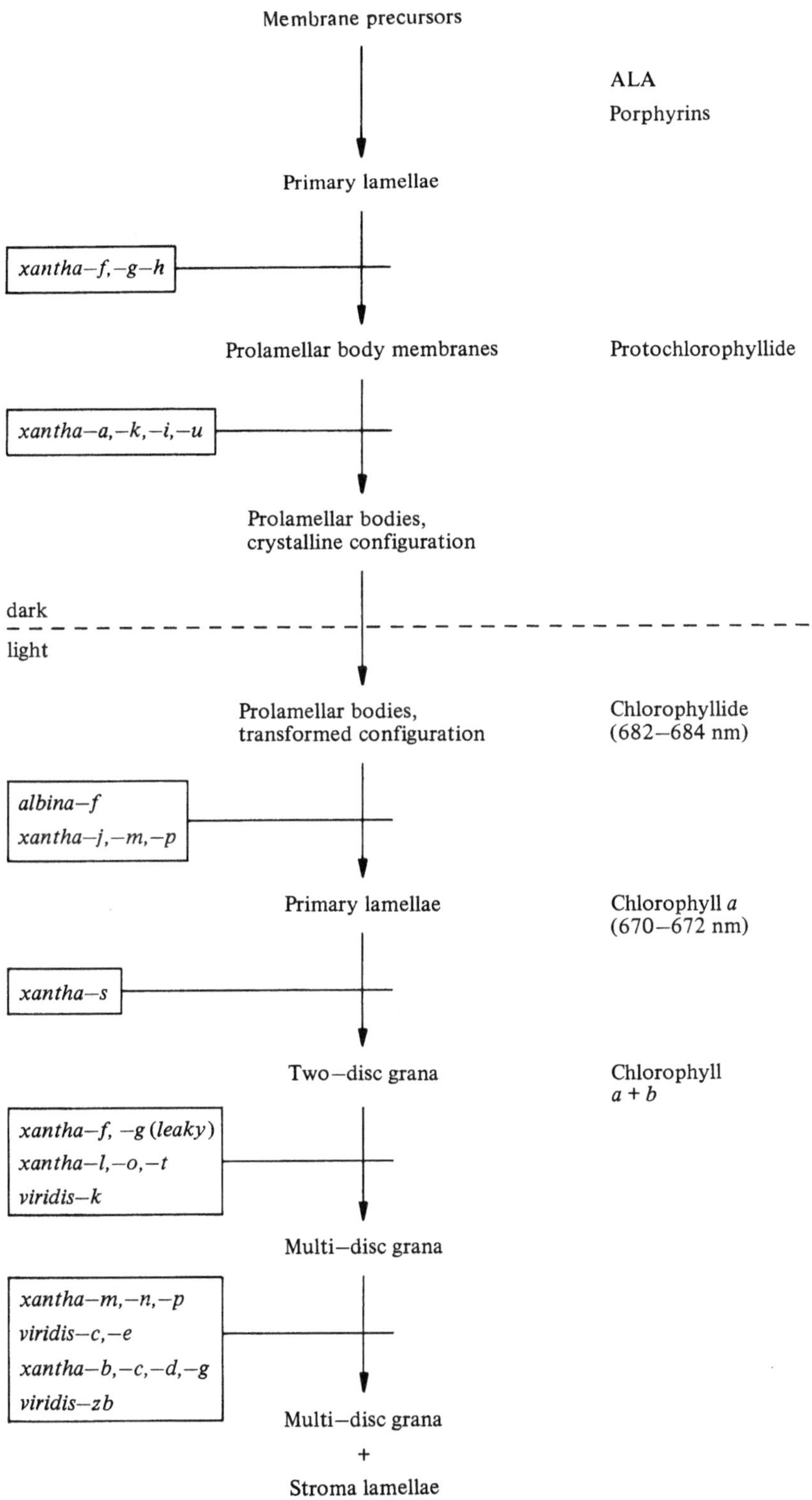

Fig. 1. Diagram illustrating steps in chloroplast development. Lesions in the developmental sequence caused by mutations in *viridis*, *xantha*, and *albina* genes of barley are indicated. (Based on data from HENNINGSEN et al. 1973)

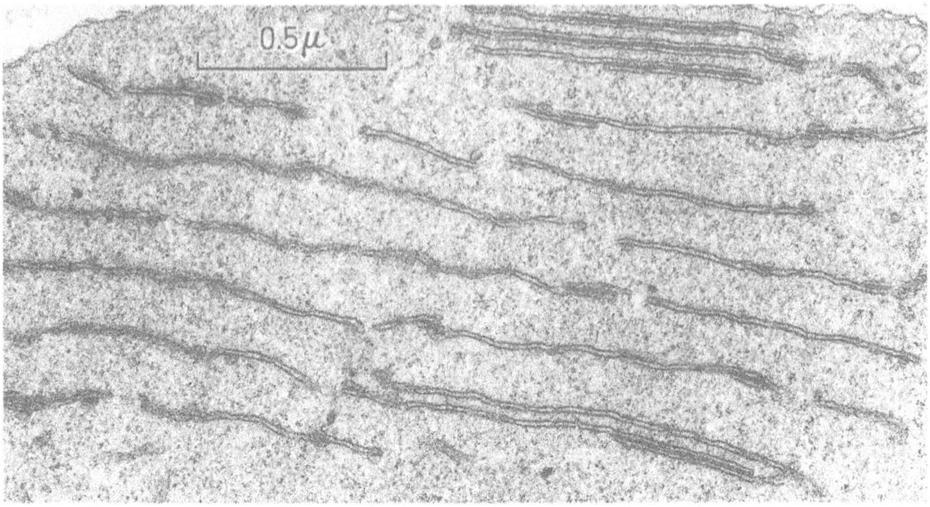

Fig. 2. Plastid section from barley mutant *xan-h*[56] developed in the dark. This mutant is deficient in protochlorophyllide and prolamellar body membranes. Primary thylakoids, largely without perforations, are present. (HENNINGSEN et al. 1973)

tids of the wild type at the early stages of greening, and grana are not formed. Conversely, mutants at the loci *xan-f*, *-g*, *-l*, and *vir-k* that are incompletely blocked in Pchl-ide synthesis form crystalline prolamellar bodies in the dark. Upon illumination the prolamellar bodies are dispersed into primary lamellar layers. However, due to the lesion in Pchl-ide synthesis, subnormal amounts of Chl accumulate in the light and the development of grana is restricted to few small diameter grana each with only two or three discs (Fig. 7).

With mutants of maize MASCIA and ROBERTSON (1978) have obtained similar results. The mutants *l-blandy-4* and *oy-1040* are completely blocked in the conversion of protoporphyrin to Mg-protoporphyrin (MASCIA 1978) and fail to assemble prolamellar bodies. The mutants *ory 1* and *ory 2* are defective in the conversion of Mg-protoporphyrin to Pchl-ide. The mutant *ory 1* appears to be completely blocked in this biosynthetic step and the etioplast membranes consist of lamellar layers and an amorphous system of tubular membranes.

Fig. 3. Plastid section from barley mutant *xan-u*[21] developed in the dark. Some protochlorophyllide is accumulated in the dark, but prolamellar body and thylakoid membranes develop into abnormal structures. (HENNINGSEN et al. 1973)

Fig. 4. Plastid section from barley mutant *xan-a*[63] developed under dim light conditions. Chlorophyll accumulates and the chloroplast membranes develop into grana, prolamellar body and honey comb-like structures. (HENNINGSEN et al. 1973)

Fig. 5. Plastid section from barley mutant *alb-f*[17] developed in the dark. Protochlorophyllide and prolamellar body membranes are present. (HENNINGSEN et al. 1973)

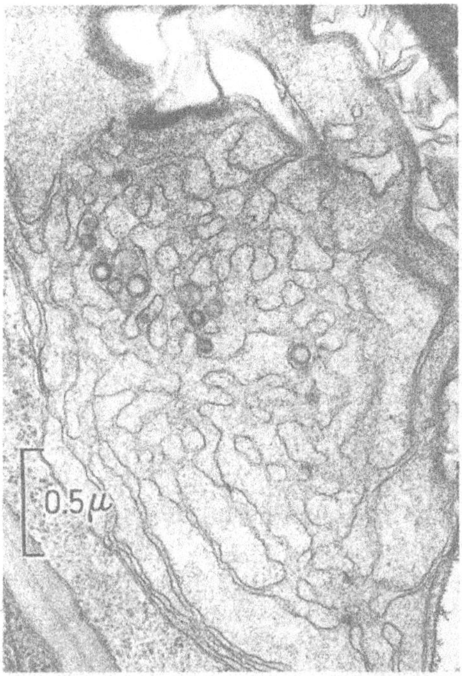

ig. 3

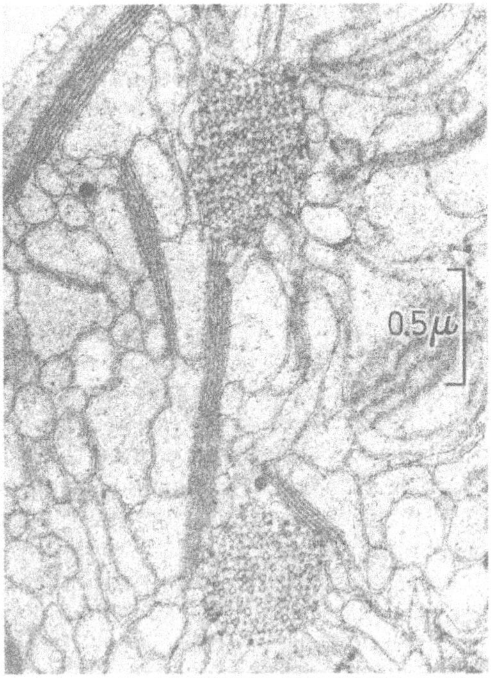

Fig. 4

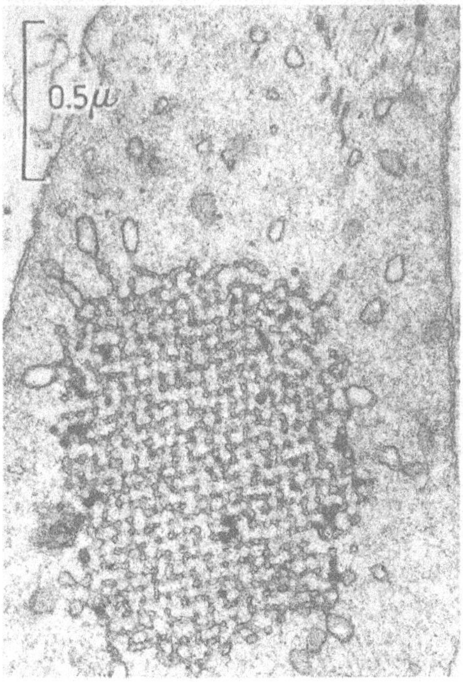

g. 5

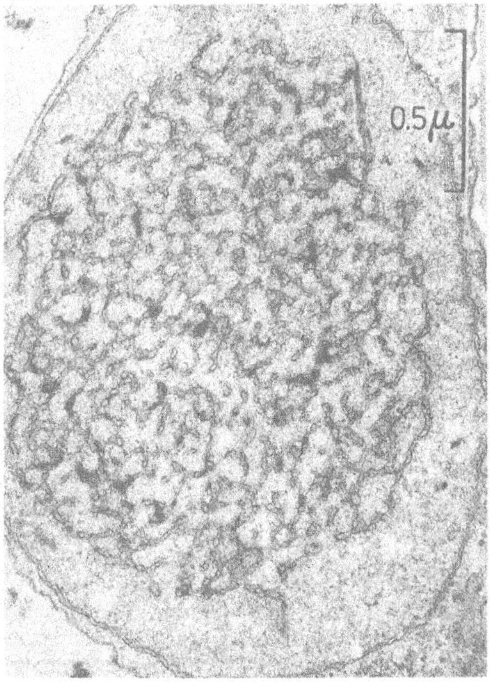

Fig. 6

Fig. 6. Plastid section from barley mutant *alb-f*[17] developed in the dark and subsequently exposed to light. Light-dependent development, in this mutant, is blocked at chlorophyllide phytolation and prolamellar body membranes remain in the transformed configuration. (Henningsen et al. 1973)

The slightly leaky mutant *ory 2* produces a small amount of Pchl-ide and some prolamellar bodies are formed.

Other mutants produce normal or almost normal amounts of Pchl-ide in the dark but exhibit defective development of the etioplast membrane system. Barley mutants at the *xantha* loci *xan-a*, *-k*, *-i*, and *-u* and at several *albina* loci (VON WETTSTEIN 1958, HENNINGSEN and BOYNTON 1967, HENNINGSEN et al. 1973), and also several mutants in maize (BACHMANN et al. 1969), are blocked in the assembly of prolamellar body membranes. The etioplast membrane system in such mutants is either confined to vesicular membranes, or an amorphous system of normally structured tubular membranes, or abnormally structured membranes can be present in prolamellar body-like organization (Fig. 3). These mutants appear to have a functional pigment biosynthesis, but lack or synthesize a defective form of a component required for the assembly of normal etioplast membranes. When developed in light of high intensity the pigments of these mutants are bleached and plastid membranes are highly deranged. Under low light intensity this group of mutants accumulates a considerable amount of leaf pigments and elaborate but abnormally structured membrane systems are assembled in the plastid (Fig. 4). In contrast to higher plants, normal cells of green algae are able to perform the synthesis of Chl a and b during dark growth. Mutants defective in Chl biosynthesis in darkness have been described in *Chlamydomonas* mutant *y-1* (SAGER 1955, WANG et al. 1977), *Chlorella,* mutant *C-10* (BRYAN et al. 1967) and *Scenedesmus,* mutant *C-2A'* (SENGER and BISHOP 1972, BISHOP and SENGER 1972). Dark-grown cells of these mutants are either completely devoid of Chl and Pchl-ide or minute amounts of Pchl-ide can accumulate. As the dark-grown cells are depleted of pigments, the plastid membranes dedifferentiate to vesicular membranes or rudimentary prolamellar body-like structure (OHAD et al. 1967a). Transfer of the cells to light results in pigment synthesis and development of normal-structured chloroplast membranes (OHAD et al. 1967b). Taken together these observations indicate that there is a close relation between the assembly of etioplast membrane structures and ability to synthesize pigments as well as other constituents of the membranes.

Fig. 7. Plastid section from barley mutant *vir-k*^{23} developed in the light. Most of the thylakoids are unpaired, but some grana, each with two or three discs, are formed. Photosystem I and II activities are high measured on a chlorophyll basis. (HENNINGSEN et al. 1974)

Fig. 8. Plastid section from barley mutant *vir-e*^{55} developed in the light. Grana are rare, but of about normal diameter. Stroma thylakoids are almost absent. This mutant is deficient in oxygen evolution from the leaves and photosystem II activity, but photosystem I activity is about normal. (HENNINGSEN, unpublished)

Fig. 9. Plastid section from barley mutant *vir-zc*^{65} developed in the light. Grana appear normal, but stroma thylakoids are strongly reduced. This mutant is deficient in oxygen evolution from the leaves, but Hill reaction and photosystem I activities are normal or subnormal. (HENNINGSEN, unpublished)

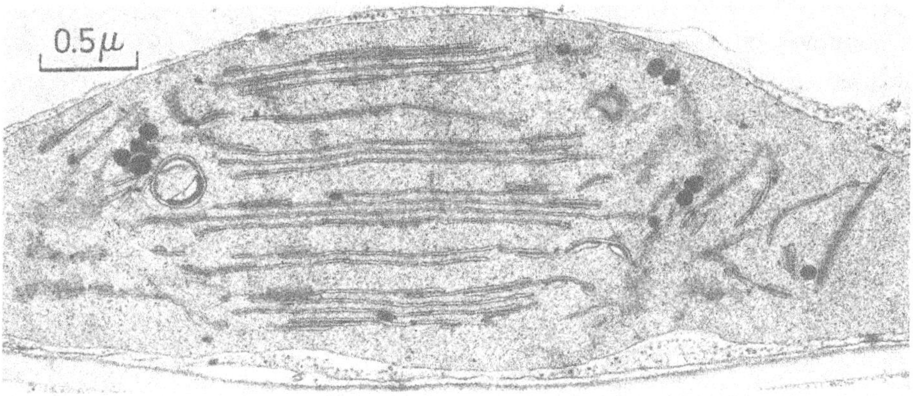

Fig. 7

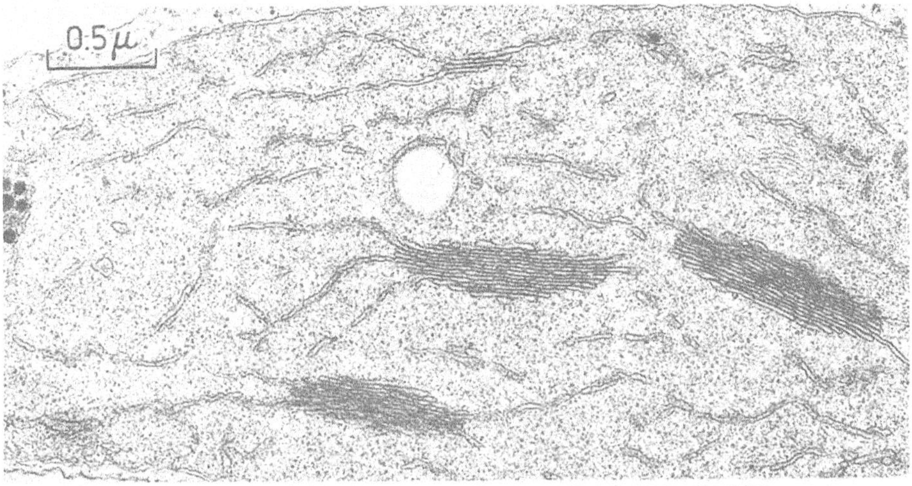

Fig. 8

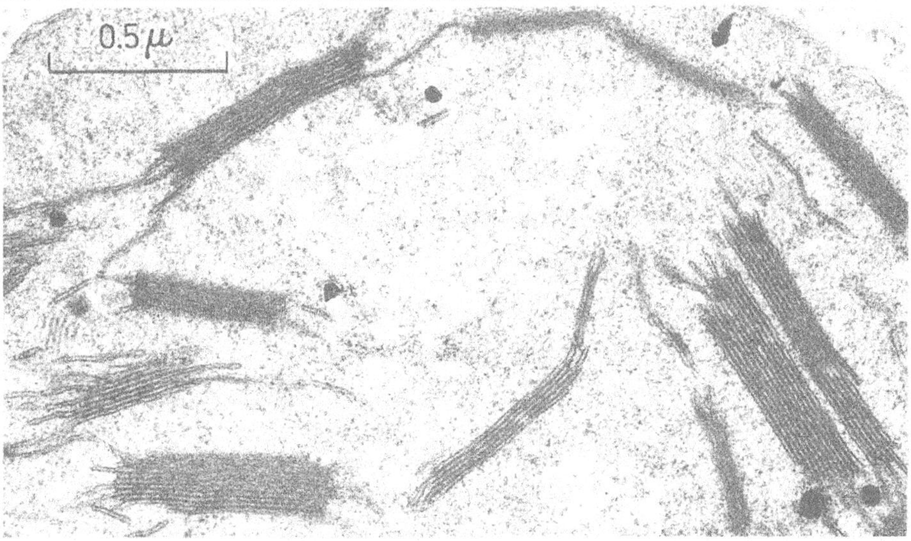

Fig. 9

3.2.2 Mutants Defective in the Initial Steps of Greening

Barley mutants of the loci *alb-f* and *xan-j* comprise a group of mutants with lesions in the initial step of greening. These mutants accumulate substantial amounts of Pchl-ide in the dark and normal crystalline prolamellar bodies are formed (Fig. 5) (HENNINGSEN and BOYNTON 1967, VON WETTSTEIN et al. 1971). Upon illumination Pchl-ide is photoconverted to Chl-ide absorbing at 682–684 nm in vivo. However, the Chl-ide absorption maximum fails to shift from 682–684 to 672 nm as observed with leaves of the wild type where it usually is rapidly completed. The mutants at loci *alb-f* and *xan-j* also fail to esterify Chl-ide and the increase in fluorescence efficiency of Chl a observed with leaves of the wild type is also missing (HENNINGSEN and THORNE 1974). These observations are interpreted as caused by a block in mutants of the gene loci *alb-f* and *xan-j* in the molecular events involving conformational relaxation of the Chl-ide holochrome and the translocation of Chl a to reaction centers of the photosystems. The lesions inferred at the molecular level are reflected at the ultrastructural level. The photoconversion of Pchl-ide to Chl-ide is accompanied by a structural transformation of the prolamellar body tubules. However, the prolamellar body remains undispersed even after the leaves have been in the light for a long period (Fig. 6). In maize, SMITH et al. (1959) found that three *albina* mutants lacked the ability for the 684 to 672 nm shift as well as for phytolation of newly formed Chl-ide. These results indicate that the aforementioned pigment changes are interconnected with the configuration of the prolamellar body membranes. The altered properties of the Chl-ide holochrome in the mutants at the loci *alb-f* and *xan-j* appears to be dependent on the location of the pigment-protein complex in the prolamellar body membranes. Pchl-ide holochrome extracted with the detergent saponin from etioplasts of the mutants $alb\text{-}f^{17}$ and $xan\text{-}j^{64}$ have spectral properties similar to the holochrome from wild type (B.M. STUMMANN, unpublished results). Therefore, the lesions in these mutants appear to be in the membrane organization.

A second type of mutant blocked at the initial stage of greening are those that upon illumination disperse the prolamellar body into lamellar layers, but fail to develop even two-disc grana (Fig. 1). In barley, *xantha* mutants at the loci *xan-m*, *-p*, and *-s*, accumulate Pchl-ide in the dark and the etioplast membranes consist of an amorphous prolamellar body organization connected with lamellar layers (HENNINGSEN et al. 1973). Photoconversion of Pchl-ide results in Chl-ide with absorption maximum at 670–672 nm, and phytolation to Chl a is much faster than in the wild type (HENNINGSEN and THORNE 1974). The tubular membranes are dispersed into perforated lamellar layers, but further structural development is blocked.

3.2.3 Mutants Defective in Differentiation of Lamellar Membranes into Grana and Stroma Regions

Barley mutants at several *xantha* and *viridis* loci accumulate substantial amounts of Pchl-ide in darkness and the structural development of etioplast membranes is similar to that of the wild type. Upon illumination of seedlings either dark-grown or developed in the light, most of these mutants accumulate substantial amounts of Chl (von Wettstein 1957, Walles 1967, von Wettstein et al. 1971, Henningsen et al. 1973, 1974). However, with respect to Chl accumulation, structural development of the plastid membranes and development of photosynthetic capacity, some of the mutants are strongly dependent on light and temperature conditions. Mutants at the *xantha* loci *xan-a*, *-e*, and *-n* are sensitive to high light intensity. Leaves of these mutants greened under high light intensity are almost devoid of Chl, lack the capacity of oxygen evolution and the plastid membranes are disorganized. When developed at low light intensity the mutant leaves accumulate Chl and develop some capacity for oxygen evolution. The plastid membrane structures in leaves developed under low light are partially normalized and consist of grana, prolamellar bodies, and honey comb-like membrane structures (Fig. 4). This type of *xantha* mutant in barley appears to be similar to carotenoid deficient mutants in maize (Robertson et al. 1966, Faludi-Dániel et al. 1968) and *Helianthus* (Walles 1967), but a further analysis might equally well show that mutation in some of the *xantha* loci causes defective synthesis of noncolored lipids or a membrane polypeptide required for a normal and light-stable binding of Chl in the membranes.

A rather common type of mutant is that with a temperature-sensitive chloroplast development. The mutant phenotype can be accentuated at either low or high temperature. The accumulation of Chl and development of chloroplast structures in barley mutants at the loci *xan-b* and *vir-zb* are dependent on growth temperature. At lower temperatures, these mutants develop slowly, the Chl content of the leaves is strongly reduced, and the plastids have abnormal membrane structures. A partial normalization of these mutants is accomplished at higher temperatures. In contrast to this are some mutants of the *chlorina* and *chlorotica* phenotypes in pea that are sensitive to high temperature, but viable at lower temperature (Stummann and Henningsen, unpublished results). It will be of interest to study further the mutants with a temperature-sensitive chloroplast development to find out whether the temperature sensitivity is caused by a thermo-labile protein-synthesizing system or due to a thermo-labile structure of an essential component in the photosynthetic membranes. Several of the mutants in higher plants that are blocked at the early stages of greening accumulate Chl in such a small amount that, in comparison with greening wild-type seedlings, one would expect at the most to find some photosystem I activity. From the study of plastid membrane structure in such mutants it is evident that membrane formation is affected in rather general fashion with only a few lamellar layers and occasionally with rudimentary two-disc grana present. It is possible that this type of mutant is defective in the assembly of ribosomes or in the translocation mechanism required for chloroplast polypeptides synthesized in the cytoplasma. A primary block caused by mutation in genes concerned with the synthesizing or translocating system for chloroplast polypeptides could be expected to cause pleiotropy of the kind observed in mutants blocked at the early stages of greening.

Other mutants accumulate substantial amounts of Chl in the light and form extensive membrane systems of various configurations, but a single or several polypeptides are missing from the chloroplast membranes and the photosynthetic capacity is impaired. Such mutants, even if completely blocked in a single essential chloroplast component, will be lethal at the seedling stage. When searching for mutants blocked in either the primary photochemical events, photosynthetic electron transport components, photophosphorylation or CO_2 fixation enzymes, the mutants of this type are the most promising to study. This implies that with mutants in higher plants, the recessive seedling lethals are the most promising mutants in spite of the labor involved in maintaining the recessive mutant alleles in heterozygous stocks.

Barley mutants at several *xantha* and *viridis* loci accumulate substantial amounts of Chl in the light. Mutants at the loci *xan-m*, *-p*, *-q* and *vir-b*, *-c*, *-e*, *-l*, *-v*, *-zb*, *-zc*, and *-zd* lack photosynthetic oxygen evolution from the leaves (HENNINGSEN et al. 1973). Plastids isolated from several of these mutants are deficient in photosystem II and/or photosystem I activities and the plastid membranes are in most instances abnormally structured (HENNINGSEN et al. 1974, NIELSEN et al. 1974, SIMPSON and VON WETTSTEIN 1980). Other of the *xantha* and *viridis* mutants of barley accumulate substantial amounts of Chl and develop a high capacity for photosynthetic oxygen evolution (HENNINGSEN et al. 1973).

Among the mutants with substantial amounts of Chl accumulated in the leaves are four major classes with respect to the structure of the chloroplast membranes. The mutants *xan-f*^{26}, *-g*^{45}, *-l*^{35}, and *vir-k*^{23} have a high ratio of Chl a to b, possess significant light-induced oxygen evolution from the leaves, and good photosystem I and II activities. The chloroplast membranes consist of lamellar layers that are mostly unpaired (Fig. 7). Only few grana, each with two to four discs are formed. The photochemical reactions require high light intensity, but under such conditions rates exceeding those of the wild type are measured on a Chl basis. In accordance with this, MACHOLD and HØYER-HANSEN (1976) found that *vir-k*^{23} chloroplast membranes lack three polypeptides, two of which are components of the light-harvesting complex.

Related to the class of mutants with long unpaired lamellar layers and few grana each with few or a reduced number of discs are homozygotes and heterozygotes of the yellow mutants in tobacco (HOMANN and SCHMID 1967, HOMANN 1968, SCHMID 1971) and in pea (HIGHKIN et al. 1969). The yellow homozygous mutant plants have very little Chl, photosystem II activity is not detectable and the lamellae are not stacked. The yellow-green plants heterozygous for these mutations have some small grana, and their photosystem II is functional at high rates when measured in high light intensity and expressed on a Chl basis. Similar characteristics have been reported for a Chl b-deficient mutant, *chlorina-f2*, in barley (BOARDMAN and HIGHKIN 1966, BOARDMAN and THORNE 1968, THORNBER and HIGHKIN 1974, BURKE et al. 1979), and a Chl b-deficient mutant in pea, *chlorina-5535*, (STUMMANN and HENNINGSEN 1980). One of the high fluorescence yield mutants, *hcf*-2*, in maize (MILES and DANIEL 1974, MILES 1980) has few and small grana. The variable part of Chl fluorescence induction kinetics is lacking, as is the activity of photosystem II, but photosystem I remains fully active. In relation to the grana-deficient type of mutants in higher plants are mutants in *Chlamydomonas* (GOODENOUGH and LEVINE 1969, LEVINE 1969, 1974). The mutants *ac-31*, *ac-115* and *ac-141* have plastids with lamellae that are mostly unpaired or with short regions of stacked lamellae.

Mutant *ac-31* is deficient in Chl and photochemical activity. The mutants *ac-115* and *ac-141* have a functional photosystem I, but the variable part of Chl fluorescence induction is not observed and photosystem II activity and cytochrome b-559 are missing. Taken together the results obtained with the grana-deficient type of mutants show that there is no relation between photochemical activity and stacking of lamellae into paired discs of grana. On the other hand it appears that components of the light-harvesting system are required for efficient functioning of photosystem II and a deficiency in one or more of these components correlates with the absence of membrane stacking.

Other mutants like those at the loci *xan-b*, *-c*, *-d*, and *vir-zb* form grana with multiple discs. The mutants are either deficient in unpaired lamellar layers of the stroma regions (Fig. 8) or unpaired stroma lamellae are lacking completely (Fig. 9). A third class of mutant chloroplast membrane system is deficient in unpaired stroma lamellae, but giant grana are formed. The giant grana have an abnormally large diameter, and the number of discs per granum is often larger than in the wild type developed under similar conditions. With many of the mutants forming giant grana the ratio of Chl a to b is often lower than the normal value of about three. The membrane polypeptide pattern of the giant grana in *xan-*b^{12} and *vir-*zd^{69} is a combination of the polypeptides from both etioplasts and chloroplasts of wild type (MACHOLD and HØYER-HANSEN 1976). Among the mutants deficient in unpaired stroma lamellae are *vir-*zb^{63} and -h^{15} that are deficient in photosystem I but have normal photosystem II activity (HILLER et al. 1980). These mutants are deficient in P-700, the reaction center complex Chl a-protein 1, as well as two low molecular weight polypeptides of the FeS-centers that are part of the photosystem I complex. On the other hand, mutants like *xan-*b^{12}, -q^{75}, *vir-*e^{64} and -zd^{69} with a chloroplast membrane system consisting of giant grana, have a defective photosystem II, while photosystem I remains fully functional. Comparable to mutants with the giant grana as found in higher plants, are the mutants *ac-80*, *ac-206*, *F1* and *F14* in *Chlamydomonas*. The mutant *ac*-206 have the lamellar layers organized in long stacks each with two discs and is deficient in cytochrome f (553) (GOODENOUGH and LEVINE 1969, LEVINE 1969, 1974). The mutants *ac-80*, *F1* and *F14* have several long, wide stacks of lamellae, and P-700 is not detectable. Membranes of these mutants lack the photosystem I reaction center complex, CP I, and a 66,000 mol.wt. apoprotein, that normally is part of this complex, is missing.

The fourth class of mutant has the lamellar membrane system differentiated into grana and unpaired lamellae in stroma regions (cf. Fig. 1). The membrane system of the mutant chloroplasts can either be indistinguishable from that of the wild type or some of the grana can be of the giant type. In mutants with multi-disc grana, the inter-thylakoid space (distance between appressed or stacked membranes) and/or intra-thylakoid space (distance between membranes within a lamellar layer) may be narrower or wider than normal. Mutants with inter-thylakoid space wider than in normal grana often have lamellae with wrinkled or folded membrane profiles (Fig. 8). This abnormal appearance of the photosynthetic membranes is typical of mutants deficient in Hill reaction activity and is likely to be caused by a deficiency in some polypeptide(s) of importance to the membrane structure. The mutant *vir-c*^{12} has lamellar layers and giant grana with wrinkled profiles. This mutant is deficient in photosystem II activity and the membranes are lacking two polypeptides that are part of the photosystem II reaction center complex. It is interesting to note that mutants with substantial amounts of Chl, but deficient in photosystem II activity, often have a reduced intra-thylakoid space, while the inter-thylakoid space appears to be wider in photosystem I-deficient mutants. Mutant chloroplasts deficient in photosystem I, but with normal differentiation of lamellae into grana and stroma region have been described in barley, mutant *vir-n*^{34} (HILLER et al. 1980) and in maize, mutant *hcf*E1481* (MILLER 1980, MILES and DANIEL 1974). These mutants are deficient in P-700 and polypeptides of the photosystem I reaction center complex, Chl a-protein 1.

The substructure of photosynthetic membranes have been studied by electron microscopy on freeze-fractured and freeze-etched preparations. The size and density of particles in or on the membranes have been related to function and composition of chloroplast mutants in barley (SIMPSON et al. 1977, VON WETTSTEIN 1981) and in maize (MILLER 1980). The barley mutants *chlorina-f2*, *vir-k*^{23} and *xan-l*^{35} are deficient in Chl b and the light harvesting complex. The inner fracture face (IFF) (cf. MÜHLETHALER 1977) of the lamellar membranes have an increased number of particles, but the size of particles is reduced. Mutants in barley like *xan-b*^{12}, *vir-c*^{12}, *-e*^{64} and *-zd*^{69} have low photosystem II activity and giant grana. The IFF of membranes from the giant grana of these mutants has a strongly reduced number of particles. Mutants reported to be deficient in photosystem I activity and polypeptides associated with the reaction center complex, CPI, can either be deficient in particles on the IFF, while the outer fracture face (OFF) has particles of normal size and density. The mutant *hcf*E1481* in maize is deficient in photosystem I activity, but photosystem II function is near normal. In this mutant the OFF of unpaired lamellae in stroma regions is without large particles, but with a large number of small particles possibly related to an incomplete photosystem I reaction center complex. In *Chlamydomonas* the mutant *F54* is lacking the 10 nm particles on the surface of the chloroplast membranes believed to be associated with the coupling factor and ATPase activity (HOWELL and MOUDRIANAKIS 1967).

Studies on the plastid membrane structures and the content of pigments accumulated are valuable in distinguishing between mutants blocked at the various stages of dark development of etioplasts or the light-dependent development of chloroplasts. Formulated in general terms, correlations between structure and composition of plastid membranes and photosynthetic capacity of leaves or isolated chloroplasts from mutants or greening wild type are difficult to make when ambiguity is to be avoided. However, studies on particles in the photosynthetic membrane by freeze-fracture or freeze-etching techniques of electron microscopy combined with electrophoretic and immunochemical analysis of pigment-protein complexes and components of electron transport and phosphorylation can yield important information on the assembly of physical units exhibiting distinct functions and this will illuminate the picture of membrane assembly in relation to acquisition of functions.

4 Synthesis of the Components of the Photosynthetic Membranes

4.1 Genetic Control of Chlorophyll Synthesis

Many mutants of higher plants and green algae that are affected in the synthesis of Chl have been described. Only those mutants for which evidence has been obtained with respect to the specific step affected will be considered (Fig. 10). GOUGH (1972) has reviewed previous studies on mutants of photosynthetic bacteria, green algae, and higher plants which accumulate or excrete precursors of Chl. The use of *Chlorella* mutants of this kind in the study of the biosynthetic pathway of Chl has been reviewed by GRANICK (1971). For *Chlorella* is is not known whether the mutations affecting Chl synthesis are located in the nuclear genome, since recombination in this organism has not been recorded. The other mutants listed in Fig. 10 are Mendelian mutants. The biosynthesis of Chl has recently been reviewed by BOGORAD (1976).

WANG et al. (1974) have isolated two mutants of *Chlamydomonas reinhardtii*, br_s*-1* and br_c*-1*, that in the dark accumulate as protoporphyrin only about 4% of the wild-type level of Chl. The mutant br_s*-1* is light-sensitive, and br_c*-1* is light-conditional, that is, it turns green when exposed to light. It was suggested that the later steps of the Chl synthesis in *Chlamydomonas* can occur via two pathways that have identical initial parts, one operating in light and another operating in the dark. The gene corresponding to the mutation br_c*-1* may function in the light-dependent pathway, while the br_s*-1* gene may function at a reaction in the common initial part of the pathways. It is also conceivable that the mutation br_c*-1* results in a gene product that can only operate efficiently under the physiological conditions prevailing in light-grown cells. WANG et al. (1975) have also isolated a mutant, *r-1*, which reduces Chl synthesis by about 40% and the level of heme by about 66%. Combination of *r-1* with br_s*-1* or br_c*-1* elevates the protoporphyrin accumulation of these mutants more than 20-fold and results in normal heme levels, presumably because the excess protoporphyrin accumulated becomes available to the Fe branch of the pathway (Fig. 10). The low levels of Chl in the mutant *r-1* can be increased by feeding δ-aminolevulinic acid (ALA). It was suggested that *r-1* may be a mutation of the gene coding for the ALA-synthesizing enzyme, which reduces the sensitivity of this enzyme to feedback inhibition by protoporphyrin or heme as well as the overall activity. The genes corresponding to the mutations br_s*-1* and br_c*-1* may code for polypeptides involved in the insertion of Mg^{2+} into protoporphyrin.

Mutants that predominantly accumulate protoporphyrin have also been described for wheat, *Chlorella*, and barley (Fig. 10). Leaves from dark-grown plants of the leaky mutant *cn-Alb* of hexaploid wheat accumulate protoporphyrin and some photoconvertible Pchl-ide when they are treated with ALA (NEWELL and RIENITS 1975). It was concluded that the mutation brings about a restriction in the rate of insertion of Mg^{2+} into protoporphyrin. Cultures of the *Chlorella vulgaris* mutant W_5B are initially pale green, but when they

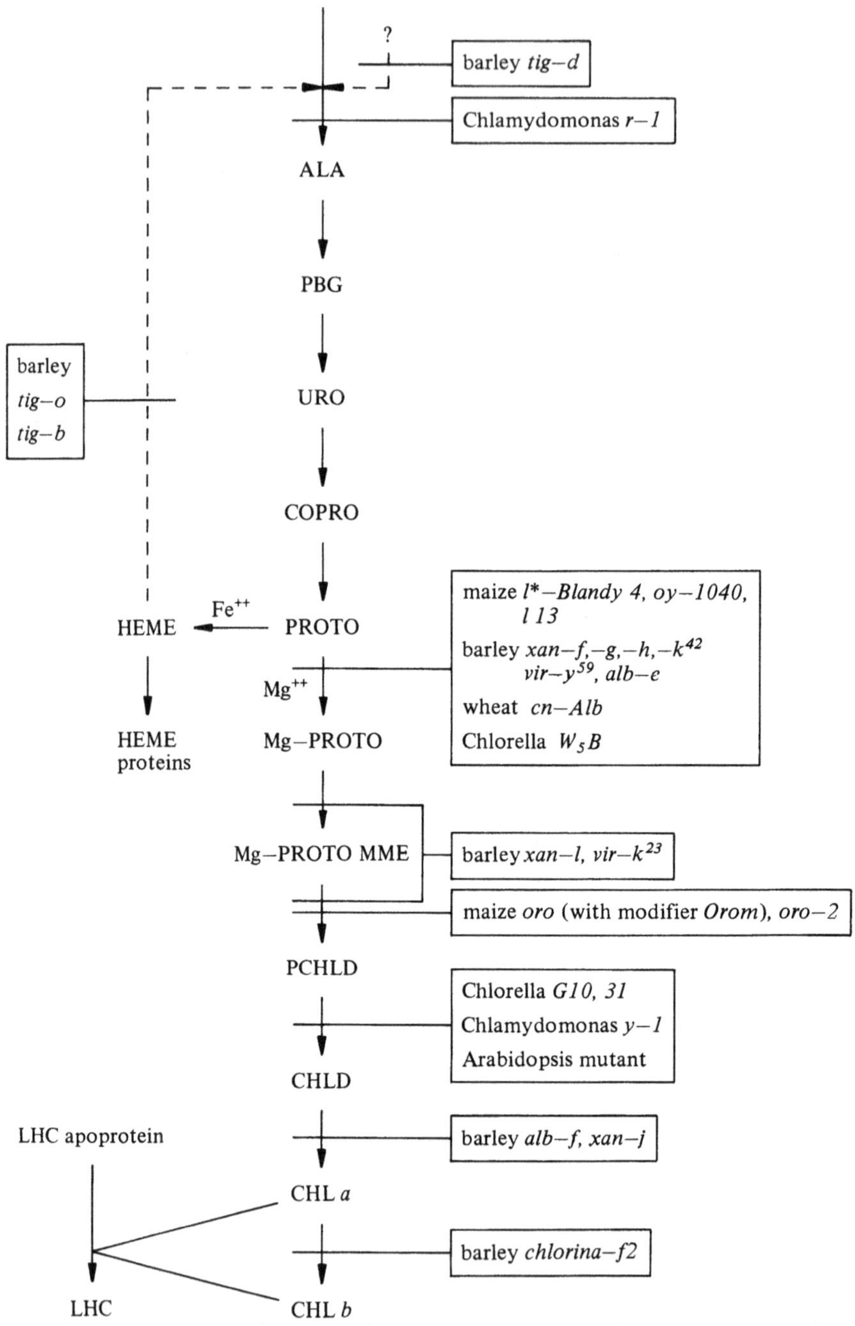

Fig. 10. Diagram of chlorophyll and heme biosynthesis and the lesions caused by mutation in genes of higher plants and algae. *Dashed lines* indicate inhibitory actions. *ALA* δ-aminolevulinic acid; *PBG* porphobilinogen; *URO* uroporphyrinogen III; *COPRO* coproporphyrinogen III; *PROTO* protoporphyrin; *MME* monomethyl ester; *PCHLD* protochlorophyllide; *CHLD* chlorophyllide; *CHL* chlorophyll; *LHC* light-harvesting complex

approach the stationary phase of growth, protoporphyrin is accumulated (GRANICK 1948). Several other mutants of *Chlorella vulgaris* have been isolated, which accumulate a mixture of the intermediates in the Chl pathway. The properties of some of these mutants have been reviewed (GOUGH 1972, GRANICK 1971). VON WETTSTEIN and coworkers have isolated and characterized several mutants of barley that affect the synthesis of Chl (HENNINGSEN and BOYNTON 1967, BOYNTON and HENNINGSEN 1967, GOUGH 1972, VON WETTSTEIN et al. 1971, 1974, KAHN et al. 1976, GOUGH et al. 1980). Some of these genes seem to be structural genes and others (*tig-d*, *-o*, *-b*, see later) have characteristics of regulatory genes. To facilitate identification of the specific step that is affected, mutant dark-grown seedlings have been incubated in solutions of ALA to sustain accumulation of the intermediates. Under such conditions the wild type accumulates predominantly Pchl-ide. Mutants at the four loci *xan-f*, *xan-g*, *xan-h* and *alb-e* accumulate mainly protoporphyrin (GOUGH 1972). Thus the products of these genes may be involved in the Mg^{2+} insertion step. The leaky mutant *xan-u*^{21} is a complex mutant that behaves as if it contains, besides the *xan* mutation, a mutation at a closely linked locus, *tig-n* (KAHN et al. 1976). When supplied with ALA in the dark it accumulates 35% of its porphyrins as uroporphyrin and/or uroporphyrinogen, 59% as protoporphyrin and 6% as protochlorophyllide (GOUGH 1972). This mutant is also deficient in carotenoid synthesis, since it forms mainly poly-cis-lycopene instead of β-carotene (NIELSEN and GOUGH 1974). Absorption spectra of ALA-treated leaves have indicated that the mutant *xan-k*^{42} accumulates coproporphyrinogen and protoporphyrin (K.W. HENNINGSEN, J.E. BOYNTON and D. VON WETTSTEIN, unpublished data), while the mutant *vir-y*^{59} seems to accumulate mainly protoporphyrin (K.W. HENNINGSEN, unpublished data). Upon ALA-feeding dark-grown seedlings of the leaky mutant *xan-l*^{35} accumulate about 12% Pchl-ide, 14% Mg protoporphyrin monomethyl ester, 34% Mg protoporphyrin and 41% protoporphyrin (GOUGH 1972). Another mutant at this locus, *xan-l*^{81}, has similar properties, but appears to be completely blocked (D. VON WETTSTEIN, personal communication). In vivo absorption spectra of the mutant *vir-k*^{23} have also provided evidence for accumulation of the Mg protoporphyrins (K.W. HENNINGSEN, unpublished data). The primary lesion of these mutants can most simply be explained as a defect in an enzyme that converts the Mg protoporphyrins to Pchl-ide. In relation to considerations about the control of porphyrin biosynthesis in plastids, it is interesting to note that leaves of *xan-l*^{35} mutant seedlings illuminated for 2–4 h contain approximately normal quantities of cytochrome f and b_6, while mutants at the loci *xan-f*, *xan-g*, and *xan-h* possess significantly increased amounts of these cytochromes (K.W. HENNINGSEN, unpublished data).

Several mutants of maize defective in Chl synthesis have been analysed (MASCIA 1978, MASCIA and ROBERTSON 1978). By ALA-feeding of dark-grown seedlings and identification of the accumulated porphyrins it was found that mutants at the three loci *l*-blandy 4*, *l 13* and *oy* are defective in conversion of protoporphyrin to Mg-protoporphyrin, while mutants at the *oro* and *oro-2* loci are defective in the conversion of Mg-protoporphyrin monomethyl ester to Pchl-ide. A dominant modifier gene, *Orom*, which allows *oro* seedlings to bypass the affected step was also described.

Dark-grown plants contain only small quantities of Pchl-ide. When they are treated with ALA larger quantities of Pchl-ide are accumulated, indicating that the formation of ALA is the rate-limiting step in the biosynthetic pathway (GRANICK 1959). Barley mutants at the loci *tig-d*, *tig-o*, and *tig-b* accumulate Pchl-ide constitutively up to 20 times the wild-type level in darkness (VON WETTSTEIN et al. 1974, NIELSEN 1974). Levulinate, an inhibitor of the enzyme ALA dehydratase which converts ALA to porphobilinogen, has been used to show that the high Pchl-ide levels in these mutants are due to higher rates of ALA formation than in the wild type (GOUGH and KANNANGARA 1979). Intact plastids from greening barley leaves can synthesize Chl from glutamate via ALA (KANNANGARA et al. 1978, GOUGH and KANNANGARA 1976, KANNANGARA and GOUGH 1977, GOUGH and KANNANGARA 1977). Some of the enzymes are located in the chloroplast membranes, and others in the stroma. The ALA-synthesizing activity is present in the soluble fraction of the etioplasts (KANNANGARA and GOUGH 1979). Combination of the *xan-l*^{35} mutation with mutations at the *tig-d*, *-b* or *-o* loci yields genotypes which accumulate Mg protoporphyrins. Some accumulation of protoporphyrin takes place upon combination of the mutation *xan-f*^{10} with *tig-b* or *tig-o* mutations. In contrast, the mutant *tig-d*^{12} does not give double mutants that accumulate protoporphyrin (VON WETTSTEIN et al. 1974, KAHN et al. 1976). These results indicate that the *tig-b* and *tig-o* genes are involved in a feedback control of the ALA-synthesizing system that involves protoporphyrin or a derivative of this compound (Fig. 10), while the *tig-d* gene product participates in another type of inhibition of this system. Partially purified preparations from greening leaves of wild-type barley of the ALA-synthesizing activity are inhibited strongly be heme and less efficiently by Mg-protoporphyrin, but are not inhibited by protoporphyrin. Feedback inhibition seems thus primarily exerted by heme (GOUGH et al. 1980). Although isolated etioplasts of the *tig-d*^{12} mutant form ALA at a higher rate than wild-type etioplasts, the ALA-synthesizing activity in plastid stroma preparations is lower for the mutant than for the wild type. As the defective control in *tig-d* is not seen with the soluble enzyme, it was hypothesized that it may require for expression components dissociating during preparation of the enzyme (GOUGH et al. 1982, GIRNTH et al. 1982).

Dark-grown seedlings of angiosperms do not contain Chl, but accumulate some Pchl-ide, which is localized in the prolamellar tubules of the etioplast (KAHN 1968). On illumination, the active Pchl-ide is reduced to Chl-ide. The photoactive Pchl-ide-protein complex – referred to as Pchl-ide holochrome – has recently been identified as an NADPH-protochlorophyllide oxidoreductase containing a single polypeptide chain with an apparent molecular weight of 36,000 (GRIFFITHS 1975, 1978, APEL et al. 1980).

RÖBBELEN (1956) has described a mutant of *Arabidopsis thaliana*, which forms Pchl-ide, but is unable to photoreduce it to Chl-ide. This mutation may affect the holochrome polypeptide or the supply of NADPH to the holochrome.

Most algae form Chl equally well in the light and when they are growing heterotrophically in darkness. Mutant strains of some of these species have been isolated that only form Chl in darkness. Several of these mutants, e.g. the *Scenedesmus obliquus* mutants *C-2A'* and *C-6D* (SENGER and BISHOP 1972,

SENGER and STRASSBERGER 1978) and the *Chlorella fusca* mutant *G10* (WILD et al. 1978) appear not to accumulate Pchl-ide in the dark. Pchl-ide accumulation indicates that the specific step affected may be the holochrome-mediated reduction of Pchl-ide to Chl-ide (Fig. 10). The *Chlorella vulgaris* mutants *31* (GRANICK 1971) and *C-10* (BRYAN et al. 1967), and the *Chlamydomonas reinhardtii* mutant *y-1* (SAGER 1955, WANG et al. 1977) accumulate Pchl-ide. The plastid of the dark-grown *y-1* mutant contains a few unpaired short thylakoids (SAGER and PALADE 1954, OHAD et al. 1967a) and some protochlorophyllide that upon illumination is converted to Chl-ide (SAGER 1961). Light-activated Chl synthesis is accompanied by the formation of thylakoid membranes (OHAD et al. 1967b). An unusual feature of mutants at this locus is that they arise at a very high frequency (SAGER and TSUBO 1962), and also revert at a high rate (HUDOCK and BART 1967, WANG 1978). In addition, it can be obtained in high yield by treatment of the cells with streptomycin (SAGER and TSUBO 1962). The *y-1* mutant and the *Chlorella* mutants *G10* and *31* may be affected in genes that directly control the ability of the holochrome complex to photoreduce Pchl-ide in the dark. It has also been suggested that the wild type has both an enzyme that does not require light and an enzyme that requires light (GRANICK 1971). In the mutants, a gene participating in the formation of the light-independent enzyme might be affected.

The barley mutants *alb-*f^{17} and *xan-*j^{59} are (as discussed in Sect. 3.2.2) unable to phytolate Chl-ide to Chl a (HENNINGSEN and THORNE 1974).

The barley mutant *chlorina-f2* lacks Chl b (BOARDMAN and THORNE 1968) and the major polypeptide of the light-harvesting Chl a/b-complex (BURKE et al. 1979). The polypeptide is probably synthesized on cytoplasmic ribosomes in the mutant (APEL and KLOPPSTECH 1978). The primary effect of mutations at this locus may therefore be inability to synthesize Chl b. This inability would result in lack of incorporation of the polypeptide normally associated with Chl b into the photosynthetic membranes. Among the higher plants mutants with no or considerably reduced quantities of Chl b have also been reported for pea (HIGHKIN et al. 1969, ANDERSON and LEVINE 1974a, b, GOTTSCHALK and MÜLLER 1964, MÜLLER 1964, STUMMANN and HENNINGSEN 1980), tobacco (SCHMID 1971) and *Arabidopsis thaliana* (RÖBBELEN 1957, HIRONO and RÉDEI 1963).

4.2 Genetic Control of Carotenoid Synthesis

Unlike Chl, carotenoids are formed in dark-grown angiosperm seedlings. However, a net synthesis of carotenoids occurs during greening, and the composition of the carotenoid fraction changes. WOLF (1963) has reported that 7–10-day-old wheat seedlings have about twice as much carotenoid when grown in light as when grown in the dark. Prolamellar bodies from etioplasts and grana from fully developed chloroplasts have nearly the same amount of carotenoid per mg of protein, while stroma lamellae have a 20 times lower carotenoid content. The ratios of β-carotene to xanthophyll are 0.05, 0.4, and 1.2 for prolamellar bodies, grana, and stroma, respectively (BAHL 1977). The rate of carotenoid

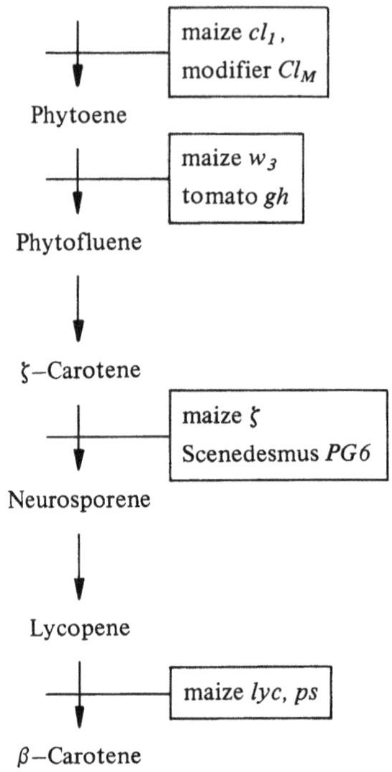

Fig. 11. Diagram of carotenoid biosynthesis and the lesions caused by mutation in genes of various plants

synthesis during greening is especially high for β-carotene and lutein (KLEUDGEN and LICHTENTHALER 1974). Only those carotenoid-deficient mutants for which some evidence has been obtained with respect to the biochemical step affected will be considered in this review (Fig. 11). Except for *Scenedesmus*, it has been shown that the mutants listed in Fig. 11 are Mendelian mutants. The biosynthesis of carotenoids has been reviewed by BRITTON (1976).

EVERETT (1949) described a recessive mutation, cl_1, in maize that in the homozygous state resulted in pale yellow or white seeds which upon germination gave albino seedlings. The mutant makes Chl, but not carotenoids. The absence of carotenoids leads to photodestruction of the Chl (ANDERSON and ROBERTSON 1960). Two other mutants at the cl_1 locus with similar properties have been found (ROBERTSON 1966). Four alleles at the locus Cl_M, Cl_2, Cl_3, Cl_4, and Cl_5, partially or completely suppress the albinism of the seedling, but do not suppress the endosperm albinism, (EVERETT 1949, ROBERTSON 1966). None of 12 other mutants of maize with properties similar to those of the mutations at the cl_1 locus were affected by the Cl_M modifiers (ROBERTSON 1966). These modifiers appear, therefore, to be highly specific.

The albina mutant w_3 in maize is also capable of forming Chl, but the Chl is bleached on continuous illumination (KOSKI and SMITH 1951). The w_3 mutant accumulates some phytoene, smaller quantities of phytofluene and ζ-carotene, and traces of neurosporene. Some other mutants affected at this locus accumulate only detectable quantities of phytoene, suggesting that this gene is involved in the conversion of phytoene to phytofluene (TREHARNE et al. 1966).

When w_3 seedlings are grown in the dark, the plastids contain prolamellar bodies and look approximately normal. When the plants are transferred to light the prolamellar bodies are transformed into primary lamellae in the usual way. However, after a few hours the lamellae start disintegrating into fragments and vesicles (BACHMANN et al. 1967). Seedlings of the tomato mutant *gh* (ghost) form Chl, but the Chl is destroyed in light. Both the chromoplasts of the mature fruit and the leaf chloroplasts contain phytoene instead of colored carotenoids (MACKINNEY et al. 1956).

The β-carotene-deficient mutants ζ and *lyc* accumulate ζ-carotene and lycopene, respectively. The Chl of the mutants is photooxidized (FALUDI-DÁNIEL et al. 1966, 1968, HORVÁTH and FALUDI-DÁNIEL 1971). Another lycopene-accumulating mutant of maize, *ps* (pink scutellum), has been described. The endosperm and the seedlings are pink (ROBERTSON 1961). The pale mutant *PG6* strain of *Scenedesmus obliquus* is extremely sensitive to photooxidation. The predominant carotenoids present are phytoene and ζ-carotene along with their epoxides. Some phytofluene is also present. This mutant appears, accordingly, to be unable to convert ζ-carotene to neurosporene (POWLS 1971).

Mutations that may affect the regulation of carotenogenesis have been described for barley, maize, and sunflower (*Helianthus annuus*). NIELSEN and GOUGH (1974) determined the composition and amount of carotenoids in etiolated seedlings of six barley mutants, comprising *xan-u*^{21} and five *tigrina* mutants with relaxed protochlorophyllide synthesis (cf. Sect. 4.1). All mutants had approximately the same total carotenoid content as the wild type, but all, except *tig-d*^{12}, deviated from the wild type with respect to carotenoid composition. Three mutants, *xan-u*^{21}, *tig*-n^{32}, and *tig-n*^{33}, contain significantly higher quantities of carotenes than the wild type. These two *tigrina* mutants, as well as *tig-b*^{23} and *tig-o*^{34}, synthesize β-carotene, but in addition accumulate various precursors of this pigment. If mutations in the *tig-b*, *tig-n*, and *tig-o* loci result in the alteration of a membrane component, it is – as suggested by NIELSEN and GOUGH (1974) – conceivable that pigment and protein molecules become associated within the membranes in a way that leads to a decrease of inhibition of the ALA-synthesizing enzyme and conservation of precursors of β-carotene. The mutant *xan-u*^{21} is unable to metabolize Mg-protoporphyrin effectively (cf. Sect. 4.1). The dark-grown seedlings do not contain β-carotene, but accumulate phytofluene, proneurosporene (poly-cis-neurosporene), and prolycopene (poly-cis-lycopene) (NIELSEN and GOUGH 1974). It may be suggested that the mutant is defective in the enzyme which converts cis-phytofluene to trans-phytofluene, and that this results in the formation of further cis-products. As pointed out by NIELSEN and GOUGH (1974) the *Chlorella vulgaris* mutant *5/520* has a strikingly similar carotene composition (CLAES 1954, 1956, 1957).

For barley plants that are homozygous for the nuclear gene *gp* (grandpa) it appears that the immature cells developing from the intercalary meristem are capable of responding to light in two ways. If the cells are permitted to develop and produce carotenoids in the dark, they will accumulate Chl upon exposure to light. Exposure to light prior to the development of carotenoids will prevent the synthesis of Chl. Thus the plants will develop a banding pattern under a light-dark cycle (MATCHETT et al. 1968). The *gp* gene may be involved in the regulation of the synthesis of carotenoids in the dark. Etiolated homozygous seedlings of the nuclear maize mutant *al* (*albescent*) are severely deficient in colored carotenoids and accumulate some phytoene and phytofluene. The plastids are destroyed and the Chl-synthesizing potential is lost when the seedlings are transferred directly to a high intensity of light. However, synthesis of plastid pigment and development of the photosynthetic apparatus appears to be normal in seedlings preilluminated with dim light prior to the transfer to high light intensity (SANDER et al. 1968, TROXLER et al. 1969). As for the *gp* mutant of barley, the mutated gene may participate in regulation of carotenoid synthesis in the dark. WALLACE and HABERMANN (1959) have described a sunflower mutant that lacks β-carotene but contains xanthophyll. Since the introduction of the hydroxyl and epoxide groups of the xanthophylls presumably occur late in the biosynthetic sequence (BRITTON 1976), it is conceivable that this mutation leads to a more effective conversion of carotenes into xanthophylls.

4.3 Genetic Control of Thylakoid Protein Synthesis

The polypeptide composition of the photosynthetic membrane is complex. By gel electrophoresis of SDS-solubilized thylakoid polypeptides at least 33 bands have been recognized in *Chlamydomonas* and pea (CHUA and BENNOUN 1975, STUMMANN and HENNINGSEN 1980), while 29 and 43 polypeptide bands have been found for maize and barley, respectively (GREBANIER et al. 1979, HØYER-HANSEN and SIMPSON 1977). In this section the efforts to characterize the function and genetic control of these polypeptides and the changes in the electrophoretic pattern during greening of wild-type seedlings will be summarized as a frame for the discussion of the mutants affected in the polypeptide composition of the thylakoids.

In barley the primary lamellar layers contain a sufficient area of membranes to support grana formation during 24 h of greening (HENNINGSEN and BOYNTON 1974). Thus the net synthesis of membrane material is limited in this period. However, the composition of the membranes changes profoundly during the first few hours of greening. The rate of Chl accumulation is maximal between 2 and 4 h of greening (HENNINGSEN and BOARDMAN 1973). HØYER-HANSEN and SIMPSON (1977) showed that at least 15 membrane polypeptides disappear and at least 23 appear during greening of dark-grown barley seedlings (Fig. 12). By purification of the NADPH-protochlorophyllide oxidoreductase (Pchl-ide holochrome) from etioplasts according to GRIFFITHS (1978) and APEL et al. (1980) followed by electrophoresis, it has recently been shown that the prominent etioplast band E_{20} (mol.wt. 36,000) represents this protein (GIRNTH et al. 1982). GREBANIER et al. (1979) demonstrated the disappearance of at least two membrane polypeptides and the appearance of at least seven during greening of etiolated maize seedlings. One of the disappearing bands has mol.wt. 37,000 and is presumably analogous to E_{20}. Both for barley and maize most of the qualitative differences have appeared after a few hours of greening, indicating that a large part of the chloroplast genome may be transcribed in this period.

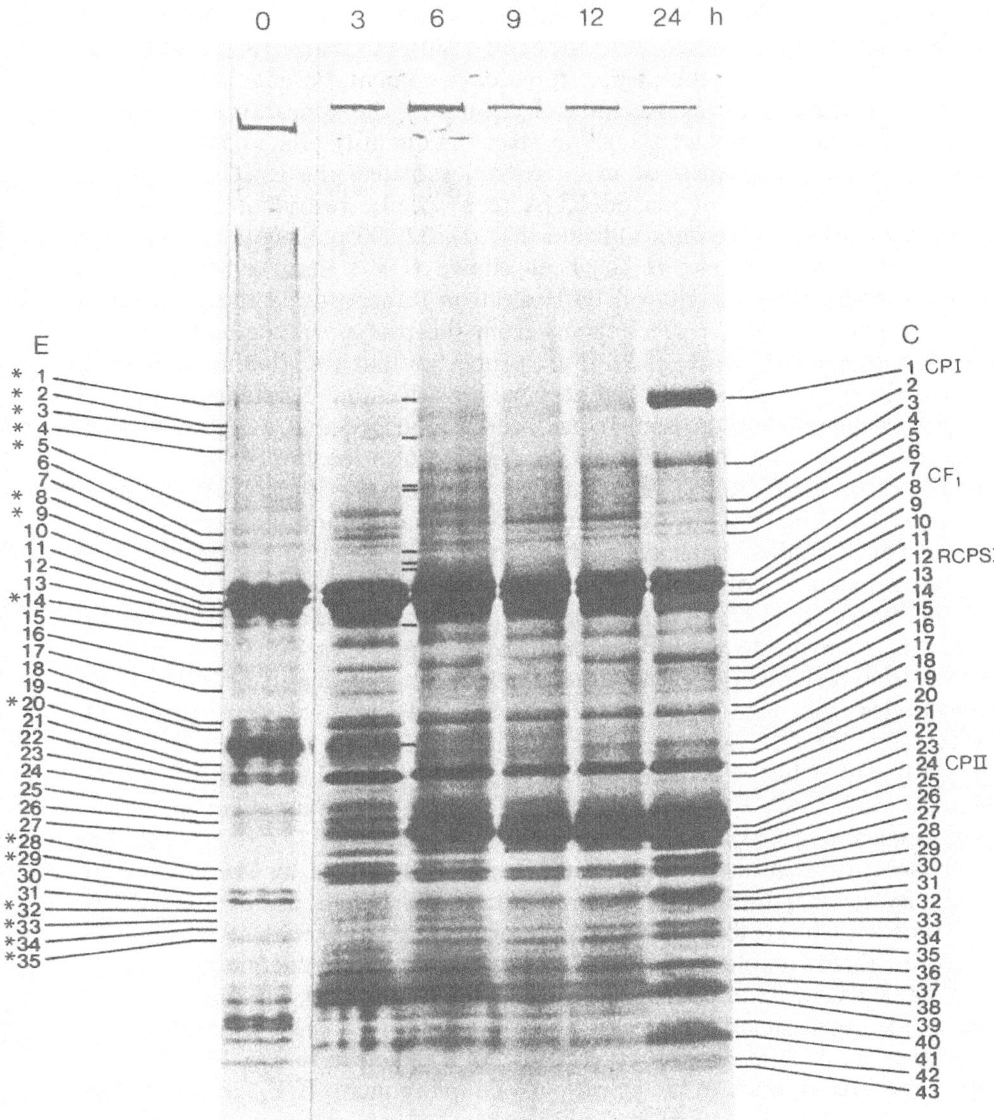

Fig. 12. Plastid membrane polypeptide patterns from etiolated (o h) and greening barley. Isolated membranes were dissociated with the detergent sodium dodecyl sulfate (SDS) and analyzed by electrophoresis in a 7.5%–15% gradient acrylamide gel containing SDS. Etioplast membrane polypeptide bands marked with an *asterisk* disappear during greening. Other polypeptide bands characteristic of membranes from fully developed chloroplasts (24 h) appear during greening. Positions are marked for some chlorophyll-protein complexes (bands *C1*, *12* and *24*) and coupling factor subunits (bands *C7* and *8*). (Høyer-Hansen and Simpson 1977)

In accordance with this notion, light-induced transcription of chloroplast DNA has been detected for pea (STUMMANN et al. 1980) and maize (BEDBROOK et al. 1978). For maize it was found that three developmental events are related in time and magnitude during greening of etioplasts: the appearance of a membrane polypeptide of mol.wt. 32,000 in vivo, the quantity of a 34,500 precursor to this polypeptide synthesized in vitro from isolated plastid RNA, and the extent of hybridization of plastid RNA to a specific restriction fragment of chloroplast DNA. These data indicate that the 32,000 polypeptide is encoded in the chloroplast genome. It has been shown (DARR et al. 1981) that lack of atrazine inhibitory activity on PS II electron transport and the presence of decreased rate of transfer of electrons from the primary acceptor (Q) to the secondary acceptor (B or R) of PS II are properties that are inherited in parallel through the maternal parent when atrazine-sensitive and -resistent biotypes of *Brassica campestris* are crossed. It has been established that a 32,000 mol.wt. thylakoid polypeptide functions as specific receptor of this PS II herbicide (PFISTER et al. 1980). It will be of interest to determine if this polypeptide is similar to the 32,000 polypeptide encoded in the maize chloroplast genome.

A series of studies that have been important in assigning a functional role to specific thylakoid polypeptides have been reviewed by BOARDMAN et al. (1978) and GILLHAM et al. (1978). We shall outline the current state of this identification process by referring to work done on barley. Some aspects of this work have recently been reviewed (VON WETTSTEIN 1980, 1981). The identity of 16 thylakoid polypeptides from wild-type barley is indicated in Fig. 13. Usually three Chl-protein complexes (Chl-P) are observed, namely Chl_a-P1 (often termed CPI), $Chl_{a/b}$-P1 and $Chl_{a/b}$-P2 (identical to the light-harvesting complex, LHC) (MØLLER et al. 1982, HILLER et al. 1980). The light-harvesting complex and its major apoprotein, $Chl_{a/b}$-AP2, have often been termed CP II (cf. Fig. 12). The nomenclature system used for the Chl-proteins is described by MACHOLD et al. (1979). By electrophoresis at low temperature and by the use of low SDS to Chl ratios ten Chl-proteins have been obtained (MACHOLD et al. 1979). In addition to the three complexes indicated in Fig. 13, two Chl a-containing polypeptides, named Chl_a-P2 and Chl_a-P3, three putative oligomers of $Chl_{a/b}$-P2, and two as yet uncharacterized green bands were observed. The apoproteins for which the electrophoretic mobility is known are indicated in Fig. 13. The position of the five subunits of the chloroplast-coupling factor (CF_1) in the electro-

Fig. 13. Chloroplast membrane polypeptide patterns from wild-type, *viridis* mutants and photosystem I particles of barley. Isolated membranes or particles were dissociated with the detergent sodium dodecyl sulfate (SDS) and analyzed by electrophoresis in a 11%–15% gradient acrylamide gel containing SDS. Mutants *vir-n*^{34}, *-zb*^{63} and *-h*^{15} are deficient in photosystem I activity and lack polypeptide bands believed to be part of photosystem I: *Chl_a-P1* chlorophyll *a*-protein; *Chl_a-AP1* apoprotein of Chl_a-P1; *FeS* polypeptides of photosystem I iron-sulfur centers. The identity (abbreviations: see text) and apparent molecular weight of several polypeptide bands are indicated. (HILLER et al. 1980)

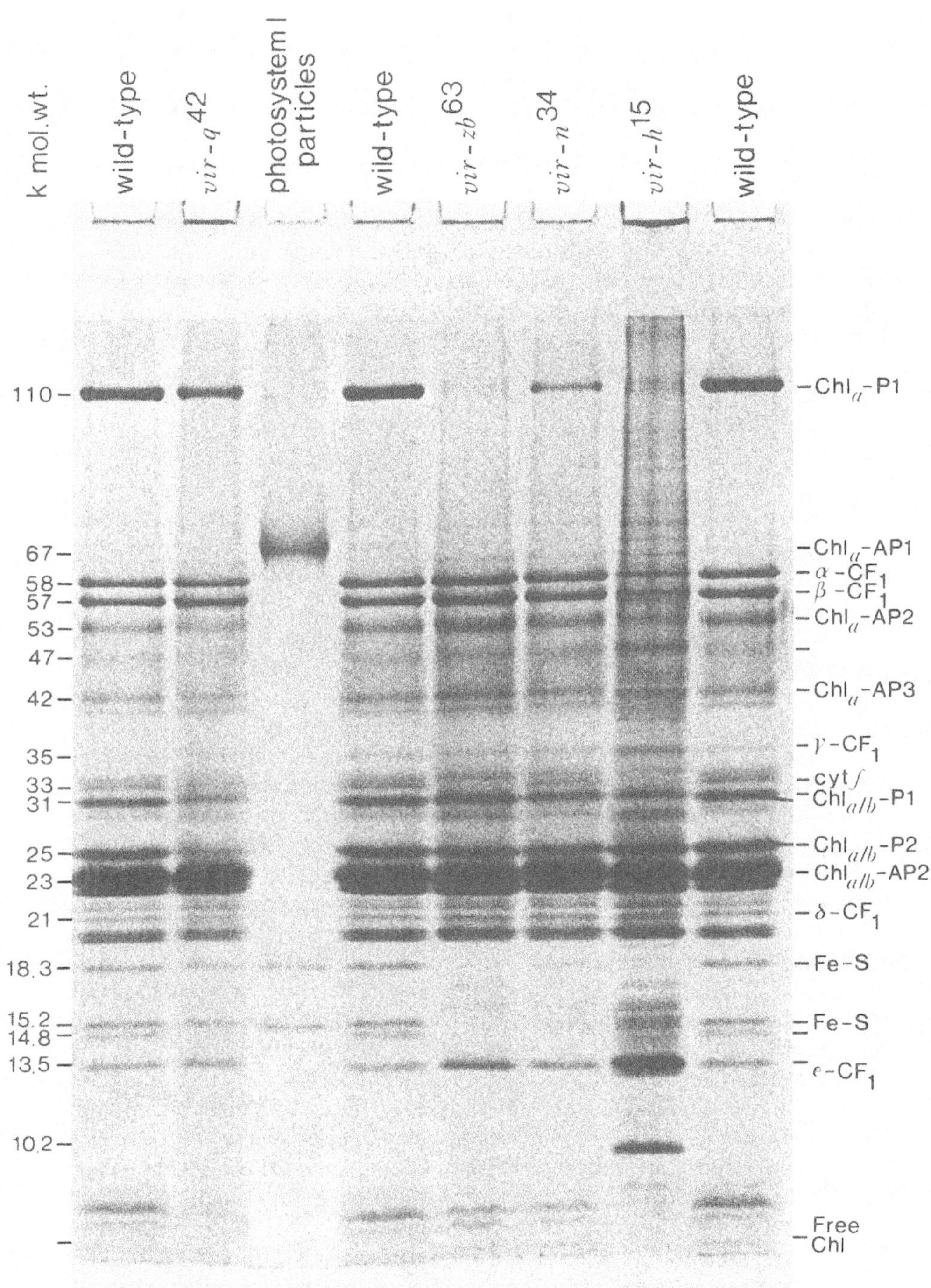
k mol.wt.
wild-type
vir-q 42
photosystem I particles
wild-type
vir-zb 63
vir-n 34
vir-h 15
wild-type
110
67
58
57
53
47
42
35
33
31
25
23
21
18.3
15.2
14.8
13.5
10.2
Chl a-P1
Chl a-AP1
α-CF1
β-CF1
Chl a-AP2
Chl a-AP3
γ-CF1
cyt f
Chl a/b-P1
Chl a/b-P2
Chl a/b-AP2
δ-CF1
Fe-S
Fe-S
ε-CF1
Free Chl

phoretic pattern has been established by comparison with electrophoretograms of isolated coupling factor and by crossed immunoelectrophoresis (HØYER-HANSEN et al. 1979, MØLLER et al. 1982). Staining of electrophoretograms according to the method of THOMAS et al. (1976) has revealed two polypeptides of molecular weight 33,000 and 20,000 with associated heme groups (HØYER-HANSEN 1980). The 33,000 polypeptide presumably represents cytochrome f (Fig. 13), while the 20,000 polypeptide may represent a subunit of cytochrome b_6 (HØYER-HANSEN 1980, WASSERMAN 1980). Particles with PS I activity – measured as O_2 consumption in the presence of methylviologen and reduced DCIP – have been isolated from barley thylakoids by treatment with the zwitterionic detergent Empigen BB of a PS I-enriched fraction obtained by the use of digitonin (MØLLER et al. 1982). The Chl to P-700 ratio of the particles was about 100, and the particles were found to comprise polypeptides of molecular weight 67,000, 18,300 and 15,200 (Fig. 13). The 67,000 band represents the apoprotein of Chl_a-P1, termed Chl_a-AP1 (MACHOLD et al. 1979). Feeding of plants with ^{59}Fe resulted in labeling of the cytochrome f band as well as two polypeptides with apparent molecular weights of about 18,300 and 15,200 (MØLLER et al. 1982). These two bands were inferred to represent FeS centers associated with PS I. The PS I particles isolated from pea by MULLET et al. (1980) contain, besides a few extra polypeptides, polypeptides with mobilities similar to those of the PS I particles from barley.

An important way of approaching or supporting some of these identifications has been through a detailed biophysical and biochemical analysis of mutants with lesions in the photosynthetic apparatus. This strategy has been used effectively for the PS I-associated polypeptides, the polypeptide associated with the reaction center of PS II, and the light-harvesting complex $Chl_{a/b}$-P2.

It has often been observed that mutants with decreased amounts of Chl_a-P1 (CP I) and its apoprotein, Chl_a-AP1, also possess impaired PS I activities. CHUA et al. (1975) have characterized two nuclear mutants of *Chlamydomonas, F1* and *F14*, which did not possess PS I activity or detectable quantities of P-700, and which specifically lacked CP I and its 66,000 mol.wt. apoprotein. BENNOUN et al. (1977) described a plastome mutant of *Chlamydomonas*, *C1*, which is PS I deficient and which only contains 10–15% of the normal amount of CP I in the thylakoids. By arsenate selection on cells previously grown in 5-fluorodeoxyuridine, SHEPHERD et al. (1979) have isolated mutants at nine chloroplast loci of *Chlamydomonas* that are affected in various photosynthetic functions. It was found that at least three chloroplast loci specifically affect CP I, including the locus of the *C1* mutation. HERRMANN (1971a) have described a plastome mutant of *Antirrhinum majus*, *en:alba-1*, which is photosynthetically inactive and lacks CP I and the corresponding apoprotein. The plastome mutant *en:gil-1* of *Pelargonium zonale* has luminiscence-properties indicating a functional PS II, but does not show the EPR signals normally associated with PS I (HERRMANN et al. 1974), and does not contain detectable quantities of CP I (HERRMANN et al. 1976). Two nuclear mutants of tomato, *sulfureapura* and *albina-2*, also lack CP I (BÖRNER et al. 1975). MILES et al. (1979) have isolated eight PS I deficient nuclear mutants of maize, which lack the CP I complex. The barley mutant *vir-q*^{42} has – on a Chl basis – about 75% of the wild-type activity in both photosystems, while the other three mutants for which the thylakoid membrane polypeptide pattern is shown in Fig. 13 have normal PS I activities, but are totally (*vir-zb*^{63}) or partly (*vir-n*^{34} and *vir-h*^{15}) deficient in PS I activity (HILLER et al. 1980). The mutants *vir-zb*^{63} and *vir-n*^{34} are clearly deficient in the 15,200 and 18,300 bands, which MØLLER et al. (1982) found as an integral part of the PS I particles isolated with the detergent Empigen BB (Fig. 13). The mutant *vir-h*^{15} also appears to be deficient in the 18,300 polypeptide. The relatively prominent bands seen at 10,500 and 13,500 for this mutant may be due to nuclear contamination (HILLER et al. 1980, HØYER-HANSEN and SIMPSON 1977). The data obtained with these mutants support the notion that Chl_a-P1 and the 18,300 and 15,200 bands are closely associated in vivo. These studies on *Chlamydomonas* and five species of higher plants, when considered together, indicate that mutations at several nuclear loci and several chloroplast loci apparently lead to specific loss of the CP I polypeptide, and possibly also of at least two other polypeptides that in vivo are associated with the CP I complex. There may exist some regulatory mechanism which ensures that the CP I polypeptide will only be synthesized and inserted into the thylakoid membrane if several other components of the membrane are also formed.

In accordance with the results described for *vir-zb*^{63}, it has been shown that this mutant lacks the EPR signals normally associated with P-700^{+} and the FeS centers of PS I (NUGENT et al. 1982). The absence of these signals made it possible to observe a signal at $g = 2.0033$ at low temperatures and under reducing conditions. The signal was similar to that observed for subchloroplast particles by KLIMOV et al. (1980), and was ascribed to reduced pheophytin. This may provide new information on the primary acceptor of PS II.

Work on mutants has played an important role in the identification of a polypeptide associated with the reaction center of PS II. CHUA and LEVINE (1969) showed that chloroplast fragments from the Mendelian *Chlamydomonas* mutant *F34* possess PS I, but not PS II activity. The fluorescence induction kinetics of this mutant consist of an abrupt rise to the maximum level of fluorescence, and this level is not affected by DCMU. These observations indicate that the primary acceptor of PS II, Q, cannot be reoxidized or is absent (CHUA and BENNOUN 1975). It was found that thylakoid polypeptide no. 6 of mol.wt. 47,000 was absent in the mutant. A suppressor mutation was obtained, that partially restored the PS II activity and the content of this polypeptide. It was inferred that polypeptide 6 is closely associated with the PS II reaction center (CHUA and BENNOUN 1975). The sites of synthesis of the major thylakoid membrane polypeptides of *Chlamydomonas* have been studied by pulse labeling of cells in the presence of inhibitors specific for chloroplast and cytoplasmic protein synthesis (CHUA and GILLHAM 1977). It was concluded that polypeptide 6 and at least eight other membrane polypeptides are made on chloroplast ribosomes. Since there is no evidence for transport of RNA across the outer membrane of plastids this result indicates that polypeptide 6 and at least eight other thylakoid polypeptides are encoded in the plastome. If this is the case the absence of polypeptide 6 from the thylakoids of the mutant *F34* does not represent the primary action of the mutation. Chloroplast preparations from the barley mutant *vir-c*^{12} lack PS II activity, but display some PS I activity (NIELSEN et al.

1974, SMILLIE et al. 1974). The polypeptide pattern of this mutant was reported by MACHOLD and HØYER-HANSEN (1976) to lack a polypeptide of mol.wt. about 46,000 as well as another band that presumably is identical to the band immediately below $Chl_{a/b}$-P1 in Fig. 13. After comparison between these results and those obtained with the mutant *F34*, it was suggested that the band at mol.wt. ca. 46,000 – which is identical to the band at 42,000 in Fig. 13 – could represent a PS II reaction center polypeptide (MACHOLD and HØYER-HANSEN 1976). It was later shown by crossed immunoelectrophoresis that antibodies raised to *Chlamydomonas* polypeptide 6 reacted with this barley polypeptide (SIMPSON et al. 1977). MACHOLD et al. (1979) prepared Chl-protein complexes from wild type and mutants of barley by the use of SDS, and separated the complexes by electrophoresis. Three Chl a-containing complexes were found, Chl_a-P1 (CP I), Chl_a-P2 (apparent mol.wt. 49,000) and Chl_a-P3 (apparent mol.wt. 41,000). Since the mutant *vir-m*^{29}, which lacks the 49,000 complex, has PS II activity, it was suggested that the 41,000 complex may represent the reaction center of PS II. In *Chlamydomonas* Chl_a-P2 and Chl_a-P3 (termed CP III and CP IV) each contain a single polypeptide chain (polypeptide 5 and 6, respectively) (DELEPELAIRE and CHUA 1979). Since the polypeptide positioned at 42,000 mol.wt. in Fig. 13 cross-reacts immunochemically with *Chlamydomonas* polypeptide 6, there seems to be ample evidence to conclude that this barley polypeptide is the apoprotein of Chl_a-P3 – termed Chl_a-AP3 – and is associated with the PS II reaction center (MACHOLD et al. 1979). It was also shown by reelectrophoresis in a dissociating system that the apoprotein of the Chl_a-P3 complex comigrated with the band, which is located at 42,000 mol.wt. in Fig. 13 (MACHOLD et al. 1979). SATOH (1979) identified 43,000 and 27,000 polypeptides from a dissociated extensively purified PS II reaction center complex. It may therefore be suggested that the band immediately below $Chl_{a/b}$-P1 in Fig. 13, which apparently is reduced or absent for the barley mutant *vir-c*^{12} (MACHOLD and HØYER-HANSEN 1976), is also closely associated with the PS II reaction center. HERRMANN (1971b) has decribed a plastome mutant of *Antirrhinum majus, en:viridis-1*, that is pale green and contains Chl b but which cannot perform the Hill reaction. A polypeptide, polypeptide no. 7, that probably is equivalent to *Chlamydomonas* polypeptide 6, is missing.

The number of polypeptides associated with the light-harvesting complex, $Chl_{a/b}$-P2, is not firmly established. MACHOLD and MEISTER (1979) have presented results which indicate that $Chl_{a/b}$-P2 from *Vicia faba* consists of two distinct Chl-proteins of mol.wt. 24,000 and 23,000. The electrophoretically purified complexes from pea, spinach, and barley have been found to contain two main polypeptides (SCHMIDT et al. 1980, ANDERSON 1980). MACHOLD et al. (1979) purified the light-harvesting complex from barley according to the method of BURKE et al. (1978). Upon electrophoresis in a dissociating system similar to the system used for the gels shown in Fig. 13, a major polypeptide of the molecular mass 25,000, and two minor bands at 26,000 and 24,000 were observed. The band of highest molecular weight has later been assumed to represent undissociated complex (VON WETTSTEIN et al. 1982, MØLLER et al. 1982) (cf. Fig. 13). BURKE et al. (1979) purified the light-harvesting complex from barley by the same procedure and found that it contained a major polypeptide of mol.wt. 25,000, a less prominent polypeptide of 23,000 and a minor band of 28,000. SCHMIDT et al. (1980) found that the two polypeptides of $Chl_{a/b}$-P2 from pea were immunochemically related. In contrast, the two low molecular weight polypeptides from the complex obtained from barley give different sets of fragments when subjected to partial digestion with papain (BURKE et al. 1979) or chymotryptic digestion (VON WETTSTEIN et al. 1982).

It has been found that when broken chloroplasts are isolated in low salt media, their thylakoids become unstacked. Addition of Mg^{2+} or other divalent cations at low concentrations – or monovalent cations at higher concentrations – induces stacking like that observed in the intact organelle (IZAWA and GOOD 1966, SMILLIE et al. 1976, BARBER 1976). It has also been reported that for preparations of broken chloroplasts Mg^{2+} increases the quantal efficiency of PS II reactions and decreases the efficiency of PS I reactions (cf. review by BARBER 1976). BARBER and CHOW (1979) have proposed a model of the thylakoid membrane that may account for these effects of cations. According to this model lateral diffusion of PS II and PS I particles can occur in the thylakoid membranes, and the driving force for unstacking is electrostatic repulsion between negatively charged PS I particles, while stacking is mediated by hydrophobic interactions between PS II particles. Screening of the coulombic repulsions by cations encourages a heterogeneous distribution of the two types of particles and stacking due to the hydrophobic forces between the PS II particles, while absence of screening causes a migration of PS I particles into the clusters of PS II particles, which leads to unstacking. If the light-harvesting complex is associated with the PS II particles the model may account for the cation-induced changes in the degree of spillover.

MULLET and ARNTZEN (1980) purified the light-harvesting complex from pea chloroplast membranes. The complex, when dialyzed to remove detergents, aggregated into planar sheets of 8 nm subunits. Addition of cations caused these sheets to become tightly apressed, indicating that $Chl_{a/b}$-P2 may possess a negative charge. Such a negative charge would represent an extra driving force for unstacking of thylakoids in low salt media. Since the light-harvesting complex appears in parallel with stacking during chloroplast development (ARGYROUDI-AKOYUNOGLOU 1977, ARMOND et al. 1976) it seems likely that this complex plays an essential role in the mediation of stacking.

The exception to these studies was the barley mutant *chlorina-f2* that has no detectable Chl b and yet has some grana (cf. review by ANDERSON 1975). It has been reported that this mutant lacks the $Chl_{a/b}$-P2 complex and its constituent polypeptides (ANDERSON and LEVINE 1974a, b). SIMPSON (1979) has published freeze-fracture replicas of wild type and *chlorina-f2* thylakoids. The most significant difference between the membranes of the mutant and the wild type is a reduction of the number of particles on the OFF face in stacked regions from ca. 6,300 per μ^2 in the wild type to ca. 1100 in the mutant. Thus the light-harvesting complex appears to be a major component of many of these particles (SIMPSON 1979). BURKE et al. (1979) have reexamined the chloroplasts from the *chlorina-f2* mutant. They found that $Chl_{a/b}$-P2 from wild-type barley comprises two major polypeptides of molecular weight 25,000 and 23,000 and a minor polypeptide of 28,000, and that the 23,000 and 28,000 polypeptides were present in the mutant. It was also demonstrated that this partial $Chl_{a/b}$-P2 complement alters the quantitative Mg^{2+} requirement for stacking, but not the overall qualitative response. MACHOLD et al. (1979) separated Chl-protein complexes from wild-type barley and from *chlorina-f2* by electrophoresis, and found a small amount of Chl a at the position corresponding to $Chl_{a/b}$-P2 in lanes of the mutant. APEL and KLOPPSTECH (1978) showed that when dark-grown seedlings of *chlorina-f2* were illuminated for 24 h, a faint protein band of mol.wt. about 25,000 appeared in the polypeptide pattern of the plastid membranes. These observations support the notion that this mutant has retained a partial light-harvesting complement. This partial complement could conceivably be responsible for the grana-formation of *chlorina-f2*. APEL and KLOPPSTECH (1978) showed that illumination of dark-grown wild type barley plants induces the appearance of a prominent cytoplasmic mRNA species coding for a 29,500 polypeptide, that is precursor for the apoprotein of the light-harvesting complex. It was also shown that illumination has the same effect on dark-grown *chlorina-f2* plants. It was concluded that the induction of mRNA for the apoprotein of the $Chl_{a/b}$-P2 complex – in contrast to the assembly of significant amounts of the complex and its insertion into the membrane – does not require the presence of Chl b.

MILES et al. (1979) separated Chl-protein complexes from wild type and a series of nuclear mutants of maize. From the wild type they obtained five different complexes, namely – in order of decreasing molecular weight – A.1, AB.1, AB.2, A.2, and AB.3. Band A.1 represents the CP I complex, and band AB.3 is the light-harvesting complex $Chl_{a/b}$-P2. Band A.2 could conceivably be identical to an unresolved mixture of Chl_a-P2 and Chl_a-P3. Alternatively, one of these complexes could be denatured in the electrophoretic system used by MILES et al. (1979). The bands AB.1 and AB.2 could be similar to two of the three Chl a and b containing bands that MACHOLD et al. (1979) interpreted as oligomers of $Chl_{a/b}$-P2.

Five of the mutants described by MILES et al. (1979) lacked specifically the CP I (A.1) complex and were PS I-deficient. Three other PS I-deficient mutants were described: The mutant *hcf*E1490* was found to have no A.1 or AB.1 complexes, while the mutants *hcf*E1266* and *hcf*E1482B* are lacking both AB.1 and AB.2 in addition to the A.1 complexes. A third series of mutants was described. These mutants have normal quantities of the A.1 and A.2 complexes, but have lost one or more of the AB-complexes. Four mutants have specifically lost the AB.2 complex. The dominant mutant *OyYg-700* also lack AB.2 when grown under low light intensity. When grown under high light intensity no or little Chl b is synthesized, and none of the three Chl b-containing complexes is seen. If the complexes AB.1 and AB.2 are interpreted as oligomeric forms of AB.3, these data indicate that mutations can eliminate the oligomers without changing the monomeric species. MILES et al. (1979) proposed that AB.1 and AB.2 could be composed of some complex-specific component associated with a fundamental Chl a- and b-containing complex, presumably the AB.3 complex.

Nuclear mutations affecting the Chl a to b ratio have also been described for *Nicotiana tabacum* and pea. The mutation *su* in tobacco shows partial dominance. Homozygous mutants are yellow and are not viable, while the heterozygotes *Su/su* are yellow-green and viable. The *Su/su* plants have a higher Chl a to b ratio than wild-type plants, and have chloroplasts with reduced pairing of thylakoids (SCHMID 1971). The light-harvesting complex is present, but in reduced amount (VERNOTTE et al. 1976). A Chl b-deficient mutant of pea with similar properties has been described (HIGHKIN et al. 1969, BOARDMAN et al. 1974, ANDERSON and LEVINE 1974a, b). STUMMANN and HENNINGSEN (1980) have described a recessive mutant of pea, *chlorina-5535*, that is deficient with respect to Chl b, the polypeptides of the light-harvesting complex, and grana formation. This mutant also has a diminished quantity of the CP I complex. SCHWARZ (1980) has decribed two mutants of pea that lack chlorophyll b and are deficient with respect to the apoprotein of the light-harvesting complex. The thylakoids of these mutants are differentiated in stroma and grana regions, but the number of discs per granum is lower than in the wild type.

MENDIOLA-MORGENTHALER et al. (1976) isolated the coupling factor CF_1 from spinach chloroplasts that had been incubated with radioactively labeled leucine or methionine in light. The five polypeptides of CF_1 were separated by SDS-gelelectrophoresis. Only the α, β, and ε subunits were labeled, indicating that they are formed in the chloroplast, while the γ and δ subunits may be formed in the cytoplasm (cf. BOTTOMLEY and BOHNERT, Chap. 16, this Vol.). Since no evidence has so far been presented for nucleic acid transport across the outer membrane of plastids, this result suggests that the α, β, and γ-subunits are encoded in the plastome. SHEPHERD et al. (1979) isolated a series of plastome mutants of *Chlamydomonas* that affected photosynthetic functions. Mutants at three loci were found to be deficient in polypeptide bands, that were identified as the α and β subunits of CF_1. It was pointed out that if the CF_1 complex must be assembled before association with the membrane, any defect that causes failure of proper assembly would result in the absence of CF_1 from thylakoid preparations.

VON WETTSTEIN et al. (1982) have devised a model of the arrangement of the components of the photosynthetic membrane, which appears to be in agreement with the studies performed on wild type and mutants of barley and with other pertinent data on the sequence of electron transfer, neighbor relationships, accessibility to antibodies as well as artificial electron donors and acceptors, and structural features revealed by freeze-etching and freeze-fracturing (Fig. 14). References to the literature on which the model is based are given by VON WETTSTEIN et al. (1982) and VON WETTSTEIN (1981). According to this model the light-energy absorbed by the light-harvesting complex is transferred, possibly via $Chl_{a/b}$-P1 and Chl_a-P2, to the reaction center of PS II, which is located in Chl_a-P3. The water-oxidizing system is represented by the three components Z_1, Z_2, and M. From the primary acceptor Q, the electrons are transferred to cytochrome f via the hypothetical two-electron acceptor R, plastoquinone, and the Rieske FeS center. From cytochrome f the electrons are transferred to the reaction center of PS I, located in CP I, via plastocyanin and the hypothetical primary donor PD. According to the model the electrons are finally transferred to ferredoxin and ferredoxin oxidoreductase via the primary acceptor A_1 and three FeS centers, A_2, A, and B. The cytochromes b_{559} and b_6 are assumed to mediate cyclic electron flow around PS II and PS I, respectively.

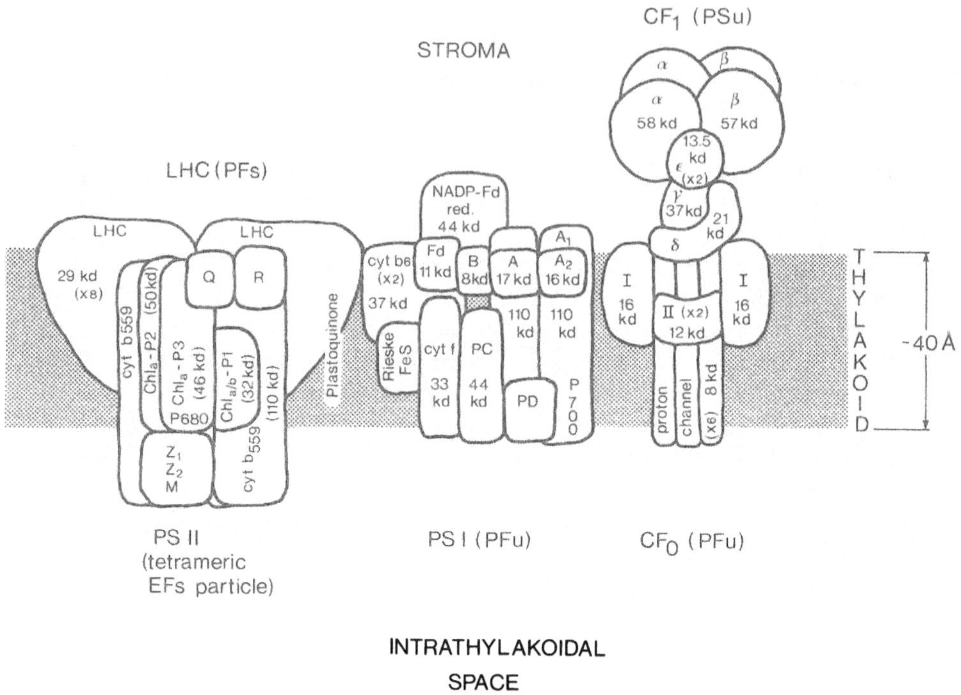

Fig. 14. Diagram representing a model for the chloroplast membrane with polypeptides arranged according to data from studies on photosynthetic electron transport, freeze-fractured membranes and accessibility to antibodies. (VON WETTSTEIN et al. 1980)

In relation to the evaluation of the coordinated expression of the nuclear and plastid genomes (Sect. 6) it is of interest to point out that the genetic information for each of the three particles in this model appears to be shared between the nucleus and the chloroplast. CHUA and GILLHAM (1977) studied the sites of synthesis of the thylakoid polypeptides of *Chlamydomonas* by the use of inhibitors specific for chloroplast and cytoplasmic protein synthesis, and found that the apoprotein of CP I as well as polypeptide 5, which is assumed to be identical to the barley polypeptide Chl_a-AP2 (VON WETTSTEIN et al. 1982, VON WETTSTEIN 1981), and polypeptide 6, which is identical to the barley polypeptide Chl_a-AP3, were among the polypeptides that were synthesized in the chloroplast. In support of the results concerning polypeptide 5, CHUA (1976) has described a uniparental mutant of *Chlamydomonas*, for which the molecular weight of polypeptide 5 appears to be 1,000 higher than the apparent molecular weight of this polypeptide from wild-type cells. The mutant did not appear to possess any change in photosynthetic capacity. There is also evidence for the synthesis of cytochrome f, cytochrome b_{559}, the α, β, and ε subunits of CF_1 and the DCCD-binding protein of CF_0 in chloroplasts (ZIELINSKI and PRICE 1977, DOHERTY and GRAY 1979a, b, MENDIOLA-MORGENTHALER et al. 1976). The following polypeptides appear to be encoded in the nuclear genome (cf. BOTTOMLEY and BOHNERT, Chap. 15, this Vol.): ferredoxin and ferredoxin oxidoreductase (ARMSTRONG et al. 1971), the apoproteins of the light-harvesting complex (APEL and KLOPPSTECH 1978, SCHMIDT et al. 1980), the δ and γ subunits of CF_1 (MENDIOLA-MORGENTHALER et al. 1976), and plastocyanin (HASLETT and CAMMACK 1974).

5 Genetic Control of the Protein-Synthesizing System of Plastids

It is now established that chloroplasts posses their own specific protein-synthesizing system. This system includes specific tRNA's and aminoacyl-tRNA synthetases that appear to be encoded in the plastome and nuclear genome, respectively (GILLHAM 1978, WEIL et al. 1976, 1982, BURKARD et al. 1980, cf. WEIL and PARTHIER, Vol. 14A, this Series, Chap. 2). It also includes a multimeric RNA polymerase with unique properties (BOGORAD et al. 1973, SMITH and BOGORAD 1974, RATCLIFF and SURZYCKI 1974, SURZYCKI and SHELLENBARGER 1976, cf. WOLLGIEHN, Chap. 5, this Vol.) and specific T and G elongation factors that seem to be encoded in the chloroplast genome (CIFERRI et al. 1980). Use of mutants and interspecific crosses has contributed significantly to our knowledge on the genetic control of chloroplast ribosomes. Several reviews of the studies performed on ribosome biogenesis have been published (HARRIS et al. 1976, GILLHAM 1978, GILLHAM et al. 1978).

The polypeptides from the two subunits of the 70S chloroplast ribosomes from *Chlamydomonas* have been separated by two-dimensional gel electrophoresis by HANSON et al. (1974), and BRÜGGER and BOSCHETTI (1975). HANSON et al. (1974) estimated the number of proteins in the large subunit and the small subunit as 26 and 22, respectively, while BRÜGGER and BOSCHETTI (1975) estimated these numbers as 35 and 25, respectively. CAPEL and BOURQUE (1977) stated that the number of polypeptides in the large subunit and the small subunit of chloroplast ribosomes from tobacco was between 34 and 38 and between 20 and 24, respectively. BOURQUE and WILDMAN (1973) detected two polypeptide differences between the large chloroplast ribosome subunits from *Nicotiana tabacum* and *N. glauca* by one-dimensional gel electrophoresis. *N. glauca* possessed a band that was absent from the polypeptide pattern of the other species, and a specific polypeptide had a higher mobility for *N. tabacum* than for *N. glauca*. The extra *N. glauca* band and the more mobile *N. tabacum* band were both present in F_1 hybrids from the two reciprocal crosses. The authors concluded that these two polypeptides were encoded in the nuclear genome. BOURQUE et al. (1977) showed that small (30S) subunits from the streptomycin-resistant and uniparentally inherited mutant *SR1* of *N. tabacum* have a ten times lower affinity for streptomycin than the small subunits from wild-type plants. By isoelectric focusing two polypeptide differences between *SR1* and wild-type 30S subunits were revealed, which suggested that the mutant has an altered ribosomal protein. If this is the cause of the streptomycin resistance, this polypeptide appears to be encoded in the chloroplast genome. YURINA et al. (1978) achieved independently similar conclusions on the nature of the *SR1* mutation.

The genetic control of chloroplast ribosome biogenesis has been most actively investigated in *Chlamydomonas reinhardtii*. Since this work has been treated thoroughly in several reviews (HARRIS et al. 1976, GILLHAM 1978, GILLHAM et al. 1978) we shall only outline the main results. As described by HARRIS et al. (1976), GILLHAM (1978), and GILLHAM et al. (1978) seven or possibly eight nuclear loci have been identified, that appear to be involved in chloroplast ribosome biogenesis. Mutants at five of these loci accumulate large (54S) subunits, while mutants at three loci do not synthesize any of the two subunits. It was stated by the authors that it could not be excluded that the ribosome-deficient mutant *ac-20* is allelic to the small subunit deficient mutant *cr-7*. SHEPHERD et al. (1979) have isolated mutants at two chloroplast loci that also appear to accumulate large subunits and be deficient with respect to the synthesis of small subunits. The mutants that are deficient in both subunits appear to affect the formation of components common to the subunits. They may, for example, affect the formation or processing of a common RNA precursor. Mutations at three nuclear and at least seven to nine chloroplast loci confer antibiotic resistance or dependence on chloroplast ribosomes (GILLHAM 1978). DAVIDSON et al. (1974) investigated four alleles at the nuclear *ery-M-1* locus that conferred erythromycin resistance on the chloroplast ribosomes. Three of the alleles changes the charge of a specific polypeptide from the large subunit, while the fourth allele leads to deletion of a large part of the same polypeptide. Thus the information for this protein is probably encoded in the nuclear genome. METS and BOGORAD (1972) reported that a mutant at another nuclear locus, *ery-M-2*, possessed an altered large subunit polypeptide. HARRIS et al. (1977) showed that 31 independently isolated streptomycin-resistant mutants that – as judged from a poly-U/isoleucine misreading test – conferred resistance on the isolated ribosomes, comprised one nuclear and four chloroplast loci. Four mutants resistant to spectinomycin, and three mutants resistant to neamine and kanamycin that also – as tested in a poly-U/phenylalanine test – conferred resistance on the isolated ribosomes, mapped at a single chloroplast locus. It appears likely that these loci code for some of the polypeptides of the chloroplast ribosomes. In conclusion, the chloroplast ribosomes are, as is the amino acid-activating system of the chloroplasts, under the control of both the plastome and the nuclear genome.

6 Coordinated Expression of the Nuclear and Plastid Genomes

As pointed out in Section 4.3 the genetic information for each of the three particles in the model presented in Fig. 14 appears to be shared between the nuclear and the chloroplast genomes. This is also the case for the chloroplast ribosomes and the amino acid-activating system (Sect. 5) and for the major stromal protein, ribulose-1,5-bisphosphate carboxylase (RuBPCase) (see BOTTOMLEY and BOHNERT, Chap. 15, this Vol.). It is evident that regulatory mechanisms must exist that imply a highly coordinated synthesis of the chloroplast proteins in the cytoplasm and in the chloroplast. Observations on the amounts and identities of chloroplast proteins that are synthesized in mutants affecting chloroplast structure and composition, and in isolated chloroplasts, are pertinent for the evaluation of these mechanisms. HAGEMANN and BÖRNER (1978) have reported that both the large and the small subunit of RuBPCase were absent – as judged from SDS gel electrophoresis and immunological studies – from a plastid ribosome-deficient plastome mutant of barley, and HALLIER et al. (1978) made similar observations on two plastome mutants of *Oenothera*. These observations indicate that the absence of the plastome-encoded large subunit from the chloroplast somehow causes a decay of synthesis of the precursor of the small subunit in the cytoplasm, or a lack of the subsequent transit through the chloroplast envelope membranes. The precursor is converted into a mature small subunit during this transit (HIGHFIELD and ELLIS 1978). In contrast to these observations, the large subunit accumulates during protein synthesis in isolated chloroplasts (ELLIS et al. 1980). Thus it could tentatively be suggested that the synthesis of the large subunit does not require the presence of small subunit.

The barley mutant *chlorina-f2* is also of interest in relation to the coordination of the chloroplast and nuclear genomes. This mutant lacks Chl b, but apparently contains a small quantity of the apoprotein of the light-harvesting complex in the thylakoid membranes (Sect. 4.3). The cytoplasmic mRNA species that codes for the precursor of the apoprotein is present in the cytoplasm (APEL and KLOPPSTECH 1978). If it is assumed that the primary action of the mutation in *chlorina-f2* is to cause the absence of Chl b, then these observations indicate that the translation of the cytoplasmic mRNA into the precursor form of the apoprotein or the translocation of the protein from the cytoplasm to the membranes do not occur. SCHMIDT et al. (1980) have resolved the apoprotein of the light-harvesting complex from pea into two polypeptides of mol.wt. about 28,000, and showed that both of these polypeptides are synthesized in the cytoplasm as precursors with a 4000–5000 higher molecular weight. The precursors are soluble and could not be detected inside the chloroplasts. The transit occurs by a post-translational mechanism that requires the presence of ATP (GROSSMAN et al. 1980). Several other cytoplasmically synthesized thylakoid membrane proteins appear to be translocated from the cytoplasm to the chloroplast by similar mechanisms (SCHMIDT et al. 1980, GROSSMAN et al. 1980). Reexamination of electron microscopic observations on greening barley plastids (unpublished data) have revealed that between 1 and 5 h of greening – when the rate of Chl accumulation is high – the plastids contain many small vesicles that are distributed throughout the stroma. In many cases the membrane of the vesicles is continuous with the inner envelope membrane or a thylakoid membrane. The inner and the outer envelope membrane are often tightly appressed at many points, and many polysomes are often observed in the region of the cytoplasm that surrounds the plastid. It may therefore be tentatively suggested that the soluble precursors of cytoplasmically synthesized thylakoid proteins are translocated through the outer envelope membrane and inserted into the inner, tightly appressed membrane, in a process that requires ATP and involves the processing of the precursor into the mature protein. For such a complicated process it appears likely that the outer envelope membrane has discrete sites for attachment of different classes of precursors to thylakoid polypeptides. In the case of the light-harvesting complex Chl a and b could conceivably be attached to their respective sites on the apoprotein during the insertion into the inner membrane. The protein might not obtain the conformation required for the correct packing of the complex in the membrane or pinching off of a vesicle in the absence of Chl b. This hypothetical model could account for the aforementioned observations on the barley mutant *chlorina-f2*. Similar models can be proposed for the assembly of structural units of the thylakoid membranes that involve polypeptides encoded in the chloroplast genome as well as polypeptides encoded in the nuclear genome.

Acknowledgments. The preparation of this paper and part of the research described results in part from support by the Danish Agricultural and Veterinary Research Council, the Danish Natural Science Research Council and the Carlsberg Foundation. We are indebted to Professor Dieter von Wettstein and staff at the Carlsberg Research Center for the communication of information and the use of illustrations. We gratefully acknowledge the technical assistance of Ulla M. Ahm in the preparation of the manuscript.

References

Anderson IC, Robertson DS (1960) Role of carotenoids in protecting chlorophyll from photodestruction. Plant Physiol 35:531–534

Anderson JM (1975) The molecular organization of chloroplast thylakoids. Biochim Biophys Acta 416:191–235

Anderson JM (1980) P-700 content and polypeptide profile of chlorophyll-protein complexes of spinach and barley thylakoids. Biochim Biophys Acta 591:113–126

Anderson JM, Levine RP (1974a) Membrane polypeptides of some higher plant chloroplasts. Biochim Biophys Acta 333:378–387

Anderson JM, Levine RP (1974b) The relationship between chlorophyll-protein complexes and chloroplast membrane polypeptides. Biochim Biophys Acta 357:118–126

Argyroudi-Akoyunoglou JH, Akoyunoglou G (1977) Correlation between cation-induced formation of heavy subchloroplast fractions and cation-induced increase in chlorophyll a fluorescence yield in tricine-washed chloroplasts. Arch Biochem Biophys 179:370–377

Armound PA, Arntzen CJ, Briantais J-M, Vernotte C (1976) Differentiation of chloroplast lamellae. I. Light harvesting efficiency and grana development. Arch Biochem Biophys 175:54–63

Apel K, Kloppstech K (1978) The plastid membranes of barley (*Hordeum vulgare*). Light-induced appearance of mRNA coding for the apoprotein of the light-harvesting chlorophyll a/b-protein. Eur J Biochem 85:581–588

Apel K, Santel H-J, Redlinger TE, Falk H (1980) The protochlorophyllide holochrome of barley (*Hordeum vulgare* L.) Isolation and characterization of the NADPH:protochlorophyllide oxidoreductase. Eur J Biochem 111:251–258

Armstrong JJ, Surzycki SJ, Moll B, Levine RP (1971) Genetic transcription and translation specifying chloroplast components in *Chlamydomonas reinhardi*. Biochemistry 10:692–701

Baalen van C (1980) Mutants of blue-green algae. In: San Pietro A (ed) Photosynthesis and nitrogen fixation. Part C. Methods in enzymology Vol 69. Academic Press, London New York, pp 38–47

Bachmann MD, Robertson DS, Bowen CC (1969) Thylakoid anomalies in relation to grana structure in pigment-deficient mutants of *Zea mays*. J Ultrastruct Res 28:435–451

Bachmann MD, Robertson DS, Bowen CC, Anderson IC (1967) Chloroplast development in pigment deficient mutants of maize. I. Structural anomalies in plastids of allelic mutants at the w_3 locus. J Ultrastruct Res 21:41–60

Bahl J (1977) Chlorophyll, carotenoid, and lipid content in *Triticum vulgare* L. Plastid envelopes, prolamellar bodies, stroma lamellae, and grana. Planta 136:21–24

Barber J (1976) Ionic regulation in intact chloroplasts and its effect on primary photosynthetic processes. In: Barber J (ed) The intact chloroplast. Elsevier/North-Holland Biomedical Press, Amsterdam, New York, pp 89–134

Barber J, Chow WS (1979) A mechanism for controlling the stacking and unstacking of chloroplast thylakoid membranes. FEBS Lett 105:5–9

Bedbrook JR, Link G, Coen DM, Bogorad L, Rich A (1978) Maize plastid gene expressed during photoregulated development. Proc Natl Acad Sci USA 75:3060–3064

Bennoun P, Girard J, Chua N-H (1977) A uniparental mutant of *Chlamydomonas reinhardtii* deficient in the chlorophyll-protein complex CPI. Mol Gen Genet 153:343–348

Bishop NI (1971) Preparation and properties of mutants: *Scenedesmus*. In: San Pietro A (ed) Photosynthesis. Part A. Methods in enzymology Vol 23. Academic Press, London, New York, pp 130–143

Bishop NI, Senger H (1972) The development of structure and function in chloroplasts of greening mutants of *Scenedesmus*. II. Development of the photosynthetic apparatus. Plant Cell Physiol 13:937–953

Blixt S (1972) Mutation genetics in *Pisum*. Agri Hort Genet 30:1–293

Boardman NK, Highkin HR (1966) Studies on a barley mutant lacking chlorophyll b. I. Photochemical activity of isolated chloroplasts. Biochim Biophys Acta 126:189–199

Boardman NK, Thorne SW (1968) Studies on a barley mutant lacking chlorophyll b. II. Fluorescence properties of isolated chloroplasts. Biochim Biophys Acta 153:448–458

Boardman NK, Anderson JM, Björkman O, Goodchild DJ, Grimme LH, Thorne SW (1974) Chloroplast differentiation in sun and shade plants: relationship between chlorophyll content, grana formation, photochemical activity and fractionation of the photosystems. In: Pais MS (ed) Proc Int Sym Plant Cell Differentiation. Port Acta Biol 14:213–236

Boardman NK, Anderson JM, Goodchild DJ (1978) Chlorophyll-protein complexes and structure of mature and developing chloroplasts. In: Rao Sanadi D, Vernon L (eds) Curr Top Bioenerg Vol VIII. Academic Press, London, New York, pp 35–109

Bogorad L (1976) Chlorophyll biosynthesis. In: Goodwin TW (ed) Chemistry and biochemistry of plant pigments, 2nd edn. Academic Press, London, New York, pp 64–148

Bogorad L, Mets LJ, Mullinex KP, Smith HJ, Strain GC (1973) Possibilities for intracellular integration: The ribonucleic acid polymerases of chloroplast and nuclei, and the genes specifying chloroplast ribosomal proteins. Biochem Soc Symp 38:17–41

Börner T, Schumann B, Krahnert S, Perchauf M, Herrmann FH, Knoth FH, Hagemann R (1975) Struktur und Funktion der genetischen Information in den Plastiden. XIII. Lamellarproteine bleicher Plastiden von Plastom- und Genmutanten von *Hordeum* und *Lycopersicon*. Biochem Physiol Pflanz 168:185–193

Bourque DP, Wildman SG (1973) Evidence that nuclear genes code for several chloroplast ribosomal proteins. Biochem Biophys Res Commun 50:532–537

Bourque DP, Horn NA, Capel MS (1977) Altered chloroplast ribosomes of a streptomycin resistant mutant of *Nicotiana tabacum*. Plant Physiol Suppl 59:110

Boynton JE, Henningsen KW (1967) The physiology and chloroplast structure of mutants at loci controlling chlorophyll synthesis in barley. Stud Biophys 5:85–88

Britton G (1976) Biosynthesis of carotenoids. In: Goodwin TW (ed) Chemistry and biochemistry of plant pigments. Vol 2 2nd edn, Academic Press, New York, London, pp 262–327

Brügger M, Boschetti A (1975) Two-dimensional gel electrophoresis of ribosomal proteins from streptomycin-sensitive and streptomycin-resistant mutants of *Chlamydomonas reinhardi*. Eur J Biochem 58:603–610

Bryan GW, Zadylak AH, Ehret CF (1967) Photoinduction of plastids and of chlorophyll in a *Chlorella* mutant. J Cell Sci 2:513–528

Burkard G, Canaday J, Crouse E, Guillemaut P, Imbault P, Kneith G, Keller M, Mubumbila M, Osorio L, Sarantoglou V, Steinmetz A, Weil JH (1980) Transfer RNAs and aminoacyl-tRNA synthetases in plant organelles. In: Leaver CJ (ed) Genome organization and expression in plants. Plenum, New York, London, pp 313–320

Burke JJ, Ditto CJ, Arntzen CJ (1978) Involvement of the light-harvesting complex in cation regulation of excitation energy distribution in chloroplasts. Arch Biochem Biophys 187:252–263

Burke JJ, Steinback KE, Arntzen CJ (1979) Analysis of the light-harvesting pigment-protein complex of the wild type and a chlorophyll-b-less mutant of barley. Plant Physiol 63:237–243

Capel MS, Bourque DP (1977) Electrophoretic characterization of chloroplast ribosomal proteins assisted by computer analysis. Plant Physiol 59, suppl:110

Chua N-H (1976) A uniparental mutant of *Chlamydomonas reinhardtii*. In: Bücher T, Neupert W, Sebald W, Werner S (eds) Genetics and biogenesis of chloroplasts and mitochondria. Elsevier North Holland, Amsterdam, London, pp 323–330

Chua N-H, Bennoun P (1975) Thylakoid membrane polypeptides of *Chlamydomonas reinhardtii*: Wildtype and mutant strains deficient in photosystem II reaction center. Proc Natl Acad Sci USA 72:2175–2179

Chua N-H, Gillham NW (1977) The sites of synthesis of the principal thylakoid membrane polypeptides in *Chlamydomonas reinhardtii*. J Cell Biol 74:441–452

Chua N-H, Levine RP (1969) The photosynthetic electron transport chain of *Chlamydomonas reinhardi*. VIII. The 520 nm light-induced absorbance change in the wild-type and mutant strains. Plant Physiol 44:1–6

Chua N-H, Matlin K, Bennoun P (1975) A chlorophyll-protein complex lacking in photosystem I mutants of *Chlamydomonas reinhardtii*. J Cell Biol 67:361–377

Ciferri O, Tiboni O, Pasquale G Di, Carbonera D (1980) Sites of synthesis and codification of chloroplast elongation factors. In: Leaver CJ (ed) Genome organization and expression in plants. Plenum, New York, London, pp 373–379

Claes H (1954) Analyse der biochemischen Synthesekette für Carotenoide mit Hilfe von *Chlorella*-Mutanten. Z Naturforsch 9b:461–470

Claes H (1956) Biosynthese von Carotinoiden bei *Chlorella*. 2. Mitt. Tetrahydrolycopin und Lycopin. Z Naturforsch 11b:260–266

Claes H (1957) Biosynthese von Carotinoiden bei *Chlorella*. III. Untersuchungen über die lichtabhängige Synthese von α- und β-Carotin und Xanthophyllen bei der *Chlorella*-Mutante 5/520. Z Naturforsch 12b:401–407

Darr S, Machado VS, Arntzen CJ (1981) Uniparantal inheritance of a chloroplast phosystem II polypeptide controlling herbicide binding. Biochim Biophys Acta 634:219–228

Davidson JN, Hanson MR, Bogorad L (1974) An altered chloroplast ribosomal protein in *ery-M1* mutants of *Chlamydomonas reinhardi*. Mol Gen Genet 132:119–129

Delepelaire P, Chua N-H (1979) Lithium dodecyl sulfate/polyacrylamide gel electrophoresis of thylakoid membranes at 4° C: Characterization of two additional chlorophyll a-protein complexes. Proc Natl Acad Sci USA 76:111–115

Doherty A, Gray JC (1979a) Synthesis of cytochrome *f* by isolated pea chloroplasts. Eur J Biochem 98:87–92

Doherty A, Gray JC (1979b) Synthesis of a dicyclohexylcarbodiimide-binding proteolipid by isolated pea chloroplasts. Biochem Soc Trans 7:1114–1115

Ellis RJ, Smith SM, Barraclough R (1980) Synthesis, transport and assembly of of chloroplast proteins. In: Leaver CJ (ed) Genome organization and expression in plants. Plenum, New York, London, pp 321–335

Everett HL (1949) Genic series controlling chloroplast pigment production in diploid *Zea mays*. Proc Natl Acad Sci USA 35:628–634

Faludi-Dániel Á, Láng F, Fradkin LI (1966) The state of chlorophyll a in leaves of carotenoid mutant maize. In: Goodwin TW (ed) Biochemistry of chloroplasts Vol I. Academic Press, London, New York, pp 269–274

Faludi-Dániel Á, Fridvalszky L, Gyurjan I (1968) Pigment composition and plastid structure in leaves of carotenoid mutants of maize. Planta 78:184–195

Gillham NW (1965) Induction of chromosomal and non-chromosomal mutations in *Chla-*

mydomonas reinhardi with N-methyl-N′-nitro-N-nitrosoguanidine. Genetics 52:529–537

Gillham NW (1978) Organelle heredity. Raven Press, New York

Gillham NW, Boynton JE, Chua N-H (1978) Genetic control of chloroplast proteins. In: Rao Sanadi D, Vernon LP (eds) Curr Top Bioenerg Vol VIII. Academic Press, London, New York, pp 211–260

Girnth C, Gough SP, Kannangara CG (1982) Δ-aminolevulinate synthesis in greening barley. 3. Characterization of regulatory mutants. In: Avron M (ed) Proc 5th Int Congr Photosynthesis. Int Sci Services, Jerusalem, in press

Goodenough UW, Levine RP (1969) Chloroplast ultrastructure in mutant strains of *Chlamydomonas reinhardi* lacking components of the photosynthetic apparatus. Plant Physiol 44:990–1000

Gottschalk W, Müller F (1964) Quantitative Pigmentuntersuchungen an strahleninduzierten Chlorophyllmutanten von *Pisum sativum*. I. Die Lethalmutanten. Planta 61:259–282

Gough S (1972) Defective synthesis of porphyrins in barley plastids caused by mutation in nuclear genes. Biochim Biophys Acta 286:36–54

Gough SP, Kannangara CG (1976) Synthesis of δ-aminolevulinic acid by isolated plastids. Carlsberg Res Commun 41:183–190

Gough SP, Kannangara CG (1977) Synthesis of δ-aminolevulinate by a chloroplast stroma preparation from greening barley leaves. Carlsberg Res Commun 42:459–464

Gough SP, Kannangara CG (1979) Biosynthesis of δ-aminolevulinate in greening barley leaves. III: The formation of δ-aminolevulinate in tigrina mutants of barley. Carlsberg Res Commun 44:403–416

Gough SP, Girnth C, Kannangara CG (1982) Δ-aminolevulinate synthesis in greening barley. 1. Regulation. In: Avron M (ed) Proc 5th Int Congr Photosynthesis. Int Sci Services, Jerusalem, in press

Granick S (1948) Magnesium protoporphyrin as a precursor of chlorophyll in *Chlorella*. J Biol Chem 175:333–342

Granick S (1959) Magnesium porphyrins formed by barley seedlings treated with δ-aminolevulinic acid. Plant Physiol 34:XVIII

Granick S (1971) Preparation and properties of *Chlorella* mutants in chlorophyll biosynthesis. In: San Pietro A (ed) Photosynthesis. Part A. Methods in enzymology Vol 23. Academic Press, London, New York, pp 162–168

Grebanier AE, Steinback KE, Bogorad L (1979) Comparison of the molecular weights of proteins synthesized by isolated chloroplasts with those which appear during greening in *Zea mays*. Plant Physiol 63:436–439

Griffiths WT (1975) Characterization of the terminal stages of chlorophyll(ide) synthesis in etioplast membrane preparations. Biochem J 152:623–635

Griffiths WT (1978) Reconstitution of chlorophyllide formation by isolated etioplast membranes. Biochem J 174:681–692

Grossman A, Bartlett S, Chua N-H (1980) Energy-dependent uptake of cytoplasmically synthesized polypeptides by chloroplasts. Nature (London) 285:625–628

Hagemann R, Börner T (1978) Plastid ribosome-deficient mutants of higher plants as a tool in studying chloroplast biogenesis. In: Akoyunoglou G, Argyroudi-Akoyunoglou JH (eds) Chloroplast development. Elsevier/North-Holland Biomedical Press, Amsterdam, New York, pp 709–719

Hallier UW, Schmitt JM, Heber U, Chaianova SS, Volodarsky AD (1978) Ribulose-1,5-bisphosphate carboxylase-deficient plastome mutants of *Oenothera*. Biochim Biophys Acta 504:67–83

Hanson MR, Davidson JN, Mets LJ, Bogorad L (1974) Characterization of chloroplast and cytoplasmic ribosomal proteins of *Chlamydomonas reinhardi* by two-dimensional gel electrophoresis. Mol Gen Genet 132:105–118

Harris EW, Boynton JE, Gillham NW (1976) Genetics of chloroplast ribosome biogenesis in *Chlamydomonas reinhardtii*. In: Lewin RA (ed) The genetics of algae. Blackwell, Oxford, London, pp 119–144

Harris EW, Boynton JE, Gillham NW, Tingle CL, Fox SB (1977) Mapping of chloroplast genes involved in chloroplast ribosome biogenesis in *Chlamydomonas reinhardtii.* Mol Gen Genet 155:249–265

Haslett BG, Cammack R (1974) The development of plastocyanin in greening bean leaves. Biochem J 144:567–572

Henningsen KW (1970) Macromolecular physiology of plastids. VI. Changes in membrane structure associated with shifts in the absorption maxima of chlorophyllous pigments. J Cell Sci 7:587–621

Henningsen KW, Boardman NK (1973) Development of photochemical activity and the appearance of the high potential form of cytochrome b-559 in greening barley seedlings. Plant Physiol 51:1117–1126

Henningsen KW, Boynton JE (1967) The physiology and ultrastructure of barley mutants at loci controlling the development of the lamellar systems in chloroplasts. Stud Biophys 5:89–90

Henningsen KW, Boynton JE (1969) Macromolecular physiology of plastids. VII. The effect of a brief illumination on plastids of dark-grown barley leaves. J Cell Sci 5:757–793

Henningsen KW, Boynton JE (1974) Macromolecular physiology of plastids. IX. Development of plastid membranes during greening of dark-grown barley seedlings. J Cell Sci 15:31–55

Henningsen KW, Thorne SW (1974) Esterification and spectral shifts of chlorophyll(ide) in wildtype and mutant seedlings developed in darkness. Physiol Plant 30:82–89

Henningsen KW, Boynton JE, Wettstein D von, Boardman NK (1973) Nuclear genes controlling chloroplast development in barley. In: Pollak JK, Lee JW (eds) The biochemistry of gene expression in higher organisms. Australia and NZ Book Co, Sydney, pp 457–478

Henningsen KW, Nielsen NC, Smillie RM (1974) The effect of nuclear mutations on the assembly of photosynthetic membranes in barley. In: Pais MS (ed) Proc Int Sym Plant Cell Differentiation. Port Acta Biol 14:323–344

Herrmann F (1971a) Genetic control of pigment-protein complexes I and Ia of the plastid mutant *en-alba-1* of *Antirrhinum majus.* FEBS Lett 19:267–269

Herrmann F (1971b) Chloroplast lamellar proteins of the plastid mutant *en-viridis-l* of *Antirrhinum majus* having impaired photosystem II. Exp Cell Res 70:452–453

Herrmann FH, Matorin D, Timofeev K, Börner T, Rubin AB, Hagemann R (1974) Structure and function of the genetic information in plastids. IX. Studies on primary reactions of photosynthesis in plastome mutants of *Antirrhinum majus* and *Pelargonium zonale* having impaired photosynthesis. Biochem Physiol Pflanz 165:393–400

Herrmann FH, Schumann B, Börner T, Knoth R (1976) Struktur and Funktion der genetischen Information in den Plastiden. XII. Die plastidalen Lamellarproteine der photosynthese-defekten Plastommutante *en:gil-1* ("Mrs. Pollock") und der Genmutante "Cloth of Gold" von *Pelargonium zonale* Arr. Photosynthetica 10:164–171

Highfield PE, Ellis RJ (1978) Synthesis and transport of the small subunit of chloroplast ribulose bisphosphate carboxylase. Nature (London) 271:420–424

Highkin HR, Boardman NK, Goodchild (1969) Photosynthetic studies on a pea-mutant deficient in chlorophyll. Plant Physiol 44:1310–1320

Hiller RG, Møller BL, Høyer-Hansen G (1980) Characterization of six putative photosystem I mutants in barley. Carlsberg Res Commun 45:315–328

Hirono Y, Rédei GP (1963) Multiple allelic control of chlorophyll *b* level in *Arabidopsis thaliana.* Nature (London) 197:1324–1325

Homann PH (1968) Fluorescence properties of chloroplasts from manganese deficient and mutant tobacco. Biochim Biophys Acta 162:545–554

Homann PH, Schmid GH (1967) Photosynthetic reactions of chloroplasts with unusual structures. Plant Physiol 42:1619–1632

Horváth G, Faludi-Dániel Á (1971) Formation and compartmentation of leaf carotenoids in normal and mutant maize. In: Forti G, Avron M, Melandri A (eds) Proc 2nd Int Congr Photosynthesis Vol III. Junk, The Hague, pp 2443–2449

Howell SH, Moudrianakis EN (1967) Function of the "quantasome" in photosynthesis:

Structure and properties of membrane-bound particle active in the dark reactions of photophosphorylation. Proc Natl Acad Sci USA 58:1261–1268
Høyer-Hansen G (1980) Identification of haem-proteins in thylakoid polypeptide patterns of barley. Carlsberg Res Commun 45:167–176
Høyer-Hansen G, Simpson DJ (1977) Changes in the polypeptide composition of internal membranes of barley plastids during greening. Carlsberg Res Commun 42:379–389
Høyer-Hansen G, Møller BL, Pan LC (1979) Identification of coupling factor subunits in thylakoid polypeptide patterns of wild-type and mutant barley thylakoids using crossed immunoelectrophoresis. Carlsberg Res Commun 44:337–351
Hudock GA, Bart C (1967) Response of a mutant strain of *Chlamydomonas reinhardi* to prolonged organotrophic growth. Plant Physiol 42:186–190
Hudock GA, Rosen H (1976) Formal genetics of *Chlamydomonas reinhardtii*. In: Lewin RA (ed) The genetics of algae. Blackwell, London, Oxford, pp 29–48
Izawa S, Good NE (1966) Effect of salts and electron transport on the conformation of isolated chloroplasts. II. Electron microscopy. Plant Physiol 41:544–552
Kahn A (1968) Developmental physiology of bean leaf plastids. II. Negative contrast electron microscopy of tubular membranes in prolamellar bodies. Plant Physiol 43:1769–1780
Kahn A, Avivi-Bleiser N, Wettstein D von (1976) Genetic regulation of chlorophyll synthesis analyzed with double mutants in barley. In: Bücher T, Neupert W, Sebald W, Werner S (eds) Genetics and biogenesis of chloroplasts and mitochondria. Elsevier/North-Holland Biomedical Press, Amsterdam, New York, pp 119–131
Kannangara CG, Gough SP (1977) Synthesis of δ-aminolevulinic acid and chlorophyll by isolated chloroplasts. Carlsberg Res Commun 42:441–457
Kannangara CG, Gough SP (1979) Biosynthesis of δ-aminolevulinate in greening barley leaves II: Induction of enzyme synthesis by light. Carlsberg Res Commun 44:11–20
Kannangara CG, Gough SP, Wettstein D von (1978) The biosynthesis of δ-aminolevulinate and its genetic regulation. In: Akoyunoglou G, Argyroudi-Akoyunoglou JH (eds) Chloroplast development. Elsevier/North-Holland Biomedical Press, Amsterdam, New York, pp 147–160
Klein S, Poljakoff-Mayber A (1961) Fine structure and pigment conversion in isolated proplastids. J Biophys Biochem Cytol 11:433–440
Klein S, Schiff JA (1972) The correlated appearance of prolamellar bodies, protochlorophyll(ide) species, and the Shibata shift during development of bean etioplasts in the dark. Plant Physiol 49:619–626
Kleudgen HK, Lichtenthaler HK (1974) Die Synthese von Lipochinonen und isoprenoiden Pigmenten bei der Ergründung von Etioplasten im Blaulicht. Z Naturforsch 29c:142–146
Klimov VV, Dolan E, Ke B (1980) EPR properties of an intermediary electron acceptor (pheophytin) in photosystem II reaction centers at cryogenic temperatures. FEBS Lett 112:97–100
Koski VM, Smith JH (1951) Chlorophyll formation in a mutant *white seedling-3*. Arch Biochim Biophys 34:189–194
Levine RP (1969) The analysis of photosynthesis using mutant strains of algae and higher plants. Annu Rev Plant Physiol 20:523–540
Levine RP (1971) Preparation and properties of mutant strains of *Chlamydomonas reinhardi*. In: San Pietro A (ed) Photosynthesis. Part A. Methods in enzymology Vol 23. Academic Press, London, New York, pp 119–129
Levine RP (1974) Mutant studies on photosynthetic electron transport. In: Stewart WDP (ed) Algal physiology and biochemistry. Bot Monogr 10. Blackwell, Oxford, London, pp 424–433
Lichtenthaler HK (1968) Verbreitung und relative Konzentration der lipophilen Plastidenchinone in grünen Pflanzen. Planta 81:140–152
Lichtenthaler HK (1969) Light stimulated synthesis of plastid quinones and pigments in etiolated barley seedlings. Biochim Biophys Acta 184:164–172

Loppes R (1968) Ethylmethanesulfonate; an effective mutagen in *Chlamydomonas reinhardi*. Mol Gen Genet 102:229–231

Machold O, Høyer-Hansen G (1976) Polypeptide composition of thylakoids from *viridis* and *xantha* mutants in barley. Carlsberg Res Commun 41:359–366

Machold O, Meister A (1980) Resolution of the light-harvesting chlorophyll a/b-protein of *Vicia faba* into two different chlorophyll-protein complexes. Biochim Biophys Acta 546:472–480

Machold O, Simpson DJ, Møller BL (1979) Chlorophyll-proteins of thylakoids from wild-type and mutants of barley (*Hordeum vulgare* L.). Carlsberg Res Commun 44:235–254

Mackinney G, Rick CM, Jenkins JA (1956) The phytoene content of tomatoes. Proc Natl Acad Sci USA 42:404–408

Mars B, Kaplan S, Shepherd W (1980) Isolation of mutants of photosynthetic bacteria. In: San Pietro A (ed) Photosynthesis and nitrogen fixation. Part C. Methods in enzymology Vol 69. Academic Press, London, New York, pp 29–37

Mascia P (1978) An analysis of precursors accumulated by several chlorophyll biosynthetic mutants of maize. Mol Gen Genet 161:237–244

Mascia PN, Robertson DS (1978) Studies of chloroplast development in four maize mutants defective in chlorophyll biosynthesis. Planta 143:207–211

Matchett RW, Pollock BM, Robertson DW (1968) The "granda" gene: A chlorophyll mutation in *Hordeum* species. J Hered 59:279–282

Mendiola-Morgenthaler LR, Morgenthaler TT, Price CA (1976) Synthesis of coupling factor CF_1 protein by isolated spinach chloroplasts. FEBS Lett 62:96–100

Mets L, Bogorad L (1972) Altered chloroplast ribosomal proteins associated with erythromycin-resistant mutants in two genetic systems of *Chlamydomonas reinhardi*. Proc Natl Acad Sci USA 69:3779–3783

Miles D (1980) Mutants of higher plants: maize. In: San Pietro A (ed) Photosynthesis and nitrogen fixation. Part C. Methods in enzymology Vol 69. Academic Press, London, New York, pp 3–23

Miles DC, Daniel CD (1974) Chloroplast reactions of photosynthetic mutants in *Zea mays*. Plant Physiol 53:589–595

Miles CD, Markwell JP, Thornber JP (1979) Effect of nuclear mutation in maize on photosynthetic activity and content of chlorophyll-protein complexes. Plant Physiol 64:690–694

Miller KR (1980) A chloroplast membrane lacking photosystem I. Changes in unstacked membrane regions. Biochim Biophys Acta 592:143–152

Møller BL, Høyer-Hansen G, Hiller RG (1982) Functional identification of barley thylakoid polypeptides resolved by SDS-polyacrylamide gel electrophoresis. In: Avron M (ed) Proc 5th Int Congr Photosynthesis. Int Sci Services, Jerusalem, in press

Mühlethaler K (1977) Introduction to structure and function of the photosynthesis apparatus. In: Trebst A, Avron M (eds) Photosynthesis I. Encyclopedia of plant physiology new ser Vol V. Springer, Berlin, Heidelberg, New York, pp 503–521

Müller F (1964) Untersuchungen über Chloroplastenfarbstoffe und Assimilationsstärke Chlorophyll b-freier durch Röntgenbestrahlung induzierter Mutanten von *Pisum sativum*. Planta 63:65–82

Mullet JE, Arntzen CJ (1980) Simulation of grana stacking in a model membrane system. Mediation by a purified pigment-protein complex from chloroplasts. Biochim Biophys Acta 589:100–117

Mullet JE, Burke JJ, Arntzen CJ (1980) Chlorophyll proteins of photosystem I. Plant Physiol 65:814–822

Newell E, Rienits KG (1975) Chlorophyll synthesis in a yellow mutant of wheat. Aust J Plant Physiol 2:543–552

Nielsen OF (1974) Macromolecular physiology of plastids. XII. *Tigrina* mutants of barley: Genetic, spectroscopic and structural characterization. Hereditas 76:269–304

Nielsen OF, Gough S (1974) Macromolecular physiology of plastids XI. Carotenes in etiolated *tigrina* and *xantha* mutants of barley. Physiol Plant 30:246–254

Nielsen NC, Henningsen KW, Smillie RM (1974) Chloroplast membrane proteins in

wild-type and mutant barley. In: Avron M (ed) Proc 3rd Int Congr Photosynthesis Vol III. Elsevier North Holland, Amsterdam, New York, pp 1603–1613
Nilan RA, Sideris EG, Kleinhofs A, Sander C, Konzak CF (1973) Azide – a potent mutagen. Mutat Res 17:142–144
Nugent JHA, Møller BL, Evans MCW (1982) EPR detection of the primary photochemistry of photosystem II in a barley mutant lacking photosystem I activity. In: Avron M (ed) Proc 5th Int Congr Photosynthesis. Int Sci Services, Jerusalem, in press
Ohad I, Siekevitz P, Palade GE (1967a) Biogenesis of chloroplast membranes. I. Plastid differentiation in a dark-grown algal mutant (*Chlamydomonas reinhardi*) J Cell Biol 35:521–552
Ohad I, Siekevitz P, Palade GE (1967b) Biogenesis of chloroplast membranes. II. Plastid differentiation during greening of a dark-grown algal mutant (*Chlamydomonas reinhardi*). J Cell Biol 35:553–584
Pfister K, Steinback KE, Gardner G, Arntzen CJ (1980) Identification of the photosystem II herbicide binding protein. Plant Physiol Suppl 65:10
Powls R (1971) A carotenoid-deficient mutant strain of *Scenedesmus obliquus* with defective chloroplast lamellae. In: Forti G, Avron M, Melandri A (eds) Proc 2nd Int Congr Photosynthesis Vol III, The Hague, pp 2611–2618
Ratcliff S, Surzycki S (1974) DNA-dependent RNA polymerases from *Chlamydomonas reinhardtii*. J Cell Biol 62:281a
Röbbelen C (1956) Über die Protochlorophyllreduktion in einer Mutante von *Arabidopsis thaliana* (L) Heynh. Planta 47:532–546
Röbbelen G (1957) Eine Blattfarbmutante ohne Chlorophyll b von *Arabidopsis thaliana* (L.) Heynh Naturwissenschaften 44:288–289
Robertson DS (1961) Linkage studies of mutant maize with pigment deficiences in endosperm and seedling. Genetics 46:649–662
Robertson DS (1966) Allelic relationships and phenotypic interactions of four dominant modifiers of the cl_1 locus in maize. Heredity 21:1–7
Robertson DS, Bachmann MD, Anderson JC (1966) Role of carotenoids in protecting chlorophyll from photodestruction. II. Studies on the effect of four modifiers of the albino cl_1 mutant of maize. Photochem Photobiol 5:797–805
Roughan PG, Boardman NK (1972) Lipid composition of the pea and bean leaves during chloroplast development. Plant Physiol 50:31–34
Sager R (1955) Inheritance in the green algae *Chlamydomonas reinhardi*. Genetics 40:476–489
Sager R (1961) Photosynthetic pigments in mutant strains of *Chlamydomonas*. Carnegie Inst Washington Yearb 60:374–376
Sager R, Palade GE (1954) Chloroplast structure in green and yellow strains of *Chlamydomonas*. Exp Cell Res 7:584–588
Sager R, Tsubo Y (1962) Mutagenic effects of streptomycin in *Chlamydomonas*. Arch Microbiol 42:159–175
Sander C, Laber LJ, Bell WD, Hamilton RH (1968) Light sensitivity of plastids and plastid pigments present in the albescent maize mutant. Plant Physiol 43:693–697
Satoh K (1979) Polypeptide composition of the purified photosystem II pigment-protein complex from spinach. Biochim Biophys Acta 546:84–92
Schiff JA (1973) The development, inheritance, and origin of the plastid in *Euglena*. Adv Morphog 10:265–312
Schiff JA, Lyman H, Russel GK (1971) Isolation of mutants from *Euglena gracilis*. In: San Pietro A (ed) Photosynthesis. Part A. Methods in enzymology Vol 23. Academic Press, London, New York, pp 143–162
Schiff JA, Lyman H, Russel GK (1980) Isolation of mutants of *Euglena gracilis*: An addendum. In: San Pietro A (ed) Photosynthesis and nitrogen fixation. Part C. Methods in enzymology Vol 69. Academic Press, London, New York, pp 23–29
Schmid GH (1971) Origin and properties of yellow plants: Yellow tobacco. In: San Pietro A (ed) Photosynthesis. Part A. Methods in enzymology Vol 23. Academic Press, London, New York, pp 171–194
Schmidt GW, Lyman H (1976) Inheritance and synthesis of chloroplasts and mitochon-

dria of *Euglena gracilis*. In: Lewin RA (ed) The genetics of algae. Blackwell, Oxford, London, pp 257–299

Schmidt GW, Matlin KS, Chua N-H (1977) A rapid procedure for selective enrichment of photosynthetic electron transport mutants. Proc Natl Acad Sci USA 74:610–614

Schmidt GW, Bartless S, Grossman AR, Cashmore AR, Chua N-H (1980) In vitro synthesis, transport, and assembly of the constituent polypeptides of the light-harvesting chlorophyll a/b protein complex. In: Leaver CJ (ed) Genome organization and expression in plants. Plenum, New York, London, pp 337–351

Schwarz HP (1980) Feinstrukturelle und biochemische Untersuchungen der Chloroplasten von Chlorophyll-Mutanten der Gartenerbse *Pisum sativum*. Thesis, Univ Bonn

Senger H, Bishop NI (1972) The development of structure and function in chloroplasts of greening mutants of *Scenedesmus* I. Formation of chlorophyll. Plant Cell Physiol 13:633–649

Senger H, Strassberger G (1978) Development of the photosystems in greening algae. In: Akoyunoglou G, Argyroudi-Akoyunoglou JH (eds) Chloroplast development. Elsevier/North-Holland Biomedical Press, Amsterdam, New York, pp 367–378

Shepherd HS, Boynton JE, Gillham NW (1979) Mutations in nine chloroplast loci of *Chlamydomonas* affecting different photosynthetic functions. Proc Natl Acad Sci USA 76:1353–1357

Shneyour A, Avron M (1975) A method for producing, selecting and isolating photosynthetic mutants of *Euglena gracilis*. Plant Physiol 55:142–144

Siefermann-Harms D, Joyard J, Douce R (1978) Light-induced changes in the carotenoid levels in chloroplast envelopes. Plant Physiol 61:530–533

Simpson DJ (1979) Freeze-fracture studies on barley plastid membranes. III. Location of the light-harvesting chlorophyll-protein. Carlsberg Res Commun 44:305–336

Simpson DJ, Wettstein von D (1980) Macromolecular physiology of plastids. XIV. *Viridis* mutants in barley: Genetic, fluoroscopic and ultrastructural characterization. Carlsberg Res Commun 45:283–314

Simpson DJ, Høyer-Hansen G, Chua N-H, Wettstein D von (1977) The use of gene mutants in barley to correlate thylakoid polypeptide composition with the structure of the photosynthetic membrane. In: Hall DO, Coombs J, Goodwin TW (eds) Proc 4th Int Congr Photosynthesis. Biochem Soc, London, pp 537–548

Smillie RM, Bishop DG, Gibbons GC, Graham D, Grieve AM, Raison JK, Reger BJ (1971) Determination of the sites of synthesis of proteins and lipids of the chloroplast using chloramphenicol and cycloheximide. In: Boardman NK, Linnane AW, Smillie RM (eds) Autonomy and biogenesis of mitochondria and chloroplasts. Elsevier North-Holland, Amsterdam, London, pp 422–433

Smillie RM, Nielsen NC, Henningsen KW, Wettstein D von (1974) Ontogeny and environmental regulation of photochemical activity in chloroplast membranes. In: Avron M (ed) Proc 3rd Int Congr Photosynthesis. Elsevier North Holland, Amsterdam, New York, pp 1841–1860

Smillie RM, Henningsen KW, Nielsen NC, Wettstein D von (1976) The influence of cations and methylamine on structure and function of thylakoid membranes from barley chloroplasts. Carlsberg Res Commun 41:27–56

Smillie RM, Nielsen NC, Henningsen KW, Wettstein D von (1977) Development of photochemical activity in chloroplast membranes. II. Studies with a mutant of barley grown under different environments. Aust J Plant Physiol 4:439–449

Smillie RM, Henningsen KW, Bain JM, Critchley C, Fester T, Wettstein D von (1978) Mutants of barley heat-sensitive for chloroplast development. Carlsberg Res Commun 43:351–364

Smith HJ, Bogorad L (1974) The polypeptide subunit structure of the DNA-dependent RNA polymerase of *Zea mays* chloroplasts. Proc Natl Acad Sci USA 71:4839–4842

Smith JHC, Durham LJ, Wurster CF (1959) Formation and bleaching of chlorophyll in albino corn seedlings. Plant Physiol 34:340–345

Stummann BM, Henningsen KW (1980) Characterization of chlorophyll deficient mutants of pea. Hereditas 93:261–275

Stummann BM, Kaufmann U, Henningsen KW (1980) Transcription products associated with developing pea chloroplast membranes. Physiol Plant 49:7–12

Surzycki SJ, Shellenbarger DL (1976) Purification of a putative sigma factor from *Chlamydomonas reinhardi*. Proc Natl Acad Sci USA 73:3961–3965

Thomas PE, Ryan D, Levin W (1976) An improved staining procedure for the detection of the peroxidase activity of cytochrome P-450 on sodium dodecyl sulfate polyacrylamide gels. Anal Biochem 75:168–176

Thornber JP, Highkin HR (1974) Composition of the photosynthetic apparatus of normal barley leaves and a mutant lacking chlorophyll b. Eur J Biochem 41:109–116

Togasaki RK, Hudock MO (1972) Effect of inorganic arsenate on the growth of *Chlamydomonas reinhardi*. Plant Physiol Suppl 49:52

Treharne KJ, Mercer EI, Goodwin TW (1966) Carotenoid biosynthesis in some maize mutants. Phytochemistry 5:581–587

Troxler RF, Lester R, Craft FO, Albright JT (1969) Plastid development in albescent maize. Plant Physiol 44:1609–1618

Vernotte C, Briantais J-M, Remy R (1976) Light-harvesting pigment protein complex requirement for spill-over changes induced by cations. Plant Sci Lett 6:135–141

Virgin HJ, Kahn A, Wettstein D von (1963) The physiology of chlorophyll formation in relation to structural changes in chloroplasts. Photochem Photobiol 2:83–91

Wallace RH, Habermann HH (1959) Genetic history and general comparisons of two albino mutations of *Helianthus annuus*. Am J Bot 46:157–162

Walles B (1967) Use of biochemical mutants in analysis of chloroplast morphogenesis. In: Goodwin TW (ed) Biochemistry of chloroplasts. Academic Press, London, New York, pp 633–653

Wang W-Y (1978) Effect of dim light on the *y-1* mutant of *Chlamydomonas reinhardtii*. Plant Physiol 61:842–846

Wang W-Y, Wang WL, Boynton JE, Gillham NW (1974) Genetic control of chlorophyll biosynthesis: Analysis of mutants at two loci mediating the conversion of protoporphyrin IX to Magnesium protoporphyrin. J Cell Biol 63:806–823

Wang W-Y, Boynton JE, Gillham NW, Gough S (1975) Genetic control of chlorophyll biosynthesis in *Chlamydomonas*: Analysis of a mutant affecting synthesis of δ-aminolevulinic acid. Cell 6:75–84

Wang W-Y, Boynton J, Gillham NW (1977) Genetic control of chlorophyll biosynthesis: effect of increased δ-aminolevulinic acid synthesis on the phenotype of the *y-1* mutant of *Chlamydomonas*. Mol Gen Genet 152:7–12

Wasserman AR (1980) Chloroplast cytochromes f, b-559, and b_6. In: San Pietro A (ed) Photosynthesis and nitrogen fixation. Part C. Methods in enzymology Vol 69. Academic Press, London, New York, pp 181–202

Weil JH, Burkard G, Guillemaut P, Jeanin G, Martin R, Steinmetz A (1976) tRNAs and aminoacyl-tRNA synthetases in plant organelles. In: Bücher T, Neupert W, Sebald W, Werner S (eds) Genetics and biogenesis of chloroplasts and mitochondria. Elsevier North-Holland, Amsterdam, New York, pp 667–675

Weil JH, Guillemaut P, Burkard G, Canaday J, Mubumbila M, Osorio ML, Keller M, Gloeckler R, Steinmetz A, Keith G, Heiser D, Crouse EJ (1982) Comparative studies on chloroplast tRNA gene localization in the rDNA units. In: Avron M (ed) Proc 5th Int Congr Photosynthesis. Int Sci Serv, Jerusalem, in press

Wettstein von D (1957) Chlorophyll-lethale und der submikroskopische Formwechsel der Plastiden. Exp Cell Res 12:427–506

Wettstein von D (1958) The formation of plastid structure. In: The photochemical apparatus, its structure and function. Brookhaven Symp Biol 11:138–159

Wettstein von D (1980) Developmental interaction between chloroplasts and the nucleus. In: Nover L, Lynen F, Mothes K (eds) Cell compartmentation and metabolic channeling. Fischer, Jena, and Elsevier/North-Holland Biomedical Press, Amsterdam, New York, pp 313–329

Wettstein von D (1981) Chloroplast and nucleus: Concerted interplay between genomes of different cell organelles. The Emil Heitz lecture. In: Schweiger G (ed) Int Cell Biol 1980–81. Springer, Berlin, Heidelberg, New York, pp 250–272

Wettstein von D, Henningsen KW, Boynton JE, Kannangara GC, Nielsen OF (1971) The genic control of chloroplast development in barley. In: Boardman NK, Linnane AW, Smillie RM (eds) Autonomy and biogenesis of mitochondria and chloroplasts. Elsevier North-Holland, Amsterdam, New York, pp 205–223

Wettstein von D, Kahn A, Nielsen OF, Gough S (1974) Genetic regulation of chlorophyll synthesis analyzed with mutants in barley. Science 184:800–802

Wettstein von D, Møller BL, Høyer-Hansen G, Simpson D (1982) Mutants in the analysis of the photosynthetic membrane polypeptides. In: Schiff JA, Stanier RY (eds) Origin of chloroplasts. Elsevier/North-Holland Biomedical Press, Amsterdam, New York, pp 245–257

Wild A, Trostmann U, Kietzmann I, Fuldner K-H (1978) Development of the photosynthetic apparatus during light-dependent greening of a mutant of *Chlorella fusca*. Planta 140:45–72

Wolf FT (1963) Effects of light on biosynthesis of carotenoid pigments in wheat seedlings. Plant Physiol 38:649–652

Yurina NP, Odintsova MS, Maliga P (1978) An altered chloroplast ribosomal protein in a streptomycin resistant tobacco mutant. Theor Appl Genet 52:125–128

Zielinski RE, Price CA (1977) Synthesis of cytochrome b_{559} by isolated spinach chloroplasts. Plant Physiol Suppl 59:8

17 Interrelationship Between Chloroplasts and the Nucleo-Cytosol Compartment in *Acetabularia*

H.-G. Schweiger

1 Introduction

In the studies on the interrelationships between the cell nucleus and cytoplasm, the unicellular and uninucleate marine green alga *Acetabularia* has become a favorite organism since Hämmerling (1932, 1953), some 50 years ago, successfully performed the merotomy experiment which had been suggested by Boveri (1918). The clear-cut results obtained with cell fragments whose nucleus had been removed, as well as with the simple and convincing method of transplanting nucleate cell fragments onto anucleate ones, opened the experimental approach to problems related to nucleocytoplasmic interrelationships (Hämmerling 1953, 1963, Brachet and Lang 1965, Schweiger 1969 Puiseux-Dao 1970, Bonotto et al. 1976, Schweiger and Berger 1979). More recently, other potent methods have been developed which again underline the exceptional suitability of this organism.

2 Acetabularia

2.1 Morphology

Acetabularia belongs to the family Dasycladaceae (Oltmanns 1922, Fritsch 1965, Valet 1969, Berger et al. 1974). It forms a subfamily to which more than 20 characterized species belong. The greater part of these species are grown in laboratory culture (Berger and Schweiger 1980). Independent of the large differences in size, all species have a similar architecture: characteristically at the end of the vegetative phase, they consist of a basal part, the rhizoid, the stalk and a highly species-specific cap-like structure which eventually bears the cysts. At a very early stage of growth of the zygote, bipolar growth results in the development of the rhizoid at the one end and the stalk at the other. With its branches, the rhizoid keeps the cell anchored to the substratum of rock, corals, wood, sea shells, or even other algae. In its apical part, the stalk carries the zone of growth where stretches of stalk and whirls are alternatingly formed. Depending on the species, a number of young whirls are kept while the old ones are lost. At the end of the vegetative phase, after the last period of stalk growth, a sporangium of a cap-like structure is formed instead of a whirl. The cap consists of rays whose number varies depending on the species. The rays of some species are interconnected and, in particular, their distal ends exhibit a pronounced morphology. Other species-specific markers are the presence or absence of a so-called corona inferior and the corona superior, their architecture and the length of the mature cells, which may vary from 2 to more than 200 mm (SCHWEIGER and BERGER 1981). The degree of calcification of the cap and of the stalk in wild-living material sometimes facilitates the identification of the species, but sometimes even changes the morphological appearance.

2.2 Life Cycle

During the entire vegetative phase, the rhizoid contains the primary nucleus in one of its branches. The vegetative phase terminates when the cap reaches its maximum diameter and the primary nucleus divides. The resulting secondary nuclei divide further as they migrate out of the rhizoid through the stalk into the cap rays (SCHULZE 1939, KOOP 1979). In the cap, each secondary nucleus forms the initiation site for one gametangium or cyst. Further divisions lead to the formation of gametes. Given that the gametes are haploid, the idea that the cell nucleus is diploid during the vegetative phase and undergoes only one reduction division at the beginning of the generative phase seems to be an oversimplification (KOOP 1979). A number of questions are raised by the observation that in the cysts a rather high percentage of secondary nuclei fuse once or perhaps more than once (REITBERGER unpublished results).

2.3 Ultrastructure

A cross-section through the stalk reveals a thick cell wall at the surface, an adjacent thin layer of plasma and a large central vacuole. The plasma layer has a bundle-like structure, where the bundles contain all the structural elements which are characteristic for a plant cell: plastids, mitochondria, Golgi apparatus and other membranaceous elements. The number of plastids and of mitochondria strongly depends on the stage of development. The number of both organelles increases during the vegetative phase and plateaus at the end of this phase. In the large species, this means that there are 10^8 or more of each of the two types of organelles (SCHWEIGER et al. 1974). The plastids are extremely heterogenous in size, varying in length from 2 to 8 μm and in width from 1 to 2 μm (PUISEUX-DAO 1970). The mitochondria are significantly smaller. The two organelles exhibit an ultrastructure which is characteristic for chloroplasts and mitochondria (WERZ 1966).

The extraordinary size of the *Acetabularia* cell has provoked the question whether *Acetabularia* is a unicellular organism at all. Alternatively, the possibility has been considered that it might instead be a multicellular organism which has lost some of its intercellular walls (BRACHET 1970). Although this is undoubtedly an interesting hypothesis, it is definitely not in agreement with the generally accepted ideas and definition of a cell, in particular since the organism has only one nucleus.

Another peculiarity of the *Acetabularia* cell is the tremendous size of the nucleus. It may reach a diameter of up to 150 μm, depending on the species. Moreover, the nucleus has an extreme localization in the rhizoid throughout the vegetative phase of the life cycle. Due to the high number of chloroplasts per cell, the total amount of plastid DNA is thought to exceed the nuclear

DNA. Among the disadvantages which limit the experimental use of *Acetabularia* is the large volume central vacuole, whose sap has a pH of 1–2.

The large size of the *Acetabularia* cell and the location of the nucleus in the rhizoid are favorable conditions for the experimenter, and also provoke consideration of the extraordinarily long distances from the nucleus to the apex and vice versa, which may be up to 200 mm or even more depending on the species. Such a consideration raises interest in this problem, which of course plays a major role in all types of cells, but is more readily studied in *Acetabularia* due to the quantitatively different dimensions.

2.4 Compartmentation

2.4.1 The Nucleo-Cytosol Compartment

The interaction between cell nucleus and cytoplasm depends on the existence of a border zone between the two compartments and is mediated by the membrane system which separates the two compartments. The basis of the interrelationships between cell nucleus and the plastids is even more complex since it includes three components: the cell nucleus with its membrane, the plastids with their envelope, and the cytosol in between. At least a superficial view of these three compartments leads to the conclusion that they are more or less independent of each other. This is true for a number of functions. With respect to genetics and gene expression, the picture can be reduced to the nucleo-cytosol compartment on the one side and the plastid compartment on the other. The two compartments are separated from each other by the plastid envelope where the interactions between the two systems occur. From this point of view, the nuclear membrane may be considered to be a regulatory site for the gene expression within the nucleo-cytosol compartment.

If the mitochondria are included in these considerations, one deals with a three-compartment system: the nucleo-cytosol compartment, the plastids, and the mitochondria. Thus, the cytosol has a dual function: to be part of the nucleo-cytosol gene expression machinery, and to serve as a transport system between the compartments.

The importance of the latter function becomes obvious if one focuses on the interrelationships between the mitochondria and chloroplasts and the role of the cytosol in these interrelationships. Both types of organelles are highly specialized in energy metabolism. The relevant exchange between the two types of organelles is mediated by molecules which carry a large amount of chemical energy or which can be easily converted into such. Here the question comes up concerning what features molecules must have in order to pass through the chloroplast and the mitochondria envelopes. Moreover, the problem arises of whether and how changed conditions might influence the permeability of these membranes and what the underlying mechanism is.

2.4.2 The Chloroplast

A major difference between the chloroplasts and the nucleo-cytosol compartment lies in the different types of genetic information which are stored in the two compartments and the essentially different mechanisms by which they are expressed. These differences are based on the structure of the two types of ribosomes, as well as on the other components of the translation system. It is generally accepted that the chloroplast gene expression machinery is related to prokaryotic systems rather than to the nucleo-cytosol system of eukaryotic cells.

A series of experiments on *Acetabularia* supported the idea of the occurence of two separate genomes and gene expression machineries in the nucleo-cytosol and the chloroplast compartment. The observation that anucleate cells of *Acetabularia* contain DNA and that this DNA is found in the chloroplast fraction was one of the early reliable pieces of evidence for the occurrence of DNA in chloroplasts.

The number of chloroplasts increases during the life cycle until the end of the vegetative phase, then it remains constant (CLAUSS et al. 1970). Ultrastructural investigations have shown that the replication of the chloroplasts is due to budding. This process, to a certain extent, is similar to replication of bacteria.

The analogy to bacteria might explain the occurrence of chloroplasts which do not contain DNA. They resemble minicells which also lack DNA (WOODCOCK and BOGORAD 1970). The number of chloroplasts increases even after removal of the cell nucleus; however, this is only for a period of up to two weeks after enucleation. This strongly indicates that the replication of the chloroplasts in some way depends on a nuclear function.

Very little is known about species specificity of chloroplasts in members of the family Dasycladaceae. Species specificity might be the basis of an incompatibility which becomes apparent in transplantation experiments (HÄMMERLING 1963, unpublished results). Heterologous transplants which were produced by grafting a rhizoid from one species to the anucleate stalk of another produce certain combinations which form cysts and can reproduce. In other combinations, fertility is reduced or lost.

The chloroplasts of an *Acetabularia* cell are heterogenous in several aspects. First of all, one has to note that the chloroplasts in the apical region have a lower content of starch than those which are in the vicinity of the nucleus (WERZ 1966). This raises the question whether these differences result from a direct influence of the nucleus on the chloroplasts or whether they arise from an indirect influence such as, for example, by means of intracellular gradients. In any case, this example shows that the structure of chloroplasts is related to cellular polarity.

3 Chloroplast Gene Expression

3.1 Chloroplast DNA

Early estimation of the amount of DNA in one chloroplast was 10^{-15} to 10^{-16} g DNA per chloroplast (GIBOR and IZAWA 1963, BALTUS and BRACHET 1963). Assuming that there is only one molecule of DNA per organelle, this would correspond to 6×10^7 to 6×10^8 molecular weight and to 2×10^5 to 2×10^6 nucleotides. This assumption, however, does not take into consideration the possibility that the amount of DNA per chloroplast is not constant. It has been shown that there are some chloroplasts which do not contain any DNA, while there are others which contain significantly more DNA than the amount given above (WOODCOCK and BOGORAD 1970, COLEMAN 1979). It remains, however, an open question whether these extreme variations reflect a real situation in a normal cell or whether they are artifacts caused by suboptimal culture conditions. This possibility cannot be excluded, since the medium used in the experiments under consideration possibly did not permit optimum growth of the cells.

Another good piece of evidence for the occurrence of DNA in chloroplasts came from electron microscopical observations on *Acetabularia* chloroplasts. On the basis of these findings, it was concluded that at least the majority of chloroplasts in *Acetabularia* contain a double-stranded DNA molecule which is about 30 μm long (WERZ and KELLNER 1968). The question remained open whether the number of DNA molecules per chloroplast is limited to one. Another report attributes lengths between 50 and 1000 μm to the DNA from *Acetabularia* chloroplasts with only 25% of the plastids containing DNA (WOODCOCK and BOGORAD 1970). More recent studies on the coding complexity of the chloroplast genome of *A. mediterranea* and *A. cliftonii* have indicated that on an average the plastids contain 2.3×10^{-15} g and 7.9×10^{-15} g DNA, respectively (PADMANABHAN and GREEN 1978, GREEN et al. 1977). From renaturation experiments, the kinetic complexities were estimated to be 1.4×10^9 and 4.7×10^9 mol. wt., which indicates that the number of genomes per plastid is 1.3 and 3 for *A. mediterranea* and *A. cliftonii* (GREEN et al. 1977).

Another quite interesting finding is that chloroplasts from the species *A. cliftonii,* in addition to the DNA molecules with a length of 35 μm, contain minicircles of DNA with a length of 4.7 μm (GREEN 1976, GREEN et al. 1977). Surprisingly enough, these minicircles could not be found in other species investigated. Nothing is known about the nature and the role of these minicircles in the cell.

3.2 Chloroplast Transcription

The first clear-cut evidence for the existence of a chloroplast-specific transcription again came from experiments on *Acetabularia* (SCHWEIGER and BERGER 1964, BERGER 1967a, b). The proof was convincing, since it was demonstrated in chloroplasts from cells in which nuclear transcription was definitely excluded by the use of anucleate fragments as a source of chloroplasts. Moreover, the transcript was identified as ribosomal RNA by density gradient centrifugation and the even distribution of labeled nucleotides after ^{32}P orthophosphate incorporation and alkaline hydrolysis.

All these early experiments appear to support the hitherto hypothesized autonomy of the chloroplasts in gene expression. At that time, there was, however, already an experiment which showed the dependence of the chloroplast transcription on the cell nucleus. This experiment was similar in design to one which had been performed on morphogenesis (SCHWEIGER and BREMER 1961). It demonstrated that anucleate cells are capable of increasing their RNA content under the following conditions. Nucleate cells were kept in darkness for ten days. After this dark period, the nucleus was removed and the cells were exposed to light. Within a few days, these anucleate cells increased their RNA content by more than twofold. This increase in RNA is supposed to be due to accumulation of chloroplast ribosomal RNA. In control cells in which the enucleation was performed prior to the dark period, no increase in cytoplasmic RNA was observed. The result indicates that the nucleus is capable of exerting some influence on chloroplast RNA synthesis.

3.3 Translation of Chloroplast Proteins as Revealed by Nuclear Exchange Experiments

The close interrelationship between the nucleo-cytosol compartment and the chloroplasts, is well known at the level of the proteins, their synthesis, and the components involved in their synthesis.

3.3.1 Malic Dehydrogenase

In this connection, an experiment can be discussed which in a very simple and convincing way may answer the question whether or not the genetic information for a specific polypeptide comes from the nuclear genome (SCHWEIGER et al. 1967, SANDAKHCHIEV et al. 1973). A prerequisite for this type of experiment is that a molecular difference for the polypeptide under investigation can be demonstrated in two species of *Acetabularia.* In this sense, a very useful marker is the behavior of the polypeptide in polyacrylamide gel electrophoresis. It has been shown that the isoenzyme banding pattern of NADP-dependent malic dehydrogenase (EC 1.1.1.37 is significantly different in *A. mediterranea* and *A. cliftonii.* While there are two bands which can be specifically stained in *A. mediterranea,* only one band was found in *A. cliftonii.* Moreover, none of the *A. mediterranea* bands comigrated with the *A. cliftonii* band. The cell nucleus was isolated from the rhizoid of one cell from each species and implanted into the anucleate fragment of the other species, resulting in heterologous combinations which contained an *A. mediterranea* nucleus in an *A. cliftonii* cytoplasm and vice versa (Fig. 1). In these hybrid cells, the isoenzyme patterns of the malic dehydrogenase were established 4 weeks after cell surgery and implantation (Fig. 2). The experiment demonstrated two important points.

a) In the *A. mediterranea* cytoplasm, the isozyme pattern characteristic for *A. cliftonii* was formed, and vice versa.
b) The appearance of the "new" isoenzyme pattern was paralleled by a disappearance of the "old" isoenzyme pattern.

Both changes obviously occurred under the influence of the implanted nucleus. The first result clearly shows that this particular polypeptide is coded by the nuclear genome. The second result is more complex, since it shows that a polypeptide disappears under the influence of an implanted nucleus. This striking result has to be compared with another one which has shown that, after removal of the nucleus, the isoenzyme pattern persists, while the total

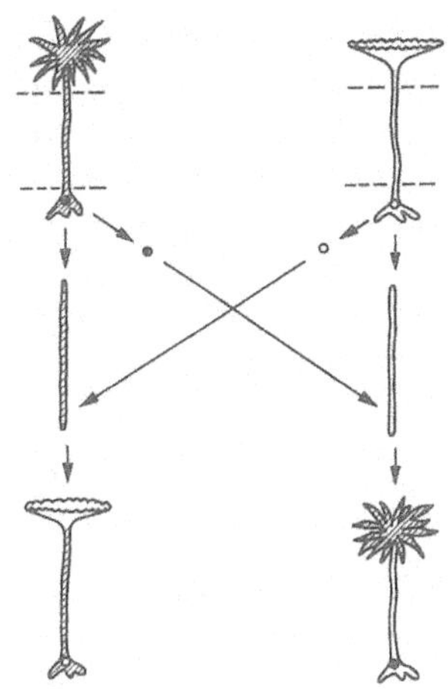

Fig. 1. Nuclear exchange between cells of *A. cliftonii* (*left*) and *A. mediterranea* (*right*). (SCHWEIGER and BERGER 1979)

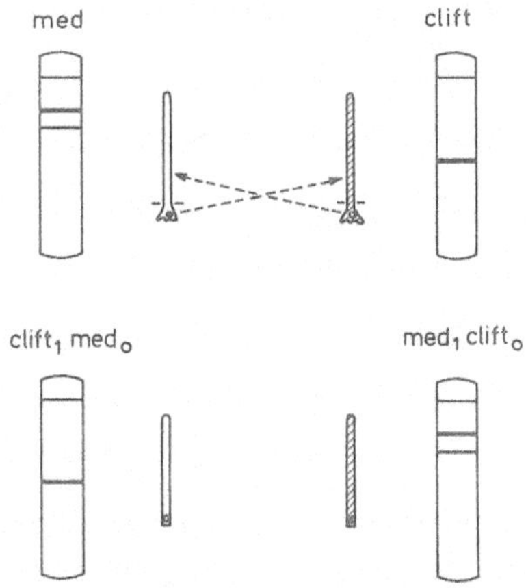

Fig. 2. Effect of nuclear exchange upon the malic dehydrogenase isozyme pattern. (SCHWEIGER et al. 1967)

enzyme activity increases over a period of 4 weeks, albeit at a slightly lower rate than in nucleate cells.

An attractive explanation for the nucleus-governed disappearance of the "old" isoenzyme pattern would be that the mRNA for the polypeptide is relatively stable in the anucleate cell, but is labile in the presence of the implanted nucleus. This might occur via a competitive situation at a site where degradation of surplus mRNA takes place. Surplus mRNA is a product of permanent transcription of the nuclear genome and is lacking in anucleate cells. A steady nonspecific degradation of a certain percentage of the total surplus mRNA would result in a slow preferential loss of the population of mRNA molecules which are not replaced by transcription. Together with a natural turnover of the polypeptides under consideration, such processes would result in an diminution of the "old" isoenzymes.

Another interesting aspect of the experiment was that at least part of the malic dehydrogenase which was shown to be nuclear-coded by the nuclear exchange is located in the chloroplasts (SCHWEIGER et al. 1967, RAHMSDORF 1977, ASTAUROVA et al. 1979). This was the first well-established example that one of the polypeptides of the chloroplasts is translated from genetic information from the nuclear genome. Such a conclusion immediately cast doubt on the then generally accepted, absolute genetic autonomy of the chloroplasts. Estimations of the encoding capacity of the genome of *Acetabularia* had already discredited the hypothesis of absolute autonomy of chloroplasts (BERGER 1967b). By a similar type of experiment, it was shown that a lactate dehydrogenase (EC 1.1.1.27) is nuclear-coded and is located in chloroplasts (REUTER and SCHWEIGER 1969).

3.3.2 Chloroplast Ribosomal Proteins

The complexity of the nucleo-cytosol-chloroplast interrelationship became evident when the coding site for the ribosomal proteins of the 70S chloroplast ribosomes was investigated. The experimental approach used was similar to that which had been successfully used for elucidating the coding site for malic dehydrogenase. The proteins of the 70S ribosomes from *A. mediterranea* and from *A. crenulata* were subjected to polyacrylamide gel electrophoresis. Differences in the pattern were established and the influence of the nuclear exchange on the species-specific pattern was followed. Under the influence of the nucleus, the species-specific patterns changed. One has to conclude that those polypeptides which differed are coded by the nuclear genome (KLOPPSTECH and SCHWEIGER 1973a, b).

The fundamental importance of this finding becomes obvious if one remembers that the RNA moiety of the 70S chloroplast ribosomes is coded by and transcribed from the chloroplast genome discussed earlier in this contribution. Within the chloroplasts, we have at least one type of structure which receives part of its genetic information from the chloroplast genome and part from the nuclear genome. This complex situation later proved to be a good model for other composite structures, not only in chloroplasts, as was shown for ribulose 1,5-bisphosphate carboxylase (KUNG 1976), but also for mitochondrial, ATPase (NELSON et al. 1980), cytochrome oxidase (TZAGOLOFF and MACINO 1979) and other less identified parts of the membrane systems.

In particular, the complexity of the interactions of the different components of the 70S ribosomes demands a very exact fitting of the components in order to retain the function of the ribosomes. A suboptimum fitting might be the basis of the incompatibility found in grafting experiments on *Acetabularia*.

All the results just discussed are in agreement with the idea that, although chloroplasts do have a certain autonomy including their own DNA and transcriptional and translational systems, they are indeed fully integrated in the cell, since a significant percentage of their proteins is coded by and transcribed in the nucleo-cytosol compartment. The question of whether there are also proteins which are coded and translated in chloroplasts is to be discussed below.

3.4 Chloroplast Translation

Besides the problem of which genome contains the genetic information for a specific protein, the question is raised where and on which type of ribosomes,

80S eukaryotic or 70S prokaryotic type, a specific protein is translated. In particular, this holds true for the proteins of the 70S ribosomes. Studies have revealed that cycloheximide, but not chloramphenicol, inhibits the incorporation of labeled amino acids into the proteins of chloroplast ribosomes (KLOPPSTECH and SCHWEIGER 1974). From such experiments, it has been concluded that these proteins are translated on 80S ribosomes of the nucleo-cytosol compartment, i.e., outside the chloroplasts. The results indicated that nuclear-coded proteins are translated only on 80S ribosomes of the same compartment and not on 70S ribosomes, and that the two gene expression systems do not become mixed.

Although this is hypothetical at present, it is difficult to see what kind of experiment could prove the hypothesis. A major piece of evidence is based on the fact that translation in the two compartments proceeds via different molecular mechanisms.

3.5 80S Ribosomes Associated with Chloroplasts

Another interesting contribution to the interrelationships between the nucleo-cytosol and the chloroplast compartment came from the following observation. Chloroplasts were carefully prepared from *Acetabularia* cells so that only very little contamination by components of the nucleocytosol compartment was left.

Such chloroplast preparations still have pronounced capability to incorporate labeled amino acids into proteins. However, a striking observation was that such highly purified chloroplast preparations were sensitive not only to chloramphenicol, but also to the 80S ribosome-specific inhibitor cycloheximide. From this result, it was concluded that these chloroplast preparations still contained some 80S ribosomes (APEL and SCHWEIGER 1973). Since a very rigorous purification procedure was used, one may be sure that these 80S ribosomes are associated with the chloroplasts in a special way. In another type of experiment, 26.5S and 18S RNA, which are characteristic for 80S ribosomes, were isolated from such chloroplast preparations. Finally, it was shown to be improbable that these 80S ribosomes, which are tightly associated with the chloroplasts, represent a simple contamination. These chloroplast-associated 80S ribosomes are involved in the synthesis of a specific fraction of the chloroplast membrane proteins which is different from those fractions which are synthesized by the other 80S ribosomes of the cytosol compartment. It will be of some interest to investigate how these 80S ribosomes are bound to the envelope of the chloroplasts and confirm that they perform specific protein synthesis.

4 Regulation of Enzyme Activity

4.1 Thymidine Kinase

The complex interrelationships between the nucleo-cytosol compartment and the cell organelles with a semiautonomous genetic apparatus were revealed by

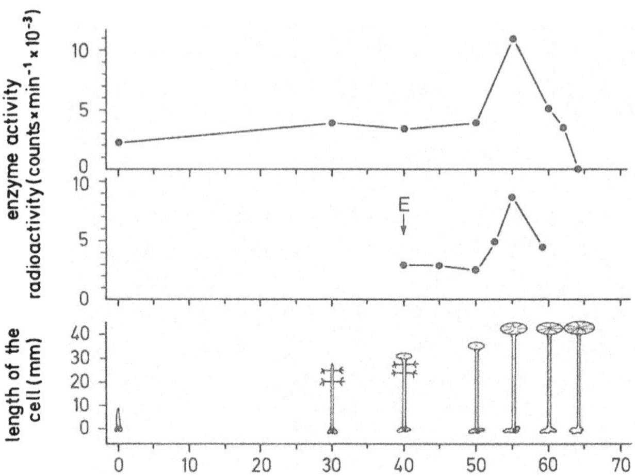

Fig. 3. Thymidine kinase activity during the life cycle of *A. mediterranea*. *Upper* nucleate cells; *middle* anucleate cells; *E* enucleation; *lower* life cycle stages correlated to time scale. (SCHWEIGER and BERGER 1979)

experiments on the behavior of the enzyme thymidine kinase (EC 2.7.1.21) during the life cycle. The activity of this enzyme slowly increases during the early vegetative phase and reaches a plateau or even decreases a little during the late vegetative phase. Just prior to the beginning of nuclear divisions, the activity of the thymidine kinase substantially increases (Fig. 3). Since at that stage of the life cycle the number of chloroplasts and mitochondria remains constant while the nucleus undergoes a series of divisions, one has to conclude that a close relationship exists between the regulation of this enzyme activity and the nuclear divisions. Thus, it was a striking observation that this regulation occurs even in the absence of the nucleus. Removal of the nucleus 14 days prior to the expected increase in enzyme activity did not suspend the regulation. This experiment clearly excluded the possibility that the regulation is directly related to a nuclear activity (BANNWARTH and SCHWEIGER 1975, BANNWARTH et al. 1977a, b).

Studies on the effect on the regulation of puromycin, a potent inhibitor of protein synthesis, led to the conclusion that protein synthesis is essential for the regulation. Another set of experiments ruled out the possibility that the regulation was mediated by the synthesis of a polypeptide activator. In these experiments, preparations from cells of the regulated and nonregulated state were mixed and in all cases the resulting enzyme activity was equal to the predicted average value. Therefore, it can be concluded that the regulation of thymidine kinase is due to de novo synthesis of the enzyme protein (BANNWARTH and SCHWEIGER 1975, BANNWARTH et al. 1977a, b). Detailed studies with a number of different inhibitors of protein synthesis like chloramphenicol, cycloheximide, and rifampicin revealed that thymidine kinase is translated on 70S ribosomes and that the enzyme is coded by the organelle genome (BANNWARTH and SCHWEIGER 1975, BANNWARTH et al. 1977a, b). The general importance of the regulation of the thymidine kinase is that we deal with a special nucleo-cytoplasmic action: an enzyme which is regulated in temporal relationship with and is necessary for nuclear divisions is coded by and translated

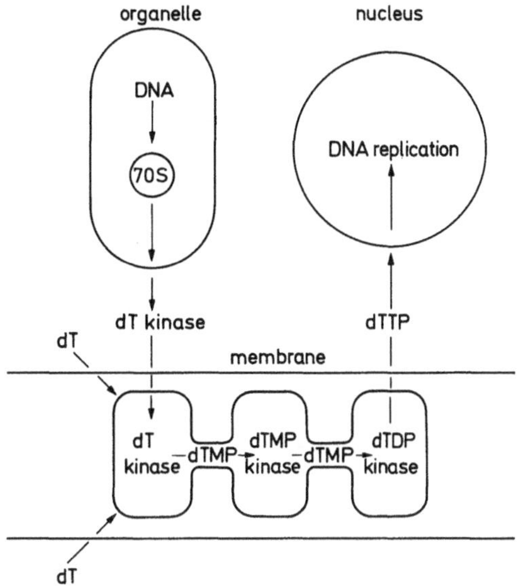

Fig. 4. Nucleocytoplasmic interactions in the regulation of the thymidine kinase. (BANNWARTH et al. 1977b)

in organelles. This necessarily means that the regulatory events must be associated with the organelles.

The enzyme thymidine kinase deserves interest from another aspect. Amputation of the basal part of the organism opens the possibility of removing, by centrifugation under cautious conditions, the greater part of the cytoplasm, including the chloroplasts and mitochondria, leaving the plasma membranes within the cell wall. On the basis of such a fractionation, it was demonstrated that at least the greater part of the enzyme activity is located in the plasma membrane. From this finding, it follows that the thymidine kinase which is coded in the organelle genome and translated on 70S ribosomes within the organelles leaves the organelles after synthesis and is integrated into the plasma membrane. Therefore, one has to recognize that besides a number of polypeptides which are transported into the organelles, there is at least one which is transported in the opposite direction (BANNWARTH et al. 1980).

The thymidine kinase is firmly fixed in the plasma membrane. If plasma membranes prepared as described above are superficially homogenized, they preferentially convert thymidine to deoxyribothymidine triphosphate and diphosphate (BANNWARTH et al. 1977a). This might mean that the thymidine kinase is in close proximity to a thymidylate kinase and a thymidine diphosphate kinase. Thorough homogenization of the plasma membranes results in preparations which catalyze the phosphorylation of thymidine to thymidine monophosphate, while no substantial label appears in the thymidine diphosphate and triphosphate (Fig. 4).

4.2 Other Enzymes

Preliminary experiments on thymidylate kinase indicate that this enzyme is regulated in a way similar to the thymidine kinase. However, this regulation takes place at a later stage of development (DE GROOT and SCHWEIGER 1980).

A pronounced regulation has been found for the enzyme which catalyzes the production of deoxyribosides in *Acetabularia*. This enzyme has been identified as a ribonucleoside reductase which has not been described before. There is good evidence that this enzyme is located in the chloroplasts (DE GROOT and SCHWEIGER in preparation).

Another enzyme which is also involved in the metabolism of DNA precursors is dCMP deaminase (EC 3.5.4.12). This enzyme is regulated in a way similar to that shown for thymidine kinase. The regulation is performed even in the absence of the nucleus and is due to de novo synthesis of the enzyme protein. The enzyme is apparently coded and synthesized in organelles (BANNWARTH et al. 1980).

Although the aspects which have been considered so far indicate that the cell organelles may play a major role in the regulatory events in the cell, this is by no means necessarily and generally valid. This may be concluded from experiments with other enzymes. Different phosphatases, in particular one with a pH optimum in the alkaline region, have been shown also to be regulated during the life cycle. However, the stage of the life cycle at which the regulation of this phosphatase takes place is much earlier than that of thymidine kinase, thymidylate kinase, and dCMP deaminase. The activity of the phosphatase increases when cap formation is initiated. This enzyme also is regulated in nucleate as well as in anucleate cells (SPENCER and HARRIS 1964).

Particular attention has been devoted to experiments in which the regulation of enzymes whose activities are attributed to distinct morphogenetic capabilities, in particular to the formation of the cap, was studied. For example, UDPG pyrophosphorylase (EC 2.7.7.9) activity increases in the final part of the vegetative phase of the life cycle. This increase in enzyme activity is observed in nucleate as well as in anucleate cells. It is, however, inhibited by cycloheximide rather than by chloramphenicol. This is in obvious contrast to the thymidine kinase. Apparently, this enzyme is translated on 80S ribosomes of the cytosol and therefore probably coded by the nuclear genome (ZETSCHE 1968). More recently, these experiments have been criticized, since the increase in enzyme activity of the UDPG pyrophosphorylase is not significantly different from the increase in protein content (DILLARD, unpublished results).

5 Circadian Rhythms

Even special features of the ultrastructure of the chloroplasts which undergo rhythmic changes are related to the cellular polarity. Estimations of the dimensions of the chloroplasts during different times of the day have revealed that the axial ratios change in a diurnal way (VANDEN DRIESSCHE 1966). These oscillating changes have been attributed to a circadian rhythm. In a similar way, the distribution of the chloroplasts within the cell is subjected to a circadian rhythm such that during the night time the chloroplasts accumulate in the vicinity of the nucleus, while they migrate into the apical regions of the cell during the day time (KOOP et al. 1978, BRODA et al. 1979). This process is highly specific for chloroplasts and differs from small vesicles, polyphosphate granules, and secondary nuclei (KOOP and KIERMAYER 1980a, b). The intracellular migration of the chloroplasts persists even if the cells are kept under constant conditions, including continuous light.

5.1 O_2 Evolution Rhythm

In particular, in *Acetabularia* it has been shown that at least one circadian rhythm is closely related to the cell nucleus. This is the photosynthesis rhythm, which is the most thoroughly studied type of oscillation in this unicellular organism. This parameter persists even under conditions of constant light and constant temperature (SWEENEY and HAXO 1961, SCHWEIGER et al. 1964a). Early experiments revealed that this diurnal rhythm of photosynthetic activity persists even after removal of the nucleus (SWEENEY and HAXO 1961, SCHWEIGER et al. 1964a, MERGENHAGEN and SCHWEIGER 1975a, KARAKASHIAN and SCHWEIGER 1976a).

Later on it was shown that this is only partially correct. It has been demonstrated that if nuclei are exchanged between two cells whose phases were different by 180°, the phase was shifted under the influence of the implanted nucleus (SCHWEIGER et al. 1964b).

This experiment led to the conclusion that, in spite of the autonomy of the cytoplasm in the absence of the nucleus, the cell nucleus is capable of exerting control over the phase of the rhythm.

The role of the nucleo-cytosol compartment in the cyclic changes of the characteristic chloroplast function of photosynthetic activity is underlined by another type of experiment. In these experiments, the influence of inhibitors of gene expression on circadian rhythms has been studied (MERGENHAGEN and SCHWEIGER 1975b, KARAKASHIAN and SCHWEIGER 1976b). These results unequivocally indicated that translation of 80S ribosomes plays an essential role in the biological clock. An important point was to exclude the possibility that the inhibitors affected the hands rather than the basic mechanism of the clock.

That this was not the case had to be concluded from experiments in which the phase was shifted in a most characteristic way by means of 8-h cycloheximide pulses. In these experiments, the circadian cycle was shown to consist of a cycloheximide-sensitive and a cycloheximide-insensitive phase (KARAKASHIAN and SCHWEIGER 1976b) (Fig. 5). The fact, however, that there is also a cycloheximide-insensitive phase in the cycle forces one to conclude that not all polypeptides which are synthesized on 80S ribosomes play an essential role in the clock.

The search for polypeptides which are essential for the central mechanism of the clock was facilitated by the observation that the cycloheximide-insensitive phase in the cycle is strongly dependent on temperature (KARAKASHIAN and SCHWEIGER 1976c). On the basis of these experiments, a search was initiated for polypeptides which fulfill the criteria just mentioned.

A candidate for such an essential polypeptide has been purified from thylakoid membranes of the chloroplasts from *Acetabularia* and has been partially characterized (LEONG and SCHWEIGER 1979). This polypeptide has a molecular

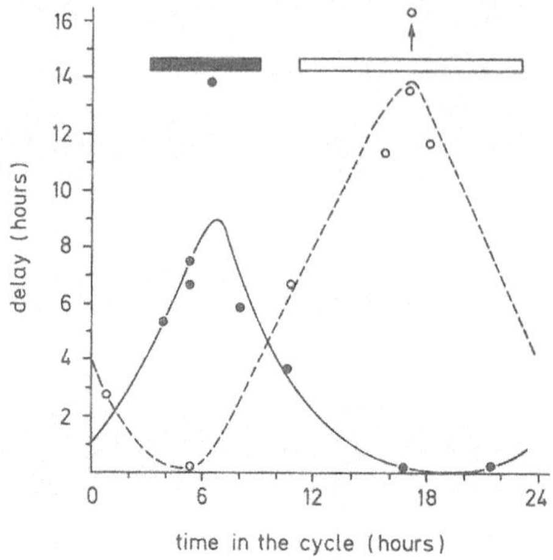

Fig. 5. Phase-shifting effect of cycloheximide in the circadian cycle at 25° C (o-----o) and 20° C (●——●). *Bars* indicate the sensitive phases. (KARAKASHIAN and SCHWEIGER 1976c)

weight of about 39,000. Although the final proof is still lacking that this polypeptide is indeed a central factor in the circadian rhythm, it is a promising starting point in the search for a function and mechanism which might explain the phenomenon of the circadian rhythm.

5.2 Coupled Translation – Membrane Model

The interrelationship between the nucleo-cytosol compartment and the chloroplast and its role in the circadian rhythm of the photosynthetic activity have led to formation of a model (SCHWEIGER and SCHWEIGER 1977). In this model, a polypeptide which is synthesized on 80S ribosomes is integrated into the thylakoid membranes, thereby changing the properties of the membranes, which results in repression of the synthesis of the essential polypeptide on the 80S ribosomes. The repression is suspended by decreasing the concentration of the essential polypeptide in the membrane below a certain threshold. This might happen via normal turnover. The derepression would then allow the cycle to repeat, beginning again with the synthesis of the essential polypeptide (Fig. 6).

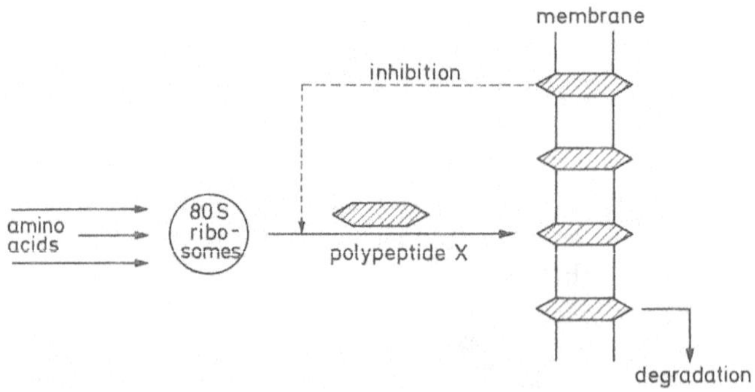

Fig. 6. The coupled-translation membrane model. (SCHWEIGER and SCHWEIGER 1977)

The expression of a photosynthesis rhythm by an individual cell of *Acetabularia* results from an oscillating behavior of a high number of chloroplasts. This, however, raises the question of how the chloroplasts are synchronized. That there is a high degree of synchronization follows from the fact that even under constant conditions, the rhythm persists over months. It is attractive to assume that the cytosol compartment which surrounds the chloroplasts plays an important role in the synchronization.

5.3 Generalization of the Model

An important question in this connection is whether the circadian rhythm of photosynthetic activity in *Acetabularia* is a special and isolated event. Recently a measuring device has been designed to monitor the oxygen evolution simultaneously with the intracellular chloroplast migration in an individual cell of *Acetabularia* over long periods (SCHWEIGER et al. 1980). Such measurements have revealed that the photosynthetic activity and chloroplast migration are subjected to related oscillations.

This observation is highly significant, since in an individual cell the period of the rhythm of photosynthetic activity changes very slowly. In the two-parameter measuring device, it could be shown that such changes in the period are observed also in the chloroplast migration and that the two parallel each other to a certain extent (SCHWEIGER et al. 1980). This experiment indicates that the two rhythms are related to each other and the results might mean that in one cell there is only one single primary oscillator, to which a number of secondary oscillating systems are coupled.

Monitoring the chloroplast migration has furthermore revealed that, similarly to the photosynthesis rhythm, this parameter can be phase-shifted by cycloheximide pulses. This again is in agreement with the assumption that in this rhythm the translation on 80S ribosomes plays an essential role. Finally, it has to be mentioned that more recently, cycloheximide and anisomycin pulses have given similar clear-cut results in a number of completely different organisms with different rhythms. This holds true for rhythms in the eye of *Aplysia* (JACKLET 1980), in *Gonyaulax* (DUNLAP et al. 1980), and in *Neurospora* (NAKASHIMA et al. 1980).

The least one can say is that there is good reason to discuss the possibility that in the different circadian rhythms a similar mechanism exists. Moreover, it seems to be attractive to assume that each cell has only one clock to which different functions are coupled. In the phenomenon of circadian rhythms, the nucleo-cytosol compartment seems to play a controlling role, while the chloroplast compartment would have only a specific function in the photosynthetic rhythm.

References

Apel K, Schweiger HG (1973) Sites of synthesis of chloroplast-membrane proteins. Evidence for three types of ribosomes engaged in chloroplast-protein synthesis. Eur J Biochem 38:373–383

Astaurova OB, Afanasova LA, Salmakha OV, Yazykow AA (1979) Isozyme composition of malate dehydrogenase in two species of *Acetabularia*, in normal conditions and after nuclear implantation. In: Bonotto S, Kefeli V, Puiseux-Dao S (eds) Developmental biology of *Acetabularia*. Elsevier/North-Holland Biomedical Press, Amsterdam New York, pp 259–268

Baltus E, Brachet J (1963) Presence of deoxyribonucleic acid in the chloroplasts of *Acetabularia mediterranea*. Biochim Biophys Acta 76:490–492

Bannwarth H, Schweiger HG (1975) Regulation of thymidine phosphorylation in nucleate and anucleate cells of *Acetabularia*. Proc R Soc London Ser B 188:203–219

Bannwarth H, Ikehara N, Schweiger HG (1977a) Thymidine phosphorylating enzymes in *Acetabularia*: evidence for the occurrence of a thymidine kinase. Proc R Soc London Ser B 198:155–176

Bannwarth H, Ikehara N, Schweiger HG (1977b) Nucleocytoplasmic interactions in the regulation of the thymidine phosphorylation in *Acetabularia*. Proc R Soc London Ser B 198:177–190

Bannwarth H, Ikehara N, Schweiger HG (1980) Regulation of dCMP deaminase during the life cycle of *Acetabularia mediterranea*. Eur J Cell Biol 22:459

Berger S (1967a) RNA-synthesis in *Acetabularia*. II RNA-synthesis in isolated chloroplasts. Protoplasma 64:13–25

Berger S (1967b) RNA-Synthese in isolierten Chloroplasten von *Acetabularia*. Inaugural Diss, Univ Köln, pp 1–71

Berger S, Schweiger HG (1980) *Acetabularia*: techniques for study of nucleo-cytoplasmic interrelationships. In: Gantt E (ed) Handbook of physiological methods. Developmental and cytological methods. Univ Press, Cambridge, pp 47–57

Berger S, Sandakhchiev L, Schweiger HG (1974) Fine structural and biochemical markers of Dasycladaceae. J Microsc 19:89–104

Bonotto S, Lurquin F, Mazza A (1976) Recent advances in research on the marine alga *Acetabularia*. Adv Mar Biol 14:123–250

Boveri T (1918) Zwei Fehlerquellen bei Merogonieversuchen und die Entwicklungsfähigkeit merogonischer, partiellmerogonischer Seeigelbastarde. Wilhelm Roux's Arch Entwicklungsmech Org 44:417–471

Brachet J (1970) Preface. In: Puiseux-Dao S (ed) *Acetabularia* and cell biology, Logos, London, IX–XII

Brachet J, Lang A (1965) The role of the nucleus and the nucleo-cytoplasmic interactions in morphogenesis In: Ruhland W (ed) Handbuch der Pflanzenphysiologie XV. Springer, Berlin Heidelberg New York, pp 1–40

Broda H, Schweiger G, Koop HU, Schmid R, Schweiger HG (1979) Chloroplast migration: a method for continuously monitoring a circadian rhythm in a single cell of *Acetabularia* In: Bonotto S, Kefeli V, Puiseux-Dao S (eds) Developmental biology of *Acetabularia*. Elsevier/North-Holland Biomedical Press, Amsterdam, New York, pp 241–253

Clauss H, Lüttke A, Hellmann F, Reinert J (1970) Chloroplastenvermehrung in kernlosen Teilstücken von *Acetabularia mediterranea* und *Acetabularia cliftonii* und ihre Abhängigkeit von inneren Faktoren. Protoplasma 69:313–329

Coleman AW (1979) Use of the fluorochrome 4′6-diamidino-2-phenylindole in genetic and developmental studies of chloroplast DNA. J Cell Biol 82:299–305

Dunlap JC, Taylor W, Hastings JW (1980) The effects of protein synthesis inhibitors on the *Gonyaulax* clock 1. Phase-shifting effects of cycloheximide. J Comp Physiol 138:1–8

Fritsch FE (1965) The family Dasycladaceae. In: Fritsch FE (ed) The structure and reproduction of the algae Vol I. Univ Press, Cambridge, pp 387–402

Gibor A, Izawa M (1963) The DNA content of the chloroplasts of *Acetabularia*. Proc Natl Acad Sci USA 50:1164–1169

Green BR (1976) Covalently closed minicircular DNA associated with *Acetabularia* chloroplasts. Biochim Biophys Acta 447:156–166

Green BR, Muir BL, Padmanabhan U (1977) The *Acetabularia* chloroplast genome: small circles and large kinetic complexity In: Woodcock CLF (ed) Progress in *Acetabularia* research. Academic Press, London, New York, pp 107–122

Groot de EJ, Schweiger HG (1980) Regulation of thymidylate kinase in *Acetabularia mediterranea* during the vegetative phase. Eur J Cell Biol 22:458

Hämmerling J (1932) Entwicklung und Formbildungsvermögen von *Acetabularia mediterranea*. II. Das Formbildungsvermögen kernhaltiger und kernloser Teilstücke. Biol Zentralbl 52:42–61

Hämmerling J (1953) Nucleo-cytoplasmic relationship in the development of *Acetabularia*. Int Rev Cytol 2:475–498

Hämmerling J (1963) Nucleo-cytoplasmic interactions in *Acetabularia* and other cells. Annu Rev Plant Physiol 14:65–92

Jacklet JW (1980) Protein synthesis requirement of the *Aplysia* circadian clock tested by active and inactive derivatives of the inhibitor anisomycin. J Exp Biol 85:33–42

Karakashian MW, Schweiger HG (1976a) Circadian properties of the rhythmic system in individual nucleated and anucleated cells of *Acetabularia mediterranea*. Exp Cell Res 97:366–377

Karakashian MW, Schweiger HG (1976b) Evidence for a cycloheximide-sensitive component in the biological clock of *Acetabularia*. Exp Cell Res 98:303–312

Karakashian MW, Schweiger HG (1976c) Temperature dependence of cycloheximide-sensitive phase of circadian cycle in *Acetabularia mediterranea*. Proc Natl Acad Sci USA 73:3216–3219

Kloppstech K, Schweiger HG (1973a) Nuclear genome codes for chloroplast ribosomal proteins in *Acetabularia*. I. Isolation and characterization of chloroplast ribosomal particles. Exp Cell Res 80:63–68

Kloppstech K, Schweiger HG (1973b) Nuclear genome codes for chloroplast ribosomal proteins in *Acetabularia*. II. Nuclear transplantation experiments. Exp Cell Res 80:69–78

Kloppstech K, Schweiger HG (1974) The site of synthesis of chloroplast ribosomal proteins. Plant Sci Lett 2:101–105

Koop HU (1979) The life cycle of *Acetabularia* (Dasycladales Chlorophyceae): a compilation of evidence for meiosis in the primary nucleus. Protoplasma 100:353–366

Koop HU, Kiermayer O (1980a) Protoplasmic streaming in the giant unicellular green alga *Acetabularia mediterranea*. I. Formation of intracellular transport systems in the course of cell differentiation. Protoplasma 102:147–166

Koop HU, Kiermayer O (1980b) Protoplasmic streaming in the giant unicellular green alga *Acetabularia mediterranea*. II. Differential sensitivity of movement systems to substances acting on microfilaments and microtubuli. Protoplasma 102:295–306

Koop HU, Schmid R, Heunert HH, Milthaler B (1978) Chloroplast migration: a new circadian rhythm in *Acetabularia*. Protoplasma 97:301–310

Kung S (1976) Tobacco fraction-1-protein: a unique genetic marker. Science 191:429–434

Leong TY, Schweiger HG (1979) The role of chloroplast-membrane-protein synthesis in the circadian clock. Purification and partial characterization of a polypeptide which is suggested to be involved in the clock. Eur J Biochem 98:187–194

Mergenhagen D, Schweiger HG (1975a) Circadian rhythm of oxygen evolution in cell fragments of *Acetabularia mediterranea*. Exp Cell Res 92:127–130

Mergenhagen D, Schweiger HG (1975b) The effect of different inhibitors of transcription and translation on the expression and control of circadian rhythm in individual cells of *Acetabularia*. Exp Cell Res 94:321–326

Nakashima H, Perlman J, Feldman JF (1980) Evidence for protein synthesis requirement for the circadian clock of *Neurospora*. Eur J Cell Biol 22:498

Nelson N, Nelson H, Schatz G (1980) Biosynthesis and assembly of the proton-translocating adenosine tri-phosphatase complex from chloroplasts. Proc Natl Acad Sci USA 77:1361–1364

Oltmanns F (1922) Morphologie und Biologie der Algen, Vol I, 2nd edn. Fischer, Jena, pp 366–382

Padmanabhan U, Green BR (1978) The kinetic complexity of *Acetabularia* chloroplast DNA. Biochim Biophys Acta 521:67–73

Puiseux-Dao S (1970) *Acetabularia* and cell biology. Logos, London

Rahmsdorf U (1977) The intracellular distribution of malate dehydrogenase isoenzymes

in *Acetabularia*. In: Woodcock CLF (ed) Progress in *Acetabularia* research. Academic Press, London, New York, pp 47–51
Reuter W, Schweiger HG (1969) Kernkontrollierte Lactatdehydrogenase in *Acetabularia*. Protoplasma 68:357–368
Sandakhchiev L, Niemann R, Schweiger HG (1973) Kinetics of changes of malic dehydrogenase isozyme patern in different regions of *Acetabularia* hybrids. Protoplasma 76:403–415
Schulze KL (1939) Cytologische Untersuchungen an *Acetabularia mediterranea* und *Acetabularia wettsteinii*. Arch Protistenkd 92:179–225
Schweiger E, Wallraff HG, Schweiger HG (1964a) Über tagesperiodische Schwankungen der Sauerstoffbilanz kernhaltiger und kernloser *Acetabularia mediterranea*. Z Naturforsch 19b:499–505
Schweiger E, Wallraff HG, Schweiger HG (1964b) Endogenous circadian rhythm in cytoplasm of *Acetabularia*: influence of the nucleus. Science 146:658–659
Schweiger HG (1969) Cell biology of *Acetabularia*. Curr Top Microbiol Immunol 50:1–36
Schweiger HG, Berger S (1964) DNA-dependent RNA synthesis in chloroplasts of *Acetabularia*. Biochim Biophys Acta 87:533–535
Schweiger HG, Berger S (1979) Nucleocytoplasmic interrelationships in *Acetabularia* and some other Dasycladaceae. Int Rev Cytol Suppl 9:12–44
Schweiger HG, Berger S (1981) Pattern formation in *Acetabularia*. In: Kiermayer O (ed) Cytomorphogenesis in plants. Springer, Wien, New York, pp 119–145
Schweiger HG, Bremer HJ (1961) Cytoplasmatische RNS-Synthese in kernlosen Acetabularien. Biochim Biophys Acta 51:50–59
Schweiger HG, Schweiger M (1977) Circadian rhythms in unicellular organisms: an endeavour to explain the molecular mechanism. Int Rev Cytol 51:315–342
Schweiger HG, Master RWP, Werz G (1967) Nuclear control of a cytoplasmic enzyme in *Acetabularia*. Nature (London) 216:554–557
Schweiger HG, Berger S, Kloppstech K, Apel K, Schweiger M (1974) Some fine structural and biochemical features of *Acetabularia major* grown in the laboratory. Phycologia 13:11–20
Schweiger HG, Broda H, Wolff D (1980) Simultaneous recording of two circadian rhythms in an individual cell of *Acetabularia*. Eur J Cell Biol 22:496
Spencer T, Harris H (1964) Regulation of enzyme synthesis in an enucleate cell. Biochem J 91:282–286
Sweeney BM, Haxo FT (1961) Persistence of a photosynthetic rhythm in anucleated *Acetabularia*. Science 134:1361–1363
Tzagoloff A, Macino G (1979) Mitochondrial genes and translation products. Annu Rev Biochem 48:419–441
Valet G (1969) Contribution à l'étude des Dasycladales II et III. Nova Hedwigia Kryptogamenkd 17:551–644
Vanden Driessche T (1966) Circadian rhythms in *Acetabularia*: Photosynthetic capacity and chloroplast shape. Exp Cell Res 42:18–30
Werz G (1966) Morphological changes in chloroplasts and mitochondria of darkened *Acetabularia* cells. Planta 68:256–268
Werz G (1974) Fine-structural aspects of morphogenesis in *Acetabularia*. Int Rev Cytol 38:319–367
Werz G, Kellner G (1968) Molecular characteristics of chloroplast DNA of *Acetabularia* cells. J Ultrastruct Res 24:109–115
Woodcock CLF, Bogorad L (1970) Evidence for variation in the quantity of DNA among plastids of *Acetabularia*. J Cell Biol 44:361–375
Zetsche K (1968) Regulation der UDPG-Pyrophosphorylase-Aktivität in *Acetabularia*. I. Morphogenese und UDPG-Pyrophosphorylase-Synthese in kernhaltigen und kernlosen Zellen. Z Naturforsch 23b:369–376

18 Use (and Misuse) of Inhibitors in Gene Expression

G. GALLING

1 Introduction

In the preceding chapters it has been stated that the genetic information of the eukaryotic plant cell is multiple. It is located in three different compartments, namely in the nucleus, the chloroplasts, and the mitochondria. Using another nomenclature, these three genetic systems are designated as genome, plastom, and chondrom.

Striking differences to the nucleo-cytoplasmic system can be noted in the prokaryotic organization of the organelles and their genetic systems. Many attempts have been made to use these differences in order to localize the genetic information for defined processes in the plant cell compartments. One of the most successful attempts was and is the antibiotic tool (AKOYUNOGLOU and ARGYROUDI-AKOYUNOGLOU 1978, ARNOLD 1978, BOULTER 1970, BÜCHER et al. 1976, ELLIS 1976).

Among the inhibitors acting on the molecular genetic level, antibacterial and antifungal antibiotics have been introduced in studies on the genetic location and on the differentiation and maintenance of many structures and pathways in the plant cell.

Nevertheless, misinterpretation of results and misleading data cannot be excluded. In some cases, in the first exuberance, unqualified conclusions have been drawn from data, resulting in simplifications or even severe errors. On the other hand, many ideas on the fundamental principles of eukaryotic organization are based on the results of an approach by means of antibiotics. This is true for the chloroplast, as well as for the mitochondrion.

Since in today's molecular biology landscape studies are performed on various levels of complexity, we have to insist that conclusions are allowed only after a combination of data from different approaches. Moreover, generalization of results obtained with one species or one group of plants is not permitted in all cases.

From the immense field of inhibitory effects on the genetic systems, only the blockage of transcription, e.g., the formation of RNA on DNA templates, and of translation, e.g., the protein synthesis with ribosomes on messenger RNA templates are described here.

This chapter is not the place to discuss the molecular action of antibiotics on RNA polymerase or on ribosomal particles, details and conceptions of which may be found in biochemical handbooks. From the large number of books and reviews, only the following are cited here: GOTTLIEB and SHAW 1967, HAHN 1979, KERSTEN and KERSTEN 1974, and PESTKA 1971.

2 Mode of Action of Various Inhibitors

2.1 General Scheme of the Molecular Biology of Plant Cells

Gene expression occurs as a two-step process of transcription from the genetic information in DNA molecules, involving the intermediary state of information as messenger RNA, and translation, resulting in polypeptides. Since in plant cells three compartments possess their own genetic information and apparatus, we can assume that interrelations between the systems exist, in especially as the organellar genetic systems are not autonomous (cf. Chaps. 15–17). Proteins can interrelate with proteins of the genetic system, with proteins, RNA, and DNA of the neighboring systems. The (hypothetical) system of interference of three equivalent genetic systems can be particularized with more details for a eukaryotic plant cell. In this case, we have to consider direct interactions of the three genetic systems, as well as modification, modulation, and transport phenomena (Fig. 1). The possible location of inhibitory action is indicated in the scheme but all secondary effects and interferences are omitted.

2.2 Inhibitors of Transcription

In Table 1, a general survey of antibiotics and inhibitors of transcription is given. Transcription can be blocked in plant cells on different levels.

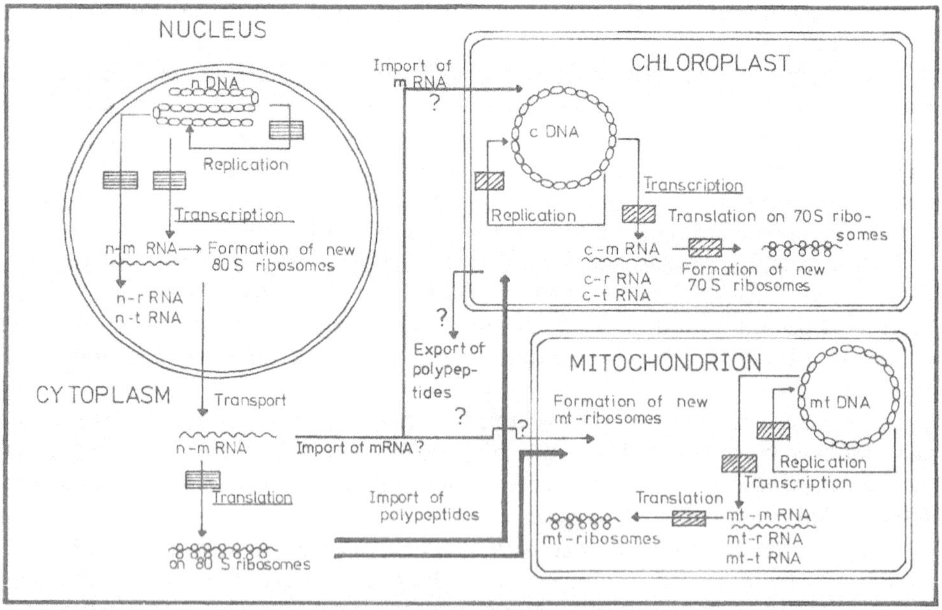

Fig. 1. Schematic diagram of possible intercommunications of the compartments of a typical green plant cell. The flow of information, e.g., messenger RNA, and of products of translation, e.g., of polypeptides, is shown as *arrows* between the different compartments of the cell. Possible blockades by antibiotic agents are symbolized by *blocks interrupting the arrows*. Only the most important pathways are drawn

Table 1. List of inhibitors of transcription. (TRÄGER 1975)

Inhibition on template level	Inhibition of RNA polymerase	Acting on prokaryotic systems	Acting on eukaryotic systems
Actinomycin		+	+
Acridines		+	+
Ethidium bromide		+	−(?)
Anthracyclines		+(GC-rich)	+
Olivomycin		+	+
Daunomycin		+	+
Distamycin		+(AT-rich)	+
Chromomycetin		+	+
Base analogs		+	+
Camptothecin		−	+
	Rifamycins	+	−
	Rifampicins	(specially viral systems)	
	Streptovaricin	+	−
	Streptolydigin	+	−
	α-Amanitin	−	+
	Heparin	+	−
	Cordycepin	−	+

In the *nucleus,* three nucleoplasmic and one nucleolar RNA-polymerases are active (cf. Chap 5, this Vol.). Only part of the RNA synthesized in the nucleus is processed and transported into the cytoplasm. Ribosomal RNA, as formed by the nucleolar polymerase, remains in the nucleus, where it is processed to form preribosomes which pass into the cytoplasm. Nuclear messenger RNA has been characterized by the fact that by post-transcriptional processing stretches of polyadenylic acid are added, and spacer parts of the original transcript are cut out (cf. Chap 6, this Vol.).

Of the antibiotics blocking nuclear transcription, only three substances act specifically on one of the several genetic systems of plant cells; most of them act on all forms of transcription, the activity being limited only by penetration conditions. The widely used inhibitor actinomycin D (Fig. 2a) has been shown to inhibit transcription in all systems tested. Plant cells, in particular algal cells, are barely influenced in vivo since the uptake is very low; even high concentrations are not very effective. The use of actinomycin D for the characterization of gene expression is useful only in studies of the life-time of messenger RNA (see STEWART and LETHAM 1973).

The drug α-amanitine (Fig. 2b) specifically blocks the synthesis of messenger RNA and short-living nuclear RNA. Neither the nucleolar polymerase nor the "prokaryotic" polymerases of chloroplasts and mitochondria seem to be influenced. The antibiotic camptothecin has been shown to inhibit the synthesis of hnRNA in *Chlamydomonas reinhardii* (HORWITZ et al. 1971).

Since in this fraction of heterogeneous nuclear RNA messenger RNA is concerned, a certain specificity is suggested. Neither chloroplast transcription nor ribosomal RNA formation seem to be affected. This effect is probably restricted to *Chlamydomonas,* since in *Chlorella* and some other algae this specificity was not observed (SSYMANK 1978).

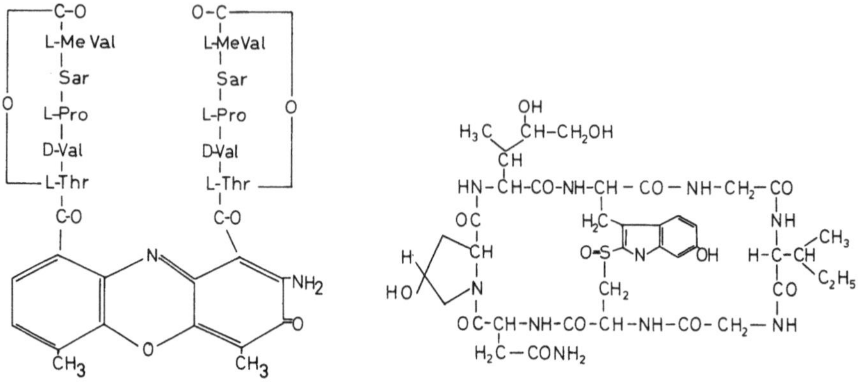

a Actinomycin D **b** α-Amanitin

Fig. 2 a, b. Formula of actinomycin D and of α-amanitin. (TRÄGER 1975)

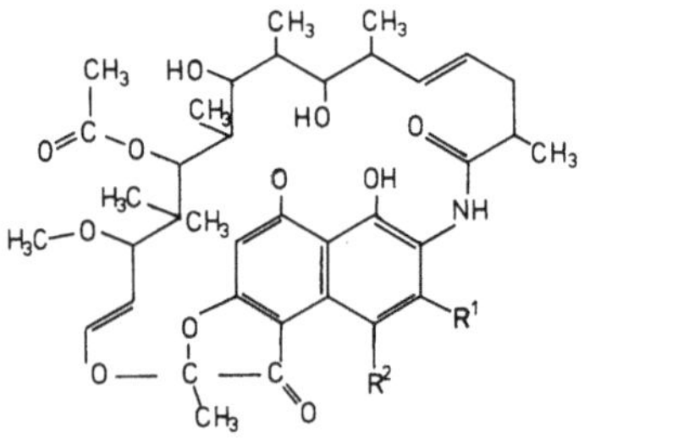

Fig. 3. Formula of some rifamycin derivatives. (TRÄGER 1975)

Cordycepin (3′deoxyadenosine) inhibits specifically polyadenylation of the 3′ end of messenger RNA and is therefore restricted in its action to eukaryotic cells. By using this inhibitor in high concentration (200–500 μg/ml) it was possible to demonstrate that new mRNA synthesis is not required for polyribosome formation in germinating embryos (SPIEGEL and MARCUS 1975, HAMMET and KATTERMAN 1975), or in light-induced etiolated bean leaves (SMITH 1976). Thus it can be used for regulation of protein synthesis at the post-transcriptional level as well as for mRNA stability measurements (ASPART et al. 1979).

Since in many cases the blockade of the nuclear RNA synthesis inhibits most of the life processes, it seems to be more promising to introduce inhibitors in the chloroplast transcription. In the chloroplast, only one RNA polymerase acts, which is responsible for the synthesis of stable and messenger RNA. The most effective inhibitors of bacterial RNA polymerases are the rifamycins (Fig. 3). This group of substances interferes with the polymerase by formation of a very stable complex (BROWN et al. 1970). Initiation of new RNA chains is then impossible, but elongation of chains initiated prior to application of the drug are terminated correctly.

Data on the action of rifamycins in the plant kingdom are contradictory. In higher plants, chloroplast RNA polymerases have been described which are resistant to the drug, and others which are sensitive (MUNSCHE and WOLLGIEHN 1973, WOLLGIEHN and PARTHIER 1980). In vivo blocking of chloroplast RNA synthesis has been observed in some cases but not throughout in higher plants (INGLE 1968). In algae also contradictory results have been published. Some strains seem to be very sensitive to the antibiotic: complete inhibition has been shown in *Chlorella* at low concentrations (GALLING 1971 1972a, b) and in *Chlamydomonas* at rather high concentrations (GOODENOUGH 1971, SURZYCKI 1969). *Euglena gracilis* does not take up the drug in sufficient amounts, and *Chlorogonium* is probably resistant (BRÄNDLE and ZETSCHE 1971, GALLING and NITSCHE 1974). When using rifampin or another rifamycin, one has to take into account that the substances are extremely sensitive to light; on the other hand, uptake sometimes may function only in the light.

In mitochondria, rifamycins act probably also but it is still unclear if this drug can be taken up by mitochondria. Purified mitochondrial RNA polymerase, which differs from other "prokaryotic" polymerases by molecular structure (BARATH and KÜNTZEL 1972), has been shown to be sensitive in some strains of fungi, in others to be resistant (WINTERSBERGER 1970). Since many algae can grow heterotrophically in the presence of rifampin over several generations, one may consider that mitochondria are not affected by the drug in vivo either by resistance or by membrane barriers at the mitochondrial level.

2.3 Inhibitors of Translation

Numerous compounds of natural and chemical provenance have been shown to inhibit translation, e.g., the specific formation of polypeptides on messenger RNA. Many of these substances are used in medical treatment as antibacterial or antifungal drugs, and some of them are known to interfere specifically within defined steps of translation (APIRION and SCHLESSINGER 1968, DAVIES et al. 1966). Some of these drugs are given in Table 2.

It is obvious that inhibitors which block translation are generally useful only in in vitro studies with translation systems. In some cases, a pseudo-specificity is suggested in vivo by simple pool effects: according to the location of an organelle near the cell surface, such an inhibitor may enter faster into the organelle than into the cytoplasm. Most of these substances, acting on any protein synthesis system, show lethal effects. This has to be mentioned for puromycin, tetracyclines, fusidic acid and many of the amino acid analogs. All these antibiotics impair the nucleo-cytoplasmic system in the same order of magnitude as the chloroplastic (and probably the mitochondrial) system. In order to discriminate between the translational systems within the cell, specific inhibitors are to be used. As listed in Table 2, many inhibitors of prokaryotic translation exist.

The 80S ribosomes of eukaryotic cells (and cell compartments), however, are rarely affected specifically. The most-used antibiotic of the cytoplasmic system is cycloheximide (formula see Fig. 4a). This compound was first described as a fungicide (WHIFFEN et al. 1946), and was introduced into plant molecular biology with *Chlorella* (MORRIS 1967). The range of concentration needed for

Table 2. Mode of action of translation inhibitors. (STEWART and LETHAM 1973)

Antibiotic	Ribosome subunit affected	Ribosome function affected			
		Initiation	Recognition	Transpeptidase	Translocase
Streptomycin	30S	++	++	+	++
Neomycin	30S	++	++	+	++
Kanamycin	30S	++	++	+	++
Spectinomycin	30S	−	−	−	++
Kasugamycin	30S	−	++		
Tetracycline	30S	++	+	+	
Edeine	30S	++	+	−	++
Pactamycin	30S, 40S	++	−	−	−
Chloramphenicol	50S	−	−	++	+
Erythromycin	50S	−	−	++	++
Lincomycin	50S		−	++	
Thiostrepton			−	−	++
Puromycin	50S, 60S	−	−	++	++
Cycloheximide	60S			−	++
Emetine	60S				++
Anisomycin	60S			++	

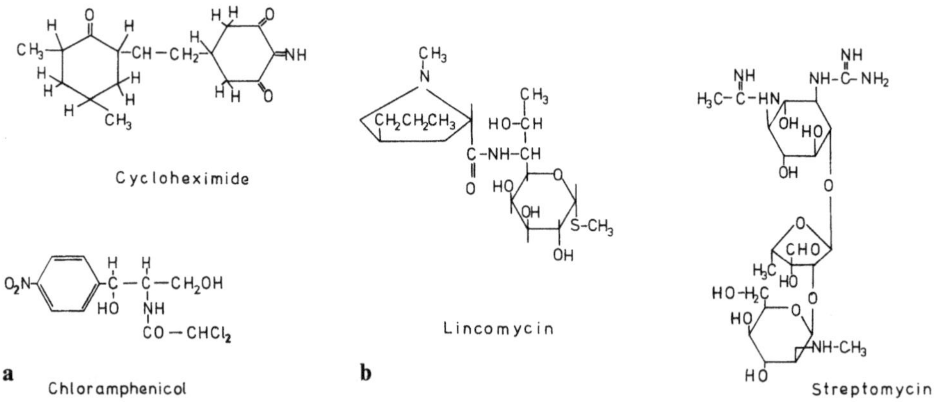

Fig. 4. Formula of inhibitors of translation: cycloheximide (actidion), chloramphenicol, lincomycin, and streptomycin. (TRÄGER 1975)

complete inhibition of cytoplasmic translation is very wide. It is not yet known if this range of concentration is due rather to high or low susceptibility of 80S ribosomes or to uptake into the cell. Cycloheximide acts on the translocase reaction of the translation, and polypeptide synthesis is blocked immediately and throughout.

Even though no more cytoplasmic proteins are formed, evidence emerged that the drug is only lethal after long-term treatment. In repair experiments, CHI can be washed off, and polypeptide synthesis starts again (GALLING 1971, GERSONI and OHAD 1980a, b, LAVINTMAN et al. 1981, MILLER and OHAD 1978).

In the presence of CHI, the cells seem to maintain a quasi-living state, and chloroplast and mitochondrial functions continue. It is possible to apply CHI together with other inhibitors; also inhibition programs can be carried out with CHI and antibacterial inhibitors (see below). In plants many side effects have been described after application of CHI. It is not yet clear if these effects are due to the depression or loss of cytoplasmic proteins. In Section 4 some of the secondary effects are discussed. In algae, however, many data on the interrelations of structural and functional processes in chloroplast development are not imaginable without CHI.

In chloroplasts and mitochondria, many antibacterial antibiotics can be introduced, provided that they penetrate into the cell and into the organelles (Boschetti et al. 1976, Bourque and Wildman 1973, Galling and Wolff 1972, Hanson and Bogorad 1978). One of the most-applied drugs is chloramphenicol (CAP) (Armstrong et al. 1971, Ellis and Hartley 1971, Ellis 1977a, b, Hoober 1970, Neumann and Parthier 1973, Liebers and Parthier 1974). This compound acts very strongly on bacteria in vivo, and it was widely used in human medical treatment prior to the detection of severe secondary effects. In plant cells, CAP was shown to penetrate into the organelles, but in vivo mitochondria are barely affected. This is not due to ineffectiveness: plant mitochondrial ribosomes are affected in vitro by blockade of transpeptidase and translocation (cf. Stewart and Letham 1973).

To inhibit protein synthesis of chloroplasts in algae, very high concentrations are needed in some species, whereas others are blocked by low concentrations.

As listed in Table 2, many other antibiotics are known to block either the small (30S) or the large (50S) subunit of chloroplast ribosomes. Some of these drugs are not active in all types of mitochondria. Since green plant mitochondria seem to have 70S ribosomes, as is well known in yeast and other fungi, most of the antibacterial compounds act in both organelles provided that they can enter (see Table 3). From the large number of antibiotics, only CAP, lincomycin (Ellis 1975a), erythromycin (Hanson and Bogorad 1978), streptomycin (Bo-

Table 3. Sensitivity of organelle protein synthesis to ribosomal inhibitors. (After Stewart and Letham 1973)

Protein-synthesizing system	CHI	CAP	Erythromycin	Neomycin	Lincomycin
Eukaryote 80S	+	−	−	−	−
Prokaryote 70S	−	+	+	+	+
Yeast mitochondria 70–80S	−	+	+	+	+
Mammalian mitochondria 50–60S	−	+	−	−	−
Chloroplast 70S	−	+	+	+	+

SCHETTI et al. 1976), and spectinomycin (GOODENOUGH 1971) are treated here (see Fig. 4b).

Prior to any use of inhibitors of chloroplast protein synthesis, the optimal and maximal concentrations have to be determined (BOULTER 1970). These can vary by orders of magnitude from one organism to the other. If, for example, spectinomycin is applied to *Chlamydomonas reinhardii,* less than 2 µg/ml kill the cells immediately (GOODENOUGH and LEVINE 1971). Other flagellates and chlorococcal algae are affected only by high doses of this drug, and vitality of the cells is not influenced.

3 Action of Inhibitors in Vivo

The main topic of this chapter is to discuss data on inhibitory effects on parts of the plant cell by inhibitors applied in vivo. It is to question if clear-cut answers can be obtained after experiments with inhibitors. From the network of intracellular interrelations (drawn in Fig. 1), we must conclude that inhibition of one pathway can have multiple effects. It is then clear that no simple answer can be expected to a complex question. Inhibition of nuclear messenger RNA formation causes not only the blockade of cytoplasmic functional proteins but probably also the nonformation of inducers, repressors, and regulators in organellar systems. Pool sizes of precursors can also be affected, resulting in an increase of organellar macromolecule synthesis. This consideration restricts the use of inhibitors to defined and clear reactions. In combination with the employment of radionuclides (FREYSSINET 1974, 1977), inhibitors can help to follow synthesis and breakdown of components within cytoplasm and organelles of plant cells. Care must be taken that the normal anabolic and catabolic metabolism is influenced as little as possible. The risks of application of antibiotics seem to be higher in vascular plants (see FEIERABEND 1966) than in unicellular eukaryotic algae: in the latter case, cell nutrition can be maintained by an external carbon source.

One of the main uses of inhibitors in plants is to elucidate the interactions and interrelations between organelles and the nucleocytoplasmic system. The question is whether we are allowed to deduce from inhibitor experiments unequivocal conclusions concerning the location of genetic expression, e.g., the location of genes within genome, plastom, and chondrom. It is so far not understood how different susceptibilities of organelles to antibiotics can occur (THOMAS and WILLIAMSON 1971). The simplest explanation is to take into account that mitochondrial ribosomes are not true 70S particles. On the other hand, in vitro studies show inhibitory effects of most of the antibiotics. The exposed membrane surface of organelles can also play a role in this different susceptibility; and another explanation takes into account that mitochondria possibly have specific membrane barriers. One of these possibilities explains the excellent (heterotrophic) growth of unicellular algae in the presence of antibacterial antibiotics. In the following, a few examples of application of antibiotics are described.

As a first example, synthesis of ribulose-1,5-bisphosphate carboxylase (RuBPCase) is described. This enzyme consists of two subunits (see Chap. 15, this Vol.) of chloroplastic and nucleo-cytoplasmic origin (ELLIS 1978, SMILLIE et al. 1967). As mentioned first in inhibitor studies, the formation of the large subunit can be blocked by inhibitors of prokaryotic translation. Also in the presence of rifampin, the formation of the large subunit is suppressed in those organisms which are susceptible to the drug. On the other hand, the small subunit is suppressed by cycloheximide in some algal strains. The heterogenous origin of the RuBPCase has been affirmed by genetic (HAGEMANN and SCHOLZE 1974) and physiological (FEIERABEND and WILDNER 1978) studies, and finally, the genetic location (and base sequence) of the gene of the large subunit has been shown, affirming the previous inhibitor data (see also BÜCHER et al. 1976, AKOYUNOGLOU and ARGYROUDI-AKOYUNOGLOU 1978). One question in this context remains open: is synthesis of the two subunits of the enzyme a highly coordinated process or not? (CHUA and SCHMIDT 1978). In some algae, the occurrence of mature molecules or precursors of the cytoplasmic part of the enzyme has been shown even after complete inhibition of chloroplastic protein synthesis. In the presence of CHI, even more holoenzyme can be found in algal cells. On the other hand, studies on the synthesis and turnover of the enzyme in *Chlamydomonas* (IWANIJ et al. 1975) or *Chlorogonium* (WESTHOFF and ZETSCHE 1981) revealed a strict correlative control existing between small and large subunits. This dependence of both syntheses has not been found in isolated soybean leaf cells (BARRACLOUGH and ELLIS 1979) or in rye after physiological blockade of 70S ribosomes by heat (FEIERABEND and MIKUS 1977). In contrast to these findings, the plastids of a 70S ribosome-free strain of *Pelargonium zonale* do not contain the cytoplasmic part of the enzyme, as proven by the antibody technique (HAGEMANN and BÖRNER 1978).

Another example is the formation of the 70S ribosome in the plastid (HANSON and BOGORAD 1978). Evidence emerged that most of the chloroplast ribosomal proteins are cytoplasmic in origin, the genes being located in the nucleus (BOURQUE and WILDMAN 1973, METS and BOGORAD 1971, NIGON and HEIZMANN 1978). After inhibition with antibacterial antibiotics, the RNA of the ribosomes is transcribed, and preribosomes are formed, which persist in an immature state. Only when the inhibitor is washed off does the maturation go on, and the RNA molecules within the preribosomes are processed to the authentic length (GALLING 1974). We can assume that only 4 or 5 ribosomal proteins (out of more than 45) are of chloroplastic origin (FREYSSINET 1974, 1977, HEIZMANN 1974). Some of these proteins have been identified by antibiotic studies. In *Euglena,* labeling of ribosomes in the presence of lincomycin revealed 4 or 5 proteins of chloroplastic origin. In *Chlamydomonas,* the maternal inheritant cleocin-resistance was shown to be linked to the proteins L-10 and L-11 of the large subunit: a non-Mendelian mutant of *Chlamydomonas* shows alteration of these two proteins in their mobility in electropherograms. Cleocin, the 7-chloro derivative of lincomycin, acts on the 50S subunit probably by inhibition of membrane binding of these particles.

A more complex example is the formation of chlorophyll-protein complexes in thylakoids (BOARDMAN et al. 1978, ELLIS 1975b, GERSONI and OHAD 1980a,

1980b). This has been studied also with antibiotics in many plant cells, in particular in algal systems (BOSCHETTI et al. 1976, GALLING et al. 1973). It was shown that both protein-synthesizing systems are necessary for the formation of active photosystems in thylakoids (FEIERABEND 1977, FEIERABEND and SCHUBERT 1978, see also PACKER et al. 1977). The intercommunication of the cytoplasmic and the chloroplastic system can be studied after application of CHI and CAP to the y-1 mutant of *Chlamydomonas reinhardi*. Data obtained in this study lead to the hypothesis that a chloroplastic aporepressor, formed on 70S ribosomes, acts together with a chlorophyll precursor in order to regulate the transcription of nuclear genes (GERSONI and OHAD 1980). The latter genes are responsible for the formation of active chlorophyll-protein complexes in thylakoids. The correlated synthesis of proteins on both cytoplasmic and chloroplastic ribosomes was shown also in *Chlorella* ts-mutant to be a prerequisite for the formation of the light-harvesting complex in photosystem II (LAVINTMAN et al. 1981).

4 Possible Errors in Interpretation

4.1 Secondary Effects of Antibiotics

The most prominent example of an antibiotic with a wide spectrum of secondary effects is cycloheximide (ELLIS 1977a). This drug has been shown to interfere with transport phenomena on membranes (EVANS and SMITH 1971, PARTHIER 1974, REILLY et al. 1970) with RNA-synthesis (TIMBERLAKE and GRIFFIN 1974) (which might be interpreted as a feedback mechanism) and with respiration (DHEIDAH and BLACK 1976). In a resistant strain of the slime mold *Physarum polycephalum,* alterations in the molecular weight and the specific activity of the glutamate dehydrogenase, and in the uptake of amino acids were described. (WENDELBERGER-SCHIEWEG et al. 1980). Since this particular strain is resistant to CHI, these effects cannot be interpreted as secondary consequences of suppressed polypeptide synthesis on 80S ribosomes. As a consequence of these data, the use of CHI-resistant strains has been suggested. Specific tests are required in order to show if this drug acts specifically on 80S ribosomes or if the inhibition of protein synthesis is by itself a secondary effect (McDONALD and ELLIS 1969).

Another source of error is the possible selectioning of resistant strains of organisms during purification of cultures from contaminating bacteria. Axenic cultures are sometimes obtained after application of antibacterial or even antifungal drugs. Since the generation time of unicellular organisms varies by orders of magnitude, selectioning may occur by chance.

During the experiments, the physical and chemical authenticity of the drugs has to be maintained. Some of the drugs, e.g., rifamycins and actinomycin D, are extremely sensitive to light; others need a highly restricted ionic environment; others again are sensitive to solvents. Possible errors in interpretation may come from inactivation or modification.

4.2 Interactions of the Genetic Systems

As mentioned in Section 2, the complexity of living systems can lead to misinterpretations of inhibitory effects. This has to be understood as "misuse" of these substances. It is clear that interpretations become more and more difficult the earlier the inhibition is located within the chain of gene expression or the metabolic pathway. Inhibition of nuclear transcription can severely affect the translation within the chloroplast or the mitochondrion, since intercommunications in the cell are modified.

If, for instance, cells of *Chlorella pyrenoidosa* are treated with cycloheximide at doses which completely block the translation on 80S ribosomes (as tested by blocked synthesis of these ribosomes), the amount and the specific activity of RuBPCase are increased. The interpretation of unexpected results such as these is very complicated, since membrane transport effects, pool sizes, and regulation effects on the modificatory level have to be kept in mind. Only the analysis of isolated subunits of the carboxylase – if possible from CHI-resistant algal strains – can give a clear-cut explanation of the result.

5 Conclusions

From the data on use and misuse of inhibitors in gene expression we cannot conclude that this approach should be abandoned. It can be a powerful method in understanding the intracellular relations between organelles and compartments. Nevertheless, it has to be supported by experiments with genetic or physiological approaches. In this case, the inhibitors are both helpful and important in increasing our understanding of living systems.

References

Akoyunoglou G, Argyroudi-Akoyunoglou JH (eds) (1978) Chloroplast development. Elsevier North Holland, New York, Oxford

Apirion D, Schlessinger D (1968) Coresistance to neomycin and kanamycin by mutations in an *Escherichia coli* locus that affects ribosomes. J Bacteriol 96:768–776

Armstrong II, Surzycki SJ, Moll B, Levine RP (1971) Genetic transcription and translation specifying chloroplast components in *Chlamydomonas reinhardi*. Biochemistry 10:692–701

Arnold CG (1978) Mitochondrien und Plastiden, extranukleare Erbträger und Zielscheiben antibakterieller Arzneistoffe. Pharm in unserer Zeit 7:151–156

Aspart L, Cooke R, Delseny M (1979) Stability of polyadenylic and polyadenylated RNAs in radish (*Raphanus sativus*) seedlings. Biochim Biophys Acta 564:43–54

Barath Z, Küntzel H (1972) Induction of mitochondrial RNA polymerase in *Neurospora crassa*. Nature (London) New Biol 240:195–197

Barraclough R, Ellis RJ (1979) The biosynthesis of ribulose bisphosphate carboxylase. Uncoupling of the synthesis of large and small subunits in isolated soybean leaf cells. Eur J Biochem 94:165–177

Boardman NK, Anderson JM, Goodchild DJ (1978) Chlorophyll protein complexes and structure of mature and developing chloroplasts. In: Current topics in bioenergetics Vol 8. Academic Press, London, New York, pp 35–109
Boschetti A, Schaffner JC, Leuenberger S (1976) Bildung der Thylakoidproteine während der Ergrünung einer vergilbten Mutante von *Chlamydomonas reinhardi*. Chimia 30:79–82
Boulter D (1970) Protein synthesis in plants. Annu Rev Plant Physiol 21:90–109
Bourque DP, Wildman SG (1973) Evidence that nuclear genes code for several chloroplast ribosomal proteins. Biochem Biophys Res Commun 50:532–537
Brändle E, Zetsche K (1971) Die Wirkung von Rifampicin auf die RNA und Proteinsynthese sowie die Morphogenese und den Chlorophyll-Gehalt kernhaltiger und kernloser *Acetabularia*-Zellen. Planta 99:46–55
Brown R, Bastia M, Haselkorn R (1970) Effect of rifampicin on transcription in chloroplasts of *Euglena*. In: I. Lepetit colloquium on biology and medicine "RNA polymerase and transcription". Elsevier North-Holland, New York, Oxford, pp 309–327
Bücher Th, Neupert W, Sebald W, Werner S (eds) (1976) Genetics and biogenesis of chloroplasts and mitochondria. Elsevier North Holland, New York, Oxford, pp 895
Chua NH, Schmidt GW (1978) In vitro synthesis, transport, and assembly of ribulose 1,5-bisphosphate carboxylase subunit. In: Siegelman HW, Hind G (eds) Photosynthetic carbon assimilation, Plenum Press, New York, pp 325–347
Davies J, Jones DS, Khorana HG (1966) A further study of misreading of codons induced by streptomycin and neomycin using ribopolynucleotides in alternating sequence as templates. J Mol Biol 18:48–57
Dheidah M, Black M (1976) The inhibitory action of cycloheximide on respiration of coleoptiles is relieved by light. Physiol Plant 37:83–88
Ellis RJ (1975a) Inhibition of chloroplast protein synthesis by lincomycin and 2-(4-methyl-2,6-dinitroanilino-N-methyl-propionamide. Phytochemistry 14:89–93
Ellis RJ (1975b) The synthesis of chloroplast membranes in *Pisum sativum*. In: Tzagoloff A (ed) Membrane biogenesis. Plenum, New York, London, pp 247–278
Ellis RJ (1976) Protein and nucleic acid synthesis by chloroplasts. In: Barber J (ed) The intact chloroplast. Elsevier/North-Holland Biomedical Press, Amsterdam, New York, pp 335–364
Ellis RJ (1977a) The synthesis of chloroplast proteins. In: Bogorad L, Weil JH (eds) Nucleic acids and protein synthesis in plants. Plenum, New York, pp 195–211
Ellis RJ (1977b) Protein synthesis in isolated chloroplasts. Biochim Biophys Acta 463:185–215
Ellis RJ (1978) Chloroplast proteins and their synthesis. In: Norton G (ed) Plant Proteins. Proc 24th Easter School in Agricult. Sci (Sutton Bonington) Butterworths, London, pp 25–40
Ellis RJ, Hartley MR (1971) Sites of synthesis of chloroplast proteins. Nature (London) 233:193–196
Evans WR, Smith C (1971) The effect of cycloheximide on membrane transport in *Euglena*. A comparative study with nigericin. J Biol Chem 246:6144–6151
Feierabend J (1966) Enzymbildung in Roggenkeimlingen während der Umstellung von heterotrophem auf autotrophes Wachstum. Planta 71:326–355
Feierabend J (1977) Capacity for chlorophyll synthesis in heat-bleached 70 S ribosome-deficient rye leaves. Planta 135:83–88
Feierabend J, Mikus M (1977) Occurrence of a high temperature sensitivity of chloroplast ribosome formation in several higher plants. Plant Physiol 59:863–867
Feierabend J, Schubert B (1978) Comparative investigation of the action of several chlorosis-inducing herbicides on the biogenesis of chloroplasts and leaf microbodies. Plant Physiol 61:1017–1022
Feierabend J, Wildner G (1978) Formation of the small subunit in the absence of the large subunit of ribulose 1,5-bisphosphate carboxylase in 70S ribosome-deficient rye leaves. Arch Biochem Biophys 186:283–291
Freyssinet G (1974) Ribosomal protein synthesis during chloroplast development of dark-

grown *Euglena*. In: Avron M (ed) Proc II. Int Congr Photosynthesis, Elsevier North Holland, Yew York, Oxford, pp 1731–1744
Freyssinet G (1977) The use of radioactive precursors for the labeling of ribosomal proteins in *Euglena*. Biologie Cellulaire 30:17–26
Galling G (1971) Der Einfluß von Rifampicin, Chloramphenicol und Cycloheximid auf den Uridin-Einbau in chloroplastidäre Ribosomenvorstufen von *Chlorella*. Planta 98:50–62
Galling G (1972a) Einbau von Uridin und Orotsäure in plastidäre ribosomale RNA von *Chlorella* nach Antibiotika-Behandlung. Arch Mikrobiol 81:245–259
Galling G (1972b) Über die Synthese ribosomaler RNA in Pflanzenzellen. Ber Dtsch Bot Ges 85:371–381
Galling G (1974) Vorstufen der plastidären ribosomalen RNA und ihre Reifung bei *Chlorella*. Planta 118:283–295
Galling G, Nitsche G (1974) Die Bedeutung von Licht und Ernährungsbedingungen für die Menge cytoplasmatischer und chloroplastidärer Ribosomen und deren RNA in *Chlorogonium elongatum*. Ber Dtsch Bot Ges 86:459–465
Galling G, Wolff E (1972) Die Wirkung von Neomycin auf Zellteilung, Wachstum und Nucleinsäuren synchronisierter Chlorellen. Arch Mikrobiol 81:295–304
Galling G, Salzmann Ch, Spiess E (1973) Synthese von Chlorophyll und Strukturelementen des Plastiden in *Chlorella* ohne Beteiligung der Chloroplasten-Ribosomen. Planta 114:269–284
Gersoni J, Ohad I (1980a) Synthesis and assembly of chlorophyll containing protein complexes of photosynthetic membranes. In: Lee CP, Schatz G, Ernster L (eds) Membrane bioenergetics, Addison-Wesley, London, pp 165–176
Gersoni J, Ohad I (1980b) Chloroplast-cytoplasmic interrelations involved in chloroplast development in *Chlamydomonas reinhardi* y-1: The effect of selective depletion of chloroplast translates. J Cell Biol 86:124–131
Goodenough U (1971) The effect of inhibitors of RNA and protein synthesis in chloroplast structure and function in wild-type *Chlamydomonas reinhardi*. J Cell Biol 50:35–49
Goodenough U, Levine RP (1971) The effect of inhibitors of RNA and protein synthesis on the recovery of chloroplast ribosomes, membrane organization, and photosynthetic electron transport in the *ac*-20 strain of *Chlamydomonas reinhardi*. J Cell Biol 50:50–62
Gottlieb D, Shaw PD (eds) (1967) Antibiotics Vol 1. Mechanism of action. Springer, Berlin, Heidelberg, New York
Hagemann R, Börner T (1978) Plastid ribosome-deficient mutants of higher plants as a tool in studying chloroplast biogenesis. In: Akoyunoglou G, Argyroudi-Akoyunoglou JH (eds) Chloroplast development. Elsevier North Holland, New York, Oxford, pp 709–720
Hagemann R, Scholze M (1974) Struktur und Funktion der genetischen Information in den Plastiden. VII. Vererbung und Entmischung genetisch unterschiedlicher Plastidensorten bei *Pelargonium zonale* Ait. Biol Zbl 93:635–648
Hahn FE (ed) (1979) Antibiotics Vol 5, Mechanism of action of antibacterial agents; mechanism of action of antieukaryotic and antiviral compounds. Springer, Berlin, Heidelberg, New York
Hammet JR, Katterman FR (1975) Storage and metabolism of poly(adenylic acid)-mRNA in germinating cotton seeds. Biochemistry 14:4375–4380
Hanson MR, Bogorad L (1978) The ery-M2 group of *Chlamydomonas reinhardii*: cold-sensitive, erythromycin-resistant mutants deficient in chloroplast ribosomes. J Gen Microbiol 105:253–262
Heizmann P (1974) Rôle des synthèses protéiques dans la formation du ribosome chloroplastique chez l'Englène. Biochimie 56:1357–1364
Hoober JK (1970) Sites for synthesis of chloroplast membrane polypeptides in *Chlamydomonas reinhardi* y-l. J Biol Chem 245:4327–4334
Horwitz SB, Chang CK, Grollmann AP (1971) Studies on camptothecin: I. Effects on nucleic acid and protein synthesis. Mol Pharmacol 7:632–644

Ingle J (1968) Synthesis and stability of chloroplast ribosomal RNAs. Plant Physiol 43:1448–1454

Iwanij V, Chua N-H, Siekewitz P (1975) Synthesis and turnover of ribulose bisphosphate carboxylase and of its subunits during the cell cycle of *Chlamydomonas reinhardi*. J Cell Biol 64:572–585

Kersten H, Kersten W (1974) Molecular biology, biochemistry and biophysics Vol 18. Inhibitors of nucleic acid synthesis. Springer, Berlin, Heidelberg, New York

Lavintman N, Galling G, Ohad I (1981) Modulation of PS II and PS I Organization during loss and repair of photosynthetic activity in a mutant of *Chlorella pyrenoidosa*. Plant Physiol 68, 1264–1272

Liebers H, Parthier B (1974) Synthese der Lamellarproteine in Chloroplasten ergrünender *Euglena gracilis*. Biochem Physiol Pflanz 165:517–530

MacDonald JR, Ellis RJ (1969) Does cycloheximide inhibit protein synthesis specifically in plant tissues? Nature (London) 222:791–792

Mets LJ, Bogorad L (1971) Mendelian and uniparental alteration in erythromycin binding by plastid ribosomes. Science 174:707–709

Miller KR, Ohad I (1978) Chloroplast membrane biogenesis in *Chlamydomonas*: Correlation between the formation of membrane components and membrane structure. Cell Biol Int Rep 2:537–549

Morris J (1967) The effect of cycloheximide (actidione) on protein and nucleic acid synthesis by *Chlorella*. J Exp Bot 18:54–64

Munsche D, Wollgiehn R (1973) Die Synthese von ribosomaler RNA in Chloroplasten von *Nicotiana rustica*. Biochim Biophys Acta 249:106–117

Neumann D, Parthier B (1973) Effects of nalidixic acid, chloramphenicol, cycloheximide, and anisomycin on structure and development of plastids and mitochondria in greening *Euglena gracilis*. Exp Cell Res 81:255–268

Nigon V, Heizmann P (1978) Morphology, biochemistry and genetics of plastid development in *Euglena gracilis*. Int Rev Cytol 53:212–189

Packer L, Papageorgiou GC, Trebst A (eds) (1977) Bioenergetics of membranes. Elsevier North Holland, New York, Oxford

Parthier B (1974) Stimulation of amino acid uptake by *Euglena gracilis* and the effect of cycloheximide. Biochem Physiol Pflanz 166:555–560

Pestka S (1971) Inhibitors of ribosome functions. Annu Rev Microbiol 25:487–562

Reilly C, Fuhrmann G, Rothstein A (1970) The inhibition of K^+ and phosphate uptake in yeast by cycloheximide. Biochim Biophys Acta 203:583–585

Smillie RM, Graham D, Dwyer MR, Grieve A, Tobin NF (1967) Evidence for the synthesis in vivo of proteins of the Calvin cycle and of the photosynthetic electron-transfer pathway on chloroplast ribosomes. Biochem Biophys Res Commun 28:604–610

Smith H (1976) Phytochrome-mediated assembly of polyribosomes in etiolated bean leaves. Evidence for post-transcriptional regulation and development. Eur J Biochem 65:161–170

Spiegel S, Marcus A (1975) Polyribosome formation in early wheat embryo germination independent of either transcription or polyadenylation. Nature (London) 256:228–230

Ssymank V (1978) Charakterisierung von Nucleinsäuren in Algenzellen und Untersuchungen über ihre Synthese. Habilitationsschrift Math nat Fakultät Univ Göttingen

Stewart PR, Letham DS (1973) The ribonucleic acids. Springer, Berlin, Heidelberg, New York

Surzycki SJ (1969) Genetic functions of the chloroplast of *Chlamydomonas reinhardi*: effect of rifampin on chloroplast DNA-dependent RNA polymerase. Proc Natl Acad Sci USA 63:1327–1334

Thomas DY, Williamson DH (1971) Products of mitochondrial protein synthesis in yeast. Nature (London) New Biol 233:196–199

Timberlake WE, Griffin DH (1974) Differential effects of analogs of cycloheximide on protein and RNA synthesis in *Achlya*. Biochim Biophys Acta 349:39–46

Träger L (1975) Einführung in die Molekularbiologie. Fischer, Stuttgart

Wendelberger-Schieweg G, Hüttermann A, Haugli FB (1980) Multiple sites of action of cycloheximide in addition to inhibition of protein synthesis in *Physarum*. Arch Microbiol 126:109–115

Westhoff P, Zetsche K (1981) Regulation of the synthesis of ribulose-1,5-bisphosphate carboxylase and its subunits in the flagellate *Chlorogonium elongatum*. Different levels of translatable messenger RNAs for the large and the small subunits in autotrophic and heterotrophic cells as determined by immunological techniques.

Whiffen AJ, Bohonos JN, Emerson RL (1946) The production of an antifungal antibiotic by *Streptomyces griseus*. J Bacteriol 52:610–611

Wintersberger E (1970) Die Bedeutung der Mitochondrien-DNA für die Organellen-Biogenese. Ber Dtsch Bot Ges 83:353–357

Wollgiehn R, Parthier B (1980) RNA and protein synthesis in plastid differentiation. In: Reinert J (ed) Results and problems in cell differentiation Vol 10. Springer, Berlin, Heidelberg, New York, pp 97–145

Author Index

Page numbers in *italics* refer to the references

Plant Name Index

Subject Index

Encyclopedia of Plant Physiology

New Series

Editors: **A. Pirson, M. H. Zimmermann**

Volume 14, Part A

Nucleic Acids and Proteins in Plants I

Structure, Biochemistry and Physiology of Proteins

Editors: **D. Boulter, B. Parthier**

With contributions by numerous experts

1982. 135 figures. XX, 768 pages. ISBN 3-540-11008-9

Contents: Introduction. – Biosynthesis and Metabolism of Protein Amino Acids and Proteins: Ammonia Assimilation and Amino Acid Metabolism. Transfer RNA and AminoacyltRNA Synthetases in Plants. Ribosomes, Polysomes and the Translation Process. Post-Translational Modifications. Protein Degradation. Physiological Aspects of Protein Turnover. Structures of Plant Proteins. Protein Types and Distribution. Cereal Storage Proteins: Structure and Role in Agriculture and Food Technology. Biochemistry and Physiology of Leaf Proteins. Microtubule Proteins and P-Proteins. Plant Peptides. Immunology. – Nucleic Acids and Proteins in Relation to Specific Plant Physiological Processes: Seed Development. Protein and Nucleic Acid Synthesis During Seed Germination and Early Seedling Growth. Leaf Senescene. Macromolecular Aspects of Cell Wall Differentiation. – Author Index. – Plant Name Index. – Subject Index.

Structure, Biochemistry and Physiology of Proteins in the first of the two-part *Nucleic Acids and Proteins in Plants.*

This volume in the **Encyclopedia of Plant Physiology** is a multiauthor work containing critical and comprehensive reviews of plant nucleic acids and proteins. It has been written to enable research workers to acquaint themselves with this rapidly expanding and important field. Plants are of growing importance not only because of the increasing problem of world food supplies, but also because they provide excellent experimental material for research into fundamental problems of biology.

In addition to its coverage of the basic aspects of plant nucleic acids and proteins, the volume deals with recent developments in genetic engineering and other applied topics, making the book a valuable reference source for agriculturalists and industrialists as well as for researchers in plant physiology and related fields.

Springer-Verlag
Berlin
Heidelberg
New York

www.ingramcontent.com/pod-product-compliance
Lightning Source LLC
Chambersburg PA
CBHW061148240926
51846CB00006B/958
* 9 7 8 3 6 4 2 6 8 3 4 9 7 *